TEACHER'S WRAPAROUND EDITION

GEOGRAPHY
THE WORLD AND ITS PEOPLE

GEOGRAPHY
THE WORLD AND ITS PEOPLE

MULTIMEDIA EDITION

TEACHER'S WRAPAROUND EDITION

NATIONAL
GEOGRAPHIC
SOCIETY

Richard G. Boehm, Ph.D.
David G. Armstrong, Ph.D.
Francis P. Hunkins, Ph.D.

GLENCOE

McGraw-Hill

New York, New York
Columbus, Ohio
Mission Hills, California
Peoria, Illinois

ABOUT THE AUTHORS

National Geographic Society

The National Geographic Society, founded in 1888 for the increase and diffusion of geographic knowledge, is the world's largest nonprofit scientific and educational organization. Since its earliest days, the Society has used sophisticated communication technologies, from color photography to holography, to convey geographic knowledge to a worldwide membership. The Educational Media Division supports the Society's mission by developing innovative educational programs—ranging from traditional print materials to multimedia programs including CD-ROMS, videodiscs, and software.

Richard G. Boehm

Richard G. Boehm, Ph.D., was one of seven authors of *Geography for Life,* national standards in geography, prepared under *Goals 2000: Educate America Act.* He was also one of the authors of the *Guidelines for Geographic Education,* in which the five themes of geography were first articulated. In 1990 Dr. Boehm was designated "Distinguished Geography Educator" by the National Geographic Society. In 1991 he received the George J. Miller award from the National Council for Geographic Education (NCGE) for distinguished service to geographic education. He was President of the NCGE and has twice won the *Journal of Geography* award for best article. He has received the NCGE's "Distinguished Teaching Achievement" award and is presently Professor of Geography at Southwest Texas State University in San Marcos, Texas.

David G. Armstrong

David G. Armstrong, Ph.D., is Dean of the School of Education at the University of North Carolina at Greensboro. A social studies education specialist with additional advanced training in geography, Dr. Armstrong was educated at Stanford University, the University of Montana, and the University of Washington. He taught at the secondary level in the state of Washington before beginning a career in higher education. Dr. Armstrong has written books for students at the secondary and university levels, as well as for teachers and university professors. He maintains an active interest in travel, teaching, and social studies education.

Francis P. Hunkins

Francis P. Hunkins, Ph.D., is Professor of Education at the University of Washington. He began his professional career as a teacher in Massachusetts. He received his masters degree in education from Boston University and his doctorate from Kent State University with a major in general curriculum and a minor in geography. Dr. Hunkins has written numerous books and articles dealing with general curriculum, social studies, and questioning and thinking for students and educators at elementary, middle school, high school, and university levels.

Glencoe/McGraw-Hill

A Division of The **McGraw-Hill** Companies

Printed in the United States of America.

Send all inquiries to:
Glencoe/McGraw-Hill, 936 Eastwind Drive, Westerville, Ohio 43081

ISBN 0-02-823291-7 (Student Edition)
ISBN 0-02-823292-5 (Teacher's Wraparound Edition)

2 3 4 5 6 7 8 9 10 11 027/043 03 02 01 00 99 98 97

CONSULTANTS

General Content Consultant

NATIONAL GEOGRAPHIC SOCIETY

Multicultural Consultants

Ricardo L. García, Ed.D.
Professor
University of Wisconsin
Stevens Point, Wisconsin

Sayyid M. Syeed, Ph.D.
Secretary General, Islamic
  Society of North America
Plainfield, Indiana

Introduction to Geography

Theodore H. Schmudde, Ph.D.
Professor of Geography
University of Tennessee
Knoxville, Tennessee

Asia

Kenji K. Oshiro, Ph.D.
Professor of Geography
Wright State University
Dayton, Ohio

United States and Canada

Frederick Isele, Ph.D.
Assistant Professor
School of Education
Indiana State University
Terre Haute, Indiana

Latin America

Frank de Varona
Region Superintendent
Dade County Public Schools
Hialeah, Florida

Europe

Sarah W. Bednarz, Ph.D.
Visiting Assistant Professor
Department of Geography
Texas A&M University
College Station, Texas

Russia and the Independent Republics

A. Naklowycz
President, Ukrainian-American
  Academic Association of
  California
Carmichael, California

Alfred Bell, Ed.S.
Social Studies Supervisor
Knox County Schools
Knoxville, Tennessee

North Africa and Southwest Asia

Mounir Farah, Ph.D.
Associate Director,
Middle East Studies Program
University of Arkansas
Fayetteville, Arkansas

Africa South of the Sahara

Mary Jane Fraser, Ph.D.
Teacher and Social Studies
  Curriculum Project Leader
Seattle High School
Seattle, Washington

Australia, Oceania, and Antarctica

Marianne Kenney
Social Studies Specialist
Colorado Department of
  Education
Denver, Colorado

TEACHER REVIEWERS

Charles M. Bateman
Social Studies Teacher and
  Department Chair
Halls Middle School
Knox County District
Knoxville, Tennessee

Rebecca A. Corley
Social Studies Teacher
Evans Junior High School
Lubbock Independent School
  District
Lubbock, Texas

Carole Mayrose
Social Studies Teacher
Northview High School
Clay Community School District
Brazil, Indiana

Don Mendenhall
Social Studies Teacher and
  Department Chair
Coleman Junior High School
Van Buren, Arkansas

Rebecca M. Revis
Social Studies Teacher
Sylvan Hills Junior High
  School
Pulaski County Special School
  District
Sherwood, Arkansas

Marita E. Sesler
Social Studies Specialist
Knoxville County School
  District
Knoxville, Tennessee

Cathy Walter
World Geography Teacher
Colony Middle School
Palmer, Alaska

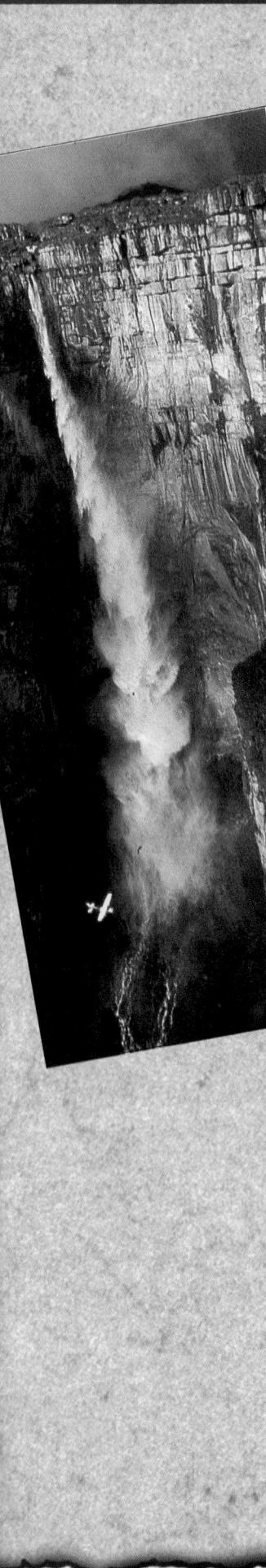

TEACHER'S WRAPAROUND EDITION

CONTENTS

BUILDING GEOGRAPHY SKILLS

MAKING CONNECTIONS

MAPS

GRAPHS, TABLES, and DIAGRAMS

NATIONAL STANDARDS IN GEOGRAPHY

Since 1994, states have been successfully implementing "standards" in a variety of school subjects, one of which is geography. The purpose of the geography standards is to provide guidance for teachers, parents, and school officials so that students can perform at internationally competitive levels as we approach the twenty-first century.

HISTORY National goals and standards in geography, published under the title *Geography for Life* (1994), were developed under *Goals 2000: Educate America Act* and were a response to evidence that students in the United States were not competitive internationally in the subject of geography. This same reality, almost a decade before, led to the preparation of the *Guidelines for Geographic Education* (1984) and the articulation of the five fundamental themes in geography: (1) location, (2) place, (3) human/environment interaction, (4) movement, and (5) region. These themes, the national standards, and current issues have become guideposts for the creative development of *Geography: The World and Its People.* The five themes appear as chapter organizers throughout the book, providing an orderly progression through geographic concepts and information, both cultural and physical.

THE 18 STANDARDS There are 18 geography standards covering the basic content, skill, and concepts of the discipline. These standards provide a framework for the geographic knowledge students should have and the skills they should be able to execute. Students will be expected to use maps, globes, and other graphic tools successfully to acquire and process geographic information. They will be expected to develop a spatial view of the world—including cultural and physical characteristics of places and regions. The standards identify, for example, critical physical and cultural processes that create patterns on the surface of the earth, such as climate, mountain-building forces, erosion, flooding, human migration, urbanization, transportation, and international trade. There is an emphasis on how people live on the earth (settlement patterns) and how people use the earth to satisfy basic needs (resources, economic development, and global interdependence). Issues-oriented geography is important in the arenas of human/environment interaction, multiculturalism, politics, and conflict and cooperation. The geography of everyday life is essential, and the well-educated student should be able to contrast how people live from one place or region to another. Finally, the standards require students to view critically the geography of past times in order to make logical decisions concerning the geography of the future.

Geography: The World and Its People is a program designed to assist students in the achievement of world-class standards in geography. The Student Edition, Teacher's Wraparound Edition, and resources have been prepared in an interesting, innovative, and realistic format that provides students with the geographic knowledge, skills, and practice they need to become informed and involved citizens in a world that is increasingly interdependent.

Physical and human phenomena are spatially distributed over Earth's surface. The outcome of *Geography for Life* is a geographically informed person (1) who sees meaning in the arrangement of things in space; (2) who sees relations between people, places, and environments; (3) who uses geographic skills; and (4) who applies spatial and ecological perspectives to life situations.

THE EIGHTEEN STANDARDS

THE WORLD IN SPATIAL TERMS

The geographically informed person knows and understands:

1 How to use maps and other geographic representations, tools, and technologies to acquire, process, and report information from a spatial perspective
2 How to use mental maps to organize information about people, places, and environments in a spatial context
3 How to analyze the spatial organization of people, places, and environments on Earth's surface

Geography studies the relationships between people, places, and environments by mapping information about them into a spatial context.

PLACES AND REGIONS

The geographically informed person knows and understands:

4 The physical and human characteristics of places
5 That people create regions to interpret Earth's complexity
6 How culture and experience influence people's perceptions of places and regions

The identities and lives of individuals and peoples are rooted in particular places and in those human constructs called regions.

PHYSICAL SYSTEMS

The geographically informed person knows and understands:

7 The physical processes that shape the patterns on Earth's surface
8 The characteristics and spatial distribution of ecosystems on Earth's surface

Physical processes shape Earth's surface and interact with plant and animal life to create, sustain, and modify ecosystems.

HUMAN SYSTEMS

The geographically informed person knows and understands:

9 The characteristics, distribution, and migration of human populations on Earth's surface
10 The characteristics, distribution, and complexity of Earth's cultural mosaics
11 The patterns and networks of economic interdependence on Earth's surface
12 The processes, patterns, and functions of human settlement
13 How the forces of cooperation and conflict among people influence the division and control of Earth's surface

People are central to geography in that human activities help shape Earth's surface, human settlements and structures are part of Earth's surface, and humans compete for control of Earth's surface.

ENVIRONMENT AND SOCIETY

The geographically informed person knows and understands:

14 How human actions modify the physical environment
15 How physical systems affect human systems
16 The changes that occur in the meaning, use, distribution, and importance of resources

The physical environment is modified by human activities, largely as a consequence of the ways in which human societies value and use Earth's natural resources, and human activities are also influenced by Earth's physical features and processes.

THE USES OF GEOGRAPHY

The geographically informed person knows and understands:

17 How to apply geography to interpret the past
18 How to apply geography to interpret the present and plan for the future

Knowledge of geography enables people to develop an understanding of the relationships between people, places, and environments over time—that is, of Earth as it was, is, and might be.

THE IMPORTANCE OF GEOGRAPHIC SKILLS

Geography for Life emphasizes what students should know and understand about geography at grades 4, 8, and 12. These national standards also suggest that students be able to master five types of skills. These skills are discussed below.

1. Asking Geographic Questions

Successful geographic inquiry involves the ability and willingness to ask, speculate on, and answer questions about why things are where they are and how they got there. Students need to be able to pose questions about their surroundings: Where is something located? Why is it there? With what is it associated? What are the consequences of its location and associations? What is this place like?

2. Acquiring Geographic Information

Geographic information is information about locations, the physical and human characteristics of those locations, and the geographic activities and conditions of the people who live in those places. To answer geographic questions, students should start by gathering information from a variety of sources in a variety of ways.

3. Organizing Geographic Information

Once collected, the geographic information should be organized and displayed in ways that help analysis and interpretation. Maps play a central role in geographic inquiry, but there are other ways to translate data into visual form, such as using different types of graphs, tables, spreadsheets, and time lines.

4. Analyzing Geographic Information

Analyzing geographic information involves seeking patterns, relationships, and connections. Students can then synthesize their observations into a coherent explanation.

Students should scrutinize maps to discover and compare spatial patterns and relationships; study tables and graphs to determine trends and relationships between and among items; and probe data through statistical methods to identify trends, sequences, correlations, and relationships.

5. Answering Geographic Questions

Successful geographic inquiry culminates in the development of generalizations and conclusions based on the data collected, organized, and analyzed. Skills associated with answering geographic questions include the ability to make inferences based on information organized in graphic form (maps, tables, graphs) and in oral and written narratives.

Adapted from *Geography for Life,* pp.42-44.

BLOCK SCHEDULING

Class scheduling is an expression of the relationship between learning and time. Traditionally, middle schools schedule six or seven 40- to 55-minute classes per day. These classes usually meet for 180 school days per school year.

Block scheduling differs from traditional scheduling in that fewer class sessions are scheduled for larger blocks of time over fewer days. For example, in block scheduling, a course might meet for 90 minutes a day for 90 days, or half a school year. Does this type of scheduling have any advantage over more traditional scheduling methods? Those schools that have tried it believe that it does.

ADVANTAGES FOR SCHOOL SYSTEMS

For the schools themselves, the greatest advantage of block scheduling is that there is a better use of resources. No additional teachers or classrooms may be needed, and more efficient use is made of those presently in school. The need for summer school is greatly reduced because the students that do not pass a course one term can take it the next term. These advantages are accompanied by an increase in the quality of teacher instruction and student's time on-task.

ADVANTAGES FOR TEACHERS

There are many advantages for teachers who are in schools that use block scheduling. Teacher-student relationships are improved. With block scheduling, teachers have responsibility for a smaller number of students at a time, so students and teachers get to know each other better. With more time, teachers are able to provide additional time and other resources for meeting the individual needs of students. Teachers can also be more focused on what they are teaching.

Block scheduling seems to result in changes in teaching approaches, classrooms that are more student centered, improved teacher morale, increased teacher effectiveness, and decreased burn-out. Teachers feel free to venture away from discussion and lecture to use more productive models of teaching.

Block scheduling cuts in half the time needed for introducing and closing classes. It also eliminates half of the time needed for class changes, which results in fewer discipline problems.

Flexibility is increased because less complex teaching schedules create more opportunities for cooperative teaching strategies such as team teaching and interdisciplinary studies. Block scheduling also increases the number of non-traditional, activity-based courses that can be offered.

ADVANTAGES FOR STUDENTS

Student success rate is found to be greater than is found with traditional scheduling because students seem to learn more and retain it better. Problem-solving skills are better developed, grades are improved, and the failure rate is lower.

Improved student-teacher relationships and a more manageable work load help students also. Students feel better about what they are learning, outside interference is reduced, and students are better able to concentrate. Generally, students feel better organized and are more aware of their progress in the class.

Many curricular advantages are also present for students. At-risk students can be scheduled for required courses during the first term. If they do not pass the course, they can repeat it during the second term instead of taking an elective. Better students can move ahead more quickly and those students who develop a late interest in certain courses can take more of them. Block scheduling has been shown to increase the number of students who take upper-level classes and earn advanced studies diplomas.

MODIFIED BLOCK SCHEDULING

Some schools use a modified form of block scheduling that combines two core classes. Under this system, students might study social studies for 90 minutes each day during the first semester and science during the second semester. Another modification has students take English and social studies blocks in one semester and science and mathematics blocks the second semester.

Block Scheduling Implementation Guide

GEOGRAPHY
THE WORLD AND ITS PEOPLE

NATIONAL
GEOGRAPHIC
SOCIETY

Boehm
Armstrong
Hunkins

ALTERNATIVE COURSE OUTLINES

1

WESTERN HEMISPHERE (9 CHAPTERS)

Unit 1	Geography of the World	Chapters 1, 2, 3
Unit 2	The United States and Canada	Chapters 4, 5
Unit 3	Latin America	Chapters 6, 7, 8, 9

2

EASTERN HEMISPHERE (21 CHAPTERS)

Unit 1	Geography of the World	Chapters 1, 2, 3
Unit 4	Europe	Chapters 10, 11, 12, 13
Unit 5	Russia and the Independent Republics	Chapters 14, 15
Unit 6	Southwest Asia and North Africa	Chapters 16, 17
Unit 7	Africa South of the Sahara	Chapters 18, 19, 20, 21
Unit 8	Asia	Chapters 22, 23, 24, 25
Unit 9	Australia, Oceania, and Antarctica	Chapters 26, 27

3

ENVIRONMENTAL GEOGRAPHY (11 CHAPTERS)

Unit 1	Geography of the World	Chapters 1, 2, 3
Unit 2	The United States and Canada	Chapter 5
Unit 3	Latin America	Chapter 9
Unit 4	Europe	Chapter 13
Unit 5	Russia and the Independent Republics	Chapter 15
Unit 6	Southwest Asia and North Africa	Chapter 17
Unit 7	Africa South of the Sahara	Chapter 21
Unit 8	Asia	Chapter 25
Unit 9	Australia, Oceania, and Antarctica	Chapter 27

TECHNOLOGY IN THE CLASSROOM

TECHNOLOGY IN THE CLASSROOM

Advances in technology are continually being made, and these advances dramatically affect all aspects of the social studies. Social studies instruction should include an awareness of advances in technology. Glencoe has developed a number of programs designed for your use in the classroom.

SOFTWARE

Testmaker The Testmaker that Glencoe has developed allows you to customize tests to fit the special needs of your students. These software programs are available in Macintosh, Apple, and IBM formats. The designers of the software have constructed it so that you are free to edit the questions available, mix questions as you wish, and add questions to meet your particular needs.

Student Self-Test: A Software Review This highly motivational program allows students to check their comprehension by answering multiple-choice questions directly at the computer. If a student chooses a wrong answer, the computer explains why the choice is incorrect. The student then may try again. This process continues until the student chooses the correct answer. It is through this process that the student receives immediate feedback to reinforce correct responses.

Vocabulary PuzzleMaker Because vocabulary development and comprehension is such an important part of social studies instruction, Glencoe has developed a special program for vocabulary. The Vocabulary PuzzleMaker allows teachers to create high-interest crossword puzzles and word searches covering the textbook's vocabulary program.

Using Software With Limited Hardware

As with many other types of instructional tools, the more equipment available, the more flexibility your program can have. The ideal situation is to have one computer for each member of the class. You also might have access several times a week to a school computer lab. If either of these situations exists, you have unlimited opportunities for providing a wide range of computer activities. It is possible, however, to provide some computer activities with limited computer hardware. This can be accomplished through student rotation and grouping.

Computer Training for Teachers

You must have a general knowledge of computer operation, but you do not have to have a knowledge of internal hardware elements or programming languages. Resources exist in every school or district for help in using troubleshooting equipment. These resources include other computer teachers, lab assistants, and technology supervisors. Concentrate on becoming familiar with the operation of your software programs. Remember that software hotlines are one of your best resources when using any type of software. Glencoe's software HOTLINE is 1-800-437-3715.

MULTIMEDIA

As a social studies teacher, you know that your students have varying learning styles. Some students are print-dominant learners; others are non-print-dominant learners. To help you accommodate these varying learning styles and to integrate listening, hands-on application, and more frequent stimulation and interactivity into the classroom, Glencoe offers a variety of multimedia.

LASERDISCS

If your school has a basic system consisting of a videodisc player and a television receiver, the laserdiscs provide an effective and interesting tool for classroom presentations. Students can see the connection between concepts and the real world. Students are also given opportunities to apply what they have learned.

USING THE *inter*NET

THE INTERNET

The Internet (or the Net) is the collective name for the connections among computers throughout the world. Computers can communicate with one another because they follow a single standard for passing information back and forth. The Internet has existed for about 25 years, originating as a military network and then expanding into the commercial realm.

Think of the Internet as similar to using the telephone. If you know the name of the computer you want to contact, you look up its number, have your computer dial it, and talk to it. Sending messages from person to person is one of the Net's most popular uses. You can reach a specific person by using E-mail, or you can join a conference call by subscribing to bulletin boards and listservers. For example, you can join the listserver for people interested in geographic education by sending the message **subscribe ge-oged Your Name** to listserv@ukcc.uky.edu. The Internet, however, is used for much more than just getting and sending messages.

WORLD WIDE WEB

Getting information through the Internet became much easier with the development of the World Wide Web. The Web is an exciting region of the Internet that contains pictures, sound, and video as well as text.

Navigating the Web

You can scan the Web using "browsers"—programs such as Netscape and Mosaic. This software can navigate the Web for you, meaning that you no longer have to know the number of the computer you want. You don't even have to know where the information you want is.

Libraries have electronic catalogues that allow you to search not only by the name of the author, but also by the name of the book, topic, keyword, and so on. Similarly, there are several ways to find what you want on the Web. You can go directly to the information if you know its **Uniform Reference Locator (URL),** or address. Every photograph, map, diagram, and text file has one of these unique addresses, and your browser can keep track of these sources as "bookmarks."

Searches

Each site mentioned below is followed by its URL. You can also use the "search engines" that are incorporated into the program you use to access the Web. Or, use another service such as Yahoo (http://www.yahoo.com/). These services work just like the electronic catalogue in a library, except that when you find the reference you want, you simply point and click and the information appears on your screen. Often, it is faster to use a search engine than to type in the URL, especially if the URL is particularly long.

The best part of the Web is that you can print and copy everything you find. If you or your school has a single computer connected to the Internet, you can print materials for your class or save files for later use. Access to a relatively inexpensive color ink jet printer allows you to print directly onto overhead transparencies.

School districts are establishing connections to the Internet at a rapid rate. One option is to join the Internet through a local university. Another option is to subscribe to a private provider such as CompuServe, America Online, or Microsoft Network. These private providers let you use the Web for a monthly fee or at an hourly rate.

SURFING THE GEOGRAPHY NET

There is a wealth of on-line resources for geography teachers. Maps, images, data, and other information can be accessed from the following list. When you type the address, you will reach the organization's "homepage." This is the best place to start if you are browsing to find what is available.

- **Library of Congress**
  http://lcweb.loc.gov/homepage/lchp.html
- **United States Geological Survey**
  http://www.usgs.gov/
- **Bureau of the Census**
  http://www.census.gov/
- **NASA**
  http://www.nasa.gov/
- **National Oceanic and Atmospheric Administration**
  http://www.noaa.gov/
- **Environmental Protection Agency**
  http://www.epa.gov/
- **CNN**
  http://www.cnn.com/
- **WorldPop** is a counter showing the world's estimated population at any given moment and is updated constantly. Depending on the software you are using, you can watch the counter tick at a brisk pace.
  http://sunsite.unc.edu/lunarbin/worldpop
- **The CIA** has the 1996 World Factbook on the Web, which is indexed by country and also contains reference maps.
  http://www.odci.gov/cia/publications/96fact/index.html
- You can get up-to-date information on **earthquakes**:
  http:www.civeng.carleton.ca/cgi-bin/quakes
- **The Jason Project** is an interactive, electronic field trip, complete with a teacher's guide.
  http:seawifs.gsfc.nasa.gov/scripts/JASON.html

- Weather images are available from several different sources. Try the **Purdue University Weather Gopher**.
  gopher://thunder.atms.purdue.edu/

  or the **University of Illinois Weather World**
  http://www.atmos.uiuc.edu/wxworld/html/detailed.html

- **The University of Texas** has an on-line map collection with maps of just about everywhere.
  http://www.lib.utexas.edu/Libs/PCL/Map_collection/Map_collection.html

Increasingly, you no longer have to look up each address individually. Many sites will provide you with links to other sites, so you can travel the world by pointing and clicking. Or you can visit several sites that are clearinghouses for information:

- **Armadillo from Rice University**
  http://chico.rice.edu/armadillo/about.html
- **ERIC Clearinghouse**
  http://www.cua.edu/www/ericae/home.html
- **Joint Education Initiative at Maryland**
  http://jei.umd.edu/

In order to reach all of your students and make their learning experiences the most rewarding for their individual needs, Geography: The World and Its People provides a wealth of instructional approaches and strategies.

ADDRESSING DIFFERENT LEARNING STYLES

VISUAL LEARNERS

Visual learners benefit the most when they can carefully look at the material to be studied. In general, visual learners retain more information if they are able to visualize what they are learning. The exercise of mental mapping used in this text is designed to heighten the learning of the visual student. These students also benefit from laserdisc presentations as well as from CD-ROMs. Visual learners benefit, too, from reading the text and studying the maps, charts, graphs, and other visuals.

AUDITORY LEARNERS

Auditory learners retain the most information when they hear what they are to learn. Oral instructions from the teacher are ideal ways to in-troduce these learners to new concepts. Technology plays an essential role in helping these learners master the course content. Audiocassette transcripts of chapters or lessons provide these students with invaluable learning aids. These students also benefit from laserdisc presentations because the soundtracks help them comprehend the information with a higher retention rate. CD-ROMs also provide students with instant auditory directions and feedback.

KINESTHETIC LEARNERS

Kinesthetic learners retain information more easily when they can actually perform basic tasks using the information. For these students, individual and group projects in which they construct models, charts, or graphs are ideal.

COOPERATIVE LEARNING

Although cooperative learning is a useful teaching strategy in many subjects, it occupies a special place in the social studies curriculum because of its success in imparting the abilities needed to work effectively in a group. Such social studies skills are beneficial for all citizens working in a democracy.

Characteristics of Cooperative Learning

Cooperative Learning requires careful monitoring and structure by the teacher if it is to be more than a simple group activity. Characteristics of Cooperative Learning include the following:

• Students work face to face in heterogeneous groups.

• The activity promotes a sense of positive interdependence.
• Each member of a group has individual accountability.
• The group has a common product or goal.

The Role of the Teacher

Although successful Cooperative Learning groups appear to work independently, this is due to the coaching of a good teacher. Students will need the teacher's help at key moments during the group's project—in agreeing upon goals, in establishing a structure of accountability, and so on.

AT-RISK STUDENTS

Most educators today agree that the nation's schools are facing an epidemic of students who are at risk of failure. It is difficult to define exactly what constitutes an at-risk student because being at risk is often linked to several environmental causes such as poverty, low self-esteem, substance abuse, or pregnancy. Current educational research has shown that certain teaching methods can help keep at-risk students from dropping out. One method is to maximize time-on-task to help students overcome distracting outside stimuli.

Another method is to establish high expectations and a school climate that supports learning. Many school activities involve parents in this process so that the expectations for success are not left inside the classroom after school is out. Many teachers give positive feedback at the end of each successfully completed assignment and include awards ceremonies for students who meet expectations.

Rather than emphasizing remedial techniques, many educators believe that at-risk students need to learn at a faster rate. Instruction emphasizes assets that at-risk students often bring to the classroom—interest in oral and artistic expression and kinesthetic learning abilities. For example, at-risk students may excel at dramatizations in which they also construct the sets.

MULTICULTURAL PERSPECTIVE

Geography: The World and Its People furnishes a wealth of material that can help students learn to appreciate the multicultural diversity of the world's peoples. By reading *Geography: The World and Its People,* students receive a balanced, broad view of the interaction of people with their physical and cultural environments, both historically and present-day in today's global village. As the world shrinks in size because of high-speed communication and transportation, it becomes increasingly important for students to see peoples different from themselves as interesting neighbors who have different ideas, customs, and languages, but who also share many of the same values. Students will be given the opportunity to read about groups of people who have

been misrepresented or omitted in the past. Inclusion of these groups will help all students develop more positive attitudes toward different cultural, racial, ethnic, religious, and gender groups.

The following five points have been identified as some of the major goals of multicultural education:

- Promoting the strength and value of cultural diversity
- Promoting human rights and respect for those who are different from oneself
- Promoting alternative life choices for people
- Promoting social justice and equal opportunity for all people
- Promoting equity in the distribution of power among groups

CRITICAL THINKING

To learn about physical and cultural geography in a way that prepares students to become thoughtful participants in this world, students must learn to think critically. They need to be able to evaluate and to question the meaning of what they see, read, and hear. The teacher plays a crucial role in this development by creating a classroom climate that actively encourages critical thinking. *Geography: The World and Its People* teaches the skills used in critical thinking.

THE CLASSROOM CLIMATE

The teacher can promote critical thinking in the classroom by verbalizing the inner thought

processes that take place. Asking questions such as "What do I want to achieve?" and "What do I already know?" models for students the importance of setting goals and of assessing current knowledge. Asking "Have I understood what I have read?" establishes the importance of checking one's progress.

CRITICAL THINKING SKILLS

Map, photograph, and graphic captions also call upon students to use critical thinking skills as applied to visual interpretation. Questions requiring critical thinking skills appear in Chapter Reviews.

Assessment is a means of identifying the degree to which students have learned the objectives and goals for the unit and duration of study.

PERFORMANCE ASSESSMENT STRATEGIES

Assessment is a means of identifying the degree to which students have learned the objectives and goals for the unit and duration of study. For assessment to be considered valid, there must be a match among the expectations of the objectives, the instructional experiences offered by the teacher, and the type of assessment item. While many objectives call for higher order thinking in predicting, analyzing, or applying, the assessment items ask only recall on the part of the student. Objectives may also call for the student to come to conceptual understanding, but only content may be measured. The instructional experiences may have met the expectations of the objectives, but the assessment did not. While factual content and some forms of thinking skills may be measured by multiple-choice or fill-in-the-blank formats, conceptual understanding and the highest levels of thinking must be measured by alternative means. These alternative means are often known as "performance assessments" since they require an evidence of "knowing" through "doing" on the part of the student. Alternative assessments often require actions by the student that may be observed by the teacher or may result in a tangible product that may be submitted for assessment.

PERFORMANCE MEASURES

The alternative measures most often utilized by classroom teachers are projects and investigations, teacher observation, performance-based essays, student interviews, and portfolios. Authentic assessment takes these means even further to place the student in a real or simulated scenario, in order to find the extent to which he/she would use the expected competencies in "real life" and the degree to which progress has been made. These scenarios, particularly in the area of geography, might include roles in the world of work, roles involving use of leisure time, roles in problematical situations, or roles demonstrated through participation in one's community or beyond. The best types of tasks are formulated around a broad or complex thinking ability—such as decision making, problem solving, creative thinking, persuasion and argumentation, and predicting or forecasting—all of which will be required of students after they leave the classroom.

Tasks may be divided into three broad types: teacher-directed tasks, student-directed tasks, and collections over time. In teacher-directed tasks students may be guided step-by-step through various phases of the task requiring differing abilities. It is often beneficial to begin with teacher-directed tasks. An example of this type of task would be one in which the teacher leads the class in brainstorming about ideas and formulating initial hypotheses, but when students realize that they do not have enough information in order to be successful, the teacher guides them to the "acquiring information" phase which takes the place of the typical lecture approach. The approach is student-centered rather than teacher-centered, with the teacher becoming the coach or facilitator. The benefits of the teacher-directed task are that the students themselves set a goal for their own learning and will actually use what they learned.

Student-directed tasks are typically used at the end of a unit of study in the place of or in concert with traditional tests. In these tasks, the teacher explains clearly what is expected of the students in process and product and makes known the criteria on which they will be graded or otherwise assessed.

Collections over time require longer periods to develop, but are helpful in allowing the student to discover a broader application for what is learned than can be accomplished in a one-time task. Collections often include news stories and articles, evidence found in products, literature, music, or other elements of culture, and samples of student applications which have been performed throughout the unit of study.

USING PERFORMANCE ASSESSMENT WITH THIS TEXT

Throughout this text, suggestions of performance tasks are supplied for each chapter. The teacher will find that they exemplify the various formats for alternative assessment and the types of tasks described above.

Assessing Performance Tasks

Student performance can be measured by several different means. The most common forms of assessing performance tasks are by rubric or percent. By using percents, the teacher is able to weight certain competencies over others. In this form, the teacher builds the total score by adding the percents gained by the student on each required quality.

If the teacher desires to assign a range of numerical scores for the product as a whole, then a rubric is required. Most rubrics include three, four, or six points beyond zero. A score of zero is almost always indicative of a task which, for one reason or another, is nonscorable. The scorable points reflect the range of success for multiple qualities and skills required of the product and the student. A three-point rubric usually ranges from a Score Point One as *unacceptable* to Score Point Three as *completely successful* with Score Point Two usually carrying terms such as *some, general, average,* or *with errors.*

In establishing a rubric, the teacher should follow five simple steps:

1. Divide the task into its component parts.
2. Decide on characteristics required of the product.
3. Determine a range of abilities for each task. Use as specific language as possible.
4. Compile in list form all of the descriptors you developed for each score point.
5. Present and explain the rubric to the class before the task is begun. Be sure students are aware that a holistic impression always exists since no one task will usually exhibit all the descriptive qualities at a given score point.

JOURNALS

The Journal approach to assessment is often used with performance-based assessment as part of an overall approach to authentic assessment. Journals contain samples of students' work collected over a period of time—often an entire grading period or even a semester.

The GeoJournal Activity, placed at the beginning of every unit of *Geography: The World and Its People,* provides students an opportunity not only to find information but also to make sense of it and put that information in context. The journal activity is an excellent way to introduce students to the unit and can be a valuable tool for them during the course of the unit. The journal may be used in several ways—you may assign a percentage of the course grade to the journal writing activity listed in the Teacher's Wraparound Edition, you may use the journal activity as a basis for class discussion, or you may wish to give students other assignments to be completed in their journals. You may give students options in how they complete their journal assignments, or in the topics they choose. Journals may also be used for students' reflections about what they are learning and how it affects them. If you wish, the journal may be used as an ongoing dialogue between teacher and student. The GeoJournal Activity and the follow-up writing assignment found in every Chapter Review may be used in a portfolio approach.

EXAMPLE RUBRIC

Score Point 4
- Complete understanding of the concept of interdependence
- No major errors in content information
- Clear, specific oral communication of ideas
- Completely plausible predictions made
- Full understanding of the problem-solving process
- Highly effective participation in group discussions
- Evidence of extensive research

Score Point 3
- Minor misunderstandings of the concept of interdependence
- Minor errors in content information
- Generally clear communication
- Some steps in problem solving less developed than others
- Generally effective participation in group discussions but minor lapses may have occurred
- Enough research to support ideas

Score Point 2
- General understanding of the concept of interdependence
- Several errors in content information
- Oral communication sometimes unclear
- Some steps in problem solving omitted
- Occasional participation in group discussion
- Some relevant research

Score Point 1
- Little or no understanding of the concept
- Major errors in content information
- Unclear, brief, or vague oral communication
- Little or no understanding of the process of problem solving
- Little or no participation in group discussions
- Brief, general or irrelevant research

For the task accompanying the chapters in this text, possible rubric features are listed for each task.

ENRICHING CLASSROOM INSTRUCTION

Quality education in the field of geography demands teaching and learning that are both innovative and participatory.

Quality education in the field of geography demands teaching and learning that are both innovative and participatory. Because we are surrounded by geography in all its forms, active teaching and learning involve the use of all senses. In this way the enrichment process is heightened.

National standards in geography suggest not only what students should know, but also what they can do with their acquired knowledge. A variety of skills enable students to observe and map geographical phenomena, ask appropriate questions, and analyze and suggest answers to problems with geographic dimensions.

An effective model for enriched geographic education should include simulations, field experiences, and the use of resource persons. While these categories differ in focus, a general model can serve as a checklist for the appropriate implementation of each. Following are some suggestions for the use of classroom extensions for geographic learning:

SIMULATIONS

Taking on different roles is often an effective means of learning by simulating a current event with spatial dimensions. For example, two- and three-student teams may represent heads of state and their staffs as they attend an international conference to discuss topics such as global warming, deforestation, or the pollution of the oceans. In another situation, students might represent a town's mayor, members of the planning and zoning commission, businesspersons, and citizens as they debate the importance of a new convention center and the resulting impact on the community. Finally, the class could assume the roles of neighborhood representatives and social workers as they debate the location of a youth services bureau.

FIELD EXPERIENCES

Begin with a walk through the neighborhood, gathering data with which to create a map of the area. Use symbols to mark the location of businesses, residences, traffic movement, parks, and unusual places. A more extensive field trip may involve traveling to a sanitary landfill, a water purification plant, an electrical utility, a newspaper office, or a supermarket (to search for foods grown and processed in various countries). To collect data about interstate travel, commerce, or tourism, students may count state license plates on automobiles at a busy intersection and then plot the appropriate states on a map. For study at home, students could work with their parents or guardians in searching for items that were made in countries other than the United States. The students could then map the countries in which these items originated.

RESOURCE PERSONS

As we become increasingly globally interdependent, more and more individuals are traveling to countries around the globe. Typically these world travelers are willing and eager to share comments about their experiences, along with pictures or slides of the areas they have seen. In addition to presenting material, these women and men may also interact with students, enhancing the experience for both.

Each of these three types of activities are valuable in and of themselves, and when implemented in conjunction with one another, provide for an integrated learning experience. As a means of enhancing this teaching and learning experience, preparation cannot be overlooked. The use of flow charts, accurately ordering the process from conception to debriefing and extension of the activity, is essential to the successful outcome of any activity.

The following general model will provide an outline of the teaching and learning enrichment activity.

MODEL FOR ENRICHMENT

Planning — Physical planning including equipment, transportation, materials, etc., necessary for the enriching event

Learning Outcomes — Goals, objectives, links to classroom activities, purposes, and geographical focus

Pre-Event Encouragement — Draws on prior knowledge of students. Could include a pretest or list of focusing ideas. Background information may include research assignment for students.

The Event — Should be accompanied by a guide of some kind including descriptive information. In the case of a simulation, roles are profiled and discussed.

Student Activity —
Note taking
Journals
Profiles of roles
Written reports

Mental mapping
Skills development
Reports (written and verbal)

Debrief —
Post-event discussion
Focus on the geographical (spatial) aspects of the event
Analyze skills used
Analyze problems

Extending Activity — Develop extending activities using the library, textbook, homework, parents, peers, and/or other enriching activities

PARTICIPATING IN THE NATIONAL GEOGRAPHY BEE

NATIONAL GEOGRAPHIC SOCIETY The National Geography Bee is a nation-wide contest for schools with any grades four through eight. It is an educational outreach program of the National Geographic Society. With a first-place prize of a $25,000 college scholarship—and other prizes in additional scholarships, cash, and classroom materials—the Bee is designed to encourage the teaching and study of geography. Students from schools in all 50 states, the District of Columbia, Guam, Puerto Rico, American Samoa, the Northern Mariana Islands, the U.S. Virgin Islands, and Department of Defense Dependents Schools take part in the National Geography Bee. Principals of eligible schools must register their schools to participate in the Bee. Principals may request registration by writing to National Geography Bee, National Geographic Society, 1145 17th Street N.W., Washington, D.C. 20036-4688, and including the $20 fee. The deadline for registering is October 15. Please check with the principal of your school to see if the school is registered.

GEO BEE, a software game, helps students ages 10 and older improve their knowledge of world geography. The game can be played alone, or with as many as four players. There are three rounds in each game. The questions are drawn from a database of over 3,000 National Geography Bee questions, and cover physical, cultural, political, and economic geography.

TEACHER'S CLASSROOM RESOURCES

The Teacher's Classroom Resources provide you with a wide variety of supplemental materials to enhance the classroom experience. These resources meet your teaching needs and your students' abilities.

Name _____ Date _____ Class _____

CHAPTER 5 Geography Skills Activity

Reading a Physical Map—British Columbia

The Rocky Mountains extends north into British Columbia. This beautiful Canadian province has many physical features that distinguish it from the rest of Canada.

DIRECTIONS: Study the map of British Columbia and then answer the questions below on a separate sheet of paper.

1. Which are higher, the Rocky Mountains or the Cariboo Mountains?

2. Name four mountain ranges in British Columbia.

3. What are the two largest island areas off the coast of British Columbia?

Critical Thinking

4. Why do you think Vancouver is British Columbia's major port city on the Pacific Ocean?

Activity

5. Choose a location in Canada that you think would be interesting to visit. Gather information and pictures from magazines. Then make a travel brochure to interest others in visiting the location you have chosen.

BRITISH COLUMBIA: Physical

Glencoe/McGraw-Hill.

135°W 130°W 125°W 120°W 115°W

YUKON TERRITORY

NORTHWEST TERRITORIES

ELEVATIONS

Feet	Meters
10,000	3,000
5,000	1,500
2,000	600
1,000	300
0	0

60°N

Churchill Peak 10,500 ft. (3,200 m)

▲ Mountain peak

Conic projection

ALASKA (U.S.)

COAST MOUNTAINS

STIKINE RANGES

R O C K Y   M O U N T A I N S

55°N

ALBERTA

Prince Rupert

Mt. Robson 12,972 ft. (3,954 m)

0   100   200 mi.
0   100   200 km

Queen Charlotte Islands

CARIBOO MTS.

PACIFIC OCEAN

Mt. Waddington 13,260 ft. (4,042 m)

Kamloops

50°N

Vancouver Island

Vancouver

Victoria

UNITED STATES

Geography: The World and Its People ⑤ **Geography Skills**

Geography Skills Activities offer practice of basic graphic skills involving maps, charts, diagrams, and graphs.

Actual size is 8 1/2" × 11"

Chapter Map Activities
provide—on a chapter basis—
an outline map, political map,
physical map, and political map
with capitals and major cities.

Actual size is 8 1/2" × 11"

Actual size is 8 ½" × 11"

Lesson 6   Living With Environmental Hazards

OBJECTIVE

Be aware of danger to life and property from environmental hazards such as hurricanes, tornadoes, and lightning

TERMS TO KNOW

hurricane (HER·uh·kayn)—largest storm in nature

hurricane warning (HER·uh·kayn WAR·ning)—notice that a hurricane is expected to strike a particular location within 24 hours

hurricane watch (HER·uh·kayn wach)—notice that a hurricane is within 24 hours of striking somewhere

earthquake (ERTH·kwayk)—strong shaking of the earth

lightning (LYT·ning)—electricity passing between a cloud and the ground

tornado (tor·NAY·doh)—most violent storm in nature

The sound may be hard to hear at first over the pounding of hail on the roof, rain lashing at the windows, or thunder and lightning. People who have heard the sound say it is like the buzzing of 10 million bees. Others say it sounds like a train running through the living room, or a jet plane landing on the roof. Once you have heard it, you will never forget it. It is the sound of a tornado about to strike.

Tornado Safety

A **tornado** is the most violent storm in nature. It can destroy anything in its path. Tornadoes can happen anywhere in the United States. However, they are most common in the United States east of the Rocky Mountains. The peak months for tornadoes are April, May, and June, but they can occur in any month. Map 3–4 shows the months of peak tornado activity in the United States.

The only safe place in a tornado is a strong underground shelter. Second best is the corner of a basement *toward the tornado*. Most injuries in tornadoes are from flying objects such as glass and lumber. If you are in the corner toward the tornado, most objects will be blown over and away from you. If the building has no basement, go to a room, hallway, or closet *in the center of the house*. Get under a piece of furniture or pull a mattress over you.

People once thought that tornadoes. It was thought th would keep the air pressure and save the house. *Opening recommended*. We now kno blow houses down.

If you are in a car outside to outrun a tornado. Drive a direction the tornado is trav best to take shelter. If you a *shelter*. If many people try to traffic jam can trap you in it most dangerous places you c

Outdoors, look for a place strong building around, hide

Lightning Safety

Tornadoes are usually bor thunderstorms. Most thunde tornadoes, but all have **light** electricity passing between is very dangerous. On averag by lightning each year in the any other weather event. Lig forest fires each year.

April-May
June-July
July
June
June
April
July
June-July
Jul-Aug
Jul-Aug
June
April
June
June
May
April
April-July
July
July
May
June
April
April
July
May
June
April
May
May
April
April-May
July-August
May-June
May
April
Feb. Apr. & Nov.
March-April
April
April-May
May
April & Nov.
June

MONTHS OF PEAK TORNADO ACTIVITY

Building Skills in Geography is a text-workbook based upon the five geographical themes.

Actual size is 8 ½" × 11"

Glencoe Social Studies Outline Map Resource Book provides a variety of updated, reproducible regional outline maps.

The **Laminated Desk Map** allows individual, hands-on practice and place location.

Facts On File: MAPS ON FILE I, MAPS ON FILE II, and GEOGRAPHY ON FILE offer excellent teaching and resource materials that include a wide selection of both regional and individual country maps, as well as special purpose maps and statistical information.

Name_____ Date _____ Class _____

CHAPTER 13 Guided Reading Activity

13-4 The Czech Republic and Slovakia

DIRECTIONS: Outlining Reading the section and completing the outline below will help you learn more about the Czech Republic and Slovakia. Use your textbook to fill in the blanks.

I. Czech Republic and Slovakia

A. In past, under _____ control as country of _____

B. In _____, peacefully ended communist rule

C. In 1992 split into Slovakia and

_____ _____

II. Czech Republic—capital city is

_____

A. East of _____; south of

_____

B. In west and north, _____

with spas and nature _____

C. In south and east, rolling hills and low

fertile _____

1. Farming and major _____ centers

2. Rivers such as the _____

and _____

D. Under communism, economy based on

_____ _____

E. Today's economy is free

_____ system; industries

F. People—most belong to

_____ ethnic group called Czechs

1. Most live in cities and _____

2. Have _____ standard of living

III. Slovakia—capital city is _____

A. To the _____ of Czech Republic

B. The _____ Mountains cover most of the north

C. The south has farms and

_____ on lowlands and plains

D. Economy mainly based on

_____; increasing service industries

E. _____ are largest ethnic

group; _____ second largest

Guided Reading Activities help students master chapter information as they read the text.

Actual size is 8 ¹/2" × 11"

Name_____ Date _____ Class _____

CHAPTER **7** Vocabulary Activity

vo•cab•u•lar•y vo-'kab-y
1: a list or collection of wo
alphabetically arranged

Central America and the West Indies: Words to Know

DIRECTIONS: Fill in the Blanks Select one of the following terms to complete each of the sentences below.

> Vocabulary Activities feature the terms and content vocabulary necessary for comprehension of chapter material.

hurricane
nutrient
literacy rate
colony
light industry

plantation
bauxite
isthmus
communism
dialect

chicle
ladino
archipelago
cooperative
commonwealth

1. Workers mine _____ for export to factories that make aluminum.

2. A high _____ reveals that most of a country's people can read.

3. In Cuba, under the system of government called _____ , farmers may work on

a government run _____ .

4. A strong _____ ripped through the islands, uprooting trees and flooding highways.

5. Crops grown on a large _____ may include tobacco, sugarcane, coffee, and bananas to sell.

6. A mineral supplying food to plants is called a _____ .

7. Rain forest workers gather _____ for making chewing gum.

8. A _____ is a Guatemalan who speaks Spanish and follows European ways.

9. In 1952 Puerto Rico became a _____ , or partly self-ruling territory; in earlier

times, it had been a _____ of Spain.

10. From the thin _____ , bordered on each side by a glistening ocean, we could

see some of the islands in the _____ .

11. The woman spoke a _____ that sounded rhythmic.

12. A _____ may produce food products, cigars, or household goods.

Geography: The World and Its People   ⑦   **Vocabulary**

Also Available in Spanish

Actual size is 8 1/2" × 11"

Name________________________ Date ____________ Class ________

CHAPTER 3 Reteaching Activity

The World's People

Geographers study Earth as the home of people. They look at similarities and differences among cultures. They analyze population patterns in different parts of the world. They also look at how people use natural resources to improve their lives.

DIRECTIONS: Making a Table Each word or phrase listed below refers to culture, population, or resources. Write each word or phrase on a line in the correct table.

Reteaching Activities help students focus on the main ideas and themes presented in each chapter.

THE WORLD'S PEOPLE

Culture	Population	Resources
________	________	________
________	________	________
________	________	________
________	________	________

- birthrate
- language service industries
- free enterprise
- subsistence farming
- urbanization
- democracy
- demographers
- solar energy
- fossil fuels
- developing country
- literacy rate

Also Available in Spanish

Actual size is 8 1/2" × 11"

EVALUATION

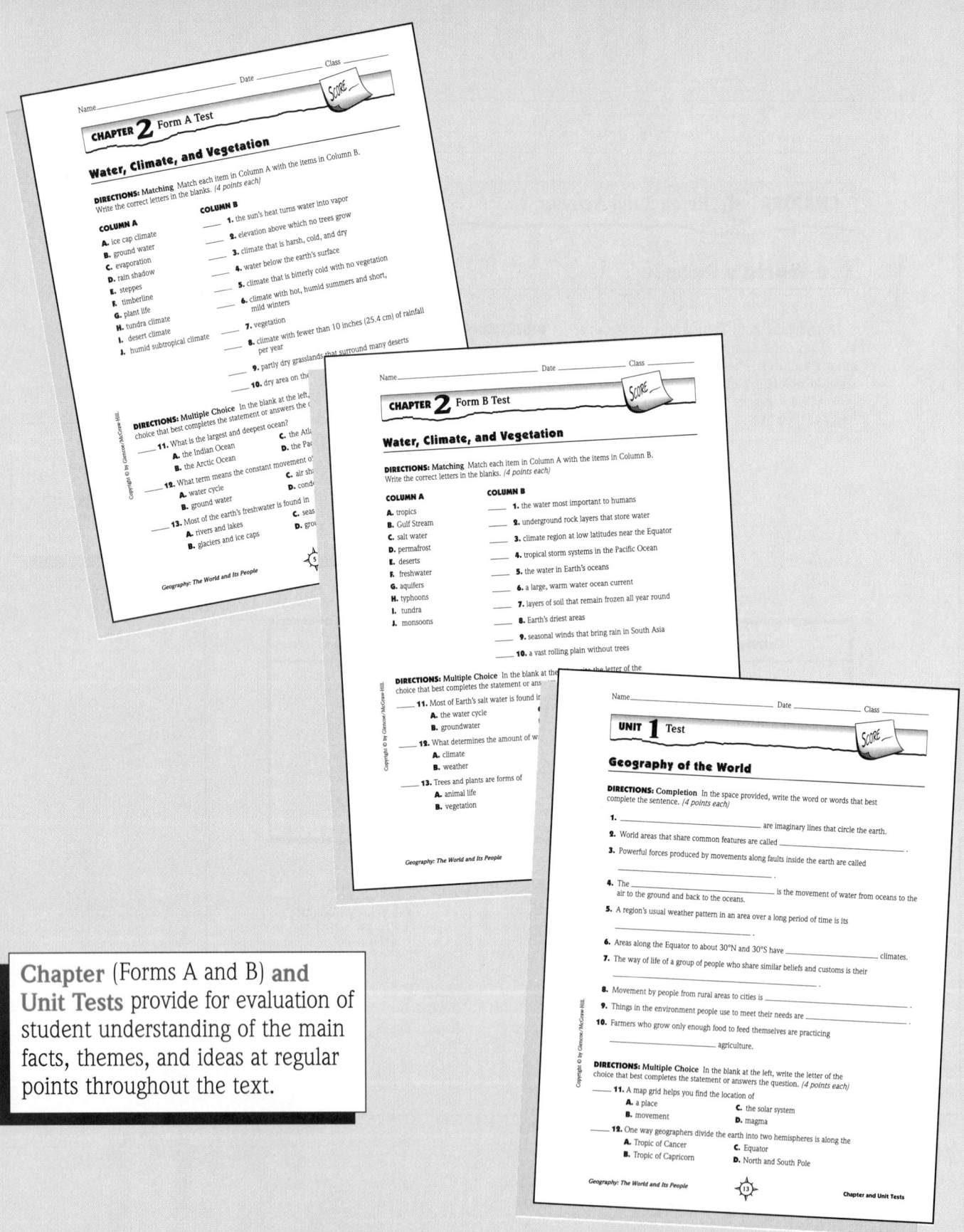

CHAPTER 2 Form A Test

Water, Climate, and Vegetation

DIRECTIONS: Matching Match each item in Column A with the items in Column B.
Write the correct letters in the blanks. *(4 points each)*

COLUMN A

A. ice cap climate
B. ground water
C. evaporation
D. rain shadow
E. steppes
F. timberline
G. plant life
H. tundra climate
I. desert climate
J. humid subtropical climate

COLUMN B

_____ **1.** the sun's heat turns water into vapor
_____ **2.** elevation above which no trees grow
_____ **3.** climate that is harsh, cold, and dry
_____ **4.** water below the earth's surface
_____ **5.** climate that is bitterly cold with no vegetation
_____ **6.** climate with hot, humid summers and short, mild winters
_____ **7.** vegetation
_____ **8.** climate with fewer than 10 inches (25.4 cm) of rainfall per year
_____ **9.** partly dry grasslands that surround many deserts
_____ **10.** dry area on the

DIRECTIONS: Multiple Choice In the blank at the left,
choice that best completes the statement or answers the

_____ **11.** What is the largest and deepest ocean?
    A. the Indian Ocean   C. the Atl
    B. the Arctic Ocean   D. the Pac

_____ **12.** What term means the constant movement of
    A. water cycle   C. air sh
    B. ground water   D. cond

_____ **13.** Most of the earth's freshwater is found in
    A. rivers and lakes   C. seas
    B. glaciers and ice caps   D. gro

Geography: The World and Its People

CHAPTER 2 Form B Test

Water, Climate, and Vegetation

DIRECTIONS: Matching Match each item in Column A with the items in Column B.
Write the correct letters in the blanks. *(4 points each)*

COLUMN A

A. tropics
B. Gulf Stream
C. salt water
D. permafrost
E. deserts
F. freshwater
G. aquifers
H. typhoons
I. tundra
J. monsoons

COLUMN B

_____ **1.** the water most important to humans
_____ **2.** underground rock layers that store water
_____ **3.** climate region at low latitudes near the Equator
_____ **4.** tropical storm systems in the Pacific Ocean
_____ **5.** the water in Earth's oceans
_____ **6.** a large, warm water ocean current
_____ **7.** layers of soil that remain frozen all year round
_____ **8.** Earth's driest areas
_____ **9.** seasonal winds that bring rain in South Asia
_____ **10.** a vast rolling plain without trees

DIRECTIONS: Multiple Choice In the blank at the   the letter of the
choice that best completes the statement or ans

_____ **11.** Most of Earth's salt water is found in
    A. the water cycle
    B. groundwater

_____ **12.** What determines the amount of w
    A. climate
    B. weather

_____ **13.** Trees and plants are forms of
    A. animal life
    B. vegetation

Geography: The World and Its People

UNIT 1 Test

Geography of the World

DIRECTIONS: Completion In the space provided, write the word or words that best
complete the sentence. *(4 points each)*

1. _____ are imaginary lines that circle the earth.

2. World areas that share common features are called _____ .

3. Powerful forces produced by movements along faults inside the earth are called
_____ .

4. The _____ is the movement of water from oceans to the
air to the ground and back to the oceans.

5. A region's usual weather pattern in an area over a long period of time is its
_____

6. Areas along the Equator to about 30°N and 30°S have _____ climates.

7. The way of life of a group of people who share similar beliefs and customs is their
_____

8. Movement by people from rural areas to cities is _____

9. Things in the environment people use to meet their needs are _____ .

10. Farmers who grow only enough food to feed themselves are practicing
_____ agriculture.

DIRECTIONS: Multiple Choice In the blank at the left, write the letter of the
choice that best completes the statement or answers the question. *(4 points each)*

_____ **11.** A map grid helps you find the location of
    A. a place   C. the solar system
    B. movement   D. magma

_____ **12.** One way geographers divide the earth into two hemispheres is along the
    A. Tropic of Cancer   C. Equator
    B. Tropic of Capricorn   D. North and South Pole

Geography: The World and Its People   13   **Chapter and Unit Tests**

Chapter (Forms A and B) **and
Unit Tests** provide for evaluation of
student understanding of the main
facts, themes, and ideas at regular
points throughout the text.

Actual size is 8 1/2" × 11"

Name_____ Date _____ Class _____

ACTIVITY 1 Performance Assessment

Use with Chapter 1
The Eye in the Sky

BACKGROUN
From the daw
series of satelli
1970s LANDS
scientists on th

TASK
A science mus
Find photograp
SAT sent back

AUDIENCE
Your audience

PURPOSE
The purpose o
satellite photog

PROCEDURE
1. Consult the
   Board Displ
2. Research to
   LANDSAT r
3. Sketch idea
   nize your m
4. Share your i
5. Revise your
   your picture

ASSESSMENT
1. Use the Asse
2. Check to se
3. Think about

Performance Assessment

Name_____ Date _____ Class _____

CHAPTER 15 Quiz 15–1

The Independent Republics

Section 1 Ukraine

DIRECTIONS: Matching Match each item in Column A with the items in Column B. Write the correct letters in the blanks. *(10 points each)*

COLUMN A
A. Donets Basin
B. steppe
C. Crimean Peninsula
D. Odessa
E. Black Sea

COLUMN B
_____ **1.** Ukraine's major seaport
_____ **2.** landmass that juts into the Black Sea
_____ **3.** vast treeless plains
_____ **4.** large body of water bordering Ukraine in the south
_____ **5.** major industrial region in eastern Ukraine

DIRECTIONS: Multiple Choice In the blank at the left, write the letter of the choice that best completes the statement or answers the question. *(10 points each)*

_____ **6.** The Ukraine's two largest rivers flow into the
   A. Sea of Azov   **C.** Black Sea
   B. Caspian Sea   **D.** Crimean Peninsula

_____ **7.** Ukraine's climate of cold winters and warm summers is called a
   A. steppe climate   **C.** high latitude climate
   B. humid continental climate   **D.** highland climate

_____ **8.** Today Ukraine's farming and industry operate in a
   A. free enterprise system   **C.** communist system
   B. command economy   **D.** socialist system

_____ **9.** The Donets Basin contains one of the world's largest deposits of
   A. oil   **C.** coal
   B. natural gas   **D.** hydroelectric power

_____ **10.** Kiev is the major transportation, industrial, and cultural center of
   kraine   G   gia
   oldova   us

Copyright © by Glencoe/McGraw-Hill.

Section Quizzes

Geography: The World and Its People

Also Available in Spanish

Actual size is 8 1/2" × 11"

Name_____ Date_____ Class_____

CHAPTER **24** Critical Thinking Skills Activity

Drawing Conclusions *(continued)*

B. Using the Skill

Use the information in t
and the graph below to
conclusions about Japanes

Japan has very little lar
able for farming. At the be
twentieth century, Japan p
80 percent of the food
Japan has grown in popu
this century. However, J
dence on imports has re
the same.

4. Which crops or food pro

_____

5. Do you think Japan will

_____

_____

6. Japan has continued to

tion. What conclusions

_____

_____

Critical Thinking Skills

Name_____ Date_____ Class_____

CHAPTER **24** Critical Thinking Skills Activity

Drawing Conclusions

Drawing conclusions from the facts you discover is a necessary step in making sense of information. You can use information you find in books, articles, and special reference materials, such as almanacs and atlases, to draw your own conclusions about a topic. Suppose you learn that a country's landscape is mostly mountainous. You can conclude that there is little land available for farming.

A. Using the Skill

Use the information below to draw some conclusions about the economy and way of life in Japan.

> **JAPAN**
>
> **Population:** 124 million
> **Population density:** 865 persons per square mile
> **Population distribution:** 77 percent urban, 23 percent rural
> **Ethnic groups:** 99.4 percent Japanese, 0.5 percent Korean, 0.1 percent Chinese
> **Literacy rate:** 99 percent
> **Land use:** 13 percent farmland
> **Principal agricultural products:** rice, vegetables, fruits, sugar beets
>
> **Principal industrial products:** automobiles, appliances, electronic consumer goods, machinery and equipment, metals and metal products
> **Principal imports:** petroleum, natural gas, coal, chemicals, textiles, metal ores and scrap, foodstuffs
> **Principal exports:** electronic goods, appliances, steel, chemicals, scientific and optical equipment, automobiles

1. What conclusions can you draw about where people work in Japan? Do you think there are more people working in factories or on farms? Explain your answer. _____

_____

2. Based on the information in the chart, how much value do you think the Japanese place on education? _____

_____

3. Look at the information about the ethnic composition of Japan. How would the social and cultural life in Japan differ from the social and cultural life in the United States? _____

_____

Copyright © by Glencoe/McGraw-Hill.

Geography: The World and Its People 47 **Critical Thinking Skills**

Critical Thinking Activities challenge students to apply what they have learned in problem-solving situations.

Actual size is 8 ¹/2" × 11"

Name_____ Date _____ Class _____

CHAPTER 2 Cooperative Learning Activities

What's the Weather Like There? *(continued)*

Cooperative Group Process

1. Decision M
brainstorm
select a favo
the daily we

2. Individual
out about th
the weather
description

3. Group Wo
decide how
columns for

4. Additional
as sun, rain
Display your

Group Proce:

- What is t
United Sta
- Did you h
pected inf
- What was
- How did y
- How was

Quick Check

1. Was the goa

_____

2. Did each te

_____

3. Were you sa

_____

Cooperative Learning

Cooperative Learning Activities offer students clear management directions for working together on a variety of activities that enrich prior learning.

Name_____ Date _____ Class _____

CHAPTER 2 Cooperative Learning Activities

What's the Weather Like There?

Background

When we talk about climate, we are looking at average conditions that exist in a location over the years. Weather, on the other hand, looks at the conditions in one location on a particular day. Both weather and climate deal with things like heat or cold, wetness or dryness, clear or cloudy skies. The continental United States has a temperate climate. The weather, however, can change dramatically from place to place in this temperate zone. Do some research to find out how daily weather is different in several parts of the temperate zone. Then track the weather in major United States cities and make a weather chart to show your findings. Share your chart with other groups to understand the different weather patterns.

Group Directions

1. Use the information in Chapter Two and library resources to find out more about the different weather conditions across the United States.

2. Each day for a week, examine the weather in cities in different regions of the country, using newspaper or TV reports. Be sure to include cities from each of these climate zones:

> **Coastal Plains**
> **Appalachian Highlands**
> **Interior Plains**
> **Interior Highlands**
> **Superior Upland**
> **Rocky Mountains**
> **Intermountain Plateaus and Basins**
> **Pacific Mountain and Valley System**

3. Make a weather chart to show your findings.

Geography: The World and Its People ③ **Cooperative Learning**

 **UNIT 9** Home Involvement Activity

Australia, Oceania, and Antarctica

 **Sharing at Home** Our class learned about the Pacific region—a world of mostly water. Australia, New Zealand, and thousands of islands make up this region. Do some of the following activities [toget]her and share the excitement of learning [about] this unusual area.

> **Home Involvement Activities** offer students geography information and activities they can share at home.

[Lear]n About Money

[Ev]ery country has its own currency, or [mone]y. Here are some dollar equivalents of [count]ries in this region:

Country	Currency	Price in US dollars
Australia	Dollar	.67
Indonesia	Rupiah	.0005
Malaysia	Ringgit	.38
New Zealand	Dollar	.54
Papua New Guinea	Kina	1.02
Philippines	Peso	.04

An exchange rate varies from day to day. It is the price of one country's currency in relation to another country's currency.

LEARN THE LANGUAGE

The version of English spoken in Australia is quite different from that of British or American English. These words and expressions are just a few of the many Australian words that seem unfamiliar to tourists.

g'day . . . . good day (the way Australians say "hello")
fair dinkum . . . . . . honest, real
arvo. . . . . afternoon
barbie . . . . barbecue
bloke. . . . . man
bush . . . . country, any place away from the city
clobber . . . . clothes
too right! . . . . absolutely!
yakka. . . . . work

Make a "Down Under" Animal Drawing

Together, find pictures in magazines of the many unique animals that live in Australia and New Zealand. Cut out the pictures—or draw them yourselves using books as guides—and paste them on a poster board. Label each animal and display your poster where your family can see it.

 Waltzing Matilda was a song sung by Australian soldiers during World Wars I and II. Since then, it has become the unofficial Australian national anthem.

Geography: The World and Its People 17 **Home Involvement**

Also Available in Spanish

Actual size is 8 1/2" × 11"

Name_____ Date _____ Class _____

CHAPTER 10 GeoLab Activity **GeoLab**

Elevation Profile: Norwegian Fjord *(continued)*

What To Do

A. Locate the So

B. Create a co
sea level–0
m); 5,000–

C. Color each

D. Glue the la

E. Cut into th
shape as th
is removed.

F. With a mar
(on large po

Lab Activity

1. Looking at t

inland out t

2. Why do you

3. Why would

4. Drawing C

of Scandina

GeoLab

Actual size is 8 1/2" × 11"

Name_____ Date _____ Class _____

CHAPTER 10 GeoLab Activity **GeoLab** ACTIVITY

Elevation Profile: Norwegian Fjord

From the classroom of Debora Bittner, John Marshall High School, San Antonio, Texas

Read more about the fjords of Scandinavia. Then build a model landscape that shows the structure of these wonders of nature.

Background

Fjords are only found in the higher latitudes of the Northern and Southern hemispheres. In Norway, on the Scandinavian Peninsula, they are located on the western and more mountainous coastlines. It is believed that the fjords were formed by ancient river glaciers pushing along to form valleys. Fjords are usually deepest at their most inland point and have steep smooth sides. The longest Norwegian fjord is the *Sognefjord.* Build a cross-section model fjord to show its features.

Believe it or NOT!

The total coastline of Norway is slightly over 12,000 miles (19,312 km). That is approximately the same as the total coastline of the United States, including Hawaii and Alaska!

Materials

- 6–$\frac{1}{4}$" sheets of polystyrene material
- glue
- markers
- knife
- large poster board (optional for mounting)

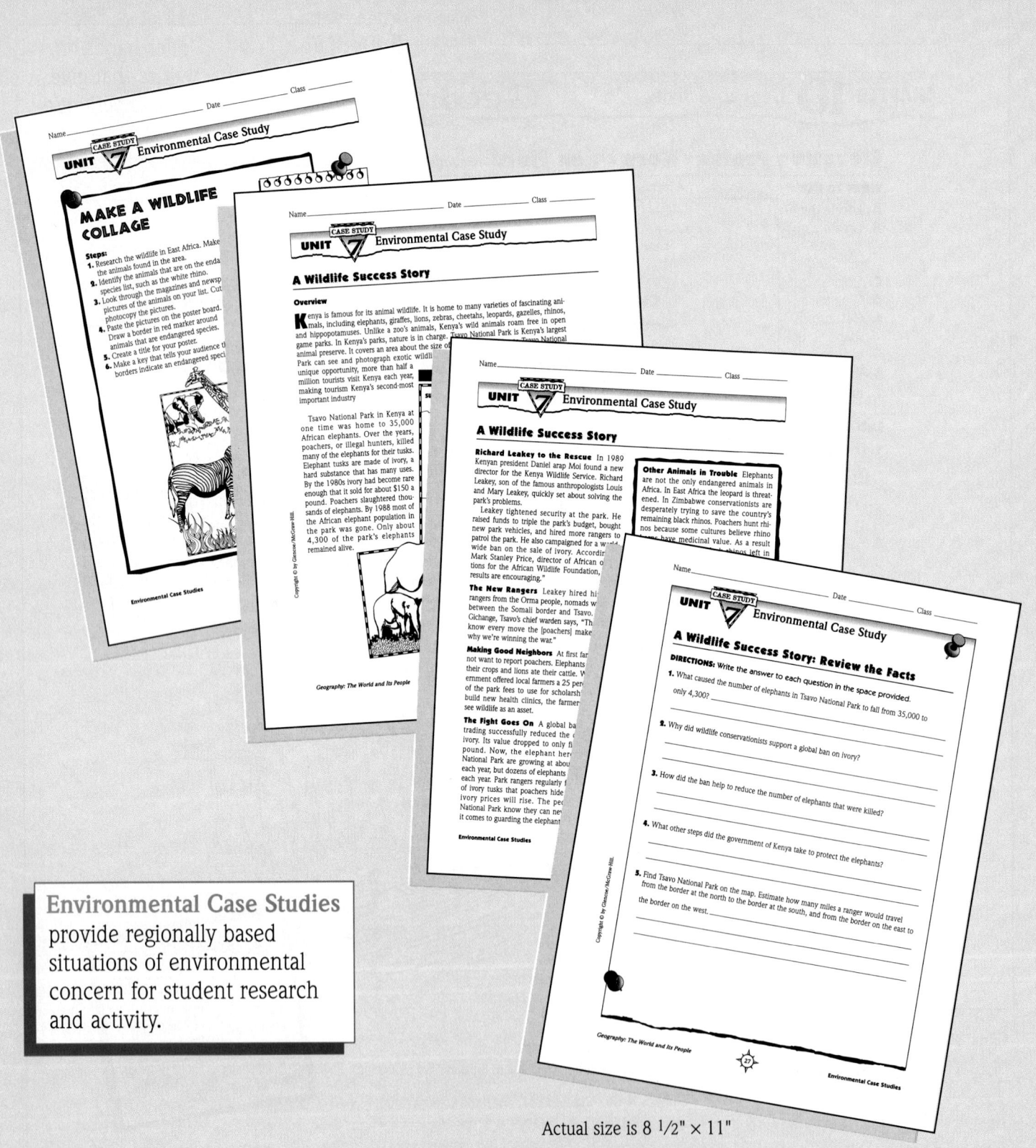

Environmental Case Studies provide regionally based situations of environmental concern for student research and activity.

Actual size is 8 1/2" × 11"

Name_____ Date _____ Class _____

The Lion City

DIRECTIONS: *Read the following article about Singapore. Then answer the questions.*

In Sanskrit the word *Singha-Pura* means "Lion City." In the early nineteenth century, the jungle-covered island of Singapore gained its name from this term. Singapore was a "lion" among the trading colonies in Southeast Asia.

In 1819 Thomas Raffles arrived by ship at the mouth of the Singapore River. Raffles was an employee of the English East India Company.

He wanted to start a company warehouse in Southeast Asia. Dutch troops kept the British traders out.

Singapore's leaders welcomed Raffles, however, and agreed to let him set up a factory, or ware-house, on the island. Raffles k[...] Dutch would challenge his co[...] ried to bring many British [...] island. Within six months [...] tants, mostly members [...] East India Compa[...] Singapore.

The Dutch were kept out, and the Lion City became a center of British trade.

Taking Another Look

1. When did Thomas Raffles arrive at Singapore? _____

2. Which company did Thomas Raffles represent? _____

3. Who were the company's rivals in Southeast Asia? _____

4. What does *Singha-Pura* mean in Sanskrit? _____

5. Making Inferences Why did Raffles rush to settle the island of Singapore before the question of

ownership was settled? _____

_____

Geography: The World and Its People ⟨29⟩ **Enrichment**

Enrichment Activities are designed to stimulate creative thinking and problem solving.

Actual size is 8 ¹/2" × 11"

Name_____ Date_____ Class_____

CHAPTER **11** Activity Workbook

Northwestern Europe: Activity 2

Identifying the Economic Products of Northwestern Europe

The countries of northwestern Europe have achieved worldwide recognition for their excellent manufactured goods and food products. These exports vary from country to country based on the resources found in each land.

Geography: The World and Its People Activity Workbook and **Annotated Teacher's Edition** give students an opportunity to expand and apply their base of knowledge through reinforcing activities.

...ed below. Note that some products are ...one country. Write each product in the ...the information in your textbook, as needed.

diamonds   steel
glass   machinery
furniture   computers
paper goods cars
clothing   watches   airplanes
lace   chemicals   electrical equipment

Copyright © by Glencoe/McGraw-Hill.

FRANCE	GERMANY
food crops, airplanes, cars, computers, chemicals, clothing, furniture	steel, machinery, cars clothing, chemicals

BELGIUM	NETHERLANDS	LUXEMBOURG
lace, diamonds, chocolate, machinery, cars, chemicals	food crops, dairy products, flower bulbs	steel

SWITZERLAND	AUSTRIA
machinery, computers, watches, chemicals, clothing, chocolate, dairy products, electrical equipment	machinery, chemicals, furniture, glass, paper goods

Geography: The World and Its People   ✦25✦   **Activity Workbook**

Actual size is 8 1/2" × 11"

Regional Recipes

Chili Con Queso
(serves 6 to 8 as an appetizer or snack)

Ingredients

1 T. vegetable oil
1 lb. Monterey Jack or Longhorn Cheddar cheese
1 cup finely chopped onions
1/2 cup finely chopped green pepper

1 garlic clove, very finely chopped
1 12-oz. can chopped tomat...
1 4- to green drain chop

Procedure Grate cheese into a larg... aside. Heat oil in large saucepan an... add onions, green pepper, and garli... stir until onions are tender. Add tom... juice) along with drained and chop... Lower heat. Add cheese. Stir and s... cheese is melted. Place in a large l... with tortilla chips for dipping.

Rice with Corn, Tomatoe...
(serves 4 to ...

Ingredients

2 T. cooking oil
1 medium onion,
1 large tomato
1 1/2 cups boiling water
1 1/2 cup long-grain rice

1/2,
1
1

Procedure Slice the onion and pe... In a heavy saucepan, heat the oil ov... carefully to coat the bottom. Add th... cook for 5 minutes, until the onion... the rice and stir for 2 to 3 minutes... brown. Add the chicken broth, bo... Return the mixture to a boil, stirri... with a tight-fitting lid and turn the... point. Cook for 12 minutes, or u... almost all of the liquid. Stir in th... heat and stir until the corn is he...

Actual size is 8 1/2" × 11"

Copyright © by Glencoe /McGraw-Hill School Publishing Company

Foods Around the World gives students a glimpse of different cultures by looking at foods from various regions of the world. Included are suggestions for planning and organizing with students a Multicultural Food Fair.

UNIT 3: LATIN AMERICA & THE CARIBBEAN

Foods of Latin America

The cuisine of Latin America contains a bounty of foods that can be traced to the earliest civilizations of the region. For example, archaeologists working in Mexico have discovered avocado seeds that date from 8000 B.C. Evidence of corn, beans, and squash has been uncovered at the sites of early Aztec and Inca civilizations. Today, corn and beans are still a major staple in Latin American cooking.

STAPLE FOODS

For early Native Americans potatoes were a vital part of the diet. Today potatoes are still grown in the mountainous regions of Argentina, Peru, Ecuador, Bolivia, and Chile. Along with corn and beans, potatoes are the most widely consumed staples in the Latin American diet. Other native foods and spices of Latin America include allspice, chocolate, sweet potatoes, tomatoes, vanilla, and, a broad variety of chili peppers, the source of the hot, spicy character of so many Latin American dishes. Native fruits include guavas, papayas, passion fruit, and pineapples.

THE INFLUENCE OF OUTSIDE CULTURES

Beginning in the late fifteenth century, Spanish explorers and conquerors came to Latin America. They took back to Europe such "unknown" foods as tomatoes and chocolate. Were it not for the early Latin American cultures, the tomato might not today be such a basis for Spanish and Italian cooking!

The Spanish also brought their own native foods to Latin America—chickens, pigs, beef, garlic, onions, olive oil, sugar cane, and rice. Latin Americans began to cook with oils and fats for the first time. They also learned how to use eggs, milk, and cheese. Once Spanish towns and cities became established in Latin America, beef quickly became a major part of the regional diet. Today Argentina leads the world in the exportation of beef.

Global Gourmet

When you are invited to someone's home in such countries as Mexico, Brazil, and Panama, you are not expected to arrive on time. In Chile, however, punctuality is insisted upon. In Venezuela, invited guests should never sit at the head of the table, and in Bolivia, guests are expected to sample all foods and eat everything on their plates.

Questions To Answer

1. What are the most widely eaten staples of the Latin American diet?
2. As you try the recipes on the following page, what foods introduced into Latin America by the Spanish are included?

Copyright © by Glencoe /McGraw-Hill School Publishing Company

Countries of the World Flash Cards are user-friendly and may be used for individual or group review, games, and self-evaluation.

Wari: a Mancala Game

3 WORLD GAMES

A Bit of History

Wari belongs to a group of related games called mancala games. Mancala games were first played in Africa and the Middle East in lands along the Red Sea. These games are very old. In fact, some mancala boards from Karnak and Luxor in Egypt are 3,500 years old.

This version of the game, called Wari, is still played by people in eastern and central Africa. The rules of the game are simple, but good Wari players play so fast that the game is exciting to watch.

How Wari is Played in Africa

Two people play Wari using a Wari board. The board has two rows of six cups, side by side. At each end is another cup for holding the pieces that each player captures. These pieces may be stones, beans, marbles, buttons, or other small objects. The goal of each player is to capture more than half of the 48 pieces.

Side 1

Africa

> World Games Activity Cards and World Crafts Activity Cards offer students an opportunity to sample elements of many cultures.

Ukrainian Easter Eggs

A Bit of History

For centuries, Christian Ukrainians have taken eggs they have decorated to their church the night before Easter to have them blessed by the priest. Then on Easter Sunday the jewel-like eggs were passed from friend to friend. These eggs were the displayed in homes and are believed by many to have the power to prevent fires and other disasters. In Ukraine, when a girl gives a boy a decorated egg on Easter she is offering her love, she knows that if the boy keeps the egg he returns her affection.

How to Blow Eggs:

Pierce a hole in both the top and bottom of the egg and insert a needle to break the yolk. Gently shake the egg to mix the yolk and white together. Place the egg over a bowl and blow through the top hole until the egg is empty. Rinse the egg by blowing a mouthful of fresh water through the egg. Bake the shells in a 200 degree oven for 10-15 minutes to dry the egg and harden any left over yolk.

Side 1

Ukraine

NATIONAL GEOGRAPHIC SOCIETY

These National Geographic Society resources will enhance learning in your classroom.

These **National Geographic Society Posters Sets**—*Images of the World* and *Eye on the Environment*— are colorful additions to your geography classroom.

National Geographic Picture Atlas of
Our World

National Geographic Society's
Picture Atlas of Our World
is an excellent resource for the
geography classroom.

National Geographic Society's
Picture Atlas of Our 50 States
is a colorful tool to aid students in
learning about the United States.

Facts On File: AFRICAN HISTORY ON FILE offers a variety of formats and topics to aid in combining history and geography.

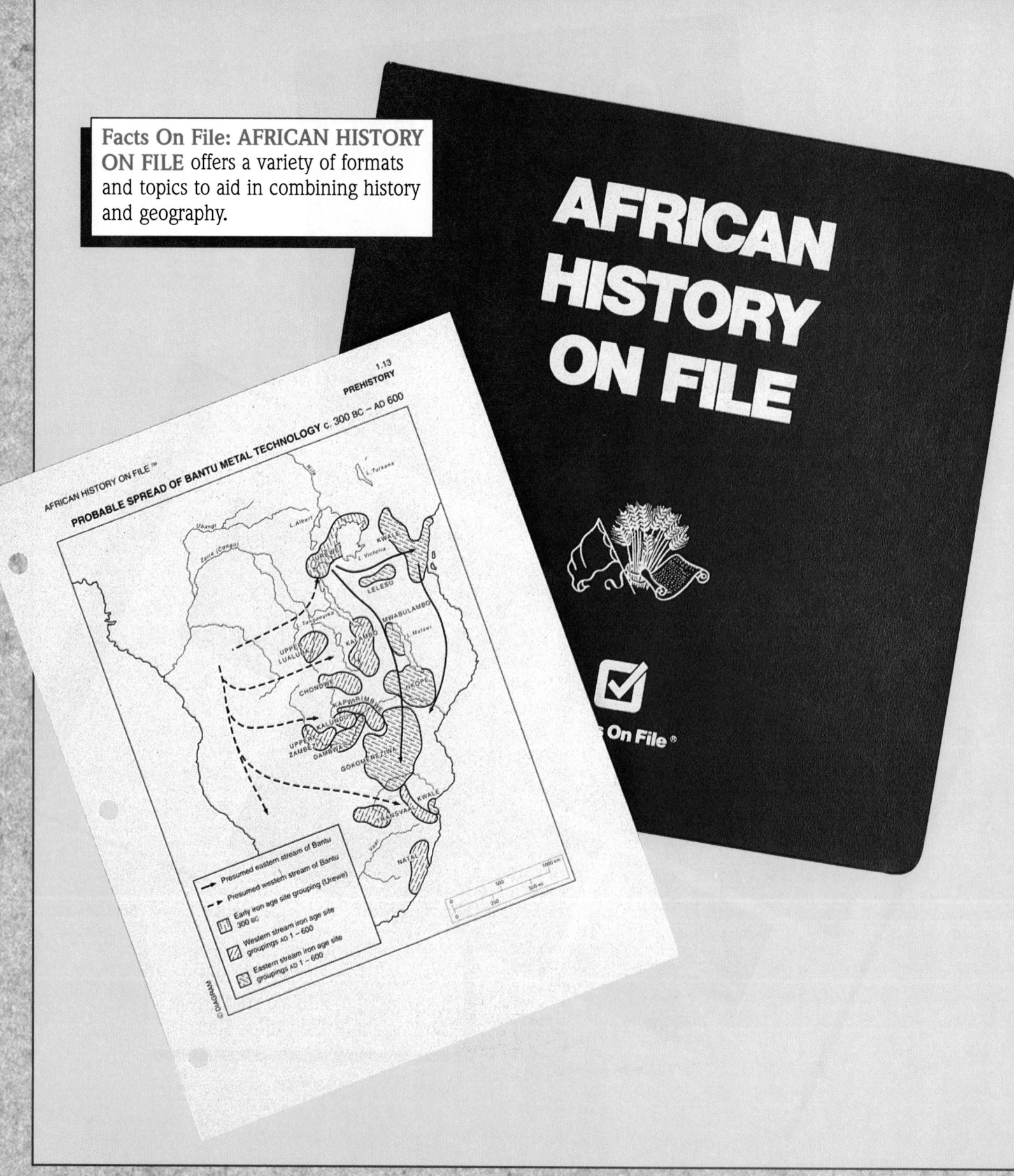

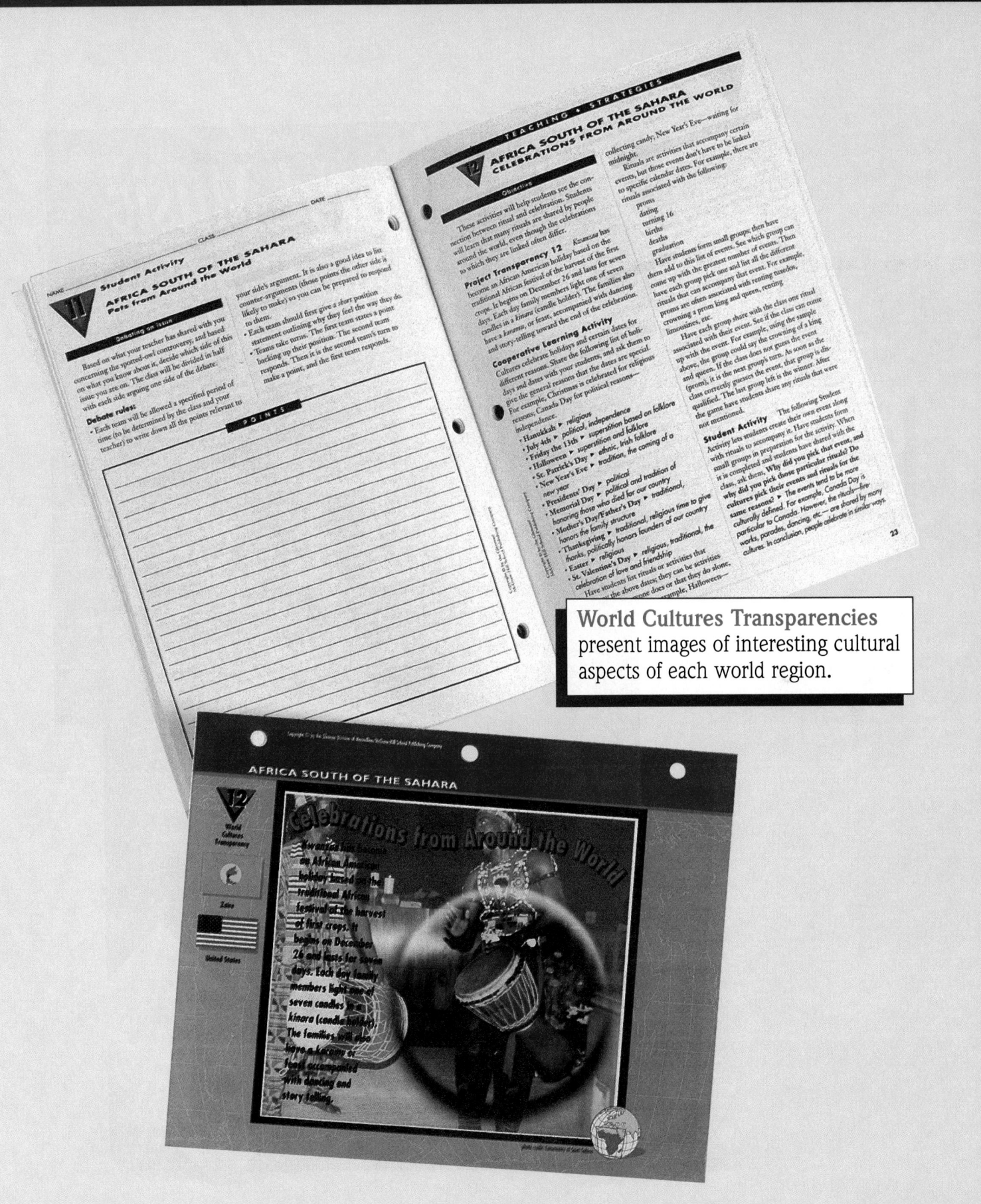

World Cultures Transparencies present images of interesting cultural aspects of each world region.

World Music: Cultural Traditions brings students into contact with music representing many cultures and regions of the world.

Focus on World Art Prints are poster-size prints that enhance student appreciation of fine arts and cultural expressions.

姿見七人化粧

Kitagawa Utamaro
Reflected Beauty from the series *Seven Women Seen in a Mirror*

c. 1790. Color woodblock print. 36.3 x 24.1 cm (14½ x 9½).
Honolulu Academy of Arts, Honolulu, Hawaii. The James A. Michener Collection.

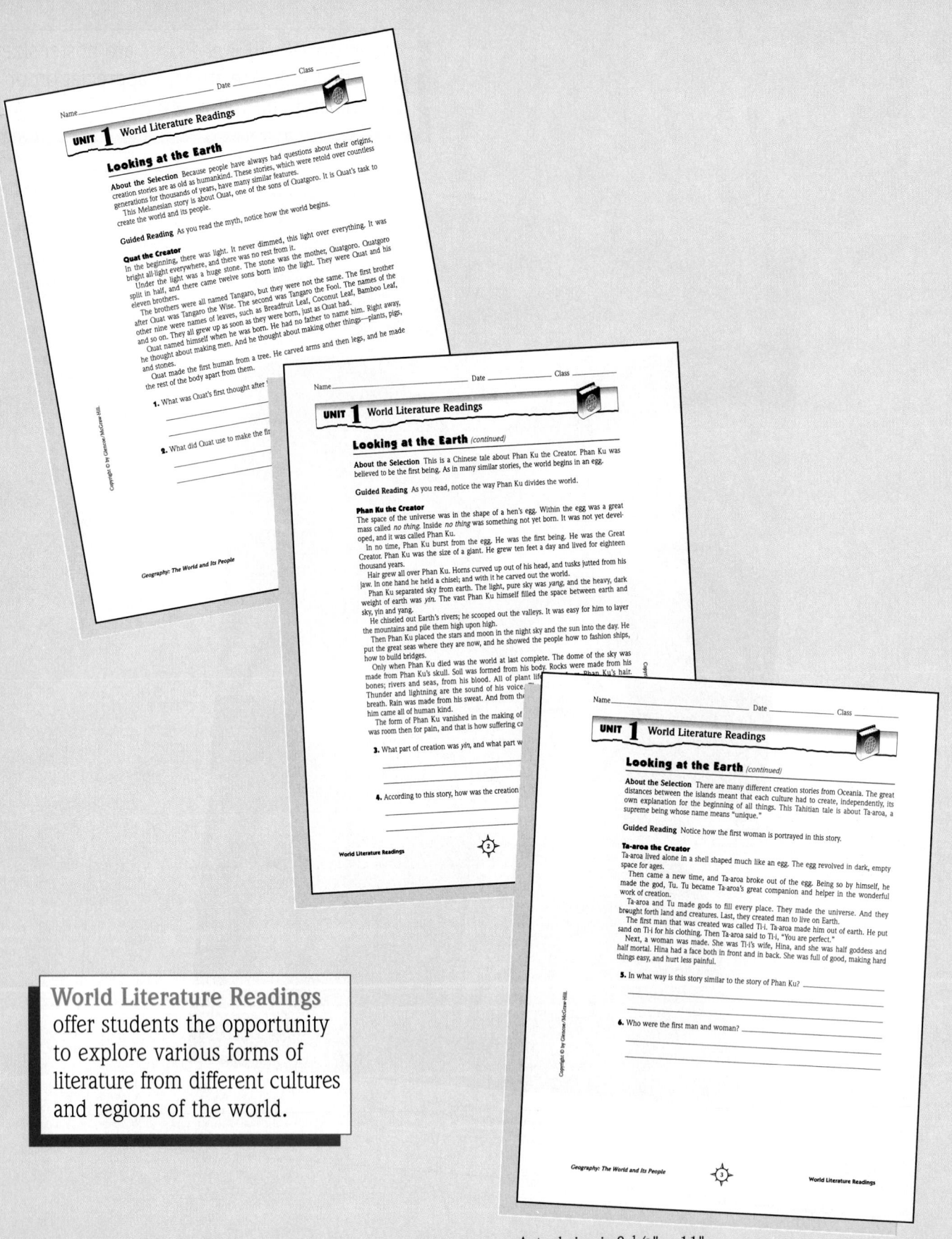

Name _____ **Date** _____ **Class** _____

UNIT 1 World Literature Readings

Looking at the Earth

About the Selection Because people have always had questions about their origins, creation stories are as old as humankind. These stories, which were retold over countless generations for thousands of years, have many similar features.

This Melanesian story is about Quat, one of the sons of Quatgoro. It is Quat's task to create the world and its people.

Guided Reading As you read the myth, notice how the world begins.

Quat the Creator
In the beginning, there was light. It never dimmed, this light over everything. It was bright all-light everywhere, and there was no rest from it.

Under the light was a huge stone. The stone was the mother, Quatgoro. Quatgoro split in half, and there came twelve sons born into the light. They were Quat and his eleven brothers.

The brothers were all named Tangaro, but they were not the same. The first brother after Quat was Tangaro the Wise. The second was Tangaro the Fool. The names of the other nine were names of leaves, such as Breadfruit Leaf, Coconut Leaf, Bamboo Leaf, and so on. They all grew up as soon as they were born, just as Quat had.

Quat named himself when he was born. He had no father to name him. Right away, he thought about making men. And he thought about making other things—plants, pigs, and stones.

Quat made the first human from a tree. He carved arms and then legs, and he made the rest of the body apart from them.

1. What was Quat's first thought after _____

2. What did Quat use to make the fir _____

Geography: The World and Its People

Name _____ **Date** _____ **Class** _____

UNIT 1 World Literature Readings

Looking at the Earth *(continued)*

About the Selection This is a Chinese tale about Phan Ku the Creator. Phan Ku was believed to be the first being. As in many similar stories, the world begins in an egg.

Guided Reading As you read, notice the way Phan Ku divides the world.

Phan Ku the Creator
The space of the universe was in the shape of a hen's egg. Within the egg was a great mass called *no thing*. Inside *no thing* was something not yet born. It was not yet developed, and it was called Phan Ku.

In no time, Phan Ku burst from the egg. He was the first being. He was the Great Creator. Phan Ku was the size of a giant. He grew ten feet a day and lived for eighteen thousand years.

Hair grew all over Phan Ku. Horns curved up out of his head, and tusks jutted from his jaw. In one hand he held a chisel; and with it he carved out the world.

Phan Ku separated sky from earth. The light, pure sky was *yang*, and the heavy, dark weight of earth was *yin*. The vast Phan Ku himself filled the space between earth and sky, yin and yang.

He chiseled out Earth's rivers; he scooped out the valleys. It was easy for him to layer the mountains and pile them high upon high.

Then Phan Ku placed the stars and moon in the night sky and the sun into the day. He put the great seas where they are now, and he showed the people how to fashion ships, how to build bridges.

Only when Phan Ku died was the world at last complete. The dome of the sky was made from Phan Ku's skull. Soil was formed from his body. Rocks were made from his bones; rivers and seas, from his blood. All of plant lif _____ Phan Ku's hair. Thunder and lightning are the sound of his voice. _____ breath. Rain was made from his sweat. And from the _____ him came all of human kind.

The form of Phan Ku vanished in the making of _____ was room then for pain, and that is how suffering ca _____

3. What part of creation is *yin*, and what part w _____

4. According to this story, how was the creation _____

World Literature Readings ②

Name _____ **Date** _____ **Class** _____

UNIT 1 World Literature Readings

Looking at the Earth *(continued)*

About the Selection There are many different creation stories from Oceania. The great distances between the islands meant that each culture had to create, independently, its own explanation for the beginning of all things. This Tahitian tale is about Ta-aroa, a supreme being whose name means "unique."

Guided Reading Notice how the first woman is portrayed in this story.

Ta-aroa the Creator
Ta-aroa lived alone in a shell shaped much like an egg. The egg revolved in dark, empty space for ages.

Then came a new time, and Ta-aroa broke out of the egg. Being so by himself, he made the god, Tu. Tu became Ta-aroa's great companion and helper in the wonderful work of creation.

Ta-aroa and Tu made gods to fill every place. They made the universe. And they brought forth land and creatures. Last, they created man to live on Earth.

The first man that was created was called Ti-i. Ta-aroa made him out of earth. He put sand on Ti-i for his clothing. Then Ta-aroa said to Ti-i, "You are perfect."

Next, a woman was made. She was Ti-i's wife, Hina, and she was half goddess and half mortal. Hina had a face both in front and in back. She was full of good, making hard things easy, and hurt less painful.

5. In what way is this story similar to the story of Phan Ku? _____

6. Who were the first man and woman? _____

Geography: The World and Its People ③ **World Literature Readings**

World Literature Readings offer students the opportunity to explore various forms of literature from different cultures and regions of the world.

Actual size is 8 ¹/₂" × 11"

Spanish Chapter Digests Audiocassette
Activities and Tests

Spanish Chapter Highlights

Spanish Reteaching Activities

Spanish Vocabulary Activities

Spanish Section Quizzes

Spanish Resources include *Spanish Vocabulary Activities, Spanish Reteaching Activities, Spanish Section Quizzes,* and *Spanish Chapter Highlights.* Spanish Chapter Digests Audiocassettes: Activities and Tests are also available.

Spanish Resources

GEOGRAPHY
THE WORLD AND ITS PEOPLE

INCLUDES
Spanish Section Quizzes
Spanish Reteaching Activities
Spanish Vocabulary Activities
Spanish Chapter Highlights

NATIONAL
GEOGRAPHIC
SOCIETY

Spanish Chapter Digests
Audiocassettes

GEOGRAPHY
THE WORLD AND ITS PEOPLE

INCLUDES
Spanish Chapter Summaries on Audiocassette
Spanish Chapter Activities
Spanish Tests

NATIONAL
GEOGRAPHIC
SOCIETY

Grade _____ Class(es) _____ Dates(s) _____ M Tu W Th F

Teacher's Name _____ Dates _____

CHAPTER 19 Reproducible Lesson Plan

Zaire *Section 1 (pp. 505-510)*

Local Objectives	TWE—Teacher's Wraparound Edition TCR—Teacher's Classroom Resources • Laser Disc Program • Software • CD-Rom

OBJECTIVES
1. Discover why the Zaire River is a highway for Zaire's people.
2. Discuss Zaire's potential wealth.
3. Explain how large movements of people have shaped Zaire's history.

FOCUS
_____ Bellringer Motivational Activity, TWE, p. 505
_____ Vocabulary Pre-check, TWE, p. 505
_____ Vocabulary PuzzleMaker Software
_____ Chapter Digests Audiocassette (English and Spanish)

TEACH
Guided Practice
_____ Activity, TWE, p. 506
_____ *Political Map Transparency 7*, TCR
_____ Cooperative Learning Activity, TWE, p. 506
_____ Meeting Special Needs Activity, TWE, p. 507
_____ Critical Thinking Activity, TWE, p. 508

Independent Practice
_____ *Guided Reading Activity 19-1*, TCR
_____ National Geographic Society: STV: World Geography, Vol. 2, • Laser Disc

ASSESS
_____ Check for Understanding, TWE, p. 508
_____ Meeting Lesson Objectives, TWE, p. 508

Evaluate
_____ *Section Quiz 19-1*, TCR (English and Spanish)
_____ Testmaker

Reteach
_____ Reteach, TWE, p. 509

Enrich
_____ Enrich, TWE, p. 509

CLOSE
_____ Close, TWE, p. 509

Geography: The World and Its People   ◆19◆   **Reproducible Lesson Plan**

> **Reproducible Lesson Plans** allow you to spend less time planning and more time teaching.

Actual size is 8 ¹/₂" × 11"

Chapter Digests Audiocassettes Activities and Tests summarize each chapter, providing review and reinforcement.

Chapter Digests Audiocassettes

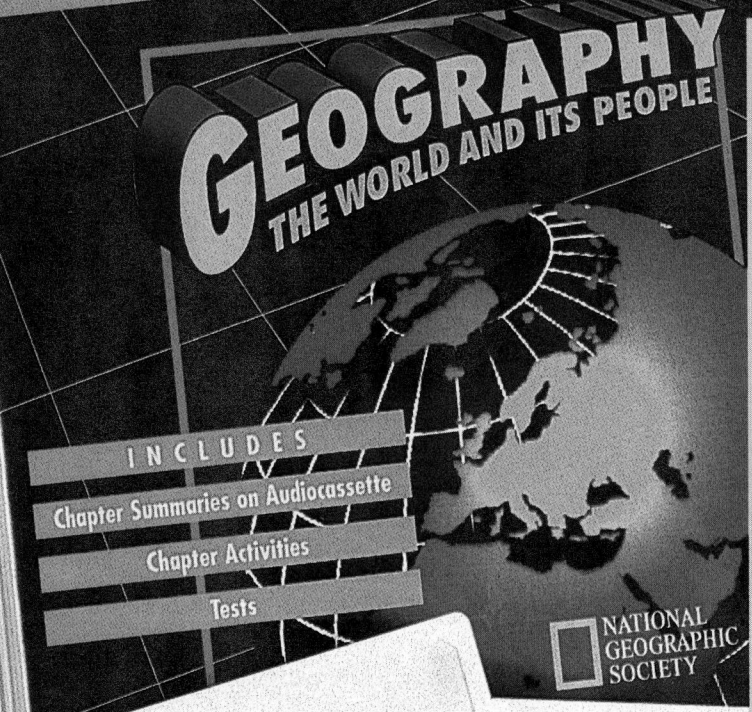

GEOGRAPHY
THE WORLD AND ITS PEOPLE

INCLUDES

Chapter Summaries on Audiocassette

Chapter Activities

Tests

NATIONAL GEOGRAPHIC SOCIETY

Name_____ Date_____ Class_____

CHAPTER 16 Audiocassette Test

Southwest Asia

DIRECTIONS: Match each item in Column A with the correct item in Column B. Write the correct letter in the blank. *(10 points each)*

COLUMN A

A. Israel
B. Saudi Arabia
C. Iran
D. Afghanistan
E. Iraq

COLUMN B

_____ 1. Twenty different ethnic groups live in this country of 18 million people.

_____ 2. This country has the most highly industrialized economy in Southwest Asia.

_____ 3. Ancient peoples called this land Persia.

_____ 4. Baghdad, once the cultural and scientific center of a Muslim empire, is the capital of this country.

_____ 5. A major share of the world's oil lies beneath the desert sands of this country.

DIRECTIONS: Write the word or phrase that best completes each sentence below. *(10 points each)*

6. Turkey's capital, Ankara, lies on the _____ _____.

7. Historians believe that one of the world's oldest cities is _____, the capital of Syria.

8. The lowest land area on earth surrounds the _____ _____.

9. Many countries in Southwest Asia are made rich by the production of _____.

10. The Empty Quarter is a _____ in the Arabian Peninsula.

Audiocassettes Activities and Tests 32 *Geography: The World and Its People*

Audiocassettes

Also Available in Spanish

These National Geographic Society PicturePack Transparencies are colorful additions to include in your lesson plans.

NATIONAL GEOGRAPHIC SOCIETY

Water

12. Glacial lakes in Canada

NGS PicturePack

The following multiple-use **Teaching Transparencies** are included in this binder:

- Unit Overlay Transparencies
- Political Map Transparencies
- World Cultures Transparencies
- Geography Handbook Transparencies

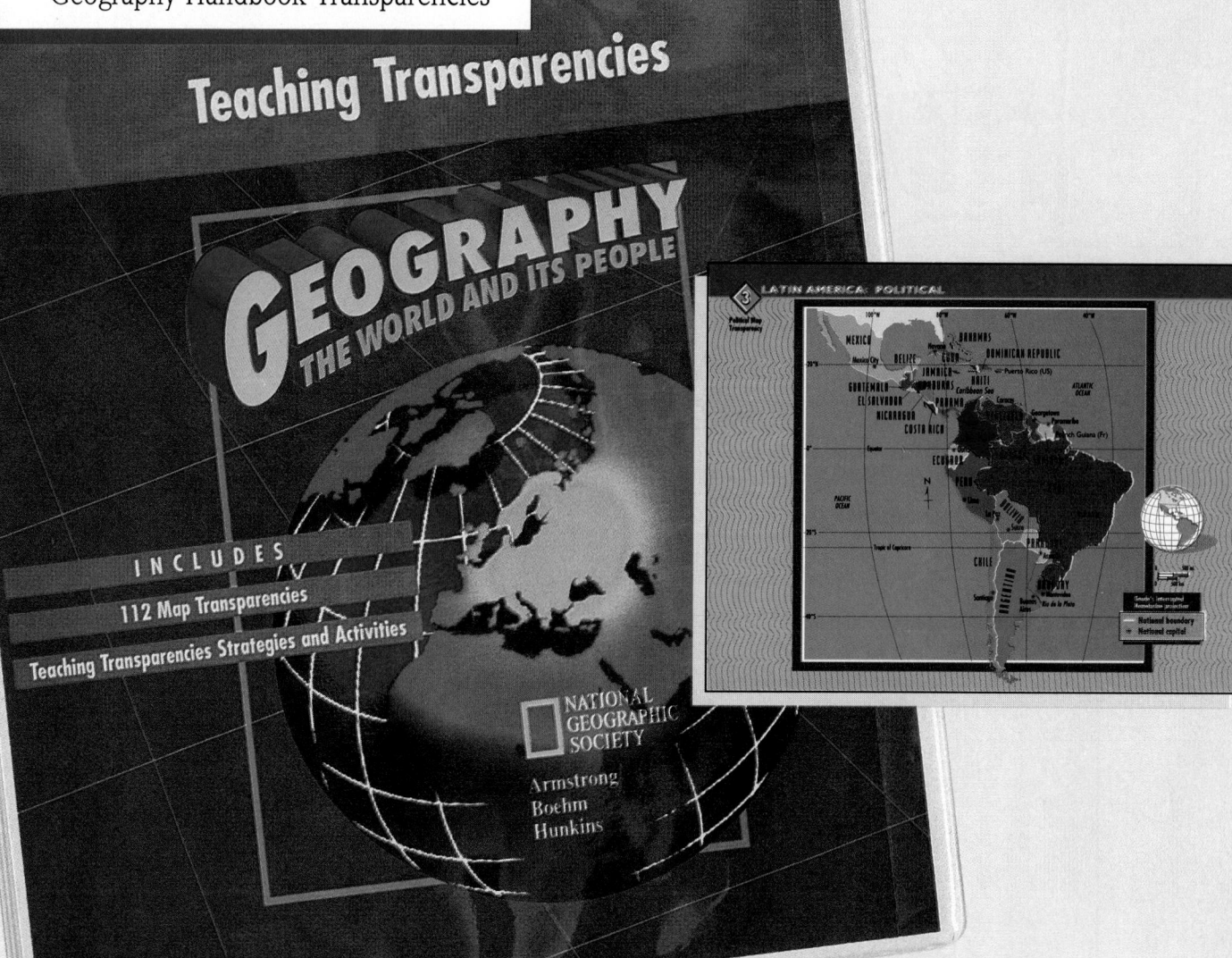

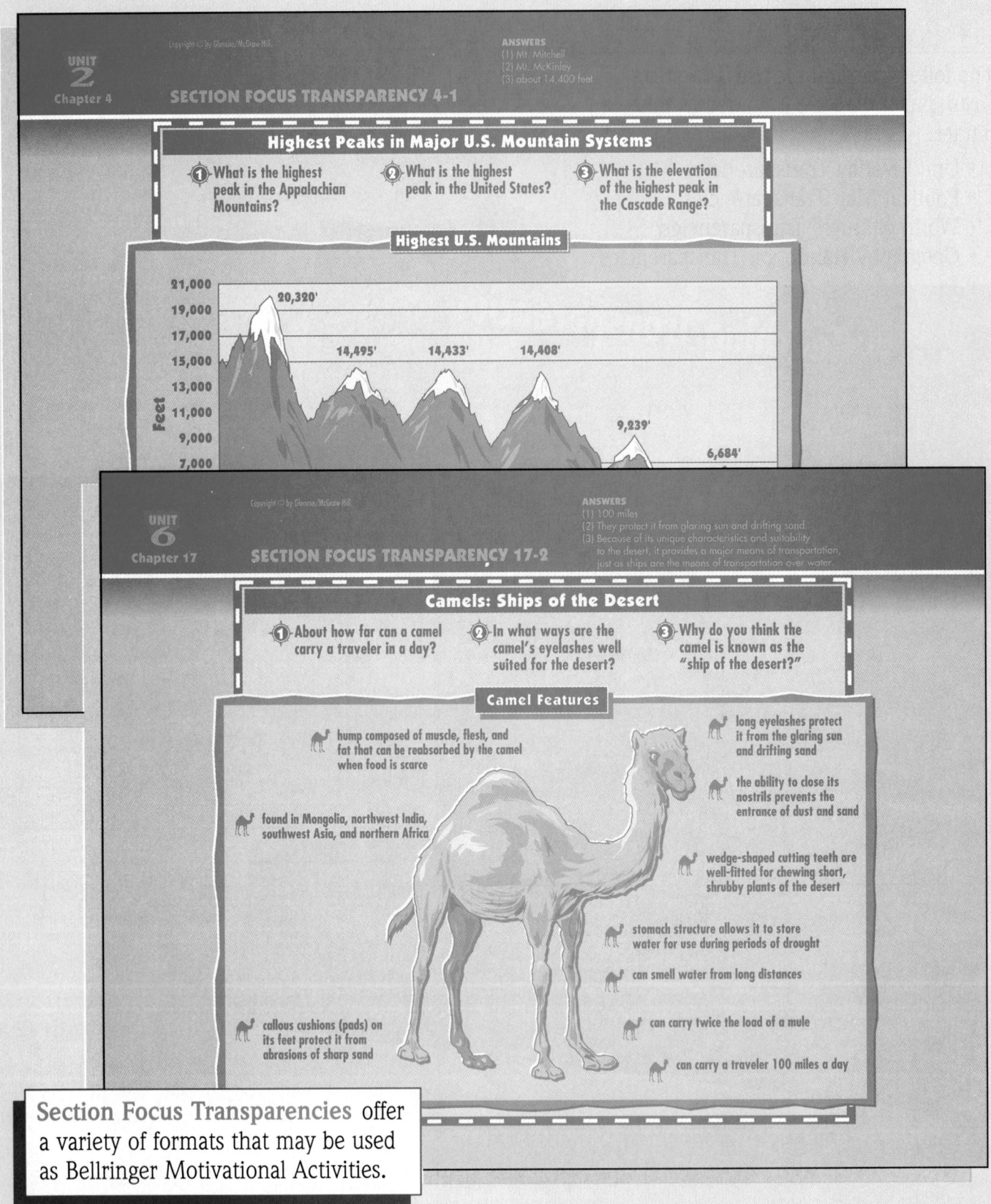

UNIT 2
Chapter 4

SECTION FOCUS TRANSPARENCY 4-1

Highest Peaks in Major U.S. Mountain Systems

1 What is the highest peak in the Appalachian Mountains?

2 What is the highest peak in the United States?

3 What is the elevation of the highest peak in the Cascade Range?

Highest U.S. Mountains

20,320'

14,495' 14,433' 14,408'

9,239'

6,684'

Feet — 21,000, 19,000, 17,000, 15,000, 13,000, 11,000, 9,000, 7,000

UNIT 6
Chapter 17

SECTION FOCUS TRANSPARENCY 17-2

Camels: Ships of the Desert

1 About how far can a camel carry a traveler in a day?

2 In what ways are the camel's eyelashes well suited for the desert?

3 Why do you think the camel is known as the "ship of the desert?"

Camel Features

- hump composed of muscle, flesh, and fat that can be reabsorbed by the camel when food is scarce
- found in Mongolia, northwest India, southwest Asia, and northern Africa
- callous cushions (pads) on its feet protect it from abrasions of sharp sand
- long eyelashes protect it from the glaring sun and drifting sand
- the ability to close its nostrils prevents the entrance of dust and sand
- wedge-shaped cutting teeth are well-fitted for chewing short, shrubby plants of the desert
- stomach structure allows it to store water for use during periods of drought
- can smell water from long distances
- can carry twice the load of a mule
- can carry a traveler 100 miles a day

Section Focus Transparencies offer a variety of formats that may be used as Bellringer Motivational Activities.

Actual size is 8 1/2" × 11"

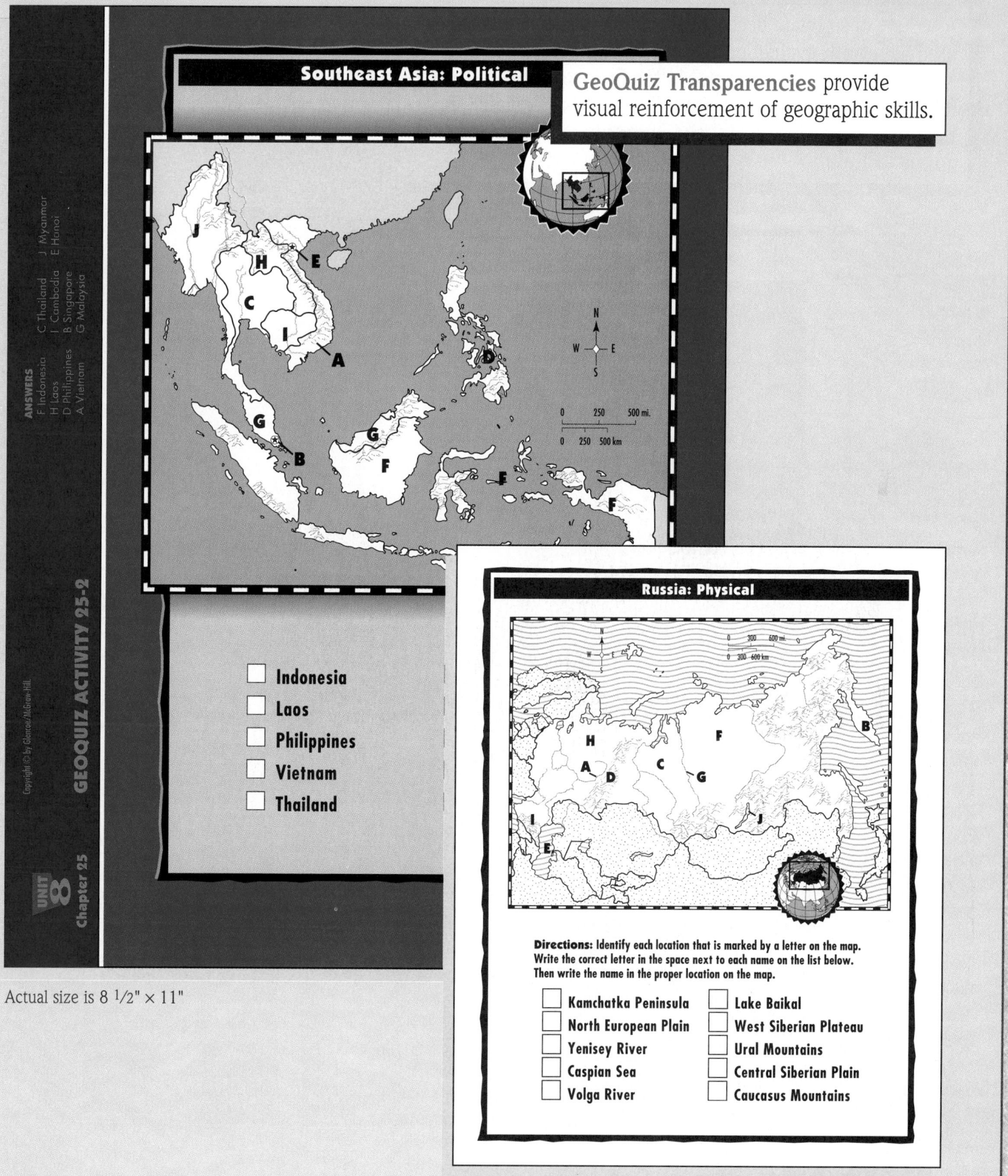

Southeast Asia: Political

GeoQuiz Transparencies provide visual reinforcement of geographic skills.

ANSWERS
F Indonesia   C Thailand   J Myanmar
H Laos   I Cambodia   E Hanoi
D Philippines B Singapore
A Vietnam   G Malaysia

Copyright © by Glencoe/McGraw-Hill.

UNIT 8
Chapter 25

GEOQUIZ ACTIVITY 25-2

☐ **Indonesia**
☐ **Laos**
☐ **Philippines**
☐ **Vietnam**
☐ **Thailand**

Actual size is 8 ¹/₂" × 11"

Russia: Physical

Directions: Identify each location that is marked by a letter on the map. Write the correct letter in the space next to each name on the list below. Then write the name in the proper location on the map.

☐ **Kamchatka Peninsula**   ☐ **Lake Baikal**
☐ **North European Plain**   ☐ **West Siberian Plateau**
☐ **Yenisey River**   ☐ **Ural Mountains**
☐ **Caspian Sea**   ☐ **Central Siberian Plain**
☐ **Volga River**   ☐ **Caucasus Mountains**

A variety of software accompanies this program and may be used as review, evaluation, or enrichment. Available software includes **Vocabulary PuzzleMaker Software, Student Self-Test: A Software Review, Testmaker** (available in IBM, Apple, or Macintosh), and **National Geographic Society ZipZapMap! World** and **ZipZapMap! USA.**

The **National Geographic Society Picture Atlas of the World** and **PictureShow** series provide colorful enrichment to the study of geography.

These **National Geographic Society** videodiscs provide visual, in-depth study of regions and environmental systems of the world. Titles available include:

- North America
- Rain Forest
- Biodiversity
- Restless Earth
- Solar System
- Water
- Atmosphere
- Planetary Manager

These **National Geographic Society** videodiscs open new worlds to your students. The World Geography series includes the following regions:

- Volume 1: Asia and Australia
- Volume 2: Africa and Europe
- Volume 3: South America and Antarctica

World Geography

LEVEL THREE
For Apple® Macintosh®

World Geography

LEVEL THREE
For Apple® Macintosh®

World Geography

LEVEL THREE
For Apple® Macintosh®

Volume 3
South America
and Antarctica

Contents
One Videodisc
Teacher's Guide
Software Diskette
Software Guide
Activity Sheets

STV

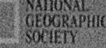
NATIONAL GEOGRAPHIC SOCIETY

Glencoe offers a variety of videodiscs. **Geography and the Environment: The Infinite Voyage** examines the earth and our relationship to it. **Reuters Issues in Geography** presents current environmental issues and challenges on a regional basis. **ABCNews InterActive™: In the Holy Land** explores issues in this region of the world. **MindJogger Videoquiz** offers preview, review, and reinforcement of chapter content in a game-show format.

Leaders in curriculum development have long favored integrative approaches to social studies instruction. Many of these experts believe that students benefit most from instructional techniques that interweave the study of social studies with that of the humanities. Such an approach allows students to study the geography of a specific region along with its accompanying art, music, and literature. This integrative curriculum is particularly useful to students with special needs. While studying the cultures of Africa South of the Sahara, for example, auditory learners would benefit from listening to the music of that region; visual learners would grasp the culture more easily by analyzing the art of the region.

To help you integrate the study of the humanities into your geography classroom, Glencoe makes available a variety of transparencies, audiotapes, and books from various disciplines. These supplemental materials are correlated to the content of *Geography: The World and Its People* on chapter interleaf pages. These materials include the following:

- World History and Art Transparencies
- Global Insights Teaching Transparencies
- World Music: Cultural Traditions
- Macmillan Literature: Introducing Literature Audiotapes
- Writer's Choice Transparencies

Photo Credits

T6, Robert W. Madden/National Geographic Society; T7, Joseph J. Scherschel/National Geographic Society; T8, Dallas & John Heaton/Westlight; T9, ©Bernard Sioberstein/FPG International; T10, James L. Stanfield/National Geographic Society; T12, James C. Simmons/Dave G. Houser; T18, ©John W. Warden/Superstock, Inc.; T28, TRIP Photographic; T30, National Geographic Society; T71, David R. Frazier.

REFERENCE

ATLAS

Contents

ATLAS KEY

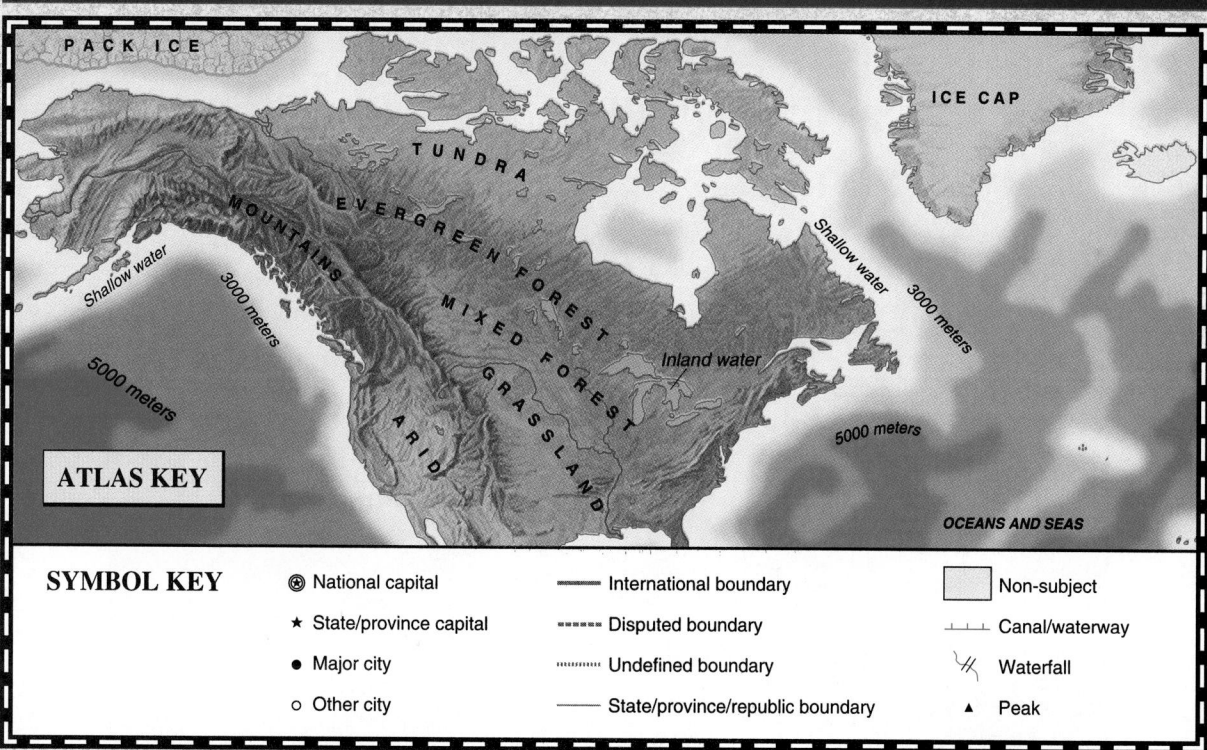

ATLAS KEY

SYMBOL KEY

⊛ National capital	▬▬▬ International boundary
★ State/province capital	▬▬▬ Disputed boundary
● Major city	▬▬▬ Undefined boundary
○ Other city	▬▬▬ State/province/republic boundary

☐ Non-subject	
┴┴┴ Canal/waterway	
⤩ Waterfall	
▲ Peak	

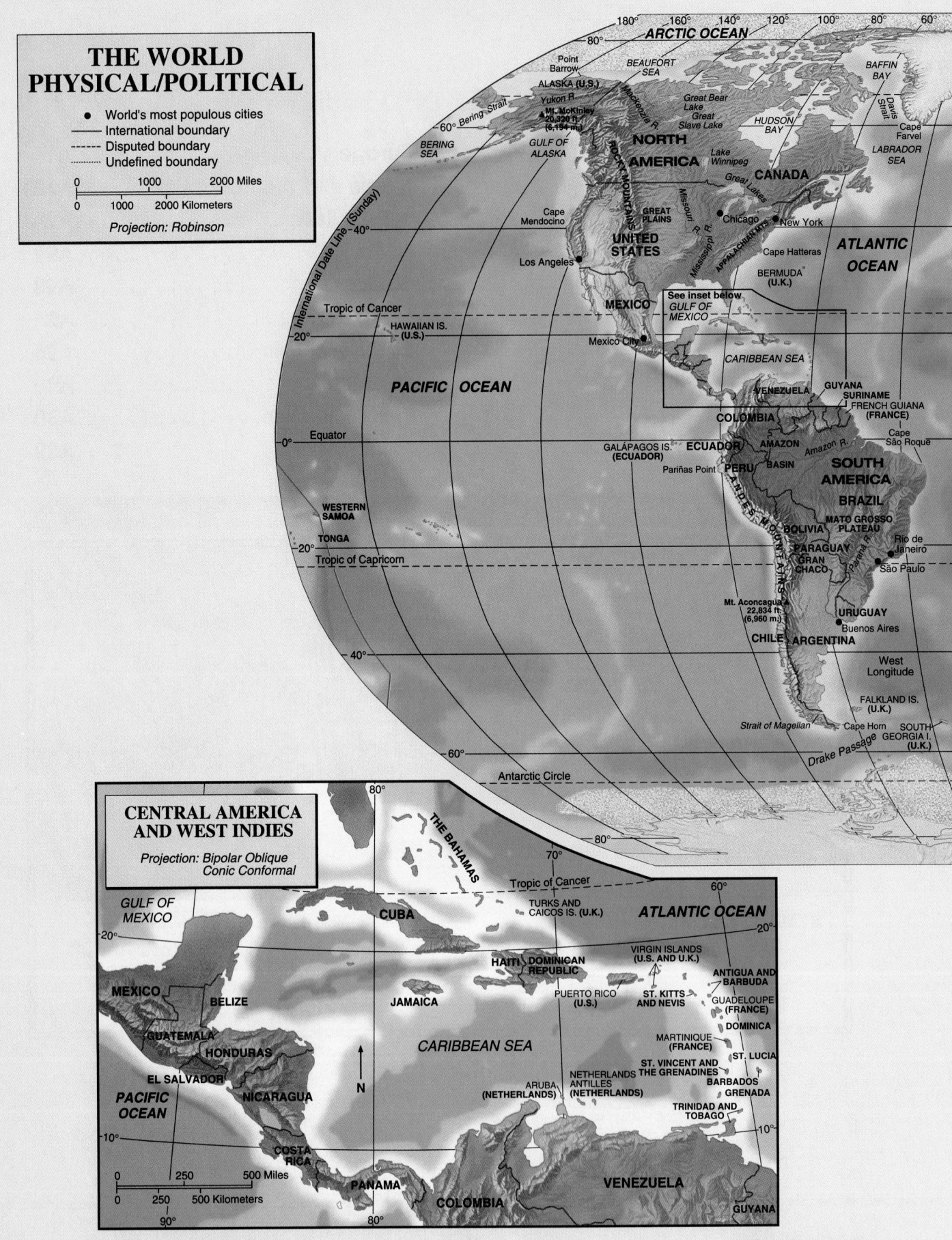

THE WORLD
PHYSICAL/POLITICAL

- • World's most populous cities
- — International boundary
- – – – Disputed boundary
- · · · · · Undefined boundary

| 0 | 1000 | 2000 Miles |
| 0 | 1000 | 2000 Kilometers |

Projection: Robinson

CENTRAL AMERICA AND WEST INDIES

Projection: Bipolar Oblique Conic Conformal

| 0 | 250 | 500 Miles |
| 0 | 250 | 500 Kilometers |

REFERENCE ATLAS

EUROPE

Projection: Azimuthal Equal Area

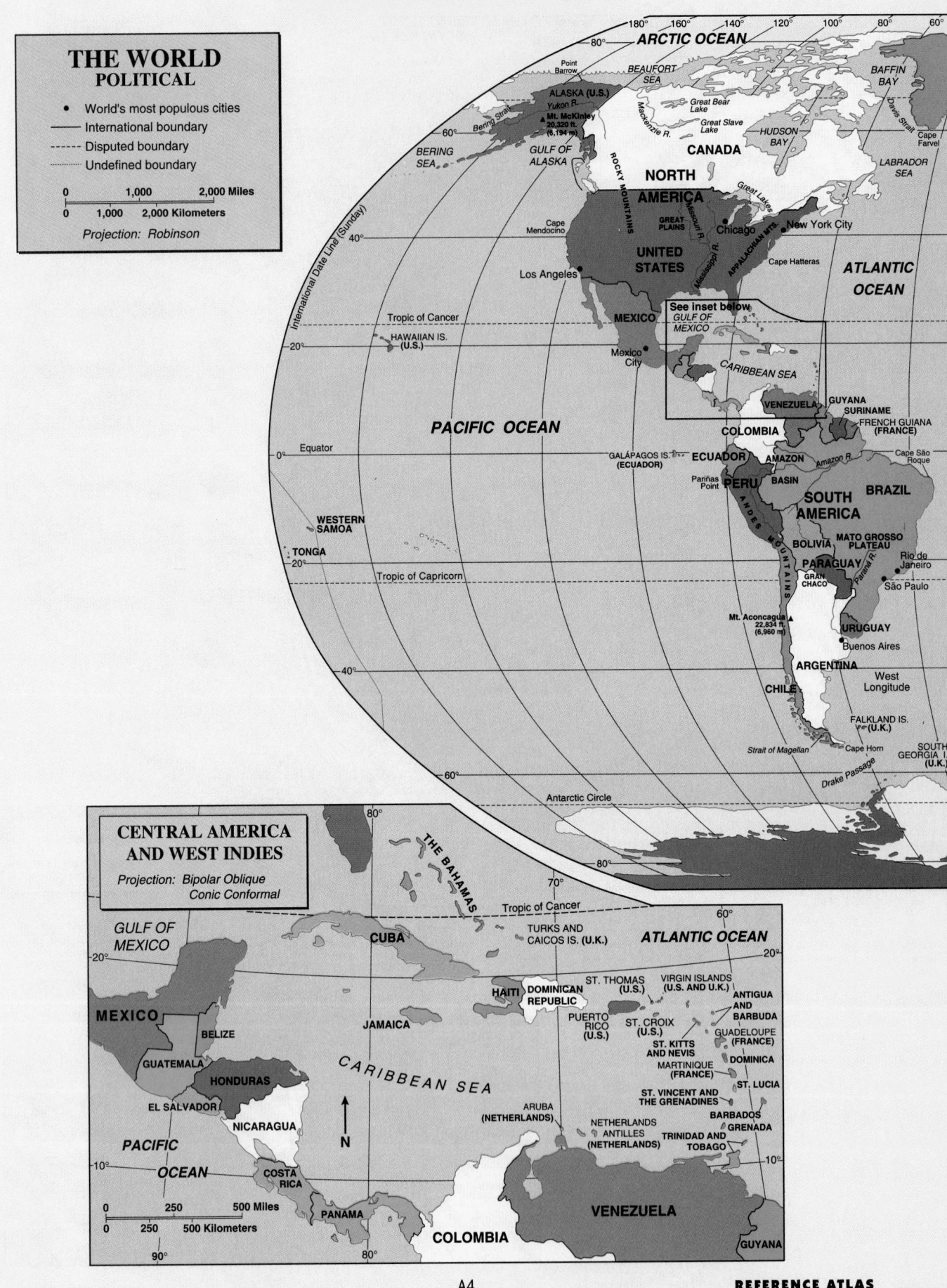

THE WORLD
POLITICAL

- World's most populous cities
— International boundary
-- Disputed boundary
··· Undefined boundary

0   1,000   2,000 Miles
0   1,000   2,000 Kilometers

Projection: Robinson

CENTRAL AMERICA AND WEST INDIES

Projection: Bipolar Oblique Conic Conformal

0   250   500 Miles
0   250   500 Kilometers

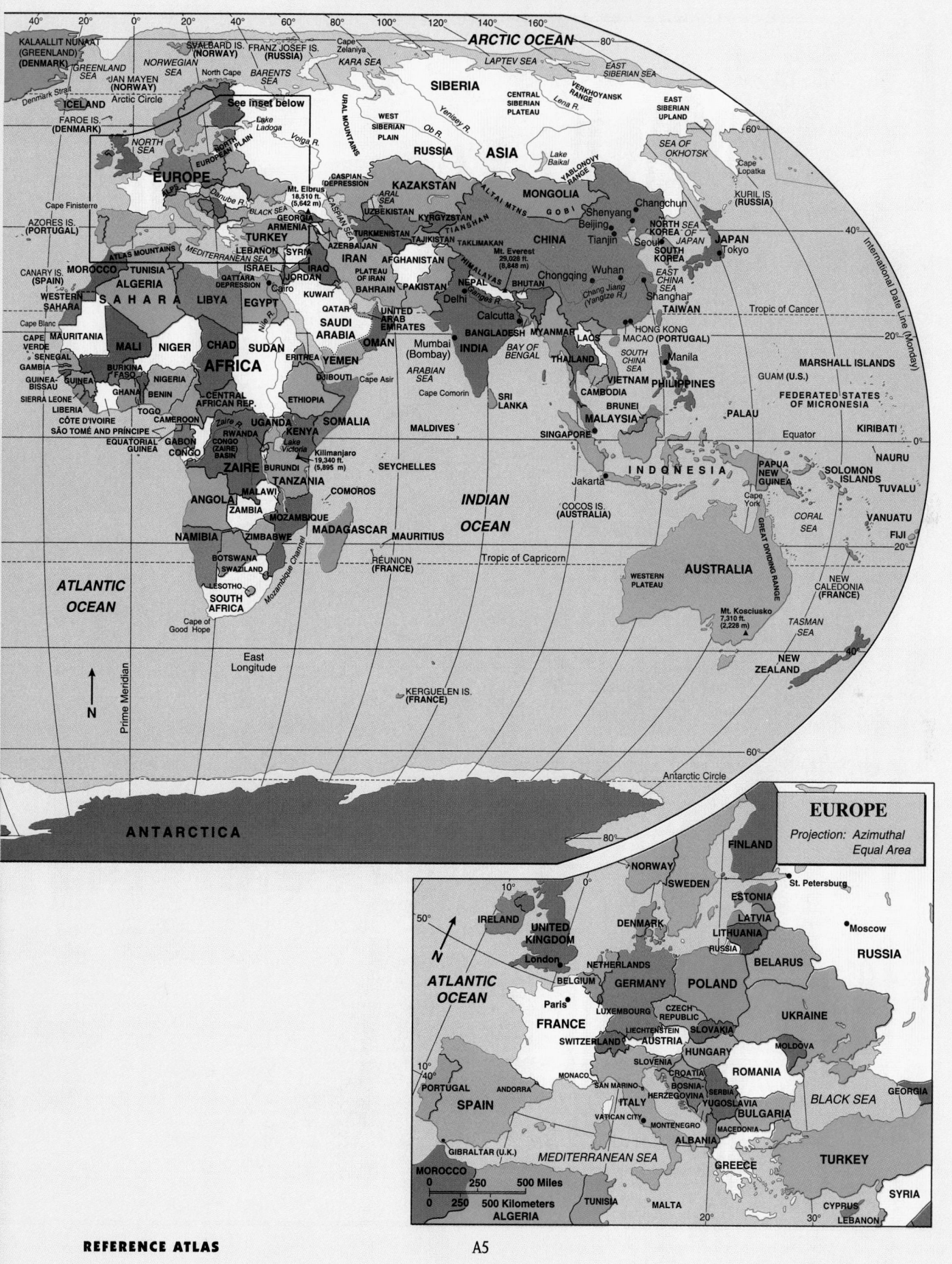

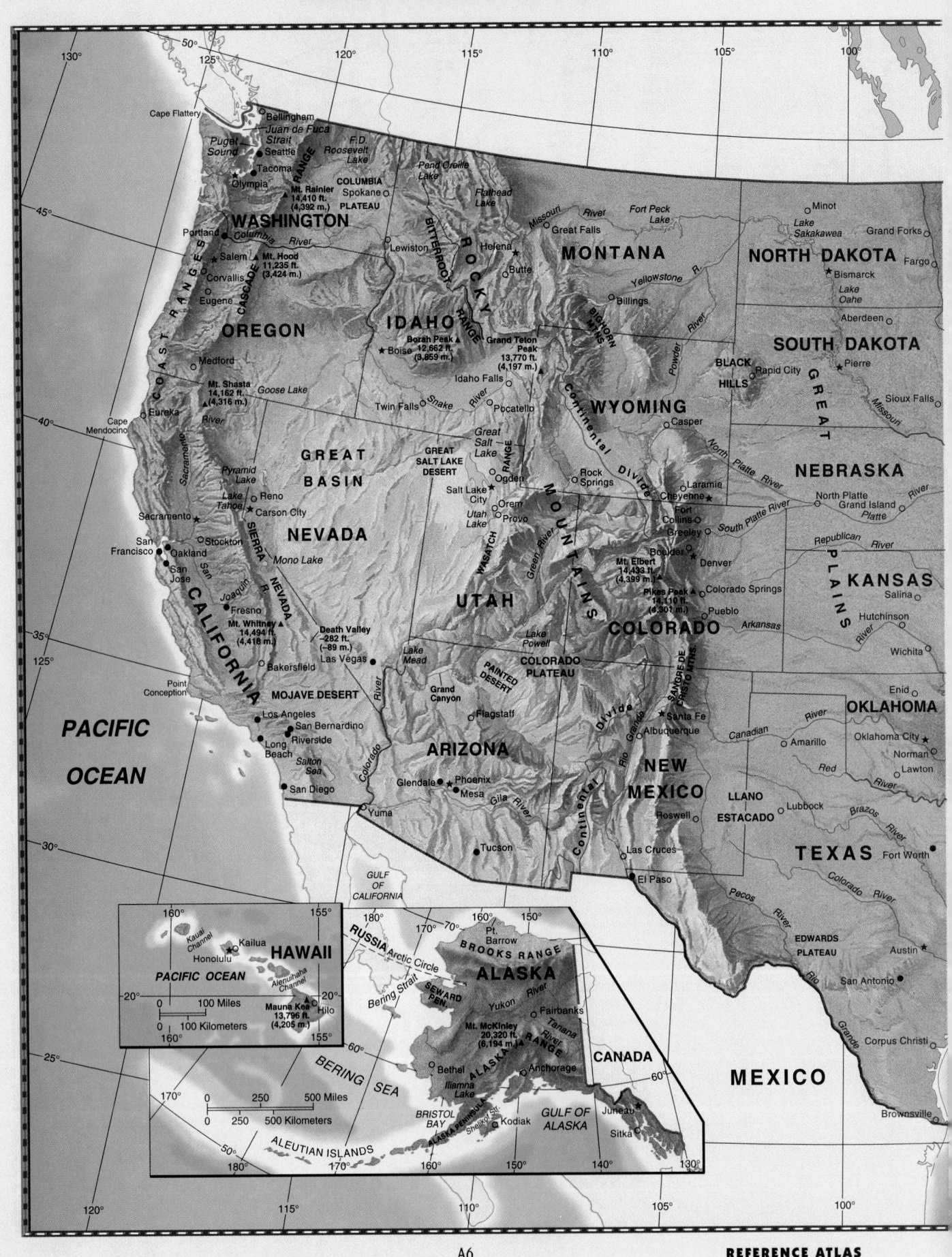

A6

UNITED STATES

- ⊛ National capital
- ★ State capital
- ● Major city
- ○ Other city
- ▬ International boundary
- — State boundary

0   150   300 Miles

0   150   300 Kilometers

Projection: Albers Equal Area

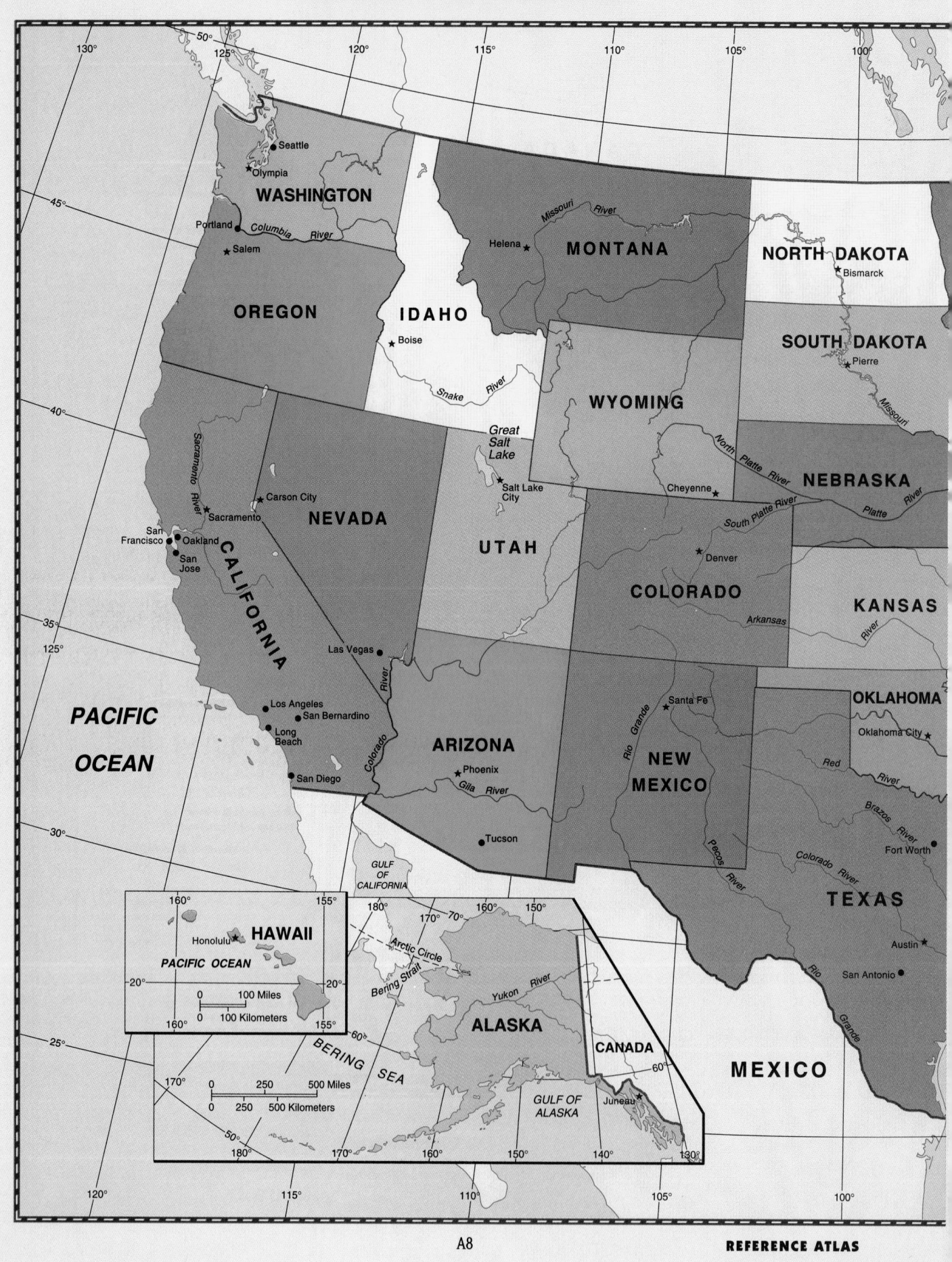

CANADA

MINNESOTA
Minneapolis · ★ St. Paul
MICHIGAN
Lake Superior
Lake Huron
WISCONSIN
Madison ★ · Milwaukee
Grand Rapids ·
Lansing ·
Detroit ·
Lake Michigan

IOWA
Des Moines ★
Omaha ·
Lincoln ★
Chicago ·
Gary ·
Hammond ·
ILLINOIS
Springfield ★

Topeka ★
Kansas City ·
Jefferson City ★
Kansas City ·
St. Louis ·
East St. Louis
MISSOURI

INDIANA
Indianapolis ★
Cincinnati ·
OHIO
Dayton ·
Columbus ★
Toledo ·
Cleveland ·
Akron · Canton ·
Youngstown ·
Pittsburgh ·

Lake Ontario
Lake Erie
Rochester ·
Buffalo ·
Syracuse ·
Albany ★
NEW YORK
Hartford ★
New York ·
Newark ·
PENNSYLVANIA
Harrisburg ★
Philadelphia ·
Camden ·
N.J.
Trenton ★
Dover ★
DEL.
MD.
Baltimore ·
Annapolis ★
Washington ⊛
D.C.

St. Lawrence River
MAINE
Augusta ★
Montpelier ★
N.H.
VT.
Concord ★
Boston ·
MASS.
Providence ★
CONN.
R.I.
Bridgeport ·
Hudson River

WEST VIRGINIA
Charleston ·
Frankfort ★
Louisville ·
KENTUCKY
Ohio River

VIRGINIA
Richmond ★
Newport News ·
Norfolk ·

ATLANTIC OCEAN

Raleigh ★
NORTH CAROLINA

ARKANSAS
Little Rock ★
Memphis ·
TENNESSEE
Nashville ★
Tennessee R.

Tulsa ·

Dallas ·

Columbia ★
SOUTH CAROLINA

Birmingham ·
ALABAMA
Montgomery ★
Jackson ★
MISSISSIPPI
GEORGIA
Atlanta ·

LOUISIANA
Baton Rouge ★
New Orleans ·
Houston ·

Jacksonville ·
Tallahassee ★
FLORIDA
Orlando ·
Tampa ·
St. Petersburg ·
Miami ·

GULF OF MEXICO

N

THE BAHAMAS

CUBA

UNITED STATES

⊛ National capital
★ State capital
· Major city
━━ International boundary
── State boundary

| 0 | | 150 | | 300 Miles |
| 0 | 150 | | 300 Kilometers | |

Projection: Albers Equal Area

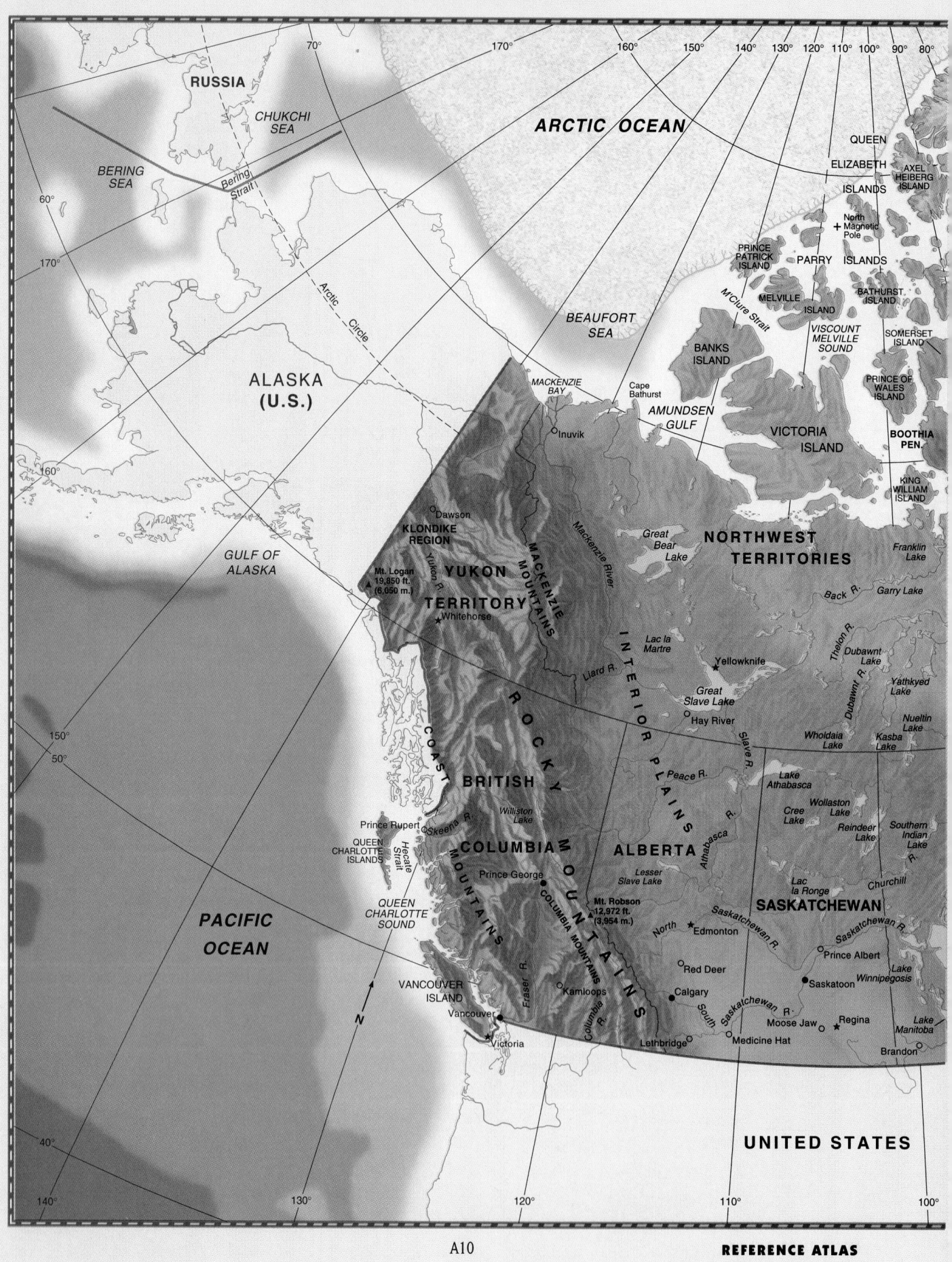

RUSSIA

CHUKCHI
SEA

BERING
SEA

Bering
Strait

ARCTIC OCEAN

QUEEN

ELIZABETH

AXEL
HEIBERG
ISLAND

ISLANDS

North
Magnetic
Pole

PRINCE
PATRICK
ISLAND

PARRY   ISLANDS

Arctic Circle

M'Clure Strait

MELVILLE
ISLAND

BATHURST
ISLAND

BEAUFORT
SEA

SOMERSET
ISLAND

ALASKA
(U.S.)

MACKENZIE
BAY

Cape
Bathurst

BANKS
ISLAND

VISCOUNT
MELVILLE
SOUND

PRINCE OF
WALES
ISLAND

Inuvik

AMUNDSEN
GULF

VICTORIA
ISLAND

BOOTHIA
PEN.

KING
WILLIAM
ISLAND

Dawson

KLONDIKE
REGION

Great
Bear
Lake

NORTHWEST
TERRITORIES

Franklin
Lake

MACKENZIE

Mackenzie River

GULF OF
ALASKA

Mt. Logan
19,850 ft.
(6,050 m.)

YUKON

Yukon R.

MOUNTAINS

TERRITORY

Whitehorse

Back R.

Garry Lake

Lac la
Martre

Yellowknife

Thelon R.

Dubawnt
Lake

Yathkyed
Lake

Liard R.

Great
Slave Lake

Dubawnt R.

INTERIOR

Hay River

Nueltin
Lake

COAST

Whoidaia
Lake

Kasba
Lake

ROCKY

BRITISH

PLAINS

Peace R.

Slave R.

Lake
Athabasca

Cree
Lake

Wollaston
Lake

Reindeer
Lake

Southern
Indian
Lake

Prince Rupert

Skeena R.

Williston
Lake

COLUMBIA

Athabasca R.

QUEEN
CHARLOTTE
ISLANDS

Hecate
Strait

MOUNTAINS

Prince George

Lesser
Slave Lake

ALBERTA

Lac
la Ronge

Churchill

PACIFIC

MOUNTAINS

Mt. Robson
12,972 ft.
(3,954 m.)

COLUMBIA MOUNTAINS

North

Saskatchewan R.

SASKATCHEWAN

QUEEN
CHARLOTTE
SOUND

Edmonton

Saskatchewan R.

Prince Albert

OCEAN

Red Deer

Lake
Winnipegosis

VANCOUVER
ISLAND

Fraser R.

Kamloops

Calgary

Saskatoon

Columbia R.

South   Saskatchewan   R.

Lake
Manitoba

Vancouver

N

Moose Jaw

Regina

Victoria

Lethbridge

Medicine Hat

Brandon

UNITED STATES

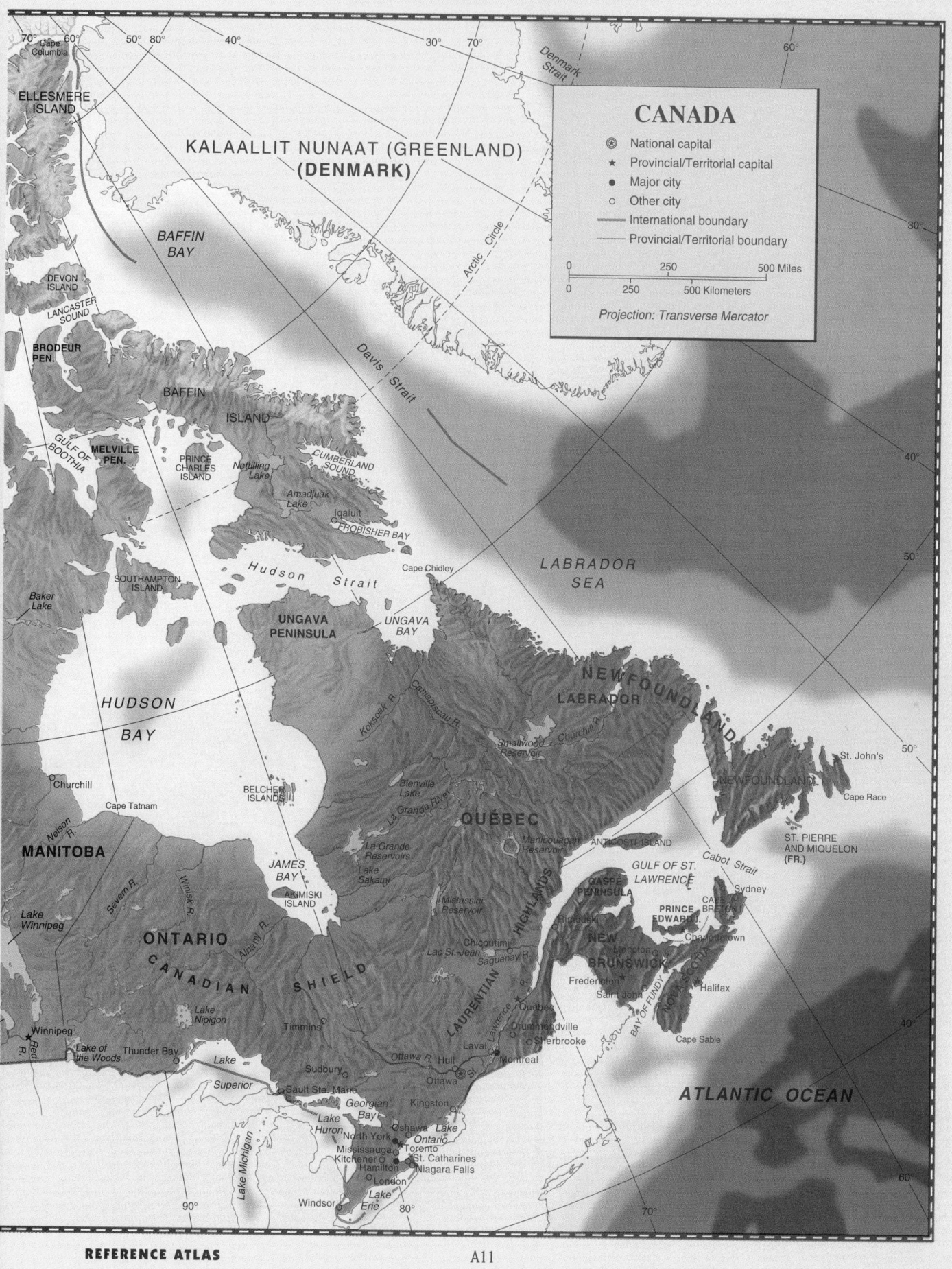

CANADA

- ◎ National capital
- ★ Provincial/Territorial capital
- ● Major city
- ○ Other city
- —— International boundary
- —— Provincial/Territorial boundary

0   250   500 Miles
0   250   500 Kilometers

Projection: Transverse Mercator

70° Cape Columbia   60°   50°   80°   40°   30°   70°   Denmark Strait   60°   30°

ELLESMERE ISLAND

KALAALLIT NUNAAT (GREENLAND) (DENMARK)

BAFFIN BAY

DEVON ISLAND
LANCASTER SOUND
BRODEUR PEN.
Arctic Circle

BAFFIN ISLAND

GULF OF BOOTHIA
MELVILLE PEN.
PRINCE CHARLES ISLAND
Nettilling Lake
CUMBERLAND SOUND

40°

Amadjuak Lake
Iqaluit
FROBISHER BAY

Davis Strait

Hudson Strait
Cape Chidley

LABRADOR SEA

50°

SOUTHAMPTON ISLAND
Baker Lake

UNGAVA PENINSULA
UNGAVA BAY

HUDSON BAY

NEWFOUNDLAND
LABRADOR

Koksoak R.
Caniapiscau R.
Smallwood Reservoir
Churchill R.

50°
St. John's
NEWFOUNDLAND
Cape Race

Churchill
Cape Tatnam

BELCHER ISLANDS

Blenville Lake
La Grande R.

QUÉBEC

ST. PIERRE AND MIQUELON (FR.)

Nelson R.

MANITOBA

JAMES BAY
AKIMISKI ISLAND

La Grande Reservoirs
Lake Sakami
Manicouagan Reservoir
ANTICOSTI ISLAND

Cabot Strait

Sydney

GASPE PENINSULA
GULF OF ST. LAWRENCE

Lake Winnipeg

Severn R.
Winisk R.

Mistassini Reservoir

CAPE BRETON I.
PRINCE EDWARD I.
Charlottetown

ONTARIO

Albany R.

Chicoutimi
Lac St. Jean
Saguenay R.

NEW BRUNSWICK

NOVA SCOTIA

Lake of the Woods
Winnipeg
Red R.

CANADIAN   SHIELD

LAURENTIAN HIGHLANDS

St. Lawrence R.
Québec
Laval
Drummondville
Sherbrooke
Montreal

Rimouski

Fredericton
Saint John
Moncton

Halifax

BAY OF FUNDY

Lake Nipigon
Timmins

Cape Sable

40°

Thunder Bay
Lake Nipigon

Sudbury

Hull
Ottawa R.
Ottawa
Kingston

ATLANTIC OCEAN

Sault Ste. Marie
Lake Superior
Georgian Bay
Lake Huron
North York
Mississauga
Kitchener
Hamilton
Oshawa
Toronto
Lake Ontario
St. Catharines
Niagara Falls
London

60°

90°
Lake Michigan
Windsor
Lake Erie
80°
70°

MEXICO, the CARIBBEAN, and CENTRAL AMERICA

⊛ National capital
• Major city
— International boundary

0   250   500 Miles
0   250   500 Kilometers

Projection: Azimuthal Equal Area

Map labels:
BAJA CALIFORNIA PENINSULA
GULF OF CALIFORNIA
SIERRA MADRE OCCIDENTAL
SIERRA MADRE ORIENTAL
MEXICAN PLATEAU
MEXICO
SIERRA MADRE DEL SUR
PACIFIC OCEAN
GULF OF MEXICO
CAMPECHE BAY
YUCATÁN PENINSULA
GULF OF HONDURAS
GUATEMALA
BELIZE
EL SALVADOR
Tropic of Cancer

Cities: Ciudad Juárez, Chihuahua, Monterrey, San Pedro River, Tampico, León, Guadalajara, Mexico City, Puebla, Veracruz, Balsas River, Rio Grande, Mérida, Belize City, Belmopan, Dolores, El Progreso, Guatemala, Tegucigalpa, Quezaltenango, Santa Ana, San Salvador

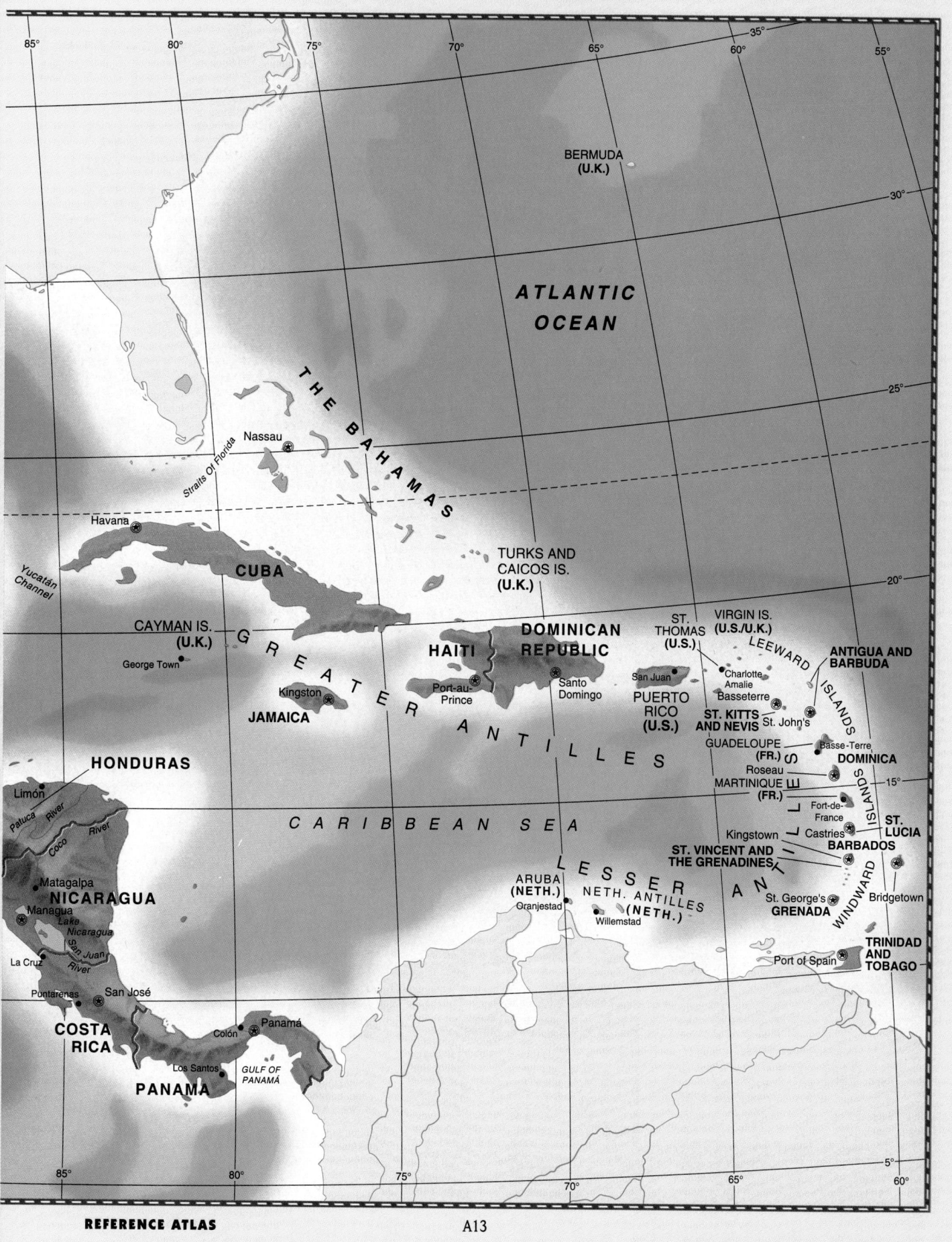

85°   80°   75°   70°   65°   35°   60°   55°

30°

**BERMUDA
(U.K.)**

*ATLANTIC
OCEAN*

25°

Nassau

Straits Of Florida

THE BAHAMAS

Havana

*Yucatán
Channel*

CUBA

**TURKS AND
CAICOS IS.
(U.K.)**

20°

**CAYMAN IS.
(U.K.)**

G R E A T E R

ST.
THOMAS
(U.S.)

**VIRGIN IS.
(U.S./U.K.)**

George Town

**DOMINICAN
REPUBLIC**

HAITI

LEEWARD

**ANTIGUA AND
BARBUDA**

Kingston

Port-au-
Prince

Santo
Domingo

San Juan

Charlotte
Amalie
Basseterre

ISLANDS

St. John's

JAMAICA

A N T I L L E S

**PUERTO
RICO
(U.S.)**

**ST. KITTS
AND NEVIS**

HONDURAS

**GUADELOUPE
(FR.)**

Basse-Terre

DOMINICA

Limón

Roseau

15°

Patuca
River

Coco
River

River

**MARTINIQUE
(FR.)**

Fort-de-
France

L
E
S
S
E
R

A
N
T
I
L
L
E
S

ISLANDS

**ST.
LUCIA**

CARIBBEAN   SEA

Kingstown

Castries

Matagalpa

**ST. VINCENT AND
THE GRENADINES**

BARBADOS

NICARAGUA

Managua

*Lake
Nicaragua*

L E S S E R

**ARUBA
(NETH.)**

NETH. ANTILLES
(NETH.)

WINDWARD

Bridgetown

*San Juan
River*

Oranjestad

St. George's
GRENADA

La Cruz

Willemstad

Puntarenas

San José

Port of Spain

**TRINIDAD
AND
TOBAGO**

**COSTA
RICA**

Colón

Panamá

Los Santos

*GULF OF
PANAMÁ*

PANAMA

5°

85°   80°   75°   70°   65°   60°

NORTH AMERICA

- ⊛ National capital
- ● Major city
- ○ Other city
- — International boundary

0   250   500   750 Miles
0   250   500   750 Kilometers

Projection: Azimuthal Equal Area

SOUTH AMERICA

- ⊛ National capital
- ● Major city
- ○ Other city
- — International boundary

0   250   500 Miles
0   250   500 Kilometers

Projection: Azimuthal Equal Area

North Cape
BARENTS SEA
Murmansk
KOLA PENINSULA
TIMAN RIDGE
Pechora R.
WHITE SEA
Arkhangel'sk
White Sea-Baltic Waterway
N. Dvina River
Vychegda River
Mt. Konzhakovskiy 5,147 ft. (1,569 m.)
URAL MOUNTAINS
FINLAND
Lake Onega
Sukhona River
Kama R.
Perm
ASIA
Tampere
Lake Saimaa
Lake Ladoga
Volga-Baltic Waterway
Turku
Helsinki
Espoo
GULF OF FINLAND
St. Petersburg
Rybinsk Reservoir
Kazan
Kama River
Ufa
Ural River
ESTONIA
Tallinn
Chudskoye Lake
Yaroslavl
Volga River
Volga-Baltic Waterway
Nizhniy Novgorod
Kuybyshev Reservoir
GULF OF RIGA
LATVIA
Riga
Dvina R.
BALTIC PLAIN
EUROPEAN PLAIN
Moscow
River
Oka
Tula
RUSSIA
Samara
Orenburg
LITHUANIA
Kaunas
Vilnius
Minsk
Smolensk
CENTRAL RUSSIAN UPLAND
Don River
Voronezh
Saratov
Volga River
Volgograd Reservoir
VOLGA UPLAND
Ural River
KAZAKSTAN
ARAL SEA
BELARUS
Pripet River
Desna R.
Kursk
Kiev
Kremenchug Reservoir
Kharkov
Lugansk
Tsimlyansk Reservoir
Volgograd
Volga River
DEPRESSION
Astrakhan
Delta of the Volga
Lvov
Dnieper
UKRAINE
DNIEPER UPLAND
Dniester
Dnepropetrovsk
Krivoy Rog
Zaporozhye
Donetsk
Rostov
Don River
CASPIAN
CASPIAN SEA
CARPATHIAN MTNS.
Debrecen
MOLDOVA
Chisinau
Prut River
DNIEPER LOWLAND
Dnieper River
Kakhovka Res.
SEA OF AZOV
Krasnodar
Grozny
Cluj-Napoca
Odessa
CRIMEA
ROMANIA
Timișoara
Brașov
Mt. Elbrus 18,510 ft. (5,642 m.)
CAUCASUS MTNS.
WALLACHIA PLAIN
Bucharest
River
Danube
Constanța
BLACK SEA
SERBIA
Ruse
Niš
BULGARIA
Varna
Sofia
Burgas
Plovdiv
Skopje
Musala Peak 9,536 ft. (2,926 m.)
Bosporus
PENINSULA
TURKEY
Salonika
Dardanelles
SEA OF MARMARA
BALKAN
Larissa
AEGEAN SEA
ASIA
GREECE
Patras
Athens
Piraeus
PELOPONNESE PEN.
RHODES
CRETE (GR.)
Iráklion

ARCTIC OCEAN

FRANZ JOSEF ISLANDS

EUROPE

BARENTS SEA

Murmansk

KOLA PENINSULA

Cape Zelaniya

NOVAYA ZEMLYA

KARA SEA

BALTIC SEA

WHITE SEA

Kara Strait

YAMAL PEN.

GYDAN PENINSULA

GULF OF FINLAND

Baltic-White Sea Canal

Lake Ladoga

Arkhangel'sk

(RUSSIA)

St. Petersburg

Volga-Baltic Waterway

VALDAI HILLS

N. Dvina R.

Vychegda

TIMAN RIDGE

River

Yenisey River

Minsk

Rybinsk Res.

Vologda

Sukhona R.

R.

Pechora

River

WEST SIBERIAN PLAIN

BELARUS

Lvov

DNIEPER UPLAND

Yaroslovl

NORTHERN HILLS

Ob

Urengoy

Dnieper

R.

Moscow

Ivanovo

Volga

URAL MOUNTAINS

Kiev

DNIEPER LOWLAND

Tula

Ryazan'

Nizhniy Novgorod

Kamsk Res.

Mt. Konzhakovsky 5,147 ft. (1,569 m.)

Ob

Vakh

R.

UKRAINE

Kazan

Izhevsk

R.

Perm

MOLDOVA

Chisinau

Kharkov

Don

River

Voronezh

Kuybyshev Res.

Ul'yanovsk

Kama

Yekaterinburg

Irtysh

River

Odessa

Nikolayev

Krivoy Rog

Penza

Saratov

VOLGA UPLAND

Tol'yatti

Ufa

Tobol

R.

Tomsk

Dnepropetrovsk

Zaporozh'ye

R.

Samara

Chelyabinsk

R.

Donetsk

Lugansk

SEA OF AZOV

Rostov

Volgograd Reservoir

Omsk

L. Chany

Kemerovo

BLACK SEA

Krasnodar

Tsimlyansk Res.

Volga

Volgograd

Ural

R.

Orenburg

KYRGYZ

STEPPE

TURGAY PLATEAU

Novosibirsk

Novosibirsk Res.

Novokuznetsk

CAUCASUS

Mt. Elbrus 18,510 ft. (5,642 m.)

CASPIAN DEPRESSION

Astrakhan

Ishim

R.

Barnaul

GEORGIA

MTNS.

Tbilisi

KAZAKSTAN

KAZAK UPLAND

Karaganda

Semipalatinsk

Mt. Belukha 14,783 ft. (4,506 m.)

ARMENIA

Yerevan

AZERBAIJAN

CASPIAN

ARAL SEA

L. Zaysan

AZERBAIJAN

Baku

SEA

USTYURT PLATEAU

BETPAK-DALA DESERT

Lake Balkhash

L. Alakol

KARA BOGAZ GOL GULF

PLAINS OF TURAN

Syr

Kzyl-Orda

Darya

Ili R.

ASIA

TURKMENISTAN

UZBEKISTAN

Almaty

KARAKUM

Amu

Tashkent

Bishkek

L. Issyk-Kul

DESERT

Ashkhabad

Samarkand

Darya

KYRGYZSTAN

ALAY MOUNTAINS

Dushanbe

Communism Pk. 24,590 ft. (7,495 m.)

TAJIKISTAN

RUSSIA AND
THE EURASIAN REPUBLICS

⊛ National capital
● Major city
○ Other city
— International boundary

| 0 | 250 | 500 Miles |
| 0 | 250 | 500 Kilometers |

Projection: Two-Point Equidistant

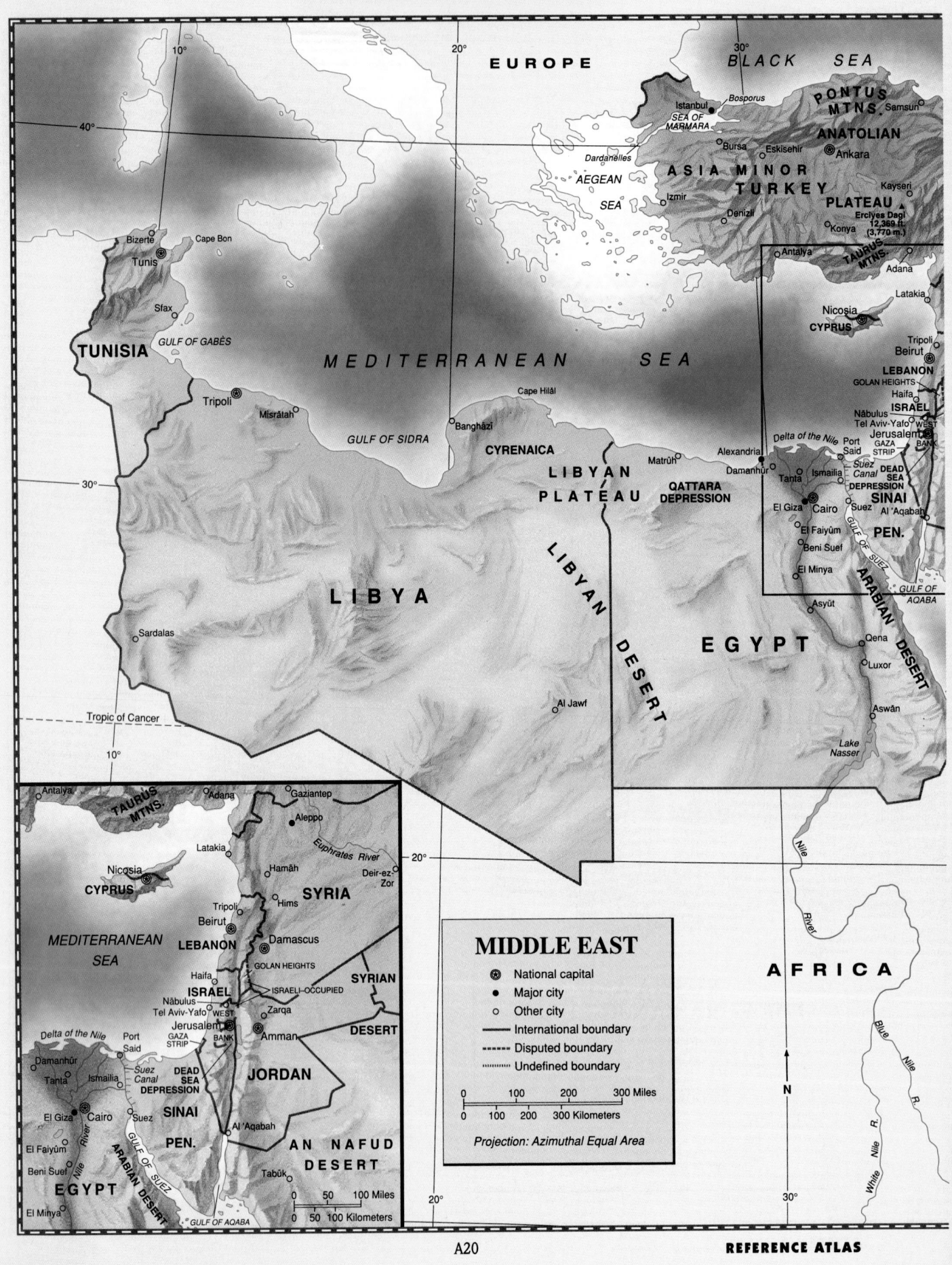

EUROPE

BLACK SEA

Istanbul · *Bosporus*
SEA OF MARMARA

Dardanelles

AEGEAN SEA

Izmir ·

PONTUS MTNS. · Samsun

ANATOLIAN

⊛ Ankara

ASIA MINOR

TURKEY

PLATEAU

Bursa · Eskisehir ·

Denizli ·

· Konya

Kayseri ·

Erciyes Dagi
12,369 ft.
(3,770 m.)

Antalya ·

TAURUS MTNS.

· Adana

Latakia ·

Nicosia ⊛
CYPRUS

Tripoli
Beirut ·
LEBANON
GOLAN HEIGHTS
Haifa ·

ISRAEL
Nâbulus ·
Tel Aviv-Yafo · WEST
Jerusalem ⊛ BANK
GAZA STRIP
DEAD SEA DEPRESSION

SINAI
PEN.
Al 'Aqabah

GULF OF AQABA

Bizerte ·
Cape Bon
Tunis ⊛

Sfax ·

TUNISIA

GULF OF GABÈS

Tripoli ·

Misrâtah ·

M E D I T E R R A N E A N   S E A

Cape Hilâl

Banghâzi ·

CYRENAICA

LIBYAN PLATEAU

QATTARA DEPRESSION

Matrûh ·

Alexandria ·
Damanhûr ·

Delta of the Nile

Tanta · Ismailia · Port Said
Suez Canal

El Giza · Cairo ⊛ Suez

El Faiyûm ·
Beni Suef ·

El Minya ·

GULF OF SUEZ

GULF OF SIDRA

LIBYA

L I B Y A N

D E S E R T

E G Y P T

Sardalas ·

Al Jawf ·

Asyût ·

Qena ·
Luxor ·

A R A B I A N   D E S E R T

Aswân ·

Lake Nasser

Tropic of Cancer

10°

20°

30°

40°

10°

20°

30°

Inset map (lower left)

Antalya ·

TAURUS MTNS.

· Adana
Gaziantep ·

· Aleppo

Nicosia ⊛
CYPRUS

Latakia ·

Hamâh ·

Euphrates River

Deir-ez-Zor ·

20°

SYRIA

MEDITERRANEAN SEA

Tripoli ·
Beirut ⊛
LEBANON

Hims ·

· Damascus ⊛

Haifa ·
GOLAN HEIGHTS

ISRAELI-OCCUPIED

SYRIAN

ISRAEL
Nâbulus ·
Tel Aviv-Yafo · WEST
Jerusalem ⊛ BANK
GAZA STRIP
Zarqa ·
· Amman ⊛

DESERT

Delta of the Nile
Damanhûr ·
Tanta · Port Said
Ismailia ·
Suez Canal
DEAD SEA DEPRESSION

JORDAN

El Giza · Cairo ⊛

El Faiyûm ·

Nile River Suez ·

SINAI
PEN.

Al 'Aqabah ·

GULF OF SUEZ

A R A B I A N   D E S E R T

AN NAFUD DESERT

Tabûk ·

Beni Suef ·

EGYPT

El Minya ·

GULF OF AQABA

20°

MIDDLE EAST

⊛ National capital
● Major city
○ Other city
— International boundary
--- Disputed boundary
···· Undefined boundary

0   100   200   300 Miles
0 100 200 300 Kilometers

Projection: Azimuthal Equal Area

0   50   100 Miles
0 50 100 Kilometers

AFRICA

Nile
River

N

Blue Nile R.

White Nile R.

30°

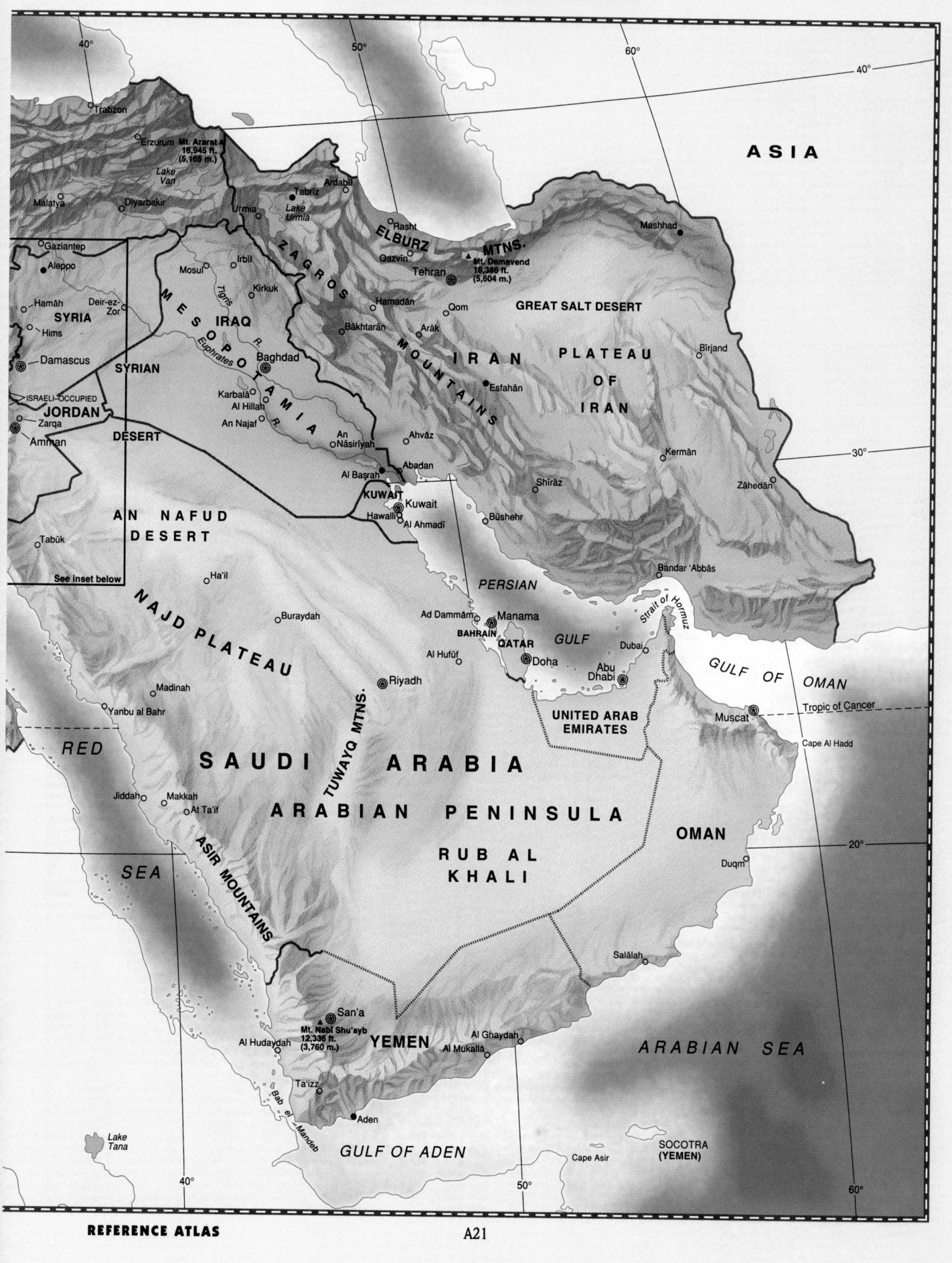

40°   50°   60°   40°

ASIA

Trabzon

Erzurum   **Mt. Ararat** ▲
16,945 ft.
(5,165 m.)

*Lake
Van*

Malatya

Diyarbakir

Urmia

Ardabil

Tabriz

*Lake
Urmia*

Rasht

Mashhad

ELBURZ

MTNS.

Qazvin

▲ **Mt. Demavend**
18,386 ft.
(5,604 m.)

Gaziantep

Aleppo

Mosul

Irbīl

Tehran

GREAT SALT DESERT

Hamāh

SYRIA

Hims

Deir-ez-
Zor

Kirkuk

Hamadān

Qom

Arāk

IRAQ

Tigris R.

Bākhtarān

Esfahān

**PLATEAU
OF
IRAN**

Bīrjand

Damascus

MESOPOTAMIA

Euphrates R.

Baghdad

**IRAN
MOUNTAINS**

SYRIAN

ISRAELI-OCCUPIED

Karbalā
Al Hillah

An
Najaf

Ahvāz

Kermān

JORDAN

Zarqa

An
ⓞNāsiriyah

Shīrāz

Zāhedān

Amman

DESERT

Abadan

30°

Al Başrah

Abadan

**AN NAFUD
DESERT**

KUWAIT

Hawalli

Kuwait

Al Ahmadī

Būshehr

Bandar 'Abbās

Tabūk

Strait of Hormuz

Ha'il

PERSIAN

See inset below

Buraydah

Ad Dammām

Manama

GULF OF OMAN

BAHRAIN

NAJD PLATEAU

Al Hufūf

QATAR

GULF

Dubai

Madinah

ⓞ Riyadh

Doha

Abu
Dhabi

Muscat

Yanbu al Bahr

TUWAYQ MTNS.

**UNITED ARAB
EMIRATES**

Tropic of Cancer

RED

Cape Al Hadd

SAUDI   ARABIA

SEA

Jiddah

Makkah

At Ta'if

ARABIAN   PENINSULA

OMAN

20°

**RUB AL
KHALI**

Duqm

ASIR MOUNTAINS

Salālah

Mt. Nabī Shu'ayb ▲
12,336 ft.
(3,760 m.)

San'a

Al Ghaydah

ARABIAN SEA

Al Hudaydah

YEMEN

Al Mukallā

*Lake
Tana*

Ta'izz

Bab el Mandeb

Aden

GULF OF ADEN

Cape Asir

**SOCOTRA
(YEMEN)**

40°   50°   60°

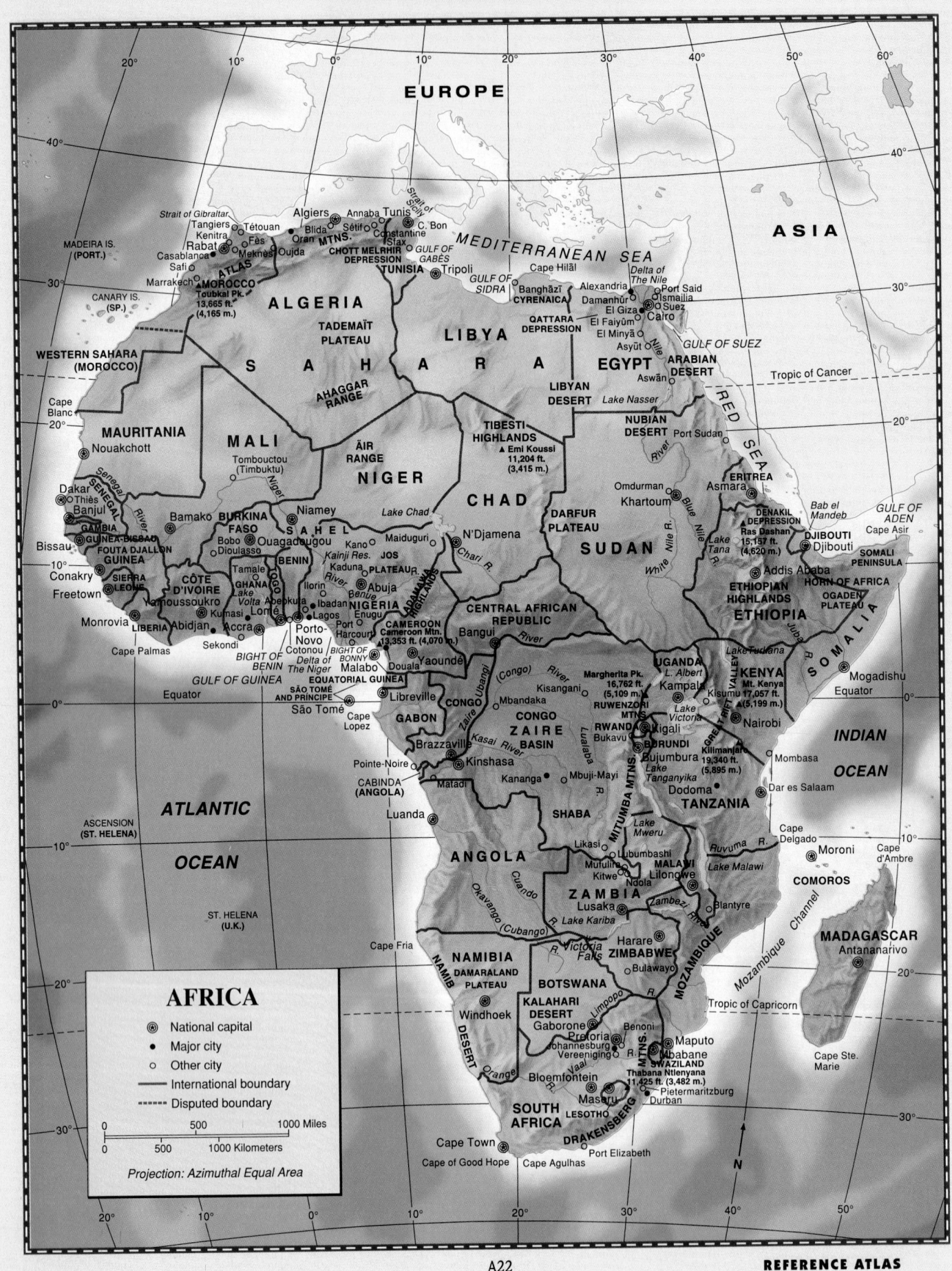

AFRICA

- ⊛ National capital
- • Major city
- ○ Other city
- —— International boundary
- ----- Disputed boundary

0   500   1000 Miles

0   500   1000 Kilometers

Projection: Azimuthal Equal Area

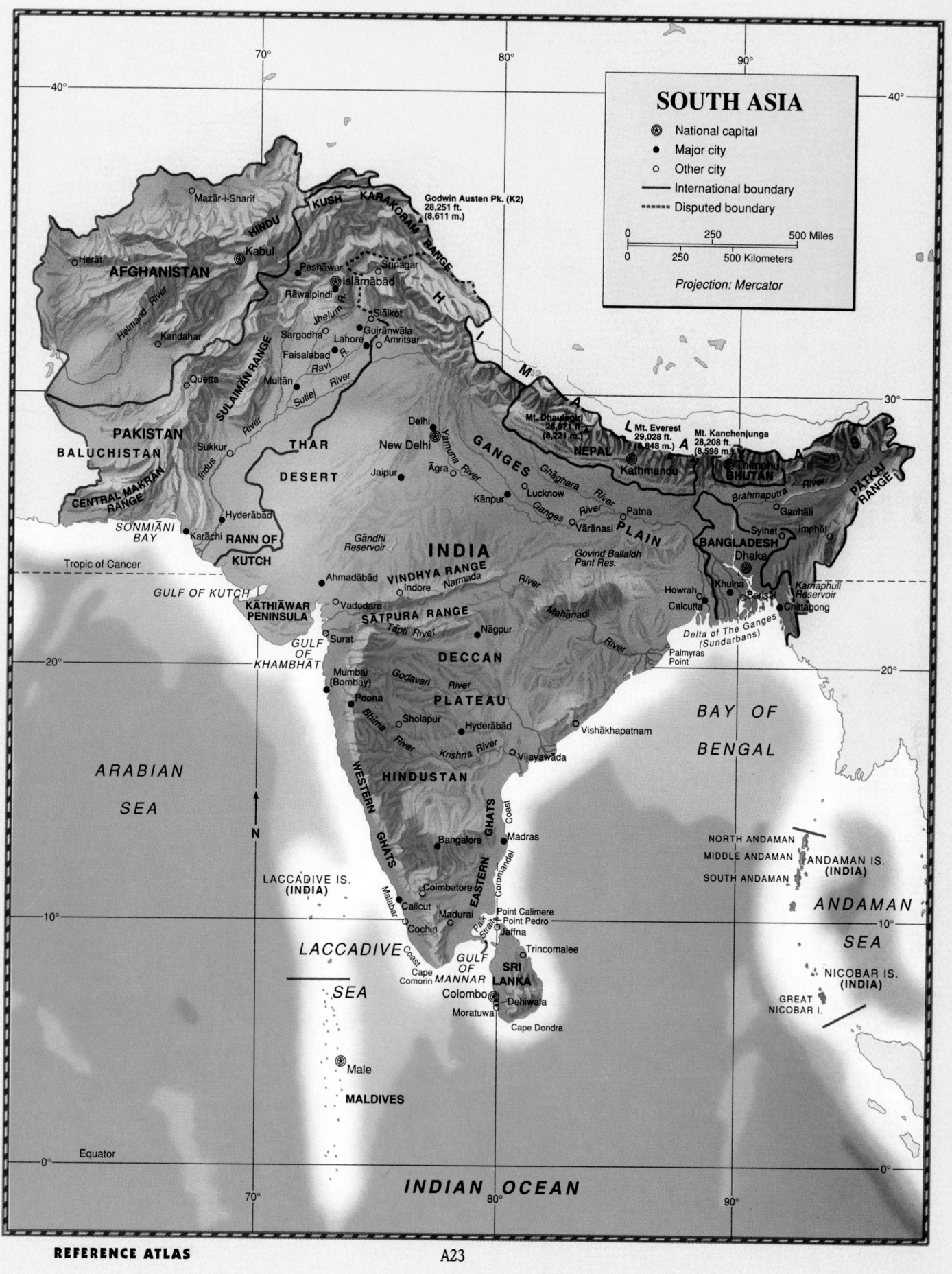

SOUTH ASIA

- ⊛ National capital
- ● Major city
- ○ Other city
- —— International boundary
- - - - Disputed boundary

| 0 | 250 | 500 Miles |

| 0 | 250 | 500 Kilometers |

Projection: Mercator

Mazār-i-Sharif

HINDU KUSH
KARAKORAM RANGE
Godwin Austen Pk. (K2)
28,251 ft.
(8,611 m.)

Herāt
Kabul
AFGHANISTAN
Peshāwar
Srinagar
Islāmābād
Rāwalpindi
Sialkot
Kandahar
Jhelum R.
Gujrānwāla
Sargodha
Lahore
Amritsar
Faisalabad
Ravi R.
Quetta
Multān
Sutlej River
Helmand River
SULAIMĀN RANGE

PAKISTAN
BALUCHISTAN
Sukkur
Indus River
THAR
DESERT
Delhi
New Delhi
Yamuna River
GANGES
CENTRAL MAKRĀN RANGE
Jaipur
Āgra
Kānpur
Lucknow
Ghāghara River
Ganges River
SONMIĀNI BAY
Hyderābād
Karāchi
RANN OF KUTCH
Gāndhi Reservoir
INDIA
Patna
Vārānasi
PLAIN
Tropic of Cancer
GULF OF KUTCH
Ahmadābād
VINDHYA RANGE
Indore
Narmada River
Govind Ballaldh Pant Res.
KĀTHIĀWAR PENINSULA
Vadodara
SĀTPURA RANGE
Tāpti River
River
Mahānadi River
GULF OF KHAMBHĀT
Surat
DECCAN
Nāgpur
Mumbai (Bombay)
Godavari River
Poona
PLATEAU
Sholapur
Hyderābād
Bhima River
Krishna River
Vijayawāda
Vishākhapatnam
WESTERN GHATS
HINDUSTAN
EASTERN GHATS
Bangalore
Madras
Coimbatore
Coast
Coromandel
Calicut
Madurai
Point Calimere
Point Pedro
Malabar
Cochin
Pak Strait
Jaffna
Trincomalee
Cape Comorin
GULF OF MANNAR
SRI LANKA
Colombo
Deniwala
Moratuwa
Cape Dondra

Mt. Dhaulāgiri
26,671 ft.
(8,221 m.)
Mt. Everest
29,028 ft.
(8,848 m.)
Mt. Kanchenjunga
28,208 ft.
(8,598 m.)
NEPAL
Kathmandu
HIMALAYA
Thimphu
BHUTAN
Brahmaputra River
Gauhāti
Sylhet
Imphāl
PATKAI RANGE
BANGLADESH
Dhaka
Khulna
Barisal
Karnaphuli Reservoir
Howrah
Calcutta
Chittagong
Delta of The Ganges (Sundarbans)
Palmyras Point

ARABIAN SEA

BAY OF BENGAL

LACCADIVE IS. (INDIA)

LACCADIVE SEA

NORTH ANDAMAN
MIDDLE ANDAMAN
SOUTH ANDAMAN
ANDAMAN IS. (INDIA)

ANDAMAN SEA

NICOBAR IS. (INDIA)

GREAT NICOBAR I.

⊛ Male
MALDIVES

Equator

INDIAN OCEAN

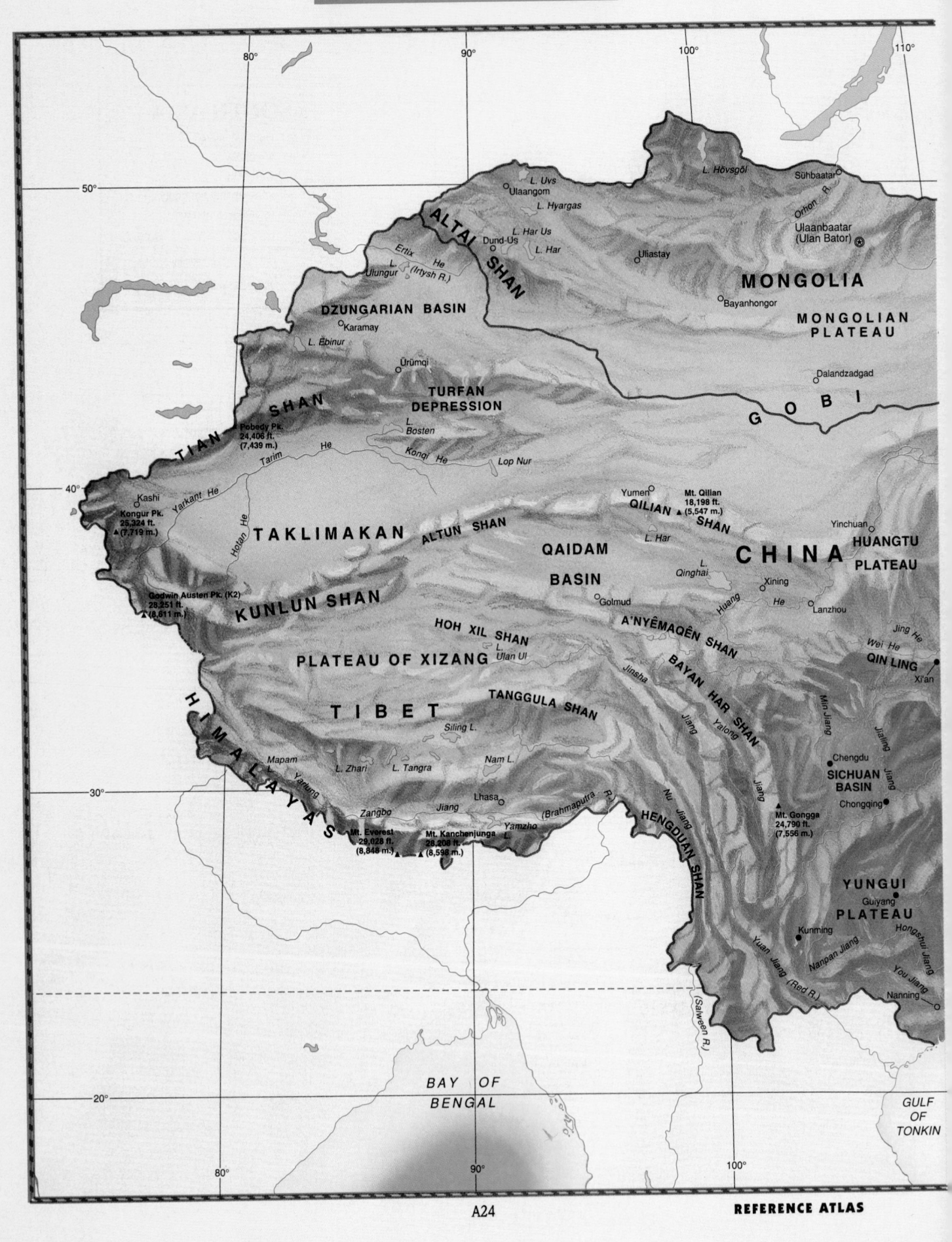

80°   90°   100°   110°

50°

L. Uvs
Ulaangom   L. Hövsgöl   Sühbaatar
L. Hyargas   Orhon R.
ALTAI SHAN   Dund-Us   L. Har Us   Ulaanbaatar
L. Har   (Ulan Bator)
Ertix He   Uliastay
(Irtysh R.)   MONGOLIA
L. Ulungur
DZUNGARIAN BASIN   MONGOLIAN
Karamay   Bayanhongor   PLATEAU
L. Ebinur
Dalandzadgad
Ürümqi
TURFAN   G O B I
DEPRESSION
L. Bosten
TIAN SHAN   He   Konqi He   Lop Nur
Pobedy Pk.   Tarim
24,406 ft.
(7,439 m.)   40°
Kashi   Yarkant He   Yumen   Mt. Qilian   Yinchuan
Kongur Pk.   QILIAN   18,198 ft.   HUANGTU
25,324 ft.   TAKLIMAKAN   ALTUN SHAN   (5,547 m.)   SHAN   CHINA   PLATEAU
(7,719 m.)   Hotan He   QAIDAM   L. Har   Xining   He   Jing He
Godwin Austen Pk. (K2)   BASIN   L.   Golmud   Huang   Lanzhou   Wei He
28,251 ft.   KUNLUN SHAN   Qinghai   A'NYÊMAQÊN SHAN   QIN LING
(8,611 m.)   HOH XIL SHAN   Xi'an
L.   BAYAN HAR SHAN   Min Jiang
PLATEAU OF XIZANG   Ulan Ul   Jinsha   Jiang
Yalong   Jiang   Chengdu   Jiang
H I M A L A Y A S   TIBET   TANGGULA SHAN   Jiang   SICHUAN
Siling L.   BASIN   Chongqing
Mapam   Nam L.   Mt. Gongga   Jiang
Yarlung   L. Zhari   L. Tangra   Lhasa   HENGDUAN SHAN   24,790 ft.
Zangbo   Jiang   (Brahmaputra R.)   (7,556 m.)   30°
Mt. Everest   Yamzho   YUNGUI
29,028 ft.   Mt. Kanchenjunga   L.   Nu Jiang   PLATEAU
(8,848 m.)   28,208 ft.   Guiyang   Hongshu Jiang
(8,598 m.)   Kunming   You Jiang
Yuan Jiang   Nanpan Jiang
(Red R.)   Nanning
(Salween R.)

20°
BAY OF   GULF
BENGAL   OF
TONKIN
80°   90°   100°

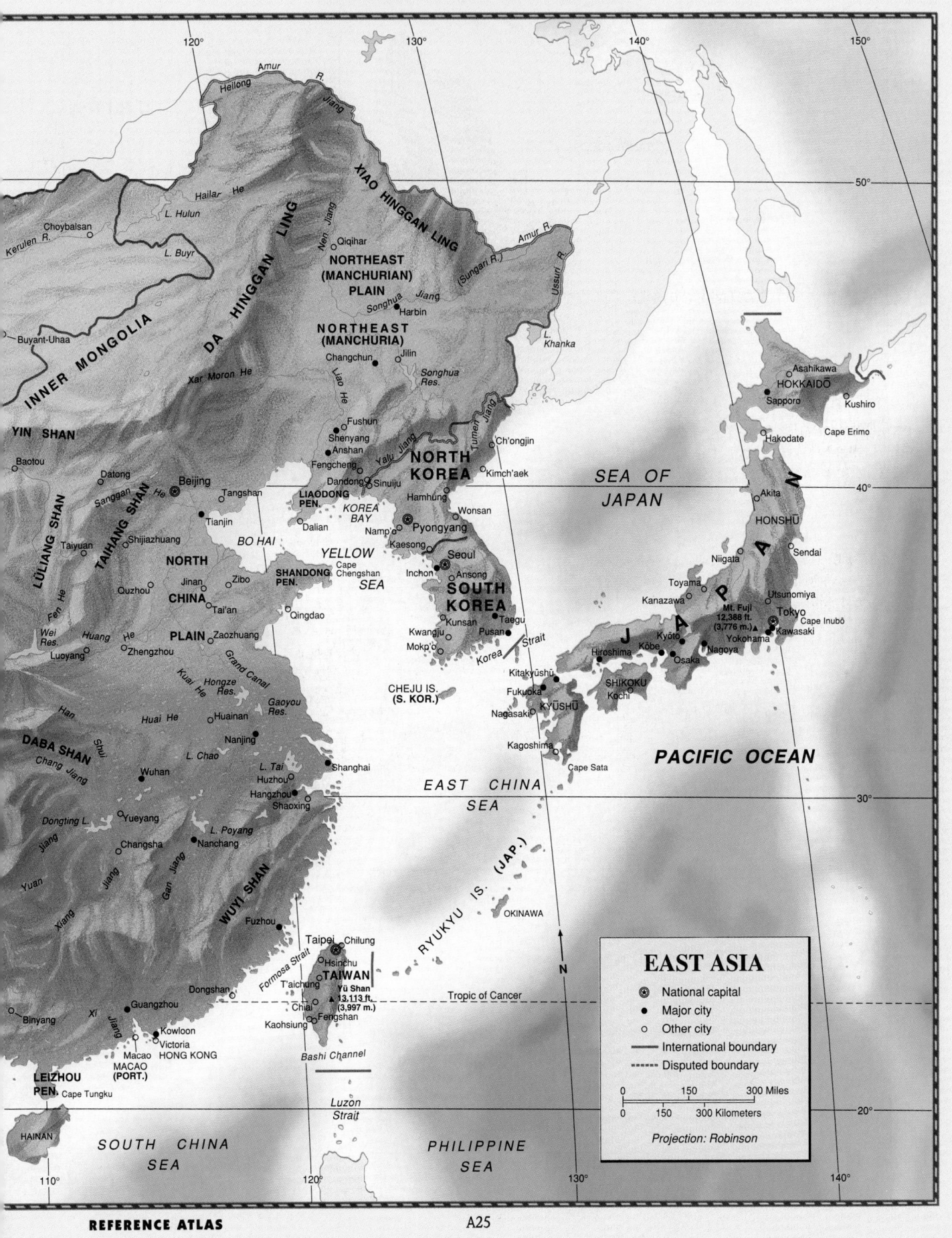

EAST ASIA

- ◉ National capital
- ● Major city
- ○ Other city
- —— International boundary
- ------ Disputed boundary

| 0 | 150 | 300 Miles |
| 0 | 150 | 300 Kilometers |

Projection: Robinson

SOUTHEAST ASIA

⊛ National capital
● Major city
○ Other city
— International boundary

| 0 | 200 | 400 Miles |

| 0 | 200 | 400 Kilometers |

Projection: Mercator

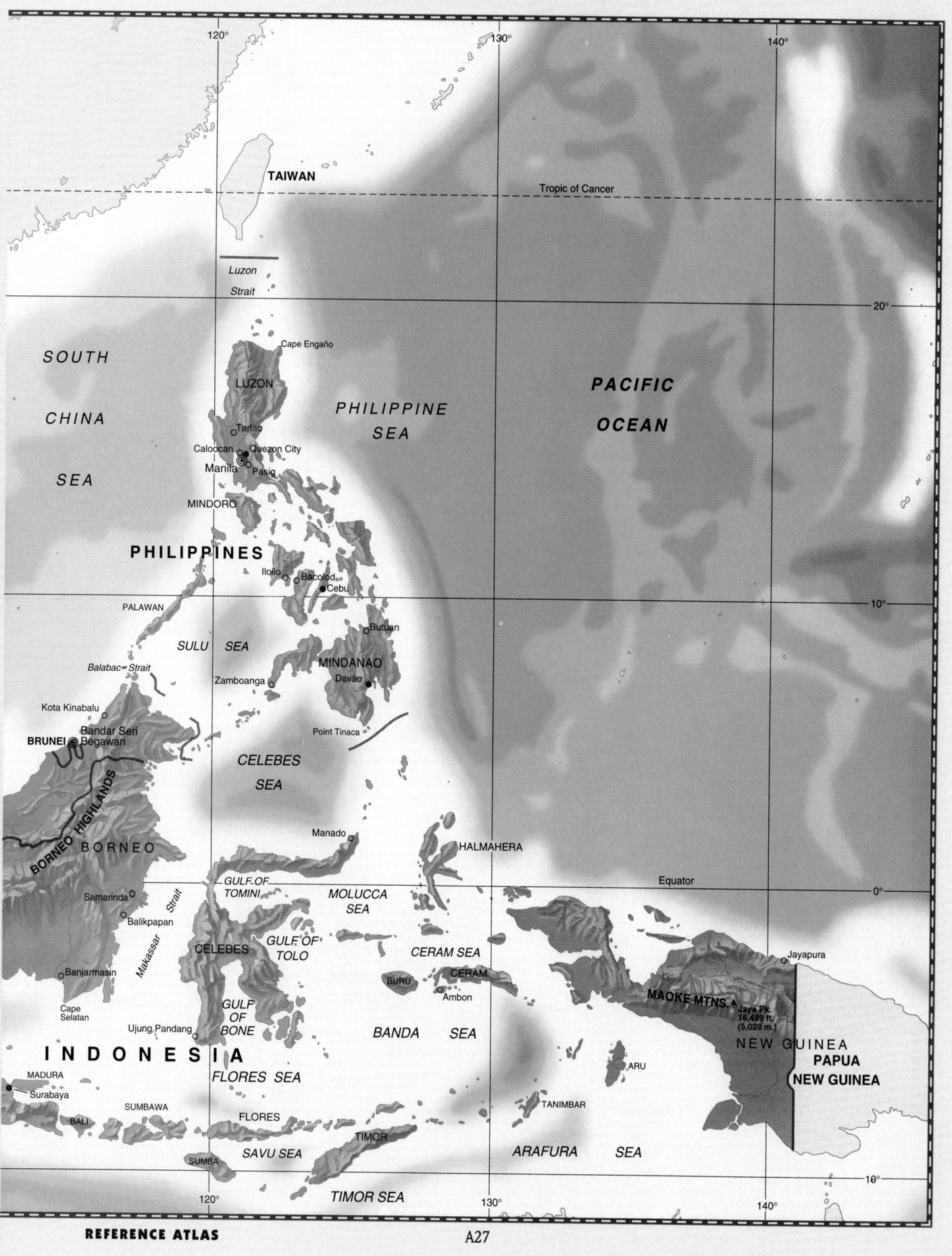

TAIWAN

Tropic of Cancer

120°   130°   140°

20°

Luzon
Strait

Cape Engaño

LUZON

SOUTH

CHINA

SEA

PHILIPPINE
SEA

PACIFIC

OCEAN

Tarlac

Caloocan ○ ● Quezon City
Manila ● ○ Pasig

MINDORO

PHILIPPINES

Iloilo ○   ○ Bacolod
● Cebu

PALAWAN

10°

Butuan

SULU   SEA

MINDANAO
Zamboanga   Davao ●

Balabac Strait

Kota Kinabalu
Bandar Seri
BRUNEI ◉ Begawan

Point Tinaca

CELEBES

SEA

BORNEO HIGHLANDS

BORNEO

Manado

HALMAHERA

Equator   0°

Samarinda

Balikpapan

GULF OF
TOMINI

MOLUCCA
SEA

CERAM SEA

Jayapura

Banjarmasin

Makassar Strait

CELEBES

GULF OF
TOLO

BURU   CERAM
Ambon

MAOKE MTNS. ▲
▲ Jaya Pk.
16,499 ft.
(5,029 m.)

Cape
Selatan

Ujung Pandang ○

GULF
OF
BONE

BANDA   SEA

NEW GUINEA
**PAPUA
NEW GUINEA**

I N D O N E S I A

MADURA
● Surabaya

FLORES SEA

ARU

BALI

SUMBAWA

FLORES

TANIMBAR

ARAFURA   SEA

SUMBA

SAVU SEA

TIMOR

10°

TIMOR SEA

120°   130°   140°

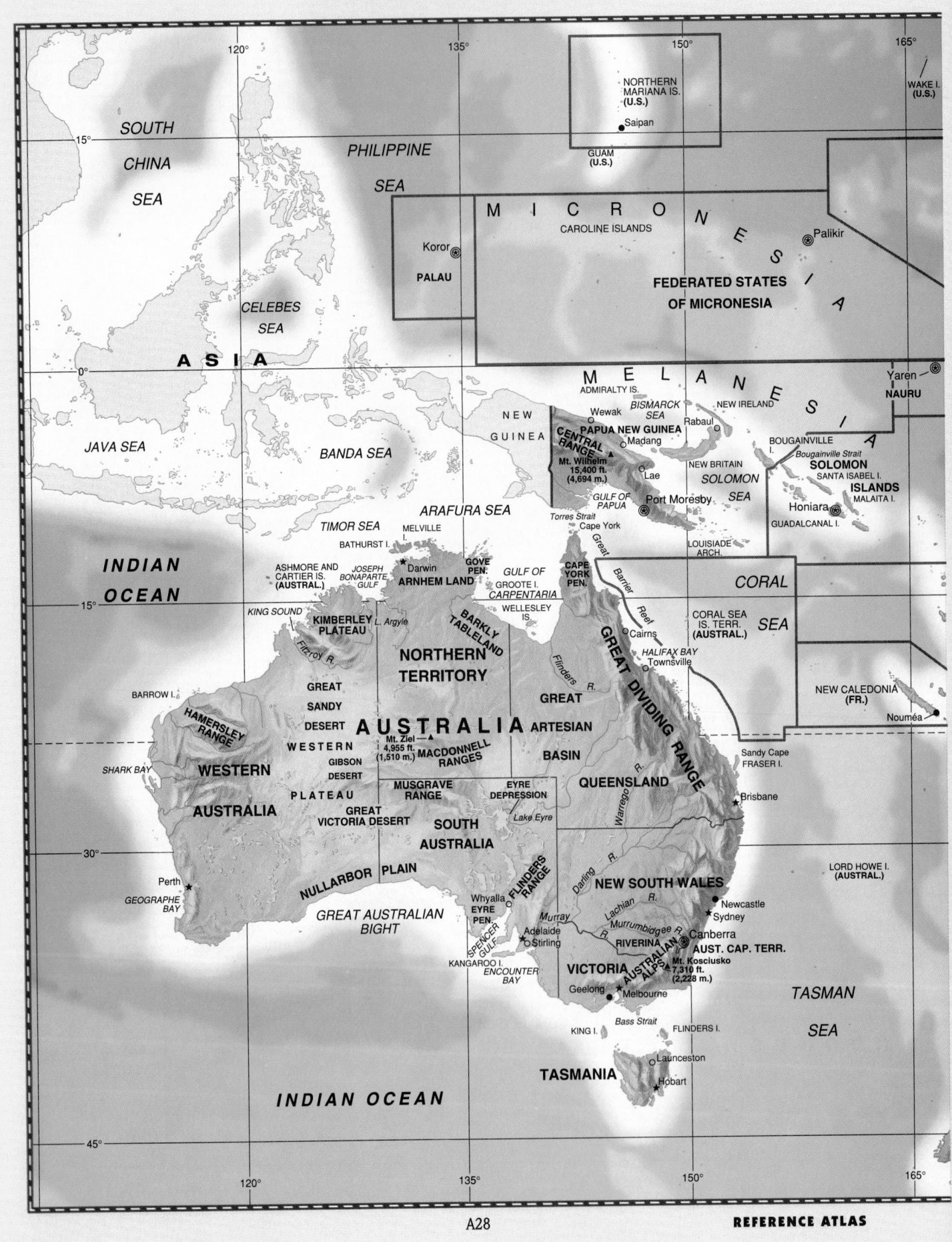

SOUTH CHINA SEA

PHILIPPINE SEA

NORTHERN MARIANA IS. (U.S.)
• Saipan

GUAM (U.S.)

WAKE I. (U.S.)

MICRONESIA

CAROLINE ISLANDS

• Palikir

Koror ⊛
PALAU

FEDERATED STATES OF MICRONESIA

CELEBES SEA

ASIA

0°

MELANESIA

Yaren ⊛
NAURU

JAVA SEA

BANDA SEA

NEW GUINEA

ADMIRALTY IS.

BISMARCK SEA

NEW IRELAND

Wewak •

PAPUA NEW GUINEA
CENTRAL RANGE ▲
Mt. Wilhelm 15,400 ft. (4,694 m.)

Madang •

Rabaul •

NEW BRITAIN

BOUGAINVILLE I.
Bougainville Strait

SOLOMON

SOLOMON ISLANDS

SANTA ISABEL I.

MALAITA I.

Lae •

ARAFURA SEA

TIMOR SEA

MELVILLE I.

GULF OF PAPUA

Port Moresby •

SEA

Honiara ⊛

GUADALCANAL I.

LOUISIADE ARCH.

INDIAN OCEAN

BATHURST I.

ASHMORE AND CARTIER IS. (AUSTRAL.)

JOSEPH BONAPARTE GULF

Torres Strait
Cape York

★ Darwin

GOVE PEN.
ARNHEM LAND

GROOTE I.
GULF OF CARPENTARIA

CAPE YORK PEN.

CORAL SEA

KING SOUND

KIMBERLEY PLATEAU

L. Argyle

WELLESLEY IS.

CORAL SEA IS. TERR. (AUSTRAL.)

15°

Fitzroy R.

BARKLY TABLELAND

Great Barrier Reef

Cairns ○

NORTHERN TERRITORY

Flinders R.

HALIFAX BAY
Townsville ○

BARROW I.

GREAT SANDY DESERT

GREAT

NEW CALEDONIA (FR.)

HAMERSLEY RANGE

AUSTRALIA

ARTESIAN

GREAT DIVIDING RANGE

Nouméa •

SHARK BAY

WESTERN

Mt. Ziel 4,955 ft. (1,510 m.) ▲
MACDONNELL RANGES

BASIN

GIBSON DESERT

Sandy Cape
FRASER I.

WESTERN AUSTRALIA

PLATEAU

MUSGRAVE RANGE

EYRE DEPRESSION

QUEENSLAND

Warrego R.

★ Brisbane

GREAT VICTORIA DESERT

Lake Eyre

Darling R.

LORD HOWE I. (AUSTRAL.)

30°

SOUTH AUSTRALIA

NULLARBOR PLAIN

Perth ★

GEOGRAPHE BAY

FLINDERS RANGE

NEW SOUTH WALES

Newcastle •

Whyalla ○
EYRE PEN.

Lachlan R.

Sydney •

GREAT AUSTRALIAN BIGHT

Murray R.

Murrumbidgee R.

SPENCER GULF

Adelaide ★
○ Stirling

RIVERINA

Canberra ⊛
AUST. CAP. TERR.

KANGAROO I.

ENCOUNTER BAY

VICTORIA

AUSTRALIAN ALPS
Mt. Kosciusko 7,310 ft. (2,228 m.) ▲

TASMAN SEA

Geelong ★ • Melbourne

KING I.

Bass Strait

FLINDERS I.

INDIAN OCEAN

45°

TASMANIA

Launceston ○

★ Hobart

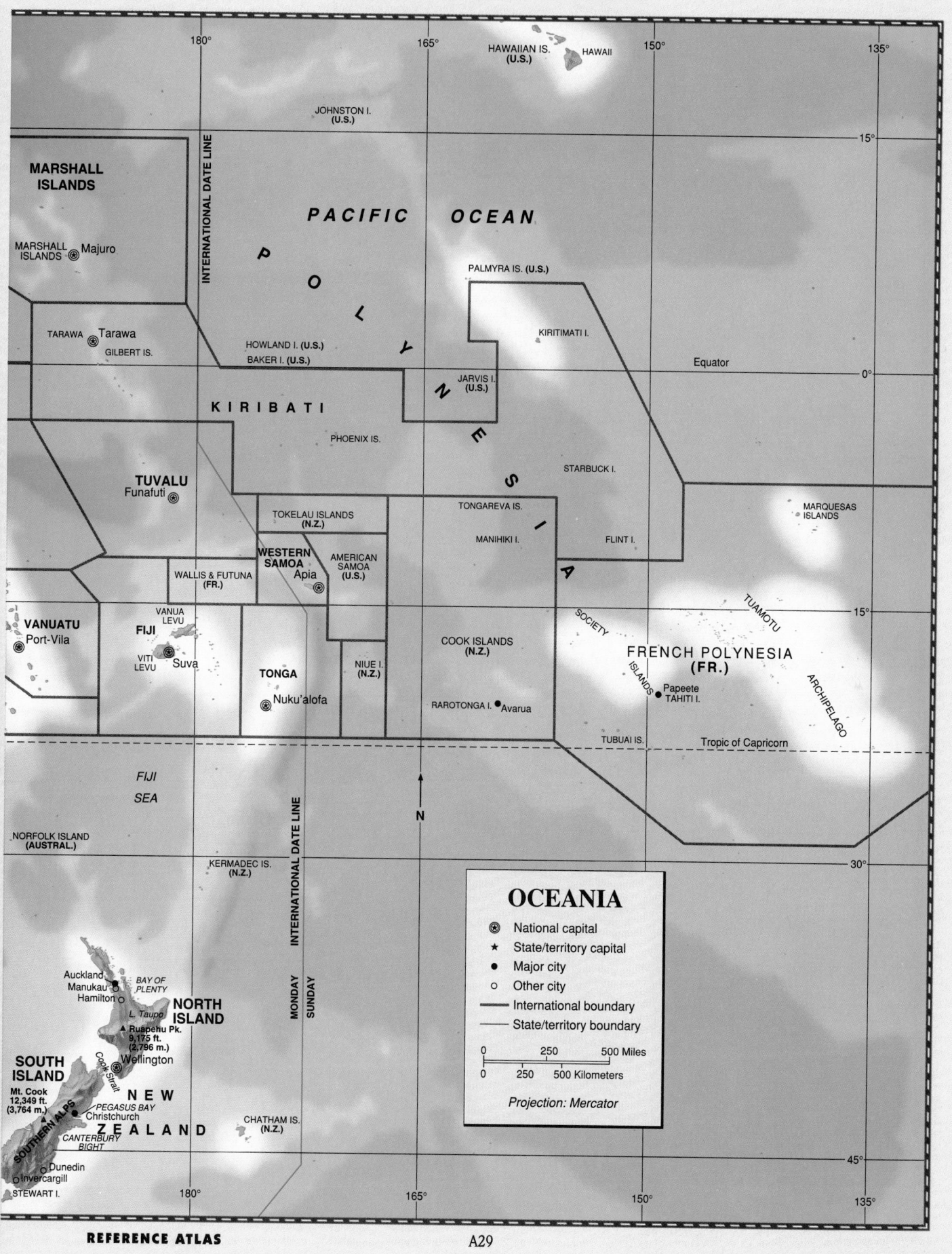

OCEANIA

⊛ National capital
★ State/territory capital
● Major city
○ Other city
— International boundary
— State/territory boundary

0   250   500 Miles
0 250 500 Kilometers

Projection: Mercator

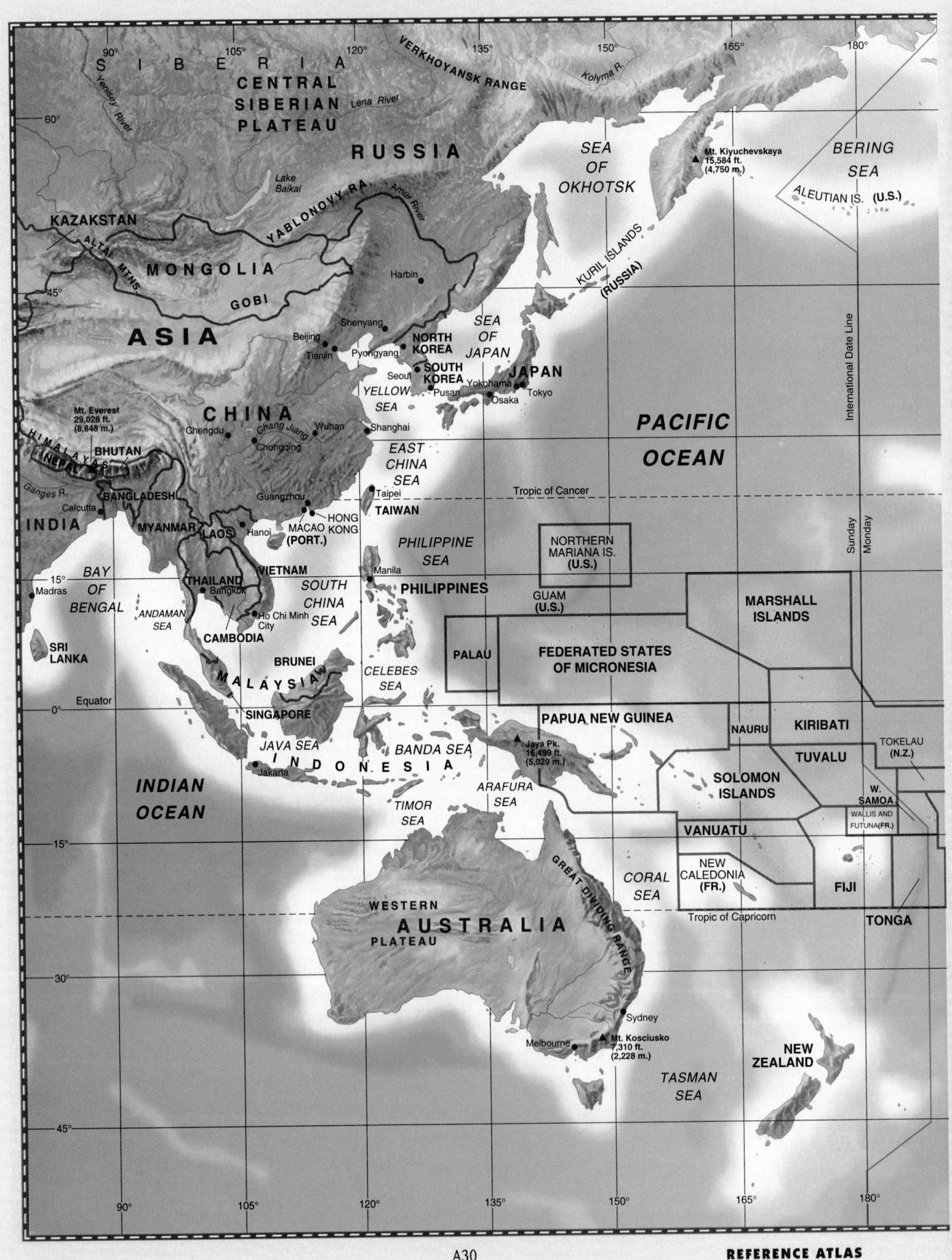

A30

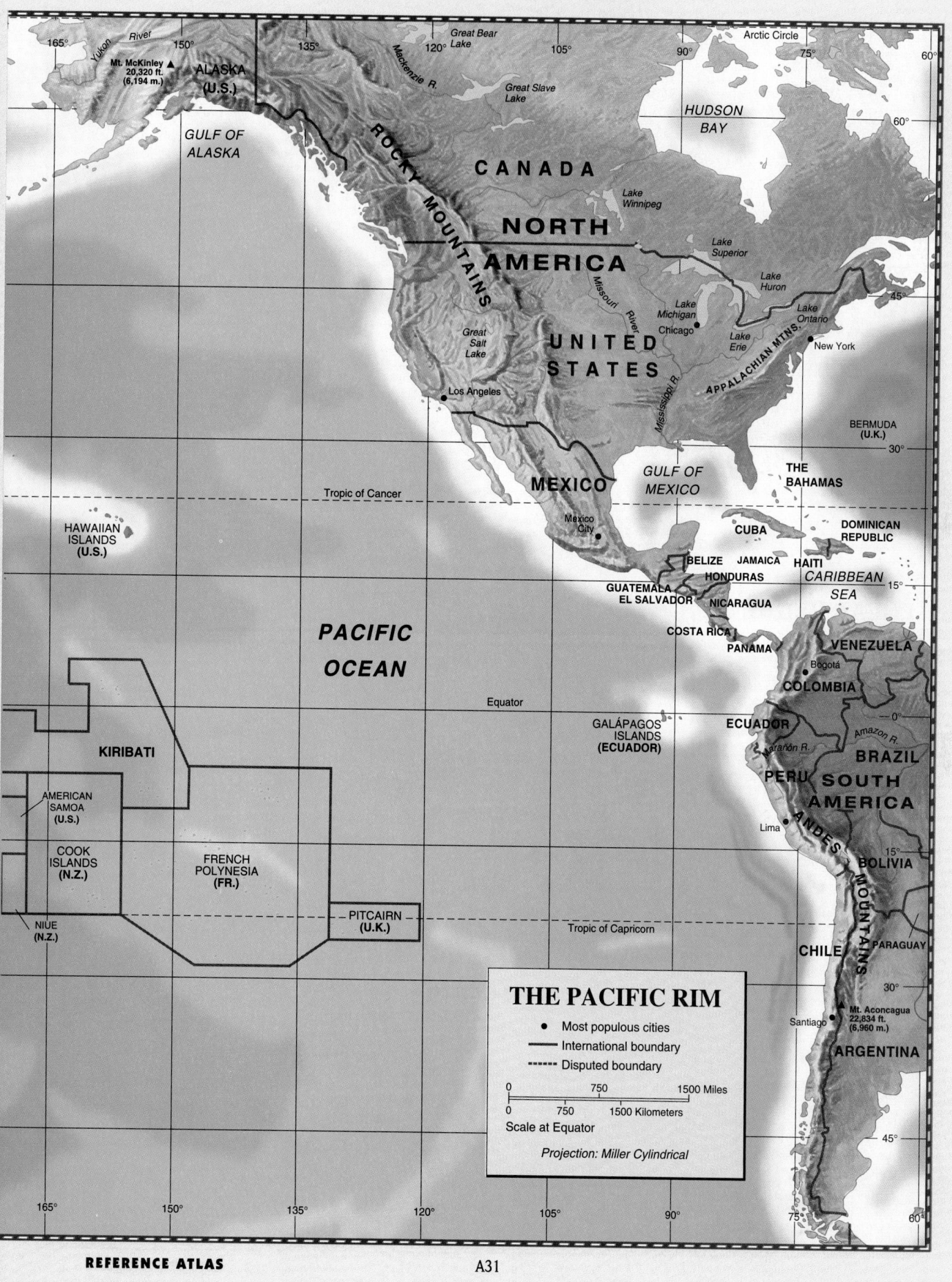

THE PACIFIC RIM

• Most populous cities
—— International boundary
----- Disputed boundary

| 0 | 750 | 1500 Miles |
| 0 | 750 | 1500 Kilometers |

Scale at Equator

Projection: Miller Cylindrical

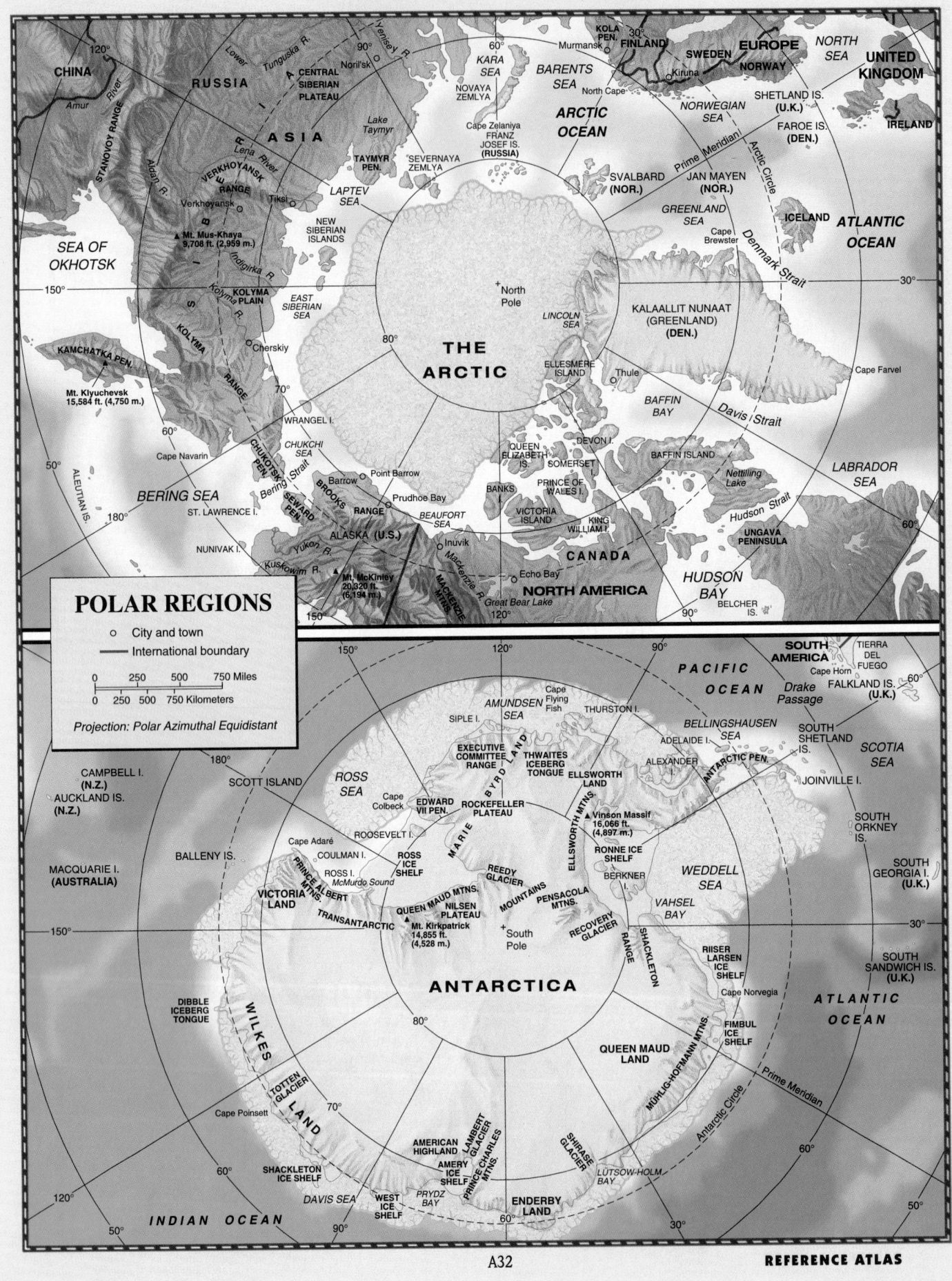

POLAR REGIONS

○ City and town
— International boundary

| 0 | 250 | 500 | 750 Miles |
| 0 | 250 | 500 | 750 Kilometers |

Projection: Polar Azimuthal Equidistant

THE ARCTIC

ANTARCTICA

Geography and Map Skills Handbook

CONTENTS

INTRODUCING Skills Handbook

Handbook Objectives

1. **Understand** the purpose and uses of globes and map projections.
2. **Describe** the parts of a map and the different types of maps geographers use.
3. **Demonstrate** how to read graphs and charts.

Cultural Kaleidoscope

The Arctic The early Inuit labeled distances on their maps with the time they took to travel rather than with miles. A map would show the distance between what is now Nome and Point Barrow as 10 days rather than 525 miles (845 km).

Map Activity

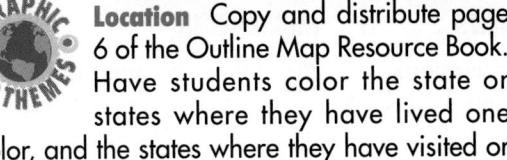

Location Copy and distribute page 6 of the Outline Map Resource Book. Have students color the state or states where they have lived one color, and the states where they have visited or driven through, another color. Then give each student an adhesive label. Tell students to put their labels in the corner of their maps and use them to make a key that explains the meanings of the colors. **L1**

Meeting National Standards

Geography for Life The following standards are highlighted in this section:
• Standards 1, 3

FOCUS

Section Objectives

1. **Understand** how mapmakers represent a round globe on a flat map.
2. **Identify** what projections are useful for different purposes.
3. **Explain** how to find an exact location.

Vocabulary Pre-check

Ask students what they associate with the word *projection.* Encourage them to think of how a movie is projected onto a screen. Tell them that the round globe is projected onto a flat surface in a map projection. **L1** **LEP**

SECTION 1 · Globes and Map Projections

PREVIEW

Words to Know
• projection
• hemisphere
• latitude
• longitude
• grid system
• absolute location
• great circle route

Read to Learn . . .
1. how mapmakers represent a round globe on a flat map.
2. what projections are useful for different purposes.
3. how to find an exact location.

Globes and the Earth

Although a globe is the most accurate way to represent the earth, using one presents some serious disadvantages. First, it is hard to carry a globe around in your pocket. Then try to imagine a globe large enough to show your community in detail. For these reasons, geographers use maps.

A globe—like the earth—is a sphere. A map, however, is a flat piece of paper. These facts explain why a flat map always distorts the surface of the earth it is showing.

Goode's Interrupted Projection

Goode's projection is called an equal-area projection. This means the projection quite accurately presents the size and shape of the continents. Distances—especially in the oceans—are less accurate. Researchers might use this type of projection to compare continent statistics according to area.

Drawing Maps

Imagine taking the whole peel from an orange and trying to flatten it on a table. You would either have to cut it or stretch parts of it. Mapmakers face a similar problem in showing the surface of the round earth on a flat map. The different ways they have found to do this are called **projections**. A map projection is a way of drawing the earth on a flat piece of paper.

Goode's Interrupted Projection

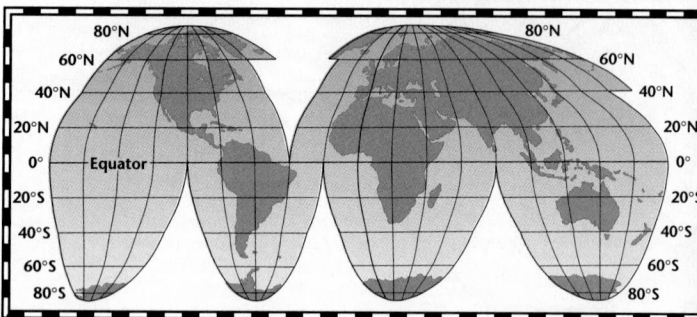

SKILLS HANDBOOK

Classroom Resources for Section 1

REPRODUCIBLE MASTERS
Geography Handbook
  Activities 1, 2, 3, 4, 5, 6
Glencoe Social Studies Outline
  Map Book

TRANSPARENCIES
Geography Handbook
  Transparencies 1, 2, 3, 4

MULTIMEDIA
National Geographic
  Society: ZipZapMap!
  USA, World

Mercator Projection

A Mercator projection was one of the earliest types of maps drawn. Mapmaker Gerardus Mercator first created this projection in 1569. This kind of projection shows land shapes fairly accurately, but not size or distance. Areas that are distant from the Equator are quite distorted on this projection. Alaska, for example, appears much larger on a Mercator map than it does on a globe. This map projection does show true directions, however, making it very useful for sea travel.

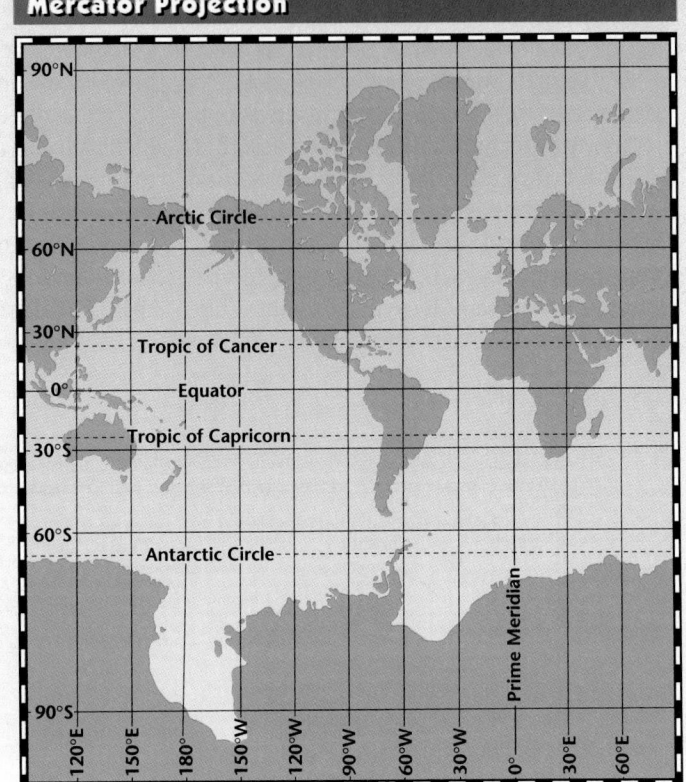

Mercator Projection

MAP STUDY

Map Skills Practice
Reading a Map What continent appears the largest on the Mercator Projection? *(Antarctica)*

CURRICULUM CONNECTION

History The Mercator projection map has been favored by sailors for more than 400 years. It is ideal for sea navigation for three reasons: 1) it shows true directions; 2) parallels and meridians are straight, not curved; and 3) parallels and meridians intersect at right angles. Thus, navigators can simply plot direct compass courses on the map and follow them at sea.

Robinson Projection

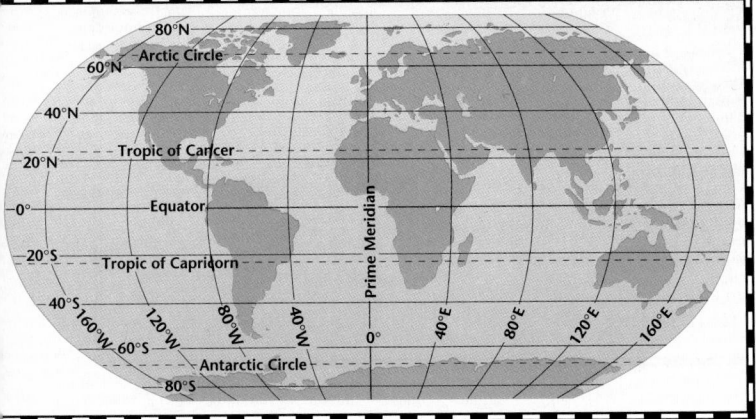

Robinson Projection

The Robinson projection shows both the size and shape of oceans and continents quite accurately. The shapes of the continents also appear much as they do on a globe. The areas most distorted on this projection are near the North and South poles. Textbook and atlas maps are often Robinson projections.

TEACH

Guided Practice

Illustrating Information Emphasize the difficulty in portraying the curved surface of the earth on a flat map by giving each student an orange. Have students draw the outlines of the continents on their oranges. Then ask students to peel the oranges and try to place the peel flat on their desks. Point out how they have to tear or stretch the peel to get it to lay flat. Tell them that they are distorting the peel to make it flat, just as mapmakers distort shapes and distances to make flat maps of the earth's curved surface. L1

Cooperative Learning Activity

Organize the class into two-person teams. Have each team record the absolute location (in degrees latitude and longitude) of 20 named places on the globe (cities, natural features, and so on). Then pit teams against each other in a round-robin Absolute Location Tournament. In each round, one team will say aloud the latitude and longitude of five places and time the other team as they locate the places on a globe. The teams will then switch roles. The team with the better time wins the round. L1

Map Skills Practice

Reading a Map What line divides the earth into Northern and Southern hemispheres? *(Equator)* What line divides the earth into Eastern and Western hemispheres? *(Prime Meridian)*

Independent Practice

Geography Handbook Activities 1, 2, 3, 4, 5, 6

There is a point on Earth with "no" latitude and "no" longitude. The absolute location of where the Prime Meridian and the Equator intersect, off the African coast in the Atlantic Ocean, is 0° N-S, 0° E-W.

Hemispheres

Remember that a globe is the most accurate way to represent the earth. To locate places on the earth, geographers set up a system of imaginary lines that crisscross the globe. One of these lines, the Equator, circles the middle of the earth like a belt. It divides the earth into "half spheres," or **hemispheres**. Everything north of the Equator is in the Northern Hemisphere. Everything south of the Equator is in the Southern Hemisphere.

Another imaginary line running from north to south divides the earth into half spheres in the other direction. Find this line—the Prime Meridian—on a globe. It is located at 0° longitude. Everything east of the Prime Meridian is in the Eastern Hemisphere. Everything west of the Prime Meridian is in the Western Hemisphere. North America lies in the Northern Hemisphere and the Western Hemisphere.

Hemispheres

NORTHERN HEMISPHERE
North Pole
NORTH AMERICA
Equator
SOUTH AMERICA
South Pole
SOUTHERN HEMISPHERE

WESTERN HEMISPHERE
North Pole
EUROPE
AFRICA
SOUTH AMERICA
Prime Meridian
South Pole
EASTERN HEMISPHERE

NORTHERN HEMISPHERE
North Pole
ASIA
Equator
AUSTRALIA
ANTARCTICA
South Pole
SOUTHERN HEMISPHERE

EASTERN HEMISPHERE
North Pole
180°
AUSTRALIA
PACIFIC OCEAN
South Pole
WESTERN HEMISPHERE

Meeting Special Needs Activity

Study Strategy The grid system is a prerequisite to understanding other aspects of geography, so ensure that all students grasp it at this point. To help students having difficulty understanding the system, draw a simplified grid of three horizontal lines labeled *A, B,* and *C,* and three vertical lines labeled *1, 2,* and *3.* Have students practice naming locations with this simple grid system. When they are comfortable with the basic concept, draw a more complex grid of 10 horizontal and 3 vertical lines labeled with degrees. Finally, guide student practice on a globe. **L1**

Latitude and Longitude

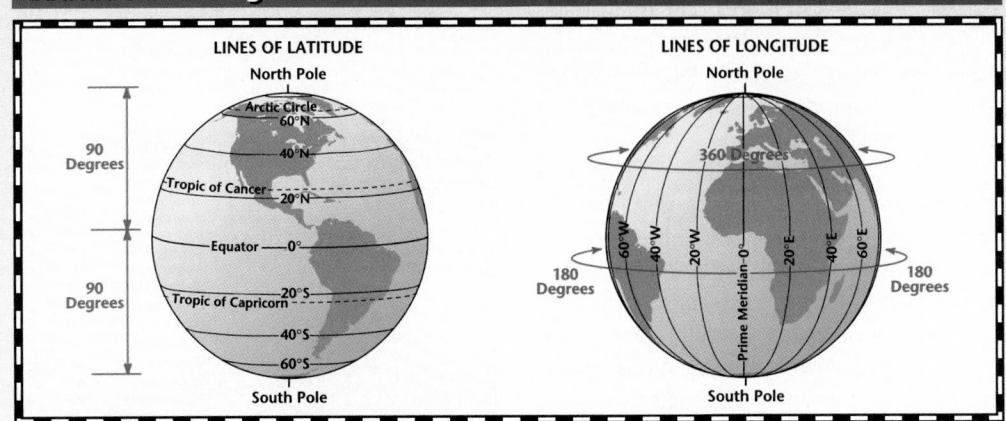

LINES OF LATITUDE

North Pole
90 Degrees
Arctic Circle 60°N
40°N
Tropic of Cancer 20°N
Equator 0°
Tropic of Capricorn 20°S
40°S
60°S
90 Degrees
South Pole

LINES OF LONGITUDE

North Pole
360 Degrees
180 Degrees
60°W 40°W 20°W Prime Meridian 0° 20°E 40°E 60°E
180 Degrees
South Pole

Latitude and Longitude

Parallels The Equator and the Prime Meridian are the starting points for two sets of lines used to find any location. Parallels circle the earth and show **latitude**, which is distance measured in degrees north and south of the Equator at 0° latitude. The letter *N* or *S* following the degree symbol tells you if the location is north or south of the Equator. The

North Pole is at 90° North *(N)* latitude, and the South Pole is at 90° South *(S)* latitude.

Two important parallels in between the poles are the Tropic of Cancer at 23½°N latitude and the Tropic of Capricorn at 23½°S latitude. You can also find the Arctic Circle at 66½°N latitude and the Antarctic Circle at 66½°S latitude.

Meridians Meridians run north to south from pole to pole. These lines signify **longitude**, which is distance measured in degrees east *(E)* or west *(W)* of the Prime Meridian at 0° longitude. On the opposite side of the earth is the International Date Line, or the 180° meridian.

Lines of latitude and longitude cross each other in the form of a **grid system**. Knowing a place's latitude and longitude allows you to locate it exactly on a map or globe. You can name an **absolute location** by naming the latitude and longitude lines that cross nearest to that location. For example, Tokyo is located at about 36°N latitude and 140°E longitude. Where those two lines cross is the absolute location of the city.

Japan

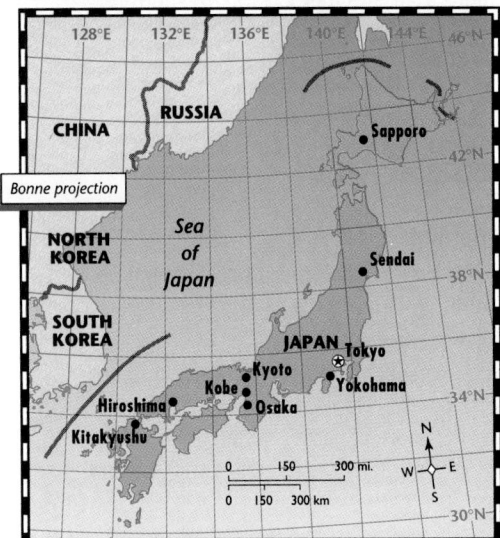

128°E 132°E 136°E 140°E 144°E 46°N
CHINA RUSSIA Sapporo
42°N
Bonne projection
NORTH KOREA
Sea of Japan Sendai
38°N
SOUTH KOREA
JAPAN Tokyo
Kyoto Yokohama 34°N
Kobe
Hiroshima Osaka
Kitakyushu

0 150 300 mi.
0 150 300 km
N W E S
30°N

MULTICULTURAL PERSPECTIVE

Culturally Speaking

The grid system provides a kind of universal language. Citizens of all countries, no matter how different their cultures, speak the same language of latitude and longitude.

MAP STUDY

Map Skills Practice
Reading a Map What line of latitude is 23½° north of the Equator? *(Tropic of Cancer)*

MAP STUDY

Map Skills Practice
Reading a Map What Japanese city is located at about 43°N and 141°E? *(Sapporo)*

Critical Thinking Activity

Identifying Central Issues Help students remember the relationships among latitude, longitude, parallel, and meridians, by writing "Parallels show latitude" and "Meridians show longitude" on the chalkboard. Then challenge students to come up with mnemonic devices to help them remember these facts. **L2**

MAP STUDY

Map Skills Practice
Reading a Map If an airplane averages 500 miles per hour (805 km per hour) on the journey between Los Angeles and Tokyo, how much time does it save by taking the great circle route? *(about 40 minutes)*

CURRICULUM CONNECTION

Language Arts The word *longitude* comes from the Latin word for "length," and *latitude* comes from the Latin word meaning "breadth."

Great Circles

The idea of a great circle illustrates one important difference between using a map and using a globe. Because both the earth and a globe are round, they accurately show great circles.

A great circle is the shortest distance between two points on earth. A great circle is any circle you can draw on the earth that divides it into two equal parts. A line drawn along the Equator around the entire earth is an example of a great circle. Traveling along a great circle is called following a **great circle route**. Airplane pilots and ship captains often use great circle routes to shorten their trips and cut down on the amount of fuel needed. These routes often take pilots over the North Pole. A trip between Moscow, Russia, and New York, New York, over the North Pole is an example of a great circle route. Other great circles lie along the routes between Hong Kong, China, and Washington, D.C., and between Cape Town, South Africa, and Washington, D.C. Find these routes on a globe.

The great circle route between two points may not appear to be the shortest distance on a flat map. On this map, the great circle route between Tokyo, Japan, and Los Angeles,

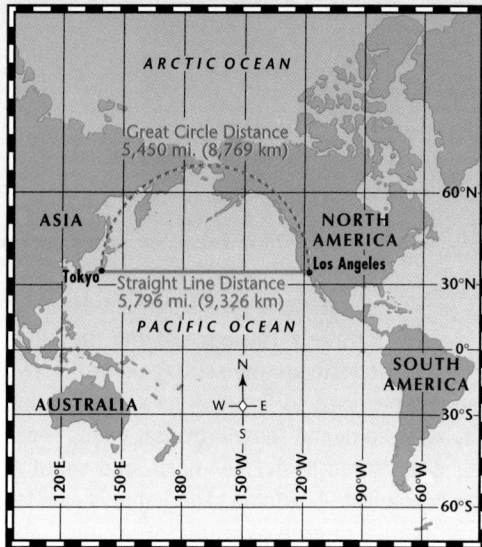

Great Circle Routes

ARCTIC OCEAN

Great Circle Distance 5,450 mi. (8,769 km)

ASIA

Tokyo

Straight Line Distance 5,796 mi. (9,326 km)

PACIFIC OCEAN

NORTH AMERICA
Los Angeles

SOUTH AMERICA

AUSTRALIA

60°N · 30°N · 0° · 30°S · 60°S

120°E · 150°E · 180° · 150°W · 120°W · 90°W · 60°W

California, appears to be far longer than the straight line route. Actually, the great circle is more than 345 miles (550 km) shorter!

SECTION 1 REVIEW

REVIEWING GLOBES AND MAP PROJECTIONS

1. **Define the following:** projection, hemisphere, latitude, longitude, grid system, absolute location, great circle route.
2. Why does every map projection distort some parts of the earth?
3. What imaginary line divides the earth into the Northern Hemisphere and Southern Hemisphere? Where is it located?
4. **ACTIVITY** Look at a globe or world map to discover more of earth's great circles. Make a list of airplane flights that could follow great circle routes between New York City and other major world cities.

6

Answers to Section 1 Review

1. All vocabulary words are defined in the Glossary.
2. because it is impossible to show a round globe completely accurately on a flat surface
3. Equator; it runs around the center of the earth
4. Answers will vary. Call on volunteers to show their routes to the rest of the class by tracing them on a globe.

SECTION 2

Building Map Skills

PREVIEW

Words to Know
- key
- cardinal direction
- compass rose
- intermediate direction
- scale bar
- scale
- relief
- elevation
- contour line
- elevation profile
- population density
- climate region

Read to Learn . . .
1. how you can find a place's exact location on a map.
2. how to recognize the special parts of a map.
3. how to identify different kinds of maps.

FOCUS

Section Objectives
1. **Demonstrate** how you can find a place's exact location on a map.
2. **Explain** how to recognize the special parts of a map.
3. **Illustrate** how to identify different kinds of maps.

Vocabulary Pre-check
Ask students what they use a key for. Then tell them that a map key "unlocks" the meaning of a map. **L1**
LEP

Parts of Maps

Maps can direct you down the street, across the country, or around the world. There are as many different kinds of maps as there are uses for them. Being able to read a map begins with learning about its parts.

Map Key The map **key** unlocks the information presented on the map. The key explains the symbols used on the map. On this map of Germany, for example, dots mark cities and towns.

On a road map, the key tells what map lines stand for paved roads, dirt roads, and interstate highways. A pine tree symbol may represent a park, while an airplane is often the symbol for an airport.

Compass Rose An important first step in reading any map is to find the direction marker. A map has a symbol that tells you where the **cardinal directions**—north, south, east, and west—are positioned. Sometimes all of these directions are shown with a **compass rose**. An intermediate direction, such as southeast, may also be on the compass rose. **Intermediate directions** fall between the cardinal directions.

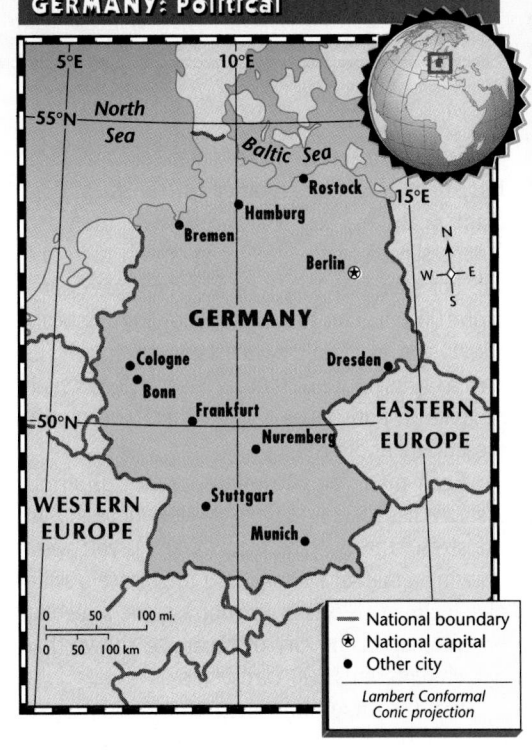

GERMANY: Political

National boundary
⊛ National capital
• Other city

Lambert Conformal Conic projection

Map Skills Practice
Reading a Map How can you tell the national capital from the other cities on this map? *(It is marked with a circled star.)*

Classroom Resources for Section 2

REPRODUCIBLE MASTERS
Geography Handbook Activities 7, 8, 9, 10, 11, 12, 13

TRANSPARENCIES
Geography Handbook Transparencies 5, 6, 7, 8

MULTIMEDIA
National Geographic Society: ZipZapMap! USA, World

MAP STUDY

Map Skills Practice

Reading a Map About what distance does an inch represent on the map of Nashville? *(about 530 yards or 485 m)*

TEACH

Guided Practice

Demonstrating Ideas Write the words *distance* and *direction* on the chalkboard. Tell students that they could give directions around the world using only these two measures. Ask students how maps show distance *(scale)* and direction *(compass rose)*. Then give students practice in determining distance by showing them how to mark a map scale on the edge of a sheet of paper. Have students practice measuring the distance between cities on the map on page 7. Next, give students practice in determining direction by having volunteers identify the direction from one city to another on the same map. Finally, have students use the map on page 8 to combine distance and direction. Suggest that they refer to the scale and compass rose to write a description of an imaginary driving tour through Tennessee. **L1**

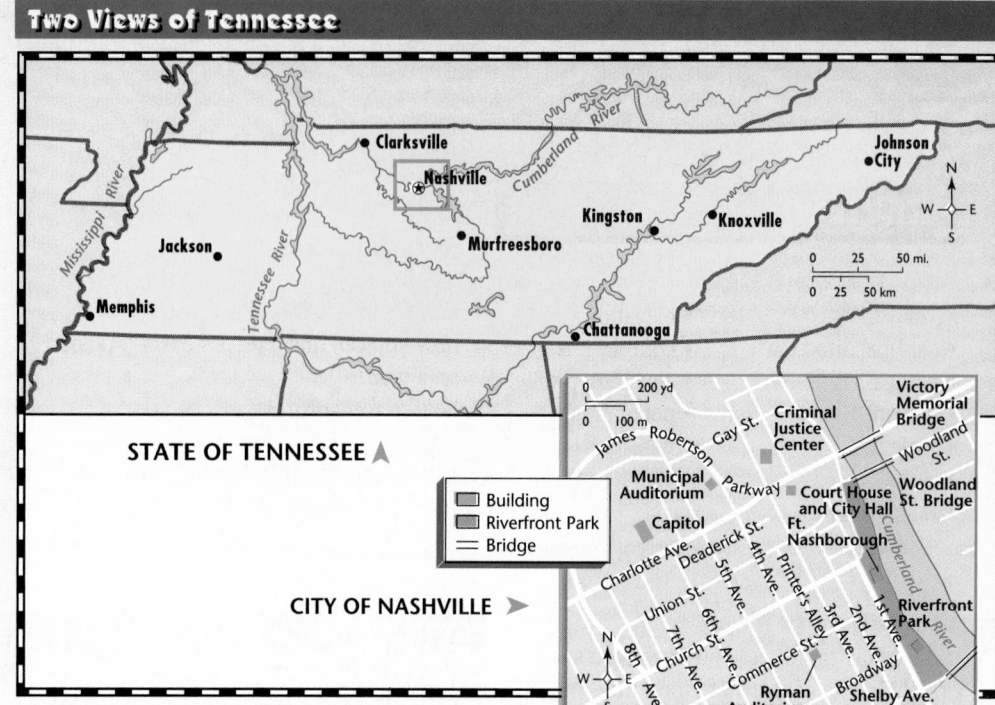

Two Views of Tennessee

STATE OF TENNESSEE ▲

CITY OF NASHVILLE ➤

Building
Riverfront Park
Bridge

Scale A measuring line, often called a **scale bar**, helps you find distance on the map. The map's **scale** tells you what distance on the earth is represented by the measurement on the scale bar. For example, 1 inch on the map may represent 100 miles on the earth. Knowing the scale allows you to visualize how large an area is, as well as to measure distances. Map scale is usually given in both miles and kilometers, a metric measurement of distance.

Each map you use might have a different scale. What scale a mapmaker uses depends on the size of the area the map shows. If you were drawing a map of a house, you might use a scale of 1 inch equals 5 feet. In contrast, the scale bar on the map of the city of Nashville shows that ⅜-inch represents 200 yards.

Scale is important when you are trying to compare the size of one area with another. You can't make a true comparison of the United States and Africa unless you are using maps drawn to the same scale.

AFRICA AND THE UNITED STATES: Land Comparison

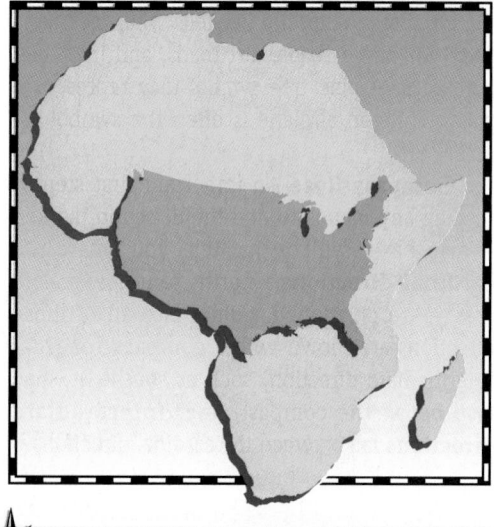

SKILLS HANDBOOK

Cooperative Learning Activity

Organize students into several groups to work as cartographers. Assign each group a particular map—a map of the classroom, the school grounds, or another small area, for example. Different group members should be responsible for the following tasks: measuring and making a scale, using a compass to determine direction and making a compass rose, creating a map key, and drawing the map itself. Group members should cooperate to coordinate their work and to display their maps. **L1**

Different Kinds of Maps

Maps that show a wide range of general information about an area are called general purpose maps. Two of the most common general purpose maps are political and physical maps.

Political Maps Political maps generally show political, or human-made, divisions of countries or regions. The political map of Germany on page 7, for example, shows boundaries between Germany and other countries. It also shows cities within Germany and bodies of water surrounding Germany.

Physical Maps A physical map shows the physical features of an area, such as its mountains and rivers. The colors used for some features on physical maps are standard—brown or green for land, blue for water. Physical maps use colors and shadings to show **relief**—how flat or rugged the land surface is. Colors also may be used to show **elevation**—the height of an area above sea level.

Mapmakers also use color to show rainfall, types of soil, or plant life. Physical maps, like political maps, have a key that explains what each color and symbol stands for.

Contour Maps One kind of physical map scientists use may not look like a map at all to some people. This map—called a contour map—shows elevation. A contour map has **contour lines**—one for each major level of elevation. All the land at the same elevation is connected by a line. These lines usually form circles or ovals—one inside the other. If contour lines come very close together, the surface is steep. If the lines are spread far apart, the land is flat or rises very gradually.

Another way to show relief is to look at the landscape from the side, or profile. This **elevation profile** is a cutaway diagram. It clearly shows level land, hills, and steeper mountains.

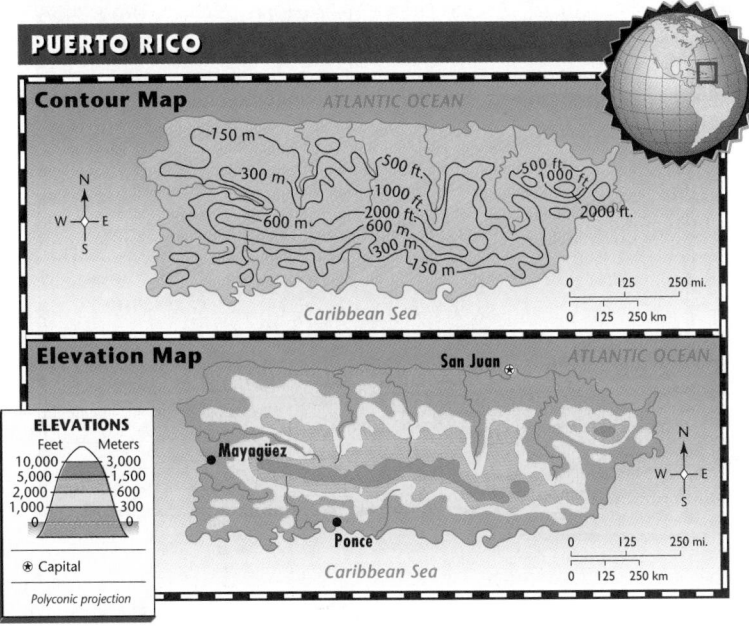

PUERTO RICO

Contour Map

ATLANTIC OCEAN

150 m
300 m
500 ft.
500 ft.
1000 ft.
600 m
1000 ft.
2000 ft.
600 m
2000 ft.
300 m
150 m

0   125   250 mi.
0   125   250 km

Caribbean Sea

Elevation Map

San Juan

ATLANTIC OCEAN

ELEVATIONS
Feet   Meters
10,000   3,000
5,000   1,500
2,000   600
1,000   300
0   0

• Mayagüez

• Ponce

⊛ Capital

Polyconic projection

Caribbean Sea

0   125   250 mi.
0   125   250 km

LESSON PLAN
Skills Handbook, Section 2

Independent Practice

Geography Handbook Activities 7, 8, 9, 10, 11, 12, 13

The political boundaries that are shown on political maps, of course, do not appear on the actual landscape. There are exceptions, however. Some of the border between the United States and Canada is marked by a wide, straight swath cut through the forest. And some borders between desert countries in Southwest Asia are marked by deep ditches dug in the sand.

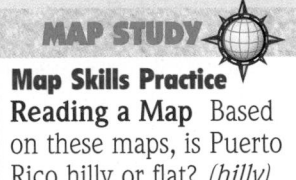

Map Skills Practice
Reading a Map Based on these maps, is Puerto Rico hilly or flat? *(hilly)*

Meeting Special Needs Activity

Study Strategy To help students remember the several different categories and examples of maps discussed in this section, guide them in creating a "Types of Maps" graphic organizer. Branching from this main cell should be two smaller cells labeled "General Purpose Maps" and "Special Purpose Maps." Have students complete the organizer by adding cells labeled with the various maps and connecting them to the correct category. **L1**

Strange but TRUE!

A special purpose map can show anything the map-maker wants. Special purpose subjects include bicycle routes, the path of killer bees, ocean currents, energy use, and pizza delivery areas. Even the treasure map that opens Robert Louis Stevenson's classic adventure story, *Treasure Island,* is a special purpose map!

MAP STUDY

Map Skills Practice
Reading a Map What metal is mined in northeastern South Africa near the Limpopo River? *(gold)*

Special Purpose Maps

Besides showing political or physical features, some maps have a special purpose. They may show cultural features, historical changes, unique physical features, population, or climates. The map's title tells what kind of specialized information it shows. Colors and symbols in the map key are especially important on this type of map.

When looking at a special purpose map, it is important to remember that you are seeing only the information that relates to the map's special purpose. For example, on a population map you will see indications of where and how many people live in a country. The map will probably not show you anything about what crops the people there raise.

Land Use and Resources Maps Look at this map of land use and resources in South Africa. Colors identify parts of the area where people follow certain ways of life, such as farming or herding. Places where manufacturing takes place are also marked with color. The symbols stand for the different kinds of natural resources found in or on the land. The number and type of symbols can vary from map to map depending upon the area shown. Diamonds, for example, are not found in all parts of the world as they are in South Africa. So diamonds might not always be in a resource map key. Resources can include minerals, timber, and important crops.

SOUTH AFRICA: Land Use and Resources

Resources
- Coal
- Copper
- Diamonds
- Fish & other seafood
- Gold
- Iron ore
- Tin
- Magnesium
- Uranium

Azimuthal Equal-Area projection

Agriculture
- Nomadic Herding
- Little or no farming
- Ranching
- Plantation farming
- Subsistence farming
- Manufacturing area

More About . . .

Geographic Information Systems (GIS) A GIS is a computer hardware and software system that is used to store, display, analyze, and map information. Geographers, urban planners, land developers, engineers, real estate agents, utility companies, and municipal officials all use these systems. GIS are vital to modern planning because they enable us to combine data and look at layers of information at the same time on a computer. One GIS, for example, may begin with a digitized base map. A retailer wishing to make an informed decision about where to build a store may combine data such as population distribution, traffic movement, land availability, and real estate prices. Using the GIS, the retailer can see and analyze all this information at the same time. GIS technology may someday help you plan your vacation!

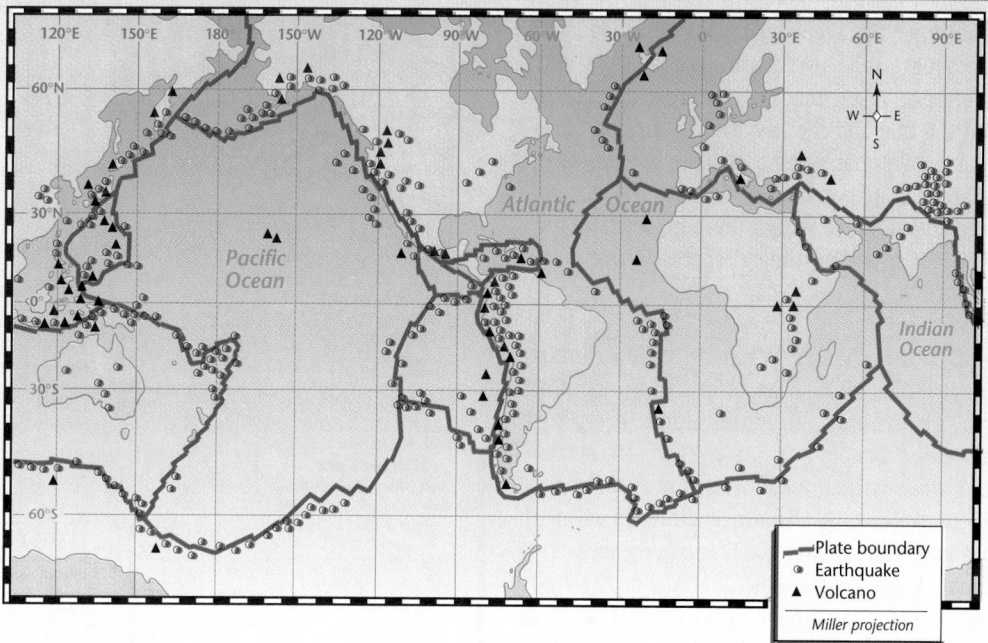

Earthquake and Volcano Zones

Plate boundary
◎ Earthquake
▲ Volcano

Miller projection

MAP STUDY

Map Skills Practice
Reading a Map Around which ocean is there the most volcanic and earthquake activity? *(Pacific Ocean)*

Cultural Kaleidoscope

Southwest Asia As with the map on this page, the direction *North* is shown as upward, or toward the top of the map. This is only a convention, however. Early Arab cartographers, who made great advances in mapmaking, put North at the bottom of their maps. This in no way affected the maps' usefulness. Marco Polo used one such map when he traveled to East Asia!

Some special purpose maps have a very limited use. You may have seen a map in your school, for example, that shows the boundaries of your school district. A map such as that would probably only be useful to people in your community and state. Other special purpose maps that show a wide area, such as the entire world, hold information that affects many more people.

Geological Maps Special purpose maps can show information about something that you cannot readily see, such as geological information. Geological maps explain the structure of the earth and how it is believed to have been formed. In addition to giving you new information, such maps might also help you make connections between events and their causes. It is often helpful to compare a special purpose map to another map to receive a complete picture of how the earth's elements work together.

Look at the Earthquake and Volcano Zones map above. It gives you a view of three things: the location of major plates that make up the earth's crust, where volcanoes have erupted, and where earthquakes have occurred most frequently. You will read in Chapter 2 about the plates that make up the earth's crust and how they move. On this map you can see—and make the connection—that earthquakes and volcanoes seem to occur where these plates meet. Geologists call the area of earthquake and volcano activity near the Pacific Ocean the Ring of Fire.

More About . . . Activity

Community Resource Person Many public works departments and local government agencies, such as the police department or the 911 emergency response system, use GIS. Contact an official of a department or agency and ask if they can supply a speaker to discuss how they use GIS with your class.

Surfing the Net Access the Internet to find out more about GIS. Go to your favorite search engine. Type in key words such as those below to find specific information:

Geographic Information Systems, urban planning, interactive maps, demographics, cartography, geospatial

The search engine should provide you with a number of links to follow. Links are "pointers" to different sites on the Internet and commonly appear as blue underlined words.

MAP STUDY

Map Skills Practice
Reading a Map What two cities in Brazil have populations in excess of five million people? *(São Paulo, Rio de Janeiro)*

Cultural Kaleidoscope

United States Maps are big business in the United States. The country's mapmaking industry has an annual turnover of about $200 million.

Strange but TRUE!

According to the *Guinness Book of World Records,* the world's largest map is a giant relief map of the state of California titled *Paradise in Panorama.* It measures 450 feet (137 m) long and 180 feet (55 m) wide and weighs 43 tons (39 t)!

Population Density Maps Another special purpose map uses colors to show **population density**, or the average number of people living in a square mile or square kilometer. As with other specialized maps, it is important to first read the title and the key. The population density map of Brazil gives a striking picture of its differences in population density. You can see large coastal cities where the population is very crowded, as well as inland spaces where fewer people live.

Climate Maps What do climate maps show? They summarize information about an area's rain, snow, and temperatures throughout the year. A climate map can help you see similarities and differences in climates between different places. A **climate region**, broad areas that have the same climate, is readily seen on a climate map. Geographers often divide the earth into four major climate regions—tropical, dry, mid-latitude, and high latitude. In which climate region do you live?

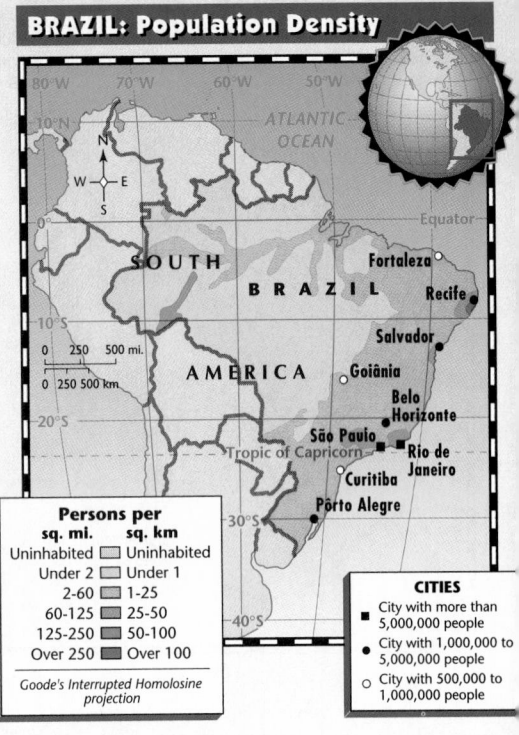

BRAZIL: Population Density

Persons per	
sq. mi.	**sq. km**
Uninhabited	Uninhabited
Under 2	Under 1
2-60	1-25
60-125	25-50
125-250	50-100
Over 250	Over 100

Goode's Interrupted Homolosine projection

CITIES
- City with more than 5,000,000 people
- City with 1,000,000 to 5,000,000 people
- City with 500,000 to 1,000,000 people

SECTION 2 REVIEW

Reviewing Map Skills
1. **Define the following:** key, cardinal direction, compass rose, intermediate direction, scale bar, scale, relief, elevation, contour line, elevation profile, population density, climate region.
2. What kinds of features does a physical map include?
3. What special purpose do geological maps serve?
4. To draw a map showing your route to school, would you use a large scale or a small scale? Why?
5. **ACTIVITY** Use the following grid system "addresses" to find the three cities located at these coordinates: 55°N, 35°E; 23°S, 43°W; 1°N, 103°E.

Answers to Section 2 Review

1. All vocabulary words are defined in the Glossary.
2. landforms, such as mountains and plains; bodies of water, such as rivers and lakes; rainfall; types of soil; plant life
3. They explain the structure of the earth and how it is believed to have been formed.
4. small scale, because the map shows a relatively small area
5. Moscow; Rio de Janeiro; Singapore

Reading Graphs and Charts

PREVIEW

Words to Know
- axis
- bar graph
- line graph
- circle graph
- pictograph
- climograph
- climate
- diagram
- flow chart
- chart
- table

Read to Learn . . .
1. how different types of graphs present information.
2. how charts, tables, and diagrams make data easier to understand.
3. how to read a flow chart.

Meeting National Standards
Geography for Life The following standards are highlighted in this section:
- Standards 1, 3

FOCUS

Section Objectives
1. **Analyze** how different types of graphs present information.
2. **Explain** how charts, tables, and diagrams make data easier to understand.
3. **Demonstrate** how to read a flow chart.

Vocabulary Pre-check

Tell students that *graph* probably is derived from two Greek words—*graphein*, meaning "to draw" or "to write"; and *gramma*, meaning "picture." Then suggest that the different kinds of graphs are "drawings" or "pictures" of information.
L1 **LEP**

Graphs

Think of the different places you see graphs and diagrams. They appear in newspapers, magazines, and even on television.

Graphs summarize and present information visually. Each part of a graph gives useful information. First read the graph's title to find out its subject. Then read the labels along the graph's **axes**—the vertical and horizontal lines along the bottom and sides of the graph. One axis will tell you what is being measured. The other axis tells what units of measurement are being used.

Bar Graphs Look carefully at this **bar graph** of the world's five most heavily populated countries. The vertical axis lists the countries. The horizontal axis gives the units of population measurement. By comparing the lengths of the bars you can quickly tell which country has the largest population. Bar graphs are especially useful for comparing quantities, and they may show the bars running across the graph or rising up from the bottom.

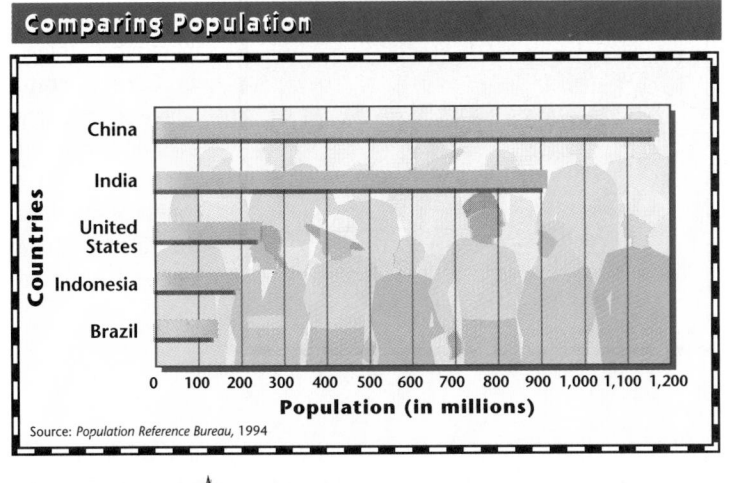

Comparing Population

Countries: China, India, United States, Indonesia, Brazil

Population (in millions): 0 100 200 300 400 500 600 700 800 900 1,000 1,100 1,200

Source: *Population Reference Bureau*, 1994

GRAPHIC STUDY

Graphic Skills Practice
Reading a Graph
Which of the countries shown on the graph is closest in size to the United States in terms of population? *(Indonesia)*

Classroom Resources for Section 3

REPRODUCIBLE MASTERS
Geography Handbook
Activities 14, 15, 16, 17, 18, 19, 20

TRANSPARENCIES
Geography Handbook
Transparencies 9, 10, 11, 12
Unit Overlay Transparencies
1-9, 1-10

13

Graphic Skills Practice
Reading a Graph
During which four-year period did India's population increase the most?
(from 1990 to 1994)

TEACH

Guided Practice

Illustrating Information
Have students identify the best way to graphically display the following information: 1) the total number of students in the class and the number of boys and girls *(circle graph)*; 2) a record of the number of students who attend class each day for a month *(line graph)*; 3) a comparison of the number of students who are of certain heights *(bar graph)*; 4) the number of books read by each student during the summer *(pictograph or bar graph)*; 5) the procedure for some classroom activity, such as taking attendance or moving to work centers *(flow chart)*. **L1**

Graphic Skills Practice
Reading a Graph If the earth's water area was expressed as a fraction, would it be closest to $\frac{1}{3}$, $\frac{2}{3}$, or $\frac{3}{4}$? *($\frac{2}{3}$)*

Line Graphs A tool that is especially good for plotting changes in something over a period of time is a **line graph**. The amounts being plotted on this graph are shown by dots. These dots are connected by a line. Line graphs sometimes have two or more lines plotted on them.

This line graph shows how the population density of India has changed since 1978. The vertical axis lists people per square mile. The horizontal axis shows the passage of time in four-year periods from 1978 to 1994.

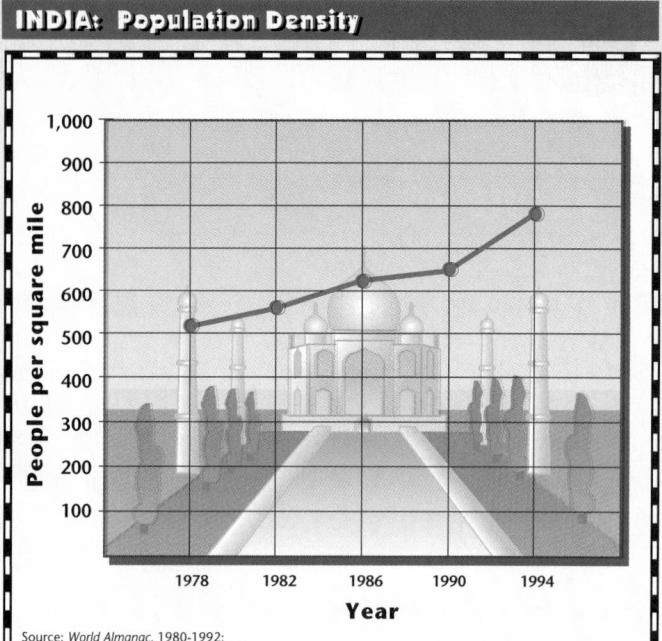

INDIA: Population Density

Source: *World Almanac*, 1980-1992;
Population Reference Bureau, 1994

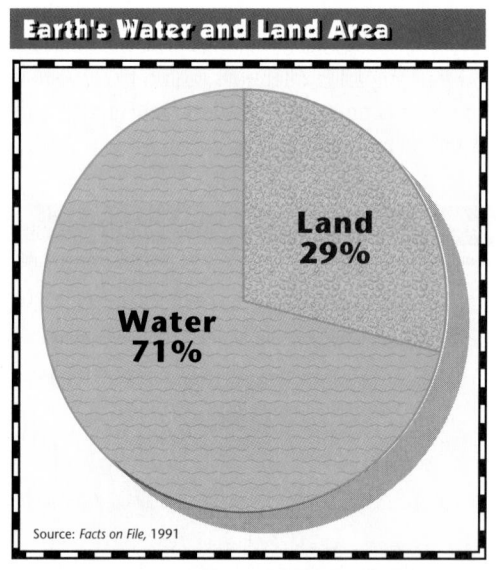

Earth's Water and Land Area

Land 29%
Water 71%

Source: *Facts on File*, 1991

Circle Graphs When you want to show how the *whole* of something is divided into its *parts*, you should use a **circle graph**. Because of their shape, circle graphs are often called pie graphs. Each "slice" represents a part or percentage of the whole "pie." Circle graphs are also good for making comparisons between two or more whole things that are broken into parts.

On this graph, the circle represents the whole surface area of the earth. Land and water each occupy a certain percentage of that whole, as shown by the way the circle is divided. You could easily compare this to another circle graph showing the total surface of another planet with land and water indicated.

Cooperative Learning Activity

Organize students into several small groups. Assign each group one or more types of graphs and charts. Group members should clip examples of their assigned graph and chart types from old newspapers and magazines. Suggest that they also log any that they see on television. Each group should then compose a headline for each of their graphic examples. Have groups cooperate in posting their materials to make a "Charts and Graphs" bulletin-board display. **L1**

Pictographs Like bar graphs, pictographs are good for making comparisons. **Pictographs** use rows of small symbols or pictures, each standing for an amount. This pictograph shows oil production in Venezuela. The key tells you that each oil rig stands for 20 million metric tons of oil. Pictographs are read like a bar graph. The total number of oil rigs in a row adds up to the production for that year.

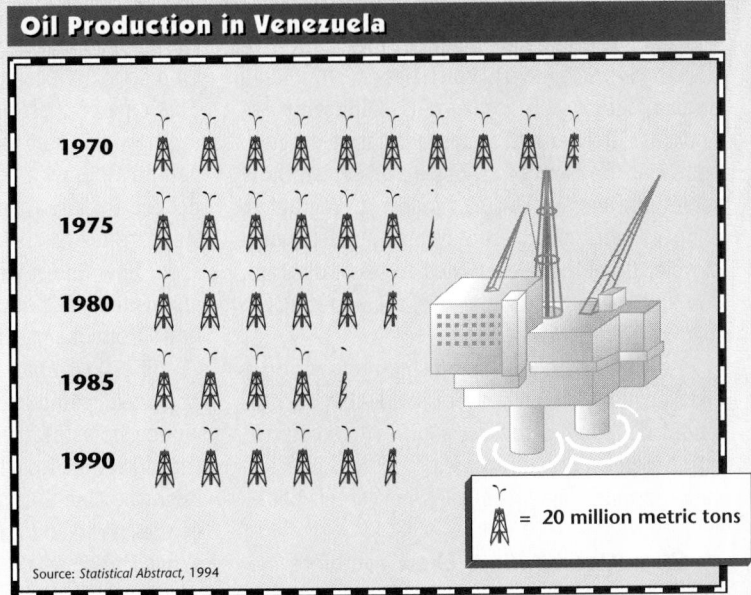

Oil Production in Venezuela

1970
1975
1980
1985
1990

= 20 million metric tons

Source: *Statistical Abstract*, 1994

Climographs A **climograph**, or climate graph, combines a line graph and a bar graph. It gives an overall picture of the **climate**— the long-term weather patterns—in a specific place. Because climographs include several kinds of information, you need to read them carefully.

Note that the vertical bars on the climograph represent average amounts of precipitation in each month of the year. To measure these bars, use the scale on the right-hand axis of the graphs.

The line above the bars, like those on other line graphs, represents changes in the average temperature based on the scale on the left. The months are abbreviated on the bottom axis of the graph.

Climograph: New York City

AVERAGE MONTHLY TEMPERATURE
°F °C
100 37.8
90 32.2
80 26.7
70 21.1
60 15.6
50 10.0
40 4.4
30 -1.1
20 -6.7
10 -12.2
0   -17.8

In. Cm
20 50.8
18 45.7
16 40.6
14 35.6
12 30.5
10 25.4
8   20.3
6   15.2
4   10.1
2   5.1
0   0
AVERAGE MONTHLY PRECIPITATION

J F M A M J J A S O N D

Source: *World Weather Guide*, 1994

GRAPHIC STUDY

Graphic Skills Practice
Reading a Graph In which year was Venezuela's oil production the lowest? *(1985)* In which year was Venezuela's oil production the highest? *(1970)*

Independent Practice

Geography Handbook Activities 14, 15, 16, 17, 18, 19, 20

GRAPHIC STUDY

Graphic Skills Practice
Reading a Graph What are the two coldest months in New York? *(January, February)* What are New York's two warmest months? *(July, August)* Is precipitation in New York heavier during winter or summer? *(slightly heavier in summer)*

Meeting Special Needs Activity

Kinesthetic Learners Students who do well with this mode of learning can benefit by translating some two-dimensional charts and graphs into three-dimensional models. For example, students can make line graphs out of tacks and string, make a bar graph out of blocks, or create a circle graph using wood or other materials. **L1**

15

Diagrams

If you enjoy science, mechanics, or working with your hands, you have probably come across diagrams. **Diagrams** are drawings that either show steps in a process, point out the parts of an object, or explain how something works. You can follow a diagram to connect parts of a stereo system or to repair a bicycle. If you have ever tried to use a diagram, you know that the simpler it is, the easier it is to follow.

Another kind of diagram can show you a cross section—or profile—of something such as landforms. Profiles can be helpful when comparing the elevations of an area. You will find elevation profile diagrams in each unit of this book.

Flow Chart A **flow chart** combines elements of a diagram and a chart. It can show the cause-and-effect process of how things happen. Or a flow chart can present the steps in a process. This type of chart often uses arrows to help you see how one step leads to another. For example, you might draw a flow chart to describe the steps needed to film a video. Arrows would link each step, from first to last. A flow chart could also show how things are interconnected. A chart of a governmental body would show how an idea flows from one part to another.

The flow chart on this page illustrates the impact of computers on different areas of life. Notice how the chart uses the arrows to show the direction in which the effects move. You might be able to think of other areas of computer use that could have been added to the chart. What are they?

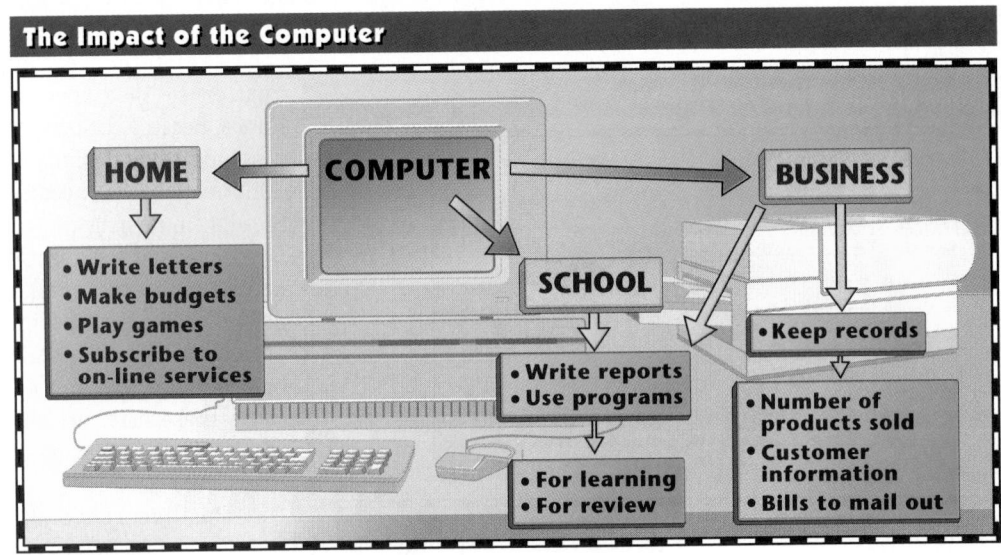

The Impact of the Computer

HOME ← COMPUTER → BUSINESS

HOME
- Write letters
- Make budgets
- Play games
- Subscribe to on-line services

SCHOOL
- Write reports
- Use programs
- For learning
- For review

BUSINESS
- Keep records
- Number of products sold
- Customer information
- Bills to mail out

Critical Thinking Activity

Recognizing Bias Point out the source lines of the graphs, charts, and diagrams in this section. Make sure students understand what they signify. Then ask students what qualities make an organization a good source of information. Ask them next what qualities make a bad source of information. Tell them that textbooks use well-established, unbiased sources for their graphs and charts, but that organizations sometimes purposely bias their information to support their point of view. Ask them what biases the following sources might have: an industry trade group, a foreign government, an environmental group. **L3**

Charts and Tables

Charts and **tables** present organized facts and statistics so they are easier to read. These graphics can also show how things are organized or how things are related. To understand a chart or table, first read the title. It tells you what information the chart or table contains. Next, read the labels at the top of each column and the left-hand side of the table or chart. They will tell you what data you are working with.

Tables are especially useful in organizing statistics in an easy-to-read way. This table shows data about the world's leading wheat-producing countries. Others might show many more columns and rows of numbers that would be difficult to understand if they were not organized in a table.

World's Leading Wheat Producers

COUNTRY	METRIC TONS PRODUCED
China	101,003,000
United States	66,920,000
India	55,087,000
Russia	46,000,000
France	32,600,000
Canada	29,871,000
Ukraine	19,473,000
Turkey	19,318,000
Kazakstan	18,500,000
Pakistan	15,684,000

Source: *The Statesman's Yearbook, 1994-1995*

GRAPHIC STUDY

Graphic Skills Practice
Reading a Chart What country is the world's third leading producer of wheat? *(India)* Which North American country produces the most wheat? *(United States)*

SECTION 3 REVIEW

Reviewing Graphs and Charts

1. **Define the following:** axis, bar graph, line graph, circle graph, pictograph, climograph, climate, diagram, flow chart, chart, table.
2. If you wanted to show how the population of Mexico City changed from 1850 to 1990, what kind of graph would you draw?
3. What two features of climate does a climograph show?
4. **ACTIVITY** Draw a flow chart showing and labeling the steps in some simple process—for example, making a sandwich or using a pay telephone. Give your flow chart to a classmate to follow it and complete the task described. Write a sentence or two about how easy your chart was to follow. See if your classmate agrees or has suggestions for improvement.

Answers to Section 3 Review

1. All vocabulary words are defined in the Glossary.
2. line graph
3. temperature and precipitation
4. Flow charts will vary. Call on volunteers to draw their charts on the chalkboard.

17

0:00 **OUT OF TIME?**

If time does not permit teaching each chapter, you may use the Chapter Highlights, Audiocassettes, and their corresponding activities and tests.

GLENCOE
TECHNOLOGY

 VIDEODISC

The Infinite Voyage:
Crisis in the Atmosphere

Chapter 3
Our Future Climate
Disc 2 Side A

interNET
CONNECTION

Resources on aspects of Earth—its atmosphere, climate, water, and landforms—are available at the following address:

World Wide Web
http://www.windows.umich.edu/earth/earth.html

To find the latest information and statistics on global population, check out the following address:

World Wide Web
http://www.albany.edu/~rws32/102-pgs2.html

Unit 1
Geography of the World

Lake Tahoe, United States

(18)

About the Illustration

Lake Tahoe is on the border of California and Nevada. The lake is perched 6,200 feet (1,900 meters) above sea level in the Sierra Nevada mountains. Because the lake is so deep, it does not freeze in winter even though area temperatures are quite low. The lake and surrounding area attract tourists year-round. The name "Tahoe" comes from a Native American word meaning "lake."

 **Location** Where is Lake Tahoe located? *(on the border of California and Nevada)*

NATIONAL
GEOGRAPHIC
SOCIETY

These materials are available from Glencoe.

 CD-ROM

Picture Atlas of the World

 VIDEODISC

STV: Solar System

The Big Picture
Side 1, Frames 00691-11815

GeoJournal Activity

Ask each student to choose one particular place to visit. Encourage students to write a paragraph or create artwork depicting the places they select. Have students arrange their work to create a "Dream Destinations" bulletin-board display. **L1**

- This journal activity provides the basis for the "GeoJournal Writing Activity" exercise in the Chapter Review.
- The GeoJournal may be used as an integral part of Performance Assessment.

GeoJournal Activity

You are about to journey to remote rain forests, bleak deserts, bustling marketplaces, and mountainous villages. You are entering the world of geography—the study of the earth and its people. Imagine that you could visit any place in the world. Where would you go? What would you want to see? Jot down your thoughts in your journal.

19

Location Activity

Display Political Map Transparency 1. Call on volunteers to answer the following questions: What covers most of the earth's surface? *(water)* Is all of the water connected? *(no, but the large bodies of water—the oceans—are)* About what percentage of the earth's surface is land? *(about 30 percent)* Is all the land connected? *(no)* On what land area is our community located? *(North America)* **L1**

PERFORMANCE ASSESSMENT ACTIVITY

WRITING SCIENCE FICTION Have students take the role of science-fiction authors. Begin the task by having students brainstorm a list of the physical attributes of the earth as discussed in the chapter. List these attributes on the left side of a T-chart. Have students complete the T-chart by assuming a change in the attribute on the left side and predicting an effect on the right. If the tilt of the earth changed, for example, then the seasons would change. Upon completion of the chart, have students choose an entry and write a science fiction story on the effects of the assumed change. Suggest that students address the five themes of geography and include maps and diagrams in their stories. Compile a class book of stories.

POSSIBLE RUBRIC FEATURES: Content information, creative thinking, composition skills, map skills, cause/effect skills, concept attainment

MENTAL MAPPING ACTIVITY

Before teaching Chapter 1, display a globe at the front of the class. Make sure that students understand that a globe is simply a model of the earth. Encourage students to spin the globe and locate various oceans, continents, and landforms around the world.

TEACHER'S CORNER

NATIONAL GEOGRAPHIC SOCIETY

INDEX TO NATIONAL GEOGRAPHIC MAGAZINE

The following articles may be used for research relating to this chapter:

- "California Desert Lands," by Michael Parfit, May 1996.
- "Into the Heart of Glaciers," Photographs by Carsten Peter, February 1996.
- "Under Our Skin," by Keay Davidson and A.R. Williams, January 1996.
- "Orion: Where Stars Are Born," by James Reston, Jr., December 1995.
- "Blueprints for Victory," by John F. Shupe, May 1995.
- "Living With California's Faults," by Rick Gore, April 1995.
- "Three Years Across the Arctic," by Ramón Hernando de Larramendi, January 1995.
- "Chile's Uncharted Cordillera Sarmiento," by Jack Miller, April 1994.
- "New Eyes on the Universe," by Bradford A. Smith, January 1994.
- "Volcanoes: Crucibles of Creation," by Noel Grove, December 1992.

NATIONAL GEOGRAPHIC SOCIETY PRODUCTS AVAILABLE FROM GLENCOE

To order the following products for use with this chapter, contact your local Glencoe sales representative or call Glencoe at 1-800-334-7344:

- *GeoBee* (Software)
- *Picture Atlas of the World* (CD-ROM)
- *Earth's Endangered Environments PictureShow* (CD-ROM)
- *Geology PictureShow* (CD-ROM)
- *GTV: Planetary Manager* (Videodisc)
- *STV: World Geography* (Videodisc)
- *STV: Solar System* (Videodisc)
- *STV: Restless Earth* (Videodisc)
- *STV: Water* (Videodisc)

ADDITIONAL NATIONAL GEOGRAPHIC SOCIETY PRODUCTS

To order the following products for use with this chapter, call National Geographic Society at 1-800-368-2728:

- *The Living Earth* (Video)
- *Atmosphere: On the Air* (Video)
- *Water: A Precious Resource* (Video)
- *Physical Geography of North America Series* (6 Videos)
- *More Than Maps: A Look at Geography:* "What is Geography?" "Geography and the World Around You," "Geography at Work." (Filmstrip)

TEACHER-TO-TEACHER
Curriculum Connection: Life Science

—from Patricia K. Bosh
Mifflin International Middle School, Columbus, OH

ANALYZING PHOTOGRAPHS *The purpose of this activity is to have students evaluate human/environment interaction by analyzing photographs of various landscapes.*

Project slides of a number of different landscapes. As each slide is projected, draw students' attention to the physical characteristics of the landscape. (Consider such features as landforms, bodies of water, vegetation, weather, and so on.) Discuss with students the possible location of the landscape. Have them find possible locations on a country or world map. Then have students note the human characteristics of the landscape. (Suggest that they look for buildings, transportation and communication systems, clothing, crops, and so on.) Encourage students to speculate on why such human/environment interactions have taken place.

Organize students into several small groups. Provide each group with photographs of different landscapes. Direct groups to note the physical characteristics of each landscape, its possible locations, its human characteristics, and the reasons for these human/environment interactions. Have groups compare their findings. Then ask students to work individually to write a few sentences on the patterns of human/environment interactions they noticed in their study of the photographs. Encourage students to apply this system of photograph analysis throughout.

BIBLIOGRAPHY

Readings for the Student

Farndon, John. *How the Earth Works.* Pleasantville, N.Y.: The Reader's Digest Association, 1992.

The Usborne Geography Encyclopedia. London, England: Usborne, 1992.

Readings for the Teacher

Davis, Kenneth C. *Don't Know Much About Geography.* New York: William Morrow, 1992.

Grillet, Donnat V. *Where on Earth?* New York: Prentice Hall, 1991.

Multimedia

Smith, David. *Mapping the World by Heart.* Watertown, Mass.: Tom Snyder Productions, 1992. Videocassette, binder, reproducible handouts, maps.

Understanding Globes. Niles, Ill.: United Learning, 1991. Videocassette, 15 minutes.

World Atlas. Novato, Calif.: The Software Toolworks, 1993. CD-ROM for Macintosh or IBM, Macintosh or IBM 3.5" disk.

KEY TO ABILITY LEVELS

Teaching strategies have been coded for varying learning styles and abilities.

L1 BASIC activities for all students

L2 AVERAGE activities for average to above-average students

L3 CHALLENGING activities for above-average students

LEP LIMITED ENGLISH PROFICIENCY activities

MEETING NATIONAL STANDARDS

Geography for Life

All of the 18 standards are demonstrated in Unit 1. The following ones are highlighted in this chapter:

• Standards 1, 2, 3, 4, 6, 7, 15, 18

Use these *Geography: The World and Its People* resources to teach, reinforce, and extend chapter content.

CHAPTER 1 RESOURCES

- *Vocabulary Activity 1
- Cooperative Learning Activity 1
- Workbook Activity 1
- Geography Skills Activity 1
- Critical Thinking Skills Activity 1
- GeoLab Activity 1
- Chapter Map Activity 1
- Performance Assessment Activity 1
- *Reteaching Activity 1
- Enrichment Activity 1
- Chapter 1 Test, Form A and Form B
- Political Map Transparency 1
- Unit Overlay Transparency 1-1
- Geoquiz Transparencies 1-1, 1-2
- *Chapter 1 Digest Audiocassette, Activity, Test
- Vocabulary PuzzleMaker Software
- Student Self-Test: A Software Review
- Testmaker Software
- MindJogger Videoquiz

If time does not permit teaching the entire chapter, summarize using the Chapter 1 Highlights on page 33, and the Chapter 1 English (or Spanish) Audiocassettes. Review students' knowledge using the Glencoe MindJogger Videoquiz. *Also available in Spanish

Use these *Geography: The World and Its People* resources to teach and reinforce section content.

SECTION 1 RESOURCES

Reproducible Lesson Plan 1-1

Section Focus Transparency 1-1

Guided Reading Activity 1-1

Section Quiz 1-1 (also available in Spanish)

SECTION 2 RESOURCES

Reproducible Lesson Plan 1-2

Section Focus Transparency 1-2

Guided Reading Activity 1-2

Section Quiz 1-2 (also available in Spanish)

SECTION 3 RESOURCES

Reproducible Lesson Plan 1-3

Section Focus Transparency 1-3

Guided Reading Activity 1-3

Section Quiz 1-3 (also available in Spanish)

ADDITIONAL RESOURCES FROM GLENCOE

 Reproducible Masters

- Glencoe Social Studies Outline Map Book, page 39

Workbook

- Building Skills in Geography, Unit 1, Lesson 7

 Transparencies

- National Geographic Society PicturePack Transparencies, Unit 1

 Laminated Desk Map

 Posters

- National Geographic Society: Eye on the Environment, Unit 1

 World Games Activity Cards Unit 1

World Crafts Activity Cards Unit 1

 Videodiscs

- Geography and the Environment: The Infinite Voyage

- National Geographic Society: STV: Atmosphere

- National Geographic Society: STV: Solar System

CD-ROM

- National Geographic Society PictureShow: Geology

Performance Assessment

Refer to the Planning Guide on page 20A for a Performance Assessment Activity for this chapter. See the *Performance Assessment Strategies and Activities* booklet for suggestions.

Chapter Objectives

1. **Discuss** the five themes of geography.
2. **Explain** how the earth moves in space and why the seasons change.
3. **Describe** the basic structure of the earth.

GLENCOE TECHNOLOGY

 VIDEODISC

Use the Chapter 1 Mind-Jogger Videoquiz to preview chapter content.

MindJogger Videoquiz

Chapter 1
Disc 1 Side A

 Also available in VHS.

Chapter

1 Looking at the Earth

World Continents and Oceans

[Map showing world continents and oceans with labels: ARCTIC OCEAN, Arctic Circle, NORTH AMERICA, EUROPE, ASIA, Tropic of Cancer, ATLANTIC OCEAN, AFRICA, Equator, PACIFIC OCEAN, SOUTH AMERICA, INDIAN OCEAN, Tropic of Capricorn, AUSTRALIA, Antarctic Circle, ANTARCTICA. Longitude lines: 150°W, 120°W, 90°W, 60°W, 30°W, 0°, 30°E, 60°E, 90°E, 120°E, 150°E. Latitude lines: 60°N, 30°N, 0°, 30°S, 60°S. Compass rose showing N, S, E, W.]

 MAP STUDY ACTIVITY

In this chapter you will learn how geographers look at the earth and its people.

1. What are the earth's seven continents?
2. What are the earth's four major oceans?

(20)

MAP STUDY ACTIVITY

Answers
1. North America, South America, Europe, Africa, Asia, Australia, Antarctica
2. Pacific Ocean, Atlantic Ocean, Indian Ocean, Arctic Ocean

Map Skills Practice
Reading a Map What is the southernmost continent? *(Antarctica)*

SECTION 1

Using the Geography Themes

PREVIEW

Words to Know
- geography
- absolute location
- hemisphere
- latitude
- longitude
- grid system
- relative location
- place
- environment
- movement
- region

Places to Locate
- Equator
- Prime Meridian

Read to Learn . . .
1. how geographers study the earth.
2. how place and location can mean different things.
3. how people relate to their environment and to each other.

Our earth is a fascinating place! Look at this photo of Tokyo, Japan, with its modern buildings. Akio

Mansai lives and goes to school in this city. He and his classmates learn about the earth and its geography—the same things you will learn as you read this book.

Geography is the study of the earth in all of its variety. When you study geography, you learn about the earth's land, water, and plant and animal life. You analyze where people are, how they live, and what they do and believe. You especially look at places people have created and try to understand *how* and *why* they are different.

Geographers study the earth as the home of people. Five geographic themes—location, place, human/environment interaction, movement, and region—are used here to help you think like a geographer.

Location

In geography, *location* means knowing where you are. Every place on the earth can be given an exact position on the globe, which is called its **absolute location**. To help geographers mark the absolute location of a place, a network or grid of imaginary lines is placed on the earth.

Equator The Equator is an imaginary line that circles the earth midway between the North Pole and the South Pole. It divides the earth into two

---

LESSON PLAN
Chapter 1, Section 1

Meeting National Standards
Geography for Life The following standards are highlighted in this section:
- Standards 1, 3

FOCUS

Section Objectives
1. **Explain** how geographers study the earth.
2. **Illustrate** how place and location can mean different things.
3. **Consider** how people relate to their environment and to each other.

Bellringer Motivational Activity

 Prior to taking roll at the beginning of the class period, project Section Focus Transparency 1-1 and have students answer the activity questions. Discuss students' responses. **L1**

Vocabulary Pre-check

The terms *latitude* and *longitude* often are confused. Encourage students to think of a phrase to help them remember the correct meanings of the two terms. **L1 LEP**

 Use the Vocabulary PuzzleMaker Software to create a crossword puzzle. **L1**

Classroom Resources for Section 1

REPRODUCIBLE MASTERS
Reproducible Lesson Plan 1-1
Guided Reading Activity 1-1
Building Skills in Geography, Unit 1, Lesson 7
Geography Skills Activity 1
Vocabulary Activity 1
GeoLab Activity 1
Section Quiz 1-1

TRANSPARENCIES
Section Focus Transparency 1-1
Geography Handbook Transparency 2
Political Map Transparency 1

MULTIMEDIA
 Vocabulary PuzzleMaker Software
Testmaker

TEACH

Guided Practice

Display Geography Handbook Transparency 2. Have students estimate the latitude and longitude of
(continued)

21

major world cities and various physical features. Call on volunteers to describe the relative location of these cities and physical features using the grid system. **L1**

Independent Practice

📁 Guided Reading Activity 1-1 **L1**

More About the Illustration

Answer to Caption
They raise animals and use tents for shelter.

 **Place** Most Tibetan nomads live on the high grasslands of northern Tibet.

ASSESS

Check for Understanding

Assign Section 1 Review as homework or an in-class activity.

Meeting Lesson Objectives

Each objective below is tested by the questions that follow it in parentheses.

1. **Explain** how geographers study the earth. (1, 2, 4)
2. **Illustrate** how place and location can mean different things. (1, 3, 5)
3. **Consider** how people relate to their environment and to each other. (1)

22

---

hemispheres, or halves. The Northern Hemisphere includes all of the land and water between the Equator and the North Pole. The Southern Hemisphere includes all of the land and water between the Equator and the South Pole.

Latitude and Longitude Other imaginary lines called lines of **latitude**, or parallels, circle the earth parallel to the Equator. They measure distance north or south of the Equator in degrees. The Equator is designated as 0°, and the poles are at 90° North and 90° South.

Lines of **longitude**, or meridians, run from the North Pole to the South Pole. They are numbered in degrees east or west of a starting line called the Prime Meridian, which is at 0° longitude. On the opposite side of the earth from the Prime Meridian is the International Date Line, or 180° longitude.

Lines of latitude and longitude cross one another in the form of a **grid system**. If you know a place's exact latitude north or south of the Equator, and its exact longitude east or west of the Prime Meridian, you can easily mark the absolute location of that place on a map or globe.

Relative Location You also can locate a place by finding out how far and in what direction it is from somewhere else. This is called **relative location** because you are learning where a place is *in relation* to another place.

Place

Place has a special meaning in geography. It means more than where a place *is*. It also describes what a place is *like*. That is, what features make this location similar to or different from another place?

These features may be physical characteristics, such as land shape, plants, animal life, or climate. They also may be characteristics of people and the things they have created, including their language, clothing, buildings, music, or ways of making a living.

The geography theme of *place* describes physical characteristics of land, such as these jagged peaks in the High Tatra mountains of Slovakia *(left)*. Place, however, also describes characteristics of cultural groups, such as the lifestyle of nomads in Tibet *(right)*.
PLACE: What can you learn about the lifestyle of the Tibetan nomads from this photo?

Characteristics of Place

22

UNIT 1

Cooperative Learning Activity

Organize the class into five groups. Assign each group one of the five themes of geography. Direct groups to apply their assigned themes to the community by 1) writing a definition of the theme; 2) writing a brief description of the community in terms of the theme; and 3) collecting newspaper articles and illustrations that show examples of that theme in the community. When they have completed their tasks, have the groups work together to create a bulletin-board display titled "The Five Themes of Geography in Our Community." **L2**

Human/Environment Interaction

Wherever humans have lived or traveled, they have changed their **environment**, or natural surroundings. People have blasted through mountains to build roads, cut down forests, built houses, and used grasslands to graze herds. Some human actions have damaged the natural environment, and some have not.

The environment influences the way people live. People adapt their lives to some environmental conditions. To live in a cold climate, for example, people must invent ways to protect themselves and make a living in the cold. To live in a place that is dry, people have to develop ways to provide water.

Movement

The theme of **movement** helps geographers understand the relationship among places. Movement describes how people in one place make contact with people from another place. People, ideas, information, and products are constantly moving around the world. They travel instantly by telephone, computer, or satellite, or go more slowly by car, train, or ship.

When people in one place want something that is not found in their area, they trade with people in areas that have what they want. People in Japan, for example, may listen to music from the United States, while people in Canada eat bananas shipped from Central America.

Region

Geographers often think about the world in **regions**, or areas that share some common characteristics. Regions can be quite small—your county, city, or neighborhood can be a region. They also can be huge—the western United States is a region.

An area can be called a region because of its physical features such as landscape or climate. A region also can be determined by human traits such as language, political boundaries, religion, or the kinds of work the people do.

High-speed Movement

This French high-speed train is one example of how people, products, and ideas move from one place to another.
MOVEMENT: Why might people board this train?

SECTION 1 REVIEW

REVIEWING TERMS AND FACTS

1. Define the following: geography, absolute location, hemisphere, latitude, longitude, grid system, relative location, place, environment, movement, region.

2. LOCATION What is the starting line for determining longitude?

3. PLACE What two kinds of features are used to describe place?

MAP STUDY ACTIVITIES

4. Turn to the world map on page 24. What large land area is both west of the Prime Meridian and entirely north of the Equator?

Answers to Section 1 Review

1. All vocabulary words are defined in the Glossary.

2. Prime Meridian

3. physical and human characteristics

4. North America

LESSON PLAN
Chapter 1, Section 1

More About the Illustration

Answer to Caption
to go on vacation, visit family, go to work

Movement
The French TGV *(train à grande vitesse),* or high-speed train, can travel at 186 miles per hour (299 km per hour).

Evaluate

Section 1 Quiz **L1**

Use the Testmaker to create a customized quiz for Section 1. **L1**

Reteach

Have students write a summary of the section using the words and places listed in the **Preview. L1**

Enrich

Have students create symbols or logos for each of the five themes of geography. **L3**

CLOSE

Call on volunteers to explain how the five themes of geography further our knowledge of the earth.

BUILDING GEOGRAPHY SKILLS

BUILDING GEOGRAPHY SKILLS

TEACH

Draw a map of the classroom on the chalkboard. Begin by drawing the perimeter of the room. Then add windows, doors, desks, bookcases, and other classroom features. Before you add each feature, call on volunteers to suggest how it might be represented on the map. *(different colors or symbols)* Record students' chosen method in a key next to the map. Complete the map by adding a compass rose.

Have students use the completed map to answer questions such as: How many windows does the classroom have? Which is closer to the door, the chalkboard or the bulletin board? Where are bookcases located in the room? Then direct students to read the skill lesson and answer the practice questions.

Additional Skills Practice

1. What mountain range lies just to the east of Los Angeles? *(Rocky Mountains)*
2. Is Sydney a national capital? *(no)*

📁 Geography Skills Activity 1

📁 Building Skills in Geography, Unit 1, Lesson 7

Using a Map Key

To understand what a map is showing, you must read the **map key**, or legend. The map key explains the meaning of special colors, symbols, and lines on the map.

On some maps, for example, colors represent different heights of land. On other maps, colors might stand for climate areas or population. Lines may stand for rivers, railroads, streets, or boundaries. The map key also explains the meaning of other symbols found on the map. To use a map key, follow these steps:

• Read the map title.
• Study the map key to find out what special information it gives.
• Find examples of each map key color, line, or symbol on the map.

The World

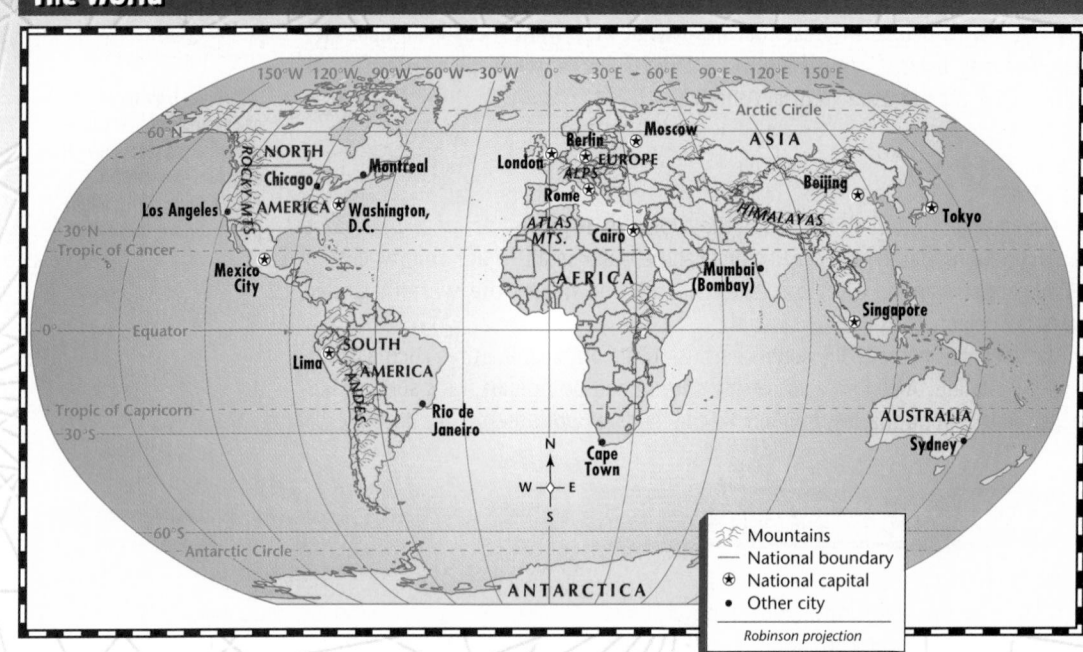

Mountains
National boundary
National capital
Other city

Robinson projection

Geography Skills Practice

1. What are the Himalayas and the Andes? How can you tell?
2. How are national boundaries shown?
3. What does the symbol ⊛ represent?

Answers to Geography Skills Practice

1. They are mountains. They are marked with a mountains symbol.
2. black lines
3. national capitals

Planet Earth

PREVIEW

Words to Know
- galaxy
- solar system
- orbit
- revolution
- leap year
- axis
- solstice
- equinox

Places to Locate
- Earth
- sun
- moon

Read to Learn . . .
1. how Earth moves in space.
2. why Earth's seasons change.

The first humans to travel in space were amazed by this view of Earth. The beautiful blue oceans and swirling clouds made the planet look like a marble against a dark sky.

O ur planet Earth is one of a group of planets revolving around the sun. The sun is just one of hundreds of millions of stars in a **galaxy**, or a huge system of stars. Earth's most important companion in space is the sun. Earth is a member of the **solar system**—planets and other bodies that revolve around our sun.

Location

The Earth in Space

Earth, eight other planets, their moons, and some smaller asteroids make up our solar system. They travel in **orbits**, or elliptical paths, around the sun. Look at the diagram of the solar system on page 26 to see Earth's position in the solar system.

Planets are sometimes classified into two types—those like Earth and those like Jupiter. The Earth-like planets are Mercury, Venus, Mars, and Pluto, which is so far away that little is known about it. These small, solid planets rotate slowly and have few or no moons. The other outer planets—Jupiter, Saturn, Neptune, and Uranus—are huge, rapidly spinning, and surrounded by gassy atmospheres and many moons.

Sun, Earth, and Moon Life on Earth could not exist without heat and light from the sun. About 93 million miles (150 million km) from Earth, the sun is made up mostly of intensely hot gases. Its great mass creates a strong pull of gravity—enough to keep the planets revolving around it.

Classroom Resources for Section 2

 REPRODUCIBLE MASTERS

Reproducible Lesson Plan 1-2
Guided Reading Activity 1-2
Cooperative Learning Activity 1
Section Quiz 1-2

 TRANSPARENCIES

Section Focus
Transparency 1-2

MULTIMEDIA

 Testmaker

Meeting National Standards
Geography for Life The following standards are highlighted in this section:
- Standards 1, 3, 15, 18

FOCUS

Section Objectives
1. **Describe** how Earth moves in space.
2. **Explain** why Earth's seasons change.

Bellringer Motivational Activity

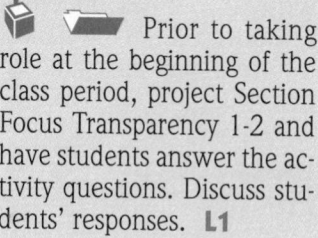 Prior to taking role at the beginning of the class period, project Section Focus Transparency 1-2 and have students answer the activity questions. Discuss students' responses. **L1**

Vocabulary Pre-check

Point out that the terms *solar* and *solstice* are derived from the Latin word for sun. Have students locate these two terms in the text and note how they are related to the sun. **L1** **LEP**

Night and day are caused by the earth's rotation on its axis. At the Equator, the speed of rotation is about 1,000 miles per hour (1,609 km per hour).

GRAPHIC STUDY

Answer
Mercury; Pluto

Graphic Skills Practice
Reading a Diagram
What is the largest planet in the solar system?
(Jupiter)

TEACH

Guided Practice

Modeling Have students work in small groups to create a model of the solar system. Encourage groups to write a brief description of the movement of the planets to accompany their models.
L2

Independent Practice

Guided Reading Activity 1-2 **L1**

ASSESS

Check for Understanding

Assign Section 2 Review as homework or an in-class activity.

Meeting Lesson Objectives

Each objective below is tested by the questions that follow it in parentheses.

1. **Describe** how Earth moves in space.
   (1, 3, 4, 5)
2. **Explain** why Earth's seasons change.
   (1, 3, 5)

The Solar System

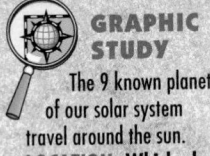

GRAPHIC STUDY

The 9 known planets of our solar system travel around the sun.
LOCATION: Which planet would be affected first if the sun suddenly died? Which would be affected last?

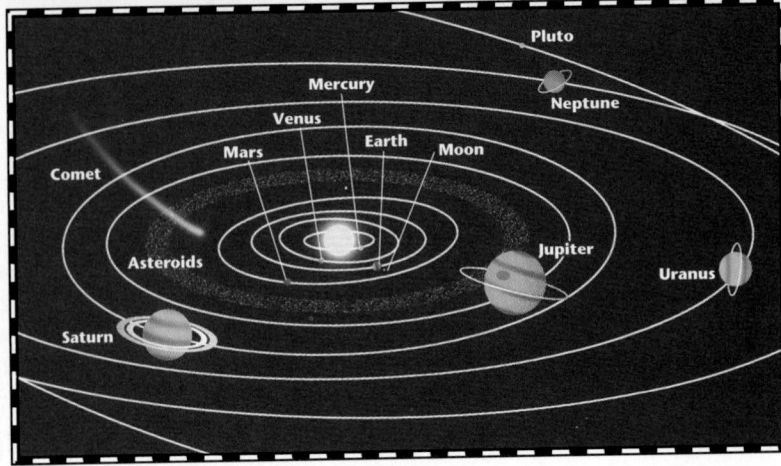

In the same way, the moon revolves around Earth—about once every 30 days. A cold, rocky sphere with no water and no atmosphere, the moon gives off no light of its own. When you see the moon shining, it is actually reflecting light from the sun. The relative positions of the sun, moon, and Earth determine whether you see a "new" or "full" moon.

Earth makes one **revolution,** or complete trip around the sun, in 365¼ days. This is what we define as one year. The extra one-fourth of a day is the reason there is a **leap year,** an extra day in the calendar every four years.

As Earth makes a revolution, it also rotates, or spins, on its axis. The **axis** is an imaginary line that runs through the earth's center between the North and South poles. As Earth turns toward and away from the sun every 24 hours, different areas are in sunlight and in darkness, causing day and night.

Place
The Seasons

Do you live in a place where the seasons change? Or does your weather stay similar year-round? Because Earth is always tilted on its axis, seasons change as Earth travels its year-long orbit around the sun. To see why this happens, look at the four globes in the diagram on page 27. Notice how sunlight falls directly on different parts of Earth at different times during the year.

On about June 21 the North Pole is tilted toward the sun. The sun appears directly overhead at the line of latitude called the Tropic of Cancer. This day is the summer **solstice,** or beginning of summer, in the Northern Hemisphere. It is the day there with the most hours of sunlight.

Six months later—about December 22—the North Pole is tilted away. The sun's direct rays strike the line of latitude known as the Tropic of Capricorn. This is the winter solstice—a time of winter in the Northern Hemisphere, but the beginning of summer in the Southern Hemisphere.

Cooperative Learning Activity

Organize the class into four groups. Assign each group one of the four seasons. Each group should create a four-page pictorial essay on its assigned season. Topics they should cover include how the movement of Earth creates seasons, differences between Northern and Southern hemispheres, and typical seasonal weather in their area. Suggest that students use maps, diagrams, and pictures cut from old newspapers and magazines as illustrations. Have groups display and discuss their essays. **L2**

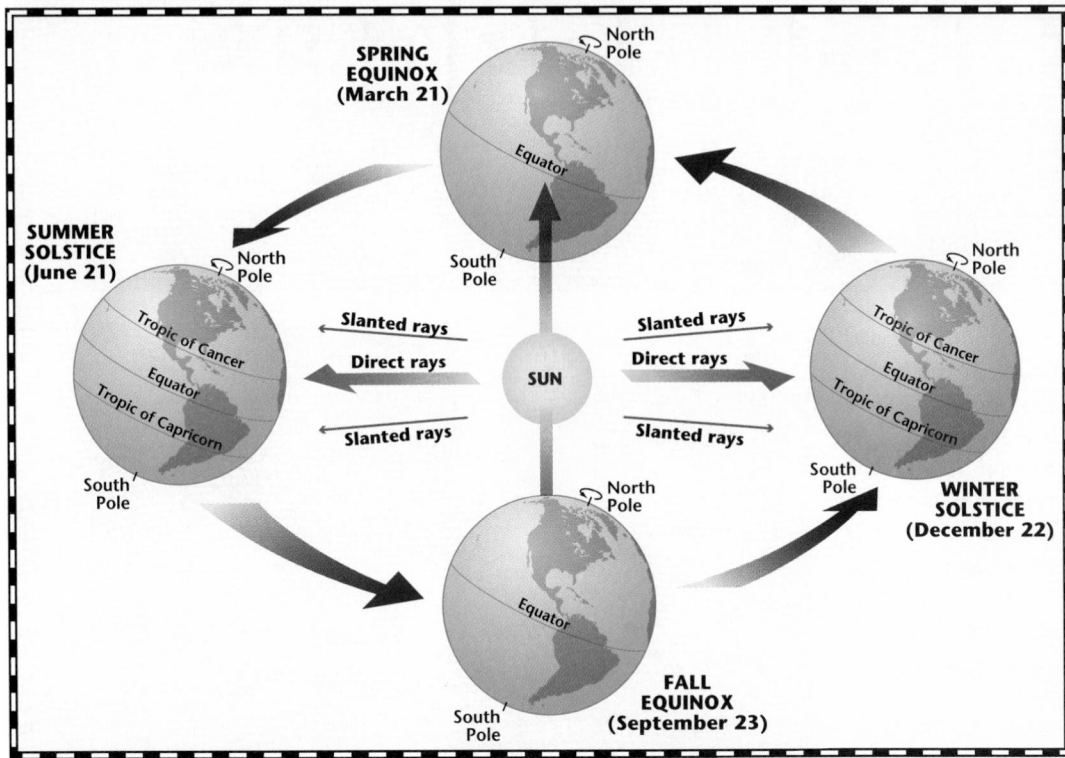

GRAPHIC STUDY
The tilt of the earth as it revolves around the sun causes the seasons to change.
LOCATION: When it is summer in the Northern Hemisphere, what season is it in the Southern Hemisphere?

GRAPHIC STUDY

Answer
winter

Graphic Skills Practice
Reading a Diagram
What does September 23 mark in the Northern Hemisphere? In the Southern Hemisphere?
(Fall, Spring)

Midway between the two solstices, about September 23 and March 21, the sun's rays are directly overhead at the Equator. These are **equinoxes,** when day and night in both hemispheres are of equal length.

SECTION 2 REVIEW

REVIEWING TERMS AND FACTS
1. Define the following: galaxy, solar system, orbit, revolution, leap year, axis, solstice, equinox.
2. PLACE What makes the moon shine?
3. MOVEMENT How do Earth's movements affect the length of the year?

GRAPHIC STUDY ACTIVITIES
4. Looking at the diagram on page 26, describe Earth's location relative to other planets.
5. Look at the diagram above. In June, where do the sun's direct rays hit Earth?

Evaluate

Section 2 Quiz **L1**

Use the Testmaker to create a customized quiz for Section 2. **L1**

Reteaching

Have students write five questions on the section content. Organize students into small groups and have group members use their questions to quiz each other on the section. **L1**

Enrich

Have students research and report on attempts to learn about our solar system through the voyages of the *Mariner, Pioneer, Viking,* and *Voyager* spacecraft. **L3**

CLOSE

Have students discuss what life in the United States would be like if the earth did not tilt at an angle.

Answers to Section 2 Review
1. All vocabulary words are defined in the Glossary.
2. reflecting light from the sun
3. A year is the length of time it takes Earth to travel around the sun. Earth rotates on its axis, turning one side toward and then away from the sun, causing day and night.
4. the third planet from the sun, between Venus and Mars
5. along the Tropic of Cancer

TEACH

Ask students who have experienced an earthquake to describe what it was like. If no one has experienced an earthquake, ask students to relate what they know about earthquakes from books, magazines, or television shows. List their responses on the chalkboard. Then have students use the information on the chalkboard to write a paragraph describing an earthquake that measures greater than 6.0 on the Richter scale. **L1**

More About . . .

Seismology Seismologists, the scientists who study earthquakes, use tools other than the Richter scale to measure the size of an earthquake. The most common is the Modified Mercalli Scale, which ranks earthquakes according to 12 levels of intensity. A Modified Mercalli Scale measurement of intensity IX (9) would be roughly equivalent to 6.0 on the Richter scale.

CURRICULUM CONNECTION

History The Chinese developed the first seismograph sometime during the later years of the Han Dynasty (around A.D. 100–200).

MAKING CONNECTIONS

MEASURING EARTHQUAKES

| MATH | SCIENCE | HISTORY | LITERATURE | TECHNOLOGY |

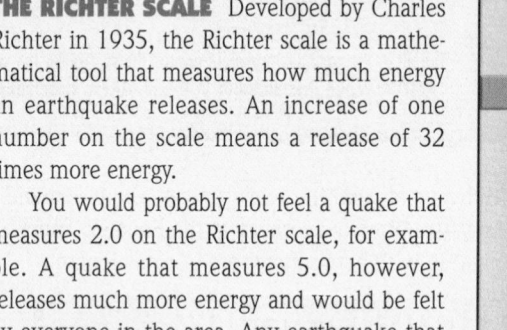

Earthquake Damage, Los Angeles, California

Imagine! The ground shook so hard that the Mississippi River actually changed its course! It was the early 1800s when one of the worst earthquakes in United States history hit New Madrid, Missouri.

Earthquakes happen when great cracks, or faults, in the earth's crust slip. To measure the energy an earthquake releases, scientists use two tools: the seismograph and the Richter scale.

SEISMOGRAPH A seismograph records the earth's tremors, or movements, during an earthquake. A sensitive writing tip draws a line on a turning drum. A tiny tremor thousands of miles away will show up as a wiggly line on the record sheet. The larger the earthquake, the larger the wiggle. The seismograph's recordings measure a quake's movements as well as help scientists locate the earthquake and determine its strength.

THE RICHTER SCALE Developed by Charles Richter in 1935, the Richter scale is a mathematical tool that measures how much energy an earthquake releases. An increase of one number on the scale means a release of 32 times more energy.

You would probably not feel a quake that measures 2.0 on the Richter scale, for example. A quake that measures 5.0, however, releases much more energy and would be felt by everyone in the area. Any earthquake that measures 6.0 or more is considered a major quake. The giant forces unleashed by earthquakes can topple buildings and bridges.

Making the Connection

1. What does a seismograph do?
2. What is the Richter scale?

28

UNIT 1

Answers to Making the Connection

1. records and measures the earth's tremors during an earthquake; helps locate the earthquake and determine its strength

2. a mathematical tool that measures how much energy an earthquake releases

PREVIEW

Words to Know
- landform
- core
- mantle
- crust
- magma
- continent
- plate tectonics
- fault
- earthquake
- tsunami
- weathering
- erosion
- plateau
- isthmus
- peninsula
- strait
- atmosphere

Places to Locate
- Mount Fuji
- Hawaii

Read to Learn . . .
1. how the earth's structure is layered.
2. how forces change landforms.
3. about the earth's major landforms.

Millions of years ago, these steep valleys in Norway were enlarged by moving ice called glaciers. Norway's landscape is a spectacular example of the forces that constantly change the earth's surface.

Humans occupy only a small percentage of the earth's surface. That surface, however, is very different from place to place. Landscapes and their **landforms,** or individual features, influence where people live and how they relate to their environment.

Place
The Changing Planet

Earth's different landscapes formed over millions of years and are still changing today. A number of forces have shaped past and present landforms. No one has been able to dig deeper than a few kilometers below the earth's crust. Still, geologists have developed a picture of the layers in the earth's structure.

Inside the Earth The inside of the earth is made up of three layers—the **core,** the **mantle,** and the **crust.** In the center of the planet is a dense core of hot metal, probably iron mixed with nickel. This core is divided into a solid inner core and an outer core of melted, liquid metal.

CHAPTER 1

29

Meeting National Standards

Geography for Life The following standards are highlighted in this section:
- Standards 1, 2, 3, 4, 7, 15

FOCUS

Section Objectives
1. **Understand** how the earth's structure is layered.
2. **Discuss** how forces change landforms.
3. **Identify** the earth's major landforms.

Bellringer Motivational Activity

Prior to taking role at the beginning of the class period, project Section Focus Transparency 1-3 and have students answer the activity questions. Discuss students' responses. **L1**

Vocabulary Pre-check

Refer students to the term *landform* in the **Words to Know** list. Have students attempt to infer the meaning of this compound word from its parts, *land* and *form.* Tell students that they can determine the meaning of other words in this manner. Point to the term *earthquake* as an example. **L1** **LEP**

Classroom Resources for Section 3

 **REPRODUCIBLE MASTERS**

Reproducible Lesson Plan 1-3
Guided Reading Activity 1-3
Workbook Activity 1
Critical Thinking Skills Activity 1
Reteaching Activity 1
Enrichment Activity 1
Section Quiz 1-3

 TRANSPARENCIES

Section Focus
    Transparency 1-3
Geoquiz Transparencies 1-1,
    1-2

MULTIMEDIA

Testmaker

More About the Illustration

Answer to Caption
Death Valley has rugged terrain and a harsh climate; Pacific atoll is too isolated and too small.

Place Parts of Death Valley are 282 feet (86 m) below sea level, the lowest elevation in the Western Hemisphere.

TEACH

Guided Practice

Organizing Information
Draw two large concentric circles on the chalkboard. Label the inner circle *Inside the Earth* and the outer circle *On the Surface of the Earth.* Direct students to copy this diagram into their notebooks. As you work through the section, draw students' attention to the boldface terms. Then have students copy these terms in the correct circle in the diagram in their notebooks. **L1**

Independent Practice

Guided Reading Activity 1-3 **L1**

The landforms in Death Valley, California *(left)*, are quite different from those found on this Pacific atoll *(right)*. **HUMAN/ENVIRONMENT INTERACTION: Why do you think it would be difficult for people to live in either of these places?**

Around the core is a thick rock layer, the mantle. Scientists calculate that the mantle is about 1,800 miles (2,900 km) thick. Mantle rock is mostly solid, but there are pockets of **magma**, or melted rock, in it that flow to the surface when a volcano erupts.

The outer layer, or the earth's crust, is the only layer scientists have studied firsthand. The crust is relatively thin—about 5 to 30 miles (8 to 50 km). It includes both the ocean floors and large land areas known as **continents**. The crust is thinnest on the ocean floor. It is thicker below the continents.

Plate Tectonics Scientists have developed a theory about the earth's structure called **plate tectonics**. This theory states that the crust is not an unbroken shell but consists of moving plates, or huge slabs of rock. These plates fit together like a jigsaw puzzle. They float—sometimes slowly, sometimes suddenly—atop soft rock in the mantle. The plates move in different directions. Some push against each other, others pull apart, and still others slide past one another. Oceans and continents ride on top of the gigantic plates.

Forces of Change Powerful forces, such as tectonic activity, have caused huge changes in the earth's surface. For millions of years these forces inside the earth have shifted rock layers and folded the crust as easily as if it were paper. Over time, such forces have built mountains and split continents. They also have caused **faults**, or cracks, in the earth's crust. Shifts along a fault can cause **earthquakes**, or violent jolts, in the area around it. In coastal areas undersea earthquakes can cause huge waves, known as **tsunamis**, which are sometimes as much as 50 feet (15 m) high!

Another force inside the earth that shapes landforms is volcanic activity. A volcano forms when magma breaks through the crust as lava. How does this

UNIT 1

The highest tsunami ever recorded measured 278 feet (85 m).

Cooperative Learning Activity

Organize the class into teams. Have teams use the Atlas on pages A1 through A32 to find and list specific examples, by name, of landforms and water bodies shown on the diagram on page 31. Allow teams a set amount of time to do their search. At the end of the time period, have teams compare their lists. Declare that the winner is the team with the most correctly listed names. **L1**

shape the land? Volcanoes may explode in a fiery burst of ashes and rock. Then cone-shaped mountains such as Japan's Mount Fuji result. Sometimes lava flows slowly, building up flat mountains like those in Hawaii.

After landforms have been created on the surface, other forces work to change them. A process called **weathering** breaks surface rocks into gravel, sand, or soil. Water, chemicals, and frost cause weathering.

Water, wind, and ice also work on landforms through **erosion**, or the wearing away of the surface. Riverbanks, for example, erode as the river water washes away soil.

Place
Types of Landforms

The earth's land surface consists of seven continents: North America, South America, Europe, Africa, Asia, Australia, and Antarctica. All have a variety of landforms—even icy Antarctica. The ocean floor also has many landforms.

Landforms and Water Bodies

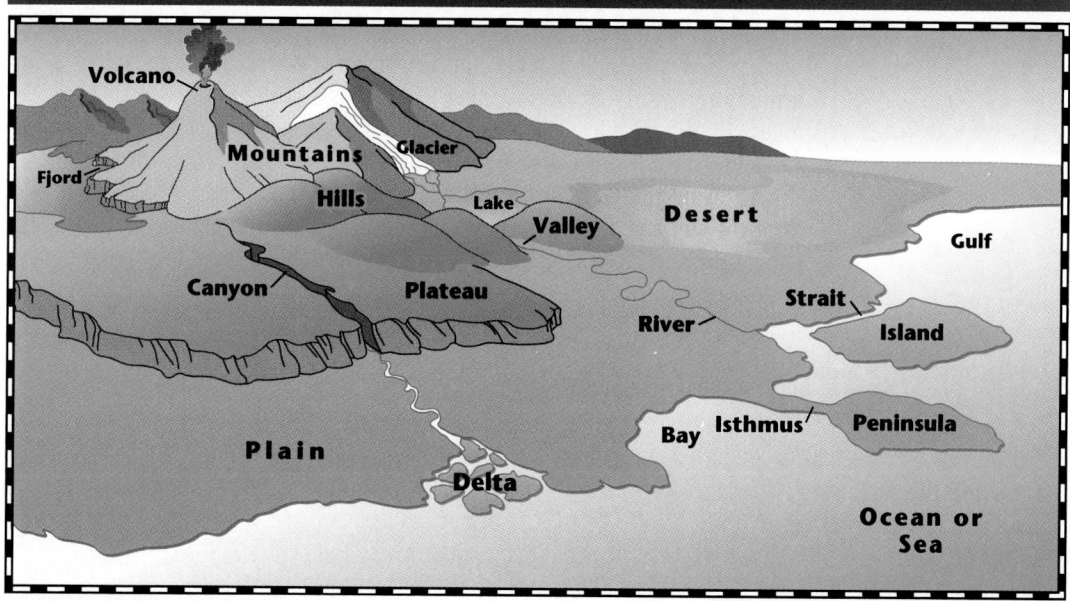

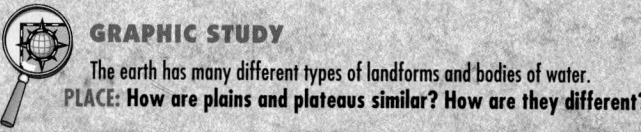

GRAPHIC STUDY

The earth has many different types of landforms and bodies of water.
PLACE: How are plains and plateaus similar? How are they different?

Meeting Special Needs Activity

Study Strategy Some students may have difficulty organizing the information they are reading. Help them by pointing out the hierarchical arrangement of the text. Note that chapters are divided into sections, and that sections are broken up by main headings and subheadings. Guide students in making chapter outlines based on section titles, headings, and subheadings. Encourage students to add notes to their outlines as they study. **L1**

ASSESS

Check for Understanding

Assign Section 3 Review as homework or an in-class activity.

Meeting Lesson Objectives

Each objective below is tested by the questions that follow it in parentheses.

1. **Understand** how the earth's structure is layered. (1, 2)
2. **Discuss** how forces change landforms. (1, 3)
3. **Identify** the earth's major landforms. (1, 3, 4, 5)

GRAPHIC STUDY

Answer
Both are flat, but a plateau is elevated.

Graphic Skills Practice
Reading a Diagram
What is the name of the narrow body of water that lies between two pieces of land? *(strait)*

MULTICULTURAL PERSPECTIVE

Culturally Speaking
To promote communication, geographers the world over use the same terminology. The terms they use come from many different languages. For example, *tsunami* is a Japanese word meaning "overflowing wave," and *fjord* is a Norwegian word meaning "long, narrow bay."

Evaluate

 Section 3 Quiz **L1**

Use the Testmaker to create a customized quiz for Section 3. **L1**

Chapter 1 Test Form A and/or Form B **L1**

Reteaching

Reteaching Activity 1 **L1**

Spanish Reteaching Activity 1 **L1**

Enrich

Enrichment Activity 1 **L3**

CLOSE

Mental Mapping Activity

Provide each student with a large sheet of white paper and map pencils. Tell students that this will not be a graded activity. Have students draw, freehand from memory, a sketch of the earth, sun, and moon in proper relation to each other. Have students add arrows to indicate the rotation of the earth, the orbit of the moon around the earth, and the orbit of the earth around the sun.

Major Landforms Major landforms include mountains, hills, plateaus, and plains. Look at the illustration on page 31 to see how these are different from one another. Mountains, which can be higher than 20,000 feet (6,100 m), have high peaks, as well as steep or rugged slopes. Hills are lower and more rounded, though they are still higher than the country around them.

Both plateaus and plains are flat, but a **plateau** rises above the land around it. A steep cliff forms at least one side of a plateau. Plains are flat or gently rolling. Smaller landforms include valleys and canyons.

Geographers describe some landforms by their relationship to larger land areas or to bodies of water. An **isthmus** is a narrow piece of land that connects two larger pieces of land. A **peninsula** is a piece of land that is surrounded by water on three sides. A body of land smaller than a continent and surrounded by water is called an island.

Water Bodies About 70 percent of the earth's surface is water. Oceans are the earth's largest bodies of water and can be more than 35,000 feet (11,000 m) deep. Smaller bodies of salt water, which are at least partly enclosed by land, are called seas, gulfs, or bays. A **strait** is a narrow body of water between two pieces of land. Lakes, streams, and rivers are freshwater landforms. You will learn more about the earth's water in Chapter 2.

Place

The Atmosphere

Earth is a living, changing planet of land, water, and atmosphere. The **atmosphere** is the air surrounding Earth. It is a cushion of gases about 1,000 miles (1,600 km) thick. About 99 percent of the atmosphere is made up of nitrogen and oxygen, with 1 percent being other gases.

All living things depend on the atmosphere. It screens out dangerous rays from the sun and reflects some heat back into space. The atmosphere holds in enough heat to make life possible, just as a greenhouse keeps in enough heat to protect plants. Without this protection, Earth would be too cold for most living things to survive.

SECTION 3 REVIEW

REVIEWING TERMS AND FACTS

1. **Define the following:** landform, core, mantle, crust, magma, continent, plate tectonics, fault, earthquake, tsunami, weathering, erosion, plateau, isthmus, peninsula, strait, atmosphere.

2. **PLACE** What materials make up the earth's inner core, outer core, and mantle?

3. **MOVEMENT** How do wind, water, and ice affect landforms on the earth's surface?

GRAPHIC STUDY ACTIVITIES

4. Look at the diagram on page 31. How are a lake and a bay similar? How are they different?

5. On the same diagram, locate a peninsula. How would you describe this landform?

(32)

UNIT 1

Answers to Section 3 Review

1. All vocabulary words are defined in the Glossary.
2. solid hot metal; molten metal; solid rock with pockets of magma
3. They weather and erode the earth's surface.
4. Both are bodies of water partly or completely surrounded by land. A bay has salt water, a lake has freshwater.
5. A peninsula is a piece of land surrounded by water on three sides.

Chapter 1 Highlights

Important Things to Know About Looking at the Earth

SECTION 1 USING THE GEOGRAPHY THEMES

- Location tells you where a place is found.
- Place tells you about the physical and human characteristics of a place.
- Human actions have changed the environment, and the environment has forced humans to adapt.
- People, ideas, information, and products move from place to place.
- Geographers divide the earth into regions based on common physical or human features.

SECTION 2 PLANET EARTH

- Earth, its moon, and other planets are part of our solar system.
- The tilt of Earth on its axis and its revolution around the sun cause the changes in seasons.
- The rotation of Earth on its axis causes areas to have day and night.

SECTION 3 LANDFORMS

- The earth has an inner and outer core, a mantle, and an outer crust.
- Forces inside the earth, such as volcanic activity, create landforms.
- Wind, water, and ice are surface forces that change landforms.
- Major types of landforms include mountains, hills, plateaus, and plains.
- The atmosphere is a cushion of gases that surrounds the earth.

LANDSAT image of Tokyo ▶

Extra Credit Project

Model Encourage students to conduct research on the Foucault pendulum, a device that provides proof that the earth rotates. Have students work in small groups to construct a Foucault pendulum. Provide the following directions: (1) suspend a large plastic soft drink bottle filled with sand from a long rope; (2) affix a pencil to the bottom of the bottle; (3) place a piece of butcher paper on the floor so that the pencil touches it. Have students set the pendulum in motion. Over time, pencil marks on the paper will make it appear that the pendulum changes direction. In reality, however, the pendulum swings in the same plane. The rotation of the earth causes the pendulum's apparent change in direction. **L2**

CHAPTER HIGHLIGHTS

Using the Chapter 1 Highlights

- Use the Chapter 1 Highlights to preview, review, condense, or reteach the chapter.

Preview/Review

 Vocabulary Puzzle-Maker Software reinforces the terms used in Chapter 1.

Condense

Have students read the Chapter 1 Highlights. Spanish Chapter Highlights also are available.

🎧 📁 Chapter 1 Digest Audiocassettes and Activity

📁 Chapter 1 Digest Audiocassette Test

Spanish Chapter Digest Audiocassettes, Activities, and Tests also are available.

📁 Guided Reading Activities

Reteach

📁 📁 Reteaching Activity 1. Spanish Reteaching Activities also are available.

MAP STUDY

Place Copy and distribute page 39 of the Outline Map Resource Book. Have students identify and label the Equator, the Prime Meridian, the Tropic of Cancer, and the Tropic of Capricorn.

33

Chapter 1 Review and Activities

ANSWERS

Reviewing Key Terms

1. C
2. B
3. D
4. G
5. E
6. H
7. F
8. A

Mental Mapping Activity

This exercise helps students to visualize the locations they have been studying and understand them in relation to one another at different times of the year. All attempts at free-hand sketching should be accepted.

REVIEWING KEY TERMS

Match the numbered terms in Column A with their definitions in Column B.

A
1. plate
2. plateau
3. latitude
4. relative location
5. geography
6. revolution
7. solar system
8. weathering

B
A. process by which water and chemicals break down rocks
B. flat, raised landform with one steep side
C. huge slab of the earth
D. imaginary lines that measure distance north and south of the Equator
E. study of the earth
F. planets and other objects that revolve around the sun
G. where a place is in relation to another place
H. complete trip around the sun by a planet

Mental Mapping Activity

On a separate piece of paper, draw a sketch that shows Earth in relation to the sun at the times of the solstices and the equinoxes. Draw four small globes in their correct positions for June 21, December 22, March 21, and September 23. Label the following terms on your globes:
• Equator
• North Pole
• South Pole
• Tropic of Cancer
• Tropic of Capricorn

REVIEWING THE MAIN IDEAS

Section 1
1. **LOCATION** What grid of lines is used to describe the exact position of a place?
2. **PLACE** What are some of the human characteristics that describe the geography of a place?
3. **REGION** What are some factors that geographers use to define a region?

Section 2
4. **MOVEMENT** How does Earth's orbit and rotation influence the length of the year and the change from day to night?
5. **LOCATION** When it is summer in the Northern Hemisphere, why is it winter in the Southern Hemisphere?

Section 3
6. **PLACE** What is the difference between the earth's mantle and crust?
7. **PLACE** What landforms are formed by volcanic activity?
8. **MOVEMENT** What forces cause weathering and erosion?
9. **PLACE** How is the atmosphere important to life on Earth?

Reviewing the Main Ideas

Section 1

1. lines of latitude and longitude
2. language, clothing, buildings, music, and so on
3. landscape, climate, language, religion, political boundaries, and so on

Section 2

4. The length of the year is based on the time it takes for Earth to make one revolution of the sun. The length of the day is based on the time it takes for Earth to rotate on its axis once. This rotation turns one side toward and then away from the sun, causing day and night.
5. Because of the tilt of the earth, the sun shines indirectly on the Southern Hemisphere.

CRITICAL THINKING ACTIVITIES

1. **Drawing Conclusions** Why is the theme of movement particularly important in geography today?
2. **Analyzing Information** What are two physical features found in the place where you live? What are two human features?

GeoJournal Writing Activity

How often do you come in contact with something or someone from another part of the world? For two or three days, use your journal to keep track of all your contacts that involve the geographic theme of movement. Start by reading the labels in your clothes. Where were your clothes made? Is your favorite TV show broadcast from another country? What about the food you had for dinner? Finish by sketching a world map that locates the sources of all your contacts.

COOPERATIVE LEARNING ACTIVITY

You are a geographer. Work with three other people to survey landforms in your area and create a landforms map. Find out the height of nearby hills, mountains, or plateaus. If exact heights are not available, record the sizes of each compared to one another. Record what water features are located nearby. Present your findings to the class in the form of a mural, or photograph local landforms for a bulletin-board display.

PLACE LOCATION ACTIVITY: THE WORLD

Match the letters on the map with the places and physical features of the world. Write your answers on a separate sheet of paper.

1. Asia
2. Australia
3. South America
4. Antarctica
5. Himalayas
6. Andes
7. Africa
8. Rocky Mountains

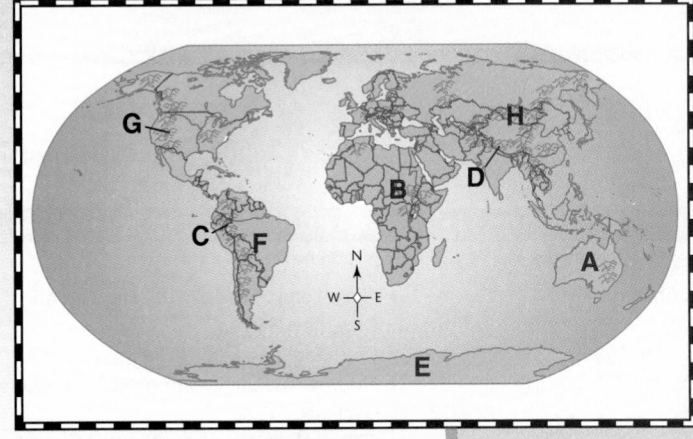

Cooperative Learning Activity

Suggest that students prepare a checklist of locations of landforms and water features before they begin this task. Display completed visuals in the classroom and invite other classes to share them.

GeoJournal Writing Activity

Encourage students to make a master map of all the world contacts made by class members during the time period.

Place Location Activity

1. H
2. A
3. F
4. E
5. D
6. C
7. B
8. G

Chapter Bonus Test Question

This question may be used for extra credit on the chapter test.

You are a rugged and steep natural landmark. You rank as a star attraction in most every country that you reside. Some earthly inhabitants put on all types of clothing and special equipment to see your apex. What are you? *(a mountain)*

Section 3

6. The mantle is solid rock with pockets of magma, the crust is the relatively thin outer layer consisting of the ocean floor and the continents.
7. cone-shaped mountains; broad, flat mountains
8. wind, flowing water, ice
9. screens out harmful rays from the sun, reflects some heat, and keeps enough heat around the earth to make life possible

Critical Thinking Activities

1. Answers may suggest that movement in the form of communication is much faster and more widespread today than it was in the past, and that there is more movement of people and goods now than in the past.
2. Answers will vary, but should show that students understand the two elements that help to describe place.

PERFORMANCE ASSESSMENT ACTIVITY

SIMULATE A CULTURE Organize students into seven groups. Assign each group a climate region or a location on the globe. Have groups use information from the text and from various maps to develop an illustrated description of a culture that might have occupied their climate region or location. Their descriptions should answer questions such as:

- What foods and other crops might they grow?
- How might these people make a living?
- What types of homes would the people have?
- What might they do for leisure activities?

Remind students that the focus of their descriptions should be the role that water, climate, and vegetation play in the development of cultures. Display completed descriptions in a classroom area or compile them into a class booklet.

POSSIBLE RUBRIC FEATURES:

Concept attainment (adaptation), content information, analysis skills, collaborative skills, ability to recognize relationships, planning and product completion skills

MENTAL MAPPING ACTIVITY

Before teaching Chapter 2, display a globe or world wall map. Encourage students to think of the earth as a natural object, with no national boundaries, cities, or other human features. Suggest that they try to visualize the earth as a natural system, which includes land, water, and air.

TEACHER'S CORNER

NATIONAL GEOGRAPHIC SOCIETY

INDEX TO NATIONAL GEOGRAPHIC MAGAZINE

The following articles may be used for research relating to this chapter:

- "Life Without Light," by Ian R. MacDonald and Charles Fisher, October 1996.
- "Exploring Antarctic Ice," by Jane Ellen Stevens, May 1996.
- "California Desert Lands," by Michael Parfit, May 1996.
- "The Amazon," by Jere Van Dyk, February 1995.
- *Water,* A National Geographic Special Edition, November 1993.
- "Lightning: Nature's High-voltage Spectacle," by William R. Newcott, July 1993.

- "Rain Forest Canopy: The High Frontier," by Edward O. Wilson, December 1991.
- "Tornado!" by Peter Miller, June 1987.
- "Monsoons: Life Breath of Half the World," by Priit J. Vesilind, December 1984.
- "El Niño's Ill Wind," by Thomas Y. Canby, February 1984.

NATIONAL GEOGRAPHIC SOCIETY PRODUCTS AVAILABLE FROM GLENCOE

To order the following products for use with this chapter, contact your local Glencoe sales representative or call Glencoe at 1-800-334-7344:

- *GeoBee* (Software)
- *Picture Atlas of the World* (CD-ROM)
- *Earth's Endangered Environments PictureShow* (CD-ROM)
- *GTV: Planetary Manager* (Videodisc)

- *STV: World Geography* (Videodisc)
- *STV: Restless Earth* (Videodisc)
- *STV: Water* (Videodisc)

ADDITIONAL NATIONAL GEOGRAPHIC SOCIETY PRODUCTS

To order the following products for use with this chapter, call National Geographic Society at 1-800-368-2728:

- *The Living Earth* (Video)
- *Nature's Fury!* (Video)
- *Volcano!* (Video)
- *Born of Fire* (Video)

- *Water: A Precious Resource* (Video)
- *The Living Ocean* (Video)
- *Ancient Forests* (Video)
- *A Swamp Ecosystem* (Video)

TEACHER-TO-TEACHER
Curriculum Connection: Earth Science

—from Charles M. Bateman
Halls Middle School, Knoxville, TN

EROSION *The purpose of this activity is to help students understand one of the primary factors responsible for the diversity of landforms and soils by demonstrating two kinds of erosion.*

Organize students into several groups. Provide half the groups with plastic jars with screw-on lids and pieces of soft rock, such as sandstone, small enough to fit in the jars. Provide the other groups with jugs of water, and paint trays or baking tins with soil piled in one end. Have group members draw a sketch of their rock or pile of soil. Direct the first groups to place rocks in the jars and half-fill the jars with water. After the jars have been sealed, have group members shake the jar vigorously for a few seconds. Have group members shake the jars at regular intervals throughout the day. Direct members of the other groups to gently pour a small amount of water down the middle of the pile of soil. Have them repeat this process at regular intervals until water jugs are empty.

During the following day's lesson, direct groups to remove the rocks from the jars. Have them compare the appearance of the rocks to the sketches they made. Have the other groups compare the sketches they made to the appearance of the piles of soil. Call on individual students to describe how the rocks or piles of soil changed and explain the process responsible for the changes.

BIBLIOGRAPHY

Readings for the Student
Bramwell, Martyn. *Weather.* New York: Franklin Watts, 1994.

Weather. New York: Alfred A. Knopf, 1993.

Readings for the Teacher
Marshall, Bruce, ed. *The Real World: Understanding the Modern World Through the New Geography.* Boston: Houghton Mifflin, 1991.

McKisson, Micki, and Linda MacRae Campbell. *The Ocean Crisis.* Tucson: Zephyr Press, 1990.

Multimedia
The Climate Factor. Washington, D.C.: National Oceanic and Atmospheric Administration. Videocassette, 25 minutes.

The Geography Tutor: Weather and Climate. Maumee, Ohio: Instructional Video, 1991. Videocassette, 18 minutes.

MEETING NATIONAL STANDARDS

Geography for Life

All of the 18 standards are demonstrated in Unit 1. The following ones are highlighted in this chapter:
• Standards 1, 2, 3, 4, 7, 8, 15

KEY TO ABILITY LEVELS
Teaching strategies have been coded for varying learning styles and abilities.

L1 **BASIC** activities for all students

L2 **AVERAGE** activities for average to above-average students

L3 **CHALLENGING** activities for above-average students

LEP **LIMITED ENGLISH PROFICIENCY** activities

Use these *Geography: The World and Its People* resources to teach, reinforce, and extend chapter content.

CHAPTER 2 RESOURCES

*Vocabulary Activity 2

Cooperative Learning Activity 2

Workbook Activity 2

Geography Skills Activity 2

GeoLab Activity 2

Critical Thinking Skills Activity 2

Chapter Map Activity 2

Performance Assessment Activity 2

*Reteaching Activity 2

Enrichment Activity 2

Chapter 2 Test, Form A and Form B

Unit Overlay Transparencies 1-2, 1-4, 1-5

Political Map Transparency 1

*Chapter 2 Digest Audiocassette, Activity, Test

Vocabulary PuzzleMaker Software

Student Self-Test: A Software Review

Testmaker Software

MindJogger Videoquiz

If time does not permit teaching the entire chapter, summarize using the Chapter 2 Highlights on page 49, and the Chapter 2 English (or Spanish) Audiocassettes. Review students' knowledge using the Glencoe MindJogger Videoquiz. *Also available in Spanish

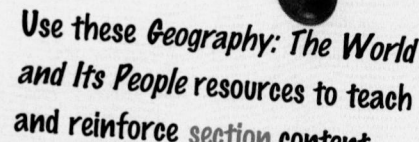

Use these *Geography: The World and Its People* resources to teach and reinforce section content.

SECTION 1 RESOURCES

Reproducible Lesson Plan 2-1

Section Focus Transparency 2-1

Guided Reading Activity 2-1

Section Quiz 2-1 (also available in Spanish)

SECTION 2 RESOURCES

Reproducible Lesson Plan 2-2

Section Focus Transparency 2-2

Guided Reading Activity 2-2

Section Quiz 2-2 (also available in Spanish)

SECTION 3 RESOURCES

Reproducible Lesson Plan 2-3

Section Focus Transparency 2-3

Guided Reading Activity 2-3

Section Quiz 2-3 (also available in Spanish)

ADDITIONAL RESOURCES FROM GLENCOE

 **Reproducible Masters**

- Glencoe Social Studies Outline Map Book, page 39
- Facts On File: GEOGRAPHY ON FILE, Climate and Weather

 **Workbook**

- Building Skills in Geography, Unit 1, Lesson 1

 Transparencies

- National Geographic Society PicturePack Transparencies, Unit 1

 Laminated Desk Map

 Videodiscs

- National Geographic Society: STV: Water

 CD-ROM

- National Geographic Society PictureShow: Earth's Endangered Environments

Performance Assessment

Refer to the Planning Guide on page 36A for a Performance Assessment Activity for this chapter. See the *Performance Assessment Strategies and Activities* booklet for suggestions.

Chapter Objectives

1. **Describe** how the water of the earth moves in a cycle.
2. **Investigate** what climate is and the factors that influence climate.
3. **Identify** each major climate region.

GLENCOE
TECHNOLOGY

 VIDEODISC

Use the Chapter 2 Mind-Jogger Videoquiz to preview chapter content.

MindJogger Videoquiz

Chapter 2
Disc 1 Side A

 Also available in VHS.

36

Chapter 2 Water, Climate, and Vegetation

World Ocean Currents

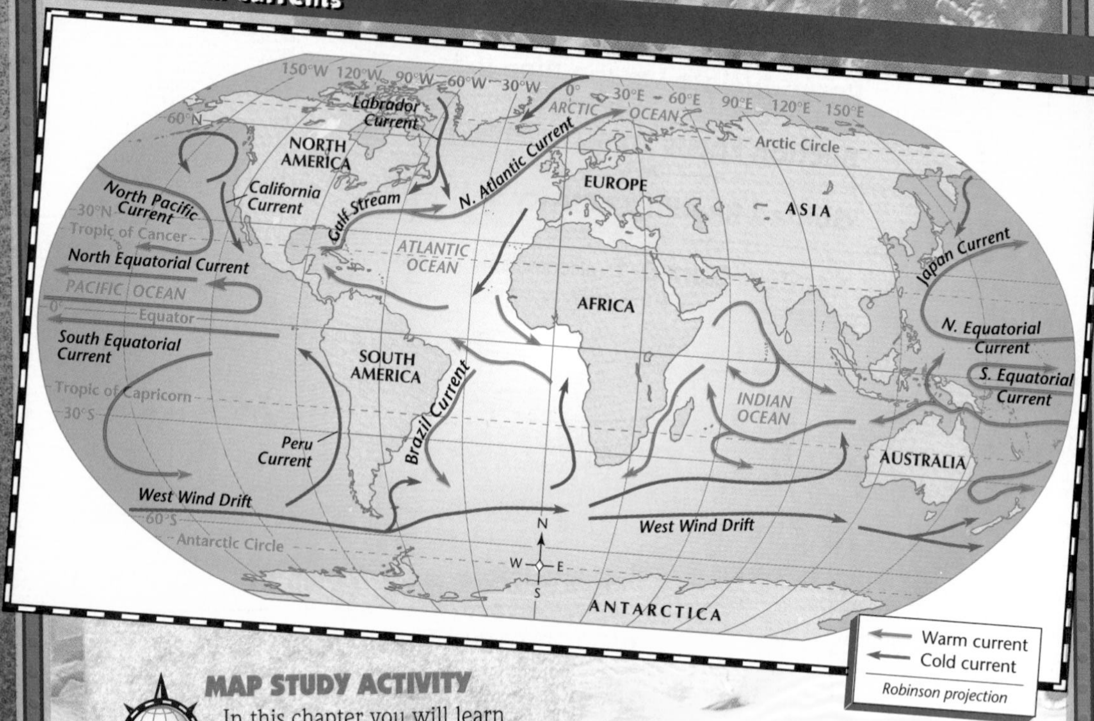

Warm current
Cold current

Robinson projection

MAP STUDY ACTIVITY

In this chapter you will learn about the ways the world's water is used.

1. What ocean current runs along North America's southeastern coast?
2. Which continent's climate is most likely to be influenced by the Peru Current?

36

MAP STUDY ACTIVITY

Answers
1. the Gulf Stream
2. South America

Map Skills Practice
Reading a Map What type of current affects the southeastern coast of Africa? *(warm)*

Earth's Water

PREVIEW

Words to Know
- vegetation
- water vapor
- water cycle
- evaporation
- condensation
- precipitation
- groundwater
- aquifer

Places to Locate
- Pacific Ocean
- Atlantic Ocean
- Indian Ocean
- Arctic Ocean
- Antarctica

Read to Learn . . .
1. how the earth's water moves in a cycle.
2. where people get freshwater.

Water, however, is vital in many other ways to life on the earth.

This fisher in Asia catches snapper for his fish market. People all over the world depend on the oceans. To them the water is a source of food.

Water covers about 70 percent of the earth's surface. Almost all of it is salt-water found in the oceans. Overall, the earth has plenty of water. Some areas, however, never have enough water to support life while other places get too much.

Why do some places get more water than others? Climate determines the amount of water a place receives, which in turn determines its **vegetation,** or plant life. Together water, climate, and vegetation influence how people in a given area live.

Movement
The Water Cycle

Water exists all around you in different forms. Rivers, lakes, and oceans contain water in liquid form. The atmosphere holds **water vapor,** or water in the form of gas. Glaciers and ice sheets are large masses of water in a frozen form. The total amount of water on the earth does not change. It is just constantly moving—from the oceans to the air to the ground and finally back to the oceans. This process is called the **water cycle.**

CHAPTER 2

Classroom Resources for Section 1

 REPRODUCIBLE MASTERS

Reproducible Lesson Plan 2-1
Guided Reading Activity 2-1
Vocabulary Activity 2
Building Skills in Geography,
  Unit 1, Lesson 1
Geography Skills Activity 2
Section Quiz 2-1

 TRANSPARENCIES

Section Focus
  Transparency 2-1

MULTIMEDIA

 Vocabulary PuzzleMaker
  Software
 Testmaker

---

Meeting National Standards
Geography for Life The following standards are highlighted in this section:
- Standards 1, 7

FOCUS

Section Objectives
1. **Describe** how the earth's water moves in a cycle.
2. **Investigate** where people get freshwater.

Bellringer Motivational Activity

Prior to taking role at the beginning of the class period, project Section Focus Transparency 2-1 and have students answer the activity questions. Discuss students' responses. **L1**

Vocabulary Pre-check

Point out the term *water cycle* to students. Ask them to speculate on how the other terms in the list are related to this key term. **L1**
LEP

Use the Vocabulary PuzzleMaker Software to create a crossword puzzle. **L1**

TEACH

Guided Practice
Making Connections Point out that there are about a trillion gallons in a cubic mile of water. Then mention that there are about 326 million cubic miles of water on Earth. Use these

(continued)

37

figures as a springboard to a discussion of how important water is to understanding geographical processes. **L1**

GRAPHIC STUDY

Answer
evaporation from oceans; because they have the largest surface area

Graphic Skills Practice
Reading a Diagram
When water evaporates, does it take the form of a gas or a liquid? *(gas)*

Independent Practice

Guided Reading Activity 2-1 **L1**

ASSESS

Check for Understanding

Assign Section 1 Review as homework or an in-class activity.

Meeting Lesson Objectives

Each objective below is tested by the questions that follow it in parentheses.

1. **Describe** how Earth's water moves in a cycle. (1, 2, 5, 6)
2. **Investigate** where people get freshwater. (3)

The Water Cycle

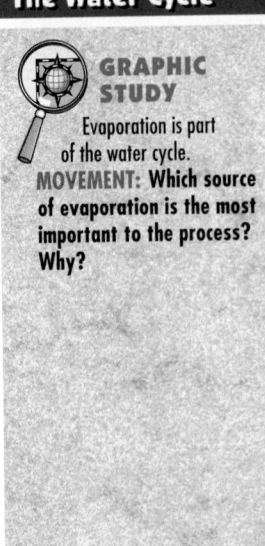

GRAPHIC STUDY

Evaporation is part of the water cycle.
MOVEMENT: Which source of evaporation is the most important to the process? Why?

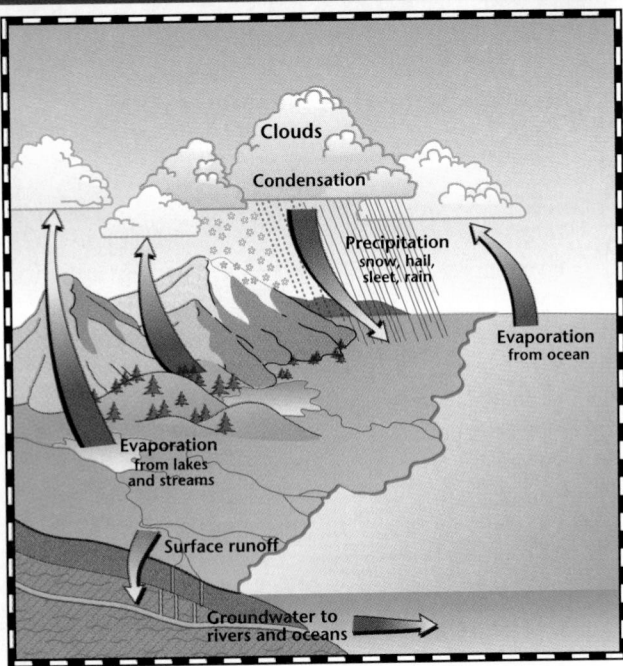

Look at the diagram above and follow the water cycle. The sun drives the cycle by evaporating water from the surface of oceans, lakes, and streams. In **evaporation** the sun's heat turns water into vapor. The amount of water vapor the air holds depends on its temperature. Warm air holds more vapor than cool air.

Warm air tends to rise and cool. As this happens, water vapor changes into a liquid—a process called **condensation**. Tiny droplets of water come together to form clouds. Eventually, the water falls as **precipitation**—rain, snow, or sleet, depending on the air temperature. This precipitation falls to the surface, where it soaks into the ground and collects in streams and lakes to return to the oceans. Soon most of it evaporates, and the cycle begins again.

Human/Environment Interaction
Types of Water

It's a hot day and you rush home for a glass of water. What if you turned on the faucet and nothing came out? All living things need water to survive. Think about the many ways you use water in just a single day—to bathe, to brush your teeth, to cook your food, to quench your thirst. The waters of the earth do more than just meet human needs, however. They also are home to millions of kinds of plants and animals. Water at the earth's surface can be freshwater or salt water. Freshwater is the most important to humans.

Cooperative Learning Activity

Organize the class into two teams. Provide one team with a map of your county and the other team with a map of your state. Assign each team the following task: Locate every body of water in your assigned area. Record your findings, including: 1) the name of each body of water; 2) its relative and/or absolute location, and; 3) the type of feature it is—lake, river, and so on. When each team has completed its task, have them create a map quiz based on their work. The teams should then trade maps and challenge each other with their quizzes. **L1**

Freshwater Only about 3 percent of the water on the earth is freshwater. Of that, almost 2 percent is frozen in glaciers and ice sheets. Lakes and rivers are the important sources of the remaining usable freshwater.

Another source of freshwater is **groundwater**, or water that fills tiny cracks and holes in the rock layers below the surface of the earth. Groundwater can be tapped by wells for people to use. An underground rock layer so rich in water that water actually flows through it is called an **aquifer**. In regions with little rainfall, both farmers and city dwellers sometimes have to depend on aquifers and other groundwater for most of their water supply.

Ocean Water Did you know that you could sail from one ocean to another without touching land? All the oceans on the earth are part of a huge, continuous body of salt water—97 percent of the planet's water. The four major oceans are the Pacific Ocean, the Atlantic Ocean, the Indian Ocean, and the Arctic Ocean. Look at the map on page 42 to see how the four oceans are really one huge body of water. Which ocean do you live closest to?

The Pacific Ocean is both the largest and the deepest ocean. It covers 64,000,000 square miles (165,760,000 sq. km)—more than all the land areas of the earth combined. If you set Mount Everest, the earth's tallest mountain, in the deepest part of the Pacific, it would still be more than one mile below the surface!

As you learned in Chapter 1, smaller bodies of salt water are called seas, gulfs, bays, or straits. Look back at page 31 to see these features again.

Fun on the Waves

This surfer demonstrates one exciting use for the world's water.
PLACE: What are the four major oceans?

SECTION 1 REVIEW

REVIEWING TERMS AND FACTS
1. **Define the following:** vegetation, water vapor, water cycle, evaporation, condensation, precipitation, groundwater, aquifer.
2. **MOVEMENT** How does water enter the air during the water cycle?
3. **HUMAN/ENVIRONMENT INTERACTION** Where do people get freshwater for drinking and growing crops?
4. **PLACE** What do the oceans of the earth have in common?

GRAPHIC STUDY ACTIVITIES
5. Turn to the diagram on page 38. What happens to water that falls on the earth's surface?
6. Using the same diagram, describe what part mountains play in the water cycle.

LESSON PLAN
Chapter 2, Section 1

More About the Illustration
Answer to Caption
Pacific, Atlantic, Indian, Arctic

Place The ocean floor has the same landforms that appear on land, including large mountain ranges and volcanoes.

Evaluate

Section 1 Quiz **L1**

Use the Testmaker to create a customized quiz for Section 1. **L1**

Reteaching

Have students create a concept map with "Water of the Earth" written in the center cell. Radiating from this cell should be branches relating to the key ideas in the lesson.

Enrich

Have students research and report on a recent news story that concerned one of the four oceans. **L3**

CLOSE

Have students brainstorm a list of the uses of water to humans.

Answers to Section 1 Review
1. All vocabulary words are defined in the Glossary.
2. sun evaporates water from surface of oceans, lakes, and other wet surfaces
3. from surface water such as lakes and rivers, from groundwater trapped in underground rock layers
4. They are part of one continuous body of salt water.
5. It runs into groundwater and rivers and back into the ocean.
6. Mountains force moist air to rise, causing condensation, cloud formation, and, eventually, precipitation.

BUILDING GEOGRAPHY SKILLS

TEACH

Use a flashlight and globe in the following demonstration. Shine the flashlight at the Equator and rotate the globe from right to left. Tell students that the flashlight beam is sunlight. Ask them if the sun rises first in New York or in Hawaii. *(New York)* Point out that because the sun rises in New York earlier than in Hawaii, New York is *east* of Hawaii. Next, have students locate the North and South poles on the globe. Explain that the directions "north" and "south" indicate the direction of the two poles. Ask students if they were facing west and wanted to go north, would they turn right or left? *(right)* Then direct students to read the skill and answer the practice questions. **L1**

Additional Skills Practice

1. You are standing on Sawyer Hill watching the sunrise, but traffic is blocking your view. On what highway is the traffic traveling? *(Interstate 495)*

2. What directions would you give a friend who wanted to travel from the junction of Linden and Barnes roads to the Town Hall taking Route 62? *(North on Barnes Road to Route 62, east on Route 62.)*

📁 Geography Skills Activity 2

📁 Building Skills in Geography, Unit 1, Lesson 1

Using Directions

To describe locations, we use the **cardinal directions** of north, south, east, and west. North and south are the directions of the North and South poles. If you stand facing north, east is the direction to your right—toward the rising sun. West is the direction on your left.

On most maps, you will find a **compass rose** showing the position of these directions. You might also see **intermediate directions**—those that fall between the cardinal directions. For example, the direction northeast falls between north and east. To use directions on a map, do the following:

• Use the compass rose to identify the four directions.
• Choose two features on the map.
• Determine whether one feature is north, south, east, or west of the other feature.

Berlin, Massachusetts

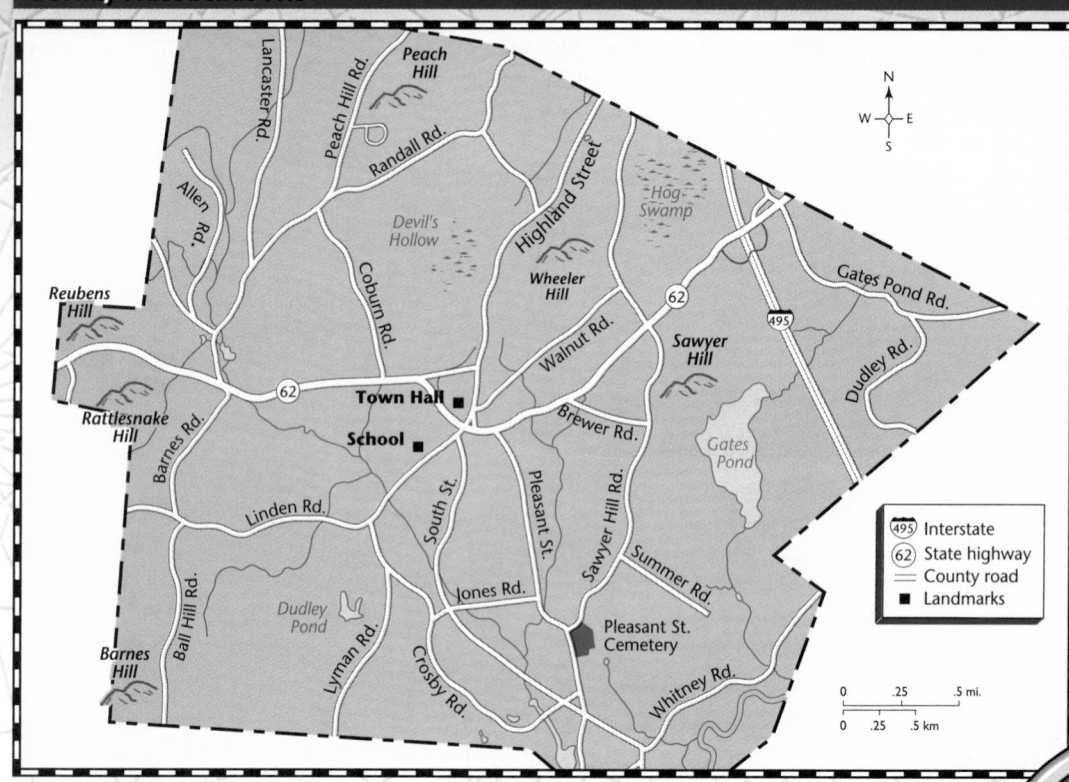

Geography Skills Practice

1. Does Hog Swamp lie north or south of Route 62?

2. From Gates Pond, what direction is Reubens Hill?

Answers to Geography Skills Practice

1. north
2. west

SECTION 2: Influences on Climate

PREVIEW

Words to Know
- weather
- climate
- tropics
- current
- hurricane
- typhoon
- monsoon
- rain shadow

Places to Locate
- Tropic of Cancer
- Tropic of Capricorn
- Equator
- Gulf Stream

Read to Learn . . .
1. what creates a particular climate.
2. how moving wind and water circulate the sun's heat.
3. what causes a rain shadow.

The power of nature's storms is awesome. The winds of a hurricane blow up to 130 miles per hour (209 km per hour), bending trees almost to the ground!

If you call a friend and say, "It's hot and rainy today," you are talking about the weather. **Weather** refers to the constant changes in the air during a short period of time. But if you say, "Summers are usually hot and rainy in my part of the country," then you have described your climate. **Climate** is the usual pattern of weather events in an area over a long period of time. Climate is determined by latitude, by location near large bodies of water, and sometimes by position near mountains.

Location
Latitude and Climate

As you learned in Chapter 1, the earth's tilt and rotation cause different regions to receive different amounts of heat and light from the sun. The sun's direct rays fall year-round at low latitudes near the Equator. This area—known as the **tropics**—lies mainly between the Tropic of Cancer and the Tropic of Capricorn. If you lived in the tropics, you would almost always enjoy a hot climate, unless you lived at a high elevation where temperatures are cooler.

Outside the tropics the sun is never directly overhead. In the high-latitude regions around the North and South poles, darkness descends for six months

CHAPTER 2

(41)

Classroom Resources for Section 2

 REPRODUCIBLE MASTERS

Reproducible Lesson Plan 2-2
Guided Reading Activity 2-2
Cooperative Learning Activity 2
Critical Thinking Skills Activity 2
Section Quiz 2-2

 TRANSPARENCIES

Section Focus
  Transparency 2-2
Geoquiz Transparencies 2-1,
  2-2

MULTIMEDIA

 Testmaker

---

Meeting National Standards

Geography for Life The following standards are highlighted in this section:
- Standards 1, 3, 4, 7, 15

FOCUS

Section Objectives
1. **Relate** what factors determine a particular climate.
2. **Describe** how moving wind and water spread the sun's heat.
3. **Identify** what causes a rain shadow.

Bellringer Motivational Activity

Prior to taking role at the beginning of the class period, project Section Focus Transparency 2-2 and have students answer the activity questions. Discuss students' responses. **L1**

Vocabulary Pre-check

Ask students to identify the terms in the **Words to Know** list that they have heard or seen before. Call on volunteers to explain where they heard or saw these terms. **L1 LEP**

TEACH

Guided Practice

Concept Web Draw a concept web on the chalkboard with the word *climate* in the central cell. Work through

(continued)

the section with students, identifying each factor that influences climate—latitude, winds, and so on—in turn. Then call on volunteers to complete the concept web by explaining how each factor relates to climate. L1

Independent Practice

📁 Guided Reading Activity 2-2 L1

ASSESS

Check for Understanding

Assign Section 2 Review as homework or an in-class activity.

Meeting Lesson Objectives

Each objective below is tested by the questions that follow it in parentheses.

1. **Relate** what factors determine a particular climate. (1, 2, 3, 4)
2. **Describe** how moving wind and water spread the sun's heat. (1)
3. **Identify** what causes a rain shadow. (1)

MAP STUDY

Answer
eastern coast of North America; western Europe

Map Skills Practice
Reading a Map What current affects the eastern coast of South America?
(Brazil Current)

---

each year. There the sun's rays hit the earth indirectly at a slant. Thus, climates in these regions are always cool or cold even in midsummer.

Movement
Wind, Water, and Currents

In addition to latitude, the movement of air and water helps create Earth's climates. Moving air and water help circulate the sun's heat around the globe. In the ocean, the moving streams of water are called **currents**.

Near the Equator, air and water are most intensely heated. Warm winds and water currents move from the tropics toward the poles, bringing warmth with them. A large, warm-water ocean current known as the Gulf Stream flows from the Gulf of Mexico through the cool Atlantic Ocean along the east coast of North America. Find the Gulf Stream on the map below. Notice how the Gulf Stream then crosses the Atlantic, bringing warm water and mild air to western Europe.

Large bodies of water influence climates in another way. Water temperatures do not change as much or as fast as land temperatures do. Thus, air over

World Ocean Currents

MAP STUDY
Warm and cold ocean currents affect the world's climates.
REGION: What areas of the world would be affected by a change in the Gulf Stream?

Cooperative Learning Activity

Organize students into several small groups. Give groups the following task: create a poster or some other form of visual that can be used to teach a lesson on the factors that affect climate in your region. Make certain all group members have an assigned task. Some might do research, others might work on the design of the visual, still others might create illustrations or write captions. Call on groups to display their finished visuals and use them to teach the rest of the class on the factors affecting climate in the region. L2

large bodies of water is warmer in winter and cooler in summer. This keeps coastal air temperatures moderate.

Storms Moving wind and water may make climates milder, but they also cause storms. **Hurricanes**, or violent tropical storm systems, form over the warm Atlantic Ocean at certain times of the year. Hurricanes bring high winds and drenching rain to the islands in the Caribbean Sea and to North America. They also rip through East Asia, although hurricanes that form in the Pacific Ocean are called **typhoons**.

Storms and heavy rains are not always destructive. In fact, during various seasons of the year, storms are an important part of some climate patterns. People in South Asia depend on **monsoons**, or seasonal winds, for the rain that waters their crops.

Rain Shadow Breezes blowing over the coast don't always move inland. Why? Natural landforms sometimes get in the way. When moist winds blow from the ocean toward a coastal mountain range, for example, the winds are forced upward over the mountains. As the winds rise, air cools and loses moisture as rain or snow. The climate on this windward, or coastal, side of mountains is moist and often foggy. Trees and vegetation are thick and green.

By the time the air reaches the inland side of the mountains, it is cool and dry. This creates a **rain shadow**, the dry area on the inland side of the mountains. The dry air of a rain shadow warms up again as it moves down mountainsides, giving the region a dry or desert climate.

The Effects of a Rain Shadow

The heavily forested Cascade Range in Washington state faces the ocean *(left)*. The inland side of these mountains is a partly dry wheat-growing area *(right)*.
LOCATION: On which side of these mountains is the rain shadow?

SECTION 2 REVIEW

REVIEWING TERMS AND FACTS

1. Define: weather, climate, tropics, current, hurricane, typhoon, monsoon, rain shadow.

2. PLACE Why are climates warm or hot year-round in places near the Equator?

3. HUMAN/ENVIRONMENT INTERACTION Why are monsoons both helpful and harmful?

MAP STUDY ACTIVITIES

4. Turn to the map on page 42. Locate the Tropic of Cancer and Tropic of Capricorn. At what latitudes are they located? What type of currents—warm or cold— is between them?

Answers to Section 2 Review

1. All vocabulary words are defined in the Glossary.

2. sun's direct rays fall year-round there

3. bring rain for watering crops, but also can cause floods and other damage

4. 23.5°N and S of the Equator; warm currents

LESSON PLAN
Chapter 2, Section 2

More About the Illustration

Answer to Caption the inland side

GEOGRAPHIC THEMES **Location** The Himalayas in Asia provide a good example of rain shadow. The ocean-facing slopes receive anywhere from 20 to 60 times more rain than the inland slopes.

Evaluate

Section 2 Quiz **L1**

Use the Testmaker to create a customized quiz for Section 2. **L1**

Reteaching

Write on the chalkboard a number of false statements on the factors that influence climate. Have students correct the statements by using information from the section. **L1**

Enrich

Encourage students to write a research report on the warm current *El Niño* and its impact on world climate. **L3**

CLOSE

Have students discuss how the local climate affects their lives.

TEACH

Before students read this feature, display a large wall map of the world. Ask students to carefully study the shape of the continents. Then tell students that, according to the theory of continental drift, all the continents once were part of one mass of land. Have students speculate on where and how the continents fit together. Encourage students to draw, freehand, the completed "puzzle" with all seven continents joined. Have students compare their drawings with the map on page 44. **L1**

More About . . .

Geology Geologists believe that some 500 million years ago six major continents existed. Over the next 250 million years, these continents slowly moved together, eventually forming Pangaea.

English philosopher Francis Bacon (1561–1626) was the first to notice the curious "jigsaw puzzle" of the continents. Scientists did not show interest in his observation, however, until the late 1700s.

MAKING CONNECTIONS

GROUND IN MOTION

| MATH | SCIENCE | HISTORY | LITERATURE | TECHNOLOGY |

Would you like to visit Japan? If the continents of North America and Asia moved together, you might be able to *walk* to Japan! That's what some geographers think could happen in about 50 million years.

You may think the earth is solid, but landmasses actually move constantly. The idea that continents have been moving around for billions of years is known as the theory of *continental drift*. The continents move, or drift, because they are parts of huge land plates that form the earth's outer crust.

THE BREAKUP In 1912 a German scientist, Alfred Wegener, proposed a theory of continental drift. He said that all the continents had at one time been part of a huge landmass called Pangaea (pan•JEE•uh). Wegener believed that about 200 million years ago this supercontinent broke into several pieces that drifted apart and became the seven continents.

Why would Wegener get this idea? He looked at a world map and wondered why the continents looked like pieces of a large puzzle. He saw that the eastern half of South America looked as if it would fit into the western half of Africa. After more research

Wegener learned that scientists had found fossils of the same kinds of animals on different continents. In addition, some rocks found in South America and Africa were alike, as were some in North America and Europe.

Continental Drift

DRIFTING CONTINENTS What caused Pangaea to break apart? Wegener believed the earth's inner energy forced the immense plates to drift to their present locations. Wegener predicted the plates will continue to drift to new locations. But don't plan to walk to Japan anytime soon—the plates move about 1 to 4 inches (2.5 cm to 10 cm) a year.

Making the Connection

1. What led Wegener to believe that the continents were once connected?
2. What did Wegener think caused Pangaea to drift apart?

44

UNIT 1

Answers to Making the Connection
1. similar fossils and rocks found on different continents
2. the earth's inner energy

SECTION 3 ·
Climate and Vegetation

PREVIEW

Words to Know
- rain forest
- savanna
- marine west coast climate
- Mediterranean climate
- humid continental climate
- humid subtropical climate
- tundra
- permafrost
- elevation
- timberline
- steppe

Places to Locate
- Amazon River basin
- Mediterranean Sea

- Arctic Circle
- Antarctica
- Greenland

Read to Learn . . .
1. what major world climate regions are like.
2. where each world climate region is located.
3. what kinds of vegetation grow in each world climate region.

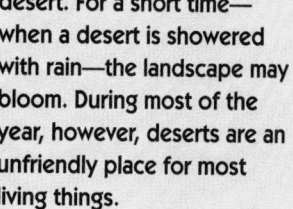

desert. For a short time—when a desert is showered with rain—the landscape may bloom. During most of the year, however, deserts are an unfriendly place for most living things.

Reds, pinks, and purples blanket a usually brown

Meeting National Standards
Geography for Life The following standards are highlighted in this section:
- Standards 1, 2, 3, 4, 8

FOCUS

Section Objectives
1. **Describe** what major world climate regions are like.
2. **Specify** where each major world climate region is located.
3. **Relate** what kinds of vegetation grow in each world climate region.

Bellringer Motivational Activity

Prior to taking role at the beginning of the class period, project Section Focus Transparency 2-3 and have students answer the activity questions. Discuss students' responses. **L1**

Vocabulary Pre-check

Have students attempt to infer the meanings of the compound words *permafrost* and *timberline* from their word parts. **L1 LEP**

The world's climates can be organized into four major regions: tropical, mid-latitude, high latitude, and dry. Some of these regions are determined by their latitude, or distance from the Equator. Others are based on the vegetation that grows in them.

Region
Tropical Climates

Tropical climate regions get their name from the tropics, the areas along the Equator reaching to about 30°N and 30°S. Temperatures here change little from season to season. The warm tropical climate region can be separated into two types. The tropical rain forest climate is wet in most months, with up to 100 inches (254 cm) of rain a year. The tropical savanna climate has two distinct seasons—one wet and one dry.

Tropical Rain Forest Climate In some parts of the tropics, the growing season lasts all year. In these areas, rain and heat produce lush vegetation and dense forests called **rain forests**. These forests are home to millions of kinds of

Classroom Resources for Section 3

 REPRODUCIBLE MASTERS
Reproducible Lesson Plan 2-3
Guided Reading Activity 2-3
GeoLab Activity 2
Reteaching Activity 2
Enrichment Activity 2
Section Quiz 2-3

 TRANSPARENCIES
Section Focus
Transparency 2-3

MULTIMEDIA
 Testmaker

Geography and the Environment: The Infinite Voyage

GLENCOE
TECHNOLOGY

 **VIDEODISC**

The Infinite Voyage: To the Edge of the Earth

Chapter 5
The Tropical Rain Forest
Disc 1 Side B

Answer
tropical rain forest

Map Skills Practice

Reading a Map What three continents have the largest areas of desert climate? *(Australia, Africa, Asia)*

TEACH

Guided Practice

Organizing Information
Draw on the chalkboard a four-column chart with the following headings: *Climate Region, Where Found, Characteristics, Typical Vegetation.* Direct students to copy the chart into their notebooks. As students work through the section, have them add information to the chart. Encourage students to retain completed charts for review purposes. **L1**

Independent Practice

Guided Reading Activity 2-3 **L1**

Climatologists agree that the United States has the most varied climate on Earth. With its tornadoes, hurricanes, thunderstorms, and blizzards, the climate of the U.S. also is one of the most violent and unpredictable.

World Climate Regions

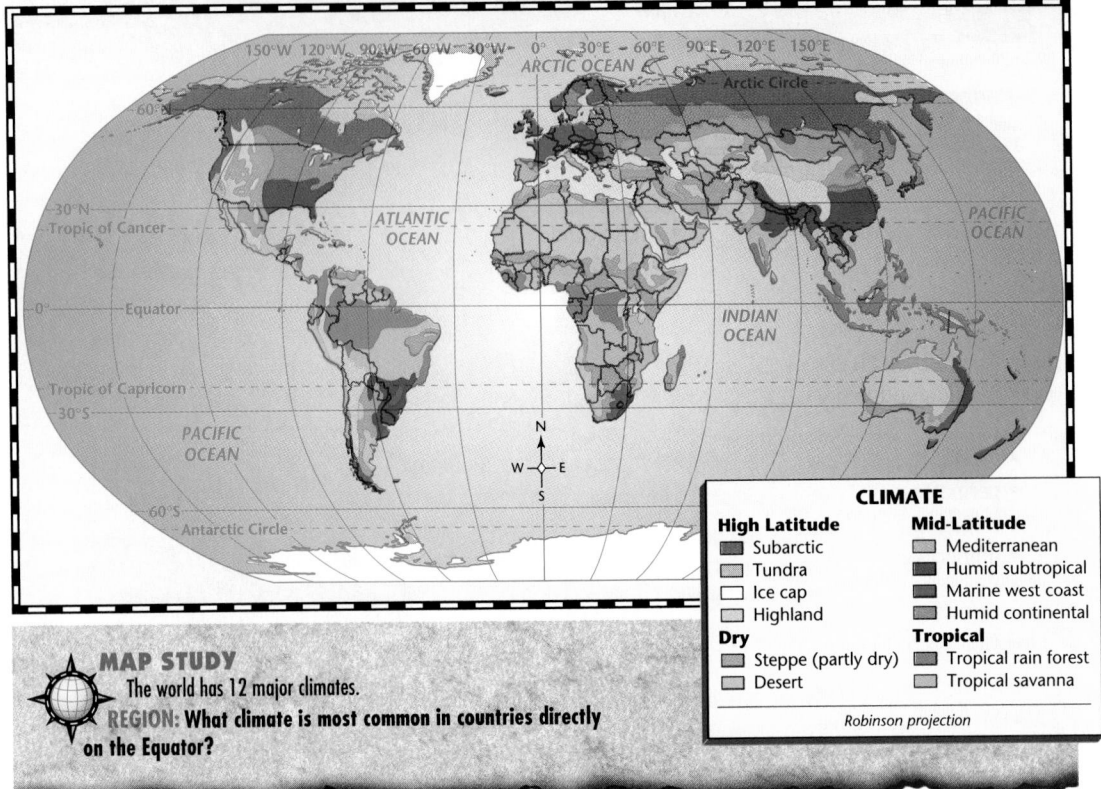

MAP STUDY
The world has 12 major climates.
REGION: What climate is most common in countries directly on the Equator?

CLIMATE

High Latitude	Mid-Latitude
Subarctic	Mediterranean
Tundra	Humid subtropical
Ice cap	Marine west coast
Highland	Humid continental
Dry	**Tropical**
Steppe (partly dry)	Tropical rain forest
Desert	Tropical savanna

Robinson projection

---

plant and animal life. Tall hardwood trees such as mahogany, teak, and ebony form the canopy, or top layer, of the forest. The vegetation at the canopy layer is so thick that little sunlight reaches the forest floor. The Amazon River basin in South America is the world's largest rain forest area.

Tropical Savanna Climate In other parts of the tropics, such as southern India and eastern Africa, most of the year's rain falls during the wet season. The rest of the year is hot and dry. **Savannas,** or broad grasslands with few trees, occur in this climate region. Find the tropical savanna climate areas on the map above.

Region
Mid-Latitude Climates

Mid-latitude, or moderate, climates are found in the middle latitudes of the Northern and Southern hemispheres. They extend from about 30°N to 60°N and from 30°S to 60°S of the Equator. Most of the world's people—and probably you, too—live in this region.

Cooperative Learning Activity

Organize students into four or more groups. Assign each group a climate type. Encourage group members to imagine that they will be stranded for one year in a remote area that has their assigned climate type. Have groups plan a survival strategy to live in that climate. Each group should identify at least the following: 1) the type of shelter they need and will be able to make; 2) the type of clothing they need and will be able to make; and 3) the way they will obtain food and water. Have each group present their survival plan to the class, and have the class challenge each plan with situations or conditions likely to arise in the climate region. **L2**

The mid-latitude region has a greater variety of climates than other regions. This variety results from a mix of air masses—warm air coming from the tropics and cool air coming from the polar regions. In most places, temperatures change with the seasons.

Marine West Coast Climate Coastal areas, where winds blow from the ocean, usually have a mild **marine west coast climate**. If you lived in one of these areas, your winters would be rainy and mild, and your summers cool.

Mediterranean Climate Another mid-latitude coastal climate is called a **Mediterranean climate** because it is similar to the climate found around the Mediterranean Sea. This climate has mild, rainy winters and hot, dry summers.

Humid Continental Climate If you live far from the oceans in inland areas of North America, Europe, or Asia, you face a harsher **humid continental climate**. In these areas, winters can be long, cold, and snowy. Summers are short but may be very hot.

Humid Subtropical Climate Mid-latitude regions close to the tropics have a **humid subtropical climate**. Rain falls throughout the year but is heaviest during the hot and humid summer months. Humid subtropical winters are generally short and mild.

Region
High Latitude Climates

High latitude climate regions lie mostly in the high latitudes of each hemisphere, from 60°N to the North Pole and 60°S to the South Pole. Climates are cold everywhere in the high latitude regions, some more severe than others.

Subarctic Climate Just below the Arctic Circle lie the subarctic areas. The few people living here face severely cold and bitter winters, but temperatures do rise above freezing during summer months. Huge evergreen forests grow in the subarctic region, especially in northern Russia.

Tundra Climate The climate of the Arctic tundra area is harsh and dry. The **tundra** is a vast rolling plain without trees. The top few inches of the ground thaw during summer months, however. This allows sturdy grasses, small berry bushes, and wildflowers to sprout. In parts of the tundra and subarctic regions, the lower layers of soil are known as **permafrost** because they stay permanently frozen.

WHAT IN THE WORLD?
A Sinking Feeling

In some Russian cities, buildings have to be built 6 feet (1.8 m) off the ground on stilts. Why? A vast area of Russia has permafrost. Heat from the buildings would melt the permafrost, creating huge marshes. Without stilts, buildings on the marshy areas might shift and topple.

47

ASSESS

Check for Understanding

Assign Section 3 Review as homework or an in-class activity.

Meeting Lesson Objectives

Each objective below is tested by the questions that follow it in parentheses.

1. **Describe** what major world climate regions are like. (1, 2, 3)
2. **Specify** where each major world climate region is located. (2, 3, 4, 5)
3. **Relate** what kinds of vegetation grow in each world climate region. (1, 3)

Evaluate

📁 Section 3 Quiz **L1**

🔘 Use the Testmaker to create a customized quiz for Section 3. **L1**

📁 Chapter 2 Test Form A and/or Form B **L1**

Meeting Special Needs Activity

Inefficient Organizers Have students combine words and graphics to visualize the information in the section. Direct them to scan the section and note the four major climate regions. Then have them find the locations of the four climate regions on the map on page 46. This activity provides students with a graphic reinforcement of the text. **L1**

More About the Illustration

Answer to Caption
elevation above which no trees grow

Location In Switzerland, many towns and villages are found on the sunny slopes of mountains. Slopes that do not receive much sun are largely undeveloped.

Reteaching

 Reteaching Activity 2 **L1**

Spanish Reteaching Activity 2 **L1**

Enrich

 Enrichment Activity 2 **L3**

CLOSE

Mental Mapping Activity

Provide each student with a copy of a world physical outline map and map pencils. Tell students that this will not be a graded activity. Have students draw, freehand from memory, approximate locations for the four major climate regions.

Village in the Highlands

This village in the Swiss Alps has cooler temperatures than areas at lower elevations.
LOCATION: What is a timberline?

Ice Cap Climate On the polar ice caps and the great ice sheets of Antarctica and Greenland, the climate is bitterly cold. Monthly temperatures average below freezing. Temperatures in Antarctica have been measured at −128°F (−103°C)! Although no vegetation grows here, some fungus-like plants can live on rocks.

Highland Climate A highland, or mountainous, climate has cool or cold temperatures year-round. The **elevation,** or height above sea level, of a place changes its climate dramatically. Higher into the mountains, the air becomes thinner. It cannot hold the heat from the sun, so the temperature drops. Even in the tropics, snow covers the peaks of high mountains.

On those peaks, you come to the **timberline,** the elevation above which no trees grow. Beyond the timberline, only small shrubs and wildflowers grow in meadows.

Region
Dry Climates

Dry climate refers to dry or partially dry areas that receive little or no rainfall. Temperatures can be extremely hot during the day and cold at night. Dry climates can also have severely cold winters.

Desert Climate You can find dry climate regions at any latitude. The driest, with less than 10 inches (25 cm) of rainfall a year, are called deserts. Only scattered plants such as cacti can survive the dry desert climate. With roots close to the surface, cacti can collect any rainfall.

Steppe Climate Many deserts are surrounded by partly dry grasslands known as **steppes.** They get more rain than deserts, averaging 10 to 20 inches (25 to 51 cm) a year. Bushes and short grasses cover the steppe landscape.

SECTION 3 REVIEW

REVIEWING TERMS AND FACTS

1. **Define:** rain forest, savanna, marine west coast climate, Mediterranean climate, humid continental climate, humid subtropical climate, tundra, permafrost, elevation, timberline, steppe.
2. **LOCATION** Where are the moderate climate regions located?
3. **PLACE** How are climates of inland areas different from those nearer coasts? Why?

MAP STUDY ACTIVITIES

4. Look at the map on page 46. In which climate region do you live?
5. Using the same map, name two countries that have high latitude climates.

Answers to Section 3 Review

1. All vocabulary words are defined in the Glossary.
2. middle latitudes of the Northern and Southern hemispheres, from about 30°N to 60°N and 30°S to 60°S
3. Inland areas tend to be much hotter in summer and, in higher latitudes, colder in winter. Oceans moderate the climate of coastal areas.
4. Answers will vary.
5. Answers will vary, but will include two of the following: Canada, Norway, Sweden, Finland, Russia

Chapter 2 Highlights

Important Things to Know About Water, Climate, and Vegetation

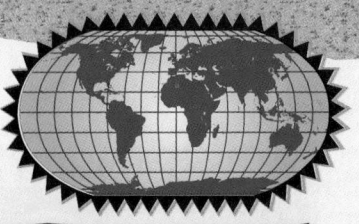

SECTION 1 EARTH'S WATER

- Water constantly moves in a cycle from oceans, to air, to land, and back to oceans.
- Rivers and lakes provide important sources of freshwater.
- Freshwater also is stored in the earth's underground layers of rock.
- The world's four oceans are the Atlantic Ocean, Pacific Ocean, Indian Ocean, and Arctic Ocean.

SECTION 2 INFLUENCES ON CLIMATE

- The kind of climate a place has depends mainly on its latitude.
- Landforms and nearby bodies of water also influence climate.
- Air and water move around the earth as warm or cold winds and currents.
- A large body of water keeps nearby land temperatures moderate.
- Areas in a rain shadow behind coastal mountains often have dry climates.

SECTION 3 CLIMATE AND VEGETATION

- Four major climate regions are tropical, mid-latitude, high latitude, and dry.
- Tropical climates can be wet, with rain forests, or dry, with savannas.
- In the mid-latitude region, climate is affected by distance from the ocean.
- Places with high latitude climates do not have much vegetation.

Ocean Life ▶

Extra Credit Project

Profile Have each student select a body of water—a river, lake, sea, or ocean. Direct students to create a profile of the body of water they have chosen. Profiles should show such features as width, depth, other physical characteristics, origin of name, and so on. Encourage students to combine their profiles in a bulletin-board collage. L1

 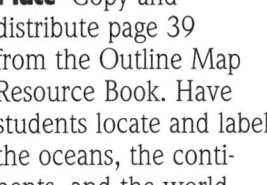
49

Chapter 2 Review and Activities

CHAPTER 2
REVIEW & ACTIVITIES

GLENCOE TECHNOLOGY

VIDEODISC

Use the Chapter 2 Mind-Jogger Videoquiz to review students' knowledge before administering the Chapter 2 Test.

MindJogger Videoquiz

Chapter 2
Disc 1 Side A

 Also available in VHS.

ANSWERS

Reviewing Key Terms

1. E
2. A
3. I
4. C
5. H
6. B
7. D
8. G
9. J
10. F

Mental Mapping Activity

This exercise helps students to visualize the locations and regions they have been studying and understand the relationships among these various points. All attempts at freehand mapping should be accepted.

REVIEWING KEY TERMS

Match the numbered terms in Column A with their definitions in Column B.

A
1. aquifer
2. climate
3. evaporation
4. hurricane
5. marine west coast
6. permafrost
7. precipitation
8. rain shadow
9. tundra
10. vegetation

B
A. long-term patterns of weather
B. layer of soil that stays frozen
C. violent storm over a tropical ocean
D. rain or snow that falls in a place
E. underground rock layer storing water
F. plants that grow naturally in a location
G. dry region inland from coastal mountains
H. mild climate with rainy winters and cool summers
I. process by which liquid water is turned into a gas
J. high latitude climate region lacking trees

Mental Mapping Activity

On a separate piece of paper, sketch a map of the world showing the continents. Then shade and label the following items on your map:
• Equator
• Atlantic Ocean
• Pacific Ocean
• Indian Ocean
• mid-latitude climate regions

REVIEWING THE MAIN IDEAS

Section 1
1. **MOVEMENT** What are the major steps in the water cycle?
2. **HUMAN/ENVIRONMENT INTERACTION** From what sources do people get usable freshwater?
3. **REGION** What are the four major oceans?

Section 2
4. **PLACE** Why are climates near the poles always cool or cold?
5. **PLACE** What effect does a large body of water have on climates in nearby places? Why?
6. **REGION** Why are desert areas often found on the inland slopes of coastal mountains?

Section 3
7. **LOCATION** At what latitudes are rain forests found?
8. **PLACE** In what climate is the vegetation limited to mostly fungus-like plants?
9. **PLACE** What creates a marine west coast climate?
10. **PLACE** What kinds of plants grow in the tundra?

Reviewing the Main Ideas

Section 1

1. Heat from the sun causes water to evaporate from the surfaces of oceans. Moist air rises and cools, causing vapor to condense into tiny droplets of water. These droplets come together to form clouds. Excess water in clouds falls as precipitation. This eventually runs back into the oceans, and the process begins again.

2. surface water—rivers, lakes, and streams—and groundwater

3. Pacific, Atlantic, Indian, Arctic

Section 2

4. sun's rays never directly strike polar areas

5. It moderates temperatures in coastal areas. Water temperatures do not change as fast or as much as air temperatures.

6. The inland sides of coastal mountains lie in a rain shadow.

CRITICAL THINKING ACTIVITIES

1. **Determining Cause and Effect** Many ancient civilizations developed near rivers. Why do you think people settled there?
2. **Synthesizing Information** Mount Kilimanjaro, the tallest mountain in Africa, is located near the Equator. What kind of climate would you find at its base? At its peak?

GeoJournal Writing Activity

Water is essential for all living things. In your journal, describe a memorable experience you had that involved water. You may want to describe an adventure you had swimming or fishing, for example. Put your ideas in the form of a poem or song.

COOPERATIVE LEARNING ACTIVITY

Organize into eight groups, with each group choosing a particular country that has one of these climates: rain forest, tropical savanna, Mediterranean, marine west coast, humid continental, tundra, desert, or highland. Have one person in your group find out more about plant life in the country, another about animal life, and a third about ways that humans have adapted to the climate. As a group, brainstorm a way to combine and display your findings.

PLACE LOCATION ACTIVITY: WORLD OCEANS AND CURRENTS

Match the letters on the map with the bodies of water and ocean currents. Write your answers on a separate sheet of paper.

1. Gulf Stream
2. Atlantic Ocean
3. Indian Ocean
4. California Current
5. Japan Current
6. Arctic Ocean

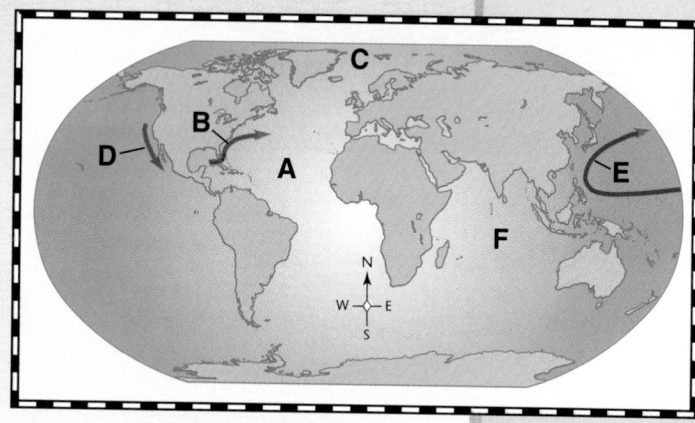

Cooperative Learning Activity

Suggest that students compare the world political map on pages A4 and A5 with the world climate map on page 46 when choosing their countries. Encourage students to illustrate their posters or displays with pictures, drawings, diagrams, and maps.

GeoJournal Writing Activity

To stimulate student thinking on the topic, offer an anecdote of your own. Interested students may wish to research poems or songs that have water as a subject.

Place Location Activity

1. B
2. A
3. F
4. D
5. E
6. C

Chapter Bonus Test Question

This question may be used for extra credit on the chapter test.

I am known as a hurricane, but when I get to the _____ , they change my name to "typhoon." Where am I? *(Pacific Ocean)*

Section 3

7. in the tropics, the low latitudes near the Equator
8. ice cap climate
9. moist winds blowing from the oceans
10. grasses, wildflowers, and small bushes

Critical Thinking Activities

1. People settled along rivers because the rivers provided a source of freshwater, which is needed for survival. Some students may add that rivers offered a means of transportation.
2. At its base, Kilimanjaro would have a tropical climate. The mountain's elevation would cause it to have a highland climate at its peak.

GeoLab
ACTIVITY

FOCUS

Ask students which of the following they think would be suitable for drinking: a glass of water from the faucet, a glass of water from a nearby river, a glass of water from a nearby lake, a glass of ocean water, a glass of water melted from a glacier. Have students explain why they believe each type of water is or is not drinkable. Then point out that water, whatever its source, may have qualities that are not apparent to the naked eye.

TEACH

Before students begin the GeoLab Activity, emphasize that it is critical to make drawings that correctly show what is on the slide. When students have completed the activity, select drawings of each sample and display them for the class. Guide students in a discussion on each sample, seeking to identify: 1) what is apparent in the drawing; 2) whether that type of water is drinkable or not; and 3) what steps can be taken to make water suitable for human consumption. **L1**

From the classroom of Eleanor Bloom, Bloom-Carroll Schools, Lithopolis, Ohio

WATER: It's not all wet!

Background

Water—the world's most precious resource—is vital to all life. It is used for drinking, irrigation, industry, transportation, and energy. But is all water the same? Let's take a look at the various kinds of water to find out what's really in this precious resource.

Moose Pond in the White Mountains, New Hampshire

Believe it or NOT!

About 97 percent of the earth's water has salt in it and is not suitable for drinking. About 2 percent of the world's water is in the form of glaciers or ice caps. Only about 1 percent of the earth's water is available for drinking.

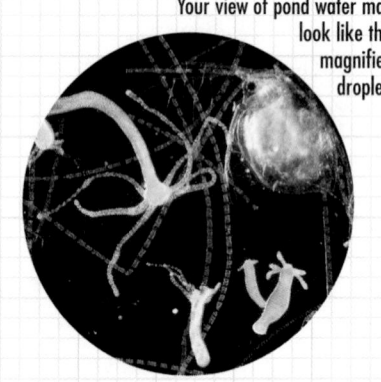
Your view of pond water may look like this magnified droplet.

Materials

- 8 empty baby-food-size jars
- microscope
- blank microscope slides and slide covers
- eyedropper
- 8 different types of water: distilled water, spring water, salt water, pond water, soapy water, faucet water, well water, filtered water

UNIT 1

Curriculum Connection Activity

Biology Depending on the magnification of the microscope and other factors, students may see living organisms in their water samples. The pond water should be especially rich with these life-forms. Tell students that they are seeing *microorganisms,* so-called because of their extremely small size. Have students extend their learning by researching the role that microorganisms play in water quality. **L1**

What To Do

A. Fill each of the jars with a different type of water. Label the jars.

B. Allow the jars to sit for 24 hours.

C. Observe and write down what you see in each jar the next day.

D. Shake each jar, then use the eyedropper to take a sample from each jar. Drip each sample onto a slide. You will end up with 8 different slides.

E. Put a slide cover over each sample and label each slide.

F. Observe each slide under the microscope.

G. Make drawings of what you see and label each drawing.

Lab Activity Report

1. After 24 hours, which two water samples differ the most?

2. Describe the differences between the water samples.

3. What did you learn about water?

4. **Drawing Conclusions** What type of water would make the best drinking water?

 Go A Step Further!

Activity

The process of desalination removes salt from water. Find out more about desalination at the library. Report to your class about this process. Use a world map to show what countries rely on desalination for the majority of their drinking water.

NATIONAL
GEOGRAPHIC
SOCIETY

These materials are available from Glencoe.

🔘 **VIDEODISC**

STV: Water

Water Quality
Side 1, Frames 00392-17833

ASSESS

Have students answer the **Lab Activity Report** questions on page 53.

CLOSE

Encourage students to complete the **Go A Step Further** writing activity. Suggest that students create a "league table" of the top 10 countries that rely most heavily on desalination for drinking water.

Answers to Lab Activity Report

1. Answers will vary, but most likely the pond water and the filtered water.

2. Answers will vary. Some water samples are clearer than others. Some of the samples have material floating in them. Some water is clear, some yellow, some bluish, and some green.

3. Answers will vary, but students should discuss how all water is not the same and the quality of the water depends on its environment.

4. Answers will vary, but students should choose the water samples that contain the least amount of foreign particles, such as distilled, filtered, or spring water.

The World's People

PERFORMANCE ASSESSMENT ACTIVITY

AN EVENING NEWSCAST Organize students into several groups and have groups act as news teams to create a 10-minute news program titled "The World's People." Groups should begin this task by brainstorming realistic headlines on such topics as population, culture, resources, agriculture, and economy. Suggest that they look for clues by studying the section previews. Groups should then select headlines that they want to cover in their news programs. Have groups assign individual members to write stories. Finally, have group members put together the stories to develop a script for the program. Encourage groups to "broadcast" their program. Encourage groups to "broadcast" their programs for the rest of the class. If video equipment is available, some groups might want to film their programs.

POSSIBLE RUBRIC FEATURES:
Concept attainment, decision-making skills, collaborative skills, oral presentation skills, composition skills

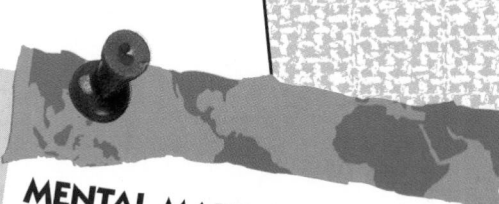

MENTAL MAPPING ACTIVITY
Before teaching Chapter 3, display a world population distribution map in the class. Have students focus on the regions that are most heavily populated. Call on volunteers to define these regions in terms of proximity to physical features. Encourage students to keep the locations of these populated areas in mind as they read about the world's people.

TEACHER'S CORNER

NATIONAL GEOGRAPHIC SOCIETY

INDEX TO NATIONAL GEOGRAPHIC MAGAZINE

The following articles may be used for research relating to this chapter:

- "The Essential Element of Fire," by Michael Parfit, September 1996.
- *Mexico,* A National Geographic Special Edition, August 1996.
- "The Three Faces of Jerusalem," by Alan Mairson, April 1996.
- "Pilgrimage to Buddhist Caves," by Reza, April 1996.
- "Our Polluted Runoff," by John G. Mitchell, February 1996.
- "Rice, the Essential Harvest," by Peter T. White, May 1994.
- "Recycling," by Noel Grove, July 1994.
- "Iraq: Crucible of Civilization," by Merle Severy, May 1991.
- "The World's Food Supply at Risk," by Robert E. Rhoades, April 1991.
- "Population, Plenty, and Poverty," by Paul L. Ehrlich and Anne H. Ehrlich, December 1988.

NATIONAL GEOGRAPHIC SOCIETY PRODUCTS AVAILABLE FROM GLENCOE

To order the following products for use with this chapter, contact your local Glencoe sales representative or call Glencoe at 1-800-334-7344:

- *GeoBee* (Software)
- *Picture Atlas of the World* (CD-ROM)
- *Earth's Endangered Environments PictureShow* (CD-ROM)
- *GTV: Planetary Manager* (Videodisc)
- *STV: World Geography* (Videodisc)

ADDITIONAL NATIONAL GEOGRAPHIC SOCIETY PRODUCTS

To order the following products for use with this chapter, call National Geographic Society at 1-800-368-2728:

- *The Living Earth* (Video)
- *Healing the Earth* (Video)
- *Ozone: Protecting the Invisible Shield* (Video)
- *Pollution: World at Risk* (Video)
- *Recycling: The Endless Circle* (Video)
- *Energy: The Problems and the Future* (Video)
- *Egypt: Quest for Eternity* (Video)
- *Living Treasures of Japan* (Video)
- *Nations of the World* (10 Videos)

TEACHER-TO-TEACHER
Curriculum Connection: Art

—from Sherry Henderson,
Northbrook School, Houston, TX

A CULTURE MOSAIC *The purpose of this activity is to have students show in what ways different peoples of the world satisfy their needs and wants.*

Call on students to identify the needs all people have in common—food, shelter, clothing, and so on. List their responses on the chalkboard. Lead the students in a discussion of how we in the United States satisfy these needs. Then encourage students to use source materials such as *Culturgram* to research the ways people in other regions of the world meet these basic needs. When they have completed their research, ask students to create mosaic-type posters depicting how people in various countries satisfy their needs. Suggest that they use a variety of materials and types of illustrations for their posters.

Display finished posters and have students discuss the following questions: How do people's needs vary from region to region? What similarities and differences do you see in the ways people meet their needs?

BIBLIOGRAPHY

Readings for the Student
Axtell, Roger E. *Do's and Taboos Around the World.* New York: John Wiley and Sons, 1993.

Steele, Philip. *The People Atlas.* Boston: Oxford University Press, 1991.

Readings for the Teacher
Comparing Cultures, revised ed. Portland, ME: J. Weston Walch, 1990. Reproducible cooperative-learning activities.

Marshall, Bruce, ed. *The Real World: Understanding the Modern World Through the New Geography.* Boston: Houghton Mifflin, 1991.

Multimedia
Culture: What Is It? Niles, Ill.: United Learning, 1993. Videocassette, 13 minutes.

Simpolicon: Simulation of Political and Economic Development. San Jose: Cross Cultural Software, 1993. Apple or Macintosh 5.25" and 3.5" disks.

MEETING NATIONAL STANDARDS

Geography for Life

All of the 18 standards are demonstrated in Unit 1. The following ones are highlighted in this chapter:
- Standards 1, 2, 3, 4, 6, 9, 10, 11, 12, 13

KEY TO ABILITY LEVELS
Teaching strategies have been coded for varying learning styles and abilities.

L1 **BASIC** activities for all students

L2 **AVERAGE** activities for average to above-average students

L3 **CHALLENGING** activities for above-average students

LEP **LIMITED ENGLISH PROFICIENCY** activities

Chapter 3 Resource Organizer
The World's People

Use these *Geography: The World and Its People* resources to teach, reinforce, and extend chapter content.

CHAPTER 3 RESOURCES

*Vocabulary Activity 3

Cooperative Learning Activity 3

Workbook Activity 3

Geography Skills Activity 3

GeoLab Activity 3

Critical Thinking Skills Activity 3

Chapter Map Activity 3

Performance Assessment Activity 3

*Reteaching Activity 3

Enrichment Activity 3

Chapter 3 Test, Form A and Form B

Unit Overlay Transparencies 1-3, 1-6, 1-7

Geography Handbook Transparency 2

Political Map Transparency 1

Geoquiz Transparencies 3-1, 3-2

*Chapter 3 Digest Audiocassette, Activity, Test

Vocabulary PuzzleMaker Software

Student Self-Test: A Software Review

Testmaker Software

MindJogger Videoquiz

*If time does not permit teaching the entire chapter, summarize using the Chapter 3 Highlights on page 69, and the Chapter 3 English (or Spanish) Audiocassettes. Review students' knowledge using the Glencoe MindJogger Videoquiz. *Also available in Spanish*

Use these *Geography: The World and Its People* resources to teach and reinforce section content.

SECTION 1 RESOURCES

Reproducible Lesson Plan 3-1

Section Focus Transparency 3-1

Guided Reading Activity 3-1

Section Quiz 3-1 (also available in Spanish)

SECTION 2 RESOURCES

Reproducible Lesson Plan 3-2

Section Focus Transparency 3-2

Guided Reading Activity 3-2

Section Quiz 3-2 (also available in Spanish)

SECTION 3 RESOURCES

Reproducible Lesson Plan 3-3

Section Focus Transparency 3-3

Guided Reading Activity 3-3

Section Quiz 3-3 (also available in Spanish)

ADDITIONAL RESOURCES FROM GLENCOE

 **Reproducible Masters**

- Glencoe Social Studies Outline Map Book, page 39

- Foods Around the World, Unit 1

 **Workbook**

- Building Skills in Geography, Unit 1, Lessons 4, 5

World Music: Cultural Traditions
Listening to the World's Music, pp. 11–12

 Transparencies

- National Geographic Society PicturePack Transparencies, Unit 1

 **Posters**

- National Geographic Society: Eye on the Environment, Unit 1

 World Games Activity Cards
Unit 1

 World Crafts Activity Cards Unit 1

Videodiscs

- Geography and the Environment: The Infinite Voyage

- National Geographic Society: STV: World Geography, Vol. 2

- National Geographic Society: GTV: Planetary Manager

 **CD-ROM**

- National Geographic Society: Picture Atlas of the World

54D

Performance Assessment

Refer to the Planning Guide on page 54A for a Performance Assessment Activity for this chapter. See the *Performance Assessment Strategies and Activities* booklet for suggestions.

Chapter Objectives

1. **Discuss** the concept of culture.
2. **Investigate** issues related to human population and to population growth.
3. **Examine** the role of renewable and nonrenewable resources in human activities.

GLENCOE TECHNOLOGY

 VIDEODISC

Use the Chapter 3 Mind-Jogger Videoquiz to preview chapter content.

MindJogger Videoquiz

Chapter 3
Disc 1 Side A

 Also available in VHS.

Chapter 3 The World's People

World Culture Regions

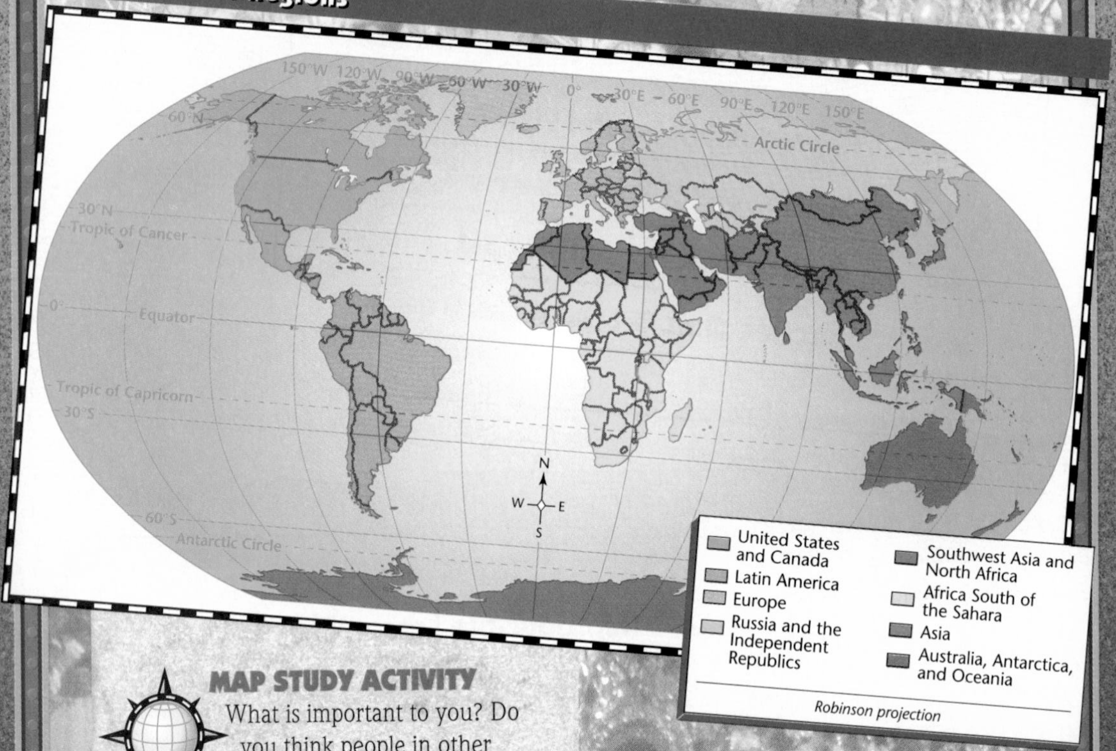

Map Legend:
- United States and Canada
- Latin America
- Europe
- Russia and the Independent Republics
- Southwest Asia and North Africa
- Africa South of the Sahara
- Asia
- Australia, Antarctica, and Oceania

Robinson projection

 MAP STUDY ACTIVITY

What is important to you? Do you think people in other countries believe as you do? In Chapter 3 you will learn about the world's people.
1. How many culture regions are shown on the map above?
2. In what culture region do you live?

54

MAP STUDY ACTIVITY

Answers
1. eight
2. Answers will vary.

Map Skills Practice
Reading a Map What culture region is indicated by the color purple?
(Australia, Antarctica, and Oceania)

SECTION 1 Culture

PREVIEW

Words to Know
- culture
- civilization
- cultural diffusion
- language family
- standard of living
- literacy rate
- free enterprise
- socialism

Places to Locate
- Huang He
- Indus River
- Tigris-Euphrates rivers
- Nile River

Read to Learn . . .
1. what *culture* means.
2. where ancient cultures began.
3. what elements make each culture unique.

This colorfully dressed dancer in South Korea reflects certain customs that are important to her. Many of her beliefs and customs have been passed down from distant ancestors. All of us hold certain beliefs and act certain ways because of what we've learned in our culture.

If you wake up to rock music, put on denim jeans, drink orange juice for breakfast, and speak English, those things are part of your culture. If you eat flat bread for breakfast, speak Arabic, and wear a long cotton robe to protect you from the hot sun, *those* things are part of your culture.

Place
What Is Culture?

When some people hear the word *culture*, they think of priceless paintings and classical symphonies. **Culture**, as used in geography, is the way of life of a group of people who share similar beliefs and customs. These people may speak the same language, follow the same religion, and dress in a certain way. The culture of a people also includes their government, their music and literature, and the ways they make a living. What things are important in your culture?

Early Cultures Some 4,000 to 5,000 years ago, at least four cultures arose in Asia and Africa. One developed in China along a river called the Huang He. Another developed near the Indus River in South Asia, a third between the

---

Classroom Resources for Section 1

REPRODUCIBLE MASTERS

Reproducible Lesson Plan 3-1
Guided Reading Activity 3-1
World Literature Reading 1
Geography Skills Activity 3
Building Skills in Geography,
   Unit 1, Lessons 4, 5
Section Quiz 3-1

TRANSPARENCIES

Section Focus
   Transparency 3-1
Geography Handbook
   Transparency 2
Geoquiz Transparency 3-1

MULTIMEDIA

- Vocabulary PuzzleMaker
   Software
- Testmaker

- National Geographic
   Society: STV: World
   Geography, Vol. 2

Meeting National Standards

Geography for Life The following standards are highlighted in this section:
- Standards 1, 2, 3, 10

FOCUS

Section Objectives
1. **Explain** what *culture* means.
2. **Locate** where ancient cultures began.
3. **Distinguish** what elements make each culture unique.

Bellringer Motivational Activity

 Prior to taking role at the beginning of the class period, project Section Focus Transparency 3-1 and have students answer the activity questions. Discuss students' responses. **L1**

Vocabulary Pre-check

Ask the class what words and ideas they associate with the term *culture*. List responses on the chalkboard. Then have students locate the definition of *culture* on page 55. Call on volunteers to circle words in the list that are relevant to this definition. **L1 LEP**

 Use the Vocabulary PuzzleMaker Software to create a crossword puzzle. **L1**

55

TEACH

Guided Practice
Making Comparisons
Make the concept of "culture" concrete for students by discussing it in the context of their own culture. For each cultural characteristic (government, economy) discussed in the lesson, have students identify examples from their culture. Work with students to provide comparative examples from other cultures with which students are familiar. **L1**

Independent Practice

Guided Reading Activity 3-1 **L1**

MULTICULTURAL PERSPECTIVE

Culturally Speaking
Human cultures are remarkable for their diversity. Despite this diversity, every single culture has certain things in common. These *cultural universals* include ways of obtaining food and providing shelter, rules about marriage and kinship, artworks, stories or folktales, and ways of settling disputes.

MAP STUDY

Answer
the United States and Canada because of the ties of colonialism, or Russia and the independent republics because of proximity

Map Skills Practice
Reading a Map The United States and what other country make up a culture region? *(Canada)*

Tigris and Euphrates rivers in Southwest Asia, and a fourth along the Nile River in North Africa.

All four river-valley cultures developed agriculture and ways of irrigating, or bringing water to the land. Why was irrigation important? Farming produced more food than hunting and gathering, which meant that larger populations could develop. People then learned trades, built cities, and made laws.

The river-valley cultures eventually became **civilizations**, or highly developed cultures. These cultures spread their knowledge and skills from one area to another, a process known as **cultural diffusion**.

Region
Culture Regions Today

Geographers today often divide the world into areas called culture regions. These regions may be based on the kind of government, social groups, economic systems, languages, or religions found in each.

Governments The kind of government, or political system, a society has reflects its culture. Until a few hundred years ago, most countries had authoritarian systems in which one person ruled with unlimited power.

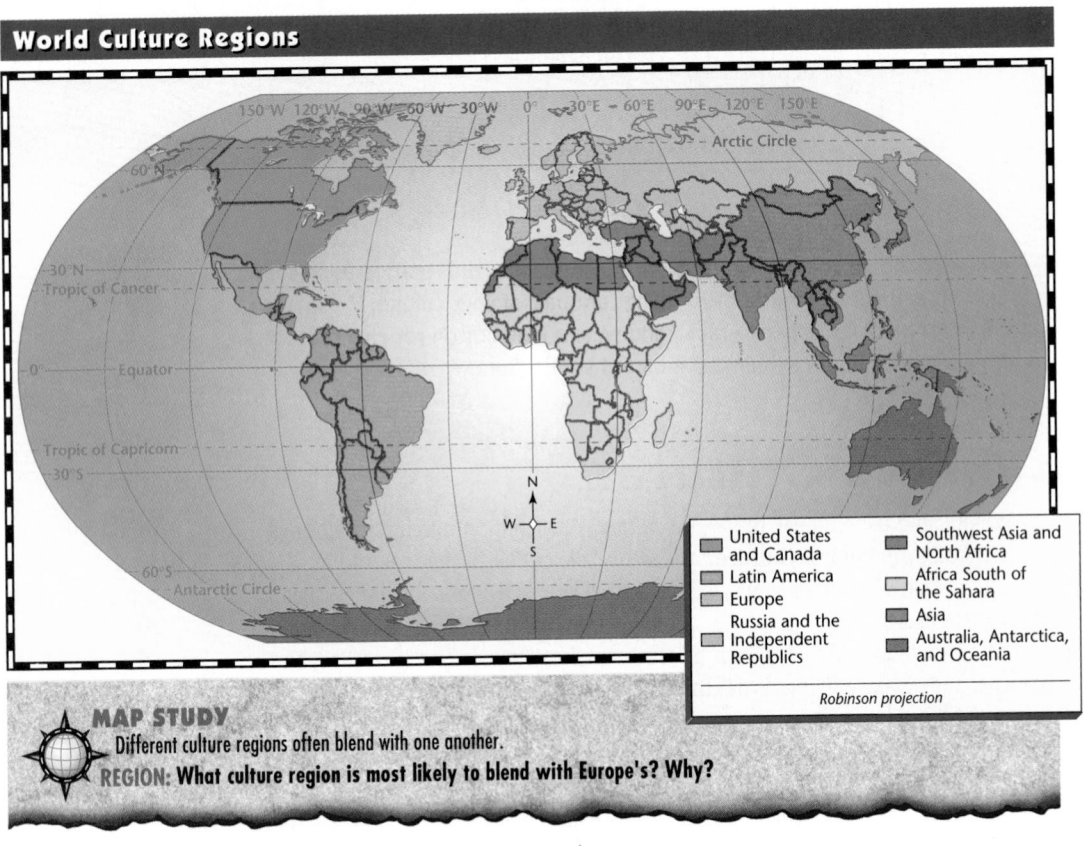

World Culture Regions

United States and Canada	Southwest Asia and North Africa
Latin America	Africa South of the Sahara
Europe	Asia
Russia and the Independent Republics	Australia, Antarctica, and Oceania

Robinson projection

MAP STUDY
Different culture regions often blend with one another.
REGION: What culture region is most likely to blend with Europe's? Why?

Cooperative Learning Activity

Organize students into five groups. Inform groups that their task is to create a five-section wall poster on culture regions. Then assign each group one of the five factors used to define culture regions—government, social groups, economic system, language, religion. Groups should develop or locate a number of visuals that illustrate their factor. They then should create their section of the poster. Some group members might do artwork or design, while others might write captions. Have groups combine their finished sections to create the wall poster. **L1**

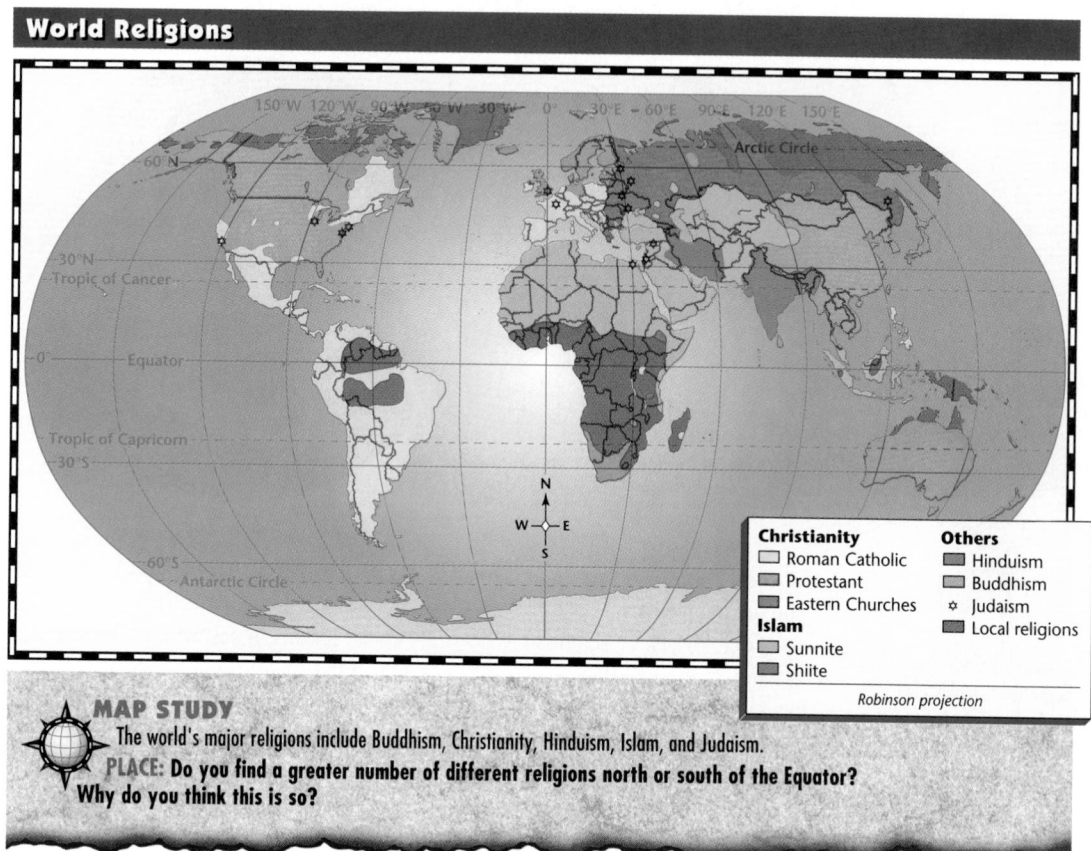

Christianity
☐ Roman Catholic
☐ Protestant
☐ Eastern Churches
Islam
☐ Sunnite
☐ Shiite

Others
☐ Hinduism
☐ Buddhism
✳ Judaism
☐ Local religions

Robinson projection

 MAP STUDY
The world's major religions include Buddhism, Christianity, Hinduism, Islam, and Judaism.
PLACE: Do you find a greater number of different religions north or south of the Equator? Why do you think this is so?

LESSON PLAN
Chapter 3, Section 1

MAP STUDY

Answer
north; students may suggest that more people live north of the Equator, people and countries are in closer proximity so cultural diffusion may be greater north of the Equator, and so on

Map Skills Practice
Reading a Map What two types of religion are most common in Africa? *(Sunnite Islam, local religions)*

When the people of a country hold the powers of government, we think of that government as a democracy. Citizens choose their leaders by voting. Once in power, leaders in a democracy are expected to obey a constitution or other long-standing traditions that require them to respect individual freedoms.

Language Language is a powerful tool, offering a way for people to share information. Sharing a language is one of the strongest unifying forces for a culture. Languages spoken in a culture region often belong to the same **language family**, or group of languages having similar beginnings. Romance languages, for example, come from Latin, the language of ancient Rome. Spanish, Portuguese, French, Italian, and Romanian are in the Romance language family.

Religion Another important part of culture is religion. Some of the major world religions are Buddhism, Christianity, Hinduism, Islam, and Judaism. The map above shows you the areas where people practice these religions.

Social Groups Another way of organizing culture regions is by social groups. A country may have many different groups of people living within its

CHAPTER 3

57

ASSESS

Check for Understanding
Assign Section 1 Review as homework or an in-class activity.

Meeting Lesson Objectives
Each objective below is tested by the questions that follow it in parentheses.
1. **Explain** what *culture* means. (1)
2. **Locate** where ancient cultures began. (2)
3. **Distinguish** what elements make each culture unique. (3, 5)

Meeting Special Needs Activity

Visual Learners Students who do well with this mode of learning can benefit by studying a variety of photographs of people, activities, and artifacts of different cultures. Bring to the class books showing photographs of various nations.

Have students find similar photographs taken of different cultures. This will help students to understand specific aspects of culture and the ways in which these aspects may vary. **L1**

More About the Illustration

Answer to Caption
Answers will vary.

Human/Environment Interaction Each country's economy is different because each has different resources.

Evaluate

 Section 1 Quiz **L1**

Use the Testmaker to create a customized quiz for Section 1. **L1**

Reteaching

Have students write a five-paragraph essay titled, "Five Ways to Define Culture Regions." **L1**

Enrich

Have students conduct research on one of the four river-valley cultures discussed in this section. Have them share their findings in a brief oral report. **L3**

CLOSE

Have students study the photographs in this section. Encourage them to identify and discuss the cultural elements illustrated in these photographs.

Industry and Agriculture in Germany

In rural Germany, grapes are harvested to make wine *(above).* In the city of Stuttgart, workers assemble Mercedes-Benz autos *(right).*
PLACE: What are two important products produced in your state?

borders. These differences are sometimes based on income, or how much money people earn.

Sociologists, or scientists who study social groups, measure how well a society meets the needs of its people. The **standard of living** is a measure of people's quality of life based on their income and the material goods they have. Another measure of a society's well-being is its **literacy rate**, or the percentage of people who can read and write.

Economies All societies have economic systems. An economic system tells how people produce goods, what goods they produce, and how those goods are then bought and sold. In most democratic societies, a market economy prevails. This is an economic system based on **free enterprise**, where people start and run businesses to make a profit with little interference from government.

Another economic system is **socialism**. In a socialist system, the government sets goals, may run some businesses, and tends to have a central role in the economy.

SECTION 1 REVIEW

REVIEWING TERMS AND FACTS

1. **Define the following:** culture, civilization, cultural diffusion, language family, standard of living, literacy rate, free enterprise, socialism.
2. **LOCATION** Along what rivers were four ancient cultures located?
3. **REGION** What three factors help identify a culture region?

MAP STUDY ACTIVITIES

4. Turn to the map on page 56. What is the name of the culture region that includes Mexico and South America?

UNIT 1

Answers to Section 1 Review

1. All vocabulary words are defined in the Glossary.
2. Huang He, Indus River, Tigris and Euphrates rivers, Nile River
3. Any three of the following: governments, social groups, economies, language, religion.
4. Latin America

BUILDING GEOGRAPHY SKILLS

Using Latitude and Longitude

To find an exact location, geographers use a set of imaginary lines. One set of lines—**latitude** lines—circles the earth's surface east to west. Each line of latitude—also called a parallel—is numbered from 1° to 90° and followed by an N or S to show it is north or south of the Equator.

A second set of **longitude** lines runs vertically from the North Pole to the South Pole. Each of these lines is called a meridian. The starting point for meridians, or 0° longitude, is called the Prime Meridian. Longitude lines are numbered from 1° to 180° followed by an E or W—east or west of the Prime Meridian. To find latitude and longitude, follow these steps:

• Choose a location on a map or globe.
• Identify the number of the nearest parallel, or line of latitude.
• Identify the number of the nearest meridian, or line of longitude, that crosses it.

The World

	National boundary
⊛	National capital
•	Other city

Robinson projection

Geography Skills Practice

1. What is the exact location of Washington, D.C.?

2. What cities on the map lie south of 0° latitude?

Answers to Geography Skills Practice

1. about 40°N and 80°W
2. Lima, Rio de Janeiro, Cape Town, Sydney

Draw on the chalkboard a simple map of the neighborhood around the school. Mark street names and include a number of local landmarks, such as the school, a park, a store used by students, and so on. Ask students to describe the locations of the various landmarks by using street names. Suggest that students use the street intersection nearest each landmark. Call on volunteers to share their answers with the rest of the class. Point out that identifying locations by using lines of latitude and longitude follows the same principle. Then have students read the skill lesson and complete the practice questions. **L1**

Additional Skills Practice

1. What is another name used for lines of longitude? for lines of latitude? *(meridians; parallels)*

2. What cities and national capitals lie between 0° and 20°N? between 20°E and 40°E? *(Mexico City, Mumbai [Bombay], Singapore; Moscow, Cairo)*

 Geography Skills Activity 3

Building Skills in Geography, Unit 1, Lessons 4, 5

SECTION 2 Population

FOCUS

Section Objectives

1. **Locate** where most people in the world live.
2. **Understand** how scientists measure population.
3. **Discuss** how the earth's population is changing.

Bellringer Motivational Activity

Prior to taking role at the beginning of the class period, project Section Focus Transparency 3-2 and have students answer the activity questions. Discuss students' responses. **L1**

Vocabulary Pre-check

Have a volunteer look up the definitions of the terms *demographer* and *demography* and share them with the class. Tell students that demography, the study of population, is closely related to geography. Have them speculate on the nature of this relationship. **L1 LEP**

PREVIEW

Words to Know
• emigrate
• refugee
• population density
• urbanization
• developed country
• developing country
• demographer
• birthrate
• death rate
• famine

Places to Locate
• China
• Vietnam
• Egypt

Read to Learn . . .
1. where most people in the world live.
2. how scientists measure population.
3. how the earth's population is changing.

Tokyo is one of the most crowded places in the world. At rush hour, thousands of Japanese workers cram the subways for the hectic ride to work or home.

Although Tokyo is crowded, some parts of the world have few people. What makes one area crowded and another empty? Climate, culture, and jobs are some of the things that help determine where people live.

Movement

Population Patterns

Some families live in the same town or on the same land for generations. Other people move frequently from place to place. In some cases people choose to leave the country in which they were born. They **emigrate**, or move, to another country. You may know of people who are forced to flee their country because of wars, food shortages, or other problems. They become **refugees**, or people who flee to another country for refuge from persecution or disaster.

Population Distribution People live on only a small part of the earth. As you learned earlier, land covers only about 30 percent of the earth's surface, and half of this land is not useful to humans. People cannot make homes, grow crops, or graze animals on land covered with ice, deserts, or high mountains.

World Population Density

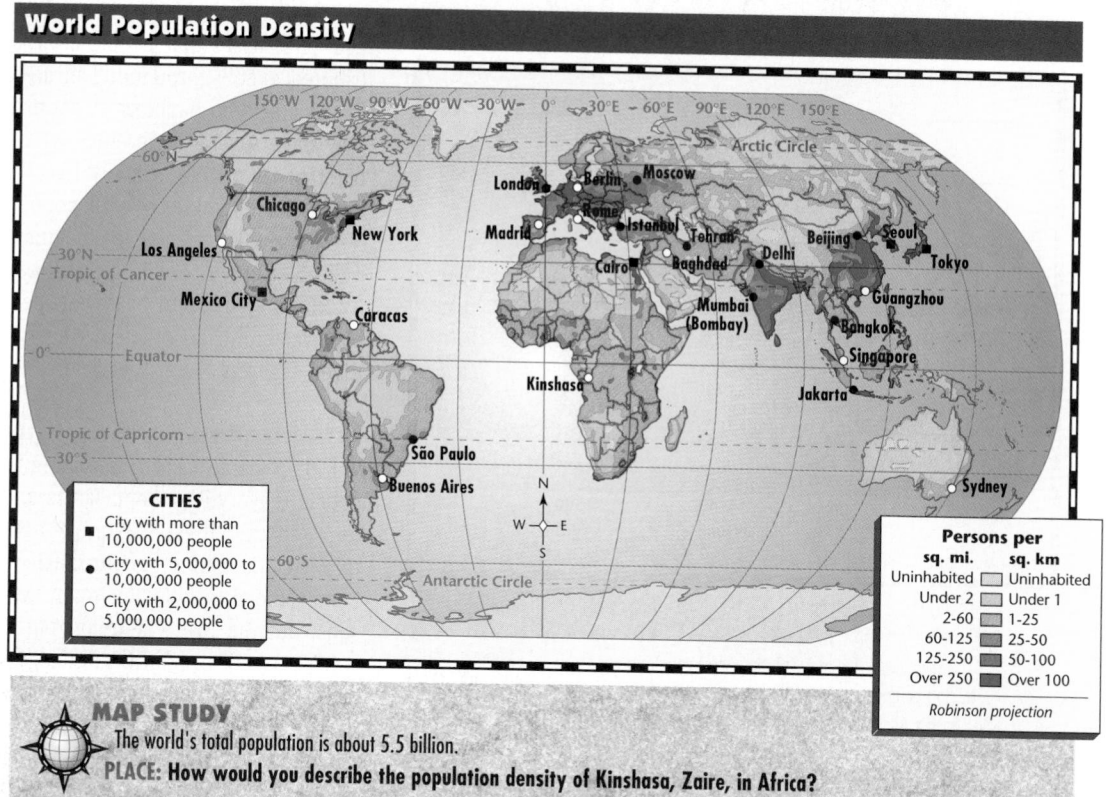

CITIES
- ■ City with more than 10,000,000 people
- ● City with 5,000,000 to 10,000,000 people
- ○ City with 2,000,000 to 5,000,000 people

Persons per	
sq. mi.	**sq. km**
Uninhabited	Uninhabited
Under 2	Under 1
2-60	1-25
60-125	25-50
125-250	50-100
Over 250	Over 100

Robinson projection

 MAP STUDY
The world's total population is about 5.5 billion.
PLACE: How would you describe the population density of Kinshasa, Zaire, in Africa?

Even on the usable land, population is not distributed, or spread, evenly. One reason is that people naturally choose to live in places with plentiful water, good land, and a favorable climate. Other reasons may lie in a people's history and culture.

Look at the map above to see how the world's population is distributed. Notice that people are most concentrated in western Europe and eastern and southern Asia. The United States also has areas of dense population.

Asia is a huge continent in land area. It also has the largest population—nearly 3.4 billion. Europe is much smaller in land area, but it has a population of about 730 million people. Africa is the second-largest continent in area and third in overall population, with 700 million people.

Population Density One way to look at population is by measuring **population density**—the average number of people living in a square mile or square kilometer. Population density gives a general idea of how crowded a country or region is. For example, the countries of Vietnam and Congo have about the same land area. The population density in Vietnam is 582 people per square mile (225 people per sq. km). Congo has an average of only 19 people per square mile (7 people per sq. km).

Cooperative Learning Activity

Organize the class into six groups and assign each group one of the following continents: North America, South America, Africa, Asia, Europe, Australia. Direct groups to locate the following information on their assigned continent: total population, population density, 10 largest cities, projected population for the year 2000 or 2010. Have each group present its findings in a list written at the top of an outline map of the continent. Groups should mark their maps to show population distribution and the 10 largest cities. Arrange the maps to form a world map of population on the bulletin board. **L1**

MAP STUDY

Answer
Kinshasa has 2-5 million people; the area surrounding it has 2-60 people per square mile.

Map Skills Practice
Reading a Map What area of the United States has the greatest population density? *(northeast)*

TEACH

Guided Practice

Cause and Effect Draw on the chalkboard a two-column chart titled "The Population Explosion." Head the two columns *Causes* and *Effects*. Guide students in identifying causes of the rapid growth in population. List their responses in the *Causes* column. Then guide students in identifying the effects of rapid population growth. List these responses in the *Effects* column. **L1**

Independent Practice

Guided Reading Activity 3-2 **L1**

Demographers project that by the year 2000 there will be 20 cities with populations of more than 10 million people. And four of them—Mexico City, Mexico; Tokyo, Japan; São Paulo, Brazil; and Seoul, South Korea—will have populations of well over 20 million.

GRAPHIC STUDY

Answer
about 5 billion people

Graphic Skills Practice
Reading a Graph In about what year did the world population reach 1 billion? *(about 1850)*

ASSESS

Check for Understanding

Assign Section 2 Review as homework or an in-class activity.

Meeting Lesson Objectives

Each objective below is tested by the questions that follow it in parentheses.

1. **Locate** where most people in the world live. (2, 3, 6)
2. **Understand** how scientists measure population. (1)
3. **Discuss** how the earth's population is changing. (1, 2, 4, 5)

World Population Growth

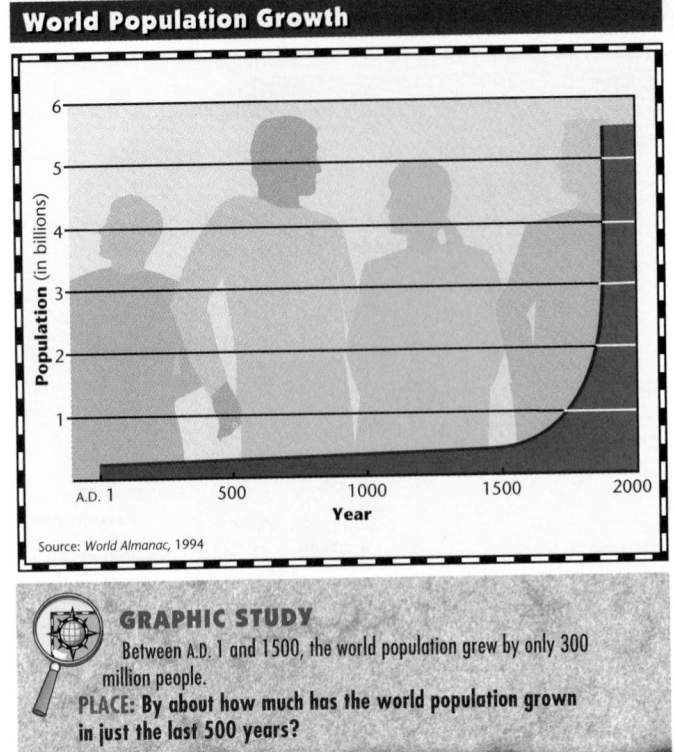

Source: *World Almanac*, 1994

GRAPHIC STUDY
Between A.D. 1 and 1500, the world population grew by only 300 million people.
PLACE: By about how much has the world population grown in just the last 500 years?

Notice that density is an *average*. It assumes that people are distributed evenly throughout an area. But this seldom happens. A country may have several large cities where most of its people actually live. In Egypt, for example, overall population density is 153 people per square mile (59 people per sq. km). In reality, about 99 percent of Egypt's people live within 20 miles of the Nile River. The rest of Egypt is desert.

Urbanization Throughout the world, populations are changing as people leave villages and farms and move to the cities. This movement to cities is called **urbanization**. People move to cities for many reasons, but the overwhelming reason is to find jobs. In South America, for example, economic hardships have caused millions of people to move to cities such as São Paulo in Brazil and Buenos Aires, Argentina.

Urban areas are usually centers of industrialization. Countries that are industrialized are called **developed countries**. Those countries that are working toward industrialization are called **developing countries**. The economies of these countries depend mainly on agriculture and developing modern industries.

Human/Environment Interaction
Population Growth

How fast has the earth's population grown? In 1800 the number of people in the world totaled about 857 million. During the next hundred years the population doubled to nearly 1.7 billion. By the mid-1990s, this figure had more than tripled to 5.5 billion people. **Demographers**, or scientists who study population, believe that the earth will have more than 7 billion people by the year 2010. Look at the graph above to see how population has grown since 1800.

Measuring Growth Populations are growing at different rates in different places. To find the growth rate in a specific area, demographers compare the birthrate with the death rate. The **birthrate** is the number of children born each year for every 1,000 people. The **death rate** is the number of deaths for every 1,000 people.

UNIT 1

Meeting Special Needs Activity

Problems With Large Numbers Discussion about the world's population necessitates the use of very large numbers that many students may have difficulty visualizing. To help students form mental pictures of large numbers of people, have them imagine "action figures," about four inches high. Tell them it would take about 10 million of these action figures to cover a football field. Using such an approach will help students visualize the population figures discussed in this section. **L1**

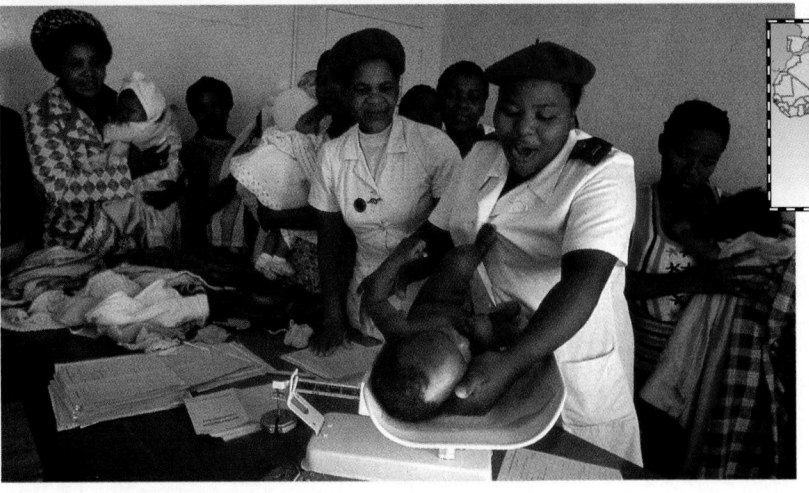

As more children receive medical care in this South African clinic, the country's death rate will drop.
PLACE: What other factor can reduce a country's death rate?

LESSON PLAN
Chapter 3, Section 2

More About the Illustration

Answer to Caption
better living conditions

GEOGRAPHIC THEMES

Place Because of a high birthrate and better medical care, the population is growing at a faster rate in Africa than in any other area of the world.

Growth rates tend to be high in developing countries. Better health care and living conditions have reduced the death rate, and people are living longer. Birthrates also may remain high because some cultures encourage large families.

Population Challenges Rapid population growth presents many challenges. A growing population requires more food. Since 1950 food production fortunately has increased faster than population on all continents except Africa. Millions of people, however, still suffer **famine**, or lack of food. Another challenge is that expanding populations use up resources more rapidly than stable populations. Some developing countries face shortages of water, housing, and jobs.

SECTION 2 REVIEW

REVIEWING TERMS AND FACTS
1. Define the following: emigrate, refugee, population density, urbanization, developed country, developing country, demographer, birthrate, death rate, famine.
2. MOVEMENT What are some reasons that people leave their home countries?
3. HUMAN/ENVIRONMENT INTERACTION Why is population distributed unevenly on the earth's land surfaces?
4. PLACE What data does a country's birthrate give?

MAP STUDY ACTIVITIES
5. Look at the graph on page 62. What major changes occurred between 1400 and 1900?
6. Turn to the map on page 61. What are the two largest cities shown south of the Equator?

Evaluate

 Section 2 Quiz **L1**

Use the Testmaker to create a customized quiz for Section 2. **L1**

Reteaching

Have each student write 10 questions on the section. Pair students and have pair members use their questions to quiz each other on section content. **L1**

Enrich

Have interested students research and report on how the population in the state has changed since 1900. **L3**

CLOSE

Encourage students to discuss how population is related to the five themes of geography.

Answers to Section 2 Review
1. All vocabulary words are defined in the Glossary.
2. to look for a better life, to escape war or some other disaster
3. because some areas of the earth are favorable for settlement while others are not
4. number of children born each year for every 1,000 people in that country
5. Population growth was slow and steady until about 1600, then population began to grow very rapidly.
6. São Paulo, Jakarta

TEACH

Encourage students to create a demographic portrait of the class. Suggest that they gather such information as total population, age at beginning of school year, month born, place of residence, and so on. Call on volunteers to report their findings to the class. Then ask students to write three generalizations on the class based on these findings. **L2**

More About . . .

Demography Englishman John Graunt initiated the study of demography as a science. He developed the first mortality table in the 1660s. Using this table he was able to estimate the number of men of military age, the number of women of child-bearing age, the total number of families, and London's population. Interest in demography grew during the 1700s, and a government-sponsored census developed around the end of the century. In 1790, the United States government began taking a census every 10 years; Great Britain began its census 11 years later.

COUNTING HEADS

| MATH | SCIENCE | HISTORY | LITERATURE | TECHNOLOGY |

Have you ever counted the number of people in a room? Geographers often rely on similar head counts of people in an area to understand the number and distribution of the many different kinds of people in our world.

WHAT DEMOGRAPHERS STUDY Demographers are the scientists who take the head counts. They collect and examine information about people—births and deaths, marriages, divorces, ages, and national backgrounds. Demographers figure out a population's *density,* or the number of people living in a given area. They also analyze *distribution,* or where people live. They even look at the movement of people from place to place.

DEMOGRAPHER'S TOOLS Two principal tools that a demographer uses are a census and a record of vital statistics. A *census* is a government's formal count of its population. Such a count keeps track of the number of people and obtains information about age, employment, income, and other characteristics. You are counted in the United States Census once every 10 years.

Vital statistics are records of basic human events—births, diseases, marriages, divorces,

Planning for Growth

and deaths. These statistics reveal what is happening or has happened to an area's population.

WHY DEMOGRAPHERS? Demographers analyze information to discover how different groups of people react to events and other people. Their findings often explain how changes in a society have affected people. For example, a rising cost of living in the United States has resulted in smaller families.

Just as often, demographic findings can predict likely events, giving people time to prepare and plan for changes. If many births are recorded, for example, cities may plan to build more schools.

Making the Connection

1. What do demographers do?
2. How can demographic findings be used?

64

Answers to Making the Connection

1. Demographers collect and examine information about people. They figure out population density and analyze population distribution.

2. to explain how changes in society have affected people; to predict future changes in populations

SECTION 3

Resources

Meeting National Standards

Geography for Life The following standards are highlighted in this section:
• Standards 1, 2, 3, 11, 14, 15, 16

PREVIEW

Words to Know
• natural resource
• renewable resource
• nonrenewable resource
• fossil fuel
• subsistence farming
• service industry
• pollution
• pesticide
• acid rain
• hydroelectric power
• solar energy

Places to Locate
• Pennsylvania
• Arabian Peninsula

Read to Learn . . .
1. what renewable and non-renewable resources are.
2. how people use resources to make a living.
3. how overusing resources may threaten the environment.

Green fields of corn have dotted North America since ancient times. This rich harvest of corn reflects the abundance of natural resources found in the region. What are natural resources, and why are they so important?

A natural resource is anything from the natural environment that people use to meet their needs. Natural resources include fertile soil, clean water, minerals, trees, and energy sources. Human skills and labor are also valuable natural resources. People use all these resources to improve their lives.

Human/Environment Interaction

Types of Resources

The value of resources changes as people discover new uses and new technology. Trees, for example, have been a valuable resource throughout history. For thousands of years people used wood to stay warm, cook food, and build homes, vehicles, and furniture.

Oil, on the other hand, was a gooey nuisance to the people of central Pennsylvania until the oil industry developed. Oil-based products then began to power cars and heat homes. Great underground pools of oil were discovered in Southwest Asia in the 1930s. Today the countries surrounding the Persian Gulf are among the richest on the earth.

FOCUS

Section Objectives
1. **Identify** what renewable and nonrenewable resources are.
2. **Explain** how people use resources to make a living.
3. **Discuss** how overusing resources may threaten the environment.

Bellringer Motivational Activity

 Prior to taking role at the beginning of the class period, project Section Focus Transparency 3-3 and have students answer the activity questions. Discuss students' responses. **L1**

Vocabulary Pre-check

Call on a volunteer to look up the term *resource* in a dictionary and read all of its definitions. Ask students to select the definition that is appropriate for this section. **L1 LEP**

Classroom Resources for Section 3

 REPRODUCIBLE MASTERS

Reproducible Lesson Plan 3-3
Vocabulary Activity 3
Guided Reading Activity 3-3
Reteaching Activity 3
Enrichment Activity 3
Section Quiz 3-3

 **TRANSPARENCIES**

Section Focus
Transparency 3-3

MULTIMEDIA

 Testmaker

 National Geographic Society: GTV: Planetary Manager

MAP STUDY

Answer
subsistence farming

Map Skills Practice
Reading a Map How is most land in China used? *(subsistence farming)*

TEACH

Guided Practice

Making Connections Lead a discussion about objects in the classroom, such as desks, chairs, and so on. Guide students into identifying: 1) the resources from which they are made; 2) whether each resource is renewable or nonrenewable; 3) how manufacturing industries, service industries, and agriculture may have contributed to its production; and 4) what environmental challenges might be associated with its production. **L2**

Independent Practice

Guided Reading Activity 3-3 **L1**

Today, forests are being cut down at an incredible rate. Rain forests alone are disappearing at the rate of more than 40,000 square miles (103,600 square km) every year—or nearly 5 square miles (13 square km) per hour!

World Land Use and Resources

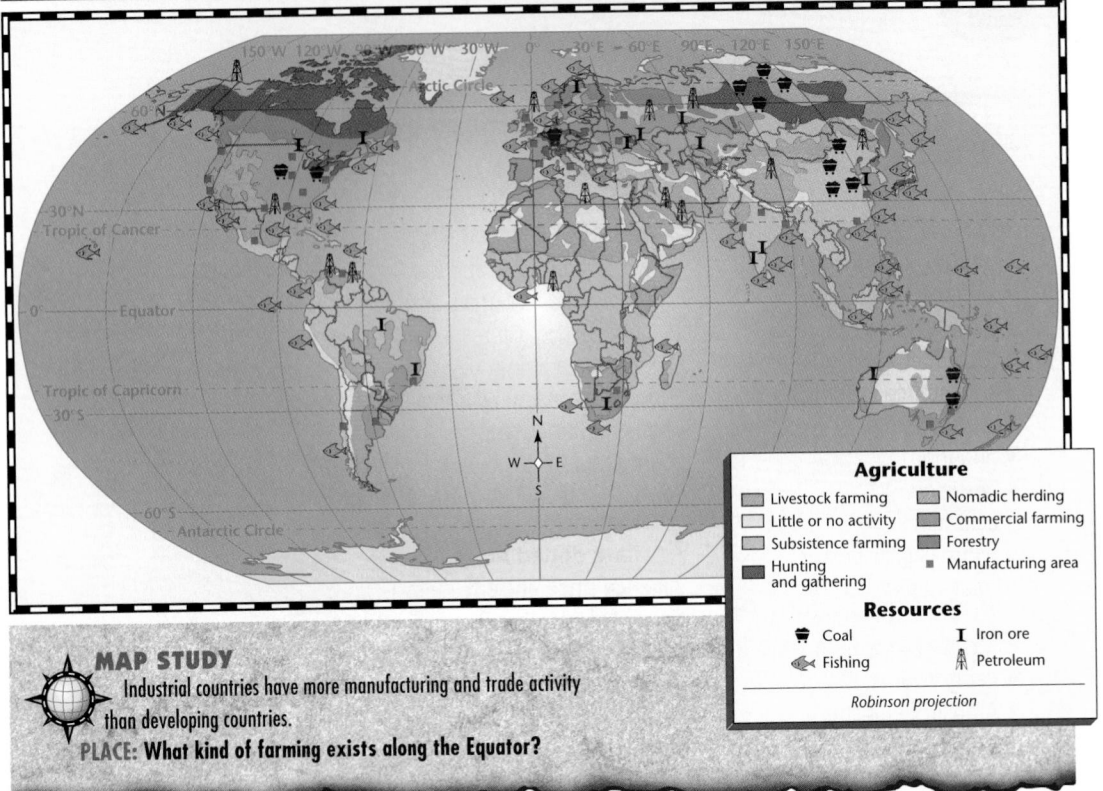

Agriculture	
▨ Livestock farming	▨ Nomadic herding
☐ Little or no activity	▨ Commercial farming
▨ Subsistence farming	▪ Forestry
■ Hunting and gathering	▪ Manufacturing area

Resources	
🛒 Coal	I Iron ore
🐟 Fishing	⚲ Petroleum

Robinson projection

MAP STUDY
Industrial countries have more manufacturing and trade activity than developing countries.
PLACE: What kind of farming exists along the Equator?

Renewable Resources Some natural resources can be replaced as they are used up. These **renewable resources** can be replaced naturally or grown fairly quickly. Forests, grasslands, plant and animal life, and rich soil all can be renewable resources if people manage them carefully. A lumber company concerned about future growth can replant as many trees as it cuts. Fishing and whaling fleets can limit the number of fish and whales they catch in certain parts of the ocean.

Nonrenewable Resources Metals and other minerals found in the earth's crust are **nonrenewable resources**. They cannot be replaced because they were formed over millions of years by geologic forces within the earth.

One important group of nonrenewable resources is **fossil fuels**—coal, oil, and natural gas. Industries and people depend on these fuels for energy and as raw materials for plastics and other goods. We also use up large amounts of other metals and minerals, such as iron, aluminum, and phosphates. Some of these can be reused, but they cannot be replaced.

66

UNIT 1

Cooperative Learning Activity

Organize students into three groups. Assign each group one of the following topics: air pollution, land pollution, water pollution. Have groups investigate their topics in their community. Groups should identify the major local threats to the environment and suggest solutions. Each group should create an illustrated poster, identifying the problems and suggesting solutions. Have groups display their posters on the bulletin board. **L2**

Human/Environment Interaction
Making a Living

People use natural resources to make the things they need and want. People also use natural resources to earn a living for themselves and their families.

Farming One of our basic needs is food. Although cities are growing through urbanization, about half the people in the world still live in rural areas. Most farmers live by **subsistence farming**, or growing only enough to feed themselves. They work in small fields with hand tools and use animals for heavy work. Sometimes they sell part of a harvest.

Commercial agriculture, on the other hand, usually operates on a large scale, with farmers producing food crops and livestock to sell. Huge plots of land may be farmed using machinery.

Industry Other working people, particularly in cities, have jobs in industry. Some industries produce goods, ranging from nails to shoes to basketballs to military jets. Other industries supply services. People in **service industries** do many different kinds of jobs. They may be computer operators, car mechanics, cooks, doctors, musicians, or taxi drivers. How many service industries have you depended on today?

Human/Environment Interaction
Environmental Challenges

When people use natural resources to make a living, they affect the environment. The careless use of resources is a threat to the environment. Many human activities can cause **pollution**—putting impure or poisonous substances into the land, water, and air.

Land and Water Only about 11 percent of the earth's surface has land good enough for farming. Chemicals that farmers use may improve their crops, but some also may damage the land. **Pesticides**, or chemicals that kill insects, can pollute rivers and groundwater.

Other human activities also pollute soil and water. Oil spills from tanker ships threaten ocean coastal areas. Illegal dumping of dangerous waste products causes problems. Untreated sewage reaching rivers pollutes lakes and groundwater as well. Salt water can also pollute both soil and groundwater.

Air Industries and vehicles that burn fossil fuels are the main sources of air pollution. Throughout the world, fumes from cars and other vehicles pollute the air. The chemicals in air pollution can seriously damage people's health.

WHAT IN THE WORLD?

All Wet

Do you waste water? The average American uses about 60 gallons (227 l) every day—maybe more. If you love long showers, remember that 6 or 7 gallons (23 to 27 l) of water are going down the drain every minute. Filling up a bathtub is not good either. That uses about 36 gallons (about 137 l).

ASSESS

Check for Understanding

Assign Section 3 Review as homework or an in-class activity.

Meeting Lesson Objectives

Each objective below is tested by the questions that follow it in parentheses.

1. **Identify** what renewable and nonrenewable resources are. (1, 2)
2. **Explain** how people use resources to make a living. (1, 3, 6)
3. **Discuss** how overusing resources may threaten the environment. (1, 4)

These materials are available from Glencoe.

 VIDEODISC

GTV: Planetary Manager

Up, Up, and Away?
Side 2, Chapter 1

Meeting Special Needs Activity

Less Proficient Readers Students having difficulty with the text can benefit by participating in partner reading. Pair students and have partners take turns reading paragraphs or sections of the text aloud. The partner not reading should follow the text as it is being read. At the end of each paragraph or section, students should discuss the reading, freely exchanging thoughts, ideas, and concerns about the text. **L1**

More About the Illustration

Answer to Caption

oil spill: more care in transporting; forest damage: reduce air pollution

 **Human/Environment Interaction** In eastern Europe, much of the acid-rain damage is caused by air pollution created by heavy industry.

Evaluate

 Section 3 Quiz **L1**

Use the Testmaker to create a customized quiz for Section 3. **L1**

Chapter 3 Test Form A and/or Form B **L1**

Reteaching

Reteaching Activity 3 **L1**

Enrich

Enrichment Activity 3 **L3**

CLOSE

 **Mental Mapping Activity**

Provide each student with a world physical outline map and map pencils. Tell students that this will not be a graded activity. Have students shade, freehand from memory, the areas of the earth where people carry on subsistence farming.

Careless use of resources has caused oil spills in Alaska *(right)* and acid rain damage to forests in eastern Europe *(above).*
HUMAN/ENVIRONMENT INTERACTION: How could the damage shown here have been prevented?

These chemicals combined with precipitation may fall as **acid rain**, or rain carrying large amounts of sulfuric acid. Acid rain eats away the surfaces of buildings, kills fish, and can destroy entire forests.

Energy Developed nations and developing nations both need safe, dependable sources of energy. Fossil fuels are most often used to generate electricity, heat buildings, run machinery, and power vehicles. Fossil fuels, however, are nonrenewable resources. In addition, they contribute to air pollution. So today many countries are trying to discover new ways of using renewable energy sources. Two of these ways are **hydroelectric power**, the energy generated by falling water, and **solar energy**, or energy produced by the heat of the sun. What other renewable energy sources are there?

SECTION 3 REVIEW

REVIEWING TERMS AND FACTS

1. **Define the following:** natural resource, renewable resource, nonrenewable resource, fossil fuel, subsistence farming, service industry, pollution, pesticide, acid rain, hydroelectric power, solar energy.
2. **HUMAN/ENVIRONMENT INTERACTION** Why are metals a nonrenewable resource?
3. **HUMAN/ENVIRONMENT INTERACTION** How are service industries different from other industries?
4. **HUMAN/ENVIRONMENT INTERACTION** How might farmers pollute the environment?

    MAP STUDY ACTIVITIES
   5. Look at the map on page 66. What areas of the world have large forests?
   6. Using the same map, find two areas where people carry on subsistence farming.

Answers to Section 3 Review

1. All vocabulary words are defined in the Glossary.
2. because metals take millions of years to form
3. other industries produce goods; service industries provide services
4. through the use of pesticides
5. Canada, far northern Europe, northern Russia
6. choices include: Central Africa, southern Africa, China, South Asia

Important Things to Know About the World's People

SECTION 1 CULTURE

- In geography, culture means a group of people who share similar beliefs and customs.
- A people's culture includes their government, economy, language, religion, and social organization.
- Four ancient cultures developed along river valleys.
- Geographers study the world in terms of culture regions.

SECTION 2 POPULATION

- People live on only about 15 percent of the world's land.
- Population is distributed very unevenly over the earth's surface.
- Population can be measured in terms of density, or the average number of people living in a square mile or square kilometer.
- Developed countries are industrialized. Developing countries are working toward that goal.
- By the mid-1990s, the world's population totaled 5.5 billion.
- Rapid population growth threatens the world's supply of food and resources.

SECTION 3 RESOURCES

- Natural resources are renewable or nonrenewable.
- Renewable resources such as forests, grasslands, and animal life can be replaced—if managed carefully.
- Nonrenewable resources such as metals and fossil fuels cannot be replaced once they are used.
- About half of the world's people live by farming—most by subsistence farming.
- People who work in industry are employed in either manufacturing or service jobs.

Rio de Janeiro, Brazil ▶

CHAPTER 3

69

CHAPTER HIGHLIGHTS

Using the Chapter 3 Highlights

- Use the Chapter 3 Highlights to preview, review, condense, or reteach the chapter.

Preview/Review

Vocabulary Puzzle-Maker Software reinforces the terms used in Chapter 3.

Condense

Have students read the Chapter 3 Highlights. Spanish Chapter Highlights also are available.

Chapter 3 Digest Audiocassettes and Activity

Chapter 3 Digest Audiocassette Test

Spanish Chapter Digest Audiocassettes, Activities, and Tests also are available.

Guided Reading Activities

Reteach

Reteaching Activity 3. Spanish Reteaching Activities also are available.

MAP STUDY

Place Copy and distribute page 39 from the Outline Map Resource Book. Direct students to shade the populated areas of the earth and locate and label the world's 10 largest cities.

Extra Credit Project

Public Opinion Survey Have students work in groups to survey other students, faculty members, family members, and neighbors. Have groups ask the following question: What makes American culture unique? Groups should collate responses and present their results to the class in a brief written report. Suggest that groups post their reports on the bulletin board for comparison purposes. **L3**

Chapter 3 Review and Activities

ANSWERS

Reviewing Key Terms

1. B
2. G
3. J
4. D
5. I
6. A
7. E
8. C
9. F
10. H

 **Mental Mapping Activity**

This exercise helps students to visualize local resources and land use. All attempts at freehand mapping should be accepted.

REVIEWING KEY TERMS

Match the numbered terms in Column A with their definitions in Column B.

A
1. birthrate
2. subsistence farming
3. famine
4. emigrate
5. fossil fuel
6. standard of living
7. renewable resource
8. urbanization
9. free enterprise
10. socialism

B
A. a measure of the quality of life
B. the number of children born each year for every 1,000 people
C. the movement of people to the cities
D. to move to another country
E. resource that can be replaced
F. people run businesses to make a profit
G. farmers grow only enough for themselves
H. economic system in which government runs some businesses
I. resource that can't be replaced
J. lack of food

Mental Mapping Activity

On a separate piece of paper, sketch a map of your state or community. Use map pencils to shade in and label the following resources or uses of land in your area. Add other resources if you need to, and don't forget to provide a map key.

- Forestry
- Commercial farming
- Manufacturing area
- Coal
- Petroleum

REVIEWING THE MAIN IDEAS

Section 1

1. PLACE What makes up the culture of a region?
2. REGION What is the difference between authoritarian and democratic rule?
3. REGION What are some measures used to judge the quality of life in a culture?

Section 2

4. MOVEMENT Why is world population distributed unevenly?
5. PLACE What does population density measure?
6. HUMAN/ENVIRONMENT INTERACTION What is a developing country?

Section 3

7. HUMAN/ENVIRONMENT INTERACTION What is the difference between renewable and nonrenewable resources?
8. MOVEMENT What are three kinds of jobs people hold in service industries?
9. HUMAN/ENVIRONMENT INTERACTION What are some human activities that contribute to land, air, and water pollution?

Reviewing the Main Ideas

Section 1

1. government, language, religion, arts and crafts, food, music, dress, ways of making a living
2. authoritarian rule: one person rules with unlimited power, often ignoring wishes of people; democratic rule: people elect government and, therefore, control government
3. standard of living, literacy rates

Section 2

4. Population is unevenly distributed because some areas of the earth are favorable for settlement while others are not.
5. the average number of people in a square mile or square kilometer
6. country in process of becoming industrialized

CRITICAL THINKING ACTIVITIES

1. **Making Comparisons** Why do population growth rates differ in developed countries and developing countries?
2. **Analyzing Information** Why are human skills and labor considered natural resources?

GeoJournal Writing Activity

Families and individuals often have personal cultural traditions. For example, your family may have holiday customs or favorite foods that come from another culture region. In your journal, describe some elements of your home culture, including influences from other culture regions.

COOPERATIVE LEARNING ACTIVITY

Working in groups of four, refer to the culture regions map on page 56. Each group should find information about the economy and population of more than one country in a culture region. Include information about literacy rate, birthrate, death rate, percentage of urban/rural dwellers, and standard of living. Brainstorm ways to present your information to the class. Make your presentation to other students.

PLACE LOCATION ACTIVITY: WORLD CULTURE REGIONS

Match the letters on the map with the culture regions of the world. Write your answers on a separate sheet of paper.

1. Africa South of the Sahara
2. United States and Canada
3. Australia, Oceania, and Antarctica
4. Latin America
5. Europe
6. Asia
7. Southwest Asia and North Africa
8. Russia and the Independent Republics

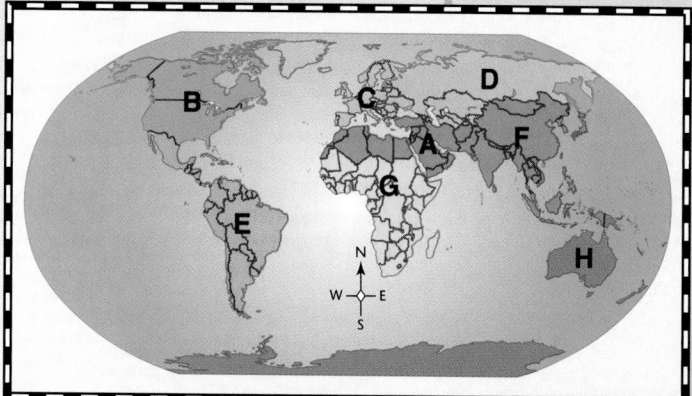

Cooperative Learning Activity

Suggest that students review all the presentations together to see if any regional similarities exist.

GeoJournal Writing Activity

Call on volunteers to share their journal entries with the rest of the class. Some students may wish to explain the significance of their families' special customs.

Place Location Activity

1. G
2. B
3. H
4. E
5. C
6. F
7. A
8. D

Chapter Bonus Test Question

This question may be used for extra credit on the chapter test.

In what culture regions are the following early civilizations located?

1. Huang He *(Asia)*
2. Indus River *(Asia)*
3. Tigris and Euphrates rivers *(Southwest Asia and North Africa)*
4. Nile River *(Southwest Asia and North Africa)*

Section 3

7. A renewable resource can be replaced naturally or grown as it is used up; a nonrenewable resource is in limited supply and cannot be replaced.
8. choices include: computer operator, car mechanic, cook, cleaner, doctor, and so on
9. burning of fossil fuels; using pesticides; oil spills; dumping dangerous waste products; releasing untreated sewage

Critical Thinking Activities

1. Birthrates are high in developing countries, because traditions encourage—and the demands of subsistence farming require—large families. At the same time, death rates are falling and people are living longer. In developed countries, because of industrialization, there is less need for large families.
2. Human skills and labor are needed to adapt other natural resources to meet needs.

FOCUS

Remind students that there would be no life on Earth without the sun's light and heat. Point out that, conversely, the sun's blinding light, searing heat, and deadly ultraviolet rays could easily destroy all life on Earth. The ozone layer protects Earth from the sun's awesome power.

TEACH

Inform students that the ozone layer lies in the upper levels of the atmosphere. Then direct students to conduct research on the atmosphere and how it mediates the energy of the sun. Suggest that they present their findings in the form of an annotated diagram. **L2**

NATIONAL GEOGRAPHIC SOCIETY
EYE ON THE ENVIRONMENT

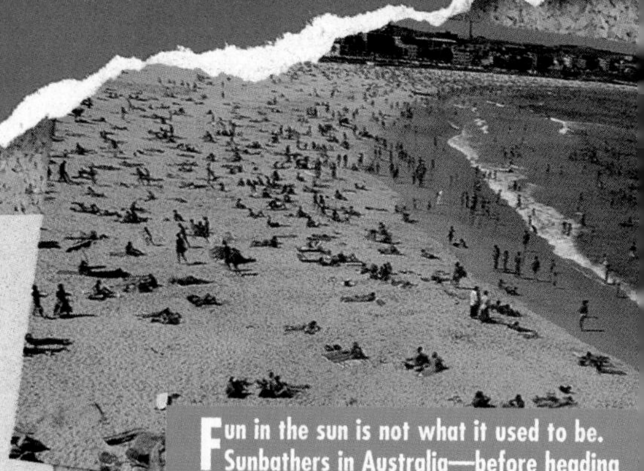

Danger: Ozone Loss

PROBLEM

Ozone: At ground level, it's a harmful gas and a major ingredient in smog. In the upper atmosphere, it's a lifesaver. Ozone acts as a sunscreen, blocking harmful ultraviolet rays that could kill all life on Earth. But manufactured chemicals, such as chlorofluorocarbons (CFCs), have thinned the ozone layer around the globe. This created an ozone hole over Antarctica—increasing the ultraviolet radiation reaching Earth. Higher levels of radiation lead to skin cancers and other health problems, and damage food chains and crops.

SOLUTIONS

● CFCs must go. An agreement approved by 74 nations calls for a ban on CFCs and other ozone-depleting chemicals by the year 2000.
● Ozone-friendly materials are replacing CFCs: Ammonia and helium are being used as coolants in refrigerators and air conditioners; propane and butane, in aerosol sprays; and water and terpenes, in cleaners.
● CFCs in existing air conditioners, refrigerators, and other products are being recycled.

Fun in the sun is not what it used to be. Sunbathers in Australia—before heading for the beach—tune in to television reports on daily ultraviolet levels. Australia has one of the world's highest rates of skin cancers.

Ozone Fact Bank

🍃 About 90 percent of all ozone in the atmosphere occurs in the stratosphere, approximately 8–30 miles (15–48 km) above Earth.

🍃 If compressed, scattered ozone molecules in the ozone layer above Earth would be about as thick as a pane of glass.

🍃 At certain times of the year, the ozone hole over Antarctica is about the size of the continental United States.

🍃 A polystyrene cup, when broken down, can add a billion CFC molecules to the atmosphere—and destroy a hundred trillion molecules of ozone. If production of CFCs stopped today, scientists estimate that it would take 50–100 years for the ozone layer to return to normal.

72

UNIT 1

More About the Issues

Human/Environment Interaction Much of what scientists know about the ozone layer comes from data obtained during space flights. Indeed, satellite and space-shuttle missions have been used to study the ozone layer and the factors that affect it. Some scientists now claim that these very missions may have contributed to ozone-layer damage. The burning of rocket fuel releases chemicals that may destroy ozone. Each space-shuttle mission, one scientist has claimed, destroys 0.25 percent of the ozone in the atmosphere.

...s of the protective ozone blanket is ...eatest over the South Pole. These images ...n by the *Nimbus 7* satellite show the ...wing ozone hole, the black areas appearing ...ctober 1983 and October 1989.

October 1989

October 1983

October 1979

TEEN TRIBUTE

Students at the George C. Soule School in Freeport, Maine, were concerned about the use of fast-food packages made with polystyrene, a plastic that contains CFCs. The students asked the town council to ban polystyrene packaging. A major fast-food chain argued against the ban, but the council sided with the students. Polystyrene packaging was banished from Freeport!

WHAT CAN YOU DO?

🌿 Stop and think...before you use any can of aerosol spray containing CFCs.

🌿 Use popcorn or newspaper for packing material instead of polystyrene pieces.

🌿 If your family owns an air conditioner, have it checked for leaks. They cause 20 percent of CFCs leaked yearly in the United States.

🌿 Organize a "Sun Alert" campaign to warn students of the danger of skin cancer from overexposure to the sun.

TAKE A CLOSER LOOK

1 Why is the ozone layer important to all life on Earth?

2 What manufactured chemicals are damaging the ozone layer? Where are these chemicals found?

good planets are hard to find

A Chinese family bikes home—carrying a new refrigerator charged with ozone-depleting CFCs.

73

ASSESS

Have students answer the **Take A Closer Look** questions on page 73.

CLOSE

Discuss the **What Can You Do?** section with students. Then have them create posters for an "Ozone Alert!" campaign. Display posters in the classroom.

*For an additional regional case study, use the following:

📁 Environmental Case Study 1 **L1**

Strange but TRUE!

Some scientists suggest that the extinction of the dinosaurs may have been caused by a loss of the ozone layer. They theorize that the ozone layer was destroyed by radiation from an exploding star. Ultraviolet radiation became so excessive on Earth that it led to the extinction of the dinosaurs.

Answers to Take A Closer Look
1. It acts as a screen, blocking out the sun's harmful ultraviolet rays.
2. chlorofluorocarbons (CFCs); in coolants in refrigerators and air conditioners, in aerosol sprays, and in cleaners

Global Issues

Interdependence When the ozone layer thins by 10 percent, the amount of harmful ultraviolet radiation reaching the earth jumps by 20 percent.

0:00 OUT OF TIME?

If time does not permit teaching each chapter, you may use the Chapter Highlights, Audiocassettes, and their corresponding activities and tests.

GLENCOE
TECHNOLOGY

 VIDEODISC

Reuters Issues in Geography

Chapter 1
The United States and Canada: The Water Supply
Disc 1 Side A

*inter*NET
CONNECTION

Canada is the world's second-largest country in land area. To find out interesting facts about Canada's geography, click on the address of the Canadian National Atlas on SchoolNet:

World Wide Web
http://www-nais.ccm.emr.ca/ schoolnet/Home.html

Interested in learning about the United States's National Parks? Contact the address of the U.S. National Park Service at:

World Wide Web
http://www.nps.gov/

Unit 2 The United States and Canada

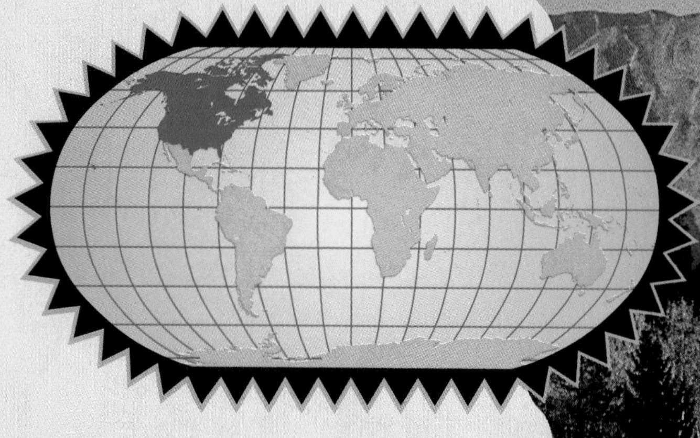

What Makes the United States and Canada a Region?

The people share . . .

- a varied physical environment.

- mostly mid-latitude climates, fertile soils, and adequate rainfall.

- a shift from a manufacturing economy to service industries.

- a fast society—fast foods, fast transportation, fast communication.

- a mixed population drawn from many world cultures.

To find out more about the United States and Canada, see the Unit Atlas on pages 76–87.

Jasper National Park, Canada

(74)

About the Illustration

The Rocky Mountains are just one of the landforms shared by the United States and Canada. This view shows the Rockies in Canada's Jasper Park in the province of Alberta. The park, which covers some 2.5 million acres (1.1 million hectares), has several peaks that rise more than 10,000 feet (3,048 m) above sea level.

 **GEOGRAPHIC THEMES**

Location Are Canada's Rocky Mountains closer to the Atlantic Ocean or the Pacific Ocean? *(Pacific Ocean)*

NATIONAL GEOGRAPHIC SOCIETY

These materials are available from Glencoe.

 CD-ROM

Picture Atlas of the World

 SOFTWARE

ZipZapMap! World

 TRANSPARENCIES

PicturePack Transparencies, Unit 2

GeoJournal Activity

Point out that the local library or historical society can provide information about ethnic groups that settled in the area. Encourage students to look for specific ways in which these immigrants influenced the community. Offer place names, museums, festivals, architectural styles, and ethnic restaurants as examples.

• This journal activity provides the basis for the "GeoJournal Writing Activity" exercise in the Chapter Review.

• The GeoJournal may be used as an integral part of Performance Assessment.

GeoJournal Activity

The United States and Canada are a cultural kaleidoscope—a mixture of the many ethnic groups who have settled in North America. Find out what groups of immigrants settled in your local area. In your journal, write an essay about the cultural influences these people have had on your town or city.

75

Location Activity

Display Political Map Transparency 2 and ask students the following questions: On what continent are the United States and Canada located? *(North America)* What body of water borders these countries on the east? *(Atlantic Ocean)* On the west? *(Pacific Ocean)*

In what hemispheres are these countries located? *(northern, western)* Where is Canada in relation to the contiguous United States? *(directly to the north)* What country borders the United States to the south? *(Mexico)* L1

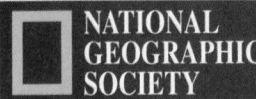

UNIT 2 ATLAS

NATIONAL GEOGRAPHIC SOCIETY

These materials are available from Glencoe.

CD-ROM

Picture Atlas of the World

POSTERS

Images of the World
Unit 2: The United States and Canada

TRANSPARENCIES

PicturePack Transparencies, Unit 2

Strange but TRUE!

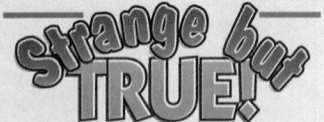

The 3,987-mile (6,416-km) border between the United States and Canada is the longest continuous—and unguarded—frontier in the world.

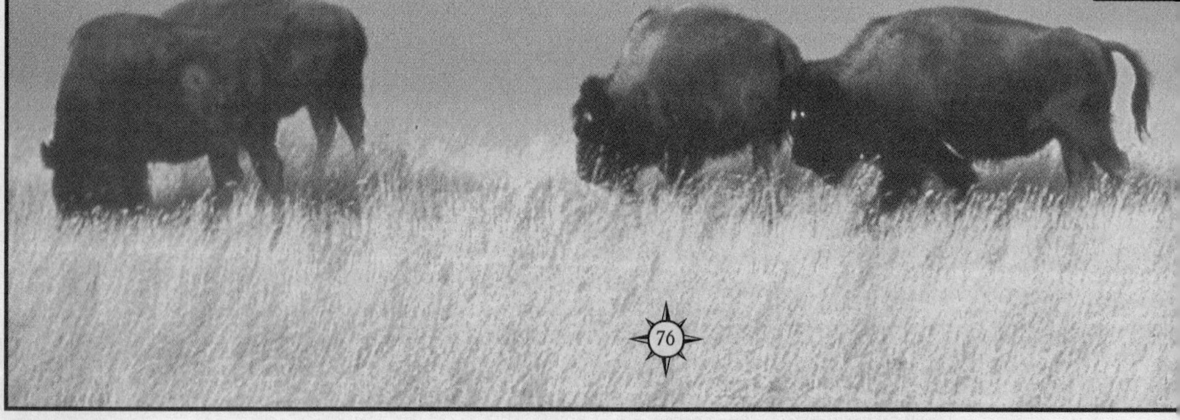

NATIONAL GEOGRAPHIC SOCIETY

IMAGES of the WORLD

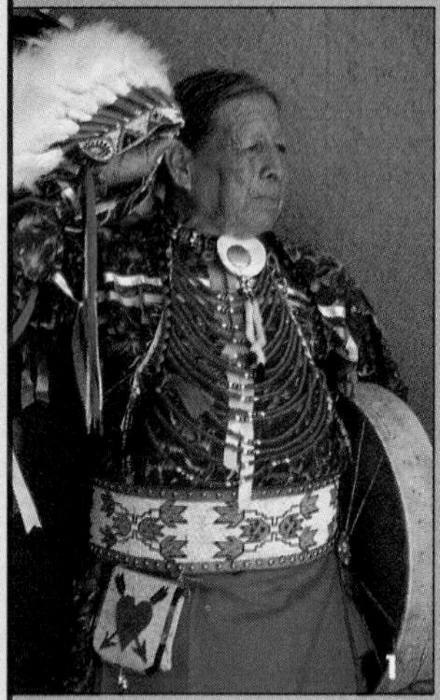

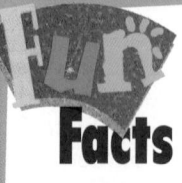

1

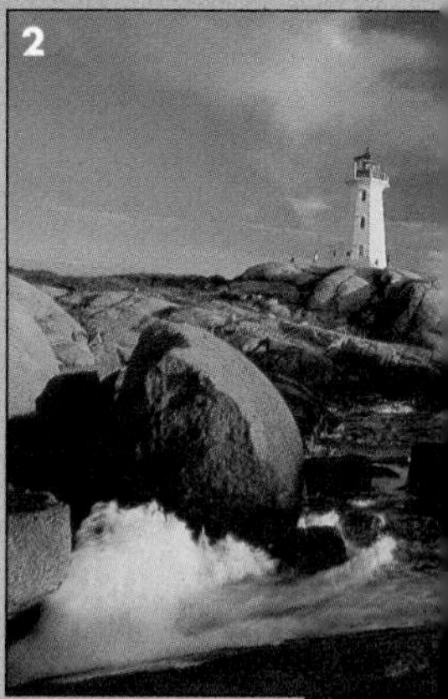

2

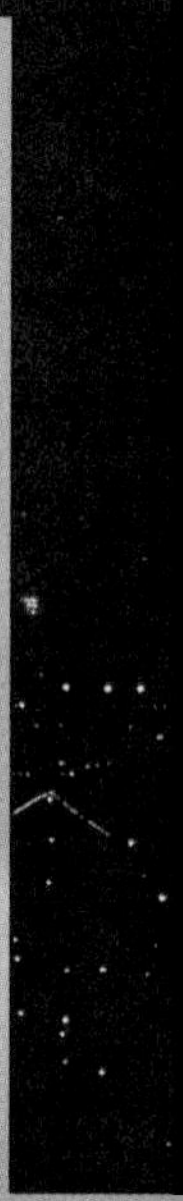

76

Fun Facts

- **Canada** The province of Prince Edward Island is famous for its potatoes. Farmers there grow 70 different kinds.
- **United States and Canada** In a time when most ships use sophisticated navigational tools, the lighthouse still serves an important purpose. Some 600 lighthouses dot the coasts of the United States and Canada, warning ships of navigational dangers.

1. Native American, New Mexico
2. Lighthouse on the Atlantic Ocean, Nova Scotia
3. Boy facing a logging truck, Oregon
4. Outfitter on a camping trip, Alberta
5. Egret in a salt marsh, Louisiana
6. Fireworks over the *Statue of Liberty*, New York City
7. Red gabled buildings, Nova Scotia

All photos viewed against bison grazing, Wyoming.

CURRICULUM CONNECTION

Literature Visitors to Prince Edward Island enjoy seeing locations associated with Lucy Maud Montgomery's novel, *Anne of Green Gables*. Montgomery based the story on her experiences as a child growing up on the island.

Geography and the Humanities

World Literature Reading 2

World Music: Cultural Traditions, Lesson 1

World Cultures Transparencies 1, 2

Focus on World Art Prints 3, 4, 24

CURRICULUM CONNECTION

Life Science The American bison is an imposing sight. A large bull can measure 12 feet (3.7 m) long, stand 6 feet (1.8 m) tall at the shoulder, and weigh as much as 3,000 pounds (1,361 kg).

Teacher Notes

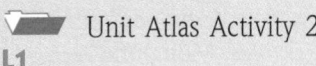

UNIT 2 ATLAS

Regional Focus

The United States and Canada are huge countries. Together they reach from the Arctic Ocean in the north to the country of Mexico in the south. West to east, they stretch from the Pacific Ocean to the Atlantic Ocean. The United States and Canada—with Mexico—form the continent of North America.

Using the Unit Atlas

These features and activities may be used as an introduction to the unit or as teaching tools throughout the course of the unit.

 Unit Atlas Activity 2
L1

 Unit Overlay Transparencies 2-0, 2-1, 2-2, 2-3, 2-4, 2-5, 2-6 L1

Home Involvement Activity 2 L1

Environmental Case Study 2 L1

Southwest Vegetation

The huge saguaro cactus grows only in the desert climates of the southwestern United States and Mexico.
PLACE: Where do you find the only deserts in the United States and Canada?

More About the Illustration

Answer to Caption
just east of the Pacific mountain ranges

GEOGRAPHIC THEMES **Place** Some saguaro grow as high as 60 feet (18 m) and weigh as much as 10 tons (10.2 t). Their trunks measure as much as 2.5 feet (0.8 m) in diameter. Desert insects, birds, and bats feed on the nectar of the saguaro's blossoms.

Region
The Land

The United States and Canada cover more than 7 million square miles (18 million sq. km). They share many of the same landscapes. Mountain ranges, basins, and plateaus run through the western part of both countries. The largest range is the massive Rocky Mountains, stretching more than 3,000 miles (4,800 km) from Alaska to New Mexico.

East of the Rockies spread the broad, rolling Great Plains. This flat area runs through the central part of the United States and Canada. The plains reach to the Canadian Shield—a huge, mineral-rich area of ancient rock—in Canada's far north. Find the Rockies and the Great Plains on the map on page 82.

Low mountains and coastal plains cover the eastern part of the United States and Canada. Here you will discover North America's second-longest mountain range, the Appalachians. East and south of these low mountains lie coastal lowlands that fan out westward to Texas.

Waterways The Mississippi River is the largest river system in North America. It flows from near the United States-Canadian border in the north to the Gulf of Mexico in the south. The Missouri River, the Ohio River, and their tributaries merge into the Mississippi River system. The largest lake system is the Great Lakes—Superior, Huron, Michigan, Erie, and Ontario. The waters of these five lakes are interconnected and eventually meet with the St. Lawrence River, which flows into the Atlantic Ocean.

Place
Climate and Vegetation

The vast size of landforms affects climate, soils, and vegetation in the United States and Canada. The forests and flatlands in far northern Canada and Alaska have short, cool summers and long, cold winters. The Pacific coast from southern Alaska to northern California has a mild, humid climate. Mist-covered forests and lush

(78)

UNIT 2

 • **Canada** About 85 percent of the water that flows over Niagara Falls flows over the Canadian section—Horseshoe Falls. During the tourist season, Canadian authorities reduce the amount of water diverted for hydroelectric power so as not to interfere with the spectacular scenery.

• **United States** The Grand Canyon extends for some 277 miles (446 km) through northwest Arizona. Parts of the canyon, which was carved by the Colorado River, are 1 mile (1.6 km) deep and 18 miles (29 km) wide.

green plants cover this area. Pacific mountain ranges keep ocean rain clouds from reaching inland areas. Just east of the mountains lie the only deserts in the region.

The Great Plains The Great Plains has a humid continental climate of cold winters and hot summers. This region receives rain from the Gulf of Mexico and the far north. Before settlers arrived in the 1800s, grasses covered much of the Great Plains. Now the land is covered with farms and cities.

Other Areas The Northeast region of the United States and Canada has the same climate as the Great Plains. Most of the southern states, however, are in a humid subtropical region that experiences mild winters. Only the tip of Florida is far enough south to have a tropical climate. The Hawaiian Islands—the only other area of the United States with a tropical climate—lie 2,400 miles (3,862 km) from the United States mainland in the Pacific Ocean.

Region
Economy

The United States and Canada base their economies on the free enterprise system. This means that individuals and groups—rather than the government—run businesses. Skilled workers, many natural resources, and the use of advanced technology have brought prosperity to the United States and Canada.

Agriculture United States and Canadian farmers raise many meats, grains, vegetables, and fruits. Up-to-date farming methods have increased the size of farms and the amount of food produced. At the same time, they have decreased the number of farms and farm workers. Fertile soil and many waterways make the Great Plains, the Pacific coast, and the Northeast major agricultural areas.

The Pacific Coast

The city of Vancouver on Canada's Pacific coast is the country's busiest port.
PLACE: On what does Canada base its economy?

(79)

NATIONAL GEOGRAPHIC SOCIETY

These materials are available from Glencoe.

 CD-ROM

Picture Atlas of the World

 SOFTWARE

ZipZapMap! USA, World

 POSTERS

Eye on the Environment, Unit 2

More About the Illustration

Answer to Caption
free enterprise system

 **Movement** Vancouver, located just 25 miles (40 km) north of the state of Washington, began as a small sawmill town in 1865. Today, it is a bustling international port. All of Canada's trade with Japan and other Asian nations flows through Vancouver.

• **United States** The Trans-Alaska pipeline carries oil some 800 miles (1,300 km) from the Prudhoe Bay oil fields on the Arctic coast to Valdez on the Pacific coast. The pipeline crosses areas of permafrost, earthquake zones, and caribou migration routes.

• **Canada** Parts of Baffin and Ellesmere islands have permanent ice caps.

Cultural Kaleidoscope

Canada Canadians call Alberta, Saskatchewan, and Manitoba, which are located on the Great Plains, the *Prairie Provinces. Prairie* is a French word that means "grassy field" or "meadow."

LESSON PLAN
Unit 2 Atlas

FactsOnFile

Use the reproducible masters from **GEOGRAPHY ON FILE,** *Northern America,* to enrich unit content.

MULTICULTURAL PERSPECTIVE

Cultural Heritage

The name *Nova Scotia* is Latin for "New Scotland." In fact, life in this Canadian province shows a strong Scottish influence. A highlight of the province's annual Highland folk arts festival is the Highland Games, which include Scottish competitions such as tossing the caber (a heavy wooden pole).

More About the Illustration

Answer to Caption

Answers will vary; see explanation below.

Region

The northern lights are also called the *aurora borealis.* Auroras are caused by electronically charged particles from the sun that are drawn toward the North and South poles by the earth's magnetic field. Energy is released in the form of light when these particles strike atoms and molecules in the earth's atmosphere. Auroras that occur in the Southern Hemisphere are called *aurora australis.*

Industry The United States and Canada are rich in such mineral resources as copper, iron ore, nickel, silver, oil, natural gas, and coal. Both countries are world leaders in manufactured goods. Their service industries—stores, banking, insurance, and entertainment—provide jobs for the largest number of people.

Trade Leaders in world trade, Americans and Canadians are linked to the rest of the globe through roads, railroads, airlines, computers, television, and telephones. In 1993 the United States, Canada, and Mexico approved the North American Free Trade Agreement (NAFTA). This agreement allows goods and money to move freely among the three countries. Other major trading partners include the countries of western Europe, Japan, Taiwan, and South Korea.

Movement
People

Throughout their history, the United States and Canada have become home to settlers from almost every part of the world. Some of these people came to find a better way of life. Still others were forced to come as enslaved labor.

Ethnic Groups Today the people of this region come from many different backgrounds. Most Americans and Canadians are European in origin. Others are African Americans, Hispanics, Asians, and Native Americans. English is the major language of the United States, although many Americans speak other languages. English is also the primary language for most Canadians, but French is the dominant language of Quebec Province.

Population About 290 million people live in the United States and Canada. In land area, Canada is the second-largest country in the world. Yet most of the nearly 29 million Canadians live in urban areas within 100 miles (161 km) of the United States border. The remaining two-thirds of Canada has very few people.

The United States is the world's third-largest country in population. Most of the 261 million Americans live in urban areas along the coasts or in the Great Lakes region. Among the largest cities are New York City, Los Angeles, Chicago, and Dallas. The area from Washington, D.C., to Boston is almost continuously urban and is called a megalopolis.

Place
Culture

Throughout their histories, Americans and Canadians have valued freedom. They have given many gifts to the world in such areas as technology, science, and the arts.

At least 12,500 years ago, Native Americans settled the area that is now the United States and Canada. Spanish, French, and British settlers arrived between the 1500s and the 1700s. As these Europeans and their American descendants

Nature's Light Show

The colorful glow of the northern lights flickers in Alaska and northern Canada.
REGION: What do you think causes the northern lights?

Fun Facts

• **Canada** The Reversing Falls of St. John is a waterfall near the mouth of the St. John River in New Brunswick. Twice each day, high tides from the Bay of Fundy force water to flow upstream up the falls!

• **United States** Mammoth Cave in central Kentucky is the world's most extensive cave system. Visitors can wander through 12 miles (19 km) of corridors on 5 levels. The lowest level is 360 feet (110 m) below the surface.

pushed westward, they wiped out many of the Native Americans living on the land.

Independence

During the late 1700s, colonists along the Atlantic coast successfully fought the British for their freedom. They founded the United States of America. In the 1800s the new Americans spread westward, built railroads, fought in a great Civil War, and then reunited. Immigrants arrived by the thousands. They helped transform the United States economy from an agricultural one to an industrial one. During the 1900s, the United States fought two world wars and became a major world power.

Canada, on the other hand, stayed apart from the United States. It gradually and peacefully won its independence from the British during the 1800s and early 1900s. As in the United States, many immigrants made their homes in Canada. Although differences between English-speaking and French-speaking Canadians threatened national unity, the people moved westward and developed their economy. In the 1900s Canadians began to play an active role in world affairs.

The Canadian West

The chuckwagon races are a popular event at the world-famous Calgary Stampede in Alberta.
REGION: Why do you think the western United States and western Canada have a similar history?

Technology Activity

Using the Internet
Search the Internet for information about settlers on the Great Plains during the 1800s and early 1900s. Using this information, write a story that describes the daily activities of a teenager living on the Great Plains about 1890 or 1900. Specify how the teenager and his or her family were affected by their new environment.

REVIEW AND ACTIVITIES

1. **REGION** What two major landforms do the United States and Canada share?
2. **PLACE** Why are deserts found in areas of the western United States?
3. **HUMAN/ENVIRONMENT INTERACTION** How has the use of modern methods affected farming in the United States and Canada?
4. **MOVEMENT** Why did people settle in the United States and Canada?
5. **LOCATION** Where do most Canadians live?

81

Answers to Review and Activities

1. mountains and plains
2. Pacific mountain ranges block rain clouds moving off the ocean.
3. They have increased the size of farms and the amount of food produced; they have decreased the number of farms and farm workers.
4. They came by choice to find a better way of life, fled bad conditions in their homelands, or were enslaved as laborers.
5. They live in urban areas close to the United States border.

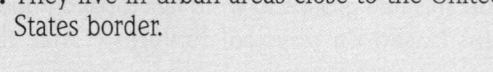

Physical Geography

Applying Geographic Themes

Place The Great Lakes contain about one-fifth of the earth's freshwater supply. Name three other large freshwater lakes located in Canada. *(Lake Winnipeg, Great Slave Lake, Great Bear Lake)*

NATIONAL GEOGRAPHIC SOCIETY

These materials are available from Glencoe.

 CD-ROM

- **Picture Atlas of the World**
- **PictureShow: Earth's Endangered Environments, Geology**

 SOFTWARE

ZipZapMap! USA, World

Answers
1. Great Plains
2. Appalachian Mountains

Map Skills Practice
Reading a Map What is the region's highest mountain? In what mountain range is it located? *(Mt. McKinley; Alaskan Range)*

THE UNITED STATES AND CANADA: Physical

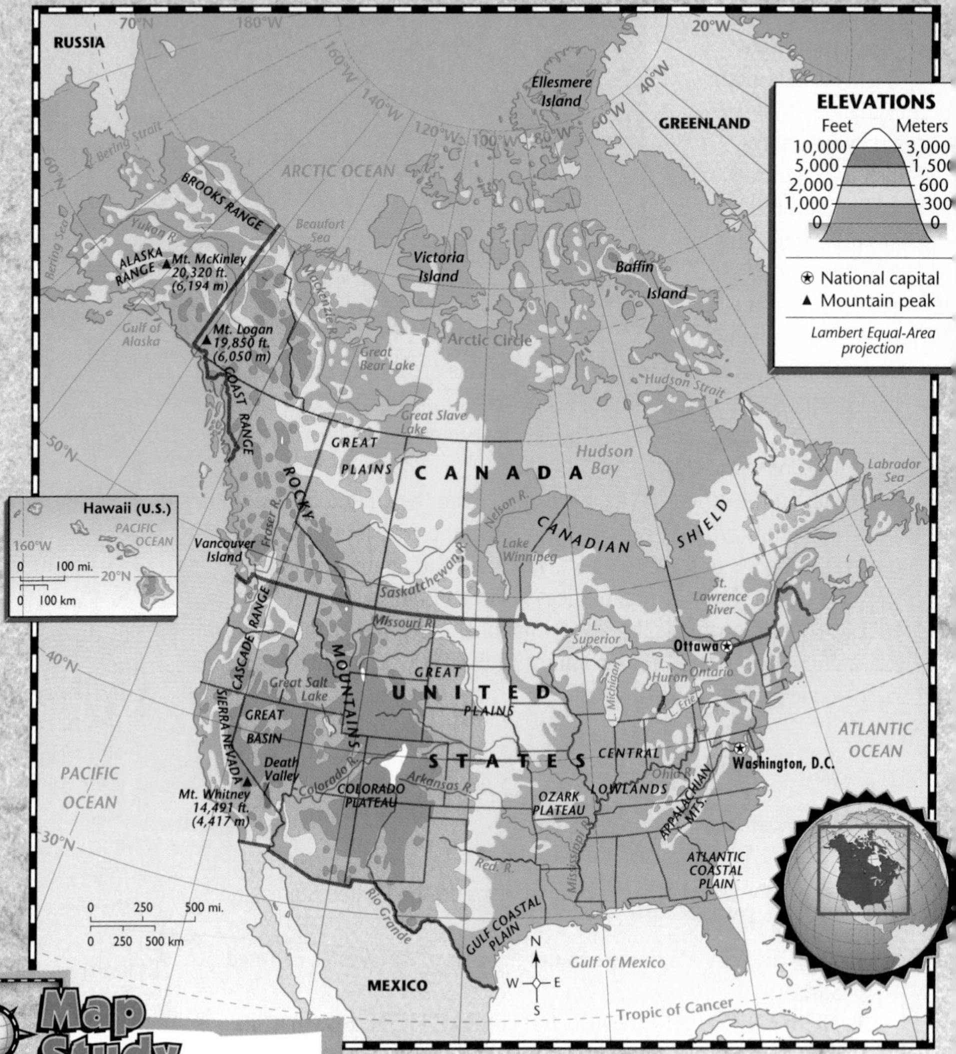

ELEVATIONS

Feet	Meters
10,000	3,000
5,000	1,500
2,000	600
1,000	300
0	0

⊛ National capital
▲ Mountain peak

Lambert Equal-Area projection

RUSSIA
GREENLAND
ARCTIC OCEAN
Ellesmere Island
BROOKS RANGE
Beaufort Sea
ALASKA RANGE ▲ Mt. McKinley 20,320 ft. (6,194 m)
▲ Mt. Logan 19,850 ft. (6,050 m)
Victoria Island
Baffin Island
Gulf of Alaska
Arctic Circle
COAST RANGE
Great Bear Lake
Hudson Strait
Great Slave Lake
GREAT PLAINS
CANADA
Hudson Bay
CANADIAN SHIELD
Labrador Sea

Hawaii (U.S.)
PACIFIC OCEAN
0 100 mi.
0 100 km
Vancouver Island
ROCKY
Nelson R.
Lake Winnipeg
St. Lawrence River
CASCADE RANGE
Saskatchewan R.
Missouri R.
L. Superior
Ottawa ⊛
SIERRA NEVADA
Great Salt Lake
MOUNTAINS
GREAT PLAINS
UNITED
Lake Michigan
L. Huron L. Ontario
ATLANTIC OCEAN
GREAT BASIN
Death Valley
STATES
CENTRAL LOWLANDS
Ohio R.
APPALACHIAN MTS.
⊛ Washington, D.C.
Mt. Whitney 14,491 ft. (4,417 m)
Colorado R.
COLORADO PLATEAU
Arkansas R.
OZARK PLATEAU
PACIFIC OCEAN
Rio Grande
Red R.
Mississippi R.
ATLANTIC COASTAL PLAIN
GULF COASTAL PLAIN
MEXICO
Gulf of Mexico
Tropic of Cancer

0 250 500 mi.
0 250 500 km

N W E S

1. **REGION** What physical region covers much of the central part of the United States and Canada?
2. **PLACE** What mountain range runs along the eastern coast of the United States and Canada?

82

Atlas Activity

Location Allow time for students to study the physical features of the United States and Canada shown on the map above. Then have them work in small groups to write five location questions based on physical features. Offer the following question as an example: What mountain range extends through the western region of both the United States and Canada? *(Rocky Mountains)* Encourage groups to challenge each other with their questions. **L1**

ELEVATION PROFILES

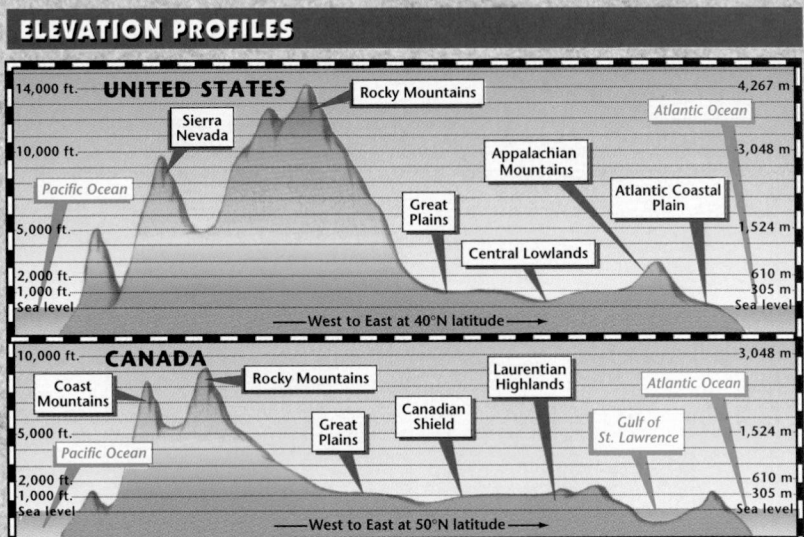

UNITED STATES

14,000 ft. — 4,267 m

Rocky Mountains

Atlantic Ocean

Sierra Nevada

10,000 ft. — 3,048 m

Appalachian Mountains

Pacific Ocean

Atlantic Coastal Plain

Great Plains

5,000 ft. — 1,524 m

Central Lowlands

2,000 ft.
1,000 ft. — 610 m
Sea level — 305 m
Sea level

← West to East at 40°N latitude →

CANADA

10,000 ft. — 3,048 m

Coast Mountains

Rocky Mountains

Laurentian Highlands

Atlantic Ocean

Great Plains

Canadian Shield

Gulf of St. Lawrence

5,000 ft. — 1,524 m

Pacific Ocean

2,000 ft.
1,000 ft. — 610 m
Sea level — 305 m
Sea level

← West to East at 50°N latitude →

Source: *Goode's World Atlas*, 19th edition

THE UNITED STATES AND CANADA: Land Comparison

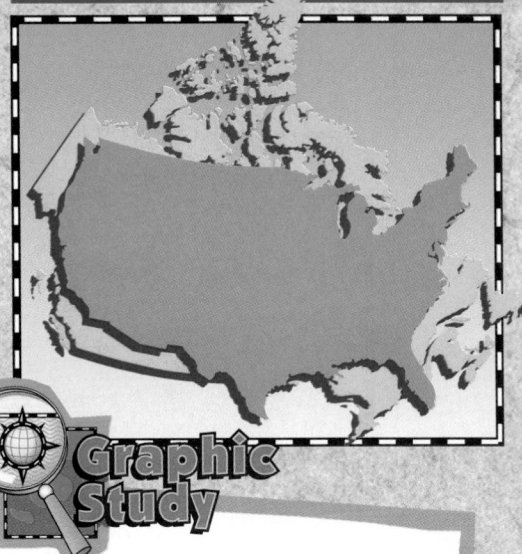

GeoFacts

Highest point: Mount McKinley (Alaska) 20,320 ft. (6,194 m) high

Lowest point: Death Valley (California) 282 ft. (86 m) below sea level

Longest river: Mackenzie River (Canada) 2,635 mi. (4,240 km) long

Largest lake: Lake Superior 31,700 sq. mi. (82,103 sq. km)

Highest waterfall: Yosemite Falls (California) 2,425 ft. (739 m) high

 Graphic Study

1. **LOCATION** In what general area—east or west—are the highest elevations in the United States and Canada found?
2. **PLACE** How would you describe the size of the United States compared with the size of Canada?

83

GLENCOE TECHNOLOGY

 VIDEODISC

The Infinite Voyage: To the Edge of the Earth

Chapter 6
Exploring Life in the Canadian Arctic
Disc 1 Side B

Chapter 7
Eskimo Survival in the Arctic
Disc 1 Side B

FactsOnFile

Use the reproducible masters from **GEOGRAPHY ON FILE**, *Northern America*, to enrich unit content.

GRAPHIC STUDY

Answers
1. west
2. The United States is slightly smaller than Canada.

Graphic Skills Practice
Reading a Graph In which country are elevations highest? *(United States)* What fertile plain does Canada share with the United States? *(Great Plains)*

Teacher Notes

LESSON PLAN
Unit 2 Atlas

Cultural Geography

Applying Geographic Themes

Location Washington, D.C., the capital of the United States, is located between the states of Maryland and Virginia. Describe the location of Canada's capital. *(in southeastern Ontario, near the border with Quebec)*

NATIONAL GEOGRAPHIC SOCIETY

These materials are available from Glencoe.

 VIDEODISC

STV: North America

|||||||||||||

The Rocky Mountains
Side 3, Frames 0020-22832

 **MAP STUDY**

Answers
1. Maine
2. Ottawa

Map Skills Practice
Reading a Map Which Canadian province lies farthest to the south? *(Ontario)* Which state in the United States lies farthest to the north? *(Alaska)*

THE UNITED STATES AND CANADA: Political

RUSSIA

Ellesmere Island

GREENLAND

ARCTIC OCEAN

National boundary
State/provincial boundary
⊛ National capital

Lambert Equal-Area projection

Alaska (U.S.)

Yukon Territory

Arctic Circle

Northwest Territories

Hudson Bay

Newfoundland

British Columbia

Alberta

C A N A D A

Manitoba

Quebec

New Brunswick

Prince Edward Is.

Saskatchewan

Ontario

Hawaii (U.S.)

Washington

Oregon

Montana

North Dakota

Minn.

Ottawa ⊛

Vt. Me.

Nova Scotia

N.H.

Idaho

Wyoming

South Dakota

Wis.

Mich.

N.Y.

Mass.
R.I.
Conn.

Nevada

Utah

Colorado

Nebraska

Iowa

Ill. Ind. Ohio

Pa.

N.J.
Md.
Del.
Washington, D.C.

ATLANTIC OCEAN

PACIFIC OCEAN

California

Kansas

Missouri

Ky.

W.Va. Va.

N.C.

Arizona

New Mexico

Oklahoma

Ark.

Tenn.

Miss. Ala. Ga.

S.C.

U N I T E D   S T A T E S

Texas

La.

Fla.

Gulf of Mexico

MEXICO

Tropic of Cancer

N W E S

0 250 500 mi.
0 250 500 km

160°W
0 100 mi.
0 100 km
20°N

 **Map Study**

1. **LOCATION** What U.S. state lies farthest east?
2. **PLACE** What is the capital of Canada?

Atlas Activity

Movement Have students use atlases and other resources to plan a cross-country trip through the United States or Canada. Direct them to choose a definite starting point and destination and plot the route between the two points. Have them calculate the approximate distance of the journey and list points of interest along the way. Allow time for students to share their itineraries. **L1**

THE UNITED STATES AND CANADA: Ethnic Groups

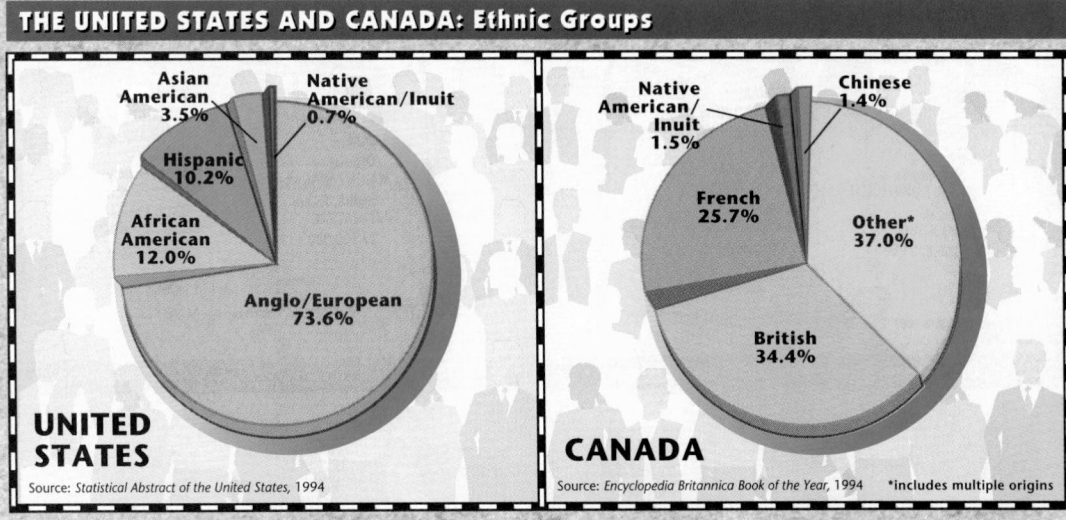

Asian American 3.5%
Native American/Inuit 0.7%
Hispanic 10.2%
African American 12.0%
Anglo/European 73.6%

UNITED STATES

Source: *Statistical Abstract of the United States*, 1994

Native American/Inuit 1.5%
Chinese 1.4%
French 25.7%
Other* 37.0%
British 34.4%

CANADA

Source: *Encyclopedia Britannica Book of the Year*, 1994   *includes multiple origins

Geography and the Humanities

📂 World Literature Reading 2

🌐 World Music: Cultural Traditions, Lesson 1

🔲 World Cultures Transparencies 1, 2

📜 Focus on World Art Prints 3, 4, 24

GeoFacts

Biggest country (land area): Canada 3,849,674 sq. mi. (9,970,610 sq. km)

Smallest country (land area): United States 3,539,230 sq. mi. (9,166,606 sq. km)

Largest city (population): New York City (1995) 14,638,000; (2000 projected) 14,648,000

Highest population density: United States 74 people per sq. mi. (28 people per sq. km)

Lowest population density: Canada 8 people per sq. mi. (3 people per sq. km)

COMPARING POPULATION: Canada and the United States

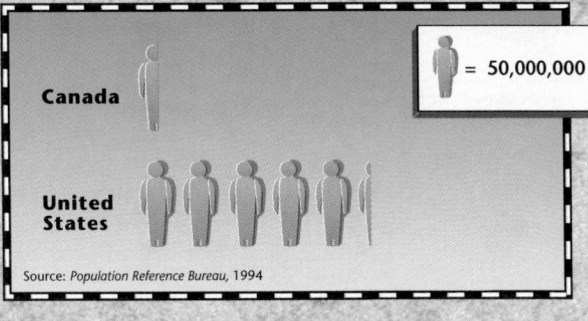

👤 = 50,000,000

Canada

United States

Source: *Population Reference Bureau*, 1994

Graphic Study

1. **PLACE** What is the population of Canada?
2. **MOVEMENT** Which country has a larger Native American population?

MULTICULTURAL PERSPECTIVE

Cultural Heritage

The United States and Canada are sometimes called *Anglo America.* This is because many of the earliest European settlers in the two countries came from England. Also, in both countries English-speakers make up a majority of the population.

GRAPHIC STUDY

Answers
1. about 29 million people
2. Canada

Graphic Skills Practice
Reading a Graph About how many more people live in the United States than in Canada? *(about 230 million)* Which is the larger ethnic group in Canada, the British or the French? *(British)*

Teacher Notes

LESSON PLAN
Unit 2 Atlas

UNIT 2 ATLAS
Countries at a Glance

Cultural Kaleidoscope

Canada Boxing Day is a holiday that Canadians celebrate on December 26. It comes from the British tradition of giving small boxed gifts to service workers.

Global Gourmet

United States The Cajuns of Louisiana enjoy many spicy dishes. Favorites include jambalaya, a rice dish cooked with ham, sausage, chicken, and shrimp or oysters; and gumbo, a thick soup of vegetables and meat or seafood.

Geography and the Humanities

📁 World Literature Reading 2

🌐 World Music: Cultural Traditions, Lesson 1

📦 World Cultures Transparencies 1, 2

📖 Focus on World Art Prints 3, 4, 24

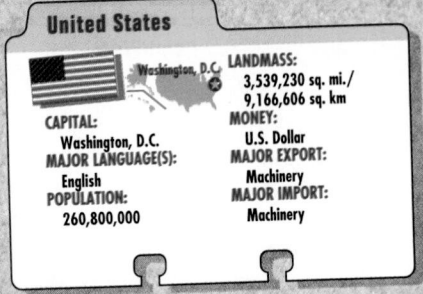

United States
CAPITAL: Washington, D.C.
MAJOR LANGUAGE(S): English
POPULATION: 260,800,000
LANDMASS: 3,539,230 sq. mi./ 9,166,606 sq. km
MONEY: U.S. Dollar
MAJOR EXPORT: Machinery
MAJOR IMPORT: Machinery

Canada
CAPITAL: Ottawa
MAJOR LANGUAGE(S): English, French
POPULATION: 29,100,000
LANDMASS: 3,849,674 sq. mi./ 9,970,610 sq. km
MONEY: Canadian Dollar
MAJOR EXPORT: Motor Vehicles
MAJOR IMPORT: Motor Vehicles

U.S. State Name Meaning and Origin

ALABAMA — Montgomery — "thicket clearers" (Choctaw)

ALASKA — Juneau — "the great land" (Aleut)

ARIZONA — Phoenix — "little spring" (Papago), or "dry land" (Spanish)

ARKANSAS — Little Rock — "downstream people" (Quapaw)

CALIFORNIA — Sacramento — unknown meaning (Spanish)

COLORADO — Denver — "red" (Spanish)

CONNECTICUT — Hartford — "beside the long tidal river" (Native American)

DELAWARE — Dover — named for Virginia's colonial governor, Baron De La Warr

FLORIDA — Tallahassee — "feast of flowers" (Spanish)

GEORGIA — Atlanta — named for England's King George II

HAWAII — Honolulu — unknown meaning (native Hawaiian)

IDAHO — Boise — unknown meaning (Native American)

ILLINOIS — Springfield — "tribe of superior men" (Native American)

INDIANA — Indianapolis — "land of Indians" (European American)

IOWA — Des Moines — unknown meaning (Native American)

KANSAS — Topeka — "people of the south wind" (Sioux)

KENTUCKY — Frankfort — "land of tomorrow" (Iroquoian)

LOUISIANA — Baton Rouge — named for France's King Louis XIV

MAINE — Augusta — named for an ancient French province

MARYLAND — Annapolis — named in honor of the wife of England's King Charles II

MASSACHUSETTS — Boston — "great mountain place" (Native American)

MICHIGAN — Lansing — "great lake" (Ojibway)

MINNESOTA — Saint Paul — "sky-tinted water" (Sioux)

MISSISSIPPI — Jackson — "father of the waters" (Native American)

MISSOURI — Jefferson City — "town of the large canoes" (Native American)

MONTANA — Helena — "mountainous" (Spanish)

NEBRASKA — Lincoln — "flat water" (Native American)

NEVADA — Carson City — "snowcapped" (Spanish)

NEW HAMPSHIRE — Concord — named for Hampshire, a county in England

NEW JERSEY — Trenton — named for Isle of Jersey off the coast of England

Countries/States/Provinces not drawn to scale.

UNIT 2

Fun Facts

- **United States** The Cajuns of Louisiana are the descendants of French Canadians who were driven from the colony of Acadia—now Nova Scotia— by the British after the French and Indian War. *Cajun* is a slurred version of the word *Acadian*.

- **Canada** The original name of Regina, the capital of Saskatchewan, was Pile o' Bones. It was so named because a large number of buffalo skeletons were found nearby.

- **United States** Giant sequoia trees once grew over much of the Northern Hemisphere. Today, these trees grow only on the western slopes of the Sierra Nevada in California.

U.S. State Name Meaning and Origin

 **NEW MEXICO** ★ Santa Fe
named for the state's former colonial ruler, Mexico

 **NEW YORK** ★ Albany
named in honor of the English Duke of York

 **NORTH CAROLINA** ★ Raleigh
named in honor of England's King Charles I

 **NORTH DAKOTA** ★ Bismarck
named for the Dakota, a Native American group

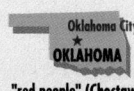 **OHIO** Columbus ★
"great river" (Native American)

 OKLAHOMA Oklahoma City ★
"red people" (Choctaw)

OREGON ★ Salem
own meaning and origin

 PENNSYLVANIA Harrisburg ★
"Penn's woodland" named for the father of Pennsylvania's founder, William Penn

RHODE ISLAND Providence ★
unknown meaning and origin

 **SOUTH CAROLINA** Columbia ★
named for England's King Charles I

 **SOUTH DAKOTA** ★ Pierre
named for the Dakota, a Native American group

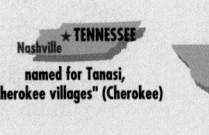 **TENNESSEE** ★ Nashville
named for Tanasi, "Cherokee villages" (Cherokee)

 **TEXAS** ★ Austin
"friends" (Tejas)

UTAH ★ Salt Lake City
"people of the ountains" (Ute)

VERMONT ★ Montpelier
"green mountain" (French)

VIRGINIA Richmond ★
named for the unmarried Queen Elizabeth I of England, known as "the Virgin Queen"

WASHINGTON ★ Olympia
named in honor of George Washington

WEST VIRGINIA Charleston ★
began as the western part of Virginia before becoming a state in 1863

WISCONSIN Madison ★
"grassy place" (Chippewa)

WYOMING Cheyenne ★
"upon the great plain" (Delaware)

Canadian Province Name Meaning and Origin

 **ALBERTA** Edmonton ★
named for the daughter of England's Queen Victoria

 **BRITISH COLUMBIA** Victoria ★
named for Christopher Columbus and the province's British heritage

 **MANITOBA** Winnipeg ★
"strait of the great spirit" (Algonquian)

 **NEW BRUNSWICK** Fredericton ★
named for English royal family of Brunswick-Luneburg

 **NEWFOUNDLAND** St. John's ★
"new found land," named by European explorer John Cabot in 1497

 **NOVA SCOTIA** Halifax ★
Latin term for "New Scotland," based on province's Scottish heritage

 **ONTARIO** Toronto ★
meaning unknown (Iroquoian)

 **PRINCE EDWARD ISLAND** Charlottetown ★
named for the son of England's King George III

 **QUEBEC** Quebec ★
"place where the river narrows" (Algonquian)

 **SASKATCHEWAN** Regina ★
"fast flowing river" (Cree)

 87

 NATIONAL GEOGRAPHIC SOCIETY

These materials are available from Glencoe.

SOFTWARE

ZipZapMap! USA, World

Cultural Kaleidoscope

United States The Iditarod Trail Dog Sled Race is an annual competition run from Anchorage to Nome in Alaska. The race commemorates a 1925 event when medicine was rushed by dog sled between the two cities.

MULTICULTURAL PERSPECTIVE

Cultural Heritage

Canada's first national game was lacrosse, a game played by Native Americans for centuries before Europeans arrived in North America. Today, ice hockey is by far the country's most popular sport.

Teacher Notes

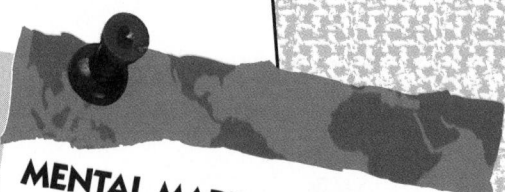

PERFORMANCE ASSESSMENT ACTIVITY

CREATING A PORTFOLIO Have students create a portfolio of materials that illustrates the geography of the United States. Suggest that students find for their portfolios at least 10 items that illustrate geographic features—natural resources, urban areas, population, landforms, climate, tourist attractions, agriculture, industry, and so on—in the United States. Encourage them to search through old newspapers and magazines and to watch television documentaries and news programs. Mention that materials may be either fiction or non-fiction, but must be clearly labeled. Newspaper and magazine articles may be clipped, but other sources may be summarized and cited. Students should write a few lines explaining how each item relates to the five themes of geography. Suggest that students arrange the materials in their portfolios according to region. Have students share and compare their portfolios.

POSSIBLE RUBRIC FEATURES:

Concept attainment, analysis skills, organization skills, comprehension skills, ability to identify relationships

MENTAL MAPPING ACTIVITY

Before teaching Chapter 4, point out the United States on a globe. Have students describe the location of the country relative to other countries in North America, relative to major bodies of water, and in terms of latitude and longitude.

TEACHER'S CORNER

NATIONAL GEOGRAPHIC SOCIETY

INDEX TO NATIONAL GEOGRAPHIC MAGAZINE

The following articles may be used for research relating to this chapter:

- "National Wildlife Refuges," by Douglas H. Chadwick, October 1996.
- "Hawk High Over Four Corners," by T.H. Watkins, September 1996.
- "California Desert Lands," by Michael Parfit, May 1996.
- "Heart of the Hudson," by Patrick Smith, March 1996.
- "Tex-Mex Border," by Richard Conniff, February 1996.
- "The Old Ones of the Southwest," by David Roberts, April 1996.
- "Utah," by Donovan Webster, January 1996.
- "A Farming Revolution," by Verlyn Klinkenborg, December 1995.
- "The Cherokee," by Geoffrey Norman, May 1995.
- *1491: America Before Columbus,* A National Geographic Special Edition, October 1991.

NATIONAL GEOGRAPHIC SOCIETY PRODUCTS AVAILABLE FROM GLENCOE

To order the following products for use with this chapter, contact your local Glencoe sales representative or call Glencoe at 1-800-334-7344:

- *GTV: A Geographic Perspective on American History* (Videodisc)
- *GTV: The American People: Fabric of a Nation* (Videodisc)
- *STV: North America* (Videodisc)
- *ZipZapMap! USA* (Software)
- *Picture Atlas of the World* (CD-ROM)
- *Native Americans, Part I* (CD-ROM)
- *Native Americans, Part II* (CD-ROM)
- *Native Americans: Eastern Woodlands and Plains* (Transparencies)
- *Native Americans: Southwest, Northwest Coast, Arctic* (Transparencies)
- *Geography of North America* (Transparencies)

ADDITIONAL NATIONAL GEOGRAPHIC SOCIETY PRODUCTS

To order the following products for use with this chapter, call National Geographic Society at 1-800-368-2728:

- *United States Geography Series* (10 Videos)
- *Physical Geography of North America Series* (6 Videos)
- *Great Lakes, Fragile Seas* (Video)
- *Chesapeake Borne* (Video)
- *More Than Maps: A Look at Geography* (Filmstrip)
- *Digging Up America's Past* (Filmstrip)

TEACHER-TO-TEACHER
Curriculum Connection: Social Studies

—from Pamela Francis
Gilbert Junior High School, Gilbert, WV

TURN THOSE LETTERS! *The purpose of this activity is to encourage students to review material for a test by having them take part in a classroom quiz game.* Prepare about 30 questions of varying difficulty on chapter content. Also prepare a number of clues and solutions. For example, one clue might be "a body of water," with the solution "Mississippi River" or "Pacific Ocean." Write solutions on large pieces of card or sheets of paper, one letter per card or sheet.

Organize the class into two teams of equal number and name a student as captain of each team. Display the first solution, face down, on a wall. Read out the clue. Toss a coin to see which team will start. Ask the winning team a question. Allow team members to confer for 30 seconds, and then have the captain offer an answer. If the team cannot answer within the time period, or offers an incorrect answer, the question passes to the other team. The team that answers the question correctly may choose a letter of the alphabet. If that letter occurs anywhere in the displayed solution, turn it face up. Continue the game by asking the other team a question. If the team gives a wrong solution, it misses a turn at answering a question. Award one point for each correct answer given, and five points for each correct solution given. At the end of the set time, count the points and award members of the winning team a bonus point on the test.

MEETING NATIONAL STANDARDS

Geography for Life

All of the 18 standards are demonstrated in Unit 2. The following ones are highlighted in this chapter:
- Standards 4, 10, 11, 14, 16

KEY TO ABILITY LEVELS

Teaching strategies have been coded for varying learning styles and abilities.

L1   **BASIC** activities for all students

L2   **AVERAGE** activities for average to above-average students

L3   **CHALLENGING** activities for above-average students

LEP   **LIMITED ENGLISH PROFICIENCY** activities

BIBLIOGRAPHY

Readings for the Student

Gordon, Patricia, and Reed C. Snow. *Kids Learn America: Bringing Geography to Life With People, Places and History.* Indianapolis: Williamson Publishing Company, 1992.

Henwood, Doug. *The State of the U.S.A. Atlas: The Changing Face of American Life in Maps and Graphics.* New York: Simon & Schuster, 1994.

Smith, Kathie Billingslea. *United States Atlas for Young People.* Mahwah, N.J.: Troll Associates, 1991.

Readings for the Teacher

Exploring Regions of the United States. Indianapolis: George H. Cram, 1992. Maps, activities, worksheets.

The United States Today: An Atlas of Reproducible Pages. Wellesley, Mass.: World Eagle, 1990.

Multimedia

The Regions of the United States. Niles, Ill.: United Learning, 1993, 1994, 1995. 5 videocassettes, 18 minutes each.

U.S. Atlas, revised ed. Novato, Calif.: The Software Toolworks, 1994. Guide; CD-ROM for Macintosh or MPC (Windows); Macintosh or IBM (Windows) 3.5" disk.

Where in the USA is Carmen Sandiego? Novato, Calif.: Broderbund, 1993. Guide; IBM, Macintosh 3.5" disk; Apple 3.5" or 5.25" disk

CHAPTER 4 RESOURCES

Use these *Geography: The World and Its People* resources to teach, reinforce, and extend chapter content.

📁 *Vocabulary Activity 4

📁 Cooperative Learning Activity 4

📁 Workbook Activity 4

📁 Geography Skills Activity 4

📁 GeoLab Activity 4

📁 Chapter Map Activity 4

📁 Performance Assessment Activity 4

📁 *Reteaching Activity 4

📁 Enrichment Activity 4

📁 Chapter 4 Test, Form A and Form B

🎧 📁 *Chapter 4 Digest Audiocassette, Activity, Test

Unit Overlay Transparencies 2-0, 2-1, 2-2, 2-3, 2-4, 2-5, 2-6

Political Map Transparency 2; World Cultures Transparency 1

Geoquiz Transparencies 4-1, 4-2

Vocabulary PuzzleMaker Software

Student Self-Test: A Software Review

Testmaker Software

MindJogger Videoquiz

Focus on World Art Prints 3, 4, 24

If time does not permit teaching the entire chapter, summarize using the Chapter 4 Highlights on page 107, and the Chapter 4 English (or Spanish) Audiocassettes. Review students' knowledge using the Glencoe MindJogger Videoquiz. *Also available in Spanish

Use these *Geography: The World and Its People* resources to teach and reinforce section content.

SECTION 1 RESOURCES

- Reproducible Lesson Plan 4-1
- Section Focus Transparency 4-1
- Guided Reading Activity 4-1
- Section Quiz 4-1 (also available in Spanish)

SECTION 2 RESOURCES

- Reproducible Lesson Plan 4-2
- Section Focus Transparency 4-2
- Guided Reading Activity 4-2
- Section Quiz 4-2 (also available in Spanish)

SECTION 3 RESOURCES

- Reproducible Lesson Plan 4-3
- Section Focus Transparency 4-3
- Guided Reading Activity 4-3
- Section Quiz 4-3 (also available in Spanish)

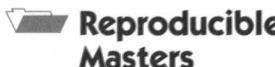

ADDITIONAL RESOURCES FROM GLENCOE

Reproducible Masters

- Glencoe Social Studies Outline Map Book, pages 5–23
- Facts On File, MAPS ON FILE II, U.S. States
- Foods Around the World, Unit 2

World Music: Cultural Traditions

Lesson 1

Transparencies

- National Geographic Society: PicturePack Transparencies, Unit 2

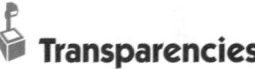

Posters

- National Geographic Society: Images of the World, Unit 2

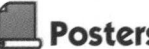

Videodiscs

- Geography and the Environment: The Infinite Voyage
- Reuters Issues in Geography
- National Geographic Society: STV: North America
- National Geographic Society: GTV: Planetary Manager

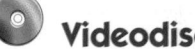

CD-ROM

- National Geographic Society: Picture Atlas of the World
- National Geographic Society PictureShow: Geology

Software

- National Geographic Society: ZipZapMap! USA

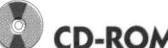

Reference Books

- Picture Atlas of Our 50 States

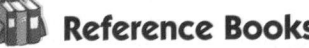

Performance Assessment

Refer to the Planning Guide on page 88A for a Performance Assessment Activity for this chapter. See the *Performance Assessment Strategies and Activities* booklet for suggestions.

Chapter Objectives

1. **Describe** the physical features and climates of the United States.
2. **Identify** economic activities in the United States.
3. **Examine** cultural influences in the United States.

GLENCOE
TECHNOLOGY

 **VIDEODISC**

Use the Chapter 4 Mind-Jogger Videoquiz to preview chapter content.

MindJogger Videoquiz

Chapter 4
Disc 1 Side A

 Also available in VHS.

Chapter 4 The United States

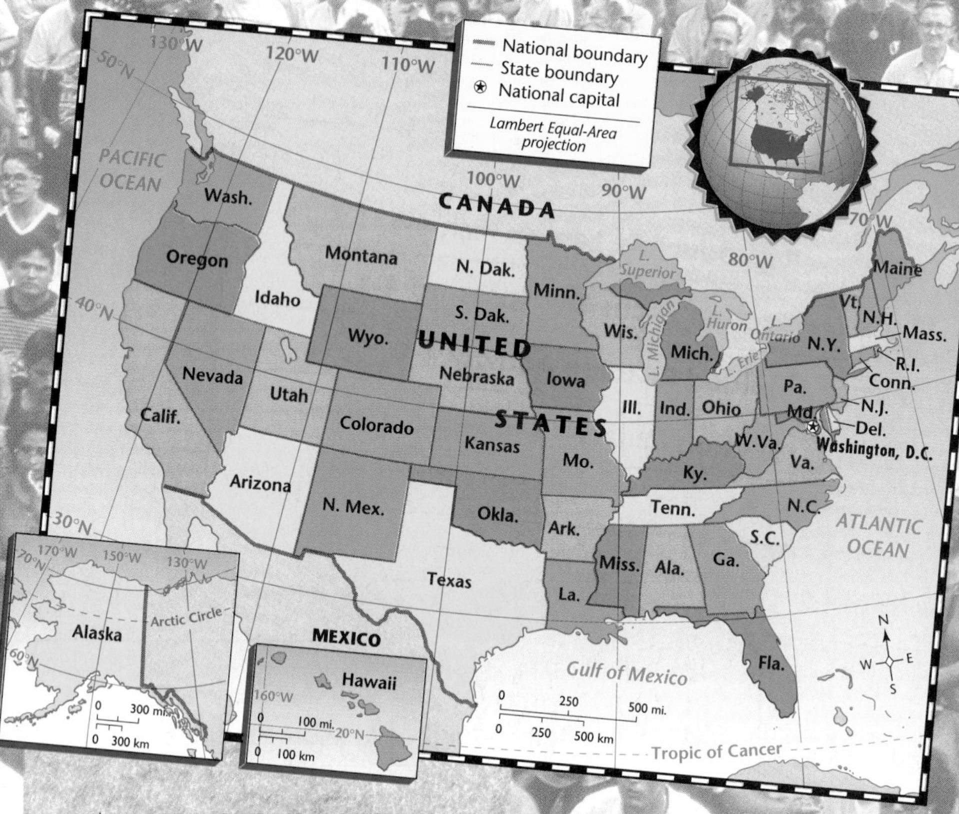

National boundary
State boundary
⊛ National capital

Lambert Equal-Area projection

MAP STUDY ACTIVITY

You may think you already know everything about the United States.

1. **Do you know what state has the largest population?**
2. **What is the distance from the country's east coast to west coast?**
3. **What is the longest river in the United States?**

88

 MAP STUDY ACTIVITY

Answers
1. California
2. 2,807 miles (4,517 km)
3. Mississippi River

Map Skills Practice
Reading a Map Which state has land located south of the Tropic of Cancer? *(Hawaii)* Which state has land located north of the Arctic Circle? *(Alaska)*

SECTION 1
The Land

PREVIEW

Words to Know
- contiguous
- urban
- megalopolis
- rural
- coral reef

Places to Locate
- Appalachian Mountains
- Mississippi River
- Great Lakes
- Great Plains
- Rocky Mountains
- Sierra Nevada Mountains
- Hawaiian Islands

Read to Learn . . .
1. what landforms are found in the United States.
2. what climates occur in the United States.

If you like to hike and camp, the Great Smoky Mountains are for you! Each year, thousands of vacationers visit the Smokies, which lie on the border between North Carolina and Tennessee. Why were they given this name? A smoky blue haze covers these high and rugged peaks in the Appalachian Mountain range.

The Great Smoky Mountains are only one of many exciting places in the United States. Look at the map on page 88 and imagine traveling through some of the 50 states. Start in the northeast. Hike along the Maine coast, and get soaked with ocean spray. Head south and clap your hands to country music in Nashville, Tennessee. Go west and take a mule ride a mile down into the Grand Canyon. End your journey on the west coast in California. All of the United States—from coast to coast—offers many sights, sounds, and adventures!

Location
A Huge Country

Most of the United States—48 of the 50 states—stretches 2,807 miles (4,517 km) across the entire middle part of North America. These 48 states are **contiguous**, or joined together inside a common boundary. The map on page 88 shows you that they touch three major bodies of water—the Atlantic Ocean, the Gulf of Mexico, and the Pacific Ocean.

CHAPTER 4

89

Classroom Resources for Section 1

REPRODUCIBLE MASTERS
Reproducible Lesson Plan 4-1
Guided Reading Activity 4-1
Vocabulary Activity 4
Workbook Activity 4
Geography Skills Activity 4
Section Quiz 4-1

TRANSPARENCIES
Section Focus
  Transparency 4-1
Political Map Transparency 2
Unit Overlay Transparencies
  2-0, 2-4

MULTIMEDIA
 Vocabulary PuzzleMaker
  Software
Testmaker

National Geographic
  Society: STV : North
  America

Meeting National Standards
Geography for Life The following standards are highlighted in this section:
- Standards 4, 5, 15

FOCUS

Section Objectives
1. **Identify** the landforms that are found in the United States.
2. **Describe** the climates that occur in the United States.

Bellringer
Motivational Activity
Prior to taking roll at the beginning of the class period, project Section Focus Transparency 4-1 and have students answer the activity questions. Discuss students' responses. **L1**

Vocabulary Pre-check
Have students skim through the text to find photographs that show rural and urban landscapes. Call on volunteers to use their photograph selections to explain the features that make a landscape either rural or urban. **L1** **LEP**

Use the Vocabulary PuzzleMaker Software to create a crossword puzzle. **L1**

TEACH

Guided Practice

 Display Political Map Transparency 2 and have students identify the approximate location of their community on the map. Then list the following on the chalkboard: Political Region (state), Physical Region, Climate Region. Direct students to use the maps on pages 88, 90, and 94 to find the political region, physical region, and climate region in which their community is located. **L1**

CURRICULUM CONNECTION

Language Arts The term *megalopolis* comes from the Greek words meaning "great city." The east coast megalopolis extends from Boston in the north to Washington, D.C., in the south. Some people refer to it as "Boswash."

MAP STUDY

Answer
about 350 miles (563 km)

Map Skills Practice
Reading a Map Where are the highest elevations found in the Great Plains? *(in the west)*

Two states lie apart from the 48 contiguous states. Alaska, the largest state, spreads over the northwestern corner of North America. Hawaii, the newest state, lies far out in the Pacific Ocean. It is about 2,400 miles (3,862 km) southwest of the California coast.

The United States—the world's fourth-largest country in size—has a total land area of 3,539,230 square miles (9,166,606 sq. km). Only Russia, Canada, and China are larger.

Region
From Sea to Shining Sea

Like a patchwork quilt, the United States consists of regional patches of different landscapes. Look at the map below to find the five physical regions of the United States.

The Coastal Plains A broad lowland runs like a quilt's border along the Atlantic Ocean and the Gulf of Mexico. Geographers divide this lowland into two parts: the Atlantic Coastal Plain and the Gulf Coastal Plain.

The Atlantic Coastal Plain borders the Atlantic coast from Massachusetts to Florida. Many of the region's deepwater ports provided excellent harbors for the first settlers' ships. Port cities such as Boston and New York City have devel-

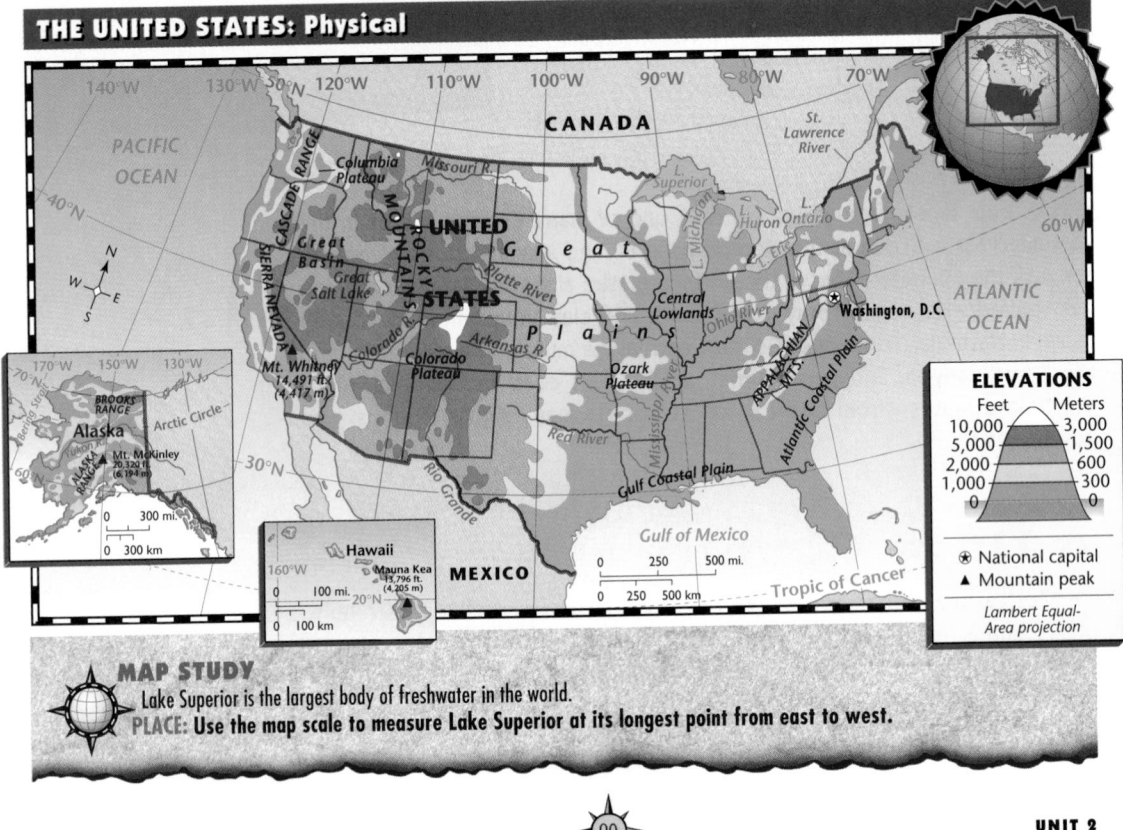

THE UNITED STATES: Physical

ELEVATIONS

Feet	Meters
10,000	3,000
5,000	1,500
2,000	600
1,000	300
0	0

⊛ National capital
▲ Mountain peak

Lambert Equal-Area projection

MAP STUDY
Lake Superior is the largest body of freshwater in the world.
PLACE: Use the map scale to measure Lake Superior at its longest point from east to west.

90

Cooperative Learning Activity

Organize students into five groups and assign each group one of the physical regions of the United States. Instruct group members to locate photographs of, or make sketches of, landscapes found in their assigned regions. Encourage groups to use their photographs and/or sketches to create a bulletin-board display titled "America the Beautiful." Have groups arrange their illustrations around a large wall map of the United States. Make certain each illustration has a caption and a lead line connecting it to its approximate location on the map. **L1**

Independent Practice

📁 Guided Reading Activity 4-1 **L1**

oped worldwide trade networks. You can drive for hundreds of miles and not always tell where one **urban**, or densely populated, area ends and another begins. Geographers refer to this urban pattern as a **megalopolis**, or super city. Farther south are more **rural**, or sparsely populated, areas. In Florida, you will even discover uninhabited marshes and swamps.

The Gulf Coastal Plain hugs the Gulf of Mexico from Florida to Texas. Look at the map on page 90. You can see that the Gulf Coastal Plain is much wider than the Atlantic Coastal Plain. The Mississippi River—the longest river in the country—drains much of the Gulf Coastal Plain. This river runs 2,340 miles (3,765 km) from the north central United States to the Gulf of Mexico. Barges carry goods to and from cities along the Mississippi—some all the way to the port city of New Orleans.

The Appalachian Mountains Curving west along the Atlantic Coastal Plain are the Appalachian (A•puh•LAY•chuhn) Mountains. The second-longest mountain range in North America, the Appalachians run almost 1,500 miles (2,400 km) from eastern Canada to western Alabama.

The Appalachians are the oldest mountain range on the continent. How do geographers know this? The rounded mountain peaks show their age. Erosion has worn them down over time. The average height of the Appalachian ranges is less than 6,000 feet (2,000 m).

The Interior Plains The central part of the country's landscape quilt is the Interior Plains. The eastern region of these plains is called the Central Lowlands. Traveling through this area, you will see thick forests, broad river valleys, rolling flatlands, and grassy hills.

The United States has five physical regions. Florida's wetland Everglades is in the Coastal Plains *(left)*. The Golden Gate Bridge, linking San Francisco to northern California, is in the Pacific Region *(right)*.
REGION: Which area looks more heavily populated?

Human/ Environment Interaction In the 1960s, the river that supplies freshwater to the Everglades was diverted into a canal. The growing number of people in the cities located close to the Everglades has diverted more water from the swampland. This reduction in the flow of water to the area has damaged plants and wildlife.

These materials are available from Glencoe.

🔴 **VIDEODISC**

STV: North America

The Central Lowlands
Side 2, Frames 00010-34860

Meeting Special Needs Activity

Spatial Disability Have students with spatial-orientation problems use Unit Overlay Transparencies 2-0 and 2-4 to practice stating locations in terms of relative location. For example, have them name states located east of the Mississippi River. Have them work in pairs to ask and answer such questions as "Which state is north of Illinois?" and "Which state is west of Colorado?" **L1**

The highest temperature ever recorded in the United States was reported in Death Valley on July 10, 1913, when the mercury reached 134° F (57° C).

St. Louis, Missouri

The Gateway Arch in St. Louis welcomes visitors to this Mississippi River port.
REGION: In what physical region is St. Louis located?

The largest group of freshwater lakes in the world—the Great Lakes—lie in the northern part of the Central Lowlands. Glaciers formed Lake Superior, Lake Michigan, Lake Huron, Lake Erie, and Lake Ontario millions of years ago. The waters of these connected lakes flow into the St. Lawrence River, which empties into the Atlantic Ocean.

Farther west the landscape, blanketed with neat fields of grain and grassy pastures, takes on a checkerboard pattern. In the Interior Plains region lie the Great Plains—a broad, high area. The Great Plains rises in elevation from about 2,500 feet (762 m) in the east to about 6,500 feet (981 m) in the west.

The Rocky Mountains The western part of the Great Plains meets the rugged Rocky Mountains. The Rockies are the largest mountain range in North America, stretching all the way from Alaska to Mexico. In the contiguous United States, the Rocky Mountains run more than 1,100 miles (1,770 km) from north to south.

Look at the physical map on page 90. The Rockies—with some peaks rising more than 14,000 feet (4,270 m)—are higher and more rugged than the Appalachians. Movements of plates under the earth's crust formed the Rockies millions of years ago. A ridge of these mountains, the Continental Divide, separates the rivers and streams flowing west to the Pacific Ocean from those flowing east toward the Mississippi River. Several important rivers—including the Colorado, Missouri, Arkansas, and Rio Grande—begin in the Rocky Mountains.

Just west of the Rockies lies an area of largely empty basins and plateaus. A large valley there, called the Great Basin, holds the Great Salt Lake. If you ever try swimming in it, you will discover that the lake's high salt levels make it easy for you to float.

To the southwest—in California—sits the lowest and hottest place in the United States. It is Death Valley, which lies about 282 feet (86 m) *below* sea level. Summer temperatures in Death Valley often climb to 125°F (52°C).

The Pacific The Pacific Ocean forms the western border of the United States's landscape quilt. Near the coast lie two major mountain ranges: the Sierra Nevadas and the Cascade Range. Like the Rockies, these Pacific ranges were formed by plate movements. Earthquakes and volcanic eruptions still shake the region. West of the Pacific mountain ranges are coastal lowlands and fertile valleys.

Alaska and Hawaii are part of the Pacific region. Glaciers, islands, and bays line Alaska's southern coastline. Central and southern mountain ranges are broken up by lowlands and plateaus. Mount McKinley, North America's highest mountain at 20,320 feet (6,194 m), towers over this area. A vast plain stretches along the coast of the Arctic Ocean in Alaska's far north.

92

UNIT 2

Critical Thinking Activity

Drawing Conclusions Create a chart on the chalkboard, using the five physical regions of the United States as horizontal column headings and *Population* and *Economic Activities* as vertical column headings. Call on volunteers to identify physical characteristics of each region. Next, ask students how they think physical characteristics might affect population in each region. Note responses on the chalkboard in the appropriate column. Then ask how physical features might influence economic activities. Note these responses on the chart. Have students copy the chart into their notebooks. Encourage them to review and adjust chart entries as they work through the chapter. **L1**

The Hawaiian Islands were formed by eruptions of volcanoes on the ocean floor. Some of the islands have **coral reefs**. These are submerged or low-lying structures formed over time from the skeletons of small sea animals.

Region
Climate

Tropical climates, mid-latitude climates, high latitude climates—all are found in the United States. Why is there so much variety? The country's huge size, changing elevations, and the flow of its ocean and wind currents create these differences.

Mid-Latitude Climates Most of the United States lies in mid-latitude climate regions. The map on page 94 shows you the parts of the country that have a humid continental climate. Winters in these areas are cold and moist; summers, long and hot. Rain falls throughout the year. Snow often blankets this area in winter, especially around the Great Lakes.

The southeastern United States has a humid subtropical climate. Winters are mild and cool, and summers are hot and humid. Severe thunderstorms, including destructive tornadoes, are common in this region during summer months.

The area along the Pacific coast from northern California to southeastern Alaska has a marine west coast climate. Temperatures here are mild year-round, and Pacific winds bring plenty of rainfall. Southern California has a Mediterranean climate of dry, warm summers and rainy, mild winters.

Dry Climate A dry climate prevails in the plateaus and basins between the Pacific mountain ranges and the Rockies. Hot, dry air gets trapped here when the Pacific ranges block humid ocean winds. This region is dotted with deserts, including the Great Salt Lake Desert, the Black Rock Desert, and Death Valley. The western part of the Great Plains also has a dry climate. Summers here tend to be hot and dry, and winters very cold.

High Latitude Climates You will find a highland climate if you visit the mountains in the western part of the contiguous United States. To the north,

Water Power Along the Colorado River

The energy of the Colorado River is harnessed by the Glen Canyon Dam in Arizona. **HUMAN/ENVIRONMENT INTERACTION: What type of energy can be created by flowing water?**

LESSON PLAN
Chapter 4, Section 1

NATIONAL GEOGRAPHIC SOCIETY

These materials are available from Glencoe.

VIDEODISC

STV: North America

The Western Dry Lands
Side 4, Frames 00004-33724

More About the Illustration

Answer to Caption
hydroelectricity

Human/Environment Interaction Although the Glen Canyon Dam is located in Arizona, it provides water and electric power for many western states.

ASSESS

Check for Understanding

Assign Section 1 Review as homework or an in-class activity.

Meeting Lesson Objectives

Each objective below is tested by the questions that follow it in parentheses.

1. **Identify** the landforms that are found in the United States. (2, 3, 5)
2. **Describe** the climates that occur in the United States. (4, 6)

More About . . .

The Appalachian Region A plateau called the Piedmont lies at the eastern edge of the Appalachians. The eastern edge of the Piedmont rises sharply from the low-lying Atlantic Coastal Plain. Where fast-moving rivers drop from the plateau to the plains, there are many rapids and waterfalls. The line formed where this drop takes place is called the fall line. Early European settlers were unable to travel beyond the fall line by boat. The waterfalls and rapids, however, later provided water power for industries. Thus, many cities—Macon, Georgia; Columbia, South Carolina; Raleigh, North Carolina; and Richmond, Virginia; for example—developed along the fall line.

MAP STUDY

Answer
two; humid continental and humid subtropical

Map Skills Practice
Reading a Map What type of climate can be found in Seattle? *(marine west coast)*

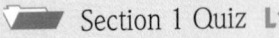

Evaluate

Section 1 Quiz **L1**

Use the Testmaker to create a customized quiz for Section 1. **L1**

Reteach

Have students locate each place listed in the Section 1 **Preview**. Then have them tell one fact about each location based on information in the section. **L1**

Enrich

Have students select one of the "natural wonders" of the United States, such as Niagara Falls or the Grand Canyon. Encourage students to create an illustrated travelogue for their selected site. **L3**

CLOSE

Remind students that each state has a nickname or popular name. Have students make a list of these names and determine which ones are related to physical geography.

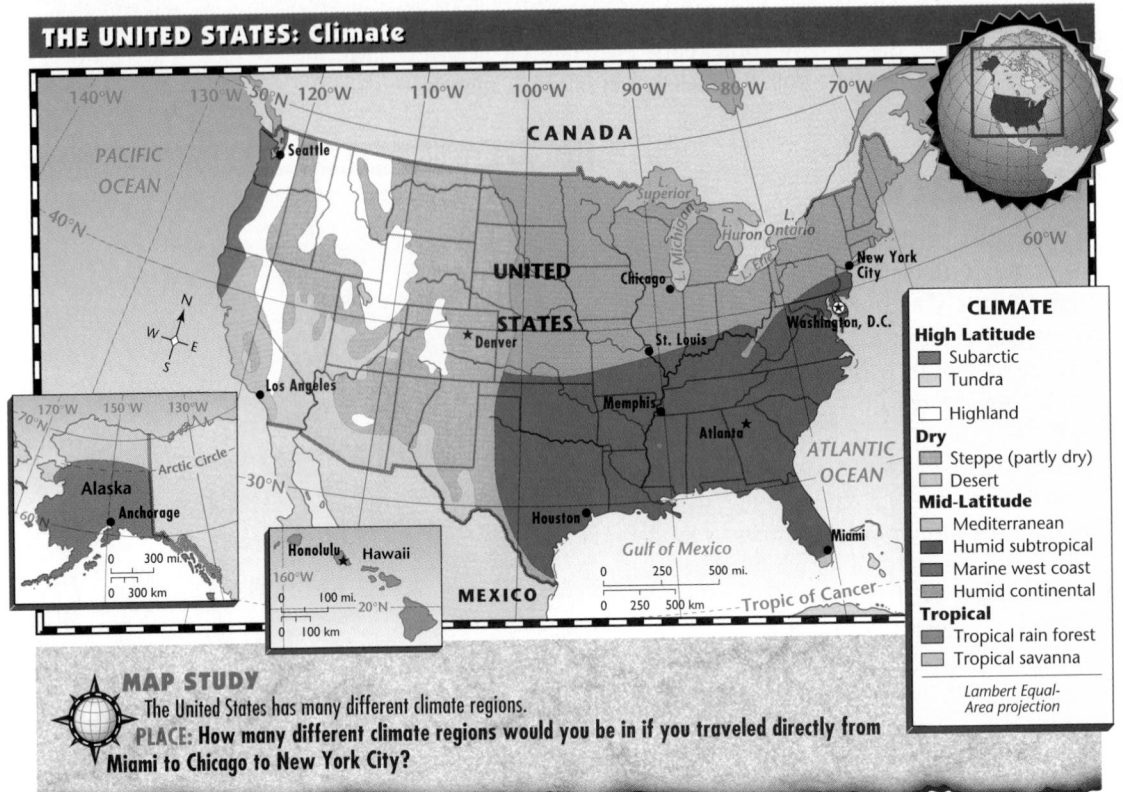

THE UNITED STATES: Climate

CLIMATE

High Latitude
- Subarctic
- Tundra
- Highland

Dry
- Steppe (partly dry)
- Desert

Mid-Latitude
- Mediterranean
- Humid subtropical
- Marine west coast
- Humid continental

Tropical
- Tropical rain forest
- Tropical savanna

Lambert Equal-Area projection

MAP STUDY
The United States has many different climate regions.
PLACE: How many different climate regions would you be in if you traveled directly from Miami to Chicago to New York City?

Alaska is the only part of the country to have subarctic and tundra climates. Winters are bitterly cold and summers are very cool along the Arctic Circle.

Tropical Climates What is the wettest state in the United States? Hawaii, which has a tropical climate, claims that distinction. You can find warm temperatures and plenty of rainfall at the southern tip of Florida, too.

SECTION 1 REVIEW

REVIEWING TERMS AND FACTS
1. **Define the following:** contiguous, urban, megalopolis, rural, coral reef.
2. **LOCATION** Where are the Great Lakes located?
3. **PLACE** What is the largest mountain range in the contiguous United States?
4. **REGION** Why does the United States have so many different types of climates?

MAP STUDY ACTIVITIES
5. Look at the physical map on page 90. What high, flat area covers much of the central United States?
6. Study the climate map above. In what climate region is Memphis, Tennessee?

Answers to Section 1 Review
1. All vocabulary words are defined in the Glossary.
2. The Great Lakes are located in the northern part of the Central Lowlands.
3. Rocky Mountains
4. because of the country's huge size, changing elevations, and the flow of its ocean and wind currents
5. Great Plains
6. humid subtropical

BUILDING GEOGRAPHY SKILLS

Using Scale

Model cars or airplanes are like the larger versions, except for size. Models are made to scale—1 inch stands for a larger measurement in the real vehicles. **Scale** is also used to represent size and distance on maps. For example, 1 inch on a map may represent 100 miles (161 km) on the earth's surface. A map scale is usually shown with a **scale bar**. This bar shows you how much real distance on the earth is shown by a measurement on the map. To use scale to find distance, follow these steps:

- On the scale bar, find the unit of measurement.
- Using this unit, measure the distance between two points.
- Multiply that number by the miles or kilometers each unit stands for.

Yellowstone National Park

Gallatin National Forest

North Entrance

MONTANA
WYOMING

Gallatin National Forest

Mammoth Hot Springs

Tower-Roosevelt

Amethyst Mt.

The Thunderer

Mt. Holmes

Norris

Canyon

Inspiration Point

Saddle Mt.

West Yellowstone

Madison R.

Sulphur Mt.

Artist Point

Castor Peak

West Entrance

Madison

Giant Geyser

Bridge Bay

Fishing Bridge

Fountain Paint Pot

Old Faithful Geyser

Natural Bridge

East Entrance

Old Faithful

West Thumb

Fishing Cone

Yellowstone Lake

IDAHO

Shoshone Lake

Lewis Lake

Heart Lake

Table Mt.

Yellowstone National Park

South Entrance

Lamar River

N W E S

Legend:
- ⚓ Park entrance
- ▲ Mountain
- ○ Point of interest
- ● City
- — Road

Conic projection

Grand Teton National Park

0   10   20 mi.

0   10   20 km

Geography Skills Practice

1. In what three states does Yellowstone National Park lie?

2. By road, about how far is the North Entrance from West Thumb?

3. About how much farther by road is it from Old Faithful Geyser to the South Entrance than from the geyser to Bridge Bay?

Answers to Geography Skills Practice

1. Wyoming, Montana, Idaho
2. about 60 miles (97 km)
3. about 50 miles (81 km)

TEACH

Encourage students to draw a map of the classroom. Discuss whether their maps accurately represent distances between objects. Point out that they could draw the map more accurately by having a unit of measurement on the map— an inch or a centimeter, for example—represent a particular distance in the classroom. Explain that this process is called *drawing to scale*. Ask students to decide on a scale. Demonstrate how to calculate distances using a scale by taking various measurements in the classroom and converting them to measurements for the map. Encourage students to use this information to redraw their maps. Then direct students to read the skill and complete the practice questions. **L1**

Additional Skills Practice

1. How can the scale of a particular map be determined? *(by looking at the scale bar)*

2. On one map, an inch represents 100 miles; on another map, an inch represents 1,000 miles. Which map shows a larger area? *(the map that has a scale of 1 inch to 1,000 miles)*

📁 Geography Skills Activity 4

FOCUS

Section Objectives

1. **Identify** how people in the United States earn their livings.
2. **Explain** why the United States ranks as a world economic leader.
3. **Discuss** the economic challenges the United States faces today.

Bellringer Motivational Activity

 Prior to taking roll at the beginning of the class period, project Section Focus Transparency 4-2 and have students answer the activity questions. Discuss students' responses. **L1**

Vocabulary Pre-check

Point out that most of the terms in the **Words to Know** consist of two words. Demonstrate how the meaning of a word changes when another is added: *farm* and *farm belt,* for example. Ask students to explain meaning changes for other terms on the list. **L1** **LEP**

SECTION 2 The Economy

PREVIEW

Words to Know
• free enterprise system
• service industry
• farm belt
• dry farming
• acid rain
• interdependent

Places to Locate
• Northeast
• South
• Midwest
• Interior West
• Pacific

Read to Learn . . .
1. how people in the United States earn their livings.
2. why the United States ranks as a world economic leader.
3. what economic challenges the United States faces today.

York City. This is the World Trade Center. From the top floor of the Trade Center, you can view the skyscrapers and harbor of New York City.

Two 110-story towers loom above the Hudson River in the Financial District of New

New York City serves as headquarters for many national and international businesses and banks. The city also is a leading center for overseas trade. New York City is one of the cities that help make the United States a world economic leader.

Place

An Economic Leader

The United States is one of the world's most developed countries. In addition to having a large land area, the country is rich in natural resources and in skilled, hardworking people. All of these benefits provide the United States with a strong, productive economy.

The American economy is based on the **free enterprise system.** Under free enterprise, people own and run businesses with limited government controls. America's economic strength was first built on agriculture, which remains important. The United States also is strong in science, technology, education, and medicine.

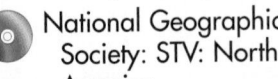

Region
Economic Regions

Geographers divide the United States into regions based on economic activity. These regions are the Northeast, the South, the Midwest, the Interior West, and the Pacific. How do the people in your community make their livings?

The Northeast The Northeast is one of the oldest manufacturing areas in the United States. By the early 1800s, New England had launched the nation's Industrial Revolution. Today you can find all types of goods and machinery manufactured here. The map below shows you that coal mining, for example, is an important economic activity in Pennsylvania and West Virginia.

The deepwater harbors and manufacturing economies of Boston, New York City, Philadelphia, and Baltimore make the Northeast a major center of world trade. New York City is also a world center of fashion, entertainment, publishing, and communications.

Many Northeastern cities employ thousands in **service industries** such as business, finance, banking, and insurance. Service industries are businesses that provide services to customers rather than producing farm or industrial products.

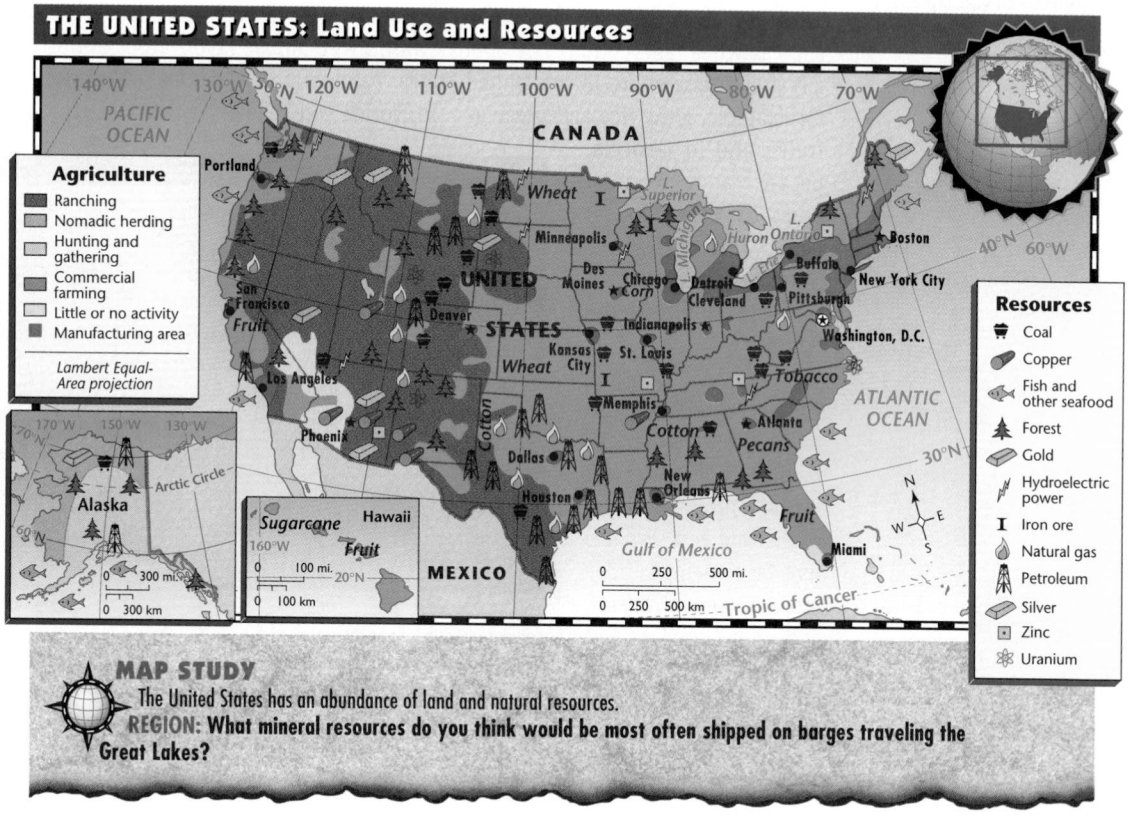

THE UNITED STATES: Land Use and Resources

Agriculture
- Ranching
- Nomadic herding
- Hunting and gathering
- Commercial farming
- Little or no activity
- Manufacturing area

Lambert Equal-Area projection

Resources
- Coal
- Copper
- Fish and other seafood
- Forest
- Gold
- Hydroelectric power
- Iron ore
- Natural gas
- Petroleum
- Silver
- Zinc
- Uranium

MAP STUDY
The United States has an abundance of land and natural resources.
REGION: What mineral resources do you think would be most often shipped on barges traveling the Great Lakes?

Cooperative Learning Activity

Organize students into five groups and assign each group one of the economic regions of the United States. Have groups prepare an annotated product map or economic activity map for their assigned region. Make certain that all group members have a task—some might conduct research, others might construct the map, and still others might write annotations. Have groups display and discuss their completed maps. **L2**

TEACH

Guided Practice
Making Connections
Have students use maps and text to identify the economic region in which they live. Direct them to make a list of the economic activities in and around their community. Then have them determine how these activities compare with those identified in the text. **L1**

Independent Practice

Guided Reading Activity 4-2 **L1**

Global Gourmet

United States Pancakes with maple syrup is a favorite breakfast across the United States. The northeastern states—most notably Vermont and New York—lead the country in maple syrup production. Some 35 to 45 gallons (132 to 170 l) of maple sap are needed to make 1 gallon (3.8 l) of syrup.

MAP STUDY

Answer
coal, iron ore, forest products, zinc
Map Skills Practice
Reading a Map Where in the United States does nomadic herding take place? *(Alaska)*

New York City has a larger population than most states.

Watching the Monitor

Computer research and development is an important service industry in Silicon Valley, California.
HUMAN/ENVIRONMENT INTERACTION: What is a service industry?

The national capital, Washington, D.C., offers service-industry jobs in government and tourism.

Agriculture is also important to the economy of the Northeast. Farmers grow fruits and vegetables and raise dairy cattle and chickens. Fishing is a leading industry along the Atlantic coast.

The South As in the Northeast, the economy in the South is varied. The people here work in manufacturing, farming, and fishing. Industry, however, has become the South's main source of income. Oil-based products, textiles, electrical equipment, and airplane parts emerge from factories and refineries here.

In recent years the South has attracted new businesses and people. Service industries have grown, as have large entertainment centers such as Seaworld and Walt Disney World in Orlando, Florida. Seaside resorts in Florida, South Carolina, and Mississippi attract tourists from all over the world. The map on page 97 shows that inland cities such as Atlanta and Dallas are major manufacturing areas. Large port cities—Houston, Miami, and New Orleans—are busy manufacturing and shipping centers.

Agriculture has always been a major economic activity in the South. Texas—with more farms than any other state—raises livestock and grows crops such as wheat and cotton. The South's warm, wet climate favors crops not usually grown elsewhere in the United States. Farmers in Louisiana and Arkansas, for example, grow rice and sugarcane. In Florida and Texas, they cultivate citrus fruits; and in Georgia, pecans and peanuts. Look at the map on page 97 to see where some agricultural products are grown in the South.

The Midwest In this region lies the American **farm belt**. Flat land and fertile soil cover much of the Midwest. Productive farms supply huge crops of corn, soybeans, and grains such as oats and wheat. In some areas of the Great Plains, farmers use **dry farming** to grow a certain kind of wheat. Dry

Waves of Grain

Contour plowing follows the curves of the land and is used to reduce water runoff.
HUMAN/ENVIRONMENT INTERACTION: Why do you think it is important to keep water from running off of farmland?

farming is a way of plowing land so that it holds rainwater. Dairy products and livestock are also important in the midwestern economy.

When you hear the term "Midwest," you probably don't think of large port cities for shipping. But the Midwest has a number of waterways and ports. The St. Lawrence Seaway links the Great Lakes to the Atlantic Ocean. The Mississippi and Ohio rivers wind through the Midwest into the Gulf Coastal Plain. These waterways are deep enough for ships to transport farm and industrial products to other parts of North America and the world.

In the late 1800s, the Midwest drew many immigrants and became a leading industrial area. Workers today produce iron, steel, heavy machinery, and cars. Industrial cities such as Chicago, Detroit, Milwaukee, and Cleveland dot the coasts of the Great Lakes. St. Louis—near the spot where the Mississippi River meets the Missouri River—is an important shipping and transportation center.

The Interior West A dry climate affects economic activity in the broad Interior West. Ranching is more common than farming in this region. A single cattle or sheep ranch may cover 2,000 acres (809 ha) or more. With limited water sources, farming depends on irrigation. Potatoes, hay, wheat, barley, and sugar beets are the region's major crops.

The Interior West has plenty of rich mineral and energy resources. Look at the map on page 97. You will see that workers in the region mine silver and copper and drill for oil.

Have you ever visited Arizona's Grand Canyon? Tourism is a rapidly growing industry in the Interior West. The cities of Albuquerque, Denver, Phoenix, and Salt Lake City are tourist and transportation centers as well as places of government and business.

The Pacific The Pacific region boasts economic activity from farming, manufacturing, and tourism. Some of the best farmland in the region is in central California. Many of the fruits and vegetables you eat every day come from here. Crops such as sugarcane, pineapples, coffee, and rice flourish in Hawaii's tropical climate and fertile volcanic soil. Fishing is also a major industry along the Pacific region's coastal waterways.

Lumber and mining are important sources of income in many states. California has gold, lead, and copper. Alaska has vast oil reserves. Industries hum in the large cities of the Pacific region—Los Angeles, San Francisco, San Diego, and Seattle. Among the leading manufactured products are airplanes and computers. Los Angeles is also the world capital of the movie industry, and San Diego is home to the United States Navy's Pacific fleet.

An estimated 47.5 million people visit the United States each year. Many of them begin their trip on America's Pacific coast, lured by snowcapped mountains, golden beaches, and crystal-clear lakes. Tourism is Hawaii's major source of income. The tropical beauty of the Hawaiian Islands attracts millions of tourist dollars annually.

The Skyline of Los Angeles, California

More than one-third of California's residents live in the sprawling city of Los Angeles. **REGION: What is Los Angeles famous for?**

Critical Thinking Activity

Synthesizing Information Have students recall the Critical Thinking Activity they undertook in Section 1, specifically their ideas on how physical geography influences economic activities. Then ask them to write a sentence for each economic region, noting how the economic activities in that region are related to physical geography. Offer the following example: Natural harbors along the Atlantic Coastal Plain make the northeast a major trade center. **L2**

LESSON PLAN
Chapter 4, Section 2

ASSESS

Check for Understanding

Assign Section 2 Review as homework or an in-class activity.

Meeting Lesson Objectives

Each objective below is tested by the questions that follow it in parentheses.

1. **Identify** how people in the United States earn their livings. (1, 2, 4)
2. **Explain** why the United States ranks as a world economic leader. (1, 2)
3. **Discuss** the economic challenges the United States faces today. (3)

More About the Illustration

Answer to Caption
movie industry

Movement Los Angeles is one of the most racially and ethnically diverse cities in the United States. According to the 1990 census, 40 percent of the people in Los Angeles are from Spanish-speaking countries, 37 percent are white, 13 percent are African American, and 10 percent are Asian. There are more people of Mexican, Chinese, Taiwanese, Korean, and Philippine backgrounds in Los Angeles than in any other city in the United States.

99

Evaluate

Section 2 Quiz **L1**

Use the Testmaker to create a customized quiz for Section 2. **L1**

Reteach

Have students work in small groups to outline Section 2. Assign one subsection to each group, then have groups put together their outlines to review the section. **L1**

Enrich

Students might investigate trade partnerships between the United States and other countries. Encourage students to present their findings in the form of an annotated map. **L3**

CLOSE

Have students work in pairs or small groups to create posters that focus attention on one of the economic challenges facing the United States today. Display the posters and lead a discussion on the steps that might be taken to meet these challenges.

Human/Environment Interaction
Today's Challenges

Americans use the latest technology to develop the country's natural resources. In some cases, however, people use resources in ways that are harmful to the environment. And some resources are being completely used up!

Use of Resources Factories in the United States use oil, gas, and coal to manufacture goods. The burning of these fuels creates pollution. When air pollution mixes with water vapor and falls to the earth in rain or snow, this is called **acid rain**. Acid rain kills fish, pollutes lakes and rivers, and damages forests in the United States and across the border in Canada.

Overusing resources is another problem. Forests are cut to provide lumber and paper products Americans want. Fishing off the coasts has led to a decline in many species of fish. In recent years, national and state governments have passed laws to protect natural resources.

Growth of Industry High-technology industries help the United States maintain its leadership role in world manufacturing. As computers and robots entered factories, however, many American workers lost their jobs. Some of the country's workforce have had to learn new skills or start new careers.

Service industries are a major part of the United States economy. Service industries such as banking, medical care, education, and government employ more workers than any other industry.

World Trade The United States leads the world in the total value of its imports and exports. Millions of jobs are linked to the exporting and importing of goods. In recent years, the economies of the United States and other countries have become more **interdependent**, or reliant on each other. In 1993 the North American Free Trade Agreement (NAFTA) was approved. It will increase trade among the United States, Canada, and Mexico.

SECTION 2 REVIEW

REVIEWING TERMS AND FACTS

1. **Define the following:** free enterprise system, service industry, farm belt, dry farming, acid rain, interdependent.
2. **REGION** Why is the Interior West different from other economic regions in the United States?
3. **HUMAN/ENVIRONMENT INTERACTION** How has acid rain affected the United States?
4. **PLACE** What form of economic activity employs the most people in the United States?

   MAP STUDY ACTIVITIES
5. Look at the land use map on page 97. What is the leading mineral resource found near the Gulf of Mexico?

UNIT 2

Answers to Section 2 Review

1. All vocabulary words are defined in the Glossary.
2. The Interior West is the driest part of the country and has limited farming. The area is, however, rich in minerals.
3. It has poisoned lakes, killing off fish. It also has destroyed trees in forested areas.
4. service industries
5. oil

MAKING CONNECTIONS

TRAVELS WITH CHARLEY   BY JOHN STEINBECK ▼ ▲

| MATH | SCIENCE | HISTORY | LITERATURE | TECHNOLOGY |

American author John Steinbeck and his dog Charley traveled the country in search of America. This excerpt describes his feelings as he drove through the Bad Lands of South Dakota.

I went into a state of flight, running to get away from the unearthly landscape. And then the late afternoon changed everything. As the sun angled, the buttes and coulees [dry streambeds], the cliffs and sculptured hills and ravines lost their burned and dreadful look and glowed with yellow and rich browns and a hundred variations of red and silver gray, all picked out by streaks of coal black. It was so beautiful that I stopped near a thicket of dwarfed and wind-warped cedars and junipers, and once stopped I was caught, trapped in color and dazzled by the clarity of the light. Against the descending sun the battlements were dark and clean-lined, while to the east, where the uninhibited light poured slantwise, the strange landscape shouted with color. And the night, far from being frightful, was lovely beyond thought, for the stars were close, and although there was no moon the starlight made a silver glow in the sky. The air cut the nostrils with dry frost. And for pure pleasure I collected a pile of dry dead cedar branches and

built a small fire just to smell the perfume of the burning wood and to hear the excited crackle of the branches. My fire made a dome of yellow light over me, and nearby I heard a screech owl hunting and a barking of coyotes, not howling but the short chuckling bark of the dark of the moon. This is one of the few places I have ever seen where the night was

The Bad Lands

friendlier than the day. And I can easily see how people are driven back to the Bad Lands.

From *Travels with Charley* by John Steinbeck. Copyright © 1961, 1962 by The Curtis Publishing Co., © 1962 by John Steinbeck, renewed © 1990 by Elaine Steinbeck, Thom Steinbeck, and John Steinbeck IV. Used by permission of Viking Penguin, a division of Penguin Books USA Inc.

Making the Connection ▼ ▲

1. How did Steinbeck's feelings toward the Bad Lands change as the day wore on?
2. How does the landscape of the Bad Lands compare with the landscape in your community?

101

Answers to Making the Connection

1. During the day, Steinbeck found the Bad Lands unearthly, burned, dreadful, and frightful. By nightfall, however, he found the area beautiful, peaceful, and friendly.
2. Answers will vary.

TEACH

Ask students to speculate on what kind of geographical features might earn a region the name of *Bad Lands.* Then inform them that this name is given to elevated areas where water has eroded the land into small, steep hills with deep gullies. **L1**

More About . . .

The Bad Lands Two areas of the Bad Lands—the Badlands National Park in South Dakota and the Theodore Roosevelt National Park in North Dakota—have been set aside as national parks. The parks' unique geographic formations are shaded by multicolored patterns of gray, blue, beige, yellow, black, red and brown. This bleak but stunning landscape has made the area a major tourist attraction. Not everyone, however, has found the area appealing. In 1864, General Alfred Sully compared the Bad Lands to "hell with the fires put out."

MULTICULTURAL PERSPECTIVE

Culturally Speaking

The term *bad lands* was officially applied to the region in 1851. The Sioux, however, had long called the area *mako sica,* or "bad land." And the first Europeans who arrived in the area, French fur traders, referred to it as *mauvaises terres á traverser,* or "bad lands to travel across."

SECTION 3

The People

Meeting National Standards

Geography for Life The following standards are highlighted in this section:
• Standards 9, 10, 17

FOCUS

Section Objectives

1. **Describe** how the United States began.
2. **Explain** why the United States is a land of many cultures.
3. **Discuss** how the arts have developed in the United States.

Bellringer Motivational Activity

Prior to taking roll at the beginning of the class period, project Section Focus Transparency 4-3 and have students answer the activity questions. Discuss students' responses. **L1**

Vocabulary Pre-check

Call on volunteers to define the terms *multicultural* and *ethnic group.* Discuss how these terms are related. Then ask students to write a sentence that pairs the two terms. Encourage them to write similar sentences for other related terms listed in **Words to Know,** such as *immigrant/colony* and *revolution/republic.* **L1 LEP**

PREVIEW

Words to Know
- immigrant
- colony
- revolution
- republic
- multicultural
- ethnic group
- mobile
- national park

Places to Locate
- Mississippi River
- Sunbelt
- Chicago

Read to Learn . . .
1. how the United States began.
2. why the United States is a land of many cultures.
3. how the arts have developed in the United States.

Every year, millions of people join in a huge, coast-to-coast birthday party on the Fourth of July. How do you celebrate this national holiday that marks the birth of the United States?

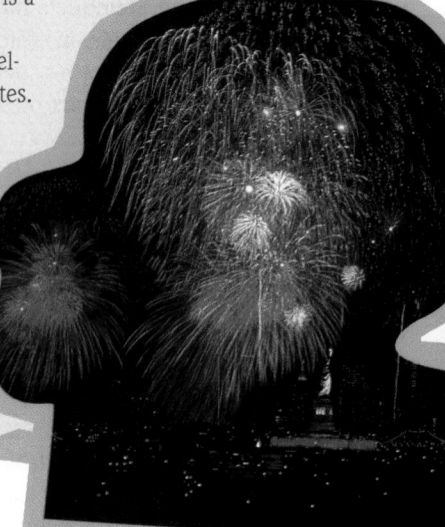

Americans have much in common, such as celebrating the birth of their nation on the Fourth of July. At the same time, they follow many different ways of life. Americans trace their roots to a variety of places around the world. Either they—or their ancestors—were **immigrants**, or people from other lands who have come to live permanently in the United States.

Place

Influences of the Past

The first Americans were nomads who followed their herds. Moving from campsite to campsite, these people from northern Asia worked their way south. They were the first immigrants to North America. These first Americans came to the country we now call the United States thousands of years ago. Over the centuries other people from Europe, Africa, and other parts of Asia and the Americas followed.

(102)

UNIT 2

Classroom Resources for Section 3

 REPRODUCIBLE MASTERS

Reproducible Lesson Plan 4-3
Guided Reading Activity 4-3
Performance Assessment Activity 4
Reteaching Activity 4
Enrichment Activity 4
Section Quiz 4-3

 **TRANSPARENCIES**

Section Focus Transparency 4-3
Political Map Transparency 2
Geoquiz Transparencies 4-1, 4-2

MULTIMEDIA

 Testmaker

National Geographic Society: GTV: Planetary Manager

Early Period The first people, some experts believe, arrived 12,500 or more years ago. They separated into different groups as they moved into almost every region of what is now the United States. The Native Americans developed cultures influenced by the environments in the areas where they settled.

In the 1600s and 1700s, Europeans settled in North America. They came seeking land, riches, and the right to live freely. European groups from Spain, France, Great Britain, and other countries set up **colonies**, or overseas settlements tied to a parent country. In settling the land, however, the Europeans and their descendants often grabbed land and killed Native Americans.

Government of the People By the late 1700s, people living in the British colonies along the Atlantic coast had started to think of themselves as Americans. They fought a war with the British that ended British rule in the American colonies. This **revolution**, or sudden political change, produced the independent country we know as the United States of America.

In 1787 a group of American leaders wrote a plan of government for the country. It became the Constitution of the United States. The Constitution created a **republic**—a form of government in which the people elect their own officials, including the leader of the country. Meanwhile, the American colonies had become states. The national government and the state governments still share the task of ruling the country.

Adobe Village in Taos, New Mexico

Some Pueblo Native Americans still live in traditional adobe villages in New Mexico.
MOVEMENT: From what continent did the first Americans migrate?

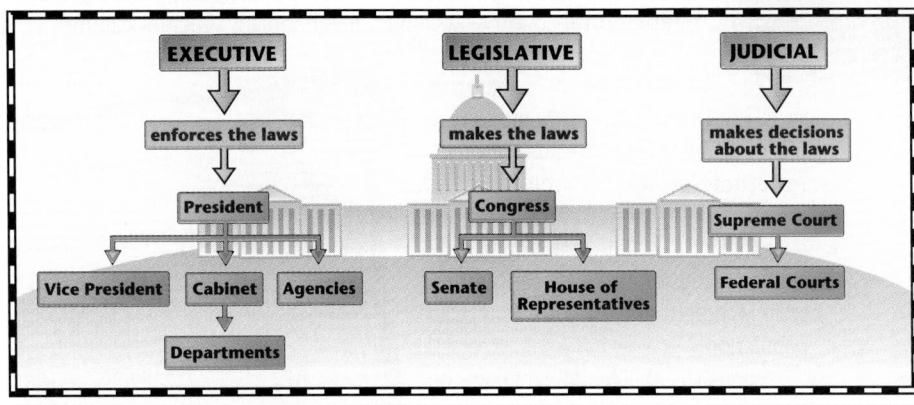
Branches of the United States Government

EXECUTIVE	LEGISLATIVE	JUDICIAL
enforces the laws	makes the laws	makes decisions about the laws
President	Congress	Supreme Court
Vice President / Cabinet / Agencies	Senate / House of Representatives	Federal Courts
Departments		

GRAPHIC STUDY
The United States government has three main branches.
PLACE: What are the two lawmaking bodies of the legislative branch?

More About the Illustration
Answer to Caption
Asia

Place Today, about 50,000 Pueblo Native Americans live along the Rio Grande in New Mexico. The Pueblo people have lived in this area for centuries. In fact, they have lived in the same area longer than any other group of people in the United States or Canada.

TEACH

Guided Practice
Analyzing Information
Write the following headings on the chalkboard: *Cultures, Homes, Standard of Living, Recreation.* Call on volunteers to identify characteristics of American people that might be entered under each heading. List responses on the chalkboard. Use this information to lead a discussion of why it is so difficult to describe a "typical American." **L1**

GRAPHIC STUDY
Answer
Senate, House of Representatives
Graphic Skills Practice
Reading a Diagram
Who heads the executive branch of government? *(president)*

Cooperative Learning Activity

Organize students into groups and assign each group a time period in American history. A possible division of time periods might be: Pre-Columbian, Colonial Times, Revolution to Early 1800s, Nineteenth Century, Twentieth Century. Using several sheets of butcher paper, construct a time line running along one or more walls of the classroom. Direct groups to research their time period and select 10 important events to enter on the time line. Have group representatives write their selections on the time line, asking them to add annotations where necessary. Retain the time line for future reference. **L1**

MULTICULTURAL PERSPECTIVE

Cultural Heritage

The greatest wave of immigration to the United States lasted from the 1880s to 1920. During that time, about 23.5 million people immigrated to the United States. Most of them came from southern and eastern Europe.

CURRICULUM CONNECTION

History The building of the transcontinental railroad in the 1860s opened up much of the West to settlement. It also encouraged immigration. The Union Pacific Railroad hired immigrant laborers, mostly from China, to lay track. Thousands of Chinese came to the United States hoping to get such jobs.

Teen Scene

Teenage Doctor

By age 7, Masoud Karkehabadi had completed high school. By age 13, he had graduated from college. By age 14, Masoud may cure Parkinson's disease. Masoud is an extremely intelligent 13-year-old who lives in California. He has been doing research to find a cure for Parkinson's disease since he was 11. "I believe my intelligence is a gift," said Masoud. "I want to use it to the best of my capacity to help society." Next year, Masoud plans to start his first year of medical school. In his free time, he plays baseball and street hockey and takes care of his pet iguana, Abi.

Industry and Expansion Over the next 200 years, the growth of industry changed the United States into an economic giant. By the mid-1800s, factories were using steam-powered machines—and later electricity—to make goods. Where did workers for these factories come from? Americans moved from farms to the cities, and thousands of immigrants poured into the United States.

Through wars, treaties, and purchases, the United States gained control of the lands west of the Mississippi River. The railroads and low prices for land drew people to this western region. Settlers there supplied minerals and agricultural products to other parts of the country.

A World Power During the 1900s, the United States became a leader in world affairs. As industry grew in the United States, foreign trade became more important. During and after World War I and World War II, the vast resources of the United States aided allies around the world. Along with weapons and manufactured goods, American culture also spread overseas and changed the way of life in many places.

Movement

One Out of Many

About 261 million people live in the United States. The variety of its people makes the United States a **multicultural** country—a country that has many different cultures.

A Variety of People

The United States is home to people of many different ethnic groups. An **ethnic group** is a group of people who share a common culture, language, or history. The graph on page 85 shows some of the different ethnic groups that make up the American population.

English is the major language of the United States, but many other languages are spoken. For example, many Spanish-speaking people live in the Southwest, in Florida, and in major cities such as New York City and Chicago.

WHAT IN THE WORLD?

An American Sound

In the late 1800s a type of music known as jazz was born in the United States. It is believed to be the only musical art form to originate in the United States. Jazz's multicultural roots blend African rhythms, African American religious music, and European harmonies.

Meeting Special Needs Activity

Visual Learners Students who learn best from visual aids may benefit from the following exercise. Have students relate illustrations to text by finding sentences or paragraphs that relate to each picture, map, or diagram in the section. Then have them compare what they learn from the illustrations with what they learn from the text. **L1**

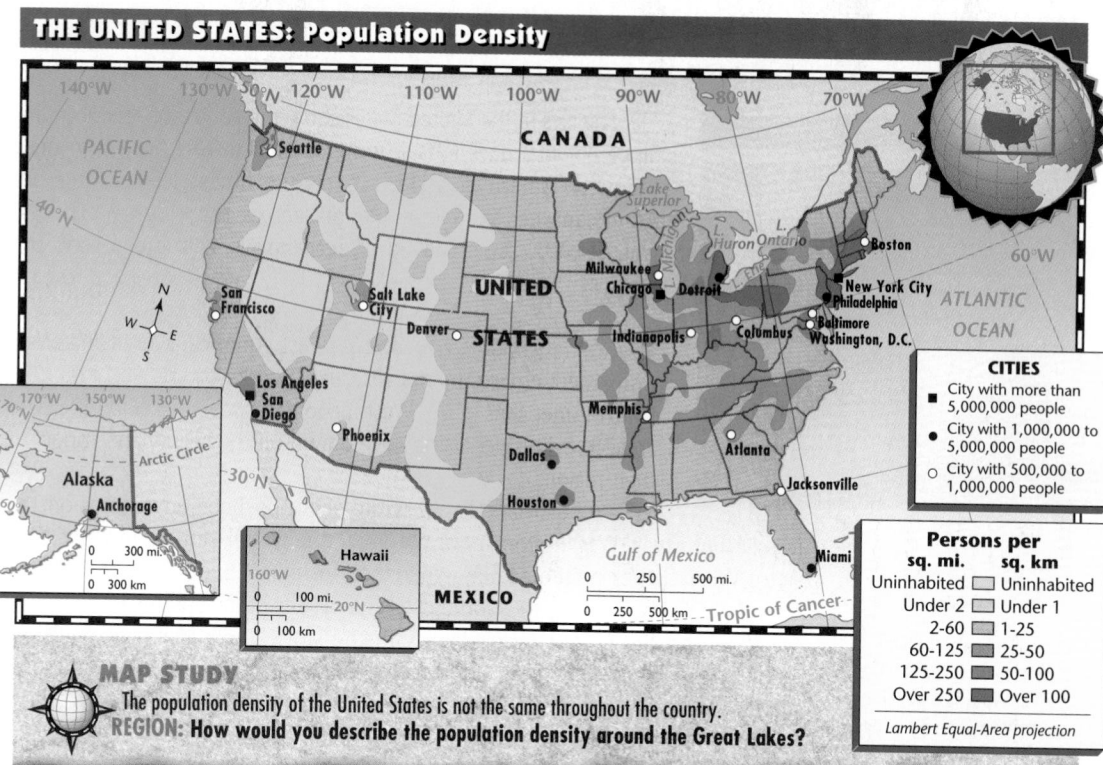

THE UNITED STATES: Population Density

CITIES
- ■ City with more than 5,000,000 people
- ● City with 1,000,000 to 5,000,000 people
- ○ City with 500,000 to 1,000,000 people

Persons per	
sq. mi.	sq. km
Uninhabited	Uninhabited
Under 2	Under 1
2-60	1-25
60-125	25-50
125-250	50-100
Over 250	Over 100

Lambert Equal-Area projection

 MAP STUDY
The population density of the United States is not the same throughout the country.
REGION: How would you describe the population density around the Great Lakes?

Religion has always been an important influence in American life. Most Americans follow some form of Christianity. Judaism, Islam, and Buddhism are also important religions in the United States.

A Nation on the Move Go, go, go! Americans have always been a **mobile** people. This means they move from place to place. Americans today move from city to city to get better jobs, better homes, or better educations.

Which city and state do you call home? You might live in Fairview—the most common city name in the country—or in Uncle Sam, Louisiana. The map above shows you that most people in the United States live in or near large cities. The Northeast, parts of the Midwest, and the Pacific coast have the greatest number of people. Since the 1970s, the fastest-growing areas in the United States have been the South and the Southwest. Industrial growth and a pleasant climate have drawn people to this area known as the Sunbelt.

How People Live Compared to most other countries, the United States enjoys a high standard of living. For most people food is abundant, not too costly, and easily available. Americans, on average, can expect to live about 76 years. Medical advances have helped bring about longer lives.

Critical Thinking Activity

Predicting Consequences Have students locate the Sunbelt region on a map of the United States. Then have them note characteristics of the region by studying the maps on pages 90, 94, 97, and 105. Ask students to use the information to predict possible geographical, eco-nomic, and political consequences of the continued migration of people to the region. Also have then consider how the loss of population may affect the Northeast and Midwest regions of the United States. **L2**

LESSON PLAN
Chapter 4, Section 3

MAP STUDY

Answer
heavy around Lake Erie and southwestern Lake Michigan; average in rest of the area

Map Skills Practice
Reading a Map In what region of the country is there the heaviest population concentration? *(Northeast)*

ASSESS

Check for Understanding
Assign Section 3 Review as homework or an in-class activity.

Meeting Lesson Objectives
Each objective below is tested by the questions that follow it in parentheses.

1. **Describe** how the United States began. (1)
2. **Explain** why the United States is a land of many cultures. (2)
3. **Discuss** how the arts have developed in the United States. (2)

Evaluate
 Section 3 Quiz **L1**

Use the Testmaker to create a customized quiz for Section 3. **L1**

Chapter 4 Test Form A and/or Form B **L1**

105

Reteach

 Reteaching Activity 4 **L1**

 Spanish Reteaching Activity 4 **L1**

Enrich

 Enrichment Activity 4 **L3**

CLOSE

 **Mental Mapping Activity**

Provide each student with a large sheet of white paper and map pencils. Inform students that this will not be a graded activity. Have students draw, freehand from memory, a map of the United States with labels for the following bodies of water and mountain ranges: Great Lakes, Mississippi River, Atlantic Ocean, Pacific Ocean, Appalachian Mountains, Rocky Mountains, Sierra Nevada.

106

Place
The Arts and Recreation

Just for Fun

Baseball is one of the most popular sports in the country. **PLACE: What is a popular recreational activity in your area?**

The arts in the United States show the variety of American life. Since colonial days artists, writers, and builders have developed uniquely American styles. In recent times, leisure time for Americans has increased. People across the country play and watch sports, camp and travel, read novels, and attend plays and movies.

Art and Literature The earliest Americans used materials from their environments to create works of art. For example, Native Americans in the West made pottery from clay found in the area. In every age, artists have drawn on the beauty and variety of their region.

Author Mark Twain wrote about boyhood adventures on the Mississippi River. Washington Irving wrote about Dutch settlers in New York state. Willa Cather presented the experiences of pioneers on the Great Plains. Georgia O'Keeffe painted the colorful cliffs and deserts of the Southwest.

American literature is also filled with the stories, folktales, and myths of many ethnic groups. Native American folktales explain the mystery of nature. African American gospel songs tell about the strength of a people in overcoming adversity.

Sports and Recreation Americans are enthusiastic sports fans and players. What sports are most popular? Baseball, football, and basketball are played by both professionals and amateurs. Americans also enjoy physical activities such as biking, skiing, golfing, and jogging. They explore the outdoors and many visit **national parks** set aside for recreation and for protection of wilderness and wildlife.

SECTION 3 REVIEW

REVIEWING TERMS AND FACTS

1. **Define the following:** immigrant, colony, revolution, republic, multicultural, ethnic group, mobile, national park.
2. **PLACE** Why is the United States known as a multicultural country?
3. **MOVEMENT** Why are Americans today so mobile?
4. **PLACE** What areas of the United States have the fastest-growing population?

 MAP STUDY ACTIVITIES
5. Turn to the map on page 105. What is the population density of Miami, Florida?
6. Using the same map, name two sparsely populated areas of the United States.

Answers to Section 3 Review

1. All vocabulary words are defined in the Glossary.
2. Americans trace their roots to many places throughout the world.
3. Americans readily move in search of better housing, education, and jobs.
4. the Sunbelt (South and Southwest)
5. 2–60 persons per sq. mi. (1–25 persons per sq. km)
6. Alaska and the interior West

Chapter 4 Highlights

Important Things to Know About the United States

SECTION 1 THE LAND

- The United States is the world's fourth-largest country in land area.
- The country consists of 48 contiguous states plus Alaska and Hawaii.
- Five physical regions are found in the country—the Coastal Plains, the Appalachian Mountains, the Interior Plains, the Rocky Mountains, and the Pacific region.
- Almost every type of climate region is found in the United States.

SECTION 2 THE ECONOMY

- The economic regions of the United States are the Northeast, South, Midwest, Interior West, and Pacific regions.
- A wealth of natural resources and skilled workers have helped to make the United States a world leader in farming and industry.
- Pollution and overuse of resources are serious environmental issues in many areas of the country.
- Service industries employ more workers than any other industry.

SECTION 3 THE PEOPLE

- Americans trace their roots to every part of the world.
- The United States became independent in the late 1700s. The Constitution established a new republic with a democratic form of government.
- By the 1900s the country had become an industrial power and a leader in world affairs.
- About 261 million people live in the United States, most in urban centers along the coasts.

Festival in Washington, D.C. ▶

Extra Credit Project

Display Have students work in small groups to create a display showing the geography of their community or state. The display might consist of captioned maps, photographs, and sketches; newspaper and magazine articles; and rock, soil, and plant samples. **L1**

CHAPTER HIGHLIGHTS

Using the Chapter 4 Highlights

- Use the Chapter 4 Highlights to preview, review, condense, or reteach the chapter.

Preview/Review

Vocabulary Puzzle-Maker Software reinforces the terms used in Chapter 4.

Condense

Have students read the Chapter 4 Highlights. Spanish Chapter Highlights also are available.

Chapter 4 Digest Audiocassettes and Activity

Chapter 4 Digest Audiocassette Test
Spanish Chapter Digest Audiocassettes, Activities, and Tests also are available.

Guided Reading Activities

Reteach

Reteaching Activity 4. Spanish Reteaching activities also are available.

MAP STUDY

Location Copy and distribute page 6 from the Outline Map Resource Book. Have students locate and label major landforms and bodies of water of the United States. Also have them shade in the country's climate regions. Remind students that they need to include a key with their maps.

107

Chapter 4 Review and Activities

REVIEWING KEY TERMS

Match the numbered terms in Column A with their definitions in Column B.

A

1. free enterprise
2. contiguous
3. national park
4. urban
5. dry farming
6. mobile
7. multicultural
8. farm belt
9. acid rain
10. immigrant

B

A. areas joined together inside a common boundary
B. a person from one country who settles permanently in another country
C. moving from place to place
D. system in which people own and run businesses with little government control
E. plowing land so that it holds rainwater
F. pollution mixed with rain or snow
G. land set aside for protection of wilderness
H. densely populated area
I. made up of many different cultures
J. an area of many productive farms

ANSWERS

Reviewing Key Terms

1. D
2. A
3. G
4. H
5. E
6. C
7. I
8. J
9. F
10. B

 Mental Mapping Activity

This exercise helps students to visualize the countries and geographic features they have been studying and understand the relationships among them. All attempts at freehand mapping should be accepted.

Mental Mapping Activity

On a separate piece of paper, draw a freehand map of the 50 states of the United States. Label the following items on your map:

- Alaska
- New England
- Great Lakes
- Mississippi River
- Rocky Mountains
- Great Plains

REVIEWING THE MAIN IDEAS

Section 1

1. **REGION** What are the five physical regions of the United States?
2. **LOCATION** In what climate region is Los Angeles located?

Section 2

3. **PLACE** Name three industrial cities of the Midwest.
4. **REGION** Why is farming limited in the Interior West?
5. **MOVEMENT** What is the purpose of NAFTA?

Section 3

6. **MOVEMENT** From which three European countries did most early settlers come?
7. **PLACE** What three branches of government does the United States have?
8. **PLACE** About how many people live in the United States?

Reviewing the Main Ideas

Section 1

1. Coastal Plains, Appalachian Mountains, Interior Plains, Rocky Mountains, Pacific Region
2. Mediterranean

Section 2

3. Chicago, Detroit, Milwaukee, Cleveland, and St. Louis are among the industrial cities of the Midwest.
4. The region is very dry and mountainous.
5. to increase trade among the United States, Canada, and Mexico

CRITICAL THINKING ACTIVITIES

1. **Determining Cause and Effect** Why does the United States have many climate regions?
2. **Analyzing Information** How does the art of Native Americans of the Southwest reflect their environment?

GeoJournal Writing Activity

Find a spot outdoors. Sit and make yourself comfortable. Facing west, draw in your journal what you see on the land and in the sky. Observe any shadows cast by the objects in your view. Then face north and draw what you see. Continue until you have drawn the eastern and southern views as well. What features are in your drawings?

COOPERATIVE LEARNING ACTIVITY

Work in a group of five. Each member of the group should choose one of the climate regions of the United States. Research and prepare a report on the vegetation and wildlife in the chosen climate zone. When research is complete, share your information with your group. As a group, prepare a poster or map that shows the group's findings.

PLACE LOCATION ACTIVITY: THE UNITED STATES

Match the letters on the map with the places and physical features of the United States. Write your answers on a separate sheet of paper.

1. Washington, D.C.
2. Appalachian Mountains
3. Detroit
4. Gulf of Mexico
5. California
6. Lake Superior
7. Great Salt Lake
8. Seattle
9. Mississippi River
10. Ohio River

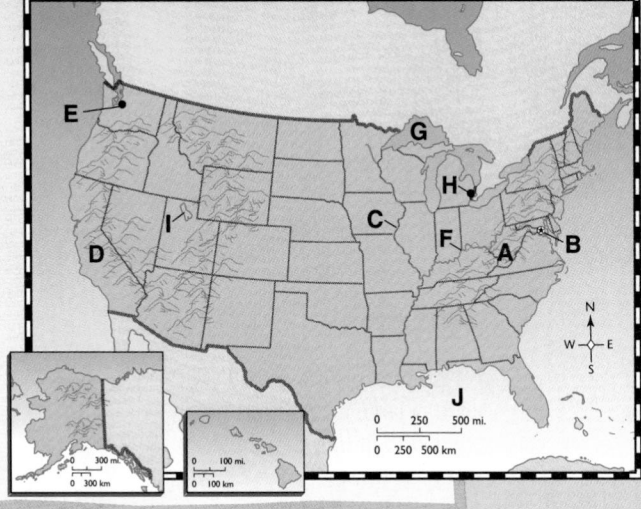

Cooperative Learning Activity

Encourage students to make their posters as colorful and dramatic as possible. Display the visuals in the classroom and invite other classes to share them.

GeoJournal Writing Activity

Encourage students to present and discuss their drawings. Have them note whose drawings are similar and why.

Place Location Activity

1. B
2. A
3. H
4. J
5. D
6. G
7. I
8. E
9. C
10. F

Chapter Bonus Test Question

This question may be used for extra credit on the chapter test.

If you were writing a story about the settlement of North America in the 1600s, which three European countries would play major roles? *(Spain, France, Great Britain)*

109

Section 3

6. Spain, France, and Britain
7. executive, legislative, and judicial branches
8. about 261 million people

Critical Thinking Activities

1. The United States has many climate regions because it is a huge country with many kinds of landforms and elevations.
2. They use materials found in their environment.

GeoLab
ACTIVITY

FOCUS

Have students identify the largest geographic features—oceans, continents, and so on. Record their responses on the chalkboard. Add the term *continental divide* to the list and remind them of its meaning. Point out that the North American continental divide runs from northern Canada through the Rockies into Mexico's Sierra Madre.

TEACH

As students create their models, point out that not all water flowing down the mountains will reach the oceans. Some streams on the western side of the Rockies, for example, flow into the deserts of the Southwest and dry up. After students have completed their models, inform them that all continents have divides. Have them locate the divides in Europe, Asia, Africa, and South America and mark them on outline maps of the world. **L1**

From the classroom of Debora Bittner, San Antonio, Texas

THE CONTINENTAL DIVIDE

Background

In North America, a high ridge of the Rockies known as the Continental Divide separates the waters flowing west to the Pacific Ocean from those flowing east toward the Mississippi River and the Atlantic Ocean. Because rivers flow downhill, whatever landforms are in their paths can direct which way water systems flow. This activity will demonstrate how mountain ranges and elevation can direct the flow of water.

Continental Divide

Believe it or NOT!

In Europe, the continental divide is found in the Alpine Mountain System, which includes the Alps. Streams and rivers drain to the Atlantic Ocean and the Arctic Ocean in one direction, and to the Mediterranean Sea and the Black Sea in the other direction.

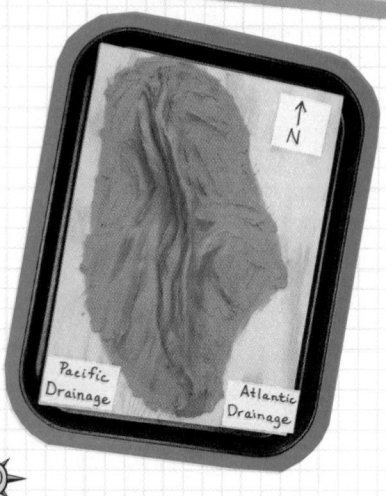
Pacific Drainage | Atlantic Drainage

Materials

- 1 piece of plywood—18" × 24"
- a tray large enough for the plywood to fit into
- 1 quart of water
- 2 pounds of modeling clay

110

UNIT 2

Curriculum Connection Activity

History Many Americans first heard of the continental divide in Mark Twain's book *Roughing It* (1872). Describing a journey in the Rockies, Twain wrote:

> We bowled along cheerily, and presently, at the very summit, we came to a spring which spent its water through two outlets and sent it in opposite directions. The conductor said one of those streams which we were

looking at, was just starting on a journey westward to . . . the Pacific Ocean. . . . He said that the other was just leaving its home among the snow-peaks on a similar journey eastward. . . .

Ask students to research the origins of the term *continental divide*. Also have them discover when and why the North American divide became known as the *Great Divide*. **L1**

What To Do

A. Choose an area in the western United States that contains a portion of the Continental Divide.

B. Build a model of the relief of this area from the modeling clay. Be sure to include different levels of elevation.

C. After the model is completed, identify the north, south, east, and west of your model.

D. Label the area to the west of the mountains "Pacific Drainage."

E. Label the area east of the mountains "Atlantic Drainage."

F. Slowly and carefully pour the water on the peaks of the mountains. Move slowly from north to south as you pour. Take note of the direction and flow of the water.

Lab Activity Report

1. In which direction did most water flow?

2. Why were there different amounts flowing in different directions?

3. How does the elevation of the mountains increase or decrease drainage?

4. *Drawing Conclusions* How will various flows affect the people living in the drainage areas?

Go A Step Further

Activity

Find out the position of the continental divide in South America. What effect does the drainage of the Amazon River have on the Brazilian rain forest? Summarize your findings in a two-paragraph report.

GeoLab
ACTIVITY

ASSESS

Have students answer the **Lab Activity Report** questions on page 111.

CLOSE

Encourage students to complete the **Go A Step Further** writing activity. Encourage students to create maps to accompany their reports. Maps should show the area drained by the Amazon River and the location of the rain forest.

A new system of trails has opened that enables cyclists to ride along the Great Divide from Montana to New Mexico.

Answers to Lab Activity Report

1. Answers will vary.

2. Because the shape of the land and its elevations direct the flow of water.

3. Answers may include: the higher the elevation, the more the drainage increases.

4. Answers may include flooding, irrigation for crops, and so on.

Cultural

HERITAGE:
THE UNITED STATES AND CANADA

FOCUS

Write the term *culture* on the chalkboard and call on students to define it. Then have them suggest things that are representative of American cultures. Remind them that culture includes ideas as well as objects. Note their responses on the chalkboard.

TEACH

Point out that one of the United States's most important exports is its culture. Ask students to discuss what they think it means to export culture. Then call on volunteers to identify some major American cultural exports. *(Answers might include: movies and television shows, music, fashion.)* **L1**

CURRICULUM CONNECTION

Music The first major jazz recordings were made in 1923 by King Oliver's Creole Jazz Band. This group included perhaps the most famous jazz musician of all time, trumpeter Louis Armstrong.

MUSIC ▶ ▶ ▶ ▶

All kinds of music are popular in the United States and Canada, but jazz is thought to be the only form of music that originated in the United States. The earliest jazz musicians were African American.

▲ ▲ ▲ ▲
FOLK ART

Since colonial times, people in the United States and Canada have made quilts as decorative covers for their beds. Quilts are made from scraps of colorful fabric sewn together in a patchwork design.

STAINED GLASS ▶ ▶ ▶ ▶

The delicate blossoms and leaves of this stained-glass lamp were created by American artist Louis Tiffany in the 1890s. Tiffany was born in New York City and created many beautiful Tiffany lamps.

112

UNIT 2

More About . . .

Painting Winslow Homer did not receive any formal art education. Rather, he developed his style on the job as a magazine illustrator. In fact, his first important works, a series of Civil War paintings, were done for the magazine *Harper's Weekly*. After Homer settled on the Maine coast in the early 1880s, his work focused more and more on the sea. These sea pictures dramatically convey the sailor's struggle against the awesome power of the ocean.

◄ ◄ ◄ ◄ ARCHITECTURE

Canada's French heritage is reflected in many of its buildings. The Chateau Frontenac, in the city of Quebec, is a large hotel patterned after a French castle.

WOOD CARVING
► ► ► ►

Totem poles were carved by Native Americans who lived in the northwestern United States and Canada. Many of the figures on totems represent clans or families. A clan's totem might include carvings of birds, fish, plants or other objects found in nature.

PAINTING ▲ ▲ ▲ ▲

Farms and country life were the subjects of many paintings by American artist Winslow Homer. Homer, born in Boston in 1836, was a realist. He tried to paint his subjects as they appeared.

APPRECIATING CULTURE

1. Quilts are called folk art because they were made by everyday people for other people to use or enjoy. They were not created as art to be put into museums. What objects in your home might be considered folk art in the future?

2. Some Native American clans or families were recognized by their totem poles. If you created a totem pole to represent the people who live in your home, what characters or objects would you use?

CHAPTER 4

PERFORMANCE ASSESSMENT ACTIVITY

STARTING A BUSINESS Organize students into groups of three or four. Have group members assign research tasks on the various provinces of Canada among themselves. As each member reports back to the group, other members should create a chart with information on the following: natural resources, landforms, water features, climate, vegetation, agriculture, and industries. Then encourage all students to take the role of a business person who would like to open a factory in Canada. Have them use the information they collected as groups to choose a product to produce and a location for the factory. Students should then write letters to prospective investors asking for investment capital. Letters should explain the geographic reasons for the selected product and the location of the factory.

POSSIBLE RUBRIC FEATURES: Persuasive writing skills, research skills, collaborative skills, accuracy of content information, ability to recognize relationships, analysis skills, decision-making skills

MENTAL MAPPING ACTIVITY

Before teaching Chapter 5, point out Canada on a globe. Have students note where it is located in relation to the United States. Also, encourage students to compare Canada, in terms of area, to the United States and to Russia.

TEACHER'S CORNER

NATIONAL GEOGRAPHIC SOCIETY

INDEX TO NATIONAL GEOGRAPHIC MAGAZINE

The following articles may be used for research relating to this chapter:

- "Toronto," by Richard Conniff, June 1996.
- "David Thompson," by Priit J. Vesilind, May 1996.
- "Rocky Times for Banff," by Jon Krakauer, July 1995.
- "Three Years Across the Arctic," by Ramón Hernando de Larramendi, January 1995.
- "Canada's Highway of Steel," by Michael Parfit, December 1994.
- "The St. Lawrence: River and Sea," by Thomas J. Abercrombie, October 1994.

- "Tatshenshini-Alsek Wilderness Park," by William R. Newcott, February 1994.
- "The Superior Way of Life," by Noel Grove, December 1993.
- "James Bay: Where Two Worlds Collide," by John G. Mitchell, November 1993.
- "Labrador, Canada's Place Apart," by Robert M. Poole, October 1993.

NATIONAL GEOGRAPHIC SOCIETY PRODUCTS AVAILABLE FROM GLENCOE

To order the following products for use with this chapter, contact your local Glencoe sales representative or call Glencoe at 1-800-334-7344:

- *GTV: A Geographic Perspective on American History* (Videodisc)
- *GTV: The American People: Fabric of a Nation* (Videodisc)
- *STV: North America* (Videodisc)

- *STV: World Geography* (Videodisc)
- *Picture Atlas of the World* (CD-ROM)
- *Geography of North America* (Transparencies)

ADDITIONAL NATIONAL GEOGRAPHIC SOCIETY PRODUCTS

To order the following products for use with this chapter, call National Geographic Society at 1-800-368-2728:

- *Physical Geography of North America Series* (6 Videos)
- *Great Lakes, Fragile Seas* (Video)
- *Yukon Passage* (Video)

- *United States Geography Series* (10 Videos)
- *Portraits of the Continents Series:* "Part I: North America; South America." (Filmstrip)

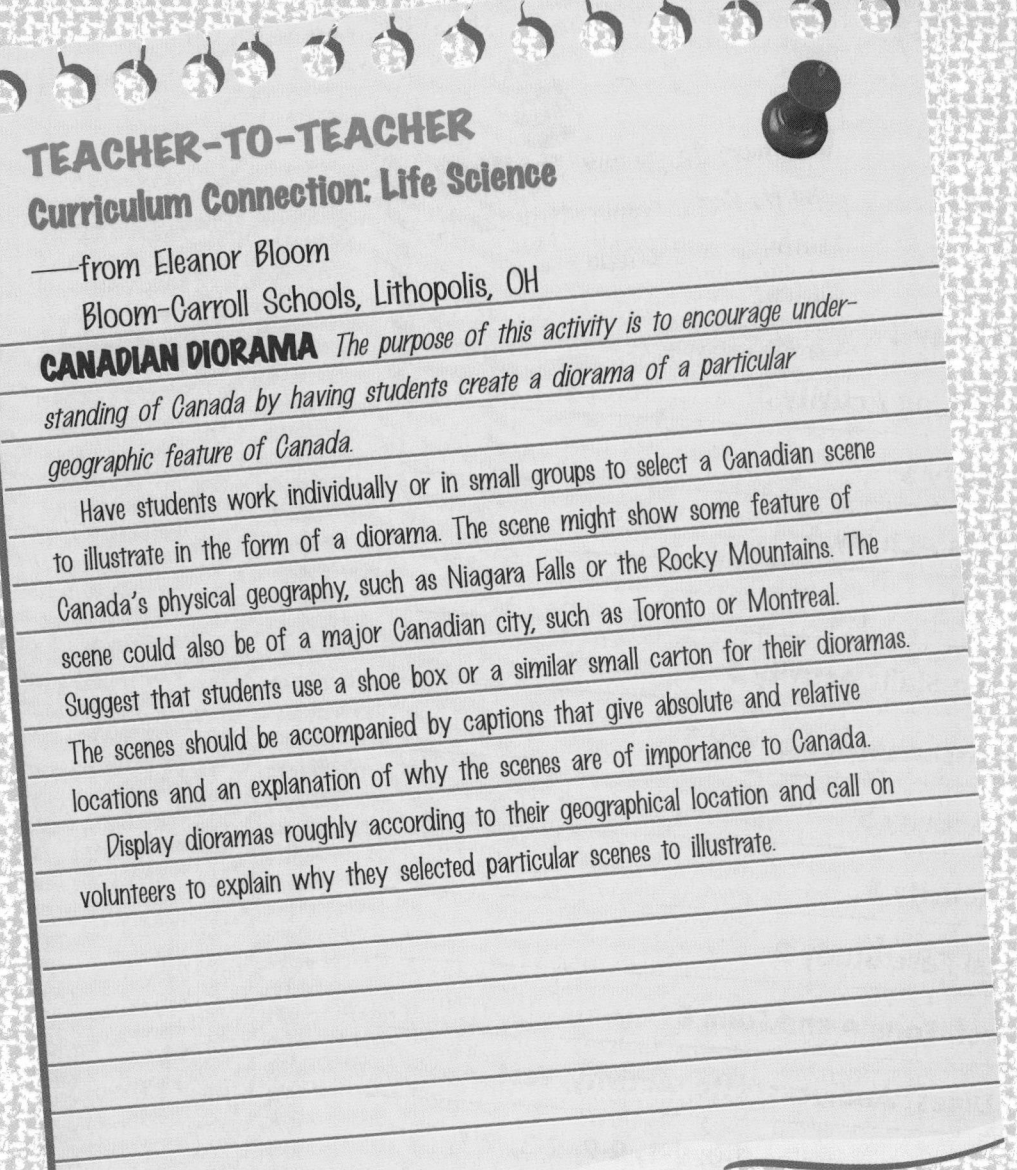

TEACHER-TO-TEACHER
Curriculum Connection: Life Science

—from Eleanor Bloom
    Bloom-Carroll Schools, Lithopolis, OH

CANADIAN DIORAMA *The purpose of this activity is to encourage understanding of Canada by having students create a diorama of a particular geographic feature of Canada.*

Have students work individually or in small groups to select a Canadian scene to illustrate in the form of a diorama. The scene might show some feature of Canada's physical geography, such as Niagara Falls or the Rocky Mountains. The scene could also be of a major Canadian city, such as Toronto or Montreal. Suggest that students use a shoe box or a similar small carton for their dioramas. The scenes should be accompanied by captions that give absolute and relative locations and an explanation of why the scenes are of importance to Canada.

Display dioramas roughly according to their geographical location and call on volunteers to explain why they selected particular scenes to illustrate.

MEETING NATIONAL STANDARDS

Geography for Life

All of the 18 standards are demonstrated in Unit 2. The following ones are highlighted in this chapter:

- Standards 4, 5, 9, 10, 12, 14

KEY TO ABILITY LEVELS

Teaching strategies have been coded for varying learning styles and abilities.

L1   **BASIC** activities for all students

L2   **AVERAGE** activities for average to above-average students

L3   **CHALLENGING** activities for above-average students

LEP **LIMITED ENGLISH PROFICIENCY** activities

BIBLIOGRAPHY

Readings for the Student

Canada: The Lands, Peoples, and Cultures Series. New York: Crabtree, 1993.

The Land and People of Canada. New York: Lippincott, 1991.

Readings for the Teacher

Discover Canada, revised ed. Milwaukee: Apple Press, 1990. Reproducible activity book.

Doran, Charles F., with Puay Tang. *Canada: Unity in Diversity.* New York: Foreign Policy Association, 1990.

Joyce, William W., ed. *Canada in the Classroom.* Washington, D.C.: National Council for the Social Studies, 1985.

Multimedia

Canada: A Nation's Quest for Identity. Madison, Wis.: Knowledge Unlimited, 1994. Videocassette, 18 minutes.

Canada Geograph II. Minneapolis: MECC, 1993. Guide with reproducible activities; Macintosh disks.

Discovering Canada. Video Visits, 1992. Videocassette, 73 minutes.

Use these *Geography: The World and Its People* resources to teach, reinforce, and extend chapter content.

CHAPTER 5 RESOURCES

- *Vocabulary Activity 5
- Cooperative Learning Activity 5
- Workbook Activity 5
- Geography Skills Activity 5
- GeoLab Activity 5
- Critical Thinking Skills Activity 5
- Performance Assessment Activity 5
- *Reteaching Activity 5
- Enrichment Activity 5
- Environmental Case Study 2
- Chapter 5 Test, Form A and Form B
- *Chapter 5 Digest Audiocassette, Activity, Test
- Unit Overlay Transparencies 2-0, 2-1, 2-2, 2-3, 2-4, 2-5, 2-6
- Political Map Transparency 2; World Cultures Transparency 2
- Geoquiz Transparencies 5-1, 5-2
- Vocabulary PuzzleMaker Software
- Student Self-Test: A Software Review
- Testmaker Software
- MindJogger Videoquiz

If time does not permit teaching the entire chapter, summarize using the Chapter 5 Highlights on page 131, and the Chapter 5 English (or Spanish) Audiocassettes. Review students' knowledge using the Glencoe MindJogger Videoquiz. *Also available in Spanish*

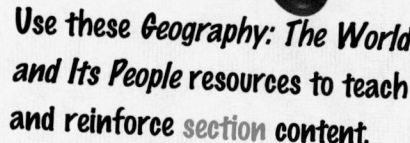

Use these *Geography: The World and Its People* resources to teach and reinforce section content.

SECTION 1 RESOURCES

Reproducible Lesson Plan 5-1

Section Focus Transparency 5-1

Guided Reading Activity 5-1

Section Quiz 5-1 (also available in Spanish)

SECTION 2 RESOURCES

Reproducible Lesson Plan 5-2

Section Focus Transparency 5-2

Guided Reading Activity 5-2

Section Quiz 5-2 (also available in Spanish)

SECTION 3 RESOURCES

Reproducible Lesson Plan 5-3

Section Focus Transparency 5-3

Guided Reading Activity 5-3

Section Quiz 5-3 (also available in Spanish)

ADDITIONAL RESOURCES FROM GLENCOE

 Reproducible Masters

- Glencoe Social Studies Outline Map Book, pages 22, 23
- Facts On File, GEOGRAPHY ON FILE, Northern America
- Foods Around the World, Unit 2

World Music: Cultural Traditions Lesson 1

 Transparencies

- National Geographic Society PicturePack Transparencies, Unit 2

Posters

- National Geographic Society: Images of the World, Unit 2
- National Geographic Society: Eye on the Environment, Unit 2

 Videodiscs

- Geography and the Environment: The Infinite Voyage
- Reuters Issues in Geography
- National Geographic Society: STV: North America
- National Geographic Society: GTV: Planetary Manager

 CD-ROM

- National Geographic Society: Picture Atlas of the World

 Software

- National Geographic Society: ZipZapMap! World

Performance Assessment

Refer to the Planning Guide on page 114A for a Performance Assessment Activity for this chapter. See the *Performance Assessment Strategies and Activities* booklet for suggestions.

Chapter Objectives

1. **Identify** the landscapes and climates of Canada.
2. **Describe** Canada's economic resources, activities, and challenges.
3. **Discuss** the history and cultural heritage of the Canadian people.

 GLENCOE
TECHNOLOGY

 **VIDEODISC**

Use the Chapter 5 Mind-Jogger Videoquiz to preview chapter content.

MindJogger Videoquiz

Chapter 5
Disc 1 Side B

Also available in VHS.

Chapter 5 Canada

MAP STUDY ACTIVITY

As you read Chapter 5, you will learn about Canada, the second-largest country in the world.

1. **What is the northernmost territory of Canada?**
2. **What is Canada's national capital?**
3. **In what province is the national capital located?**

114

 MAP STUDY ACTIVITY

Answers
1. Northwest Territories
2. Ottawa
3. Ontario

Map Skills Practice
Reading a Map Which Canadian province borders on the Pacific Ocean?

(British Columbia) New Brunswick, Nova Scotia, and Prince Edward Island are sometimes called the Maritime Provinces. Why do you think this is so? *(because they all border on the sea)*

LESSON PLAN
Chapter 5, Section 1

PREVIEW

Words to Know
- prairie
- cordillera

Places to Locate
- Appalachian Highlands
- St. Lawrence River
- Canadian Shield
- Hudson Bay
- Rocky Mountains
- Coast Mountains

Read to Learn . . .
1. how Canada's landscapes differ from region to region.

2. where the oldest rock formations in North America are found.
3. how climate affects where Canadians live.

Imagine yourself in this photograph. Do you hear the waves of the Atlantic crashing against the rocky coast? This lovely place is Cape

Breton Island, one of Canada's many scenic areas.

Meeting National Standards

Geography for Life The following standards are highlighted in this section:
- Standards 4, 7, 15

FOCUS

Section Objectives
1. **Describe** how Canada's landscapes differ from region to region.
2. **Identify** where the oldest rock formations in North America are found.
3. **Explain** how climate affects where Canadians live.

Bellringer Motivational Activity

 Prior to taking roll at the beginning of the class period, project Section Focus Transparency 5-1 and have students answer the activity questions. Discuss students' responses. **L1**

Vocabulary Pre-check

Have students define the terms *prairie* and *cordillera*. Then ask students to contrast these two physical features based on physical appearance, natural vegetation, and location. **L1** **LEP**

 Use the Vocabulary PuzzleMaker Software to create a crossword puzzle. **L1**

Vikings first landed their longboats on its eastern coast around A.D.1000. Niagara Falls thunders in the southeast. Grizzly bears roam its western territories. What country are we describing? It is Canada.

Location
A Northern Land

The map on page 114 shows you that Canada lies north of the contiguous United States. Between the two countries lies the longest undefended border in the world. Like the United States, Canada has the Atlantic Ocean on its eastern coast and the Pacific Ocean on its western coast. The Arctic Ocean lies north of Canada.

The second-largest country in the world in area, Canada covers 3,849,674 square miles (9,970,610 sq. km). Only Russia is larger. Canada contains 10 provinces and 2 territories. Look at the map on page 114 to find the provinces of Newfoundland, Nova Scotia, New Brunswick, Prince Edward Island, Quebec (kwih•BEHK), Ontario, Manitoba, Saskatchewan (suh•SKA•chuh•wuhn), Alberta, and British Columbia.

CHAPTER 5

(115)

Classroom Resources for Section 1

 REPRODUCIBLE MASTERS

Reproducible Lesson Plan 5-1
Guided Reading Activity 5-1
Vocabulary Activity 5
Chapter Map Activity 5
Geography Skills Activity 5
Section Quiz 5-1

TRANSPARENCIES

Section Focus
   Transparency 5-1
Political Map Transparency 2

MULTIMEDIA

 Vocabulary PuzzleMaker
   Software

Testmaker

National Geographic
   Society: STV:
   North America

115

TEACH

Guided Practice

Display Political Map Transparency 2. Have students compare Canada's landforms and bodies of water with those of the United States. Then have them identify the physical features that the two countries share. *(Appalachians, Great Lakes, Rocky Mountains, Great Plains)* L1

More About the Illustration

Answer to Caption
Great Plains

Place
Alberta's farmers grow huge amounts of oats, barley, and rye, as well as wheat. They also raise beef cattle. Despite all this agricultural activity, only about 7 percent of Alberta's workers are involved in farming. The service industries are the province's biggest employers, accounting for 37 percent of the workforce.

MULTICULTURAL PERSPECTIVE

Culturally Speaking
The name *Canada* probably comes from the Iroquois word *Kanata-Kon,* which means "to the village" or "to the small houses."

Loading Grain in Stavely, Alberta

The wide-open prairies of Canada's Interior Plains are dotted with grain elevators filled with wheat and other grains.
REGION: What physical region in the United States is an extension of Canada's Interior Plains?

Now look for the Northwest Territories and the Yukon Territory. In 1999 a third territory—Nunavut (NOO•nuh•VUHT)—will be carved out of part of the Northwest Territories. This area is claimed by the Inuit.

Regions
Landforms

Canada—from an Iroquois word for "village"—can be divided into six major physical regions: the Appalachian Highlands, the St. Lawrence and the Great Lakes Lowlands, the Canadian Shield and the Arctic Islands, the Interior Plains, the Rocky Mountains, and the Pacific Coast. From border to border, Canada has more lakes and inland waterways than any other country in the world.

The Appalachian Highlands Along Canada's southeastern Atlantic coast stretch the Appalachian Highlands. They continue through the eastern United States as the Appalachian Mountains. Traveling around this part of Canada, you will see rolling hills and low mountains. The valleys between are dotted with farms. Forests blanket much of the Appalachian Highlands. Many deepwater harbors nestle along the region's jagged, rocky coast.

The St. Lawrence and Great Lakes Lowlands Sprawling urban areas and slow-moving barges are a common sight in this region. The St. Lawrence River and the Great Lakes form the major waterways leading

WHAT IN THE WORLD

Nature's Flowerpots
On the shoreline of Ontario's Georgian Bay are strange-looking landforms called flowerpots. Wind and water have worn away the bases of these tall rock formations. Their wide tops are covered with flowers, plants, and small shrubs—just like a flowerpot!

116

UNIT 2

Cooperative Learning Activity

Organize students into six groups and assign each group one of Canada's six major physical regions. Inform groups that their task is to create a puzzle map of Canada. Display a large wall map of Canada and have group representatives trace the outline of their region from it. Instruct group members to mark and label their regions' most important physical features. Suggest that groups use shading to show elevation. Have groups paste their finished map sections onto cardboard. Then have group representatives assemble the map puzzle. L1

into central Canada. Find the St. Lawrence River valley on the map below. Now find the peninsula in Ontario that is bordered by Lake Erie, Lake Ontario, and Lake Huron. These areas are lowlands with rich soil and vital transportation routes. You can find most of Canada's people, farms, and industries here. The world-famous Niagara Falls—where the waters of Lake Erie flow into Lake Ontario—also impresses visitors to the region.

The Canadian Shield and the Arctic Islands Picture a huge, abandoned parking lot. The concrete slabs have split and buckled, and wildflowers poke through the cracks and puddles. The Canadian Shield looks a little like this, except that ice and snow cover its 2 million square miles (5,180,000 sq. km)—more than half of Canada's entire land area!

Wrapped around Hudson Bay is the huge, horseshoe-shaped Canadian Shield. Hills worn down by erosion and hundreds of lakes carved by glaciers cover much of the Shield. Scientists claim this region holds some of the oldest rock formations—some more than 3 billion years old—in North America.

What plants can thrive in the cold northern areas of the Canadian Shield? Mosses and small shrubs grow here. In the southern part, evergreen forests provide shelter and food to deer, elk, and moose. Although the soil here is not good for farming, it is rich in mineral resources such as iron ore, copper, nickel, and gold.

CANADA: Physical

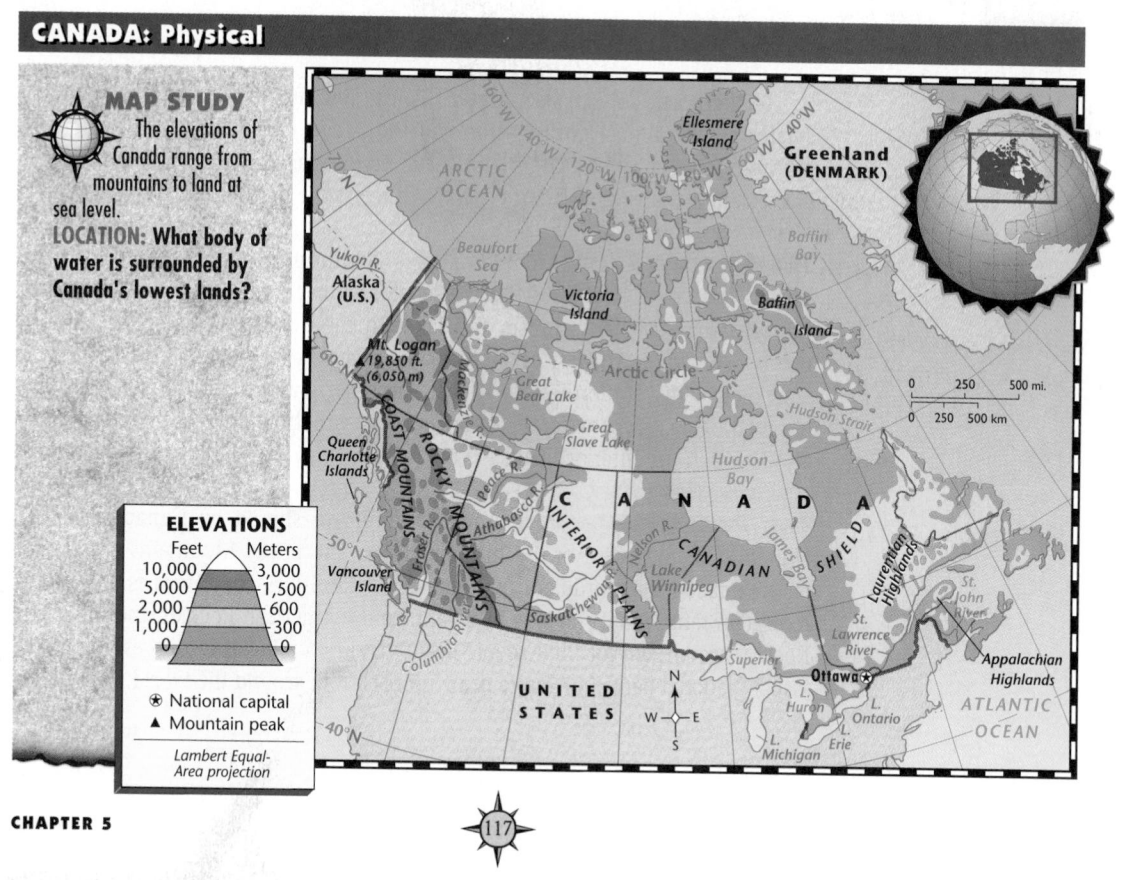

MAP STUDY The elevations of Canada range from mountains to land at sea level.

LOCATION: What body of water is surrounded by Canada's lowest lands?

ELEVATIONS

Feet	Meters
10,000	3,000
5,000	1,500
2,000	600
1,000	300
0	0

✯ National capital
▲ Mountain peak

Lambert Equal-Area projection

CHAPTER 5

117

Independent Practice

Guided Reading Activity 5-1 **L1**

In the roughly 180 miles (290 km) from Lake Ontario to the city of Montreal, the St. Lawrence River falls some 220 feet (67 m) through a series of rapids.

NATIONAL GEOGRAPHIC SOCIETY

These materials are available from Glencoe.

 VIDEODISC

STV: North America

Canadian Shield
Any side, Frame 53629

 **MAP STUDY**

Answer
Hudson Bay

Map Skills Practice
Reading a Map Where are the highest elevations found in Canada? *(in the west)*

Meeting Special Needs Activity

Writing Disability Help students with writing problems obtain information from the text by creating questions for each subhead. Offer these examples for the subhead "The Canadian Shield and the Arctic Islands": "Where is the Canadian Shield located?" "What types of plants grow in the Canadian Shield?" "What mineral resources are found in the Canadian Shield?" Ask students to read the subhead and write brief answers to these questions. Then have them write their own questions and answers for the subheads titled "The Interior Plains" and "The Rocky Mountains." **L1**

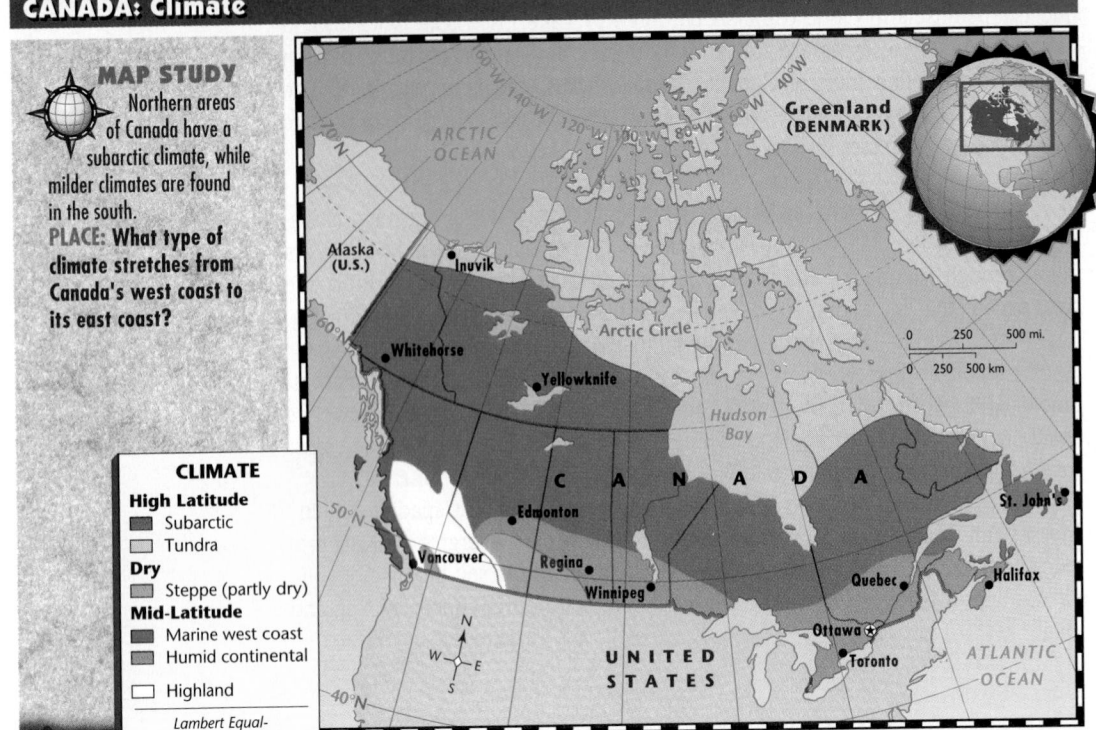

CANADA: Climate

MAP STUDY Northern areas of Canada have a subarctic climate, while milder climates are found in the south.
PLACE: What type of climate stretches from Canada's west coast to its east coast?

CLIMATE
High Latitude
  Subarctic
  Tundra
Dry
  Steppe (partly dry)
Mid-Latitude
  Marine west coast
  Humid continental
  Highland

Lambert Equal-Area projection

To the north lie Canada's Arctic Islands. Ten large islands and hundreds of small ones lie almost entirely north of the Arctic Circle. Because of their far northern location, the Arctic Islands have no forests. Only tundra plants such as grasses and lichens grow here. Glacial ice covers the northernmost islands.

The Interior Plains Canada shares landforms with its neighbor to the south. The Great Plains of the United States become the Interior Plains in Canada. You see many forests and lakes in the northern part of the Interior Plains. The southern part of these plains is a huge rolling **prairie,** or inland grassland, with very fertile soil. Wheat and other grainfields echo with the rumble of tractors. Rich mineral resources such as petroleum and natural gas are also found in the Interior Plains.

The Rocky Mountains Another landform shared by Canada and the United States lies west of the Interior Plains. The Rocky Mountains are part of an area called the Cordillera (KAW•duhl•YEHR•uh). A **cordillera** is a group of mountain ranges that run side by side. Like the American Rockies, the Canadian Rockies are known for their scenic beauty and rich mineral resources. Banff and Jasper National parks in Alberta draw tourists from around the world.

118

Critical Thinking Activity

Making Comparisons Encourage students to compare Canada's climate regions with those of the United States. Then ask them to determine what geographic features contribute to the differences in climate in the two countries. **L1**

The Pacific Coast Mount Logan—Canada's highest peak—towers 19,850 feet (6,050 m) over the Pacific coastal region. A major group of the Cordillera—the Coast Mountains—rise above this western part of Canada. Like the Rockies, the Coast Mountains cross into the United States. Near Canada's border with Alaska, several mountains soar more than 15,000 feet (4,572 m).

Place
Climate

One thing Canada does not share with the United States is climate. The map on page 118 shows you that Canada generally has a cool or cold climate because it lies in the high latitudes of the Western Hemisphere. As you read in Chapter 2, however, differences in climate can be caused by a place's location or its nearness to oceans.

In northern Canada people shiver in the cold polar climate. Farther south, between 50° and 70° North latitude, you find a subarctic climate with short, cool summers and long, cold winters. The southeastern part of Canada has a humid continental climate. Most Canadians live in this area.

The southwestern Pacific coast is Canada's only area of wet, mild winter climate. There the Coast Mountains cause warm winds from the Pacific Ocean to release moisture. The western, or windward, side of the mountains gets more rain and warmer temperatures than any other part of Canada. On the eastern side of the mountains, however, the climate is dry or partly dry in the rain shadow.

Traveling Along the Trans-Canada Highway

Canada's ten provinces are linked by a highway that extends from the Atlantic to the Pacific coasts.
MOVEMENT: What two areas of Canada are not linked by the Trans-Canada Highway?

SECTION 1 REVIEW

REVIEWING TERMS AND FACTS
1. Define the following: prairie, cordillera.
2. PLACE What is the landscape of the Appalachian Highlands like?
3. PLACE In what physical region do most Canadians live?
4. REGION What area of Canada enjoys a mild climate and plenty of rainfall?

MAP STUDY ACTIVITIES
5. Turn to the physical map on page 117. Which of the Great Lakes lies farthest north?
6. Look at the climate map on page 118. In which climate region do you find Ottawa?

Answers to Section 1 Review

1. All vocabulary words are defined in the Glossary.
2. The Appalachian Highlands have rolling hills and low mountains. Forests cover much of the area. The rocky coastline has many deepwater harbors.
3. St. Lawrence and Great Lakes Lowlands
4. the southwest Pacific coast of British Columbia
5. Lake Superior
6. humid continental

LESSON PLAN
Chapter 5, Section 1

More About the Illustration
Answer to Caption
Yukon and Northwest Territories

Movement
The Trans-Canada Highway stretches some 4,860 miles (7,825 km) from St. John's, Newfoundland, to Victoria, British Columbia. Opened in 1962, it is the world's longest national highway.

Evaluate
Section 1 Quiz **L1**

Use the Testmaker to create a customized quiz for Section 1. **L1**

Reteach
Have students work in pairs to quiz each other on the content of Section 1. **L1**

Enrich
Have students create posters illustrating plants and animals of Canada. Display and discuss posters, asking students to identify which species, if any, are unique to Canada. **L3**

CLOSE
Encourage students to imagine they are taking a trip west on the Trans-Canada Highway. Have them write a description of the landscapes they see on their journey.

119

BUILDING GEOGRAPHY SKILLS

TEACH

Display a blank outline map and a physical map of the same region. Have students note the information shown on the physical map that does not appear on the outline map. *(physical features—rivers, lakes, oceans, mountains, for example)* Encourage students to imagine they are planning to take a vacation in an unfamiliar region. Ask them how a physical map would help them determine the activities they could take part in on the vacation. *(Possible responses: rivers, lakes and oceans would indicate water sports; mountains would indicate climbing or skiing.)* Then direct students to read the skill and complete the practice questions. **L1**

Additional Skills Practice

1. How would you describe British Columbia in terms of elevation. *(Much of the province is above 2,000 feet—600 m—in elevation.)*
2. Based on the physical geography of British Columbia, what economic activities do you think take place in the province? *(Possible answers: fishing, mining, logging, tourism)*

📁 Geography Skills Activity 5

Reading a Physical Map

It is difficult to imagine the shape of land you have not seen. Physical features such as mountains, plains, oceans, and rivers are shown on maps. A map that shows these natural features is called a *physical map.* A physical map often uses color to show the land's elevation, or height above sea level. To read a physical map, apply these steps:

- Identify the region shown on the map.
- Use the map key to find the meaning of colors and symbols.
- Find important physical features including elevation, rivers, and coastlines. Imagine the actual shape of the land.

BRITISH COLUMBIA: Physical

ELEVATIONS

Feet	Meters
10,000	3,000
5,000	1,500
2,000	600
1,000	300
0	0

▲ Mountain peak

Conic projection

Geography Skills Practice

1. Name two major mountain ranges that cross British Columbia. In which directions do they run?

2. Why are many lakes in British Columbia so narrow?

Answers to Geography Skills Practice

1. Rocky Mountains, Coast Mountains; both ranges run northwest to southeast.
2. They lie in narrow valleys between the mountain ranges.

SECTION 2
The Economy

PREVIEW

Words to Know
- fossil fuel
- newsprint

Places to Locate
- Newfoundland
- Maritime Provinces
- Quebec
- Ontario
- Prairie Provinces
- British Columbia

Read to Learn . . .
1. what natural resources Canada has.
2. how the Canadian people earn their livings.
3. what challenges face Canada and its economy today.

Hundreds of bays and fishing villages dot the jagged Atlantic coast of Newfoundland. Not far from the village shown here lie the Grand Banks—one of the great fishing areas of the world. Fleets from Canada and many other countries harvest fish in the Grand Banks.

Fishing is Canada's oldest industry. It is only one of many economic activities that Canadians carry on today. Productive farms, mines, and businesses help make Canada one of the world's most economically developed countries.

Region
Economic Specialties

Canada's economy is very similar to that of the United States. The country is known for its rich farmland, many natural resources, and skilled workers. Service industries, manufacturing, and farming are the country's major economic activities. Canada, too, has an economy based on free enterprise. The Canadian government, however, plays an active part in some economic activities, such as broadcasting, transportation, and health care.

Like the economy of the United States, Canada's economy differs from region to region. Geographers group Canada's provinces and territories into six economic regions: Newfoundland and the Maritime Provinces, Quebec, Ontario, the Prairie Provinces, British Columbia, and the North.

CHAPTER 5 — 121 —

Meeting National Standards
Geography for Life The following standards are highlighted in this section:
- Standards 11, 14, 16

FOCUS

Section Objectives
1. **Identify** Canada's natural resources.
2. **Discuss** how Canadians earn a living.
3. **Describe** the challenges Canada and its economy face today.

Bellringer Motivational Activity
Prior to taking roll at the beginning of the class period, project Section Focus Transparency 5-2 and have students answer the activity questions. Discuss students' responses. **L1**

Vocabulary Pre-check
Have students define the terms *fossil fuel* and *newsprint*. Then have them determine how each term is related to Canada's economy. **L1 LEP**

Classroom Resources for Section 2

REPRODUCIBLE MASTERS
Reproducible Lesson Plan 5-2
Guided Reading Activity 5-2
Vocabulary Activity 5
Workbook Activity 5
Cooperative Learning Activity 5
Section Quiz 5-2

TRANSPARENCIES
Section Focus
Transparency 5-2
Unit Overlay Transparencies 2-0, 2-5
Political Map Transparency 2

MULTIMEDIA
 Testmaker

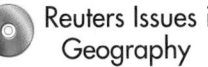 Reuters Issues in Geography

TEACH

Guided Practice

Display Unit Overlay Transparencies 2-0 and 2-5. Have students speculate on the following questions: Where might farming take place in Canada? Where in Canada might forestry be a major economic activity? Which locations in Canada might be manufacturing centers? Direct students to record their speculations in their notebooks. Encourage them to check and adjust their speculations as they work through the section. **L2**

MAP STUDY

Answer
Possible answers: St. John's, Corner Brook, Sydney, Halifax, Saint John, Victoria, Vancouver

Map Skills Practice
Reading a Map Which economic region do you think is the center of Canada's oil industry? *(Prairie Provinces)*

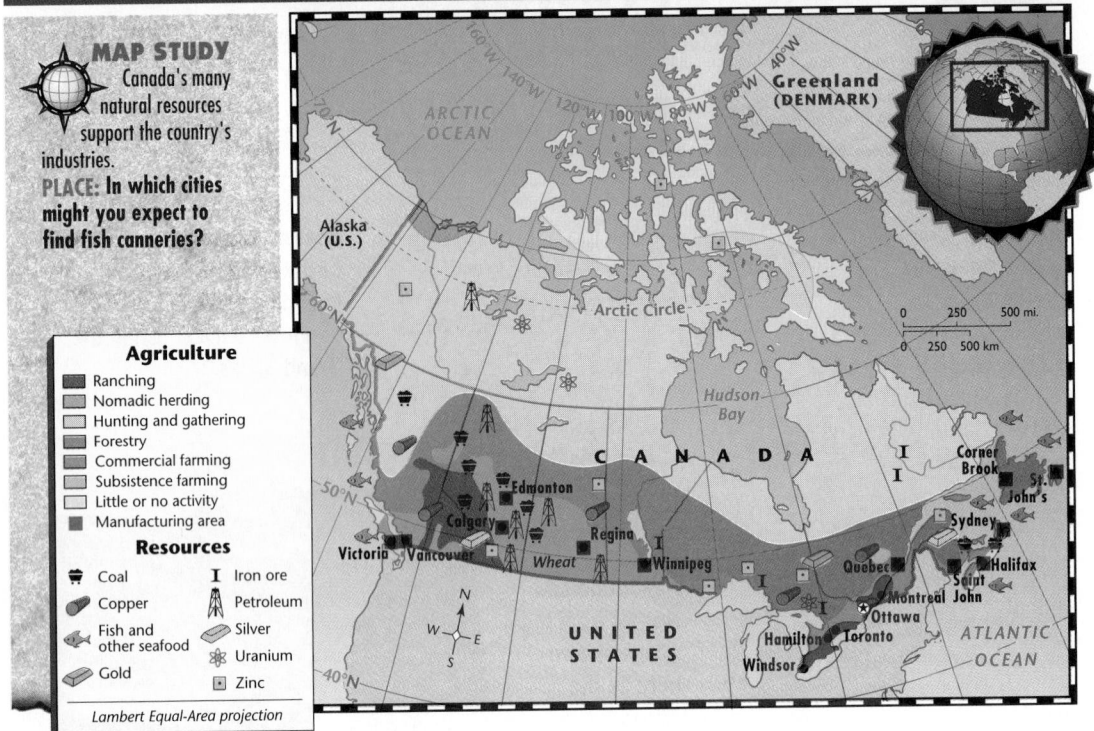

CANADA: Land Use and Resources

MAP STUDY
Canada's many natural resources support the country's industries.
PLACE: In which cities might you expect to find fish canneries?

Agriculture
- Ranching
- Nomadic herding
- Hunting and gathering
- Forestry
- Commercial farming
- Subsistence farming
- Little or no activity
- Manufacturing area

Resources
- Coal
- Copper
- Fish and other seafood
- Gold
- Iron ore
- Petroleum
- Silver
- Uranium
- Zinc

Lambert Equal-Area projection

Newfoundland and the Maritime Provinces Canada's easternmost economic region is formed by Newfoundland and the Maritime Provinces. These provinces include Nova Scotia, New Brunswick, and Prince Edward Island. Fishing and shipbuilding were major economic activities here. Because of overfishing, however, only 3 percent of the region's workers now make a living from the sea.

By looking at the map above, you can see why farming is limited in this region. A short growing season and thin, rocky soil discourage large-scale farming. Small farms throughout the Maritime Provinces, however, grow crops such as potatoes and apples.

Manufacturing and mining provide jobs for most people in this area of Canada. Halifax, the capital of Nova Scotia, is a major shipping center. Its harbor remains open in winter when ice closes most other eastern Canadian ports. Tourism is also important to the economy.

Quebec About 25 percent of Canada's people live in Quebec, the largest province in area. Manufacturing and service industries are major economic activities here. Montreal on the St. Lawrence River is Canada's second-largest city and the center of Quebec's economic and social life. To the north, Quebec's mines provide iron ore, copper, and gold. Other economic activities in Quebec are agriculture and fishing.

Cooperative Learning Activity

Organize students into 12 groups and assign each group one of Canada's provinces or territories. Have groups brainstorm a list of all the economic activities in their assigned province or territory. Direct groups to use their lists to create a symbol and motto that represents their province's or territory's economy. Call on groups to display and explain their symbols and mottoes. **L1**

Ontario Ontario is Canada's second-largest province, but it has the most people and the greatest wealth. The map on page 122 shows you that Ontario is rich in natural resources. This is one reason Ontario is the leading manufacturing region of Canada. The region produces more than half of Canada's manufactured goods.

Toronto, the capital of Ontario, is Canada's largest city. It is also the country's chief manufacturing, financial, and communications center. As a center of industry, Ontario has faced industrial pollution and other threats to the environment.

Ontario borders the Great Lakes and the St. Lawrence River. To open the Great Lakes to ocean shipping, the United States and Canada built the St. Lawrence Seaway. Look at the map and diagram below. You see how this system of locks and canals allows ships to travel between the Great Lakes and the Atlantic Ocean. The St. Lawrence Seaway has helped make Ontario a major shipping area.

Farmers in southern Ontario grow grains, fruits, and vegetables and raise beef and dairy cattle. Canada's capital city, Ottawa, lies in Ontario near the Quebec border. Many Canadians work in government offices in Ottawa.

The Prairie Provinces The heartland of Canada overflows with resources. The Prairie Provinces are Manitoba, Saskatchewan, and Alberta. The map on page 122 shows you that farming and raising cattle are major economic activities in this region. The Prairie Provinces are important wheat producers. Ranchers in Alberta also raise most of Canada's cattle.

The St. Lawrence Seaway

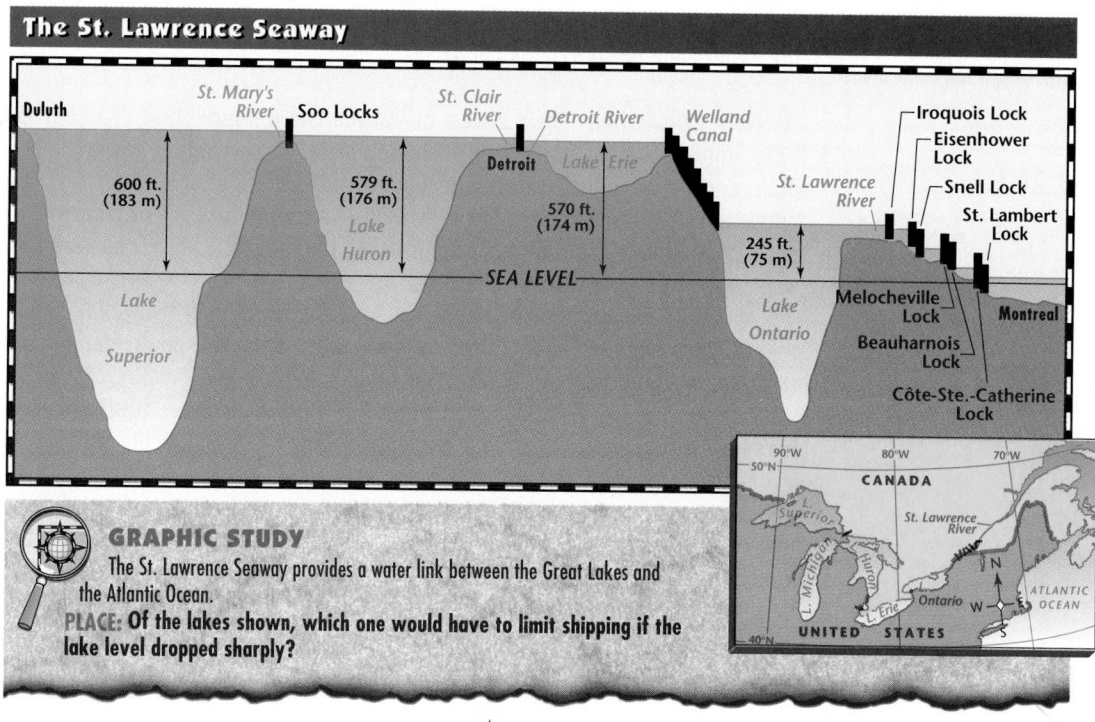

 GRAPHIC STUDY
The St. Lawrence Seaway provides a water link between the Great Lakes and the Atlantic Ocean.
PLACE: Of the lakes shown, which one would have to limit shipping if the lake level dropped sharply?

LESSON PLAN
Chapter 5, Section 2

Independent Practice

Guided Reading Activity 5-2 **L1**

CURRICULUM CONNECTION

Life Science Some Canadians favor draining wetlands in farming areas of the Prairie Provinces. They argue that this would create additional farmland and lead to increased crop production. Conservationists oppose draining the wetlands because these areas are the habitat of many migratory birds.

GRAPHIC STUDY

Answer
Lake Erie

Graphic Skills Practice
Reading a Diagram Between which two lakes is there the greatest drop? *(Lake Erie and Lake Ontario)*

Meeting Special Needs Activity

Attention Deficiency Help students with attention problems by making them more aware of their behavior during study periods. Have students divide a piece of paper into two columns with the headings *On-task* and *Off-task*. Direct students to silently read Section 2. At intervals of two to three minutes, call out "Check." Each time the call is made, students should enter a check mark in the *On-task* column if they are paying close attention to their reading. If they are not paying close attention, they should enter a check mark in the *Off-task* column. When students have completed the reading, encourage them to honestly evaluate how well they attended during the exercise. **L1**

123

MULTICULTURAL PERSPECTIVE

Cultural Heritage

Life in Vancouver, British Columbia, has a strong Asian flavor. About 15 percent of the population is Chinese, and many people of Japanese heritage also live in the city.

ASSESS

Check for Understanding

Assign Section 2 Review as homework or an in-class activity.

Meeting Lesson Objectives

Each objective below is tested by the questions that follow it in parentheses.

1. **Identify** Canada's natural resources. (6)
2. **Discuss** how Canadians earn a living. (2, 5)
3. **Describe** the challenges Canada and its economy face today. (4)

More About the Illustration

Answer to Caption
Quebec

Movement Almost half the people who live in Toronto are of English descent. Large numbers of people from Italy and Portugal immigrated to the city after World War II. Toronto also has large populations of people of German, French, Chinese, and Caribbean heritage.

A vast prairie surrounds the rapidly growing city of Calgary, Alberta *(left)*. A skywalk links business and shopping areas in Toronto, Ontario *(right)*.
PLACE: Which of Canada's provinces is largest in area?

Some of the world's largest reserves of fossil fuels—oil and natural gas—were discovered in Alberta and Saskatchewan. A **fossil fuel** is a fuel that is formed from the remains of prehistoric plants and animals.

The Prairie Provinces also have some of the fastest-growing cities in Canada. Calgary and Edmonton, both in Alberta, are leading centers for the oil industry. Winnipeg, in Manitoba, is a major transportation and business center that links the eastern and western parts of Canada.

British Columbia The far west of Canada is a treasure chest of resources. Thick forests blanket much of British Columbia. As you might guess, timber, pulp, and paper industries provide much of British Columbia's income. The province helps to make Canada the world's leading producer of **newsprint,** or paper that is made into newspapers. The mining of coal, copper, and lead also adds to the wealth of British Columbia.

Agriculture and fishing are strong economic activities in British Columbia. In river valleys between mountains, farmers raise cattle and poultry and grow many fruits and vegetables. Fishing fleets sailing out into the Pacific Ocean bring their catch of salmon, halibut, and herring to port. Vancouver, British Columbia's largest city, is a bustling trade center and Canada's major Pacific port.

The North Geese, moose, bear, and reindeer lay claim to northern regions that cover 40 percent of Canada's land. The North economic region includes the Yukon Territory and the Northwest Territories. The far north is Arctic wilderness, and the south produces forest products and minerals. The Yukon Territory and the Northwest Territories are still largely undeveloped.

Critical Thinking Activity

Analyzing Information Have students work in pairs to draw up a list of social and economic challenges that face Canada today. Then ask pairs to rank the challenges, from the most pressing to the least important. Call on pairs to present and justify their rankings. **L2**

Place
Today's Challenges

The differences that give Canada a rich, diverse economy also provide challenges to the country. One issue concerns two of Canada's cultures. Others involve trade and environmental issues.

Regional Differences Not only physical and economic differences but also cultural differences divide the regions of Canada. About 80 percent of Quebec's population is French-speaking. The rest of Canada, however, is largely English-speaking. Many people in Quebec want freedom for their province. In 1995 voters there narrowly defeated a proposal for independence.

NAFTA Since World War II, Canada has strengthened its economic ties with the United States. Today about 70 percent of Canada's trade is with the United States. In contrast, only about 20 percent of United States trade is with Canada. Some Canadians want even closer economic ties with the United States. Others fear loss of Canadian jobs to the United States.

In 1993 the three major countries of North America—Canada, the United States, and Mexico—created a new trading partnership. They approved the North American Free Trade Agreement (NAFTA). This agreement aims to do away with trade barriers among the three countries by 1999.

Pollution The protection of its natural resources is a major challenge facing Canada. The national government works with the provinces to deal with air and water pollution. Environmental agreements between Canada and the United States aimed at reducing pollution have also been made.

SECTION 2 REVIEW

REVIEWING TERMS AND FACTS
1. **Define the following:** fossil fuel, newsprint.
2. **PLACE** Why is Ontario a major shipping area?
3. **REGION** How are the Prairie Provinces like the Great Plains of the United States?
4. **MOVEMENT** How do Canadians view close economic ties between Canada and the United States?

 MAP STUDY ACTIVITIES
5. Look at the map on page 122. In which provinces is fishing a major economic activity?
6. Using the same map, name the mineral found north of the Arctic Circle.

LESSON PLAN
Chapter 5, Section 2

GLENCOE
TECHNOLOGY

 VIDEODISC

Reuters Issues in Geography

Chapter 1
The United States and Canada: The Water Supply
Disc 1 Side A

Evaluate

 Section 2 Quiz **L1**

Use the Testmaker to create a customized quiz for Section 2. **L1**

Reteach
Have students create a chart of Canada's economic regions showing ways in which people make a living in each region. **L1**

Enrich
Have students research and report on the building of the St. Lawrence Seaway. Reports should discuss how the seaway reflects the interdependence of Canada and the United States. **L3**

CLOSE

Have students write "want ads" for jobs in Canada. Call on volunteers to read their ads and have the rest of the class identify in which regions the various employment opportunities might exist.

Answers to Section 2 Review
1. All vocabulary words are defined in the Glossary.
2. It borders the Great Lakes and the St. Lawrence River, which are linked to the Atlantic by the St. Lawrence Seaway.
3. Both the Prairie Provinces and the Great Plains have flat, fertile land that is highly suitable for growing grain.
4. Some Canadians believe that closer ties between the United States and Canada will open up American markets to Canadian goods. Others fear that these ties will result in the loss of jobs to the United States.
5. Newfoundland, Nova Scotia, New Brunswick, Prince Edward Island, British Columbia
6. zinc

125

MAKING CONNECTIONS

TEACH

Have students use atlases to locate the various places named in the feature. Then have them study the maps on pages 94 and 118 to discover the kinds of climates the gold seekers found on their travels to, and when they arrived in, the Klondike. Call on volunteers to discuss how climate added to the problems that the gold seekers had to overcome. **L1**

More About . . .

History All of the following are strange-but-true stories of the Klondike gold rush.

- The mayor and a quarter of the police force of Seattle resigned and went in search of Klondike gold.
- A group of enterprising people hacked out a 1,500-step staircase, called the *Golden Stairs,* up the mountain face to Chilkoot Pass. They then charged a toll to use the steps.
- Dawson laundry workers could scoop as much as $20 worth of gold dust from the bottoms of their tubs after washing miners' clothing.
- On some claims, gold nuggets lay on top of the ground where they could easily be picked up. One miner's wife collected $10,000 worth of gold in a year on strolls through her husband's claim.

GOLD! GOLD! GOLD!

| MATH | SCIENCE | HISTORY | LITERATURE | TECHNOLOGY |

"Gold! Gold! Gold!" shouted the headline of the *Seattle Post-Intelligencer* on July 17, 1897. Gold seekers from Canada's Klondike region in the Yukon Territory had found gold dust and nuggets. Within 10 days, ships began landing excited gold seekers in the Alaskan ports of Skagway and Dyea. Getting to the goldfields, however, was a difficult problem. Miners had to brave mountains, glaciers, and the Yukon River.

NO SAFE WAY Three routes led to the Klondike's goldfields in Dawson, Canada. One crossed a glacier; the second and third went through mountain passes. None were safe.

The Malspina Glacier, near the Yukon Territory's southern border, was a dangerous, slippery ice field. Huge cracks lay hidden near its surface. Many people died or were injured in falls. Others were blinded by the sunlight on the ice.

Another route, the 45-mile-long (73-km) Skagway Trail, began with an easy climb. The sight facing the miners from the top of the trail, however, was row upon row of boulder-covered hills and deep, muck-filled swamps.

The third route, Dyea Trail, was only 33 miles (54 km) long, but it led to Chilkoot

Pass. Crossing this pass meant making a very steep climb thousands of feet up a snow-covered mountain face. Most miners were not experienced in this type of rugged travel.

Chilkoot Pass, Alaska

THE YUKON RIVER If they made it past the coastal mountains, gold seekers still had to build boats and sail down the Yukon River. They faced rapids, a deadly whirlpool, and huge boulders before reaching the goldfields at Dawson.

Nearly 40,000 people made it to Dawson during the Klondike's three-year gold rush. Only 200 to 300 miners actually found gold. Was it worth the trip?

Making the Connection

1. Why did so many people head for the Klondike in 1897?
2. What geographical barriers did gold seekers face?

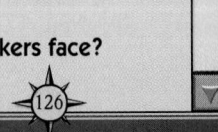 126

Answers to Making the Connection

1. to search for gold
2. They had to cross high mountains, muck-filled swamps, and dangerous glaciers. They also had to sail down the Yukon River, which was dotted with rapids, whirlpools, and huge boulders.

SECTION 3 — The People

PREVIEW

Words to Know
- parliamentary democracy
- prime minister
- bilingual

Places to Locate
- Quebec
- Montreal

Read to Learn . . .
1. how Canada gained its independence.
2. what groups make up the Canadian people.
3. where most Canadians live.

Founded by French explorers in 1608, Quebec is one of the oldest cities in North America. It has many picturesque buildings, churches, and even a fortress overlooking the St. Lawrence River. In Quebec, you can see signs in French and taste mouth-watering French cooking in the restaurants.

Like the United States, Canada is a multicultural country. The circle graph on page 85 shows you the groups that make up Canada's population. Native Americans and Inuit form a small but growing part of the population. Most Canadians, however, share a European background.

Place
Influences of the Past

The great migration of the first Asians to North America brought Canada its earliest people some 12,500 years ago. They later became the Native Americans and the Inuit. Many other groups—most of them European—later settled in Canada.

The French In 1534 French explorer Jacques Cartier sailed into the St. Lawrence River valley and claimed the area for France. French explorers, settlers, and missionaries later founded cities such as Quebec and Montreal. For almost 230 years, France ruled the area around the St. Lawrence and the Great Lakes as New France.

CHAPTER 5

127

Meeting National Standards

Geography for Life The following standards are highlighted in this section:
- Standards 6, 10, 17

FOCUS

Section Objectives
1. **Explain** how Canada gained its independence.
2. **Identify** the groups that make up the Canadian people.
3. **Locate** where most Canadians live.

Bellringer Motivational Activity

Prior to taking roll at the beginning of the class period, project Section Focus Transparency 5-3 and have students answer the activity questions. Discuss students' responses. **L1**

Vocabulary Pre-check

Point out that the prefix *bi* means "two" and that *lingua* means "language." Have students use these clues to define the term *bilingual*. **L1 LEP**

Classroom Resources for Section 3

 REPRODUCIBLE MASTERS

Reproducible Lesson Plan 5-3
Guided Reading Activity 5-3
GeoLab Activity 5
Performance Assessment Activity 5
Reteaching Activity 5
Enrichment Activity 5
Section Quiz 5-3

 TRANSPARENCIES

Section Focus Transparency 5-3
World Cultures Transparency 2
Geoquiz Transparencies 5-1, 5-2

MULTIMEDIA

 Testmaker

 National Geographic Society: Picture Atlas of the World

TEACH

Guided Practice

Making Comparisons
Have students create a chart that compares the governments of Canada and the United States. Suggest that they use bases of comparison such as head of state, lawmaking body, capital, and regional governments. **L1**

Independent Practice

📁 Guided Reading Activity 5-3 **L1**

Quebec's annual winter carnival includes ice sculpting, fireworks, and canoe races *(left)*. Canada's national police force is the Royal Canadian Mounted Police *(right)*.
PLACE: What are the two official languages of Canada?

The British In 1497 John Cabot landed on the Atlantic coast near Newfoundland and claimed the area for England. In the 1600s and 1700s, the British and the French fought over land in North America. The British won control of New France. After the American Revolution, many Americans loyal to the British settled in Canada.

The Birth of Canada Several cultures emerged in Canada in the 1700s. Both English-speaking and French-speaking settlers arrived. The western part—Ontario—became a colony for English-speaking settlers. The eastern part with its large French-speaking population became the colony of Quebec. In 1867 the British united the provinces of Ontario, Quebec, Nova Scotia, and New Brunswick into one nation known as the Dominion of Canada.

Prince Edward Island, the western territories, and Newfoundland eventually joined the Dominion as provinces. In 1982 Canadians claimed the right to change their constitution without British approval. The British king or queen reigns, but the British government has no real power in Canada.

Canada's Government Features borrowed from both the United Kingdom and the United States formed Canada's government. The Canadians have a British-style **parliamentary democracy.** Voters elect representatives to a lawmaking body called Parliament. The head of the largest political party in Parliament is named the **prime minister,** or leader. Over the years, power has shifted from the national government to the provinces.

Movement

A Changing Population

Although Canada is large in land area, it has only about 29 million people. The population has grown rapidly since World War II, however, because of

Cooperative Learning Activity

Organize students into several groups and assign each group a time period in Canadian history. Determine reasonable time periods—perhaps 50- or 100-year increments from 1600 to the present. Using several sheets of butcher paper, construct a time line running parallel to the one created for the Cooperative Learning Activity for Chapter 4, Section 3 (see page 103). Direct groups to research their time period and select 10 important events to enter on the time line. Have group representatives write their selections on the time line, asking them to add annotations where necessary. Then have students analyze the two time lines and compare the historical development of Canada and the United States. **L1**

heavy immigration and a high birthrate. Look at the population density map below. It shows you that most Canadians live in cities near the Canadian-American border. This heavily populated area covers only about one-tenth of the country!

Canada is a **bilingual** country, or a country with two official languages—English and French. But other European languages, as well as Native American languages, are also spoken.

Canada has a long heritage of religious freedom. Although most Canadians are Roman Catholics or Protestants, other Canadians follow Judaism, Buddhism, or Islam.

Place
The Arts and Recreation

Canada's culture is a blend of many heritages. Each province holds on to its ethnic traditions with pride.

Literature, Art, Music From poetry to novels to drama, Canadian authors write in either English or French about many subjects. The works of Canadian painters and sculptors are as varied as Canada's people. In the early 1900s, a group of artists known as the Group of Seven painted the northern Canadian

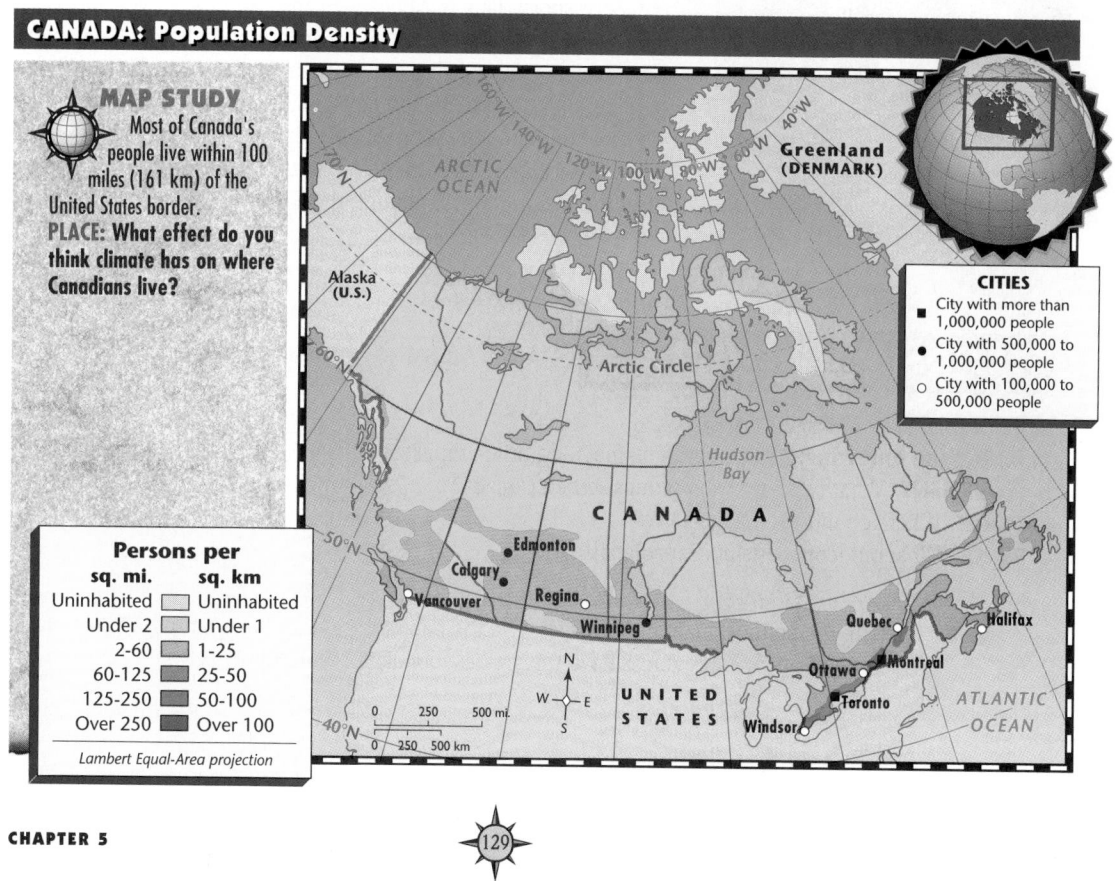

CANADA: Population Density

MAP STUDY
Most of Canada's people live within 100 miles (161 km) of the United States border.
PLACE: What effect do you think climate has on where Canadians live?

CITIES
■ City with more than 1,000,000 people
● City with 500,000 to 1,000,000 people
○ City with 100,000 to 500,000 people

Persons per	
sq. mi.	**sq. km**
Uninhabited	Uninhabited
Under 2	Under 1
2-60	1-25
60-125	25-50
125-250	50-100
Over 250	Over 100

Lambert Equal-Area projection

ASSESS

Check for Understanding

Assign Section 3 Review as homework or an in-class activity.

Meeting Lesson Objectives

Each objective below is tested by the questions that follow it in parentheses.

1. **Explain** how Canada gained its independence. (1)
2. **Identify** the groups that make up the Canadian people. (2)
3. **Locate** where most Canadians live. (3, 5)

Global  **Gourmet**

Canada *Poutine* is a popular fast food in Quebec. It consists of french fries covered with spicy gravy and cheese curds.

MAP STUDY

Answer
Areas close to the United States border, where most Canadians live, have a milder climate. The cold northern areas, however, are sparsely populated.

Map Skills Practice
Reading a Map Name Canada's two largest cities. *(Toronto, Montreal)*

Meeting Special Needs Activity

Memory Disability Note taking especially benefits students with memory problems. Point out that taking notes helps students to focus on important information, to organize facts, and to review for tests. Direct students to take notes on the subsection titled "The Arts and Recreation." When students have completed the task, review their note-taking procedures and, if necessary, suggest ways to improve. **L1**

Evaluate

 Section 3 Quiz **L1**

 Use the Testmaker to create a customized quiz for Section 3. **L1**

 Chapter 5 Test Form A and/or Form B **L1**

Reteach

 Reteaching Activity 5 **L1**

 Spanish Reteaching Activity 5 **L1**

Enrich

 Enrichment Activity 5 **L3**

CLOSE

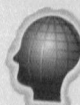 **Mental Mapping Activity**

Provide each student with a large sheet of white paper and map pencils. Inform students that this will not be a graded activity. Have students draw, freehand from memory, a map of Canada with labels for the following provinces and territories: Newfoundland, Nova Scotia, New Brunswick, Prince Edward Island, Quebec, Ontario, Manitoba, Saskatchewan, Alberta, British Columbia, Northwest Territories, Yukon Territory.

Teen Scene

Let's See the Puck!

Anton Lafonte has been playing hockey since he was four. As soon as the water freezes on the lake near his home, he and his friends choose up teams for the season. Because winters are so long in northern Quebec, Anton has had plenty of ice time to fine-tune his game. This year Anton has recruited two new players for his team—his younger sisters Yvette and Jennifer.

wilderness in brilliant colors. British Columbia's green forests and colorful Native American totem poles appear in Emily Carr's works.

Canadians share a love for all kinds of music—from jazz to classical. Music of different cultures—such as Scottish folk music, reggae from the Caribbean area, and African American rap—is also popular. Canadians of Native American, Ukrainian, and African ancestry host many festivals where their ethnic dances are performed.

Food and Recreation Canadians enjoy the same foods as their neighbors in the United States. A variety of meats and seafood—beef, chicken, and fish—are served with vegetables and potatoes. Ethnic foods are popular from coast to coast.

French Canadians create their own special dishes. Thirteen-year-old Marie Villeneuve and her family live in Quebec. At Christmas they enjoy hearty pea soup, a meat pie called a *tourtière,* and a dessert made of maple syrup.

Canadians enjoy a variety of activities, especially outdoor sports. You will find local parks or national parks crowded with sports enthusiasts. One of these sports fans is Michael Evans. He is a 15-year-old Canadian who lives in Toronto. Michael, like many Canadians, enjoys playing and watching ice hockey. He also takes part in winter sports, including skiing, skating, and snowshoeing. During the summer Michael goes sailing on Lake Ontario and plays baseball. He tries to attend all the baseball games played at Toronto's large indoor stadium, the Skydome.

SECTION 3 REVIEW

REVIEWING TERMS AND FACTS
1. **Define the following:** parliamentary democracy, prime minister, bilingual.
2. **MOVEMENT** What two European groups settled Canada?
3. **PLACE** Where do most Canadians live?
4. **PLACE** What outdoor sports do Canadians enjoy?

 MAP STUDY ACTIVITIES
5. Compare the political map on page 114 with the population density map on page 129. Which two Canadian provinces have the greatest population densities?

130

Answers to Section 3 Review

1. All vocabulary words are defined in the Glossary.
2. British and French
3. in cities near the Canadian-American border
4. ice hockey, skiing, skating, snowshoeing in winter; sailing in summer
5. Quebec, Ontario

Chapter 5 Highlights

Important Things to Know About Canada

SECTION 1 THE LAND

- Canada covers most of the northern part of North America.
- Canada has a variety of landforms, including mountains, lowlands, prairies, and Arctic wilderness.
- Most of Canada has a cool or cold climate. Milder temperatures are found in the southern part of the country.

SECTION 2 THE ECONOMY

- Canada has one of the world's most developed economies. It is rich in natural resources, farmland, and skilled workers.
- Workers in Canada's cities fill jobs in banking, communication, and other service industries. Rural Canadians mostly grow grain crops and raise livestock.
- Canada faces three major challenges: holding its separate regions together, working out its trade relationship with the United States, and dealing with environmental issues.

SECTION 3 THE PEOPLE

- Native Americans and Inuit were the first Canadians. Europeans, mainly British and French, later settled Canadian lands and founded the modern country of Canada.
- Cultural differences exist between French-speaking Quebec and the rest of Canada, which is largely English-speaking.
- Canada today is a multicultural country with many different peoples from throughout the world.
- Many people in Quebec want freedom for their province. In 1995 voters there narrowly defeated a proposal for independence.

Canadians cheering in Toronto's Skydome ▶

Extra Credit Project

Calendar Have students locate or draw 12 pictures suitable for a calendar titled "Canada Through the Year." Pictures should show landscapes or events appropriate for each month. Encourage students to add captions identifying the landscapes or events shown. Call on volunteers to display their calendars and explain why they selected particular pictures. **L1**

Using the Chapter 5 Highlights

- Use the Chapter 5 Highlights to preview, review, condense, or reteach the chapter.

Preview/Review

 Vocabulary Puzzle-Maker Software reinforces the terms used in Chapter 5.

Condense

Have students read the Chapter 5 Highlights. Spanish Chapter Highlights also are available.

🎧 📁 Chapter 5 Digest Audiocassettes and Activity

📁 Chapter 5 Digest Audiocassette Test

Spanish Chapter Digest Audiocassettes, Activities, and Tests also are available.

📁 Guided Reading Activities

Reteach

📁 📁 Reteaching Activity 5. Spanish Reteaching activities also are available.

MAP STUDY

Location Copy and distribute page 23 from the Outline Map Resource Book. Have students locate and label each of the places listed in this chapter's **Section Previews**.

131

Chapter 5 Review and Activities

ANSWERS
Reviewing Key Terms

1. E
2. A
3. F
4. B
5. D
6. G
7. C

 **Mental Mapping Activity**

This exercise helps students to visualize the countries and geographic features they have been studying and understand the relationships among them. All attempts at freehand mapping should be accepted.

REVIEWING KEY TERMS

Match the numbered terms in Column A with their definitions in Column B.

A
1. cordillera
2. parliamentary democracy
3. prairie
4. newsprint
5. prime minister
6. fossil fuel
7. bilingual

B
A. voters elect representatives to a law-making body called Parliament
B. paper that is made into newspapers
C. area in which two languages are spoken
D. head of government chosen from the largest political party in Parliament
E. group of mountain ranges that run side by side
F. inland area of grasslands
G. product formed from the remains of prehistoric plants and animals

Mental Mapping Activity

On a separate piece of paper, draw a freehand map of Canada. Label the following items on your map:
- Nova Scotia
- British Columbia
- Quebec
- Coast Mountains
- Montreal
- Ontario
- Arctic Ocean

REVIEWING THE MAIN IDEAS

Section 1
1. **LOCATION** What three oceans surround Canada?
2. **REGION** Which physical region is mostly rolling hills and low mountains?
3. **PLACE** What climate is found in the southeastern part of Canada?

Section 2
4. **REGION** In which economic region do you find wheat and oil?
5. **HUMAN/ENVIRONMENT INTERACTION** What are two important economic activities in British Columbia?
6. **PLACE** Which province is mostly French in culture and language?

Section 3
7. **MOVEMENT** Who were the first people to settle Canada?
8. **PLACE** What are Canada's two official languages?

Reviewing the Main Ideas

Section 1
1. Atlantic, Pacific, Arctic
2. Appalachian Highlands
3. humid continental

Section 2
4. Prairie Provinces
5. any two of the following: timber, pulp, and paper industries; mining; agriculture; fishing
6. Quebec

CRITICAL THINKING ACTIVITIES

1. **Making Comparisons** Compare the overall climate of Canada with the overall climate of the United States.
2. **Analyzing Information** Why do most Canadians live near the border of the United States?

GeoJournal Writing Activity

Imagine that you and your family are moving to Canada. Describe which province or territory you would like to live in. Why did you choose this area? How would your life change if you moved there?

COOPERATIVE LEARNING ACTIVITY

Work in a group of four to learn more about the major cities in Canada. Each group will choose one of the following cities: Montreal, Ottawa, Toronto, Winnipeg, or Vancouver. Each member will research one of the following topics: (a) major buildings and historic places; (b) cultural activities; (c) sports and recreation; or (d) economic activities. After your research is complete, share your information with your group. Prepare a commercial that encourages people to visit or live in your city. Use charts, posters, or music in your commercial.

PLACE LOCATION ACTIVITY: CANADA

Match the letters on the map with the places and physical features of Canada. Write your answers on a separate sheet of paper.

1. Ontario
2. St. Lawrence River
3. Winnipeg
4. Hudson Bay
5. Vancouver
6. Yukon Territory
7. Newfoundland
8. Rocky Mountains
9. Ottawa

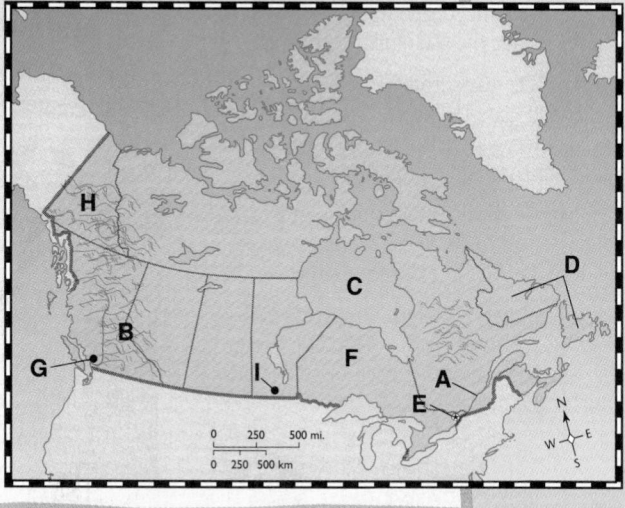

Cooperative Learning Activity

Encourage groups to make their commercials in the form of a tour guide to their selected city. Call on groups to present their commercials to the rest of the class.

GeoJournal Writing Activity

Encourage students who selected the same province or territory to compare their journal entries.

Place Location Activity

1. F
2. A
3. I
4. C
5. G
6. H
7. D
8. B
9. E

Chapter Bonus Test Question

This question may be used for extra credit on the chapter test.

Identify the Canadian province described below.

1. It is a leading oil producer.
2. Wheat is one of its agricultural products.
3. Edmonton is the provincial capital.
(Alberta)

Section 3

7. Asians, who later became Native Americans and Inuit
8. English and French

Critical Thinking Activities

1. Much of Canada has a cold climate—either subarctic or tundra. The United States, in contrast, has a variety of climates—from tropical to tundra.
2. Areas near the border have the mildest climates, the most fertile land, and the easiest access to navigable waterways.

133

FOCUS

Mention that the United States generates more trash per person than any other country in the world. Then ask students what kinds of things they think Americans throw away. Note their responses on the chalkboard.

TEACH

Next to the list of student suggestions, write the following:

Paper Products	37%
Yard Waste	18%
Plastics	8%
Glass	7%
Aluminum	1%
Other Metals	7%
Miscellaneous	22%

Have students present this information in a circle graph. Suggest that they highlight categories that are relatively easy to recycle—paper, yard waste, glass, and aluminum, for example—with shading or a different color. Encourage students to write a caption noting what percentage of American trash these recyclable materials represent.

L1

NATIONAL GEOGRAPHIC SOCIETY
EYE ON THE ENVIRONMENT

United States and Canada
TRASH

PROBLEM

The United States and Canada generate a lot of trash—approximately 200 million tons a year. That's about 4 pounds per person per day. Most of this garbage is burned in incinerators or dumped in landfills. But incinerators can pollute the air, and landfills are running out of space. Nearly half of all landfills in the United States will be land*fulls* by the year 2000.

SOLUTIONS

● Environmentalists want paper manufacturers to pay a tax on each ton of new paper produced—which should encourage the use of recycled paper.

● The list of products made from recycled materials includes crushed glass mixed with asphalt, a paving material called glasphalt, and recycled plastic made into park benches and clothing.

● A firm in Eugene, Oregon, makes building panels called Enviroboards out of recycled plastic, newspaper, cardboard, rubber, and polystyrene.

● A hospital in Burlington, Vermont, recycles 600–900 pounds of kitchen waste a day. The food waste is composted, and the compost is used on vegetables grown nearby for the hospital.

Printed on recycled paper.

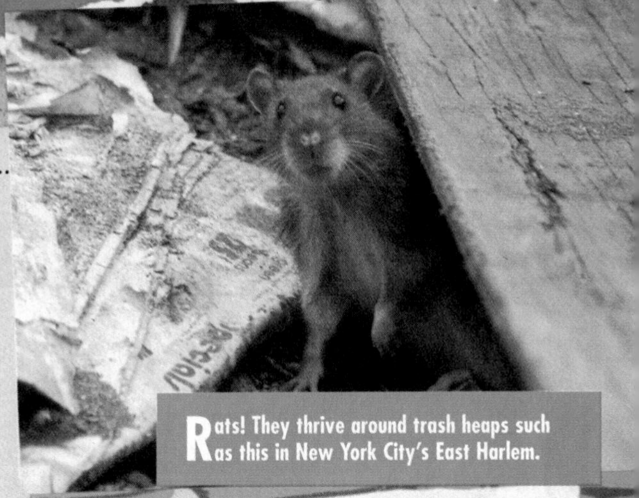

Rats! They thrive around trash heaps such as this in New York City's East Harlem.

TRASH FACT BANK

🍂 In the United States, 19 percent of trash is recycled, 17 percent incinerated, and 64 percent dumped in landfills.

🍂 Every $1,000 spent in fast-food stores creates 200 pounds of trash.

🍂 The garbage thrown away each year in the United States could fill garbage trucks lined up bumper-to-bumper on a four-lane highway circling the earth.

🍂 Paper, construction debris, and yard cuttings take up nearly 90 percent of the space in landfills.

🍂 A ton of recycled paper saves more than 4,000 kilowatt-hours of energy, 7,000 gallons of water, and 17 trees.

(134)

More About the Issues

GEOGRAPHIC THEMES

Location *Hazardous wastes are* materials, such as toxic chemicals, that present an exceptional threat to the environment. Disposal of these materials is closely regulated by the Environmental Protection Agency (EPA). In some places, however, businesses have dumped hazardous wastes without due care and attention. The EPA has identified more than 1,200 such hazardous waste sites throughout the country—an average of about 25 sites per state.

This landfill near Washington, D.C., is part of the solution for getting rid of trash—and part of the problem. Landfills in the United States are overflowing, and it's difficult to open new ones.

NATIONAL GEOGRAPHIC SOCIETY

These materials are available from Glencoe.

VIDEODISC

GTV: Planetary Manager

Tidy World
Side 2, Chapter 5

ASSESS

Have students answer the **Take A Closer Look** questions on page 135.

CLOSE

Discuss with students the **What Can You Do?** question. Encourage students to research and report on trash disposal in their community. Consider inviting a representative of a local waste utility or an environmental group to give a brief presentation and answer students' questions.

*For an additional case study, use the following:

Environmental Case Study 2 **L1**

TEEN TRIBUTE

As part of an ecology club project, Taz Guishard tends red worms living in a bin in his classroom at Franko Middle School in Mount Vernon, New York. The worms recycle organic waste in a process called vermicomposting. One pound of worms can chew through one pound of waste and change it into rich soil in one month. "A lot of people talk about saving the earth," says Taz, "but I'm doing something about it."

environmental activities

TAKE A CLOSER LOOK

1 What happens to most garbage in the United States and Canada?

2 What is expected to happen to landfills in the United States by the year 2000?

REDUCE · RECYCLE · REUSE · RESTORE ·

This boy takes a step to solve the trash problem by picking up litter. Much of the trash taken to disposal areas can be recycled and reused.

WHAT CAN YOU DO?

🍃 Stop and think before you throw ANYTHING away!

🍃 Recycle aluminum—you can save 90 percent of the energy needed to make new cans by reusing the old.

🍃 Organize a recycling program for your lunchroom. Could leftover food go to a homeless shelter?

🍃 Boycott fast-food restaurants that do not use recycled materials for packaging.

Answers to Take A Closer Look

1. It is burned in incinerators or dumped in landfills.
2. Nearly half of all the landfills in the United States will be full by the year 2000.

Global Issues

Pollution The most popular form of trash disposal worldwide is the open dump. Such dumps provide a breeding ground for rats and disease, foul the air with disgusting odors and smoke from burning garbage, and pollute nearby groundwater.

0:00 OUT OF TIME?

If time does not permit teaching each chapter, you may use the Chapter Highlights, Audiocassettes, and their corresponding activities and tests.

GLENCOE
TECHNOLOGY

 VIDEODISC

Reuters Issues in Geography

Chapter 2
Latin America:
The Disappearing
Rain Forests
Disc 1 Side 1

CONNECTION

An electronic reference desk on all of the countries of Latin America (including Mexico and the nations of Central America and the Caribbean) is located at the following address:

World Wide Web
http://lanic.utexas.edu/las.html

Latin America is home to many indigenous peoples. Explore the diverse cultures of these "first Americans" by checking out:

World Wide Web
http://www.maxwell.
syr.edu/nativeweb/
geography/latinam/
latinam.html

Unit 3 Latin America

What Makes Latin America a Region?

The people share . . .

- a strong Spanish or Portuguese influence on culture and language.
- an African and Native American heritage.
- a mostly tropical or subtropical climate.
- the world's largest zone of tropical rain forest climate.

To find out more about Latin America, see the Unit Atlas on pages 138–151.

Iguaçú Falls, Brazil

136

About the Illustration

The Iguaçú Falls, located near the junction of the Iguaçú and Paraná rivers, are 269 feet (82 m) high and about 2 miles (3 km) wide, four times the width of Niagara Falls. The horseshoe-shaped falls are formed by 275 separate waterfalls, which cascade over lava formations and basalt rock. Spectacular rainbows are created from the rising spray and from the water deflected by protruding rocks. Brazil and Argentina, both bordering the falls, have established national parks to protect the vegetation, wildlife, and scenic beauty of the falls.

 **Human/Environment Interaction** If the national parks surrounding Iguaçú Falls had not been established, what changes might farmers have made to this area? *(cleared forests for farmland, dug irrigation ditches, built homes, apartments, roads, dams, and so on)*

NATIONAL GEOGRAPHIC SOCIETY

These materials are available from Glencoe.

 **CD-ROM**

Picture Atlas of the World

 SOFTWARE

ZipZapMap! World

TRANSPARENCIES

PicturePack Transparencies, Unit 3

GeoJournal Activity

Ask students to choose one specific aspect of Latin American culture or a physical feature that interests them. Have students describe it in terms of taste, smell, feel, look, or sound. Have students illustrate their ideas using poster board and markers or paint. Display students' work on a class bulletin board. **L2**

• This journal activity provides the basis for the "GeoJournal Writing Activity" exercise in the Chapter Review.

• The GeoJournal may be used as an integral part of Performance Assessment.

GeoJournal Activity

As you read about this region, use each of your senses. Record your ideas about the tastes, smells, feel, look, and sounds that make up Latin America.

137

Location Activity

Display Political Map Transparency 3 and ask students the following questions: What continents form Latin America? *(South America and part of North America)* Why is part of Latin America called Central America? *(It is geographically in the center between North America and South America.)* What countries border Argentina? *(Bolivia, Paraguay, Brazil, Uruguay, and Chile)* What part of Latin America contains groups of islands? *(Caribbean)* What bodies of water border Latin America to the east and west? *(Gulf of Mexico, Caribbean Sea, Atlantic Ocean, Pacific Ocean)* **L1**

IMAGES OF THE WORLD

NATIONAL GEOGRAPHIC SOCIETY

These materials are available from Glencoe.

CD-ROM

Picture Atlas of the World

POSTERS

Images of the World
Unit 3: Latin America

TRANSPARENCIES

PicturePack Transparencies, Unit 3

Strange but TRUE!

The Amazon River reaches its broadest width where it meets the Atlantic Ocean. There, it is about 40 miles (64 km) wide. Seasonal tides push the Amazon's waters upriver in large waves at 22 miles (35 km) an hour.

NATIONAL GEOGRAPHIC SOCIETY

IMAGES of the WORLD

1. **Girl in traditional dress, Guatemala**
2. **Crowds in movie district, Buenos Aires**
3. **Gaucho swinging a rope, Argentina**
4. **Maya potter, Mexico**
5. **Men drilling for oil, Brazil**
6. **Rio de Janeiro at night**
7. **Desert landscape, Mexico**

All photos viewed against a panorama of Iguaçú Falls, Brazil.

138

Fun Facts

• **Ecuador** This country takes its name from its location on the Equator. A monument near Quito, Ecuador's capital, marks the line of the Equator.

• **Colombia** Tutunendo, in the rain forests of Colombia, has the highest annual average rainfall in the world.

Geography and the Humanities

World Literature Reading 3

World Music: Cultural Traditions, Lesson 2

World Cultures Transparencies 3, 4

Focus on World Art Print 11

Teacher Notes

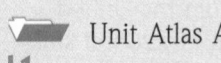

LESSON PLAN
Unit 3 Atlas

Using the Unit Atlas

These features and activities may be used as an introduction to the unit or as teaching tools throughout the course of the unit.

📁 Unit Atlas Activity 3
L1

🔧 📁 Unit Overlay Transparencies 3-0, 3-1, 3-2, 3-3, 3-4, 3-5, 3-6

📁 Home Involvement Activity 3

📁 Environmental Case Study 3

🗂 Countries of the World Flashcards, Unit 3

🧩 World Games Activity Cards, Unit 3

More About the Illustration

Answer to Caption
Orinoco River, Río de la Plata, São Francisco River, Amazon River

Location
Rising in the Andes in Colombia, the Magdalena River flows about 995 miles (1,600 km) to the Caribbean Sea. Shallow-draft vessels, such as the one shown in the photograph, can sail some 149 miles (240 km) upstream.

Regional Focus

Latin America is the name given to the vast region that lies south of the United States. The map on page 144 shows you that Latin America begins at the Rio Grande, a river dividing the United States and Mexico. It ends at the southern tip of the continent of South America.

Region
The Land

Latin America covers some 7,900,000 square miles (20,400,000 sq. km)—about 16 percent of the earth's surface. This region includes Mexico and the countries of Central America and South America. It also takes in the West Indies island countries in the Caribbean Sea.

Cruising the Magdalena River in Colombia

A river ferry carries passengers and their luggage down the Magdalena River.
LOCATION: What other major rivers are found in South America?

Mountains High mountains can be found in much of Latin America. In Mexico, three mountain ranges meet to form what looks like the letter *Y*. Mountains also run south through much of Central America. Most of the West Indies islands are the tops of volcanic mountains rising up from the Caribbean Sea. Along the west coast of South America, you will find the Andes—the world's longest mountain chain. Volcanic activity and earthquakes are common in these mountainous areas of Latin America.

Plains Vast plains cover other areas of Latin America. You will find plains along the coasts of Mexico and Central America. Broad inland plains also can be found in parts of South America. The two major South American plains are the *pampas* of Argentina and the *llanos* of Colombia and Venezuela. These plains areas are made up of rolling grassland with few trees.

Rivers The map on page 144 shows you that South America has five major rivers: the Magdalena, the Orinoco, the Río de la Plata, the São Francisco, and the Amazon. The Amazon is the longest river in the Western Hemisphere. It stretches for about 4,000 miles (6,436 km) across South America from the Andes eastward to the Atlantic Ocean. It carries 20 percent of the world's unfrozen freshwater and drains a basin of more than 3 million square miles (7.8 million sq. km).

(140)

Fun Facts

• **Brazil** The Amazon River was named for the Amazons, the female warriors of Greek mythology. Francisco de Orellana, the first European to explore the river, supposedly chose the name after being attacked by a group of female warriors.

• **Venezuela** *Por puesto* is a popular system of transportation in Venezuela. Taxi-like automobiles travel a regular route throughout a city, picking up and dropping off passengers at any point. The fare is more than a bus but less than a taxi.

Natural Resources The countries of Mexico and Venezuela are among the world's leading producers of oil and natural gas. Latin America in general is rich in natural resources such as copper, iron ore, silver, tin, lead, oil, and natural gas. Rivers and waterfalls provide electric power to many countries. Farmers use the rich soil in many areas of Latin America to grow grains, fruits, and coffee.

Region
Climate and Vegetation

Latin America has a variety of climates and different types of vegetation. Traveling through the region, you can journey through deserts and rain forests and from grassy plains to empty plateaus.

Elevation Most of Latin America lies in the tropics and has some form of tropical climate. Elevation, however, affects climate throughout the region. Land at low elevations is hot and humid with green tropical vegetation. Crops such as sugarcane, bananas, and cacao beans grow well here. At higher elevations, you will notice that the climate becomes milder and temperatures cooler. Crops such as coffee, corn, and wheat grow in this area. In the highest areas, you will see very little plant life and may even spot frost and snow.

Rain Forests Rain forests cover lowland areas of Latin America. They are rainy and hot all the time. Rain forests have more kinds of plants and animals than anywhere else on the earth. The world's largest tropical rain forest is in Brazil in the Amazon basin. Other rain forests grow along the eastern coast of Central America and in some of the Caribbean islands.

Amazon Rain Forest

Plants in the Amazon basin provide sources for one-fourth of the world's medicines.
LOCATION: Where are other rain forests in Latin America?

(141)

NATIONAL GEOGRAPHIC SOCIETY

These materials are available from Glencoe.

 CD-ROM

Picture Atlas of the World

 SOFTWARE

ZipZapMap! World

📙 **POSTERS**

Eye on the Environment, Unit 3

More than 2 million species of insects live in the Amazon basin.

More About the Illustration

Answer to Caption
along the eastern coast of Central America and in some of the Caribbean islands

 **Place** The Amazon basin has huge reserves of resources such as bauxite, copper, gold, lumber, manganese, nickel, and tin.

Fun Facts

• **Paraguay** When Paraguayans visit one another, they make their presence known by clapping at the gate. It is impolite for guests to enter a yard until the host invites them to do so.

• **Andean Countries** Llamas are the largest South American members of the camel family. They are useful pack animals because they can carry as much as 130 pounds (60 kg) and are sure-footed on mountain trails.

LESSON PLAN
Unit 3 Atlas

FactsOnFile

Use the reproducible masters from **GEOGRAPHY ON FILE,** *Latin America,* to enrich unit content.

More About the Illustration

Answer to Caption

about 70 percent

Place Many people in Ecuador live in poverty, and malnutrition and disease is widespread among the poor. The Ecuadorian government has established programs to provide better housing and medical care for those in need. With the average pay for many workers standing at about a dollar a day, however, the government faces a formidable task.

CURRICULUM CONNECTION

Life Science Mexico has a great variety of plants and animals. Some species have become popular in other parts of the world. One of these is the chihuahua—the world's smallest dog. The brilliant red poinsettia plant, a popular holiday decoration, is native to Mexico.

Goods to Market

Workers load cases of bananas onto a waiting ship in the port of Guayaquil *(top).* A woman from Ecuador weaves hats that she will sell in the city's market *(bottom).*
PLACE: What percentage of Latin Americans live in cities?

Region
Economy

In the past, the economies of most Latin American countries were based largely on agriculture. Today the region is developing manufacturing and service industries.

Agriculture Fertile soil and a warm, wet climate allow farmers to grow tropical crops in many areas of Latin America. These crops—coffee, bananas, and sugarcane—are sold to other countries. In addition to farming, Latin Americans also make use of grasslands to raise livestock. Argentina, Mexico, and Brazil are among the world's leaders in cattle raising and meat production.

Industrial Growth Service industries and manufacturing have grown rapidly in Latin America. Venezuela, Mexico, and Brazil are among the industrial leaders in the region. Workers in industrial areas of Latin America produce iron and steel, cars, textiles, cement, chemicals, and electrical goods.

Not all countries have industrial economies. Some have difficulty feeding, clothing, and housing their people. They lack the money, skilled labor, and the transportation systems needed to build industries. Rugged mountains and thick tropical vegetation provide physical barriers to movement.

Movement
The People

Latin America has about 470 million people, about 12 percent of the world's population. In recent decades, Latin America's population has grown rapidly. Latin Americans come from many different ethnic groups. They include Native Americans, Europeans, Africans, and Asians.

Latin Americans occupy only about one-third of the region's land. You will find a great number of them living along the coasts of South America. Many also live in a broad strip of land reaching from central Mexico into Central America.

In recent years, many Latin Americans have moved from the countryside to the cities. About 70 percent of the region's people are now city dwellers. Four Latin American cities rank among the 10 largest urban areas in the world. They are Mexico City, Mexico; São Paulo and Rio de Janeiro, Brazil; and Buenos Aires, Argentina. Latin American governments are trying to improve health care, education, and standards of living for their people. Urban slums, air pollution, and sputtering economies are handicaps to economic development.

History and Culture The culture of Latin America is rooted in its history. It reflects the influences of the many ethnic groups that settled the region. Native Americans were the first people to settle in Latin America. The Maya, Aztec, and Inca all had advanced civilizations long before Europeans arrived in the Western Hemisphere.

Fun Facts

- **Bolivia** Potatoes are a main part of the Bolivian diet. *Saltenas*—meat or chicken pies with potatoes, olives, and raisins—are popular among Bolivian city dwellers.

- **Barbados** Some families on this island live in *chattel houses.* These small wooden homes can be taken apart and moved. They are made of such good-quality materials that they can be passed from one generation to the next.

During the 1500s Europeans enslaved and destroyed the Native American civilizations. They later enslaved Africans and brought them in to work the plantations they had set up. Spain and Portugal ruled most of Latin America from the 1500s to the early 1800s. Spanish and Portuguese settlers brought the Roman Catholic faith and the Spanish and Portuguese languages to the region. Because these languages are based on Latin, the region became known as Latin America. Most Latin Americans today are Roman Catholics.

Many Latin American countries became independent of colonial rule in the early 1800s. Small groups of wealthy landowners and military officers, however, controlled the governments of these nations. These ruling groups often ignored demands for change that would benefit poor farmers and workers. By the mid-1900s, the leaders were unable to meet challenges created by the rise of cities and industries. With wide public support, new democratic governments are beginning to replace the old, harsh political systems.

The Arts Native Americans, Europeans, Africans, and Asians—all have left their mark on the arts of Latin America. In marketplaces you can see colorful cloth, pottery, and metalwork based on Native American styles. Many churches in Latin America are built in European styles. Latin American Africans have contributed to the music and dance of the region.

Lima, Peru

Peruvian soldiers stand guard at Lima's Presidential Palace. **PLACE: What European countries ruled most of Latin America until the 1800s?**

Technology Activity

Using a Computerized Card Catalog Use an online card catalog to find two books and a video or CD-ROM about the Aztec, Inca, and Maya. Use these sources to create an illustrated time line that identifies when and where each group settled and what they contributed to Latin America.

REVIEW AND ACTIVITIES

1. **REGION** What separate areas make up Latin America?
2. **LOCATION** Where are the Andes located?
3. **REGION** What is Latin America's most common climate?
4. **HUMAN/ENVIRONMENT INTERACTION** Name three tropical crops grown in Latin America.
5. **PLACE** What percentage of Latin Americans live in cities?
6. **MOVEMENT** What peoples have settled in Latin America?

(143)

Answers to Review and Activities

1. Mexico, Central America, Caribbean islands, South America
2. along the western coast of South America
3. tropical
4. bananas, coffee, sugarcane
5. about 70 percent
6. Native Americans, Europeans, Africans, Asians

UNIT 3 ATLAS

Physical Geography

Applying Geographic Themes

Place The Amazon River basin of Brazil, fed by many rivers, is reported to have 50,000 types of plants and more than 3,000 species of fish. What other major rivers feed into the Amazon Basin? *(Rio Negro, Rio Madeira)*

NATIONAL GEOGRAPHIC SOCIETY

These materials are available from Glencoe.

CD-Rom

- **Picture Atlas of the World**
- **PictureShow:** Earth's Endangered Environments, *Rain Forests*

Software

ZipZapMap! World

MAP STUDY

Answers
1. the Andes
2. the Caribbean Sea

Map Skills Practice
Reading a Map What area of Mexico has the highest elevation? *(central Mexico)*

LATIN AMERICA: Physical

120°W 100°W UNITED STATES 80°W 60°W 40°W

ATLANTIC OCEAN

Rio Grande
Baja California
Plateau of Mexico
SIERRA MADRE OCCIDENTAL
SIERRA MADRE ORIENTAL
SIERRA MADRE DEL SUR
MEXICO
Gulf of Mexico
BAHAMAS
Tropic of Cancer
WEST INDIES
Yucatán Peninsula
CUBA
Greater
HAITI
DOM. REP.
BELIZE
JAMAICA
Antilles
Puerto Rico (U.S.)
HONDURAS
Lesser Antilles
GUATEMALA
EL SALVADOR
NICARAGUA
COSTA RICA
Caribbean Sea
PANAMA
Isthmus of Panama
Lake Maracaibo
GUYANA
SURINAME
French Guiana (Fr.)
LLANOS
VENEZUELA
Orinoco River
Guiana Highlands
COLOMBIA
Magdalena R.
Rio Negro
AMAZON
Amazon River
Equator
Galápagos Islands (Ecuador)
ECUADOR
BASIN
PERU
Rio Madeira
BRAZIL
Brazilian
Rio São Francisco
ANDES MOUNTAINS
Lake Titicaca
BOLIVIA
Altiplano
Highlands
Paraná R.
GRAN CHACO
PARAGUAY
20°S
Tropic of Capricorn
Atacama Desert
Paraguay R.
CHILE
PATAGONIA

PACIFIC OCEAN

Aconcagua 22,834 ft. (6,960 m)
ARGENTINA
URUGUAY
PAMPAS
Rio de la Plata

ELEVATIONS
Feet	Meters
10,000	3,000
5,000	1,500
2,000	600
1,000	300
0	0

▲ Mountain peak

0   300   600 mi.
0   300   600 km

Goode's Interrupted Homolosine projection

40°S

Falkland (Malvinas) Is. (U.K.)
Tierra del Fuego
Cape Horn
Strait of Magellan

Map Study

1. **PLACE** What mountain chain runs along the western coast of South America?
2. **LOCATION** What body of water lies south of the West Indies?

(144)

UNIT 3

Atlas Activity

Region Organize students into small groups to write at least five questions about the physical geography of Latin America with which to challenge another group. Tell students to include direction, key, scale, or physical map questions, such as: In what direction would you travel from Mexico to the West Indies island with the highest elevation? *(east to Hispaniola)* Allow time for groups to write questions and challenge one another. You may also want to repeat this activity at the end of the unit. **L1**

ELEVATION PROFILE: South America

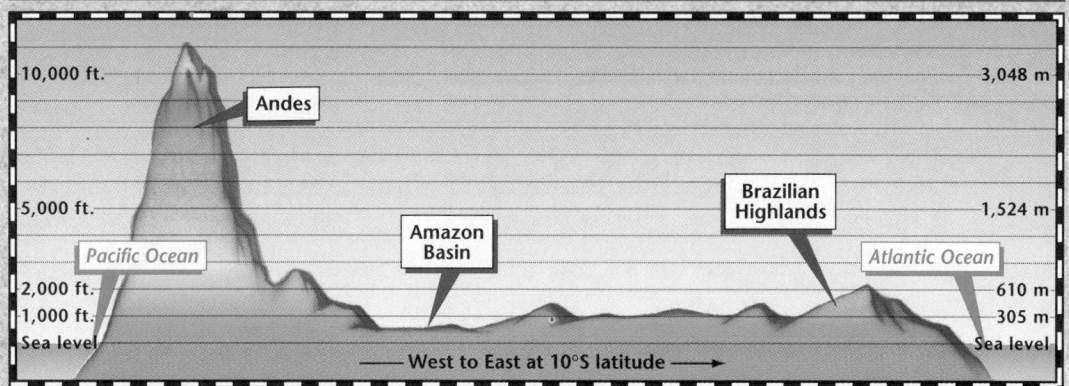

10,000 ft. — 3,048 m

Andes

5,000 ft. — 1,524 m

Brazilian Highlands

Pacific Ocean

Amazon Basin

Atlantic Ocean

2,000 ft. — 610 m
1,000 ft. — 305 m
Sea level — Sea level

← West to East at 10°S latitude →

Source: *Goode's World Atlas*, 19th edition

GeoFacts

Highest point: Mt. Aconcagua (Argentina) 22,834 ft. (6,960 m)

Lowest point: Valdes Peninsula (Argentina) 131 ft. (40 m) below sea level

Longest river: Amazon River (Brazil) 4,000 mi. (6,436 km) long

Largest lake: Lake Maracaibo (Venezuela) 5,217 sq. mi. (13,512 sq. km)

Highest waterfall: Angel Falls (Venezuela) 3,212 ft. (979 m)

Largest rain forest: Amazon rain forest (Brazil) 2,700,000 sq. mi. (6,993,000 sq. km)

Highest lake: Lake Titicaca (Peru and Bolivia) 12,500 ft. (3,811 m) in altitude

LATIN AMERICA AND THE UNITED STATES: Land Comparison

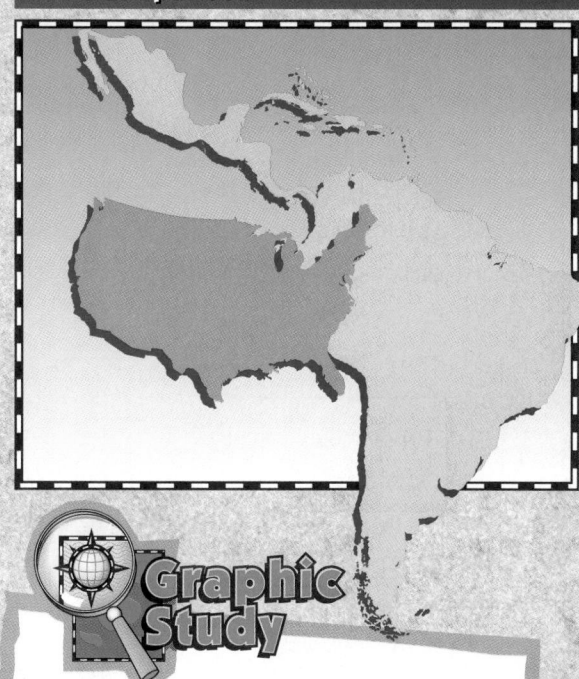

Graphic Study

1. **LOCATION** What is the highest point in South America?
2. **PLACE** Using the land comparison map, compare the land area of the United States to that of Latin America.

145

GLENCOE
TECHNOLOGY

VIDEODISC

The Infinite Voyage: To the Edge of the Earth

Chapter 4
Exploring the Galápagos Islands
Disc 1 Side B

Chapter 5
The Tropical Rain Forest
Disc 1 Side B

FactsOnFile

Use the reproducible masters from **GEOGRAPHY ON FILE,** *Latin America,* to enrich unit content.

GRAPHIC STUDY

Answers
1. Mt. Aconcagua in Argentina
2. The continental U.S. is about one-third the size of Latin America.

Graphic Skills Practice
Reading a Chart How many feet (meters) lower is the lowest point in Argentina compared to the highest point? *(22,965 ft or 7,000 m)*

Teacher Notes

Cultural Geography

LESSON PLAN
Unit 3 Atlas

Applying Geographic Themes

Location Latin America has the greatest latitudinal span of any world region. Have students estimate South America's greatest east-west distance. *(about 3,200 miles or 5,000 km)* Then have them identify the countries located along the Equator. *(Ecuador, Colombia, and Brazil)*

NATIONAL GEOGRAPHIC SOCIETY

These materials are available from Glencoe.

VIDEODISC

STV: World Geography Vol. 3: South America and Antarctica

South America
Side 1, Frames 00001-
50061

SOFTWARE

ZipZapMap! World

MAP STUDY

Answers
1. Mexico
2. Buenos Aires

Map Skills Practice
Reading a Map Which South American countries do not share a border with Brazil? *(Ecuador, Chile)*

LATIN AMERICA: Political

(map of Latin America showing countries, capitals, and surrounding oceans)

- National boundary
- ✪ National capital
- *Goode's Interrupted Homolosine projection*

0   300   600 mi.
0   300   600 km

Map Study

1. **LOCATION** What is the northernmost Latin American country?
2. **PLACE** What is the capital of Argentina?

UNIT 3

Atlas Activity

Location Assign a Latin American country to each student. Have students make up riddles about their assigned country, based on its location relative to other countries or to bodies of water. Then have students take turns reading their riddles and calling on classmates to identify the country. **L1**

LATIN AMERICA: Largest Urban Areas

Mexico City, Mexico

São Paulo, Brazil

Rio de Janeiro, Brazil

Buenos Aires, Argentina

Lima, Peru

Bogotá, Colombia

Guadalajara, Mexico

Caracas, Venezuela

Havana, Cuba

= 3,000,000

*estimated populations in mid-1995

Source: *The 1994 Information Please Almanac*

Geography and the Humanities

📁 World Literature Reading 3

🎵 World Music: Cultural Traditions, Lesson 2

📦 World Cultures Transparencies 3, 4

📕 Focus on World Art Print 11

COMPARING POPULATION: Latin America and the United States

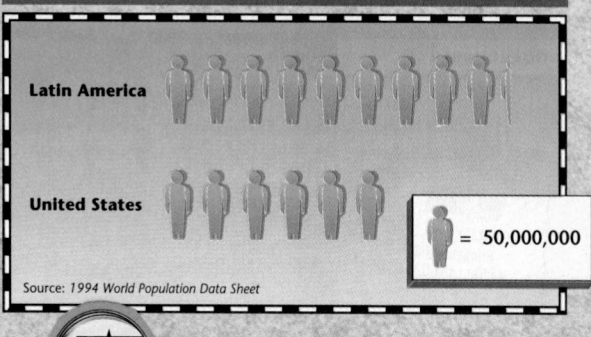

Latin America

United States

= 50,000,000

Source: *1994 World Population Data Sheet*

Cultural Kaleidoscope

Haiti Most of Haiti is mountainous. Its name comes from a Native American word meaning "high ground."

 GeoFacts

Biggest country (land area): Brazil 3,265,060 sq. mi. (8,456,505 sq. km)

Smallest country (land area): Grenada 130 sq. mi. (337 sq. km)

Largest city (population): Mexico City (1995) 23,913,000; (2000 projected) 27,872,000

Highest population density: Barbados 1,765 people per sq. mi. (682 people per sq. km)

Lowest population density: French Guiana 4 people per sq. mi. (1 person per sq. km)

Graphic Study

1. **PLACE** How does the population of Latin America compare to the population of the United States?
2. **PLACE** What are the three most populated urban areas in Latin America?

(147)

GRAPHIC STUDY

Answers
1. The population of the United States is roughly two-thirds the size of the population of Latin America.
2. Mexico City, Mexico; São Paulo, Brazil; Rio de Janeiro, Brazil

Graphic Skills Practice
Reading a Graph About how many people live in Buenos Aires, Argentina? *(nearly 13 million)*

Teacher Notes

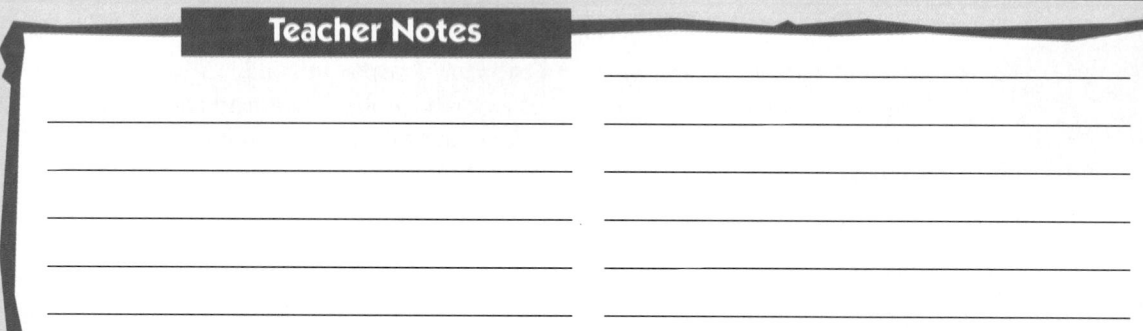

Cultural Kaleidoscope

Venezuela Flowers are very important in Venezuelan festivities. During each holiday, statues of Simón Bolívar, the founder of Venezuela, are decorated with colorful wreaths.

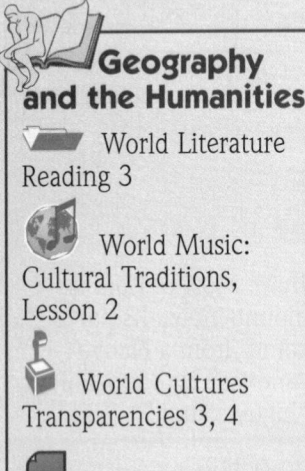

Geography and the Humanities

World Literature Reading 3

World Music: Cultural Traditions, Lesson 2

World Cultures Transparencies 3, 4

Focus on World Art Print 11

UNIT 3 ATLAS

Countries at a Glance

Antigua and Barbuda
St. Johns
CAPITAL: St. Johns
MAJOR LANGUAGE(S): English
POPULATION: 100,000
LANDMASS: 170 sq. mi./ 440 sq. km
MONEY: Eastern Caribbean dollar
MAJOR EXPORT: Petroleum Products
MAJOR IMPORT: Food and Live Animals

Bolivia
La Paz, Sucre
CAPITALS: La Paz, Sucre
MAJOR LANGUAGE(S): Spanish, Quechua, Aymara
POPULATION: 8,200,000
LANDMASS: 418,680 sq. mi./ 1,084,381 sq. km
MONEY: Boliviano
MAJOR EXPORT: Zinc
MAJOR IMPORT: Raw Materials

Argentina
Buenos Aires
CAPITAL: Buenos Aires
MAJOR LANGUAGE(S): Spanish
POPULATION: 33,900,000
LANDMASS: 1,056,640 sq. mi./ 2,736,698 sq. km
MONEY: Peso
MAJOR EXPORT: Food and Live Animals
MAJOR IMPORT: Machinery

Brazil
Brasília
CAPITAL: Brasília
MAJOR LANGUAGE(S): Portuguese
POPULATION: 155,300,000
LANDMASS: 3,265,060 sq. mi./ 8,456,505 sq. km
MONEY: Cruzeiro
MAJOR EXPORT: Machinery
MAJOR IMPORT: Petroleum

Bahamas
Nassau
CAPITAL: Nassau
MAJOR LANGUAGE(S): English
POPULATION: 300,000
LANDMASS: 3,860 sq. mi./ 9,997 sq. km
MONEY: Dollar
MAJOR EXPORT: Petroleum
MAJOR IMPORT: Fuels

Chile
Santiago
CAPITAL: Santiago
MAJOR LANGUAGE(S): Spanish
POPULATION: 14,000,000
LANDMASS: 289,110 sq. mi./ 748,795 sq. km
MONEY: Peso
MAJOR EXPORT: Paper Products
MAJOR IMPORT: Intermediate Goods

Barbados
Bridgetown
CAPITAL: Bridgetown
MAJOR LANGUAGE(S): English
POPULATION: 300,000
LANDMASS: 170 sq. mi./ 440 sq. km
MONEY: Dollar
MAJOR EXPORT: Sugar
MAJOR IMPORT: Machinery

Colombia
Bogotá
CAPITAL: Bogotá
MAJOR LANGUAGE(S): Spanish
POPULATION: 35,600,000
LANDMASS: 401,040 sq. mi./ 1,038,694 sq. km
MONEY: Peso
MAJOR EXPORT: Petroleum
MAJOR IMPORT: Machinery

Belize
Belmopan
CAPITAL: Belmopan
MAJOR LANGUAGE(S): English, Spanish, native Creole dialects
POPULATION: 200,000
LANDMASS: 8,800 sq. mi./ 22,792 sq. km
MONEY: Belize Dollar
MAJOR EXPORT: Sugar
MAJOR IMPORT: Machinery

Costa Rica
San José
CAPITAL: San José
MAJOR LANGUAGE(S): Spanish
POPULATION: 3,200,000
LANDMASS: 19,710 sq. mi./ 51,049 sq. km
MONEY: Colón
MAJOR EXPORT: Clothing
MAJOR IMPORT: Petroleum

Countries not drawn to scale.

Fun Facts

- **Tierra del Fuego** As the explorer Ferdinand Magellan sailed round the southern tip of South America, he noticed large fires on land. He named the area *Tierra del Fuego,* or "Land of Fire."
- **Ecuador** People in each rural region of Ecuador have their own traditional clothing, colors, and fabrics. For example, people from coastal regions generally wear white, yellow, or red. Those from highland regions prefer blue, brown, or black.

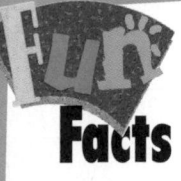

Cuba

Havana

CAPITAL: Havana
MAJOR LANGUAGE(S): Spanish
POPULATION: 11,100,000
LANDMASS: 42,400 sq. mi./ 109,816 sq. km
MONEY: Peso
MAJOR EXPORT: Sugar
MAJOR IMPORT: Fuels

Dominica

Roseau

CAPITAL: Roseau
MAJOR LANGUAGE(S): English, French, Creole
POPULATION: 100,000
LANDMASS: 290 sq. mi./ 751 sq. km
MONEY: Eastern Caribbean Dollar
MAJOR EXPORT: Bananas
MAJOR IMPORT: Machinery

Dominican Republic

Santo Domingo

CAPITAL: Santo Domingo
MAJOR LANGUAGE(S): Spanish
POPULATION: 7,800,000
LANDMASS: 18,680 sq. mi./ 48,381 sq. km
MONEY: Peso
MAJOR EXPORT: Ferronickel
MAJOR IMPORT: Petroleum

Ecuador

Quito

CAPITAL: Quito
MAJOR LANGUAGE(S): Spanish, Quechua, Jivaroan
POPULATION: 10,600,000
LANDMASS: 106,890 sq. mi./ 276,845 sq. km
MONEY: Sucre
MAJOR EXPORT: Petroleum
MAJOR IMPORT: Raw Materials

El Salvador

San Salvador

CAPITAL: San Salvador
MAJOR LANGUAGE(S): Spanish
POPULATION: 5,200,000
LANDMASS: 8,000 sq. mi./ 20,720 sq. km
MONEY: Colón
MAJOR EXPORT: Coffee
MAJOR IMPORT: Chemicals

French Guiana*

Cayenne

CAPITAL: Cayenne
MAJOR LANGUAGE(S): French
POPULATION: 127,505
LANDMASS: 35,126 sq. mi./ 91,000 sq. km
MONEY: French Franc
MAJOR EXPORT: Shrimp
MAJOR IMPORT: Food

Grenada

St. George's

CAPITAL: St. George's
MAJOR LANGUAGE(S): English, French patois
POPULATION: 100,000
LANDMASS: 130 sq. mi./ 337 sq. km
MONEY: Eastern Caribbean Dollar
MAJOR EXPORT: Bananas
MAJOR IMPORT: Machinery

Guatemala

Guatemala

CAPITAL: Guatemala
MAJOR LANGUAGE(S): Spanish, Mayan languages
POPULATION: 10,300,000
LANDMASS: 41,860 sq. mi./ 108,417 sq. km
MONEY: Quetzal
MAJOR EXPORT: Coffee
MAJOR IMPORT: Petroleum

Guyana

Georgetown

CAPITAL: Georgetown
MAJOR LANGUAGE(S): English, Amerindian dialects
POPULATION: 800,000
LANDMASS: 76,000 sq. mi./ 196,840 sq. km
MONEY: Dollar
MAJOR EXPORT: Sugar
MAJOR IMPORT: Fuels

Haiti

Port-au-Prince

CAPITAL: Port-au-Prince
MAJOR LANGUAGE(S): French, Creole
POPULATION: 7,000,000
LANDMASS: 10,640 sq. mi./ 27,558 sq. km
MONEY: Gourde
MAJOR EXPORT: Textiles
MAJOR IMPORT: Food and Live Animals

*Territory of France

149

LESSON PLAN
Unit 3 Atlas

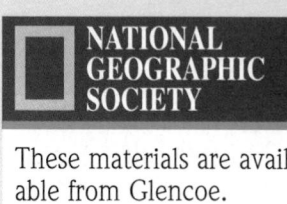

NATIONAL GEOGRAPHIC SOCIETY

These materials are available from Glencoe.

VIDEODISC

STV: World Geography
Vol. 3: South America and Antarctica

South America
Side 1, Frames 00001-50061

SOFTWARE

ZipZapMap! World

Cultural Kaleidoscope

Chile The *abrazo* is the most common greeting among relatives and friends in Chile. It consists of a handshake and a hug, sometimes followed by a kiss to the right cheek for women or family members.

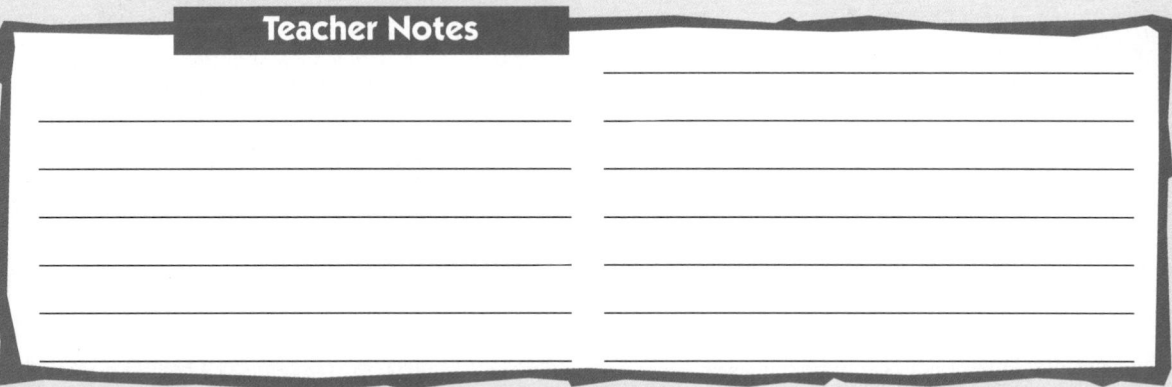

Teacher Notes

UNIT 3 ATLAS

Countries at a Glance

CURRICULUM CONNECTION

History Inca builders constructed cities of stone that clung to steep mountainsides. The stones for the Inca buildings were cut so accurately that they fit together without cement.

Cultural Kaleidoscope

Brazil On New Year's Eve, some Brazilians traditionally honor *Iemanja*, the sea goddess. They dress in blue and white and place flowers and candles on beaches.

Honduras

CAPITAL:
Tegucigalpa
MAJOR LANGUAGE(S):
Spanish
POPULATION:
5,300,000
LANDMASS:
43,200 sq. mi./
111,888 sq. km
MONEY:
Lempira
MAJOR EXPORT:
Bananas
MAJOR IMPORT:
Machinery

Jamaica

CAPITAL:
Kingston
MAJOR LANGUAGE(S):
English, Jamaican Creole
POPULATION:
2,500,000
LANDMASS:
4,180 sq. mi./
10,826 sq. km
MONEY:
Dollar
MAJOR EXPORT:
Alumina
MAJOR IMPORT:
Fuels

Mexico

CAPITAL:
Mexico City
MAJOR LANGUAGE(S):
Spanish, Amerindian languages
POPULATION:
91,800,000
LANDMASS:
736,950 sq. mi./
1,908,700 sq. km
MONEY:
New Peso
MAJOR EXPORT:
Machinery
MAJOR IMPORT:
Machinery

Nicaragua

CAPITAL:
Managua
MAJOR LANGUAGE(S):
Spanish
POPULATION:
4,300,000
LANDMASS:
45,850 sq. mi./
118,751 sq. km
MONEY:
Cordoba Oro
MAJOR EXPORT:
Coffee
MAJOR IMPORT:
Petroleum

Panama

CAPITAL:
Panamá
MAJOR LANGUAGE(S):
Spanish, English
POPULATION:
2,500,000
LANDMASS:
29,340 sq. mi./
75,991 sq. km
MONEY:
Balboa
MAJOR EXPORT:
Bananas
MAJOR IMPORT:
Fuels

Paraguay

CAPITAL:
Asunción
MAJOR LANGUAGE(S):
Spanish, Guarani
POPULATION:
4,800,000
LANDMASS:
153,400 sq. mi./
397,306 sq. km
MONEY:
Guarani
MAJOR EXPORT:
Cotton Products
MAJOR IMPORT:
Machinery

Peru

CAPITAL:
Lima
MAJOR LANGUAGE(S):
Spanish, Quechua, Aymara
POPULATION:
22,900,000
LANDMASS:
494,210 sq. mi./
1,280,004 sq. km
MONEY:
Sol
MAJOR EXPORT:
Copper
MAJOR IMPORT:
Machinery

Puerto Rico*

CAPITAL:
San Juan
MAJOR LANGUAGE(S):
Spanish
POPULATION:
3,600,000
LANDMASS:
3,420 sq. mi./
8,858 sq. km
MONEY:
Dollar
MAJOR EXPORT:
Chemical Products
MAJOR IMPORT:
Chemicals

St. Kitts-Nevis

CAPITAL:
Basseterre
MAJOR LANGUAGE(S):
English
POPULATION:
40,000
LANDMASS:
140 sq. mi./
363 sq. km
MONEY:
Eastern Caribbean Dollar
MAJOR EXPORT:
Sugar and Molasses
MAJOR IMPORT:
Food and Live Animals

St. Lucia

CAPITAL:
Castries
MAJOR LANGUAGE(S):
English, French patois
POPULATION:
100,000
LANDMASS:
240 sq. mi./
622 sq. km
MONEY:
Eastern Caribbean Dollar
MAJOR EXPORT:
Bananas
MAJOR IMPORT:
Machinery

Countries not drawn to scale.

150

*U.S. Commonwealth

Fun Facts

- **Colombia** This country is the world's leading producer of emeralds. The value of the rich green gemstones depends on their color and lack of flaws. Perfect emeralds are more expensive than diamonds.

- **Peru** The Central Railway of Peru climbs from sea level to 15,800 feet (4,800 m). On its journey from Lima to Huancayo, the train crosses 59 bridges and goes through 66 tunnels.

Ocho Rios, Jamaica

St. Vincent and the Grenadines

Kingstown

CAPITAL:
Kingstown
MAJOR LANGUAGE(S):
English
POPULATION:
100,000

LANDMASS:
150 sq. mi./
389 sq. km
MONEY:
Eastern Caribbean Dollar
MAJOR EXPORT:
Bananas
MAJOR IMPORT:
Food Products

Beautiful beaches and a
warm climate make tourism
a main source of income for
Jamaica and other islands in
Latin America.

Suriname

Paramaribo

CAPITAL:
Paramaribo
MAJOR LANGUAGE(S):
Dutch, Sranantonga, English
POPULATION:
400,000

LANDMASS:
60,230 sq. mi./
155,996 sq. km
MONEY:
Guilder
MAJOR EXPORT:
Alumina
MAJOR IMPORT:
Machinery

Uruguay

Montevideo

CAPITAL:
Montevideo
MAJOR LANGUAGE(S):
Spanish
POPULATION:
3,200,000

LANDMASS:
67,490 sq. mi./
174,799 sq. km
MONEY:
New Peso
MAJOR EXPORT:
Textiles
MAJOR IMPORT:
Machinery

Trinidad and Tobago

Port of Spain

CAPITAL:
Port of Spain
MAJOR LANGUAGE(S):
English
POPULATION:
1,300,000

LANDMASS:
1,980 sq. mi./
5,128 sq. km
MONEY:
Dollar
MAJOR EXPORT:
Petroleum
MAJOR IMPORT:
Food

Venezuela

Caracas

CAPITAL:
Caracas
MAJOR LANGUAGE(S):
Spanish
POPULATION:
21,300,000

LANDMASS:
340,560 sq. mi./
882,050 sq. km
MONEY:
Bolívar
MAJOR EXPORT:
Petroleum
MAJOR IMPORT:
Machinery

151

Cultural Kaleidoscope

Dominican Republic Dominicans point with puckered lips rather than with a finger.

MULTICULTURAL PERSPECTIVE

Cultural Heritage

In 1991, the Venezuelan government set aside land in the far south for the country's Yanomami people. The reservation covers more than 30,000 square miles (80,000 sq km). It is off-limits to all but Yanomami settlers.

Teacher Notes

Chapter 6 Planning Guide
Mexico

PERFORMANCE ASSESSMENT ACTIVITY

DEVELOP A PLAY Have students take the roles of set designers for a play that is intended to inform students in the United States about their closest neighbor to the south. Have students work individually or in groups to develop a title for the play, distinct titles and a list of essential content for each of three acts in the play, and an illustrated description of the scenery and set design. These materials should be submitted with a cover letter of introduction to a panel of "producers" who will decide whether to hire the designers. The panel may be comprised of students, parents, or school staff.

POSSIBLE RUBRIC FEATURES: Content information, classificatory skills, creative thinking, composition skills, planning and product completion skills, decision-making skills

MENTAL MAPPING ACTIVITY

Before teaching Chapter 6, use a globe to show students the location of Mexico. Have students identify the hemisphere in which Mexico is located and where the country is located relative to the United States.

TEACHER'S CORNER

NATIONAL GEOGRAPHIC SOCIETY

INDEX TO NATIONAL GEOGRAPHIC MAGAZINE

The following articles may be used for research relating to this chapter:

- *Mexico,* A National Geographic Special Edition, August 1996.
- "Tex-Mex Border," by Richard Conniff, February 1996.
- "The Timeless Vision of Teotihuacan," by George E. Stuart, December 1995.
- "Mexico's Desert Aquarium," by George Grall, October 1995.
- "Huautla Cave Quest," by William C. Stone, September 1995.

- "Maya Masterpiece Revealed," by Mary Miller, February 1995.
- "The Song of Oaxaca," by Sandra Dibble, November 1994.
- "New Light on the Olmec," by George E. Stuart, November 1993.
- "Mural Masterpieces of Ancient Cacaxtla," by George E. Stuart, September 1992.
- "Mexico's Bajío - The Heartland," by Charles E. Cobb, Jr., December 1990.

NATIONAL GEOGRAPHIC SOCIETY PRODUCTS AVAILABLE FROM GLENCOE

To order the following products for use with this chapter, contact your local Glencoe sales representative or call Glencoe at 1-800-334-7344:

- *STV: North America* (Videodisc)
- *STV: World Geography* (Videodisc)
- *STV: Rain Forest* (Videodisc)
- *Picture Atlas of the World* (CD-ROM)
- *Geography of North America* (Transparencies)

- *ZipZapMap! World* (Software)
- *GeoBee* (Software)
- *Images of the World* (Posters)
- *Eye on the Environment* (Posters)

ADDITIONAL NATIONAL GEOGRAPHIC SOCIETY PRODUCTS

To order the following products for use with this chapter, call National Geographic Society at 1-800-368-2728:

- *Nations of the World Series:* "Mexico." (Video)
- *Lost Kingdoms of the Maya* (Video)

- *The Mexicans: Through Their Eyes* (Video)
- *Rain Forest* (Video)

—from Cathy Salter,
Educational Consultant, Hartsburg, MO

THE TREASURES OF MEXICO *The purpose of this activity is to involve students in Mexico's history, geography, and culture by conducting research and taking part in a scavenger hunt titled "Searching for the Treasures of Mexico."*

Organize students into teams of three or four. Assign each team a number of topics from the following list: geography, history, economy, politics, cultures, archaeological sites, cities, music, art, writers and their works. Each team should research its topics, using resources such as textbooks, atlases, almanacs, encyclopedias, and, if available, videocassettes and computer software. Have each team submit four scavenger-hunt questions based on their research. Suggest that they write their questions on index cards, noting the answer on the back of each card.

Make copies of the questions and give them to each team, making certain teams do not receive questions they wrote. Launch teams on a scavenger hunt to find answers to the questions they received. When all teams have completed their hunt, ask students to note the most surprising and unusual points of information that they learned about Mexico.

BIBLIOGRAPHY

Readings for the Student

Grayson, George W. *The North American Free Trade Agreement.* New York: Foreign Policy Association, 1993.

Irizarry, Carmen. *Passport to Mexico.* New York: Franklin Watts, 1994.

Readings for the Teacher

Creative Activities for Teaching About Mexico. Stevens and Shea, 1990.

Latin America Today: An Atlas of Reproducible Pages. Wellesley, Mass.: World Eagle, 1992.

Multimedia

Mexico: Our Neighbor to the South. United Learning, 1993. Two videocassettes, 40 minutes total.

Toward a Better World, Kit 4: Tackling Poverty in Rural Mexico. Washington, D.C.: World Bank, 1992. Student books, pamphlets, sound filmstrip, guide.

MEETING NATIONAL STANDARDS

Geography for Life

All of the 18 standards are demonstrated in Unit 3. The following ones are highlighted in this chapter:

- Standards 3, 4, 8, 10, 16, 18

KEY TO ABILITY LEVELS

Teaching strategies have been coded for varying learning styles and abilities.

L1 **BASIC** activities for all students

L2 **AVERAGE** activities for average to above-average students

L3 **CHALLENGING** activities for above-average students

LEP **LIMITED ENGLISH PROFICIENCY** activities

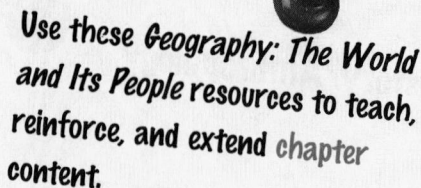

Use these Geography: The World and Its People resources to teach, reinforce, and extend chapter content.

CHAPTER 6 RESOURCES

📁 *Vocabulary Activity 6

📁 Cooperative Learning Activity 6

📁 Workbook Activity 6

📁 Geography Skills Activity 6

📁 GeoLab Activity 6

📁 Chapter Map Activity 6

📁 Performance Assessment Activity 6

📁 *Reteaching Activity 6

📁 Enrichment Activity 6

📁 Chapter 6 Test, Form A and Form B

🎧 📁 *Chapter 6 Digest Audiocassette, Activity, Test

Unit Overlay Transparencies 3-0, 3-1, 3-2, 3-3, 3-4, 3-5, 3-6

Geoquiz Transparencies 6-1, 6-2

Political Map Transparency 3; World Cultures Transparency 3

Vocabulary PuzzleMaker Software

Student Self-Test: A Software Review

Testmaker Software

MindJogger Videoquiz

Focus on World Art Print 11

If time does not permit teaching the entire chapter, summarize using the Chapter 6 Highlights on page 169, and the Chapter 6 English (or Spanish) Audiocassettes. Review students' knowledge using the Glencoe MindJogger Videoquiz. *Also available in Spanish

Use these *Geography: The World and Its People* resources to teach and reinforce **section** content.

ADDITIONAL RESOURCES FROM GLENCOE

 Reproducible Masters

- Glencoe Social Studies Outline Map Book, pages 22, 23
- Facts On File: MAPS ON FILE I, North and Central America and the Caribbean
- Foods Around the World, Unit 3

 **Workbook**

- Building Skills in Geography, Unit 2, Lesson 2

World Music: Cultural Traditions Lesson 2

 Transparencies

- National Geographic Society PicturePack Transparencies, Unit 3

 Laminated Desk Map

Videodiscs

- Geography and the Environment: The Infinite Voyage
- National Geographic Society: STV: Restless Earth
- National Geographic Society: GTV: Planetary Manager

 **CD-ROM**

- National Geographic Society: Picture Atlas of the World

Software

- National Geographic Society: ZipZapMap! World

SECTION 1 RESOURCES

- Reproducible Lesson Plan 6-1
- Section Focus Transparency 6-1
- Guided Reading Activity 6-1
- Section Quiz 6-1 (also available in Spanish)

SECTION 2 RESOURCES

- Reproducible Lesson Plan 6-2
- Section Focus Transparency 6-2
- Guided Reading Activity 6-2
- Section Quiz 6-2 (also available in Spanish)

SECTION 3 RESOURCES

- Reproducible Lesson Plan 6-3
- Section Focus Transparency 6-3
- Guided Reading Activity 6-3
- Section Quiz 6-3 (also available in Spanish)

Performance Assessment

Refer to the Planning Guide on page 152A for a Performance Assessment Activity for this chapter. See the *Performance Assessment Strategies and Activities* booklet for suggestions.

Chapter Objectives

1. **Examine** the location of Mexico and its physical and climatic features.
2. **Specify** some economic challenges facing the three regions of Mexico.
3. **Describe** Mexican culture and the groups that have influenced it.

GLENCOE
TECHNOLOGY

VIDEODISC

Use the Chapter 6 Mind-Jogger Videoquiz to preview chapter content.

MindJogger Videoquiz

Chapter 6
Disc 1 Side B

Also available in VHS.

Chapter 6 Mexico

National boundary
State boundary
⊛ National capital
● Other city

Polyconic projection

0 — 125 — 250 mi.
0 — 125 — 250 km

MAP STUDY ACTIVITY

As you read Chapter 6, you will learn about Mexico, the most northern country in Latin America.

1. **How many states make up the country of Mexico?**
2. **What is the capital of Mexico?**

MAP STUDY ACTIVITY

Answers
1. 31 states, one Federal District
2. Mexico City

Map Skills Practice
Reading a Map Locate the following cities and name the state where each is located: Acapulco *(Guerrero)*, Monterrey *(Nuevo León)*, Veracruz *(Veracruz)*, Guadalajara *(Jalisco)*.

The Land

PREVIEW

Words to Know
- land bridge
- peninsula
- basin
- latitude
- altitude

Places to Locate
- Isthmus of Tehuantepec
- Baja California
- Yucatán Peninsula
- Sierra Madre Occidental
- Sierra Madre Oriental
- Sierra Madre del Sur
- Plateau of Mexico

- Mexico City
- Rio Grande

Read to Learn . . .
1. where Mexico is located.
2. why Mexico is called the "land of the shaking earth."
3. what climates are found in Mexico.

Clara Ponce is a teenager who lives in Mazatlán, a large port city and tourist resort on the coast of the Pacific Ocean. When Clara's parents take her and her sisters on vacation, they have some spectacular choices—Mexico's sunny beaches, hot deserts, misty rain forests, or snowy mountains.

Clara is proud of the Spanish and Native American heritage that Mexico shares with its Latin American neighbors to the south. Mexico also has strong links with its northern neighbor, the United States.

Location
Bridging Two Continents

Mexico forms part of a **land bridge**, or narrow strip of land that joins two larger landmasses. This land bridge connects North America and South America. Look at the map on page 154. You can see that Mexico sweeps to the southeast like a crooked funnel. The widest part of Mexico borders the United States in the north. Farther south, Mexico reaches its narrowest point at the Isthmus of Tehuantepec (tuh•WAHN•tuh•PEHK). Here, only 140 miles (225.3 km) separates the Pacific Ocean from the Gulf of Mexico.

The Pacific Ocean borders Mexico on the west. Extending south along this west coast is Baja (BAH•hah) California. It is a long, thin **peninsula**, or piece of land surrounded by water on three sides. On the eastern side, the Gulf of

Classroom Resources for Section 1

 **REPRODUCIBLE MASTERS**

Reproducible Lesson Plan 6-1
Guided Reading Activity 6-1
GeoLab Activity 6
Geography Skills Activity 6
Building Skills in Geography,
   Unit 2, Lesson 2
Section Quiz 6-1

 TRANSPARENCIES

Section Focus
   Transparency 6-1
Political Map Transparency 3

 MULTIMEDIA

 National Geographic
   Society: STV: Restless
   Earth
Vocabulary PuzzleMaker
   Software
Testmaker

Meeting National Standards

Geography for Life The following standards are highlighted in this section:
- Standards 3, 8, 10

FOCUS

Section Objectives
1. **Locate** Mexico on a globe, map, or atlas.
2. **Explain** why Mexico is called the "land of the shaking earth."
3. **Identify** climates found in Mexico.

Bellringer
Motivational Activity

 Prior to taking roll at the beginning of the class period, project Section Focus Transparency 6-1 and have students answer the activity questions. Discuss students' responses. **L1**

Vocabulary Pre-check

Point out to students that the terms *latitude* and *altitude* often are confused. Have students look up these words in a dictionary and then use them in separate sentences. **L1** **LEP**

Use the Vocabulary PuzzleMaker Software to create a crossword puzzle. **L1**

153

TEACH

Guided Practice

Compare Have students construct a chart comparing Mexico's altitude zones. Bases of comparison might include location, temperature range, and vegetation. Use the completed chart to analyze the relationship between altitude and climate.
L1

These materials are available from Glencoe.

  VIDEODISC

STV: Restless Earth

Earthquake destruction, 1985, Mexico
Any side, Frame 52170

Earthquake rescue, 1985; Mexico City
Any side, Frame 52271

 **MAP STUDY**

Answer
5,000 feet to 10,000 feet (1,500 m to 3,000 m)

Map Skills Practice
Reading a Map Where are Mexico's major plains areas located? *(along the coast of the Gulf of Mexico)*

Mexico and the Caribbean Sea wash Mexico's shores. Jutting far out into the Gulf is another large peninsula—the Yucatán Peninsula. Find Baja California and the Yucatán Peninsula on the map below.

Place

Land of the Shaking Earth

From spectacular mountains and volcanoes to deep valleys, Mexico has a rugged landscape. Why? The country sits where some plates of the earth's crust have collided for billions of years. These collisions formed mountains and volcanoes and caused terrifying earthquakes. One famous volcano was named Popocatepetl (POH•puh•KAH•tuh•PEH•tuhl), or "smoky mountain," by ancient Native Americans. The Aztec, one of the early peoples of Mexico, called their country the "land of the shaking earth." The ground is still moving! Mexico City, the capital, has been shaken by many earthquakes in the past 50 years.

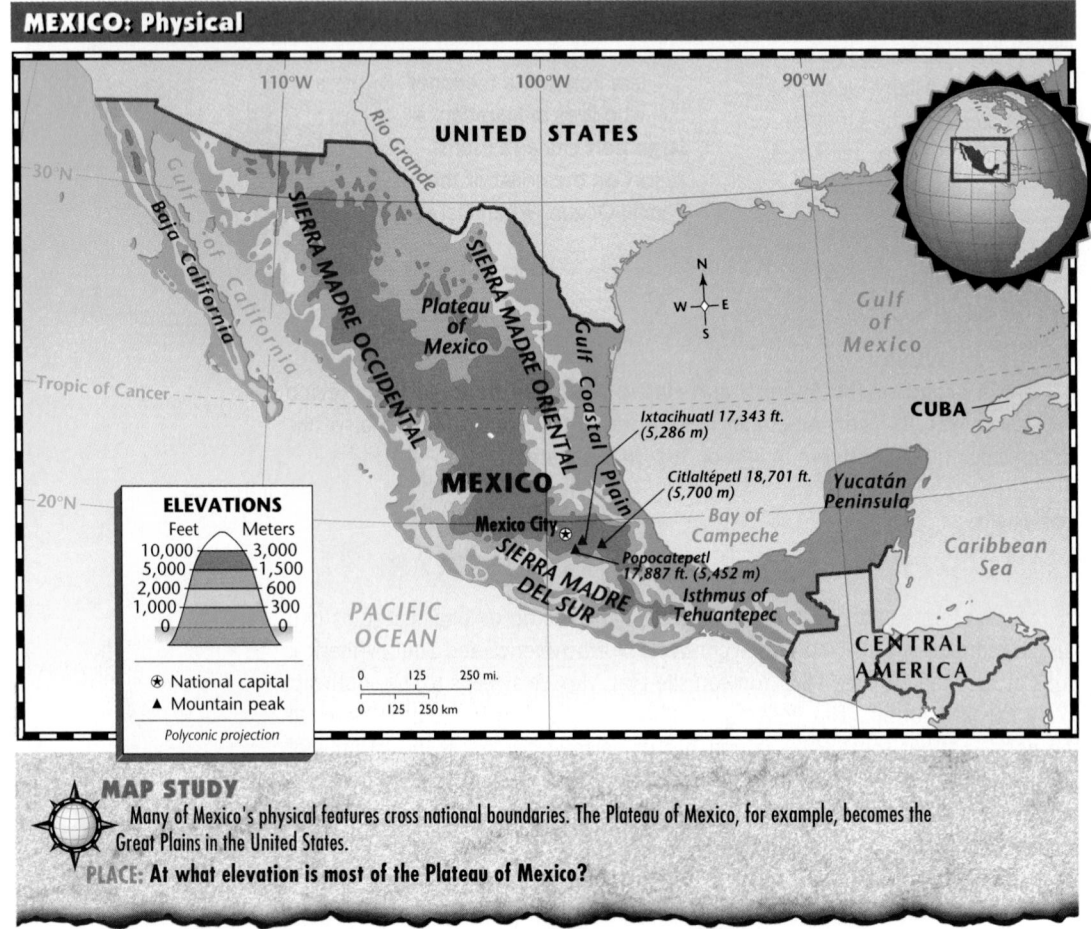

MAP STUDY
Many of Mexico's physical features cross national boundaries. The Plateau of Mexico, for example, becomes the Great Plains in the United States.
PLACE: At what elevation is most of the Plateau of Mexico?

154

UNIT 3

Cooperative Learning Activity

Have students work in small groups to create a relief map of Mexico. Assign regions or specific landforms to each group. Provide clay or crumpled paper for groups to use to form mountains. Students should create a map key and paint the landforms appropriately to represent their elevation. **L1**

A one-lane road crisscrosses the jagged peaks of the Sierra Madre Occidental.
LOCATION: Along which coast of Mexico does the Occidental range extend?

LESSON PLAN
Chapter 6, Section 1

Independent Practice

⬦ Guided Reading Activity 6-1 **L1**

Strange but TRUE!

Paricutín is the most recently formed volcano in the Western Hemisphere. It appeared in a Mexican cornfield in 1943. Volcanic material began to erupt through a crack in the earth. By the end of a week, the volcanic cone was 450 feet (140 m) high. After 8 months, it had reached 1,500 feet (457 m). When the eruptions finally ended in 1952, Paricutín stood 9,210 feet (2,807 m) high.

Mountains and Plateau Three major mountain ranges tower over Mexico. One range—the Sierra Madre Occidental—runs north and south along western Mexico near the Pacific Ocean. Clara's village lies in this lightly populated area. Another major range, the Sierra Madre Oriental, runs along Mexico's eastern side. The steep ridges and deep canyons of yet another range—the Sierra Madre del Sur—rise in southwestern Mexico.

The Sierra Madres surround the large, flat center of the country—the Plateau of Mexico—which covers 40 percent of the country. You find mostly deserts and grassy plains in the northern part of the Plateau. As you move south, the Plateau steadily rises in elevation with occasional **basins**, or broad, flat valleys. Snowcapped mountains, some of which are volcanoes, rise above the basins. Volcanic eruptions have left fertile soil in the basins. The map on page 154 shows you the location of the mountain ranges and plateau.

Coastal Lowlands Mexico's coastal plains stretch along the Pacific Ocean and the Gulf of Mexico. Many of the country's longest rivers flow through these coastal plains. The Rio Grande, one of Mexico's great rivers, empties into the Gulf of Mexico. The Rio Grande forms about 1,300 miles (2,090 km) of Mexico's border with the United States.

Region
Climate

By looking at the map on page 156, you see that Mexico has many different climates. What causes such differences? As you read in Chapter 2, **latitude**—or location north or south of the Equator—affects temperatures. The Tropic of Cancer runs through the center of Mexico. It marks the northern

Meeting Special Needs Activity

Language Delayed Students with language difficulties may have problems formulating questions. Form small groups and ask members to practice developing questions based on the text. In order to limit the task, assign each group a specific heading in Section 1. Praise groups for producing many, varied questions. **L1**

MAP STUDY

Answer
20°N latitude

Map Skills Practice
Reading a Map Mexico City lies within the tropics but does not have a tropical climate. Why? (*because of its high altitude*)

CURRICULUM CONNECTION

Life Science Marlin, swordfish, and tarpon are among the game fish caught off Mexico's coasts. Many deep-sea anglers prefer the Cape San Lucas area at the southern tip of Baja California.

ASSESS

Check for Understanding

Assign Section 1 Review as homework or an in-class activity.

Meeting Lesson Objectives

Each objective below is tested by the questions that follow it in parentheses.

1. **Locate** Mexico on a globe, map, or atlas. (2, 5)
2. **Explain** why Mexico is called the "land of the shaking earth." (3, 4)
3. **Identify** the climates found in Mexico. (6)

156

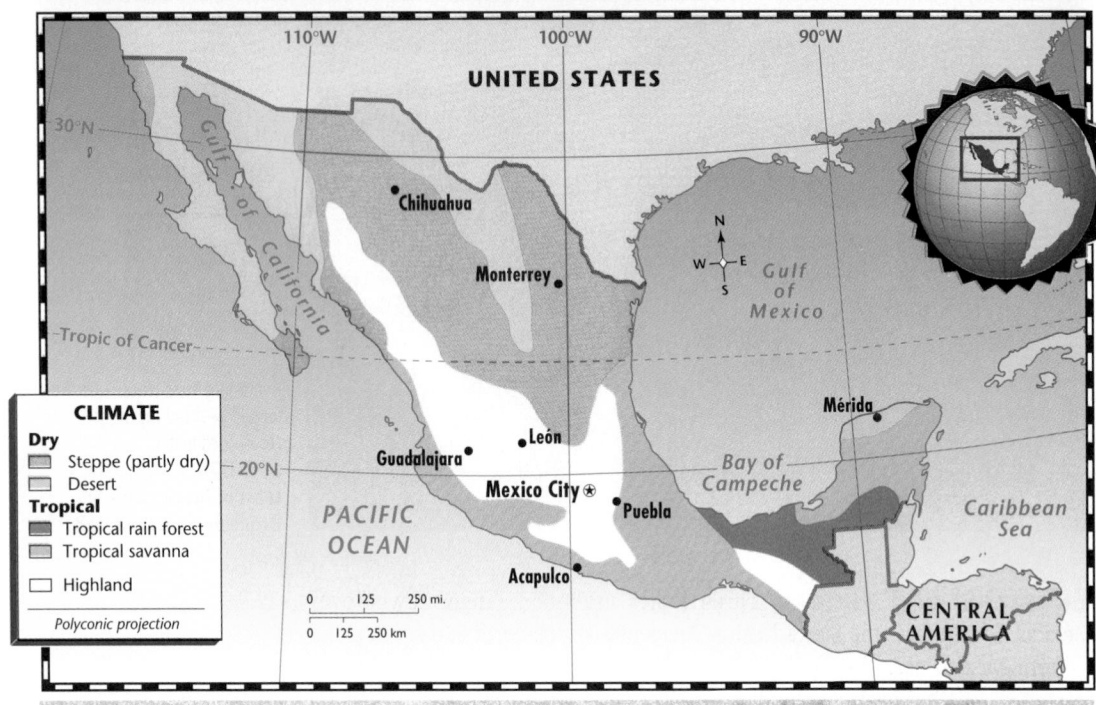

MEXICO: Climate

CLIMATE
Dry
- Steppe (partly dry)
- Desert

Tropical
- Tropical rain forest
- Tropical savanna

- Highland

Polyconic projection

0   125   250 mi.
0   125   250 km

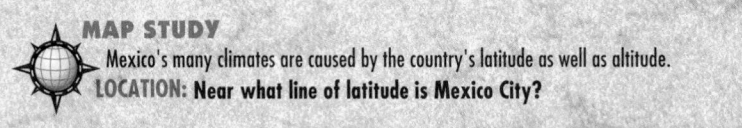

MAP STUDY
Mexico's many climates are caused by the country's latitude as well as altitude.
LOCATION: Near what line of latitude is Mexico City?

A Rugged Land

WHAT IN THE WORLD?

In 1521, a Spanish soldier named Hernán Cortés conquered Mexico. When he returned to Spain, the Spanish king asked him to describe Mexico. Legend says that Cortés crumpled a piece of paper and threw it on a table, saying, "This, your Majesty, is the land of Mexico." In this way he showed that Mexico is a very mountainous country.

edge of the tropics, where temperatures are generally warm year-round.

Altitude Zones Mexico's mountains also affect its temperatures. The diagram on page 157 shows that the country's mountains create three **altitude** zones. You could travel through each of these zones in a day's trip across the Sierra Madres.

Hot Land Your trip begins as you stand on the coastal plain near the sea in the *tierra caliente* (tee•AY•rah kah•lee•AYN•teh), which means "hot land" in Spanish. Hot and humid all year, the *tierra caliente* temperatures average 77°F (25°C) to 82°F (28°C). Dense rain forests or tall grasses blanket much of this zone, although bananas, rice, sugarcane, and oranges also grow here.

156

Critical Thinking Activity

Making Generalizations Help students practice the skill of making generalizations based on information in this section. For example, point out that many of the place names in Mexico are Spanish. The introduction to the section refers to the Spanish heritage that Mexico shares with other Latin American countries. Students might generalize that people from Spain colonized Mexico. Challenge students to formulate additional generalizations. They might test the accuracy of their statements after completing the chapter. **L2**

Mexico's Altitude Zones

SIERRA MADRE OCCIDENTAL — Mexico City — SIERRA MADRE ORIENTAL

PACIFIC OCEAN

Caribbean Sea

Tierra fría   6,500 feet   (2,000 m)

Tierra templada   2,500 feet   (760 m)

Tierra caliente   0 feet   (0 m)

Sea Level   Sea Level

Source: *Mexico,* Amanda Hopkinson

GRAPHIC STUDY

In the diagram, you see that the *tierra caliente* includes the low coastal plains. Hot, humid climates prevail along the coasts, with temperatures ranging from 60° to 120°F (15.6° to 48.9°C).

PLACE: What kinds of vegetation are shown growing in the *tierra caliente*?

Temperate Land As you climb into the mountains, the temperatures become a little cooler. The trees are larger and have more leaves. You are in the middle elevation—the *tierra templada* (tehm•PLAH•dah), or "temperate land" where temperatures hover around 70°F (21°C). Farmers can grow a wide variety of crops here.

Cold Land As you reach the top of the mountains, the climate becomes very cold. You are traveling through the *tierra fría* (FREE•ah), or "cold land." During the year, temperatures here average below 68°F (20°C). Short, stunted trees and a few wildflowers cling to rocky soil. As you travel higher, plant life becomes very scarce and then disappears under snow-topped peaks.

SECTION 1 REVIEW

REVIEWING TERMS AND FACTS

1. **Define the following:** land bridge, peninsula, basin, latitude, altitude.
2. **LOCATION** What country borders Mexico to the north?
3. **PLACE** How does the landscape of the Plateau of Mexico differ from north to south?
4. **PLACE** What three altitude zones would you go through if you crossed Mexico's mountains?

MAP STUDY ACTIVITIES

5. Turn to the map of Mexico on page 154. What bodies of water surround Mexico?
6. Look at the map on page 156. What climates are found in northern Mexico?

Answers to Section 1 Review

1. All vocabulary words are defined in the Glossary.
2. United States
3. Deserts and grassy plains are mostly found in the northern part of the Mexican Plateau. To the south, the Plateau rises in elevation, with occasional basins.
4. *tierra caliente, tierra templada, tierra fría*
5. Pacific Ocean, Gulf of California, Gulf of Mexico, Caribbean Sea
6. steppe, desert, and highland

GRAPHIC STUDY

Answer
tropical vegetation

Graphic Skills Practice
Reading a Diagram In which altitude zone is Mexico City located?
(Tierra fría)

Evaluate

Section 1 Quiz **L1**

Use the Testmaker to create a customized quiz for Section 1. **L1**

Reteach

Write each of the terms and places listed in the **Preview** on a slip of paper. Have students take turns drawing a slip of paper and telling what they know about the term or place. **L1**

Enrich

Have students research and report on the Yucatán Peninsula. Topics they should cover include geology, vegetation, history, and cultures. **L3**

CLOSE

Refer students to the **Preview** on page 153. Direct them to write a brief letter to Clara Ponce telling her where in Mexico they would like to vacation and why.

BUILDING GEOGRAPHY SKILLS

TEACH

Allow students five minutes to prepare an answer to the following: Compare how people live in lowland areas to how people live in mountainous areas. Consider the type of land, the climate, and the ways in which people work and enjoy their leisure time. Write as many differences as possible.

At the end of the time period, call on volunteers to read their answers. Point out that a relief map can offer clues as to where the two ways of life might be practiced.

Have students read the skill lesson. Then ask: What do relief maps show? *(differences in elevation)* How do the maps show relief? *(Different colors or shading represent different elevations.)* **L2**

Additional Skills Practice

1. What two mountain ranges are shown on the map? *(Sierra de Juárez, Sierra San Pedro Mártur)*
2. Where are the lowest elevations in Baja California found? *(along the coasts)*

 Geography Skills Activity 6

 Building Skills in Geography, Unit 2, Lesson 2

Reading a Relief Map

Differences in the elevation, or height, of land areas are called *relief*. In a relief map, colors or shadings show areas of different elevations.

Relief maps are an important source of geographic information. To read a relief map, follow these steps:

• Read the map title to identify the land area shown on the map.
• Use the map key to determine what elevations are shown on the map.
• Identify the areas of highest and lowest elevations on the map.

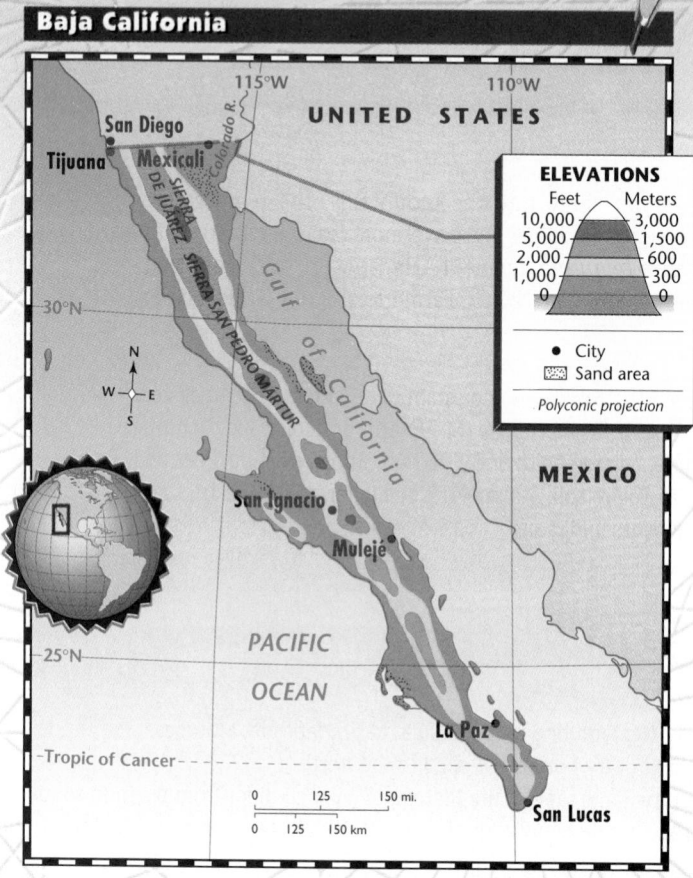

Geography Skills Practice

1. What area of Mexico is shown on the map?
2. At what elevation is San Ignacio?
3. What is the elevation of most of the area along the west coast of Baja?

158

Answers to Geography Skills Practice

1. Baja California
2. 1,000-2,000 feet (300-600 m)
3. 0-1,000 feet (0-300 m)

SECTION 2
The Economy

PREVIEW

Words to Know
- service industry
- *maquiladora*
- subsistence farm
- plantation
- industrialized
- smog

Places to Locate
- Guadalajara
- León
- Puebla

Read to Learn . . .
1. how Mexicans earn a living.
2. what the three economic regions of Mexico are.
3. what economic challenges face modern Mexico.

Oil rigs stand offshore in the Gulf of Mexico as reminders of Mexico's hopes for a better economy. Every time workers strike oil underneath the Gulf, Mexico improves its position as one of the world's top oil producers.

Oil fuels Mexican businesses and homes. It is also exported, which brings much money into the country. The map on page 160 shows that oil is only one of many minerals found in Mexico.

Region
Economic Regions

Manufacturing and mining are vital to the Mexican economy. Almost one-fifth of the world's silver is mined in Mexico. **Service industries** also strengthen the country's economy. A service industry, you remember, is a business that provides services to people instead of making goods. Banking and tourism are Mexico's main service industries. The geography of Mexico influences its economy. Many geographers divide Mexico into three economic regions: central Mexico, the north, and the south.

Although rich in minerals, Mexico is poor in fertile land. The country's many mountains, deserts, and rain forests limit the land that can be farmed to only 11 percent of the total area. Farmers make good use of this land by growing coffee, corn, cotton, oranges, and sugarcane.

CHAPTER 6

159

Meeting National Standards

Geography for Life The following standards are highlighted in this section:
- Standards 11, 14, 16

FOCUS

Section Objectives
1. **Explain** how Mexicans earn a living.
2. **Identify** the three economic regions of Mexico.
3. **Discuss** the economic challenges that face modern Mexico.

Bellringer
Motivational Activity

 Prior to taking roll at the beginning of the class period, project Section Focus Transparency 6-2 and have students answer the activity questions. Discuss students' responses. **L1**

Vocabulary Pre-check

Have students compare subsistence farming with plantation farming. How many similarities and differences can they list? **L1 LEP**

Classroom Resources for Section 2

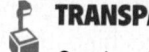 **REPRODUCIBLE MASTERS**

Reproducible Lesson Plan 6-2
Guided Reading Activity 6-2
Cooperative Learning Activity 6
Workbook Activity 6
Section Quiz 6-2

 TRANSPARENCIES

Section Focus Transparency 6-2
Unit Overlay Transparencies 3-0, 3-1, 3-3, 3-5

MULTIMEDIA

 National Geographic Society: GTV: Planetary Manager

Testmaker

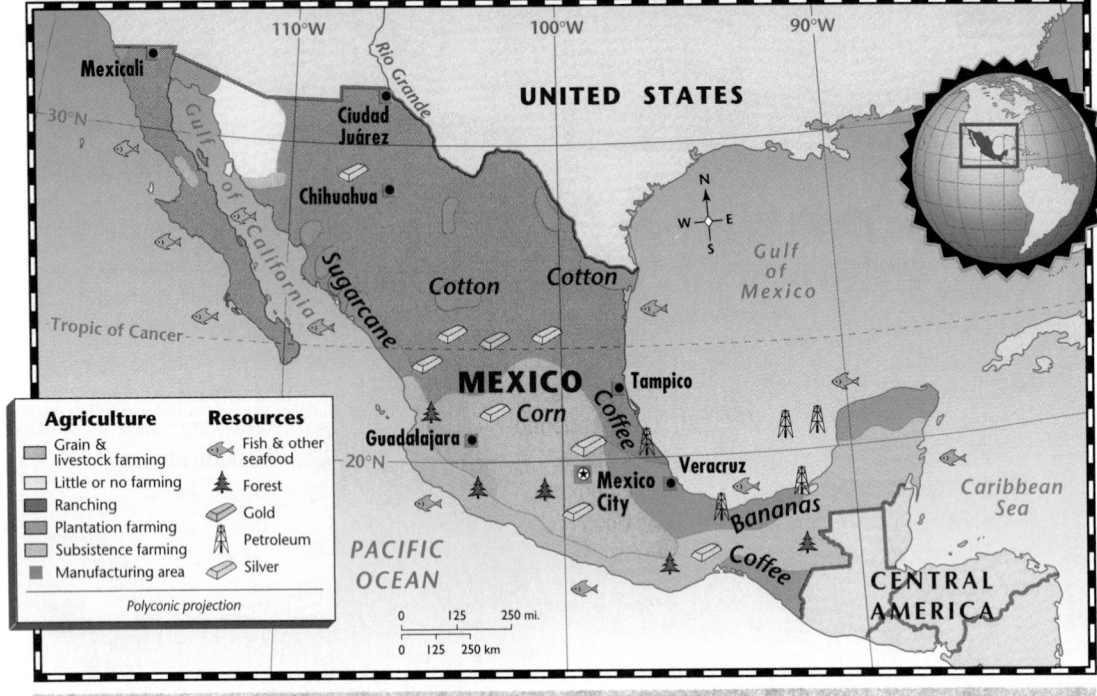

LESSON PLAN
Chapter 6, Section 2

MAP STUDY

Answer

north: fish, gold, silver; south: petroleum, fish, silver, forests

Map Skills Practice

Reading a Map Which would be a better location for a wood processing mill—northern or southern Mexico? Why? *(southern, because of forests there)*

TEACH

Guided Practice

Display Unit Overlay Transparencies 3-0, 3-1, 3-3, and 3-5. Have students use markers to indicate areas of Mexico best suited for service industries, *maquiladoras,* small farms, and plantations. Students should consider the physical geography, population settlement, and climate of each region. **L2**

Independent Practice

Guided Reading Activity 6-2 **L1**

MEXICO: Land Use and Resources

MAP STUDY

Mexico's resources differ greatly as you move from the north to the south.

PLACE: What resources are found in the north? In the south?

Central Mexico Central Mexico is the economic heart of the country. This area is home to more than half of the population and offers favorable conditions for agriculture. Large industrial cities, such as Mexico City and Guadalajara (GWAH•duhl•uh•HAHR•uh), also flourish here. Other cities in central Mexico include León (lay•OHN) and Puebla (PWEH•bluh). Find these cities on the map on page 161.

Mexico City is the giant hub of central Mexico. More than 23 million people live in Mexico City and its suburbs, making it one of the largest urban areas in the world. It is also the center of Mexico's government.

The North The northern economic region includes Baja California and the northern part of the Plateau of Mexico. Much of the land in the north is too dry to farm, but farmers irrigate in some places to grow cotton, fruits, cereals, and vegetables. In hilly areas ranchers raise cattle, sheep, goats, and pigs. Ranching, as we know it today, originated in Mexico. *Vaqueros,* or cowhands, developed the tools to herd, rope, and brand cattle.

160

Cooperative Learning Activity

Organize students into three groups and assign one of Mexico's economic regions to each group. Have group members read and discuss the text relating to their assigned region. Then form new groups of three students, consisting of one member from each of the original groups.

Members of the new groups are responsible for teaching one another about the economic region they studied. At the end of the activity, all students should be able to compare the three regions. **L1**

The largest city in this region is Monterrey, which leads the country in steel production. Workers in the region mine copper, silver, lead, and zinc. Other workers are employed in **maquiladoras** (mah•KEE•luh•DOH•rahz), factories that assemble parts shipped from other countries. Workers in *maquiladoras* assemble automobiles, stereo systems, computers, and other electronic products. These assembled goods are then sent to foreign countries, especially the United States.

The South The southern region stretches from Mexico City to the Yucatán Peninsula. People have lived here since at least 2000 B.C. Those living in this area are among the poorest in Mexico. Farmers in many mountain villages are subsistence farmers. **Subsistence farms**, small plots where farmers grow only enough food to feed their families, are common in this area. In the valleys, wealthy farmers grow coffee or sugarcane on **plantations**, large farms that raise a single crop for money. Tourism is also important in the south. The region's beautiful beaches and historic monuments draw visitors from all over the world.

Human/Environment Interaction
Economic Challenges

During the past 50 years, Mexico has **industrialized**. What does this mean? It means that Mexico has become less a country of farms and villages and more a country of factories and cities. Many challenges arise with industrial growth, however. They include conserving land, controlling pollution, creating new jobs, and increasing trade with other countries.

Pollution Industrial growth in Mexico affects the surrounding environment. As you read earlier, mountains encircle Mexico City, blocking the flow of air. Mexico City's many factories and cars pollute the air, leaving a thick haze of

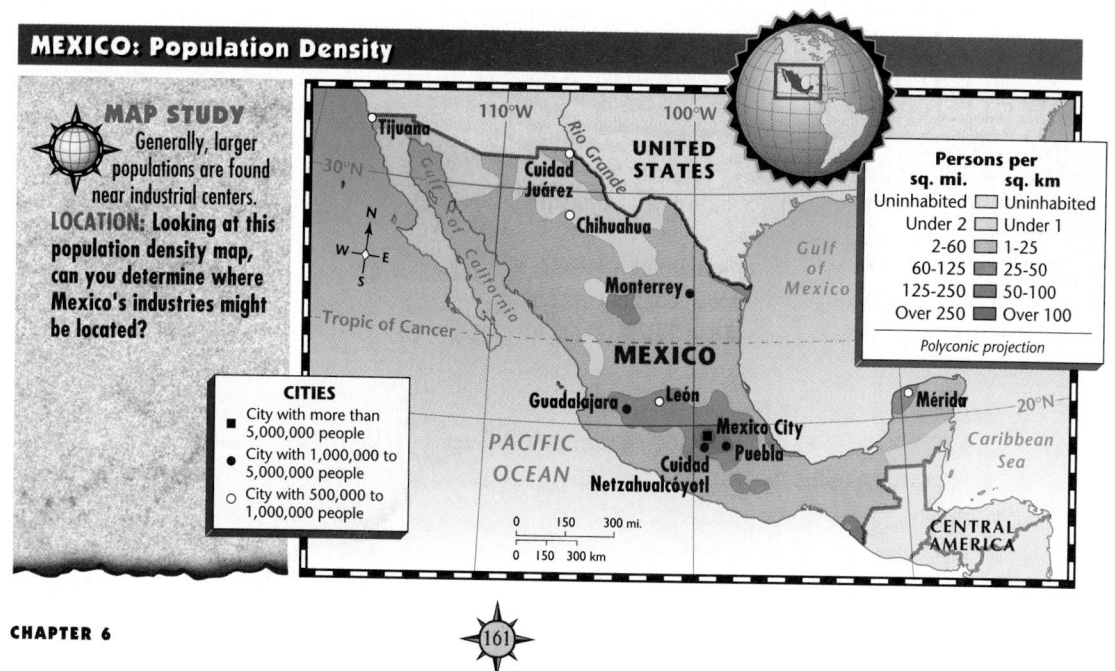

MEXICO: Population Density

MAP STUDY
Generally, larger populations are found near industrial centers.
LOCATION: Looking at this population density map, can you determine where Mexico's industries might be located?

Persons per	
sq. mi.	**sq. km**
Uninhabited	Uninhabited
Under 2	Under 1
2-60	1-25
60-125	25-50
125-250	50-100
Over 250	Over 100

Polyconic projection

CITIES
■ City with more than 5,000,000 people
● City with 1,000,000 to 5,000,000 people
○ City with 500,000 to 1,000,000 people

NATIONAL GEOGRAPHIC SOCIETY

These materials are available from Glencoe.

VIDEODISC

GTV: Planetary Manager

Mexico City smog
Any side, Frame 47839

ASSESS

Check for Understanding
Assign Section 2 Review as homework or an in-class activity.

Meeting Lesson Objectives

Each objective below is tested by the questions that follow it in parentheses.

1. **Explain** how Mexicans earn a living. (3, 6)
2. **Identify** the three economic regions of Mexico. (3, 6)
3. **Discuss** the economic challenges that face modern Mexico. (4, 5)

MAP STUDY

Answer
central southern Mexico, near the United States border

Map Skills Practice
Reading a Map Which city has the larger population, Guadalajara or Mérida? *(Guadalajara)*

Meeting Special Needs Activity

Writing Disability Students with writing problems can obtain essential information from the text by making questions from each subhead. For example, "How many economic regions are in Mexico?" and "Which economic region is too dry to farm?" can be derived from the subhead "Economic Regions." Ask students to read about the economic regions and to write brief responses for each question. Then have them write their own set of questions and answers for the same subhead. L1

161

Evaluate

Section 2 Quiz **L1**

Use the Testmaker to create a customized quiz for Section 2. **L1**

Reteach

Have students work in pairs to outline the section. Encourage students to retain the outlines for review purposes.

Enrich

Ask interested students to investigate the economic crisis that confronted Mexico in 1995. Have them detail the steps the international community took to aid Mexico.

CLOSE

Have students complete a circle graph comparing economic activities in Mexico. Each part of the circle should indicate what percentage of the labor force is employed in the economic activity it represents.

A Floating Village?

On the outskirts of Mexico City lies the floating village of Xochimilco (SOH•chih•MEEL•koh). When it was settled in the 1200s, crops were planted on mud-covered rafts held together by reeds. Roots from the plants eventually anchored the floating gardens to the bottom of the lake surrounding Mexico City. Xochimilco's farmland is still a sort of raft, and visitors travel on canals instead of roads.

smog, or fog mixed with smoke and chemicals, to settle over the city. Schoolchildren wear masks at recess to filter out the pollution, and sometimes the city completely shuts down because people must stay indoors.

In the countryside, farmers often burn wild vegetation to make room for crops, destroying thousands of acres of forests. Mexico must find ways to use and conserve its resources without spoiling the countryside.

Population Changes Mexico's population is growing twice as fast as the population of the United States. Because health care and diet have improved, Mexico's birthrate is rising and more people are living longer. With this increase in population, Mexico cannot provide jobs for all the people who want to work.

Look at the map on page 161. It shows that most of Mexico's 92 million people live in the southern part of the Plateau of Mexico. The many people living in this one area strain resources. Crowded conditions and unemployment are common. Many Mexican people in search of work move to the United States. Some enter the United States illegally, without proper travel papers or official permission to work.

Free Trade In 1993 Mexico, the United States, and Canada approved the North American Free Trade Agreement, or NAFTA. This agreement eventually will allow goods and money to move freely among these three countries. It has already begun to create many new jobs in Mexico. Those who support NAFTA hope that increased trade will help Mexico's economy grow and improve life for all Mexicans.

SECTION 2 REVIEW

REVIEWING TERMS AND FACTS

1. **Define the following:** service industry, *maquiladora,* subsistence farm, plantation, industrialized, smog.
2. **PLACE** What is Mexico's most important mineral?
3. **REGION** What are Mexico's three economic regions?
4. **HUMAN/ENVIRONMENT INTERACTION** What environmental problems face those who live in Mexico City?

MAP STUDY ACTIVITIES

5. Look at the map on page 160. What minerals, besides oil, are found in Mexico?
6. According to the map on page 160, what kinds of farming take place on the Yucatán Peninsula?

Answers to Section 2 Review

1. All vocabulary words are defined in the Glossary.
2. oil
3. central Mexico, the north, the south
4. severe pollution
5. gold, silver
6. subsistence and plantation farming

MAKING CONNECTIONS

THE ANCIENT MAYA

| MATH | SCIENCE | HISTORY | LITERATURE | TECHNOLOGY |

The Maya (MY•uh) were Native Americans who lived in the Yucatán Peninsula and Central America. From about A.D. 250 to A.D. 900, their civilization flourished, making advances in astronomy, architecture, and mathematics. They also invented a sophisticated form of writing hundreds of years before Europeans came to the Americas.

RELIGION Religion was the strongest force in Maya society. The Maya worshipped many gods and goddesses, including a corn god, a rain god, and a moon goddess. To keep accurate records for their religious festivals, the Maya became skilled astronomers. They used their knowledge of the sun and stars to predict eclipses and to develop a calendar. The Maya calendar had 360 days and 5 "unlucky" days to equal 365 days.

ARCHITECTURE Ruins of Maya cities lie hidden in the steamy rain forests of the Yucatán Peninsula. One of the greatest Maya cities was Chichén Itzá (CHIH•chehn EET•suh). Most Maya cities had religious temples built as huge pyramids. The main pyramid at Chichén Itzá reveals the organized thinking of Maya architects. The pyramid has 4 sides, each with a stairway of 91 steps running to

the top of a platform—a total of 364 steps. The platform at the top brings the total to 365, the number of days in a year.

The Main Pyramid at Chichén Itzá

MATHEMATICS The Maya were also skilled in mathematics. They created a number system based on the number 20 that allowed them to count into the millions. Hundreds of years before the Arabs, the Maya developed the concept of zero.

Maya civilization mysteriously collapsed sometime after A.D. 800. No one knows why the Maya abandoned their cities. Descendants of the Maya, however, today live in southern Mexico and nearby areas of Guatemala.

Making the Connection

1. Where in Mexico did the Maya live?

2. What were some Maya achievements in architecture? In mathematics?

163

SECTION 3
The People

FOCUS

Section Objectives

1. **Identify** groups that influenced Mexican culture.
2. **Contrast** city life with country life in Mexico.
3. **Explain** what makes up Mexican culture today.

Bellringer Motivational Activity

Prior to taking roll at the beginning of the class period, project Section Focus Transparency 6-3 and have students answer the activity questions. Discuss students' responses. **L1**

Vocabulary Pre-check

Have students find the meaning of the term *adobe*. Then have them make a list of other kinds of materials used to build houses. Discuss how each material is suited to a particular climate. **L1 LEP**

PREVIEW

Words to Know
• colony
• *mestizo*
• adobe

Places to Locate
• Tenochtitlán

Read to Learn . . .
1. what groups influenced Mexican culture.
2. how city life differs from country life in Mexico.
3. what makes up Mexican culture today.

Thousands of tourists visit Mexico City every year. They often visit the unique Plaza of Three Cultures. Here, in a single place, you can see the ruins of an ancient Native American monument, a stone church built by the Spaniards, and modern glass and steel apartment buildings!

Mexico City displays the mix of cultures—old and new—that make Mexico a fascinating country. Two groups—Native Americans and Spaniards—have made important contributions to Mexico's growth as a nation.

Place
Influences of the Past

Historians usually divide Mexican history into three periods. The first is the age of great Native American civilizations. The second is the time of Spanish rule, and the third is the age of modern Mexico.

Native Americans The first people to live in Mexico came from Asia. Later known as Native Americans, these people traveled south through North America and entered Mexico thousands of years ago. Look at the map on page 165 to locate Mexico's major Native American civilizations.

One group, the Maya, flourished in the Yucatán area between A.D. 250 and 900. They built cities around towering stone temples in thick rain forests. These temples honored Maya gods and rulers. Turn to page 163 to read about the ancient Maya.

164

UNIT 3

MEXICO: Native American Civilizations

MAP STUDY Some of the most advanced civilizations of the Western Hemisphere lived in Mexico.
LOCATION: Which civilization lived mostly in the Yucatán Peninsula?

110°W   100°W   90°W

NORTH AMERICA

30°N

Gulf of California

Rio Grande

Gulf of Mexico

Tropic of Cancer

Chichén Itzá

20°N

PACIFIC OCEAN

Tenochtitlán

Caribbean Sea

Maya land (300 B.C.–A.D. 900)
Aztec land (A.D. 1450–1521)
• City

Polyconic projection

0   150   300 mi.
0   150   300 km

Around A.D. 1200 the Mexica, whom the Spanish called the Aztec, built a city named Tenochtitlán (tay•NOHCH•teet•LAHN) in central Mexico. Mexico City is located in this area today. The Aztec were fierce warriors as well as builders and traders. Merchants in Tenochtitlán set up marketplaces, which were filled with pottery, woven baskets, cloth, gold, and silver.

The Spanish Heritage In 1519 a Spanish army led by Hernán Cortés arrived on Mexican soil, and, 2 years later, conquered the Aztec. Mexico remained a Spanish **colony,** or overseas territory, for about 300 years. During that time many Spaniards controlled the lives of Native Americans, forcing them to work on farms and in mines. As a result, the Spanish and Native American cultures mixed. The term *mestizo* (meh•STEE•zoh) refers to a person of mixed Native American and European heritage. Today about 60 percent of Mexico's people are *mestizos.* Another 30 percent are Native American.

You can also see the influence of the Spanish on religion in Mexico. Most Mexicans today—about 90 percent—are Roman Catholic. Festivals honor the country's patron saint, the Virgin of Guadalupe (GWA•duhl•OO•pay).

The Zócalo, Mexico City

Each year thousands of people gather in the Zócalo, or Constitution Square, in remembrance of the Mexican revolution.
PLACE: How many years did the Mexican revolution last?

CHAPTER 6

165

TEACH

Guided Practice

Multicultural Point out that NAFTA has brought about a greater exchange of cultural ideas between Canada, the United States, and Mexico. Have students suggest ways that Mexico's culture might be affected by NAFTA. Students should consider how the visual arts, music, literature, cuisine, fashion, politics, and architecture might be affected by the interchange that occurs between these nations. Note student responses on the chalkboard. Then have students use information on the chalkboard to write a paragraph on the cultural impact of NAFTA. **L2**

More About the Illustration

Answer to Caption
over 10 years

Place The Zócalo is the main plaza in Mexico City. It is the site of the National Palace, the National Cathedral, and the Supreme Court of Justice.

Cooperative Learning Activity

Organize students into three groups to research the cultural heritage of Mexico. Assign each group one of the following cultures: Maya, Aztec, or Spanish. Group members should report on the history of their assigned Native American or European culture. Then they should find illustrations to indicate how that culture influences Mexico today. **L2**

Modern Mexico Mexico gained its independence from Spain in 1821. During much of the 1800s, a few wealthy families, the army, and church leaders controlled the Mexican government. They did little to improve life for Mexico's many poor people who were farmers. In 1910 the discontent of the people finally exploded in a revolution.

The Mexican revolution lasted until the early 1920s and greatly influenced the type of government that Mexico has today. Like the United States, Mexico is a federal republic. Powers are shared by a national government and the 31 state governments. Since the late 1920s, Mexico has been ruled by one political party. Look at the map on page 152 to see the states that make up modern Mexico.

Place
City Life

Bustling cities dot Mexico, providing homes for three out of four Mexicans. These cities have both modern and older sections. Guadalajara, in central Mexico, has broad streets lined with palm trees. The city's tall modern buildings and colorful shops convey a sense of excitement and progress. Guadalajara's citizens also enjoy historic areas with parks, fountains, and splendid old churches.

Areas of beautifully preserved old homes are found in many Mexican cities. These homes, usually made of **adobe** (uh•DOH•bee), or sun-dried clay brick, often are built around courtyards with fountains and pots of blooming plants. Houses in poor urban areas are made of scraps of wood, metal, or whatever materials can be found. Most of these lack electricity and running water.

Guadalajara, Mexico

Nicknamed the "Pearl of the West," Guadalajara has more than three million people.
MOVEMENT: Do most people in Mexico live in rural or urban areas?

166

UNIT 3

Place
Country Life

Country life is very different from city life in Mexico. Most Mexican villages are very poor. Some village homes are built of cement blocks with a flat, red-tiled roof, and others are made of sheet metal, straw, or clay. Homes line narrow streets that lead to a central plaza, which may include a small church, a few shops, and the local government building.

Almost every village has a marketplace where men, women, and children sell or trade clothes, food, baskets, or pottery. Some show their goods in a large building that resembles a warehouse. Others display their wares outside on brightly colored shawls spread upon the street.

Place
The Arts and Recreation

Mexican art explores and represents the pride of the people in their achievements and their heritage. Early Native Americans gave Mexico beautiful temples and dramatic sculptures. Spaniards and Native Americans later built and decorated many of the country's churches. Modern composers, artists, and writers have also expressed the spirit of Mexico. Craftworkers in Acapulco, for example, combine ancient and modern skills to make silver jewelry, colorful pottery, and handblown glass.

Painters and Writers Mexican artists and writers have created many national treasures. In the early 1900s, Mexican artists produced beautiful murals, or wall paintings. Among the most famous mural painters were José Clemente Orozco and Diego Rivera, who painted scenes from Mexican history. Much of Mexican literature, such as works by modern writers Carlos Fuentes and Octavio Paz, describes daily life.

Music and Dance Mexicans listen to and compose all kinds of music—from modern rock to classical. Traditional Mexican music is played by mariachi (mahr•ee•AH•chee) bands. If you listen to a typical mariachi band, you will hear a singer, two violinists, two guitarists, two horn players, and a bass player. These musicians dress in colorful outfits and wear wide-brimmed hats called sombreros.

Teen Scene

In Touch With the Past

Like most of his friends, Benito Carrea loves to listen to Mexican rock music. On Sundays, however, Benito leaves his radio at home and goes to the park with his family. He takes the flute and practices the traditional music his father and grandfather taught him. Benito's family are descendants of the Maya. More than 1,700 years ago, the Maya played flutes similar to Benito's along with drums, shells, and rattles made from gourds.

Critical Thinking Activity

Determining Cause and Effect Have students determine how the region's physical features and climate have contributed to Mexico's cultural diversity. Discuss how Mexico's diverse vegetation and farm production provide a diet rich in corn, beef, and grains. Students could point out that the region's harsh physical features and warm climate have led to the building of adobe structures, fountains, red-tiled roofs, and central plazas. Craftworkers have also used the region's resources to create silver jewelry, colorful pottery, baskets, and hand-blown glass. Have students summarize the findings in a chart. **L1**

LESSON PLAN
Chapter 6, Section 3

ASSESS

Check for Understanding

Assign Section 3 Review as homework or an in-class activity.

Meeting Lesson Objectives

Each objective below is tested by the questions that follow it in parentheses.
1. **Identify** groups that influenced Mexican culture. (2, 5, 6)
2. **Contrast** city life with country life in Mexico. (3)
3. **Explain** what makes up Mexican culture today. (3, 4)

Evaluate

▱ Section 3 Quiz **L1**

◉ Use the Testmaker to create a customized quiz for Section 3. **L1**

▱ Chapter 6 Test Form A and/or Form B **L1**

Reteach

▱ Reteaching Activity 6 **L1**

▱ Spanish Reteaching Activity 6 **L1**

Enrich

▱ Enrichment Activity 6 **L3**

More About the Illustration

Answer to Caption
guitar, violin, trumpet

Movement The guitar was introduced to Mexico by the Spanish in the 1500s. It was introduced to Spain some 800 years earlier by Muslim invaders from North Africa.

CURRICULUM CONNECTION

Language Arts The term *mariachi* may date from the mid-1800s, when the French army occupied Mexico. Many French soldiers married Mexican women. They hired small bands to play at their weddings. The bands later were known as mariachis, from *mariage,* the French word for wedding or marriage.

CLOSE

Mental Mapping Activity

Provide each student with a large sheet of white paper and map pencils. Inform students that this will not be a graded activity. Have students draw and label, freehand from memory, a map of Mexico that includes the following: Gulf of Mexico, Caribbean Sea, Gulf of California, Sierra Madre Occidental, Sierra Madre Oriental, and Sierra Madre Del Sur.

168

Mexican Music

Mariachi bands play lively Mexican music.
PLACE: What musical instruments do you recognize in this photo?

Celebrations Listen to the excited voices and music in the plaza! What is going on? Throughout the year, Mexicans enjoy special celebrations called fiestas. The word *fiesta* means "feast day" in Spanish. Fiestas are festivals that include parades, fireworks, music, and dancing. Independence Day (September 15-16) and Cinco de Mayo (May 5) are patriotic days of celebration.

Most Mexicans also celebrate Christmas. Actors perform religious plays every evening from mid-December until Christmas Day. After each play, children gather around a piñata (peen•YAH•tuh), a hollow figure of stiff paper filled with toys and candy. Wearing blindfolds, the children try to break the piñata open with a stick.

Foods Traditional Mexican food combines Spanish and Native American cooking. Corn, a crop grown since ancient times, has always been the most important food in Mexico. Mexicans make tortillas, a thin flat cornmeal bread shaped by hand and cooked on a griddle. A folded tortilla filled with vegetables, cheese, beans, or meat becomes a taco. As you probably know, tacos are now as popular in the United States as they are in Mexico.

Sports If you like soccer, visit Mexico and join the crowds. Soccer is the country's most popular sport. Major soccer games are played in Mexico City's Azteca Stadium, which holds about 100,000 people. Baseball, which came to Mexico from the United States more than 50 years ago, also draws large crowds.

Bullfighting, introduced by Spaniards, is a favorite spectator sport for tourists. Many Mexicans admire expert bullfighters for their bravery. Other Mexicans, however, think that bullfighting is cruel because of the way that bulls are killed during the event.

SECTION 3 REVIEW

REVIEWING TERMS AND FACTS

1. **Define the following:** colony, *mestizo,* adobe.
2. **PLACE** Name two groups that influenced modern Mexican culture.
3. **PLACE** How does city life differ from country life in Mexico?
4. **REGION** What celebrations are common throughout Mexico?

MAP STUDY ACTIVITIES

5. Turn to the map on page 165. Name an ancient civilization that lived in central Mexico.
6. Using the same map, name the major city of the Aztec Empire.

(168)

UNIT 3

Answers to Section 3 Review

1. All vocabulary words are defined in the Glossary.
2. Native Americans and Spanish
3. Cities have modern conveniences, are crowded and very active, and are inhabited by both the wealthy and the poor. Country towns and villages have fewer people, fewer modern buildings, people are poorer, and things move at a slower pace.
4. religious celebrations and traditional fiestas
5. Aztec
6. Tenochtitlán

Chapter 6 Highlights

Important Things to Know About Mexico

SECTION 1 THE LAND

- Mexico is part of the land bridge connecting North America and South America.
- Major landforms in Mexico are deserts, mountains, plateaus, and lowlands.
- Much of Mexico's climate is tropical or desert.

SECTION 2 THE ECONOMY

- Central Mexico has the largest cities and the greatest industrial development of Mexico's three economic regions.
- Oil is Mexico's most important product.
- Coffee, rice, sugarcane, and fruit are grown on large and small farms.
- Challenges facing Mexico include encouraging industry, creating jobs, and protecting the environment.

SECTION 3 THE PEOPLE

- Native Americans were the earliest people to live in Mexico. Spanish explorers and settlers arrived in the 1500s.
- The Roman Catholic religion remains an important influence on the people of Mexico.
- Mexico is a federal republic. Its 31 states share power with the national government.
- The culture of modern Mexico mixes old and new art, music, and literature grown from Native American, Spanish, and Mexican roots.

A market in Guanajuato, Mexico ▶

Extra Credit Project

Poster Have each student investigate the plant and animal life of Mexico. Instruct them to research how plants and animals adapt to the country's climate. Have students prepare illustrated posters to share with the class. **L1**

169

Chapter 6 Review and Activities

ANSWERS

Reviewing Key Terms

1. J
2. C
3. I
4. D
5. A
6. B
7. E
8. G
9. F
10. H

Mental Mapping Activity

This exercise helps students to visualize the geographical features they have been studying and understand the relationships among them. All attempts at freehand mapping should be accepted.

REVIEWING KEY TERMS

Match the numbered terms in Column A with their definitions in Column B.

A
1. smog
2. basin
3. subsistence farm
4. plantation
5. industrialized
6. colony
7. *maquiladora*
8. adobe
9. *mestizo*
10. land bridge

B
A. growth of factories and cities
B. overseas territory
C. broad valley surrounded by mountains
D. large farm that grows a single crop
E. factory that assembles parts from other countries
F. person of mixed Native American and European ancestry
G. sun-dried clay brick
H. narrow strip of land that connects two larger land masses
I. to grow only enough to feed your family
J. fog mixed with smoke and chemicals

Mental Mapping Activity

On a separate piece of paper, draw a freehand map of Mexico. Label the following items on your map:
• Yucatán Peninsula
• Gulf of Mexico
• Baja California
• Pacific Ocean
• Gulf of California
• Mexico City

REVIEWING THE MAIN IDEAS

Section 1
1. REGION What continents does Mexico connect?
2. PLACE Name Mexico's three major mountain ranges.
3. LOCATION How does Mexico's latitude affect its climate?
4. PLACE What plants grow in the *tierra fría*?

Section 2
5. LOCATION Where are most of Mexico's oil fields located?
6. PLACE What are Mexico's main service industries?
7. HUMAN/ENVIRONMENT INTERACTION What economic activities take place in the north?
8. MOVEMENT What does Mexico hope NAFTA will do for its foreign trade?

Section 3
9. MOVEMENT Name two Native American civilizations that influenced Mexico's history.
10. MOVEMENT What European nation once ruled Mexico?
11. PLACE What does the term *mestizo* mean?
12. PLACE What are Mexico's most popular sports?

Reviewing the Main Ideas

Section 1

1. North America and South America
2. Sierra Madre Occidental, Sierra Madre Oriental, and Sierra Madre Del Sur
3. Mexico is located primarily in the tropical latitudes, creating a warm, tropical climate over much of the country.
4. short, stunted trees and a few wildflowers

Section 2

5. offshore in the Gulf of Mexico and along the eastern coast
6. banking and tourism
7. farming, raising of livestock, mining
8. create more jobs

CRITICAL THINKING ACTIVITIES

1. **Determining Cause and Effect** Why are temperatures in Mexico City—which is south of the Tropic of Cancer—usually not as hot in the summer as temperatures in northern Mexico?

2. **Evaluating Information** What examples of Mexican culture do you find in your own community? What conclusions can you draw from this knowledge?

GeoJournal Writing Activity

In the GeoJournal Activity on page 137, you were asked to use your five senses—sight, sound, smell, feel, and taste—to describe Latin American countries. Look at your notes about Mexico. Imagine you are a European immigrant who came to Mexico during the days of Spanish colonial settlement. Write a letter to a friend or relative in Europe describing your new home.

COOPERATIVE LEARNING ACTIVITY

Work in groups of 3 to learn more about 1 of the 31 states of Mexico. Each group will choose a state, then each member will select 1 of the following topics to research: (a) the people; (b) the climate; or (c) the vegetation of your chosen state. After your research is complete, share your information with your group. As a group, prepare a written report, poster, or map that illustrates the group's findings.

PLACE LOCATION ACTIVITY: MEXICO

Match the letters on the map with the places and physical features of Mexico. Write your answers on a separate sheet of paper.

1. Sierra Madre Occidental
2. Sierra Madre Oriental
3. Sierra Madre del Sur
4. Plateau of Mexico
5. Mexico City
6. Isthmus of Tehuantepec
7. Acapulco
8. Guadalajara
9. Gulf of Mexico
10. Rio Grande

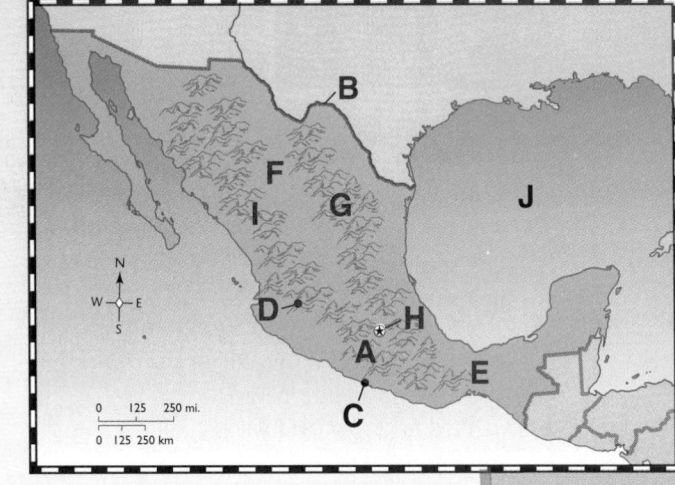

Cooperative Learning Activity

Remind students not to ignore the history of the people in their chosen state. Encourage students to make their presentations as dramatic and colorful as possible, suggesting that they use other mediums besides paper and markers. Find space in the classroom to display the visuals and invite other classes to share them.

GeoJournal Writing Activity

Encourage students to share their letters with the rest of the class.

Place Location Activity

1. I
2. G
3. A
4. F
5. H
6. E
7. C
8. D
9. J
10. B

Chapter Bonus Test Question

This question may be used for extra credit on the chapter test.

What geographic factors have helped promote economic development in Mexico? *(Mexico has large deposits of oil and silver, has rich farmland in the central and southern sections, and has spectacular physical and cultural features essential to tourism.)*

Section 3
9. Maya and Aztec
10. Spain
11. a person of mixed Native American and European heritage
12. soccer, baseball, bullfighting

Critical Thinking Activities
1. Mexico City is at a much higher altitude than northern Mexico.
2. Answers will vary. Conclusions should focus on the theme of movement.

171

GeoLab
ACTIVITY

From the classroom of Marita Sesler, Knoxville, Tennessee

VOLCANOES: Powerful Giants

FOCUS

Read aloud the following quote from Keith Ronnholm, a survivor of the volcanic eruption at Mount St. Helens, Washington, in 1980: *"I stopped and looked and saw a boiling mass of clouds pursuing me. The cloud was like a wall about three miles across, extending straight up . . ."* Then ask students how they think they might have felt if they had been confronted by such a sight.

TEACH

To help students visualize the movement of magma within a volcano, have them perform the following experiment. Take a closed bottle of maple or corn syrup that has been refrigerated and turn it upside down a few times. Ask students what they see. *(A bubble of air floats slowly to the top each time the bottle is turned.)* Place the syrup bottle in hot water for a few minutes. Then repeat the experiment. Ask students to note what happens and compare it to the first results. *(The bubbles of air in the warmed syrup should rise faster.)* Tell students that the air bubble in the warmed syrup is like magma—it moves faster as it becomes hotter. **L1**

Background

Fierce giants named Irazú, Cotopaxi, and Lascar threaten Latin America. Why don't the people try to destroy these giants? They can't—the giants are all volcanoes. In Chapter 1 you learned that volcanoes occur where hot, fluid lava forces its way up to the surface of the earth by extreme pressure. Find out more about erupting volcanoes by making your own volcanic eruption.

Mount Kilauea Erupts!

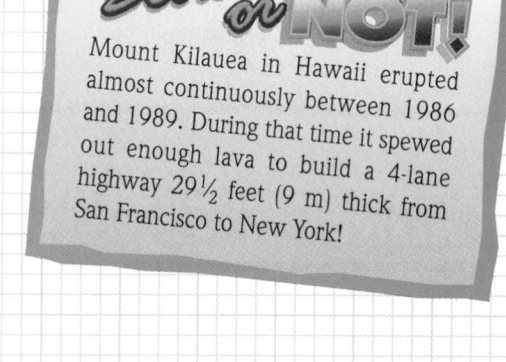

Believe it or NOT!

Mount Kilauea in Hawaii erupted almost continuously between 1986 and 1989. During that time it spewed out enough lava to build a 4-lane highway 29½ feet (9 m) thick from San Francisco to New York!

Materials

- sand or dirt, slightly damp
- 1/2 cup water
- small, empty tin can
- empty quart bottle
- 4 tablespoons baking soda
- 1/4 cup dishwashing liquid
- 1/4 cup white vinegar
- red food coloring (optional)
- long-handled spoon
- large baking pan

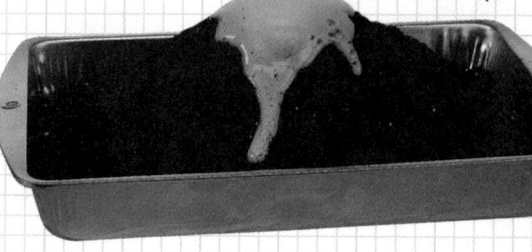

Build your mountain carefully to leave the mouth of the volcano open.

UNIT 3

Curriculum Connection Activity

Geology Ask students to think about what happens inside the earth to cause volcanoes. Then tell them that rocks deep inside the earth melt to form magma, and that when magma flows out onto the earth's surface it is called *lava*. Point out that magma is less dense than the rock around it, so it very slowly rises toward the surface. Display some lava rock, which can be found at most garden centers or plant nurseries. Point out the porous nature of the rock and have students determine what might have caused this. *(the escape of gas bubbles during the rapid cooling of the lava)* **L1**

What To Do

A. Place the baking pan on a level surface.

B. Put the tin can in the center of the pan and build a mountain of sand around it, leaving the top of the can uncovered. (This makes the crater of your volcano.)

C. Pour the baking soda into the can.

D. In the quart bottle mix the water, dishwashing liquid, vinegar, and a few drops of red coloring.

E. Pour only half of the mixture into the tin can. (You can repeat your eruption with the rest later.)

F. Step back and watch your volcano erupt. If the lava does not flow from your volcano right away, carefully stir the can's mixture with the spoon.

Lab Activity Report

1. Did the lava flow more to one side of your volcano, or evenly all around it?

2. Describe how fast the lava flowed at the beginning compared with the end of the eruption.

3. What did you learn about volcanoes from this experiment?

4. Drawing Conclusions
How do you think an erupting volcano might affect the wildlife that lives around it?

 Go A Step Further

Activity
Find out what destructive volcano erupted in the West Indies in 1902. What happened to the people and the land around it? Write a newspaper article describing the event as if you had been there.

Answers to Lab Activity Report

1. Answers will vary. Students should accurately describe the lava flow.
2. Answers will vary. Most students will note that the lava flowed more quickly at the beginning of the eruption than at the end.
3. Answers will vary. Some students may note that they have had experience, on a smaller scale, with the type of energy involved in a volcanic eruption.
4. Answers will vary. Most students will note that an eruption will threaten wildlife, plant life, and human life in the vicinity.

 **GLENCOE**
TECHNOLOGY

VIDEODISC

The Infinite Voyage: To the Edge of the Earth

Chapter 3
Exploring Volcanoes: To the Center of the Earth
Disc 1 Side B

ASSESS

Have students answer the **Lab Activity Report** questions on page 173.

CLOSE

Encourage students to complete the **Go A Step Further** writing activity. Suggest that those who wish to illustrate their newspaper stories do so.

Go A Step Further Answer

In 1902 Mt. Pelee on Martinique in the West Indies erupted, wiping out the city of St. Pierre and killing 40,000 people.

In 1982 El Chichón, located in the Chiapas region of Mexico, erupted. The cloud of smoke and ash was so enormous that it caused two days of darkness in the area. Within a week, it had turned the skies over Hawaii a milky white.

Central America and the West Indies

PERFORMANCE ASSESSMENT ACTIVITY

RECOMMEND A VACATION Organize students into several small groups. Have groups act as travel agents who have been asked to recommend a vacation spot in Central America and the West Indies. Have groups develop an annotated list of the countries discussed in the chapter, noting geographic advantages and disadvantages of each country. Have each group select the three countries it would most recommend as vacation destinations. Groups should then use information on their three countries to create a travel brochure on full-sized poster board. Information should be illustrated with maps, charts, and photographs cut from old magazines. Call on groups to present and discuss their brochures.

POSSIBLE RUBRIC FEATURES:
Decision-making skills, content information, analysis and inference skills, collaborative skills, visual presentation skills

MENTAL MAPPING ACTIVITY
Before teaching Chapter 7, point out Central America on a world map. Have students note that this region is located on a land bridge that connects North America and South America.

TEACHER'S CORNER

NATIONAL GEOGRAPHIC SOCIETY

INDEX TO NATIONAL GEOGRAPHIC MAGAZINE

The following articles may be used for research relating to this chapter:

- "Treasure From the Silver Bank," by Tracy Bowden, July 1996.
- "Feast of the Tarpon," by David Doubilet, January 1996.
- "El Salvador," by Mike Edwards, September 1995.
- "Poison-Dart Frogs," by Mark W. Moffett, May 1995.
- "Trinidad and Tobago," by A.R. Williams, March 1994.
- "New Sensors Eye the Rain Forest," by Thomas O'Neill, September 1993.
- "The Violent Saga of a Maya Kingdom," by Arthur A. Demarest, February 1993.

- "La Isabela, Europe's First Foothold in the New World," by Kathleen A. Deagan, January 1992.
- "Rain Forest Canopy: The High Frontier," by Edward O. Wilson, December 1991.
- "Cuba at a Crossroads," by Peter T. White, August 1991.

NATIONAL GEOGRAPHIC SOCIETY PRODUCTS AVAILABLE FROM GLENCOE

To order the following products for use with this chapter, contact your local Glencoe sales representative or call Glencoe at 1-800-334-7344:

- *STV: World Geography* (Videodisc)
- *STV: Rain Forest* (Videodisc)
- *Picture Atlas of the World* (CD-ROM)
- *Physical Geography of the World* (Transparencies)

- *ZipZapMap! World* (Software)
- *GeoBee* (Software)
- *Images of the World* (Posters)
- *Eye on the Environment* (Posters)

ADDITIONAL NATIONAL GEOGRAPHIC SOCIETY PRODUCTS

To order the following products for use with this chapter, call National Geographic Society at 1-800-368-2728:

- *Tropical Kingdom of Belize* (Video)
- *Jewels of the Caribbean Sea* (Video)
- *Lost Kingdoms of the Maya* (Video)
- *Rain Forest* (Video)

- *Changing Faces of Communism Series:* "Cuba." (Video)
- *Nations of the World Series:* "Central America." (Video)
- *Central America:* "Geography," "Everyday Life," "Cultural and Ethnic Heritage," "Conflict in the Region." (Filmstrip)

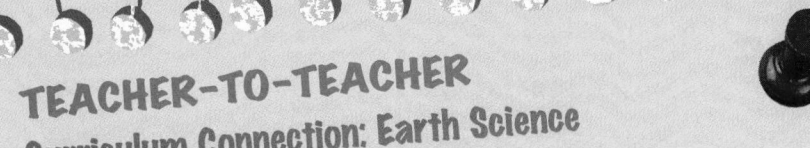

TEACHER-TO-TEACHER
Curriculum Connection: Earth Science

—from Janis Coffman,
Bellmont Middle School, Decatur, IN

STRINGING THE CARIBBEAN *The purpose of this activity is to have students assimilate information about the geographical location and history of the islands of the West Indies by making group presentations and a "string map."*

Organize students into pairs or groups of three. Assign one or more of the islands of the West Indies to each group. Ask groups to develop a two- to three-minute oral presentation that tells about the history of their assigned islands. Emphasize that research on their islands' colonial history is a vital part of this activity.

Display a large wall map of the world. As groups make their presentations, have them use "push pins" to run lengths of string between their islands and the countries that have influenced their islands.

Ask students what conclusions they can draw about the development of the West Indies from the presentations and the "string map." Some groups might want to undertake a follow-up activity by looking at present-day influences in the West Indies. Suggest that they construct another "string map" showing their islands' chief trading partners and the countries with which they have treaty ties.

BIBLIOGRAPHY

Readings for the Student

Caribbean Connections: Overview of Regional History. Washington, D.C.: EPICA/NECA, 1991.

Goodwin, Paul B., ed. *Latin America: Global Studies.* Guilford, Conn.: Dushkin, 1992.

Readings for the Teacher

The Cambridge Encyclopedia of Latin America and the Caribbean. New York: Cambridge University Press, 1992.

Dingle, James, and Jon Esler. *Geography, Culture, History, Politics of Latin America.* Denver: Center for Teaching International Relations, 1991.

Multimedia

Cesar's Story: My Mountain Home in Guatemala. Educational Design, 1992. Videocassette, 22 minutes.

Giving Up the Canal. Alexandria, Va.: PBS Video, 1990. Videocassette, 58 minutes.

MEETING NATIONAL STANDARDS

Geography for Life

All of the 18 standards are demonstrated in Unit 3. The following ones are highlighted in this chapter:

• Standards 3, 4, 6, 9, 10

KEY TO ABILITY LEVELS

Teaching strategies have been coded for varying learning styles and abilities.

L1 **BASIC** activities for all students

L2 **AVERAGE** activities for average to above-average students

L3 **CHALLENGING** activities for above-average students

LEP **LIMITED ENGLISH PROFICIENCY** activities

CHAPTER 7 RESOURCES

- *Vocabulary Activity 7
- Cooperative Learning Activity 7
- Workbook Activity 7
- Performance Assessment Activity 7
- Geography Skills Activity 7
- Chapter Map Activity 7
- GeoLab Activity 7
- Critical Thinking Skills Activity 7
- *Reteaching Activity 7
- Enrichment Activity 7
- Chapter 7 Test, Form A and Form B
- *Chapter 7 Digest Audiocassette, Activity, Test
- Unit Overlay Transparencies 3-0, 3-1, 3-2, 3-3, 3-4, 3-5, 3-6
- Political Map Transparency 3
- Geoquiz Transparencies 7-1, 7-2
- MindJogger Videoquiz
- Vocabulary PuzzleMaker Software
- Student Self-Test: A Software Review
- Testmaker Software
- Focus on World Art Print 11

> Use these *Geography: The World and Its People* resources to teach, reinforce, and extend *chapter* content.

If time does not permit teaching the entire chapter, summarize using the Chapter 7 Highlights on page 191, and the Chapter 7 English (or Spanish) Audiocassettes. Review students' knowledge using the Glencoe MindJogger Videoquiz. * Also available in Spanish

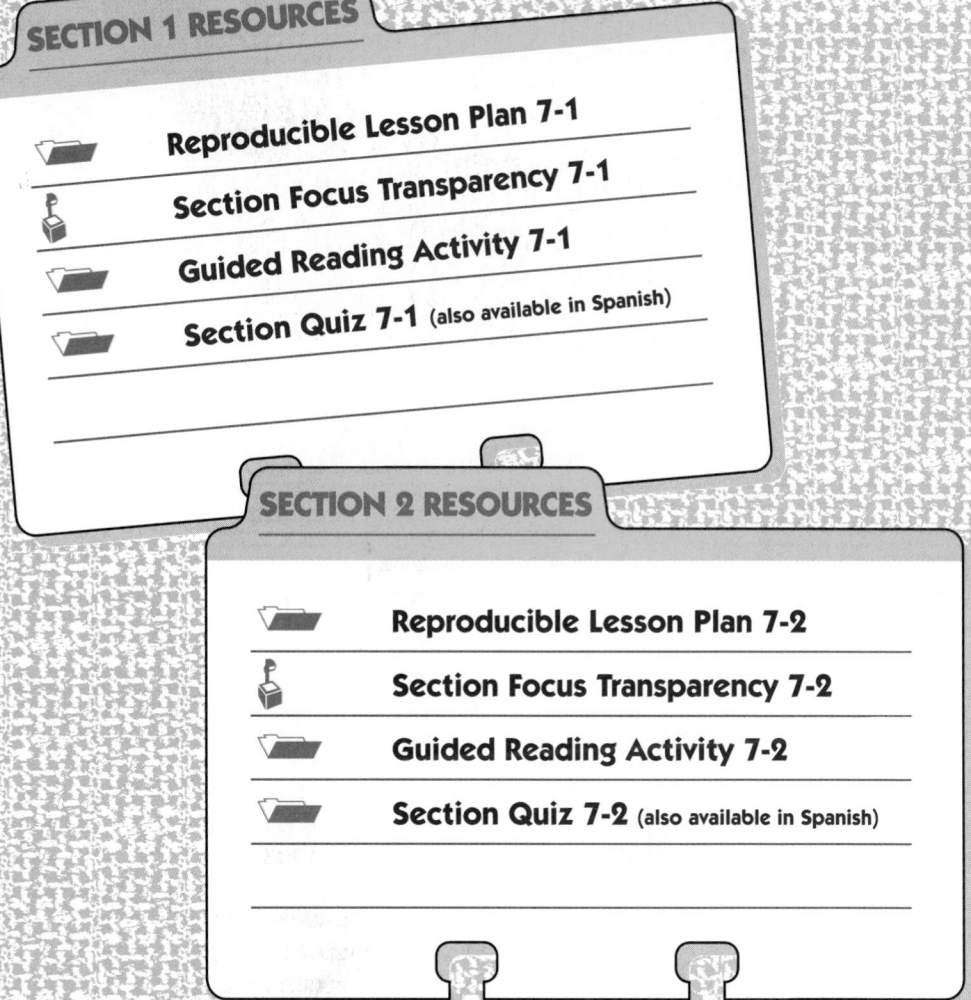

Use these Geography: The World and Its People resources to teach and reinforce section content.

SECTION 1 RESOURCES

Reproducible Lesson Plan 7-1

Section Focus Transparency 7-1

Guided Reading Activity 7-1

Section Quiz 7-1 (also available in Spanish)

SECTION 2 RESOURCES

Reproducible Lesson Plan 7-2

Section Focus Transparency 7-2

Guided Reading Activity 7-2

Section Quiz 7-2 (also available in Spanish)

ADDITIONAL RESOURCES FROM GLENCOE

 **Reproducible Masters**

- Facts On File: MAPS ON FILE I, North and Central America and the Caribbean

- Foods Around the World, Unit 3

 **World Music: Cultural Traditions**
Lesson 2

 **Posters**

- National Geographic Society: Eye on the Environment, Unit 3

 **Countries of the World Flashcards**
Unit 3

World Games Activity Cards
Unit 3

World Crafts Activity Cards Unit 3

 **Videodiscs**

- Geography and the Environment: The Infinite Voyage

- National Geographic Society: STV: Rain Forest

- National Geographic Society: STV: North America

CD-ROM

- National Geographic Society: Picture Atlas of the World

Software

- National Geographic Society: ZipZapMap! World

Performance Assessment

Refer to the Planning Guide on page 174A for a Performance Assessment Activity for this chapter. See the *Performance Assessment Strategies and Activities* booklet for suggestions.

Chapter Objectives

1. **Examine** the physical and cultural geography of Central America.
2. **Describe** the major physical and cultural features of the West Indies.

GLENCOE TECHNOLOGY

 VIDEODISC

Use the Chapter 7 Mind-Jogger Videoquiz to preview chapter content.

MindJogger Videoquiz

Chapter 7
Disc 1 Side B

 Also available in VHS.

Chapter 7 Central America and the West Indies

MAP STUDY ACTIVITY

As you read Chapter 7, you will learn about Central America and its Caribbean neighbors in the West Indies.

1. What is the southernmost country in Central America?
2. What is its capital?

174

MAP STUDY ACTIVITY

Answers
1. Panama
2. Panamá

Map Skills Practice
Reading a Map Using the map scale, find the shortest distance between Cuba and the southern tip of mainland Florida. *(about 90 miles or 145 km)*

SECTION 1
Central America

PREVIEW

Words to Know
- hurricane
- plantation
- chicle
- nutrient
- bauxite
- *ladino*
- literacy rate
- isthmus

Places to Locate
- Guatemala
- Belize
- El Salvador
- Honduras
- Nicaragua
- Costa Rica
- Panama
- Central Highlands
- Pacific Lowlands
- Caribbean Lowlands
- Panama Canal

Read to Learn . . .
1. where Central America is located.
2. how farming supports the economy of Central America.
3. what groups of people settled Central America.

Imagine a fiery red sea of lava rolling toward you. For those who lived in the shadow of Mount Izalco, this was not just a daydream. Today, however, Mount Izalco is an inactive volcano in Central America.

Central America is part of the land bridge that lies between the continents of North America and South America. The map on page 174 shows that Central America includes seven countries: Belize, Guatemala, Honduras, El Salvador, Nicaragua, Costa Rica, and Panama.

These countries all have mountainous inland areas, and most have two coastlines—one on the Caribbean Sea and the other on the Pacific Ocean. All except Belize share a mostly Native American and Spanish cultural heritage. Belize, once ruled by the British, has strong British and African influences.

Place
The Land

Central America extends more than 1,000 miles (1,600 km) north to south and measures 300 miles (483 km) at its widest point. Look at the map on page 174. You see that the region stretches from Mexico southward to South America. If you travel across it from east to west, you will journey from the Caribbean Sea to the Pacific Ocean.

CHAPTER 7

(175)

FOCUS

Section Objectives
1. **Locate** Central America on a globe or map.
2. **Explain** how farming supports the economy of Central America.
3. **Identify** groups of people who settled Central America.

Bellringer Motivational Activity

 Prior to taking roll at the beginning of the class period, project Section Focus Transparency 7-1 and have students answer the activity questions. Discuss students' responses. **L1**

Vocabulary Pre-check

Have students find the pronunciation and meaning of the word *isthmus*. Then direct them to locate examples of this physical feature on a world map. **L1 LEP**

 Use the Vocabulary PuzzleMaker Software to create a crossword puzzle. **L1**

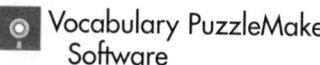

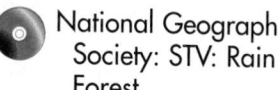
175

Answer
Tajumulco

Map Skills Practice
Reading a Map In which country is Central America's highest mountain peak? *(Guatemala)*

TEACH

Guided Practice

Identifying Information Have each student make up seven riddles, one for each of the countries of Central America. The riddle might deal with location, landforms, climate, and/or population. Each riddle must be specific enough so that the country can be identified. Allow time for students to share and solve the riddles. **L2**

CENTRAL AMERICA AND THE WEST INDIES: Physical

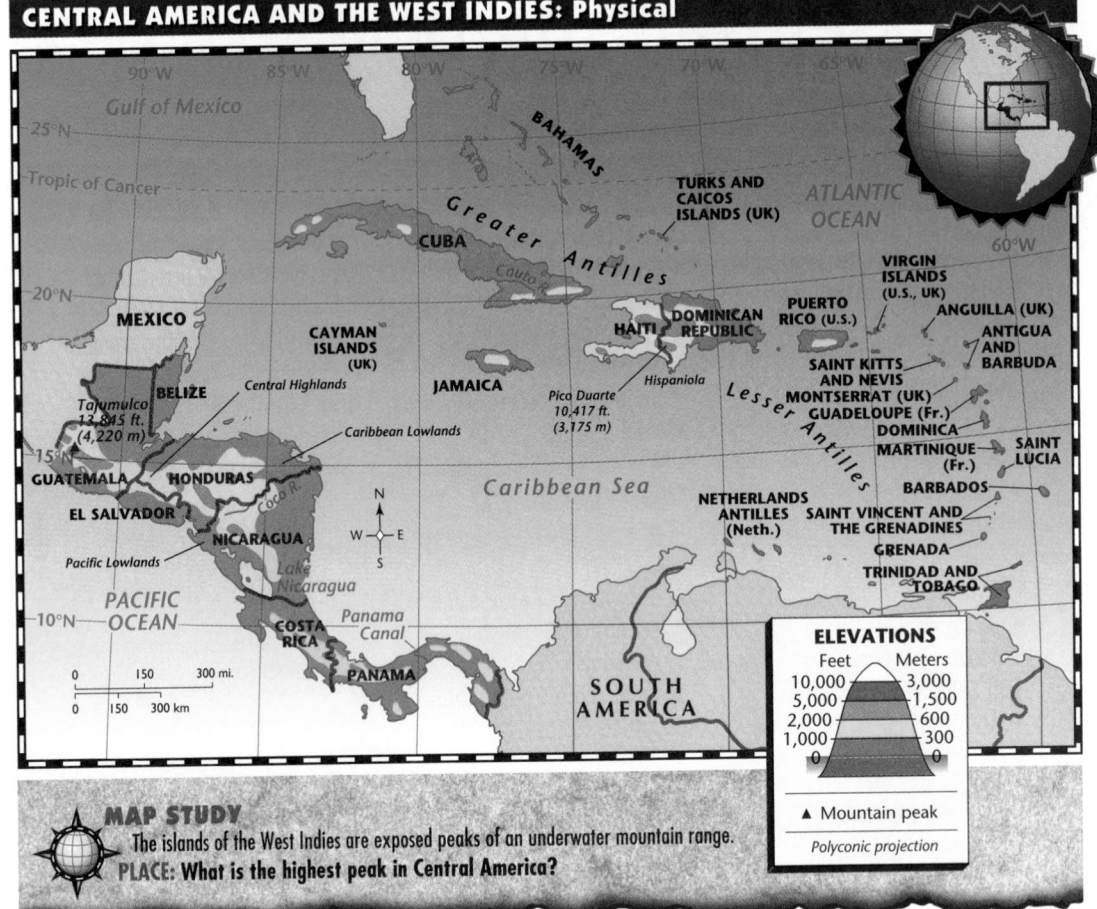

MAP STUDY
The islands of the West Indies are exposed peaks of an underwater mountain range.
PLACE: What is the highest peak in Central America?

Landforms Volcanic eruptions are part of life in Central America. Like Mexico, these countries have a rugged landscape and many active volcanoes. A chain of volcanic mountains called the Central Highlands rises like a backbone along most of the region. Volcanic material that has broken down over the centuries has left rich, fertile soil. Throughout Central America, farmers grow coffee, bananas, sugarcane, and other crops for export.

The map above shows you that on either side of the Central Highlands lie coastal plains. The Pacific Lowlands extend along the Pacific side. The Caribbean Lowlands lie on the Caribbean side.

Climate Central America's climate is mostly tropical, but there are some differences from country to country. What causes these differences? In general, altitude and location on the continent determine climate. Mountains can block the movement of winds and moisture.

UNIT 3

Cooperative Learning Activity

Organize students into seven groups and assign one of the countries of Central America to each group. Instruct groups to use the text and other references to find out about their assigned country's political history. They should choose five or more important events to share with the class. Then have groups use these events to create annotated time lines on poster boards or butcher paper. Call on groups to display and discuss their time lines. **L1**

Look at the map on page 184. The central part of the region has mountains and highland areas that are dry and cool year-round. In the Pacific Lowlands on the west coast, a tropical savanna climate prevails. Temperatures are warm and rain is plentiful from May through November. From December through April, the climate is hot and dry.

If you don't mind a daily rain shower, you would be happy living in the Caribbean Lowlands. These eastern lowlands have a hot, tropical rain forest climate year-round. Here you could expect about 100 inches (254 cm) of rainfall each year and cool breezes from the Caribbean Sea. Those breezes, however, can turn into deadly hurricanes during the summer and autumn months. **Hurricanes** are fierce storms with winds of more than 74 miles (119 km) per hour. They pose a threat for those who live on the coastal plains.

WHAT IN THE WORLD?

Hurricane Hugo

Hurricanes can be deadly, and Hurricane Hugo certainly was. In 1989 its 220-mile-per-hour (about 350-km-per-hour) winds stripped the Caribbean of vegetation and devastated almost every island in its path. Hugo finally blew itself out after smashing into the United States. More than 500 lives were lost, and property damage totaled between $2 billion and $4 billion.

Region
The Economy

Fertile soil and rain forests rank as Central America's most important resources. The economies of different countries depend on the exporting of agricultural products from the farms and wood products from the rain forests to world markets.

Farming Two kinds of farmers form the base of Central America's economy. Owners of **plantations**—large farms that grow produce for sale—raise coffee, bananas, and sugarcane. They export their harvest to the United States, western Europe, and other parts of the world. International trade is not as important to the other type of farmer—the subsistence farmer. Most raise small crops of corn, beans, rice, and livestock to feed their families. At nearby village marketplaces, they sell extra food for other supplies.

Rain Forests Under the green canopy of the Central American rain forests, great treasures can be found. The dense forests offer valuable woods— mahogany, balsa, and teak—which are exported. Workers also tap the sapodilla tree for **chicle** (CHIH•kuhl), a substance used in making chewing gum.

Scientists use trees and plants from the rain forest in medical research or to make new medicines. Unusual animals found nowhere else on earth also roam among the rain forest plants and trees.

In the Caribbean Lowlands, farmers have cleared rain forest areas to raise crops. Because of heavy rains throughout the year, the soil erodes and loses many of its **nutrients,** or minerals that supply food to plants. Farmers in this area now earn most of their living from raising livestock.

Many Central Americans are worried about the rapid clearing of so many acres of rain forest. What are they doing about it? Costa Rica has turned some forest areas into national parks. Other countries control all logging operations

LESSON PLAN
Chapter 7, Section 1

Independent Practice

Guided Reading Activity 7-1 **L1**

Workbook Activity 7 **L1**

MULTICULTURAL PERSPECTIVE
Culturally Speaking
The word *Honduras* means "depths." Christopher Columbus named the area for the deep waters off its northern coast.

NATIONAL GEOGRAPHIC SOCIETY

These materials are available from Glencoe.

 VIDEODISC

STV: Rain Forest

What Makes a Rain Forest?
Side 1, Frames 05181-06511

Meeting Special Needs Activity

Mixed Learners Some students learn best visually, while others rely on auditory skills. Mixed learners use a combination of visual and auditory skills. Have students compare the information that they get from the pictures and graphics in Section 1 with the information they get from the text. Help them determine how each kind of information aids their learning. **L1**

and assist workers in replanting cleared areas. In this way, Central American governments hope to save some of the rain forest for future generations. You can read more about the problems facing the rain forests on page 236.

Industry Missing from the skylines of most major Central American cities are the towers and chimneys of industry. Although the region has a number of small industries, not much manufacturing has been developed. Most of the countries are poor in mineral resources, especially fuels.

Guatemala and Costa Rica are exceptions. Guatemala sends crude, or unrefined, oil to overseas markets. Costa Rica exports **bauxite,** a mineral used to make aluminum. The map below shows other resources found in Central America.

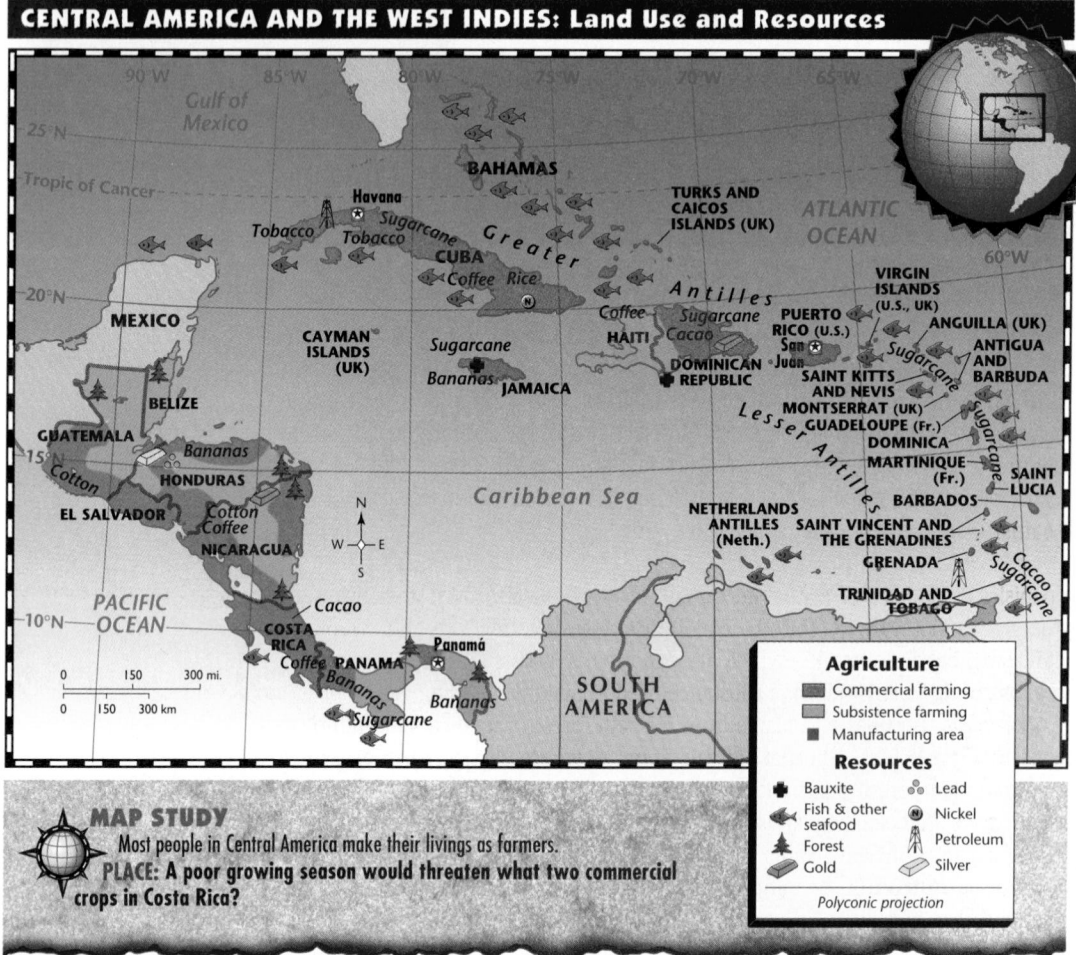

CENTRAL AMERICA AND THE WEST INDIES: Land Use and Resources

Agriculture
- Commercial farming
- Subsistence farming
- Manufacturing area

Resources
- Bauxite
- Fish & other seafood
- Forest
- Gold
- Lead
- Nickel
- Petroleum
- Silver

Polyconic projection

MAP STUDY
Most people in Central America make their livings as farmers.
PLACE: A poor growing season would threaten what two commercial crops in Costa Rica?

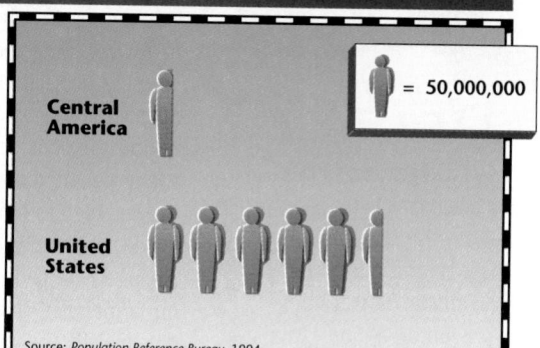

GRAPHIC STUDY

Seven countries make up the region known as Central America.
PLACE: What is the approximate population of Central America?

Central America

United States

= 50,000,000

Source: *Population Reference Bureau, 1994*

GRAPHIC STUDY

Answer
about 32 million people

Graphic Skills Practice
Reading a Graph How many more people live in the United States than in Central America? *(about 229 million people)*

Movement
The People

In Guatemala the cooking stones of the ancient Maya stand cold—overgrown with creeping vines. Steep stairways lead to empty temples. Water canals lay dry and cracked. The ancient Maya have long since disappeared—but their descendents live on in Central America.

Influences of the Past Like Mexico, Central America is a mix of cultures. The Maya settled throughout Central America about 250 to 400 B.C. As you learned in Chapter 6, the Maya developed an advanced civilization. After A.D. 800 the Maya mysteriously left their cities and scattered. Today some of their descendents live in Guatemala as well as in parts of Mexico.

In the late 1400s, the Spanish settled in the Central American region. In the 1500s they claimed territory along the Caribbean coast. For the next 300 years Spanish landowners forced Native Americans to work on plantations. Eventually, the cultures blended. Native Americans gradually adopted the Spanish language and many adopted the Roman Catholic religion.

Settled by the British in the 1600s, the area that is now the country of Belize eventually came under British rule. The British enslaved Africans and brought them to work in Belize's forests. Africans are the largest ethnic group in Belize.

Independence Most Central American countries gained independence by 1821. With help from the United States, Panama won independence from Colombia in 1903. The last Central American country to gain independence was Belize, which ceased to be a colony of the United Kingdom in 1981.

Revolutions have rocked most of the countries of Central America from the mid-1800s to the present. In many of these countries the people continue to fight for the government that best meets their needs.

Panajachel, Guatemala

Lake Atitlán in the highlands of Guatemala is surrounded by volcanic peaks.
HUMAN/ENVIRONMENT INTERACTION: How does the lake provide a living for this Guatemalan?

More About the Illustration

Answer to Caption
fishing

Place Lake Atitlán is 11 miles (17.7 km) long, 8 miles (12.9 km) wide, and as much as 1,000 feet (304.8 m) deep. It was probably created when mounds of ash from nearby volcanoes dammed the valley. A popular resort area, Lake Atitlán attracts many tourists each year.

Global Issues Activity

Panama Canal Have students research the Panama Canal treaties of 1977. Direct students to pay close attention to opinions, both for and against, on the transfer of control of the canal to Panama. Have volunteers debate the issue. Then poll the class to determine which viewpoint students favor. L2

ASSESS

Check for Understanding

Assign Section 1 Review as homework or an in-class activity.

Meeting Lesson Objectives

Each objective below is tested by the questions that follow it in parentheses.

1. **Locate** Central America on a globe or map. (2)
2. **Explain** how farming supports the economy of Central America. (5)
3. **Identify** groups of people who settled Central America. (3)

Cultural Kaleidoscope

Guatemala Rigoberta Menchú, who is descended from the Quiché Maya, won the Nobel Peace Prize in 1992 for her efforts to end discrimination against the Maya.

Evaluate

📁 Section 1 Quiz **L1**

💿 Use the Testmaker to create a customized quiz for Section 1. **L1**

Teen Scene

On the Job

It's market day in the village, and Queta Aponte is waiting for her next customer. Spread out before her are the beautiful handcrafted items that she and other members of her family have made. The brightly patterned cloth was woven on a loom. It is used to make blankets, shawls, bags, and small stuffed animals. Queta started selling her family's goods at the weekly market when she was seven years old. The money she earns will help to buy food and other things that her family needs.

The Population Today Nearly 32 million people live in Central America. More than 10 million people live in Guatemala, the most heavily populated country in the region. Only about 200,000 people live in Belize, the most sparsely populated Central American country. Spanish is the official language throughout the region, except for English-speaking Belize. Many people also speak Native American languages, such as Mayan.

Rural living is common in Central America. About 52 percent of all Central Americans live on farms or in small towns. At least one major city, usually the capital, is densely populated in each country. Guatemala's capital, also called Guatemala, ranks highest in population among Central American cities with about 1,675,000 people.

People in urban areas hold manufacturing or service industry jobs, or they work on farms outside the cities. Those living in coastal areas harvest shrimp, lobster, and other seafood to sell in city markets or for export.

Celebrations During the steamy month of August, Nicaraguan teenager Juan Ruiz and his friends celebrate the festival of Santo Domingo. They join the crowds overflowing the streets of Managua, Nicaragua's capital. Special foods, music, and dancing highlight the festivities. When there are no festivals to attend, Juan and his friends like to play baseball and soccer.

Place
National Profiles

To better understand the variety of people and places in Central America, look closely at Guatemala, Costa Rica, and Panama.

Guatemala The beautiful lakes of Guatemala reflect its majestic volcanoes. Most Guatemalans live in the southern Central Highlands area. The culture of Guatemalans comes from both Native American and Spanish influences. More than 40 percent of the people follow a rural way of life similar to that of their Native American ancestors. These people live in villages and usually do not travel beyond their country's borders. At an area market, you can tell exactly where villagers come from. How? Look at their clothing—each Native American village has its own special colors and styles of dress.

Guatemalans who speak Spanish and practice European ways are called *ladinos.* Most ladinos live in cities in modern houses or apartments and work

UNIT 3

More About . . .

Central American Celebrations A blending of Native American, Spanish, and African cultures adds to the richness of Central American traditions. In each country, common foods such as corn, beans, and rice, are mixed with spe-cial spices resulting in a variety of dishes. And while each country has its own national music, common rhythms can be heard throughout the region.

as laborers or businesspeople. For more than 34 years, a civil war has been fought between the government and groups living in the highlands. About 150,000 Guatemalans have died in this war.

Costa Rica Costa Ricans like to say, *"Darse buena vida,"* which is Spanish for "Enjoy the great life!" Why? Costa Rica offers one of the highest standards of living in the world. It also has one of the highest **literacy rates**, or percentage of people who can read and write. Most Costa Ricans are of Spanish ancestry and only a small number claim Native American heritage.

Wars have rarely occurred in this country. Costa Rica has enjoyed good relations with its neighbors. A well-developed democratic government rules, supported by a police force. Today Costa Ricans boast that their country has more schools than police barracks.

Most of the country's 3.2 million people live in the cool Central Highlands. The capital, San José, is in this area, as are most of Costa Rica's other cities. Near the highlands, farmers grow the country's major export—coffee.

Panama Panama is one of the major crossroads of the world. Across it stretches the Panama Canal. This southernmost Central American country lies on an **isthmus.** Linking Central and South America, the isthmus of Panama also separates the Caribbean Sea and the Pacific Ocean.

In 1903 the United States helped Panama win its independence from Colombia. Then the United States proposed building a canal across Panama to shorten shipping time between the Atlantic and Pacific oceans. On the map on page 182 you can see how this route crosses Panama. Engineers completed the Panama Canal—one of the engineering wonders of the world—in 1914. Since then the United States has held ownership of the canal and the land on either side of it. In the year 2000 Panama will gain control of the canal zone.

Nearly one-half of Panama's 2.5 million people live and work in the area near the canal. Most are of mixed Native American and Spanish ancestry and speak Spanish, the official language. English, however, is also widely spoken. The national capital, Panamá—located on the Pacific side of the isthmus—is the leading center of culture and industry.

SECTION 1 REVIEW

REVIEWING TERMS AND FACTS
1. **Define:** hurricane, plantation, chicle, nutrient, bauxite, *ladino,* literacy rate, isthmus.
2. **LOCATION** What continents does Central America link?
3. **MOVEMENT** What groups of people have influenced the development of Central America?

 MAP STUDY ACTIVITIES
4. Turn to the physical map on page 176. Which country in Central America has the highest elevation?
5. Look at the land use map on page 178. What resources are found in the region?

Answers to Section 1 Review

1. All vocabulary words are defined in the Glossary.
2. North America and South America
3. Native Americans, Europeans (Spanish and British), Africans
4. Guatemala
5. bauxite, fish and other seafood, forests, gold, lead, nickel, silver, petroleum

 **NATIONAL GEOGRAPHIC SOCIETY**

The following materials are available from Glencoe.

 VIDEODISC

GTV: A Geographic Perspective on American History

Panama Canal
Any side, Frame 53487

Reteach

Have students work in pairs to review Section 1 material. Partners should take turns asking and answering questions based on the text. **L1**

Enrich

Encourage students to research the plants and animals of Central American rain forests. Have them create a poster to share with the class. **L3**

CLOSE

Refer students to the discussion of celebrations on page 180. Direct students to write a letter to Juan Ruiz that compares celebrations in the United States with those in Central America.

TEACH

Have students trace on an outline map of the Western Hemisphere (page 27 of the Outline Map Resource Book) several possible routes from the Atlantic to the Pacific Ocean. Suggest the following: by water through northern Canada; around the southern tip of South America; overland across the United States. Then have students locate and label Panama on the map. Encourage them to discuss why a canal through Panama would be more beneficial for trade and travel than the other routes they marked on the map. **L1**

More About . . .

History Earlier efforts had been made to build a canal across Panama. In 1881, a French company run by Ferdinand de Lesseps—who had built the Suez Canal in Egypt—began work on a canal. Difficult terrain and disease among workers, however, made the task much slower and more expensive than de Lesseps had expected. By 1887, the company had gone bankrupt and de Lesseps had abandoned the project.

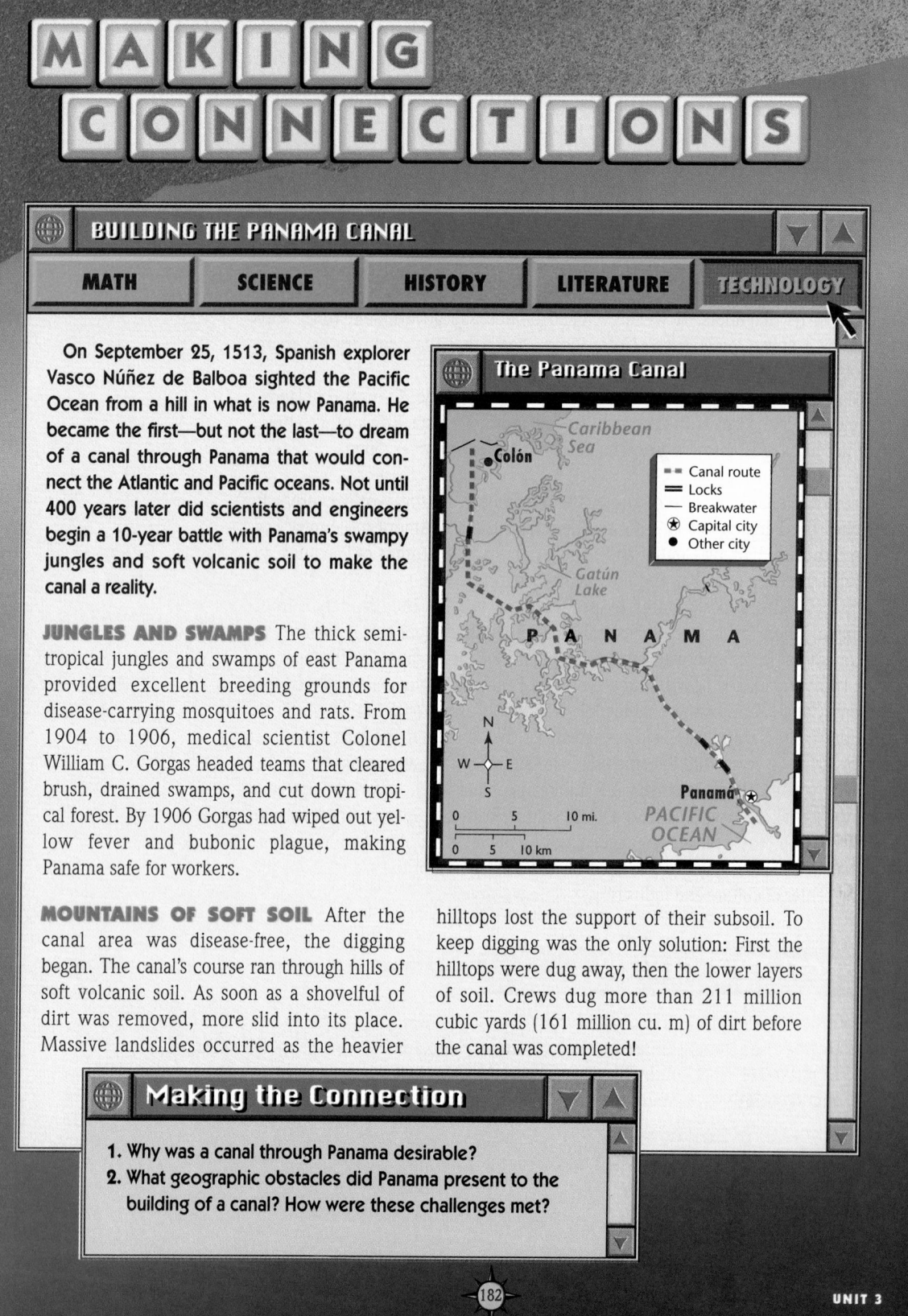

MAKING CONNECTIONS

BUILDING THE PANAMA CANAL

MATH   SCIENCE   HISTORY   LITERATURE   **TECHNOLOGY**

On September 25, 1513, Spanish explorer Vasco Núñez de Balboa sighted the Pacific Ocean from a hill in what is now Panama. He became the first—but not the last—to dream of a canal through Panama that would connect the Atlantic and Pacific oceans. Not until 400 years later did scientists and engineers begin a 10-year battle with Panama's swampy jungles and soft volcanic soil to make the canal a reality.

JUNGLES AND SWAMPS The thick semi-tropical jungles and swamps of east Panama provided excellent breeding grounds for disease-carrying mosquitoes and rats. From 1904 to 1906, medical scientist Colonel William C. Gorgas headed teams that cleared brush, drained swamps, and cut down tropical forest. By 1906 Gorgas had wiped out yellow fever and bubonic plague, making Panama safe for workers.

The Panama Canal

Colón — Caribbean Sea

Gatún Lake

P A N A M A

- - - Canal route
═ Locks
— Breakwater
✪ Capital city
● Other city

N / S / W / E

0 5 10 mi.
0 5 10 km

Panamá ✪

PACIFIC OCEAN

MOUNTAINS OF SOFT SOIL After the canal area was disease-free, the digging began. The canal's course ran through hills of soft volcanic soil. As soon as a shovelful of dirt was removed, more slid into its place. Massive landslides occurred as the heavier hilltops lost the support of their subsoil. To keep digging was the only solution: First the hilltops were dug away, then the lower layers of soil. Crews dug more than 211 million cubic yards (161 million cu. m) of dirt before the canal was completed!

Making the Connection

1. Why was a canal through Panama desirable?
2. What geographic obstacles did Panama present to the building of a canal? How were these challenges met?

UNIT 3

Answers to Making the Connection

1. to connect the Atlantic and Pacific oceans
2. jungles and swamps that bred disease-carrying mosquitoes, soft volcanic soil that continuously slid down into the canal; swamps were drained, brush cleared, and forests cut down; crews kept digging

The West Indies

PREVIEW

Words to Know
• archipelago
• colony
• communism
• cooperative
• light industry
• dialect
• commonwealth

Places to Locate
• Caribbean Sea
• Greater Antilles
• Lesser Antilles
• Cuba
• Dominican Republic
• Haiti
• Hispaniola
• Jamaica
• Puerto Rico

Read to Learn . . .
1. what landforms and climates are common to the West Indies.
2. how people in the West Indies earn a living.
3. what cultures are found in the West Indies.

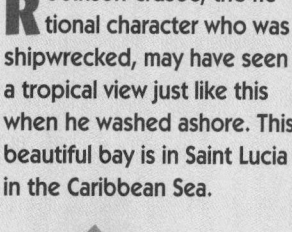

Robinson Crusoe, the fictional character who was shipwrecked, may have seen a tropical view just like this when he washed ashore. This beautiful bay is in Saint Lucia in the Caribbean Sea.

FOCUS

Section Objectives
1. **Describe** the landforms and climates of the West Indies.
2. **Explain** how people in the West Indies earn a living.
3. **Identify** the cultures found in the West Indies.

Bellringer Motivational Activity

 Prior to taking roll at the beginning of the class period, project Section Focus Transparency 7-2 and have students answer the activity questions. Discuss students' responses. **L1**

Vocabulary Pre-check

Have students look up the meaning of the vocabulary words. Then have them categorize the terms under the headings *physical geography, politics and government, economy,* and *cultural heritage.* Discuss and resolve differences regarding placement of terms. **L1** **LEP**

The Caribbean **archipelago,** or group of islands, is called the West Indies. It sweeps, like the tail of a cat, from Florida to the northeast coast of South America.

The islands of the Bahamas lie southeast of Florida. The rest of the northern Caribbean area, or the Greater Antilles, includes the large islands of Jamaica, Cuba, Hispaniola (HIHS•puh•NYOH•luh), and Puerto Rico. The Lesser Antilles includes a number of smaller islands rising from the blue waters of the southern Caribbean.

Place
The Land

If you flew above the clouds over a mountain range, what would you see? Probably only the tops of mountains would be visible. That is what you see when you're looking at the islands of the West Indies—the tops of mountains.

Many islands are really an underwater chain of mountains, some of which are active volcanoes. Others are limestone mountains pushed up from the ocean floor by pressures under the earth's crust.

Classroom Resources for Section 2

REPRODUCIBLE MASTERS
Reproducible Lesson Plan 7-2
Guided Reading Activity 7-2
Geography Skills Activity 7
Critical Thinking Skills Activity 7
Reteaching Activity 7
Enrichment Activity 7
Section Quiz 7-2

TRANSPARENCIES
Section Focus
    Transparency 7-2
Unit Overlay Transparencies
    3-0, 3-1, 3-2, 3-3, 3-4, 3-5,
    3-6
Geoquiz Transparencies 7-1,
    7-2

MULTIMEDIA
 Testmaker

National Geographic
    Society: ZipZapMap!
    World

TEACH

Guided Practice

Identifying Information
After students have read the section, organize them into two teams for a quiz. Offer the first team a clue that will help them identify a West Indian country. If the team gives a wrong answer, or does not know the answer, offer a second clue to the second team. If that team is unable to give the correct answer, throw the competition open to both teams and offer a third clue. If teams still are unable to give the correct answer, reveal the identity of the country. Continue the quiz with clues on other countries. The team with the most correct answers wins the quiz. **L1**

MAP STUDY

Answer
Both have mostly tropical climates. Central America has an area with a highland climate.

Map Skills Practice
Reading a Map What type of climate is found in Jamaica? *(tropical savanna)*

Landforms The islands of the West Indies cover an area of about 90,699 square miles (234,909 sq. km). Most Caribbean islands have central highlands that slope down to the sea. Strips of fertile coastal plain line these coasts.

Look at the map on page 174 to see how the West Indies islands vary in size. Cuba, the largest island, covering about 42,400 square miles (109,816 sq. km), is the size of the state of Ohio. Anguilla (an•GWIH•luh) is one of the smallest islands, just 34 square miles (88 sq. km). It is about half the size of Salt Lake City, Utah.

Climate The picture-postcard view of coconut palms and sunny beaches in the West Indies can be enjoyed year-round. The islands of the West Indies lie in the tropics, and most enjoy a fairly constant tropical savanna climate. Cool northeast breezes sweep across the Caribbean Sea, keeping temperatures between 70°F (20°C) and 85°F (30°C) and bringing gentle rains.

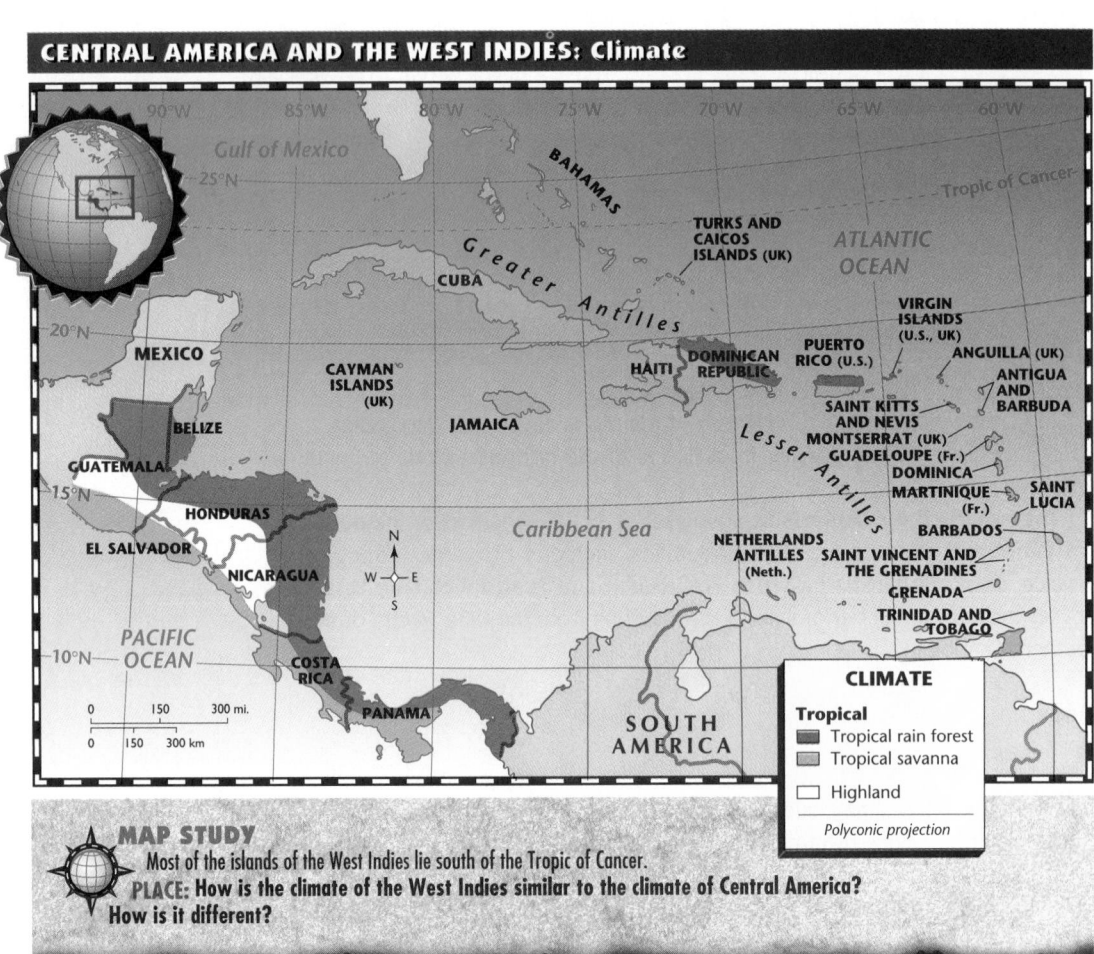

CENTRAL AMERICA AND THE WEST INDIES: Climate

CLIMATE

Tropical
- Tropical rain forest
- Tropical savanna
- Highland

Polyconic projection

MAP STUDY
Most of the islands of the West Indies lie south of the Tropic of Cancer.
PLACE: How is the climate of the West Indies similar to the climate of Central America? How is it different?

184

Cooperative Learning Activity

Organize students into four groups and assign one of the following crops to each group: sugarcane, coffee, bananas, tobacco. Have group members find out how their assigned crop is grown and why it is well suited to the region. Groups should present their findings in illustrated reports. Suggest that groups include in their reports a discussion on how the plantation system influenced the history of the West Indies. **L2**

Caribbean rains are not *always* gentle, however. For more than half of the year, hurricanes batter the West Indies. The word "hurricane" comes from the Taíno people of Cuba, who worshiped a god of storms named Hurakan.

Region
The Economy

Farming and tourism are the most important economic activities in the West Indies. Many people, however, cannot find jobs or have to work for low pay. Governments in the West Indies are trying to develop new industries that will bring more jobs to the region.

One common economic problem is that many islands have a farm economy based on only one crop. Sugar, for example, makes up 75 percent of the exports of the islands of Saint Kitts and Nevis. If an island's major crop fails, or if too much sugar is produced and prices fall, the economy is in serious trouble.

Farming Wealthy West Indian plantation owners grow crops such as sugarcane, bananas, coffee, and tobacco for export. Many full-time laborers are needed to work the plantations. Look at the map on page 178 to see where plantation crops are grown. The map also shows that some West Indians are subsistence farmers who own or rent small plots of land.

Tourism A warm climate, beautiful beaches, and friendly people draw large numbers of tourists to the West Indies. Airlines and cruise ships from the United States and Europe make regular stops in the Caribbean region. The service industries created by tourism provide jobs and a strong base for the economies of many island nations.

Mining, Manufacturing, and Trade Mining and manufacturing generally play a small role in the economies of the West Indies. Some island countries, however, do contain mineral resources. Find these areas on the map on page 178. Jamaica, for example, is a leading world producer of bauxite, used to make aluminum, and Trinidad exports crude oil.

Trading ships from Europe and the rest of Latin America have sailed to the West Indies since the late 1500s. The building of the Panama Canal also brought trade and commerce to the region.

Island Life

Workers harvest sugarcane in the Dominican Republic *(left)*. Trinidad's annual carnival attracts many tourists *(right)*. **HUMAN/ENVIRONMENT INTERACTION: What are the two most important economic activities in the West Indies?**

MAP STUDY

Answer
Haiti

Map Skills Practice
Reading a Map How do the population densities of Jamaica and Puerto Rico compare? *(They are the same.)*

Global Gourmet

Barbados The national dish in Barbados is flying fish and *cou cou* (okra and cornmeal). Another popular local dish, called *conkit,* consists of cornmeal, coconut, pumpkin, raisins, sweet potatoes, and spices steamed in a banana leaf.

MULTICULTURAL PERSPECTIVE

Cultural Diversity

Religion plays a major role in life in Jamaica. While the majority of Jamaicans—some 60 percent—are Christian, many different faiths are practiced on the island. The Jewish faith is especially well established in Jamaica, tracing its roots to members of Columbus's crews. A strong bond exists between all the faiths, and people of one religion may attend the services of another religion. Christians, for example, often share in Hindu and Muslim ceremonies.

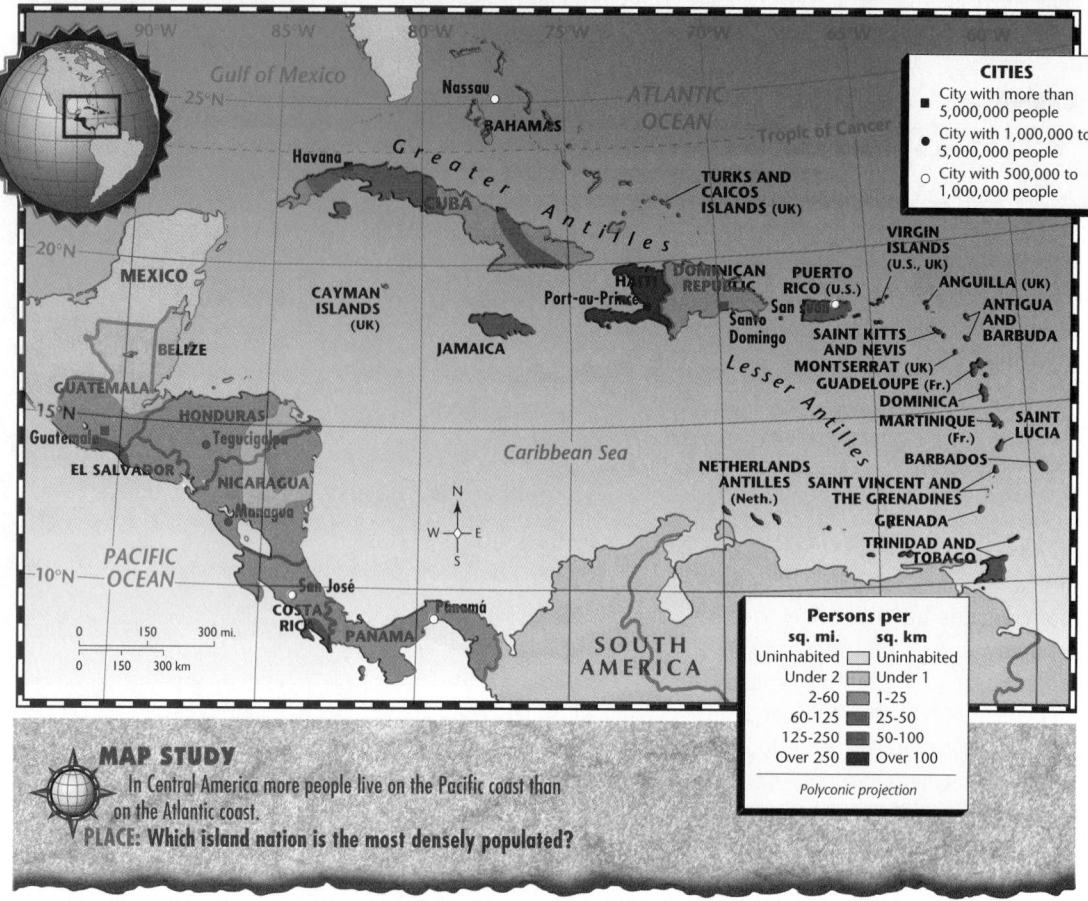

CENTRAL AMERICA AND THE WEST INDIES: Population Density

MAP STUDY
In Central America more people live on the Pacific coast than on the Atlantic coast.
PLACE: Which island nation is the most densely populated?

Movement

The People

The cultures of Africans, Native Americans, Europeans, and Asians have mixed and mingled to create the cultures of the West Indies. Why do you find so many groups represented here? Think about the location of the islands. Over the centuries, a variety of peoples have passed through and settled in this region.

Influences of the Past When Christopher Columbus sailed into the harbor of Hispaniola in 1492, who met him? As you might think, it was a Native American group—the Taíno. At their first meal together, a dish made of vegetables and meat called pepper pot stew was probably served.

The first permanent European settlement in the Western Hemisphere was established in 1498 by the Spanish in the present-day Dominican Republic.

186

UNIT 3

Critical Thinking Activity

Analyzing Information Pose the following question for students to consider: Is tourism beneficial or harmful to the West Indies? Why? Provide time for students to reflect on the question and discuss it among themselves. Then ask several volunteers to present opposing viewpoints. *(Some may feel that tourism is beneficial because it helps local economies. Others may argue that countries that rely on tourism fail to develop other industries.)* To conclude, have students determine which viewpoints are supported by information in the text and/or in other sources. **L2**

During the next 200 years, the Spanish, the English, the French, and the Dutch also founded **colonies,** or overseas settlements, on many of the islands.

Forced Labor The Europeans forced the peoples of the islands to farm the land and to work in the mines. After most Native Americans died from European diseases or harsh treatment, the Europeans turned to Africa for workers. They enslaved Africans and shipped them to the islands to work the sugar plantations. Most of the Caribbean population today traces its ancestry to Africa.

An upset in the Caribbean plantation system took place in the mid-1800s. World sugar prices collapsed, and many plantations closed. After that, the ruling European governments passed laws that ended the enslavement of workers.

Plantation owners still in need of workers brought them from Asia, particularly India. The Asians agreed to work a set number of years in return for free travel to the West Indies. Today large Asian populations live in Trinidad, Jamaica, and other islands.

Independence During this century most of the smaller Caribbean islands won their freedom from colonial rule. Today some countries of the West Indies are republics. Others are constitutional monarchies.

Cuba is the only country in the Western Hemisphere with a government based on the ideas of **communism.** Communism favors strong government control of the society and the economy. Some Caribbean islands still have close links to European nations or to the United States.

A Mix of People About 36 million people live in the West Indies. Find the island of Cuba on the map on page 186. With more than 11 million people, Cuba is the most heavily populated island in the West Indies. Five of the Caribbean's smallest countries—Antigua and Barbuda, Dominica, Grenada, Saint Lucia, and Saint Vincent and the Grenadines—have only about 100,000 people each.

Most West Indians have added some aspects of European cultures to their own. Many have adopted European languages and the Christian religion. Some have mixed them with their Native American and African cultures.

Culture Music and dance are important to the cultures of the Caribbean islands. The bell-like tones of the steel drum, a musical instrument developed in Trinidad, are part of the rich musical heritage of the region. Enslaved Africans created a kind of music called calypso. Its strong rhythms have spread around the world.

Daily Life Rajo Hassanali, a young woman of Asian descent, lives in Trinidad. When Rajo is not in school, she works on a nearby cocoa plantation. Many of her friends travel to the cities to work in hotels, restaurants, or shops. During their free time the young people enjoy baseball, basketball, and soccer. Urban areas of the West Indies are home to about 60 percent of the people. The other 40 percent of the population live and work in the countryside.

CHAPTER 7 187

CURRICULUM CONNECTION

History In the late 1800s, mongooses were brought from India to the West Indies to kill the rats that infested the sugarcane fields. Within a few years, the rat problem had been brought under control. The mongooses soon found a new source of food—the chickens raised by West Indian farmers. By 1900, the mongoose had become such a threat to the farming economy that West Indian governments paid hunters to kill them.

Cultural Kaleidoscope

Haiti Some Haitians follow the customs their ancestors brought from Africa. For example, people plant and harvest crops to the sound of music and singing. This practice of combining work and play is called a *combite*.

Global Issues Activity

Developing Countries List the following characteristics of developing countries on the chalkboard: shortages of products to meet basic needs, few resources, and low gross domestic product (GDP)—the value of finished goods and services produced within a country in a year. Have students research reasons why West Indian countries, such as Haiti, have struggled to develop strong economies. Suggest that they consider such issues as one-crop economies, political turmoil, and rapidly growing populations. L3

 **Place** The walls of Christophe's fortress were 100 feet (30.5 m) high and 20 feet (6.1 m) thick. Workers were forced to carry the heavy materials and guns up the steep mountainside. Many died during construction of the fortress.

ASSESS

Check for Understanding

Assign Section Review as homework or an in-class activity.

Meeting Lesson Objectives

Each objective below is tested by the questions that follow it in parentheses.

1. **Describe** the landforms and climates of the West Indies. (2, 5)
2. **Explain** how people in the West Indies earn a living. (5)
3. **Identify** the cultures found in the West Indies. (4)

Place
Caribbean Islands

The islands of the West Indies share many similarities and differences. Some of these differences can be seen in Cuba, Haiti, Jamaica, and Puerto Rico.

Historic Landmark in Haiti

Haitian ruler Henri Christophe had this fortress built in the early 1800s to protect the island from invaders.
PLACE: What country ruled Haiti until 1804?

Cuba One of the world's top sugar producers, Cuba lies about 90 miles (about 140 km) south of Florida. About 50 percent of the Cuban population are black, and another 25 percent are of mixed African and Spanish ancestry. Most Cuban farmers work on **cooperatives,** or farms owned and operated by the government. In addition to sugarcane, they grow coffee, tobacco, rice, and fruit. In Havana, Cuba's capital, workers in **light industries** produce food products, cigars, and household goods.

Currently, Cuba is a communist state led by dictator Fidel Castro. Cuba won its independence from Spain in 1898. A revolution led by Castro in 1959 set up a communist government. Opposed by the United States, Castro turned to the Soviet Union for support. When the Soviet Union broke up in 1989, it withdrew its support from Castro. Since then the Cuban economy has collapsed, and many Cubans live in poverty.

Haiti Haiti shares the island of Hispaniola with the Dominican Republic. More than 90 percent of Haiti's 7 million people are of African ancestry. Power struggles and civil war have left Haiti's economy in ruins. Haiti's people are generally poor and live in rural areas. Coffee, the major export crop, is shipped through Port-au-Prince, the country's capital.

Haiti won its independence from France in 1804. It was the second independent republic in the Western Hemisphere (after the United States). During most of the 1900s, dictators ruled Haiti. In 1990 Haitian voters elected Jean-Bertrand Aristide (zhawn behr•trahn ah•rih•STEED) as a democratic president. Military leaders, however, forced Aristide out of the country. With the help of the United States, Aristide finally returned to power in 1994. Conflict among political groups, however, has slowed progress toward a stable government.

Jamaica What languages would you expect to hear if you visited Jamaica? If you said English, you are correct. But other languages such as Creole, a mix of English and French, are also spoken there. Some Jamaicans also speak **dialects,** or local forms of a language, such as patois.

188

UNIT 3

More About . . .

Jamaica At least 100,000 Jamaicans are *Rastafarians.* They take their name from Ras Tafari, the original title held by the late Emperor Haile Selassie I of Ethiopia. *Rastafarians* believe that Haile Selassie is the one true god. They also believe that the true location of the holy land of Zion is Ethiopia, not Israel. *Rastafarians* call Jamaica *Babylon*—a reference to the Babylonian Captivity of the 580s B.C., when the Jews were driven from their homeland into exile in Babylon.

More than 90 percent of Jamaica's 2.5 million people are of African or mixed African and European backgrounds. The other 10 percent are of Asian ancestry.

Once a British colony, Jamaica became independent in 1962. About 52 percent of the Jamaican population live in urban areas. Kingston is the country's capital and largest city.

This island country is known for its misty blue mountains and sunny tropical beaches. Jamaica's economy relies mainly on tourism. One of the few Caribbean islands with mining interests, the country exports bauxite in addition to bananas, sugar, and coffee.

Puerto Rico To be or not to be . . . a state? That is a question Puerto Ricans ask themselves every few years. Should they officially become a state in the United States? Their last vote on the matter was no.

Today about 2.8 million Americans are of Puerto Rican ancestry. Puerto Rico was a Spanish colony from 1508 to 1898. After the Spanish-American War in 1898, Puerto Rico came under the control of the United States. Since 1952 the island has been a **commonwealth**, a partly self-governing territory, under American protection.

Most of Puerto Rico's 3.6 million people are of Spanish ancestry. Many of them, however, have mixed European, African, and Native American backgrounds. About 75 percent of the Puerto Rican population live in towns and cities. San Juan, the capital and largest city, is home to about 1.8 million people.

Puerto Rico boasts more industry than any other island in the West Indies. Factories there make chemicals, clothing, medicines, and metal products. Puerto Rico is an island of mountains, rain forests, and coastal lowlands. Agriculture and tourism form the basis of Puerto Rico's economy. Sugarcane and coffee rank as its major crops.

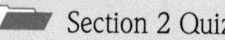

WHAT IN THE WORLD?

Reggae Rhythms

One of Jamaica's most well-known exports is reggae music, a style of music that combines American pop music with African rhythms. Bob Marley, the most famous performer of reggae, lived in Jamaica all of his life. He is a national hero today.

SECTION 2 REVIEW

REVIEWING TERMS AND FACTS
1. **Define:** archipelago, colony, communism, cooperative, light industry, dialect, commonwealth.
2. **PLACE** What landforms are typical in the West Indies? How do they influence climate?
3. **PLACE** What is the capital of Puerto Rico?
4. **MOVEMENT** What different groups have settled in the West Indies in the last 400 years?

MAP STUDY ACTIVITIES
5. Turn to the climate map on page 184. Why do the West Indies have a tropical climate?
6. Look at the population density map on page 186. Name two of the most densely populated islands in the Greater Antilles.

Evaluate

Section 2 Quiz **L1**

Use the Testmaker to create a customized quiz for Section 2. **L1**

Chapter 7 Test Form A and/or Form B **L1**

Reteach

Reteaching Activity 7 **L1**

Spanish Reteaching Activity 7 **L1**

Enrich

Enrichment Activity 7 **L3**

CLOSE

Mental Mapping Activity

Provide each student with a large sheet of white paper and map pencils. Inform students that this will not be a graded activity. Have students draw, freehand from memory, a map of Central America and the West Indies. Have them label each of the following: Guatemala, Belize, El Salvador, Honduras, Nicaragua, Costa Rica, Panama, Cuba, Dominican Republic, Haiti, Hispaniola, Jamaica, and Puerto Rico.

Answers to Section 2 Review

1. All vocabulary words are defined in the Glossary.
2. coastal plains and inland highland areas; the coastal plains experience tropical climates, whereas the inland mountains are dry and cool and block the movement of winds and moisture
3. San Juan
4. Native Americans, Europeans, Africans, Asians
5. because they lie within the tropics
6. any two of Puerto Rico, Jamaica, Hispaniola (Haiti and the Dominican Republic)

BUILDING GEOGRAPHY SKILLS

BUILDING GEOGRAPHY SKILLS

TEACH

Direct students to study the physical map of Central America and the West Indies on page 176. Point out that a physical map is similar to an elevation profile in that they both provide information about relief and landforms. Ask students to try to translate information on the map into an elevation profile. Suggest that they draw a cross section from the Pacific coast of Guatemala through the Gulf of Honduras and Cuba to the Bahamas. Call on volunteers to use their profiles to locate and identify the highest elevations and lowest elevations on the cross section. Then have students read the skill and complete the practice questions. **L2**

Additional Skills Practice

1. What sea surrounds Jamaica? *(Caribbean Sea)*

2. What are the lowest mountains? *(Don Figuerero Mountains)*

📁 Geography Skills Activity 7

Interpreting an Elevation Profile

You've learned that differences in land elevation are often shown on relief maps. Another way to show elevation is on **elevation profiles**. An elevation profile is a diagram that shows a side view of the landforms in an area.

Suppose you could slice right through a country from top to bottom and could look at the inside, or *cross section.* The *cross section,* or elevation profile, below pictures the island of Jamaica. It shows how far Jamaica landforms extend below or above sea level.

Use these steps to understand an elevation profile:

• Read the profile title.

• Read the labels to identify the different landforms shown.

• Compare the highest and lowest points in the profile.

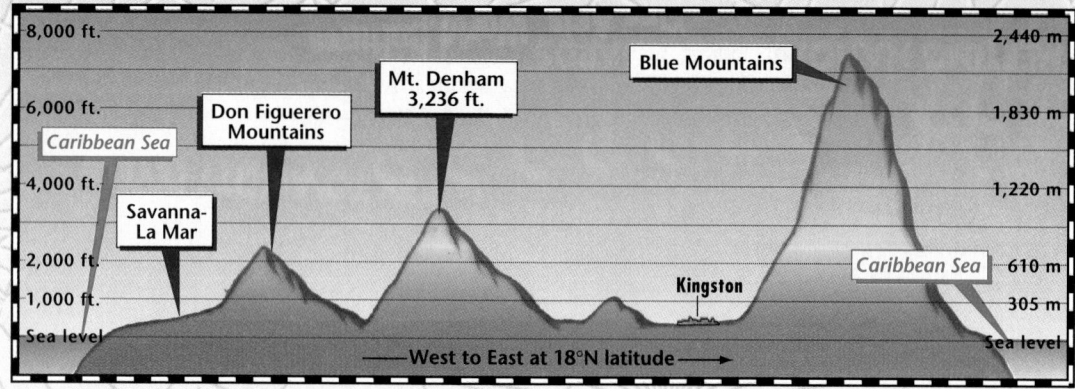

JAMAICA: Elevation Profile

Source: *Goode's World Atlas,* 19th edition

Geography Skills Practice

1. At about what elevation is Kingston?

2. What are the highest mountains, and where are they located?

3. Where are the lowest regions?

4. Along what line of latitude was this cross section taken?

190

Answers to Geography Skills Practice

1. approximately 500 feet (153 m)

2. Blue Mountains, in eastern Jamaica

3. along the coasts

4. 18°N latitude

Chapter 7 Highlights

Important Things to Know About Central America and the West Indies

SECTION 1 CENTRAL AMERICA

- Central America extends from Mexico south to South America.
- Central America includes seven countries: Belize, Guatemala, Honduras, El Salvador, Nicaragua, Costa Rica, and Panama.
- The landscape of Central America drops from inland highlands to coastal plains.
- Central America's climate is mostly tropical savanna or tropical rain forest.
- Central America's economy relies on farm products and rain forest resources.
- About 32 million people live in Central America. Guatemala, with 10 million people, is the most heavily populated country in the region.

SECTION 2 THE WEST INDIES

- The West Indies consists of two major island groups: the Greater Antilles and the Lesser Antilles.
- The islands of the Bahamas are also part of the West Indies. They lie southeast of Florida.
- Most of the islands of the West Indies have a tropical savanna climate with wet and dry seasons.
- Farming and tourism are the major economic activities of the West Indies.
- The cultures of the West Indies mix African, Native American, and European influences.

Market in Chichicastenango, Guatemala ▶

Using the Chapter 7 Highlights

- Use the Chapter 7 Highlights to preview, review, condense, or reteach the chapter.

Preview/Review

Vocabulary Puzzle-Maker Software reinforces the terms used in Chapter 7.

Condense

Have students read the Chapter 7 Highlights. Spanish Chapter Highlights are also available.

Chapter 7 Digest Audiocassettes and Activity

Chapter 7 Digest Audiocassette Test
Spanish Chapter Digest Audiocassettes, Activities, and Tests also are available.

Guided Reading Activities

Reteach

Reteaching Activity 7. Spanish Reteaching activities also are available.

MAP STUDY

Place Display Political Map Transparency 3 and ask students to identify the location of each of the photographs in the chapter. Have students write captions for each photograph, relating photograph content to the main ideas presented in the chapter.

Extra Credit Project

Letter Encourage students to write an "update" letter to Christopher Columbus. Suggest that they discuss the positive and negative changes that have taken place in Central America and the West Indies since Columbus first visited the region 500 years ago. Call on volunteers to share their letters with the class. **L2**

Chapter 7 Review and Activities

CHAPTER 7 REVIEW & ACTIVITIES

GLENCOE TECHNOLOGY

 **VIDEODISC**

Use the Chapter 7 Mind-Jogger Videoquiz to review students' knowledge before administering the Chapter 7 Test.

MindJogger Videoquiz

Chapter 7
Disc 1 Side B

Also available in VHS.

ANSWERS

Reviewing Key Terms

1. C
2. E
3. G
4. H
5. D
6. F
7. A
8. I
9. B

 Mental Mapping Activity

This exercise helps students to visualize the countries and geographic features they have been studying and understand the relationships among them. All attempts at freehand mapping should be accepted.

REVIEWING KEY TERMS

Match the numbered terms in Column A with their definitions in Column B.

A
1. hurricane
2. bauxite
3. *ladino*
4. nutrients
5. archipelago
6. light industry
7. chicle
8. isthmus
9. dialect

B
A. substance used to make chewing gum
B. local form of a language
C. fierce storm
D. group of islands
E. mineral used to make aluminum
F. producing household goods
G. Guatemalan who speaks Spanish and follows European ways
H. minerals providing food for plants
I. narrow piece of land connecting two large landmasses

Mental Mapping Activity

On a separate sheet of paper, draw a freehand map of Central America and the West Indies. Label the following items on your map:

• Central Highlands
• Pacific Ocean
• Caribbean Sea
• Cuba
• Greater Antilles
• Lesser Antilles
• Isthmus of Panama

REVIEWING THE MAIN IDEAS

Section 1

1. **PLACE** What countries make up Central America?
2. **LOCATION** What two bodies of water border Central America?
3. **REGION** How have rain forests boosted the economy of Central America?
4. **MOVEMENT** What European people conquered most of Central America in the 1500s?
5. **HUMAN/ENVIRONMENT INTERACTION** How did the United States change the landscape of Panama?

Section 2

6. **LOCATION** Where are the Bahamas located?
7. **REGION** What is the largest island in the West Indies?
8. **HUMAN/ENVIRONMENT INTERACTION** What two kinds of farms are found in the West Indies?
9. **PLACE** What two islands have large Asian populations?
10. **PLACE** What is the capital of Haiti?

Reviewing the Main Ideas

Section 1

1. Guatemala, Belize, El Salvador, Honduras, Nicaragua, Costa Rica, Panama
2. the Pacific Ocean and Caribbean Sea
3. Rain forests have important resources: valuable kinds of woods, such as teak, balsa, and mahogany; chicle, the main ingredient of chewing gum; and rare plant and animal species that are used in medicine.
4. Spanish
5. The United States built the Panama Canal across the Isthmus of Panama.

CRITICAL THINKING ACTIVITIES

1. **Determining Cause and Effect** Why was Panama chosen as a good location to build a canal?
2. **Evaluating Information** How have early groups had a lasting influence on the West Indies?

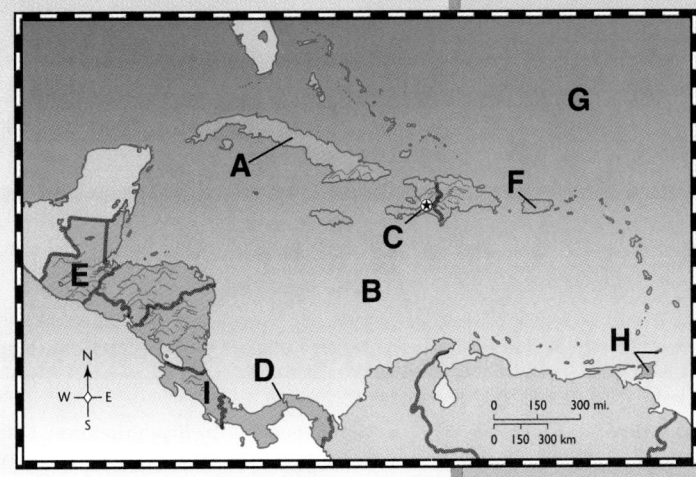

GeoJournal Writing Activity

Imagine you are a Costa Rican teenager with a pen pal in the United States. Write a letter to your pen pal describing what you would most like to know about a teenager's life in the United States.

COOPERATIVE LEARNING ACTIVITY

In groups of three, investigate which Caribbean nations were at one time colonies; what European country ruled these colonies; and when the colonies gained their independence. Each member of the group can research a different group of islands. Someone may also want to find out more about how these countries celebrate their independence days. When you have gathered your information, brainstorm a way to present it to the rest of the class. Try to include at least one visual element.

PLACE LOCATION ACTIVITY: CENTRAL AMERICA AND THE WEST INDIES

Match the letters on the map with the places of Central America and the West Indies. Write your answers on a separate sheet of paper.

1. Costa Rica
2. Trinidad and Tobago
3. Panama Canal
4. Cuba
5. Atlantic Ocean
6. Caribbean Sea
7. Port-au-Prince
8. Guatemala
9. Puerto Rico

Cooperative Learning Activity

Suggest that students also research the relations between Caribbean countries and their former colonial rulers since independence. Encourage students to make their visuals as colorful and dramatic as possible.

GeoJournal Writing Activity

Select students to read their letters to the class. Then call on volunteers to compose answers to the questions contained in the letters.

Place Location Activity

1. I
2. H
3. D
4. A
5. G
6. B
7. C
8. E
9. F

Chapter Bonus Test Question

This question may be used for extra credit on the chapter test.

Soft-drink canning companies might be very interested in Costa Rica for their products. Why? *(Costa Rica has bauxite, which is the principal source of aluminum, which is used to make soft-drink cans.)*

Section 2

6. southeast of Florida and northeast of Cuba
7. Cuba
8. commercial plantations, subsistence farms
9. Trinidad, Jamaica
10. Port-au-Prince

Critical Thinking Activities

1. Because it lies on the Isthmus of Panama, the narrowest piece of land between the Atlantic and Pacific oceans.
2. Most of the islands have adopted the language and many of the customs of their former European colonial rulers. The culture of the area has also been influenced by the enslaved Africans brought to work on plantations.

193

PERFORMANCE ASSESSMENT ACTIVITY

CREATE A COUNTRY Have students work in groups to develop a new country for the area. The country should have characteristics borrowed from each of the countries actually studied in Chapter 8. Groups should determine a location for the country in relation to the others and sketch a revised map of the area. The various characteristics groups choose for their country should be geographically sound. For example, the country may have a fishing industry only if it is located on the coast. Some characteristics, such as cultural and historical features, may be more creatively borrowed, however. When groups have made all their selections, they should incorporate this information on a full-sized poster map of the country. The map should be illustrated with pictures, symbols, charts, and graphs. Call on groups to present their maps and explain why they selected the various characteristics.

POSSIBLE RUBRIC FEATURES:
Accuracy of content information, identification of relationships, quality of maps and graphics, communication skills

MENTAL MAPPING ACTIVITY
Before teaching Chapter 8, point out Brazil on a map of South America. Have students note its size in comparison to other countries on the continent. Then have them note the course of the Amazon River.

TEACHER'S CORNER

NATIONAL GEOGRAPHIC SOCIETY

INDEX TO NATIONAL GEOGRAPHIC MAGAZINE

The following articles may be used for research relating to this chapter:

- "Feast of the Tarpon," by David Doubilet, January 1996.
- "Leafcutter Ants," by Mark W. Moffett, July 1995.
- "Poison-Dart Frogs," by Mark W. Moffett, May 1995.
- "The Amazon," by Jere Van Dyk, February 1995.
- "Remote World of the Harpy Eagle," by Neil Retting, February 1995.

- "Simón Bolívar," by Bryan Hodgson, March 1994.
- "New Sensors Eye the Rain Forest," by Thomas O'Neill, September 1993.
- "Plotting a New Course," by Sandra Dibble, August 1992.
- "Rain Forest Canopy: The High Frontier," by Edward O. Wilson, December 1991.

NATIONAL GEOGRAPHIC SOCIETY PRODUCTS AVAILABLE FROM GLENCOE

To order the following products for use with this chapter, contact your local Glencoe sales representative or call Glencoe at 1-800-334-7344:

- *STV: World Geography* (Videodisc)
- *STV: Rain Forest* (Videodisc)
- *Picture Atlas of the World* (CD-ROM)
- *Physical Geography of the World* (Transparencies)

- *ZipZapMap! World* (Software)
- *GeoBee* (Software)
- *Images of the World* (Posters)
- *Eye on the Environment* (Posters)

ADDITIONAL NATIONAL GEOGRAPHIC SOCIETY PRODUCTS

To order the following products for use with this chapter, call National Geographic Society at 1-800-368-2728:

- *Jewels of the Caribbean Sea* (Video)
- *Amazon: Land of the Flooded Forest* (Video)
- *Rain Forest* (Video)

- *Physical Geography of the Continents Series:* "South America." (Video)
- *Portraits of the Continents Series:* "Part I: North America; South America." (Filmstrip)

TEACHER-TO-TEACHER
Curriculum Connection: Fine Arts, World Cultures

—from Charles M. Bateman,
Halls Middle School, Knoxville, TN

CARNIVAL The purpose of this activity is to allow students to experience the spirit of celebration of the pre-Lenten festival known as Carnival in Brazil and much of South America.

Before starting the activity, discuss Lent, the Christian ritual of denial for 40 days prior to the celebration of Easter. Then provide students with materials such as colored construction paper, colored tissue paper, glue and glue sticks, glitter, crayons, scissors, and so on. Have students work individually, using these materials to make brightly colored and decorated masks. On a selected day, have students don their masks and parade around the school yard, playing various percussion instruments. Play recordings of samba music to add to the festivities.

After a set period of time, bring the celebration to an end. Mention that in Brazil, people may spend months—and a great deal of money—making their costumes and floats for Carnival. Also mention that the celebration lasts the whole day, not just a few minutes. Then have students discuss why a period of personal denial might be preceded by festivals and celebration.

BIBLIOGRAPHY

Readings for the Student

Portraits of the Nations: The Land and the People Series. Venezuela. Philadelphia: Lippincott, 1992.

The Rainforests. Charleston: Cambridge Research Group, Ltd., 1991.

Readings for the Teacher

Exploring the Developing World: Life in Africa and Latin America. Denver: Center for Teaching International Relations, 1993.

Latin America Today: An Atlas of Reproducible Pages, revised ed. Wellesley, Mass.: World Eagle, 1992.

Multimedia

Amazon: Paradise Lost? Boston: Christian Science Monitor, 1991. Videocassette, 60 minutes.

Children of Brazil. Chicago: Coronet/MTI, 1990. Videocassette, 25 minutes.

MEETING NATIONAL STANDARDS

Geography for Life

All of the 18 standards are demonstrated in Unit 3. The following ones are highlighted in this chapter:
• Standards 3, 4, 6, 7, 8

KEY TO ABILITY LEVELS

Teaching strategies have been coded for varying learning styles and abilities.

L1   **BASIC** activities for all students

L2   **AVERAGE** activities for average to above-average students

L3   **CHALLENGING** activities for above-average students

LEP   **LIMITED ENGLISH PROFICIENCY** activities

Use these *Geography: The World and Its People* resources to teach, reinforce, and extend chapter content.

CHAPTER 8 RESOURCES

- *Vocabulary Activity 8
- Cooperative Learning Activity 8
- Workbook Activity 8
- Geography Skills Activity 8
- GeoLab Activity 8
- Chapter Map Activity 8
- Critical Thinking Skills Activity 8
- Performance Assessment Activity 8
- *Reteaching Activity 8
- Enrichment Activity 8
- Chapter 8 Test, Form A and Form B
- *Chapter 8 Digest Audiocassette, Activity, Test
- Unit Overlay Transparencies 3-0, 3-1, 3-2, 3-3, 3-4, 3-5, 3-6
- Geography Handbook Transparency 7
- Political Map Transparency 3
- Geoquiz Transparencies 8-1, 8-2
- Vocabulary PuzzleMaker
- Student Self-Test: A Software Review
- Testmaker Software
- MindJogger Videoquiz

If time does not permit teaching the entire chapter, summarize using the Chapter 8 Highlights on page 211, and the Chapter 8 English (or Spanish) Audiocassettes. Review students' knowledge using the Glencoe MindJogger Videoquiz. *Also available in Spanish*

Use these Geography: The World and Its People resources to teach and reinforce section content.

SECTION 1 RESOURCES

- Reproducible Lesson Plan 8-1
- Section Focus Transparency 8-1
- Guided Reading Activity 8-1
- Section Quiz 8-1 (also available in Spanish)

SECTION 2 RESOURCES

- Reproducible Lesson Plan 8-2
- Section Focus Transparency 8-2
- Guided Reading Activity 8-2
- Section Quiz 8-2 (also available in Spanish)

SECTION 3 RESOURCES

- Reproducible Lesson Plan 8-3
- Section Focus Transparency 8-3
- Guided Reading Activity 8-3
- Section Quiz 8-3 (also available in Spanish)

ADDITIONAL RESOURCES FROM GLENCOE

Reproducible Masters

- Glencoe Social Studies Outline Map Book, pages 24, 25
- Foods Around the World, Unit 3

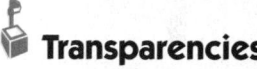

Workbook

- Building Skills in Geography, Unit 5, Lesson 3

World Music: Cultural Traditions

Lesson 2

Transparencies

- National Geographic Society PicturePack Transparencies, Unit 3

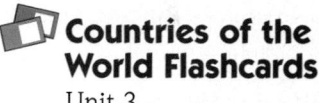

Posters

- National Geographic Society: Images of the World, Unit 3

Countries of the World Flashcards

Unit 3

Videodiscs

- Reuters Issues in Geography
- National Geographic Society: GTV: Planetary Manager

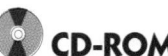

CD-ROM

- National Geographic Society PictureShow: Earth's Endangered Environments

Software

- National Geographic Society: ZipZapMap! World

Performance Assessment

Refer to the Planning Guide on page 194A for a Performance Assessment Activity for this chapter. See the *Performance Assessment Strategies and Activities* booklet for suggestions.

Chapter Objectives

1. **Discuss** the major geographical features of Brazil.
2. **Compare** and contrast the geography of the countries of Caribbean South America.
3. **Describe** the physical, economic, and cultural geography of Uruguay and Paraguay.

GLENCOE
TECHNOLOGY

VIDEODISC

Use the Chapter 8 Mind-Jogger Videoquiz to preview chapter content.

MindJogger Videoquiz

‖‖‖‖‖‖‖‖‖‖‖‖

Chapter 8
Disc 1 Side B

Also available in VHS.

Chapter
8 Brazil and Its Neighbors

MAP STUDY ACTIVITY
In this chapter you will read about Brazil, the largest country in South America, and its neighbors.

1. **What is the capital of Brazil?**
2. **What other countries border this huge country?**

194

 MAP STUDY ACTIVITY

Answers
1. Brasília
2. Venezuela, Guyana, Suriname, French Guiana, Paraguay, Uruguay are shown on the map

Map Skills Practice
Reading a Map What sets Paraguay off from the other countries in this region? *(It is landlocked, while all the other countries have coastlines.)*

Meeting National Standards
Geography for Life The following standards are highlighted in this section:
• Standards 8, 14, 16

PREVIEW

Words to Know
• basin
• *selva*
• escarpment
• sisal
• *favela*

Places to Locate
• Amazon River
• Paraná River
• São Francisco River
• Brazilian Highlands
• Great Escarpment
• São Paulo
• Rio de Janeiro
• Brasília

Read to Learn . . .
1. where Brazil is located.
2. what landforms and climates are found in Brazil.
3. what natural resources Brazil's economy depends on.

I magine standing inside this photograph of the Amazon rain forest—the largest rain forest in the world. Can you hear this parrot chatter? Do you feel the stickiness of the steamy, tropical air? Lush flowers and trees surround you so thickly that you can't see the sky!

FOCUS

Section Objectives
1. **Locate** Brazil on a map or globe.
2. **Describe** Brazil's landforms and climates.
3. **Identify** the natural resources Brazil's economy depends on.

Bellringer Motivational Activity

 Prior to taking roll at the beginning of the class period, project Section Focus Transparency 8-1 and have students answer the activity questions. Discuss students' responses. **L1**

Vocabulary Pre-check

In **Words to Know,** have students identify the two terms that refer to landforms. *(basin, escarpment)* Then have them find or draw illustrations of these physical features. **L1 LEP**

 Use the Vocabulary PuzzleMaker Software to create a crossword puzzle. **L1**

T he Amazon rain forest covers the northern half of South America. Flowing through this vast, flat region is the mighty Amazon River. What country lays claim to this tropical beauty? The Amazon River and its surrounding rain forest are located mostly in the country of Brazil.

Region
The Land

The largest country in South America, Brazil has a land area of 3,265,060 square miles (8,456,505 sq. km). Because it covers such a large area, Brazil encompasses many types of landforms. The map on page 196 shows you that Brazil has lowland river valleys as well as highland areas and coastal plains.

Rivers and Lowlands The Amazon is the world's second longest river, winding almost 4,000 miles (6,436 km) from the Andes to the Atlantic Ocean. On its journey to the ocean, the Amazon passes through a wide, flat basin. A **basin** is a low area entirely surrounded by higher land. Most of the Amazon basin is covered with thick tropical forests, which Brazilians call the *selva*.

CHAPTER 8

195

Classroom Resources for Section 1

REPRODUCIBLE MASTERS
Reproducible Lesson Plan 8-1
Guided Reading Activity 8-1
Vocabulary Activity 8
Geography Skills Activity 8
Building Skills in Geography,
  Unit 5, Lesson 3
Section Quiz 8-1

TRANSPARENCIES
Section Focus
  Transparency 8-1
Unit Overlay Transparency 3-5
Geography Handbook
  Transparency 7

MULTIMEDIA
Vocabulary PuzzleMaker
  Software
Testmaker

National Geographic
  Society: GTV: Planetary
  Manager

TEACH

Guided Practice

 Project Unit Overlay Transparency 3-5. Have students identify the **Places to Locate** listed in the **Preview** on page 195. Point out the extent of the Amazon rain forest. Then have students locate and identify other landscapes found in Brazil. **L1**

Independent Practice

 Guided Reading Activity 8-1 **L1**

NATIONAL GEOGRAPHIC SOCIETY

These materials are available from Glencoe.

 **VIDEODISC**

STV: World Geography

Vol. 3: South America and Antarctica

Amazon Basin
Side 1, Frames 32121-43167

MAP STUDY

Answer
0–1,000 feet (0–300 m)

Map Skills Practice
Reading a Map What upland area is located in east central Brazil?
(*Brazilian Highlands*)

In addition to the Amazon basin, Brazil has two other lower land areas. They are located along the Paraná River and the São Francisco River. Trace these rivers on the map below. Both rivers begin in the center of Brazil, but they flow in opposite directions. The Paraná flows to the southwest, and the São Francisco flows to the northeast.

Highlands Don't picture Brazil as one big rain forest. Highland areas, in fact, cover more than one-half of the country. The Brazilian Highlands—the largest highland area—begin south of the Amazon River basin and cover much of east central Brazil. Low mountain ranges in the highlands drop sharply to the Atlantic Ocean, a drop called the Great Escarpment. An **escarpment** is a steep cliff between a higher and lower surface.

The Climate Brazil's climates are as varied as its landforms. If you like steam baths, you'd love living in the Amazon basin. With its tropical rain forest climate, the basin has hot, steamy temperatures and torrents of rain year-round. In most of Brazil's highland areas, you find a tropical savanna climate, with wet

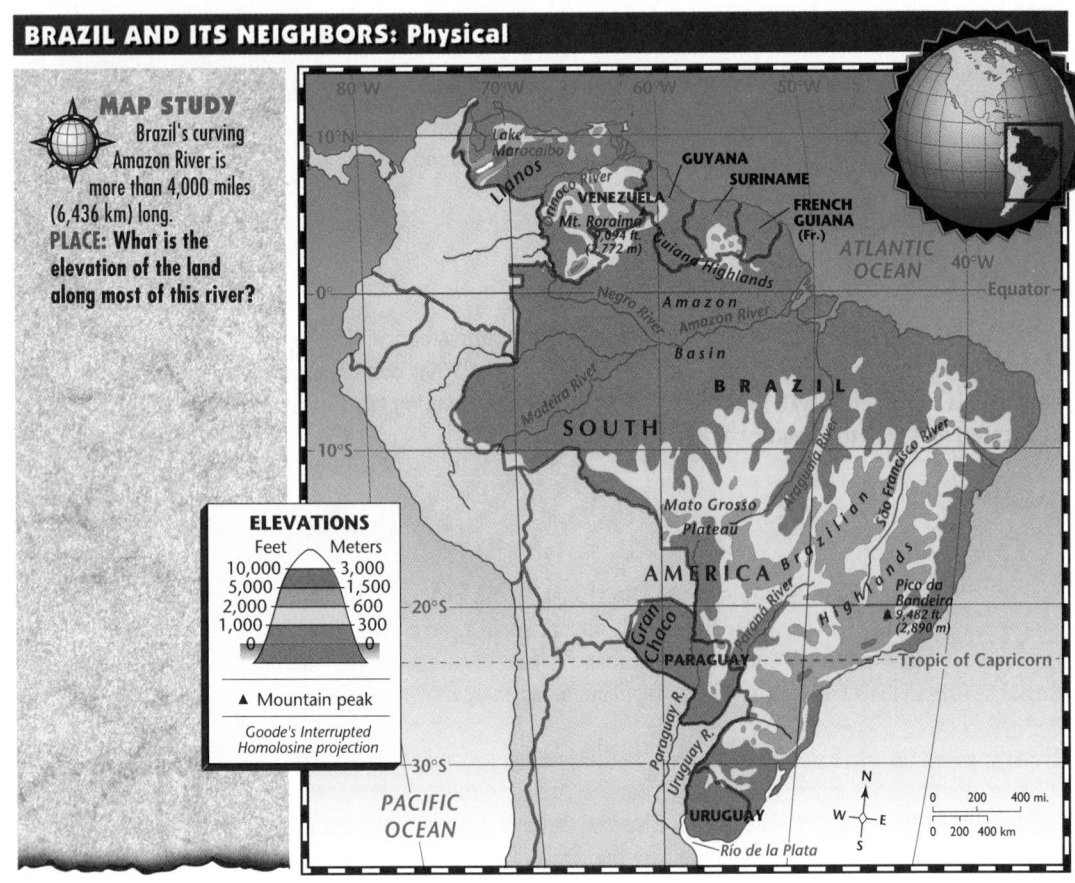

BRAZIL AND ITS NEIGHBORS: Physical

MAP STUDY
Brazil's curving Amazon River is more than 4,000 miles (6,436 km) long.
PLACE: What is the elevation of the land along most of this river?

ELEVATIONS
Feet	Meters
10,000	3,000
5,000	1,500
2,000	600
1,000	300
0	0

▲ Mountain peak

Goode's Interrupted Homolosine projection

Cooperative Learning Activity

Organize students into small groups to create a diorama of the Amazon rain forest. Assign specific topics to each group. These might include trees, plants, birds, reptiles, mammals, or amphibians. Instruct members of each group to research their assigned topic. Then have them choose representative species to illustrate for the diorama. Have groups work together to assemble the display on a table or in a large box. Upon completion of the diorama, have students discuss the challenge of portraying the incredible variety of plant and animal life in this region. **L2**

and dry seasons. For those who prefer a drier climate, moderate temperatures prevail in the southern part of the country year-round.

Region
Economic Regions

How do Brazilians earn a living? Brazil's diverse economy includes agriculture, manufacturing, and service industries. Agriculture has been important in the country for centuries, and rapid industrial growth during this century has made Brazil the leading manufacturing nation in South America.

The North For many years the Amazon rain forest was a mysterious region whose secrets were guarded by the Native Americans living there. In recent years, however, new roads have allowed more people to settle in the north. The map on page 204 reveals that the north is rich in mahogany, teak, and other forest woods. The area also contains minerals such as bauxite and iron ore. As a result, the Brazilian government has encouraged logging and mining in the north.

Many people worry that the increasing use of forest resources is destroying the rain forest and its wildlife. They also fear that economic development threatens the survival of Native Americans still living in the area. Turn to page 236 to learn more about the destruction of the rain forests.

The Northeast In the northeast region, farmers and ranchers have cleared coastal rain forests to grow crops and raise cattle. Sugarcane, cotton, cacao, and **sisal**—a plant fiber used to make rope—are this region's major farm products. Overgrazing in this dry area has ruined much of the land, however.

The Southeast The southeast is rich in mineral resources and fertile farmland. This region boasts one of the largest iron-ore deposits in the world, and the area also supplies coffee drinkers with their favorite beverage. The graph on page 198 shows that Brazil is the world's leading producer of coffee.

The towering forests of the north are matched by the towering skyscrapers of the southeast. Here stand Brazil's major cities and centers of industry. São Paulo (sown POW•loo) is home to more than 18 million people, making it one of the fastest-growing urban areas in the world. It is also Brazil's leading trade and industrial center. Tourists flock to Rio de Janeiro (REE•oh day zhuh•NEHR•oh) for its beautiful coastline, sandy beaches, and colorful festivals.

The South The south's vast plains support huge herds of cattle. Brazilian beef is exported all over the world. Another product of the south is *yerba maté,* a tea-like drink that is popular throughout the southern part of South America. *Yerba maté* is made from the leaves of holly trees that grow in the region.

WHAT IN THE WORLD?
Setting the Style
In Brazil traditional clothing styles differ from region to region. In the northeast, women wear colorful skirts, bright blouses, and much jewelry. In the south, cowhands wear baggy pants, felt hats with wide brims, and ponchos, or sleeveless coverings.

CURRICULUM CONNECTION

Earth Science Brazilians are harnessing the power of their water resources to supply their energy needs. The Itaipu Dam power plant on the Paraná River is a joint $20-billion project between Brazil and Paraguay. It is designed to supply one-fifth of Brazil's electricity.

NATIONAL GEOGRAPHIC SOCIETY

These materials are available from Glencoe.

VIDEODISC

GTV: Planetary Manager

Steel mill in Brazil
Any side, Frame 47884

MULTICULTURAL PERSPECTIVE

Cultural Heritage
The name *Brazil* comes from a Portuguese word meaning "glowing ember." When Portuguese sailors first saw the trees that grow along the Brazilian coast, they thought the color of the wood looked like the glowing embers of a fire.

Meeting Special Needs Activity

Inefficient Readers Provide structure for students with reading problems by helping them locate specific information in each subhead. Have students work in pairs. Both partners should write one question based on each subhead. Then they should exchange papers and write an answer to each of their partner's questions. **L1**

Answer to Caption
fútbol

Place Soccer is the favorite sport of Brazilians, but auto racing, basketball, and horse racing are also popular.

ASSESS

Check for Understanding

Assign Section 1 Review as homework or an in-class activity.

Meeting Lesson Objectives

Each objective below is tested by the questions that follow it in parentheses.

1. **Locate** Brazil on a map or globe. (4, 5)
2. **Describe** Brazil's landforms and climates. (2, 5)
3. **Identify** the natural resources Brazil's economy depends on. (1, 2)

GRAPHIC STUDY

Answer
Colombia, Mexico, Guatemala

Graphic Skills Practice
Reading a Graph About how much more coffee does Brazil produce than the second-ranking country? *(about 1.6 billion pounds, 0.6 billion kilograms)*

Soccer Game in Rio de Janeiro, Brazil

Maracana Stadium in Rio de Janeiro is the world's largest stadium.
PLACE: What is the Brazilian word for soccer?

West Central Region Inland highlands and plateaus cover the west central region, which is very isolated and has few settlers. The soil here is not fertile, and the land has been overgrazed. The Brazilian government hopes to encourage better farming practices that will boost the economy of this region.

Movement
People

With more than 155 million people, Brazil has the largest and fastest-growing population of any Latin American nation. Unlike most of the countries in South America, Brazil has a culture that is largely Portuguese rather than Spanish. The Portuguese were the first and largest European group to colonize Brazil. Today Brazilians are of European, African, Native American, or mixed ancestry. They also speak their own Brazilian variety of the Portuguese language.

Influences of the Past Native Americans were the first people to live in Brazil. In the 1500s the Portuguese took control of the northeast and other coastal areas. They forced Native Americans to work on sugar plantations or in mines. Many Native Americans died from disease or overwork. Needing workers, the Portuguese traders enslaved Africans and then shipped them to

Leading Coffee-Producing Countries

GRAPHIC STUDY

Brazil's warm, moist climate and fertile volcanic soil make it the world's leading coffee producer.
PLACE: What other Latin American countries are among the top seven coffee-producing countries?

Coffee Beans Produced in One Year

= .25 billion pounds

Billions of pounds — Billions of kilograms

Brazil, Colombia, Indonesia, Mexico, Vietnam, Côte d'Ivoire, Guatemala

Source: *World Book,* 1994

Critical Thinking Activity

Determining Cause and Effect Have students create a chart to illustrate the cause-and-effect relationships that have influenced settlement patterns in Brazil. For example: moving the capital from Rio de Janeiro to Brasília encouraged settlement of inland areas. **L2**

Brazil. Enslavement finally ended in 1888, but many Africans remained. Today about 6 percent of Brazil's population are of purely African ancestry. African traditions and customs have influenced Brazil's religions, music, dance, and foods.

In 1822 Brazil became a monarchy after winning its independence from Portugal. About 70 years later it became a republic. During the twentieth century Brazil's government has seesawed between dictatorship and democracy.

Brazilians Today Look at the map on page 209. You see that most of Brazil's people live along the country's Atlantic coast. The Brazilian government has tried to encourage people to move from these crowded coastal areas to less populated inland areas. In the early 1960s Brazil moved its capital from the coastal city of Rio de Janeiro to the newly built inland city of Brasília. With more than 1.5 million people, Brasília is a modern and rapidly growing city.

Most rural Brazilians live in villages of one- or two-room houses made of stone or brick. Some work on plantations or ranches, and others cultivate subsistence farms. About 76 percent of Brazil's people live in cities. Many city dwellers are very poor and live in *favelas*, or slum areas. Others have good jobs in industry and government and live in city apartments or suburban houses.

Antonio Vargas is a Brazilian teenager who lives in Brasília. He enjoys showing visitors Brasília's modern government buildings and university. But Antonio often visits his cousins in Rio de Janeiro. Rio, as it is usually called, is said to be one of the world's most beautiful cities. Buildings from Brazil's early history stand side by side with modern skyscrapers. Busy downtown streets lead to white, sandy beaches along Guanabara Bay.

Brazilian Sports and Celebrations It is safe to say that *fútbol,* or soccer, is a way of life in Brazil. Every village has some kind of soccer field, and the larger cities have stadiums. In Rio de Janeiro, the stadium can seat 220,000 fans! The most famous soccer player in the world, Pelé, is from Brazil, and the country frequently competes for the World Cup.

Brazilians enjoy many celebrations. The festival known as Carnival is celebrated just before the beginning of Lent, the Christian holy season that comes before Easter. The most spectacular Carnival is held each year in Rio de Janeiro.

SECTION 1 REVIEW

REVIEWING TERMS AND FACTS

1. Define the following: basin, *selva,* escarpment, sisal, *favela.*
2. REGION What kinds of landforms and climates does Brazil have?
3. PLACE How does city life in Brazil differ from rural life?

MAP STUDY ACTIVITIES

4. Look at the political map on page 194. How far from Rio de Janeiro is Brasília?
5. Study the physical map on page 196. How does northern Brazil's landscape differ from southern Brazil's?

199

LESSON PLAN
Chapter 8, Section 1

MULTICULTURAL PERSPECTIVE

Cultural Heritage

The Opera House of Manaus, built in 1896, is one of Brazil's architectural masterpieces. It serves as a reminder of the wealth from the Amazon rubber trade that helped build the city.

Evaluate

 Section 1 Quiz **L1**

Use the Testmaker to create a customized quiz for Section 1. **L1**

Reteach

Project Unit Overlay Transparency 3-5. As you point to a region of Brazil, call on students to describe the major landforms, climate, and vegetation. **L1**

Enrich

Have students choose one of the plant or animal species they depicted in the rain forest diorama. Ask them to report on the plant's or animal's place in the rain forest ecosystem. **L3**

CLOSE

Have each student list five facts he or she learned about Brazil from this lesson. Then have students work in groups to create a summary of the lesson based on their members' lists.

Answers to Section 1 Review

1. All vocabulary words are defined in the Glossary.
2. lowlands with river basins, highland areas with mountains, plateaus, and coastal plains; mostly tropical—rain forest and savanna.
3. Most rural Brazilians live in villages of one- or two-room houses, some work on ranches or plantations; others depend on subsistence farming. Many city dwellers are very poor and live in slum areas. Others have good jobs and live in apartments, mansions, or houses.
4. approximately 600 miles (965 km)
5. northern: covered by the Amazon rain forest; southern: covered by highlands.

BUILDING GEOGRAPHY SKILLS

Organize the classroom into four sections. Tell students that each section represents a country. Select two students from each section to represent the country's capital city and its major industrial city. Ask students how they might show this information on a map. List their suggestions on the chalkboard. Then have students use the list to create a map showing the four countries and their cities. Point out that the map they have drawn is a political map. Then have students read the skill and complete the practice questions. **L1**

Additional Skills Practice

1. How do political maps show boundaries between states or countries? *(with lines)*
2. How does the use of color differ on physical and political maps? *(On a physical map, colors show elevations; on political maps, colors show political units.)*

📁 Geography Skills Activity 8

📁 Building Skills in Geography, Unit 5, Lesson 3

Reading a Political Map

Think about the last sports event you participated in or watched. Boundaries—the outer limits of the playing area—are set up by the rules of the sport. In the same way, political areas such as cities, states, and countries have boundaries. People living within these borders follow a common set of rules or laws.

Political maps show these boundaries. Lines represent the boundaries between political areas. These areas are often shown in contrasting colors. Political maps also include cities and important physical features. To read a political map, apply these steps:

- Read the map title.
- Identify the political areas on the map.
- Locate the major cities.

VENEZUELA: Political

National boundary
⊛ National capital
• Other city

Azimuthal Equal-Area projection

0   75   150 mi.
0   75   150 km

Geography Skills Practice

1. What countries are shown on this map?
2. Name the capital city and three other major cities of Venezuela.
3. Where are most of the cities located?
4. How would you explain why cities are located in these areas?

Answers to Geography Skills Practice

1. Venezuela, Colombia, Brazil, Guyana
2. The capital of Venezuela is Caracas; other major cities include Maracaibo, Valencia, Maracay, Barquisimeto, Cumaná, Maturín, Ciudad Bolívar, Barinas, San Cristóbal.
3. along the Caribbean coast
4. Coastal areas promote trade and industry, which leads to the growth of cities.

Caribbean South America

PREVIEW

Words to Know
- *llanos*
- hydroelectric power
- altitude
- *caudillo*

Places to Locate
- Venezuela
- Caribbean Sea
- Andean Highlands
- Caracas
- Lake Maracaibo
- Orinoco River
- Guyana
- Suriname
- French Guiana

Read to Learn . . .
1. what landscapes and climates are found in Caribbean South America.
2. what early groups influenced Caribbean South America.
3. how countries in Caribbean South America use their resources.

Angel Falls—the highest waterfall in the world— roars down a cliff in a highland area of the country of Venezuela. If you stood at the bottom of the waterfall, you'd strain your neck as you gazed 3,212 feet (979 m) up to the top!

Meeting National Standards

Geography for Life The following standards are highlighted in this section:
- Standards 4, 6, 17

FOCUS

Section Objectives
1. **Examine** the landscapes and climates found in Caribbean South America.
2. **Identify** the early groups that influenced Caribbean South America.
3. **Explain** how countries in Caribbean South America use their resources.

Bellringer Motivational Activity

 Prior to taking roll at the beginning of the class period, project Section Focus Transparency 8-2 and have students answer the activity questions. Discuss students' responses. **L1**

Vocabulary Pre-check

Have students look up the pronunciation of the word *caudillo*. Note that the double "l" at the end of the word sounds like a "y." **L1**
LEP

Venezuela is located in Caribbean South America. This region extends across the northern part of South America along the Caribbean Sea. Besides Venezuela, two other countries and a territory border the Caribbean Sea: Guyana, Suriname, and French Guiana.

Place

Venezuela

Venezuela is the largest country in Caribbean South America. Its boundaries touch the countries of Brazil, Colombia, and Guyana. The waters of the Caribbean Sea and the Atlantic Ocean wash its shores.

The Land What types of landforms would you see in Venezuela if you could look down on it from space? Looking northwest you would see the Maracaibo (MAR•uh•KY•boh) basin. This lowland coastal area surrounds Lake Maracaibo, the largest lake in South America. To the east you would see a hilly area known as the Andean Highlands—home to most of Venezuela's people.

CHAPTER 8

(201)

Classroom Resources for Section 2

 REPRODUCIBLE MASTERS

Reproducible Lesson Plan 8-2
Guided Reading Activity 8-2
Workbook Activity 8
Cooperative Learning Activity 8
Section Quiz 8-2

 TRANSPARENCIES

Section Focus
    Transparency 8-2
Unit Overlay Transparency 3-5

MULTIMEDIA

 Testmaker

National Geographic
    Society: STV: World
    Geography, Vol. 3

TEACH

Guided Practice

 Project Unit Overlay Transparency 3-5. Have students make predictions about climate in Caribbean South America. They should consider location relative to the Equator and to large bodies of water. In addition, they should take note of the relationship between altitude and temperature. Have students list their predictions and verify them as they read this section. **L2**

NATIONAL GEOGRAPHIC SOCIETY

MAP STUDY

Answer
tropical rain forest, tropical savanna, steppe, desert

Map Skills Practice
Reading a Map What word best describes the climate of Caribbean South America? *(tropical)*

In the south you would see the Guiana Highlands with dense rain forests and very few people. Grassy plains known as the ***llanos*** lie between the Andean and Guiana highlands. The *llanos* have many ranches, farms, and oil fields. Venezuela's most important river—the Orinoco—flows across the *llanos*. It is a valuable source of **hydroelectric power**, or water-generated electricity, for Venezuela's cities.

The Climate Because it is close to the Equator, Venezuela has mostly a tropical rain forest climate. Cool, dry winds from the northeast, however, keep temperatures moderate. As in Mexico, temperatures in Venezuela also differ with **altitude**, or height above sea level. The lowland Maracaibo basin and inland river valleys are hot and rainy. As you travel up into the highland areas, you are usually warm in the daytime but cool at night.

The Economy Venezuelans once depended on crops such as coffee and cacao to earn their livings. Since the 1920s, though, petroleum has changed all that. Today Venezuela, a world leader in oil production, is one of the wealthiest countries in South America. Bauxite, coal, diamonds, and gold also are mined there.

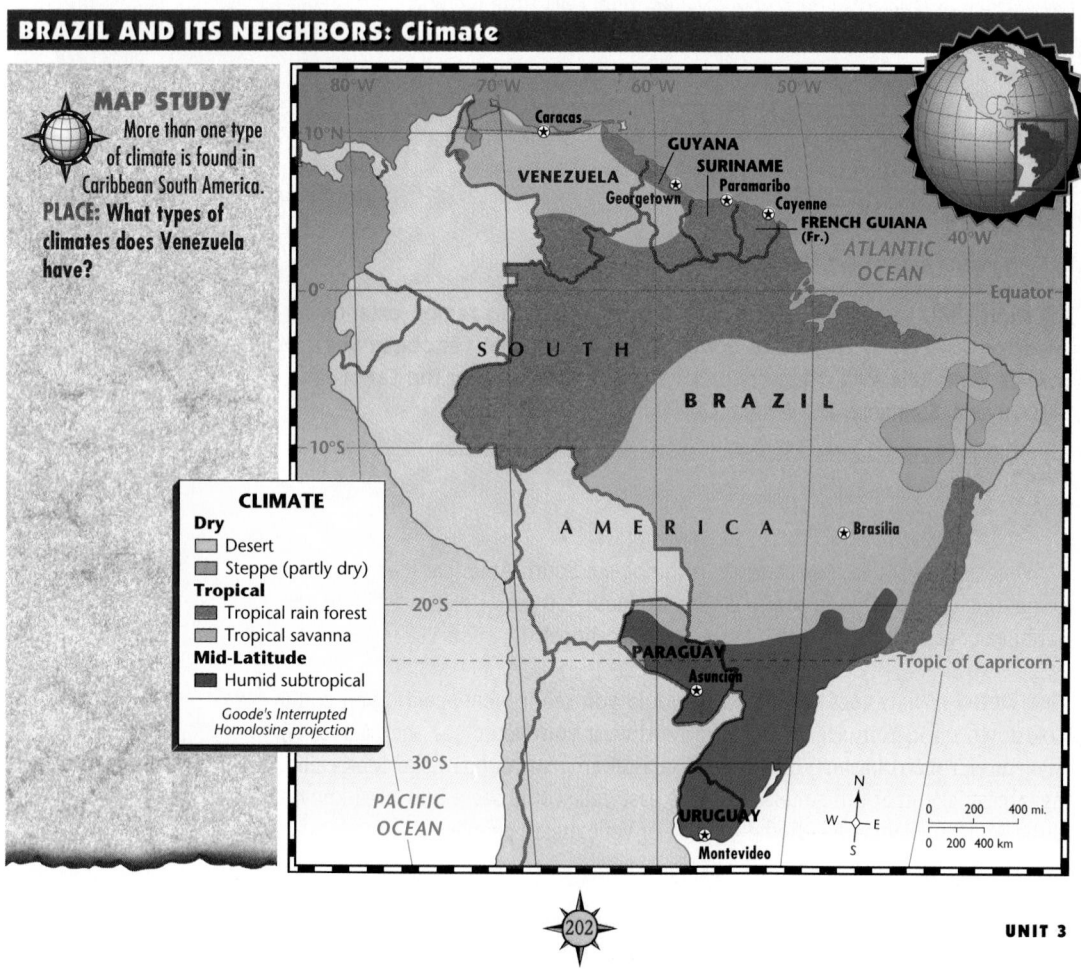

BRAZIL AND ITS NEIGHBORS: Climate

MAP STUDY
More than one type of climate is found in Caribbean South America.
PLACE: What types of climates does Venezuela have?

CLIMATE
Dry
- Desert
- Steppe (partly dry)
Tropical
- Tropical rain forest
- Tropical savanna
Mid-Latitude
- Humid subtropical

Goode's Interrupted Homolosine projection

Cooperative Learning Activity

Organize students into four groups and assign one of the countries of Caribbean South America to each group. Direct groups to research the history and culture of their assigned countries. Have groups present their findings in an illustrated report. After all reports have been presented, have students discuss similarities and differences among the countries of Caribbean South America. **L2**

Caracas, Venezuela

Place The population of Caracas tripled between 1961 and 1981—the years that marked the height of Venezuela's oil boom. Today, the city of Caracas has a population of more than 3 million and a population density greater than New York City.

About 90 percent of Venezuela's people work in service industries or manufacturing. Agriculture, however, remains an important part of the country's economy. Landowners and ranchers export beef and dairy cattle. Smaller farms grow bananas, coffee, corn, and rice.

The Past Spanish explorers gave Venezuela its name. With its rivers and natural canals, the country reminded them of Venice, Italy. Like many other Latin American nations, Venezuela was a Spanish colony from the early 1500s to the early 1800s. In 1821 South Americans rejoiced as Simón Bolívar (see•MOHN buh•LEE•VAHR) and his soldiers freed the northern part of the continent from Spanish rule. Nine years later Venezuela became an independent republic. During most of the 1800s and 1900s, the country was governed by harsh military rulers known as **caudillos.** But since the 1960s Venezuela has developed into a stable democracy.

The People Most of Venezuela's 22 million people are a mix of European, African, and Native American backgrounds. Spanish is the major language of the country; the major religion is Roman Catholicism.

A remarkable 90 percent of the population of Venezuela live in cities. María González, a teenager, is among the 3.2 million people who live in Caracas (kuh•RAH•kuhs), Venezuela's capital and largest city. Traveling on freeways, her school bus passes towering skyscrapers on its daily route. Like many people in Caracas, María and her family live in a high-rise apartment. Because of oil, the people of Venezuela enjoy one of the highest standards of living in South America.

Venezuela's modern capital is nestled in a valley surrounded by mountains.
PLACE: What natural resource gives Venezuelans one of the highest standards of living in South America?

MULTICULTURAL PERSPECTIVE

Cultural Heritage
Most Venezuelan cities have a *Plaza Bolívar,*—a public square honoring the South American liberator, Simón Bolívar. Venezuelans consider it rude to behave disrespectfully in the square. Further, they take negative comments about Bolívar as an insult.

CHAPTER 8

 203

Meeting Special Needs Activity

Study Strategy Model the study strategy SQ3R (Survey, Question, Read, Recite, Review) for students with reading disabilities. First, skim the section, reading the heads and subheads. Then, rephrase each heading as a question and have students read the text under each heading to answer the question. Next, call on a volunteer to recite what he or she remembers from reading. Finally, have students review the material to check for errors or omissions from the recitation. **L1**

MAP STUDY

Answer
bauxite, gold
Map Skills Practice
Reading a Map What are the major crops grown in Caribbean South America? *(coffee, sugarcane, tobacco, corn, rice, bananas)*

ASSESS

Check for Understanding

Assign Section 2 Review as homework or an in-class activity.

Meeting Lesson Objectives

Each objective below is tested by the questions that follow it in parentheses.

1. **Examine** the landscapes and climates found in Caribbean South America. (2)
2. **Identify** the early groups that influenced Caribbean South America. (4)
3. **Explain** how countries in Caribbean South America use their resources. (3, 5)

Evaluate

 Section 2 Quiz **L1**

Use the Testmaker to create a customized quiz for Section 2. **L1**

BRAZIL AND ITS NEIGHBORS: Land Use and Resources

MAP STUDY
The lumber from rain forests is important to Brazil's economy, but harvesting the lumber can harm the environment.
REGION: What two minerals are mined by most of the countries in this region?

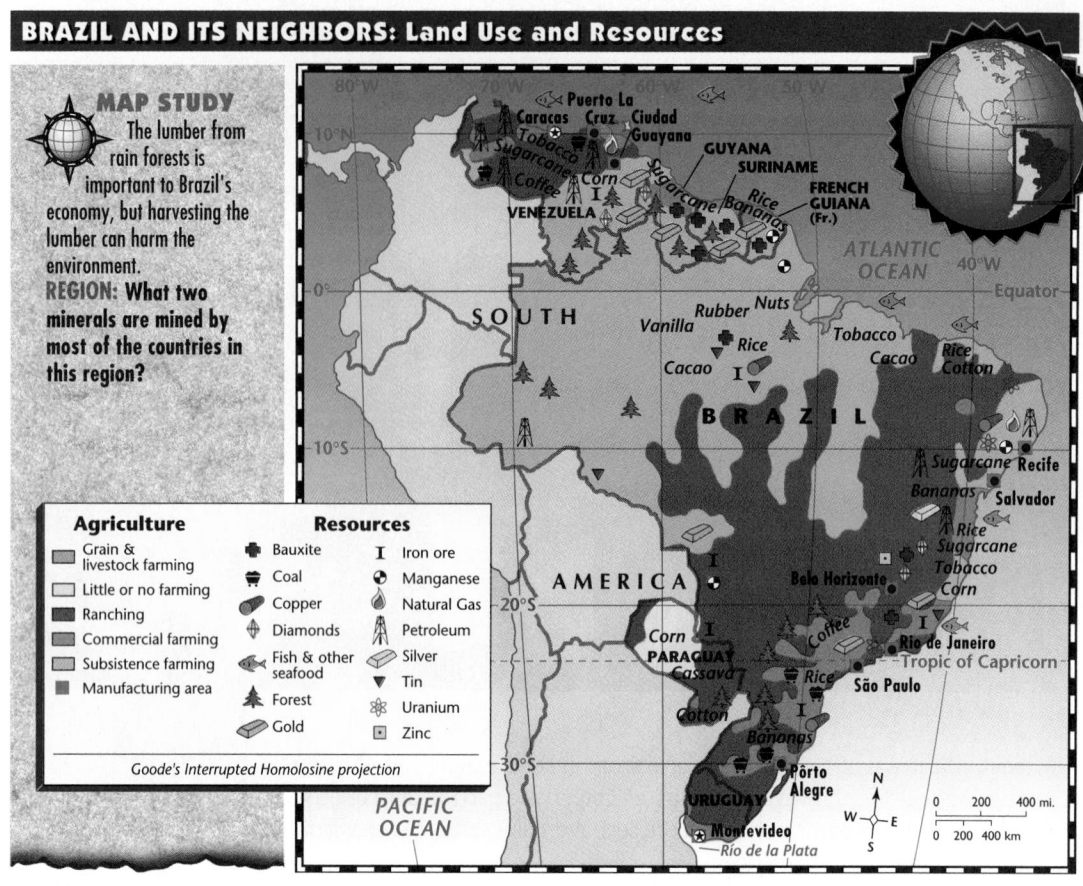

Agriculture
- Grain & livestock farming
- Little or no farming
- Ranching
- Commercial farming
- Subsistence farming
- Manufacturing area

Resources
- Bauxite
- Coal
- Copper
- Diamonds
- Fish & other seafood
- Forest
- Gold
- Iron ore
- Manganese
- Natural Gas
- Petroleum
- Silver
- Tin
- Uranium
- Zinc

Goode's Interrupted Homolosine projection

Place
The Guianas

In addition to Venezuela, Caribbean South America includes the countries of Guyana (gy•A•nuh), Suriname (sur•uh•NAH•muh) and the territory of French Guiana (gee•A•nuh).

Guyana Guyana lies just north of the Equator. About the size of the state of South Dakota, Guyana's high inland plateau is covered by thick rain forests. Closer to the coast the land gradually descends to a low flatland. The climate of Guyana is largely tropical rain forest.

When the British arrived in Guyana in the early 1800s, they forced the Native Americans and, later, Africans to work on sugarcane plantations. Asians came and worked voluntarily. Guyana finally won its independence in 1966. Today about one-half of Guyana's population are of Asian ancestry. Another one-third are of African ancestry. The rest of Guyana's people are Native Americans and Europeans. Most live along the Caribbean coast where Georgetown, the capital, is the major city.

Critical Thinking Activity

Drawing Conclusions Have students review the pros and cons of Venezuela's petroleum industry. Suggest that they consider the following questions: How has the country benefited from its status as a world leader in oil production? What are the risks of depending on income from oil? Call on volunteers to present and defend their opinions. **L2**

Guyana's past still influences its economy. The map on page 204 shows you that one of Guyana's major products is sugar. Other people earn their living mining gold and bauxite, or working in urban service industries.

Suriname Suriname ranks as the smallest independent country in South America in both area and population. Native Americans were the first settlers. In the late 1600s the region came under Dutch rule. The Dutch enslaved African people—and later brought in Asians—to work on their large plantations. In 1975 Suriname won its freedom from the Dutch. Most of Suriname's people live along the coast. Paramaribo (PAR•uh•MAR•uh•BOH) is the capital and chief port.

Suriname's economy depends on agriculture and mining. Rice is the country's largest crop, followed by bananas, coconuts, and sugar. Suriname also mines bauxite and gold. The ring of buzz saws throughout the thick rain forests signals a thriving lumber industry.

French Guiana To the east lies French Guiana with a low coastal plain and inland rain forests. The climate inland is hot and humid. Cooling ocean winds however, keep coastal areas comfortable at about 80°F (27°C).

The French settled this area during the early 1600s. They, too, founded a colony of sugarcane plantations worked by enslaved Africans. The colony also served as a place for French prisoners from the 1790s to the 1940s.

Today French Guiana is considered part of France. The French government builds hospitals and schools in the capital, Cayenne (ky•EHN), and other cities. It also provides government jobs for many of French Guiana's people. Most are of African or mixed African and European ancestry.

Teen Scene

Catch the Wind

This is Luis Torres's favorite way to spend a Saturday. "I love living in Caracas," Luis says. "I'm only one subway ride and a short bus ride from the beach. The weather is always warm here, so I can windsurf almost year-round. I like to go with my friend Emilio. We bring enough money to buy lunch at the food stand on the beach. I usually eat an *arepa*—a type of corn-meal bread with meat or cheese in the middle."

SECTION 2 REVIEW

REVIEWING TERMS AND FACTS

1. **Define the following:** *llanos,* hydroelectric power, altitude, *caudillo.*
2. **LOCATION** Why are Brazil's northern neighbors known as Caribbean South America?
3. **HUMAN/ENVIRONMENT INTERACTION** What resource has changed Venezuela's economy?
4. **PLACE** What European countries have influenced the Guianas?

MAP STUDY ACTIVITIES
5. Look at the land use map on page 204. What are the major resources of Caribbean South America?

Reteach

Have students draw up a chart comparing the countries of Caribbean South America. Suggest they use bases of comparison such as culture, economic activities, landscapes, and so on. **L1**

Enrich

Three of the world's highest waterfalls are in Caribbean South America— Angel Falls (Venezuela), Cuquenan (Venezuela), and King George VI (Guyana). Have students draw annotated diagrams of these waterfalls. **L3**

CURRICULUM CONNECTION

Life Science Guyana's wildlife includes *hoatzins*—brightly colored birds that live in marshy areas. At birth, *hoatzins* have claws on their wings. They use the claws to climb on tree branches until they learn to fly. As the birds mature, the claws fall off.

CLOSE

Challenge students to list a person, a city, a country, a crop, natural resource, or an activity from the section for each letter in the term *Caribbean South America.*

Answers to Section 2 Review

1. All vocabulary words are defined in the Glossary.
2. because these countries border the Caribbean Sea
3. petroleum
4. The British ruled Guyana, the Dutch ruled Suriname, and the French still control French Guiana.
5. petroleum, natural gas, forests, coal, iron ore, bauxite, manganese, gold, diamonds

MAKING CONNECTIONS

TEACH

Ask students to compare the landscape shown in the photograph to the terrain in their area, giving differences and similarities. Have them locate northern Mexico and the southwest United States on a map. Point out that the area is part of the North American land region known as the Western Plateaus, Basin, and Ranges. **L1**

More About . . .

Geography Lupe's route went through the Sierra Madre Occidental, one of Mexico's most rugged regions. The beautiful view which so moved Lupe took in the western edge of the Plateau of Mexico. Here, fast-running streams have cut canyons deeper than a mile through the slopes. The Barranca del Cobre is so rugged and wild that parts of it are still unexplored.

NATIONAL GEOGRAPHIC SOCIETY

These materials are available from Glencoe.

VIDEODISC

STV: Biodiversity

Jaguar, South American forest
Any side, Frame 46142

RAIN OF GOLD   BY VICTOR VILLASEÑOR

| MATH | SCIENCE | HISTORY | LITERATURE | TECHNOLOGY |

Latin American author Victor Villaseñor's grandparents grew up in Mexico during the early 1900s. He wrote his family history in a novel called *Rain of Gold*. In this passage, Villaseñor tells about the girlhood journey of his grandmother, Lupe, through the countryside of northern Mexico.

Copper Canyon

Lupe came to the short new pines that had grown after the white pine forest had been burnt by the meteorite. The sound of the waterfalls was devastating. Carefully, Lupe continued through the young white pines and went out through the loose rock break on the north rim of the canyon.

. . . Suddenly, passing through the break, the whole world opened up. She was above the mountain peaks and flat mesas piled up for hundreds of miles in every direction. There was beauty to the left, beauty to the right. Beauty surrounded Lupe. This was the high country of northwest Mexico, unmapped and uncharted. Some fingers of the canyon of La Barranca del Cobre [Copper Canyon] ran deeper than the Grand Canyon of Arizona.

Following one of the little waterways, Lupe crossed the meadows of tightly woven wildflowers of blue and yellow and red and pink. It was quiet up here without the waterfall's terrible roar. Her brother's little dog flushed out deer and mountain quail as they went along.

Crossing a tiny creek, Lupe saw the fresh tracks in the cold mud of the terrible jaguar. The little dog's hair came up on his back. Lupe petted him, looking around carefully, but saw nothing. Jaguars, after all, were fairly common, and so people were more respectful than afraid of them, just as they were of any natural force.

Excerpt from *Rain of Gold* by Victor Villaseñor is reprinted with permission from the publisher (Houston: Arte Publico Press—University of Houston, 1991)

Making the Connection

1. Describe the countryside that Lupe sees on her journey.
2. The jaguar was an animal that early Native Americans prized highly. Why do you think that was so?

206

UNIT 3

Answers to Making the Connection

1. rugged land of mountains, waterfalls, and mesas; on the Plateau of Mexico there were meadows with wildflowers.

2. Answers will vary. Some students might suggest that the Native Americans prized it for its skin; others might note that since it was a natural force, the Native Americans might have looked on it as a god.

Uruguay and Paraguay

PREVIEW

Words to Know
- landlocked
- buffer state
- welfare state
- *gaucho*
- cassava

Places to Locate
- Uruguay
- Montevideo
- Río de la Plata
- Paraguay
- Gran Chaco
- Asunción

Read to Learn . . .
1. where Uruguay and Paraguay are located.
2. how the people of Uruguay and Paraguay earn a living.
3. what cultures have influenced Uruguay and Paraguay.

Imagine strolling through this elegant park in Montevideo, the capital of Uruguay. Through the trees, you see modern office buildings. Seagulls from the nearby beaches swoop by.

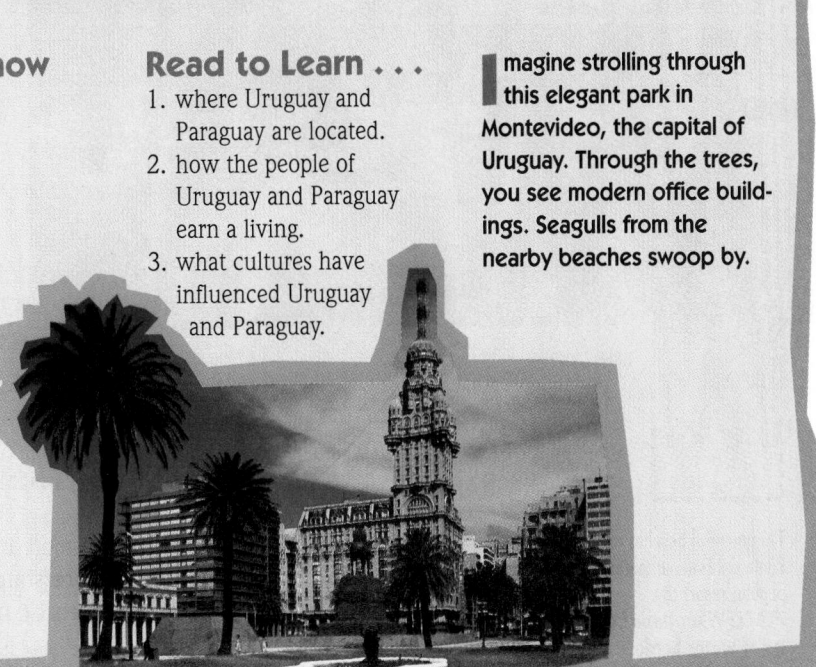

On Brazil's southern border lie two small countries—Uruguay (UR•uh•GWY) and Paraguay (PAR•uh•GWY). Although they both border Brazil, Uruguay and Paraguay are geographically very different. Uruguay has a coastline that links it to other parts of the world. Landlocked Paraguay has been more isolated. A **landlocked** country is one without a seacoast.

Place

Uruguay

Find Uruguay on the map on page 194. See how Brazil borders it on the north? The rest of Uruguay faces water: the Atlantic Ocean, the Río de la Plata, and the Uruguay River.

Uruguay's motto should be "Moderation in All Things." The country's landscape is undramatic, with low-lying interior grasslands and narrow, fertile coastal plains. Its climate is also generally moderate. Most of the country experiences a humid subtropical climate. Uruguay's grasslands, fertile soil, and mild climate help farmers to grow wheat and ranchers to raise livestock.

CHAPTER 8

(207)

Classroom Resources for Section 3

 **REPRODUCIBLE MASTERS**

Reproducible Lesson Plan 8-3
Guided Reading Activity 8-3
Reteaching Activity 8
Enrichment Activity 8
Section Quiz 8-3

 TRANSPARENCIES

Section Focus
    Transparency 8-3
Geoquiz Transparencies 8-1,
    8-2

 **MULTIMEDIA**

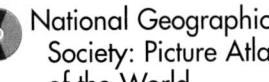 Testmaker

National Geographic
    Society: Picture Atlas
    of the World

FOCUS

Section Objectives
1. **Specify** the locations of Uruguay and Paraguay.
2. **Explain** how the people of Uruguay and Paraguay make a living.
3. **Identify** the cultures that have influenced Uruguay and Paraguay.

Bellringer Motivational Activity

Prior to taking roll at the beginning of the class period, project Section Focus Transparency 8-3 and have students answer the activity questions. Discuss students' responses. **L1**

Vocabulary Pre-check

Point out that one definition of the word *buffer* is "something that acts as a protective barrier." Have students use this information to speculate on the meaning of the term *buffer state*. **L1**
LEP

TEACH

Guided Practice

Making Connections
Have students locate Paraguay and Uruguay on a physical map of South America. Ask them to hypothesize about how the differences in location and place might influence economic activities in these two countries. Have students test their hypotheses based on information in this section. **L2**

Independent Practice

Guided Reading Activity 8-3 **L1**

Cultural Kaleidoscope

Uruguay Women in Uruguay received the right to vote in 1905. Women in the United States did not gain voting rights until 1920.

Punta del Este, Uruguay

The beautiful beaches of Punta del Este make it a popular resort city.
PLACE: What three bodies of water border southern Uruguay?

The Economy The map on page 204 shows you that Uruguay's major economic activities are raising sheep and cattle. The country's leading exports are primarily animal products—wool, beef, and hides. Uruguay also has factories that make tires, textiles, and leather goods. Industry, however, is limited because of Uruguay's lack of mineral resources.

Influences of the Past Uruguay won its independence from Spain in 1811. Moderation guides its politics. For most of its history, the country has been a buffer state between Brazil and Argentina. A **buffer state** is a small country located between larger, often hostile, neighbors. Uruguay maintains good relations with Argentina and Brazil.

Today the government of Uruguay spends more on education than it does on defense. It has turned Uruguay into a **welfare state**, a country that uses tax money to provide its people with help if they are sick, needy, or out of work.

The People Uruguay's 3.2 million people are in many ways more European than Latin American. Most of them are European in ancestry. They are descended from Spanish and Italian settlers who came to Uruguay during the 1800s and early 1900s. Only a small number of people have Native American or African ancestry. Spanish is the official language, and the Roman Catholic faith is the major religion. About 1.2 million Uruguayans live in the coastal city of Montevideo (MAHN•tuh•vuh•DAY•OH), the country's capital.

Most of Uruguay's people enjoy a comfortable standard of living. City dwellers hold jobs in government, business, and industry. They live in apartments or single-family houses. Unskilled workers, however, often live in tiny shacks edging the cities.

Only 11 percent of Uruguay's population live in rural areas. Many of them own or rent small farms. Others work as *gauchos*, or cowhands, on huge ranches. Legends about the *gauchos* have inspired Uruguay's folk music and literature.

If you lived in Uruguay you would eat a lot of meat, especially beef. You would enjoy sipping *yerba maté*, a tealike drink, through a straw from a gourd or the dried shell of a fruit. In your spare time, you and other Uruguayans might enjoy soccer and *gaucho* rodeos.

208

UNIT 3

Cooperative Learning Activity

Organize students into four groups and assign groups the following task: create a four-page illustrated report titled: "Uruguay and Paraguay—A Comparison." Suggest that students focus on topics such as physical geography, economy, people, and history. Make certain that each member of a group has a specific task—research, artwork and design, caption writing, and so on. Have groups share their reports with the rest of the class. **L2**

Place
Paraguay

Landlocked Paraguay is located near the center of South America. Two contrasting geographic areas make up Paraguay. The Paraguay River divides Paraguay into east and west regions. Rolling hills, forests, streams, rivers, and a humid subtropical climate make the eastern region a pleasant place to live. Most of Paraguay's people live here.

West of the Paraguay River is a plains region called the Gran Chaco. It covers three-fifths of Paraguay but holds less than 5 percent of its people. Winter droughts and summer floods often affect this area. Grasses, palm trees, and quebracho (kay•BRAH•choh) trees thrive in a tropical savanna climate. Quebracho trees are a source of tannin, a chemical used to process leather.

The Economy Forestry and farming are the major economic activities in Paraguay. Large cattle ranches cover much of the country. Most farmers, however, own or settle on small, fertile plots of land and grow corn, cotton, or cassava.

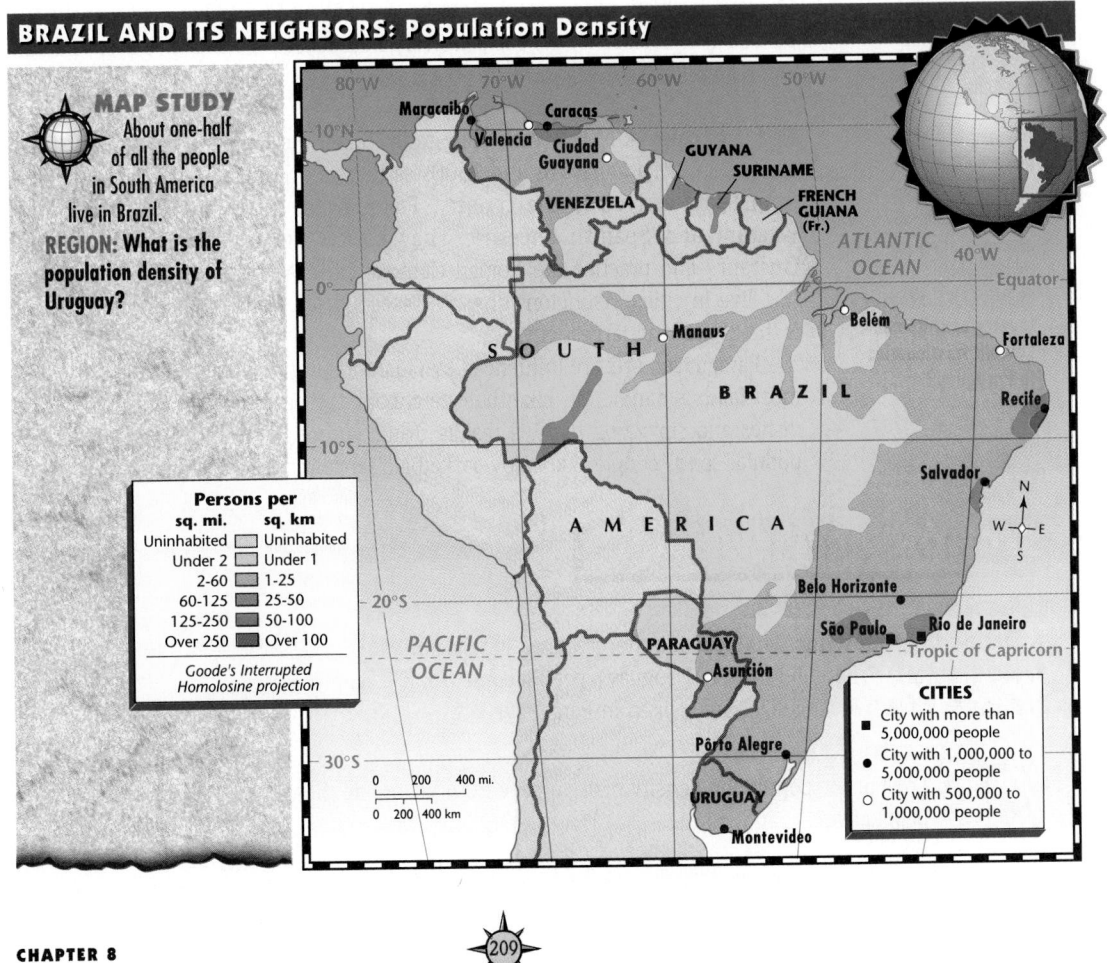

BRAZIL AND ITS NEIGHBORS: Population Density

MAP STUDY
About one-half of all the people in South America live in Brazil.

REGION: What is the population density of Uruguay?

Persons per	
sq. mi.	sq. km
Uninhabited	Uninhabited
Under 2	Under 1
2-60	1-25
60-125	25-50
125-250	50-100
Over 250	Over 100

Goode's Interrupted Homolosine projection

0   200   400 mi.
0 200 400 km

CITIES
■ City with more than 5,000,000 people
● City with 1,000,000 to 5,000,000 people
○ City with 500,000 to 1,000,000 people

ASSESS

Check for Understanding

Assign Section 3 Review as homework or an in-class activity.

Meeting Lesson Objectives

Each objective below is tested by the questions that follow it in parentheses.

1. **Specify** the locations of Uruguay and Paraguay. (1)
2. **Explain** how the people of Uruguay and Paraguay earn a living. (1)
3. **Identify** the cultures that have influenced Uruguay and Paraguay. (2, 3)

Global  Gourmet

Paraguay The *empanada*, a deep-fried dough pocket stuffed with ground meat or vegetables, is a staple of the Paraguayan diet.

MAP STUDY

Answer
2–60 persons per sq. mi. (1–25 persons per sq. km)

Map Skills Practice
Reading a Map Which city is larger, Asunción or Montevideo? *(Montevideo)*

Meeting Special Needs Activity

Study Strategy Model the "Question" component of study strategies for students who are not fluent in asking questions. Read aloud the first sentence of the first paragraph under the heading "The People" on page 210. Then ask, "How is the Guaraní heritage reflected in the Paraguayan population today?" Have students form questions by using the first sentence of the next paragraph under the heading "The People." **L1**

209

Paraguay—Past and Present

More About the Illustration

Answer to Caption
Spain

GEOGRAPHIC THEMES

Place The *gauchos* of Paraguay work on huge ranches called *estancias.* Many wear the traditional baggy trousers called *bombachas* and usually ride barefoot.

The ruins of a mission mark the Roman Catholic influence on Paraguay *(left).* A modern-day gaucho drinks his *yerba maté (right).*
MOVEMENT: What country sent Roman Catholic missionaries to Paraguay?

Cassava is a plant with roots that can be ground up and eaten or used to make pudding such as tapioca. These roots are very nutritious. You may have eaten pudding made from Paraguay's cassava.

The People A Native American group called the Guaraní were the first people to live in Paraguay. By the 1500s settlers and Roman Catholic missionaries from Spain arrived in the country. Paraguayans today are mostly of mixed Guaraní and Spanish ancestry. They speak two languages—Spanish and Guaraní—and practice the Roman Catholic faith. About one-half of the population live in cities. Asunción (uh•SOONT•see•OHN), with about 600,000 people, is the capital and largest city.

Paraguayan arts are influenced by Guaraní culture. Guaraní lace is Paraguay's most famous handicraft. Like their neighbors, the people of Paraguay enjoy meat dishes and sip *yerba maté,* a tealike drink. Although soccer is Paraguay's most popular sport, people also enjoy basketball, horse racing, and swimming.

Evaluate

 Section 3 Quiz **L1**

Use the Testmaker to create a customized quiz for Section 3. **L1**

 Chapter 8 Test Form A and/or Form B **L1**

Reteach

 Reteaching Activity 8 **L1**

Enrich

Enrichment Activity 8 **L1**

SECTION 3 REVIEW

REVIEWING TERMS AND FACTS

1. Define the following: landlocked, buffer state, welfare state, *gaucho,* cassava.

2. MOVEMENT Where did most of Uruguay's people come from?

3. PLACE What two languages are spoken in Paraguay?

 MAP STUDY ACTIVITIES

4. Look at the population density map on page 209. Where do most of Paraguay's people live?

CLOSE

 **Mental Mapping Activity**

Provide students with a large sheet of white paper and map pencils. Inform students that this will not be a graded activity. Have students draw, freehand from memory, a map of South America that includes the following labeled countries: Brazil, Venezuela, Guyana, Suriname, French Guiana, Uruguay, and Paraguay.

(210) **UNIT 3**

Answers to Section 3 Review

1. All vocabulary words are defined in the Glossary.

2. Europe, mainly Spain and Italy

3. Spanish and Guaraní

4. in the eastern half of the country

Important Things to Know About Brazil and Its Neighbors

SECTION 1 BRAZIL

- Brazil is South America's largest country in size and population.
- The Amazon River basin holds the world's largest rain forest.
- Rapid industrial growth during the 1900s has made Brazil the leading manufacturing nation in South America.
- Many people fear that economic development in the Amazon rain forest threatens the forest, its wildlife, and the Native Americans living there.
- The people of Brazil are of European, African, Native American, or mixed backgrounds.

SECTION 2 CARIBBEAN SOUTH AMERICA

- Caribbean South America includes Venezuela, Guyana, Suriname, and French Guiana.
- Oil has made Venezuela one of the wealthiest countries in South America.
- The people of the Guianas trace their heritages to Asian, African, European, and early Native American cultures.

SECTION 3 URUGUAY AND PARAGUAY

- Most of Uruguay's people are of European descent. They live along the coast.
- Paraguay is a landlocked country. Its people speak two languages: Spanish and Guaraní.
- Uruguay's economy depends on livestock raising. Paraguay relies on forestry and farming.

Carnival in Rio de Janeiro, Brazil ▶

CHAPTER HIGHLIGHTS

Using the Chapter 8 Highlights

- Use the Chapter 8 Highlights to preview, review, condense, or reteach the chapter.

Preview/Review

Vocabulary Puzzle-Maker Software reinforces the terms used in Chapter 8.

Condense

Have students read the Chapter 8 Highlights. Spanish Chapter Highlights also are available.

Chapter 8 Digest Audiocassettes and Activity

Chapter 8 Digest Audiocassette Test

Spanish Chapter Digest Audiocassettes, Activities, and Tests also are available.

Guided Reading Activities

Reteach

Reteaching Activity 8. Spanish Reteaching activities are also available.

MAP STUDY

Human/ Environment Interaction

Copy and distribute page 25 of the Outline Map Resource Book. Have students label the countries discussed in this chapter. Then have them locate and mark with symbols and/or colors the main resources and forms of economic activity in these countries. Remind students to include a key.

Extra Credit Project

Picture Postcards Have students choose one of the countries studied in this chapter and imagine that they are visiting there. Ask them to create a picture postcard to send to a friend in the United States. The front of the postcard should depict a landscape, a city, a place of interest, or a local culture. The back of the postcard should have a brief explanation of the picture and a message from the student about his or her visit to the country. **L1**

Chapter 8 Review and Activities

CHAPTER 8 REVIEW & ACTIVITIES

GLENCOE TECHNOLOGY

VIDEODISC

Use the Chapter 8 MindJogger Videoquiz to review students' knowledge before administering the Chapter 8 Test.

MindJogger Videoquiz

Chapter 8
Disc 1 Side B

 Also available in VHS.

ANSWERS

Reviewing Key Terms

1. I
2. F
3. A
4. J
5. G
6. C
7. H
8. B
9. D
10. E

Mental Mapping Activity

This exercise helps students to visualize the countries and geographic features that they have been studying and to understand the relationships among these various points. All attempts at freehand mapping should be accepted.

REVIEWING KEY TERMS

Match the numbered terms in Column A with their definitions in Column B.

A
1. sisal
2. *favela*
3. landlocked
4. *yerba maté*
5. *gaucho*
6. *selva*
7. basin
8. *llanos*
9. buffer state
10. *caudillo*

B
A. having no seacoast
B. plains area in Caribbean South America
C. Amazon rain forest
D. small country between two larger, hostile countries
E. military leader
F. slum area in Brazil
G. South American cowhand
H. lowland area surrounded by higher land
I. plant fiber used to make rope
J. tealike drink

Mental Mapping Activity

On a separate sheet of paper, draw a freehand map of Brazil, Caribbean South America, Uruguay, and Paraguay. Label the following items on your map:

- Caribbean Sea
- Atlantic Ocean
- Amazon River
- Brasília
- Caracas
- Asunción
- Montevideo

REVIEWING THE MAIN IDEAS

Section 1

1. **PLACE** About how much of Brazil is highlands?
2. **REGION** What kind of economy does Brazil have?
3. **HUMAN/ENVIRONMENT INTERACTION** How has the Brazilian government tried to develop the country's inland areas?

Section 2

4. **PLACE** What large lake is found in Venezuela?
5. **REGION** Where do most of Venezuela's people live?
6. **HUMAN/ENVIRONMENT INTERACTION** What two resources are mined in the Guianas?

Section 3

7. **LOCATION** What body of water lies directly south of Uruguay?
8. **REGION** What two landscapes are found in Paraguay?
9. **PLACE** In what major way do Paraguay and Uruguay differ in their geography?

Reviewing the Main Ideas

Section 1

1. more than one-half of the country
2. a mixed economy of farming and industry
3. It built a new capital, Brasília, and roads to encourage people to move to the interior.

Section 2

4. Lake Maracaibo
5. in cities
6. gold, bauxite

CRITICAL THINKING ACTIVITIES

1. **Analyzing Information** Why are some people concerned about Brazil's rain forests?
2. **Drawing Conclusions** Find the Paraguay River on the map on page 196. Why do you think the Paraguay River is important to the people of landlocked Paraguay?

GeoJournal Writing Activity

Imagine what it would be like to be a rural Brazilian moving to the city of São Paulo. Write a letter home about your experiences in the city.

COOPERATIVE LEARNING ACTIVITY

Working in groups of three, learn more about daily life in Brazil, Caribbean South America, Uruguay, and Paraguay. Each group will choose one country, then each person will select one of the following topics to research: (a) holidays and festivals; (b) the arts; or (c) sports and leisure activities. Share your information with the rest of your group. As a group, prepare a travel brochure to encourage tourists to visit your country.

PLACE LOCATION ACTIVITY: BRAZIL AND ITS NEIGHBORS

Match the letters on the map with the places of Brazil and its neighbors. Write your answers on a separate sheet of paper.

1. Río de la Plata
2. Caracas
3. Suriname
4. Brasília
5. Uruguay
6. Rio de Janeiro
7. Orinoco River
8. Brazilian Highlands
9. Amazon River

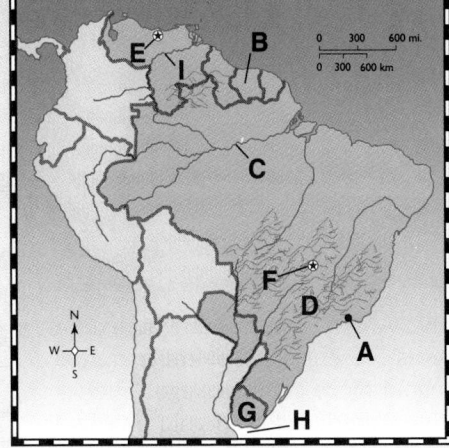

Cooperative Learning Activity

Obtain a number of travel brochures from a travel agency. Have groups study them to gain ideas on how they might present their findings.

GeoJournal Writing Activity

Encourage students to make a comparison of city life and rural life. Select students to read their letters to the class.

Place Location Activity

1. H
2. E
3. B
4. F
5. G
6. A
7. I
8. D
9. C

Chapter Bonus Test Question

This question may be used for extra credit on the chapter test.

When I was young and living in Brazil I knew my family was going to move because my dad was with a top government agency. What caused this move? *(The capital of Brazil changed from Rio de Janeiro to Brasília.)*

Section 3

7. Río de la Plata
8. the Gran Chaco plains area in the west and an area of rolling hills in the east
9. Paraguay is landlocked, while Uruguay has an extensive coastline.

Critical Thinking Activities

1. They are concerned that too many trees have been cut down to clear the land for farming. The destruction of so many trees endangers the wildlife and the Native American peoples that occupy the rain forest. Indeed, tree clearance puts the continued survival of the rain forest itself in question.
2. Because it provides a transportation route to the Río de la Plata and the Atlantic Ocean.

Cultural
HERITAGE:
LATIN AMERICA

FOCUS

Before students study this feature, ask them what image they would choose to illustrate Latin American culture. Call on volunteers to explain their choices. Then have students read the feature.

TEACH

Review with students the various entries in the feature. Call on volunteers to describe in their own words what is shown in each illustration. Then ask students what, based on the feature entries, appear to be the major influences on Latin American culture. *(Most will note that entries reflect Native American and/or Spanish influences.)* **L1**

CURRICULUM CONNECTION

Music The music of the panpipes is most closely related to the Andean region of Latin America. Today, a group that performs traditional music of the Andes typically plays instruments such as the panpipes; various flutes and recorders; the *bombo,* a kind of kettle drum; the *charango,* a 10-stringed guitar; and the harp.

Cultural HERITAGE: LATIN AMERICA

FABRIC ART ▶ ▶ ▶ ▶

This colorful cloth animal is a *mola,* a type of folk art made by the Cuna Indians of Panama. Designs often include abstract patterns and images of native plants and animals.

◀ ◀ ◀ ◀ MUSIC

Native American and mestizo musicians play many instruments, including the panpipes pictured here. These are similar to instruments used by their ancestors long ago.

METALWORKING ▶ ▶ ▶ ▶

Native American artisans who lived in what is present-day Colombia crafted this gold piece centuries ago. The Native Americans used gold objects for decoration and religious purposes. Today museums house artifacts such as these.

214

UNIT 3

More About . . .

Metalworking Latin America is rich in metal resources. Copper, gold, silver, and tin, for example, are found in abundance. For centuries, the people of Latin America have worked these metals into a variety of practical, ornamental, and religious objects. The metalworkers shaped the metal by various means, including casting, filigreeing, embossing, and repoussé. Perhaps the most often used technique is also the simplest and the oldest: hammering.

◀ ◀ ◀ ◀ PAINTING

Diego Rivera was a Mexican artist famous for bold murals with large, simplified figures. His works illustrate the culture and history of Mexico. Some of his murals are in the National Palace in Mexico City.

DANCE ▶ ▶ ▶ ▶

Many Latin American countries sponsor national dance companies that preserve traditional dances. Dancers such as this with the Ballet Folklórico of Mexico perform colorful productions of Native American and Spanish traditional dances.

ARCHITECTURE ▲ ▲ ▲ ▲

In the late 1600s Europeans built many cathedrals in the baroque style in Latin America. The baroque style features carved columns, ornamental sculptures, colored tile, gold, and silver.

CHAPTER 8 (215)

APPRECIATING CULTURE

1. Which type of art shown here appeals most to you? Why?

2. Latin America has a rich and varied cultural history. In what ways have Latin Americans tried to preserve this heritage?

Answers to Appreciating Culture

1. Answers will vary. Encourage students to make a list of pros and cons for each type of art before they answer the question.

2. Answers will vary. Students may mention such ways as housing artifacts in museums, establishing national dance companies that preserve traditional dance styles, adopting instruments and approaches used by their ancestors, and telling the story of their past through art.

PERFORMANCE ASSESSMENT ACTIVITY

A SUMMER COLLEGE COURSE Many colleges offer immersion courses in which students may travel for an extended time to learn not only the language but the culture of a foreign land. Have students take the roles of college students who have been offered a travel-course opportunity to any one of the Andean countries. Students should decide which country they would like to visit and submit proposals or persuasive letters in which they explain why they should be selected for the program. Proposals and letters should reflect what students know and appreciate about the geography of the country and also include an outline of the topics they would like to learn more about. Students will need to use chapter information in order to decide where they would like to study. They may also choose to research further about the country of their choice before submitting their proposals or letters.

POSSIBLE RUBRIC FEATURES:
Decision-making skills, accuracy of content information, persuasive writing skills, research skills

MENTAL MAPPING ACTIVITY
Before teaching Chapter 9, point out the Andes mountain range on a map of South America. Have students identify the countries that share this landform.

TEACHER'S CORNER

NATIONAL GEOGRAPHIC SOCIETY

INDEX TO NATIONAL GEOGRAPHIC MAGAZINE

The following articles may be used for research relating to this chapter:

- "Peru's Ice Maidens," by Johan Reinhard, June 1996.
- "Peru Begins Again," by John McCarry, May 1996.
- "Leafcutter Ants," by Mark W. Moffett, July 1995.
- "Poison-Dart Frogs," by Mark W. Moffett, May 1995.
- "Chincorro Mummies," by Bernardo Arriaza, March 1995.

- "The Amazon," by Jere Van Dyk, February 1995.
- "Remote World of the Harpy Eagle," by Neil Retting, February 1995.
- "Buenos Aires," by John J. Putnam, December 1994.
- "Chile's Uncharted Cordillera Sarmiento," by Jack Miller, April 1994.
- "Simón Bolívar," by Bryan Hodgson, March 1994.

NATIONAL GEOGRAPHIC SOCIETY PRODUCTS AVAILABLE FROM GLENCOE

To order the following products for use with this chapter, contact your local Glencoe sales representative or call Glencoe at 1-800-334-7344:

- *STV: World Geography* (Videodisc)
- *STV: Rain Forest* (Videodisc)
- *Picture Atlas of the World* (CD-ROM)
- *Physical Geography of the World* (Transparencies)

- *ZipZapMap! World* (Software)
- *GeoBee* (Software)
- *Images of the World* (Posters)
- *Eye on the Environment* (Posters)

ADDITIONAL NATIONAL GEOGRAPHIC SOCIETY PRODUCTS

To order the following products for use with this chapter, call National Geographic Society at 1-800-368-2728:

- *Amazon: Land of the Flooded Forest* (Video)
- *Rain Forest* (Video)
- *Physical Geography of the Continents Series:* "South America." (Video)

- *Portraits of the Continents Series:* "Part I: North America; South America." (Filmstrip)

TEACHER-TO-TEACHER
Curriculum Connection: Language Arts, Social Studies

—from Rodney Zeisig
Boston Middle School, LaPorte, IN

ANDEAN COLLAGE The purpose of this activity is to increase students' awareness and understanding of the Andean region of South America by creating a regional collage.

Organize the class into three groups and tell them that their task is to create a three-panel collage on the Andean region of South America. Assign one of the following topics to each group: the region's physical geography, the region's cultural geography, regional issues in the news. Have groups search through books, newspapers, and magazines and watch television news programs to find suitable materials for the collage. Remind students that they should clip articles and pictures only from old newspapers and magazines that are no longer in use. Materials from books and recent periodicals—and from television programs—should be summarized and accompanied by maps, sketches, or diagrams. After they have completed the research, have each group select 10 to 12 items to display in the collage. Then have groups construct the collage. Direct each group to develop a title for its panel in the display.

As a follow-up activity, encourage students to select the most distinctive physical and cultural features of the region and the most pressing issue the region faces at this time. Call on volunteers to explain their selections.

BIBLIOGRAPHY

Readings for the Student

Portraits of the Nations: The Land and the People Series. Argentina; Bolivia. Philadelphia: Lippincott, 1992.

A Survey of World Cultures: Latin America. Culver City, Ca.: Media Materials/SSSS, 1991.

Readings for the Teacher

The Cambridge Encyclopedia of Latin America and the Caribbean. New York: Cambridge University Press, 1992.

Latin America Today: An Atlas of Reproducible Pages, revised ed. Wellesley, Mass.: World Eagle, 1992.

Multimedia

Festivals and Holidays in Latin America. Video Knowledge, 1988–1989. Videocassette, 50 minutes.

Global Studies with Country Databases. Program 2: Latin America. Worldview Software, 1993. Apple 5.25 in. disk, IBM 3.5 in. disk, guide.

KEY TO ABILITY LEVELS

Teaching strategies have been coded for varying learning styles and abilities.

L1 **BASIC** activities for all students

L2 **AVERAGE** activities for average to above-average students

L3 **CHALLENGING** activities for above-average students

LEP **LIMITED ENGLISH PROFICIENCY** activities

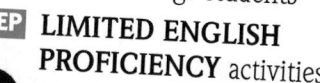

MEETING NATIONAL STANDARDS

Geography for Life

All of the 18 standards are demonstrated in Unit 3. The following ones are highlighted in this chapter:
• Standards 5, 6, 10, 15, 17

Chapter 9

Resource Organizer

The Andean Countries

Use these *Geography: The World and Its People* resources to teach, reinforce, and extend *chapter* content.

CHAPTER 9 RESOURCES

- *Vocabulary Activity 9
- Cooperative Learning Activity 9
- Workbook Activity 9
- Geography Skills Activity 9
- GeoLab Activity 9
- Critical Thinking Skills Activity 9
- Performance Assessment Activity 9
- *Reteaching Activity 9
- Enrichment Activity 9
- Environmental Case Study 3
- Chapter 9 Test, Form A and Form B
- Chapter Map Activity 9
- *Chapter 9 Digest Audiocassette, Activity, Test
- Unit Overlay Transparencies 3-0, 3-1, 3-2, 3-3, 3-4, 3-5, 3-6
- Political Map Transparency 3; World Cultures Transparency 3
- Geoquiz Transparencies 9-1, 9-2
- Vocabulary PuzzleMaker
- Student Self-Test: A Software Review
- Testmaker Software
- MindJogger Videoquiz

If time does not permit teaching the entire chapter, summarize using the Chapter 9 Highlights on page 233, and the Chapter 9 English (or Spanish) Audiocassettes. Review students' knowledge using the Glencoe MindJogger Videoquiz. *Also available in Spanish

Use these *Geography: The World and Its People* resources to teach and reinforce section content.

SECTION 1 RESOURCES

- Reproducible Lesson Plan 9-1
- Section Focus Transparency 9-1
- Guided Reading Activity 9-1
- Section Quiz 9-1 (also available in Spanish)

SECTION 2 RESOURCES

- Reproducible Lesson Plan 9-2
- Section Focus Transparency 9-2
- Guided Reading Activity 9-2
- Section Quiz 9-2 (also available in Spanish)

SECTION 3 RESOURCES

- Reproducible Lesson Plan 9-3
- Section Focus Transparency 9-3
- Guided Reading Activity 9-3
- Section Quiz 9-3 (also available in Spanish)

SECTION 4 RESOURCES

- Reproducible Lesson Plan 9-4
- Section Focus Transparency 9-4
- Guided Reading Activity 9-4
- Section Quiz 9-4 (also available in Spanish)

ADDITIONAL RESOURCES FROM GLENCOE

Reproducible Masters

- Facts On File: MAPS ON FILE I, South America; GEOGRAPHY ON FILE, Latin America
- Foods Around the World, Unit 3

World Music: Cultural Traditions

Lesson 2

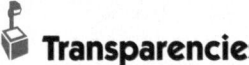

Transparencies

- National Geographic Society PicturePack Transparencies, Unit 3

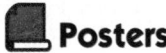

Posters

- National Geographic Society: Eye on the Environment, Unit 3

Countries of the World Flashcards

Unit 3

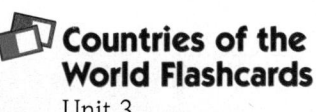

Videodiscs

- Geography and the Environment: The Infinite Voyage
- Reuters Issues in Geography
- National Geographic Society: STV: World Geography, Vol. 3
- National Geographic Society: GTV: Planetary Manager

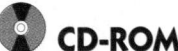

CD-ROM

- National Geographic Society: Picture Atlas of the World
- National Geographic Society PictureShow: Earth's Endangered Environments

Performance Assessment

Refer to the Planning Guide on page 216A for a Performance Assessment Activity for this chapter. See the *Performance Assessment Strategies and Activities* booklet for suggestions.

Chapter Objectives

1. **Examine** the major geographical features of Colombia.
2. **Compare** the geography of Peru to that of Ecuador.
3. **Contrast** Bolivia with Chile.
4. **Prepare** a geographic portrait of Argentina.

GLENCOE
TECHNOLOGY

 VIDEODISC

Use the Chapter 9 Mind-Jogger Videoquiz to preview chapter content.

MindJogger Videoquiz

Chapter 9
Disc 2 Side A

 Also available in VHS.

Chapter 9 The Andean Countries

MAP STUDY ACTIVITY

 In Chapter 9 you will learn about the countries that the Andes run through.

1. **What is the capital of Argentina?**
2. **What is unusual about Bolivia's capital?**

(216)

MAP STUDY ACTIVITY

Answers
1. Buenos Aires
2. There are two—La Paz and Sucre

Map Skills Practice
Reading a Map What three countries border Chile? (*Peru, Bolivia, Argentina*)

Colombia

PREVIEW

Words to Know
- cordillera
- *llanos*
- cash crop
- *campesino*

Places to Locate
- Colombia
- Andes
- Magdalena River
- Bogotá

Read to Learn . . .
1. where Colombia is located.
2. what products Colombia exports.
3. how Colombia became independent.

Towering peaks of the Andes stand guard over the city of Bogotá, the capital of Colombia. The 6 million people of Bogotá are nestled in a high basin of the Andes—the world's longest mountain range. It stretches more than 4,500 miles (7,240 km) along the western length of South America!

Colombia shares the majestic Andes with the countries of Ecuador, Peru, Bolivia, Chile, and Argentina. Together these six South American lands are known as the Andean countries.

Region
The Land

Colombia is the only country in South America that borders both the Caribbean Sea and the Pacific Ocean. Its land area totals 401,040 square miles (1,038,694 sq. km)—about three-fourths the size of Alaska.

Colombia's Landforms Narrow ribbons of coastal lowlands stretch along the Caribbean Sea and the Pacific Ocean. Bananas, cotton, and sugarcane are grown on large plantations along the Caribbean. Busy coastal ports handle Colombia's trade. In the west, thick forests spread over the Pacific lowlands. Few people live here.

The Andes spread outward through the center of Colombia. The Andes form a **cordillera**, or a group of mountain chains that run side by side. Colombia's

Meeting National Standards

Geography for Life The following standards are highlighted in this section:
- Standards 4, 11, 17

FOCUS

Section Objectives
1. **Locate** Colombia on a map or globe.
2. **List** the products that Colombia exports.
3. **Explain** how Colombia became independent.

Bellringer Motivational Activity

Prior to taking roll at the beginning of the class period, project Section Focus Transparency 9-1 and have students answer the activity questions. Discuss students' responses. **L1**

Vocabulary Pre-check

Have students check the meaning of the term *cordillera*. Then have them locate this feature on a map of Colombia. **L1 LEP**

Use the Vocabulary PuzzleMaker Software to create a crossword puzzle. **L1**

 **REPRODUCIBLE MASTERS**

Reproducible Lesson Plan 9-1
Guided Reading Activity 9-1
Vocabulary Activity 9
Workbook Activity 9
Section Quiz 9-1

 TRANSPARENCIES

Section Focus
  Transparency 9-1
Political Map Transparency 3

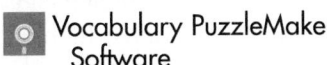 **MULTIMEDIA**

 Vocabulary PuzzleMaker Software

Testmaker

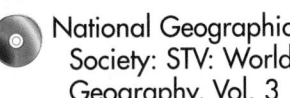 National Geographic Society: STV: World Geography, Vol. 3

Classroom Resources for Section 1

217

MAP STUDY

Answer
Argentina, Chile

Map Skills Practice
Reading a Map Do the Argentine Pampas lie east or west of the Andes?
(east)

TEACH

Guided Practice

Reading Maps Have students use the maps in Chapter 9 to write a brief geographic description of Colombia. Guide students' observations by asking questions such as: What physical features influence Colombia's climate? Where do most of Colombia's people live? What resources influence economic activities in Colombia? **L2**

Independent Practice

 Guided Reading Activity 9-1 **L1**

NATIONAL GEOGRAPHIC SOCIETY

These materials are available from Glencoe.

 VIDEODISC

STV: World Geography
Vol. 3: South America and Antarctica

South America
Side 1, Frames 00001-50061

218

THE ANDEAN COUNTRIES: Physical

MAP STUDY
The Andes extend along the entire west coast of South America.
PLACE: South America's highest peak is found near the border between what two countries?

major river, the Magdalena, winds between the central and eastern Andean chains to the Caribbean Sea. Most of Colombia's cities—including its capital, Bogotá (BOH•guh•TAH)—lie in the valleys of this mountainous area.

Vast plains cover the remaining 60 percent of Colombia. Tropical rain forests spread across the southeast plain into the Amazon River basin in neighboring Brazil. Colorful birds such as toucans, parrots, and parakeets screech from the trees of the tropical forests. Only a few Native American groups live in this hot, steamy region.

In the northeast you find hot grasslands called the *llanos.* Here on large estates ranchers drive their cattle across the rolling plains.

The Climate Look at the climate map on page 223. You can see that Colombia lies entirely within the tropics. Yet Colombia does not have a totally tropical climate.

UNIT 3

Cooperative Learning Activity

Organize students into three groups and assign each group one of the following regions of Colombia: lowlands, highlands, plains. Have members of each group study the way of life in their assigned region. Then instruct groups to prepare a presentation about their region to share with the class. Presentations may consist of skits, panel discussions, or graphics. **L2**

In the high altitudes of the Andes, temperatures are very cool for a tropical area. Bogotá, for example, lies at 8,355 feet (2,547 m) above sea level. High temperatures average only 67°F (20°C). Low temperatures average about 50°F (10°C).

Human/Environment Interaction
The Economy

Did you know that the average American eats 27.3 pounds (12.4 kg) of bananas a year? Colombia ranks as the second-largest Latin American supplier of this fruit. Most Colombians earn a living as farmers, factory workers, or miners.

Colombian factories produce automobiles, machinery, clothing, and food products. These industries are supported by the rich minerals found in the country, including coal, iron ore, petroleum, and natural gas. A leading supplier of gold, Colombia also produces 90 percent of the world's emeralds.

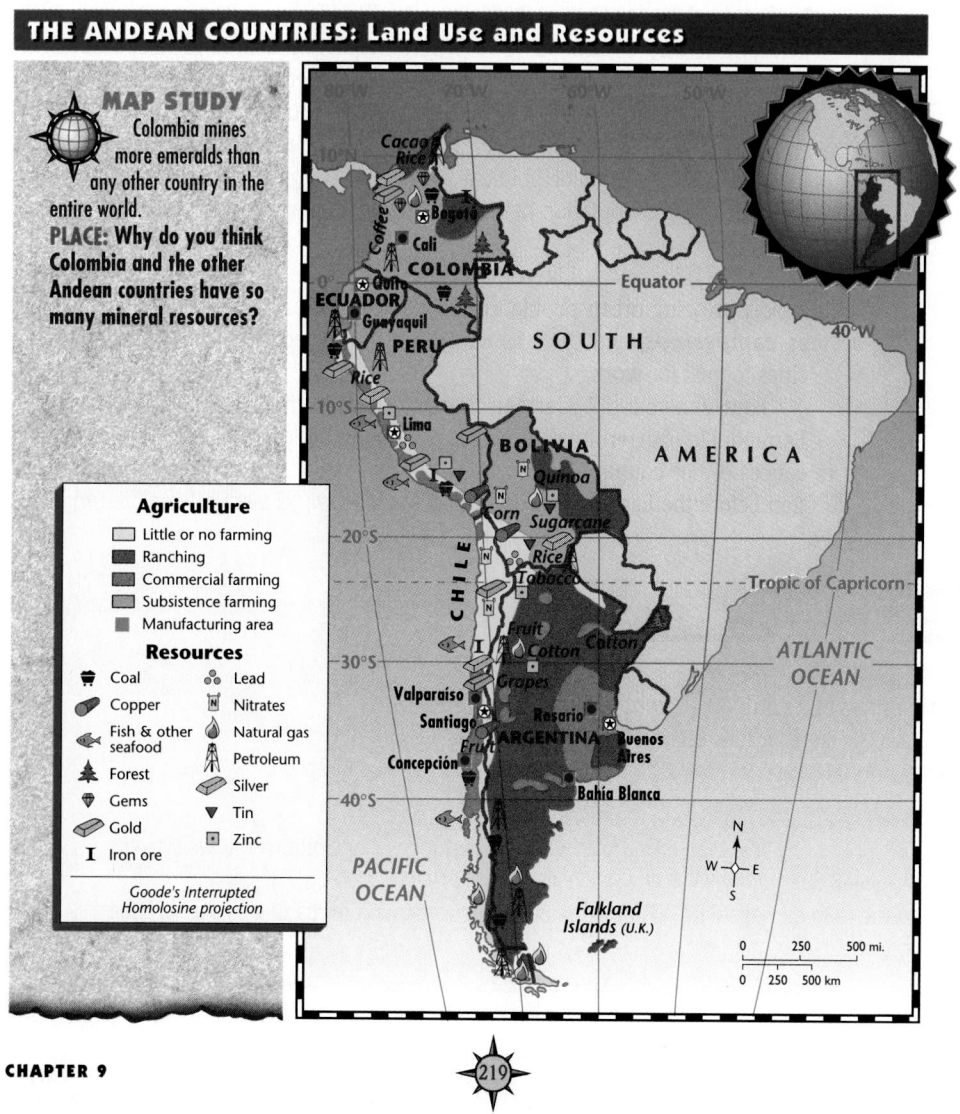

THE ANDEAN COUNTRIES: Land Use and Resources

MAP STUDY Colombia mines more emeralds than any other country in the entire world.
PLACE: Why do you think Colombia and the other Andean countries have so many mineral resources?

Agriculture
- Little or no farming
- Ranching
- Commercial farming
- Subsistence farming
- Manufacturing area

Resources
- Coal
- Copper
- Fish & other seafood
- Forest
- Gems
- Gold
- Iron ore
- Lead
- Nitrates
- Natural gas
- Petroleum
- Silver
- Tin
- Zinc

Goode's Interrupted Homolosine projection

Meeting Special Needs Activity

Study Strategy Model the OH-RATS (Overview, Headings, Read, Answer, Test, Study) strategy for reading. "Overview" consists of reading the title and subheads, looking at illustrations, reading the introduction and summary. "Headings" involves writing possible questions based on each heading. "Read" refers to reading the text to answer the questions. "Answer" requires writing answers to the questions. "Test-Study" provides a review of the material. Have students use the OH-RATS strategy for the material in Section 1. **L1**

Cultural Kaleidoscope

Colombia The legend of *El Dorado*—the kingdom of gold—probably grew out of a coronation ceremony conducted by the Chibcha people, who lived around what today is Bogotá. According to tradition, the Chibcha covered their new ruler with gold dust. The ruler then dived into a lake to wash off the gold as an offering to the gods.

ASSESS

Check for Understanding

Assign Section 1 Review as homework or an in-class activity.

Meeting Lesson Objectives

Each objective below is tested by the questions that follow it in parentheses.

1. **Locate** Colombia on a map or globe. (2)
2. **List** the products Colombia exports. (3)
3. **Explain** how Colombia became independent. (2)

MAP STUDY

Answer
because the mountains contain a wealth of minerals

Map Skills Practice
Reading a Map What is the main form of agriculture practiced in Argentina? *(ranching)*

219

Evaluate

 Section Quiz 9-1 **L1**

Use the Testmaker to create a customized quiz for Section 1. **L1**

Reteach

Have students use the headings in this section to outline the content. Then ask them to write a summary statement for each heading. **L1**

Enrich

The Chibcha people of Colombia were skilled goldsmiths. Have students research and report on the *Museo del Oro* (Gold Museum) in Bogotá, where many Chibcha artifacts are displayed. **L3**

CLOSE

Have students review the geographic description they wrote for the **Guided Practice** activity. Direct them to make adjustments based on their reading of the section.

Agriculture Differences in land elevation allow Colombians to grow a variety of crops. Coffee is the country's major **cash crop**—a crop that is usually sold for export. Colombian coffee—grown on large plantations—is known all over the world for its rich flavor. Colombia also exports cotton, tobacco, and cut flowers. Huge herds of cattle roam large ranches in the *llanos*. The rain forests of the eastern plains also supply a valuable resource—lumber.

Movement

The People

The population of Colombia totals about 36 million. Most Colombians live in the valleys of the Andes. Nearly all Colombians speak Spanish, but they are of mixed European, African, and Native American backgrounds. The Roman Catholic faith is Colombia's major religion.

Independence Colombia, Venezuela, Ecuador, and Panama became independent as one country in 1819. Símon Bolívar, whom you read about in Chapter 8, led this struggle for independence. Over time Colombia's neighbors broke away from Colombia and became separate countries.

During the 1800s and 1900s, political disagreements divided Colombia. In the 1950s the two major political parties finally agreed to govern the country together. Their cooperation has led to a stable democratic government.

Growth of Cities Like other South American countries, Colombia has a rapidly growing urban population. Since the 1940s Colombian farmers, known as ***campesinos,*** and their families have journeyed from their villages to the cities to look for work.

Fourteen-year-old Carmen Rivera lives in the Caribbean port city of Barranquilla (BAHR•ruhn•KEE•yuh). She enjoys performing many of the regional dances of her country. Her brother, Tomás, likes to join in the Carnival procession before the Easter season. In the parade, he wears a brightly colored mask.

SECTION 1 REVIEW

REVIEWING TERMS AND FACTS

1. **Define the following:** cordillera, *llanos,* cash crop, *campesino.*
2. **PLACE** Where is Colombia located?
3. **HUMAN/ENVIRONMENT INTERACTION** What is Colombia's main export?

 MAP STUDY ACTIVITIES

4. Look at the physical map on page 218. What major river of Colombia flows into the Caribbean Sea?
5. Turn to the map on page 219. What Andean countries depend on fishing as part of their economies?

Answers to Section 1 Review

1. All vocabulary words are defined in the Glossary.
2. northwestern South America between the Caribbean Sea and the Pacific Ocean
3. coffee
4. Magdalena River
5. Peru, Chile

MAKING CONNECTIONS

INCA ENGINEERING

MATH | SCIENCE | HISTORY | LITERATURE | TECHNOLOGY

They built thousands of miles of roads. They guaranteed one-day delivery for messages delivered up to 140 miles (225 km) away. They built a city in the sky and the first true suspension bridges. No, we're not talking about modern architects. The people who accomplished these feats—and more—lived during the 1400s and 1500s. They were the Inca. Surprisingly, these Native Americans built their structures without the aid of the wheel, wagons, or pulleys.

ROADS To control their huge empire that stretched about 2,500 miles (4,023 km) along the west coast of South America, the Inca needed roads. Eventually, they carved 14,260 miles (22,944 km) of roads throughout the Andes. To conquer the steep slopes, the Inca laid roads in a zigzag pattern. They also built stairways of long, shallow steps. On the steepest slopes, roads were just a series of stone steps thrust into the mountainside. A system of way stations provided stops for the Inca army or messengers to rest. Across deep gorges, the Inca built sturdy suspension bridges made of rope and wood.

CITY IN THE SKY Probably the most spectacular feat of ancient engineering in the

Western Hemisphere is the Inca city, Machu Picchu (MAH•choo PEE•choo), built high in

Machu Picchu, Peru

the Andes. You can see its ruins today if you visit the country of Peru. Jagged Andean cliffs hid and protected Machu Picchu, which rose 2,000 feet (about 610 m) above a narrow valley. Using only stone hammers and a polish made of wet sand, the Inca erected Machu Picchu's 143 buildings from granite blocks cut from the mountains. They transformed the steep slopes below the city into gardens by building terraces, or stepped ridges of land suitable for farming.

Making the Connection

1. How did the Inca overcome the rugged Andes in their road building?
2. What features make Machu Picchu a remarkable city?

221

Answers to Making the Connection

1. laid roads in a zigzag pattern; built stairways of long, shallow steps; on steepest hills, made roads as a series of stepping stones

2. Answers will vary, but many students will note that the Inca engineers shaped the stones for the buildings using only stone hammers and a wet-sand polish.

TEACH

Ask students to study the photo of Machu Picchu. Then ask them to suggest the problems that engineers would have to overcome to build such a city in such a location. Emphasize that the Inca overcame these problems with—by today's standards—very limited technology. **L1**

More About . . .

Engineering The stones that comprise Machu Picchu's walls and buildings were cut and polished to fit so well that no cement was needed. A joint between any two stones looks like a hairline, but it cannot be felt with the fingertips. To connect the various buildings and parts of the city, engineers carved stone stairways into the faces of the surrounding mountain slopes.

More About . . .

Communications Relay teams of messengers carried messages and packages back and forth along Inca roads. One team had a special task—carrying fresh fish from the Pacific to the emperor's palace high in the Andes. They completed the journey in less than a day!

FOCUS

Section Objectives

1. **Compare** landforms found in Peru and Ecuador.
2. **Identify** mineral resources mined in Peru and Ecuador.
3. **Discuss** how people live in Peru and Ecuador.

Bellringer Motivational Activity

Prior to taking roll at the beginning of the class period, project Section Focus Transparency 9-2 and have students answer the activity questions. Discuss students' responses. **L1**

Vocabulary Pre-check

Have students check definitions of the terms listed in **Words to Know.** Direct students to use each term correctly in a sentence. **L1** **LEP**

NATIONAL GEOGRAPHIC SOCIETY

These materials are available from Glencoe.

 **VIDEODISC**

GTV: Planetary Manager

Peru, desert ecosystem
Any side, Frame 47491

SECTION 2
Peru and Ecuador

Words to Know
• *altiplano*
• navigable
• empire

Places to Locate
• Peru
• Lima
• Lake Titicaca
• Ecuador
• Quito
• Guayaquil
• Galápagos Islands

Read to Learn . . .
1. what landforms are found in Peru and Ecuador.
2. what mineral resources are mined in Peru and Ecuador.
3. how people live in Peru and Ecuador.

in the Pacific Ocean about 600 miles (965 km) west of South America? Since 1832 the Galápagos Islands have belonged to Ecuador, one of the Andean countries.

What do you suppose this lizard is thinking as it suns itself on the rocky coast of the Galápagos Islands? Is it aware that its home rises up

Ecuador and its larger neighbor, Peru, lie in western South America along the Pacific Ocean. They are lands of enormous contrasts in landscape and climate, but they share a Native American and Spanish heritage.

Place
Peru

Peru is South America's third-largest country in area. Peru—a Native American word that means "land of abundance"—is rich in mineral resources.

The Land and Economy Dry deserts, snowtopped mountains, and hot, humid rain forests greet you in Peru. On a narrow coastal strip of plains and deserts lies Peru's many farms and cities. The cold Peru Current in the Pacific Ocean keeps coastal temperatures fairly mild.

The Andes, with their broad valleys and plateaus, sweep through the center of Peru. The southern part of Peru's Andean region contains a large plateau known as the *altiplano*. If you travel on the *altiplano*, you will see Lake Titicaca (TIH•tih•KAH•kuh), the highest navigable lake in the world. **Navigable** means that a body of water is wide and deep enough to allow the passage of ships.

222

Classroom Resources for Section 2

 **REPRODUCIBLE MASTERS**

Reproducible Lesson Plan 9-2
Guided Reading Activity 9-2
Cooperative Learning Activity 9
Section Quiz 9-2

 TRANSPARENCIES

Section Focus Transparency 9-2
Unit Overlay Transparencies 3-1, 3-5

MULTIMEDIA

 Testmaker

National Geographic Society: GTV: Planetary Manager

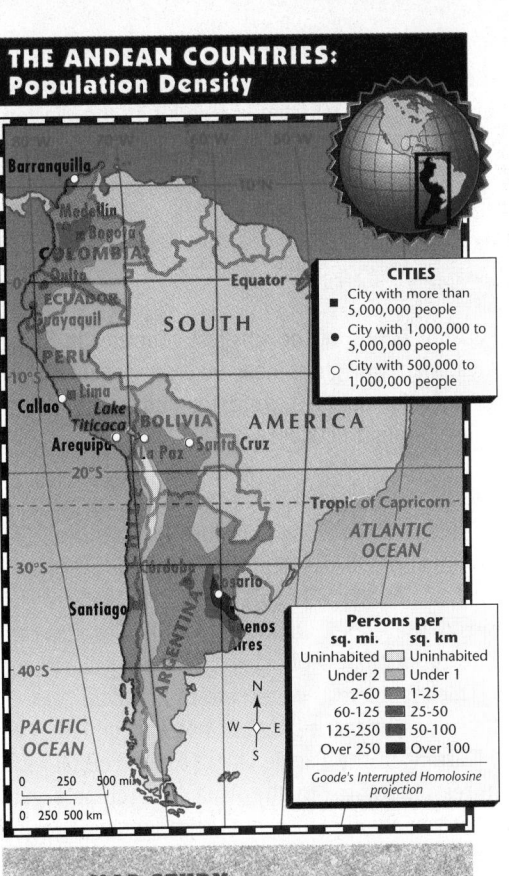

THE ANDEAN COUNTRIES: Population Density

CITIES
- City with more than 5,000,000 people
- City with 1,000,000 to 5,000,000 people
- City with 500,000 to 1,000,000 people

Persons per
sq. mi.	sq. km
Uninhabited	Uninhabited
Under 2	Under 1
2-60	1-25
60-125	25-50
125-250	50-100
Over 250	Over 100

Goode's Interrupted Homolosine projection

0   250   500 miles
0 250 500 km

MAP STUDY The most rugged highlands of the Andes are uninhabited.
HUMAN/ENVIRONMENT INTERACTION: What is the population density around Lake Titicaca?

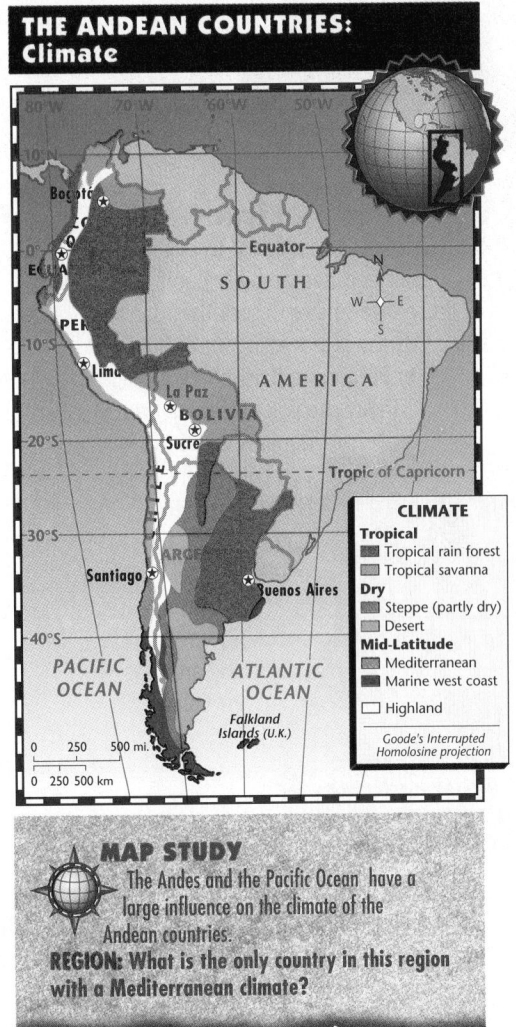

THE ANDEAN COUNTRIES: Climate

CLIMATE
Tropical
- Tropical rain forest
- Tropical savanna
Dry
- Steppe (partly dry)
- Desert
Mid-Latitude
- Mediterranean
- Marine west coast
- Highland

Goode's Interrupted Homolosine projection

0   250   500 mi.
0 250 500 km

MAP STUDY The Andes and the Pacific Ocean have a large influence on the climate of the Andean countries.
REGION: What is the only country in this region with a Mediterranean climate?

East of the Andes you run into low foothills and flat plains. Thick rain forests cover almost all of the plains area, where the mighty Amazon River begins. Rainfall is plentiful, and temperatures remain high throughout the year.

The map on page 219 shows you that farming is the major economic activity of Peru. Coffee, cotton, and sugarcane are Peru's major export crops. Peru also ranks among the world's leading producers of copper, lead, silver, and zinc. These minerals come from mines in the Andean highlands.

The People During the 1400s the Inca had a powerful civilization in the area that is now Peru. Their **empire,** or group of lands under one ruler, stretched more than 2,500 miles (4,023 km). In 1530 the Spaniards defeated the Inca and made Peru a Spanish territory. Peru gained independence from Spain in 1821.

CHAPTER 9

223

Cooperative Learning Activity

Have students work in small groups to create an illustrated dictionary of Latin American geographic terms. Review geographic terms from this chapter that refer to physical features in Latin America. *(altiplano, llanos, cordillera)* Then have students skim the unit and list other terms to include in their dictionary. Assign a specific task to each group member, such as alphabetizing, writing definitions, illustrating terms, photocopying, and collating pages. Make certain all group members are issued a copy of the dictionary. **L1**

MAP STUDY

Answers
(left map) 2–60 persons per sq. mi./1–25 persons per sq. km
(right map) Chile

Map Skills Practice
Reading Maps What type of climate is found in the city of Rosario? *(tropical rain forest)*

TEACH

Guided Practice

Project Unit Overlay Transparency 3-5. Direct students' attention to Peru and Ecuador. Have them speculate which areas of these countries might be densely populated. Then project Unit Overlay Transparency 3-1 and have students verify their speculations. **L1**

Independent Practice

Guided Reading Activity 9-2 **L1**

ASSESS

Check for Understanding

Assign Section 2 Review as homework or an in-class activity.

Meeting Lesson Objectives

Each objective below is tested by the questions that follow it in parentheses.

1. **Compare** landforms found in Peru and Ecuador. (2)
2. **Identify** mineral resources mined in Peru and Ecuador. (3)
3. **Discuss** how people live in Peru and Ecuador. (5)

Evaluate

 Section 2 Quiz **L1**

Use the Testmaker to create a customized quiz for Section 2. **L1**

Reteach

Organize students into two groups. Have members of one group outline the material about Peru. Have members of the other group outline the material about Ecuador. Allow time for students to share their outlines. **L1**

Enrich

Have students research and report on the animals of the Galápagos Islands. **L3**

CLOSE

Have students choose a destination in Peru or Ecuador that they would like to visit. Direct students to provide geographic information to explain their choices.

224

Teen Scene

Inca Celebration

The harvest festival is about to begin. Mitena Alcala and her family will join others in the village square to socialize, dance, and sing. Mitena is a descendant of the Inca who lived in Peru more than 700 years ago. Today she will observe a traditional Inca religious day. Mitena made her brightly colored shawl and hat from alpaca and llama wool. The colors and patterns in her handwoven clothing are those of her village.

Peru's 23 million people live mostly in cities and towns. Lima (LEE•muh), with a population of 6 million, is the capital and largest city. Peru has the largest Native American population in the Western Hemisphere. People of *mestizo* and European ancestry make up the rest of Peru's population.

Place

Ecuador

Ecuador is one of the smallest countries in South America. Can you guess how Ecuador got its name? *Ecuador* is the Spanish word for "equator," which runs right through Ecuador.

The Land and Economy Swamps, deserts, and fertile plains stretch along Ecuador's Pacific coast. The Andes run through the center of the country. Nearly one-half of Ecuador's people live in the valleys and plateaus of the Andes. Quito (KEE•toh), Ecuador's capital, lies more than 9,000 feet (2,700 m) above sea level. Thick rain forests cover eastern Ecuador.

Ecuador's inland climate is hot and humid. The Peru Current in the Pacific Ocean keeps coastal temperatures moderate. In the Andes the climate gets colder as you climb higher into the mountains.

Agriculture is Ecuador's most important economic activity. Bananas, cacao, sugarcane, and other export crops grow in the coastal lowlands. Here you will find Guayaquil (gwy•uh•KEEL), Ecuador's most important port city. Farther inland, farms in the Andean highlands grow coffee, beans, corn, potatoes, and wheat. Ecuador's major mineral export is petroleum.

The People Of Ecuador's 11 million people, most claim Native American or *mestizo* ancestry. A small number are of African background. About 57 percent of Ecuadorians live in urban areas.

SECTION 2 REVIEW

REVIEWING TERMS AND FACTS

1. **Define the following:** *altiplano,* navigable, empire.
2. **REGION** What landforms do Peru and Ecuador share?
3. **HUMAN/ENVIRONMENT INTERACTION** What is Ecuador's major mineral export?

 MAP STUDY ACTIVITIES

4. Look at the climate map on page 223. What climate do you find in the eastern lowlands of Peru and Ecuador?

(224)

UNIT 3

Answers to Section 2 Review

1. All vocabulary words are defined in the Glossary.
2. Both countries have coastal lowlands, the Andes highlands, and eastern lowlands covered with rain forests.
3. petroleum
4. tropical rain forest

SECTION 3
Bolivia and Chile

PREVIEW

Words to Know
- *quinoa*
- sodium nitrate

Places to Locate
- Bolivia
- La Paz
- Sucre
- Chile
- Tierra del Fuego
- Santiago

Read to Learn . . .
1. where Bolivia and Chile are located.
2. what landforms and climates are found in Bolivia and Chile.
3. how the economies of Bolivia and Chile differ.

Do you think you would want to be a farmer in the high Andean plateau of Bolivia? You would tend a flock of llamas, which are raised for wool and carrying goods. Native Americans in Bolivia live much as their ancestors did hundreds of years ago. Their way of life is common throughout much of rural Bolivia.

At first glance, Bolivia and its neighbor Chile seem very different. Bolivia lacks a seacoast, while Chile has a long, narrow coastline. The Andes, however, affect the climate and cultures of both countries.

Place
Bolivia

Bolivia lies near the center of South America, cut off from the sea. The Andes dominate Bolivia's landscape. Breathtaking scenery draws visitors from all over the world. The mountains, however, present hardships to Bolivia's people. The many Bolivians who live at high elevations must work hard to farm and extract minerals from the earth.

Land and Climate Bolivia has mountains, lowland plains, and tropical rain forests. Look at the map on page 218. You see that in western Bolivia, the Andes surround a high plateau called the *altiplano*. This region has a cool climate, kept moderate by Lake Titicaca. Few trees grow in this region, and the land is too dry to farm. Farms are more common in south central Bolivia, an area of gently sloping hills and broad valleys.

CHAPTER 9

(225)

Meeting National Standards

Geography for Life The following standards are highlighted in this section:
• Standards 8, 11, 12

FOCUS

Section Objectives
1. **Examine** the locations of Bolivia and Chile.
2. **Describe** landforms and climates found in Bolivia and Chile.
3. **Contrast** the economies of Bolivia and Chile.

Bellringer Motivational Activity

 Prior to taking roll at the beginning of the class period, project Section Focus Transparency 9-3 and have students answer the activity questions. Discuss students' responses. **L1**

Vocabulary Pre-check

Have students find the meaning of the word *quinoa*. Ask them to list other types of grains. **L1** **LEP**

NATIONAL GEOGRAPHIC SOCIETY

These materials are available from Glencoe.

 VIDEODISC

STV: World Geography Vol. 3: South America and Antarctica

Lake Titicaca
Side 1, Frames 15915-19520

Classroom Resources for Section 3

REPRODUCIBLE MASTERS
Reproducible Lesson Plan 9-3
Guided Reading Activity 9-3
Performance Assessment Activity 9
Section Quiz 9-3

TRANSPARENCIES
Section Focus Transparency 9-3
Political Map Transparency 3

MULTIMEDIA
 Testmaker

National Geographic Society: STV: World Geography, Vol. 3

225

TEACH

Guided Practice

Comparing Draw a Venn diagram on the chalkboard. Call on volunteers to complete the diagram by making a comparison of Bolivia and Chile. In one circle, have volunteers list unique features of Bolivia. In the other circle, have volunteers list unique features of Chile. In the intersecting part of the circles, have them list common features of the two countries. **L2**

Independent Practice

📁 Guided Reading Activity 9-3 **L1**

ASSESS

Check for Understanding

Assign Section 3 Review as homework or an in-class activity.

Meeting Lesson Objectives

Each objective below is tested by the questions that follow it in parentheses.

1. **Examine** the locations of Bolivia and Chile. (2, 5)
2. **Describe** landforms, climates, and resources found in Bolivia and Chile. (3, 5)
3. **Contrast** the economies of Bolivia and Chile. (3, 4, 5)

A vast lowland plain spreads over northern and eastern Bolivia. Tropical rain forests cover the northern end. Grasslands and swamps sweep across the rest of the plain. Most of this area has a hot, humid climate.

The Economy Bolivia's economy relies partly on farming. Bolivian farmers struggle to grow corn, potatoes, wheat, and a cereal grain called *quinoa.* They also raise cattle for beef and llamas for wool.

Abundant mineral resources balance Bolivia's economy. The country is rich in tin, copper, and lead. Miners extract these minerals from mines high in the Andes. The country ranks as one of the world's leading tin producers. The eastern lowlands provide gold, petroleum, and natural gas.

The People Bolivians face a problem that no other Latin American people face. How do they direct someone to their capital? They have two of them! Bolivia's two capital cities—La Paz (luh•PAHZ) and Sucre (SOO•kray)—are located in the *altiplano.* La Paz—the highest capital in the world—lies at an elevation of 12,000 feet (3,660 m).

Most of Bolivia's 8.2 million people live in the Andean highlands. The majority claim Native American or mixed Spanish and Native American ancestry. In the cities, most people observe European or North American customs. Most people in the countryside are Native Americans and follow the traditions of their ancestors. The men carry on subsistence farming, and the women weave textiles or make pottery to earn money.

Place
Chile

Find Chile on the map on page 216. Do the words "l-o-n-g" and "narrow" come to mind? Chile runs about 2,650 miles (4,265 km) along the Pacific coast of western South America. The country's average width, however, is only 100 miles (160 km).

The Land and Climate Chile contains many types of landforms. In the north lies the Atacama Desert, one of the world's driest places. Farther east, the Andes run along Chile's border with Argentina. Fanning out from the mountains in central Chile is the Central Valley. With its fertile soil and mild climate, the Central Valley supports most of Chile's people, industries, and farms.

The southern part of Chile is a stormy, windswept region of snow-tipped volcanoes, thick forests, and huge glaciers. In the far south, the Strait of Magellan separates mainland Chile from a group of islands known as Tierra del Fuego (fu•AY•goh)—or Land of Fire. Cold ocean waters batter the rugged coast around Cape Horn, the southernmost point of South America.

The Economy Chile has the fastest-growing economy in Latin America. It relies on the mining and export of valuable minerals. Copper is the country's most important resource and major export. Chile ranks as the world's leading

UNIT 3

Cooperative Learning Activity

Organize students into three groups and assign one of the following topics to each group: landforms, climate, or resources. Group members should work together to prepare a map of Bolivia and Chile depicting their assigned topic. Display the completed maps and have students discuss how economic activities in these two countries are related to the landforms, climates, and resources. **L1**

Chile has a wide variety of climates and landforms. The capital city of Santiago in central Chile *(left)* contrasts sharply with the icy southern region with massive glaciers *(right)*. **PLACE: What chain of islands lies at the southern tip of Chile?**

copper producer. Chile also mines and exports gold, silver, iron ore, and **sodium nitrate.** Sodium nitrate is used as a fertilizer and in explosives.

Agriculture and manufacturing are also major economic activities in Chile. Farmers produce wheat, corn, barley, rice, and oats and raise cattle, sheep, and other livestock. Chilean factory workers produce clothing, wood products, chemicals, and transportation equipment. In Chile's cities, service industries such as banking and tourism thrive.

The People Of the 14 million people in Chile, about 75 percent are mestizos. About 20 percent are of pure European ancestry. Nearly all the people speak Spanish, and most are Roman Catholics. About 85 percent of Chile's population live in urban areas. Santiago, the capital, has about 5.3 million people.

Antonio Pérez, a teenager from central Chile, attends rodeos to watch Chilean cowboys perform remarkable feats. He—and many other Chileans— also enjoy sports such as soccer, skiing, and volleyball. Antonio likes to vacation on the beautiful beaches along Chile's Pacific coast.

SECTION 3 REVIEW

REVIEWING TERMS AND FACTS

1. Define the following: *quinoa,* sodium nitrate.

2. PLACE What are the two capitals of Bolivia? Where are they located?

3. HUMAN/ENVIRONMENT INTERACTION What is Chile's most important mineral resource?

 MAP STUDY ACTIVITIES

4. Look at the land use map on page 219. What region holds most of Chile's farms?

5. Using the same map, name the mineral resources found in Bolivia.

Answers to Section 3 Review

1. All vocabulary words are defined in the Glossary.

2. La Paz, Sucre; in the *altiplano*

3. copper

4. Central Chile, in the Central Valley

5. tin, copper, lead, nitrates, zinc, natural gas, silver

LESSON PLAN
Chapter 9, Section 3

More About the Illustration

Answer to Caption
Tierra del Fuego

 GEOGRAPHIC THEMES

Human/ Environment Interaction Air pollution has become a major problem in Santiago. Clouds of exhaust fumes from cars and smoke from factories hover in the air, trapped by the high mountains that surround the city.

Evaluate

Section 3 Quiz **L1**

Use the Testmaker to create a customized quiz for Section 3. **L1**

Reteach

Name a physical feature, type of climate, resource, economic activity, or culture group. Have students determine whether the characteristic refers to Bolivia, Chile, or both countries. **L1**

Enrich

Have students investigate traditional clothing in Bolivia and Chile. Ask them to make annotated drawings to illustrate different styles. **L3**

CLOSE

Call on a volunteer to summarize the information in this section. Have the rest of the class note what, if anything, they think should be added to the summary.

227

FOCUS

Section Objectives

1. **Identify** the physical regions that make up Argentina.
2. **List** the products that come from Argentina.
3. **Describe** where Argentina's people live.

Bellringer Motivational Activity

 Prior to taking roll at the beginning of the class period, project Section Focus Transparency 9-4 and have students answer the activity questions. Discuss students' responses. **L1**

Vocabulary Pre-check

Have students define the word *gaucho*. Discuss reasons why the *gaucho* is the national symbol of Argentina. **L1 LEP**

NATIONAL GEOGRAPHIC SOCIETY

These materials are available from Glencoe.

 **VIDEODISC**

STV: World Geography Vol. 3: South America and Antarctica

Iguaçú Falls
Any side, Frame 50487

228

PREVIEW

Words to Know
• tannin
• *estancia*
• *gaucho*

Places to Locate
• Argentina
• Gran Chaco
• Patagonia
• Pampas
• Buenos Aires
• Río de la Plata

Read to Learn . . .
1. what physical regions make up Argentina.
2. what products come from Argentina.
3. where Argentina's people live.

The North American song "Home on the Range" could have been written about this vast, treeless landscape in the country of Argentina. Here cowhands drive large herds of cattle across this plains area. Beef—one of Argentina's major products—is known throughout the world for its quality and flavor.

Argentina occupies most of the southern part of South America. Uruguay, Brazil, Paraguay, and Bolivia lie on its northern borders. Argentina's eastern coastline is washed by the Atlantic Ocean. Its southern tip reaches almost to the continent of Antarctica.

Region
The Land

Argentina's land area of 1,056,640 square miles (2,736,698 sq. km) makes it South America's second-largest country, after Brazil. Within Argentina's vast spaces are a variety of landscapes: mountains, forests, plains, and deserts.

The North A roaring waterfall greets visitors to Argentina's north region. The spectacular Iguaçú (EE•gwuh•SOO) Falls, on Argentina's border with Brazil, is one of the scenic wonders of the world.

Lowland areas stretch across northern Argentina. To the west, great forests cover an area known as the Gran Chaco. As you learned in Chapter 8, the Gran

Highest Peaks of the Americas

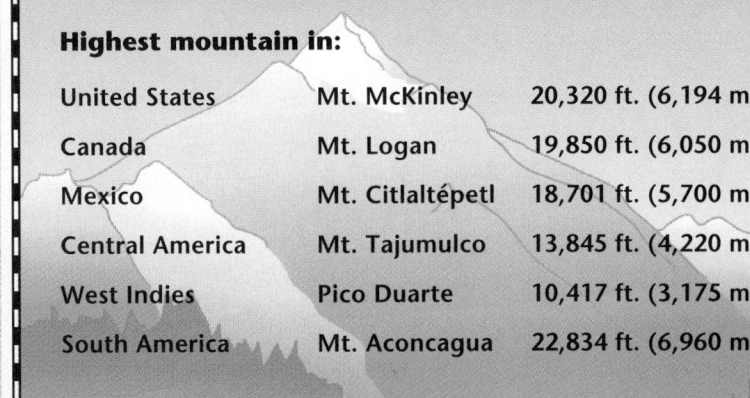

GRAPHIC STUDY

The highest peak in the United States is Mt. McKinley in Alaska.
PLACE: What is the highest peak in the Americas?

Highest mountain in:

United States	Mt. McKinley	20,320 ft. (6,194 m)
Canada	Mt. Logan	19,850 ft. (6,050 m)
Mexico	Mt. Citlaltépetl	18,701 ft. (5,700 m)
Central America	Mt. Tajumulco	13,845 ft. (4,220 m)
West Indies	Pico Duarte	10,417 ft. (3,175 m)
South America	Mt. Aconcagua	22,834 ft. (6,960 m)

Source: *World Almanac,1994;
Goode's World Atlas,* 19th edition

Answer
Mt. Aconcagua

Graphic Skills Practice
Reading a Graph What is the elevation of the highest peak in the West Indies? *(10,417 feet, or 3,175 m)*

Chaco also extends into neighboring Paraguay. This region, dry for most of the year, experiences sudden downpours during the summer months. The few people that live here harvest quebracho trees and practice subsistence agriculture. These hardwood trees are a source of **tannin,** a substance used in making leather.

To the east, hot, humid grasslands lie between the Paraná and Paraguay rivers. Farmers graze livestock and grow crops on this fertile soil. Along this area's northern border, the Iguaçú River empties into the Paraná River. Near where the two rivers join, about 275 waterfalls form Iguaçú Falls.

The Andes The Andes tower over the western part of Argentina. Snow-capped peaks and clear blue lakes draw tourists who come to ski and hike. Mount Aconcagua (A•kuhn•KAH•gwuh), soaring to a height of almost 23,000 feet (7,000 m), is the highest mountain in the Western Hemisphere. The chart above shows how Aconcagua compares in height with mountains in other parts of the Americas.

East of the Andes is a region of rolling hills and desert valleys. Mountain streams flow through this area. Farmers use these waters to grow sugarcane, corn, cotton, and grapes. Cities in this area are the oldest Spanish settlements in Argentina.

The Pampas In the center of Argentina are treeless plains known as the Pampas. The Pampas spread almost 500 miles (804 km) from the Atlantic coast to the Andes. Argentina's economy depends on this region's fertile soil and mild climate. Similar to the Great Plains of the United States, the Pampas are home to farmers growing grains and ranchers raising livestock. Most of Argentina's urban areas are here, where more than two-thirds of the country's people live.

Buenos Aires, Argentina's capital and largest city, lies in the area where the Pampas meet the Río de la Plata. The Río de la Plata is a funnel-shaped bay that enters the Atlantic Ocean.

CHAPTER 9

 229

TEACH

Guided Practice
Analyzing Information
Direct students to read the section. Have them write a paragraph that answers the following question: Based on your analysis of this material, what do you think life is like in Argentina? Why? Call on volunteers to read their paragraphs to the class. Then lead a discussion on how life in Argentina compares to life in other Andean countries. **L2**

CURRICULUM CONNECTION

Earth Science Argentina's varied geography includes the Perito Moreno glacier on Lake Argentino in the southern Andes. This wall of ice is one of the few glaciers in the world that is still advancing.

Cooperative Learning Activity

Organize students into four groups and assign each group one of the major land regions of Argentina: the north, the Andes, the Pampas, or Patagonia. Have groups create a mural representing their assigned region. Suggest that groups focus on topics such as landforms, climate, resources, economic activities, major cities, everyday life, and so on. Have groups display and discuss their completed murals. **L1**

Independent Practice

📁 Guided Reading Activity 9-4 **L1**

📁 Critical Thinking Skills Activity 9 **L1**

More About the Illustration

Answer to Caption
because it is dry and windswept and has poor soil

Movement The name *Patagonia* means "big feet." The name was chosen by the explorer, Ferdinand Magellan. When he first saw the Native Americans of the area, they were wearing huge boots.

ASSESS

Check for Understanding

Assign Section 4 Review as homework or an in-class activity.

Meeting Lesson Objectives

Each objective below is tested by the questions that follow it in parentheses.

1. **Identify** the physical regions that make up Argentina. (4, 5, 6)
2. **List** products that come from Argentina. (3)
3. **Describe** where Argentina's people live. (2)

The Andes tower over the Lake of Horns in southern Patagonia *(left)*. The "pink house" in Buenos Aires is the office of Argentina's president *(right)*. **HUMAN/ENVIRONMENT INTERACTION: Why do few people live in Patagonia?**

Patagonia South of the Pampas is a dry, windswept plateau called Patagonia. Most of Patagonia gets little rain and has poor soil. Sheep raising is the region's major economic activity.

Region
The Economy

Although Argentina today has one of the most developed economies in Latin America, Argentines struggled for much of the past 50 years under unstable governments. Argentina's economy depends on agriculture, manufacturing, and service industries.

Agriculture Argentina's major farm products are beef, corn, and wheat. Most of these products come from the Pampas and the northeastern part of the country. Huge *estancias,* or ranches, cover the Pampas. Owners of these estates hire *gauchos,* or cowhands, to take care of the livestock. The *gaucho* is the national symbol of Argentina. Many people in Argentina admire the *gauchos* for their independence and their horse-riding skills.

Manufacturing Argentina is one of the most industrialized countries in South America. The map on page 219 shows you that most of the country's factories are in or near Buenos Aires. Argentina's leading manufactured goods are food products, leather goods, electrical equipment, and textiles.

Petroleum is Argentina's most valuable mineral resource. The country's major oil fields are in Patagonia and the Andes. Other minerals, such as iron ore, lead, zinc, and uranium, also lie in the Andean region.

230

UNIT 3

Meeting Special Needs Activity

Writing Disability Help students who have difficulty writing sentences and paragraphs learn the skill of note taking. Direct students to write key words or phrases from each of the section's subheads. Then help students use their notes to summarize main ideas orally. **L1**

Movement

The People

The population density map on page 223 reveals that Argentina's 34 million people live mostly in certain areas of the country. Because of the harsh climate and landscape of the Andean region and Patagonia, settlement in these areas is sparse. More than one-third of the country's people live in or around Buenos Aires. Another one-fourth live on the Pampas.

Influences of the Past The region that is now Argentina had very few Native Americans before the arrival of the Europeans. The Spanish arrived in the 1500s from Chile and later founded a colony centered around Buenos Aires. Most of the Native Americans in the region died from disease or were killed by Europeans.

In 1816 a general named José de San Martín helped Argentina win its independence from Spain. During the late 1800s and early 1900s, some of Argentina's people grew wealthy from livestock raising and industry. Military leaders controlled Argentina's government.

In 1982 Argentina suffered defeat in a war with the United Kingdom for control of the Falkland Islands. The Falklands, known in Argentina as the Malvinas, are located in the Atlantic Ocean off the coast of Argentina. Argentina's loss led to the overthrow of its military leaders and the creation of a democracy.

Argentina Today Argentines are mostly of European ancestry. Settlers came to Argentina from Spain and Italy during the 1800s. Native Americans and mestizos form only a small part of Argentina's population. The official language of Argentina is Spanish, and the Roman Catholic faith is the major religion. About 86 percent of Argentina's people live in cities and towns. Thirteen-year-old Gabriella Marcelli is among the 11.7 million Argentines who live in Buenos Aires. She admires the city's attractive parks and elegant government buildings.

WHAT IN THE WORLD?

Let's Play Pato

Many Argentines play a sport called *pato*. *Pato* combines polo and basketball. Players on horseback form two teams. Each team tries to keep control of a six-handled ball and toss it into the opposing team's basket.

SECTION 4 REVIEW

REVIEWING TERMS AND FACTS

1. **Define the following:** tannin, *estancia, gaucho.*
2. **PLACE** Where is Buenos Aires located?
3. **PLACE** What are the most important agricultural products of Argentina?
4. **REGION** How is the Andean region important to Argentina's economy?

GRAPHIC STUDY ACTIVITIES

5. Look at the chart on page 229. What is the highest mountain in Central America?
6. How much taller is Mt. Aconcagua than the highest mountain in the United States?

Answers to Section 4 Review

1. All vocabulary words are defined in the Glossary.
2. where the Pampas meets the Río de la Plata
3. beef, corn, wheat
4. the Andes region has many minerals
5. Mt. Tajumulco
6. 2,514 feet (766 m)

Global Gourmet

Argentina Argentina's per-person consumption of beef is greater than any other nation in the world. Most Argentines eat beef every day.

Evaluate

Section 4 Quiz **L1**

Use the Testmaker to create a customized quiz for Section 4. **L1**

Chapter 9 Test Form A and/or Form B **L1**

Reteach

Reteaching Activity 9 **L1**

Spanish Reteaching Activity 9 **L1**

Enrich

Enrichment Activity 9 **L3**

CLOSE

Mental Mapping Activity

Provide each student with a large sheet of white paper and map pencils. Inform students that this will not be a graded activity. Have students draw, freehand from memory, a map of South America that includes the following labeled countries: Colombia, Peru, Ecuador, Chile, Bolivia, and Argentina.

BUILDING GEOGRAPHY SKILLS

TEACH

Display a road map with insets of major cities. If possible, use an opaque projector for easier viewing. Have students imagine that they are traveling by car along an interstate highway shown on the map. Trace the highway with a pointer. Point to a particular city and mention that when they reach this point, they must leave the highway to find a store in the center of the city. Ask students what part of the map would help them to find the street on which the store is located. *(Students should identify the inset map for the relevant city. If they do not, indicate the inset maps along the margin and have them find the correct one.)* Ask students to describe how the inset map differs from the larger map. *(different scale; gives detailed information about a smaller area)* Then have students draw rough maps of their neighborhood with an inset map of the streets on which they live.

Have students read the skill and answer the practice questions. **L1**

Additional Skills Practice

1. What map would be useful in finding the location of airports in and around Buenos Aires? *(inset map)*
2. What map would be useful in finding the location of Buenos Aires relative to the city of Rosario? *(main map)*

▼ Geography Skills Activity 9

Using an Inset Map

Have you ever used a magnifying lens to observe something in greater detail? An *inset map* serves the same purpose. It magnifies, or enlarges, one section of a map. If a map contains an inset, study both maps together. Each will be drawn to a different scale. The map that covers the larger area gives an overall view. A square, circle, or dot on that area marks the section you see in the inset map. This inset shows a smaller region in more detail. To use an inset map, apply the following steps:

• Identify the areas covered by each map.

• Study the scale of each map. Look for details about the area shown in the inset.

• Notice the relationship between the two maps.

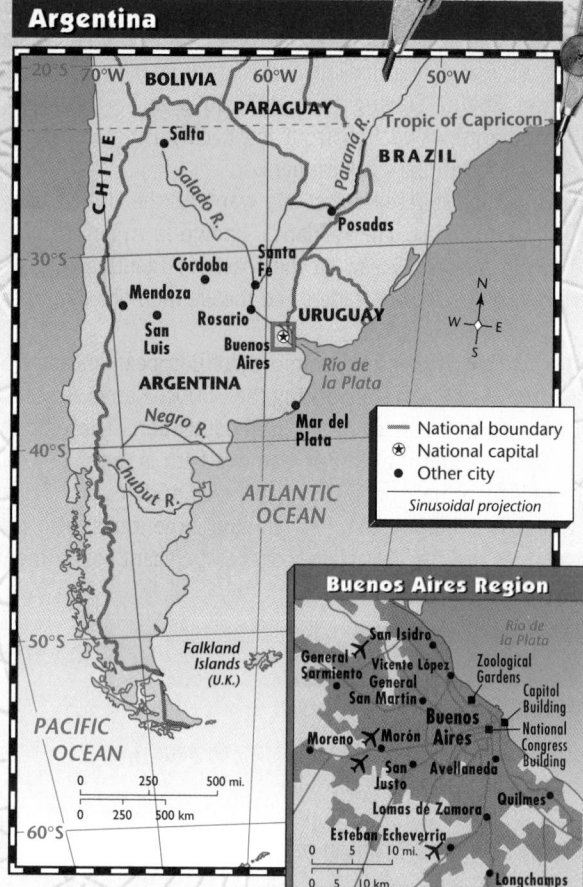

Geography Skills Practice

1. What areas are shown in each map?
2. What is the scale of each map?
3. Describe the location of Buenos Aires compared to the rest of Argentina.
4. What details are revealed in the inset that do not appear on the other map?

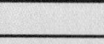

Answers to Geography Skills Practice

1. Argentina; the Buenos Aires region
2. 7/8" = 500 miles; 1/2" = 10 miles
3. Buenos Aires is located in eastern Argentina, on the Río de la Plata.
4. location of airports and local landmarks, major roads, names and locations of suburbs and other nearby communities

Chapter 9 Highlights

Important Things to Know About the Andean Countries

SECTION 1 COLOMBIA

- Colombia has coastlines on the Caribbean Sea and the Pacific Ocean.
- Coastal lowlands, central highlands, and inland rain forests make up Colombia's landscape.
- Coffee is Colombia's major farm product.

SECTION 2 PERU AND ECUADOR

- Peru's long Pacific coastline is mostly desert and plains.
- The land that is now Peru was the center of a great Native American civilization—the Inca Empire.
- Most of the people of Peru live in coastal cities and in mountain valleys.

SECTION 3 CHILE AND BOLIVIA

- Bolivia is a landlocked country near the center of South America. Chile is a long, narrow country along the Pacific coast.
- Most of Bolivia's people are Native Americans.
- The majority of Chile's population live in the fertile central valley between the mountains.
- Copper is the major mineral export of Chile; tin is Bolivia's major mineral export.

SECTION 4 ARGENTINA

- Argentina is the second-largest country in South America, after Brazil.
- A vast, treeless plain called the Pampas is home to Argentina's beef cattle industry.
- About one-third of Argentina's people live in Buenos Aires, the nation's capital.

Inca terraces, Pisac, Peru ▶

Extra Credit Project

Distance Tables Locate a distance table in a road atlas and display it for students. Direct students to create a similar table showing distances between major cities in the Andean countries. For information, students might refer to distance scales on the maps in the text or to various atlases. Distances should be expressed in miles and kilometers. Have students compare their distance tables. **L1**

CHAPTER HIGHLIGHTS

Using the Chapter 9 Highlights

- Use the Chapter 9 Highlights to preview, review, condense, or reteach the chapter.

Preview/Review

Vocabulary Puzzle-Maker Software reinforces the terms used in Chapter 9.

Condense

Have students read the Chapter 9 Highlights. Spanish Chapter Highlights also are available.

Chapter 9 Digest Audiocassettes and Activity

Chapter 9 Digest Audiocassette Test

Spanish Chapter Digest Audiocassettes, Activities, and Tests also are available.

Guided Reading Activities

Reteach

Reteaching Activity 9. Spanish Reteaching activities also are available.

MAP STUDY

Human/Environment Interaction
Have students return to the resources/economic activities map they developed for the **Chapter 8 Highlights** Map Study activity. Ask them to complete the map by adding information for the Andean countries.

Chapter 9 Review and Activities

REVIEWING KEY TERMS

Match the numbered terms in Column A with their definitions in Column B.

A
1. *quinoa*
2. cordillera
3. sodium nitrate
4. tannin
5. *estancia*
6. *altiplano*
7. *campesino*
8. *gaucho*
9. empire
10. *llanos*

B
A. Argentine ranch
B. substance used in making leather
C. collection of territories under one ruler
D. a cereal grain
E. side-by-side mountain chains
F. mineral used for fertilizer and explosives
G. farmer
H. treeless plains in Colombia
I. high, dry plateau in the Andes
J. cowhand in Argentina

Mental Mapping Activity

On a separate sheet of paper, draw a freehand map of the Andean countries. Label the following items on your map:
- Buenos Aires
- Atacama Desert
- Cape Horn
- Andes
- Lake Titicaca
- Bogotá

REVIEWING THE MAIN IDEAS

Section 1
1. **PLACE** How do the Andes affect Colombia's climate?
2. **HUMAN/ENVIRONMENT INTERACTION** What part of Colombia has a lumber industry?

Section 2
3. **PLACE** What is the capital of Peru?
4. **PLACE** What body of water is the highest navigable lake in the world?

Section 3
5. **REGION** Describe the climate of northern Chile.
6. **HUMAN/ENVIRONMENT INTERACTION** What minerals are mined in Bolivia?

Section 4
7. **LOCATION** Where is Patagonia located?
8. **PLACE** What are the major economic activities in the Pampas?

ANSWERS
Reviewing Key Terms
1. D
2. E
3. F
4. B
5. A
6. I
7. G
8. J
9. C
10. H

Mental Mapping Activity

This exercise helps students to visualize the countries and geographic features they have been studying and understand the relationships among them. All attempts at freehand mapping should be accepted.

Reviewing the Main Ideas

Section 1

1. Because of the Andes, some areas of Colombia have mild climates, even though the country lies within the tropics.
2. eastern lowlands

Section 2
3. Lima
4. Lake Titicaca

CRITICAL THINKING ACTIVITIES

1. **Determining Cause and Effect** What effect does the Peru Current have on the coastal areas of Peru and Ecuador?
2. **Evaluating Information** How did the arrival of Europeans affect the Native Americans living in the Andean region of South America?

GeoJournal Writing Activity

Imagine that you are working on an *estancia*, a large cattle ranch in Argentina. Write a letter to a family member or a friend describing how you begin your day, your work on the ranch, and how you end your day.

COOPERATIVE LEARNING ACTIVITY

Work in a group of three to learn more about daily life in one of the capital cities of the Andean countries. Choose one city and select one of the following topics to research: (a) where and how people live; (b) what people do for recreation; or (c) what holidays and festivals they celebrate. After your research is complete, share your information with the group. As a group prepare a script for a travel show about these countries. Share your information with the class.

PLACE LOCATION ACTIVITY: THE ANDEAN COUNTRIES

Match the letters on the map with the places and physical features of the Andean countries. Write your answers on a separate sheet of paper.

1. Paraguay River
2. Bogotá
3. Lima
4. Strait of Magellan
5. Pampas
6. Río de la Plata
7. Bolivia
8. Ecuador
9. Lake Titicaca
10. Andes

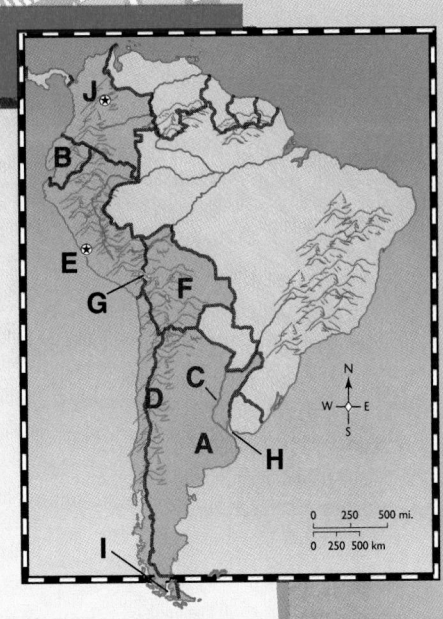

Cooperative Learning Activity

Have groups present their scripts to the class. Then call on students to discuss the following: Which of these cities would you most like to visit? Why?

GeoJournal Writing Activity

Remind students to discuss such subjects as the kinds of clothes they wear and the different foods they eat. Select students to read their letters to the class.

Place Location Activity

1. C
2. J
3. E
4. I
5. A
6. H
7. F
8. B
9. G
10. D

Chapter Bonus Test Question

This question may be used for extra credit on the chapter test.

Is Ecuador in the Southern Hemisphere or Northern Hemisphere? *(both)*

Section 3
5. desert (dry)
6. nitrates, zinc, tin, copper, silver, lead

Section 4
7. southern Argentina
8. agriculture, raising livestock

Critical Thinking Activities

1. cools temperatures, which, because these countries lie in the tropics, would normally be hot
2. Many Native Americans died from overwork and the diseases Europeans brought to South America. Even so, Native American culture survived and plays an important role in life in the Andean countries today.

235

NATIONAL GEOGRAPHIC SOCIETY

The following materials are available from Glencoe.

 **POSTERS**

Eye on the Environment, Unit 3

FOCUS

Ask students to imagine the following situation: A lake near their town is about to be filled in to allow a shopping mall to be built. The lake is the only place where gilded grumpfish are known to live. Would students vote for the mall plan to proceed, or would they vote to stop the project to protect the gilded grumpfish? After discussion, tell students that governments controlling rain forest areas face a similar problem.

TEACH

Ask students to research animals that live in the rain forests of Brazil. Tell them to choose one animal to represent all endangered species there. Suggest that they use this animal as the basis for the slogan and design of a "Save the Rain Forest Wildlife" campaign button. Have students display their button designs. **L1**

NATIONAL GEOGRAPHIC SOCIETY
EYE ON THE ENVIRONMENT

The Disappearing Rain Forest

PROBLEM

The rain forests of the Amazon basin in South America are being cut or burned at a very fast rate. Why? The trees are being cut for export to countries that will pay high prices for the fine hardwoods. Farmers are burning the forest to clear the land for other profitable crops. Many people are worried that these practices—which are sometimes good for the economies of South America—will cause the rain forests to completely disappear.

SOLUTIONS

- Farmers who settle on newly cleared land are learning soil conservation.
- Governments in Latin America have set up land preserves.
- Researchers supported by a group of Brazil's largest companies are looking for less harmful ways to develop land.
- Latin America is promoting markets for goods that can be harvested without destroying trees.

This brilliant tree frog is one of about 4,000 species of frogs living in the forests of Central America whose habitats are threatened.

Rain Forest Fact Bank

- Every minute of every day, between 30 and 50 acres (12 to 20 ha) of rain forest disappear.
- Every year, 27,000 species of forest plants and animals are destroyed.
- Rain forests supply one-fifth of the world's oxygen supply.
- Plants in the Amazon basin provide sources for one-fourth of the world's medicines.
- At least one-half of all plant and animal species on Earth live in the rain forests.
- Destroying the rain forests wipes out an ecosystem that does not exist anywhere else on Earth.

236

UNIT 3

More About the Issues

Region One of the many problems resulting from the development of the rain forests is the displacement of indigenous peoples. These native peoples, recognized by countries as an important resource, are as endangered as the plants and animals in the region. Throughout the Americas and beyond, local, national, and international organizations have begun to protest the displacement of these peoples and to work to secure their land and resource rights.

A family of farmers winnows dry-land rice from a plot in western Brazil. Slash-and-burn agriculture is the single largest cause of tropical forest loss around the world.

NATIONAL GEOGRAPHIC SOCIETY

These materials are available from Glencoe.

  VIDEODISC

STV: Rain Forest

Upward and Onward
Side 2, Frames 33538-49359

ASSESS

Have students answer the **Take A Closer Look** questions on page 237.

CLOSE

Discuss with students the **What Can You Do?** question. Encourage them to find out more about the groups that are working to save the environment. Students might want to start their own local organization.

*For an additional regional case study, use the following:

📁 Environmental Case Study Unit 3 **L1**

TEEN TRIBUTE

Roland Tiensuu, 13, of Sorunda, Sweden, wanted to help save the rain forests. In the early 1990s, he and fellow teenage students raised about $2 million to buy and protect 23,500 acres (9,185 ha) of forests in Costa Rica.

WHAT CAN YOU DO?

environmental activities

TAKE A CLOSER LOOK

1 What are some of the ways the rain forests affect the whole world?

2 What are the major reasons the rain forests are being cut down?

🍃 Plan and carry out an awareness campaign in your school or community to let people know about the destruction of the rain forests.

🍃 Help groups who are trying to protect the forests by raising money.

🍃 Plant a tree.

🍃 Write to your government representatives and encourage them to support plans that help the rain forest.

🍃 Find out what natural lands might be threatened in your community.

SAVE THE RAINFORESTS

237

Botanist David Neill collects plant specimens from the tops of trees felled by road builders in Ecuador. Roughly 25 per cent of all prescription medicines in the U.S. are derived from rain forest plants.

Global Issues

Interdependence Many scientists believe that the medicinal applications of rain forest plants and animals have been only partially explored. Further experimentation might also help to preserve the rain forest.

Answers to Take A Closer Look

1. Rain forests provide one-fifth of the world's oxygen supply, a home for rare animals and plants found in no other environment, and are the source of many important medicines.

2. Rain forests are being cut down for their valuable lumber resources and to clear land for farming and other types of development.

237

Unit 4 Europe

What Makes Europe a Region?

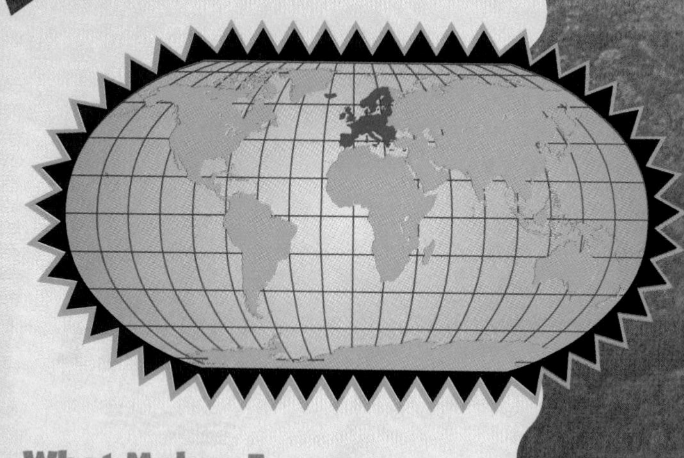

The people share . . .

• high productivity in agriculture and industry.

• densely populated cities.

• global connections in trade, communications, travel, and tourism.

• multinational cooperation since World War II.

To find out more about Europe, see the Unit Atlas on pages 240–253.

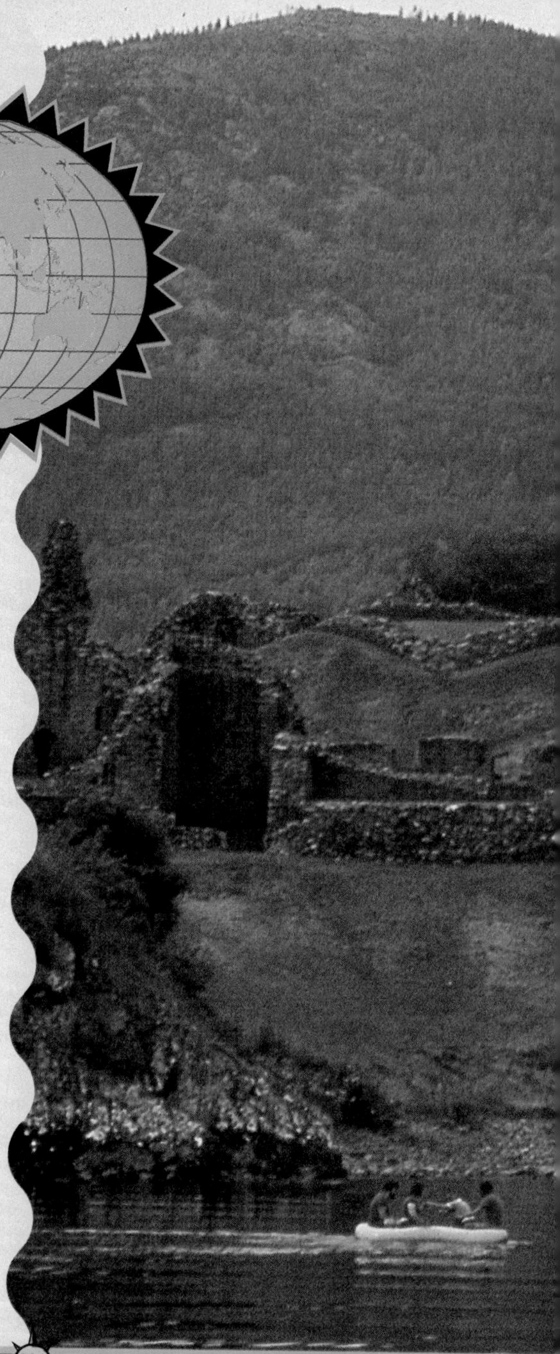

Loch Ness, Scotland

About the Illustration

The ruins of Urquhart Castle overlook Loch Ness in the Scottish Highlands. The mention of "Loch Ness" conjures up images of the dinosaur-like "monster" that thousands of people have reported seeing in the lake. The first recorded sighting of "Nessie" was made as early as A.D. 565. Since then, many people have tried to prove that the strange beast really exists. Scientists and tourists observe the lake, hoping to catch a glimpse of the elusive creature.

 **Human/Environment Interaction** Why do you think some scientists are interested in discovering if the Loch Ness monster really exists? *(Students might suggest that it could be an animal unlike any other, or that it might provide information about extinct animals.)*

NATIONAL GEOGRAPHIC SOCIETY

These materials are available from Glencoe.

 CD-ROM

Picture Atlas of the World

SOFTWARE

ZipZapMap! World

GeoJournal Activity

Draw a chart on the chalkboard with *Color, Subject,* and *Style* as horizontal headings and *Climate, Features,* and *Water Bodies* as vertical headings. Have students copy the chart into their notebooks. As students review photographs in the unit, have them check the geographic features of the region that they think influenced the subject, color, or style of the piece. Then ask students to sum up how environment influences art. **L2**

- This journal activity provides the basis for the "GeoJournal Writing Activity" exercise in the Chapter Review.
- The GeoJournal may be used as an integral part of Performance Assessment.

GeoJournal Activity

Much of the world's great art comes from Europe. As you read about this region, imagine that you are an art critic. Choose at least five pieces of art or architecture shown in this unit's photographs. Then analyze each piece and comment on its style, its color, and how it reflects the environment in which it was created.

Location Activity

Display Political Map Transparency 4 and ask students the following questions: What continent adjoins Europe? *(Asia)* What European country lies closest to Africa? *(Spain)* Why are Spain, France, Italy, and Greece known as Mediterranean countries? *(They border the Mediterranean Sea.)* Why is Norway generally cooler than Italy? *(Norway lies nearer the Arctic.)* What European country lies farthest west? *(Iceland)* What bodies of water separate the United Kingdom from northern Europe? *(North Sea, English Channel)* What countries border the Czech Republic? *(Austria, Germany, Poland, Slovakia, and Hungary)*

IMAGES OF THE WORLD

UNIT 4 ATLAS

Global Gourmet

France French etiquette requires that a diner never cut his or her lettuce but fold it into small pieces with a fork.

NATIONAL GEOGRAPHIC SOCIETY

IMAGES *of the* WORLD

1. Mowing at Stonehenge, England
2. Open-air barber shop, Greece
3. Snow-covered peaks, Lofoten Islands, Norway
4. Children and pony returning from a horse fair, Dublin
5. Grand Canal, Venice
6. Veteran wearing his medals of honor, Paris
7. Parade of bullfighters, Madrid

All photos viewed against an aerial photo of fields and hedgerows, England.

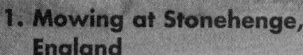

240

Fun Facts

• **United Kingdom** Europe's longest underwater train tunnel runs 30.6 miles (49.2 km) under the seabed of the English Channel and links Great Britain with France and the rest of Europe. Regular passenger service through the "Chunnel" began in November 1994.

• **Bulgaria** Bulgarians shake their head from side to side to denote "yes" and nod up and down to denote "no."

• **Romania** Transylvania, home of the vampire in Bram Stoker's *Dracula,* is a real place located in central and northwestern Romania. The character of Dracula was based on Transylvania's Prince Vlad, who was infamous for his cruelty.

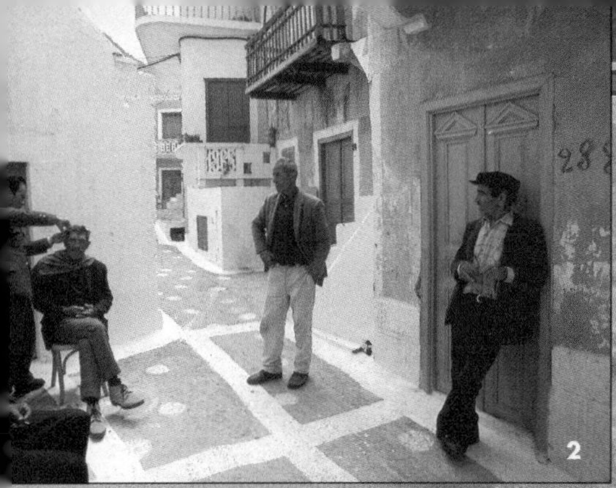

2

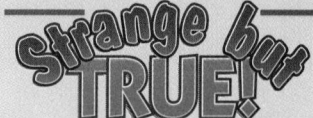

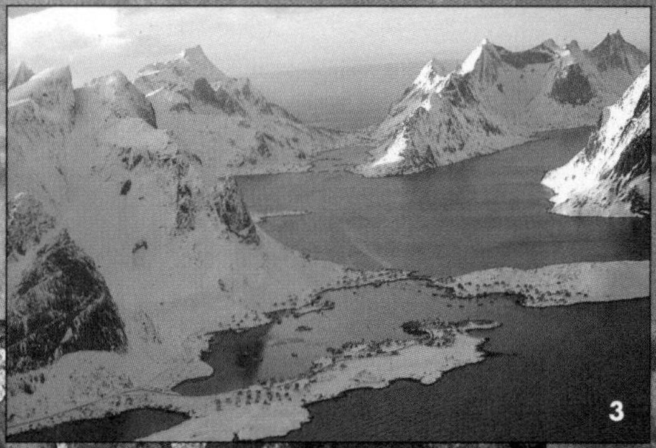

3

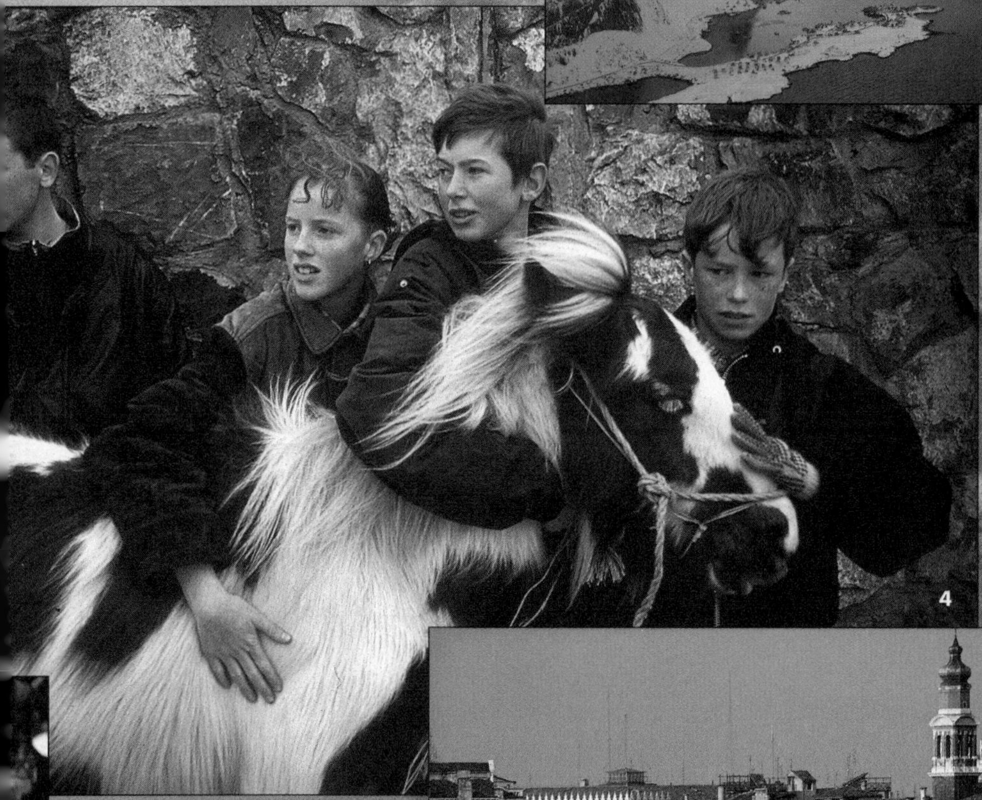

4

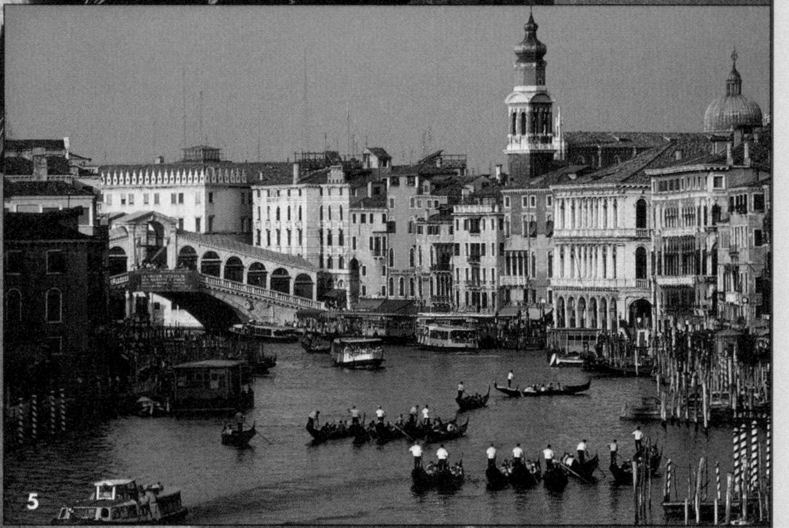

5

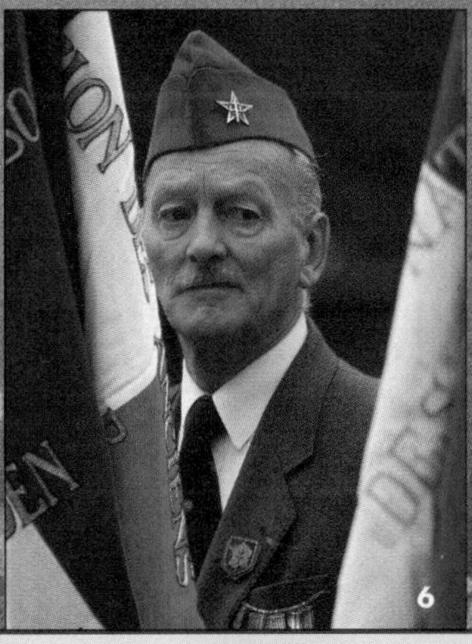

6

241

Teacher Notes

IMAGES OF THE WORLD

Strange but TRUE!

Although Spain is surrounded on three sides by water, it has no large lakes.

Cultural Kaleidoscope

Denmark Trolls originated in Danish legend. Folklore explains that trolls spend their nights burying treasure and their days guarding their loot.

Geography and the Humanities

World Literature Reading 4

World Music: Cultural Traditions, Lesson 3

World Cultures Transparencies 5, 6

Focus on World Art Prints 1, 2, 5, 6, 8, 9, 14, 15, 17, 18, 19, 20, 21, 22, 23

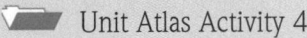

LESSON PLAN
Unit 4 Atlas

Using the Unit Atlas

These features and activities may be used as an introduction to the unit or as teaching tools throughout the course of the unit.

📁 Unit Atlas Activity 4

📁 Unit Overlay Transparencies 4-0, 4-1, 4-2, 4-3, 4-4, 4-5, 4-6

📁 Home Involvement Activity 4

📁 Environmental Case Study 4

📁 Countries of the World Flashcards, Unit 4

✂ World Crafts Activity Cards, Unit 4

🧩 World Games Activity Cards, Unit 4

More About the Illustration
Answer to Caption
the Rhine

Movement
The canals of Amsterdam not only help to drain water from low-lying land, but also create a transportation network that links the rivers of the Netherlands. In winter, Amsterdam's canals often freeze over. Many people ice-skate to work, and schools close early to allow children more skating time.

242

Regional Focus

Europe is both a continent and a region. It is a huge peninsula that runs westward from the landmass of Asia. More than 30 countries make up Europe. Their peoples speak about 50 different languages.

Region
The Land

Europe covers about 1.9 million square miles (5.1 million sq. km). The map on page 246 shows you that several bodies of water touch Europe. The largest of these are the Arctic Ocean, the Mediterranean Sea, the Baltic Sea, and the Atlantic Ocean.

The Canals of Amsterdam in the Netherlands

Houseboats and tall, narrow apartment buildings line the banks of Amsterdam's many canals.
PLACE: What is western Europe's major river?

Seas and Coasts Europe's long, jagged coastline has many peninsulas and offshore islands. Deep bays, narrow seas, and well-protected inlets shelter fine harbors. Closeness to the sea has enabled Europeans to trade with other lands. Many Europeans also depend on the sea for food.

Mountains Europe is a continent almost completely covered by mountains. You will find low mountains in the British Isles, Scandinavia, and parts of France, Germany, and eastern Europe. Mountains elsewhere on the continent are higher and more rugged. They include the Pyrenees, which form the border between France and Spain, and the Alps. Europe's highest mountain range, the Alps sweep across southern and central parts of the continent. Joined to the Alps are the Carpathians, a chain of mountains that towers over the landscape of eastern Europe.

Plains Broad, fertile plains curve around Europe's mountains. The North European Plain stretches from the British Isles to Russia. Farms, towns, and cities dot its rolling land. You will find other plains in the peninsulas of southern and eastern Europe.

Rivers Europe's rivers provide trade links between inland areas and coastal ports. The Rhine—western Europe's major river—begins in the Alps, flows through northwestern Europe, and empties into the North Sea. Europe's other important waterway—the Danube River—flows eastward from central Europe to the Black Sea.

(242)

UNIT 4

Fun Facts

• **Liechtenstein** The government of Liechtenstein earns one-tenth of its income from the sale of postage stamps.

• **Greece** The people of Greece are well known for their hospitality. Ancient Greek myths tell of mortals rewarded for kindness to strangers and of those punished for treating strangers inhospitably. The tradition of hospitality may stem from the ancient belief that one should treat strangers kindly in case they were gods in disguise.

Region
Climate and Vegetation

Europe's small land area offers many contrasts in its climate. Europe's northern location and closeness to the sea are key to this variety of climates.

A Mild Climate Europe lies far north, but many areas of the continent enjoy mild temperatures year-round. This happens because of an ocean current known as the North Atlantic Current. It brings warm Atlantic Ocean waters and winds, which provide rainfall, to Europe's western coast.

Europe's northern location also influences the climate of the region. Northern Europe has longer, colder winters and shorter, cooler summers than southern Europe. Eastern Europe lies farther inland from the Atlantic and its mild current. Therefore, its winters are longer and colder—and its summers shorter and hotter—than those in western Europe.

Vegetation In the far north of Europe, mosses and small shrubs blanket the land. Mixed forests and grasslands are found in the milder climate regions of northwestern and eastern Europe. Air pollution from Europe's factories and cars, however, has harmed many woodlands throughout the continent. In the south the hot, dry summers of the Mediterranean area produce shrubs and short trees.

Region
The Economy

Agriculture, manufacturing, and service industries are Europe's leading economic activities. Skilled workers, a few key natural resources, and closeness to waterways have made Europe one of the economic giants of the world.

Agriculture Europe has some of the world's most productive farmland. European farmers use modern equipment to produce huge yields of grains, fruits, and vegetables on small areas of land. They also raise some of the world's finest breeds of cattle and sheep.

Industry Iron ore, coal, and other minerals are found throughout the North European Plain. Vast reserves of oil and natural gas lie beneath the North Sea and in southeastern Europe. Most of Europe's industrial centers have developed near major mineral deposits.

Many European countries are among the world's leading manufacturing centers. They produce steel, machinery, cars, textiles, electronic equipment, food products, and household goods. Service industries such as banking, insurance, and tourism are also important to Europe's economy.

Moving Products Along the Danube

Barges line the banks of the Danube River in Bratislava, Slovakia.
REGION: What factors have made Europe an economic giant?

(243)

NATIONAL GEOGRAPHIC SOCIETY

 CD-ROM

Picture Atlas of the World

 SOFTWARE

ZipZapMap! World

 POSTERS

Eye on the Environment, Unit 4

Global Gourmet

The Netherlands Favorite foods of the Dutch are chocolate spread on bread for breakfast and smoked eel for the main meal.

More About the Illustration

Answer to Caption skilled workers, a few key natural resources, and closeness to waterways

Place Bratislava's location on the Danube has made it an important trade center for Central Europe. It also is a major industrial city, with textiles, chemicals, oil refining, and food processing among its industries.

Facts

• **Czech Republic** Czechs give each other marzipan candies shaped like pigs for good luck in the New Year.

• **Norway** Courtesy and respect are central to Norwegian life. An important custom involves thanking hosts for preparing or providing a meal. Norwegian children are taught to say *Takk for maten* ("Thank you for the food") before getting up from the table.

FactsOnFile

Use the reproducible masters from **GEOGRAPHY ON FILE** and **MAPS ON FILE I**, *Europe,* to enrich unit content.

Global  **Gourmet**

Slovakia The national dish of Slovakia consists of small dumplings with processed sheep cheese.

More About the Illustration

Answer to Caption
about 510 million people

Place During the 1700s and 1800s, Vienna was an important musical center. Many great composers, including Beethoven, Brahms, Haydn, Mozart, Schubert, and the Strauss family, lived and worked there.

UNIT 4 ATLAS

Region
The People

Some European countries have one leading ethnic group; others are made up of two or more ethnic groups. Differences among Europe's many peoples have often led to conflicts such as civil war in the former Yugoslav republics.

Population The region of Europe has about 510 million people—a large number for a small space. It is one of the world's most densely populated areas. Yet not all of the region is crowded. Most Europeans live in major cities or their suburbs, while large stretches of countryside have few people.

A Street in Old Vienna, Austria

People gather at sidewalk cafes in Vienna's historic Inner City to socialize and enjoy pastries and coffee.
REGION: What is the population of Europe?

City and Countryside Many of the world's leading cities are in Europe. They include London, Paris, Berlin, Madrid, Rome, Warsaw, and Athens. Europe's cities have palaces, churches, and monuments that recall great moments in history. Their skylines boast skyscrapers, high-rise apartment blocks, and modern shopping malls. In rural areas life still centers on villages.

History Europe has had a powerful influence on the world. Europeans made many advances in learning and the arts as well as in science and technology. They explored, settled, and conquered other lands. In doing so, they brought their ideas and ways of life to every part of the globe.

The ancient Greeks developed one of the first civilizations in Europe. They laid the foundation of European government, art, literature, theater, science, and philosophy. The Romans later built an empire that united much of Europe under a common system of law. In western Europe several Germanic kingdoms, influenced by Christianity, replaced the Roman Empire. In eastern Europe the Roman Empire survived for another thousand years as the Christian Byzantine Empire.

From A.D. 500 to 1500, Christianity and the ideas of ancient Europeans merged to form a new European civilization. In the 1400s and 1500s, the Renaissance—or "rebirth" of art and scholarly activities—advanced learning and the value of the individual person. Western Europeans also sailed on overseas voyages, conquered foreign civilizations, and pushed for trade.

Political changes in western Europe after the 1600s increased freedom for the common people. An interest in science and the invention of machines during the Industrial Revolution changed the economy and, in the long run, raised

Fun Facts

• **France** The French take an almost chauvinistic pride in the beauty of their language. Some want the purity of the language closely protected. For example, they want to ban the use of such foreign adaptations as *le weekend* and *le sandwich.*

• **Luxembourg** Going to the polls is compulsory for voters in this constitutional monarchy.

standards of living. In eastern Europe once-powerful empires faced growing challenges from ethnic groups that wanted independence.

Competition for land and trade among Europe's nations led to World War I and World War II. These global conflicts put an end to western Europe's worldwide power. After World War II, Europe was divided into communist eastern Europe and noncommunist western Europe. The fall of communism in the late 1980s brought a new era for Europeans. Most western European countries joined the European Union to bring their economies and governments closer together.

The Arts The arts of Europe developed over many centuries. The ancient Greeks and Romans left behind graceful temples, such as the Parthenon in Athens and the Pantheon in Rome. Notre Dame Cathedral in Paris reflects Europe's Christian heritage. The Gothic style of many of Europe's cathedrals allowed builders to replace large portions of walls with stained-glass windows.

European artists of the Renaissance focused on everyday subjects as well as religious themes. After the Renaissance, new music forms—such as opera and the symphony—developed. During the 1900s, artists experimented with shapes and colors instead of showing realistic scenes.

Tiranë, Albania

The main square of Albania's capital is home to the country's national museum.
PLACE: How did the ancient Greeks influence Europe?

REVIEW AND ACTIVITIES

1. **PLACE** What is Europe's highest mountain range?
2. **REGION** Why does much of Europe enjoy mild temperatures year-round?
3. **HUMAN/ENVIRONMENT INTERACTION** How have Europe's industries affected many of its woodland areas?
4. **LOCATION** Where do most Europeans live?
5. **REGION** What changes have come to Europe since the late 1980s?
6. **REGION** Why was the Renaissance important to Europe?

245

Technology Activity

Developing a Multimedia Presentation You are planning a tour to three European countries. Use the Collector's button in National Geographic's "Picture Atlas of the World" CD-ROM to explore the countries. Incorporate and organize images and text from the CD-ROM as the basis for a multimedia presentation that will interest students.

Answers to Review and Activities

1. Alps
2. The North Atlantic Current brings warm waters to Europe's west coast.
3. Air pollution from European factories has harmed many woodland areas.
4. in the major cities and their suburbs
5. the fall of communism in eastern Europe and the formation of the European Union that brought many western European governments and economies closer together
6. The Renaissance advanced learning and the value of the individual.

LESSON PLAN
Unit 4 Atlas

UNIT 4 ATLAS

Physical Geography

Applying Geographic Themes

Human/Environment Interaction Much of the Netherlands lies below sea level. Find this area on the map. In 1995 rivers swollen by heavy rains and melting snow threatened to flood this lowland and cause the Netherlands' greatest natural disaster since 1953. In that year, the sea broke through the country's extensive system of dams and dikes, killing more than 1,800 people. What sea borders the Netherlands? *(North Sea)*

NATIONAL GEOGRAPHIC SOCIETY

These materials are available from Glencoe.

 CD-ROM

Picture Atlas of the World

MAP STUDY

Answers
1. Spain and Portugal
2. Baltic Sea

Map Skills Practice

Reading a Map What river forms part of the border between Romania and Bulgaria? *(Danube)* What body of water joins the Atlantic Ocean and the Mediterranean Sea? *(Strait of Gibraltar)*

EUROPE: Physical

ELEVATIONS

Feet	Meters
10,000	3,000
5,000	1,500
2,000	600
1,000	300
0	0

ARCTIC OCEAN

ICELAND

Arctic Circle

Norwegian Sea

SCANDINAVIAN PENINSULA

NORWAY

FINLAND

SWEDEN

Shetland Is.

ESTONIA

Orkney Is.

LATVIA

North Sea

Jutland Peninsula DENMARK

Baltic Sea

LITHUANIA

UNITED KINGDOM

IRELAND

NETHERLANDS

Elbe R.

EUROPEAN PLAIN

Vistula R.

GERMANY

POLAND

English Channel

Thames R.

ATLANTIC OCEAN

BELGIUM

NORTH

CZECH REPUBLIC

CARPATHIAN MTS.

SLOVAKIA

LUXEMBOURG

Loire R.

Seine R.

LIECHTENSTEIN

AUSTRIA

Hungarian Plain

HUNGARY

ROMANIA

FRANCE

Mt. Blanc 15,771 ft. (4,807 m)

SWITZERLAND

Rhone R.

SLOVENIA

Gran Paradiso 13,323 ft. (4,061 m)

Po R.

CROATIA

YUGOSLAVIA

Danube R.

IBERIAN

PYRENEES

ANDORRA

SAN MARINO

BOSNIA HERZEGOVINA

BULGARIA

BALKAN MTS.

PORTUGAL

Douro R.

Ebro R.

MONACO

APENNINES

ITALY

Adriatic Sea

MACEDONIA

ALBANIA

BALKAN

SPAIN

Corsica

GREECE

PENINSULA

Sardinia

Aegean Sea

Strait of Gibraltar

Sicily

Mediterranean Sea

MALTA

Crete

▲ Mountain peak

Lambert Conformal Conic projection

0 125 250 mi.
0 125 250 km

ASIA

CYPRUS

Mediterranean Sea

(globe inset)

Map Study

1. **LOCATION** What two countries are located on the Iberian Peninsula?
2. **PLACE** What sea separates the Scandinavian Peninsula from Poland?

246

Atlas Activity

Location On the chalkboard, copy this puzzle, omitting the letters above the blanks.

A E G E A N   S E A
 E L B E   R I V E R
P O   R I V E R
 G R A N   P A R A D I S O
C A R P A T H I A N   M O U N T A I N S
 A L P S

A P E N N I N E S
R H O N E   R I V E R
P Y R E N E E S

Have students complete the acrostic with names of physical features from the map. Provide hints, such as "Mountain range in Italy." **L1**

ELEVATION PROFILE: Europe

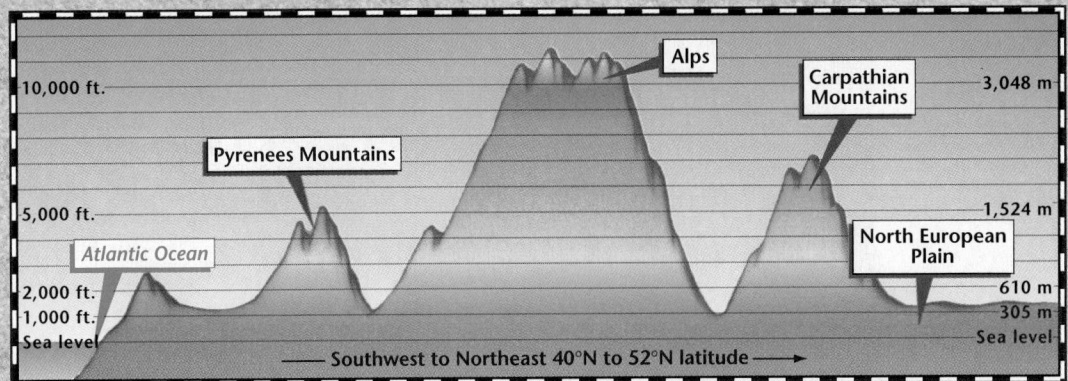

- 10,000 ft.
- 5,000 ft.
- 2,000 ft.
- 1,000 ft.
- Sea level

Atlantic Ocean

Alps

Carpathian Mountains

Pyrenees Mountains

North European Plain

- 3,048 m
- 1,524 m
- 610 m
- 305 m
- Sea level

◄— Southwest to Northeast 40°N to 52°N latitude —►

 VIDEODISC

The Infinite Voyage: Crisis in the Atmosphere

Chapter 9
Energy Conservation: Changes in Switzerland
Disc 2 Side A

Reuters Issues in Geography

Chapter 3
Europe: United or Divided
Disc 1 Side A

GeoFacts

Highest mountain peak: Mont Blanc (France-Italy) 15,771 ft. (4,807 m) high

Longest river: Danube (central Europe) 1,776 mi. (2,858 km) long

Largest lake: Vänern (Sweden) 2,156 sq. mi. (5,584 sq. km)

Highest waterfall: Mardalsfossen [Southern] (Norway) 2,149 ft. (655 m) high

EUROPE AND THE UNITED STATES: Land Comparison

Graphic Study

1. **LOCATION** What are the highest mountains in Europe?
2. **REGION** Look at the land comparison map of Europe and the United States. Knowing that Europe's population is almost double that of the United States, what can you infer about Europe's population density?

247

GRAPHIC STUDY

Answers
1. the Alps
2. Europe's population density is greater than that of the United States.

Graphic Skills Practice
Reading a Graph What is the elevation of the Pyrenees? *(just over 5,000 feet or 1,524 m)*

More About . . .

Region Remind students that the *continent* of Europe extends to the Ural Mountains in Russia. The *region* of Europe—as discussed in Unit 4—includes the countries shown on the map on page 246. This distinction will help explain why, for example, Mont Blanc is the highest mountain in the *region* of Europe and Mount Elbrus (in Russia) is the highest mountain on the *continent* of Europe.

FactsOnFile

Use the reproducible masters from **GEOGRAPHY ON FILE**, *Europe,* to enrich unit content.

LESSON PLAN
Unit 4 Atlas

Cultural Geography

UNIT 4 ATLAS

Applying Geographic Themes

Location San Marino is the world's smallest republic and Europe's oldest existing state. Its origins as an independent state reach back about 1,600 years. Its inhabitants, however, do not have their own language. Find San Marino on the map. Based on the country's location, what language do its people probably speak? *(Italian)*

These materials are available from Glencoe.

VIDEODISC

STV: World Geography Vol. 2: Africa and Europe

Europe
Side 2, Frames 00001-47934

SOFTWARE

ZipZapMap! World

EUROPE: Political

National boundary
⊛ National capital

Lambert Conformal Conic projection

ARCTIC OCEAN

Arctic Circle

ICELAND
⊛ Reykjavik

Norwegian Sea

SWEDEN   FINLAND

NORWAY   Helsinki

Oslo   Stockholm   ⊛ Tallinn
          Baltic   ESTONIA
          Sea
                    ⊛ Riga   LATVIA
North   LITHUANIA
Sea   DENMARK   ⊛ Vilnius
UNITED   Copenhagen
KINGDOM
⊛ Dublin   NETHERLANDS
IRELAND   Warsaw
         London   ⊛ Amsterdam   Berlin   ⊛
                              GERMANY   POLAND
ATLANTIC   ⊛ Brussels
                BELGIUM   Prague
OCEAN   ⊛ Luxembourg   ⊛
         Paris   LUXEMBOURG   CZECH   SLOVAKIA
                    LIECHTEN-   REPUBLIC ⊛ Bratislava
                    STEIN   Vienna   ⊛ Budapest
                    AUSTRIA   HUNGARY   ROMANIA
         FRANCE   Bern ⊛   Ljubljana ⊛   Zagreb   Bucharest
                    SWITZER-   SLOVENIA ⊛   ⊛ Belgrade   ⊛
                    LAND   CROATIA   YUGOSLAVIA
                    SAN   BOSNIA-
                    MARINO   HERZEGOVINA   BULGARIA
                    MONACO   ⊛ Sarajevo   SERBIA ⊛ Sofia
PORTUGAL   ANDORRA   ITALY   ⊛ MONT.   ⊛ Skopje
         ⊛ Madrid   ⊛ Rome   Tiranë ⊛   MACEDONIA
⊛ Lisbon   SPAIN   ALBANIA
                                        GREECE
                              Mediterranean   ⊛ Athens
                                   Sea
                    MALTA
                    ⊛ Valletta

ASIA
Nicosia
CYPRUS
Mediterranean Sea

0   125   250 mi.
0   125   250 km

Map Study

1. **LOCATION** What large island country lies about 600 miles northwest of Norway?
2. **PLACE** What is the capital of Bosnia-Herzegovina?

248

UNIT 4

MAP STUDY

Answers
1. Iceland
2. Sarajevo

Map Skills Practice
Reading a Map What two countries are located on the southern border of Poland? *(Czech Republic, Slovakia)*

Atlas Activity

Place Assign each student a European country. Have students divide their assigned country's population by its area. (See **Countries at a Glance** for statistics.) Explain that the answer gives the country's population density. Have students use their answers to list on the chalkboard the countries in the order of population density. Then have students compare the list on the chalkboard with the population and area statistics in the **Countries at a Glance** entries. Ask them to draw conclusions from their comparison. *(Answers will vary. Students may say that the most populous countries are not necessarily the most densely populated or that many small countries have high population densities.)* **L1**

248

COMPARING POPULATION: Selected European Countries

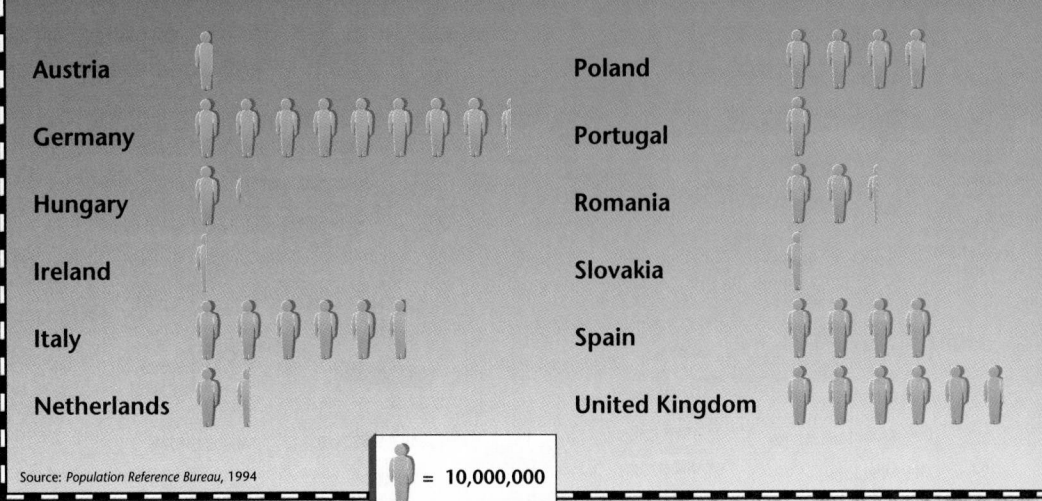

Austria

Germany

Hungary

Ireland

Italy

Netherlands

Poland

Portugal

Romania

Slovakia

Spain

United Kingdom

Source: *Population Reference Bureau, 1994*

= 10,000,000

GeoFacts

Biggest country (land area): France
212,390 sq. mi. (550,090 sq. km)

Smallest country (land area): Vatican
City .17 sq. mi. (.44 sq. km)

Largest city (population): London
(1995) 8,897,000; (2000 projected) 8,574,000

Highest population density:
Monaco 49,520 people per sq.
mi. (18,570 people per sq. km)

Lowest population density: Iceland
8 people per sq. mi. (3 people
per sq. km)

COMPARING POPULATION:
Europe and the United States

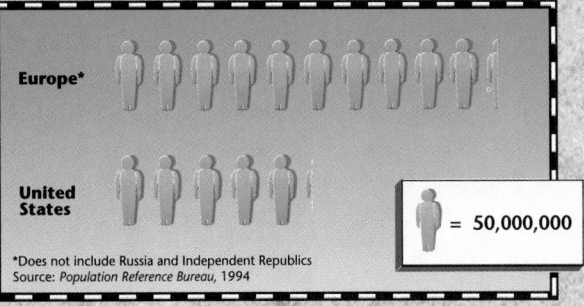

Europe*

United
States

= 50,000,000

*Does not include Russia and Independent Republics
Source: *Population Reference Bureau, 1994*

Graphic Study

1. **PLACE** About how many people live in
Europe?
2. **REGION** How much difference is there
between the most heavily populated and most
sparsely populated countries on the graph?

249

Teacher Notes

LESSON PLAN
Unit 4 Atlas

Countries at a Glance

UNIT 4 ATLAS

Global Gourmet

Italy Spaghetti and meatballs is not an authentic Italian dish. Cooks in Italy serve meat and pasta as two separate courses.

Cultural Kaleidoscope

Austria Dinner guests in Austria bring their host an odd number of flowers because an even number is considered unlucky. They also avoid roses, which symbolize romantic love.

Albania
CAPITAL:
Tiranë
MAJOR LANGUAGE(S):
Albanian, Greek
POPULATION:
3,400,000
LANDMASS:
10,580 sq. mi./
27,402 sq. km
MONEY:
Lek
MAJOR EXPORT:
Fuels
MAJOR IMPORT:
Machinery

Andorra
CAPITAL:
Andorra la Vella
MAJOR LANGUAGE(S):
Catalan
POPULATION:
54,000
LANDMASS:
185 sq. mi./
479 sq. km
MONEY:
French franc,
Spanish peseta
MAJOR EXPORT:
Clothing
MAJOR IMPORT:
Electronic Equipment

Austria
CAPITAL:
Vienna
MAJOR LANGUAGE(S):
German
POPULATION:
8,000,000
LANDMASS:
31,940 sq. mi./
82,725 sq. km
MONEY:
Schilling
MAJOR EXPORT:
Machinery
MAJOR IMPORT:
Machinery

Belgium
CAPITAL:
Brussels
MAJOR LANGUAGE(S):
Flemish, French,
Italian, German
POPULATION:
10,100,000
LANDMASS:
11,750 sq. mi./
30,433 sq. km
MONEY:
Franc
MAJOR EXPORT:
Machinery
MAJOR IMPORT:
Machinery

Bosnia and Herzegovina
CAPITAL:
Sarajevo
MAJOR LANGUAGE(S):
Serbo-Croatian
POPULATION:
4,600,000
LANDMASS:
19,740 sq. mi./
51,127 sq. km
MONEY:
New Yugoslav dinar
MAJOR EXPORT:
Machinery
MAJOR IMPORT:
Fuels

Bulgaria
CAPITAL:
Sofia
MAJOR LANGUAGE(S):
Bulgarian, Turkish
POPULATION:
8,400,000
LANDMASS:
42,680 sq. mi./
110,541 sq. km
MONEY:
Leva
MAJOR EXPORT:
Machinery
MAJOR IMPORT:
Machinery

Croatia
CAPITAL:
Zagreb
MAJOR LANGUAGE(S):
Croatian
POPULATION:
4,800,000
LANDMASS:
21,830 sq. mi./
56,540 sq. km
MONEY:
Croatian dinar
MAJOR EXPORT:
Machinery
MAJOR IMPORT:
Machinery

Cyprus
CAPITAL:
Nicosia
MAJOR LANGUAGE(S):
Greek, Turkish, English
POPULATION:
700,000
LANDMASS:
3,570 sq. mi./
9,246 sq. km
MONEY:
Pound
MAJOR EXPORT:
Clothing
MAJOR IMPORT:
Transport Equipment

Czech Republic
CAPITAL:
Prague
MAJOR LANGUAGE(S):
Czech
POPULATION:
10,300,000
LANDMASS:
30,590 sq. mi./
79,228 sq. km
MONEY:
Koruna
MAJOR EXPORT:
Chemicals
MAJOR IMPORT:
Chemicals

Denmark
CAPITAL:
Copenhagen
MAJOR LANGUAGE(S):
Danish
POPULATION:
5,200,000
LANDMASS:
16,360 sq. mi./
42,372 sq. km
MONEY:
Krone
MAJOR EXPORT:
Machinery
MAJOR IMPORT:
Machinery

Countries not drawn to scale.

UNIT 4

Fun Facts

- **Andorra** The tiny state of Andorra has two heads of state—the French president and a Spanish bishop. It does not however, have an army, a constitution, or an income tax.

- **France** About 30,000 years ago, bison and rhinoceroses wandered what is now France. How do we know? In December 1994, a French official discovered cave paintings of these animals that date back to the Stone Age.

- **Italy** Rome was home to the first professional barbers, who set up shop in 303 B.C. Early Romans prized dark hair as much as a good cut, and they used dyes made by boiling walnut shells and leeks. These dyes were so harsh that they often made the Romans' hair fall out.

Estonia

CAPITAL:
Tallinn
MAJOR LANGUAGE(S):
Estonian, Russian
POPULATION:
1,500,000
LANDMASS:
17,410 sq. mi./
45,092 sq. km
MONEY:
Kroon
MAJOR EXPORT:
Textiles and Clothing
MAJOR IMPORT:
Textiles and Clothing

Finland

CAPITAL:
Helsinki
MAJOR LANGUAGE(S):
Finnish, Swedish
POPULATION:
5,100,000
LANDMASS:
117,610 sq. mi./
304,610 sq. km
MONEY:
Markka
MAJOR EXPORT:
Metal Products
MAJOR IMPORT:
Fuels

France

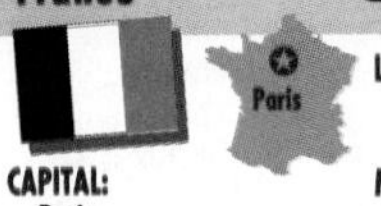

CAPITAL:
Paris
MAJOR LANGUAGE(S):
French
POPULATION:
58,000,000
LANDMASS:
212,390 sq. mi./
550,090 sq. km
MONEY:
Franc
MAJOR EXPORT:
Machinery
MAJOR IMPORT:
Machinery

Germany

CAPITAL:
Berlin
MAJOR LANGUAGE(S):
German
POPULATION:
81,200,000
LANDMASS:
134,930 sq. mi./
349,469 sq. km
MONEY:
Mark
MAJOR EXPORT:
Machinery
MAJOR IMPORT:
Machinery

Greece

CAPITAL:
Athens
MAJOR LANGUAGE(S):
Greek
POPULATION:
10,400,000
LANDMASS:
50,520 sq. mi./
130,847 sq. km
MONEY:
Drachma
MAJOR EXPORT:
Food and Beverages
MAJOR IMPORT:
Machinery

Hungary

CAPITAL:
Budapest
MAJOR LANGUAGE(S):
Hungarian
POPULATION:
10,300,000
LANDMASS:
35,650 sq. mi./
92,334 sq. km
MONEY:
Forint
MAJOR EXPORT:
Food and Live Animals
MAJOR IMPORT:
Machinery

Iceland

CAPITAL:
Reykjavik
MAJOR LANGUAGE(S):
Icelandic
POPULATION:
300,000
LANDMASS:
38,710 sq. mi./
100,259 sq. km
MONEY:
Kronur
MAJOR EXPORT:
Seafood
MAJOR IMPORT:
Ships

Ireland

CAPITAL:
Dublin
MAJOR LANGUAGE(S):
English, Irish
POPULATION:
3,600,000
LANDMASS:
26,600 sq. mi./
68,894 sq. km
MONEY:
Punt
MAJOR EXPORT:
Machinery
MAJOR IMPORT:
Machinery

Italy

CAPITAL:
Rome
MAJOR LANGUAGE(S):
Italian
POPULATION:
57,200,000
LANDMASS:
113,540 sq. mi./
294,069 sq. km
MONEY:
Lira
MAJOR EXPORT:
Machinery
MAJOR IMPORT:
Machinery

Latvia

CAPITAL:
Riga
MAJOR LANGUAGE(S):
Latvian
POPULATION:
2,500,000
LANDMASS:
24,900 sq. mi./
64,491 sq. km
MONEY:
Latvian ruble
MAJOR EXPORT:
Machinery
MAJOR IMPORT:
Textiles

Cultural Kaleidoscope

Hungary Hungarians usually greet each other by shaking hands. If their hands are dirty, however, they offer their elbows.

Strange but TRUE!

Finland has about 60,000 lakes. Together, these lakes occupy about one-tenth of the country's total area.

NATIONAL GEOGRAPHIC SOCIETY

These materials are available from Glencoe.

 SOFTWARE

ZipZapMap! World

 REFERENCE BOOKS

Picture Atlas of Our World

Teacher Notes

LESSON PLAN
Unit 4 Atlas

Countries at a Glance

Liechtenstein
CAPITAL: Vaduz
MAJOR LANGUAGE(S): German, Alemannic dialect
POPULATION: 30,000
LANDMASS: 60 sq. mi./ 155 sq. km
MONEY: Swiss franc
MAJOR EXPORT: Machinery
MAJOR IMPORT: Machinery

Lithuania
CAPITAL: Vilnius
MAJOR LANGUAGE(S): Lithuanian
POPULATION: 3,700,000
LANDMASS: 25,210 sq. mi./ 65,294 sq. km
MONEY: Lit
MAJOR EXPORT: Machinery
MAJOR IMPORT: Petroleum and Gas

Luxembourg
CAPITAL: Luxembourg
MAJOR LANGUAGE(S): French, German, Letzeburgesch
POPULATION: 400,000
LANDMASS: 990 sq. mi./ 2,564 sq. km
MONEY: Franc
MAJOR EXPORT: Metal Products
MAJOR IMPORT: Metal Products

Macedonia
CAPITAL: Skopje
MAJOR LANGUAGE(S): Macedonian, Albanian, Bulgarian, Greek
POPULATION: 2,100,000
LANDMASS: 9,930 sq. mi. 25,719 sq. km
MONEY: Denar
MAJOR EXPORT: Clothing and Footwear
MAJOR IMPORT: Food and Beverages

Malta
CAPITAL: Valletta
MAJOR LANGUAGE(S): Maltese, English
POPULATION: 400,000
LANDMASS: 120 sq. mi./ 311 sq. km
MONEY: Maltese lira
MAJOR EXPORT: Machinery
MAJOR IMPORT: Machinery

Monaco
CAPITAL: Monaco
MAJOR LANGUAGE(S): French
POPULATION: 29,712
LANDMASS: 0.6 sq. mi./ 1.6 sq. km
MONEY: French franc or Monégasque franc

Netherlands
CAPITAL: Amsterdam
MAJOR LANGUAGE(S): Dutch
POPULATION: 15,400,000
LANDMASS: 13,100 sq. mi./ 33,929 sq. km
MONEY: Guilder
MAJOR EXPORT: Machinery
MAJOR IMPORT: Machinery

Norway
CAPITAL: Oslo
MAJOR LANGUAGE(S): Norwegian
POPULATION: 4,300,000
LANDMASS: 118,470 sq. mi./ 306,837 sq. km
MONEY: Kroner
MAJOR EXPORT: Fuels
MAJOR IMPORT: Machinery

Poland
CAPITAL: Warsaw
MAJOR LANGUAGE(S): Polish
POPULATION: 38,600,000
LANDMASS: 117,550 sq. mi./ 304,455 sq. km
MONEY: Zloty
MAJOR EXPORT: Machinery
MAJOR IMPORT: Machinery

Portugal
CAPITAL: Lisbon
MAJOR LANGUAGE(S): Portuguese
POPULATION: 9,900,000
LANDMASS: 35,500 sq. mi./ 91,945 sq. km
MONEY: Escudo
MAJOR EXPORT: Textiles and Apparel
MAJOR IMPORT: Motor Vehicles

Countries not drawn to scale.

UNIT 4

Fun Facts

- **Luxembourg** A recent survey revealed that "sleeping" and "resting" are the favorite activities of people in Luxembourg.

- **Sweden** Swedish Easter is similar to American Halloween. Children dress as witches—brooms and all—and go from house to house collecting candy.

- **Vatican City** The Vatican Library, established in the 1400s, is the oldest public library in the world.

Romania

CAPITAL:
Bucharest
MAJOR LANGUAGE(S):
Romanian, Hungarian, German
POPULATION:
22,700,000
LANDMASS:
88,930 sq. mi./
230,329 sq. km
MONEY:
Leu
MAJOR EXPORT:
Mineral Fuels
MAJOR IMPORT:
Raw Materials

San Marino

CAPITAL:
San Marino
MAJOR LANGUAGE(S):
Italian
POPULATION:
20,000
LANDMASS:
20 sq. mi./
52 sq. km
MONEY:
Italian lira
MAJOR EXPORT:
Wine
MAJOR IMPORT:
Oil

Slovakia

CAPITAL:
Bratislava
MAJOR LANGUAGE(S):
Slovak
POPULATION:
5,300,000
LANDMASS:
18,790 sq. mi./
48,666 sq. km
MONEY:
New koruna
MAJOR EXPORT:
Machinery
MAJOR IMPORT:
Petroleum

Slovenia

CAPITAL:
Ljubljana
MAJOR LANGUAGE(S):
Slovenian, Serbo-Croatian
POPULATION:
2,000,000
LANDMASS:
7,820 sq. mi./
20,254 sq. km
MONEY:
Tolar
MAJOR EXPORT:
Machinery
MAJOR IMPORT:
Machinery

Spain

CAPITAL:
Madrid
MAJOR LANGUAGE(S):
Spanish, Catalan, Galician, Basque
POPULATION:
39,200,000
LANDMASS:
192,830 sq. mi./
499,430 sq. km
MONEY:
Peseta
MAJOR EXPORT:
Transport Equipment
MAJOR IMPORT:
Machinery

Sweden

CAPITAL:
Stockholm
MAJOR LANGUAGE(S):
Swedish
POPULATION:
8,800,000
LANDMASS:
158,930 sq. mi./
411,629 sq. km
MONEY:
Krona
MAJOR EXPORT:
Machinery
MAJOR IMPORT:
Machinery

Switzerland

CAPITAL:
Bern
MAJOR LANGUAGE(S):
German, French, Italian
POPULATION:
7,000,000
LANDMASS:
15,360 sq. mi./
39,782 sq. km
MONEY:
Franc
MAJOR EXPORT:
Chemical Products
MAJOR IMPORT:
Machinery and Electronics

United Kingdom

CAPITAL:
London
MAJOR LANGUAGE(S):
English, Welsh, Gaelic
POPULATION:
58,400,000
LANDMASS:
93,280 sq. mi./
241,595 sq. km
MONEY:
Pound
MAJOR EXPORT:
Machinery
MAJOR IMPORT:
Machinery

Vatican City

MAJOR LANGUAGE(S):
Italian, Latin
POPULATION:
802
LANDMASS:
.17 sq. mi./
.44 sq. km
MONEY:
Lira

Yugoslavia

CAPITAL:
Belgrade
MAJOR LANGUAGE(S):
Serbian, Macedonian, Hungarian, Albanian
POPULATION:
10,500,000
LANDMASS:
26,940 sq. mi./
69,775 sq. km
MONEY:
Dinar
MAJOR EXPORT:
Clothing
MAJOR IMPORT:
Machinery

Although Norway's Atlantic coast stretches only about 1,500 miles (2,414 km), its many fjords and islands add another 10,000 miles (16,090 km) of shoreline.

Geography and the Humanities

📁 World Literature Reading 4

🌐 World Music: Cultural Traditions, Lesson 3

📦 World Cultures Transparencies 5, 6

📕 Focus on World Art Prints 1, 2, 5, 6, 8, 9, 14, 15, 17, 18, 19, 20, 21, 22, 23

253

Teacher Notes

The British Isles and Scandinavia

PERFORMANCE ASSESSMENT ACTIVITY

POSSIBLE RUBRIC FEATURES: Organization skills, inferencing skills, concept attainment, accuracy of content information, ability to identify relationships

MAKING AN ITINERARY Many Americans travel to the United Kingdom, Ireland, and Scandinavia every year. Have students imagine that they are able to make several trips to the area each year. Ask students to make four itineraries, each to a different country and at different times of the year. The itineraries should include information such as the clothes they would take, the places of interest they would visit, the activities they might take part in, the transportation they might use, and items they might want to remember to bring back for family members and friends. These itineraries should demonstrate students' knowledge of the geographic features of the region.

MENTAL MAPPING ACTIVITY

Before teaching Chapter 10, use a globe to point out the British Isles and Scandinavia to students. Have them identify the hemisphere in which the British Isles and Scandinavia are located and tell where the countries are located relative to the United States.

TEACHER'S CORNER

NATIONAL GEOGRAPHIC SOCIETY

INDEX TO NATIONAL GEOGRAPHIC MAGAZINE

The following articles may be used for research relating to this chapter:

- "Scotland," by Andrew Ward, September 1996.
- "The Aran Islands," by Lisa Moore LaRoe, April 1996.
- "Oxford," by Bill Bryson, November 1995.
- "Saving Britain's Shore," by Alan Mairson, October 1995.
- "Ireland on Fast-Forward," by Richard Conniff, September 1994.

- "England's Lake District," by Bill Bryson, August 1994.
- "English Channel Tunnel," by Thomas B. Allen, May 1994.
- "The Wings of War," by Thomas B. Allen, March 1994.
- "Britain's Hedgerows," by Bill Bryson, September 1993.
- "Sweden in Search of a New Model," by Don Belt, August 1993.

NATIONAL GEOGRAPHIC SOCIETY PRODUCTS AVAILABLE FROM GLENCOE

To order the following products for use with this chapter, contact your local Glencoe sales representative or call Glencoe at 1-800-334-7344:

- *STV: World Geography* (Videodisc)
- *Picture Atlas of the World* (CD-ROM)
- *Physical Geography of the World* (Transparencies)
- *ZipZapMap! World* (Software)

- *GeoBee* (Software)
- *Images of the World* (Posters)
- *Eye on the Environment* (Posters)

ADDITIONAL NATIONAL GEOGRAPHIC SOCIETY PRODUCTS

To order the following products for use with this chapter, call National Geographic Society at 1-800-368-2728:

- *Democratic Governments Series:* "United Kingdom." (Video)
- *1914-1918: World War I* (Video)
- *Physical Geography of the Continents Series:* "Europe." (Video)
- *Ballad of the Irish Horse* (Video)

- *Portraits of the Continents Series:* "Part III: Europe; Asia; Africa." (Filmstrip)
- *Geography of Europe Series:* "Pt. I: Northern Europe;" "Pt. II: Western Europe;" "Pt. III: Central Europe;" "Pt. IV: Southern Europe." (Filmstrip)

TEACHER-TO-TEACHER
Curriculum Connection: Earth Science, History

—from Charles M. Bateman,
Halls Middle School, Knoxville, TN

STONEHENGE *The purpose of this activity is to help students understand the significance and function of ancient monuments.*

Display a model sundial and, using a flashlight, demonstrate how it can be employed to tell time. Explain how the revolution of the earth around the sun causes the sun to change its position in the sky throughout the year. Point out that to have a fairly accurate idea of the time, the sundial user must know the approximate day of the year so that adjustments can be made for the sun's position in the sky. Then display a large picture or diagram of Stonehenge. Explain how the builders of this monument used it to mark the summer solstice.

Organize students into groups and provide each group with small rocks and stones of various sizes, and a flashlight. Direct groups to use the rocks and stones to build their own model of Stonehenge. Then have each group, in turn, demonstrate the purpose of Stonehenge, using the flashlight to simulate the rising sun. Some groups may wish to use their models to teach other classes about Stonehenge.

As a wrap-up activity, have students identify and discuss other ancient monuments that are used for astronomical purposes.

MEETING NATIONAL STANDARDS

Geography for Life

All of the 18 standards are demonstrated in Unit 4. The following ones are highlighted in this chapter:
- Standards 3, 4, 6, 9, 10, 11, 12, 13, 15, 17

KEY TO ABILITY LEVELS

Teaching strategies have been coded for varying learning styles and abilities.

L1   **BASIC** activities for all students

L2   **AVERAGE** activities for average to above-average students

L3   **CHALLENGING** activities for above-average students

LEP   **LIMITED ENGLISH PROFICIENCY** activities

BIBLIOGRAPHY

Readings for the Student
Lander, Patricia S., and Claudette Charbonneau. *The Land and People of Finland.* New York: Harper-Collins, 1990.

Langley, Andrew. *Passport to Great Britain,* revised ed. New York: Franklin Watts, 1994.

Tames, Richard, and Sheila Tames. *Great Britain.* New York: Franklin Watts, 1994.

Readings for the Teacher
Europe Today: An Atlas of Reproducible Pages, revised ed. Wellesley, Mass.: World Eagle, 1993.

Mapping Europe: A Curriculum Unit for Grades 6-10. Stanford, Calif.: SPICE, 1992. Lesson plans with reproducible pages.

Multimedia
Europe. Washington, D.C.: National Geographic Society, 1991. Videocassette, 26 minutes.

European Geography: Insular Region. Chicago: Encyclopedia Britannica, 1993. Videocassette, 20 minutes.

European Geography: Northern Region. Chicago: Encyclopedia Britannica, 1993. Videocassette, 20 minutes.

The British Isles and Scandinavia

CHAPTER 10 RESOURCES

- *Vocabulary Activity 10
- Cooperative Learning Activity 10
- Workbook Activity 10
- Geography Skills Activity 10
- GeoLab Activity 10
- Chapter Map Activity 10
- Performance Assessment Activity 10
- *Reteaching Activity 10
- Enrichment Activity 10
- Chapter 10 Test, Form A and Form B
- *Chapter 10 Digest Audiocassette, Activity, Test
- Unit Overlay Transparencies 4-0, 4-1, 4-2, 4-3, 4-4, 4-5, 4-6
- Geography Handbook Transparency 14
- Political Map Transparency 4; World Cultures Transparency 5
- Geoquiz Transparencies 10-1, 10-2
- Vocabulary PuzzleMaker Software
- Student Self-Test: A Software Review
- Testmaker Software
- MindJogger Videoquiz
- Focus on World Art Print 15

Use these *Geography: The World and Its People* resources to teach, reinforce, and extend chapter content.

If time does not permit teaching the entire chapter, summarize using the Chapter 10 Highlights on page 271, and the Chapter 10 English (or Spanish) Audiocassettes. Review students' knowledge using the Glencoe MindJogger Videoquiz. *Also available in Spanish

Use these Geography: The World and Its People resources to teach and reinforce section content.

SECTION 1 RESOURCES

- Reproducible Lesson Plan 10-1
- Section Focus Transparency 10-1
- Guided Reading Activity 10-1
- Section Quiz 10-1 (also available in Spanish)

SECTION 2 RESOURCES

- Reproducible Lesson Plan 10-2
- Section Focus Transparency 10-2
- Guided Reading Activity 10-2
- Section Quiz 10-2 (also available in Spanish)

SECTION 3 RESOURCES

- Reproducible Lesson Plan 10-3
- Section Focus Transparency 10-3
- Guided Reading Activity 10-3
- Section Quiz 10-3 (also available in Spanish)

ADDITIONAL RESOURCES FROM GLENCOE

Reproducible Masters

- Glencoe Social Studies Outline Map Book, pages 29, 30
- Facts On File: GEOGRAPHY ON FILE, Europe
- Foods Around the World, Unit 4

World Music: Cultural Traditions

Lesson 3

Transparencies

- National Geographic Society PicturePack Transparencies, Unit 4

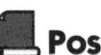

Laminated Desk Map

Posters

- National Geographic Society: Images of the World, Unit 4

Videodiscs

- National Geographic Society: STV: World Geography, Vol. 2
- National Geographic Society: GTV: Planetary Manager

CD-ROM

- National Geographic Society: Picture Atlas of the World

Software

- National Geographic Society: ZipZapMap! World

Reference Books

- Picture Atlas of Our World

**Performance
Assessment**

Refer to the Planning Guide on page 254A for a Performance Assessment Activity for this chapter. See the *Performance Assessment Strategies and Activities* booklet for suggestions.

Chapter Objectives

1. **Identify** the major geographic features of the United Kingdom.
2. **Describe** the geography of Ireland.
3. **Point out** geographic similarities and differences among the Scandinavian countries.

GLENCOE
TECHNOLOGY

 VIDEODISC

Use the Chapter 10 MindJogger Videoquiz to preview chapter content.

MindJogger Videoquiz

Chapter 10
Disc 2 Side A

Also available in VHS.

Chapter
10 The British Isles and Scandinavia

MAP STUDY ACTIVITY

As you read Chapter 10, you will learn about the British Isles and Scandinavia.

1. What countries in this region are islands?
2. What country lies farthest north?
3. What capital is farthest south?

254

MAP STUDY ACTIVITY

Answers
1. United Kingdom, Ireland, Iceland
2. Norway
3. London

Map Skills Practice
Reading a Map What city located in Ireland is actually a part of the United Kingdom? *(Belfast)*

SECTION 1
The United Kingdom

PREVIEW

Words to Know
- moor
- loch
- parliamentary democracy
- constitutional monarchy

Places to Locate
- England
- London
- Wales
- Scotland
- Northern Ireland

Read to Learn . . .
1. what landscapes and climate are found in the United Kingdom.
2. how the British earn their livings.
3. how the United Kingdom has influenced other countries of the world.

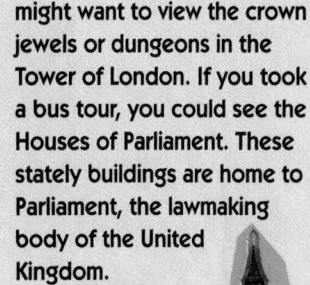

What sights would you want to see if you visited London, England? You might want to view the crown jewels or dungeons in the Tower of London. If you took a bus tour, you could see the Houses of Parliament. These stately buildings are home to Parliament, the lawmaking body of the United Kingdom.

What do you think of when you hear the words "British Isles"? Perhaps you have visions of the legendary King Arthur and his Knights of the Round Table. Maybe you hear a bagpipe echoing across the foggy highlands of Scotland. Or you might picture rock groups such as the Rolling Stones or U2.

The British Isles lie just off the Atlantic coast of western Europe. The islands form two countries: the United Kingdom and the Republic of Ireland. The United Kingdom is made up of four regions: England, Scotland, Wales, and Northern Ireland.

Region
The Land

Slightly smaller in area than the state of Oregon, the United Kingdom depends on the sea for many things. The country also lies close to the mainland of Europe. This closeness to both sea and mainland helped make the United Kingdom one of the world's major trading nations.

CHAPTER 10

(255)

FOCUS

Section Objectives
1. **Identify** the landscapes and climate that are found in the United Kingdom.
2. **Summarize** how the British earn their livings.
3. **Discuss** how the United Kingdom has influenced other countries of the world.

Bellringer Motivational Activity
 Prior to taking roll at the beginning of the class period, project Section Focus Transparency 10-1 and have students answer the activity questions. Discuss students' responses. **L1**

Vocabulary Pre-check
Ask students to look up the words *democracy* and *monarchy* in a dictionary. Then have them explain the difference between the two kinds of government. **L1** **LEP**

 Use the Vocabulary PuzzleMaker Software to create a crossword puzzle. **L1**

Meeting National Standards
Geography for Life The following standards are highlighted in this section:
- Standards 3, 6, 9, 11, 17

Classroom Resources for Section 1

 REPRODUCIBLE MASTERS

Reproducible Lesson Plan 10-1
Guided Reading Activity 10-1
Chapter Map Activity 10
Vocabulary Activity 10
Workbook Activity 10
Section Quiz 10-1

 **TRANSPARENCIES**

Section Focus Transparency 10-1
Unit Overlay Transparencies 4-0, 4-4
Political Map Transparency 4

MULTIMEDIA

 Vocabulary PuzzleMaker Software

Testmaker

National Geographic Society: GTV: Planetary Manager

MAP STUDY

Answer
North Sea

Map Skills Practice
Reading a Map What mountain range is located in the northern United Kingdom? *(Grampian Mountains)*

TEACH

Guided Practice

Display Unit Overlay Transparencies 4-0, 4-4. Have a student draw a line across the United Kingdom from the southwest to the northeast. After students read about the landscape of the United Kingdom, have them identify the side of the line that has plains and meadows and the side that has highlands. Ask a student to point out the lochs along Scotland's coast. **L1**

More About the Illustration

Answer to Caption
the warm North Atlantic Current influences climate

Human/ Environment Interaction

The British have many words to describe the different landforms in their country. *Downs*, for example, are hilly grassland areas; *fens* are marshy areas, and *wolds* are low chalk hills.

THE BRITISH ISLES AND SCANDINAVIA: Physical

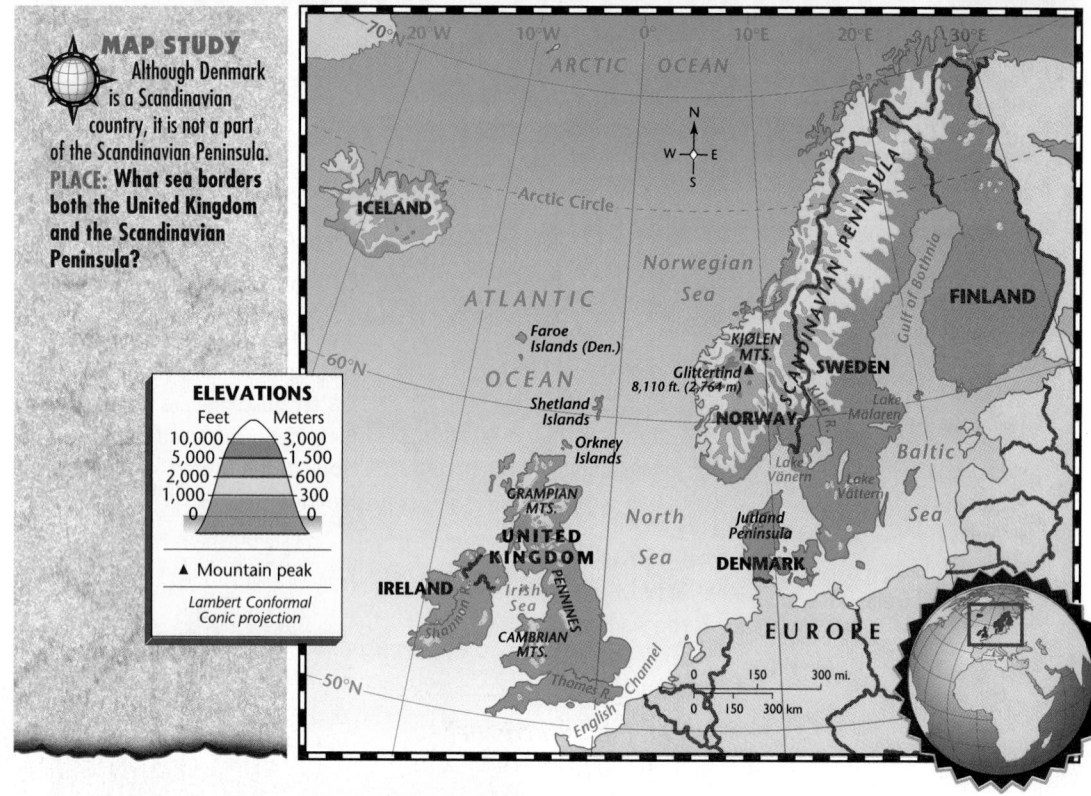

MAP STUDY
Although Denmark is a Scandinavian country, it is not a part of the Scandinavian Peninsula.
PLACE: What sea borders both the United Kingdom and the Scandinavian Peninsula?

ELEVATIONS

Feet	Meters
10,000	3,000
5,000	1,500
2,000	600
1,000	300
0	0

▲ Mountain peak

Lambert Conformal Conic projection

English Countryside

Rolling meadows surround this village in southern England.
PLACE: Why is the United Kingdom's climate mild compared to other countries at that latitude?

The Landscape If you drew an imaginary line across the United Kingdom from the southwest to the northeast, you would find lowlands east of this line. From the air you would see rolling plains covered by a patchwork of fertile fields and meadows.

West of this imaginary line lies a region of highland areas, most of which are **moors**—treeless landscapes often swept by strong winds. In Scotland narrow bays called **lochs** cut into the highland coasts. They reach far inland and are nearly surrounded by steep mountain slopes.

The Climate The map on page 258 shows you that the United Kingdom has a mild climate even though it lies as far north as Canada. What causes this? Winds blowing over the North Atlantic Current—an extension of the Gulf Stream—warm the United Kingdom in winter and cool it in summer. The sea winds also bring plenty of rain. If you visit the United Kingdom, take your raincoat! Foggy, damp weather is common throughout the country.

256

Cooperative Learning Activity

Organize students into small groups. Assign each group a different letter of the alphabet. Have groups research the United Kingdom, using encyclopedias and magazines such as *Europe* and *National Geographic,* to find a plant, animal, landform, and body of water for their assigned letters. Ask groups to draw the items they find and to place a box in the lower right corner of each drawing. In the box, have groups supply the following: letter, category, name, and where found. Display the drawings on the bulletin board in alphabetical order. **L1**

Human/Environment Interaction
The Economy

More than 200 years ago, inventors and scientists in the United Kingdom sparked the Industrial Revolution. Fuel-powered machines in factories began producing a greater variety and supply of goods. The Industrial Revolution made the United Kingdom the world's leading economic power during the 1800s. Although its economic influence declined in the latter half of the 1900s, the United Kingdom is still a major industrial and trading country.

Resources and Manufacturing The United Kingdom's leading natural resource is oil, pumped from under the North Sea. Other important energy resources are coal and natural gas. All three fuels help drive industries in the United Kingdom and are exported to other countries. The United Kingdom is Europe's leading oil producer.

Heavy machinery, ships, textiles, and cars were once the United Kingdom's major industrial products. Today most of the industries making these goods have declined because of stiff competition from other countries. New high-technology industries, especially computers and electronic equipment, are replacing the older industries.

Farming In recent years farmers in the United Kingdom have adopted modern agricultural methods. Yields are high, but the United Kingdom still must import about one-third of its food supply. Why? The country does not have enough farmland to support its large population. In addition, other countries in Europe produce food at lower costs than the United Kingdom does.

The main crops grown in the United Kingdom are wheat, barley, potatoes, fruits, and vegetables. British farmers also raise cattle and sheep, and the dairy industry thrives.

Place
The People

About 59 million people live in the United Kingdom. The capital, London, ranks as the most heavily populated city in Europe. Many people in the United States and other parts of the world trace their ancestry to the United Kingdom. The British people speak English, and most are Protestants.

Let the Games Begin!

Once a year, the sound of bagpipes signals the beginning of the Highland Games in Braemar, Scotland. Brian Dunbar and Patty McClory perform traditional Highland dances for the huge crowds that attend the games. The plaid, or tartan, on their kilts represents their clan—a group of related families with the same name. Patty's favorite competition is the tossing of the caber—a 100-pound pole.

Independent Practice

📁 Guided Reading Activity 10-1 **L1**

📁 Chapter Map Activity 10 **L1**

MULTICULTURAL PERSPECTIVE

Cultural Heritage
The nursery rhyme "London Bridge" refers to the actual destruction of an early wooden bridge that spanned the river Thames in London. Norway's King Olaf and his Viking raiders destroyed the bridge in the eleventh century.

NATIONAL GEOGRAPHIC SOCIETY

These materials are available from Glencoe.

 VIDEODISC

GTV: Planetary Manager

Oil drilling under North Sea
Any side, Frame 47568

Oil refining in Wales
Any side, Frame 47963

Meeting Special Needs Activity

Language Delayed A troublesome challenge for students with language difficulties is asking their teachers questions. Form small groups and ask group members to practice developing questions, using the text under "The Economy." Praise groups for producing varied questions. **L1**

MAP STUDY

Answer
Denmark
Map Skills Practice
Reading a Map Which countries have regions with subarctic climates? *(Norway, Sweden, Finland)*

CURRICULUM CONNECTION

Language Arts The plays of England's William Shakespeare are rich not only in poetry, but also in insults, such as "tickle-brain," "viper vile," and "purple-hued maltworms."

ASSESS

Check for Understanding

Assign Section 1 Review as homework or an in-class activity.

Meeting Lesson Objectives

Each objective below is tested by the questions that follow it in parentheses.

1. **Identify** the landscapes and climate that are found in the United Kingdom. (1, 2, 5)
2. **Summarize** how the British earn their livings. (3)
3. **Discuss** how the United Kingdom has influenced other countries of the world. (4)

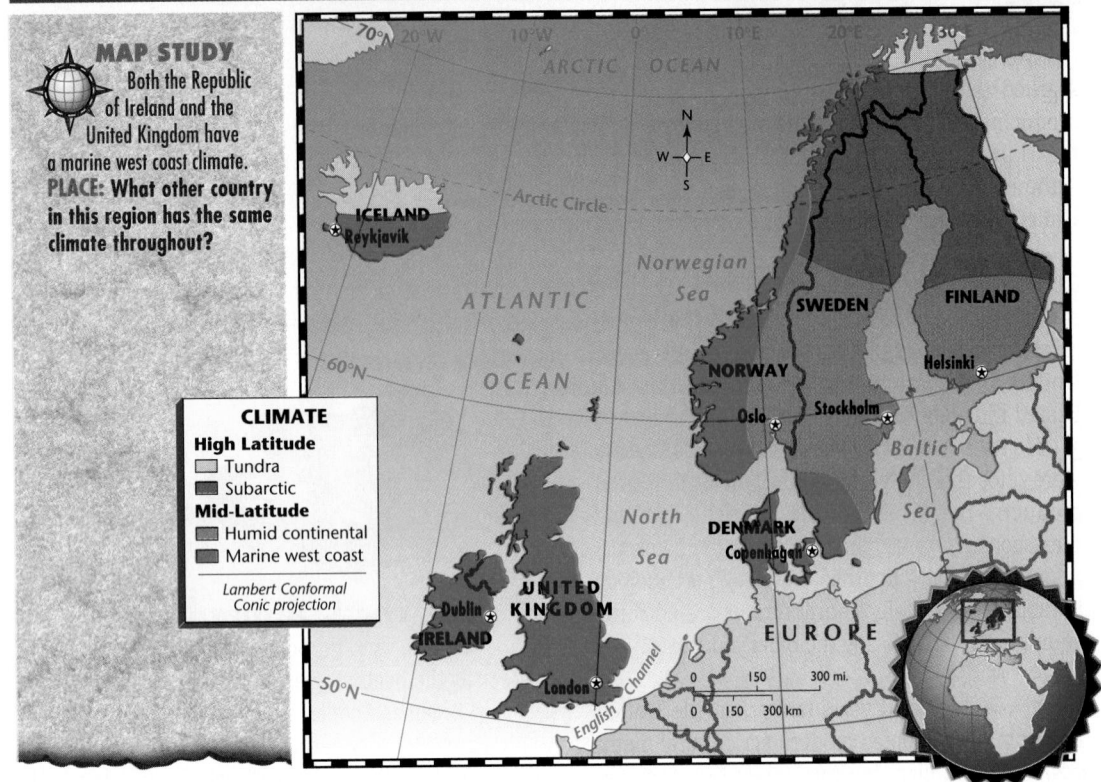

THE BRITISH ISLES AND SCANDINAVIA: Climate

MAP STUDY Both the Republic of Ireland and the United Kingdom have a marine west coast climate.
PLACE: What other country in this region has the same climate throughout?

CLIMATE
High Latitude
☐ Tundra
■ Subarctic
Mid-Latitude
■ Humid continental
■ Marine west coast

Lambert Conformal Conic projection

From Past to Present The British are descendants of various groups. Among early arrivals were the Celts (KEHLTS), who sailed from the European mainland around 500 B.C. Their descendants live today in Scotland, Wales, and Ireland. From the A.D. 400s to 1000s, various Anglo-Saxon groups from northern and western Europe settled England. Their descendants became known as the English.

England became a major European power in the late 1500s. In 1707 England and Wales united with Scotland to form the United Kingdom. A seafaring people, British traders, soldiers, and settlers won control of large areas of land throughout the world. By the mid-1800s, the United Kingdom governed the world's largest overseas empire. This empire was so vast that, according to a popular saying, "the sun never sets on the British Empire."

In the 1900s, two world wars weakened the United Kingdom. Nearly all of the British Empire broke into independent countries. Some of the former colonies joined the United Kingdom in a new organization called the Commonwealth of Nations. Recently the United Kingdom has joined other European countries to form the European Union. The goal of this organization is to unify and strengthen the economies of member countries.

258

Critical Thinking Activity

Making Comparisons Point out Oregon on a map and tell students that the United Kingdom is slightly smaller than Oregon. Then display a transparency with the following Oregon data: **Area:** 97,073 square miles; **Population:** 2,977,331; **Climate:** mild, rainy coast and dry interior; **Manufactured goods:** lumber and wood products, foods, machinery, fabricated metals, paper, printing and publishing, primary metals; **Crops:** hay, wheat, potatoes, grass seed, pears, and onions. Refer students to the section and **Countries at a Glance** for similar data on the United Kingdom. Ask students to list three similarities and three differences between the United Kingdom and Oregon. *(Answers may include that both places border oceans and that the United Kingdom has a greater population density than Oregon.)* **L2**

Gifts to the World The literature of the United Kingdom ranks as one of its greatest gifts to the world. Its writers include William Shakespeare, Robert Burns, Charles Dickens, and Charlotte and Emily Brontë. Today the United Kingdom remains a center for literature, theater, and the arts.

When you vote in your first election in the United States, you will be sharing another British gift. The British form of government—**parliamentary democracy**—was copied by the Framers of the U.S. Constitution. In a parliamentary democracy, voters elect representatives to a lawmaking body called Parliament. The largest political party in Parliament chooses the government's leader, the prime minister.

The United Kingdom is also a **constitutional monarchy,** in which a king or queen represents the country at public events. The prime minister and other elected officials, however, actually hold the powers of government.

A Crowded Country Look at the map on page 269. Like other European countries, the United Kingdom is densely populated. About 90 percent of the British live in cities and towns. The city of London in southern England is a world center of trade, business, and banking. More than 8 million people live in London and surrounding suburbs.

The ancient and the modern meet in London. Thirteen-year-old Pamela Wilson lives with her family in a small house on the outskirts of London. She enjoys going to the West End, an area of London famous for its shops, museums, theaters, and restaurants. Not far from the West End are the church of Westminster Abbey, where Britain's kings and queens are crowned, and Buckingham Palace, the London home of the British royal family. Pamela and her friends also enjoy watching soccer matches, as do many other British people.

WHAT IN THE WORLD?

The Chunnel

In 1994 England and France, separated for thousands of years by the English Channel, were once again joined. The Channel Tunnel, nicknamed the Chunnel, linked the island country to mainland Europe. By the year 2003 the Chunnel is expected to carry more than 120,000 people between England and France each day. At night, tons of freight will move through the Chunnel.

SECTION 1 REVIEW

REVIEWING TERMS AND FACTS

1. **Define the following:** moor, loch, parliamentary democracy, constitutional monarchy.
2. **REGION** What landforms are found in the United Kingdom?
3. **HUMAN/ENVIRONMENT INTERACTION** What natural resources do the British export?
4. **MOVEMENT** In what ways have the British influenced other parts of the world?

MAP STUDY ACTIVITIES

5. Look at the climate map on page 258. What type of climate does the United Kingdom have?
6. Turn to the political map on page 254. What is the distance between the cities of London and Glasgow?

Answers to Section 1 Review

1. All vocabulary words are defined in the Glossary.
2. lowlands and highlands
3. coal, oil, natural gas
4. The British spread their form of government and their culture to other parts of the world.
5. marine west coast
6. about 350 miles (563 km)

Evaluate

 Section 1 Quiz **L1**

Use the Testmaker to create a customized quiz for Section 1. **L1**

Reteach

Have students write the headings *PEOPLE, PLACES,* and *PRODUCTS* at the top of a sheet of paper. Then tell them to reread the section and to copy nouns from their reading under the appropriate headings. **L1**

Enrich

Have students research the Industrial Revolution in the United Kingdom. Encourage them to prepare a report on the impact of the Industrial Revolution on life in rural and urban areas. **L3**

CLOSE

Tell students to imagine that they are Pamela Wilson. Ask them to plan a day of sight-seeing in London with a friend from the United States.

MAKING CONNECTIONS

MAKING
CONNECTIONS

TEACH

Project views of moorland landscape, or show moorland scenes from the movie, *The Secret Garden.* Then have students speculate on why people might find moorland scenery both beautiful and desolate. **L1**

More About . . .

Earth Science One of the most common plants on the moors is the low evergreen shrub called *heather.* This plant has grayish hairy stalks, broomlike branches, and needlelike leaves. Some moors are covered by a layer of wet marshy peat, a substance consisting of partially decayed plants, herbs and mosses. The topmost layer of peat is a pale gray-green; lower layers look like mud.

THE SECRET GARDEN by Frances Hodgson Burnett

| MATH | SCIENCE | HISTORY | LITERATURE | TECHNOLOGY |

The Secret Garden is a novel set in England in the early 1900s. The following excerpt gives us a view of the English countryside through the eyes of a young traveler.

Heather Blooming on the Moors

"What is a moor?" [Mary] said suddenly to Mrs. Medlock.

"Look out of the window . . . you'll see," the woman answered. . . .

The carriage lamps shed a yellow light on a rough-looking road which seemed to be cut through bushes and low-growing things which ended in the great expanse of dark apparently spread out before and around them. A wind was rising and making a singular, wild, low, rushing sound.

"It's—it's not the sea, is it?" asked Mary, looking round at her companion.

"No, not it," answered Mrs. Medlock. "Nor it isn't fields nor mountains, it's just miles and miles and miles of wild land that nothing grows on but heather and gorse and broom, and nothing lives on but wild ponies and sheep."

"I feel as if it might be the sea, if there were water on it," said Mary. "It sounds like the sea just now."

"That's the wind blowing through the bushes," Mrs. Medlock said. "It's a wild, dreary enough place to my mind, though there's plenty that likes it—particularly when the heather's in bloom."

. . . Two days later . . . the rainstorm had ended and the gray mist and clouds had been swept away in the night by the wind. The wind itself had ceased and a brilliant, deep blue sky arched high over the moorland. Never, never had Mary dreamed of a sky so blue. . . . The far-reaching world of the moor itself looked softly blue instead of gloomy purple-black or awful dreary gray.

From *The Secret Garden* by Frances Hodgson Burnett. Copyright © 1938 by Verity Constance Burnett, J. B. Lippincott Company.

Making the Connection

1. How did Mrs. Medlock describe the view from the carriage?
2. To what geographic feature did Mary compare the landscape?

260

UNIT 4

Answers to Making the Connection

1. miles and miles of wild land that nothing grows on but heather and gorse and broom
2. the sea

SECTION 2
The Republic of Ireland

PREVIEW

Words to Know
- peat
- bog

Places to Locate
- Republic of Ireland
- Dublin
- Killarney

Read to Learn . . .
1. why Ireland is called the Emerald Isle.
2. how the Irish struggled to win their independence.
3. how urban and rural Irish live.

Who lives next door to you? Are they just like you, or are they different? Ireland's way of life is very different from that of its neighbor, the United Kingdom. One of the ways in which Ireland differs is in its religion—more than 90 percent of Ireland's people are Roman Catholics. Churches such as this one dot Irish lands.

Besides religion, other key differences distinguish Ireland and the United Kingdom. Only 3.6 million people live in Ireland, compared with the United Kingdom's 59 million people. And while nearly all the British live in urban areas, only a little more than one-half of the Irish live in cities.

Place
The Emerald Isle

If you could fly over Ireland on a summer day, you would see lush green meadows and tree-covered hills. Surrounded on three sides by the Atlantic Ocean, Ireland's green color is so striking that it was named the Emerald Isle.

The Landscape The map on page 256 shows you Ireland's plains and highlands. At Ireland's center lies a wide, rolling plain dotted with low hills. Forests and farmland cover this central lowland. Much of the area is rich in **peat,** or wet ground with decaying plants that can be used for fuel. Peat is dug from **bogs,** or low swampy lands.

CHAPTER 10

261

FOCUS

Meeting National Standards
Geography for Life The following standards are highlighted in this section:
- Standards 3, 4, 9, 10, 13

Section Objectives
1. **Explain** why Ireland is called the Emerald Isle.
2. **Describe** how the Irish struggled to win their independence.
3. **Examine** how urban and rural Irish live.

Bellringer Motivational Activity

 Prior to taking roll at the beginning of the class period, project Section Focus Transparency 10-2 and have students answer the activity questions. Discuss students' responses. **L1**

Vocabulary Pre-check
Have students speculate on whether *peat* is animal, mineral, or vegetable. Then have them read the term's meaning in the section to find out if their speculations are correct. **L1 LEP**

Classroom Resources for Section 2

REPRODUCIBLE MASTERS
Reproducible Lesson Plan 10-2
Guided Reading Activity 10-2
Cooperative Learning Activity 10
GeoLab Activity 10
Geography Skills Activity 10
Section Quiz 10-2

TRANSPARENCIES
Section Focus Transparency 10-2
Geography Handbook Transparency 14
Political Map Transparency 4

MULTIMEDIA
Testmaker

National Geographic Society: ZipZapMap! World

TEACH

Guided Practice

Graphic Skills On the chalkboard, write the following statistics for Ireland's labor force:

1,334,000 workers;
15% agricultural;
29% manufacturing and construction;
51% services;
5% other.

Tell students to present this information in graphic form. Suggest that they use a circle graph or a one-line bar graph. Remind them to add a title and use labels where necessary. L1

Independent Practice

▼ Guided Reading Activity 10-2 L1

▼ Cooperative Learning Activity 10 L1

Harvesting Peat

Two Irish workers cut slabs of peat, which will be used for heating homes.
HUMAN/ENVIRONMENT INTERACTION: On what type of land is peat usually found?

Along the Irish coast, the land rises in rocky highlands. In some places, however, the central plain spreads all the way to the sea. On the map on page 254, you will find Dublin—Ireland's capital—on an eastern stretch of the plain.

The Climate Whether plain or highland, no part of Ireland is more than 70 miles (113 km) from the sea. This nearness to the sea gives Ireland a uniform climate. Like the United Kingdom, Ireland is warmed by moist winds blowing over the North Atlantic Current. The mild weather, along with frequent rain and mist, makes Ireland's landscape green year-round.

The Economy If you look at the map on page 263, you see that Ireland has few mineral resources. The country, however, does have rich soil and pastureland.

The mild and rainy climate favors farming. In the mid-1800s, Irish farmers grew potatoes as their main food crop. When too much rain and a blight caused the potatoes to rot in the fields, famine struck, bringing hardship to the Irish. This disaster forced many Irish to emigrate to other countries, especially to the United States.

Although farming is still important to Ireland, industry now also contributes to economic development. The economy depends on the manufacturing of machinery and transportation equipment exported to the United Kingdom and the European mainland. Ships bringing mineral and energy resources to Ireland dock at the country's many ports, including Dublin and Cork.

Cooperative Learning Activity

Attach a number of pieces of butcher paper to the bulletin board. Draw a long line across the butcher paper, label the beginning of the line A.D. 1000 and the end of the line A.D. 2000. Add interval labels for each century. Organize the class into 10 groups and assign each group one century. Have groups research their period in Irish history to find five major events to record on the bulletin-board time line. After they have made their entries on the time line, call on groups to explain why they selected particular events. L1

Place
The People

Most of the Irish trace their ancestry to groups of people who settled Ireland more than 7,000 years ago. The Celts and the British made the biggest impact. Their languages—Gaelic and English—are Ireland's two official languages today. Most Irish, however, speak English as their everyday language.

Influences of the Past Stormy politics mark Ireland's history. From the 1100s to the early 1900s, the British governed Ireland. Religion and government controls mixed to cause disagreement. The Irish people resisted British rule and demands that the Roman Catholic country become Protestant. British officials seized land in Ireland and gave it to English and Scottish Protestants. At one time the British drove out Irish Catholics to make room for the new settlers.

WHAT IN THE WORLD?

What's in a Name?

Many Irish and Scottish last names begin with *Mac, Mc,* or *O'*, such as MacDermott, McCormack, and O'Casey. Why? *Mac, Mc,* and *O'* mean "descendant of." *MacDermott*, then, means descendant of Dermott; *McCormack,* descendant of Cormack; and *O'Casey,* descendant of Casey.

Cultural Kaleidoscope

Ireland Barefoot pilgrims have been climbing rocky Croagh Patrick—a mountain in County Mayo—for centuries. According to legend, St. Patrick fasted there in A.D. 441 for 40 days before driving all of the snakes from Ireland.

MAP STUDY

Answer
zinc, lead, peat, hydroelectric power

Map Skills Practice
Reading a Map In what country are reindeer raised? *(Norway)*

ASSESS

Check for Understanding
Assign Section 2 Review as homework or an in-class activity.

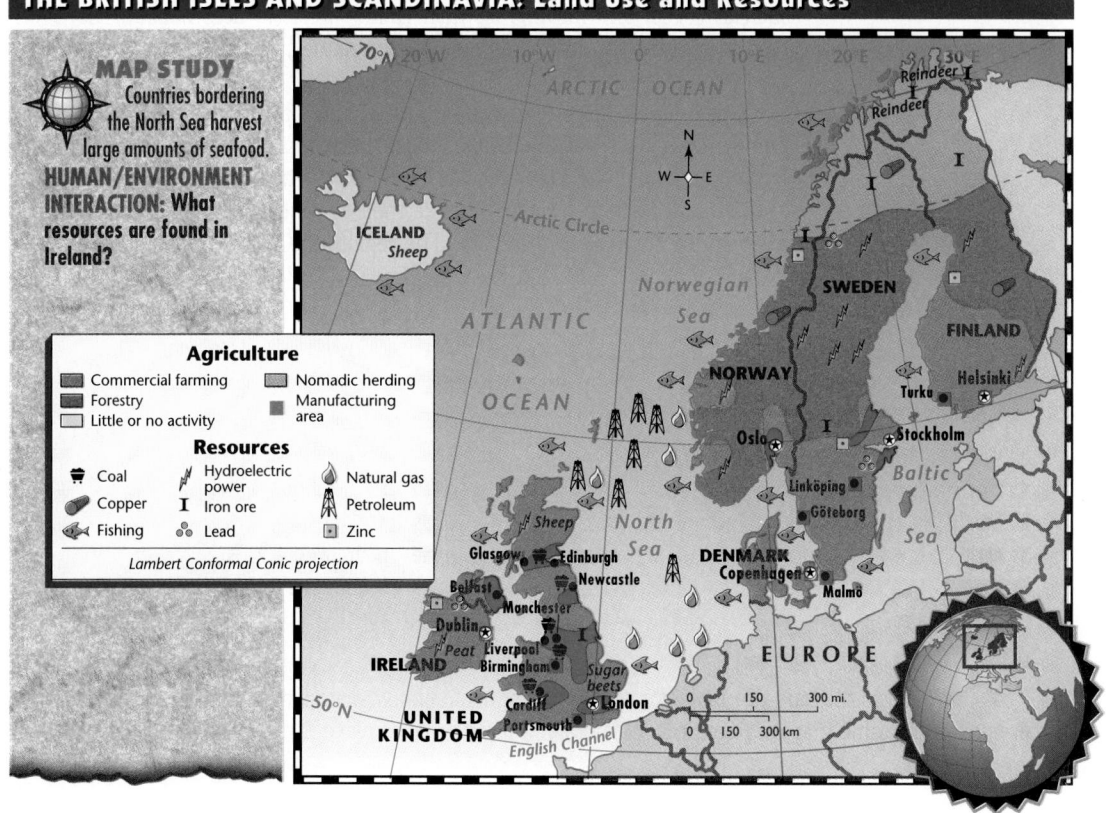

THE BRITISH ISLES AND SCANDINAVIA: Land Use and Resources

MAP STUDY Countries bordering the North Sea harvest large amounts of seafood. **HUMAN/ENVIRONMENT INTERACTION:** What resources are found in Ireland?

Agriculture
- Commercial farming
- Forestry
- Little or no activity
- Nomadic herding
- Manufacturing area

Resources
- Coal
- Copper
- Fishing
- Hydroelectric power
- Iron ore
- Lead
- Natural gas
- Petroleum
- Zinc

Lambert Conformal Conic projection

Meeting Special Needs Activity

Memory Disability Note taking benefits all students but especially those with memory problems. Explain that notes help students focus on important information in the text, organize the information, and review for tests. Have students take notes on the material under the headings "Continuing Changes" and "Daily Life." Then discuss students' note-taking procedures and, if necessary, offer ideas on how these procedures might be improved. **L1**

Meeting Lesson Objectives

Each objective below is tested by the questions that follow it in parentheses.

1. **Explain** why Ireland is called the Emerald Isle. (2)
2. **Describe** how the Irish struggled to win their independence. (3)
3. **Examine** how urban and rural Irish live. (4, 5)

Evaluate

Section 2 Quiz **L1**

Use the Testmaker to create a customized quiz for Section 2. **L1**

Reteach

Give students copies of the bold-faced headings from the section. Direct students to list important facts from the section under each heading. **L1**

Enrich

Ireland is renowned for its great literature. Have students work in small groups to develop exhibits for a museum display titled "The Literature of Ireland." **L3**

CLOSE

Encourage students to write a letter to Sean Carroll telling him how life in Dublin compares to life in their community.

In 1921 the southern part of Ireland finally won its independence from the United Kingdom and became the Republic of Ireland. The northern part remained within the United Kingdom as Northern Ireland. Friction among the three regions continued—especially between Catholics and Protestants.

Disagreement over Northern Ireland exploded into violence in the 1960s. The violence continued into the 1990s. In 1994 British and Irish officials agreed to seek peace, but new violence in Northern Ireland threatened this effort.

Continuing Changes The Republic of Ireland continues to make changes to benefit its people. It is a member of the European Union (EU), which increases the number of markets for its products. In 1990 the Irish elected their first female president.

Daily Life Nearly 56 percent of Ireland's 3.6 million people live in cities and towns. Nearly one-third of its people live in the city of Dublin alone. Rural population has decreased. Life in both urban and rural areas often centers on the neighborhood or village church, where people take part in social gatherings and other activities.

Fifteen-year-old Sean Carroll lives in Dublin, Ireland's capital and largest city. He rides his bicycle to school across one of the bridges that span the River Liffey, which runs through the heart of Dublin. Sean, like most Irish boys, likes sports such as soccer and hurling, an Irish game similar to field hockey. His favorite food is Irish stew, a dish made by boiling potatoes, onions, and mutton in a covered pot.

Sean's father teaches Irish literature at the University of Dublin. Ireland is known for its writers, such as George Bernard Shaw and James Joyce. His mother works at the National Museum, which houses famous artworks from Celtic times. Sean and his family take their summer holiday in Killarney, a little town in the southwest of Ireland. The region of Killarney is popular because of its beautiful lakes, hills, and woods.

SECTION 2 REVIEW

REVIEWING TERMS AND FACTS
1. **Define the following:** peat, bog.
2. **LOCATION** Where in Ireland are plains and highlands located?
3. **MOVEMENT** Why did many Irish emigrate to other countries in the mid-1800s?
4. **HUMAN/ENVIRONMENT INTERACTION** How does Irish stew reflect the rural landscape of Ireland?

MAP STUDY ACTIVITIES
5. Turn to the land use map on page 263. What is the main manufacturing area in Ireland?
6. Look at the same map on page 263. What type of farming takes place throughout Ireland?

264

Answers to Section 2 Review

1. All vocabulary words are defined in the Glossary.
2. The plains are mostly in the center of the country; the highlands lie between the plains and the sea.
3. famine and lack of work
4. It reflects the agricultural crops and livestock raised there.
5. around the city of Dublin
6. commercial farming

BUILDING GEOGRAPHY SKILLS

Reading a Climate Map

Before you dress for school in the morning, you probably check the weather. You want to know how warm or cold or rainy it is today. If you're packing to go on vacation to another area, you might check a climate map. You want to know what the weather usually is like at this time of year.

A climate map reveals the usual weather patterns of a region. A climate region can be defined by its temperature and amount of precipitation. Near the Equator, for example, the climate is warm year-round. Some areas, however, have *rain forest climates* that are warm and very wet. Others have *desert climates* that are warm and very dry.

On a climate map, colors represent climate regions. The map key explains what the colors mean. To read a climate map, apply these steps:

- Identify the area shown in the map.
- Study the map key to identify the climate regions on the map.
- Locate the countries or areas in each climate region.

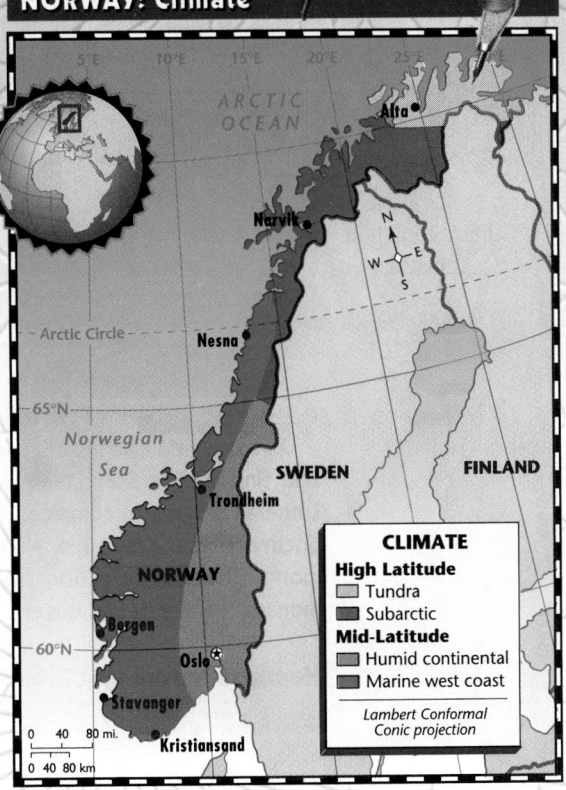

NORWAY: Climate

CLIMATE
High Latitude
- Tundra
- Subarctic
Mid-Latitude
- Humid continental
- Marine west coast

Lambert Conformal Conic projection

0   40   80 mi.
0   40   80 km

Geography Skills Practice

1. What two major kinds of climates are found in Norway?
2. Does the west coast of Norway have more or less snow than inland areas of the country? Why?
3. How is the climate of northern Norway different from that of southern Norway?

TEACH

Ask students to volunteer information on their own climatic region by answering the following: What seasons does the area have? What are average high and low temperatures in these seasons? How much precipitation does the area receive? When does the region receive the most precipitation? When does the region receive the least precipitation? Ask students to identify how climate affects the economy, recreation, dress, food, and housing. Then have students speculate on how their lives might change if they lived in another climatic region.

Direct students to read the skill and complete the practice questions. **L1**

Additional Skills Practice

1. How do cartographers show climate on a map? *(by using colors to represent climatic regions)*
2. How could the map of Norway help in the planning of a trip to that country? *(when to go, what activities will be available, what clothes to bring, and so on)*

 Geography Skills Activity 10

Answers to Geography Skills Practice

1. high latitude and mid-latitude
2. The west coast has less snow because the ocean moderates the land temperature, producing a marine west coast climate; it has cool summers and mild damp winters.
3. Northern Norway has a subarctic climate with colder winters and shorter, cooler summers than the southern part of the country.

265

FOCUS

Section Objectives

1. **Analyze** how the Atlantic Ocean affects climate in Scandinavia.
2. **Locate** where most Scandinavians live.
3. **Compare** how Scandinavians work and enjoy leisure time.

Bellringer Motivational Activity

Prior to taking roll at the beginning of the class period, project Section Focus Transparency 10-3 and have students answer the activity questions. Discuss students' responses. **L1**

Vocabulary Pre-check

Explain that *geyser* is Icelandic for "to gush" and refers to a naturally occurring, hot-water gusher. Ask: What park in the United States is famous for its geyser named *Old Faithful?* (Yellowstone) **L1** **LEP**

SECTION 3
Scandinavia

PREVIEW

Words to Know
- fjord
- welfare state
- geyser

Places to Locate
- Norway
- Sweden
- Denmark
- Iceland
- Finland

Read to Learn . . .
1. how the Atlantic Ocean affects climate in Scandinavia.
2. where most Scandinavians live.
3. how Scandinavians work and enjoy leisure time.

Does the sun ever shine at midnight? If you ask the Sami—reindeer herders who live in northern Scandinavia—they will say, "Of course it does." Between April and August, the sun shines 24 hours a day in the northern area of Norway.

Norway is part of Scandinavia, a large region in northern Europe. Scandinavia also includes four other countries: Sweden, Denmark, Iceland, and Finland. The Scandinavian countries have democracy and the Protestant Lutheran religion in common. All share similar cultures, but Finland has a distinctive language. Sweden, Norway, and Denmark also have something in common with their neighbor, the United Kingdom. They have constitutional monarchies—kings and queens who have no real power.

Human/Environment Interaction

Norway

The Scandinavian Peninsula sweeps from the Arctic Ocean to the North Sea. Norway is a narrow kingdom that runs the length of the peninsula. The map on page 256 shows you Norway's jagged Atlantic coastline. Its many **fjords,** or steep-sided valleys, are inlets of the sea. What caused these deep valleys? Thousands of years ago glaciers scoured the land. The fjords now provide Norway with sheltered harbors.

266

UNIT 4

Landscape and Climate The snowcapped Kjølen Mountains tower over most of Norway. Limited areas of land are available for farming. Rivers rushing down from the mountains provide hydroelectric power to Norway's farms, factories, and homes.

Warm winds from the North Atlantic Current give Norway's southern and western coasts a mild climate. Most of the country's 4.3 million people live in these areas. Oslo, the capital and largest city, lies at the end of a fjord on the southern coast.

People of the Sea From the icy oceans lapping its long coastline, Norway built its economy. About 1,000 years ago, the Vikings—ancestors of most modern Scandinavians—sailed west. They founded settlements in Iceland, Greenland, and even northern Canada. Today their Norwegian descendants harvest fish from the sea. The map on page 263 shows that the Norwegians also benefit from oil and natural gas found beneath the North Sea waters.

Love of Sports Sonja Hamsun, an Oslo teenager, likes to travel to the mountains of Norway's interior. This part of the country has a colder climate because the mountains block the warm coastal winds. Snow covers the ground at least three months of the year. Like most Norwegians, Sonja loves outdoor sports, especially skiing. Before modern transportation linked Norway's cities, Norwegians traveled over the snow on skis. Today skiing is Norway's national sport.

Place
Sweden

What would you see from an airplane flying east from Norway? The many mountains of Sweden rise up to greet you. Then the shimmering peaks and valleys end abruptly at the seacoast. Sweden's long coastline of sandy beaches and rocky cliffs touches the Baltic Sea, the Gulf of Bothnia, and a narrow arm of the North Sea.

Landscape and Climate The mountains along Sweden's border with Norway block the warm winds of the North Atlantic Current. This causes northern Sweden to have cool summers and cold winters. As you descend from the snow-covered mountains, you come to forested highlands and then fertile lowlands. Many lakes, hills, and farms dot the lowland coast of southern Sweden. The North Atlantic winds reach this area, giving it a mild climate.

Norwegians Bring in Their Catch

A Norwegian fishing crew prepares its catch for a local fish market in Bergen, Norway. **PLACE: How were most of Norway's harbors formed?**

TEACH

Guided Practice

Display Unit Overlay Transparencies 4-0 and 4-3. Read aloud the paragraphs from the section about the land and climate of each Scandinavian country. Ask students to use information in these readings to identify the areas of Scandinavia that probably are most populous. Have volunteers shade these areas on the transparency with markers. Then tell students to look for passages that affirm their answers as they read the rest of the section. **L1**

More About the Illustration

Answer to Caption
by glacial movement

Human/Environment Interaction Norway has one of the largest fishing fleets in the world. Most of its ships work the waters off the Norwegian coast, catching haddock, herring, cod, mackerel, and many other kinds of fish.

Global Gourmet

Norway Although fish is a staple on Norwegian menus, reindeer dishes are also popular. For instance, a favorite tourist stop near Trondheim called the Tavern serves reindeer meat in a spinach sauce for lunch.

Cooperative Learning Activity

Organize students into five groups and assign a Scandinavian country to each group. Have groups research the folktales of their assigned countries and write summaries of their favorite stories. Suggest *Favorite Folktales Around the World*, edited by Jane Yolen, as a possible source. Have groups combine their summaries into a "book." Make copies of the "book" and give one to each student. **L1**

Independent Practice

 Guided Reading Activity 10-3

Place Sweden is a model welfare state. Every worker is guaranteed a minimum income and an annual four-week paid vacation. And, in most cases, medical care and medicines are free or are available for a minimal charge.

Cultural Kaleidoscope

Denmark A 20-acre park in Copenhagen inspired Disneyland. Before Walt Disney built his amusement park, he spent two months in Tivoli Gardens. This park, which is more than 100 years old, has a lake full of swans, a concert hall, a children's theater, a flying-carpet ride, several five-star restaurants, and many other attractions.

Celebrating St. Lucia Day in Sweden

These people are part of a pageant on St. Lucia Day. Young people often awaken their families with a traditional song and a breakfast of buns and coffee.
PLACE: What other holiday is popular in Sweden?

Prosperous Economy Sweden has one of Europe's most prosperous economies. The Swedes have developed profitable industries based on timber, iron ore, and water power. With this economic wealth, Sweden has become a **welfare state.** In a welfare state, the government uses tax money to provide education, medical services, health insurance, and other services for its citizens.

The People The map on page 269 shows you that most of Sweden's 8.8 million people live in cities in the southern lowlands. Stockholm, on the Baltic coast, is the country's capital and largest city. Stockholm's churches, palaces, and modern buildings spread out over several islands connected by more than 40 bridges. Another city, Göteborg (YUHR•tuh•BAWR•ee), is a major port.

Like the Norwegians, the Swedes enjoy the countryside and outdoor sports. One of their favorite holidays, Midsummer's Eve, comes in late June and celebrates summer. On this holiday people dance and feast into the early morning hours—with the sky still light from the midnight sun.

Location
Denmark

Picture the Scandinavian Peninsula as the head of a huge dragon, its jaws open wide. Its morsel of food is the country of Denmark. Denmark is a small country south of Sweden between the North and Baltic seas.

Landscape and Climate The map on page 256 shows you that most of Denmark is made up of the Jutland Peninsula, which connects the country to the European mainland. Denmark also includes hundreds of nearby small islands. In addition, tiny Denmark rules the gigantic island of Greenland, 1,300 miles (2,090 km) away off the northeastern coast of Canada. Rolling, green hills and low, fertile plains run through most of Denmark. Warm North Atlantic winds give the country a mild, damp climate.

Farm Country Denmark is poor in natural resources, but it has some of the richest farmland in Scandinavia. Danish farm products include butter, cheese, bacon, and ham. Foods make up much of Denmark's exports. Food exports help pay for the fuels and metals the Danish people must import for their industries. Danish workers produce beautifully designed furniture, porcelain, and silver.

The People Like the Swedes, the 5.2 million Danes have a high standard of living and a welfare state. More than half of them live on the islands near the Jutland Peninsula. Copenhagen, Denmark's capital and largest city, is on the

UNIT 4

Meeting Special Needs Activity

Inefficient Readers Students often profit by using a form of rapid "reading" called scanning to locate specific information. Model this procedure for students by sliding your finger down the middle of a column rapidly. Demonstrate scanning as a way to find all the products of Denmark mentioned under "Farm Country." Then time students as they scan the rest of Section 3 for the products of Norway, Sweden, Iceland, and Finland. **L1**

largest island. This city is known for its old church spires and red tile rooftops as well as its modern buildings and parks. It is also a major port. Denmark is the most densely populated Scandinavian country.

Place
Iceland

When and where was the first government legislature in the world established? In A.D. 930 in Iceland—a small island republic that lies just south of the Arctic Circle in the Atlantic Ocean. Once part of Denmark, Iceland declared its independence in 1944.

Land of Fire and Ice Most of Iceland is a rugged plateau that sits on top of a fault line, or break in the earth's crust. This makes Iceland a land of volcanoes, hot springs, and **geysers**—springs that spout hot water and steam. Icelanders use the naturally heated water to heat their buildings.

Because Iceland is so far north, it is a land of glaciers. Fast-flowing rivers, some formed by melting snow and glaciers, provide a good source of hydroelectric power. The North Atlantic Current warms most of Iceland's coast and keeps temperatures from getting too cold.

THE BRITISH ISLES AND SCANDINAVIA: Population Density

MAP STUDY
London has more people than any other city in Europe.
PLACE: What city in Denmark has more than 1 million people?

Persons per

sq. mi.	sq. km
Uninhabited	Uninhabited
Under 2	Under 1
2-60	1-25
60-125	25-50
125-250	50-100
Over 250	Over 100

CITIES
■ City with more than 1,000,000 people
● City with 500,000 to 1,000,000 people
○ City with 100,000 to 500,000 people

Lambert Conformal Conic projection

CHAPTER 10

ASSESS

Check for Understanding
Assign Section 3 Review as homework or an in-class activity.

Meeting Lesson Objectives
Each objective below is tested by the questions that follow it in parentheses.
1. **Analyze** how the Atlantic Ocean affects climate in Scandinavia. (2, 4)
2. **Locate** where most Scandinavians live. (5)
3. **Compare** how Scandinavians work and enjoy leisure time. (1, 2)

MAP STUDY

Answer
Copenhagen
Map Skills Practice
Reading a Map Which Scandinavian country is most densely populated? *(Denmark)*

Critical Thinking Activity

Predicting Consequences Give students the following information: The Faroe Islands are semi-independent possessions of Denmark that lie halfway between Norway and Iceland. Most Faroese earn their livings from fishing and whaling. In fact, meat from the pilot whale is a staple of the Faroese diet. In 1981 Faroese scientists warned the people to reduce their intake of whale meat because it contains high levels of mercury. Yet each year, the Faroese kill about 2,000 pilot whales and eat about 66 pounds (30 kg) of whale meat per person. Ask students to predict the consequences if the Faroese continue to hunt and eat whales. *(People might get sick from the mercury; if the Faroese continue to kill whales at the present rate, pilot whales will soon become very scarce.)* **L2**

269

Evaluate

 Section 3 Quiz **L1**

Use the Testmaker to create a customized quiz for Section 3. **L1**

Chapter 10 Test Form A and/or Form B **L1**

Reteach

Reteaching Activity 10 **L1**

Spanish Reteaching Activity 10 **L1**

Enrich

Enrichment Activity 10 **L3**

CLOSE

Mental Mapping Activity

Provide each student with a large sheet of white paper and map pencils. Inform students that this will not be a graded activity. Have students draw, freehand from memory, a map of Scandinavia that shows its climate regions in different colors and a key that explains the colors.

The Economy and People Iceland has few mineral resources and areas of farmland. The country's economy depends on fishing. Fish exports provide the money for Iceland to buy food and consumer goods from other countries.

Icelanders trace their heritage to Viking explorers who came from mainland Scandinavia. Today nearly all 300,000 Icelanders live near the coast. About half live in Reykjavík (RAY•kyuh•vihk), the most northern capital city in the world. Icelanders, with a 100 percent literacy rate, like reading and sports. Iceland's most famous works of literature are sagas, or long tales, written during the 1100s about Vikings.

Location

Finland

Finland is a long, narrow country tucked among Sweden, Norway, and Russia. Ruled first by Sweden and later by Russia, Finland became an independent republic in 1917. Finland lies on a flat plateau broken by small hills and valleys. Thick green forests and thousands of blue lakes cover most of the country. Finland has humid continental and subarctic climates because of its great distance from the warm North Atlantic Current.

The Economy Most of Finland's wealth comes from its huge forests. Paper and wood production are the country's major industries. Because of poor soil and a short growing season, Finland's farming areas are in the south. There, farmers raise wheat, rye, and livestock, including dairy cattle.

The People Most of Finland's people belong to an ethnic group known as Finns. The Finns' culture and language are very different from those of the other Scandinavian peoples. The rest of Finland's people are Swedes, Russians, or Sami. The Sami are a small ethnic group who live in northern Scandinavia and have their own separate culture and language. Their traditional nomadic way of life is based on herding reindeer. Most of Finland's 5.1 million people live in towns and cities on the southern coast. Helsinki, the capital and largest city, is known for its scenic harbor and modern buildings.

SECTION 3 REVIEW

REVIEWING TERMS AND FACTS
1. Define the following: fjord, welfare state, geyser.
2. HUMAN/ENVIRONMENT INTERACTION How have Norwegians adapted to their environment?
3. LOCATION How does Iceland's location affect its landscape?
4. PLACE In what ways is Finland different from other Scandinavian countries?

 MAP STUDY ACTIVITIES
5. Look at the population density map on page 269. What area of Norway is most sparsely populated?

270

Answers to Section 3 Review
1. All vocabulary words are defined in the Glossary.
2. Vikings from what now is Norway ventured forth on the Atlantic seeking new lands. Today, Norwegians fish in the ocean and pump oil and natural gas from beneath it.
3. Iceland's location on a fault line makes it a land of volcanoes, hot springs, and geysers.
4. Finland's climate is less affected by the North Atlantic Current. Also, Finland's language and culture are different from those of the other Scandinavian countries.
5. in southern central Norway

Chapter 10 Highlights

Important Things to Know About the British Isles and Scandinavia

SECTION 1 THE UNITED KINGDOM

- The United Kingdom is made up of England, Scotland, Wales, and Northern Ireland.
- The landscape of the United Kingdom consists of highlands and lowlands.
- Winds blowing over the warm North Atlantic Current provide the United Kingdom with a mild climate.
- The United Kingdom is a major industrial and trading nation.
- Literature and parliamentary democracy are two of the United Kingdom's greatest gifts to the world.

SECTION 2 THE REPUBLIC OF IRELAND

- Ireland is called the Emerald Isle because of its lush vegetation that stays green year-round.
- The Irish were once dependent on the land. Now they are developing an industrial economy.
- Ireland's stormy political history has caused frequent changes in government policies.

SECTION 3 SCANDINAVIA

- The region of Scandinavia is made up of the countries of Norway, Sweden, Denmark, Iceland, and Finland.
- The warm North Atlantic Current, an extension of the Gulf Stream, keeps temperatures mild in many parts of Scandinavia.
- The economies of the Scandinavian countries depend on produce from the land and the sea.

Royal Guards in London, England ▶

Using the Chapter 10 Highlights

- Use the Chapter 10 Highlights to preview, review, condense, or reteach the chapter.

Preview/Review

 Vocabulary Puzzle-Maker Software reinforces the terms used in Chapter 10.

Condense

Have students read the Chapter 10 Highlights. Spanish Chapter Highlights also are available.

Chapter 10 Digest Audiocassettes and Activity

Chapter 10 Digest Audiocassette Test
Spanish Chapter Digest Audiocassettes, Activities, and Tests also are available.

Guided Reading Activities

Reteach

Reteaching Activity 10. Spanish Reteaching activities also are available.

MAP STUDY

Place Copy and distribute page 30 from the Outline Map Resource Book. Have students label the British Isles, Ireland, and the countries of Scandinavia. Then direct them to locate and label the major cities and to shade the most populous areas of each country.

Extra Credit Project

Book Jacket Have each student choose a country from the British Isles or Scandinavia and research its tourist attractions, such as dramatic coastal scenery or imposing castles and cathedrals. Have students use their research findings to design book jackets for hypothetical travel guides. Direct them to include a catchy title along with pictures of major tourist attractions on the front and brief "teasers" describing the highlights of their country on the back. Wrap the jackets around real books, and display them in the classroom. **L1**

Chapter 10 Review and Activities

REVIEWING KEY TERMS

Match the numbered terms in Column A with their definitions in Column B.

A
1. moor
2. parliamentary democracy
3. fjord
4. geyser
5. peat
6. constitutional monarchy
7. loch
8. welfare state

B
A. government in which a king or queen shares power with elected officials
B. a type of decayed plant material
C. spring that spouts hot water and steam
D. steep-sided valley filled with seawater
E. form of government in which the people rule through Parliament
F. wild, treeless land in the United Kingdom
G. system under which the government provides major services to citizens
H. narrow bay cut into the highland coasts of Scotland

ANSWERS

Reviewing Key Terms

1. F
2. E
3. D
4. C
5. B
6. A
7. H
8. G

Mental Mapping Activity

On a separate piece of paper, draw a freehand map of the British Isles and Scandinavia. Label the following items on your map:
• Reykjavík
• Dublin
• Scotland
• North Sea
• Finland

Mental Mapping Activity

This exercise helps students to visualize the countries and geographic features they have been studying and understand the relationships among them. All attempts at freehand mapping should be accepted.

REVIEWING THE MAIN IDEAS

Section 1
1. **LOCATION** Where are the lowlands of the United Kingdom located?
2. **HUMAN/ENVIRONMENT INTERACTION** What is the United Kingdom's major natural resource?
3. **PLACE** What kind of government does the United Kingdom have?

Section 2
4. **PLACE** What is the landscape of Ireland like?
5. **MOVEMENT** How does the North Atlantic Current affect Ireland's climate?
6. **MOVEMENT** How has membership in the European Union helped Ireland?

Section 3
7. **HUMAN/ENVIRONMENT INTERACTION** How has Scandinavia's northern location affected people's daily lives?
8. **LOCATION** What is the capital of Finland?

272

Reviewing the Main Ideas

Section 1

1. east of an imaginary line running southwest to northeast across Great Britain
2. oil
3. a parliamentary democracy and a constitutional monarchy

Section 2

4. central area of lowlands ringed by highlands
5. The Atlantic's Gulf Stream Current gives Ireland a warm, moist climate.
6. Ireland has been able to sell more goods in Europe. At the same time, its industries have grown.

CRITICAL THINKING ACTIVITIES

1. **Evaluating Information** How has the Industrial Revolution that began in the United Kingdom affected your life?
2. **Making Comparisons** What comparisons can you make between the landscapes of Norway and Denmark? Which country's land would affect their tourist industry the most?

GeoJournal Writing Activity

Imagine you are a Viking who has traveled through time to the present-day British Isles and Scandinavia. In your journal describe how these lands have changed over the centuries. Suggest reasons for these changes.

COOPERATIVE LEARNING ACTIVITY

Discover how the movement of peoples into England affected the English language. Organize into three groups. Each group will look through a dictionary to list at least 15 commonly known words that came from the language of (a) the Angles, Saxons, or Jutes (Old English); (b) Norman French (Old French); or (c) Old Norse. Write your group's complete list on a chart. Then share your words with the class through a game of charades.

PLACE LOCATION ACTIVITY: BRITISH ISLES AND SCANDINAVIA

Match the letters on the map with the places listed below. Write your answers on a separate sheet of paper.

1. Reykjavík
2. Baltic Sea
3. London
4. Stockholm
5. North Sea
6. Northern Ireland
7. Denmark
8. English Channel
9. Republic of Ireland
10. Kjølen Mountains

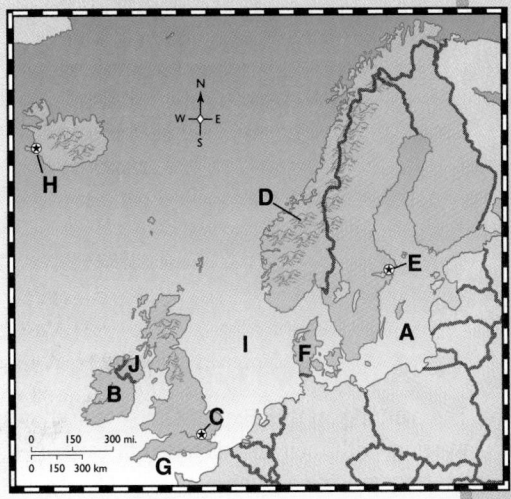

Cooperative Learning Activity

Direct group members to work together to come up with ideas for charades. Before the students play charades, go over the standard signals, such as fingers on the forearm to denote the number of syllables and the thumb and index finger close together to denote "small word."

GeoJournal Writing Activity

Encourage students to illustrate their journal entries with sketches. Call on volunteers to share their entries with the rest of the class.

Place Location Activity

1. H
2. A
3. C
4. E
5. I
6. J
7. F
8. G
9. B
10. D

Chapter Bonus Test Question

This question may be used for extra credit on the chapter test.

Which of the countries in this part of Europe would be most affected by a sharp drop in oil prices? Why? *(United Kingdom; it is Europe's leading oil producer.)*

Section 3

7. Skiing is a national sport, people make their living by fishing because farming land is limited, and so on.
8. Helsinki

Critical Thinking Activities

1. Answers should emphasize the changes brought about by the shift from agricultural to industrial economies, specifically the introduction of modern technologies.
2. Norway has a long, jagged seacoast and inland mountains with little farmland. Denmark, consisting of a peninsula and many islands, is mostly flat with a great deal of farmland. Many Norwegians look to the sea for their livelihood; Norway

273

Cultural HERITAGE:
EUROPE

Cultural **HERITAGE**

FOCUS

Ask students to name school subjects in which European people or ideas play a prominent role. They should see that there is at least a European influence in most areas. Tell students that, in fact, Europe has had a great influence on world history and culture.

TEACH

Display Political Map Transparency 1 and circle Europe in one color. Have students suggest regions that have been powerfully influenced by European settlement, technology, and ideas, and circle them with a second color. Guide students into identifying the following regions: the Western Hemisphere; Australia and Oceania; Northern Asia; the Indian subcontinent; the southern Arabian Peninsula; Africa. **L1**

ARCHITECTURE ▶ ▶ ▶ ▶

The wooden stave churches in Scandinavia were built during the early 1100s. They were the first Christian churches in Scandinavia. Their steeply angled roofs are covered with scalelike shingles and are often decorated with dragons and other Viking symbols.

◀ ◀ ◀ ◀ SCULPTURE

The European Renaissance was a time of invention and creativity. Michelangelo, a brilliant Italian sculptor and painter, created the Pietá in 1500. This marble sculpture of the Virgin Mary and Christ is housed in St. Peter's Basilica in Rome.

ANCIENT POTTERY ▶ ▶ ▶ ▶

The painting on this ancient Greek vase shows warriors with their shields raised in battle. Artifacts provide a glimpse of what life was like more than 2,000 years ago in Europe.

274

UNIT 4

More About . . .

Painting The French Impressionists, who painted during the last third of the nineteenth century, remain popular today. Their use of color and light has wide appeal, and their unique style is instantly recognizable. French Impressionism marks a watershed in artistic history, when artists turned from painting formal interior scenes to portraying outdoor scenes of everyday life. Thus, many people consider Impressionist works to be the first modern art. Famous Impressionists include Edouard Manet, Camille Pissarro, Edgar Degas, Alfred Sisley, and Pierre-Auguste Renoir.

◄ ◄ ◄ ◄ ROYAL JEWELS

This diamond-studded gold crown is part of the collection of the British crown jewels. The precious ornaments and jewels of Great Britain's kings and queens are stored in an underground vault in the Tower of London.

PAINTING ► ► ► ►

The French Impressionists were a group of painters who used short brush strokes and brilliant color to show the play of light on outdoor subjects. Their goal was to give an *impression* rather than to show reality. Claude Monet painted this scene in Giverny, France, in the late 1800s.

PORCELAIN ▲ ▲ ▲ ▲

These porcelain figures were made in Germany in the 1700s. The tiny sculptures were used as table decorations. A world-famous porcelain factory still operates near Dresden, Germany.

275

APPRECIATING CULTURE

1. Describe what you like or do not like about Monet's Impressionist painting.

2. How is the sculpture by Michelangelo different from the miniature porcelain sculptures made in Germany?

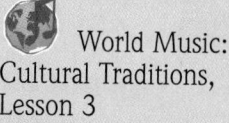

Cultural
HERITAGE

Geography and the Humanities

📁 World Literature Reading 4

🌎 World Music: Cultural Traditions, Lesson 3

📦 World Cultures Transparencies 5, 6

📕 Focus on World Art Prints 1, 2, 5, 6, 8, 9, 14, 15, 17, 18, 19, 20, 21, 22, 23

ASSESS

Have students answer the **Appreciating Culture** questions on page 275.

CLOSE

Have students debate whether the global influence of European culture has been a positive or negative development.

Answers to Appreciating Culture

1. Answers will vary. Suggest that students focus their criticism on such features as the color and light, the subject matter, or the atmosphere the painting creates.

2. Answers will vary. Differences students might note include: size, materials used, process by which works created, subject matter, reasons why artists created works, period when works created.

PERFORMANCE ASSESSMENT ACTIVITY

AN IMMIGRANT'S DIARY Many immigrants have come to the United States from northwestern Europe. Have each student play the role of an immigrant from one of the countries in the region. Students should complete a "Before/After" chart about the geography of their country of origin and the area in which they settle in the United States. Have students use these charts to write a series of diary entries from their departure from Europe through their arrival in the United States. Suggest that they focus on the adaptations they make to life in America. Encourage students to illustrate their diaries with maps, sketches, and so on.

Suggest that they "bind" their diaries with appropriate covers.

POSSIBLE RUBRIC FEATURES: Accuracy of content information, concept attainment for adaptation, classification skills, inference skills, composition skills, visual quality of product

MENTAL MAPPING ACTIVITY

Before teaching Chapter 11, point out the countries of northwestern Europe on a globe to students. Have them identify the water bodies surrounding the countries and estimate their distance from the United States.

TEACHER'S CORNER

NATIONAL GEOGRAPHIC SOCIETY

INDEX TO NATIONAL GEOGRAPHIC MAGAZINE

The following articles may be used for research relating to this chapter:

- "The Basques," by Thomas J. Abercrombie, November 1995.
- "Essence of Provence," by Bill Bryson, September 1995.
- "Blueprints for Victory," by John F. Shupe, May 1995.
- "Hanseatic League," by Edward Von der Porten, October 1994.
- "English Channel Tunnel," by Cathy Newman, May 1994.
- "Europe Faces an Immigrant Tide," by Peter Ross Range, May 1993.
- "Main-Danube Canal: Linking Europe's Waterways," by Bill Bryson, August 1992.
- "Are the Swiss Forests in Peril?" by Christian Mehr, May 1989.
- "The Dutch Touch," by Bart McDowell, October 1986.
- "Those Eternal Austrians," by John J. Putman, April 1985.

NATIONAL GEOGRAPHIC SOCIETY PRODUCTS AVAILABLE FROM GLENCOE

To order the following products for use with this chapter, contact your local Glencoe sales representative or call Glencoe at 1-800-334-7344:

- *STV: World Geography* (Videodisc)
- *Picture Atlas of the World* (CD-ROM)
- *Physical Geography of the World* (Transparencies)
- *ZipZapMap! World* (Software)
- *GeoBee* (Software)
- *Images of the World* (Posters)
- *Eye on the Environment* (Posters)

ADDITIONAL NATIONAL GEOGRAPHIC SOCIETY PRODUCTS

To order the following products for use with this chapter, call National Geographic Society at 1-800-368-2728:

- *Democratic Governments Series:* "France," "Germany." (Video)
- *1914-1918: World War I* (Video)
- *Europe: The Road to Unity* (Video)
- *Physical Geography of the Continents Series:* "Europe." (Video)
- *Portraits of the Continents Series:* "Part III: Europe; Asia; Africa." (Filmstrip)
- *Geography of Europe Series:* "Pt. I: Northern Europe;" "Pt. II: Western Europe;" "Pt. III: Central Europe;" "Pt. IV: Southern Europe." (Filmstrip)

TEACHER-TO-TEACHER
Curriculum Connection: History, Political Science

—from Cathy Salter,
Educational Consultant, Hartsburg, MO

A PROFILE OF GERMANY, 1940s-1990s *The purpose of this activity is to have students understand the changes that have taken place in Germany since World War II by having them prepare and teach a lesson on the subject.*

Briefly review the changes that have occurred in Germany since 1945, focusing on the Berlin Wall—its construction, purpose, and final collapse. Then organize students into four groups. Inform groups that their task is to develop a 15-minute lesson on Germany from the 1940s to the 1990s. Suggest that their lessons cover topics such as Germany at the end of World War II, the division of Germany, communism versus democracy in Germany, the Berlin Wall, the collapse of communism, reunification, and Germany today. Encourage groups to illustrate their lessons with maps, charts, and other appropriate visuals and to develop useful handouts such as fact sheets, time lines, and so on.

Have groups present their lessons to the class or to other social studies classes. As a follow-up activity, have students discuss the following: Suppose your city or town had been divided by a wall, with people on either side adopting different forms of government. What would everyday life be like? What would your reactions be if the wall came down? What problems would you face in adjusting to life after the collapse of the wall?

MEETING NATIONAL STANDARDS

Geography for Life

All of the 18 standards are demonstrated in Unit 4. The following ones are highlighted in this chapter:

- Standards 1, 2, 3, 5, 7, 8, 10, 11, 13, 14, 15

KEY TO ABILITY LEVELS

Teaching strategies have been coded for varying learning styles and abilities.

L1   **BASIC** activities for all students

L2   **AVERAGE** activities for average to above-average students

L3   **CHALLENGING** activities for above-average students

LEP   **LIMITED ENGLISH PROFICIENCY** activities

BIBLIOGRAPHY

Readings for the Student
Dunnan, Nancy. *One Europe.* Brookfield, Conn.: Millbrook Press, 1992.

Ganeri, Anita, and Rachel Wright. *France.* New York: Franklin Watts, 1993.

James, Ian. *Inside the Netherlands.* New York: Franklin Watts, 1990.

Spencer, William. *Germany Then and Now.* New York: Franklin Watts, 1994.

Readings for the Teacher
Benelux Red Guide. Greenville, S.C.: Michelin, 1993.

Europe Today: An Atlas of Reproducible Pages, revised ed. Wellesley, Mass.: World Eagle, 1993.

Timeline: Teaching Project Germany. Coral Gables, Fla.: Timeline U.S.A., 1993. Overheads, reproducible fact sheets, student pamphlets.

Multimedia
European Geography: Central Region. Chicago: Encyclopedia Britannica, 1993. Videocassette, 20 minutes.

European Geography: Western Region. Chicago: Encyclopedia Britannica, 1993. Videocassette, 20 minutes.

Germany: Video Visits. San Ramon, Calif.: International Video Network, 1991. Videocassette, 90 minutes.

Paris. Chicago: Questar/Travel-Network, 1989. Videocassette, 30 minutes.

Use these *Geography: The World and Its People* resources to teach, reinforce, and extend chapter content.

CHAPTER 11 RESOURCES

*Vocabulary Activity 11

Cooperative Learning Activity 11

Workbook Activity 11

Geography Skills Activity 11

Chapter Map Activity 11

GeoLab Activity 11

Critical Thinking Skills Activity 11

Performance Assessment Activity 11

*Reteaching Activity 11

Enrichment Activity 11

Chapter 11 Test, Form A and Form B

*Chapter 11 Digest Audiocassette, Activity, Test

Unit Overlay Transparencies 4-0, 4-1, 4-2, 4-3, 4-4, 4-5, 4-6

Political Map Transparency 4; World Cultures Transparency 6

Geoquiz Transparencies 11-1, 11-2

Vocabulary PuzzleMaker Software

Student Self-Test: A Software Review

Testmaker Software

MindJogger Videoquiz

Focus on World Art Prints 1, 2, 5, 17, 22

*If time does not permit teaching the entire chapter, summarize using the Chapter 11 Highlights on page 293, and the Chapter 11 English (or Spanish) Audiocassettes. Review students' knowledge using the Glencoe MindJogger Videoquiz. *Also available in Spanish*

Use these *Geography: The World and Its People* resources to teach and reinforce section content.

SECTION 1 RESOURCES

Reproducible Lesson Plan 11-1

Section Focus Transparency 11-1

Guided Reading Activity 11-1

Section Quiz 11-1 (also available in Spanish)

SECTION 2 RESOURCES

Reproducible Lesson Plan 11-2

Section Focus Transparency 11-2

Guided Reading Activity 11-2

Section Quiz 11-2 (also available in Spanish)

SECTION 3 RESOURCES

Reproducible Lesson Plan 11-3

Section Focus Transparency 11-3

Guided Reading Activity 11-3

Section Quiz 11-3 (also available in Spanish)

SECTION 4 RESOURCES

Reproducible Lesson Plan 11-4

Section Focus Transparency 11-4

Guided Reading Activity 11-4

Section Quiz 11-4 (also available in Spanish)

ADDITIONAL RESOURCES FROM GLENCOE

Reproducible Masters

- Glencoe Social Studies Outline Map Book, page 30

- Foods Around the World, Unit 4

World Music: Cultural Traditions
Lesson 3

Transparencies

- National Geographic Society PicturePack Transparencies, Unit 4

Posters

- National Geographic Society: Images of the World, Unit 4

World Games Activity Cards
Unit 4

Videodiscs

- Geography and the Environment: The Infinite Voyage

- Reuters Issues in Geography

- National Geographic Society: STV: World Geography, Vol. 2

CD-ROM

- National Geographic Society: Picture Atlas of the World

Software

- National Geographic Society: ZipZapMap! World

Reference Books

- Picture Atlas of Our World

Performance Assessment

Refer to the Planning Guide on page 276A for a Performance Assessment Activity for this chapter. See the *Performance Assessment Strategies and Activities* booklet for suggestions.

Chapter Objectives

1. **Identify** the major geographic features of France.
2. **Specify** the link between history and geography in Germany.
3. **Compare** the geographic characteristics of the Benelux countries.
4. **Name** landforms, languages, and population centers of the Alpine countries.

GLENCOE
TECHNOLOGY

 VIDEODISC

Use the Chapter 11 MindJogger Videoquiz to preview chapter content.

MindJogger Videoquiz

Chapter 11
Disc 2 Side A

 Also available in VHS.

Chapter 11 Northwestern Europe

MAP STUDY ACTIVITY

 As you read Chapter 11, you will learn about countries in northwestern Europe.

1. What northwest European countries have borders on the North Sea?
2. What is the capital of Germany?
3. What countries border Liechtenstein?

276

MAP STUDY ACTIVITY

Answers
1. Belgium, the Netherlands, Germany
2. Berlin
3. Austria, Switzerland

Map Skills Practice
Reading a Map What is the capital of the Netherlands? *(Amsterdam)*

France

LESSON PLAN
Chapter 11, Section 1

PREVIEW

Words to Know
- navigable
- republic

Places to Locate
- France
- Paris
- Loire River
- Seine River
- Pyrenees
- Alps

Read to Learn . . .
1. what landforms are found in France.
2. why France is able to produce huge amounts of food.
3. what French culture offers the rest of the world.

Imagine yourself on a river barge gliding between fairy-tale castles and peaceful villages. You ask the captain about this beautiful river, the Loire (LWAHR), and find out you are cruising on the longest river in France. Along its banks lie rich farmlands and prosperous manufacturing centers. France welcomes you.

France, Germany, and their smaller neighbors rank as major economic and cultural centers of Europe. In the first half of this century, warfare among these nations tore Europe apart. Since the end of World War II, however, they have put aside their differences. Joined in economic partnership in the European Union, they look forward to a peaceful and prosperous twenty-first century.

Region
The Land

Home to one of the oldest civilizations in the world, France's land area covers an area of 212,390 square miles (550,090 sq. km). Slightly smaller in area than the state of Texas, about one-half of France's border is coastline. The Mediterranean Sea lies along its southeast coast. Along the country's west and northwest shores are the Atlantic Ocean and the English Channel. A new tunnel—the Chunnel—built under this channel connects France with Great Britain. For the first time, citizens and freight can move directly between these two nations by truck, rail, and automobile.

Classroom Resources for Section 1

 REPRODUCIBLE MASTERS

Reproducible Lesson Plan 11-1
Guided Reading Activity 11-1
Vocabulary Activity 11
Geography Skills Activity 11
Section Quiz 11-1

 TRANSPARENCIES

Section Focus
   Transparency 11-1
Political Map Transparency 4
Unit Overlay Transparency 4-0

MULTIMEDIA

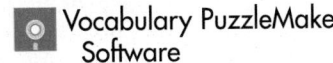 Vocabulary PuzzleMaker
Software

Testmaker

 National Geographic
Society: STV: World
Geography, Vol. 2

FOCUS

Section Objectives
1. **Identify** the landforms that are found in France.
2. **Explain** why France is able to produce huge amounts of food.
3. **Appreciate** what French culture offers to the rest of the world.

Bellringer Motivational Activity

Prior to taking roll at the beginning of the class period, project Section Focus Transparency 11-1 and have students answer the activity questions. Discuss students' responses. **L1**

Vocabulary Pre-check

Explain that the suffix *-or* means "one who," the suffix *-ation* means "act of," and the suffix *-able* means "can be." Point out that a navigator is "one who sails" and navigation is "the act of sailing." Then ask what *navigable* means. *(can be sailed)* **L1 LEP**

Use the Vocabulary PuzzleMaker Software to create a crossword puzzle. **L1**

Meeting National Standards
Geography for Life The following standards are highlighted in this section:
- Standards 2, 4, 10, 11, 14

TEACH

Guided Practice

Graphic Skills On the chalkboard, write the following facts: Everest in Tibet and Nepal, 29,028 feet (8,848 m); Changtse in Tibet, 24,780 feet (7,553 m); El Muerto in Argentina and Chile, 21,456 feet (6,540 m); and McKinley in Alaska, 20,320 feet (6,194 m). Have students copy the information into their notebooks, adding statistics for Mont Blanc in France. Then have them present this information in the form of a graph. Call on volunteers to display their graphs. **L1**

Independent Practice

 Guided Reading Activity 11-1 **L1**

MAP STUDY

Answer
France, Switzerland

Map Skills Practice
Reading a Map What bodies of water separate the United Kingdom from northwestern Europe?
(English Channel, North Sea)

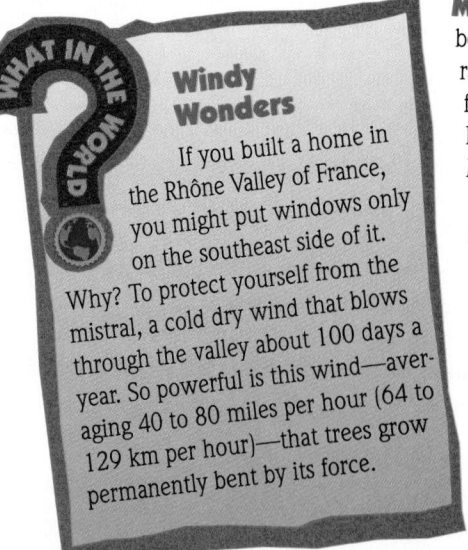

Windy Wonders

If you built a home in the Rhône Valley of France, you might put windows only on the southeast side of it. Why? To protect yourself from the mistral, a cold dry wind that blows through the valley about 100 days a year. So powerful is this wind—averaging 40 to 80 miles per hour (64 to 129 km per hour)—that trees grow permanently bent by its force.

Mountains Great mountain ranges also form part of France's borders. In the southwest the Pyrenees (PIHR•uh•NEEZ) separate France from Spain. In the southeast the Alps divide France from Italy and Switzerland. The soaring, icy peaks of the French Alps include Mont Blanc. At 15,771 feet (4,807 m), Mont Blanc is one of the highest mountains in Europe.

Plains and Rivers The map on page 246 shows you that most of northern France is part of the vast North European Plain. It stretches from the British Isles east to Russia. In France a variety of crops thrive in the rich soil of the plain. Livestock are raised here, too, and the area is home to many industrial cities.

An important transportation network of rivers connects major manufacturing areas within France. Most of these rivers are **navigable,** or usable by large ships. The Seine (SAYN) River flows through the capital city of Paris. The Rhône (ROHN) River lies in the southeast. The Rhine River—Europe's most important inland waterway—forms part of France's eastern border.

The Climate How many different climate regions of France do you see on the map on page 284? Note that France has three types of climate. Along the Mediterranean coast, summers are hot and winters are mild. The climate of

NORTHWESTERN EUROPE: Physical

MAP STUDY
Mont Blanc is one of Europe's highest mountains.
MOVEMENT: A trip from Mont Blanc to the Matterhorn would start and end in what countries?

ELEVATIONS
Feet	Meters
10,000	3,000
5,000	1,500
2,000	600
1,000	300
0	0

▲ Mountain peak

Lambert Conformal Conic projection

278

Cooperative Learning Activity

Print on poster boards or sheets of butcher paper the following French proverbs about wine: "Good wine makes the horse go"; "Good wine needs no sign"; "Since the wine is drawn, it must be drunk"; "Wine poured out is not wine swallowed"; "Wine will not keep in a foul vessel." Organize students into five groups and assign one of the proverbs to each group. Have groups discuss the literal and metaphorical meanings of their proverbs. Call on groups to share the results of their discussions with the class. Conclude by asking students why they think wine features in so many French sayings. *(France is famous for its wines.)* **L1**

Scenes of Paris, France

More About the Illustration

Answer to Caption
red, white, and blue

Place Bastille Day commemorates the storming of the Bastille prison by Parisians in 1789, the action which set off the French Revolution.

western France, influenced by Atlantic winds, has cool summers and mild winters with plenty of rainfall. As you travel farther east—away from the ocean—winters are colder, summers warmer, and there is less rain.

Place

The Economy

France's well-developed economy relies on agriculture, industry, and commerce. France is the largest food producer in western Europe. French farmers grow fruits, vegetables, and grains. They also raise beef and dairy cattle. Grapes rank as one of the most important French crops. Why? The grapes are used to make famous French wines.

France is also one of the world's leading manufacturing countries, with Paris as its chief industrial center. France's natural resources include coal, iron ore, and bauxite. A variety of industrial goods—including airplanes, cars, computers, chemicals, clothing, and furniture—are produced by French workers. France is also a leading center of commerce, with an international reputation in fashion.

Place

The People

"Liberté . . . Egalité . . . Fraternité" (Liberty, Equality, Fraternity)—France's national motto—describes the spirit of the people. Although they have regional differences, the French share a strong national loyalty. Most French trace their ancestry to the Celts and Romans of early Europe. They speak French, and about 75 percent of them are Roman Catholics.

CHAPTER 11

The Seine River and Eiffel Tower are two of Paris's most famous landmarks *(left)*. On Bastille Day, soldiers parade in Paris to commemorate the French Revolution *(right)*.
PLACE: Judging from the parade, what are France's national colors?

ASSESS

Check for Understanding

Assign Section 1 Review as homework or an in-class activity.

Meeting Lesson Objectives

Each objective below is tested by the questions that follow it in parentheses.

1. **Identify** the landforms that are found in France. (2, 5)
2. **Explain** why France is able to produce huge amounts of food. (4)
3. **Appreciate** what French culture offers to the rest of the world. (3)

Meeting Special Needs Activity

Reading Disability Students with reading problems often are helped by having a specific purpose for reading. Focused reading exercises present a limited goal. They force the reader to make judgments about the applicability of the material in question. Have students read "The Economy" to find out why grapes rank as one of France's most important crops. **L1**

Evaluate

 Section 1 Quiz **L1**

Use the Testmaker to create a customized quiz for Section 1. **L1**

CURRICULUM CONNECTION

The Arts In 1874 Louis Leroy, a French art critic, coined the term *Impressionist* while reviewing Claude Monet's painting *Impression, Sunrise.* Leroy compared the painting unfavorably to wallpaper.

Reteach

On the chalkboard, copy and number the section's bold-faced heads in outline form. Have students complete the outline with subheads and details.

Enrich

Encourage students to read Eve Titus's books about the French mouse Anatole. Have them write similar stories to young children about the geography of France. **L3**

CLOSE

Have students design postcards showing the Eiffel Tower, Notre Dame, the Tour de France, or other French tourist attractions and write appropriate messages from Hervé Lefort on the back.

Influences of the Past Kings ruled France for nearly 800 years, establishing France as a great European power. In 1792 the French people overthrew their last king. They set up a **republic,** a government in which the people elect all important government officials. A few years later, a general named Napoleon Bonaparte seized power. He conquered much of Europe before an alliance of European countries defeated him. In the first half of the 1900s, France was a battleground during the two world wars. As strong as its motto, France has survived and holds an important role in modern world affairs.

City Life If you drew a line connecting the five largest French cities from south to north, you would form a triangle. You can see on the map on page 292 that the areas around Marseille, Bordeaux, and Paris hold most of France's people. A superb system of railroads connects these and other cities. French passenger trains are some of the fastest in the world, many averaging 132 miles per hour (212 km per hour). The population of France totals about 58 million people. About 80 percent of the French people live in cities and large towns.

Paris Paris itself is home to about 8.7 million people. France's literacy rate of 99 percent shows that the French prize education. Paris, a world center of art and learning, is home to many universities, museums, and other cultural sites. It is also France's leading center of industry, transport, and communications. Millions of tourists from all over the world visit Paris annually.

Thirteen-year-old Hervé Lefort lives in Paris and can see the Eiffel Tower every day. He likes to visit the tower to enjoy the wonderful view of the whole city. Among the most famous sights in Paris is the beautiful cathedral of Notre Dame, which sits on an island in the Seine River. The nearby Louvre (LOOV) is one of the greatest art museums in the world.

Hervé and his friends look forward every year to the Tour de France—Europe's premier bicycle race. Held since 1903, this three-week competition matches cyclists from all over the world.

SECTION 1 REVIEW

REVIEWING TERMS AND FACTS

1. **Define the following:** navigable, republic.
2. **REGION** What is important about the region of the North European Plain?
3. **PLACE** Why is Paris an important city?
4. **HUMAN/ENVIRONMENT INTERACTION** What are France's major agricultural products?

MAP STUDY ACTIVITIES

5. Look at the physical map on page 278. What mountain ranges are located along France's borders?
6. Turn to the political map on page 276. What is France's smallest neighbor along its eastern border?

UNIT 4

Answers to Section 1 Review

1. All vocabulary words are defined in the Glossary.
2. In France, the plain's rich soil supports a great variety of crops and livestock. The plain is also the location of many of France's industrial cities.
3. Paris is France's capital and largest city and a major world cultural center.
4. grapes, grains, fruits, vegetables, beef, and dairy products
5. Alps and Pyrenees
6. Monaco

BUILDING GEOGRAPHY SKILLS

Reading a Vegetation Map

Environmental groups challenge us to "Save a Tree!" rather than "Save the Vegetation!" But that is exactly what a tree is—vegetation. A vegetation map helps you see whole regions that have the same type of plant life. These areas often share other natural elements, too, such as climate and soil type. For example, in rain forests, broad-leaved evergreens grow consistently throughout the region. Rain forests also have the same climate and soil types. To read a vegetation map, apply these steps:

- Use the map key to match colors with vegetation regions.
- Use a dictionary to define unfamiliar vegetation terms.
- Draw conclusions about the climate and soil of the vegetation region.

FRANCE: Vegetation

Vegetation
- Deciduous forest
- Coniferous forest
- Mixed forest (coniferous and deciduous)
- Mediterranean vegetation
- Alpine vegetation

Lambert Conformal Conic projection

UNITED KINGDOM · BELGIUM · LUXEMBOURG · GERMANY · Paris · Strasbourg · Nantes · FRANCE · SWITZERLAND · Bordeaux · Lyon · ITALY · Toulouse · Montpellier · Nice · MONACO · Marseille · Mediterranean Sea · SPAIN · ANDORRA

ATLANTIC OCEAN · English Channel · Seine R. · Marne R. · Loire R. · Garonne R. · Rhône

0   50   100 mi.
0   50 100 km

Geography Skills Practice

1. What vegetation region covers most of France?

2. What type of vegetation is found only along France's Mediterranean coast?

3. From the map, what conclusions can you draw about the amount of rain the regions of France receive?

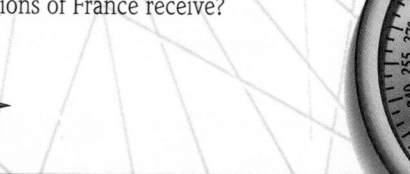

CHAPTER 11

281

FOCUS

Section Objectives

1. **Describe** the landscape of Germany.
2. **Explain** why the German economy is so strong.
3. **Consider** how historical events affected the geography of Germany.

Bellringer Motivational Activity

 Prior to taking roll at the beginning of the class period, project Section Focus Transparency 11-2 and have students answer the activity questions. Discuss students' responses. **L1**

Vocabulary Pre-check

On the chalkboard, write *soil, pollution, form of socialism, highway, language,* and *mass murder.* Have students scan the section to find out which term on the chalkboard comes closest in meaning to each term in the **Words to Know** list. **L1**
LEP

Global 🍴 Gourmet

Germany Germans prefer soft drinks without ice because they consider cold drinks unhealthy.

SECTION 2 — Germany

PREVIEW

Words to Know
• communism
• acid rain
• loess
• *autobahn*
• dialect
• Holocaust

Places to Locate
• Germany
• Berlin
• Alps
• Black Forest
• the Ruhr

Read to Learn . . .
1. what the landscape is like in Germany.
2. why the German economy is so strong.
3. how historical events affected the geography of Germany.

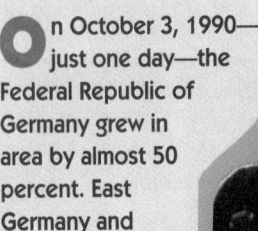 On October 3, 1990—in just one day—the Federal Republic of Germany grew in area by almost 50 percent. East Germany and West Germany had officially reunited. Germans hurried to chip away at the Berlin Wall—the symbol of the once-divided country.

In the early 1900s, Germany ranked as one of the strongest countries in the world. Following World War I, however, it needed rebuilding. Then after World War II, it became a divided country. One part—West Germany—was based on democracy and closely tied to the countries of the West. The other part—East Germany—was tied to the Soviet Union and based on communism. **Communism** is a form of socialism in which the government, usually under one strong ruler, controls the economy and the society in general.

In 1990 communism in Europe collapsed and the two parts of Germany came together. They formed one democratic country. Modern, united Germany has a strong, free market economy.

Region
The Land

Now a much larger country, Germany covers about 134,930 square miles (349,469 sq. km). It lies in the heart of Europe, south of Scandinavia and northeast of France. Germany's northern border touches Denmark and the Baltic and North seas.

 282

Classroom Resources for Section 2

 **REPRODUCIBLE MASTERS**
Reproducible Lesson Plan 11-2
Guided Reading Activity 11-2
Cooperative Learning Activity 11
Workbook Activity 11
Section Quiz 11-2

 **TRANSPARENCIES**
Section Focus Transparency 11-2
Political Map Transparency 4
World Cultures Transparency 6

MULTIMEDIA
 Testmaker

National Geographic Society: GTV: World Geography, Vol. 2

The Landscape Mountains, plateaus, and plains form the landscape of Germany. In the south the Alps rise over the German border with Switzerland and Austria. The lower slopes of these mountains—a favorite area of snow skiers—are covered with forests. Beyond the forests lie open meadows. Above the slopes are the mountain peaks, many capped with snow year-round.

Have you ever heard a cuckoo clock announce the hour? That clock probably came from Germany's Black Forest. Lying north of the Alps, this part of the Central Uplands is famous for its wood products. Is the forest really black? No, but the trees grow so close together the forest appears to be black.

In recent decades, the Black Forest has suffered severe damage from **acid rain,** or rainfall with high amounts of chemical pollution. Growing numbers of European factories and automobiles are responsible. Germans are making efforts to conserve their forests. A solution to the acid rain problem, however, has yet to be found.

From Germany's Central Uplands flows one of Europe's most important rivers—the Danube. It winds eastward for about 400 miles (650 km) across southern Germany. The Rhine, the Elbe, and the Weser rivers stretch across the North European Plain.

The Climate Westerly Atlantic Ocean winds crossing Europe help warm Germany in winter and cool it in summer. The **loess** (LEHS), or soil deposited by winds, produces good crops. Most of Germany has a marine west coast climate. Southern areas more distant from the ocean have colder winters and warmer summers.

Place

The Economy

The product label "Made in Germany" means superior quality and construction to consumers all over the world. Why? Germany's manufacturing economy has two major strengths: highly skilled workers and an abundance of key industrial resources.

The Ruhr Rich deposits of coal and iron ore found in the Ruhr help make this region in western Germany the industrial heart of Europe. Battles have been fought over the control of this productive area. Factories here produce machinery, cars, clothing, and chemical products. In the Ruhr, often called a "smokestack region," cities and factories stand as close together as teeth on a comb. They cover almost every square mile of the Ruhr.

Farming Germany imports about one-third of its food, although it is a leading producer of beer, wine, and cheese. German farmers raise livestock and crops

The Bavarian Alps in Germany

The beauty of the snowcapped Bavarian Alps attracts many tourists.
LOCATION: Besides Germany, through what two countries do the Bavarian Alps extend?

More About the Illustration

Answer to Caption
Switzerland, Austria

Location The Zugspitze, located in the Bavarian Alps, is Germany's highest peak. It rises to a height of 9,721 feet (2,963 m).

TEACH

Guided Practice

History Project a transparency showing these dates: late 1800s, 1930s, following 1945, and 1990. Ask volunteers to match an event discussed in the section with each date. *(late 1800s—Bismarck unites Germany, 1930s—Hitler gains control, following 1945—Germany is divided, 1990—Germany is united)* L1

MULTICULTURAL PERSPECTIVE

Cultural Heritage
Robert Browning's poem "The Pied Piper of Hamelin" is based on a true story. In 1984 the German town of Hamelin commemorated the 700th anniversary of the disappearance of 130 children. No one really believes a piper lured them away. Some think the children were kidnapped and made serfs by local landowners. Others believe the 130 joined the Children's Crusade to the Holy Land.

Cooperative Learning Activity

Organize students into three groups and assign one of the following headings to each group: "The Land," "The Economy," "The People." Have group members read and discuss the text under their assigned heading. Then form new groups of three students, consisting of one member from each of the original groups. Members of the new groups are responsible for teaching one another about the text they studied. At the end of the activity, all students should have a full picture of Germany's geography. L1

Independent Practice

 Guided Reading Activity 11-2 **L1**

There are no speed limits on some sections of the *autobahns*.

Answer
marine west coast

Map Skills Practice
Reading a Map What body of water affects the climate of Southern France? *(Mediterranean Sea)*

ASSESS

Check for Understanding

Assign Section 2 Review as homework or an in-class activity.

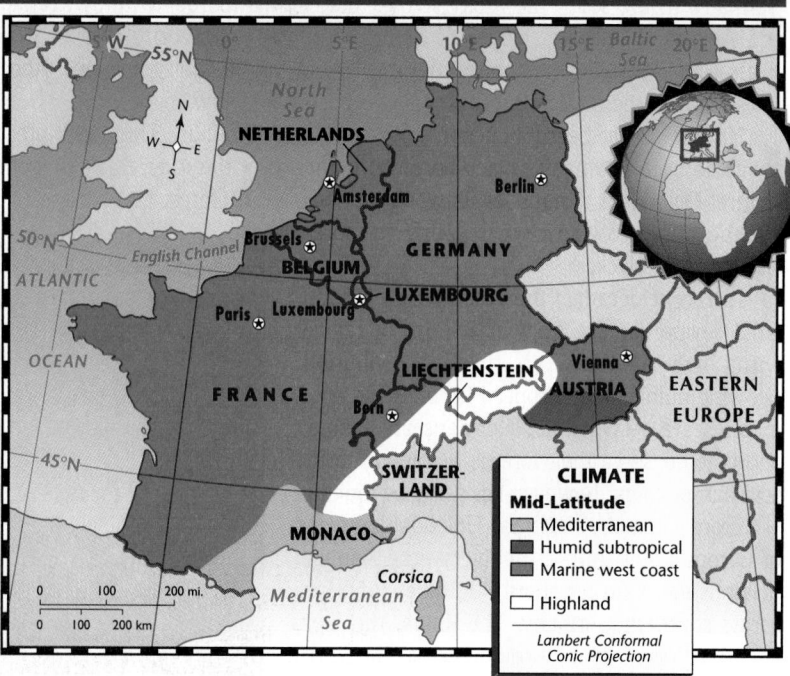

NORTHWESTERN EUROPE: Climate

MAP STUDY The central European Alps have a highland climate.
PLACE: What type of climate is found in the European countries that border the Atlantic Ocean?

NETHERLANDS
Amsterdam
Berlin
English Channel
Brussels
BELGIUM
GERMANY
ATLANTIC
LUXEMBOURG
Paris
Luxembourg
OCEAN
FRANCE
LIECHTENSTEIN
Vienna
AUSTRIA
EASTERN EUROPE
Bern
SWITZER-LAND
MONACO
Corsica
Mediterranean Sea

CLIMATE
Mid-Latitude
Mediterranean
Humid subtropical
Marine west coast
Highland

Lambert Conformal Conic Projection

such as grains, vegetables, and fruits. German workers and products travel to market on superhighways called *autobahns.* These highways, along with canals and railroads, link Germany's cities.

Movement
The People

If a German said to you, *"Guten tag,"* how would you reply? You would return this greeting of "Good day" with a similar one. The German language commonly spoken within Germany has two major **dialects,** or local forms of a language. Most of Germany's 82 million people trace their ancestry to groups of people who settled in Europe from about the A.D. 100s to the 400s. Roman Catholics and Protestants make up most of the population and are fairly evenly represented.

Influences of the Past For hundreds of years, Germany was a collection of small territories ruled by princes. During the late 1800s, a leader named Otto von Bismarck united the territories into a single nation—Germany. The country's efforts to become a world power, in part, caused World War I.

During the 1930s dictator Adolf Hitler gained control of Germany. Under his leadership Germany increased its military power and invaded neighboring countries, setting off World War II.

One of the horrors of World War II was the **Holocaust,** the systematic murder of 6 million European Jews and 6 million others by Hitler and his followers.

Meeting Special Needs Activity

Learning Disabilities Students with learning problems often need help organizing and classifying material. Have them create a large chart with the following headings: *Life in the Past, Life Today.* Direct students to reread the section and list the facts that they consider important under the appropriate heading. Then have students choose one of the headings and write a paragraph using the facts listed on the chart. **L1**

The Allies—led by the United States, Great Britain, and the Soviet Union—defeated Germany and ended the Holocaust.

Following World War II, Germany was divided into democratic West Germany and communist East Germany. West Germany built a powerful free enterprise economy. East Germany made slower progress under communism. East Germany and West Germany reunited in 1990. An expanded free enterprise system and greater European unity are goals of the new German government.

Daily Life About 85 percent of Germans live in cities and towns. Berlin, with a population of 3.5 million, is the largest city and the official capital. Hamburg, in the north, is Germany's major port. Many government offices are located in Bonn and Cologne, cities on the Rhine River. The southern cities of Munich and Stuttgart rank as the fastest-growing cities in Germany. Industrial expansion and technology have spurred this growth.

Fourteen-year-old Eva Reinhardt lives in Munich. Her favorite subjects in school are music, German literature, and mathematics. Eva is proud of what her country has contributed to the worlds of music, literature, science, and good food.

Germans excel at sports and physical activities. Eva often joins friends for a hike or a ski trip in the Bavarian Alps near Munich. Oktoberfest, a 16-day fall festival, offers an opportunity for Eva and her friends to wear traditional costumes and march in a parade. The festival includes amusement park rides, bands, folk dancers, and plenty of food—especially frankfurters.

Teen Scene

German Rockers

By day, Sebastian Haussen and Ava Rottermich are students at a music academy in Berlin. By night, they are students of rock 'n roll! Sebastian and Ava started their own band two years ago. Someday they hope to play in one of Berlin's many cabarets. Cabarets are cafes with live entertainment. Sebastian lives in East Berlin. Now that the Berlin Wall is gone, he can travel freely in Germany.

SECTION 2 REVIEW

REVIEWING TERMS AND FACTS

1. **Define the following:** communism, acid rain, loess, *autobahn,* dialect, Holocaust.
2. **PLACE** What are the three major landforms of Germany?
3. **PLACE** What natural resources are found in the Ruhr?
4. **MOVEMENT** How are goods transported within Germany?

 MAP STUDY ACTIVITIES

5. Study the climate map on page 284. What is the major climate of Germany?

Answers to Section 2 Review

1. All vocabulary words are defined in the Glossary.
2. the Alps, the Central Uplands, the North European Plain
3. coal, iron
4. highways, rivers and canals, railroads
5. marine west coast

Meeting Lesson Objectives

Each objective below is tested by the questions that follow it in parentheses.

1. **Describe** the landscape of Germany. (2)
2. **Explain** why the German economy is so strong. (3, 4)
3. **Consider** how historical events affected the geography of Germany. (1)

Evaluate

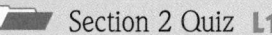 Section 2 Quiz **L1**

Use the Testmaker to create a customized quiz for Section 2. **L1**

Reteach

Have students sum up the section by expressing the main idea of each paragraph in one or two sentences. **L1**

Enrich

Have students listen to recordings of music by Bach, Beethoven, and other German composers. Encourage them to write a critical appreciation of the music. **L3**

CLOSE

Encourage students to discuss the following statement: You cannot understand Germany's geography without understanding the country's history.

285

TEACH

Read the following lines from Winston Churchill's 1946 speech, made at Westminster College in Fulton, Missouri: "From Stettin in the Baltic, to Trieste, in the Adriatic, an Iron Curtain has descended across the Continent. Behind that line lie all the capitals of the ancient states of Central and Eastern Europe. . . . These famous cities . . . lie in what I must call the Soviet sphere. . . ." Then point out that for many people, the Berlin Wall was Churchill's "Iron Curtain" made real. Have students discuss why people felt this way. **L1**

More About . . .

History For the leaders of the Communist East German government, the Berlin Wall was an economic necessity. In the summer of 1961, as many as 1,000 East Berliners fled to the West each day. East Germany's work force was literally disappearing. The Wall served its purpose—it stemmed the flow of people to the West. Only about 5,000 East Germans managed to escape over the Berlin Wall in the 28 years that it stood.

MAKING CONNECTIONS

THE BERLIN WALL COMES CRASHING DOWN

| MATH | SCIENCE | **HISTORY** | LITERATURE | TECHNOLOGY |

Yelling—shouting—cheering voices rang through the streets of Berlin, Germany. It was midnight on November 9, 1989. The joyful crowd had gathered to tear down the hated Berlin Wall.

A BARRIER AND A SYMBOL The Berlin Wall—named *Schandmauer* (wall of shame) by Berliners—divided Berlin from 1961 to 1989. About 27½ miles (43 km) of concrete and barbed wire, it created East Berlin and West Berlin, dividing friends and families. The wall also became a symbol of the separation between democracy and communism. Why was the wall built?

After World War II, defeated Germany and Berlin, its capital, were divided among the victorious Allies. West Germany and West Berlin were allied to the Western democracies. They remained democratic. Communism and the Soviet Union ruled East Germany and East Berlin.

THE WALL GOES UP During the 1950s, many people escaped from East Germany by crossing into West Berlin. To stop the escapes, the East Germans built the Berlin Wall. As the world watched in shock, German soldiers laid bricks and barbed wire to seal off East Berlin. Along the wall appeared guard towers and searchlights. East German soldiers, armed with machine guns and German shepherd dogs, patrolled nearby.

Hello, Neighbor!

Crossing from East Berlin to West Berlin—without government permission—was now almost impossible. Yet hundreds of East Germans risked their lives to try.

THE WALL COMES DOWN Communism weakened in the 1980s. In 1989 East Germany announced many government changes. Included was the opening of the Berlin Wall. All that remains of the wall today are a few sections kept as reminders of a divided Germany.

Making the Connection

1. In what way was the Berlin Wall a geographical barrier?
2. In what ways was the Berlin Wall a cultural barrier?

286

UNIT 4

Answers to Making the Connections

1. The wall, built through the middle of Berlin, divided the city, and blocked the passage of people from one part to the other.

2. Answers may vary. Most students will point out that a culture based on democracy and individual freedom grew up on one side of the wall, while on the other side a culture based on communist totalitarianism developed.

SECTION 3
The Benelux Countries

PREVIEW

Words to Know
- polder
- multinational firm

Places to Locate
- Netherlands
- Amsterdam
- Belgium
- Brussels
- Luxembourg

Read to Learn . . .
1. what two ethnic groups live in Belgium.
2. how the people of the Netherlands have changed their environment.
3. why tiny Luxembourg attracts many businesses.

Tulips blossom as far as the eye can see! Tulips are not only beautiful but, in the Netherlands, they are also a major commercial crop.

The name *Benelux* comes from combining the first letters of *Bel*gium, the *Ne*therlands, and *Lux*embourg. The three small Benelux countries of northwestern Europe have much in common. Much of their land area is low, flat, and densely populated. Most of their people live in cities, work in businesses or factories, and enjoy a high standard of living. Governments in the Benelux countries are parliamentary democracies with monarchs.

Location
Belgium

Belgian lace . . . Belgian diamonds . . . Belgian chocolate . . . all enjoy a worldwide reputation for excellence. Belgium is bordered by France, Germany, and the Netherlands. Because Belgium is centrally located, European wars have often been fought on its lands. Its central location has also contributed to trade and industry and a strong economy. The Belgian people import metals and fuels to produce manufactured goods for export. Rolling hills stretch through Belgium's southeast, and flat plains spread through the northwest. Traveling

CHAPTER 11

(287)

Meeting National Standards
Geography for Life The following standards are highlighted in this section:
- Standards 1, 4, 14

FOCUS

Section Objectives
1. **Describe** the two ethnic groups that live in Belgium.
2. **Recall** how the people of the Netherlands have changed their environment.
3. **Understand** why tiny Luxembourg attracts many businesses.

Bellringer Motivational Activity

Prior to taking roll at the beginning of the class period, project Section Focus Transparency 11-3 and have students answer the activity questions. Discuss students' responses. L1

Vocabulary Pre-check
Tell students that in this section, the word *firm* refers to business. Ask students where they think a multinational firm does business. Offer the following hint: the prefix *multi-* means "many."
L1 LEP

Classroom Resources for Section 3

 REPRODUCIBLE MASTERS

Reproducible Lesson Plan 11-3
Guided Reading Activity 11-3
Critical Thinking Skills
  Activity 11
Chapter Map Activity 11
Section Quiz 11-3

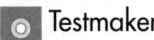 **TRANSPARENCIES**

Section Focus
  Transparency 11-3
Unit Overlay Transparencies
  4-0, 4-1, 4-3

MULTIMEDIA

 Testmaker

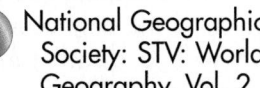 Reuters Issues in
  Geography

National Geographic
  Society: STV: World
  Geography, Vol. 2

287

TEACH

Guided Practice

Map Skills Draw an outline map of the Benelux countries on the chalkboard. As students read the section, call on volunteers to add symbols for resources and products. When the map is complete, have another volunteer create a key to explain the symbols. **L1**

Independent Practice

Guided Reading Activity 11-3 **L1**

MAP STUDY

Answer
Belgium and Luxembourg are mostly industrial, while the Netherlands is mostly agricultural.

Map Skills Practice
Reading a Map What is Belgium's chief resource?
(coal)

ASSESS

Check for Understanding

Assign Section 3 Review as homework or an in-class activity.

---

along Belgium's excellent road system, you are never very far from a town or city. Look at the map on page 292 to see the population density of Belgium.

The People The Walloons and the Flemings dominate the culture of Belgium. These two ethnic groups have lived in Belgium for hundreds of years. The French-speaking Walloons are the largest group of Belgians. The Flemings speak Dutch and often prefer to be called Flemish. Most of Belgium's people are Roman Catholics.

Brussels, the nation's centrally located capital city, has an international air. Many world organizations locate their headquarters there. Banks and other businesses use Brussels as a trade and commerce center.

Human/Environment Interaction
The Netherlands

The Netherlands—a country about half the size of the state of Maine—is located on the North Sea north of Belgium. At its elevation, about 40 percent of the country should be under water. That is how much of its land lies below sea level.

NORTHWESTERN EUROPE: Land Use and Resources

Agriculture
- Commercial farming
- Manufacturing area

Resources
- Bauxite
- Coal
- Copper
- Fish & other seafood
- Forest
- Iron ore
- Lead
- Natural gas
- Petroleum
- Potash
- Silver
- Uranium
- Hydroelectric power
- Zinc

Lambert Conformal Conic projection

MAP STUDY
The Benelux countries are members of the European Union.
PLACE: How do Belgium and Luxembourg differ from the Netherlands in terms of land use?

Cooperative Learning Activity

Explain the following issues facing Benelux countries: Dutch shopping hours are among the shortest in the world. Shops are open until 6 P.M. on weekdays, until 5 P.M. on Saturdays, and not at all on Sundays. Should laws regulating Dutch shopping hours remain the same?

Belgium is the only country in the European Union with a death penalty for murder, although it was last carried out in 1918. Members of the Belgian parliament have tried to abolish the death penalty, but public opinion favors keeping the guillotine. Should the death penalty be abolished?

Assign two groups to each issue, and have them debate the pros and cons before the class. **L1**

The Land and People The Dutch, or the people of the Netherlands, have been reclaiming their land from the sea since the A.D. 1100s. Land is valuable to the Dutch because the Netherlands is one of the most densely populated countries in the world. The Dutch method of reclaiming land is simple, although it takes hard work. First a dam is built across a coastal bay to hold back the sea. Then workers pump out the seawater to create a **polder**—an area of reclaimed land. Today, about 3,000 square miles (7,770 sq. km) of the Netherlands is made up of polders. In addition to growing tulips and other bulbs, the Dutch raise food crops and dairy cattle on polders. Amsterdam, the capital and largest city, is located on a polder.

About 90 percent of the Dutch live in towns and cities. They speak Dutch. About half of the population are Protestants and half Catholics. Because of their high standard of living and their efforts in reclaiming land from the sea, the Dutch have a reputation for hard work.

Location

Luxembourg

If you travel southeast of Belgium, you can visit Luxembourg, Luxembourg. The city of Luxembourg is the capital and largest city of one of Europe's smallest countries. The country is only about 55 miles (89 km) in length and about 35 miles (56 km) in width.

The Land and People In spite of its small size, Luxembourg is a land of contrasts. Its scenic areas of rolling hills, dense forests, and charming villages draw tourists from all over the world. Many **multinational firms,** or companies that do business in several countries, have their headquarters in Luxembourg. Ranked as one of Europe's leading steel producers, this Benelux country has a strong economy.

The people of Luxembourg have close cultural ties to both France and Germany. The country has three official languages: French, German, and Letzeburgesch—a dialect of German. The Luxembourgers, however, proudly cling to their independence, which they have enjoyed for more than 1,000 years. More than 95 percent of the people are Roman Catholics.

SECTION 3 REVIEW

REVIEWING TERMS AND FACTS
1. Define the following: polder, multinational firm.
2. LOCATION How has Belgium's location affected its economic development?
3. HUMAN/ENVIRONMENT INTERACTION Why do the Dutch reclaim land from the sea?

MAP STUDY ACTIVITIES
4. Study the land use map on page 288. Which of the Benelux countries is landlocked?
5. Look again at the map on page 288. What resources are found in Belgium?

Answers to Section 3 Review

1. All vocabulary words are defined in the Glossary.
2. Belgium's location near several large countries has made it a center of trade. On occasions, it also has made Belgium a battleground.
3. Since the Netherlands is so densely populated, the Dutch need to find more land.
4. Luxembourg
5. coal

Meeting Lesson Objectives

Each objective below is tested by the questions that follow it in parentheses.

1. **Describe** the two ethnic groups that live in Belgium. (2)
2. **Recall** how the people of the Netherlands have changed their environment. (1, 3)
3. **Understand** why tiny Luxembourg attracts many businesses. (1)

Evaluate

Section 3 Quiz **L1**

 Use the Testmaker to create a customized quiz for Section 3. **L1**

Reteach

Have each student write 10 questions on the section. Pair students and have pair members use their questions to quiz each other. **L1**

Enrich

Have students prepare an illustrated report on the Dutch port city of Rotterdam. **L3**

CLOSE

Ask students to imagine they live in a Benelux country. Have them draw the view from a window in their house. Their drawings should give clues to where they live. Display drawings and have the class identify the countries.

SECTION 4
The Alpine Countries

FOCUS

Section Objectives

1. **Classify** the landscape in the Alpine countries.
2. **Name** the languages spoken in Switzerland and Austria.
3. **Find** where most people in Switzerland and Austria live.

Bellringer
Motivational Activity

Prior to taking roll at the beginning of the class period, project Section Focus Transparency 11-4 and have students answer the activity questions. Discuss students' responses. **L1**

Vocabulary Pre-check

Have students use each term in the **Words to Know** list correctly in a sentence. **L1** **LEP**

TEACH

Guided Practice

Before students read Section 4, display Unit Overlay Transparency 4-5. Point out the Alps. Have students hypothesize about how the Alps affect travel and communications in nearby countries. **L1**

PREVIEW

Words to Know
• neutrality
• watershed

Places to Locate
• Switzerland
• Jura Mountains
• Bern
• Austria
• Vienna
• Danube River
• Liechtenstein

Read to Learn . . .
1. what the landscape is like in the Alpine countries.
2. what languages are spoken in Switzerland and Austria.
3. where most people in Switzerland and Austria live.

A breathtaking sight, the Matterhorn soars 14,691 feet (4,478 m) into the sky! This dramatic peak has become a symbol of the Alpine countries.

The Alps form most of the landscape in Switzerland, Austria, and Liechtenstein. That is why they are called the Alpine countries. Switzerland and Austria together cover about 47,300 square miles (122,507 sq. km), an area about the size of the state of Alabama. Sandwiched between them is tiny Liechtenstein, covering only 60 square miles (156 sq. km).

Location

Switzerland

What do you think of when you imagine the country of Switzerland? Do mounds of Swiss chocolate or soaring mountaintops come to mind? The rugged Swiss Alps have always created a barrier to travel between northern and southern Europe. For centuries Switzerland guarded the few routes that cut through this barrier. Because of its location, Switzerland practiced **neutrality**—refusing to take sides in disagreements and wars between countries.

The Land and Economy The Alps make Switzerland the **watershed** of central Europe. A watershed is a high place from which rivers flow in different directions. Several rivers, such as the Rhine and the Rhône, begin in the Swiss

---

Classroom Resources for Section 4

 REPRODUCIBLE MASTERS

Reproducible Lesson Plan 11-4
Guided Reading Activity 11-4
GeoLab Activity 11
Reteaching Activity 11
Enrichment Activity 11
Section Quiz 11-4

 TRANSPARENCIES

Section Focus Transparency 11-4
Geoquiz Transparencies 11-1, 11-2
Unit Overlay Transparency 4-5

 **MULTIMEDIA**

Testmaker

National Geographic Society: STV: World Geography, Vol. 2

Alps. Dams harness many of Switzerland's rivers, producing great amounts of hydroelectricity.

Another mountain range, the Jura, runs across northwest Switzerland. Between the Jura Mountains and the Alps lies a plateau called the *Mittelland,* or "Middle Land." Most of Switzerland's industries and its richest farmlands are found here. Bern, Switzerland's capital, and Zurich, its largest city, are also located here.

Switzerland's climate is strongly influenced by the Alps. Though temperatures differ with elevation, in most parts of the country winters are cold and summers are warm.

With no coastline and few natural resources, you might not expect Switzerland to be industrialized. It is, however, a thriving industrial nation. Using imported materials, Swiss workers make high-quality goods such as machine tools, electrical equipment, computers, and watches. They also produce chemicals, chocolate, cheese, and other dairy products. Zurich and Geneva are important international banking centers.

The People Multicultural describes the people of Switzerland. Switzerland has many different ethnic groups and religions. If you went to school there, you might have to learn all three of the official languages—German, French, and Italian. About half of the Swiss are Protestants and half Catholics. The Swiss have enjoyed a stable democratic government for more than 700 years.

Region
Austria

Austria's landscape looks like the setting for a fairy tale. Small when compared to some other European countries—31,940 square miles (82,725 sq. km)—Austria is slightly larger than the state of Maryland.

At one time, Austria ranked as one of the largest and most powerful countries in Europe. Like many European nations, it was once ruled by a strong monarchy that later lost power and territory. The Austrian Empire dominated politics and culture in central Europe from around 1280 until 1918.

The Land and Economy Mountains cover three-fourths of Austria. It is one of the most mountainous countries in the world. An important European river, the Danube, flows 217 miles (350 km) from east to west across the country's northern region.

Austria mines a variety of mineral resources. Iron ore, coal, copper, lead, and graphite fuel its many industries. Austrian factories produce machinery, chemical products, clothing, furniture, glass, and paper goods. The mountains cover much of the land. Austrian farmers, however, still manage to raise dairy cattle and other livestock, potatoes, sugar beets, and rye.

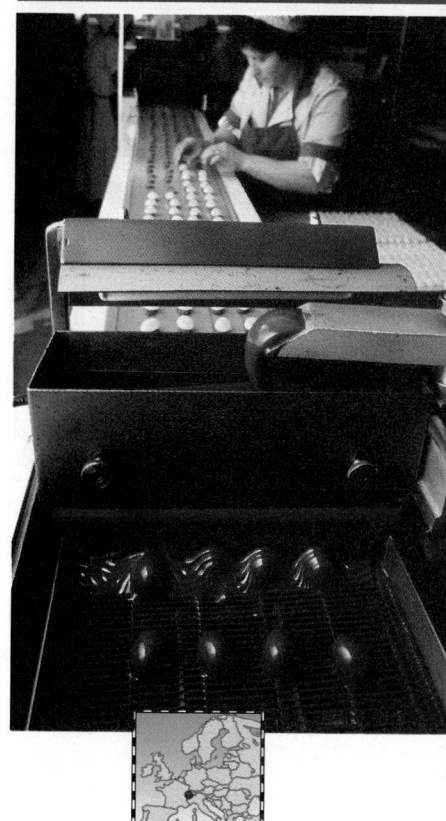

Swiss Chocolate Factory

Switzerland's factories produce some of the best chocolate in the world.
PLACE: What other products are made in Switzerland?

MAP STUDY

Answer
over 250 people per square mile (over 100 per sq. km)

Map Skills Practice
Reading a Map Which part of Austria is more densely populated, east or west? *(east)*

Evaluate

Section 4 Quiz **L1**

Use the Testmaker to create a customized quiz for Section 4. **L1**

Chapter 11 Test Form A and/or Form B **L1**

Reteach

Reteaching Activity 11 **L1**

Enrich

Enrichment Activity 11 **L3**

CLOSE

Mental Mapping Activity

Have students draw, freehand from memory, a map of Austria, Switzerland, and Liechtenstein that shows the Matterhorn, Naafkopf Peak, the Swiss Alps, the Jura Mountains, and the Rhine, Rhône, and Danube rivers.

NORTHWESTERN EUROPE: Population Density

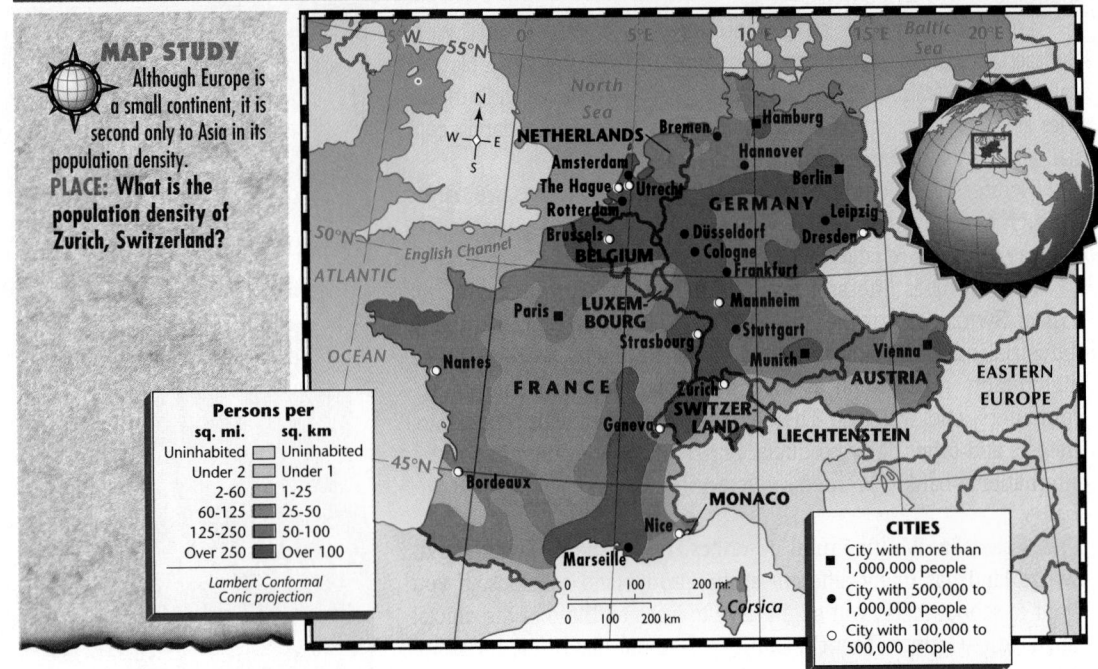

MAP STUDY
Although Europe is a small continent, it is second only to Asia in its population density.
PLACE: What is the population density of Zurich, Switzerland?

Persons per
sq. mi.	sq. km
Uninhabited	Uninhabited
Under 2	Under 1
2-60	1-25
60-125	25-50
125-250	50-100
Over 250	Over 100

Lambert Conformal Conic projection

CITIES
■ City with more than 1,000,000 people
● City with 500,000 to 1,000,000 people
○ City with 100,000 to 500,000 people

The People Austria, a republic, is home to about 8 million people. More than half of its population live in cities and towns. Most speak German and share cultural traditions with Switzerland. About 90 percent of Austrians are Roman Catholics. Austria is well known for its concert halls, historic palaces and churches, and its architecture.

The music of Vienna, Austria's capital, has entertained the world. Some of the world's greatest composers—Joseph Haydn, Wolfgang Amadeus Mozart, Franz Schubert—lived or performed in Vienna, the largest city in Austria.

SECTION 4 REVIEW

REVIEWING TERMS AND FACTS
1. **Define the following:** neutrality, watershed.
2. **REGION** How has Switzerland's location affected its relations with neighbors?
3. **PLACE** What makes Switzerland a multicultural country?
4. **HUMAN/ENVIRONMENT INTERACTION** What minerals are found in Austria?

 MAP STUDY ACTIVITIES
5. Compare the population density map above with the physical map on page 278. Where do most of the people in Switzerland live? What led them to live there?

(292)

UNIT 4

Answers to Section 4 Review

1. All vocabulary words are defined in the Glossary.
2. Switzerland guards the main routes through the Alps that connect northern and southern Europe; Switzerland's location has won European recognition of its policy of neutrality.
3. Switzerland's population is made up of different ethnic groups who speak different languages and follow different religions.
4. iron ore, coal, copper, lead, graphite
5. Central Switzerland; the other areas of the country are too mountainous for settlement.

Chapter 11 Highlights

Important Things to Know About Northwestern Europe

SECTION 1 FRANCE

- France contains mountains and fertile northern plains.
- France's well-developed economy balances agriculture and industry.
- Paris, the capital of France, is one of the world's leading cultural centers.

SECTION 2 GERMANY

- Germany is hilly and mountainous in the south and flat in the north.
- The German economy is strong because of its skilled workers and natural resources.
- Germany reunited as one country in 1990.

SECTION 3 THE BENELUX COUNTRIES

- Belgium has two major ethnic groups: the Walloons and the Flemings.
- The Dutch have reclaimed much of their land from the sea.
- Luxembourg's capital is home to the headquarters of many multi-national companies.

SECTION 4 THE ALPINE COUNTRIES

- The Alpine countries include Switzerland, Austria, and Liechtenstein.
- The Alps cover much of Switzerland and Austria.
- Switzerland has three official languages: German, French, and Italian.
- Most Austrians speak German, share cultural customs with Switzerland, and are Roman Catholic.

Berlin business district ▶

CHAPTER HIGHLIGHTS

Using the Chapter 11 Highlights

- Use the Chapter 11 Highlights to preview, review, condense, or reteach the chapter.

Preview/Review

Vocabulary Puzzle-Maker Software reinforces the terms used in Chapter 11.

Condense

Have students read the Chapter 11 Highlights. Spanish Chapter Highlights also are available.

Chapter 11 Digest Audiocassettes and Activity

Chapter 11 Digest Audiocassette Test
Spanish Chapter Digest Audiocassettes, Activities, and Tests also are available.

Guided Reading Activities

Reteach

Reteaching Activity 11. Spanish Reteaching activities also are available.

MAP STUDY

Place Copy and distribute page 30 from the Outline Map Resource Book. Have students label France, Germany, the Netherlands, Belgium, Luxembourg, Switzerland, Austria, and Liechtenstein. Then direct them to locate and label the following urban centers: Paris, Berlin, Amsterdam, Brussels, Bern, and Vienna.

Extra Credit Project

Diorama Have each student choose a highlight from the chapter, such as the Matterhorn or the Eiffel Tower, and reproduce it in miniature in a shoe box. Have the student print the highlight's name, location, and background information on an index card. Display the cards alongside the dioramas in the classroom. **L1**

Chapter 11 Review and Activities

ANSWERS

Reviewing Key Terms

1. B
2. H
3. G
4. C
5. I
6. F
7. A
8. D
9. E

 **Mental Mapping Activity**

This exercise helps students to visualize the countries and geographic features they have been studying and understand the relationships among them. All attempts at freehand mapping should be accepted.

REVIEWING KEY TERMS

Match the numbered terms in Column A with their definitions in Column B.

A
1. navigable
2. multinational firm
3. republic
4. *autobahn*
5. communism
6. polder
7. acid rain
8. watershed
9. Holocaust

B
A. chemical pollution that destroys plants
B. usable by ships
C. German superhighway
D. a high place from which rivers begin
E. the organized killing of 6 million European Jews and 6 million others
F. an area of reclaimed land
G. government in which the people elect their officials
H. a company that does business in several countries
I. form of government that controls society

Mental Mapping Activity

On a separate piece of paper, draw a freehand map of Europe. Label the following items on your map:
• Mediterranean Sea
• Alps
• Danube River
• North European Plain
• Belgium
• Rhine River
• Vienna

REVIEWING THE MAIN IDEAS

Section 1
1. **PLACE** What river forms France's border on the east?
2. **HUMAN/ENVIRONMENT INTERACTION** What makes France's land good for agriculture?

Section 2
3. **REGION** What part of Germany's Central Uplands is famous for its wood products?
4. **LOCATION** What is Germany's leading industrial region?
5. **REGION** What four major rivers flow through Germany?

Section 3
6. **REGION** Why are Belgium, the Netherlands, and Luxembourg grouped together into one region called Benelux?
7. **LOCATION** Where do most of the people of the Netherlands live?

Section 4
8. **PLACE** What is the capital of Switzerland?
9. **PLACE** What river runs through northern Austria?

Reviewing the Main Ideas

Section 1
1. Rhine River
2. the rich soil of the North European Plain and a mild climate

Section 2
3. Black Forest
4. the Ruhr in western Germany
5. Rhine, Danube, Weser, Elbe

Section 3
6. They are all small countries located in northwestern Europe; the name *Benelux* is made up of the first two or three letters of each country.
7. in towns and cities

Section 4
8. Bern
9. Danube

CRITICAL THINKING ACTIVITIES

1. **Making Generalizations** Why do you think so many historic castles are found along Europe's rivers?
2. **Making Comparisons** What similarities are there among the people of Germany and Austria?

GeoJournal Writing Activity

Imagine that you used to live in communist East Germany before 1990. You have made a daring escape to democratic West Germany. Write a letter home to a friend describing your escape and your new life.

COOPERATIVE LEARNING ACTIVITY

Work in a group of three or four to write a play about one of the countries discussed in the chapter. Choose a topic for your play that relates to the people, economy, or geography of the country. Each person can research and write a scene of the play. Review each scene as a group and share your play with the rest of the class. You may want to choose a cast for your play and present it to another class.

PLACE LOCATION ACTIVITY: NORTHWESTERN EUROPE

Match the letters on the map with the places of northwestern Europe. Write your answers on a separate sheet of paper.

1. Netherlands
2. Austria
3. Mediterranean Sea
4. English Channel
5. Danube River
6. Paris
7. Alps
8. Berlin
9. Seine River
10. Luxembourg

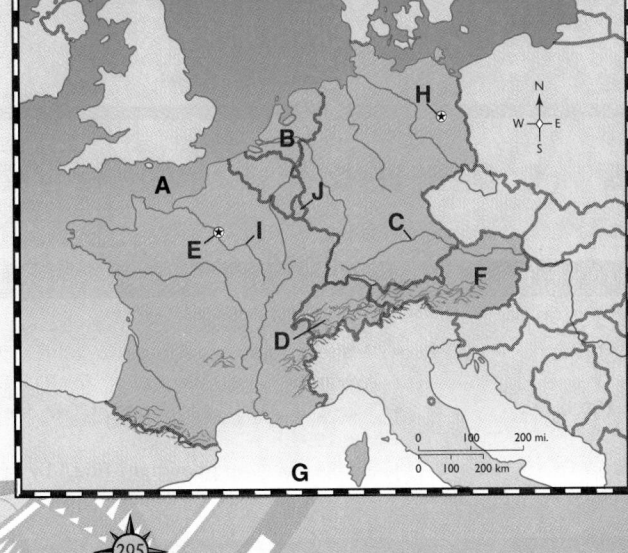

GeoJournal Writing Activity

Suggest that students focus their letters on the new freedom they have found in the West. Call on volunteers to read their letters to the class.

Place Location Activity

1. B
2. F
3. G
4. A
5. C
6. E
7. D
8. H
9. I
10. J

Chapter Bonus Test Question

This question may be used for extra credit on the chapter test.

Your travel agent tells you the following about the country where you plan to take a vacation:

1. Its southern coastal areas have a Mediterranean climate.
2. It is famous for its wines.
3. Its capital is a world center of art and learning.
4. Its railroad runs some of the fastest passenger trains in the world.

Where will you be taking a vacation? *(France)*

Critical Thinking Activities

1. Rivers were critical to the movement of goods and people and, therefore, needed to be protected.
2. Both peoples speak German and have a culture that is based on German heritage.

Cooperative Learning Activity

Remind students that when they are scripting their plays they will need to include lists of characters, stage directions, descriptions of costumes and scenery, and so on.

Southern Europe

PERFORMANCE ASSESSMENT ACTIVITY

CHOOSING AN OLYMPIC SITE The first modern Olympic Games took place in Greece. Other countries in southern Europe also have hosted the Olympics. In order for a site to be selected to host the games, it must meet certain criteria. A winter Olympics site, for example, must be near mountains for the skiing competitions. Organize students into groups of three. Have groups examine almanacs or other sources to find where the Olympic Games have been held during the last 25 years. Direct groups to draw up a list of conclusions on the characteristics required of sites for either the winter or summer games. Have groups use their lists to select two sites in southern Europe—one winter and one summer—where the Olympic Games might be held in the future. Groups should prepare an illustrated display intended to persuade the Olympic Committee to award the games to their selected sites. Displays may be arranged in a special exhibit area or presented, with an oral explanation, to the class.

POSSIBLE RUBRIC FEATURES:

Research skills, inference skills, ability to identify relationships, accuracy of content information, decision-making skills, visual quality of product

MENTAL MAPPING ACTIVITY

Before teaching Chapter 12, point out the countries of southern Europe on a globe. Have students identify the water bodies surrounding the countries and determine the direction that they would have to travel from their home to reach southern Europe.

TEACHER'S CORNER

NATIONAL GEOGRAPHIC SOCIETY

INDEX TO NATIONAL GEOGRAPHIC MAGAZINE

The following articles may be used for research relating to this chapter:

- "Treasure From the Silver Bank," by Tracy Bowden, July 1996.
- "Monaco," by Richard Conniff, May 1996.
- "The Basques," by Thomas J. Abercrombie, November 1995.
- "Sicily," by Jane Vessels, August 1995.
- "Brindisi Bronzes," by O. Louis Mazzatenta, April 1995.
- "Venice," by Erla Zwingle, February 1995.
- "Madeira Toasts the Future," by John McCarry, November 1994.
- "When the Greeks Went West," by Rick Gore, November 1994.
- "Michelangelo's 'Last Judgment,'" by Meg Nottingham Walsh, May 1994.
- "Europe Faces an Immigrant Tide," by Peter Ross Range, May 1993.

NATIONAL GEOGRAPHIC SOCIETY PRODUCTS AVAILABLE FROM GLENCOE

To order the following products for use with this chapter, contact your local Glencoe sales representative or call Glencoe at 1-800-334-7344:

- *STV: World Geography* (Videodisc)
- *Picture Atlas of the World* (CD-ROM)
- *Physical Geography of the World* (Transparencies)
- *ZipZapMap! World* (Software)
- *GeoBee* (Software)
- *Images of the World* (Posters)
- *Eye on the Environment* (Posters)

ADDITIONAL NATIONAL GEOGRAPHIC SOCIETY PRODUCTS

To order the following products for use with this chapter, call National Geographic Society at 1-800-368-2728:

- *In The Shadow of Vesuvius* (Video)
- *Soul of Spain* (Video)
- *1914-1918: World War I* (Video)
- *Physical Geography of the Continents Series:* "Europe." (Video)
- *Portraits of the Continents Series:* "Part III: Europe; Asia; Africa." (Filmstrip)
- *Geography of Europe Series:* "Pt. I: Northern Europe;" "Pt. II: Western Europe;" "Pt. III: Central Europe;" "Pt. IV: Southern Europe." (Filmstrip)

BIBLIOGRAPHY

Readings for the Student

Lye, Keith. *Passport to Spain,* revised ed. New York: Franklin Watts, 1994.

Mariella, Cinzia. *Passport to Italy,* revised ed. New York: Franklin Watts, 1994.

Readings for the Teacher

Lister, Charles. *Between Two Seas: A Walk Down the Appian Way.* North Pomfret, Vt.: Trafalgar, 1992.

Taggie, Benjamin F., et al, eds. *Spain and the Mediterranean.* Kirksville, Mo.: TJU Press, 1992.

Multimedia

European Geography: Southern Region. Chicago: Encyclopedia Britannica, 1993. Videocassette, 20 minutes.

Portugal and the Sea. Maumee, Ohio: Instructional Video. Videocassette, 29 minutes.

TEACHER-TO-TEACHER
Curriculum Connection: Social Studies

—from Pamela Francis
Gilbert Junior High School, Gilbert WV

REVIEW VOLLEYBALL *The purpose of this activity is to encourage students to review material for a test in a classroom volleyball game.*

Prepare 20 to 30 questions of varying difficulty on the content of this chapter. Organize the class into two teams of equal number and name a student as captain of each team. Have teams face each other. Inform teams that they will play volleyball with a balloon. Point out that regular volleyball rules will be followed except for the following: students may not rise from their seats to strike the balloon; a balloon bouncing outside of the area bounded by the desks is out of bounds; as is a balloon that strikes the ceiling or light fixtures; an imaginary line about halfway between the front rows of the teams will act as the net.

Toss a coin to see which team will serve. When a point is won, ask the winning team a question. Allow team members to confer and then have the captain offer an answer. If the team cannot answer within the time period, or offers an incorrect answer, the question passes to the other team. The team that answers the question correctly serves the next point. If neither team offers a correct answer, the team that won the previous point retains serve. Award one point for each correct answer. At the end of the set time, count the points and award members of the winning team a bonus point on the test.

MEETING NATIONAL STANDARDS

Geography for Life

All of the 18 standards are demonstrated in Unit 4. The following ones are highlighted in this chapter:

- Standards 1, 2, 4, 9, 10, 11, 12, 14, 15, 16, 17, 18

KEY TO ABILITY LEVELS

Teaching strategies have been coded for varying learning styles and abilities.

L1 **BASIC** activities for all students

L2 **AVERAGE** activities for average to above-average students

L3 **CHALLENGING** activities for above-average students

LEP **LIMITED ENGLISH PROFICIENCY** activities

Chapter 12

Resource Organizer

Southern Europe

Use these *Geography: The World and Its People* resources to teach, reinforce, and extend chapter content.

CHAPTER 12 RESOURCES

- *Vocabulary Activity 12
- Cooperative Learning Activity 12
- Workbook Activity 12
- Geography Skills Activity 12
- GeoLab Activity 12
- Chapter Map Activity 12
- Performance Assessment Activity 12
- *Reteaching Activity 12
- Enrichment Activity 12
- Chapter 12 Test, Form A and Form B
- *Chapter 12 Digest Audiocassette, Activity, Test
- Unit Overlay Transparencies 4-0, 4-1, 4-2, 4-3, 4-4, 4-5, 4-6
- Geography Handbook Transparency 13
- Political Map Transparency 4
- Geoquiz Transparencies 12-1, 12-2
- Vocabulary PuzzleMaker Software
- Student Self-Test: A Software Review
- Testmaker Software
- MindJogger Videoquiz
- Focus on World Art Print 6

*If time does not permit teaching the entire chapter, summarize using the Chapter 12 Highlights on page 313, and the Chapter 12 English (or Spanish) Audiocassettes. Review students' knowledge using the Glencoe MindJogger Videoquiz. *Also available in Spanish*

Use these *Geography: The World and Its People* resources to teach and reinforce section content.

ADDITIONAL RESOURCES FROM GLENCOE

SECTION 1 RESOURCES

Reproducible Lesson Plan 12-1

Section Focus Transparency 12-1

Guided Reading Activity 12-1

Section Quiz 12-1 (also available in Spanish)

SECTION 2 RESOURCES

Reproducible Lesson Plan 12-2

Section Focus Transparency 12-2

Guided Reading Activity 12-2

Section Quiz 12-2 (also available in Spanish)

SECTION 3 RESOURCES

Reproducible Lesson Plan 12-3

Section Focus Transparency 12-3

Guided Reading Activity 12-3

Section Quiz 12-3 (also available in Spanish)

 Reproducible Masters

- Facts On File: GEOGRAPHY ON FILE, Europe
- Foods Around the World, Unit 4

Workbook

- Building Skills in Geography, Unit 2, Lesson 4

 **World Music: Cultural Traditions** Unit 4

Transparencies

- National Geographic Society PicturePack Transparencies, Unit 4

 **Posters**

- National Geographic Society: Images of the World, Unit 4

 World Games Activity Cards Unit 4

 Videodiscs

- Reuters Issues in Geography
- National Geographic Society: STV: World Geography, Vol. 2
- National Geographic Society: STV: Rain Forest

CD-ROM

- National Geographic Society: Picture Atlas of the World

 **Software**

- National Geographic Society: ZipZapMap! World

Performance Assessment

Refer to the Planning Guide on page 296A for a Performance Assessment Activity for this chapter. See the *Performance Assessment Strategies and Activities* booklet for suggestions.

Chapter Objectives

1. **Compare** the geographic features of Spain and Portugal.
2. **Summarize** the physical features and cultural contributions of Italy.
3. **Outline** the physical geography and cultural contributions of Greece.

GLENCOE °
TECHNOLOGY

 VIDEODISC

Use the Chapter 12 MindJogger Videoquiz to preview chapter content.

MindJogger Videoquiz

Chapter 12
Disc 2 Side A

Also available in VHS.

Chapter 12 Southern Europe

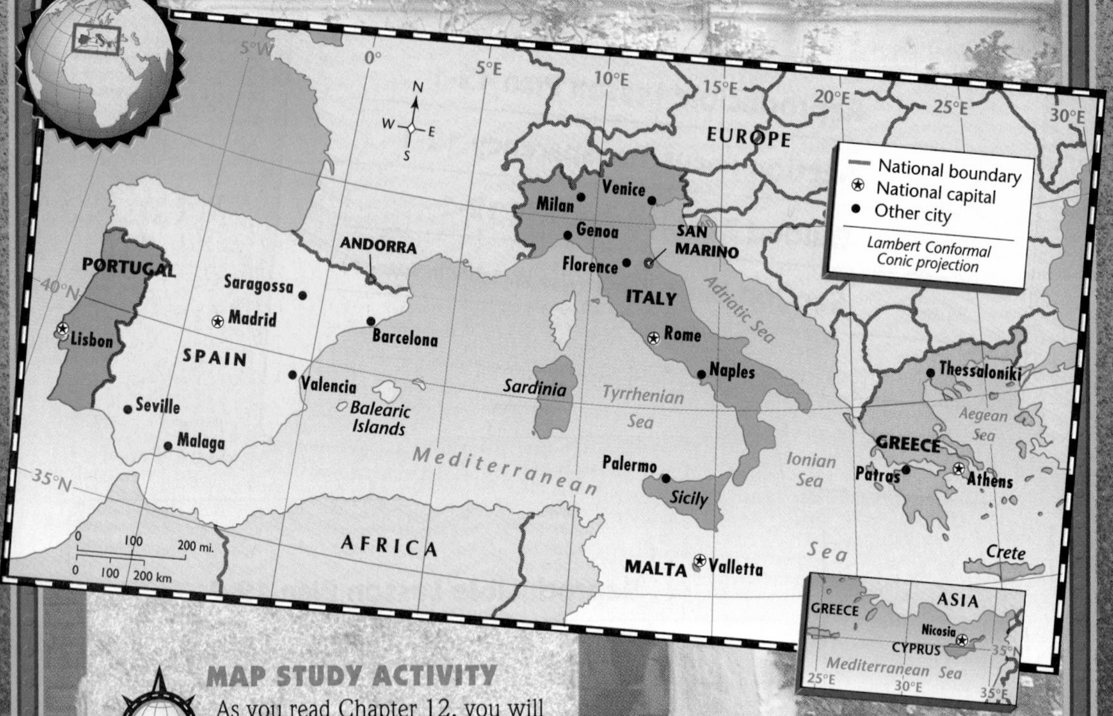

National boundary
⊛ National capital
● Other city

Lambert Conformal Conic projection

EUROPE

PORTUGAL · Lisbon · Seville · Malaga

SPAIN · Madrid · Barcelona · Valencia

ANDORRA · Saragossa

Balearic Islands

Milan · Venice · Genoa · Florence · SAN MARINO · ITALY · Rome · Naples

Sardinia · Tyrrhenian Sea

Palermo · Sicily

Adriatic Sea

GREECE · Patras · Athens · Thessaloniki

Aegean Sea · Ionian Sea · Crete

Mediterranean Sea

AFRICA

MALTA ⊛ Valletta

GREECE · Nicosia · CYPRUS · Mediterranean Sea · ASIA

MAP STUDY ACTIVITY

As you read Chapter 12, you will learn about the countries in southern Europe.

1. What southern European countries are islands?
2. What is the capital of Greece?
3. What sea do all of these countries except Portugal border?

296

MAP STUDY ACTIVITY

Answers
1. Malta, Cyprus
2. Athens
3. Mediterranean Sea

Map Skills Practice
Reading a Map What is the capital of Spain? *(Madrid)*

SECTION 1

Spain and Portugal

PREVIEW

Words to Know
- plateau
- textile
- dialect

Places to Locate
- Iberian Peninsula
- Spain
- Madrid
- Meseta
- Portugal
- Lisbon

Read to Learn . . .
1. about the landscape of Spain and Portugal.
2. how the people of Spain and Portugal earn their livings.
3. what cultural groups are found in Spain.

Past and present meet in Spain. This scene of plateaus and windmills

seems untouched by the passage of time. Yet just a few hours by train over the dusty horizon is Spain's capital—Madrid—a thriving, modern urban center.

Three countries—Spain, Portugal, and Andorra—are found on the Iberian Peninsula. The map on page 298 shows you the square-shaped Iberian Peninsula. Spain takes up five-sixths of the peninsula's land area. The remaining one-sixth, on the peninsula's western edge, is Portugal. Andorra, a tiny nation of 185 square miles (479 sq. km), perches high in the Pyrenees next to France.

Region
The Land

Most of the Iberian Peninsula is high and rugged. In the northern region the Pyrenees form a wall dividing the peninsula from the rest of Europe. In the southern region another mountain chain—the Sierra Nevada—parallels the Mediterranean coast.

The Meseta Between these mountain ranges in the heart of the peninsula lies a huge central plateau called the Meseta (muh•SAY•tuh), or "tableland." A **plateau**, you will recall, is a flat landmass higher than the surrounding land.

Meeting National Standards

Geography for Life The following standards are highlighted in this section:
- Standards 4, 5, 9, 10, 14, 15, 18

FOCUS

Section Objectives
1. **Describe** the landscape of Spain and Portugal.
2. **Compare** how the people of Spain and Portugal earn their livings.
3. **Identify** the cultural groups found in Spain.

Bellringer Motivational Activity

Prior to taking roll at the beginning of the class period, project Section Focus Transparency 12-1 and have students answer the activity questions. Discuss students' responses. **L1**

Vocabulary Pre-check

Have students define the term *plateau*. Then have them locate examples of this landform on a world physical map. **L1 LEP**

Use the Vocabulary PuzzleMaker Software to create a crossword puzzle. **L1**

Classroom Resources for Section 1

 **REPRODUCIBLE MASTERS**

Reproducible Lesson Plan 12-1
Guided Reading Activity 12-1
Vocabulary Activity 12
Chapter Map Activity 12
Geography Skills Activity 12
Building Skills in Geography Unit 2, Lesson 4
Section Quiz 12-1

 TRANSPARENCIES

Section Focus Transparency 12-1

MULTIMEDIA

 Vocabulary PuzzleMaker Software

 Testmaker

National Geographic Society: STV: Rain Forest

 National Geographic Society: STV: World Geography, Vol. 2

MAP STUDY

Answer
Spain, Portugal

Map Skills Practice

Reading a Map What mountains form Spain's northern border?
(Pyrenees)

TEACH

Guided Practice

Map Skills Draw an outline map of the Iberian Peninsula on the chalkboard. After students have read about climates in Spain and Portugal, call on volunteers to shade the climate areas on the map in different colors of chalk and to explain the colors in a key. **L1**

In November 1755, a series of earthquakes flattened Lisbon, raised parts of the coast as much as 20 feet (6 m), and killed between 10,000 and 60,000 people.

298

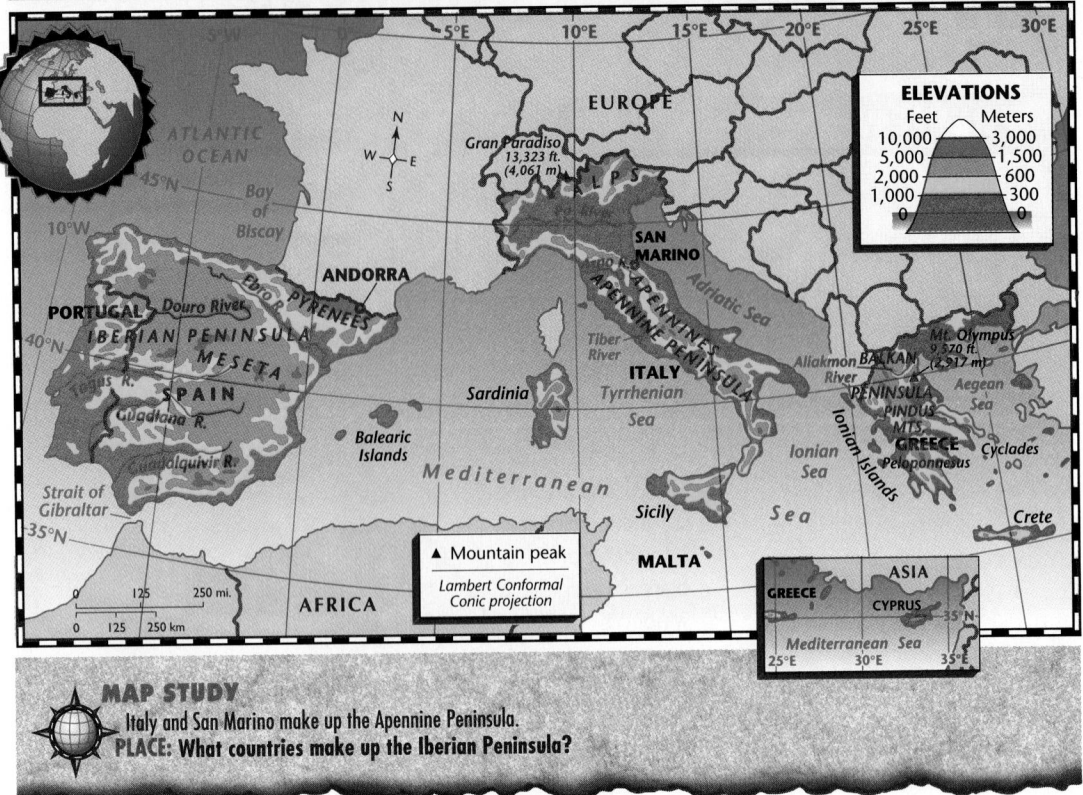

SOUTHERN EUROPE: Physical

MAP STUDY
Italy and San Marino make up the Apennine Peninsula.
PLACE: What countries make up the Iberian Peninsula?

The Meseta covers about two-thirds of the Iberian Peninsula. Its dry climate and poor, reddish-yellow soil must be irrigated for successful farming. Farmers of the Meseta raise wheat and vegetables, and they herd sheep and goats.

Many of the Iberian Peninsula's major rivers begin in the Meseta. For example, the Tagus (TAY•guhs)—Spain's longest river—and the Guadalquivir (GWAH•duhl•kih•VIHR) River flow from the Meseta to the Atlantic Ocean. Most of the peninsula's rivers are too narrow and shallow for large ships, however. The map above shows you that several fertile basins and coastal plains reach far into the Meseta. The largest is the Guadalquivir Basin in southern Spain.

The Climate To see how climate differs throughout the Iberian Peninsula, look at the map on page 305. Coastal areas of Spain and Portugal have a Mediterranean climate with mild winters and hot summers. Northern areas, however, enjoy a marine west coast climate. There, warm winds from the Atlantic Ocean bring much rain. The ocean's influence on the land diminishes as you travel inland. Parts of the Meseta have cold winters, hot summers, and very little rain.

(298)

Cooperative Learning Activity

Organize students into five or six groups and supply each group with 10 large index cards. Inform groups that their task is to create a set of 10 geography information cards on the Iberian Peninsula. First, have group members spend time in discussion on what they think are the 10 most important geographic features of the region. Then, direct groups to produce their cards. Some group members should design visuals for each feature. These should be drawn on the front of the cards. Other group members should research the features and record their findings on the back of the cards. Call on groups to present their cards to the class. Challenge them to explain why they selected certain features. **L1**

Place
The Economy

Tour buses filled with travelers from all over the world regularly crisscross Spain. In recent years tourism has brought much money to the Spanish. Visitors enjoy the sunny climate, beautiful beaches, and storybook castles. Spain also has a rapidly growing industrial and service economy. Spanish workers make machinery, trucks, cars, and clothing.

Farming and fishing form an important part of the Spanish economy. Acres and acres of olive groves make Spain the world's leading producer of olive oil. Spain also exports citrus fruits, cork, and wine.

Portugal lacks the economic development of Spain. Although Portugal benefits from **textile**, or clothing, and tourist industries, it is still mainly a subsistence farming economy. Portuguese farmers grow wine grapes, citrus fruits, olives, rice, cork, and wheat. Because of the slow growth of industry, Portugal's standard of living is one of the lowest in Europe.

Region
The People

The people of Spain and Portugal share similar histories and cultures. Many historians believe Iberians migrated to these countries from North Africa more than 5,000 years ago. They were later followed by Romans from Italy and Moors from North Africa. These earlier civilizations influenced Spanish and Portuguese culture, religion, and architecture.

Fountains and flower vendors brighten the main square of Lisbon, Portugal's capital *(left)*. Cork is an important product in both Portugal and Spain *(right)*.
MOVEMENT: Which country—Spain or Portugal—has a more rural population?

Urban and Rural Life in Portugal

Independent Practice

 Guided Reading Activity 12-1 **L1**

 **NATIONAL GEOGRAPHIC SOCIETY**

These materials are available from Glencoe.

 **VIDEODISC**

STV: Rain Forest

Crowded beach, Catalonia
Any side, Frame 51829

STV: World Geography Vol. 2: Africa and Europe

Women packing anchovies, Spain
Any side, Frame 50354

More About the Illustration

Answer to Caption
Portugal

 **Movement** Lisbon grew to prominence during the 1500s, when Portugal gained control of the east-west oceanic trade routes.

Meeting Special Needs Activity

Inefficient Readers Students with problems decoding unfamiliar words often skip over them. They often are successful, however, in defining the words based on the context of the sentence. Ask students to scan Section 1 for words that are unfamiliar to them. Have them write the words in their notebooks and interpret the meanings from context. Some words that may need definition or review are: *irrigated, exports, empires,* and *dictators.* **L1**

ASSESS

Check for Understanding

Assign Section 1 Review as homework or an in-class activity.

More About the Illustration

Answer to Caption
Roman Catholicism

Place The *Alhambra,* which means "Red Castle," was built by the Moors between 1248 and 1354. Following tradition, each new Moorish ruler built on his own rooms rather than occupying the rooms of his predecessor. When the Moors were driven from Spain in 1492, the Alhambra became a Christian palace.

By the late 1400s, Portugal and Spain had become independent, powerful, and Roman Catholic. A long Atlantic coastline and good harbors encouraged a spirit of adventure in both countries. Portugal and Spain sponsored explorers, such as Christopher Columbus and Vasco da Gama, who sailed abroad and set up empires in Africa, Asia, and the Americas.

The Court of Lions in the Alhambra

The Alhambra, a palace for Moorish royalty, was built when Spain was under Islamic rule.
PLACE: What is the primary religion of Spain today?

Cultural Regions After the decline of their overseas empires in the 1800s and 1900s, Spain and Portugal turned their attention to local affairs. Portugal developed a unified culture based on the Portuguese language. Spain remained in many ways a "country of different countries." Today Spain is still made up of several unique cultural regions.

The region of Castile, in the Meseta, dictated the culture of Spain for centuries. The **dialect** of the Castilian people became the major language of Spain. A dialect is a local form of a language. Today Castilian Spanish is spoken by most Spaniards.

People in other regions of Spain, however, regard themselves as separate groups. In the Mediterranean coastal region of Catalonia, the people speak Catalan, a language related to French. Galicia in northwest Spain and Andalusia in the south also have their unique cultural traditions.

The Basque people of northern Spain speak a language that is unlike any other language spoken in Europe. These people think of themselves as separate from the rest of Spain. Many Basques want complete independence from Spain in order to preserve their way of life.

A Rich Cultural Heritage The royal courts of early Spain and Portugal echoed with poetry and music. Spain and Portugal have a rich artistic heritage. In the 1500s Luiz Vaz de Camões (kuh•MOHNSH) of Portugal wrote *Os Lusídas* (OHS loo•SEE•dahs), a long poem about his country's history and heroes. In the 1600s Spain's Miguel de Cervantes (suhr•VAN•TEEZ) wrote *Don Quixote* (kee•HOH•tee), one of the world's first great novels. One of the most famous painters of the 1900s was Spain's Pablo Picasso, whose works had a major influence on modern art.

The cultures of Spain and Portugal have gained new freedoms in the past 20 years. From the 1930s until the mid-1970s, both countries were ruled by dictators. When Spanish dictator Francisco Franco died in 1975, democracy returned. Today Spain is a constitutional monarchy, and Portugal, too, has become a more open and democratic society.

Rural and City Life If you lived in Portugal today, you would most likely live in a rural village. About 65 percent of Portugal's 10 million people are villagers. Lisbon, with a population of about 2 million is Portugal's capital and major city.

UNIT 4

Critical Thinking Activity

Drawing Conclusions Tell students that once the outside world knew little about Andorra. Most Americans learned about the country for the first time from the travel writings of Richard Halliburton in the early 1900s. Direct students to the description of Andorra on page 297. Then ask them to draw conclusions about why Andorra remained so isolated through much of its history. (*Answers will vary but may include: because Andorra is located in the mountains, it has been difficult to reach.*) **L2**

COMPARING POPULATION:
Spain and the United States

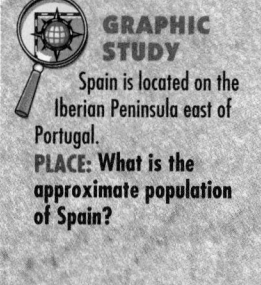

GRAPHIC STUDY

Spain is located on the Iberian Peninsula east of Portugal.
PLACE: What is the approximate population of Spain?

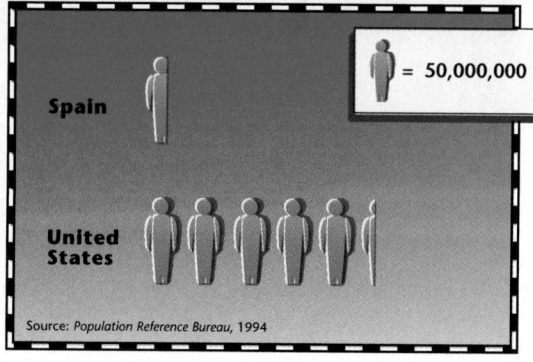

Spain

United States

 = 50,000,000

Source: *Population Reference Bureau, 1994*

In contrast, almost 80 percent of Spain's nearly 40 million people live in cities and towns. A city of 4.5 million people, Madrid ranks as one of Europe's leading cultural centers. Barcelona, along the Mediterranean coast, is Spain's leading seaport and industrial center. About 4 million people live there.

Juan García, a Madrid teenager, enjoys the fast pace of life in his city. Like most Spaniards, he and his family usually do not sit down to dinner until 9 or 10 o'clock at night. The meal may not end until midnight. On special occasions, the García family enjoys *paella*, a traditional dish of shrimp, lobster, chicken, ham, and vegetables mixed with seasoned rice.

Although Juan enjoys rock music and jazz, his parents prefer folk singing and dancing. In Spain the people of each region enjoy their own special songs and dances. Musicians often accompany singers and dancers on guitars, castanets, and tambourines. Spanish dances such as the bolero and flamenco have spread throughout the world.

SECTION 1 REVIEW

REVIEWING TERMS AND FACTS
1. **Define the following:** plateau, textile, dialect.
2. **REGION** What is the major landform of Spain?
3. **REGION** How are the economies of Spain and Portugal different?
4. **LOCATION** What is the capital of Spain?

MAP STUDY ACTIVITIES
5. Study the physical map on page 298. What two major rivers flow through both Spain and Portugal?
6. Look at the political map on page 296. About how far is it from Lisbon to Seville?

Evaluate

📁 Section Quiz 12-1
L1

💿 Use the Testmaker to create a customized quiz for Section 1. **L1**

Reteach

Write on the chalkboard a number of false statements on the geography of the Iberian Peninsula. Have students correct the statements by using information from the section. **L1**

Enrich

Have students prepare illustrated reports on the rise and fall of the Portuguese and Spanish empires of the 1500s and 1600s. **L3**

CLOSE

Have students compare their dinner times, favorite foods, and favorite music with Juan García's.

CHAPTER 12

301

Answers to Section 1 Review
1. All vocabulary words are defined in the Glossary.
2. the Meseta, a rugged, high plateau
3. They both are becoming more industrialized, but Spain's economy has developed faster than that of Portugal.
4. Madrid
5. Tagus and Douro rivers
6. about 200 miles (about 322 km)

BUILDING GEOGRAPHY SKILLS

TEACH

Display a map of your city or county. Have students quickly draw a freehand copy of the map. Then ask them to mark on the map areas where they think a lot of people live—the city center or high-rise developments, for example—and areas where they think few people live—beaches, parks, empty lots, and so. Call on volunteers to share their maps with the class. Point out that although people live all over the earth, some places are very crowded, while others are quite empty. Geographers express these differences in terms of population density. Conclude by having students discuss what impact the population density of their city or county has on their lives.

Have students read the skill and complete the practice questions. **L1**

Additional Skills Practice

1. How is population density expressed? *(as the number of people living in a square mile or kilometer)*

2. How is population density shown on maps?" *(Colors represent various ranges of population density.)*

📂 Geography Skills Activity 12

📂 Building Skills in Geography, Unit 2, Lesson 4

Reading a Population Map

At a baseball game, most of the filled seats are in the infield. Outfield seats have fewer people, and the upper decks have few people or none at all. The infield seats, you might say, have the greatest population density in the stadium.

Population density is the average number of people living in a square mile or square kilometer. Population density maps often use color to show differences in population. Dots or squares represent cities of different population sizes. To read a population density map, follow these steps:

• Study the map key to identify the population densities on the map.

• On the map, find areas of lowest and highest population density.

• Compare the map with other information to explain the region's population pattern.

SPAIN AND PORTUGAL: Population Density

CITIES
■ City with more than 1,000,000 people
• City with 500,000 to 1,000,000 people
○ City with 100,000 to 500,000 people

Persons per
sq. mi.	sq. km
2-60	1-25
60-125	25-50
125-250	50-100
Over 250	Over 100

Conic projection

Geography Skills Practice

1. What color represents areas with 125–250 people per square mile (50–100 per sq. km)?

2. Which cities have more than 1 million people?

3. Which areas have the lowest population density? Why?

4. What color stands for over 250 people per square mile (over 100 per sq. km)?

Answers to Geography Skills Practice

1. red
2. Barcelona and Madrid
3. The Pyrenees has the lowest population density; this is probably due to rugged terrain.
4. purple

SECTION 2 · Italy

PREVIEW

Words to Know
- sirocco
- city-state

Places to Locate
- Italy
- Apennines
- San Marino
- Rome
- Vatican City
- Po River
- Sicily
- Sardinia
- Adriatic Sea

Read to Learn . . .
1. about Italy's physical regions.
2. how the northern and southern parts of Italy differ.
3. how Italy's rich history has influenced the world.

The glistening dome of St. Peter's Basilica rises above the skyline of Rome. This historic city attracts people to Italy from all over the world.

Meeting National Standards

Geography for Life The following standards are highlighted in this section:
- Standards 1, 9, 12, 15, 16, 17

FOCUS

Section Objectives
1. **Describe** Italy's physical regions.
2. **Differentiate** between the northern and southern parts of Italy.
3. **Appreciate** how Italy's rich history has influenced Europe.

Bellringer Motivational Activity

 Prior to taking roll at the beginning of the class period, project Section Focus Transparency 12-2 and have students answer the activity questions. Discuss students' responses. **L1**

Vocabulary Pre-check

Read aloud the following sentence: "Siroccos blow across the Apennine Peninsula from North Africa, bringing hot, dry weather." Then ask students to define *sirocco* based on its context. *(hot, dry wind)* **L1 LEP**

Italy is a land of seacoasts. Look at the map on page 296. You see that Italy is a peninsula jutting out from southern Europe about 750 miles (1,207 km) into the Mediterranean Sea. The Italian Peninsula looks like a boot about to kick a triangle-shaped football. The "football" is the Italian island of Sicily. Another large Mediterranean island, Sardinia, is also part of Italy. Between Sicily and Africa lies Malta, an independent island country that has close ties to the United Kingdom as well as to Italy.

Where can you find the smallest nation in the world? Two tiny countries—San Marino and Vatican City—lie within the Italian "boot." San Marino is near the northwestern coast. Vatican City—the smallest nation—lies completely within Italy's capital, Rome. Its area totals only 0.17 square miles (0.44 sq. km).

Region
The Land

The land area of Italy—113,540 square miles (294,069 sq. km)—equals the combined areas of the states of Florida and Georgia. Mountains and highlands run down Italy's length. Plains and lowlands hug its coasts.

CHAPTER 12

303

Classroom Resources for Section 2

 REPRODUCIBLE MASTERS

Reproducible Lesson Plan 12-2
Guided Reading Activity 12-2
Workbook Activity 12
Cooperative Learning
  Activity 12
GeoLab Activity 12
Section Quiz 12-2

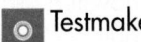 **TRANSPARENCIES**

Section Focus
  Transparency 12-2
Political Map Transparency 4

MULTIMEDIA

 Testmaker

National Geographic
  Society: STV: World
  Geography, Vol. 2

TEACH

Guided Practice

Categorizing On the chalkboard, list Italy's products, natural resources, cities, and physical features—natural gas, grapes, Milan, the Po River, for example. As students read the section, have them mark each item on the list with *S* if it is connected with southern Italy, *N* if it is connected with northern Italy, or *S/N* if it is connected with both sections of the country. **L1**

Independent Practice

Guided Reading Activity 12-2 **L1**

The Landscape Locate Italy's mountain regions on the map on page 298. The Alps tower over northern Italy, separating the country from France, Switzerland, and Austria. Another mountain range, the Apennines (A•puh•NYNZ), runs down the center of the Italian Peninsula to the toe of the boot. These mountains descend to low hills in the east and west.

The Amalfi Coast, Italy

Cliffside towns hug the jagged shoreline of Italy's southwestern coast.
PLACE: What mountain range extends north to south through Italy?

The rumbling of volcanic mountains echoes through the southern part of the peninsula and the island of Sicily. Mount Etna, one of the world's most famous volcanoes, rises 11,122 feet (3,390 m) on Sicily's eastern coast.

Although mountains and hills cover most of Italy, you can also find lowland regions. In the north the largest lowland is the Po River valley, the most densely populated part of the country. The Po River is Italy's longest river, flowing 400 miles (644 km) west to east. It stretches from the French border to the Adriatic Sea. You will learn later how important the Po River valley is to Italy.

The Climate Nearly all of Italy has a mild Mediterranean climate with sunny summers and rainy winters. In spring and summer, hot dry winds called **siroccos** blow across the Italian Peninsula from North Africa. In fall and winter, cool moist air from the Atlantic Ocean replaces this hot air. Most of Italy receives enough rain to grow crops. Parts of Sicily and Sardinia, however, record little rainfall.

Human/Environment Interaction
The Economy

The economy of Italy is as unevenly balanced as a seesaw. It ranks as the most prosperous country in southern Europe. The country's wealth, however, is not evenly distributed. The lives of wealthy bankers in Milan contrast sharply with farmers who carve out a living on rocky slopes in southern Italy.

The Prosperous North Fertile soil and steady, year-round rainfall have made northern Italy very productive. Most of Italy's agriculture centers on the Po River valley. Farmers there grow wheat, corn, rice, and sugar beets. In hilly areas of northern and central Italy, farmers grow grapes for wine. Italy is one of the world's major wine-producing countries.

Most manufacturing takes place in the northern Po River valley in the cities of Milan and Turin. Skilled workers there produce cars, machinery, chemicals, clothing, and leather goods. More industry is located in Genoa, an important port city. Italy has few mineral resources, but valuable natural gas deposits lie in the Po River valley. Factories in the northern region also use hydroelectric power from water sources in the nearby Alps.

UNIT 4

Cooperative Learning Activity

Point out that Italy's fashion and food industries are world-renowned. Then organize the class into an even number of groups. Have half the groups create an annotated collage on the Italian fashion industry. The other groups should do the same for the Italian food industry.

Suggest that groups look for pictures in old copies of such magazines as *Vogue, Vanity Fair, Gourmet,* and *Bon Appétit.* Have groups display their finished collages on the bulletin board. **L1**

CLIMATE

Dry
- ▢ Steppe (partly dry)

Mid-Latitude
- ▢ Mediterranean
- ▢ Humid subtropical
- ▢ Marine west coast
- ▢ Highland

Lambert Conformal Conic projection

EUROPE

ATLANTIC OCEAN

ANDORRA

SAN MARINO

ITALY

PORTUGAL

Lisbon ⊛ Madrid

SPAIN

Sardinia

Adriatic Sea

Rome ⊛

Tyrrhenian Sea

Balearic Islands

Mediterranean

Sicily

GREECE

Aegean Sea

Athens ⊛

Ionian Sea

Sea

Crete

Valletta ⊛
MALTA

AFRICA

0   125   250 mi
0   125   250 km

GREECE

ASIA

Nicosia ⊛

Crete   CYPRUS

Mediterranean Sea

MAP STUDY
Almost all of southern Europe has hot, dry summers and mild, rainy winters.
PLACE: What climate type does this describe?

LESSON PLAN
Chapter 12, Section 2

MAP STUDY

Answer
Mediterranean

Map Skills Practice
Reading a Map Which is the only country in the region with steppe climate areas? *(Spain)*

ASSESS

Check for Understanding

Assign Section 2 Review as homework or an in-class activity.

Meeting Lesson Objectives

Each objective below is tested by the questions that follow it in parentheses.

1. **Describe** Italy's physical regions. (2)
2. **Differentiate** between the northern and southern parts of Italy. (3)
3. **Appreciate** how Italy's rich history has influenced Europe. (1, 4)

The Developing South Southern Italy is poorer and less developed than northern Italy. From the map on page 309, you can tell that southern Italy lacks hydroelectric power. Southern Italy also lacks the natural gas deposits and rich soil found in northern Italy. The dry, rugged landscape of the southern region is used mainly for pastureland. The region's volcanic and clay soils are also good for growing citrus fruits, olives, and grapes. Southern Italy benefits from a steady stream of tourists.

Place
The People

Have you been on a crowded bus or in a crowded store recently? Then you know how some Italians feel living in their crowded country. About two-thirds of Italy's nearly 58 million people live on only one-fourth of the land. Why? Mountains take up so much space in the country. Many Italians from the poorer southern region have moved to northern Italian cities to find jobs. Others have emigrated to northern Europe and the United States.

CHAPTER 12

305

Meeting Special Needs Activity

Poor Learners Most students are familiar with the study system known as "SQ3R" (Survey, Question, Read, Recite, Review), but those with reading and learning problems may have difficulty surveying information. One task involved in surveying is predicting what kind of information will be given in a piece of text. Ask students to consider whether the text under each subhead in this section will 1) describe the physical geography of Italy, 2) examine geography's effect on Italian life, or 3) discuss Italian history. **L1**

Evaluate

 Section 2 Quiz **L1**

 Use the Testmaker to create a customized quiz for Section 2. **L1**

More About the Illustration

Answer to Caption

less pollution from cars, less noise, slower pace of life

Human/ Environment Interaction Venice is located on a group of about 120 islands that lie some 2.5 miles (4 km) off the Italian mainland. The islands are connected by 150 canals and 400 bridges.

Reteach

Distribute copies of a list of words, places, and phrases from the section, such as *city-states, Venice,* and *volcanic and clay soils.* Tell students that these are answers. Direct students to write an appropriate question for each "answer." **L1**

Enrich

Have students write word histories for 12 English words with Latin roots. Encourage students to combine their work in an illustrated display titled "Where Does That Word Come From?" **L3**

CLOSE

Have each student share with the rest of the class one fact he or she finds fascinating about Italy.

Cruising Along Venice's Grand Canal

Canals serve as streets for the many gondolas and motorboats in Venice, Italy.
HUMAN/ENVIRONMENT INTERACTION: What do you think it would be like to live in a city where boats are used more than cars?

Italy's Heritage For hundreds of years, Italy was the heart of Western civilization. The Roman Empire, based in Italy, influenced the government, arts, and architecture of Europe. After the fall of the Roman Empire in the A.D. 400s, Italy was divided into many small territories and **city-states**. Each city-state included an independent city and its surrounding countryside. The Renaissance developed in Italy's city-states during the 1300s and spread throughout Europe. It was a period of great achievement in the arts.

Italy was finally united as an independent country in 1861. From the 1920s to the early 1940s, dictator Benito Mussolini ruled Italy. He also became a supporter of Germany's Adolf Hitler and pulled Italy into World War II. Italy was defeated in the war.

In 1946 Italy became a democratic republic. Democracy, however, did not bring a stable government. Since the late 1940s, political power in Italy has constantly changed hands. Rivalry between the wealthy north and the poorer south has caused political tensions.

Daily Life To the rest of the world, Italy's capital—Rome—is the Eternal City. It is the keeper of historic ruins, ancient monuments, and beautiful churches and palaces. About 70 percent of Italy's population live in towns and cities. Three cities in Italy—Rome, Milan, and Naples—have populations of more than 3 million each.

The people of Italy speak Italian, which developed from Latin, the language of ancient Rome. Italian is closely related to French and Spanish. Do you know any Italian words? Many have been adopted into the English language. Pasta, made from flour and water, is the basic dish in Italy. Some pasta dishes are spaghetti, lasagna, and ravioli.

Celebrating the religious festivals of the Roman Catholic Church is a widely shared part of Italian life. More than 95 percent of Italy is Roman Catholic. Vatican City in Rome is the headquarters of the Roman Catholic Church.

SECTION 2 REVIEW

REVIEWING TERMS AND FACTS
1. **Define the following:** sirocco, city-state.
2. **REGION** What are two major mountain regions of Italy?
3. **HUMAN/ENVIRONMENT INTERACTION** How is the economy of northern Italy different from the economy of southern Italy?
4. **PLACE** Why is Rome so important to Italy?

MAP STUDY ACTIVITIES
5. Using the climate map on page 305, identify the climate of Sicily.

Answers to Section 2 Review
1. All vocabulary words are defined in the Glossary.
2. Alps, Apennines
3. north: fertile soil and a mild climate have helped farming to flourish, natural gas and abundant hydroelectric power have encouraged the development of industry; south: few resources to encourage industry, the terrain/climate have made farming difficult; soils do support citrus fruits, grapes, and olives. Tourism is a major industry.
4. Rome is Italy's capital and largest city; it was the center of the ancient Roman civilization; it is the location of the Vatican, the center of the Roman Catholic Church.
5. Mediterranean

MAKING CONNECTIONS

ROMAN BUILDERS

MATH   SCIENCE   HISTORY   LITERATURE   TECHNOLOGY

Paved roads and running water—how would you live without them? You may believe that they are inventions of the modern world. They had their origins, however, centuries ago in Italy. Among other achievements, the Romans built good roads and reliable water supplies.

ROMAN ROADS Road building played a key role in Roman military conquest. It allowed Roman legions to move quickly from one trouble spot to another in the empire. Wherever Roman soldiers went, they built roads.

Engineers first analyzed the soil and marked the road edges. Then, under the watchful eyes of supervisors, teams of soldiers dug down several feet to prepare the roadbed. A flattened layer of sand was further layered with rocks, small stones, and then gravel. These layers were topped off with flat paving stones carefully cut to fit closely together.

This top pavement was raised slightly in the middle so rainwater drained off to the sides. The roads were then usable in all kinds of weather. This approach is still used today to build highways!

AQUEDUCTS The Romans also built aqueducts, or stone troughs held aloft by supports.

Towering Aqueducts

The aqueducts were a way to move water from its source in the mountains to communities in lower regions.

To keep the water flowing, all aqueducts were built at a constant slope from beginning to end. Each aqueduct stood about 50 feet (15.2 m) off the ground, on top of its own row of supports and foundations. The outside of the supports and foundations was made of close-fitting stones. The inside was coated with layers of concrete, which the Romans invented by mixing volcanic stone with stone rubble.

Making the Connection

1. What building materials did Romans use?
2. What is an aqueduct?

307

TEACH

Ask students to speculate on the origin of the saying: "All roads lead to Rome." Then inform them that the network of roads built by the Romans extended to the furthest corners of the empire. Because these roads interlocked with, and ran into, each other, people said that no matter on which road you began your journey, you eventually would end up in Rome. **L1**

More About . . .

Roman Engineering By about A.D. 100, the city of Rome had a system of 11 aqueducts running some 300 miles (483 km). Pipes made of clay, lead, or wood, as well as stone and concrete, carried water to the city from as far away as 50 miles (80 km). The aqueducts were built high above the ground to prevent people from stealing or poisoning the water!

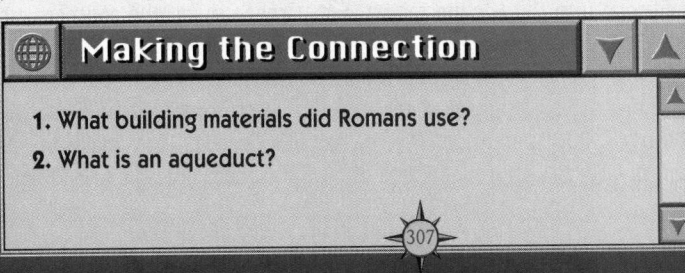

Answers to Making the Connection

1. sand, gravel, stones, concrete
2. an elevated stone trough used to carry water from its source in the mountains to communities in the lower regions

SECTION 3
Greece

Meeting National Standards

Geography for Life The following standards are highlighted in this section:
• Standards 2, 11, 15, 16, 17, 18

FOCUS

Section Objectives

1. **Illustrate** how mountains and seas divide Greece.
2. **Report** on how Greeks earn their livings.
3. **Cite** contributions ancient Greece made to Western civilization.

Bellringer Motivational Activity

Prior to taking roll at the beginning of the class period, project Section Focus Transparency 12-3 and have students answer the activity questions. Discuss students' responses. **L1**

Vocabulary Pre-check

Direct students to a physical map and explain that its key shows the elevation of land features. Have students make up a definition of *elevation* based on this hint and then compare it to the definition in the section. **L1 LEP**

PREVIEW

Words to Know
• mainland
• elevation
• service industry
• suburb
• emigrate

Places to Locate
• Greece
• Athens
• Pindus Mountains
• Plain of Thessaly
• Peloponnesus
• Crete
• Aegean Sea

Read to Learn . . .
1. how mountains and seas divide Greece.
2. how Greeks earn their livings.
3. what contributions ancient Greece made to Western civilization.

Were you named after someone—your grandmother or grandfather perhaps? The people of Athens, Greece, know how their city was named. It was called Athens in honor of Athena, the Greek goddess of wisdom. This graceful temple—the Parthenon—stands high above the city of Athens.

In ancient times Athens was a small town. Today the city is a modern urban center that is growing rapidly in size and population. The ancient Greeks would be amazed to see the changes that have taken place in Athens over the centuries. Yet they would find many other places in Greece that have changed very little. Outside the large cities, much of Greek life still follows traditional ways.

Region
The Land

Like Spain and Italy, much of Greece lies on a large peninsula that juts out from Europe into the Mediterranean Sea. Greece sits at the southern tip of the Balkan Peninsula.

Most of Greece is rocky and mountainous. High peaks separate valleys and plains. Long arms of the sea reach into the coast, forming many small peninsulas. Nearly 2,000 islands dot the sea around Greece. They make up about 20 percent of Greece's entire land area of 50,520 square miles (130,847 sq. km).

308

UNIT 4

Classroom Resources for Section 3

 REPRODUCIBLE MASTERS
Reproducible Lesson Plan 12-3
Guided Reading Activity 12-3
Reteaching Activity 12
Performance Assessment
 Activity 12
Enrichment Activity 12
Section Quiz 12-3

 **TRANSPARENCIES**
Section Focus
 Transparency 12-3
Geoquiz Transparencies 12-1,
 12-2
Political Map Transparency 4

 **MULTIMEDIA**
 Testmaker

National Geographic Society: STV: World Geography, Vol. 2

Agriculture

⬛ Commercial farming   ⬛ Manufacturing area

Resources

♣ Bauxite	⚡ Hydroelectric power	▣ Potash
⬛ Coal	I Iron ore	▣ Petroleum
▬ Copper	⚬ Lead	✳ Uranium
🐟 Fish & other seafood	💧 Natural gas	▣ Zinc
🌲 Forest		

Lambert Conformal Conic projection

MAP STUDY

Answer
commercial farming
Map Skills Practice
Reading a Map What are the major resources and crops of Portugal? *(uranium, forests, fish and other seafood, cork)*

MAP STUDY

Southern Europe has many types of agriculture because of its mild climate.
PLACE: What is the major form of agriculture in Cyprus?

TEACH

Guided Practice

Constructing Diagrams
Describe one effect of elevation on the climate of Greece by reading the following sentence to the class: "The north and west slopes of the mountains receive more rain than the east and south slopes because of the prevailing westerly winds." Ask students to draw a diagram to illustrate this point. Suggest that they start with an elevation profile of Greece. Then have them add direction arrows for prevailing winds, and show where rain falls and where a rain shadow exists. Have them complete the diagram by labeling highland areas and bodies of water and adding a title. **L1**

Highlands As you look at the map on page 298, you can see that the Pindus Mountains run through the center of the Greek **mainland**. This part of Greece connects to the European continent. The Pindus and other smaller ranges divide Greece into many separate regions. Historically, this has kept people in one region isolated and cut off from people in other regions. Because of the poor, stony soil, most people living in the highlands graze sheep and goats.

Lowlands and Peninsulas To the east of the Pindus Mountains lie two fertile lowlands—the Plain of Thessaly and the Macedonia-Thrace Plain. These two plains form Greece's major farming areas.

At the southern end of the Pindus range is another lowland, the Plain of Attica. About one-third of the Greek population live on this plain. Athens, the Greek capital, spreads out in all directions on the Plain of Attica.

Southwest of the Plain of Attica lies the Peloponnesus (PEH•luh•puh•NEE•suhs), a large peninsula of rugged mountains and deep valleys. If you traveled through this area thousands of years ago, you might have heard the cheering crowds attending the ancient Olympic games. Among other ruins on the Peloponnesus stands Olympia. Here the first Olympic games were held more than 2,600 years ago.

Cooperative Learning Activity

Organize students into three groups and assign each group one of the following headings from the section: "The Land," "The Economy," "The People." Tell groups their task is to teach the material under their headings to the rest of the class. Encourage groups to develop unique ways of presenting the material, such as role playing, cartoons, sketches, or stories. Have groups present their lessons in the order their headings appear in the section. **L1**

MULTICULTURAL PERSPECTIVE

Culturally Speaking

The name *Greece* comes from the Latin word *Graeci*, the name that the Romans gave to the people who lived in what today is northern Greece.

Independent Practice

 Guided Reading Activity 12-3

More About the Illustration

Answer to Caption
because most people want to be close to the shoreline, which is more densely populated

 **Place**
Mikonos is one of the more than 430 islands that make up the Grecian Archipelago. These islands comprise about one-fifth of Greece's total land area.

Global Gourmet

Greece At midnight on New Year's Eve, the Greeks serve *vasilopitta*—a cake with a coin in it. Whoever gets the piece with the coin supposedly has good luck for the rest of the year.

The Aegean Island of Mikonos

Gleaming white buildings line the shore of the Greek island of Mikonos.
HUMAN/ENVIRONMENT INTERACTION: Why do you think most of the buildings on Mikonos are so close together?

Islands A large number of islands are scattered on all sides of the Greek mainland. Some islands are mere dots of land less than an acre (0.4 ha) in area. Others, like Crete, are large—more than 3,000 square miles (7,770 sq. km)—in area. The islands are named after the branch of the Mediterranean Sea in which they are located. The Ionian Islands lie in the Ionian Sea. The Aegean (ih•JEE•uhn) Islands spread across the Aegean Sea. Farther east in the Mediterranean is the independent island country of Cyprus. It has close cultural and historic ties to both Greece and Turkey.

The Climate Greece has a Mediterranean climate with hot, dry summers and mild, rainy winters. As you learned in Chapter 2, however, climate may vary, depending on **elevation**, or height above sea level. High-elevation areas in Greece are often cooler and wetter than lowlands.

Movement

The Economy

Greece has one of the least developed economies in Europe. With few mineral resources, Greece must import many manufactured goods. In recent years, however, manufacturing has grown rapidly. Greek workers produce cement, tobacco products, and clothing. Tourism and **service industries** also have grown. Service industries provide a service—such as banking, insurance, or transport—instead of making goods.

Agriculture Greece's rocky soil limits the land available for farming. Yet many Greeks manage to earn their living by tilling the soil. Small in size, most Greek farms lie in valleys where the land is fertile. There farmers cultivate citrus fruits, olives, wine grapes, and tobacco.

Harvest Time

Melina Stavros and her mother have spent the day gathering ripe olives from the family's olive grove. Melina lives on a small farm with her mother, father, and three brothers. "My family is very traditional," she says, "but nothing like my grandparents' family!" Melina said her grandparents chose who her mother would marry.

 310

Meeting Special Needs Activity

Inefficient Readers Students often profit by using a form of rapid "reading" called scanning to locate specific information. Tell the students to use scanning to find and list all the ways in Section 3 that Greeks make a living. Model the scanning procedure by rapidly sliding your finger down the middle of the column under "The Land" and starting the list with "raising goats and sheep." Have students scan under "The Economy" and "The People" to find items to finish the list. **L1**

Shipping and Tourism The sea has long been important to the Greeks. No part of Greece is more than 85 miles (137 km) from the sea. In ancient times the Greeks depended on the sea for fishing and trade. Many of their legends describe heroes taking long sea voyages. Today the sea still provides fish and trade. Greece has one of the largest merchant fleets in the world, and Greek shipping is vital to the economy.

Greek ships not only transport goods, but also carry passengers. Each year millions of tourists come to Greece to sunbathe and relax. Many visit the sunny Aegean islands and historic sites such as the Parthenon in Athens. Tourism, which has increased rapidly since the 1960s, is now one of Greece's most important industries.

Place
The People

About 70 percent of Greece's 10.5 million people live in urban areas. Athens, the capital and largest city, is home to about 700,000 people. Another 2.8 million live in its **suburbs**, or surrounding areas.

SOUTHERN EUROPE: Population Density

Persons per

sq. mi.	sq. km
Uninhabited	Uninhabited
Under 2	Under 1
2-60	1-25
60-125	25-50
125-250	50-100
Over 250	Over 100

Lambert Conformal Conic projection

CITIES
- City with more than 1,000,000 people
- City with 500,000 to 1,000,000 people
- City with 100,000 to 500,000 people

MAP STUDY
Italy has more people than any other southern European country.
PLACE: Of the countries in this chapter, which appears to have the second-highest number of people?

CHAPTER 12   311

Critical Thinking Activity

Making Comparisons Reproduce the following chart, omitting the answers in parentheses:

	Greece	USA
Type of Government	(presidential/ parliamentary)	federal republic
Head of State	(president)	president
Head of Government	(prime minister)	president
Lawmaking Body	(parliament)	Congress
# of Lawmakers	(300)	535

Distribute copies of the chart and have students fill in the blanks. Refer them to the text, almanacs, and encyclopedias for help. **L2**

ASSESS

Check for Understanding

Assign Section 3 Review as homework or an in-class activity.

Meeting Lesson Objectives

Each objective below is tested by the questions that follow it in parentheses.

1. **Illustrate** how mountains and seas divide Greece. (5)
2. **Report** on how Greeks earn their livings. (2, 3)
3. **Cite** contributions ancient Greece made to Western civilization. (4)

MAP STUDY

Answer
Spain
Map Skills Practice
Reading a Map What part of Greece has the highest population density? *(southwestern coast)*

311

Evaluate

📁 Section Quiz 12-3
L1

⊙ Use the Testmaker to create a customized quiz for Section 3. **L1**

📁 Chapter 12 Test Form A and/or Form B **L1**

Reteach

📁 Reteaching Activity 12 **L1**

📁 Spanish Reteaching Activity 12 **L1**

Enrich

📁 Enrichment Activity 12 **L3**

CLOSE

Mental Mapping Activity

Have students draw, free-hand from memory, a map of the Greek mainland and Crete. Also have them label the Aegean Sea, the Ionian Sea, Mediterranean Sea, and the Pindus Mountains.

WHAT IN THE WORLD?

A Lost Continent?

Centuries ago, the Greek thinker Plato wrote about a continent called Atlantis that had disappeared in a spectacular natural disaster. According to Plato, the continent sank under the sea in an earthquake. Some historians now think the lost "continent" was actually Thera, one of Greece's Aegean Islands. In 1500 B.C. a volcano erupted and destroyed life on the island.

The population of Athens has mushroomed because many Greek farmers have left their villages to look for jobs in the city. Some Greeks have **emigrated**, or moved to live in other countries. Today more than 3 million people of Greek descent make their home in the United States, Australia, and western Europe.

The Ancient Greeks Many firsts mark the history of Greece. The first theories of geometry, medical science, astronomy, and physics were advanced by Greeks. The ancient Greeks developed the first European civilization.

Greek civilization reached its height in Athens during the mid-400s B.C.—the Golden Age of Greece. During this period, the city-state of Athens set up one of the world's first democratic governments. The ancient Greeks prized freedom and stressed the importance of the individual. Greek thinkers, such as Plato and Aristotle, laid the foundations of Western science and philosophy. They sought logical explanations for what happened in the world around them. Greek writers, such as Sophocles (SAH•fuh•KLEEZ), created dramas that explored human thoughts and feelings. Following the Golden Age, Greek civilization declined. Greece came under foreign rule for about 2,000 years. The Greeks did not regain their freedom until 1829.

The Modern Greeks The Greeks of today have much in common with their ancestors. They debate political issues with great enthusiasm. They love their sunny, rugged land and spend much time outdoors.

About 95 percent of the Greek population are Eastern Orthodox Christians. Religion influences much of Greek life, especially in rural areas. Easter is the most important Greek holiday. Traditional holiday foods include lamb, fish, and feta cheese—made from sheep's or goat's milk.

SECTION 3 REVIEW

REVIEWING TERMS AND FACTS

1. **Define the following:** mainland, elevation, service industry, suburb, emigrate.
2. **REGION** Why is so little of Greece's land good for farming?
3. **HUMAN/ENVIRONMENT INTERACTION** How does the sea help Greece's economy?
4. **PLACE** What great contributions did ancient Greece make to Western civilization?

MAP STUDY ACTIVITIES

5. Study the land use map on page 309. What sea would you cross if you moved products by ship from Greece to Italy?
6. Using the population density map on page 311, locate Greece's main population centers.

Answers to Section 3 Review

1. All vocabulary words are defined in the Glossary.
2. because much of Greece is mountainous
3. Fishing is a major economic activity in Greece. Also, the Greeks have a fleet of merchant ships that transports goods for other countries.
4. The ancient Greeks created great art, architecture, literature, and philosophy. Among the important ideas the ancient Greeks passed on to the modern world were the value of the individual and the right of citizens to run their own government.
5. Ionian Sea
6. Athens, Piraeus, and Patras

Chapter 12 Highlights

Important Things to Know About Southern Europe

SECTION 1 SPAIN AND PORTUGAL

- Spain's dry central highland slopes down to fertile coastal lowlands. Mountains rise on its northern and southern coasts.
- Both Spain and Portugal have a mostly Mediterranean climate.
- Spain has developed an industrial economy. Portugal is still mostly agricultural.
- Portugal is more culturally uniform than Spain, which has several distinct cultural regions.

SECTION 2 ITALY

- Italy's major landform is the Apennine mountain chain that runs through the center of the Italian Peninsula.
- Fertile soil, a mild climate, and hydroelectric power help make northern Italy a productive region.
- The ancient Romans made important contributions to Western civilization in language, government, and architecture.
- Rome, Italy's capital, encompasses Vatican City, the World's smallest nation and headquarters of the Roman Catholic Church.

SECTION 3 GREECE

- Greece consists of a mountainous mainland and nearly 2,000 offshore islands.
- The Mediterranean climate found throughout Greece varies only slightly with changes in elevation.
- The Greek economy relies on tourism and shipping.
- Ancient Greece laid the foundations of Western science, art, philosophy, government, and drama.

Cordoba, Spain ▶

Using the Chapter 12 Highlights

- Use the Chapter 12 Highlights to preview, review, condense, or reteach the chapter.

Preview/Review

 Vocabulary Puzzle-Maker Software reinforces the terms used in Chapter 12.

Condense

Have students read the Chapter 12 Highlights. Spanish Chapter Highlights also are available.

🎧 📁 Chapter 12 Digest Audiocassettes and Activity

📁 Chapter 12 Digest Audiocassette Test

Spanish Chapter Digest Audiocassettes, Activities, and Tests also are available.

📁 Guided Reading Activities

Reteach

📁 📁 Reteaching Activity 12. Spanish Reteaching activities also are available.

MAP STUDY

Region Copy and distribute page 30 from the Outline Map Resource Book. Have students label Portugal, Spain, Italy, and Greece on the map and the following cities: Lisbon, Madrid, Barcelona, Milan, Genoa, Turin, Rome, Naples, and Athens.

Extra Credit Project

Time Line Have interested students choose a country from the chapter and draw an illustrated time line of the country's history stretching from earliest times to the present. Refer the students to the text, encyclopedias, and other sources for dates and events. L1

313

Chapter 12 Review and Activities

REVIEWING KEY TERMS

Match the numbered terms in Column A with their definitions in Column B.

A
1. suburb
2. elevation
3. city-state
4. dialect
5. mainland
6. sirocco
7. service industry
8. emigrate

B
A. business that does not produce a product
B. an ancient city with an independent government that rules the countryside around it
C. height above sea level
D. local form of a language
E. land that is part of another large landmass
F. an area that surrounds or is next to a city center
G. move from one country to settle in another country
H. hot, dry wind from North Africa

ANSWERS

Reviewing Key Terms

1. F
2. C
3. B
4. D
5. E
6. H
7. A
8. G

Mental Mapping Activity

On a separate piece of paper, draw a freehand map of southern Europe. Label the following items on your map:
- Meseta
- Lisbon
- Tagus River
- Adriatic Sea
- Peloponnesus
- Athens
- Ionian Islands

Mental Mapping Activity

This exercise helps students to visualize the geographic features of the countries they have been studying and understand the relationships among them. All attempts at freehand mapping should be accepted.

REVIEWING THE MAIN IDEAS

Section 1
1. REGION   What areas of Spain and Portugal enjoy a Mediterranean climate?
2. PLACE   What agricultural products are grown in Portugal?
3. REGION   What are the cultural divisions of Spain?

Section 2
4. REGION   What is the most productive region in Italy?
5. HUMAN/ENVIRONMENT INTERACTION   How does industry in northern Italy depend on the environment?
6. PLACE   What form of government does Italy have today?

Section 3
7. REGION   What two plains form Greece's major farming areas?
8. MOVEMENT   Where have Greeks emigrated to find jobs?
9. PLACE What religion is practiced by about 95 percent of Greeks?

Reviewing the Main Ideas

Section 1
1. the coastal and lowland areas, especially in the south
2. grapes for wine, citrus fruits, olives, rice, wheat
3. Castile, Catalonia, Galicia, Andalusia, the Basque Provinces

Section 2
4. around the Po River in the north
5. it uses hydroelectric power generated by waterfalls in the Alps
6. democratic republic

CRITICAL THINKING ACTIVITIES

1. **Drawing Conclusions** If you had the chance to live in southern Europe, in which country would you prefer to live? Give reasons for your choice.
2. **Making Generalizations** How did the ancient Greeks and Romans contribute to your world today?

GeoJournal Writing Activity

Imagine you are writing an article for *NATIONAL GEOGRAPHIC* magazine. Describe your travels through southern Europe by focusing on the region's physical features.

COOPERATIVE LEARNING ACTIVITY

Organize into small groups, with each group researching one of the following Greek heroes, gods, or goddesses: Odysseus, Hercules, Zeus, Aphrodite, Apollo, Ares, Artemis, Athena, Demeter, Hephaestus, Hera, Hermes, or Poseidon. After your research is complete, write an adventure involving your subject, and decide how it will be presented to the class. Some possibilities might include acting out a myth or making a videotape.

PLACE LOCATION ACTIVITY: SOUTHERN EUROPE

Match the letters on the map with the places and physical features of southern Europe. Write your answers on a separate sheet of paper.

1. Adriatic Sea
2. Aegean Sea
3. Athens
4. Madrid
5. Portugal
6. Sicily
7. Sardinia
8. Tagus River
9. Pindus Mountains
10. Apennines

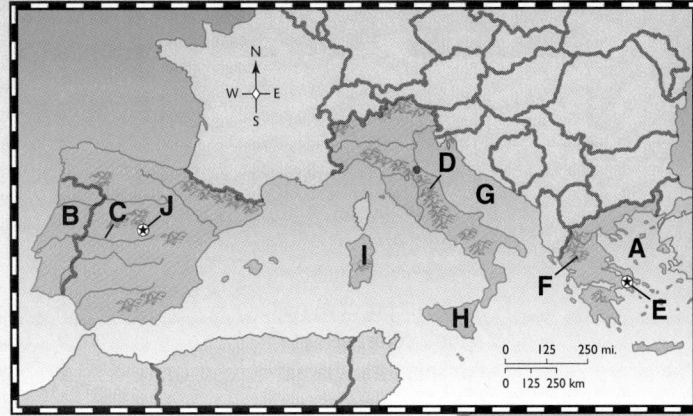

Cooperative Learning Activity

Refer students to *Myths and Mythology,* retold by Anthony Horowitz, or other books about Greek mythology for stories to use as models for their adventures.

GeoJournal Writing Activity

Suggest that students illustrate their journal entries with pictures from travel brochures and old magazines. Call on volunteers to present their entries to the class.

Place Location Activity

1. G
2. A
3. E
4. J
5. B
6. H
7. I
8. C
9. F
10. D

Chapter Bonus Test Question

This question may be used for extra credit on the chapter test.

Why is agriculture a difficult occupation in much of Southern Europe? *(because of unsuitable soils and lack of rainfall)*

Section 3

7. Plain of Thessaly, Macedonia-Thrace Plain
8. from rural to urban areas, to the United States, Australia, and western Europe
9. Eastern Orthodoxy

Critical Thinking Activities

1. Answers may include: Portugal—pleasant Mediterranean climate in the south, relaxed pace of life, rich culture; Spain—mild Mediterranean climate in the south, booming industrial country with modern city life, rich culture; Italy—prosperous industrial economy in the north, fertile agricultural area in Po River Valley, rich culture; Greece—beautiful island beaches, many historic places to visit.
2. Greeks: political ideas, plays and other literature, architectural styles, the Olympic Games; Romans: architectural styles, law and government.

GeoLab
ACTIVITY

FOCUS

Ask students what they understand by the phrase "the population of a country." *(total number of people who live in that country)* Next, have them state what they understand by the phrase "the area of a country." *(measure of space occupied by that country)* Finally, ask students how the population and area of a country can be used to calculate other important population statistics. Make sure students in their answers define *population density* as the average number of people living in a square mile or square kilometer.

TEACH

Point out that while *population density* shows the average number of people in a given area, *population distribution* shows the actual pattern of population. For example, while a population density map might show that an average of 250 people live in every square mile in a city, a population distribution map might show that there are heavier concentrations of people in the central areas than on the outskirts. Then ask students to speculate how population distribution might be affected by natural and social forces.

(continued)

From the classroom of Rebecca A. Corley, Evans Junior High School, Lubbock, Texas.

POPULATION: A Corny Map

Background

Western Europe has some of the most densely populated areas of the world. The density and distribution of Europe's population sometimes cause problems for the governments and environments of the region. To better understand Europe's population distribution, use popcorn to create your own population map of Europe.

Shoppers Crowd London's Portobello Market

Believe it or NOT!

One of Europe's most heavily populated areas—the lowlands of Belgium, the Netherlands, and sections of Germany—holds more than 900 people per square mile (347 people per sq. km). The interior of Iceland is the most sparsely populated area in Europe with about 9 people per square mile (3 people per sq. km).

Materials

- population density map of Europe
- 1 large poster board
- newspaper
- popcorn kernels
- hot-air popcorn popper
- food coloring
- 6 small transparent cups or jars
- 1 glue stick
- 1 pencil
- 6 large mixing bowls
- 6 spoons

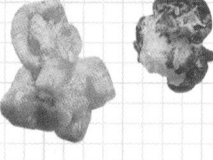

316

UNIT 4

Curriculum Connection Activity

Civics Give students firsthand knowledge of their community by having them calculate its population density. Students should first define a region for their study (a neighborhood, community, or city). They should then determine the area of the region in square miles by using a map or published figures. Next, have students estimate or use published figures to determine the population of their region. By dividing this figure by the number of square miles, they will have calculated the population density. **L2**

What To Do

A. Using a population density map of Europe as a reference, trace Europe's outline onto the poster board.

B. Add the lines marking the various levels of population density for the different European regions that you see on the map.

C. Most maps include 6 density levels, from uninhabited to more than 250 people per square mile. You will need 6 shades of colored popped corn, one for each level.

D. Pop your popcorn. Divide it into 6 bowls.

E. Color your popcorn. Mix food coloring drops with a small amount of water in a cup or jar. Make the color darker by adding more coloring. Then stir the colored water into a bowl of popped corn.

F. Spread the popcorn on newspaper to dry.

G. Using a glue stick, glue the colored popcorn onto your map to represent the different population densities throughout Europe.

Lab Activity Report

1. What areas of Europe are the most densely populated?

2. What areas are the most sparsely populated?

3. What inference can you make between where most people live and Europe's rivers?

4. **Drawing Conclusions** What are two physical features in Europe that contribute to low population densities in some regions?

Go A Step Further

Activity

High population density can cause serious problems for a region. Imagine that you live in a densely populated area of Europe. Write a letter to the editor of a local newspaper in which you describe your concerns for the environment.

Guide students by mentioning such natural forces as landforms and climate and by pointing out that social forces include cultural, historical, technological, and economic factors. **L2**

ASSESS

Have students answer the **Lab Activity Report** questions on page 317.

CLOSE

Encourage students to complete the **"Go A Step Further"** writing activity. In addition to describing their concerns, students may propose various solutions to the problems they identify.

Europe is home to about 1 out of every 8 people in the world.

Answers to Lab Activity Report

1. lowland areas, along the coasts, and along rivers

2. highland areas, areas where climate is inhospitable

3. Answers may vary, but most students will suggest that Europe's rivers tend to coincide with areas of high population density.

4. mountains and dry lands

PERFORMANCE ASSESSMENT ACTIVITY

WORKING ON A DOCUMENTARY FILM

Have students assume that a filmmaker is planning a major documentary film about the countries of eastern Europe during this century. Ask students to take the role of an aspiring scriptwriter, assistant director, photographer, or announcer who would like to take part in the project. Have each student write a persuasive letter to the filmmaker asking to take part in the project. Point out to students that one method of convincing the filmmaker would be to demonstrate their knowledge of the region. Suggest that they do this by writing a synopsis of the geographic information that they believe should be included in the film. Remind them that the information should be based on the five themes of geography. Call on volunteers to share their letters with the class.

POSSIBLE RUBRIC FEATURES:

Accuracy of content information, concept attainment for the themes of geography, persuasive writing skills, analytical skills, inferencing and evaluation skills

MENTAL MAPPING ACTIVITY

Before teaching Chapter 13, point out the countries of eastern Europe on a map. Have students note the other European countries that border eastern Europe. Also ask them to list the countries that are landlocked and those that have coasts.

TEACHER'S CORNER

NATIONAL GEOGRAPHIC SOCIETY

INDEX TO NATIONAL GEOGRAPHIC MAGAZINE

The following articles may be used for research relating to this chapter:

- "In Focus: Bosnia," (no author) June 1996.
- "Macedonia," by Priit J. Vesilind, March 1996.
- "Albania Opens the Door," by Dusko Doder, July 1992.
- "East Europe's Dark Dawn," by Jon Thompson, June 1991.
- "Dispatches from Eastern Europe," by Tad Szulc, March 1991.
- "The Baltic Nations," by Priit J. Vesilind, November 1990.
- "Yugoslavia: A House Much Divided," by Kenneth C. Danforth, August 1990.
- "The Baltic: Arena of Power," by Priit J. Vesilind, May 1989.
- "Hungary: A Static Society," by Paul R. Ehrlich and Anne H. Ehrlich, December 1988.
- "Poland: The Hope That Never Dies," by Tad Szulc, January 1988.

NATIONAL GEOGRAPHIC SOCIETY PRODUCTS AVAILABLE FROM GLENCOE

To order the following products for use with this chapter, contact your local Glencoe sales representative or call Glencoe at 1-800-334-7344:

- *STV: World Geography* (Videodisc)
- *Picture Atlas of the World* (CD-ROM)
- *Physical Geography of the World* (Transparencies)
- *ZipZapMap! World* (Software)
- *GeoBee* (Software)
- *Images of the World* (Posters)
- *Eye on the Environment* (Posters)

ADDITIONAL NATIONAL GEOGRAPHIC SOCIETY PRODUCTS

To order the following products for use with this chapter, call National Geographic Society at 1-800-368-2728:

- *Changing Faces of Communism Series:* Poland (Video)
- *Physical Geography of the Continents Series:* "Europe." (Video)
- *Portraits of the Continents Series:* "Part III: Europe; Asia; Africa." (Filmstrip)
- *Geography of Europe Series:* "Pt. I: Northern Europe;" "Pt. II: Western Europe;" "Pt. III: Central Europe;" "Pt. IV: Southern Europe." (Filmstrip)

TEACHER-TO-TEACHER
Curriculum Connection: Social Studies

—from Debora F. Bittner
John Marshall School, San Antonio, TX

MAPPING ETHNIC AND POLITICAL BOUNDARIES *The purpose of this activity is to help students understand the ethnic diversity of eastern Europe by having them examine the number, size, and location of ethnic groups in the area.*

Have students work in small groups to draw up a list of countries in eastern Europe. Direct students to use almanacs, cultural atlases, and other resources to discover information on the ethnic groups that live in each of the countries on the list. For each country, have students note the ethnic groups and the percentage of the total population each group constitutes. Have groups draw a large map of eastern Europe on a piece of butcher paper or poster board. Direct them to present their information on ethnic groups on their maps. Suggest that they use bar or circle graphs, or color or shading for the ethnic groups within each country.

Display the maps. Then have students discuss the following questions: Which countries have the greatest ethnic diversity? Which countries, at present, are experiencing ethnic conflicts? How do you think political boundaries that do not recognize ethnic differences—or, conversely, fail to respect ethnic similarities— have contributed to these conflicts?

MEETING NATIONAL STANDARDS

Geography for Life

All of the 18 standards are demonstrated in Unit 4. The following ones are highlighted in this chapter:

• Standards 1, 2, 3, 4, 5, 7, 9, 10, 11, 12, 14, 15, 16, 17, 18

KEY TO ABILITY LEVELS

Teaching strategies have been coded for varying learning styles and abilities.

L1 **BASIC** activities for all students

L2 **AVERAGE** activities for average to above-average students

L3 **CHALLENGING** activities for above-average students

LEP **LIMITED ENGLISH PROFICIENCY** activities

BIBLIOGRAPHY

Readings for the Student
Kronenwetter, Michael. *The New Eastern Europe.* New York: Franklin Watts, 1991.

Lithuania: Then and Now. Minneapolis: Lerner, 1993. (This series also includes books on Latvia and Estonia.)

Readings for the Teacher
CIS and Eastern Europe on File. New York: Facts on File, 1993. Reproducible maps, charts, fact sheets.

Eastern Europe. Stanford, Calif.: SPICE, 1992. Reproducible activity books.

Haynes, Jim, ed. *Poland: People to People.* Somerville, Mass.: Zephyr Press, 1992.

Multimedia
Eastern Europe. Stanford, Calif.: SPICE, 1992. 3 videocassettes.

European Geography: Balkan Region. Chicago: Encyclopedia Britannica, 1993. Videocassette, 20 minutes.

Chapter 13

Resource Organizer

Eastern Europe

Use these *Geography: The World and Its People* resources to teach, reinforce, and extend chapter content.

CHAPTER 13 RESOURCES

*Vocabulary Activity 13

Cooperative Learning Activity 13

Workbook Activity 13

Chapter Map Activity 13

Geography Skills Activity 13

GeoLab Activity 13

Critical Thinking Skills Activity 13

Performance Assessment Activity 13

*Reteaching Activity 13

Enrichment Activity 13

Environmental Case Study 4

Chapter 13 Test, Form A and Form B

*Chapter 13 Digest Audiocassette, Activity, Test

Unit Overlay Transparencies 4-0, 4-1, 4-2, 4-3, 4-4, 4-5, 4-6

Political Map Transparency 4

Geoquiz Transparencies 13-1, 13-2

Vocabulary PuzzleMaker Software

Student Self-Test: A Software Review

Testmaker Software

MindJogger Videoquiz

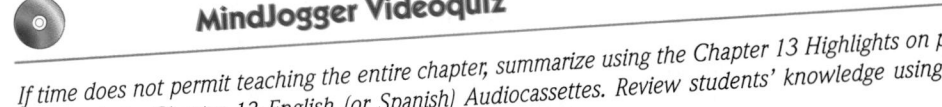

*If time does not permit teaching the entire chapter, summarize using the Chapter 13 Highlights on page 339, and the Chapter 13 English (or Spanish) Audiocassettes. Review students' knowledge using the Glencoe MindJogger Videoquiz. *Also available in Spanish*

> *Use these Geography: The World and Its People resources to teach and reinforce section content.*

SECTION 1 RESOURCES

📁 **Guided Reading Activity 13-1**

🖨 **Section Focus Transparency 13-1**

📁 **Section Quiz 13-1** (also available in Spanish)

SECTION 2 RESOURCES

📁 **Guided Reading Activity 13-2**

🖨 **Section Focus Transparency 13-2**

📁 **Section Quiz 13-2** (also available in Spanish)

SECTION 3 RESOURCES

📁 **Guided Reading Activity 13-3**

🖨 **Section Focus Transparency 13-3**

📁 **Section Quiz 13-3** (also available in Spanish)

SECTION 4 RESOURCES

📁 **Guided Reading Activity 13-4**

🖨 **Section Focus Transparency 13-4**

📁 **Section Quiz 13-4** (also available in Spanish)

SECTION 5 RESOURCES

📁 **Guided Reading Activity 13-5**

🖨 **Section Focus Transparency 13-5**

📁 **Section Quiz 13-5** (also available in Spanish)

ADDITIONAL RESOURCES FROM GLENCOE

📁 **Reproducible Masters**

- Glencoe Social Studies Outline Map Book, page 30

- Facts On File: MAPS ON FILE I, Europe

- Foods Around the World, Unit 4

 World Music: Cultural Traditions
Lesson 3

 **Transparencies**

- National Geographic Society PicturePack Transparencies, Unit 4

 Laminated Desk Map

 **Posters**

- National Geographic Society: Eye on the Environment, Unit 4

 **Countries of the World Flashcards**
Unit 4

 **Videodiscs**

- Reuters Issues in Geography

- National Geographic Society: STV: World Geography, Vol. 2

 **CD-ROM**

- National Geographic Society: Picture Atlas of the World

🖥 **Software**

- National Geographic Society: ZipZapMap! World

 Reference Books

- Picture Atlas of Our World

Performance Assessment

Refer to the Planning Guide on page 318A for a Performance Assessment Activity for this chapter. See the *Performance Assessment Strategies and Activities* booklet for suggestions.

Chapter Objectives

1. **Examine** the physical geography, politics, and culture of the Baltic republics.

2. **Discuss** the land, economy, and culture of Poland.

3. **Describe** the geography of Hungary.

4. **Compare** the geographies of the Czech Republic and Slovakia.

5. **Note** geographical similarities and differences among the Balkan countries.

GLENCOE TECHNOLOGY

 VIDEODISC

Use the Chapter 13 MindJogger Videoquiz to preview chapter content.

MindJogger Videoquiz

Chapter 13
Disc 2 Side B

Also available in VHS.

Chapter 13 Eastern Europe

MAP STUDY ACTIVITY

As you read Chapter 13, you will learn about the countries of eastern Europe.
1. Which eastern European country has the largest area?
2. What two countries border on the Black Sea?
3. What is the capital of Poland?

 MAP STUDY ACTIVITY

Answers
1. Poland
2. Romania, Bulgaria
3. Warsaw

Map Skills Practice
Reading a Map What is the capital of the Czech Republic? *(Prague)* of Slovakia? *(Bratislava)*

SECTION 1

The Baltic Republics

PREVIEW

Words to Know
- communism
- oil shale

Places to Locate
- Baltic Sea
- Estonia
- Latvia
- Lithuania

Read to Learn . . .
1. how the sea affects climate in the Baltic republics.
2. what political changes occurred in the Baltic republics.
3. where most people in the Baltic republics live.

Today—with great effort—this farmer in the Baltic country of Lithuania clears land. The people of the Baltic republics are working hard to develop their economies.

Along the eastern shore of the Baltic Sea are three small republics—Estonia, Latvia, and Lithuania. In 1940 the Baltic republics came under the rule of the neighboring Soviet Union. The leaders of the Soviet Union practiced **communism,** a political and economic system in which the government controls how its citizens live and produce goods. In the early 1990s, the Soviet Union collapsed. The Baltic republics soon won back their independence. Today they are democracies with free enterprise economies.

Place
Estonia

The map on page 318 shows you that Estonia is the northernmost Baltic republic. With 17,410 square miles (45,092 sq. km), Estonia is also the smallest of the three Baltic countries.

The Land Low plains, forests, and swamps cover the land. Sandy beaches sweep along the Baltic coast. For a northern country, Estonia has a relatively mild climate. Mild Baltic Sea winds keep the weather from becoming too hot or too cold.

CHAPTER 13

319

Classroom Resources for Section 1

 REPRODUCIBLE MASTERS

Reproducible Lesson Plan 13-1
Guided Reading Activity 13-1
Vocabulary Activity 13
Geography Skills Activity 13
Section Quiz 13-1

 **TRANSPARENCIES**

Section Focus
  Transparency 13-1
Political Map Transparency 4

MULTIMEDIA

Vocabulary PuzzleMaker
  Software

Testmaker

Meeting National Standards
Geography for Life The following standards are highlighted in this section:
- Standards 1, 7, 9, 11, 12, 14, 15, 18

FOCUS

Section Objectives
1. **Identify** how the sea affects climate in the Baltic republics.
2. **Analyze** the political changes that occurred in the Baltic republics.
3. **Note** where most people in the Baltic republics live.

Bellringer Motivational Activity

Prior to taking roll at the beginning of the class period, project Section Focus Transparency 13-1 and have students answer the activity questions. Discuss students' responses. **L1**

Vocabulary Pre-check

Ask students to define the term *communism* by comparing this political system to that of the U.S. **L1 LEP**

Use the Vocabulary PuzzleMaker Software to create a crossword puzzle. **L1**

TEACH

Guided Practice

Map Skills Refer students to the "Countries at a Glance" section of the Unit 4 Atlas. Have them use the area and

(continued)

319

population statistics to calculate the population density of each of the Baltic republics. Then ask students to use their results to draw a simple population density map of the area. **L1**

Independent Practice

 Guided Reading Activity 13-1 **L1**

MAP STUDY

Answer
mostly below 1,000 feet (300 m)

Map Skills Practice
Reading a Map What two countries lie south of Poland's border? *(Czech Republic and Slovakia)*

ASSESS

Check for Understanding

Assign Section 1 Review as homework or an in-class activity.

Meeting Lesson Objectives

Each objective below is tested by the questions that follow it in parentheses.

1. **Identify** how the sea affects climate in the Baltic republics. (5)
2. **Analyze** the political changes that occurred in the Baltic republics. (2)
3. **Note** where most people in the Baltic republics live. (3)

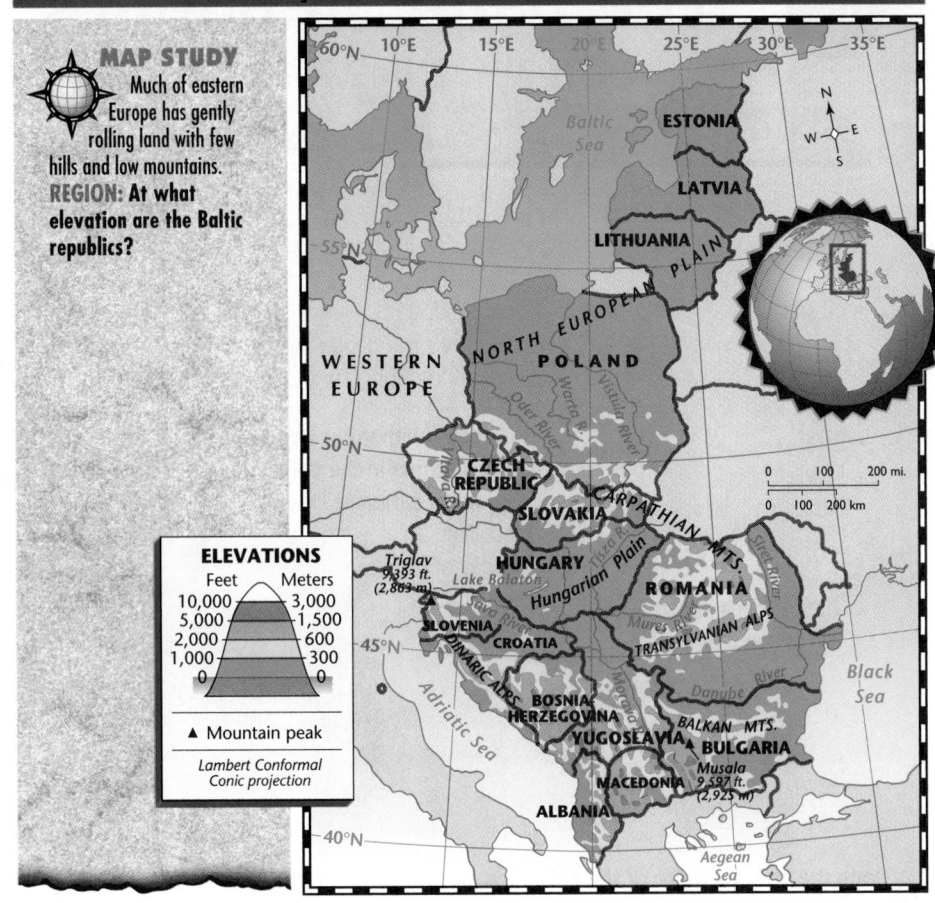

EASTERN EUROPE: Physical

MAP STUDY Much of eastern Europe has gently rolling land with few hills and low mountains.
REGION: At what elevation are the Baltic republics?

Estonian farmers raise livestock and grow crops such as apples, potatoes, and sugar beets. Industry, however, provides jobs for most of Estonia's people. The country has large deposits of **oil shale**, a rock that contains oil.

The People About 64 percent of the country's people are Estonians who have close ties to the Finns. After nearly 50 years of Soviet rule, about 30 percent of Estonia's people are Russian. Most of Estonia's 1.5 million people live in cities and towns. Tallinn, the capital, is on the Baltic coast. German and Estonian builders created Tallinn's beautiful churches and castles hundreds of years ago.

Location
Latvia

South of Estonia lies the Baltic republic of Latvia. Latvia covers 24,900 square miles (64,491 sq. km)—about the area of West Virginia. Latvia's location on the Baltic Sea has made it a trading center.

UNIT 4

Cooperative Learning Activity

After students read the section, organize them into teams. Explain that you will give one- or two-word clues and that team members must try to identify the Baltic republic to which each clue refers. Tell the students that "all three" is also an option. For each correct answer, a team earns one point. After all the students have had turns, the team with the most points wins. Use clues such as "oil shale" *(Estonia)*, "mild climate" *(all three)*, "Riga" *(Latvia)*, and "largest" *(Lithuania)*. **L1**

The Land and Economy If you visited Latvia, you would find a landscape of coastal plains, low hills, and forests. The mild climate of Latvia is similar to that of Estonia.

Latvia's industrial development is greater than that of other Baltic republics. Latvia's factory workers produce electronic equipment, machinery, and household goods. Latvian farmers raise dairy cattle and grow barley, flax, oats, potatoes, and rye.

The People Only about 53 percent of Latvia's people are Latvian. Under Soviet rule, many Russians settled in Latvia. Russians make up about 33 percent of the population. The Latvians and the Russians, who have a history of conflict, often clash today. Most of Latvia's people are city dwellers. Riga, the capital and largest city, is an important shipping and industrial center.

Place
Lithuania

Lithuania covers 25,210 square miles (65,294 sq. km). About 3.7 million people—at a density of 141 people per square mile (55 people per sq. km)—fill its cities and villages. In both size and population, Lithuania is the largest of the Baltic republics. It, too, struggles to develop a strong economy.

The Land and Economy Flat plains and forested hills cover most of Lithuania. Lithuania enjoys mild winter and summer temperatures. The country's economy is largely industrial with some farming. Lithuanian factory workers produce chemicals, electronic products, and machinery. Farmers here raise dairy cattle and other livestock.

The People About 80 percent of the country's people are Lithuanian, and most are city dwellers. Vilnius, the capital and largest city, is a major cultural and industrial center. Although the other Baltic republics are Protestant, Lithuania is largely Roman Catholic.

Praying in Lithuania

Worshipers gather at a Catholic church in Vilnius, Lithuania. **PLACE: Which Baltic republic is the largest in size?**

SECTION 1 REVIEW

REVIEWING TERMS AND FACTS
1. **Define the following:** communism, oil shale.
2. **REGION** What country ruled the Baltic countries from 1940 to 1991?
3. **PLACE** Where do most people live in the Baltic republics?
4. **PLACE** How does Lithuania compare in population to the other republics?

MAP STUDY ACTIVITIES
 5. Study the physical map on page 320. What large body of water lies to the west of Latvia?

Answers to Section 1 Review
1. All vocabulary words are defined in the Glossary.
2. Soviet Union
3. in cities
4. Lithuania is the largest in population.
5. Baltic Sea

LESSON PLAN
Chapter 13, Section 1

More About the Illustration
Answer to Caption
Lithuania

Place Before World War II and the Holocaust, Vilnius had a large Jewish population that supported some 96 synagogues.

Evaluate

Section 1 Quiz **L1**

Use the Testmaker to create a customized quiz for Section 1. **L1**

Reteach

Help students organize the details from the section into outline form. **L1**

Enrich

Ask students to make and display circle graphs showing the ethnic make-up of people in the Baltic republics. **L3**

CLOSE

Have students list geographic similarities and differences among the Baltic republics.

BUILDING GEOGRAPHY SKILLS

BUILDING GEOGRAPHY SKILLS

TEACH

Ask students to imagine that they have invited a new friend to visit them. Encourage them to draw a map of the route from school to their homes. Allow students a short time to draw sketch maps. Then ask how many students made a mental picture of the route in order to draw the map. Explain that the process of visualizing places and routes is called mental mapping. Direct students to read the skill and complete the practice questions. **L1**

Additional Skills Practice

1. Look at a map of North America. Then draw a mental map of the continent, labeling Canada, the United States, Mexico, the Great Lakes, and the Pacific and Atlantic oceans.

2. Draw a mental map of a road trip from Chicago to New Orleans. Label the states the route passes through.

Geography Skills Activity 13

Mental Mapping

Think about how you get from place to place each day. In your mind you have a picture—or mental map—of your route. If necessary, you could probably draw sketch maps of many familiar places like the one shown below.

In the same way you can develop mental maps of places in the world. When studying a world region, picture its shape in your mind. Think about its important features and cities. Also try to imagine where in the world it is located. To develop your mental mapping skills, follow these steps:

• Picture a place in your mind. See its shape and most important features.

• Draw a sketch map of it.

• Compare your sketch to the actual place or a map. Revise your mental map and sketch.

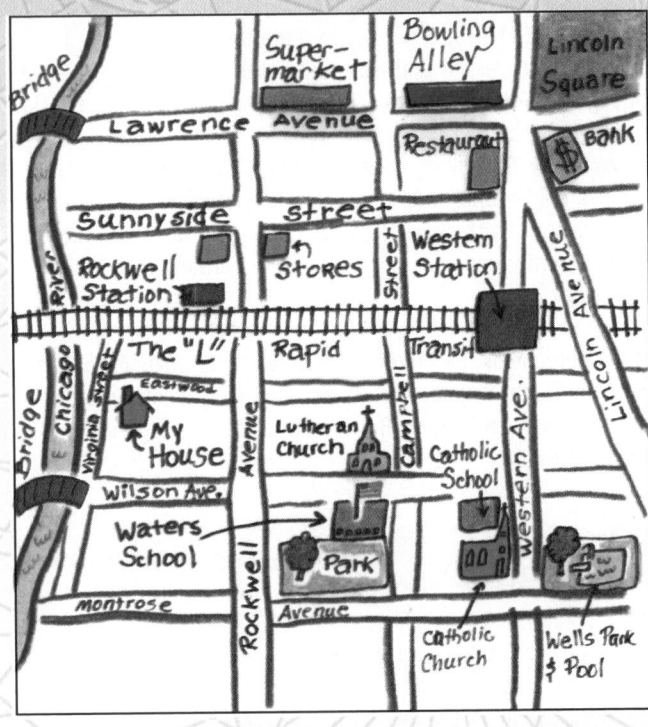

Geography Skills Practice

Study the sketch map above. On a separate sheet of paper, imagine your own neighborhood. Draw a sketch map of it from your mental map and answer these questions:

1. Which neighborhood streets or roads did you include?

2. What are the three most important features on your map?

Answers to Geography Skills Practice

1. Answers will vary.
2. Answers will vary.

PREVIEW

Words to Know
- bog
- pope

Places to Locate
- Poland
- Carpathian Mountains
- Warsaw

Read to Learn . . .
1. how Poland's landscape differs from north to south.
2. how Poland's economy has changed in recent years.
3. what customs and beliefs the Polish people value.

Today Poland is a vigorous and visible country facing new challenges. But that was not the case in the late 1700s when Poland disappeared from world maps. This cafe in Kraków, the third-largest city in Poland, shows that the country is alive and well today!

Meeting National Standards

Geography for Life The following standards are highlighted in this section:
- Standards 1, 2, 4, 5, 10, 14, 17, 18

FOCUS

Section Objectives
1. **Point out** how Poland's landscape differs from north to south.
2. **Evaluate** how Poland's economy has changed in recent years.
3. **Summarize** the customs and beliefs the Polish people value.

Bellringer Motivational Activity

Prior to taking roll at the beginning of the class period, project Section Focus Transparency 13-2 and have students answer the activity questions. Discuss students' responses. **L1**

Vocabulary Pre-check

Encourage students to use dictionaries to discover the origins of the word *pope*. **L1** **LEP**

The map on page 318 shows you that Poland lies south of the Baltic countries and east of Western Europe. In the late 1700s, Poland was swallowed up by Austria, Russia, and Germany. Poland later regained independence, but then communist rule was imposed on it after World War II.

Region
The Land

Today Poland is an independent nation. Covering 117,550 square miles (304,455 sq. km), Poland is slightly smaller than the state of New Mexico. On the north its coastline stretches for 326 miles (525 km) along the Baltic Sea. Sandy beaches and busy ports line the coast. Most of Poland lies in the North European Plain. Northern Poland also includes a hilly area with thousands of small lakes. Forests and **bogs**, or small swamps of spongy ground, run through the lakes region.

Plains and Highlands A vast plain stretches through central Poland. It forms part of the fertile North European Plain that spreads across Europe. Most of the Polish people live in this plains area.

Classroom Resources for Section 2

 **REPRODUCIBLE MASTERS**
Reproducible Lesson Plan 13-2
Guided Reading Activity 13-2
Workbook Activity 13
Section Quiz 13-2

 **TRANSPARENCIES**
Section Focus Transparency 13-2
Political Map Transparency 4

MULTIMEDIA
 Testmaker

National Geographic Society: ZipZapMap! World

 **NATIONAL GEOGRAPHIC SOCIETY**

These materials are available from Glencoe.

 SOFTWARE
ZipZapMap! World

MAP STUDY

Answer

Poland: humid continental; Hungary: marine west coast; Albania: Mediterranean

Map Skills Practice

Reading a Map What type of climate is found on the coast of Croatia? *(Mediterranean)*

TEACH

Guided Practice

Graphic Skills On the chalkboard, write the heading *Polish Workers in the Early 1990s* and list the following data underneath it: Industry and construction—35 percent; Agriculture—27 percent; Transport and communications—16 percent; Government and other—22 percent. Encourage students to present this information in a pictogram. Have students present and compare their pictograms. **L1**

Independent Practice

Guided Reading Activity 13-2 **L1**

CURRICULUM CONNECTION

Life Science The Bialowieza (or Belovezha) Forest, on Poland's eastern border, is one of the few places where the European bison, or wisent, can be found in the wild.

324

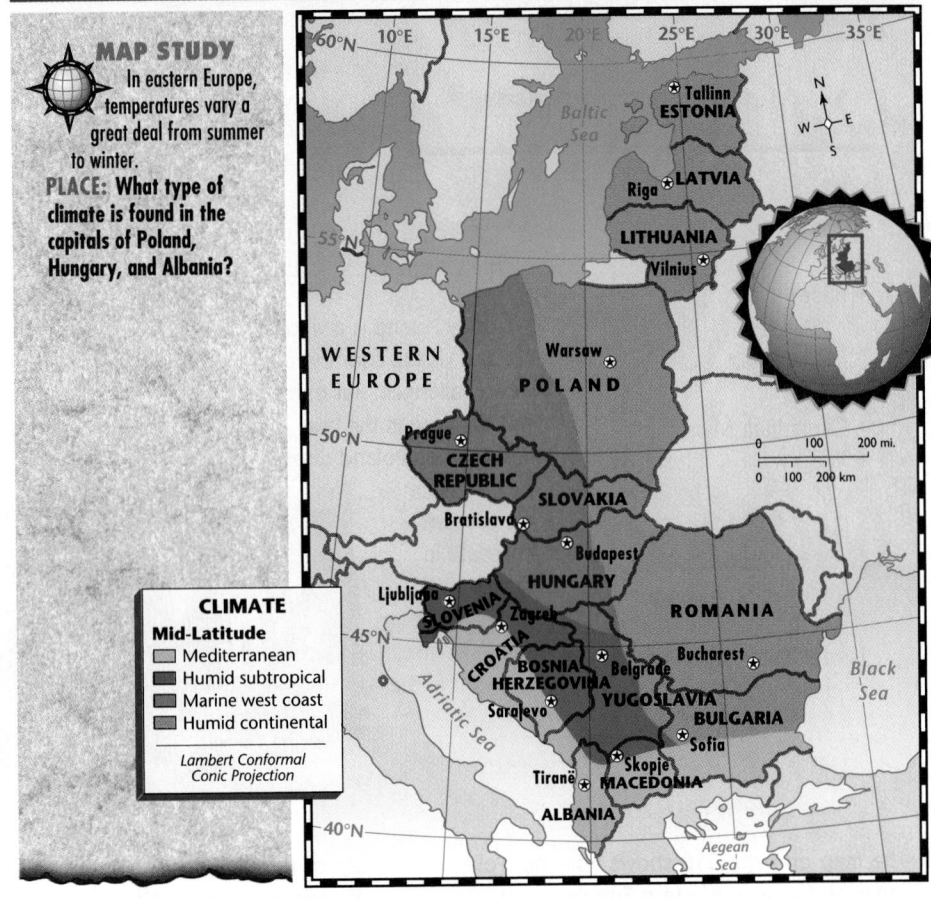

EASTERN EUROPE: Climate

MAP STUDY
In eastern Europe, temperatures vary a great deal from summer to winter.
PLACE: What type of climate is found in the capitals of Poland, Hungary, and Albania?

CLIMATE
Mid-Latitude
☐ Mediterranean
☐ Humid subtropical
☐ Marine west coast
☐ Humid continental

Lambert Conformal Conic Projection

In southern Poland the plains gradually rise in elevation to meet the low Sudeten (soo•DAY•tuhn) Mountains and the towering Carpathian (kahr•PAY•thee•uhn) Mountains. These two mountain areas form Poland's southern border.

The source of the mighty Vistula (VIHS•chuh•luh)—the main river of Poland—lies in the Carpathians. Winding more than 680 miles (1,094 km), the Vistula River carves a huge letter *S* through the country before it empties into the Baltic Sea.

The Climate Look at the map above. You see that the climate differs from one part of Poland to another. Western Poland has a marine west coast climate. Warm winds blowing across Europe from the Atlantic Ocean bring mild weather year-round to this part of the country. Farther from the ocean, eastern Poland experiences a humid continental climate with hot summers and cold winters.

UNIT 4

Cooperative Learning Activity

Inform students that coal is Poland's chief source of energy and the principal cause of its air and water pollution. Then have students work in small groups to research the connection between using coal as fuel and the following topics: acid rain, forest damage, respiratory ailments, and the greenhouse effect. Ask groups to report their findings to the class. Encourage students to use this information as the basis for suggestions as to how the Poles might improve their environment. **L2**

Human/Environment Interaction
The Economy

Some people joke that going to work is going "back to the salt mines." In Wieliczka (vyeh•LEECH•kah), Poland, workers say that and mean it! The world's oldest working salt mines lie beneath this city. The map on page 330 shows that the Polish economy depends on agriculture as well as industry. Polish farmers raise livestock and grow potatoes, sugar beets, wheat, and rye on the central plains. About 30 percent of the Polish population work on farms.

An Industrial Land Mining and manufacturing take place mostly in Poland's central and southern regions. Polish miners dig up coal, copper, and iron ore. Coal mining is Poland's most important industry and is concentrated near the Czech Republic. Factory workers produce machinery, cars, and trucks. Shippers at ports on the Baltic, such as Gdansk (guh•DAHNSK), send many of these products to other lands. Solidarity—the Polish labor union—struggles to protect the rights of these and other workers.

A New Economy Until the early 1990s, Poland had a communist economy in which the government ran industry and set prices and wages. Since then Poland has been moving toward free enterprise.

The transition from communism to free enterprise has brought Poland new challenges. Under the communist system, many Poles had lifelong jobs in industries that were often inefficient. These jobs have disappeared as Polish companies try to compete on the world market. Even though some have lost jobs, most Poles prefer free enterprise to communism.

Environmental Challenges Poland has begun to make a serious attempt to correct its environmental problems. For years Polish factories poured wastes into the air, water, and soil. Pollution in Poland still causes severe health problems for humans and animals. The Poles are now trying to improve the environment without further loss of jobs.

Place
The People

Poland has a population of 38.6 million. Most of Poland's citizens are Poles who belong to a larger ethnic group known as Slavs. Neighboring groups such as the Russians, Ukrainians, and Czechs are also Slavs.

CHAPTER 13 · 325 ·

A Potato Harvest in Poland

Families work together to harvest the year's potato crop. **HUMAN/ENVIRONMENT INTERACTION: What has caused the pollution of the soil, air, and water in Poland?**

More About the Illustration

Answer to Caption factory wastes

Place Before the Communists took control of Poland, farming was the country's most important economic activity. Today, about 34 percent of the labor force is involved in industry, while 29 percent works in farming.

ASSESS

Check for Understanding

Assign Section 2 Review as homework or an in-class activity.

Meeting Lesson Objectives

Each objective below is tested by the questions that follow it in parentheses.

1. **Point out** how Poland's landscape differs from north to south. (1, 2)
2. **Evaluate** how Poland's economy has changed in recent years. (3, 4)
3. **Summarize** the customs and beliefs that the Polish people value. (5)

Meeting Special Needs Activity

Inefficient Readers Students with reading problems often have difficulty differentiating main ideas from details. Teach them to skim for a main idea in the subsection titled "The Economy." Have students tell in their own words what the subsection is about. *(the work Poles do and the products they produce)* Next, have them read the first paragraph of the subsection. Point out that the main idea of a piece of writing often is stated in the first paragraph. Then have students state the main idea of the subsection. *(Poles work and produce in agriculture and industry.)* **L1**

Evaluate

Section Quiz 13-2
L1

Use the Testmaker to create a customized quiz for Section 2. **L1**

Reteach

List names of physical features, people, and places from the section and tell students to briefly identify each item on the list. **L1**

Enrich

Ask interested students to research Lech Walesa or Karol Wojtyla. Have them use their findings to create a poster listing highlights from the life of their selected subject. **L3**

CLOSE

Have students use information in the text to write a profile of a typical Polish school student.

A Traditional Polish Wedding

Many Polish couples dress in traditional clothes on their wedding day.
PLACE: To what ethnic group do most of Poland's people belong?

The Poles speak Polish, a Slavic language similar to Russian. Poles use the Latin alphabet, which English speakers also use. Some other Slavic groups, such as the Russians, use the Cyrillic (suh•RIH•lihk) alphabet. Look at the chart on page 334 to see the Slavic and other major language families of Europe.

Influences of the Past Situated on a vast plain, Poland has no natural defenses on its eastern and western borders. Over the centuries, it has been an easy target for invading armies. The ethnic term *Slav* comes from the word *slave*. Polish people have often been enslaved throughout their history. The Poles' love of country has helped them face these trials.

Daily Life About 62 percent of the Polish people live in cities and towns. Warsaw, the capital of Poland, has been an important urban center for hundreds of years. Poles still value their rural heritage and customs, however. They sometimes wear traditional costumes and often perform folk dances at weddings and other special occasions. At the same time, Polish young people enjoy rock and jazz as well as sports such as soccer, skiing, and basketball.

Religion has a strong influence on Polish life. Social life often centers around the local Roman Catholic church. Poles celebrate many religious holidays, especially Christmas and Easter. In 1978 Karol Wojtyla (voy•TEE•wah), a Polish church leader, became **pope,** or head of the Roman Catholic Church. The first Polish pope in history, Wojtyla took the name John Paul II.

SECTION 2 REVIEW

REVIEWING TERMS AND FACTS

1. **Define the following:** bog, pope.
2. **REGION** What part of Poland consists of a vast plain?
3. **HUMAN/ENVIRONMENT INTERACTION** How has industry affected the environment in Poland?
4. **HUMAN/ENVIRONMENT INTERACTION** How has Poland's geography affected its history?
5. **PLACE** What traditions do most Poles have in common?

MAP STUDY ACTIVITIES

6. Look at the climate map on page 324. What two climates are found in Poland?

Answers to Section 2 Review

1. All vocabulary words are defined in the Glossary.
2. central area of Poland
3. Factories have poured pollution into the air, water, and soil, causing considerable damage to the environment.
4. Poland has no natural defenses on its eastern and western borders. Therefore, it has often been overrun and occupied by invading armies.
5. love of country, attachment to rural customs, loyalty to the Roman Catholic Church
6. humid continental, marine west coast

COPERNICUS ▼ ▲

| MATH | SCIENCE | HISTORY | LITERATURE | TECHNOLOGY |

How many times have you stared up at the sky and wondered about the stars? A young Polish man named Nicolaus Copernicus (koh•PUHR•nih•kuhs) also wondered about the night sky. The sun, stars, and planets fascinated him throughout his life. Copernicus (1473–1543) became a scientist and astronomer and changed people's view of the solar system.

GREEK INFLUENCE The people of Copernicus's time believed, as the ancient Greeks had, that Earth was the center of the solar system. They thought that Earth remained still, while the sun, moon, planets, and stars circled it, with the planets "wandering" from place to place.

THE VIEW OF COPERNICUS Copernicus regularly watched the heavens. He carefully noted movements and measurements of stars. In 1543 he published his findings in a book called *On the Revolution of the Celestial Spheres.* Using mathematical proof, Copernicus established the sun as the center of the solar system. He showed that Earth did not stay in one place but rotated on its axis, as did all the other planets. He also proved that the

moon revolved around Earth, while Earth and the other planets revolved around the sun.

1543 Map of the Universe

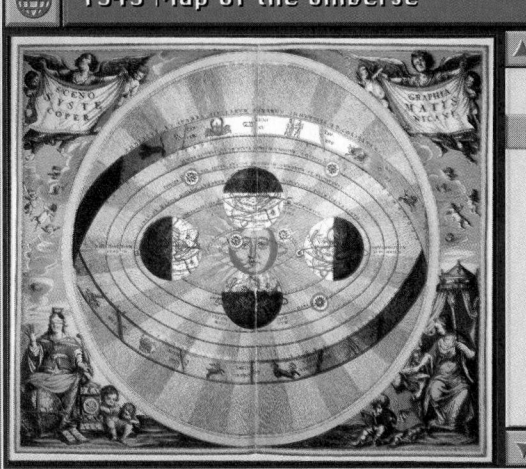

MODERN ASTRONOMY Copernicus did what no other scientist had done. He proved that scientific findings and mathematical equations could predict and compare the movements of stars and planets. For the first time, he explained the solar system. For these contributions, later scientists called Copernicus the Father of Modern Astronomy.

Making the Connection ▼ ▲

1. How did Copernicus change people's view of the solar system?
2. Why is Copernicus called the Father of Modern Astronomy?

CHAPTER 13

(327)

FOCUS

Section Objectives

1. **Explain** why the Danube River is important to Hungary.
2. **Relate** how Hungary's economy changed after the fall of communism.
3. **Describe** the types of food Hungarians enjoy.

Bellringer
Motivational Activity

Prior to taking roll at the beginning of the class period, project Section Focus Transparency 13-3 and have students answer the activity questions. Discuss students' responses. **L1**

Vocabulary Pre-check

Explain that Alix asked a number of her friends, Michael, Kate, and John, to invest in her bake sale business by buying flour, sugar, and other ingredients. In return, Alix promised to give them a share of the money made from the sale. Then ask students to define *invest* based on its context in your explanation. **L1** **LEP**

SECTION 3 Hungary

PREVIEW

Words to Know
• invest

Places to Locate
• Hungary
• Budapest
• Danube River
• Great Hungarian Plain

Read to Learn . . .
1. why the Danube River is important to Hungary.
2. how Hungary's economy changed after the fall of communism.
3. what types of foods Hungarians enjoy.

Have you ever ridden a horse or taken care of one? In Hungary raising horses has been a favorite activity in rural areas for centuries. Raised on the broad, fertile Hungarian Plain, Hungary's horses are famous throughout Europe.

Hungary lies east of Austria. Until 1918 Hungary and Austria were partners in a large central European empire. Like Austria, Hungary lost much territory after its defeat in World War I. Today it is a much smaller, landlocked country of 35,650 square miles (92,334 sq. km).

Region
The Land

Because it has no coastline, Hungary depends on the Danube River for transportation and trade. Look at the map on page 320. You see that the Danube twists and turns through several European countries, including Hungary. This river flows through so many countries that it has seven different names: Danube, Donau, Duna, Donaj, Dunarea, Dunav, and Dunai. It winds across 1,776 miles (2,858 km) before emptying into the Black Sea.

The Landscape The land of Hungary is mostly plains and highlands. The Great Hungarian Plain runs through eastern Hungary. Rivers and small hills wind across it. Dotted with farms, the plain is known for its excellent farming and grazing land.

328

UNIT 4

Classroom Resources for Section 3

 REPRODUCIBLE MASTERS

Reproducible Lesson Plan 13-3
Guided Reading Activity 13-3
Cooperative Learning
  Activity 13
GeoLab Activity 13
Section Quiz 13-3

TRANSPARENCIES

Section Focus
  Transparency 13-3
Unit Overlay Transparencies
  4-0, 4-4, 4-5

MULTIMEDIA

 Testmaker

National Geographic
  Society: ZipZapMap!
  World

The Danube River separates the Great Hungarian Plain from a very different region to the west. This region is called Transdanubia because it lies "across the Danube." Rolling hills, wide valleys, and forests cover Transdanubia's landscape. Lake Balaton, one of Europe's largest lakes, is in this region. Many Hungarians spend their vacations there sailing, swimming, and hiking.

In northern Hungary the Carpathian Mountains tower above the horizon. In this scenic part of Hungary you can wander through thick forests, find strange rock formations, and explore underground caves.

The Climate The map on page 324 shows you that Hungary is located far from any large body of water. As you might expect, it has a humid continental climate with cold winters and hot summers. Temperatures climb as high as 106°F (41°C) and drop as low as -29°F (-34°C). Western Hungary receives more rainfall than the eastern part of the country.

Movement

The Economy

Hungary's most important economic activities are agriculture and manufacturing. Hungarian farmers grow corn, potatoes, sugar beets, wheat, and wine grapes. They also raise dairy cattle, horses, and sheep. The country's factory workers produce chemicals, steel, textiles, and buses.

In the late 1940s, Hungary's economy came under communist control. When communism collapsed in the early 1990s, free enterprise returned. To encourage the growth of Hungarian companies, many foreign businesses now are investing in Hungary's free enterprise system. To **invest** means to put money into a company in return for a share of the profits.

Place

The People

The majority of Hungary's 10.3 million people are Hungarians, or Magyars. Unlike other eastern Europeans, Hungarians do not belong to the Slavic ethnic group. The ancestors of present-day Hungarians, the Magyars, came to eastern Europe from central Asia more than 1,000 years ago.

Hungarians are similar to other Europeans in their religious choices. About 67 percent of Hungary's people are Roman Catholic. Another 25 percent are Protestant.

CHAPTER 13

329

Window-Shopping in Budapest

Window-shoppers look at the merchandise in Budapest's fashionable shops.
PLACE: How has Hungary's economic system changed since the collapse of communism?

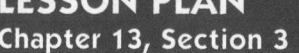
More About the Illustration

Answer to Caption
It now has a free enterprise economic system, which encourages the development of privately owned businesses.

GEOGRAPHIC THEMES **Place** Hungary had followed a policy of greater economic freedom and better standards of living since the 1960s. Some people referred to this approach as "goulash communism."

TEACH

Guided Practice

Project Unit Overlay Transparencies 4-0, 4-4, and 4-5, and ask a student to trace with a pointer the Danube River. Call on students to point out features and places along the river as they read about them in the section. **L1**

Independent Practice

Guided Reading Activity 13-3 **L1**

Cooperative Learning Activity

Organize the students into groups, and assign each group one of the following topics in Hungarian history: Hungary as part of the Roman empire, Magyar rule in Hungary, Hungary during the Renaissance, Hungary under the Hapsburgs, Hungary before and during World War I, Hungary during World War II, the Hungarian uprising of 1956, political changes in Hungary between 1988 and 1990. Encourage groups to research their topics and to create comic books based on the research. Distribute comics about history for groups to use as models. Allow time for groups to exchange and read one another's completed comics. **L2**

MAP STUDY

Answer
manufacturing

Map Skills Practice
Reading a Map What resources are found in Hungary? *(petroleum, natural gas, bauxite)*

ASSESS

Check for Understanding

Assign Section 3 Review as homework or an in-class activity.

Meeting Lesson Objectives

Each objective below is tested by the questions that follow it in parentheses.

1. **Explain** why the Danube River is important to Hungary. (2, 4)
2. **Relate** how Hungary's economy changed after the fall of communism. (3)

Evaluate

 Section 3 Quiz **L1**

Use the Testmaker to create a customized quiz for Section 3. **L1**

EASTERN EUROPE: Land Use and Resources

MAP STUDY

Coal mining is an important industry in Poland.
PLACE: What other economic activity takes place in Poland in the areas where coal is found?

Agriculture
- Commercial farming
- Subsistence farming
- Manufacturing area

Resources
- Bauxite
- Coal
- Copper
- Fish and other seafood
- Hydroelectric power
- Iron ore
- Lead
- Natural gas
- Petroleum
- Zinc

Lambert Conformal Conic projection

City on the Danube About 63 percent of Hungarians are city dwellers. Nearly 2.5 million people live in Budapest (BOO•duh•PEHST), the capital and largest city. Fifteen-year-old Mátyás Kemény and his family live in Budapest. Mátyás always explains to visitors that Budapest is really two cities that stand on opposite banks of the Danube River. Buda, on the west bank, is a historic city of churches and palaces. Across the river, Pest is more modern and the site of many factories.

Past and Present Mátyás's father remembers when the Hungarians rose up against strict communist rule in 1956. Soviet troops crushed the uprising, and many Hungarians died. At that time, about 200,000 people left the country. Many of them moved to the United States.

Today Mátyás understands why his father does not have a job. His factory, which made parts for railroad cars, closed down because it could not compete successfully in the new free enterprise market. The family is hopeful for the future. Whatever happens, they know they do not want to return to communism.

Meeting Special Needs Activity

Inefficient Readers Scanning a passage before reading can help students develop a time frame for events or ideas that are presented. Have students scan material under "The Economy" for time markers such as years. "In the late 1940s . . ." and "in the early 1990s" in the second paragraph tells them that the time frame is less than 50 years. **L1**

Hungary's capital extends along both banks of the Danube River.
PLACE: Why is the Danube so important to Hungary's economy?

Daily Life Neither the grim communist past nor the present shortage of jobs can dim the Hungarians' enjoyment of life. Mátyás and his family like to celebrate. On special occasions they go to a restaurant where they can hear an orchestra play fast-paced Hungarian music.

Like most Hungarians, Mátyás enjoys good food, especially a thick stew called goulash. It is made of beef, gravy, onions, and potatoes. He also likes to flavor his food with a seasoning of red peppers called paprika. You may have seen paprika on potato salad or other foods that your family has prepared.

SECTION 3 REVIEW

REVIEWING TERMS AND FACTS

1. Define the following: invest.
2. REGION What does Transdanubia look like?
3. MOVEMENT How are foreign businesses helping Hungary's economy?
4. PLACE What river divides Budapest into two parts?

MAP STUDY ACTIVITIES

5. Study the land use map on page 330. Where would be a good place to build an oil refining plant?
6. Look again at the map on page 330. What is Hungary's major manufacturing center?

331

Answers to Section 3 Review

1. All vocabulary words are defined in the Glossary.
2. Transdanubia has rolling hills, wide valleys, and forests.
3. They are investing in Hungarian businesses to help them grow.
4. Danube
5. in southern Hungary
6. Budapest

More About the Illustration

Answer to Caption
Hungary depends on the river for transportation and trade links.

GEOGRAPHIC THEMES **Human/ Environment Interaction** Many Europeans travel to Budapest to soak in the city's many baths. These baths are fed by more than 120 hot springs around the city. Some say that the mineral-rich waters from the springs have healing powers.

Reteach

Have students work in small groups to write 10 questions on the section content. Then have groups use their questions to quiz other groups. **L1**

Enrich

Encourage students to tape interviews with Hungarian Americans about the traditions they follow. Have students play the taped interviews for the class. **L3**

CLOSE

Tell students to imagine that they are Mátyás Kemény. Ask them to prepare a brief travelogue on Hungary for friends who are visiting from the United States.

331

FOCUS

Section Objectives

1. **Examine** how people earn their livings in the Czech Republic and Slovakia.
2. **Describe** the physical features that dominate the Czech Republic and Slovakia.
3. **Understand** why the Czech Republic and Slovakia became separate, independent countries.

Bellringer Motivational Activity

Prior to taking roll at the beginning of the class period, project Section Focus Transparency 13-4 and have students answer the activity questions. Discuss students' responses. **L1**

Vocabulary Pre-check

Ask students to think of a synonym for the word *preserve. (save, protect, maintain, care for)* Then have them use the synonym to complete this sentence: "A nature preserve is land set aside to . . ." **L1 LEP**

SECTION 4

The Czech Republic and Slovakia

PREVIEW

Words to Know
• nature preserve
• service industry

Places to Locate
• Czech Republic
• Prague
• Slovakia
• Bratislava

Read to Learn . . .
1. how people earn their livings in the Czech Republic and Slovakia.
2. what physical features dominate the Czech Republic and Slovakia.
3. why the Czech Republic and Slovakia became separate, independent countries.

Charles Bridge is a charming spot in the city of Prague, the capital of the Czech Republic. From the bridge, you can see the domes and towers of Prague. Prague is one of the best-preserved historic cities in Europe. It is also a modern city whose people look forward to a bright future.

The Czech (CHEHK) Republic and its neighbor Slovakia (sloh•VAH•kee•uh) once formed a larger country called Czechoslovakia. Like Poland and Hungary, Czechoslovakia was under communist control and closely linked to the Soviet Union. In 1989 the Czech and Slovak peoples peacefully ended communist rule. On January 1, 1993, they also peacefully split into two separate countries.

Place
The Czech Republic

The Czech Republic lies south of Poland and east of Germany. It is a landlocked country, lying deep within the continent. This interior location gives the republic warm summers and cold winters.

The Land The map on page 320 shows that mountains run through the western and northern parts of the Czech Republic. In these areas, spas famous for their healthful waters attract bathers to hot mineral springs. Tourists can also explore **nature preserves,** or lands set aside for plant and animal wildlife. The

332

natural beauty, however, cannot hide the vast areas of bare trees ruined by industrial pollution.

Between low mountain ranges, the Czech Republic is a land of rolling hills and low fertile plains. These areas boast the nation's best farmland and major industrial centers. Rivers such as the Elbe (EHL•buh) and Vltava (VUHL•tuh•vuh) flow gracefully through the region.

The Economy Under communism the Czech economy was based on heavy industry. Today there is a free enterprise economy, and many Czechs have set up new **service industries.** They now own hotels, repair shops, and stores that sell clothing and household goods. Farming provides jobs for only about 10 percent of the population. Major crops include barley, corn, fruits, oats, and potatoes.

The People Most of the Czech Republic's 10.3 million people belong to a Slavic ethnic group called Czechs. Most Czechs live in cities and towns. Prague (PRAHG), with about 1 million people, is the capital and largest city. The Czechs have one of the highest standards of living in eastern Europe. Many of them own cars and household appliances. City dwellers, however, often have to live in crowded high-rise apartment buildings.

Teen Scene

Bells of Freedom

It was an election that brought a smile to the face of Ilsa Blazhova. The man who led the fight to overturn communism in Czechoslovakia was now the head of the Czech Republic. In 1990 the Czech people made Vaclav Havel president. Ilsa and hundreds of others showed their support in Prague's main square by ringing bells for him.

Place
Slovakia

Newly separated from the Czech Republic, Slovakia lies to the east. The Slovak people and the Czechs share a common Slavic heritage. The two groups, however, have different languages and cultures.

WHAT IN THE WORLD?

Castle Craze

Are you a victim of castle-mania? Do you find castles thrilling and fascinating? If so, you should travel to the Czech Republic and Slovakia. There you can visit a different castle every day—for seven years! Some 2,500 castles still stand in the two countries.

The Land and Climate A range of the Carpathian Mountains towers over most of northern Slovakia. Rugged gray peaks, thick forests, and blue lakes make this area a popular vacation spot. Farther south,

TEACH

Guided Practice

Making Comparisons As students read the section, have them list the kinds of physical features, industries, and crops that the Czech Republic and Slovakia have in common. **L1**

Independent Practice

📁 Guided Reading Activity 13-4 **L1**

ASSESS

Check for Understanding

Assign Section 4 Review as homework or an in-class activity.

Meeting Lesson Objectives

Each objective below is tested by the questions that follow it in parentheses.

1. **Examine** how people earn their livings in the Czech Republic and Slovakia. (1)
2. **Describe** the physical features that dominate the Czech Republic and Slovakia. (3)

Cooperative Learning Activity

Organize students into several small groups. Assign half the groups the Czech Republic and the other groups Slovakia. Have groups create a mural representing their assigned country. Suggest that groups focus on topics such as landforms, climate, resources, economic activities, major cities, everyday life, and so on. Have groups display and discuss their completed murals. **L1**

Evaluate

 Section 4 Quiz **L1**

Use the Testmaker to create a customized quiz for Section 4. **L1**

Reteach

Have students reread the part of the section about the Czech Republic and write a two-paragraph summary. Then have them do the same for the part of the section about Slovakia. **L1**

Enrich

Ask students to write a magazine article about Prague. Topics they should cover include: history, economy, population, and famous landmarks. **L3**

CLOSE

Have students imagine they are Ilsa Blazhova. Encourage them to express their feelings about being part of the fight to overturn communism.

Language Families of Europe

GRAPHIC STUDY
Europe has seven main language families.
MOVEMENT: From what language family did English develop?

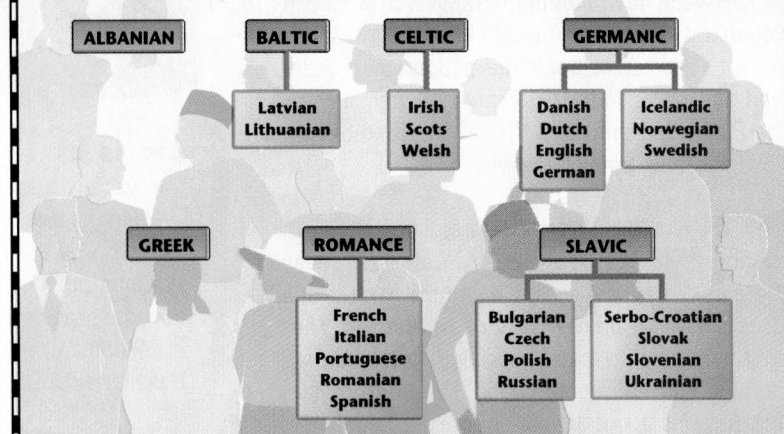

ALBANIAN	BALTIC	CELTIC	GERMANIC	
	Latvian Lithuanian	Irish Scots Welsh	Danish Dutch English German	Icelandic Norwegian Swedish

GREEK	ROMANCE	SLAVIC	
	French Italian Portuguese Romanian Spanish	Bulgarian Czech Polish Russian	Serbo-Croatian Slovak Slovenian Ukrainian

vineyards and farms spread across the fertile lowlands and plains that stretch to the Danube River. Elevation affects Slovakia's climate, but most of the country experiences cold winters and warm summers.

The Economy Slovakia has been a farming country throughout most of its history. Slovak farmers grow barley, corn, potatoes, sugar beets, and wine grapes. Under communist rule Slovakia began building factories for heavy industry. Slovakia now has a free enterprise economy with a growing number of service industries. Many inefficient factories have shut down, causing job losses.

The People An ethnic group known as Slovaks make up the majority of the country's population. People of Hungarian descent form the second-largest group. About 60 percent of Slovakia's 5.3 million people live in urban areas. Bratislava (BRA•tuh•SLAH•vuh), a port on the Danube, is Slovakia's capital and largest city. Most Slovaks are Catholics.

SECTION 4 REVIEW

REVIEWING TERMS AND FACTS
1. **Define the following:** nature preserve, service industry.
2. **LOCATION** Where are the Czech Republic's major farmlands and industrial centers located?
3. **PLACE** What mountain range runs through northern Slovakia?

GRAPHIC STUDY ACTIVITIES

4. Study the chart of language families above. From which family does the Czech language come?

 334

Answers to Section 4 Review
1. All vocabulary words are defined in the Glossary.
2. in the rolling hills and plains to the east and south of the mountains
3. Carpathians
4. Slavic

SECTION 5
The Balkan Countries

PREVIEW

Words to Know
- consumer goods
- mosque

Places to Locate
- Slovenia
- Croatia
- Yugoslavia (Serbia and Montenegro)
- Bosnia-Herzegovina
- Macedonia
- Romania
- Bulgaria
- Albania

Read to Learn . . .
1. why Yugoslavia broke up into separate countries.
2. how Romanians are like western Europeans.
3. what Albania needs to build its economy.

Have you ever seen such a towering wall of rock? It is one of the many mountain ranges that runs through the Balkan Peninsula in the southeast corner of Europe. The word *balkan* means "mountain" in one of the peninsula's local languages. The rugged Balkan Peninsula is home to a rich variety of cultures.

The Balkan Peninsula lies between the Adriatic Sea and the Black Sea. It also stretches into the Mediterranean Sea. The map on page 318 shows you that several countries make up this Balkan region. They are the former Yugoslav republics, plus Romania, Bulgaria, and Albania.

Place
Former Yugoslav Republics

Rugged mountains cover most of the republics that formerly made up Yugoslavia. Branches of the Alps reach into northwestern and coastal areas. Highlands at the center of the republics flatten into northern plains. The Danube River flows through the plains region.

Breakup of the Region This region once formed a large communist country called Yugoslavia. After communism ended in the region, Yugoslavia broke apart because of cultural differences among its many ethnic groups. Today the region is made up of five independent republics: Slovenia, Croatia, Yugoslavia (Serbia and Montenegro), Bosnia-Herzegovina (BAHZ•nee•uh HERT•suh•goh•VEE•nuh), and Macedonia.

CHAPTER 13

(335)

Meeting National Standards
Geography for Life The following standards are highlighted in this section:
- Standards 1, 2, 3, 10, 12, 16

FOCUS

Section Objectives
1. **Comprehend** why Yugoslavia broke up into separate countries.
2. **Note** how Romanians are like western Europeans.
3. **Identify** what Albania needs to build its economy.

Bellringer Motivational Activity

 Prior to taking roll at the beginning of the class period, project Section Focus Transparency 13-5 and have students answer the activity questions. Discuss students' responses. **L1**

Vocabulary Pre-check
Tell students that when they shop, they act as consumers. Ask students to list the kinds of goods that they and their families buy regularly. After the students finish, tell them that the goods they listed are known as *consumer goods*. **L1 LEP**

Classroom Resources for Section 5

REPRODUCIBLE MASTERS
Reproducible Lesson Plan 13-5
Guided Reading Activity 13-5
Reteaching Activity 13
Enrichment Activity 13
Section Quiz 13-5

 **TRANSPARENCIES**
Section Focus
   Transparency 13-5
Political Map Transparency 4

MULTIMEDIA
 Testmaker

National Geographic Society: Picture Atlas of the World

TEACH

Guided Practice

Drawing Conclusions On the chalkboard, write the following heads: *COUNTRY, ETHNIC GROUPS, RELIGION.* Call on volunteers to come to the chalkboard and list details from the section under the appropriate headings. Then have students use information on the chalkboard to determine which countries are most alike in their ethnic and religious make-up. **L1**

Independent Practice

Guided Reading Activity 13-5 **L1**

MULTICULTURAL PERSPECTIVE

Cultural Diffusion

The Croats began the tradition of wearing neckties. Croatian soldiers of the seventeenth century wore loosely-knotted scarves around their necks. The French adopted the fashion, calling the scarf a *cravate,* their name for the Croatian soldiers. Over time, other people adopted the fashion, which eventually evolved into the necktie.

Slovenia Slovenia lies in the mountains of the northwest. It has more factories and service industries than the other republics. It also has the region's highest standard of living. Most of the 2 million Slovenians are Slavs who practice the Roman Catholic religion.

Croatia Beyond Croatia's island-studded Adriatic coast, mountains rise to a fertile, inland plain. An industrialized republic, Croatia supports an agricultural economy as well. Like the Slovenians, the 4.8 million Croats (KROH•atz) are Slavs and Roman Catholics.

Serbia and Montenegro (Yugoslavia) Serbia and Montenegro spread over inland plains and mountains. United as Yugoslavia, the economies of the two republics are based on agriculture and industry. The major urban center is Belgrade, the region's largest city. The 10.5 million Serbs and Montenegrins belong to the Eastern Orthodox Church. While the Slovenians and Croats use the Latin alphabet, the Serbs and Montenegrins write in the Cyrillic alphabet.

Bosnia-Herzegovina Bosnia-Herzegovina lies west of Serbia. It consists of mountains, thick forests, and fertile river valleys. Many of the Bosnian people are Muslims. Others are Eastern Orthodox Serbs or Roman Catholic Croats.

Civil war ravaged this area in the early 1990s. Bosnian Serbs tried to force out Bosnian Muslims. In 1995 the Dayton Peace Treaty ended the fighting. It also divided Bosnia into two separate regions under one government.

Macedonia Macedonia is the most southern of the republics. It has a developing economy largely based on agriculture. The 2 million Macedonians are a mixture of many different Balkan peoples.

Place

Romania

Romania spreads eastward from Yugoslavia to the Black Sea. It has hot summers and cold winters. The scenic Carpathian Mountains curve through northern and central Romania. Between the mountain ranges stretch a vast plateau and plains. Quaint villages and modern urban centers dot these flatlands. Transylvania, the setting for many horror novels, lies in this region.

The Economy Romania's major economic activities include farming, manufacturing, and mining. The oil industry is important in the southeastern part of the country. Under communism Romania's factories produced machinery but few **consumer goods**—clothing, shoes, and other products made for people. Romania now has a free enterprise economy with factories that turn out more consumer goods.

The People What does Romania's name tell you about its history? If you guessed that the Romans once ruled this region, you are correct. Most of Romania's 22.7 million people are descended from the Romans. The Romanian

---

Cooperative Learning Activity

Organize students into several groups. Have groups create a board game that people in the Balkan countries can use to learn about free enterprise. Suggest that groups use Monopoly as a model but that they name the squares on the playing board after streets in Balkan cities. Refer groups to maps of cities in travel guides to find the names. Help groups make up cards that reflect situations in a free enterprise economy, such as: "You bought a convertible without first comparing prices. Pay the bank 300 *kuna* in extra interest on your auto loan." Have classroom artists design paper money similar to Balkan currency—Albanian *lek,* Romanian *leu,* Bulgarian *lea,* Croatian *dinar,* Macedonian *denar,* and Slovenian *tolar.* Copy enough money for several players. Allow class time for students to play the game. **L1**

MAP STUDY
Of the region, the Czech Republic has the greatest population density.
PLACE: What is the population density of land along the Adriatic Sea?

CITIES
- City with more than 1,000,000 people
- City with 500,000 to 1,000,000 people
- City with 100,000 to 500,000 people

Persons per

sq. mi.	sq. km
Uninhabited	Uninhabited
Under 2	Under 1
2–60	1–25
60–125	25–50
125–250	50–100
Over 250	Over 100

Lambert Conformal Conic projection

LESSON PLAN
Chapter 13, Section 5

MAP STUDY

Answer
60–125 per sq. mi. (25–50 per sq. km)

Map Skills Practice
Reading a Map What is the population density in the areas around the cities of Zagreb and Belgrade? *(125–250 per sq. mi., or 50–100 per sq. km)*

ASSESS

Check for Understanding

Assign Section 5 Review as homework or an in-class activity.

Meeting Lesson Objectives

Each objective below is tested by the questions that follow it in parentheses.

1. **Comprehend** why Yugoslavia broke up into separate countries. (2)
2. **Note** how Romanians are like western Europeans. (1)
3. **Identify** what Albania needs to build its economy. (4)

---

language comes from the Roman language, Latin. Romanian is closer to French, Italian, and Spanish than it is to other eastern European languages. In other ways, the Romanians are more like their Slavic neighbors. For example, many Romanians are Eastern Orthodox Christians. Half of Romania's population live in cities and towns. Bucharest, the capital and largest city, has wide streets and modern buildings.

Place

Bulgaria

Bulgaria lies south of Romania. Like Romania, Bulgaria has a coastline on the Black Sea. Two mountain ranges—the Balkan Mountains and the Rhodope (RAH•duh•pee) Mountains—span most of Bulgaria. Fertile valleys and plains separate the mountains in many areas. Bulgaria's climate varies considerably. The Black Sea coast has warmer year-round temperatures than the mountainous inland areas.

Meeting Special Needs Activity

Inefficient Organizers Have students scan the section for names of crops and manufactured goods from Bulgaria and Albania. Then have them copy and add to the following chart:

L1

Country	Crops	Manufactured goods
Bulgaria	fruits	chemicals

Evaluate

 Section Quiz 13-5
L1

Use the Testmaker to create a customized quiz for Section 5. **L1**

337

📁 Chapter 13 Test Form A and/or Form B **L1**

Reteach

📁 Reteaching Activity 13 **L1**

More About the Illustration

Answer to Caption

 as factory workers and farmers

Human/ Environment Interaction

Bulgaria's Valley of the Roses provides 70 percent of the attar, or rose oil, used in making the world's perfumes.

Enrich

📁 Enrichment Activity 13 **L1**

CLOSE

Mental Mapping Activity

Provide each student with a large sheet of paper and map pencils. Inform students that this will not be a graded activity. Have students draw, freehand from memory, the Balkan Peninsula and the borders of the Balkan countries. Tell them to label the countries and to shade the former Yugoslav republics in red.

338

A Sweet-Smelling Job

A Bulgarian woman handpicks roses to be used in making perfume.
PLACE: In what other ways do Bulgarians earn their livings?

The Economy Bulgaria is making the transition from a communist to a free enterprise economy. Bulgarian factories produce chemicals, machinery, and textiles. The country's farms grow fruits, vegetables, and grains. Roses are grown for their sweet-smelling oil, which is used in perfumes.

The People Most of Bulgaria's 8.4 million people trace their ancestry to the Slavs and other groups from central Asia. The Bulgarian language is similar to Russian and uses the Cyrillic alphabet. Most Bulgarians live in cities and towns. Sofia (SOH•fee•uh) is the country's capital and largest city.

Place
Albania

Tucked southwest of Yugoslavia lies the small country of Albania. Bordering the Adriatic coast of the Balkan Peninsula, Albania is slightly larger than the state of Maryland. Mountains cover most of Albania, contributing to Albania's isolation from neighboring countries. A small coastal plain runs along the Adriatic Sea. Most of Albania has a mild climate. Temperatures in coastal areas, however, are warmer than in mountainous inland areas.

The Economy Albania has one of Europe's least developed economies. The country is rich in mineral resources but lacks the technology to develop them. Under communist rule Albania began to develop heavy industry. Yet today, under free enterprise, most Albanians still make their living from farming. They grow corn, grapes, olives, potatoes, sugar beets, and wheat in mountain valleys.

The People About 64 percent of Albania's 3.4 million people live in the countryside. Most Albanians are Muslims. Since the fall of communism, many **mosques**, or Muslim houses of worship, have opened in Albania.

SECTION 5 REVIEW

REVIEWING TERMS AND FACTS
1. **Define the following:** consumer goods, mosque.
2. **PLACE** What former Yugoslav republic is torn by warfare?
3. **LOCATION** What Balkan countries have coastlines on the Black Sea?
4. **HUMAN/ENVIRONMENT INTERACTION** Why is Albania unable to develop its resources?

MAP STUDY ACTIVITIES
5. Look at the population density map on page 337. What city in the Yugoslav region has more than 1 million people?

338

UNIT 4

Answers to Section 5 Review
1. All vocabulary words are defined in the Glossary.
2. Bosnia-Herzegovina
3. Romania, Bulgaria
4. Albania lacks the technology to develop its resources.
5. Belgrade

Chapter 13 Highlights

Important Things to Know About Eastern Europe

SECTION 1 THE BALTIC REPUBLICS

- The Baltic republics of Estonia, Latvia, and Lithuania share a Baltic Sea coastline.
- After decades of communist rule, Estonia, Latvia, and Lithuania are now independent democracies.

SECTION 2 POLAND

- Most of Poland's farms, factories, and cities lie on a vast plain.
- Poland's economy changed from communism to free enterprise.
- Most Poles belong to the Roman Catholic Church.

SECTION 3 HUNGARY

- Landlocked Hungary depends on the Danube River for trade.
- Most Hungarians belong to an ethnic group known as Magyars.
- Hungarian businesses grow through foreign investments.

SECTION 4 THE CZECH REPUBLIC AND SLOVAKIA

- Cultural differences led Czechs and Slovaks to form two separate nations in 1993.
- Mountains, rolling hills, and plains sweep across the Czech Republic and Slovakia.
- Both republics are industrialized, with large urban populations.

SECTION 5 THE BALKAN COUNTRIES

- The Balkan Peninsula has many different ethnic groups speaking several different languages.
- The breakup of Yugoslavia has led to civil war in the region.

Celebrating independence in Tallinn, Estonia ▶

Using the Chapter 13 Highlights

- Use the Chapter 13 Highlights to preview, review, condense, or reteach the chapter.

Preview/Review

 Vocabulary Puzzle-Maker Software reinforces the terms used in Chapter 13.

Condense

Have students read the Chapter 13 Highlights. Spanish Chapter Highlights are also available.

Chapter 13 Digest Audiocassettes and Activity

Chapter 13 Digest Audiocassette Test

Spanish Chapter Digest Audiocassettes, Activities, and Tests are also available.

Guided Reading Activities

Reteach

Reteaching Activity 13. Spanish Reteaching activities are also available.

MAP STUDY

Movement Copy and distribute page 30 from the Outline Map Resource Book. Have students label the countries of eastern Europe on the map and locate and label the region's major river and sea ports. Suggest that students refer to the text and atlases for help.

Extra Credit Project

Picture Essay Encourage students to research domestic animals of eastern Europe, such as dairy cattle in Latvia and horses in Hungary, and to report on the animals' history, care, and feeding. Suggest that students present their reports in the form of a pictorial essay. Display the reports in the classroom. **L1**

339

Chapter 13 Review and Activities

ANSWERS

Reviewing Key Terms

1. H
2. F
3. D
4. E
5. A
6. I
7. C
8. B
9. G

Mental Mapping Activity

This exercise helps students to visualize the countries and geographic features they have been studying and understand the relationships among them. All attempts at freehand mapping should be accepted.

REVIEWING KEY TERMS

Match the numbered terms in Column A with their definitions in Column B.

A
1. consumer goods
2. service industry
3. oil shale
4. bog
5. pope
6. invest
7. nature preserve
8. mosque
9. communism

B
A. leader of the Roman Catholic Church
B. Muslim house of worship
C. land set aside to protect plants and animal wildlife
D. a type of rock that contains petroleum
E. small swamp of spongy ground
F. an industry such as banking or tourism
G. the government controls how citizens live and produce goods
H. goods made for people's use, such as shoes and clothing
I. to spend money on a business, hoping to make a profit

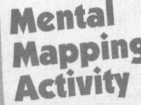

Mental Mapping Activity

On a separate piece of paper, draw a freehand map of Eastern Europe. Label the following items on your map:

• Slovakia
• Sofia
• Great Hungarian Plain
• Czech Republic
• Albania
• Poland

REVIEWING THE MAIN IDEAS

Section 1
1. PLACE What two features make Estonia stand out among the Baltic countries?
2. LOCATION How does Latvia's location affect its economy?

Section 2
3. REGION What are the main landforms of Poland?
4. PLACE What religion do most Poles practice?

Section 3
5. MOVEMENT Where did the original Magyars come from?
6. PLACE Name a typical Hungarian food dish.

Section 4
7. REGION What two rivers flow through the Czech Republic?
8. REGION How has free enterprise changed Slovakia's economy?

Section 5
9. PLACE What is the major landform of the Balkan Peninsula?
10. PLACE What is the capital of Romania?

Reviewing the Main Ideas

Section 1

1. Estonia is the smallest of the Baltic Republics; it has large deposits of oil shale; its people have a unique culture that is closely related to that of the Finns.
2. Latvia's location on the Baltic coast has made it an important trading center.

Section 2

3. sandy coastlines, plains, hilly areas with bogs and forests, mountains
4. Roman Catholicism

Section 3

5. central Asia
6. the thick stew called *goulash*

CRITICAL THINKING ACTIVITIES

1. **Making Comparisons** How do people earn a living in the Czech Republic? In Albania?
2. **Making Generalizations** How have mountains affected ways of life in the Balkan Peninsula?

GeoJournal Writing Activity

Write a journal entry describing a trip down the Danube River from southern Germany to the Black Sea. Describe the cities and the countryside through which you would pass.

COOPERATIVE LEARNING ACTIVITY

Work in a group of three to learn more about recent conflicts in the former Yugoslav republics. Research the various ethnic groups of the region. What does each group want? What land are they fighting over? Draw a map of the Balkan region with labels showing where different groups live. Propose a peace settlement that seems fair. Then share your plan with the rest of the class.

PLACE LOCATION ACTIVITY: EASTERN EUROPE

Match the letters on the map with the places and physical features of eastern Europe. Write your answers on a separate sheet of paper.

1. Poland
2. Budapest
3. Bosnia-Herzegovina
4. Latvia
5. Balkan Mountains
6. Danube River
7. Adriatic Sea
8. Warsaw
9. Carpathian Mountains
10. Baltic Sea

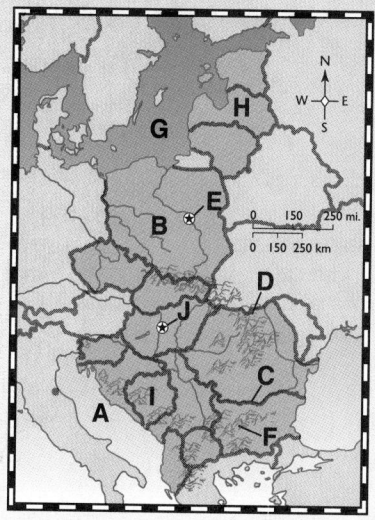

Cooperative Learning Activity

Suggest that groups add other visual aids, such as pictures of traditional costumes, to their maps. Then allow time for groups to report their findings in class.

GeoJournal Writing Activity

Encourage students to sketch maps to illustrate their description. Call on volunteers to share their descriptions with the rest of the class.

Place Location Activity

1. B	6. C
2. J	7. A
3. I	8. E
4. H	9. D
5. F	10. G

Chapter Bonus Test Question

This question may be used for extra credit on the chapter test.

The communist government of Albania discouraged contact with the outside world. How has this created economic problems for Albania's new government? *(Answers will vary. Most students will point out that lack of contact with the outside world has meant that Albania has not shared in technology developed elsewhere. Also, lack of established trade ties makes it harder for Albania to function in an interdependent world economy.)*

Section 4

7. Elbe and Vltava rivers
8. number of service industries has increased; many inefficient factories have closed, causing job losses

Section 5

9. rugged mountains
10. Bucharest

Critical Thinking Activities

1. The Czech Republic has an economy based on industry. Most Albanians are involved in farming.
2. Most students will point out that mountains have made travel and communication difficult; mountains have isolated one group of people from another, keeping them from appreciating each other's culture.

FOCUS

Ask students what comes to mind when they here the word *Europe*. Encourage them to identify the romantic images often associated with the continent—castles, monarchs, and so on. Record their responses on the chalkboard. Without calling attention to it, intersperse student responses with terms like *acid rain* and *nuclear contamination*. When students object, point out that just as your terms spoiled the students' images, pollution is spoiling Europe.

TEACH

Organize students into three "environmental research groups" and assign each group one of the following topics: European land pollution, European air pollution, European water pollution. Have groups conduct research to identify the major threats in their areas to Europe's environment. Each group should present its findings by compiling a report, creating a special-purpose map, or giving an oral presentation. **L2**

NATIONAL GEOGRAPHIC SOCIETY
EYE ON THE ENVIRONMENT

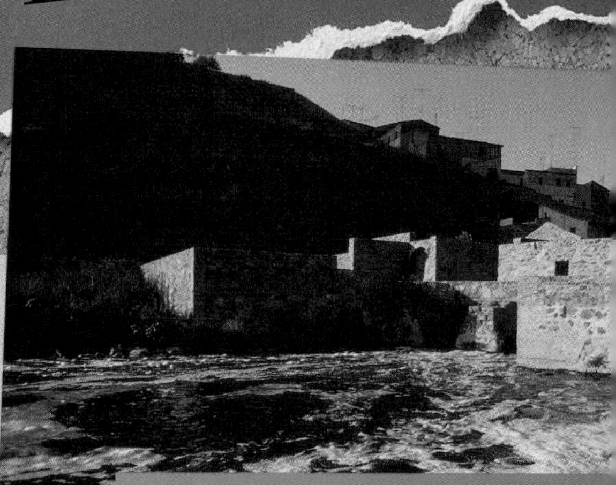

A layer of suds pollutes the Tagus River, running beneath a stone bridge in Toledo, Spain.

Europe
POLLUTION

PROBLEM

Air, water, and soil pollution do not respect national boundaries. Smoke belching from one country's smokestacks becomes another country's acid rain or smog. Europe is one of the world's most polluted continents. Smoke from British factories kills forests in Scandinavia. Raw sewage and industrial toxins in the Baltic Sea spoil beaches and kill fish, birds, and other wildlife. Pollutants dumped into the Danube River end up in the drinking water of eight nations.

SOLUTIONS

- European Union members have pledged to reduce air pollution by 90 percent by the year 2005.
- Britain has established "protection zones" limiting the use of pesticides in certain areas.
- In the Czech Republic, scientists have developed a "clean coal" that burns without releasing sulfur dioxide—the cause of acid rain.
- In France, taxes on industries are being used to install machines that sniff the air. They sound a warning when pollution is detected.
- In Greece and Italy, some cities allow cars and buses to operate only on certain days.

POLLUTION FACT BANK

- In 100 European cities, air pollution is 10 times greater than the standards set by the World Health Organization.

- Air pollution and acid rain have laid waste to many German forests. In the Czech Republic and Slovakia, nearly a million acres of woodland have been damaged by acid rain.

- Most of the water in Poland's rivers is undrinkable, and 50 percent of river water is so toxic it corrodes industrial machinery.

- In Hungary, air pollution causes 1 in 17 deaths. An hour's stroll through Budapest's polluted streets is as bad for the lungs as smoking 20 cigarettes!

342

UNIT 4

More About the Issues

Human/Environment Interaction By far, the most frighteningly polluted region of Europe is eastern Europe. The fall of the Iron Curtain revealed a nightmare of land, air, and water pollution—a legacy of Communist rulers who valued economic progress over environmental protection. The extent of the damage is incredible: In one Bulgarian town, snow fell black with pollution. In one east German city, 80 percent of children under the age of seven have lung or heart diseases. In many areas of eastern Europe, drivers use headlights during the day to see through the thick pollution. Eastern Europe is afflicted with nearly every known kind of pollution. Unfortunately, environmental cleanup activities are not a high priority in many eastern European countries.

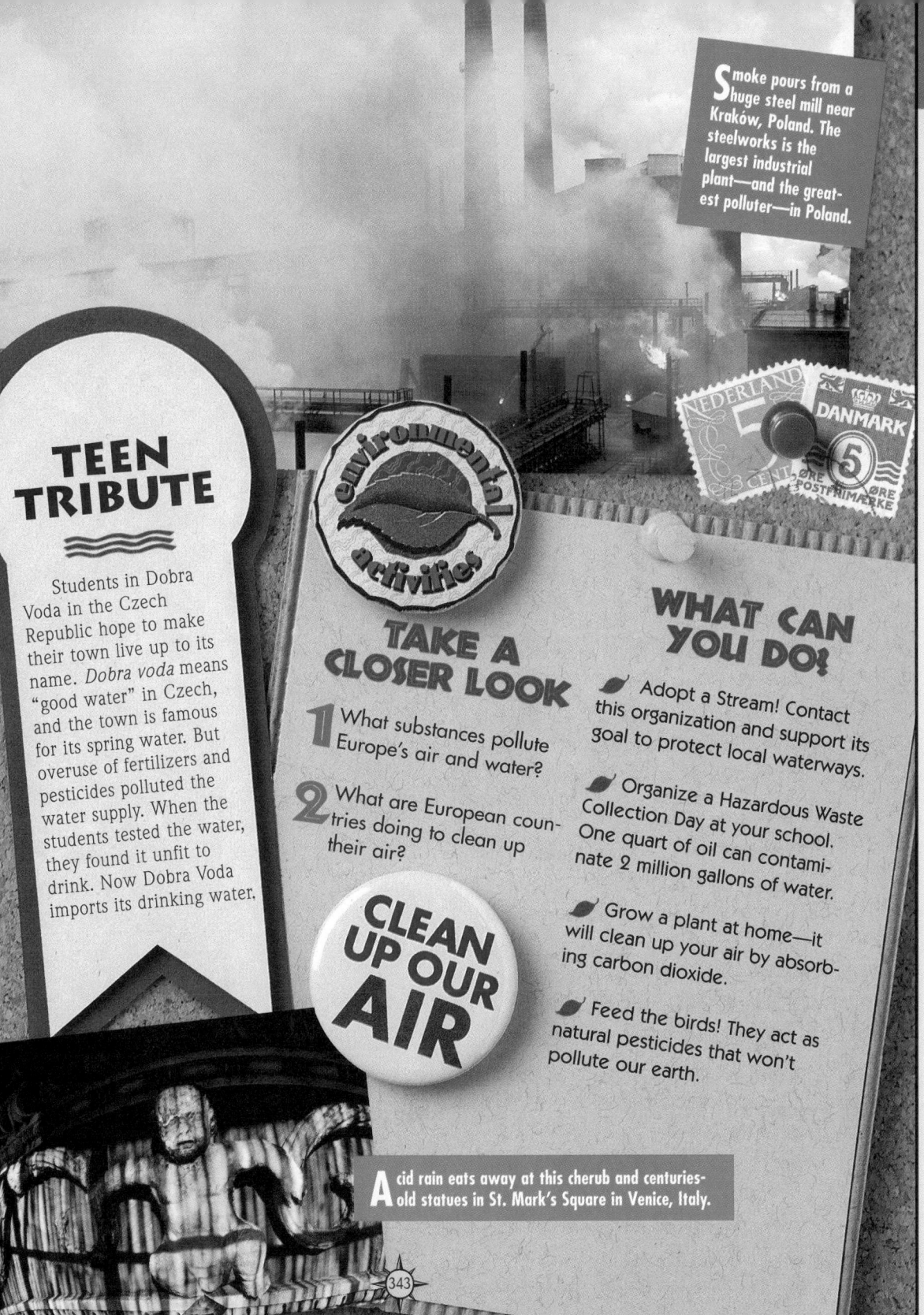

Smoke pours from a huge steel mill near Kraków, Poland. The steelworks is the largest industrial plant—and the greatest polluter—in Poland.

TEEN TRIBUTE

Students in Dobra Voda in the Czech Republic hope to make their town live up to its name. *Dobra voda* means "good water" in Czech, and the town is famous for its spring water. But overuse of fertilizers and pesticides polluted the water supply. When the students tested the water, they found it unfit to drink. Now Dobra Voda imports its drinking water.

environmental activities

TAKE A CLOSER LOOK

1 What substances pollute Europe's air and water?

2 What are European countries doing to clean up their air?

CLEAN UP OUR AIR

WHAT CAN YOU DO?

🍃 Adopt a Stream! Contact this organization and support its goal to protect local waterways.

🍃 Organize a Hazardous Waste Collection Day at your school. One quart of oil can contaminate 2 million gallons of water.

🍃 Grow a plant at home—it will clean up your air by absorbing carbon dioxide.

🍃 Feed the birds! They act as natural pesticides that won't pollute our earth.

Acid rain eats away at this cherub and centuries-old statues in St. Mark's Square in Venice, Italy.

343

Answers to Take A Closer Look

1. smoke, raw sewage, industrial toxins

2. EU members have pledged to reduce air pollution by 90 percent; Britain has established "protection zones" where the use of pesticides is limited; Czech scientists have developed coal that does not release sulfur dioxide when burned; France uses taxes on industries to install machines that monitor pollution levels; some cities in Greece and Italy allow cars and buses to operate only on certain days.

0:00 OUT OF TIME?

If time does not permit teaching each chapter, you may use the Chapter Highlights, Audiocassettes, and their corresponding activities and tests.

GLENCOE
TECHNOLOGY

 VIDEODISC

Reuters Issues in Geography

Chapter 4
Russia: Managing Resources
Disc 1 Side A

*inter***NET**
CONNECTION

Russia is the world's largest country in area. Use "The Russian Page" at the following address to locate exciting information about this immense land:

World Wide Web
http://www.public.iastate.edu/~andrei/R/index.html

What is it like to live in Ukraine, the Caucasus, or Central Asia? Locate resources for these areas at:

World Wide Web
http://www.city.net/regions/europe/commonwealth_of_independent_sta/

Unit 5 Russia and the Independent Republics

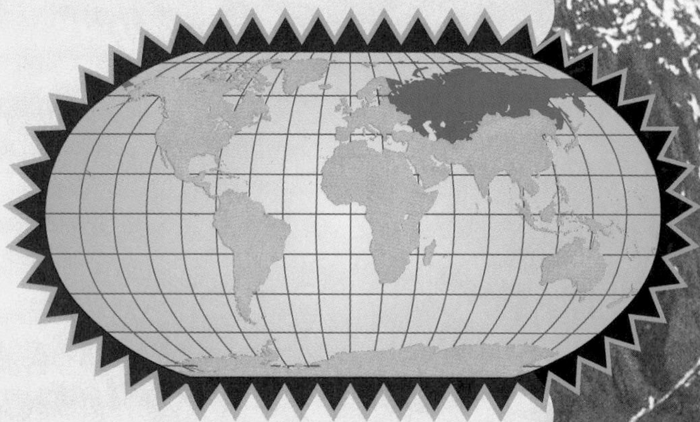

What Makes Russia and the Independent Republics a Region?

The people share . . .

- a vast land—8.5 million square miles (22.1 million sq. km)!

- very diverse cultures within republics.

- geographic features such as permafrost, mountains, and long distances between towns and cities.

- the change to a capitalist economy.

To find out more about Russia and the independent republics, see the Unit Atlas on pages 346–357.

Caucasus Mountains, Kazbeji, Georgia

About the Illustration

The Caucasus Mountains extend through Russia, Georgia, and Azerbaijan, forming a natural boundary between Europe and Asia. There are more than 2,000 glaciers in the Caucasus range. Many are larger than those found in the Alps. Mount Elbrus, Europe's highest peak, is located in the Caucasus Mountains.

 **MOVEMENT** What effect do you think the Caucasus Mountains would have had on the movement of people into and out of this area? *(They would serve as a natural barrier to migration.)*

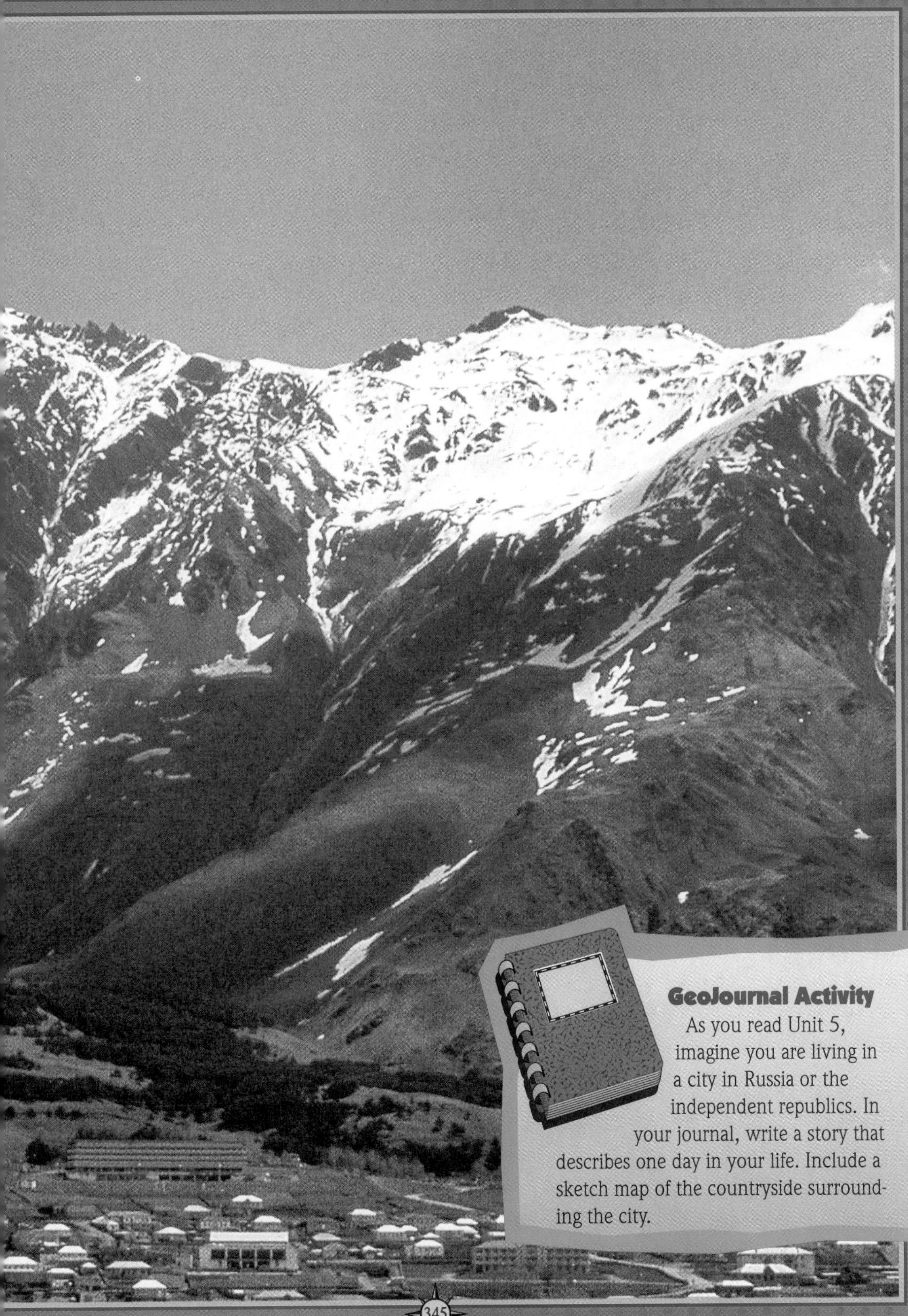

GeoJournal Activity

As you read Unit 5, imagine you are living in a city in Russia or the independent republics. In your journal, write a story that describes one day in your life. Include a sketch map of the countryside surrounding the city.

345

NATIONAL GEOGRAPHIC SOCIETY

These materials are available from Glencoe.

CD-ROM

Picture Atlas of the World

SOFTWARE

ZipZapMap! World

TRANSPARENCIES

PicturePack Transparencies, Unit 5

GeoJournal Activity

Suggest that students consult reference books, travel guides, or travel brochures to find pictures of cities in Russia or the independent republics. Students might then incorporate descriptions of buildings or monuments in their stories. Point out that the mention of actual places and/or street names will make their stories interesting and realistic. L2

- This journal activity provides the basis for the "GeoJournal Writing Activity" exercise in the Chapter Review.
- The GeoJournal may be used as an integral part of Performance Assessment.

Location Activity

Have students look at Atlas pages A18 and A19 in their texts and then ask the following questions: Which two continents does Russia occupy? *(Europe and Asia)* Which independent republic is largest in area? *(Kazakstan)* Which independent republics border Russia on the west? *(Belarus and Ukraine)* What body of water borders Russia on the north? *(Arctic Ocean)* What body of water borders Russia, Georgia, and Ukraine? *(Black Sea)* What body of water borders all of the following: Russia, Kazakstan, Turkmenistan, and Azerbaijan? *(Caspian Sea)* L1

IMAGES OF THE WORLD

UNIT 5 ATLAS

Cultural Kaleidoscope

Uzbekistan According to tradition, adult women wear their hair in 2 thick braids. Girls, however, may wear their hair in as many as 40 thin braids.

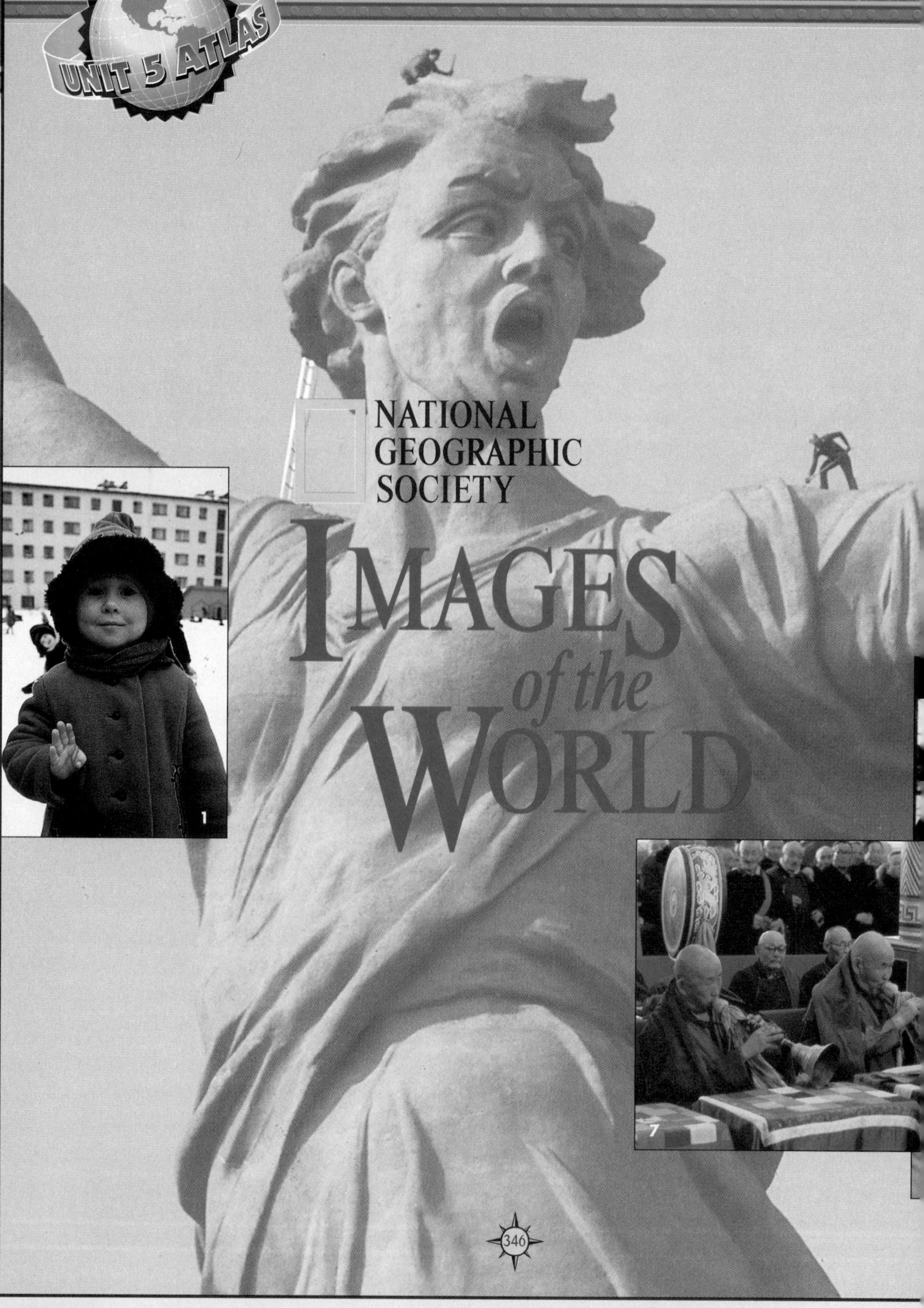

NATIONAL GEOGRAPHIC SOCIETY

IMAGES of the WORLD

Fun Facts

- **Armenia** Noah's Ark supposedly settled on Mount Ararat, which was once within the borders of Armenia. According to legend, Armenians are descended from Noah's great-great-grandson, Haik.

- **Georgia** Average life expectancy in Georgia—about 72 years—is among the highest in the region. Many Georgians live well beyond the average. In the early 1990s, for example, there were more than 20,000 Georgians aged 90 or older. Georgians attribute their good health and longevity to their country's beneficial climate.

1. **Child using sign language, Murmansk**
2. **Gold domes, St. Petersburg**
3. **Gas field workers, Siberia**
4. **Village, western Ukraine**
5. **Castle on the Black Sea, Ukraine**
6. **Open-air market, Kazan**
7. **Buddhist monks, Ulan-Ude**

All photos viewed against the statue *Motherland Calls*, Volgograd, Russia.

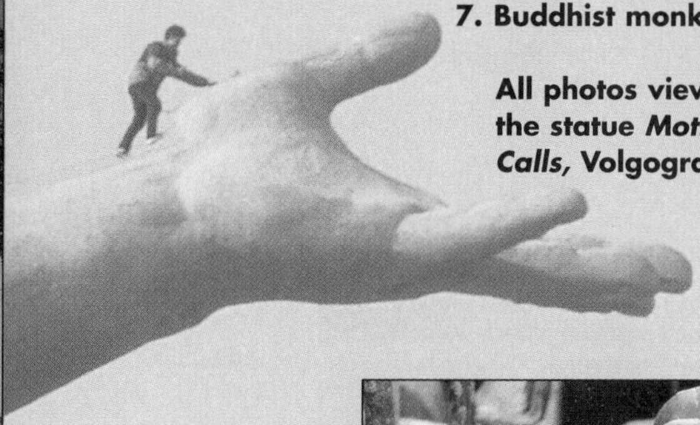

Global Gourmet

The potato is a staple of the Russian diet. It was not always so popular, however. When the government ordered peasants to plant potatoes on common ground in 1840, "anti-potato" riots broke out in 10 Russian provinces.

Geography and the Humanities

📁 World Literature Reading 5

🌐 World Music: Cultural Traditions, Lesson 4

📦 World Cultures Transparencies 7, 8

Teacher Notes

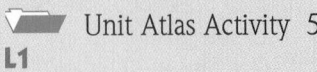

Using the Unit Atlas

These features and activities may be used as an introduction to the unit or as teaching tools throughout the course of the unit.

Unit Atlas Activity 5
L1

Unit Overlay Transparencies 5-0, 5-1, 5-2, 5-3, 5-4, 5-5, 5-6

Home Involvement Activity 5 **L1**

Environmental Case Study 5 **L1**

Countries of the World Flashcards, Unit 5

World Games Activity Cards, Unit 5

More About the Illustration

Answer to Caption
because Siberia has a harsh climate

Human/ Environment Interaction
The Russians began to develop the Urals in the eighteenth century by exploiting the area's natural resources. The area possesses incredible mineral wealth, including vast deposits of bauxite, copper, iron ore, and zinc.

Regional Focus

Russia and the other independent republics span the continents of Europe and Asia. Russia is the largest republic in area and population. The other republics in the region are Ukraine, Belarus, Moldova, Armenia, Georgia, Azerbaijan, Kazakstan, Uzbekistan, Turkmenistan, Kyrgyzstan, and Tajikistan.

Region
The Land

The region of Russia and the independent republics covers about 8.5 million square miles (22.1 million sq. km). This land area is greater than the size of Canada, the United States, and Mexico combined.

Overlooking the Ural Mountains

The Ural Mountains separate European Russia from the Asian part of Russia known as Siberia.
HUMAN/ENVIRONMENT INTERACTION: Why do you think most Russians live west of the Urals?

The North The far north of Russia is made up of low hills. Snow and ice cover the land for most of the year. The few people who live in the far north either fish, hunt, or herd reindeer. Farther south you will find deep, thick forests. In some areas farmers have cleared the woods to grow crops.

South of the forests is a vast open plain. It stretches more than 3,000 miles (4,800 km) from eastern Europe to Asia. For centuries routes across this plain brought invading armies. Today it is an important agricultural and industrial area for Russia, Ukraine, Moldova, and Belarus. The plain's fertile soil is among the best in the world for farming.

Mountains Mountains cover many areas in Russia and the independent republics. You will find two mountain ranges in areas where the continents of Europe and Asia meet. The Ural Mountains divide Russia into European and Asian parts. To the south, the Caucasus range rises along the borders of Russia, Georgia, and Azerbaijan. Find these mountains on the map on page 352.

Mountains also cross through several republics in central Asia. The Pamirs in Tajikistan have some of the region's highest peaks. The Tian Shan range in Kyrgyzstan holds some of the world's largest glaciers.

Bodies of Water Russia and the independent republics contain many bodies of water. You will find two of the world's largest lakes here. One is the Caspian Sea, which is really a salt lake. The other is Lake Baikal, the deepest lake in the world. The region also has long rivers. Some of the rivers flow eastward, like the Amur, which forms Russia's border with China. Others, like the Volga and the Dnieper, flow south through plains. Many that flow north, such as the Lena, Ob, and Yenisey, are frozen much of the year.

348

UNIT 5

Fun Facts

- **Kazakstan** Some nomads of Kazakstan live in *yurts*—round felt tents built on a hoop frame with about 60 interlocking poles. Even with all these parts, the yurt can be quickly taken apart or put together.

- **Russia** Moscow's Metro carries more than 7 million passengers a day. Its 140 miles (225 km) of track link about 140 stations. These subway stations, which are the fanciest in the world, are decorated with chandeliers, marble panels, stained glass, paintings, and statues.

Region
Climate and Vegetation

The far northern location of Russia and the independent republics affects their climates. Most of the republics lie far from great bodies of water that keep land temperatures mild. Also, the region's flat lands do not provide shelter from hot summer winds and freezing winter storms.

Climate Regions Far northern areas have tundra and subarctic climates with cold temperatures most of the year. The plains have a humid continental climate, with short, hot summers and long, cold winters. Farther south, desert regions in central Asia have long, hot summers and short, cold winters. In the area along the Black Sea, you will find humid subtropical and Mediterranean climates. Winters there are short and rainy, while the summers are hot and dry.

Vegetation The region has as many kinds of vegetation as it has climates. In the north, most of the deep soil is always frozen. Only low grasses, reeds, and mosses grow there. Farther south, forests stretch to the plains area. Grasses cover much of the plains. In the central Asian deserts, you will notice very little plant life, except for small shrubs. Along the Black Sea, however, subtropical plants and citrus fruits grow well.

Vineyards in Moldova

Enough wine is produced from Moldova's vineyards to export to Russia and other independent republics.
PLACE: What crops grow well along the Black Sea?

Region
The Economy

For many years, Russia and the independent republics formed one country called the Soviet Union. It had an economy planned and run by the communist leaders. In 1991 each republic became independent and responsible for its own economy. The republics now allow people to run their own businesses and farms. This change to free enterprise has been a challenge for the people in the region.

349

Strange but TRUE!

In northern Siberia the permafrost layer is about 5,250 feet (1,600 m) thick.

More About the Illustration

Answer to Caption subtropical plants and citrus fruits

Place As well as wine grapes, Moldovan farmers grow grain, sugar beets, tobacco, and various vegetables. Much of the country's industry is agriculture-related. Food processing, sugar refining, and winemaking, for example, are among Moldova's leading economic activities.

Fun Facts

- **Kyrgyzstan** The folklore of Kyrgyzstan is preserved in the *Manas*, the longest oral chronicle of its kind. It tells the adventures of the folk hero Manas the Strong, including how he defended his people against many invaders.

- **Russia** Fully one-third of the world's forests lie across northern Russia in a huge expanse known as the taiga. It covers an area roughly 33 times larger than the Amazon rain forest.

LESSON PLAN
Unit 5 Atlas

CURRICULUM CONNECTION

Life Science Sturgeon eggs are used to make caviar, which is one of Russia's most valuable products. Dams on the Volga River prevented sturgeon from migrating to their normal spawning ground to deposit their eggs. As a result, the government set up fish hatcheries to replenish the supply of sturgeon.

More About the Illustration

Answer to Caption
the Slavs

Place
Samarkand, an important industrial city in Uzbekistan, has been a Muslim city since the A.D. 800s, when it was overrun by Arab invaders. Tashkent, Uzbekistan's capital, is considered the center of the Islamic faith in Central Asia.

Shopping in Moscow

The GUM department store—Russia's largest shopping mall—sells everything from coats to caviar *(above).* Blue-tiled mosques line the central square of the historic city of Samarkand *(right).*
REGION: What is the largest ethnic group in Russia and the independent republics?

Agriculture Farmers in Russia, Ukraine, Belarus, and Moldova grow mainly wheat, rye, oats, barley, and sugar beets. Cotton is an important agricultural crop in the central Asian republics. Climate, however, makes farming difficult in many areas of Russia and the independent republics. Some areas are too cold for growing crops. Others do not get enough rain.

Industry Russia and the independent republics are rich in minerals such as coal, oil, natural gas, copper, silver, manganese, and gold. Dams on rivers in Russia and Ukraine supply large amounts of electric power to factories and cities. Many industrial areas are linked by rivers, roads, and railroads stretching through Russia and Ukraine. Industry and growth under Soviet rule, however, led to widespread pollution of the air and water.

Place
The People

About 285 million people live in Russia and the independent republics. Russia has the region's largest population with 148 million people.

Ethnic Groups Each of the republics has a major ethnic group, language, and culture. There are also many smaller groups in each republic. More than 100 different ethnic groups live throughout the entire region.

The largest ethnic group in the region is the Slavs. Russians, Ukrainians, and Belarussians belong to this group. Most Slavs practice Eastern Orthodox Christianity but speak Slavic languages. The peoples of central Asia—Uzbeks, Kazaks, and Turkmenis—belong to the Turkic ethnic group. They practice the religion of Islam and have their own languages and cultures.

Mosques in Samarkand, Uzbekistan

Facts

- **Kazakstan** The Kazaks preserve many traditions. Families sit on carpets and eat from low tables. Almost everyone can play the *domras*—a two-stringed musical instrument.

- **Uzbekistan** Bukhara and Samarkand were once stopping places for caravans traveling along the Silk Road. Today they are important industrial cities in Uzbekistan.

The people of Armenia and Georgia belong to various ethnic groups known as Caucasian. They practice their own forms of Orthodox Christianity.

Population Climate and landscape affect where people live in Russia and the independent republics. You will find that most of the population lives west of the Urals, where the climates are mild and the land is fertile. In the past most people lived in the countryside. Today most are city dwellers. The three largest cities are Moscow, St. Petersburg, and Kiev.

History Russia and the independent republics have a rich history. Centuries ago groups of Slavs set up city-states along rivers in present-day Ukraine and Russia. By the 1400s powerful rulers called czars ruled Russia. Over several centuries, the Russians took over all of the region that is today Russia and the independent republics. Their territory was called the Russian Empire.

In 1917 the rule of the czars came to an end. A communist dictatorship emerged and the Russian Empire became the Union of Soviet Socialist Republics. The Communists made the Soviet Union an industrial power but denied the people basic freedoms. In 1991 the Communists fell from power, and the separate parts of the Soviet Union became independent republics. Since 1994, the Chechen people have tried to separate from Russia.

The Arts The arts of Russia and the independent republics include architecture, painting, music, and dance. Each republic has its own rich heritage. You probably have seen Russia's onion-domed churches and heard the classical music of Peter Tchaikovsky and other Russian composers. Ukraine is known for its lively folk music and colorfully decorated Easter eggs. Ancient churches with drumlike tops and bells dot the rugged countryside of Armenia and Georgia. In the central Asian republics, you will see beautiful tiles in swirling patterns decorate Islamic mosques.

Technology Activity

Using a Database
Create a database with a separate index card form for each of the republics. For each republic, create fields for four questions: one for land area, one for population and major ethnic groups, one for major cities, and one for major bodies of water. Make a list of the information in the forms. Using the list, reorganize the information in the database to prepare five categories of questions to ask a classmate.

REVIEW AND ACTIVITIES

1. **REGION** How large is Russia and the independent republics?
2. **PLACE** What climate is found in the plains area of Russia and Ukraine?
3. **HUMAN/ENVIRONMENT INTERACTION** Name three natural resources mined in Russia and the independent republics.
4. **LOCATION** Where do most of the region's people live?
5. **REGION** What religion is practiced in the central Asian republics?
6. **PLACE** What political group ruled the region from 1917 to 1991?

351

Russia covers such a great distance from east to west that it stretches across 11 time zones. When it is 11:00 P.M. in Moscow and people are going to bed, in easternmost Siberia it is 9:00 A.M. the following morning and people have already begun their work or school day!

Answers to Review and Activities

1. about 8.5 million square miles (22.1 million sq. km)
2. humid continental, with short, hot summers, and long, cold winters
3. Answers may include coal, oil, natural gas, copper, silver, manganese, and gold.
4. in the area west of the Urals, where the climates are mild and the land is fertile
5. Islam
6. the Communists

Physical Geography

LESSON PLAN
Unit 5 Atlas

Applying Geographic Themes

Region Russia has more land than any other country on Earth. Have students use the map on this page and maps in the atlas to discover which landform separates Russia's European and Asian sections. *(Ural Mountains)*

GLENCOE
TECHNOLOGY

 VIDEODISC

Reuters Issues in Geography

**Chapter 4
Russia: Managing Resources**
Disc 1 Side A

 **MAP STUDY**

Answers
1. Sayan and Yablonovy
2. Kara Kum

Map Skills Practice
Reading a Map What peninsula is bordered by the Bering Sea on the north and the Sea of Okhotsk on the south? *(Kamchatka)* Into what sea do the Ural and Volga rivers flow? *(Caspian)*

RUSSIA AND THE INDEPENDENT REPUBLICS: Physical

 Map Study

1. **LOCATION** What two mountain ranges border the Central Siberian Plateau on the south?
2. **LOCATION** What large desert is found in Turkmenistan?

UNIT 5

Atlas Activity

Location Have students write a description of the location of each feature listed under "GeoFacts" on page 353. Instruct students that their descriptions should be based on relative location and should be as specific as possible.

Offer the following example: Communism Peak is located in northeastern Tajikistan, near the border with Kyrgyzstan. Have students take turns locating the features on a map based on a classmate's description. **L1**

352

ELEVATION PROFILE: Russia

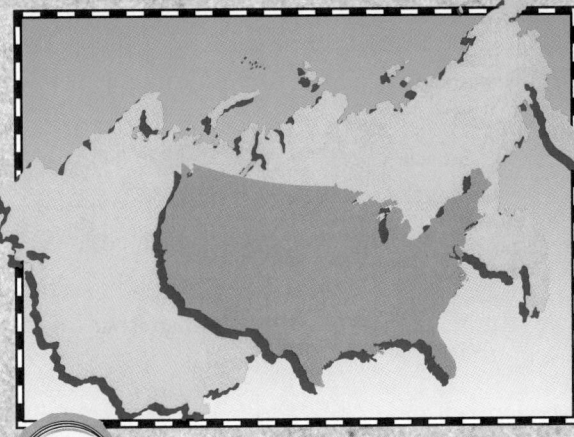

10,000 ft. — 3,048 m

North European Plain

West Siberian Plain

East Siberian Uplands

Ural Mountains

Central Siberian Plateau

Sea of Okhotsk

5,000 ft. — 1,524 m

Pacific Ocean

2,000 ft. — 610 m
1,000 ft. — 305 m
Sea level — Sea level

← West to East at 60°N latitude →

Source: *Goode's World Atlas*, 19th edition

GeoFacts

Highest point: Communism Peak (Tajikistan) 24,590 ft. (7,495 m) high

Lowest point: Caspian Sea (Russia, Azerbaijan) 92 ft. (28 m) below sea level

Longest river: Ob-Irtysh 3,362 mi. (5,409 km) long

Largest lake: Caspian Sea (Eurasia) 143,244 sq. mi. (371,002 sq. km)

Deepest lake: Lake Baikal (Russia) 5,315 ft. (1,620 m) deep

Largest desert: Kara Kum (Turkmenistan) 120,000 sq. mi. (310,800 sq. km)

RUSSIA/INDEPENDENT REPUBLICS AND THE UNITED STATES: Land Comparison

 Graphic Study

1. **LOCATION** Where are the highest elevations found in Russia?
2. **REGION** About how much larger in area are Russia and the independent republics than the continental United States?

 353

 **NATIONAL GEOGRAPHIC SOCIETY**

These materials are available from Glencoe.

 VIDEODISC

STV: World Geography Vol. 1: Asia and Australia

Caspian Sea
Side 1, Frames 05266-05811

 **SOFTWARE**

ZipZapMap! World

POSTERS

Eye on the Environment Posters, Unit 5

FactsOnFile

Use the reproducible masters from **GEOGRAPHY ON FILE**, *Commonwealth of Independent States,* to enrich unit content.

GRAPHIC STUDY

Answers
1. Ural Mountains
2. Russia and the independent republics are around three times larger than the continental United States.

Graphic Skills Practice
Reading a Graph What two Russian plains have the same elevation?
(North European Plain, West Siberian Plain)

Teacher Notes

_____   _____
_____   _____
_____   _____
_____   _____
_____   _____
_____   _____

353

LESSON PLAN
Unit 5 Atlas

UNIT 5 ATLAS

Cultural Geography

Applying Geographic Themes

Location Refer students to the physical map on page 352 and remind them that the Ural Mountains and the Caucasus Mountains serve as the dividing line between Europe and Asia. Then ask students to name the national capitals that are European cities. *(Minsk, Chisinau, Kiev, Moscow, Tbilisi, Yerevan, Baku, and Ashkhabad)*

NATIONAL GEOGRAPHIC SOCIETY

These materials are available from Glencoe.

 VIDEODISC

STV: World Geography

Vol. 1: Asia and Australia

Interior and North Asia
Side 1, Frames 13631-17712

 SOFTWARE

ZipZapMap! World

RUSSIA AND THE INDEPENDENT REPUBLICS: Political

ARCTIC OCEAN

EUROPE

Barents Sea

Bering Sea

Arctic Circle

Minsk • BELARUS
Chisinau • Kiev
Moscow

UKRAINE
MOLDOVA

R U S S I A

Sea of Okhotsk

Black Sea

GEORGIA

Lake Baikal

Tbilisi
Yerevan
AZERBAIJAN
ARMENIA Baku

KAZAKSTAN
Aral Sea
Lake Balkhash
UZBEKISTAN

A S I A

Sea of Japan

Caspian Sea

Bishkek • Almaty

Ashkhabad
TURKMENISTAN
Tashkent
Dushanbe
KYRGYZSTAN
TAJIKISTAN

— National boundary
✷ National capital

Lambert Equal-Area projection

PACIFIC OCEAN

0 250 500 mi.
0 250 500 km

 Map Study

1. **PLACE** What is the largest independent republic?
2. **PLACE** What is the capital of Belarus?

354

UNIT 5

MAP STUDY

Answers
1. Kazakstan
2. Minsk

Map Skills Practice
Reading a Map Which countries border the Black Sea? *(Russia, Ukraine, Georgia)*

Atlas Activity

Place Have students work in pairs or small groups to research the cultural diversity of Russia and the independent republics. Assign one country to each pair or group of students. Instruct students to use almanacs and/or world fact books to find and list the major cultural groups in their assigned country. Have students compare their lists and note which countries have the greatest diversity and/or which cultural groups are represented in the most countries. **L2**

Population: Russia and the Independent Republics

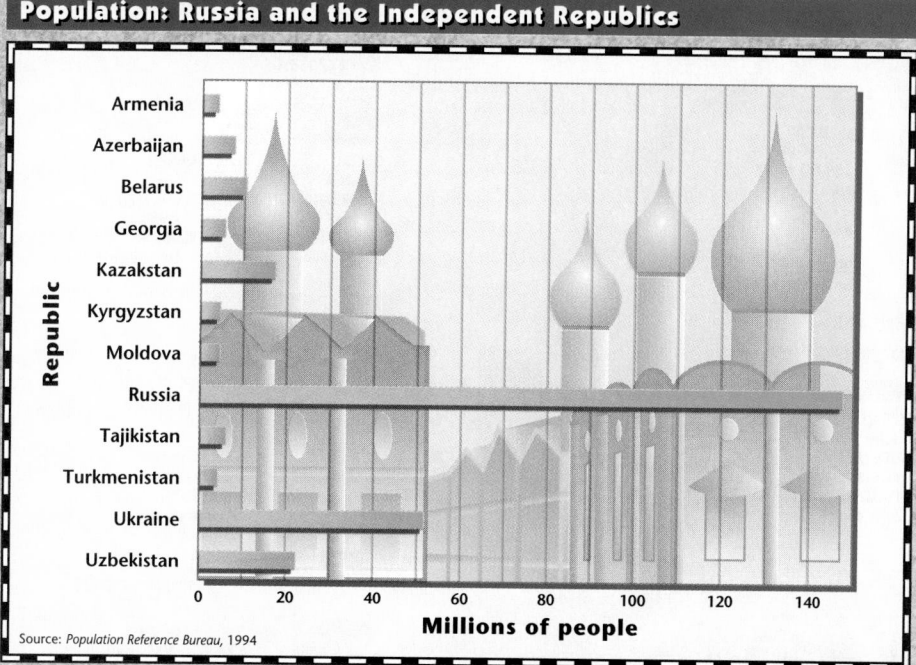

Source: *Population Reference Bureau, 1994*

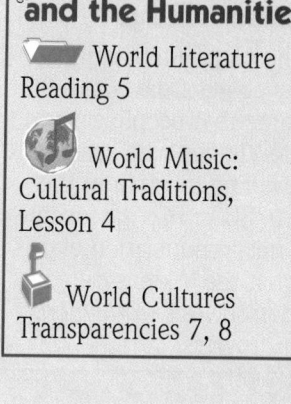

Geography and the Humanities

📁 World Literature Reading 5

🌎 World Music: Cultural Traditions, Lesson 4

🗂 World Cultures Transparencies 7, 8

COMPARING POPULATION:
Russia/Independent Republics and the U.S.

Russia/Independent Republics

United States

= 50,000,000

Source: *Population Reference Bureau, 1994*

GeoFacts

Biggest country (land area):
Russia 6,592,800 sq. mi.
(17,075,352 sq. km)

Smallest country (land area):
Armenia 11,500 sq. mi.
(29,785 sq. km)

Largest city (population):
Moscow (1995) 10,769,000
(2000 projected) 11,121,000

Highest population density:
Moldova 350 people per sq. mi.
(135 per sq. km)

Lowest population density:
Kazakstan 16 people per sq. mi.
(6 people per sq. km)

Graphic Study

1. **PLACE** What is the combined population of Russia and the independent republics?
2. **PLACE** Which country in the chart above has the second-highest population?

355

GRAPHIC STUDY

Answers
1. about 285 million people
2. Ukraine

Graphic Skills Practice
Reading a Graph What is the difference in population between Russia and the independent republics and the United States? *(Russia and the independent republics have about 25 million more people.)* What are the four smallest countries in terms of population? *(Armenia, Kyrgyzstan, Moldova, Turkmenistan)*

Teacher Notes

LESSON PLAN
Unit 5 Atlas

UNIT 5 ATLAS
Countries at a Glance

MULTICULTURAL PERSPECTIVE

Cultural Heritage

The name *Russia* probably comes from *Rus*—the name that people gave to the Vikings who started invading the area in the A.D. 800s. *Rus* is probably a mispronunciation of what the Vikings called themselves—*rothsmen,* or "rowers."

Geography and the Humanities

▭ World Literature Reading 5

◉ World Music: Cultural Traditions, Lesson 4

▮ World Cultures Transparencies 7, 8

Armenia
LANDMASS:
11,500 sq. mi./
29,785 sq. km
CAPITAL:
Yerevan
MONEY:
Ruble
MAJOR LANGUAGE(S):
Armenian
MAJOR EXPORT:
Machinery
POPULATION:
3,700,000
MAJOR IMPORT:
Machinery

Azerbaijan
LANDMASS:
33,400 sq. mi./
86,506 sq. km
CAPITAL:
Baku
MONEY:
Manat
MAJOR LANGUAGE(S):
Azeri, Turkish, Russian
MAJOR EXPORT:
Food Products
POPULATION:
7,400,000
MAJOR IMPORT:
Food Products

Belarus
LANDMASS:
80,200 sq. mi./
207,718 sq. km
CAPITAL:
Minsk
MONEY:
Belarus Ruble
MAJOR LANGUAGE(S):
Belorussian, Russian
MAJOR EXPORT:
Machinery
POPULATION:
10,300,000
MAJOR IMPORT:
Machinery

Georgia
LANDMASS:
26,900 sq. mi./
69,671 sq. km
CAPITAL:
Tbilisi
MONEY:
Ruble
MAJOR LANGUAGE(S):
Georgian, Russian
MAJOR EXPORT:
Food
POPULATION:
5,500,000
MAJOR IMPORT:
Machinery

Kazakstan
LANDMASS:
1,049,200 sq. mi./
2,717,428 sq. km
CAPITAL:
Almaty
MONEY:
Ruble
MAJOR LANGUAGE(S):
Kazak, Russian, German
MAJOR EXPORT:
Raw Materials
POPULATION:
17,100,000
MAJOR IMPORT:
Raw Materials

Kyrgyzstan
LANDMASS:
76,600 sq. mi./
198,394 sq. km
CAPITAL:
Bishkek
MONEY:
Som
MAJOR LANGUAGE(S):
Kyrgyz
MAJOR EXPORT:
Machinery
POPULATION:
4,500,000
MAJOR IMPORT:
Light-industrial Products

Moldova
LANDMASS:
14,170 sq. mi./
36,700 sq. km
CAPITAL:
Chisinau
MONEY:
Ruble
MAJOR LANGUAGE(S):
Romanian, Ukrainian
MAJOR EXPORT:
Food Products
POPULATION:
4,400,000
MAJOR IMPORT:
Machinery

Russia
LANDMASS:
6,592,800 sq. mi./
17,075,352 sq. km
CAPITAL:
Moscow
MONEY:
Ruble
MAJOR LANGUAGE(S):
Russian, Ukrainian,
Belorussian, Uzbek
MAJOR EXPORT:
Fuels
POPULATION:
147,800,000
MAJOR IMPORT:
Machinery

Tajikistan
LANDMASS:
55,300 sq. mi./
143,227 sq. km
CAPITAL:
Dushanbe
MONEY:
Ruble
MAJOR LANGUAGE(S):
Tadzhik, Russian
MAJOR EXPORT:
Aluminum
POPULATION:
5,900,000
MAJOR IMPORT:
Chemicals

Turkmenistan
LANDMASS:
188,500 sq. mi./
488,215 sq. km
CAPITAL:
Ashkhabad
MONEY:
Ruble
MAJOR LANGUAGE(S):
Turkmen, Russian
MAJOR EXPORT:
Natural gas
POPULATION:
4,100,000
MAJOR IMPORT:
Machinery

Countries not drawn to scale.

356

UNIT 5

Fun Facts

• **Kazakstan** Centuries of breeding produced the small rugged Kazak horses. They weigh about 700–800 pounds (320–360 kg). The horses' great speed made them valuable for military use.

• **Russia** Guests often bring gifts of flowers to their hosts. An odd number of flowers is given, however, since an even number is for funerals.

Ukraine

CAPITAL:
Kiev
MAJOR LANGUAGE(S):
Ukrainian
POPULATION:
51,500,000

LANDMASS:
233,100 sq. mi./
603,729 sq. km
MONEY:
Karbovanet
MAJOR EXPORT:
Machinery
MAJOR IMPORT:
Machinery

Uzbekistan

CAPITAL:
Tashkent
MAJOR LANGUAGE(S):
Uzbek
POPULATION:
22,100,000

LANDMASS:
172,700 sq. mi./
447,293 sq. km
MONEY:
Ruble
MAJOR EXPORT:
Cotton
MAJOR IMPORT:
Food

Lunch in Moscow

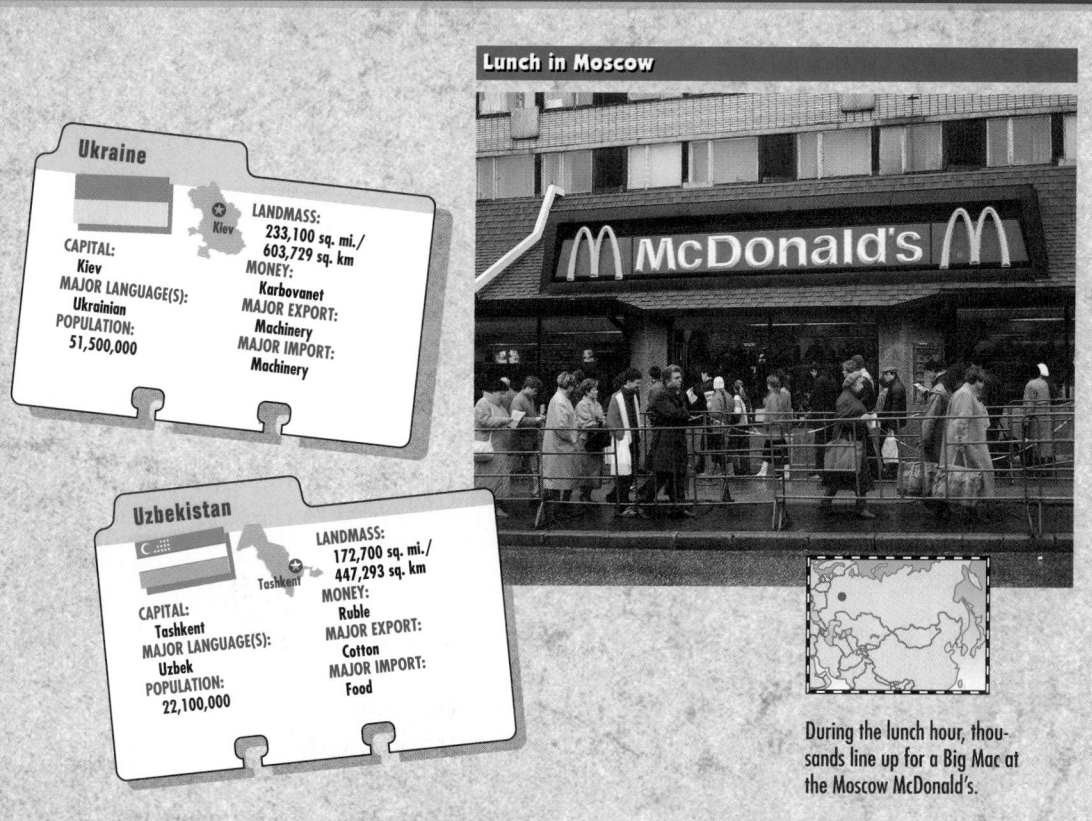

During the lunch hour, thousands line up for a Big Mac at the Moscow McDonald's.

Street Scene in Lvov, Ukraine

The city of Lvov has large squares and narrow streets much like the cities of Europe.

NATIONAL GEOGRAPHIC SOCIETY

These materials are available from Glencoe.

CD-Rom

Picture Atlas of the World

MULTICULTURAL PERSPECTIVE

Cultural Heritage

In the 1950s and 1960s, the Soviet "Virgin Lands" project converted the vast steppes of Kazakstan into cropland. Thousands of people from other parts of the region migrated there to work in the fields. Many settled there permanently.

Orchards around Almaty in Kazakstan produce apples that can weigh as much as 18 ounces (500 grams) a piece!

357

POSSIBLE RUBRIC FEATURES: Accuracy of content information, research skills, composition skills, creative thinking, analytical thinking skills, concept attainment

PERFORMANCE ASSESSMENT ACTIVITY

WRITING RUSSIAN FOLKTALES Have students work individually or in groups to find and read a number of Russian folktales. Suggest that they note the format and any common themes in the tales. Then have students write a folktale of their own that, in telling a story, describes the geographic characteristics of Russia. Students may use information from the chapter or from resources in libraries. Encourage students to create an illustrated title page for their folktales. Compile completed folktales into an anthology.

MENTAL MAPPING ACTIVITY

Before teaching Chapter 14, point out Russia on a world map or globe. Have students identify the bodies of water and countries that border Russia. Then have them locate the Ural Mountains, which separate European Russia from Asian Russia.

TEACHER'S CORNER

NATIONAL GEOGRAPHIC SOCIETY

INDEX TO NATIONAL GEOGRAPHIC MAGAZINE

The following articles may be used for research relating to this chapter:

- "Kuril Islands," by Charles E. Cobb, Jr., October 1996.
- "Searching for the Scythians," by Mike Edwards, September 1996.
- "Siberian Mummy Unearthed," by Natalya Polosmak, October 1994.
- "Crimea: Pearl of a Fallen Empire," by Peter T. White, September 1994.
- "Chornobyl," by Mike Edwards, August 1994.

- "Soviet Pollution," by Mike Edwards, August 1994.
- "A Russian Voyage," by Miles Clark, June 1994.
- "Siberian Cranes," by George Archibald, May 1994.
- "Kamchatka," by Bryan Hodgson, April 1994.
- "Russia: Playing by New Rules," by Mike Edwards, March 1993.

NATIONAL GEOGRAPHIC SOCIETY PRODUCTS AVAILABLE FROM GLENCOE

To order the following products for use with this chapter, contact your local Glencoe sales representative or call Glencoe at 1-800-334-7344:

- *STV: World Geography* (Videodisc)
- *Picture Atlas of the World* (CD-ROM)
- *Physical Geography of the World* (Transparencies)
- *ZipZapMap! World* (Software)

- *GeoBee* (Software)
- *Images of the World* (Posters)
- *Eye on the Environment* (Posters)

ADDITIONAL NATIONAL GEOGRAPHIC SOCIETY PRODUCTS

To order the following products for use with this chapter, call National Geographic Society at 1-800-368-2728:

- *Rise and Fall of the Soviet Union* (Video)
- *Russia: After the U.S.S.R.* (Video)
- *Voices of Leningrad* (Video)

- *1917: Revolution in Russia* (Video)
- *Physical Geography of the Continents Series:* "Asia," "Europe." (Video)
- *Soviet World in Transition:* "Pt. I: Geography;" "Pt. II: History." (Filmstrip)

TEACHER-TO-TEACHER
Curriculum Connection: Global Studies, Home Economics

—from Rodney Zeisig,
Boston Middle School, LaPorte, IN

RUSSIAN FOOD DAY The purpose of this activity is to have students learn about some aspects of Russian culture by having them plan and prepare a typical Russian meal.

Organize students into four groups and inform them that their task is to prepare a Russian meal. Assign one of the following dishes to each group: an appetizer, a main course, a side dish, and a dessert. Direct groups to use library resources to find a recipe for their assigned dish. Review the recipes to make sure they are appropriate. Ingredients should be relatively cheap and preparation and cooking time should be short. When all groups have an approved recipe, set a date for "Russian Food Day." Also, direct groups to provide items such as paper plates, plastic cutlery, napkins, and so on.

On Russian Food Day, have groups use facilities in the Home Economics classroom to prepare their dishes. Make sure there is enough of each dish for every student to have a small portion. As students are sampling the food, call on a representative of each group to describe the ingredients and explain the preparation techniques of his or her dish. In the next class period, lead students in a discussion of Russian foods and cooking methods.

MEETING NATIONAL STANDARDS

Geography for Life

All of the 18 standards are demonstrated in Unit 5. The following ones are highlighted in this chapter:
- Standards 4, 11, 13, 16, 18

KEY TO ABILITY LEVELS

Teaching strategies have been coded for varying learning styles and abilities.

L1   **BASIC** activities for all students

L2   **AVERAGE** activities for average to above-average students

L3   **CHALLENGING** activities for above-average students

LEP   **LIMITED ENGLISH PROFICIENCY** activities

BIBLIOGRAPHY

Readings for the Student

Diller, Daniel C. *Russia and the Independent States,* revised ed. Washington, D.C.: Congressional Quarterly, 1993.

Kort, Michael. *The Rise and Fall of the Soviet Union.* New York: Franklin Watts, 1992.

Russia: Then and Now. Minneapolis: Lerner, 1993.

Readings for the Teacher

CIS and Eastern Europe on File. New York: Facts on File, 1993.

Smith, Hedrick. *The New Russians.* New York: Avon, 1991.

Winpenny, Patricia, et al. *Teaching About the Former Soviet Union: History, Language, Culture, Art,* revised ed. Denver: Center for Teaching International Relations, 1994.

Multimedia

The Rise and Fall of the Soviet Union. Tapeworm, 1992. 2 videocassettes, 124 minutes.

Russia: After the U.S.S.R. Washington, D.C.: National Geographic Society, 1994. Videocassette, 25 minutes.

Russia and Neighboring Countries. Maplewood, N.J.: Hammond, 1992. Political wall map.

Use these *Geography: The World and Its People* resources to teach, reinforce, and extend chapter content.

CHAPTER 14 RESOURCES

*Vocabulary Activity 14

Cooperative Learning Activity 14

Workbook Activity 14

Geography Skills Activity 14

GeoLab Activity 14

Critical Thinking Skills Activity 14

Performance Assessment Activity 14

*Reteaching Activity 14

Enrichment Activity 14

World Literature Reading 5

Chapter 14 Test, Form A and Form B

*Chapter 14 Digest Audiocassette, Activity, Test

Unit Overlay Transparencies 5-0, 5-1, 5-2, 5-3, 5-4, 5-5, 5-6

Political Map Transparency 5

World Cultures Transparencies 7, 8

Geoquiz Transparencies 14-1, 14-2

Vocabulary PuzzleMaker Software

Student Self-Test: A Software Review

Testmaker Software

MindJogger Videoquiz

If time does not permit teaching the entire chapter, summarize using the Chapter 14 Highlights on page 375, and the Chapter 14 English (or Spanish) Audiocassettes. Review students' knowledge using the Glencoe MindJogger Videoquiz. *Also available in Spanish*

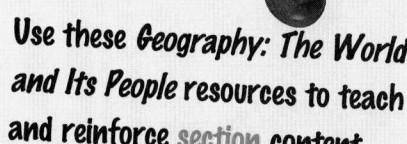

Use these *Geography: The World and Its People* resources to teach and reinforce section content.

SECTION 1 RESOURCES

- Reproducible Lesson Plan 14-1
- Section Focus Transparency 14-1
- Guided Reading Activity 14-1
- Section Quiz 14-1 (also available in Spanish)

SECTION 2 RESOURCES

- Reproducible Lesson Plan 14-2
- Section Focus Transparency 14-2
- Guided Reading Activity 14-2
- Section Quiz 14-2 (also available in Spanish)

SECTION 3 RESOURCES

- Reproducible Lesson Plan 14-3
- Section Focus Transparency 14-3
- Guided Reading Activity 14-3
- Section Quiz 14-3 (also available in Spanish)

ADDITIONAL RESOURCES FROM GLENCOE

Reproducible Masters

- Facts On File: MAPS ON FILE I, Asia: GEOGRAPHY ON FILE, Commonwealth of Independent States
- Foods Around the World, Unit 5

 ### Workbook

- Building Skills in Geography, Unit 4, Lesson 6

 ### World Music: Cultural Traditions
Lesson 4

 ### Transparencies

- National Geographic Society PicturePack Transparencies, Unit 5

 ### Posters

- National Geographic Society: Images of the World, Unit 5

World Crafts Activity Cards Unit 5

 ### Videodiscs

- Geography and the Environment: The Infinite Voyage
- Reuters Issues in Geography
- National Geographic Society: STV: World Geography, Vol. 1

CD-ROM

- National Geographic Society: Picture Atlas of the World

Software

- National Geographic Society: ZipZapMap! World

Performance Assessment

Refer to the Planning Guide on page 358A for a Performance Assessment Activity for this chapter. See the *Performance Assessment Strategies and Activities* booklet for suggestions.

Chapter Objectives

1. **Examine** the location, landforms, and climates found in Russia.
2. **Compare** Russia's economy under communism with Russia's economy today.
3. **Describe** cultural influences in Russia.

GLENCOE TECHNOLOGY

VIDEODISC

Use the Chapter 14 MindJogger Videoquiz to preview chapter content.

MindJogger Videoquiz

Chapter 14
Disc 2 Side B

 Also available in VHS.

Chapter 14 Russia

MAP STUDY ACTIVITY

As you read Chapter 14, you will learn about the land, people, and history of Russia.

1. In what part of Russia is its capital located?
2. What sea borders Vladivostok?
3. About how many miles (km) separate Moscow and St. Petersburg?

Map legend:
- National boundary
- ⊛ National capital
- • Other city

Lambert Equal-Area projection

358

MAP STUDY ACTIVITY

Answers
1. western part
2. Sea of Japan
3. about 450 miles (725 km)

Map Skills Practice
Reading a Map What is the westernmost Russian city marked on the map? *(St. Petersburg)*

SECTION 1
The Land

PREVIEW

Words to Know
• tundra
• taiga
• permafrost

Places to Locate
• North European Plain
• West Siberian Plain
• Central Siberian Plateau
• East Siberian Uplands
• Ural Mountains
• Volga River
• Caspian Sea
• Lake Baikal
• Kamchatka Peninsula

Read to Learn . . .
1. where Russia is located.
2. what landforms are found in Russia.
3. what two major climates Russia has.

Brown bear, elk, and fox roam this "green ocean"—a vast forest area that stretches across the northern part of Russia. Few people visit or live here, however. Why? Although summers are warm, snow covers the ground for as long as eight months a year!

If you had to describe Russia in one word, that word would be "BIG!" Russia is the largest country in the world in area. Its almost 6.6 million square miles (17,075,400 sq. km) are spread across two continents—Europe and Asia. If you crossed Russia by train, it would take about one full week of travel and you would pass through 11 time zones!

Location
A Northern Country

The map on page 358 shows you two important facts about Russia—its far northern location and its isolation. Russia's longest coastline faces the Arctic Ocean. Along the Arctic coast, ice makes shipping difficult or impossible most of the year. Although Russia does have ports on the Baltic Sea, the Black Sea, and the Pacific Ocean, many of these are frozen for several weeks each winter.

CHAPTER 14

359

Classroom Resources for Section 1

 REPRODUCIBLE MASTERS

Reproducible Lesson Plan 14-1
Vocabulary Activity 14
GeoLab Activity 14
Guided Reading Activity 14-1
Geography Skills Activity 14
Building Skills in Geography, Unit 4, Lesson 6
Section Quiz 14-1

 **TRANSPARENCIES**

Section Focus Transparency 14-1
Unit Overlay Transparencies 5-0, 5-3
Political Map Transparency 5

MULTIMEDIA

 Vocabulary PuzzleMaker Software

Testmaker

National Geographic Society: STV: World Geography, Vol. 1

LESSON PLAN
Chapter 14, Section 1

Meeting National Standards

Geography for Life The following standards are highlighted in this section:
• Standards 3, 4, 7

FOCUS

Section Objectives

1. **Locate** Russia on a map or globe.
2. **Describe** the landforms found in Russia.
3. **Identify** the major climates Russia has.

Bellringer Motivational Activity

Prior to taking roll at the beginning of the class period, project Section Focus Transparency 14-1 and have students answer the activity questions. Discuss students' responses. **L1**

Vocabulary Pre-check

Have students study the term *permafrost*. Ask them to speculate on its meaning. Then have them find the term in text to check the accuracy of the definitions. **L1** **LEP**

Use the Vocabulary PuzzleMaker Software to create a crossword puzzle. **L1**

Global Gourmet

Many Russian dishes have become popular in other parts of the world. These include *blinis* (thin pancakes filled with smoked salmon and sour cream) and *beef stroganoff* (beef strips cooked with onions and mushrooms in sour cream sauce).

359

TEACH

Guided Practice

Project Unit Overlay Transparency 5-3. Have students identify the kinds of climate found in Russia. *(subarctic and humid continental)* Ask students to speculate on how each type of climate might affect ways of life in Russia. Encourage students to support their ideas. Call on volunteers to share their ideas with the rest of the class. **L1**

Independent Practice

Guided Reading Activity 14-1 **L1**

Cultural Kaleidoscope

Russia In the past, Russian and Soviet governments banished criminals and political prisoners to the frigid lands of Siberia. Prisoners were forced to work in factories and mines. Today, the Russian government offers high salaries and long vacations to attract workers to Siberia.

RUSSIA: Physical

ELEVATIONS

Feet	Meters
10,000	3,000
5,000	1,500
2,000	600
1,000	300
0	0

⊛ National capital
▲ Mountain peak

Lambert Equal-Area projection

MAP STUDY
Most of Russia's people live on the North European Plain.
LOCATION: On what rivers do these people depend for water and transportation?

Ural Mountains Find the Ural Mountains on the map above. Can you see why geographers use them to mark the boundary between the continents of Europe and Asia? The Ural Mountains extend for 1,500 miles (2,414 km) from the Arctic Ocean in the north to near the Aral Sea in the south. The Urals—worn by erosion—are old mountains, rising only a few thousand feet.

Plains and Highlands West of the Urals lies a vast area known as the North European Plain. About 75 percent of the Russian people live on this plain. Russia's largest cities are located here, including its capital, Moscow, and St. Petersburg, a leading industrial and cultural center. Most of Russia's industries and farms dot this region, too.

Now look east of the Urals. This is the Asian part of Russia known as Siberia. Another plain lies on this side of the mountains—the West Siberian Plain. It is the world's largest area of flat land.

As you move farther east of the West Siberian Plain, you come to the Central Siberian Plateau. It rises in elevation from the Arctic Ocean in the north to the Sayan Mountains in the south. Another highland area—the East Siberian Uplands—forms the largest region of Siberia. This wilderness of forests,

360

Cooperative Learning Activity

Have students work in small groups to create a display titled "Russia's Geographic Hall of Fame." Assign each group a characteristic of Russia that qualifies for the hall of fame. These might include: largest land area, largest area of flat land, tallest mountain in Europe, deepest lake, and coldest winters. Have groups create an appropriate plaque or monument to represent their assigned characteristics. Allow time for groups to share and discuss their work. **L1**

mountains, and plateaus runs from the Central Siberian Plateau to Russia's eastern coast. The endangered Siberian tiger, among other animals, makes its home here.

In Siberia's far north, large areas of **tundra,** or cold treeless plains, stretch along the Arctic shore. South of the tundra is a vast area of evergreen forests known as the **taiga.** It has so many trees that it is called a "green ocean." On its far eastern edges, a finger of land called the Kamchatka Peninsula stretches south between the Bering Sea and Sea of Okhotsk.

Caucasus Mountains In southwest Russia you can see the mighty Caucasus (KAW•kuh•suhs) Mountains. An extension of the Alps, the Caucasus are thickly covered with pines and other trees. Mount Elbrus in the Caucasus rises 18,510 feet (5,642 m). It is the tallest mountain on the continent of Europe.

Inland Water Areas Russia's landscape also includes many inland bodies of water. In the south Russia borders on the Caspian Sea—a saltwater lake. About the size of the state of California, it is the largest inland body of water in the world. The Caspian Sea lies 92 feet (28 m) *below* sea level.

RUSSIA: Climate

CLIMATE
High latitude
☐ Subarctic
☐ Tundra
Mid-latitude
☐ Humid continental
Dry
☐ Steppe (partly dry)

Lambert Equal-Area projection

MAP STUDY
The Arctic Circle passes through northern Russia.
PLACE: What two types of climates are found between the Arctic Circle and 80°N?

LESSON PLAN
Chapter 14, Section 1

NATIONAL GEOGRAPHIC SOCIETY

These materials are available from Glencoe.

  VIDEODISC

STV: World Geography Vol. 1: Asia and Australia

Interior and North Asia
Side 1, Frames 13631-17712

ASSESS

Check for Understanding

Assign Section 1 Review as homework or an in-class activity.

Meeting Lesson Objectives

Each objective below is tested by the questions that follow it in parentheses.

1. **Locate** Russia on a map or globe. (2)
2. **Describe** the landforms found in Russia. (4)
3. **Identify** the major climates Russia has. (6)

Meeting Special Needs Activity

Slow Readers Help slow readers work more efficiently by teaching them to scan for the location of specific information. Demonstrate the scanning procedure. Then ask students to scan the subhead "Inland Water Areas" to find out in what part of Russia the Caspian Sea is found *(southern)* and what kind of water it contains *(saltwater).* **L1**

MAP STUDY

Answer
subarctic, tundra
Map Skills Practice
Reading a Map What climate type is found in Moscow? *(humid continental)*

Evaluate

▢ Section 1 Quiz L1

◉ Use the Testmaker to create a customized quiz for Section 1. L1

Reteach

Have students brainstorm words that describe Russia. (Choices might include northern, big, cold, forested.) List words on the chalkboard. L1

Enrich

Have students make an illustrated report about animals of the tundra of Russia. L3

CLOSE

Ask students to write a geographical description of a trip across Russia from east to west. Direct them to choose a specific starting point and destination.

Now That's Cold!

In Yakutsk, a city in northeastern Siberia, temperatures have plunged as low as −108°F (−77°C).
REGION: What kind of climate prevails in northern Siberia?

Farther east is Siberia's Lake Baikal (by•KAHL). It is the deepest lake in the world—about 5,315 feet (1,620 m) deep. It also holds more water than any other freshwater lake—about 20 percent of the world's unfrozen freshwater.

Many rivers fan across Russia's landscape. Rivers in Siberia, such as the Lena (LEE•nuh), the Ob (AHB), and the Yenisey (YEH•nuh•SAY), are among the longest in the world. Most of them flow south to north into the Arctic Ocean.

Rivers in European Russia, on the western side of the Urals, are connected to one another and to the Caspian Sea by canals. This flow of river traffic has helped unify the country. European Russia's longest river—the Volga—flows 2,193 miles (3,528 km) southward from the Moscow area to the Caspian Sea.

Region
Climate

Russia's northern location near the cold Arctic Ocean results in short, warm summers and long, cold winters for most of the country. Imagine traveling west across the forested southern parts of Siberia. These areas have a subarctic climate. Summers are cool, and snow is on the ground eight to nine months each year. Average temperatures in January plunge to -45°F (-42.8°C).

The region north of the Arctic Circle has a tundra climate with perhaps the coldest winters in the world. Across much of Siberia, a permanently frozen layer of soil called **permafrost** lies beneath the ground's surface.

Temperatures warm on the North European Plain. This area has a humid continental climate. Winters are cold, but not so brutally cold as farther north and east. Summers are rainy and warm. In the city of Moscow, snow falls only five months each year, and rainfall averages 20 inches (50 cm) in the summer.

Farmers in some areas of southern Russia are not as fortunate regarding rainfall. Here rainfall is scarce. These areas have a steppe, or partly dry, climate. The natural vegetation in this area is largely grasses.

SECTION 1 REVIEW

REVIEWING TERMS AND FACTS
1. **Define the following:** tundra, taiga, permafrost.
2. **LOCATION** How does Russia's location affect its climate?
3. **REGION** In what area do most of Russia's people live?

MAP STUDY ACTIVITIES

4. Look at the physical map on page 360. What oceans and seas border Russia?
5. Turn to the climate map on page 361. How would you describe the climate in the far north of Siberia?

Answers to Section 1 Review
1. All vocabulary words are defined in the Glossary.
2. Russia is a large land mass located in the northern latitudes. As a result, it has continental and polar climates.
3. on the North European Plain
4. Arctic Ocean, Pacific Ocean, Baltic Sea, Black Sea
5. tundra climate—extremely cold for much of the year

BUILDING GEOGRAPHY SKILLS

Reading a Transportation Map

Transportation maps show how people and goods move from place to place. Lines and colors represent different types of transportation. Areas with many transportation routes through them are *accessible,* or easy to reach. These areas usually have more trade, industry, and population. When reading a transportation map, apply these steps:

- Read the title and the map key to find the region and the kinds of transportation shown on the map.
- Find the areas that are most and least accessible.
- Determine what kinds of transportation are most important to this region.
- Use the map to conclude how transportation affects the region.

RUSSIA: Transportation Routes

Map Key
- ——— National boundary
- - - - Sea route
- +—+ Railroad
- ——— Trans-Siberian Railroad
- ⊛ National capital
- ● Other city
- ⚓ Seaport

0   250   500 mi.
0   250 500 km

Lambert Equal-Area projection

Geography Skills Practice

1. What region and kinds of transportation are shown on this map?
2. Which major railroad extends to the far eastern part of Russia?
3. Which parts of Russia are most accessible? Least accessible?

TEACH

Have students identify transportation routes that link your city or town to the rest of the country and beyond. Tell students to consider highways and roads, railroads, seaports, inland waterways, and airports. Discuss your community's accessibility and compare it to accessibility of other towns in your region. Then ask students to explain how accessibility affects a region's economy. *(Producers can bring goods to consumers, buyers can get access to goods, services develop to support trade.)* Have students read the skill and complete the practice questions. **L1**

Additional Skills Practice

1. What kinds of transportation are found in central and northern Russia? *(rivers)*
2. What is the major seaport on the Sea of Japan? *(Vladivostok)*

▭ Geography Skills Activity 14

▭ Building Skills in Geography, Unit 4, Lesson 6

Answers to Geography Skills Practice

1. Russia; rivers, railroads, sea routes
2. Trans-Siberian Railroad
3. Western and southern Russia are most accessible; eastern and northern Russia are least accessible.

363

SECTION 2
The Economy

FOCUS

Section Objectives

1. **Explain** what Russia's economy was like under communism.
2. **Describe** how Russia's economy has changed in recent years.
3. **Identify** the economic regions of Russia.

Bellringer Motivational Activity

Prior to taking roll at the beginning of the class period, project Section Focus Transparency 14-2 and have students answer the activity questions. Discuss students' responses. **L1**

Vocabulary Pre-check

Write the terms *heavy industry* and *light industry* on the chalkboard and call on volunteers to offer definitions. Then name a number of products and have students classify each one under the proper heading. **L1** **LEP**

PREVIEW

Words to Know
• communism
• command economy
• heavy industry
• light industry
• free enterprise system
• consumer goods

Places to Locate
• Moscow
• St. Petersburg
• Siberia

Read to Learn . . .
1. what Russia's economy was like under communism.
2. how Russia's economy has changed in recent years.
3. what economic regions make up Russia.

You may not be able to tell by looking at it, but there is something special about this electronics factory in Russia. What is it? In the past, this same factory made weapons for the military. For years the communist Russian government told workers what to produce. Today Russians are making their own decisions about what to make and sell. They are entering the growing world economy.

If you travel across Russia, you will see people farming the land and working in factories. Others take resources from the earth, forests, and rivers. Russia is rich in natural and human resources, but its economy has gone through many ups and downs in this century.

Place
Today's Challenges

From 1922 to 1991, Russia was part of an even larger empire called the Soviet Union. The Soviet Union practiced **communism,** an authoritarian political system in which the government controls the economy. Private ownership of property is not allowed.

Government Controls The Soviet Union's Communist leaders created the world's largest command economy. In a **command economy,** the national government owns all land, resources, industries, farms, and railroads. The government also makes all economic decisions.

364

Communist leaders used five-year plans to run the Soviet economy. These plans set production goals, prices of goods, and workers' wages. Plans emphasized **heavy industry,** or the making of goods such as machinery, mining equipment, and military weapons. They gave less attention to **light industry,** or the making of such goods as clothing, shoes, furniture, and household products.

Collapse of Communism Many workers in the Soviet Union made the change from farmers to factory workers under communism. In the 1970s, however, inefficiencies in the centrally controlled communist system gradually weakened the economy. In 1985 the Soviet leader Mikhail Gorbachev (GOR•buh•CHAWF) pushed to rebuild the economy. He reduced some of the national government's rigid controls and allowed more decisions to be made at the local level. Before his plans could bring about change, however, the Soviet Union collapsed. In 1991 Russia and the other parts of the Soviet Union became independent countries.

A New Beginning Independence turned Russia and its economy upside down. The Russian people are making the painfully slow change to a **free enterprise system** like that of the United States. In a free enterprise system, businesses are privately owned, operating for profit with little government control. Today Russia's economy is still struggling. Goods are often scarce, and prices have steadily increased.

The colorful domes of St. Basil's Cathedral stand just outside the Kremlin in Moscow's Red Square.
PLACE: Why do you think the Kremlin was surrounded by a wall?

TEACH

Guided Practice

Comparing Draw a two-column chart on the chalkboard with *Command Economy* and *Free Enterprise Economy* as column headings. Ask students to make comparisons between these two economic systems. Suggest that they consider bases of comparison such as ownership of resources and industries, power to make economic decisions, and impact on workers. Enter their responses under the appropriate heading. Then have students use the chart to speculate why the Soviet Union's command economy failed. **L2**

Independent Practice

📂 Guided Reading Activity 14-2 **L1**

CHAPTER 14

 365

Cooperative Learning Activity

Organize students into four groups and assign each group one of the following economic regions: Moscow, St. Petersburg, the Volga and Urals, Siberia. Instruct groups to create a marketing plan to attract new businesses to their assigned regions. Plans should be based on the resources and economic opportunities in the region. Call on groups to present their plans. Then have the class determine which economic region has the greatest opportunity for growth. **L2**

MAP STUDY

Answer
commercial farming
Map Skills Practice
Reading a Map What resources are found in the area between the Caspian and Black seas? *(zinc, lead, coal, petroleum, natural gas)*

ASSESS

Check for Understanding

Assign Section 2 Review as homework or an in-class activity.

Meeting Lesson Objectives

Each objective below is tested by the questions that follow it in parentheses.

1. **Explain** what Russia's economy was like under communism. (2)
2. **Describe** how Russia's economy has changed in recent years. (3, 4)
3. **Identify** the economic regions of Russia. (3, 4, 5)

RUSSIA: Land Use and Resources

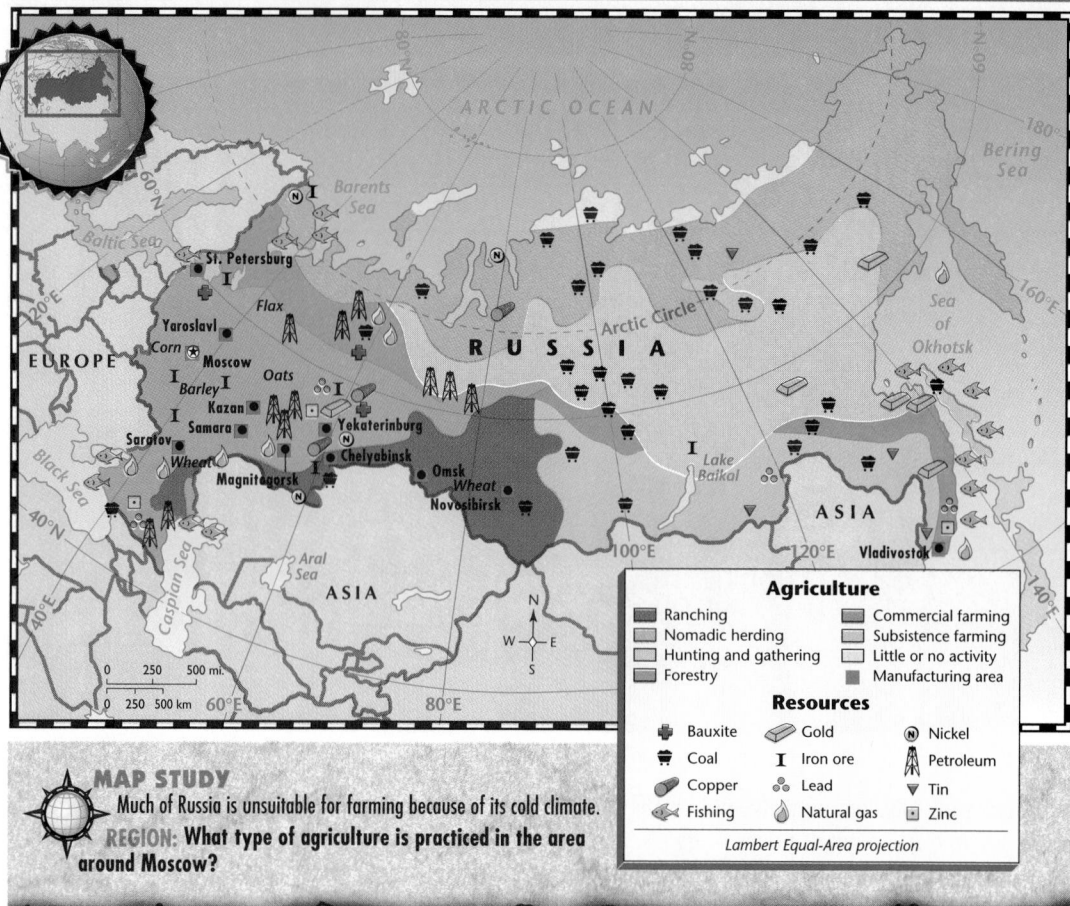

MAP STUDY
Much of Russia is unsuitable for farming because of its cold climate.
REGION: What type of agriculture is practiced in the area around Moscow?

Region
Economic Regions

Look at the map above. You see that most of Russia's industries are located west of 60°E in the European part of the country. Most farms are in this area of Russia, too. Although the country has a short growing season, its farmland is extremely productive. A dark, fertile soil spreads across part of the North European Plain into southwestern Siberia. Farmers in this area grow wheat, barley, rye, and sugar beets.

Moscow Russia's rich history shines in the city of Moscow located in the western part of Russia. About 800 years old, Moscow is the political, economic, and cultural center of Russia. This major industrial city produces textiles, electrical equipment, and automobiles. Moscow's factories recently have started to make more **consumer goods,** or household and electronics products.

Ivan Borisov, a Moscow teenager, enjoys greater freedom of choice than his parents or grandparents did. Ivan and his family—among Moscow's 10.8 million people—live in one of the city's many tall apartment houses. Ivan likes to take foreign visitors on a trip through the Metro, the Moscow subway that links every part of the city. His favorite stop is near Red Square in the center of Moscow.

St. Petersburg Northwest of Moscow lies St. Petersburg, Russia's second-largest city. Located on the Neva River near the Gulf of Finland, St. Petersburg is noted for shipbuilding. Its factories produce light machinery, textiles, and scientific and medical equipment.

St. Petersburg's 4.5 million residents take pride in their beautiful city. Built in the early 1700s, St. Petersburg spreads over about 100 islands. A network of graceful bridges and streets joins the islands.

The Volga and Urals Tucked between European Russia and Siberia lies the industrial region of the Volga River and the Ural Mountains. The Volga carries almost one-half of Russia's river traffic. It also provides water for hydroelectric power and irrigation. Both the Volga and the Urals contain large deposits of oil and natural gas. The Ural Mountains are rich in other minerals, too, such as tungsten, zinc, iron ore, and nickel.

Siberia East of the Volga and the Urals spreads Siberia, a promising economic region. Siberia has the largest supply of mineral resources in the country, including iron ore, uranium, gold, diamonds, and coal. In the Arctic region, huge deposits of oil and natural gas lie below the permafrost. About two-thirds of Siberia is covered by forests that provide lumber. In spite of this wealth in natural resources, Siberia is mostly undeveloped because of its climate and isolation. Since the late 1800s, the governments ruling Russia have built roads and railroads there to encourage development.

"Chizburger," Anyone?

The familiar hamburger chain, McDonald's, has introduced the people of Moscow to the "chizburger." Russians, like Americans, are enthusiastic burger fans. Daily, more than 50,000 people pack into the Moscow McDonald's to get lunch.

SECTION 2 REVIEW

REVIEWING TERMS AND FACTS
1. **Define the following:** communism, command economy, heavy industry, light industry, free enterprise system, consumer goods.
2. **PLACE** Before 1991, how was the Soviet Union's economy run?
3. **REGION** What are Russia's four major economic regions?
4. **HUMAN/ENVIRONMENT INTERACTION** What region has the most promising future? Why?

MAP STUDY ACTIVITIES
5. Look at the land use map on page 366. What resources are found in Siberia? What kind of industry might locate there?

Answers to Section 2 Review
1. All vocabulary words are defined in the Glossary.
2. From the Russian Revolution to 1991, the communist government ran the economy, setting production goals, prices, and wages.
3. Moscow, St. Petersburg, the Volga and Urals, Siberia
4. Siberia, because of its wealth in minerals and timber resources
5. petroleum, coal, tin, gold, nickel, copper, timber; mining, oil refining, lumber industry

Evaluate

Section 2 Quiz **L1**

Use the Testmaker to create a customized quiz for Section 2. **L1**

Reteach

Have students use the terms from the **Words to Know** list to write a paragraph explaining how Russia's economy has changed in the twentieth century. **L1**

Enrich

Have students work in groups to research Russia's major cities. Then have groups prepare visual aids for a "tour" of places of interest. **L3**

MULTICULTURAL PERSPECTIVE

Cultural Heritage
St. Petersburg's main shopping street is Nevsky Prospekt. It is named for Alexander Nevsky, a Russian prince who, in 1240, won a great victory against Sweden on the banks of the Neva River.

CLOSE

Ask volunteers to role-play Russian workers and farmers before and after the collapse of communism. Discuss how economic changes have affected their lives.

MAKING CONNECTIONS

TEACH

Ask students to name highways or industrial complexes in the area that are named for a product or industry. List suggestions on the chalkboard. Then point out that one of the world's oldest and most famous trade routes was named for a product—the Silk Road. **L1**

More About...

History One of the most important stops on the Silk Road in Central Asia was the city of Samarkand. Founded in the sixth century B.C., Samarkand quickly became an important trade center because of its location on the Silk Road. Its great wealth attracted many invading armies. Alexander the Great's soldiers destroyed Samarkand in 329 B.C. In A.D. 1219, the great Mongol leader Genghis Khan laid waste to much of the city. Another Mongol leader, Timur the Lame, or Tamerlane, invaded Samarkand in the late 1300s. He, however, did not destroy it. Rather, he made the city the capital of his empire. Samarkand continued to flourish until European explorers opened up sea routes to China. As the Silk Road's importance faded, so too did Samarkand's.

THE SILK ROAD

| MATH | SCIENCE | HISTORY | LITERATURE | TECHNOLOGY |

A road that is 2,000 years old? It sounds unbelievable. Yet stretching 7,000 miles (11,263 km) across central Asia and Russia lies the ancient Silk Road. For about 14 centuries, this road was a vital trade route between Asia and Europe. Armies, merchants, missionaries, and adventurers from both eastern and western worlds traveled this road.

THE ROUTE Beginning in northern China about 100 B.C., the Silk Road wound its way from oasis to oasis across Asia's great deserts. To the south, always in plain view, were majestic snowcapped mountains. Beyond the deserts, travelers began the terrifying climb through the Himalayas. On the other side of these mountains, the road split. One branch continued westward toward Europe, winding through what is today Russia and the independent republics. Another branch turned south into India.

THE REASON Again and again, travelers on the Silk Road braved fierce desert sandstorms, howling Himalaya blizzards, and murderous thieves. Why? The only land route between

Asia and Europe, the Silk Road made it possible for costly Chinese goods to reach Europe. The most valuable of these goods was silk—the soft, shiny fabric that gave the road its name. Wealthy Europeans were willing to pay handsomely for this luxurious cloth.

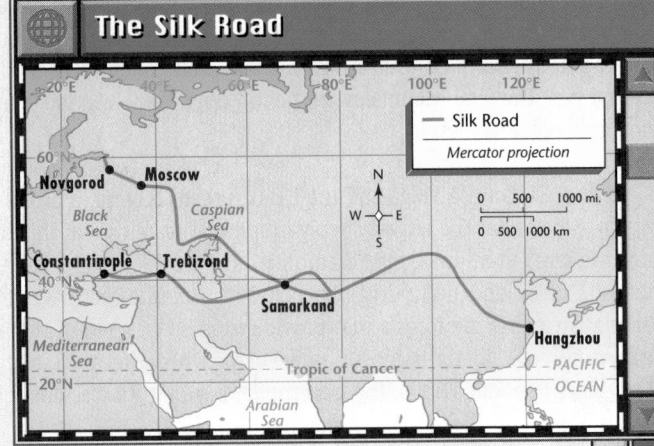
The Silk Road

THE RESULT Heavily laden camels formed caravans that transported the silk. They also carried gemstones, perfumes, tea, fine china, and gunpowder. More importantly, travelers on the Silk Road traded ideas and customs. This cultural exchange greatly enriched the peoples of central Asia, China, India, Southwest Asia, and eventually, western Europe.

Making the Connection

1. What and where was the Silk Road?
2. What products did Silk Road travelers carry from China?

368

Answers to Making the Connection

1. The Silk Road was the major overland trade route between China and Europe. Beginning in northern China, it wound its way across Asia's great deserts and through the Himalayas. The road then split. One route went west through Central Asia to Europe; the other went south into India.

2. silk, gemstones, perfumes, tea, fine china, gunpowder, as well as ideas and various aspects of culture

The People

PREVIEW

Words to Know

- czar
- serf
- cold war
- *glasnost*
- ethnic group

Places to Locate

- Muscovy
- Union of Soviet Socialist Republics (Soviet Union)
- Commonwealth of Independent States (CIS)

Read to Learn . . .

1. what groups influenced Russia's culture.
2. how city life differs from country life in Russia.
3. what makes up Russia's culture today.

These Russian teenagers are enjoying a weekend outing in Moscow. After eating a "chizburger" at a fast-food restaurant, they will explore the city. In recent years, they and other Russians have seen two amazing changes in their country—the fall of communism and the first stirrings of democracy. Like teenagers around the world, these young people hope for a better future.

Meeting National Standards

Geography for Life The following standards are highlighted in this section:
- Standards 6, 9, 13

FOCUS

Section Objectives

1. **Identify** groups that influenced Russia's culture.
2. **Contrast** city life and country life in Russia.
3. **Discuss** Russia's culture today.

Bellringer Motivational Activity

Prior to taking roll at the beginning of the class period, project Section Focus Transparency 14-3 and have students answer the activity questions. Discuss students' responses. **L1**

Vocabulary Pre-check

Have students find the meanings of the words *czar* and *serf*. Then have them identify terms with similar meanings. (For example: emperor, kaiser; servant, slave) **L1** **LEP**

Russia and the other countries that were once part of the Soviet Union have a dramatic history. Today these countries are independent republics, trying to build democracies and a better way of life for their people.

Movement

A Dramatic Past

To understand the challenges facing Russia today, let's look back through Russia's history. As an early Russian you would have belonged to a loose union of people called Eastern Slavs who dominated the area from the 900s to the 1200s. Then Mongols swept in from central Asia in the 1200s and ruled the Slavic territories in eastern Europe for 200 years. During this time, what is now Moscow became the center of a territory called Muscovy (MUHS•kuh•vee). As a citizen in the late 1400s, you would have rejoiced when Ivan III, a prince of Muscovy, drove out the Mongols and made the territory independent.

Classroom Resources for Section 3

REPRODUCIBLE MASTERS

Reproducible Lesson Plan 14-3
Guided Reading Activity 14-3
Workbook Activity 14
Critical Thinking Skills Activity 14
Reteaching Activity 14
Enrichment Activity 14
Section Quiz 14-3

TRANSPARENCIES

Section Focus Transparency 14-3
World Cultures Transparencies 7, 8

MULTIMEDIA

Testmaker

TEACH

Guided Practice

Making Comparisons
Write the headings *Country Life* and *City Life* on the chalkboard. Ask students to offer information on urban and rural life in Russia and note their responses under the appropriate heading. Then have students use the information on the chalkboard to write a script for a brief television news feature on life in Russia. Call on volunteers to "broadcast" their features for the rest of the class. **L2**

The Trans-Siberian Railroad is the longest continuous rail line in the world. Begun in the 1880s, the 5,750-mile (9,250-km) line was completed in 1916.

CURRICULUM CONNECTION

Literature In the 1970s, Russian novelist Aleksandr Solzhenitsyn wrote a study of the Soviet prison system titled *The Gulag Archipelago.* He explained the title thus: *GULAG* was the Russian acronym for the agency that ran the system, and the hundreds of prison camps stretched across the Soviet Union resembled a great chain of islands.

Teen Scene

Out With the Old

Nadia Abramov often visits the park near her family's apartment in Moscow. Sometimes she climbs on the toppled statues of former Communist leaders. The fallen statues remind her of the many changes that Russians have experienced since the breakup of the Soviet Union in 1991. Nadia is happy with most of the changes, especially those at school. She still goes to school six days a week, but she doesn't have to wear uniforms anymore. The classes on communism have been dropped, and her new history books are filled with information that was once forbidden by the former government. For the first time, she can also take religion courses if she wants to.

Strong Rulers Muscovy slowly developed into the country we know today as Russia. The Muscovite rulers extended their power and began calling themselves **czars,** or emperors. As a Muscovite, you would have feared Czar Ivan IV, who used a secret police force to keep his people under control. He also expanded Russia's territory by conquest eastward into the Volga region and Siberia.

Over the centuries the country grew as czars conquered other lands. Many non-Russian peoples came into the Russian Empire. Czars such as Peter I and Catherine II pushed the empire's borders westward and southward. They also tried to make Russia more like Europe. A new capital—St. Petersburg—was built in the early 1700s to look like a European city. If you had been a Russian noble at this time, you would have worn European clothes and spoken French instead of Russian.

The actions of the czars, however, had little effect on ordinary citizens. Most Russians were **serfs,** or laborers who were bound to the land. They were too poor to be interested in European ways and kept their Russian culture.

Toward Revolution In 1861 Czar Alexander II freed the serfs. At about the same time, Russia began to set up industries like those of other European countries. Railroads, including the famous Trans-Siberian Railroad, spread across the country. It linked Moscow in the west with Vladivostok on Russia's Pacific coast.

Russia, however, did not progress politically. The czars clung to their power and rejected democracy. Revolution brewed. In 1917 the political leaders and workers forced Czar Nicholas II to give up the throne. At the end of the year, a group of Communists led by Vladimir Ilyich Lenin came to power. They set up a Communist government and soon moved its capital to Moscow.

The Soviet Union In 1922 the Communists formed the Union of Soviet Socialist Republics, or the Soviet Union. The new country included Russia and most of the conquered territories of the old Russian Empire. During the late 1920s, Joseph Stalin became the ruler of the Soviet Union and set out to make it a great industrial power. To reach this goal, the government took control of all industry and farming. Stalin, a cruel dictator, put down any opposition to his rule. Millions of people were either killed or sent to prison labor camps.

370

Cooperative Learning Activity

Organize students into five groups and assign each group one of the following historical periods: 900s–1400s, 1400s–1800s, 1800s–1917, 1920s–1980s, 1980s–present. Have each group research its assigned time period and choose events to include on an illustrated time line of Russian history. Attach to a classroom wall a long strip of paper on which dates are indicated. Have groups place annotated illustrations of their selected events in appropriate places on the strip of paper. Suggest that they use yarn to connect each illustration with its correct date on the time line. **L1**

After World War II the Soviet Union further expanded its territory and extended communism to eastern Europe. From the late 1940s to the late 1980s, the Soviet Union and the United States waged a **cold war.** They competed for world influence without actually waging war on one another.

A New Beginning In 1985 Mikhail Gorbachev came to power in the Soviet Union. In addition to economic changes, Gorbachev supported a policy of *glasnost,* or openness. He wanted people to speak freely about the Soviet Union's problems. Gorbachev's efforts, however, failed to stop the collapse of the Soviet Union. Many of the non-Russian nations had long resented Russian rule and now wanted independence. By late 1991 the Soviet Union had broken apart. Russia and the newly independent countries formed a loose union called the Commonwealth of Independent States (CIS).

Place
City and Country Life

The teens you read about earlier are part of Russia's 148 million people. Look at the graph below to see how Russia's population compares with that of the United States. Russian Slavs make up about 80 percent of the population, and Russian is the official language. The country, however, has hundreds of smaller **ethnic groups,** or people who have a common language, culture, and history.

City Life Nearly three-fourths of Russia's people live in cities and towns. The map on page 372 shows that a large part of Russia's population lives around Moscow. Many cities are crowded, and residents face severe housing shortages. Whole families often live in one- or two-room apartments. City residents in Russia also have to deal with other problems such as food shortages, high prices, crime, and pollution from factories. The government is trying to correct such concerns and is also working to improve medical care. When people in cities relax, they spend time with family and friends, taking walks through parks, or attending concerts, movies, and the circus.

COMPARING POPULATION:
Russia and the United States

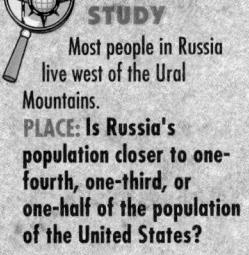

GRAPHIC STUDY
Most people in Russia live west of the Ural Mountains.
PLACE: Is Russia's population closer to one-fourth, one-third, or one-half of the population of the United States?

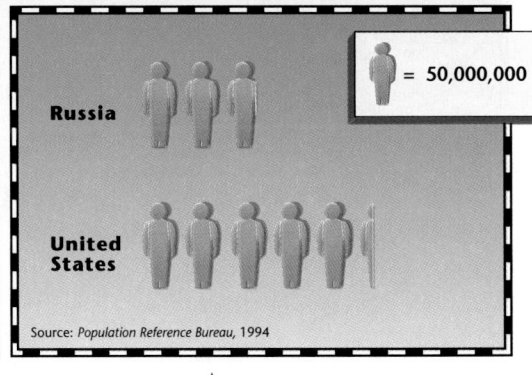

 = 50,000,000

Russia

United States

Source: *Population Reference Bureau, 1994*

Independent Practice

📁 Guided Reading Activity 14-3 L1

📁 Critical Thinking Skills Activity 14 L1

📁 Workbook Activity 14 L1

MULTICULTURAL PERSPECTIVE
Culturally Speaking
Gift-giving is an important part of life in Russia. As in other cultures, holidays and special events are accompanied by gifts. Russians, however, will mark even the most casual visit to friends with some kind of present.

GRAPHIC STUDY
Answer
one-half
Graphic Skills Practice
Reading a Graph What is the difference in population between Russia and the United States?
(United States has about 110 million more people.)

Meeting Special Needs Activity

Inefficient Readers Help students learn how to scan a passage before reading so that they may develop a time frame for the ideas that are presented. Have students scan the subhead "Toward Revolution." Instruct them to look for time markers. Examples include the following: "in 1861," "at about the same time," "in 1917," and "by the end of the year." Then have students follow the same procedure for the next subhead. L1

Cultural
Kaleidoscope

Russia Russian is the third most widely spoken European language. It is one of the six official languages of the United Nations. Russian uses the Cyrillic alphabet, which has 33 letters. Many of these letters are unlike any in the Roman alphabet, which the English language uses.

Geography and the Humanities

📁 World Literature
Reading 5

🌐 World Music:
Cultural Traditions,
Lesson 4

📼 World Cultures
Transparencies 7, 8

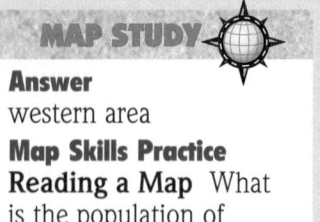

MAP STUDY

Answer
western area

Map Skills Practice
Reading a Map What is the population of St. Petersburg? *(more than 1 million)*

Country Life If you visit a village in Russia today, you will find that most people live in houses built of wood. In some rural areas, people live without heat, electricity, or plumbing. Consumer goods are scarce in the countryside, but food is more plentiful than in some cities. The quality of health care, education, and cultural activities is lower in rural areas than in urban areas. Over the years, many people have left the villages to find work in the cities.

Place

The Arts and Recreation

Russia has a rich tradition of art, music, and literature. The Russian people view their country's cultural achievements with pride. Some of their finest works of art depict religious and historical themes. Others reflect the daily lives of ordinary people.

Religion Communist rulers of the Soviet Union banned religion. They often closed houses of worship and persecuted religious people. Since the fall of communism, many Russians have returned to their religious traditions. Russians are Jews, Protestants, and Roman Catholics. Most Russians, however, follow Eastern Orthodox Christianity. Orthodox churches in Russia that have survived are excellent examples of Russian architecture. Many contain religious works of art that are hundreds of years old.

Arts and Music If you want to see Russian culture at its best, where do you go? If you enjoy dance, you might attend a ballet performed at Moscow's Bolshoi Theater. Russia has long been a world leader in ballet.

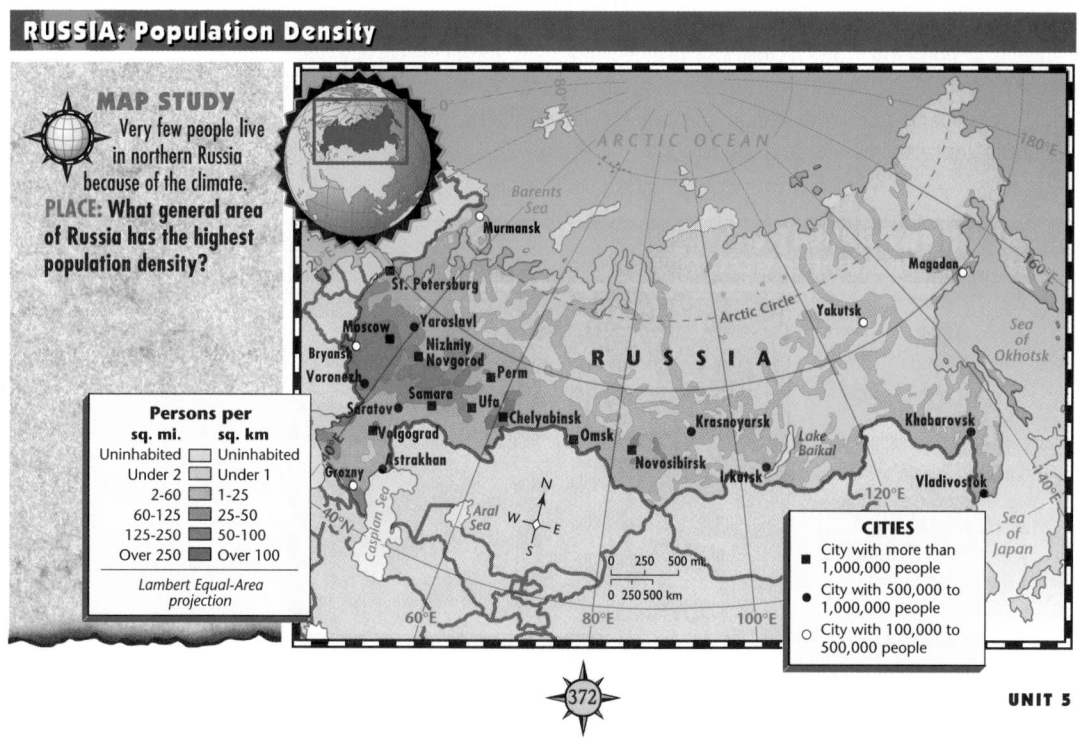

RUSSIA: Population Density

MAP STUDY
Very few people live in northern Russia because of the climate.
PLACE: What general area of Russia has the highest population density?

Persons per
sq. mi.	sq. km
Uninhabited	Uninhabited
Under 2	Under 1
2-60	1-25
60-125	25-50
125-250	50-100
Over 250	Over 100

Lambert Equal-Area projection

CITIES
■ City with more than 1,000,000 people
● City with 500,000 to 1,000,000 people
○ City with 100,000 to 500,000 people

Critical Thinking Activity

Predicting Consequences Point out that under communism, Russians were accustomed to obeying the government without question. Ask students to predict how freedom might affect people's attitudes and behavior. Have students consider the following question: What skills might Russians need to develop to handle this new-found freedom? *(Answers might include: the ability to negotiate, to take risks, to make economic decisions)* **L2**

The wooden houses in rural Russia *(left)* have much more room than the cramped apart-ment buildings in Russia's cities *(right).*
PLACE: About how many of Russia's people live in cities?

If you enjoy painting, you might stroll through the Hermitage, St. Petersburg's most famous museum. Works of noted Russian as well as western European painters hang in the Hermitage. In the early 1900s, Russia led the modern art movement. Two of the most famous Russian-born painters of this movement are Kazimir Malevich and Wassily Kandinsky.

Perhaps you prefer music. Classical music fills concert halls throughout Russia. Many of the world's greatest composers were Russian. They include Peter Tchaikovsky and Nikolay Rimsky-Korsakov. You may be familiar with Tchaikovsky's *Nutcracker Suite* and his *1812 Overture.*

The 1800s and 1900s saw a great flowering of Russian literature. Among the best Russian writers of the nineteenth century were Alexander Pushkin, Leo Tolstoy, and Fyodor Dostoyevsky. Their works vividly describe the lives of nobles, merchants, serfs, and city workers. In this century, the leading Russian writers include Maxim Gorky and Alexander Solzhenitsyn. They and other mod-ern writers have discussed the lack of individual freedom in Russia and the hopes for a better society.

Celebrations Russians enjoy small family get-togethers as well as national celebrations. New Year's Eve is the most festive nonreligious holiday for the Russian people. On this day children decorate a fir tree and exchange presents with their families. Another important holiday is May Day. In parades and speeches, May Day celebrates Russian workers and their contributions to their country.

Food If you have dinner with a Russian family, you might begin with a big bowl of *borsch,* a soup made from beets, or *shchi,* a soup made from cabbage. Next you might be served meat turnovers called *pirozhki.* For the main course, you would have meat, poultry, or fish with boiled potatoes. If you were having a

373

Global Issues Activity

Have students research how the end of the cold war has affected relations between Russia and the United States. Encourage them to bring in articles from newspapers and news maga-zines. Have students study and write an analy-sis of their articles. Then ask them to create a bulletin-board display with their articles and analyses. **L2**

More About the Illustration

Answer to Caption
Bolshoi Theater

Place Ballet is a passion for most Russians. The leading dancers are highly respected in Russian society. Dancers begin their training at a very early age, and many of their instructors were once great dancers themselves.

Reteach

 Reteaching Activity
14 **L1**

Enrich

 Enrichment Activity
14 **L1**

CLOSE

 **Mental Mapping Activity**

Provide each student with a large sheet of white paper and map pencils. Inform students that this will not be a graded activity. Have students draw, freehand from memory, a map of Russia that includes the following labeled bodies of water and mountain ranges: Volga River, Caspian Sea, Lake Baikal, Arctic Ocean, Ural Mountains, Caucasus Mountains.

374

Ballet Practice

Dedicated young dancers practice at a ballet school in St. Petersburg.
LOCATION: What famous ballet theater is located in Moscow?

special holiday dinner, you might have caviar, or fish eggs, fresh from the Caspian Sea.

Sports Russians love soccer, tennis, and ice hockey. On the lakes and in the mountains, you will find Russians skiing, hiking, camping, mountain climbing, and hunting. Reindeer races and dog-team competitions are popular sports in the cold lands of Siberia. And if you watch the Olympics, you know that Russians are major competitors in the games.

SECTION 3 REVIEW

REVIEWING TERMS AND FACTS
1. **Define the following:** czar, serf, cold war, *glasnost,* ethnic group.
2. **PLACE** What was life like for most Russians under the czars?
3. **PLACE** How does city life differ from country life in Russia?
4. **HUMAN/ENVIRONMENT INTERACTION** How would you compare the practice of religion in Russia before and after 1991?

 MAP STUDY ACTIVITIES
5. Turn to the population density map on page 372. Which region of Russia is sparsely populated? Why is this so?
6. Look again at the same map. About how many people live in Murmansk?

374

UNIT 5

Answers to Section 3 Review

1. All vocabulary words are defined in the Glossary.
2. Most Russians were serfs bound to the land and had little interest in European ways.
3. Cities are crowded and often have shortages. Health care, education, and cultural activities tend to be better than in rural areas. Food is more plentiful in rural areas.
4. The communist government discouraged the practice of religion and closed many houses of worship and placed hardships on religious people. Since 1991, many Russians have begun to practice religion openly.
5. Siberia; it has a harsh climate
6. between 100,000 and 500,000

Important Things to Know About Russia

SECTION 1 THE LAND

- Russia is the largest country in the world in area.
- Russia extends across both Europe and Asia.
- Russia is a vast lowland divided or surrounded by mountains.
- Inland waterways help connect Russia's widespread regions.
- Much of Russia's climate is subarctic or humid continental.

SECTION 2 THE ECONOMY

- Russia's economy is moving from government control to a free enterprise system.
- Russia's leading industrial centers are located in Moscow, St. Petersburg, the Volga and the Urals, and Siberia.
- Oil and gas are important resources in Russia.
- Siberia, the Asian part of Russia, has vast timber resources.

SECTION 3 THE PEOPLE

- Russia was the center of a powerful empire under rulers known as czars.
- A revolution in 1917 overthrew the czars; later a communist empire called the Union of Soviet Socialist Republics, or Soviet Union, was established.
- Russia and other parts of the Soviet Union have become independent republics since the fall of communism in the early 1990s.
- Most Russians live in urban areas in the European part of Russia.
- Russia has a rich tradition of art, music, literature, and dance.

St. Petersburg, Russia ▶

CHAPTER HIGHLIGHTS

Using the Chapter 14 Highlights

- Use the Chapter 14 Highlights to preview, review, condense, or reteach the chapter.

Preview/Review

Vocabulary Puzzle-Maker Software reinforces the terms used in Chapter 14.

Condense

Have students read the Chapter 14 Highlights. Spanish Chapter Highlights also are available.

Chapter 14 Digest Audiocassettes and Activity

Chapter 14 Digest Audiocassette Test
Spanish Chapter Digest Audiocassettes, Activities, and Tests also are available.

Guided Reading Activities

Reteach

Reteaching Activity 14. Spanish Reteaching activities also are available.

MAP STUDY

Location Copy and distribute page 34 from the Outline Map Resource Book. Have students locate and label Russia's major landforms and bodies of water. Also have them shade in the country's climate regions. Remind students that they need to include a key with their maps.

Extra Credit Project

Advertisement Have students work individually or in pairs to write an advertisement to attract workers to Siberia. Direct students to review the text to determine what kinds of jobs are available in this region and what the pros and cons of working in the area might be. Then have them use this information to create their advertisements. **L2**

Chapter 14 Review and Activities

ANSWERS
Reviewing Key Terms

1. I
2. C
3. B
4. H
5. F
6. G
7. A
8. J
9. D
10. E

Mental Mapping Activity

This exercise helps students to visualize the countries and geographic features they have been studying and understand the relationships among them. All attempts at freehand mapping should be accepted.

REVIEWING KEY TERMS

Match the numbered terms in Column A with their definitions in Column B.

A
1. taiga
2. command economy
3. czar
4. heavy industry
5. tundra
6. serf
7. permafrost
8. ethnic group
9. light industry
10. free enterprise system

B
A. permanent layer of frozen soil
B. Russian ruler
C. economy planned and directed by the communist government
D. production of household products
E. privately owned businesses with little government control
F. cold, treeless plain
G. farmers bound to the land
H. production of machinery, mining equipment, and military weapons
I. vast forest area of Siberia
J. people having a common culture

Mental Mapping Activity

On a separate sheet of paper, draw a freehand map of Russia. Label the following items on your map:
- Black Sea
- Volga River
- Moscow
- Caspian Sea
- Central Siberian Plateau
- St. Petersburg

REVIEWING THE MAIN IDEAS

Section 1
1. **PLACE** How large, in area, is Russia?
2. **REGION** What is the world's largest region of flat land?
3. **REGION** What Russian lake is the deepest in the world?

Section 2
4. **HUMAN/ENVIRONMENT INTERACTION** What changes have come to Russia's economy since the fall of communism?
5. **HUMAN/ENVIRONMENT INTERACTION** Why is the Volga River important to Russia?
6. **REGION** Where are Russia's oil and natural gas deposits located?

Section 3
7. **MOVEMENT** What railroad links Moscow with Russia's eastern coast?
8. **PLACE** What effect did Mikhail Gorbachev's changes have on the Soviet Union?
9. **PLACE** What ethnic group makes up more than 80 percent of Russia's population?
10. **PLACE** What sports do Russians enjoy?

Reviewing the Main Ideas

Section 1

1. 6.6 million square miles (17,075,400 sq. km)—the largest country in the world
2. West Siberian Plain
3. Lake Baikal

Section 2

4. Russia has moved to a free enterprise economy; some people own their own small businesses and some people farm their own land. Economic problems continue, however, most notably high prices and food shortages.
5. carries most of Russia's river freight; source of hydroelectric power, irrigation
6. in the Volga and Urals region

CRITICAL THINKING ACTIVITIES

1. **Determining Cause and Effect** How do you think climate affects the way Russians make a living?
2. **Evaluating Information** What problems did Russia face after becoming independent in the early 1990s?

GeoJournal Writing Activity

Imagine you are a friend of Ivan Borisov. He has asked you to join his family on a trip to Siberia. Write about your visit to this region of Russia. You may use the maps in your text or a reference book to help you.

COOPERATIVE LEARNING ACTIVITY

Work in a group of three or more to plan a trip from Moscow to one of the following areas: (1) the Volga River; (2) the Ural Mountains; (3) Lake Baikal; or (4) an area of your choice. Each member of the group will do one of the following: (a) find out historic places to visit; (b) draw the route to follow on a map; or (c) make a list of provisions and clothes to bring. Combine your findings in a travel information packet. Share your trip with the rest of the class.

PLACE LOCATION ACTIVITY: RUSSIA

Match the letters on the map with the places and physical features of Russia. Write your answers on a separate sheet of paper.

1. Vladivostok
2. West Siberian Plain
3. Ural Mountains
4. Caspian Sea
5. Lena River
6. Kamchatka Peninsula
7. Lake Baikal
8. Caucasus Mountains
9. Moscow
10. Sea of Okhotsk

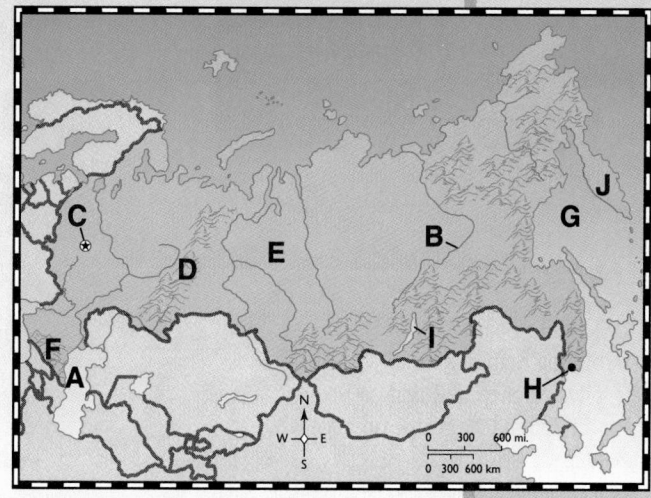

Cooperative Learning Activity

Encourage students to illustrate their plans with maps, sketches, and photographs clipped from old magazines or travel brochures. Display the finished information packets in the classroom and invite other classes to share them.

GeoJournal Writing Activity

Encourage students to share the descriptions of their visit with the rest of the class.

Place Location Activity

1. H
2. E
3. D
4. A
5. B
6. J
7. I
8. F
9. C
10. G

Chapter Bonus Test Question

This question may be used for extra credit on the chapter test.

You are the longest river in European Russia. You left Moscow and are flowing southward for 2,193 miles (3,528 km). What will be your final destination? *(the Caspian Sea)*

Section 3

7. Trans-Siberian Railroad
8. Gorbachev loosened rigid government control of the economy and allowed people more freedom in the hope of reforming the Soviet system. His changes were too little and too late, however, and the Soviet Union collapsed.
9. Russian Slavs
10. soccer, tennis, ice hockey, skiing, hiking, camping, mountain climbing, hunting; reindeer and dog-team competitions in Siberia

Critical Thinking Activities

1. Answers will vary. Most students will note that the long, cold winters limit the time that can be spent outside as well as the amount of work that can be done effectively.
2. high prices and shortages of food, consumer goods, and housing

GeoLab
ACTIVITY

FOCUS

Pose the following problem: Suppose you wanted to give a friend from another community an idea of the geography of your school. Would a photograph serve this purpose? Or would a small-scale model of the school be better? Direct the discussion so that students identify the following advantages of a model: it provides for three-dimensional viewing; it provides the chance to use another sense, touch, in addition to sight; and, it accurately shows the relationship among the various parts of the school. Tell students that these are similar to the advantages that a relief map has over a regular map.

TEACH

Emphasize the concept of *scale* as students create their maps. Point out that small-scale relief maps always exaggerate relief in order to show landforms. For example, the highest mountain in Russia is Mt. Elbrus at 18,510 feet (5,642 m). If students were to mold this peak exactly to scale on a three-foot wide map of Russia, the mountain would only be a few hundredths of an inch tall. Ask students why, given this distortion, relief maps are still a valuable tool for geographers. Guide them into understanding the maps' value in showing regional divisions and differences and in providing an idea of the real appearance of the terrain. **L1**

From the classroom of Dana Moseley, Carl Stuart Middle School, Conway, Arkansas

MAPMAKING: Dough It!

Background

Russia can be divided into five major geographic regions: the North European Plain, the West Siberian Plain, the Central Siberian Plateau, the East Siberian Uplands, and the lowland region east of the Caspian Sea. In this activity, you will make a salt-dough relief map of these regions.

The Ural Mountains in Russia

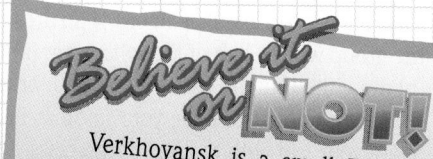

Verkhoyansk is a small Russian town in northeast Siberia. It has one of the world's most extreme climates. In January, the coldest month, the average temperature plunges to -58.2°F.

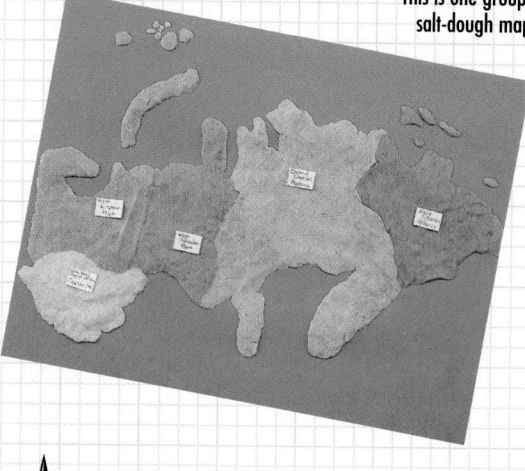

This is one group's salt-dough map.

Materials

- 1 poster board
- outline map of Russia
- physical map of Russia
- salt dough (see recipe on page 379)
- glue
- food coloring

378

Curriculum Connection Activity

Mathematics Assist students in making graphic scales for their maps. They should determine the distance between two points on a published map of Russia in miles and the distance in inches between these points on their own maps.

By comparing the number of inches to the number of miles, students will have the information needed to create a graphic ruler scale. Have them add the scale to their maps. **L2**

What To Do

A. Draw an outline map of Russia onto the poster board.
B. Mix the salt dough using this recipe:
- *1 cup flour*
- *1 tablespoon vegetable oil*
- *1/2 cup salt*
- *2 teaspoons cream of tartar*
- *1 cup water*

Heat all ingredients in a saucepan over medium heat, stirring about 4 minutes or until mixture forms a ball. Remove from saucepan and let dough cool on counter for 5 minutes. Knead dough about 1 minute or until it is smooth and blended. Let cool completely before using on map.

C. Separate the dough into five sections and use the food coloring to make each region a different color—brown, green, red, yellow, and blue.
D. Press the colored salt dough into the appropriate places on your outline map of Russia. Make sure you recreate the correct geographic landforms and regions.
E. Label each region.

Lab Activity Report

1. Which region covers the greatest area?

2. Which region has the highest relief overall?

3. Describe the relief of each region.

4. Drawing Conclusions
What adjustments in your life would you have to make if you lived in each region?

Go A Step Further!

Activity
Research and create a time line depicting Russia's history. Include important people, places, and events. Illustrate your time line with small, colorful pictures that focus on important events in Russia's history.

GeoLab
ACTIVITY

ASSESS

Have students answer the **Lab Activity Report** questions on page 379.

CLOSE

Encourage students to complete the **Go A Step Further** writing activity. You may wish to assign different themes of Russian history, such as political developments, artistic movements, and so on, to different students. Combine these individual time lines into a single multi-banded one.

CURRICULUM CONNECTION

History In this activity, students are asked to make accurate maps of Russia. During the Cold War, Soviet mapmakers deliberately falsified maps of their country. These maps misrepresented location and distorted distances. This practice, which was designed to confuse western military planners, was discontinued when satellite photography made such deception fruitless.

Answers to Lab Activity Report

1. East Siberian Uplands
2. East Siberian Uplands
3. The North European Plain is a vast plain with elevations of 0–1,000 ft. On its east lie the Ural Mountains. The West Siberian Plain is the world's largest area of flat land. Its elevation also ranges from 0–1,000 ft. The Central Siberian Plateau rises in elevation as you go from the Arctic Ocean in the north to the Sayan Mountains in the south (0–10,000 ft.) The East Siberian Uplands is a highland area with elevations mostly of 2,000 to 10,000 ft. The Caspian Sea area has low elevations, with some land lying below sea level.
4. Answers will vary.

FOCUS

Read the following words aloud to the class: *Russian, Ukrainian, Byelorussian, Latvian, Lithuanian, Armenian, Georgian, Moldavian, Tajik, Azerbaijani, Bashkir, Chuvash, Kazak, Kirghiz, Tartar, Turkmen, Uzbek, Yakut, Estonian, Finnish, Kaelian, Komi, Mari, Morodavian,* and *Udmurt.* Tell students that these are just *some* of the dozens of languages spoken by the more than 120 different ethnic groups in Russia and the independent republics. Write the word *Diversity* on the chalkboard and emphasize that, despite having many cultural similarities, the peoples of this region are very distinct.

TEACH

Display a map of the former Soviet Union and point out areas occupied by various ethnic groups. Have students discuss how political boundaries that fail to respect language and cultural differences can cause problems. Then have students speculate on how ethnic diversity in this region may have contributed to the collapse of the Soviet Union. **L1**

Cultural HERITAGE:
RUSSIA AND THE INDEPENDENT REPUBLICS

RELIGIOUS ART ▶ ▶ ▶ ▶

Icons hang on the walls of many homes and churches in Russia and the independent republics. Icons are paintings of Christian religious figures on wood panels. Many are hundreds of years old.

◀ ◀ ◀ ◀ MUSIC

Folk music is very popular in all the independent republics, especially Ukraine. This Ukrainian student plays the *bandura*, a multistringed instrument that is plucked like a guitar. Her brightly embroidered blouse and skirt are traditional Ukrainian clothing.

PAINTING ▶ ▶ ▶ ▶

Kazimir Malevich was one of the first painters to introduce modern art to Russia. This painting from the early 1900s does not show real objects. Instead Malevich used shapes and pure colors on plain backgrounds to represent objects.

380

More About . . .

Religious Art The Eastern Orthodox Church has more followers in Russia and the neighboring republics than any other branch of Christianity. Icons are an integral part of Eastern Orthodox Christianity. In Russia, and elsewhere in the region, Orthodox churches are filled with icons. Rigid rules specify how these images should be painted and displayed. In churches, icons are displayed on a screen called the *iconostasis,* which separates the sanctuary—the area around the altar—from the main part of the church. The faithful also display icons in special areas of their homes. They light candles and pray before the icons.

◄ ◄ ◄ ◄ WEAVING

The people of Uzbekistan take great pride in their beautiful wool carpets and rugs. Each year, factories in the area weave more than 3 million square miles (7.9 million sq. km) of carpet. The bright colors and bold patterns of the rugs reflect the influence of Islamic ancestors.

ARCHITECTURE ► ► ► ►

Churches, mosques, and synagogues reflect the many types of architecture found in Russia and the independent republics. This old church with a round tower in Msteka, Georgia, is made of stone. Others have elaborate onion-shaped domes and colorful mosaics.

FABERGÉ EGG ▲ ▲ ▲ ▲

Russia's most famous jeweler, Carl Fabergé, designed this beautiful egg for a Russian czar. He crafted a series of these elaborate eggs from gold, silver, enamel, and precious stones.

APPRECIATING CULTURE

1. Russia and the independent republics have a rich and varied history. How has some of that history been recorded in artwork from this area?

2. Which type of art shown here most appeals to you? Why?

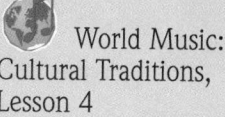
Geography and the Humanities

📁 World Literature Reading 5

🌐 World Music: Cultural Traditions, Lesson 4

📦 World Cultures Transparencies 7, 8

CURRICULUM CONNECTION

Art One of Fabergé's most remarkable works was a tiny gold clockwork model of the Trans-Siberian Express. The passenger cars had crystal windows and the engine's headlight was a ruby. The whole train was housed in a gold egg engraved with a map of the railroad.

ASSESS

Have students answer the **Appreciating Culture** questions on page 381.

CLOSE

Ask students to cite images and information in this feature to support the use of the word *diversity* as a description of the culture of the region.

Answers to Appreciating Culture

1. Answers will vary. Most students will note that different forms of architecture, weaving designs, and traditional music and dress all offer ample evidence of a rich and varied history.

2. Answers will vary.

The Independent Republics

PERFORMANCE ASSESSMENT ACTIVITY

COMPARING THE INDEPENDENT
REPUBLICS Organize students in groups of three or four. Have groups create charts of the geographic characteristics of each country in the area. Charts should include characteristics such as physical and water features, climate, agriculture, industry, population, major cities, religion, language, and so on. Upon completion of the charts, groups should compare and contrast the countries. Based on this comparison, groups should make recommendations for alliances among countries that have common cultural heritage, interests, or challenges. Call on groups to make oral presentations giving explicit reasons why the various countries in the alliances would benefit from working together.

POSSIBLE RUBRIC FEATURES: Problem-solving skills, persuasive thinking, accuracy of content information, collaborative skills, comparing and contrasting skills

MENTAL MAPPING ACTIVITY

Before teaching Chapter 15, point out the independent republics on a map. Have students note the locations of Ukraine and the other independent republics relative to Russia.

TEACHER'S CORNER

NATIONAL GEOGRAPHIC SOCIETY

INDEX TO NATIONAL GEOGRAPHIC MAGAZINE

The following articles may be used for research relating to this chapter:

- "The Fractured Caucasus," by Mike Edwards, February 1996.
- "Crimea: Pearl of a Fallen Empire," by Peter T. White, September 1994.
- "Chornobyl," by Mike Edwards, August 1994.
- "Soviet Pollution," by Mike Edwards, August 1994.
- "Russia: Playing by New Rules," by Mike Edwards, March 1993.
- "Ukraine: Running on Empty," by Mike Edwards, March 1993.
- "Mother Russia on a New Course," by Mike Edwards, February 1991.

NATIONAL GEOGRAPHIC SOCIETY PRODUCTS AVAILABLE FROM GLENCOE

To order the following products for use with this chapter, contact your local Glencoe sales representative or call Glencoe at 1-800-334-7344:

- *STV: World Geography* (Videodisc)
- *Picture Atlas of the World* (CD-ROM)
- *Physical Geography of the World* (Transparencies)
- *ZipZapMap! World* (Software)
- *GeoBee* (Software)
- *Images of the World* (Posters)
- *Eye on the Environment* (Posters)

ADDITIONAL NATIONAL GEOGRAPHIC SOCIETY PRODUCTS

To order the following products for use with this chapter, call National Geographic Society at 1-800-368-2728:

- *Rise and Fall of the Soviet Union* (Video)
- *Russia: After the U.S.S.R.* (Video)
- *Physical Geography of the Continents Series:* "Asia," "Europe." (Video)
- *Soviet World in Transition:* "Pt. I: Geography;" "Pt. II: History." (Filmstrip)

TEACHER-TO-TEACHER
Curriculum Connection: Art
—from Janis Coffman,
Belmont Middle School, Decatur, IN

EGG-CITING FOLK ART *The purpose of this activity is to have students learn about the folk art, traditions, and customs of the region.*

Have students research the work of Carl Fabergé. In preparation for this research, tell students that Fabergé used a traditional Ukrainian art style, the psyanky egg. The egg had some form of decoration on the shell and a "surprise" inside. One of the most notable of Fabergé's eggs was made of gold and had a map of the Trans-Siberian Railroad on the shell. It contained a gold train decorated with crystal and precious stones.

Provide each student with two egg-shaped pieces of construction paper. Have students use these to create their own psyanky eggs. Direct them to draw the shell decorations on one oval and the "surprise" the egg contains on the other. Encourage students to make their designs colorful. Designs might represent some aspect of the geography or culture of the region. Suggest that students use resources in museums and libraries for help with design ideas. Mention that *psyanky* means "to write," and direct students to write explanations of the symbols they used.

Follow up by displaying the psyanky eggs and their explanations on the bulletin board. Then lead students in a discussion of the following questions: What legends grew up around the psyanky eggs? Why?

BIBLIOGRAPHY

Readings for the Student
Diller, Daniel C. *Russia and the Independent States,* revised ed. Washington, D.C.: Congressional Quarterly, 1993.

Kronenwetter, Michael. *The New Eastern Europe.* New York: Franklin Watts, 1991.

Ukraine: Then and Now. Minneapolis: Lerner, 1993. (This series also includes books on Armenia, Azerbaijan, Belarus, Georgia, Kazakstan, Kyrgyzstan, Moldova, Tajikistan, Turkmenistan, and Uzbekistan.)

Readings for the Teacher
CIS and Eastern Europe on File. New York: Facts on File, 1993.

The Soviet Union. Columbus, Ohio: Good Apple, 1993. Reproducible activity book.

Multimedia
The Changing Face of Russia and the New Commonwealth. Rockford, Ill: Giant Photos, 1992. Laminated write-on/wipe-off wall map.

The Rise and Fall of the Soviet Union. Tapeworm, 1992. 2 videocassettes, 124 minutes.

Russia: After the U.S.S.R. Washington, D.C.: National Geographic Society, 1994. Videocassette, 25 minutes.

KEY TO ABILITY LEVELS
Teaching strategies have been coded for varying learning styles and abilities.

L1   BASIC activities for all students

L2   AVERAGE activities for average to above-average students

L3   CHALLENGING activities for above-average students

LEP   LIMITED ENGLISH PROFICIENCY activities

MEETING NATIONAL STANDARDS

Geography for Life

All of the 18 standards are demonstrated in Unit 5. The following ones are highlighted in this chapter:

• Standards 3, 4, 10, 13, 16

Use these *Geography: The World and Its People* resources to teach, reinforce, and extend chapter content.

CHAPTER 15 RESOURCES

- *Vocabulary Activity 15
- Cooperative Learning Activity 15
- Workbook Activity 15
- Geography Skills Activity 15
- GeoLab Activity 15
- Critical Thinking Skills Activity 15
- Performance Assessment Activity 15
- *Reteaching Activity 15
- Enrichment Activity 15
- World Literature Reading 5
- Environmental Case Study 5
- Chapter 15 Test, Form A and Form B
- *Chapter 15 Digest Audiocassette, Activity, Test
- Unit Overlay Transparencies 5-0, 5-1, 5-2, 5-3, 5-4, 5-5, 5-6
- Political Map Transparency 5
- Geoquiz Transparencies 15-1, 15-2
- Vocabulary PuzzleMaker Software
- Student Self-Test: A Software Review
- Testmaker Software
- MindJogger Videoquiz

*If time does not permit teaching the entire chapter, summarize using the Chapter 15 Highlights on page 399, and the Chapter 15 English (or Spanish) Audiocassettes. Review students' knowledge using the Glencoe MindJogger Videoquiz. *Also available in Spanish*

Use these *Geography: The World and Its People* resources to teach and reinforce section content.

SECTION 1 RESOURCES

Reproducible Lesson Plan 15-1

Section Focus Transparency 15-1

Guided Reading Activity 15-1

Section Quiz 15-1 (also available in Spanish)

SECTION 2 RESOURCES

Reproducible Lesson Plan 15-2

Section Focus Transparency 15-2

Guided Reading Activity 15-2

Section Quiz 15-2 (also available in Spanish)

SECTION 3 RESOURCES

Reproducible Lesson Plan 15-3

Section Focus Transparency 15-3

Guided Reading Activity 15-3

Section Quiz 15-3 (also available in Spanish)

SECTION 4 RESOURCES

Reproducible Lesson Plan 15-4

Section Focus Transparency 15-4

Guided Reading Activity 15-4

Section Quiz 15-4 (also available in Spanish)

ADDITIONAL RESOURCES FROM GLENCOE

Reproducible Masters

- Facts On File: GEOGRAPHY ON FILE, Commonwealth of Independent States

- Foods Around the World, Unit 5

Workbook

- Building Skills in Geography, Unit 1, Lesson 13

 World Music: Cultural Traditions Lesson 4

Transparencies

- National Geographic Society PicturePack Transparencies, Unit 5

Posters

- National Geographic Society: Eye on the Environment, Unit 5

World Crafts Activity Cards Unit 5

Videodiscs

- National Geographic Society: STV: World Geography, Vol. 1

CD-ROM

- National Geographic Society: Picture Atlas of the World

Software

- National Geographic Society: ZipZapMap! World

Performance Assessment

Refer to the Planning Guide on page 382A for a Performance Assessment Activity for this chapter. See the *Performance Assessment Strategies and Activities* booklet for suggestions.

Chapter Objectives

1. **Describe** the geography of Ukraine.
2. **Compare** the geography of Belarus and Moldova.
3. **List** the major geographic features of the Caucasus republics.
4. **Identify** the unique geographical characteristics of the Central Asian republics.

GLENCOE
TECHNOLOGY

 VIDEODISC

Use the Chapter 15 MindJogger Videoquiz to preview chapter content.

MindJogger Videoquiz

Chapter 15
Disc 2 Side B

Also available in VHS.

Chapter 15 The Independent Republics

MAP STUDY ACTIVITY

As you read Chapter 15, you will learn about the independent republics that, like Russia, were once part of the former Soviet Union.

1. **How many independent republics are there?**
2. **What republics share a border with Russia?**

382

MAP STUDY ACTIVITY

Answers
1. 11
2. Belarus, Ukraine, Georgia, Azerbaijan, Kazakstan

Map Skills Practice
Reading a Map What is the capital of the largest of the independent republics? *(Almaty)*

SECTION 1 Ukraine

PREVIEW

Words to Know
- steppe

Places to Locate
- Ukraine
- Black Sea
- Dnieper River
- Dniester River
- Crimean Peninsula
- Kiev
- Odessa
- Donets Basin

Read to Learn . . .
1. what landforms are found in Ukraine.
2. why Ukraine is known as the breadbasket of Europe.
3. what challenges Ukraine faces as an independent country.

The sputtering of this tractor and the "whoosh" of harvested wheat echo across the broad plains of Ukraine. Farmers grow a variety of grains in Ukraine's rich soil and ship them abroad to foreign markets. Because of the abundance of its grain harvests, Ukraine has been called the breadbasket of Europe.

Ukraine and many of its neighbors were once part of the Soviet Union. Since the breakup of the Soviet Union in 1991, they have been independent republics. To help their economies grow, these new countries and Russia have formed the Commonwealth of Independent States (CIS).

Human/Environment Interaction
Land, Climate, Economy

Ukraine—a word meaning "frontier" in a Slavic language—was named for its location. It once marked the far western edge of the Russian empire. With 233,100 square miles (603,729 sq. km), Ukraine is about the size of Texas.

Landforms Find Ukraine on the map on page 384. Lowlands dotted with farms and forests spread across the north. Eastern highlands stretch toward the Black Sea. A coastal plain curves along the Black Sea. From the plain, a landmass juts into the water. This landmass, the Crimean Peninsula, is one of the Black Sea's most scenic areas.

CHAPTER 15

383

Classroom Resources for Section 1

REPRODUCIBLE MASTERS
Reproducible Lesson Plan 15-1
Guided Reading Activity 15-1
Cooperative Learning
  Activity 15
Geagraphy Skills Activity 15
Building Skills in Geography
  Unit 1, Lesson 13
Section Quiz 15-1

TRANSPARENCIES
Section Focus
  Transparency 15-1
Unit Overlay Transparencies
  5-4, 5-5
Political Map Transparency 5

MULTIMEDIA
 Vocabulary PuzzleMaker
  Software
 Testmaker

LESSON PLAN
Chapter 15, Section 1

Meeting National Standards
Geography for Life The following standards are highlighted in this section:
- Standards 11, 14, 18

FOCUS

Section Objectives
1. **Describe** the landforms that are found in Ukraine.
2. **Explain** why Ukraine is known as the breadbasket of Europe.
3. **Identify** challenges that Ukraine faces as an independent country.

Bellringer Motivational Activity
Prior to taking roll at the beginning of the class period, project Section Focus Transparency 15-1 and have students answer the activity questions. Discuss students' responses. **L1**

Vocabulary Pre-check
Have students find the definition of *steppe*. Then have them compare steppe and tundra regions. **L1** **LEP**

Use the Vocabulary PuzzleMaker Software to create a crossword puzzle. **L1**

TEACH

Guided Practice

Project Unit Overlay Transparencies 5-4 and 5-5. Have students take turns stating one fact about Ukraine. List facts on the chalkboard. Challenge students to use the facts to write a paragraph about Ukraine. **L1**

Independent Practice

Guided Reading Activity 15-1 **L1**

ASSESS

Check for Understanding

Assign Section 1 Review as homework or an in-class activity.

Meeting Lesson Objectives

Each objective below is tested by the questions that follow it in parentheses.

1. **Describe** the landforms that are found in Ukraine. (1)
2. **Explain** why Ukraine is known as the breadbasket of Europe. (2)
3. **Identify** challenges Ukraine faces as an independent country. (3)

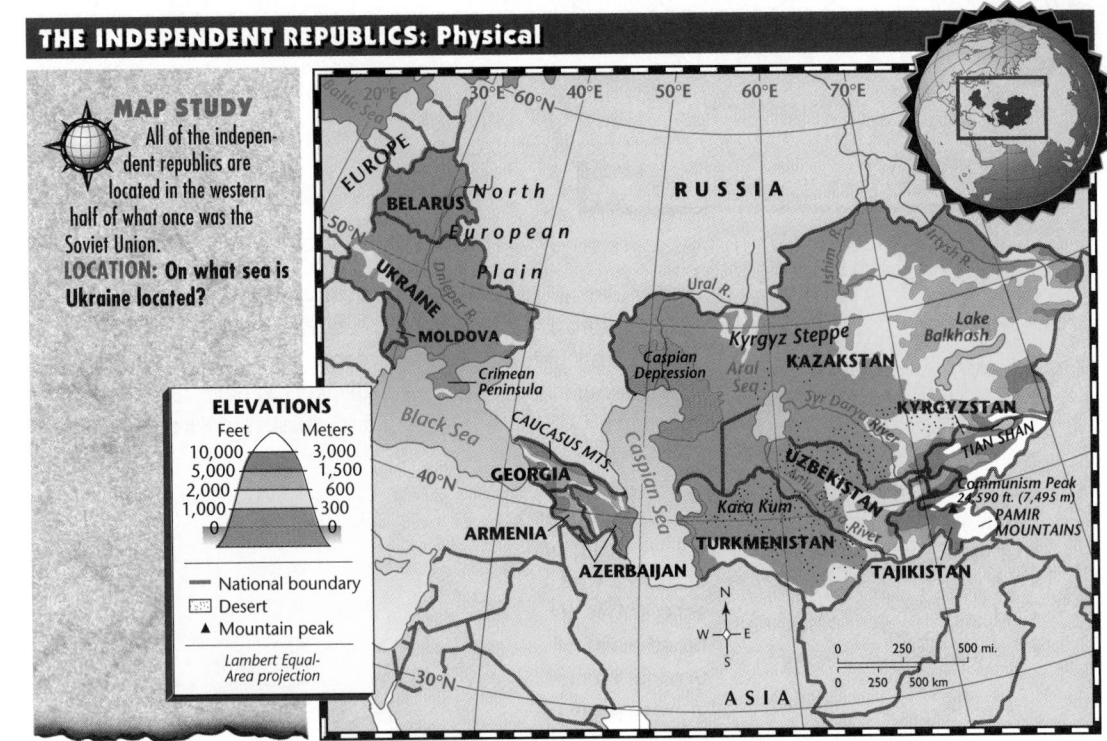

THE INDEPENDENT REPUBLICS: Physical

MAP STUDY
All of the independent republics are located in the western half of what once was the Soviet Union.
LOCATION: On what sea is Ukraine located?

ELEVATIONS

Feet	Meters
10,000	3,000
5,000	1,500
2,000	600
1,000	300
0	0

— National boundary
▨ Desert
▲ Mountain peak

Lambert Equal-Area projection

Vast treeless plains called **steppes** sweep through the center of Ukraine. Rich, fertile soil and plentiful rainfall make these steppes the most productive farmland in the region. Through the steppes to the Black Sea flows Ukraine's important river—the Dnieper (NEE•puhr). The Dniester (NEES•tuhr) also flows through the area. These rivers and other waterways link Ukraine's industrial areas with the Black Sea and world markets. The city of Odessa is Ukraine's major port.

The Climate Most of Ukraine has a humid continental climate of cold winters and warm summers. Rainfall is more plentiful in the northern and central parts of the country than in the south. On the Crimean Peninsula, mountains shield the Black Sea coast from the cold northern winds.

The Economy Under Soviet rule, what Ukraine produced was controlled by the government. Today Ukraine is changing its economy to one based on free enterprise. Farming and industry are the mainstays of Ukraine's economy. Farmers harvest grains, potatoes, sugar beets, and dairy products.

Ukraine is more than a breadbasket, however. The country is rich in coal, iron ore, manganese, potash, and natural gas. The Donets Basin in eastern Ukraine contains one of the largest coal deposits in the world. It also is the country's major industrial area. A gigantic dam on the Dnieper River supplies electricity to the region's factories.

384

Cooperative Learning Activity

Organize students into four groups. Assign each group one of the following topics: landforms, agricultural products, mineral resources, industrial centers. Provide a large outline map of Ukraine. Have groups create pictures and/or symbols representing their topic. Then have groups place the pictures and symbols in appropriate locations on the map. Also have groups create a key to explain the meaning of their pictures and symbols. **L1**

Place
People

Ukraine has a population of about 52 million. About 75 percent of the people—including Anna Lyashenko, a 13-year-old citizen of Ukraine's capital—are ethnic Ukrainians. They have their own language and culture. The rest of Ukraine's population are mostly Russian. Eastern Orthodox Christianity is the country's main religion.

From Past to Present About two-thirds of Ukraine's people live in urban areas. Kiev (KEE•EHF), the capital and largest city, has about 2.6 million people. Located on the Dnieper River, Kiev is a transportation, industrial, and cultural center. In history class, Anna has learned that, from A.D. 900 to the 1200s, Kiev was the center of a group of powerful Slavic city-states. In past centuries, outsiders such as the Poles and the Russians ruled Anna's country, and Ukrainians suffered. Independence, Anna admits, has not solved all problems. Ukraine's economy is shaky. Many people have lost their jobs as factories change ways of producing goods. Prices are high in the stores, and some goods are scarce. In addition, the growth of heavy industry during the years of Soviet rule has seriously harmed Ukraine's environment. Factory smoke and other wastes have polluted the air and water.

Culture and Celebrations The people of Ukraine take pride in a rich cultural heritage. Ukrainians are famous for their folktales and proverbs. The chart above lists several favorite proverbs of Ukraine and Russia.

During the Easter season, Ukrainian families observe the ancient tradition of decorating eggs with colorful designs. Folk music is popular, and many Ukrainian musicians perform on a stringed instrument called the *bandura.* In a popular dance—the *hopak*—male dancers make astounding acrobatic leaps!

Ukrainian and Russian Proverbs

- **Where the road is straight, don't look for a short cut.**
- **A bad peace is better than a good war.**
- **There is plenty of sound in an empty barrel.**
- **A near neighbor is better than a distant relative.**
- **When the Czar has a cold, all Russia coughs.**

Source: *The Prentice-Hall Encyclopedia of World Proverbs*

GRAPHIC STUDY
A proverb is a saying that describes a truth.
HUMAN/ENVIRONMENT INTERACTION: On this chart, which proverb might be rewritten as "don't take the easy way out"?

SECTION 1 REVIEW

REVIEWING TERMS AND FACTS
1. **Define the following:** steppe.
2. **PLACE** How does the climate of Ukraine differ from north to south?
3. **HUMAN/ENVIRONMENT INTERACTION** How is the Donets Basin important to Ukraine's industrial growth?

MAP STUDY ACTIVITIES
4. Turn to the physical map on page 384. What major body of water lies south of Ukraine?

Answers to Section 1 Review
1. All vocabulary words are defined in the Glossary.
2. north: humid continental with cold winters and warm summers; south: steppe, or partly dry, without the cold northern winds
3. It contains one of the largest coal deposits in the world and is located in the country's major industrial area.
4. Black Sea

GRAPHIC STUDY

Answer
Where the road is straight, don't look for a short cut.

Graphic Skills Practice
Reading a Chart What do you think is meant by "A bad peace is better than a good war"? *(Peace is always a better solution than war.)*

Evaluate

Section 1 Quiz **L1**

Use the Testmaker to create a customized quiz for Section 1. **L1**

Reteach

Have students review the **Places to Locate** list on page 383. Call on volunteers to locate listed places on a map and explain each one's importance. **L1**

Enrich

Have students research and report on Ukraine's struggle for independence from the Soviet Union. **L3**

CLOSE

Ask students to adopt the role of Anna Lyashenko. Have them write about Ukraine from her viewpoint.

TEACH

Find an old map of your town or county. Make copies and distribute them to the class. Have students carefully study the map. Then copy and distribute a recent map of the area. Direct students to compare this map with the first one. Have them note the changes that have taken place in the area over time. Also, encourage them to point out the features that have remained the same. Then ask: How would you categorize the nature of the things that have changed—are they mostly physical features or political features? *(political)* Then have students read the skill and complete the practice questions. **L1**

Additional Skills Practice

1. Maps of Russia and the independent republics have changed greatly since the late 1980s. What caused these changes? *(breakup of the Soviet Union)*

2. What events might cause changes in the map shown in the skill feature? *(invasions, wars, treaties involving exchange of land)*

📁 Geography Skills Activity 15

📁 Building Skills in Geography, Unit 1, Lesson 13

BUILDING GEOGRAPHY SKILLS

Reading a Historical Map

Suppose you tried to draw a map of your town as it looked 100 years ago. You would not include roads or bridges as they are today. You certainly wouldn't find any airports. Perhaps there was no town at all!

A historical map shows the cultural and political features of an area in an earlier period. Physical features of a landscape may stay much the same. Cultural features, however, change dramatically over time. To read a historical map, apply these steps:

- Read the title to identify the region and time period.
- Use the key to locate political units.
- Compare the map with a recent physical map to observe changes.
- Compare the map with a recent political map to observe changes.

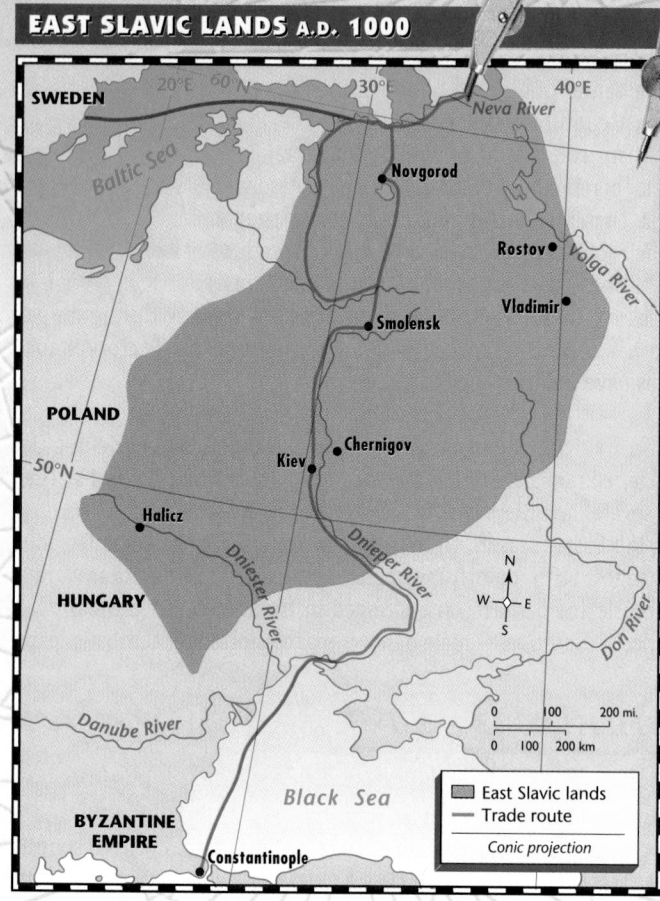

EAST SLAVIC LANDS A.D. 1000

■ East Slavic lands
— Trade route
Conic projection

Geography Skills Practice

1. What region and time period are shown in this map?
2. Near what rivers did city-states develop?
3. What present-day independent republics were once part of the East Slavic lands?

Answers to Geography Skills Practice

1. East Slavic lands, about A.D. 1000
2. Dnieper, Dniester, and Volga rivers
3. Belarus, Ukraine, Moldova

SECTION 2

Belarus and Moldova

LESSON PLAN
Chapter 15, Section 2

PREVIEW

Words to Know
- nature preserve

Places to Locate
- Belarus
- Minsk
- Moldova
- Chisinau

Read to Learn . . .
1. where Belarus and Moldova are located.
2. what Belarus and Moldova produce.
3. what groups of people live in Belarus and Moldova.

What makes this farmer in Belarus so happy? Is it a good crop of potatoes, flax, or wheat? Or is he feeling successful because he tends his own land? While most rural families in Belarus work on government-owned farms, many have been encouraged to run their own farms and businesses.

Belarus (BEE•luh•ROOS) lies north of Ukraine, and the country of Moldova (mahl•DOH•vuh) lies south. Like Ukraine, Belarus and Moldova—both land-locked countries—were once part of the Soviet Union. Today they are independent but have joined other former Soviet republics in the CIS.

Place
Belarus

Belarus covers 80,200 square miles (207,718 sq. km)—roughly the area of the state of Kansas. The people of Belarus, or Belarussians, are Slavs who are closely related to the Ukrainians and Russians. They also have close cultural ties to Poland, another neighboring Slavic country.

The Land and Climate Dense green undergrowth, towering trees, and wide marshes mark an area Belarus and Poland also share—Belovezha (BEHL•oh•VYEH•zah) Forest. This unique stretch of land is a famous **nature preserve,** or land set aside by the government to protect plants and wildlife. Rare European bison, similar to the American buffalo, graze in the woods and

CHAPTER 15

387

Meeting National Standards

Geography for Life The following standards are highlighted in this section:
- Standards 10, 13, 17

FOCUS

Section Objectives
1. **Describe** the locations of Belarus and Moldova.
2. **List** the products of Belarus and Moldova.
3. **Identify** groups of people that live in Belarus and Moldova.

Bellringer Motivational Activity

  Prior to taking roll at the beginning of the class period, project Section Focus Transparency 15-2 and have students answer the activity questions. Discuss students' responses. **L1**

Vocabulary Pre-check

Have students define the term *nature preserve.* Then have them describe a nature preserve located near their community. **L1 LEP**

MULTICULTURAL PERSPECTIVE

Culturally Speaking

The name *Belarus* comes from two Russian words *Belaya Rus,* which mean "White Russia."

Classroom Resources for Section 2

REPRODUCIBLE MASTERS
Reproducible Lesson Plan 15-2
Guided Reading Activity 15-2
Workbook Activity 15
Performance Assessment Activity 15
Section Quiz 15-2

TRANSPARENCIES
Section Focus Transparency 15-2
Unit Overlay Transparencies 5-4, 5-5
Geoquiz Transparencies 15-1, 15-2

MULTIMEDIA
Testmaker

National Geographic Society: ZipZapMap! World

MAP STUDY

Answer
humid continental

Map Skills Practice
Reading a Map What climates affect Uzbekistan? *(desert and steppe)*

TEACH

Guided Practice

Project Unit Overlay Transparencies 5-4 and 5-5. Have students compare Belarus and Moldova based on the maps. Bases of comparison might include area, location, borders, landforms, and bodies of water. **L1**

Independent Practice

Guided Reading Activity 15-2 **L1**

ASSESS

Check for Understanding

Assign Section 2 Review as homework or an in-class activity.

Meeting Lesson Objectives

Each objective below is tested by the questions that follow it in parentheses.

1. **Describe** the locations of Belarus and Moldova. (4)
2. **List** the products of Belarus and Moldova. (3)
3. **Identify** groups of people that live in Belarus and Moldova. (3)

meadows of the forest. Low flatlands cover the rest of Belarus in the north. Marshes, swamps, and forests are found in the south.

The map below shows that Belarus has a humid continental climate with cold winters and warm summers. The temperature averages about 22°F (-6°C) in January, the coldest month. In July—the hottest month—the temperature averages about 65°F (18°C).

The Economy Most Belarussians earn their livings as factory workers and farmers. They also work in service industries. Factory workers in Belarus make trucks, tractors, engineering equipment, and consumer goods such as television sets and refrigerators. The country's farmers produce barley, flax, potatoes, rye, sugar beets, dairy products, and livestock. Forests supply wood products that include furniture, matches, and paper goods.

The People Belarus's 10.3 million people are mostly Slavs who are proud of their separate language and culture. Smaller ethnic groups in the country include Russians, Ukrainians, and Poles. Two-thirds of the population live in cities. Minsk, the capital and largest city, has a population of 1.5 million. Belarus became independent in 1991. Minsk was chosen as the headquarters of the CIS.

Nina Muzychenko lives in Neglyubka—a village known for its textiles woven in elaborate patterns. Like most Belarussians, Nina follows the Eastern Orthodox Christian religion. Her family lives in a wooden home decorated with roof carvings.

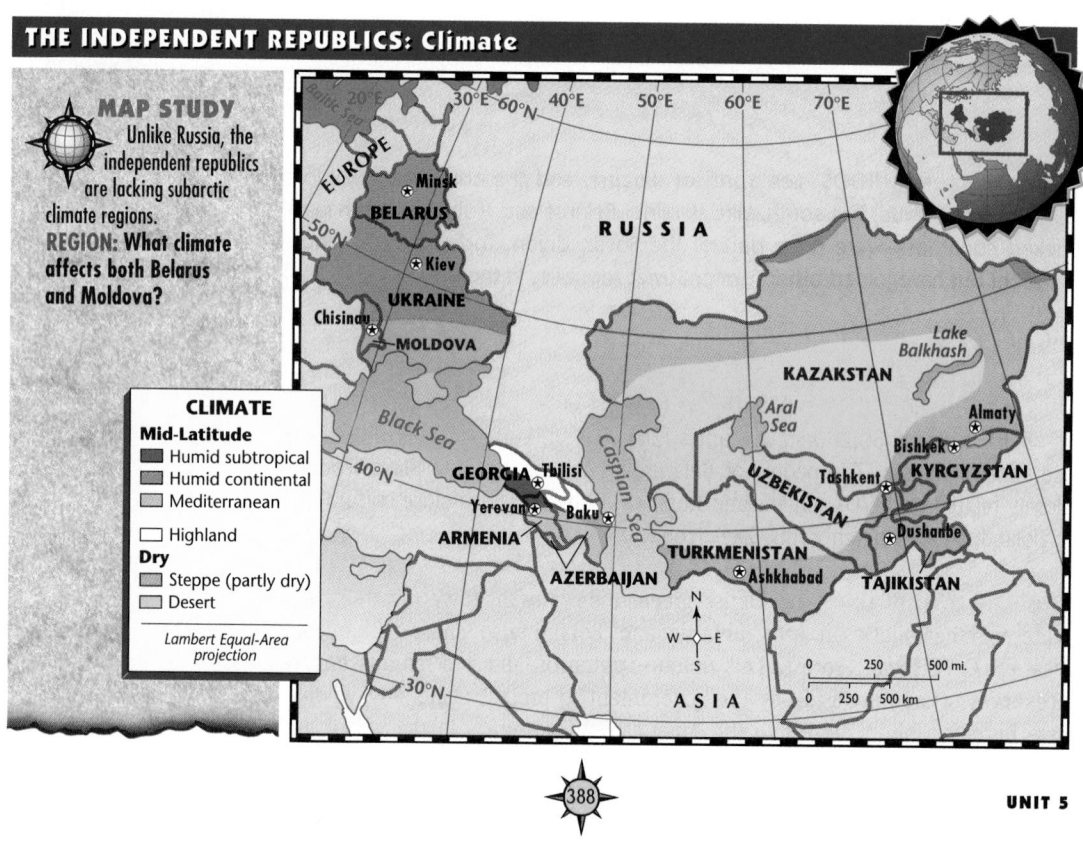

THE INDEPENDENT REPUBLICS: Climate

MAP STUDY
Unlike Russia, the independent republics are lacking subarctic climate regions.
REGION: What climate affects both Belarus and Moldova?

CLIMATE
Mid-Latitude
- Humid subtropical
- Humid continental
- Mediterranean
- Highland

Dry
- Steppe (partly dry)
- Desert

Lambert Equal-Area projection

UNIT 5

Cooperative Learning Activity

Organize the class into two groups and assign Belarus to one group and Moldova to the other. Instruct groups to subdivide into four teams, with each team researching one of the following topics related to their assigned country: climate, landforms, economy, and people. Encourage groups to present their findings in a four-section, illustrated wallchart. **L1**

Place
Moldova

A small country of 14,170 square miles (36,700 sq. km), landlocked Moldova broke away from the Soviet Union and became independent in 1991. Most of northern and central Moldova is hilly. Many rivers and streams flow through the rolling countryside. The river valleys of Moldova boast dark, fertile soil. Grassy plains sweep across much of the country's southern area. The map on page 388 shows you that northern Moldova has a humid continental climate of long, cold winters and short, hot summers. The southern part of the country has a steppe climate of cold, snowy winters and hot, dry summers.

Human/Environment Interaction Moldova relies on farming as its major economic activity. Farmers pluck grapes to make the country's popular wines. On the plains of the south, farmers harvest corn and wheat. In addition to farming, Moldova's light industries produce a variety of consumer and household goods. Like the other countries of the former Soviet Union, however, Moldova is struggling to move from a government-run economy to a privately owned economy with an expanded service sector.

The population of Moldova totals about 4.5 million people. It is one of the most densely populated countries in this region. Most of Moldova's people trace their language and culture to neighboring Romania. The rest of Moldova's population are Ukrainians, Russians, and Turks. Such diverse ethnic and national backgrounds often lead to conflict.

Moldova's capital and largest city is Chisinau (KEE•shih•NOW). More than half of Moldova's people, however, live in the countryside. Much of Moldova's culture is based on rural ways of life. Villagers celebrate special occasions with folk music and good food. Moldovan feasts include lamb, cornmeal pudding, and cheese made from goat's milk.

Bustling Minsk

Minsk, the capital of Belarus, is an important center of science, industry, and culture.
PLACE: In what year did Moldova and Belarus gain their independence from the Soviet Union?

SECTION 2 REVIEW

REVIEWING TERMS AND FACTS
1. **Define the following:** nature preserve.
2. **PLACE** Why is the Belovezha Forest unique?
3. **PLACE** How do the economies of Belarus and Moldova differ?

MAP STUDY ACTIVITIES
4. Turn to the map on page 388. What republics share a border with Moldova?
5. Look at the map on page 388. In what climate region is the city of Minsk?

More About the Illustration
Answer to Caption
1991

Place An old city, Minsk was an important trade center as far back as the eleventh century.

Evaluate

Section 2 Quiz **L1**

Use the Testmaker to create a customized quiz for Section 2. **L1**

Reteach

Have students complete a chart comparing Belarus and Moldova. Suggest that they include both physical and human characteristics in their comparison. **L1**

Enrich

Ask students to find out more about the Belovezha Forest. Have them create an illustrated report on the forest's plants and animals. **L3**

CLOSE

Have students imagine they are visiting Belarus and Moldova. Have them write a postcard to friends in the United States briefly describing the geography of these two countries.

Answers to Section 2 Review
1. All vocabulary words are defined in the Glossary.
2. It is the largest of Europe's ancient forest lands that once covered much of the continent. The forest contains rare animals, such as the European bison.
3. Belarus's economy is based on farming and industry. Crops include flax, barley, wheat, sugar beets, and potatoes. Its industries manufacture trucks, engineering equipment, and consumer goods. Moldova's major economic activity is farming. Important crops include grapes, corn, and wheat. Moldova also has some light industries.
4. Ukraine and Romania
5. humid continental

TEACH

Ask students to volunteer thoughts they have when they hear the words "buried treasure." Note appropriate feelings and descriptive terms on the board. Encourage students to compare their feelings about such treasure with the narrator's as they read the excerpt. **L1**

More About . . .

Literature Russian writer Mikhail Sholokhov discussed the history and ways of the Don Cossacks in such works as *Don Tales* (1925), *The Quiet Don* (1928–1940) and *Virgin Soil Upturned* (1941, 1960). Long passages describing the stark beauty of the region appear throughout Sholokhov's work.

MULTICULTURAL PERSPECTIVE

Culturally Speaking

The Turks gave the Black Sea its name. They considered it a dark and threatening body of water because violent storms often arose on it without any warning.

MAKING CONNECTIONS

THE FIRST SONG — BY SEMYON E. ROSENFELD

MATH | **SCIENCE** | **HISTORY** | **LITERATURE** | **TECHNOLOGY**

In this excerpt from *The First Song*, a Russian boy tells about moving to a new home. He and his family are leaving Mariopol on the Azov Sea to live in the port city of Odessa on the Black Sea.

I had spent most of my life here in Mariopol. In the summer, I would almost live on the beach, baking myself on the hot light-yellow sand, and when I was overcome with the heat, I would throw myself into the green transparent water of the Azov Sea, swim and dive. Here, also, [about two-thirds of a mile] from the city, on the Kalka—a river associated with many legends about the fighting between the Russians and the invading Mongols, centuries ago—we boys would often make our way across in an old flat-bottomed rowboat. Tall reeds grew thickly on the opposite shore—this was almost the beginning of the Don Cossack Region. On this spot we had made a momentous decision: to explore this land of the Cossacks for buried Tartar [Mongol] treasure, hoping to find a cauldron of gold, a trunkful of daggers, and a chestful of ornamented saddles and harness. Yes, I was leaving behind me a great deal, and my spirit was heavy.

. . . We soon came to the end of the Azov Sea and entered the Kerchinsky Strait, and after passing one more floating beacon, we were on the Black Sea. But I experienced a

Port of Odessa

real disappointment now—this sea was not at all black. And no one was able to explain why it was called black. I thought even the Azov was darker. However, when I found out that the Black Sea was bigger and deeper, and that its water was bitter and salty and never froze over, while the water of the Azov Sea was sweet and fresh and froze over for several months in the year, I unwillingly felt respect for its vastness and depth.

From *The First Song* by Semyon E. Rosenfeld, in *A Harvest of Russian Children's Literature*, University of California Press, 1967. Reprinted by permission of the Estate of Miriam Morton.

Making the Connection

1. What activities with friends was this boy leaving behind?
2. How is the Black Sea different from the Azov Sea?

390

UNIT 5

Answers to Making the Connection

1. searching for buried treasure taken by Cossacks from invading Mongols in past centuries

2. Black Sea is bigger, deeper, saltier, and never freezes

SECTION 3
The Caucasus Republics

PREVIEW

Words to Know
- fault
- food processing
- ethnic group

Places to Locate
- Caucasus Mountains
- Armenia
- Yerevan
- Georgia
- Tbilisi
- Azerbaijan

Read to Learn . . .
1. what major landform is found in the Caucasus republics.
2. what farm products are grown in this region.
3. how people live in the republics of the Caucasus.

Imagine living alongside a mountain range that seems to touch the sky! You can enjoy such a view in the country of Armenia. There snow-topped peaks form part of an enormous chain of mountains called the Caucasus, which stretches about 750 miles (1,210 km) between the Black and Caspian seas.

Meeting National Standards
Geography for Life The following standards are highlighted in this section:
- Standards 7, 10, 15

FOCUS

Section Objectives
1. **Identify** the major landform found in the Caucasus republics.
2. **List** the farm products grown in this region.
3. **Discuss** the ways people live in the republics of the Caucasus.

Bellringer Motivational Activity

 Prior to taking roll at the beginning of the class period, project Section Focus Transparency 15-3 and have students answer the activity questions. Discuss students' responses. **L1**

Vocabulary Pre-check
Have students define the term *food processing*. Then have them identify types of food-processing industries. **L1 LEP**

Armenia shares the Caucasus Mountains with two neighboring countries: Georgia and Azerbaijan (A•zuhr•BY•JAHN). Together these three lands are known as the Caucasus republics. All of them lie south of Russia. Once part of the Soviet Union, they are now independent countries.

Place
Armenia

Tiny landlocked Armenia takes pride in a history that reaches back to the ancient world. The map on page 382 shows that today Armenia is without a coastline. Centuries ago—as early as 800 B.C.—Armenia ruled a large empire that reached from the Caspian Sea to the Mediterranean Sea.

Rugged Land Most of Armenia is a rugged plateau of ridges, narrow valleys, and lakes. Formed millions of years ago by volcanic eruptions, this plateau today sits across many **faults,** or openings in the earth's crust. What occurs near faults? If you guessed earthquakes, you are correct. Earthquakes occur quite

CHAPTER 15

391

Classroom Resources for Section 3

 REPRODUCIBLE MASTERS
Reproducible Lesson Plan 15-3
Guided Reading Activity 15-3
Critical Thinking Skills
  Activity 15
GeoLab Activity 15
Section Quiz 15-3

 **TRANSPARENCIES**
Section Focus
  Transparency 15-3

MULTIMEDIA
 Testmaker

National Geographic
  Society: Picture Atlas of
  the World

More About the Illustration

Answer to Caption
It is on the Black Sea coast and has a mild climate.

Region Georgia has been controlled by many different peoples. The Byzantine Empire and Persia fought over this land, and in A.D. 654 it was conquered by the Arabs. Later, Georgia was a part of the Mongol and Ottoman empires.

TEACH

Guided Practice

Display a physical and political map of the independent republics. Have students locate the Caucasus Mountains and identify the countries through which this landform extends. *(Armenia, Georgia, Azerbaijan)* Then ask students what aspects of a country's geography might be affected by mountains. *(climate, economic activities, ways of life)* Encourage students to support their answers with examples from the Caucasus republics. **L2**

Independent Practice

📁 Guided Reading Activity 15-3 **L1**

---

frequently in Armenia. In 1988 a severe quake shook the country, killing thousands of people and destroying many villages and towns.

The Economy How do people in Armenia earn a living? In years past, many Armenians left their own country to work in the oil fields of neighboring Azerbaijan. Today they work in their own factories, making chemicals, electronic goods, machinery, and textiles. Armenia's miners take copper, gold, lead, and zinc out of the earth. Farmers—although few in number—grow wheat, barley, grapes, and other fruits in the hot, dry climate of the valleys.

Overlooking Tbilisi, Georgia

The ruins of an ancient castle and fort stand guard over the Georgian capital of Tbilisi.
PLACE: Why is Georgia popular with tourists?

The People Armenia's 3.7 million people are mostly ethnic Armenians who share a unique language, culture, and the Christian religion. In recent centuries Turkey and Russia ruled Armenia. The Armenians suffered under foreign rule, and many settled in other countries.

During the years of foreign rule, the Armenians protected their culture. They now value their newly won independence. Today about 70 percent of Armenia's people live in urban areas. Yerevan (YEHR•uh•VAHN), the capital, has broad streets, attractive fountains, and colorful buildings made of volcanic stone. Founded in 782 B.C., Yerevan is one of the ancient cities of the world.

Victor Basmajian is a teenager who lives in Yerevan. Like most Armenians, Victor and his family share close ties and like to entertain friends at home. Victor enjoys foods such as shish kebab, or meat and vegetables cooked on an iron skewer, and cabbage leaves stuffed with rice and meat. He also likes to play chess and a board game called backgammon.

Place
Georgia

A Georgian poet once said, "Three divine gifts have been bestowed on us by our ancestors: language, homeland, faith." The homeland of Georgia, as you can see on the map on page 384, lies north of Armenia. Georgia's west coast touches the Black Sea.

Mountains and Highlands Sandwiched between two mountain ranges, Georgia is a rugged land said to have been named for St. George, a Christian saint. Fertile mountain valleys and lowlands along the Black Sea coast are home to most of Georgia's people.

Cooperative Learning Activity

Have the class formulate a list of questions about the Caucasus republics. Write the questions on the chalkboard. Organize students into three groups and assign each group one of the Caucasus republics. Groups should answer the listed questions with information about their assigned country. Then form new groups consisting of three students, one student from each of the original groups. Students in the new groups should share their answers to the questions. **L1**

MAP STUDY

Azerbaijan is a major oil-producing country.
PLACE: What city in Azerbaijan might make use of that oil in manufacturing?

Agriculture

- Ranching
- Commercial farming
- Subsistence farming
- Little or no activity
- Manufacturing area

Resources

Coal	Iron ore	Nickel
Copper	Lead	Petroleum
Fishing	Manganese	Potash
Gold	Natural gas	Zinc

Lambert Equal-Area projection

LESSON PLAN
Chapter 15, Section 3

MAP STUDY

Answer
Baku

Map Skills Practice
Reading a Map What kinds of crops are grown in the Caucasus republics?
(tea, citrus fruits)

ASSESS

Check for Understanding

Assign Section 3 Review as homework or an in-class activity.

Meeting Lesson Objectives

Each objective below is tested by the questions that follow it in parentheses.

1. **Identify** the major landform found in the Caucasus republics. (2, 3)
2. **List** the farm products grown in this region. (3)
3. **Discuss** the ways people live in the republics of the Caucasus. (4)

Farms and Industries The map above shows you that farming ranks as Georgia's major economic activity. What would you grow if you were a Georgian farmer? You would probably harvest citrus fruits, tea, grapes, or wheat.

Georgia's major industries are wine making and **food processing,** or the preparing of foods as products for sale. Another leading industry is tourism. Resorts along the mild, scenic Black Sea coast draw thousands of visitors each year.

The People About 70 percent of Georgia's 5.5 million people are ethnic Georgians who are proud of their language, unique alphabet, and Christian heritage. The rest of Georgia's people belong to a number of other **ethnic groups**, or people having a common language, culture, and history. Since independence in 1991, conflict has broken out between Georgians and other ethnic groups. These groups want to separate from Georgia and create their own countries. Georgia first rejected membership in the Commonwealth of Independent States, but later joined in 1993.

More than half of Georgia's people live in cities and towns. Tbilisi (tuh•BEE•luh•see), the capital and largest city, prizes its warm mineral springs and scenic location. Georgians are known for their business skills, love of festivals, and friendliness to foreign visitors. They also maintain close family ties. Popular Georgian foods at family gatherings include shish kebab, pressed fried chicken, cheese, olives, and fruit.

CHAPTER 15

393

Meeting Special Needs Activity

Inefficient Readers Students with reading problems may have difficulty stating main ideas. Ask them to skim for a main idea under the subheading "The People" on page 393. Ask students to tell in their own words what the subheading is about. *(ethnic groups and ways of life in Georgia)* Then direct students' attention to the first sentence of the subheading. Point out that the main idea of a paragraph generally is stated in the first sentence. Have students state the main idea. *(Most of Georgia's people are ethnic Georgians.)* **L1**

Global Gourmet

Azerbaijan Azeris often serve *dovga,* a mixture of yogurt, rice, and herbs, after the main meal to improve digestion.

More About the Illustration

Answer to Caption
petroleum

Region The government of Azerbaijan claims that 44 of every 100,000 people in the country live past the age of 100. The Azeris attribute their longevity to diet and an active lifestyle.

Evaluate

Section 3 Quiz **L1**

Use the Testmaker to create a customized quiz for Section 3. **L1**

Reteach

Direct students to write sentences that summarize the main ideas of each subheading in Section 3. Have students read and compare their sentences. **L1**

Enrich

Have students research Baku, the capital of Azerbaijan. Ask them to create an illustrated report on its landmarks, major industries, and other points of interest. **L3**

CLOSE

Direct students' attention to the Georgian poet's saying on page 392. Have students create a chart with the headings *Language, Homeland,* and *Faith* across the top. Direct them to make chart entries for each of the ethnic groups from the Caucasus republics.

Preparing Lunch in Azerbaijan

An Azerbaijani family prepares vegetables for an outdoor lunch.
PLACE: What is Azerbaijan's most important natural resource?

Place
Azerbaijan

Also bordering Armenia is Azerbaijan, a country about the size of the state of Maine. It lies on the Caspian Sea. A narrow piece of Azerbaijan is separated from the rest of the country by Armenia. Since winning independence in 1991, Azerbaijan and Armenia have battled for control of territory and people.

The Land and Climate Like Armenia and Georgia, Azerbaijan is a mountainous country. The Caucasus ranges, having a highland climate, spread through the northern and western parts of Azerbaijan. A Mediterranean climate runs along the Caspian Sea. A dry coastal plain stretches to the north. Most of Azerbaijan is partly dry, with hot summers and mild winters.

The Economy Azerbaijan is largely agricultural. The country's farmers grow cotton, wheat, rice, fruits, and tobacco. Yet, along the shore of the Caspian Sea, countless oil-drilling rigs rise above the water. As you can see on the map on page 393, petroleum, lumber, and gold are Azerbaijan's main natural resources.

The People The 7.4 million people of Azerbaijan are mostly Azeris, a people distantly related to the people of Turkey in Southwest Asia. Armenians and Russians make up a small part of the population. The major faith of Azerbaijan is the religion of Islam.

About 55 percent of Azerbaijan's people live in cities. Baku (bah•KOO), the capital and largest city, has a population of 1.7 million. Founded more than 1,000 years ago, Baku got its name from Persian words meaning "windy town." Baku lies in one of the world's major oil-producing areas.

SECTION 3 REVIEW

REVIEWING TERMS AND FACTS
1. **Define the following:** fault, food processing, ethnic group.
2. **REGION** What two bodies of water lie on either side of Armenia, Georgia, and Azerbaijan?
3. **PLACE** What kind of landscape is found in Armenia?
4. **PLACE** What is Azerbaijan's major religion?

MAP STUDY ACTIVITIES
5. Look at the land use map on page 393. What is Georgia's major economic activity?

(394)

UNIT 5

Answers to Section 3 Review

1. All vocabulary words are defined in the Glossary.
2. Black Sea, Caspian Sea
3. rugged plateau
4. Islam
5. farming

The Central Asian Republics

PREVIEW

Words to Know
- nomad
- oasis

Places to Locate
- Kazakstan
- Almaty
- Kyrgyzstan
- Bishkek
- Uzbekistan
- Tashkent
- Tajikistan
- Dushanbe
- Turkmenistan
- Ashkhabad
- Aral Sea
- Kara Kum

Read to Learn . . .
1. what landforms and climates are found in the Central Asian republics.
2. how people earn a living in these republics.
3. how cultures here blend traditions with modern ways.

H ave you ever bargained for goods at a market? This seller displays tomatoes and dill at a market in Almaty, the capital of Kazakstan. Taking its first steps toward capitalism, the country is building a free market economy. Many people in Kazakstan's cities and towns now own small businesses.

Meeting National Standards

Geography for Life The following standards are highlighted in this section:
- Standards 4, 10, 15

FOCUS

Section Objectives
1. **Identify** the landforms and climates found in the Central Asian republics.
2. **Describe** how people earn a living in these republics.
3. **Discuss** how cultures in the Central Asian republics blend traditions with modern ways.

Bellringer Motivational Activity

Prior to taking roll at the beginning of the class period, project Section Focus Transparency 15-4 and have students answer the activity questions. Discuss students' responses. **L1**

Vocabulary Pre-check
Review the definition of *oasis.* Point out that this word is Greek in origin and takes a special plural form— *oases.* **L1** **LEP**

T he Central Asian republics include Kazakstan (KA•zak•STAN), Kyrgyzstan (KIHR•gih•STAN), Tajikistan (tah•JIH•kih•STAN), Turkmenistan (turk•MEH•nuh•STAN), and Uzbekistan (uz•BEH•kih•STAN). Reaching from east of the Caspian Sea to the borders of China, these republics cover an immense area. Once part of the Soviet Union, these countries became independent in 1991.

Place
Kazakstan

Kazakstan is a large country covering 1,049,200 square miles (2,717,428 sq. km). About one-third the size of the United States, this country lies south of Russia in central Asia.

A Variety of Landscapes Kazakstan has lowlands, steppes, deserts, and mountains. The map on page 384 shows you that the Tian Shan (tee•AHN SHAHN) mountain range extends through the eastern part of Kazakstan. Major rivers, such as the Irtysh (ihr•TISH), provide water to farms in dry areas.

CHAPTER 15

 395

Classroom Resources for Section 4

 **REPRODUCIBLE MASTERS**

Reproducible Lesson Plan 15-4
Guided Reading Activity 15-4
Reteaching Activity 15
Enrichment Activity 15
Section Quiz 15-4

 **TRANSPARENCIES**

Section Focus
    Transparency 15-4
Unit Overlay Transparency 5-4

MULTIMEDIA

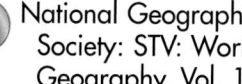 Testmaker

National Geographic
    Society: STV: World
    Geography, Vol. 1

TEACH

Guided Practice

Display Unit Overlay Transparency 5-4. Have students locate the Central Asian republics and draw conclusions about why they are so called. *(They are located near the center of Asia.)* Then guide students in using the respellings on page 395 to pronounce the names of these countries. **L1**

NATIONAL GEOGRAPHIC SOCIETY

These materials are available from Glencoe.

VIDEODISC

STV: World Geography

Vol. 1: Asia and Australia

Interior and North Asia
Side 1, Frames 13631-17712

MAP STUDY

Answer
under 2 persons per sq. mi./under 1 person per sq. km

Map Skills Practice
Reading a Map Is population density greater around the Black Sea or the Caspian Sea? *(Black Sea)*

Kazakstan as a whole receives very little rainfall. Its people endure very cold winters and long, hot summers. Wheat farming and sheep raising are the country's major economic activities. Kazakstan is also rich in minerals such as coal and oil. Some 9 billion barrels of oil may lie beneath its soil. The map on page 393 shows natural resources of Kazakstan. The government recently opened its border with China, hoping to increase trade with Asia.

The People Many Kazaks and other people of the region trace their ancestry to the Mongols. Kazaks make up about 40 percent of Kazakstan's population. Most follow Islam and live in rural areas. Russians, who form the second-largest group, live in the major cities. Most of the 1.5 million people of Almaty (al•MAH•tee), Kazakstan's capital, are Russians.

Kazak culture is rich in folk songs and legends. At one time Kazaks were **nomads,** or people who moved with their flocks in search of pasture. Today Kazaks live in rural houses or city apartments.

Place
Kyrgyzstan

The country of Kyrgyzstan—one of the most southern republics—lies southeast of Kazakstan. Kyrgyzstan's 76,600 square miles (198,394 sq. km) make it about the size of the state of Minnesota.

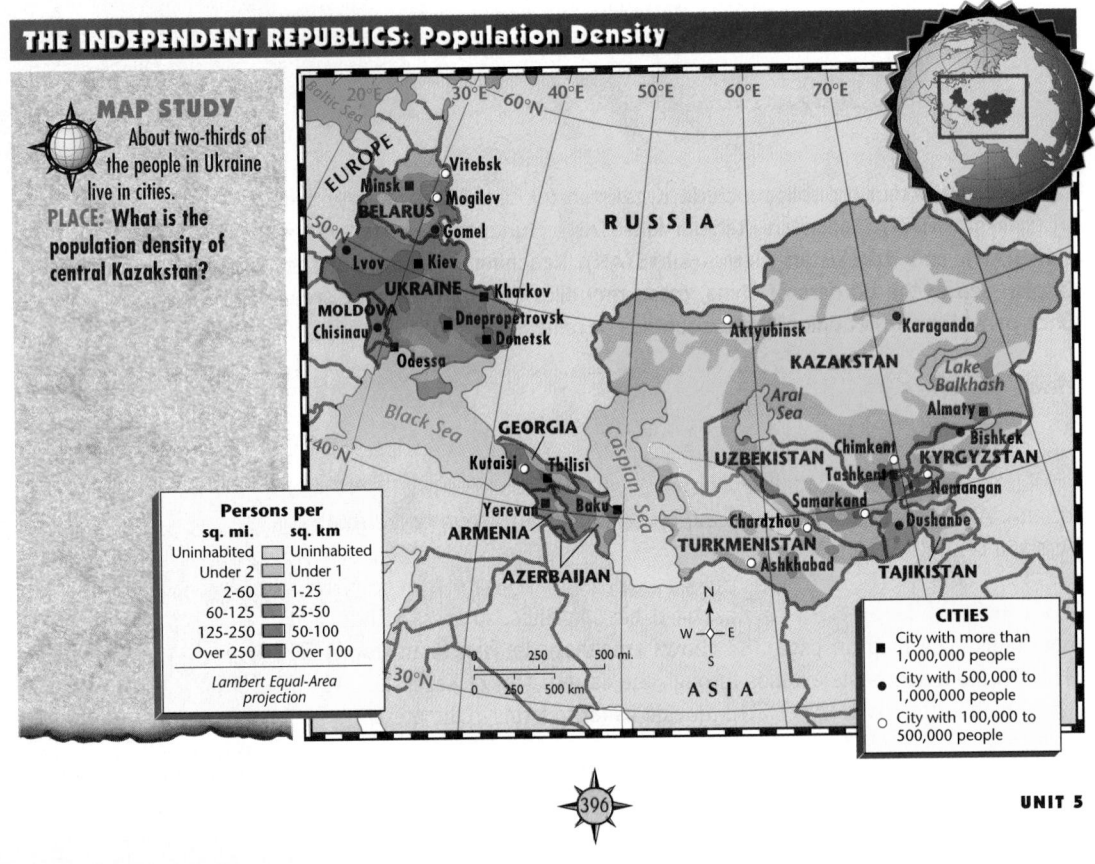

THE INDEPENDENT REPUBLICS: Population Density

MAP STUDY
About two-thirds of the people in Ukraine live in cities.
PLACE: What is the population density of central Kazakstan?

Persons per
sq. mi.	sq. km
Uninhabited	Uninhabited
Under 2	Under 1
2-60	1-25
60-125	25-50
125-250	50-100
Over 250	Over 100

Lambert Equal-Area projection

CITIES
■ City with more than 1,000,000 people
● City with 500,000 to 1,000,000 people
○ City with 100,000 to 500,000 people

Cooperative Learning Activity

Have students work in groups of five to make puzzle maps of the Central Asian republics. Have members of each group trace the outlines of these countries on poster board. Direct other group members to label and then cut out the countries along the borders. Use the map puzzles to play a game in which groups compete by answering questions on the Central Asian republics. For each correct answer, a group may put one piece of the puzzle in place. The first group to assemble its puzzle wins. **L1**

A Land of Mountains The Tian Shan mountain range covers a large area of Kyrgyzstan. Nestled between the mountains lie high grassy plains and low river valleys. As in other highland areas, the climate of Kyrgyzstan differs with elevation. Lower valleys and plains have warm, dry summers and chilly winters. Highland areas have cool summers and bitterly cold winters. Most of Kyrgyzstan's people raise livestock or grow cotton, vegetables, and fruits.

The People More than half of Kyrgyzstan's 4.5 million people belong to the Kyrgyz (kihr•GEEZ) ethnic group. Russians make up the rest of the country's population. The Kyrgyz tend to live in rural areas, and the Russians live in the cities. Bishkek, with a population of about 616,000, is the capital and largest city.

Place
Tajikistan

Southwest of Kyrgyzstan lies the mountainous country of Tajikistan. Covering an area of 55,300 square miles (143,227 sq. km), Tajikistan is about the size of the state of Iowa.

Mountains and Valleys "Roof of the World" is what some people call Tajikistan's mountains. In one spot there, four major Asian mountain ranges come together. The map on page 384 shows that the Pamir Mountains rise in southeastern Tajikistan. Here Communism Peak reaches 24,590 feet (7,495 m).

Tajikistan is a farming country that depends on irrigation. In the river valleys, Tajik farmers grow cotton, rice, and fruits. In highland areas, coal, lead, and zinc are mined.

The People Most Tajiks are Muslims who speak a language similar to that spoken in Iran. About two-thirds of Tajikistan's people live in villages along the rivers. The largest city is Dushanbe (doo•SHAM•buh), the capital.

Place
Turkmenistan

Turkmenistan has some of the most isolated areas in the Central Asian republics. Covering 188,500 square miles (488,215 sq. km), it is slightly larger than the state of California.

Desert Country More than 85 percent of Turkmenistan is a vast desert known as the Kara Kum (KAHR•uh•KOOM), which means "black sand." For their water supply, the people of Turkmenistan rely on a few rivers and **oases**, or fertile desert areas watered by underground springs.

CHAPTER 15 397

Teen Scene
Making His Mark

Kasim Ben Amir has found a way to be creative and earn money at the same time. Fifteen-year-old Kasim lives near Khodzent, Tajikistan. Much of the cotton and silk produced on Tajikistan farms is made into textiles here. Kasim buys plain shirts for a few rubles and then stamps them with designs that he creates. He sells his creations at a nearby market. Kasim gives some of the money he earns to his family. He's saving the rest to pay for a trip to Iran to visit his cousins.

More About the Illustration

Answer to Caption
Uzbekistan and Kazakstan

Human/Environment Interaction Estimates suggest that irrigation would have to be halted for 30 years to bring the Aral Sea back to its 1960 level.

Evaluate

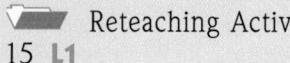 Section 4 Quiz **L1**

🔘 Use the Testmaker to create a customized quiz for Section 4. **L1**

📁 Chapter 15 Test Form A and/or Form B **L1**

Reteach

📁 Reteaching Activity 15 **L1**

Enrich

📁 Enrichment Activity 15 **L3**

CLOSE

Mental Mapping Activity

Provide each student with a large sheet of white paper and map pencils. Inform students that this will not be a graded activity. Have students draw and label, freehand from memory, a map of the 11 independent republics.

The Shrinking Aral Sea

Stranded ships lie aground on the shoreline of the shrinking Aral Sea.
HUMAN/ENVIRONMENT INTERACTION: What two countries share the Aral Sea?

Cotton, Turkmenistan's major crop, covers acres of irrigated land. Other farm products include grains, grapes, and potatoes. Turkmenistan's natural resources include natural gas, petroleum, copper, gold, and lead.

The People Most of Turkmenistan's 4.1 million people belong to the Turkmen ethnic group. Islam is Turkmenistan's major religion. The population is fairly evenly divided between rural and urban dwellers. Ashkhabad (ASH•kuh•BAD), with about 400,000 people, is the capital and largest city.

Place
Uzbekistan

Like Turkmenistan, Uzbekistan is slightly larger than California—about 172,700 square miles (447,293 sq. km). But Uzbekistan has about 8 times as many people as Turkmenistan.

Cotton Country Deserts and high plains cover much of Uzbekistan. In the west are dry lowlands around the Aral Sea, a large inland salt lake. In the east are fertile valleys that receive plenty of water from the Tian Shan mountain range. Uzbekistan's economy relies on cotton, grains, fruits, and vegetables that are grown in irrigated areas.

Cotton growing in Uzbekistan has taken its toll on the environment. Under the Soviets, farmers were ordered to irrigate crops with water drained from the rivers flowing into the Aral Sea. Since the 1960s the Aral Sea has shrunk drastically in size, and its salt content has increased. Fish and other wildlife have disappeared. Salt particles carried by wind from the lake pollute the air and soil.

The People Most of Uzbekistan's 22.1 million people are Uzbeks. More than 2 million people live in the capital city of Tashkent—the largest city in all of central Asia. Most Uzbeks practice Islam and live in rural areas.

SECTION 4 REVIEW

REVIEWING TERMS AND FACTS
1. Define the following: nomad, oasis.
2. REGION What type of landscape covers most of Turkmenistan?
3. PLACE What is the capital of Uzbekistan?
4. HUMAN/ENVIRONMENT INTERACTION What caused the Aral Sea to shrink?

 MAP STUDY ACTIVITIES
5. Look at the population density map on page 396. What Central Asian republic is the most densely populated?

Answers to Section 4 Review
1. All vocabulary words are defined in the Glossary.
2. desert
3. Tashkent
4. irrigation from rivers that flow into the Aral Sea
5. Uzbekistan

Chapter 15 Highlights

Important Things to Know About the Independent Republics

SECTION 1 UKRAINE

- Ukraine's landscape consists of lowlands, highlands, a coastal plain, and a central plains area.
- Ukraine is called the breadbasket of Europe because of its fertile farmland.
- Ukraine's capital, Kiev, was the center of an early Slavic civilization.

SECTION 2 BELARUS AND MOLDOVA

- The major economic activities of Belarus are farming, manufacturing, and service industries.
- Moldovans have strong ties to Romanians in language and culture.

SECTION 3 THE CAUCASUS REPUBLICS

- The Caucasus Mountains cover most of Armenia, Georgia, and Azerbaijan.
- The people of Armenia and Georgia have separate cultures, languages, and heritages.
- Located on the Caspian Sea, Azerbaijan is a major oil-producing country.

SECTION 4 THE CENTRAL ASIAN REPUBLICS

- Mountains, deserts, and plains form the landscapes of the Central Asian republics.
- Most of the peoples of the Central Asian republics practice the Islamic religion.
- Farmers in these republics depend on irrigation to grow cotton, grains, fruits, and vegetables.

Ukrainian folk dancers ▶

Extra Credit Project

Drawing Graphs Have students draw various types of graphs to compare features of the countries studied in this chapter. Bases of comparison might include area, population, major cities, urban and rural population, and economic activities. Display completed graphs in the classroom. **L2**

MAP STUDY

Human/ Environment Interaction

Copy and distribute outline maps of the independent republics. Have students label the republics and mark the major economic activities and natural resources in the region. Remind students to include a key with their map.

399

Chapter 15 Review and Activities

ANSWERS
Reviewing Key Terms

1. G
2. F
3. E
4. A
5. C
6. D
7. H
8. B

 Mental Mapping Activity

This exercise helps students to visualize the countries and geographic features they have been studying and understand the relationships among them. All attempts at freehand mapping should be accepted.

REVIEWING KEY TERMS

Match the numbered terms in Column A with their definitions in Column B.

A
1. nomad
2. steppe
3. nature preserve
4. landlocked
5. ethnic group
6. oasis
7. food processing
8. fault

B
A. having no coastline
B. break in the earth's crust
C. has a common culture, language, and history
D. fertile place in a desert area supplied by underground water
E. land set aside for the protection of plants and wildlife
F. vast treeless plain
G. person who travels in search of pasture for grazing animals
H. the preparing of food for sale as products

Mental Mapping Activity

On a separate sheet of paper, draw a freehand map of the independent republics. Label the following items on your map:
- Kiev
- Black Sea
- Tian Shan Mountains
- Dnieper River
- Aral Sea
- Georgia

REVIEWING THE MAIN IDEAS

Section 1
1. **LOCATION** Where is Ukraine located?
2. **REGION** Why is Ukraine called a breadbasket?

Section 2
3. **REGION** What landforms are found in Belarus?
4. **HUMAN/ENVIRONMENT INTERACTION** What agricultural products come from Moldova?

Section 3
5. **MOVEMENT** Why did many Armenians migrate to other countries?
6. **HUMAN/ENVIRONMENT INTERACTION** What are the major industries of Georgia?
7. **REGION** What are the two major religions of the Caucasus republics?

Section 4
8. **REGION** What countries make up the Central Asian republics?
9. **HUMAN/ENVIRONMENT INTERACTION** What is the role of rivers in the Central Asian republics?

Reviewing the Main Ideas
Section 1
1. south and east of Russia along the Black Sea
2. It produces wheat that is exported to other parts of the world.

Section 2
3. lowlands in the north; marshes, swamps, and forests in the south
4. wine from grapes, wheat, corn

Section 3
5. to escape harsh rule of their homeland by foreigners
6. wine making, food processing, and tourism
7. Christianity and Islam

Section 4
8. Kazakstan, Kyrgyzstan, Tajikistan, Turkmenistan, Uzbekistan

CRITICAL THINKING ACTIVITIES

1. **Drawing Conclusions** Why do you think Moldova is a densely populated country?
2. **Evaluating Information** Did years of Soviet rule help or harm the development of the independent republics? Explain.

GeoJournal Writing Activity

Imagine you are vacationing in Ukraine and the other independent republics. Write a letter home describing what you are seeing and where you are staying. Use travel books and your text to describe in detail each of your stops. In your letter, describe the scenery, climate, and the activities of the people you see.

COOPERATIVE LEARNING ACTIVITY

Organize into 11 groups, and then choose one of the republics discussed in this chapter. Each group will make a diorama of a republic, using a mixture of flour, salt, and water to indicate land features and rivers. Refer to the Geolab activity on page 378 for the salt-dough recipe. Paint your diorama and exhibit it in your classroom.

PLACE LOCATION ACTIVITY: THE INDEPENDENT REPUBLICS

Match the letters on the map with the places of the independent republics. Write your answers on a separate sheet of paper.

1. Black Sea
2. Caucasus Mountains
3. Ukraine
4. Crimean Peninsula
5. Aral Sea
6. Dnieper River
7. Kazakstan
8. Kara Kum
9. Baku
10. Minsk

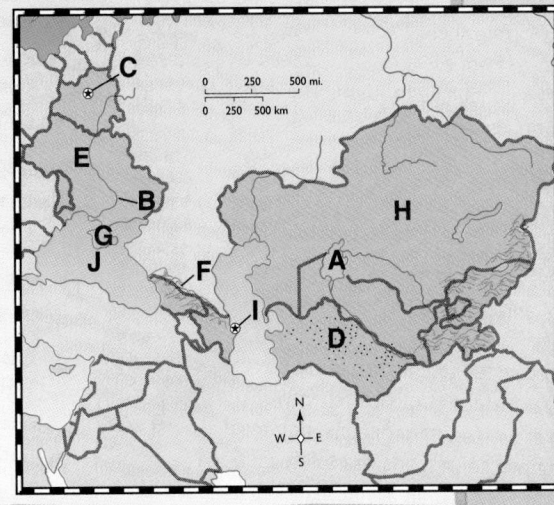

Cooperative Learning Activity

Challenge students to develop colorful and imaginative ways to label land features and bodies of water on their dioramas. Display the finished dioramas in the classroom and invite other classes to share them.

GeoJournal Writing Activity

Encourage students to share their letters with the rest of the class. Then have students discuss which countries in the region they would like to visit and why.

Place Location Activity

1. J
2. F
3. E
4. G
5. A
6. B
7. H
8. D
9. I
10. C

Chapter Bonus Test Question

This question may be used for extra credit on the chapter test.

How is Ukraine of Europe similar to Iowa in the United States?
(They are both known as "breadbasket" because of their grain harvest.)

9. Rivers provide water to irrigate farmland in this mostly dry region.

Critical Thinking Activities

1. Answers will vary, but should mention the country's mild climate and fertile soil.

2. Answers will vary. Students might mention that the Soviet rule helped the independent republics by bringing modern ways and industrialization. They may also mention that Soviet rule harmed the area by curtailing freedom, causing tremendous loss of life, and pollution of the environment. Most students will conclude that Soviet rule did more harm than good.

FOCUS

Have students imagine the following situation: the sound of a distant explosion is immediately followed by the lights going out. Sirens wail, and frightened people begin congregating in the streets. Rumors of an explosion at the nearby nuclear power plant begin to spread. How would students feel? What actions would they take? Who would they turn to for help?

TEACH

Remind students that Ukraine was still part of the Soviet Union when the Chernobyl accident occurred. Point out that the Soviet state was extremely secretive and security-conscious. Ask students to speculate on how such practices may have worsened the Chernobyl disaster. Then have students write a brief paragraph showing how the Chernobyl accident illustrates the following statement: Nature does not respect political boundaries. **L2**

NATIONAL GEOGRAPHIC SOCIETY

EYE ON THE ENVIRONMENT

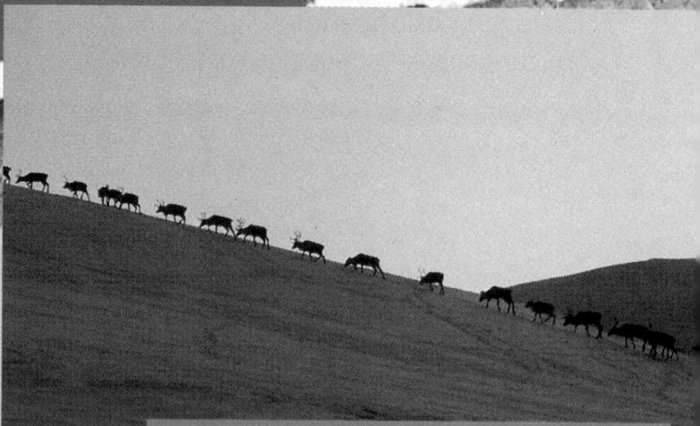

CHERNOBYL
NUCLEAR DISASTER

PROBLEM

On April 26, 1986, Reactor Number 4 of the Chernobyl Nuclear Power Plant exploded. Radioactive material flew more than 3 miles into the air. Blown by the wind, radioactive clouds contaminated thousands of square miles. Nearly 5 million people in Ukraine, Belarus, and Russia were affected.

Authorities quickly encased the reactor in a steel-and-concrete shell. They evacuated 116,000 people from the area and scraped away tons of contaminated topsoil. The cleanup continues to this day.

SOLUTIONS

● Ukraine authorities have agreed to shut down the 3 nuclear reactors remaining in Chernobyl and to work with the United States to replace electricity generated by the plants.
● Russia has agreed to encourage the use of natural gas instead of nuclear energy.
● Better worker-training programs, improved operating procedures, and fire safety measures have been started at nuclear plants.

Reindeer move up a ridge in Lapland. Radiation from the Chernobyl explosion contaminated reindeer herds, making the meat and milk unsafe for human consumption.

CHERNOBYL FACT BANK

🌶 The Chernobyl accident was the worst nuclear-power-plant disaster of all time. An estimated 5,000 people died and 30,000 were disabled.

🌶 More than 30,000 square miles of farmland were contaminated. Nothing can be grown there again for 100 years.

🌶 About 23 percent of the land in Belarus was contaminated.

🌶 Cases of thyroid cancer in children and birth defects have increased in the area near the reactor and in the path of the radioactive cloud.

🌶 Penguins in Antarctica, reindeer in Lapland, and countless other animals have suffered the effects of the radioactive clouds that encircled the planet.

402

More About the Issues

Human/Environment Interaction

The potential for catastrophic accidents is a common criticism of nuclear power. Another serious problem is what to do with the by-products that result from the creation of nuclear energy. Nuclear power plants generate a variety of radioactive waste materials that remain extremely dangerous for thousands of years. Currently, most of this waste is stored at the power plants in pools of water. The building of long-term underground storage vaults is proceeding. Critics charge, however, that seepage from such permanent-storage facilities poses as great a threat to the environment as accidents like that at Chernobyl.

At 1:23 A.M. on April 26, 1986, a test on emergency systems in the Chernobyl Nuclear Power Plant went wrong—and led to the explosion of Reactor Number 4. As pictured in this illustration, the explosion ripped apart the reactor and sent radioactive material high into the atmosphere on a plume of intense heat.

go by bike!

environmental activities

TEEN TRIBUTE

In April 1986, a group from Ramapo High School in Spring Valley, New York, was visiting Leningrad (now St. Petersburg) when the Chernobyl reactor blew up. The visitors returned home—their trip cut short by the disaster. Since then, Ramapo students and their teacher, working with community leaders and drug companies, have provided more than $11 million worth of medicine, vitamins, and toys to hospitals in Belarus and Russia.

WHAT CAN YOU DO?

TAKE A CLOSER LOOK

1 What disaster occurred in Ukraine at the Chernobyl Nuclear Power Plant?

2 What were the lasting effects of the Chernobyl disaster?

🍃 SAVE ENERGY–ride a bike!

🍃 Support a group working for safe disposal of toxic and nuclear wastes.

🍃 Don't add to toxic waste–use fewer batteries and buy only rechargeable ones.

🍃 SAVE ENERGY– encourage your parents to lower the thermostat 6 degrees in winter. If all Americans did, we'd save more than 500,000 barrels of oil a day!

403

This Ukrainian couple had to resettle after the accident, with only their colorful wall hangings to remind them of their lost home.

ASSESS

Have students answer the **Take A Closer Look** questions on page 403.

CLOSE

Discuss with students the **What Can You Do?** question. Have students find out how electricity for their community is generated and what impact that type of power plant has on the environment.

* For an additional regional case study, use the following:

📁 Environmental Case Study 5

Answers to Take A Closer Look

1. The power plant exploded, spewing radioactive material into the atmosphere.
2. The accident resulted in the contamination of 30,000 square miles of farmland. Nothing can be grown on this land for 100 years. Also, the incidence of cancers and birth defects increased in contaminated areas.

Wildlife hundreds of miles from the accident site has been affected by radioactivity.

0:00 OUT OF TIME?

If time does not permit teaching each chapter, you may use the Chapter Highlights, Audiocassettes, and their corresponding activities and tests.

GLENCOE
TECHNOLOGY

 **VIDEODISC**

Reuters Issues in Geography

Chapter 5
North Africa and Southwest Asia:
The Influence of Oil and Water
Disc 1 Side A

interNET
CONNECTION

Resources on the countries of Southwest Asia and North Africa are available at the following address:

World Wide Web
http://www.columbia.edu/cu/libraries/indiv/area/MiddleEast/region.html

Interested in the latest news from Southwest Asia and North Africa? Contact the web sites of the Arab countries at:

World Wide Web
http://www.liii.com/~hajeri/arab.html

Contact an Israeli web site at:

World Wide Web
http://ucsu.colorado.edu/~jsu/information.html#isnf

Unit 6 Southwest Asia and North Africa

What Makes Southwest Asia and North Africa a Region?

The people share . . .

• a constant quest for water.

• Mediterranean, steppe, and desert climates.

• an abundance of oil and natural gas reserves.

• the historical beginnings of the Islamic, Jewish, and Christian religions.

To find out more about Southwest Asia and North Africa, see the Unit Atlas on pages 406–417.

The Great Pyramids near Cairo, Egypt

404

About the Illustration

One's most vivid image of Egyptian life probably includes the Pyramids located at Giza. To appreciate their massive size, consider that the Pyramid of Khufu covers an area of 13 acres (5.3 ha). The five largest cathedrals in the world could be placed within its base with room to spare! The pyramid was made by piling 2.3 million blocks of limestone, each averaging 2.5 tons (2.3 metric tons) to a height of 480 feet (146.3 m). This makes the pyramid about as high as the Washington Monument.

 **Movement** Ask students how they think these pyramids were built. *(Scholars believe that thousands of workers quarried and dragged/slid stone to the construction site. Stones may have been dragged up ramps of earth and sand that were raised with each level of the structure.)*

NATIONAL GEOGRAPHIC SOCIETY

These materials are available from Glencoe.

 CD-ROM

Picture Atlas of the World

 SOFTWARE

ZipZapMap! World

GeoJournal Activity

As students write their descriptions of Southwest Asia and North Africa, have them include information about cultural differences. For example, suggest that they note how gestures and body language in this region differ from those in Western cultures. Call on volunteers to read their newspaper articles to the class. **L2**

- This journal activity provides the basis for the "GeoJournal Writing Activity" exercise in the Chapter Review.
- The GeoJournal may be used as an integral part of Performance Assessment.

GeoJournal Activity

As you read about this region, imagine that you are a journalist writing a book about Southwest Asia and North Africa. Write descriptions of at least five cities or areas. Then use the descriptions to write an article for your local newspaper.

405

Location Activity

Display Political Map Transparency 6 and ask students the following questions: What body of water borders most of the countries of North Africa and the western countries of Southwest Asia? *(Mediterranean Sea)* Which countries border the Persian Gulf? *(Iran, Iraq, Kuwait, Saudi Arabia, Qatar, Bahrain, United Arab Emirates, and Oman)* Which country forms a link between Europe and Asia? *(Turkey)* Which country in this region has no coastline? *(Afghanistan)* Which countries border on more than one body of water? *(Morocco, Egypt, Saudi Arabia, Yemen, Oman, Iran, Turkey)* **L1**

IMAGES OF THE WORLD

NATIONAL GEOGRAPHIC SOCIETY

These materials are available from Glencoe.

CD-ROM

Picture Atlas of the World

POSTERS

Images of the World
Unit 6: Southwest Asia and North Africa

TRANSPARENCIES

PicturePack Transparencies, Unit 6

Cultural Kaleidoscope

Southwest Asia and North Africa In many countries in the region, it is considered an insult to point the soles of the feet at another person. For this reason, people generally avoid sitting cross-legged.

UNIT 6 ATLAS

NATIONAL GEOGRAPHIC SOCIETY

IMAGES of the WORLD

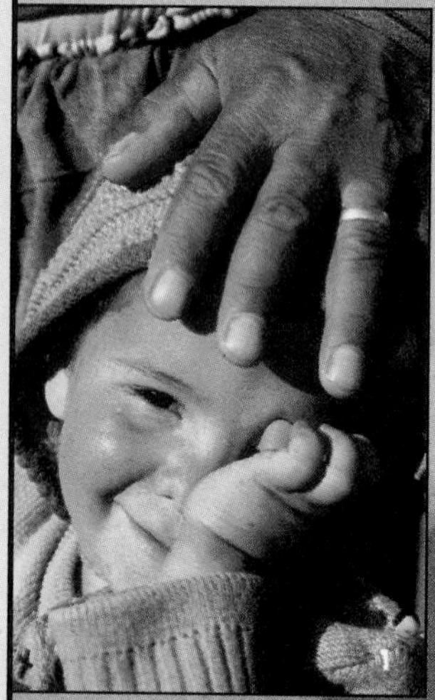

1. Portrait of a child, Turkey
2. Boy in a poppy field, Tunisia
3. Dome of the Rock, Jerusalem
4. Hobbled camels, Egypt
5. Woman and children, Saudi Arabia
6. Pyramid, Egypt
7. Fishermen with their catch, Yemen

All photos viewed against a desert landscape, Tunisia.

406

Fun Facts

• **Oman** This country is one of the hottest in the world. Temperatures often reach 130°F (54.4°C).

• **Saudi Arabia** Frankincense and myrrh are fragrant materials that come from trees that grow in this country. Since ancient times, people have burned these materials as incense or used them in perfumes.

Geography and the Humanities

 World Literature Reading 6

World Music: Cultural Traditions, Lesson 5

World Cultures Transparencies 16, 17

Focus on World Art Print 7

Global Gourmet

Southwest Asia The pomegranate has been popular in Southwest Asia since ancient times. A picture of the fruit, for example, appeared on the pillars of Solomon's temple in Jerusalem. People in the region use the fruit's crimson-colored pulp to make drinks.

Teacher Notes

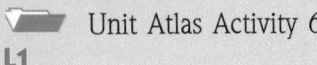

Using the Unit Atlas

These features and activities may be used as an introduction to the unit or as teaching tools throughout the course of the unit.

📁 Unit Atlas Activity 6
L1

📁 Unit Overlay Transparencies 6-0, 6-1, 6-2, 6-3, 6-4, 6-5, 6-6

📁 Home Involvement Activity 6

📁 Environmental Case Study 6

🗂 Countries of the World Flashcards, Unit 6

🧩 World Games Activity Cards, Unit 6

More About the Illustration

Answer to Caption
The land near the rivers is more fertile and can be used as farmland.

Region The fertile plain between Iraq's Tigris and Euphrates rivers is the site of the first known civilization. The area, once known as Mesopotamia, supported the civilization of Sumer around 3500 B.C.

Regional Focus

Southwest Asia and North Africa lie where the continents of Asia, Africa, and Europe meet. You will find the world's largest desert and part of the world's longest river in this region. Some of the world's earliest civilizations and three great world religions started here. Today this region is one of the world's major suppliers of oil.

Region
The Land

The region of Southwest Asia and North Africa covers about 5.5 million square miles (14.2 million sq. km)—about 10 percent of the earth's total land area. Look at the physical map on page 412. The region stretches from sandy beaches along the Atlantic Ocean on the west to the towering Hindu Kush on the east. North to south, the region reaches from the deep blue waters of the Mediterranean Sea to the Sahara.

Mountains Towering mountains rise over much of Southwest Asia and North Africa. The Taurus Mountains run east and west through the southern edge of the country of Turkey. Another range—the Zagros—stretches between Iran and Iraq. Other mountain chains are the Elburz Mountains of Iran and the Asir Mountains of Saudi Arabia. Farther west, in North Africa, stretch the Atlas Mountains—Africa's longest mountain range.

Deserts Mountains and prevailing winds have created vast deserts in many parts of Southwest Asia and North Africa. The Rubal Khali, or Empty Quarter, of the Arabian Peninsula is made up of shifting sand dunes. The Sahara—the largest desert in the world—covers much of North Africa.

Water Several bodies of water border the region of Southwest Asia and North Africa. These include the Mediterranean Sea, the Black and Caspian seas, the Persian Gulf, and the Arabian Sea. Freshwater, however, is scarce. Land near rivers and underground springs is green and fertile. With the help of irrigation, crops grow in these areas.

The Nile River, more than 4,000 miles (6,400 km) long, flows north through Egypt to the Mediterranean Sea. It is the world's longest river. Two other major rivers—the Tigris and the Euphrates—flow through southeastern Turkey, Syria, and Iraq into the Persian Gulf.

Euphrates River in Iraq

A thick stand of palm trees lines the banks of the Euphrates River.
REGION: Why do you think most people in Iraq live near the Tigris and Euphrates rivers?

Facts

• **Egypt** Ancient Egyptian women dyed their hair, wore red lip powder, and painted their fingernails. They used gray, black, or green paint to outline their eyes and color their eyebrows. Men often wore as much makeup as women.

• **Iraq** Baghdad's *suq,* or marketplace, is one part of the city that has escaped modernization. For centuries, the covered bazaar has provided shoppers with goods ranging from spices to brass pots.

Natural Resources Oil and natural gas are the most important natural resources in Southwest Asia and North Africa. You will find these minerals in the countries of North Africa and those bordering the Persian Gulf. Southwest Asia and North Africa are also rich in iron ore, copper, lead, manganese, zinc, and phosphate. Phosphate is used for making fertilizers.

Region
Climate and Vegetation

The many mountains of Southwest Asia and North Africa block the flow of rain clouds to the inland side of the region. Only coastal areas enjoy regular and adequate rainfall. Rainfall averages about 10 inches (25 cm) or less each year. This amount is inadequate to grow most food crops, and irrigation is necessary.

Desert and steppe climates cover much of Southwest Asia and North Africa. During the day, temperatures in the desert may soar above 100°F (38°C). In steppe areas, enough rain falls for grasses to grow. This allows farmers to raise livestock. A Mediterranean climate of hot, dry summers and cool, rainy winters is found on coastal plains.

Human/Environment Interaction
Economy

Economies differ from country to country in Southwest Asia and North Africa. The people living in countries that have oil, manufacturing, and trade generally enjoy a high standard of living. People living in countries where the economies are based on farming have a low standard of living.

Agriculture Most people in Southwest Asia and North Africa are farmers or herders. Cereals, citrus fruits, grapes, and dates grow in river valleys and coastal areas. Another source of food for the people of the region is livestock—mostly cattle and sheep. Farmers in river valleys or in irrigated areas also grow cotton, one of the region's major exports.

Industrial Growth Oil and natural gas are important sources of wealth for many Southwest Asian and North African countries. These resources are major exports and have allowed industries to develop. Factories in the region produce textiles, fertilizers, medicines, plastics, and paints. Service industries such as banking, insurance, and tourism have also grown in recent years.

Transportation and Communication In the past, mountains and deserts made transportation costly in the region. Today roads and railroads link cities, oil fields, and seaports. Inland waterways such as the Nile River and the Suez Canal also move people and goods.

409

Damascus, Syria

People gather in marketplaces to shop for items made by local workers.
PLACE: What mineral is an important source of wealth for many countries in this region?

NATIONAL GEOGRAPHIC SOCIETY

These materials are available from Glencoe.

CD-ROM

Picture Atlas of the World

SOFTWARE

ZipZapMap! World

POSTERS

Eye on the Environment, Unit 6

More About the Illustration
Answer to Caption
oil

Place Many of Damascus's marketplaces are in the old section of the city on narrow, winding streets. Merchants sell food and objects made of leather, brass, silver, and gold. People go to the marketplaces not only to shop, but also to discuss business and to socialize.

Fun Facts
• **Egypt** Ancient Egyptians buried their kings in a secret chamber inside or beneath a pyramid. They filled the chamber with gold and other treasures as well as practical, everyday items. Egyptians believed that the king would need these things in the afterlife.

• **North Africa** Couscous, a kind of wheat meal, is a staple of the diet throughout North Africa. As a main course, couscous is steamed and served with vegetables and meats or fish. Sometimes a spicy sauce is added. Sweetened couscous, mixed with raisins, pomegranates, or dates, is served as a dessert.

Global Gourmet
Lebanon A lettuce scoop is used to eat *tabbouleh*—a popular Lebanese salad made with cracked wheat, parsley, mint, minced onions, diced tomatoes, lemon juice, and olive oil. Lettuce is not part of the salad.

UNIT 6 ATLAS

FactsOnFile

Use the reproducible masters from **GEOGRAPHY ON FILE**, *North Africa and Southwest Asia,* to enrich unit content.

CURRICULUM CONNECTION

Life Science Camels have been used as pack animals in Southwest Asia and North Africa for centuries. These "ships of the desert" can carry as much as 1,000 pounds (454 kg), although their usual load is about one-third of that weight. Working camels generally travel about 25 miles (40 km) a day.

More About the Illustration

Answer to Caption
Egypt

Location The Temple of Luxor is about 400 miles (644 km) south of Cairo on the Nile River. The pharaoh Amenhotep III built the temple at the center of the city of Thebes at the beginning of his reign in 1417 B.C.

Place
The People

For centuries Southwest Asia and North Africa have served as the meeting place for the peoples of Asia, Africa, and Europe. The region today has people from many different ethnic backgrounds.

History and Culture Southwest Asia and North Africa have a rich culture and history. Centuries ago the region's farmers were the first to raise many of the grains, vegetables, and animals still used as basic foods in much of the world. Some of the world's oldest civilizations developed in the Nile and Tigris-Euphrates river valleys.

Three great religions—Judaism, Christianity, and Islam—began in Southwest Asia. All of them share the belief in one God. In the mid-600s, Islam became the region's leading religion as a result of the teachings of Muhammad. An Islamic state soon arose and became a powerful empire. The study of sciences and the arts flourished under Islamic rule. Islamic thinkers passed on much of their knowledge to western Europe and Asia.

From the 800s to the 1500s, a number of Islamic empires flourished in Southwest Asia and North Africa. By the 1800s, these empires had declined and Europeans ruled most of the region. After World Wars I and II, direct European rule ended. By the 1960s most of the Arab peoples in Southwest Asia and North Africa had set up independent states. These include a mix of monarchies, democratic republics, and dictatorships.

Ethnic Groups Most of the people of Southwest Asia and North Africa are Arabs. About 90 percent are Muslims who practice the religion of Islam. Most of the remaining 10 percent are Christians. The Arab culture, especially the Arabic language, has had a lasting influence on the region.

Two other groups—the Iranians and the Turks—once ruled over empires in this region. Neither group is Arab, but many in each group are Muslims. Living among Turks, Iranians, and Arabs are two smaller ethnic groups—Armenians and Kurds.

In the Southwest Asian country of Israel, most of the people are Jewish. Israel was founded in 1948 as a Jewish state. Israeli Jews trace their ancestry to

Ancient Thebes

Giant stone figures guard the 3,000-year-old Temple of Luxor.
LOCATION: In what country can you see ancient pyramids?

410

Fun Facts

• **Iran** Islam requires all Muslims to dress modestly. After the Iranian Revolution in 1979, religious leaders passed laws enforcing this requirement. Women, for example, must wear *chadors*, or head scarves, and long, loose-fitting dresses when in public.

• **Egypt** In traditional Egyptian families, a girl is protected by her brothers. They may even accompany her in public. Indeed, a man's honor is judged by his ability to protect the women in his family.

the Hebrews who settled the region in ancient times and believed that God had given them the land as a permanent home.

For many years, conflicts have taken place between Jews and Arabs. Palestinians—Arabs living in Israeli-ruled territory—especially opposed the Jewish state. Today, however, peace agreements between Israeli and Arab leaders have led to greater self-rule for Palestinians in the area.

Population About 350 million people live in Southwest Asia and North Africa. Because of the vast deserts, most people live along seacoasts or rivers, or near highlands. The environment in these areas makes it possible to grow crops and raise animals. Turkey, Egypt, and Iran have the largest populations.

For centuries the people of Southwest Asia and North Africa lived in rural areas. Today many of them are moving to large cities such as Cairo, Tehran, Baghdad, Istanbul, Algiers, Tunis, and Casablanca. Because of the increasing population, food and housing shortages have become a problem in some cities. Air and water pollution are also major concerns.

The Arts The arts of Southwest Asia and North Africa reflect the region's early civilizations and its three great religions. In Egypt you can see the huge pyramids and temples left by the ancient Egyptians. Many Muslim places of worship are covered by glistening ceramic tiles. These tiles are decorated with geometric patterns and verses from the Quran, the Muslim holy book. In Israel you can see modern sculpture depicting events in both ancient and modern Jewish history.

Tangier, Morocco

From the hillsides of Tangier, you can overlook the city's many beaches.
PLACE: What types of climate are found in much of Southwest Asia and North Africa?

Technology Activity

Using a Word Processor Create a two-column format newsletter that includes at least one brief article, story, or poem that you have written about either the physical or cultural geography of each country of Southwest Asia and North Africa. Title the newsletter, and use the options in your program to help you design it.

REVIEW AND ACTIVITIES

1. **REGION** What two major landforms are found in Southwest Asia and North Africa?
2. **LOCATION** Where do farmers raise crops in the region?
3. **MOVEMENT** What two major movements of people have taken place in Southwest Asia and North Africa in recent years?
4. **HUMAN/ENVIRONMENT INTERACTION** How has oil affected standards of living in the region?
5. **LOCATION** Where in the region did two of the world's oldest civilizations develop?
6. **REGION** What three world religions began in Southwest Asia?

411

LESSON PLAN
Unit 6 Atlas

Physical Geography

Applying Geographic Themes

Movement The Khyber Pass is a passage through the Hindu Kush mountain range. Over the centuries, invading armies took the pass to reach India. Traders leading camel caravans traveled through the pass on their way to the Mediterranean. Today, the Khyber Pass remains an important travel route, with a paved highway and a railroad. According to the map on page 412, in what country is the Hindu Kush mountain range located? *(Afghanistan)*

NATIONAL GEOGRAPHIC SOCIETY

These materials are available from Glencoe.

VIDEODISC

STV: World Geography
Vol. 1: Asia and Australia

Middle East
Side 1, Frames 06887-07190

SOFTWARE

ZipZapMap! World

MAP STUDY

Answers
1. Sahara
2. Persian Gulf

Map Skills Practice
Reading a Map What mountain range is found in northern Morocco and Algeria? *(Atlas Mountains)*

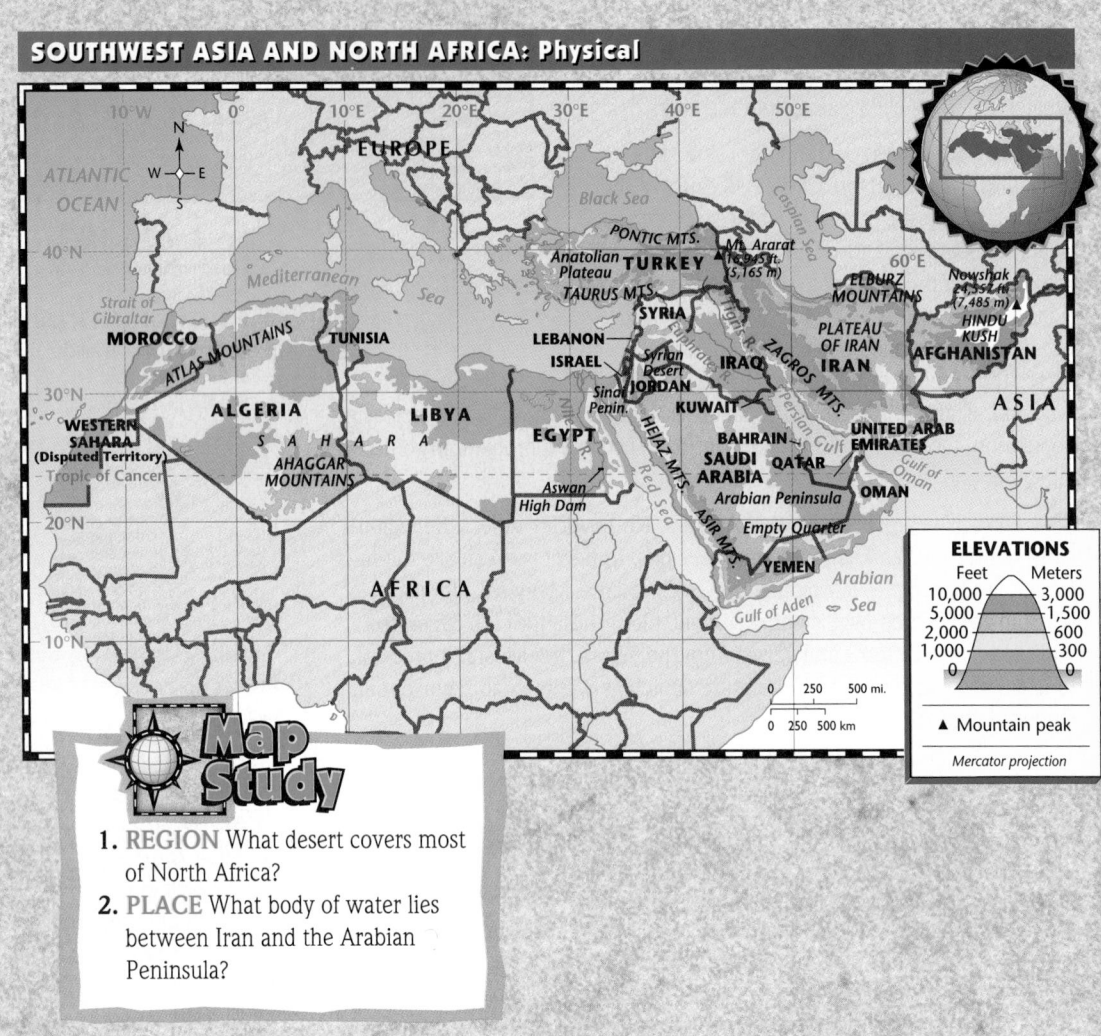

SOUTHWEST ASIA AND NORTH AFRICA: Physical

Map Study

1. **REGION** What desert covers most of North Africa?
2. **PLACE** What body of water lies between Iran and the Arabian Peninsula?

UNIT 6

Atlas Activity

Place Have students study and identify the physical features of Southwest Asia and North Africa. Then have them predict which parts of the region might be used for farming. Ask students to verify their predictions by consulting other maps and the text. Finally, have students write generalizations about the relationships among physical geography and land use. **L2**

ELEVATION PROFILE: Southwest Asia and North Africa

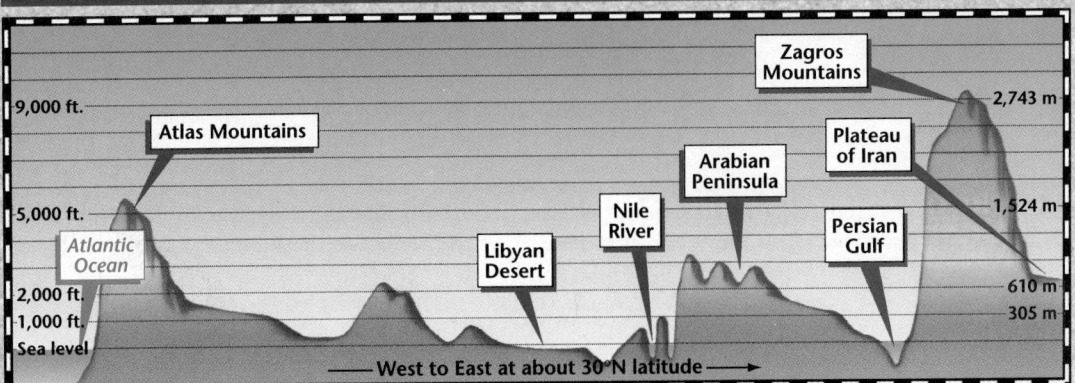

9,000 ft.

Atlas Mountains

5,000 ft.

Atlantic Ocean

Libyan Desert

Nile River

Arabian Peninsula

Plateau of Iran

Zagros Mountains

2,743 m

1,524 m

Persian Gulf

610 m

305 m

2,000 ft.
1,000 ft.
Sea level

◄— West to East at about 30°N latitude —►

Source: *Goode's World Atlas,* 19th edition

SOUTHWEST ASIA/NORTH AFRICA and the UNITED STATES: Land Comparison

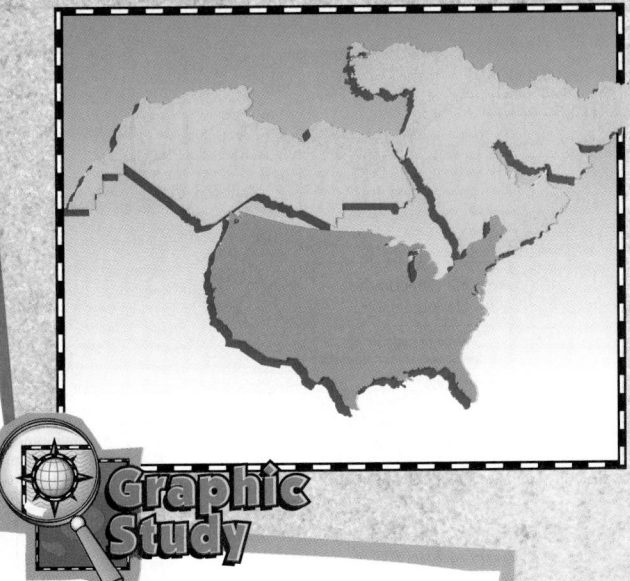

GeoFacts

Highest point: Nowshak (Afghanistan) 24,557 ft. (7,485 m) high

Lowest point: Dead Sea (Israel-Jordan) 1,312 ft. (400 m) below sea level

Longest river: Nile River (Africa) 4,160 mi. (6,693 km) long

Largest lake: Caspian Sea (Europe-Asia) 143,244 sq. mi. (371,002 sq. km)

Largest desert: Sahara (Northern Africa) 3,500,000 sq. mi. (9,065,000 sq. km)

Graphic Study

1. **HUMAN/ENVIRONMENT INTERACTION** How might elevation affect the population in the Zagros and Atlas mountains?
2. **PLACE** How would you describe the physical size of the United States compared to the combined sizes of Southwest Asia and North Africa?

(413)

GRAPHIC STUDY

Answers
1. population would be low
2. Southwest Asia and North Africa combined are much larger than the United States

Graphic Skills Practice
Reading a Graph What is the elevation of the Libyan Desert? *(sea level)* Which mountains are higher—the Atlas or the Zagros? *(Zagros)*

FactsOnFile

Use the reproducible masters from **GEOGRAPHY ON FILE,** *North Africa and Southwest Asia,* to enrich unit content.

Teacher Notes

413

LESSON PLAN
Unit 6 Atlas

UNIT 6 ATLAS
Cultural Geography

Applying Geographic Themes

Place The greatest population densities in Southwest Asia and North Africa are found around bodies of water. The Nile River Valley, for example, is one of the world's most heavily populated areas. What three capital cities of North Africa are located near the Mediterranean Sea? *(Algiers, Tunis, and Tripoli)*

Geography and the Humanities

📁 World Literature Reading 6

💿 World Music: Cultural Traditions, Lesson 5

📱 World Cultures Transparencies 16, 17

📕 Focus on World Art Print 7

MAP STUDY

Answers
1. Turkey
2. Algiers

Map Skills Practice

Reading a Map What two countries form the border of Kuwait? *(Iraq, Saudi Arabia)* On what sea is Lebanon's capital located? *(Mediterranean Sea)*

414

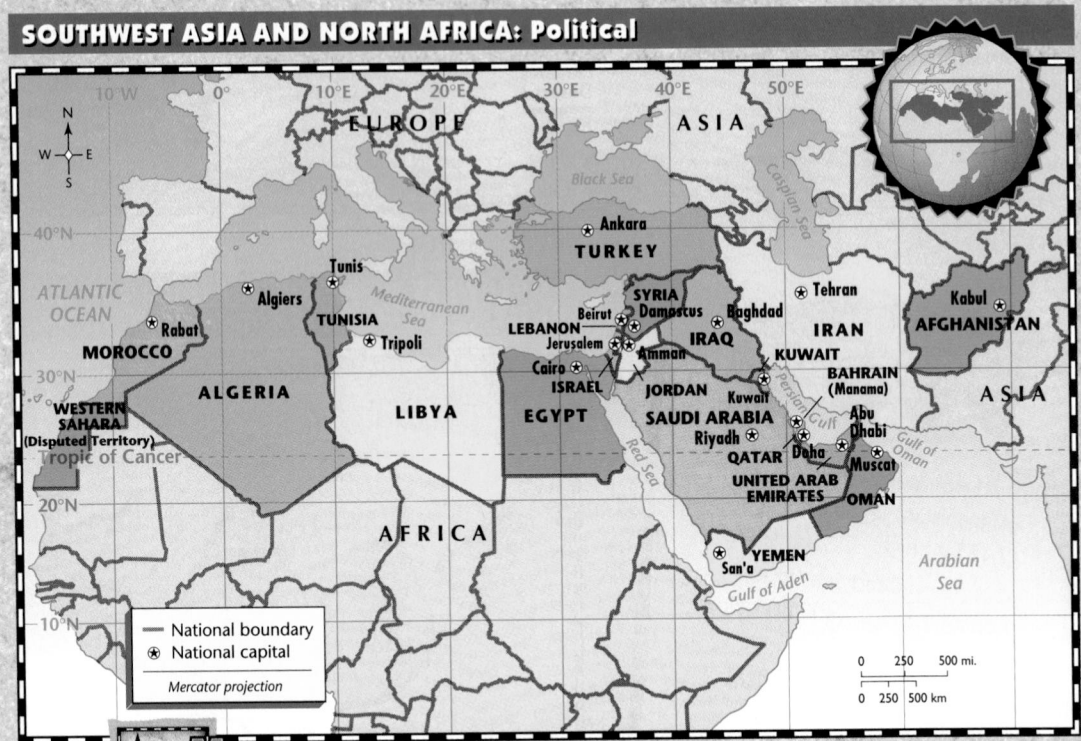

SOUTHWEST ASIA AND NORTH AFRICA: Political

EUROPE   ASIA

Black Sea

Caspian Sea

Ankara

ATLANTIC OCEAN

TURKEY

Tunis   Tehran   Kabul

Algiers   SYRIA   AFGHANISTAN

Rabat   Mediterranean Sea   Beirut Damascus Baghdad

TUNISIA   LEBANON   IRAN   ASIA

MOROCCO   Tripoli   Jerusalem Amman IRAQ   KUWAIT

Cairo   BAHRAIN (Manama)

WESTERN SAHARA (Disputed Territory)   ISRAEL   JORDAN   Kuwait

Tropic of Cancer   ALGERIA   LIBYA   EGYPT   SAUDI ARABIA   Abu Dhabi

Riyadh   QATAR Doha Muscat

Red Sea   UNITED ARAB EMIRATES   OMAN

AFRICA   Gulf of Oman

San'a   YEMEN   Arabian Sea

Gulf of Aden

— National boundary
⊛ National capital

Mercator projection

0   250   500 mi.
0   250   500 km

Map Study

1. **LOCATION** What country lies north of Syria?
2. **PLACE** What is the capital of Algeria?

Atlas Activity

Movement Assign a country to each student or pair of students. Direct students to use library resources to discover their assigned country's major imports, exports, and trading partners. Create a wall chart by pinning a sheet of butcher paper to the bulletin board. Have students enter their findings on the chart. Then call on volunteers to use information on the wall chart to make generalizations about the region's imports, exports, and trading partners. **L2**

LARGEST URBAN AREAS: Southwest Asia and North Africa

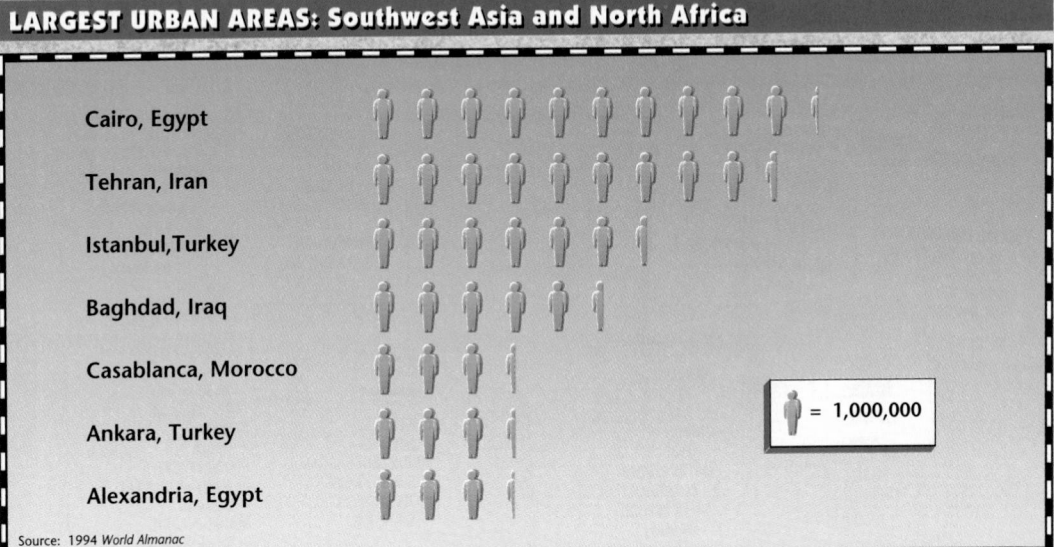

Cairo, Egypt

Tehran, Iran

Istanbul, Turkey

Baghdad, Iraq

Casablanca, Morocco

Ankara, Turkey

Alexandria, Egypt

= 1,000,000

Source: 1994 *World Almanac*

COMPARING POPULATION:
Southwest Asia/North Africa and the United States

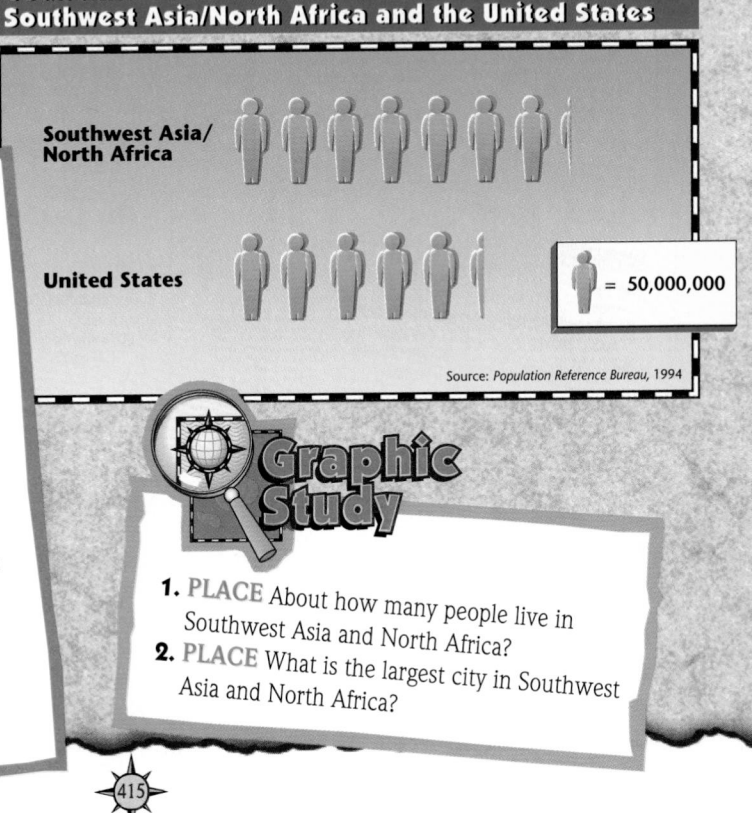

Southwest Asia/
North Africa

United States

= 50,000,000

Source: *Population Reference Bureau, 1994*

GeoFacts

Biggest country (land area): Algeria
919,590 sq. mi.
(2,381,738 sq. km)

Smallest country (land area): Bahrain
260 sq. mi. (673 sq. km)

Largest city (population): Cairo
(1995) 11,155,000
(2000 projected) 12,512,000

Highest population density: Bahrain
2,308 people per sq. mi. (892
people per sq. km)

Lowest population density: Libya
8 people per sq. mi. (3 people
per sq. km)

Graphic Study

1. **PLACE** About how many people live in
Southwest Asia and North Africa?
2. **PLACE** What is the largest city in Southwest
Asia and North Africa?

415

ABC NEWS INTERACTIVE™

These materials are available from Glencoe.

VIDEODISC

In the Holy Land

The Issues
Side 2 Chapter 8

Leaders and People
Side 2 Chapter 18

GRAPHIC STUDY

Answers
1. about 360 million people
2. Cairo

Graphic Skills Practice
Reading a Graph About how many more people live in Southwest Asia and North Africa than in the United States? *(about 100 million more)* What three cities have about the same population? *(Casablanca, Ankara, and Alexandria)*

Teacher Notes

UNIT 6 ATLAS

Countries at a Glance

Afghanistan 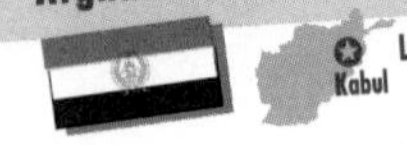
Kabul
LANDMASS: 251,770 sq. mi./ 652,084 sq. km
CAPITAL: Kabul
MAJOR LANGUAGE(S): Pushtu, Dari Persian, Uzbek
POPULATION: 17,800,000
MONEY: Afghani
MAJOR EXPORT: Dried Fruits and Nuts
MAJOR IMPORT: Machinery

Algeria 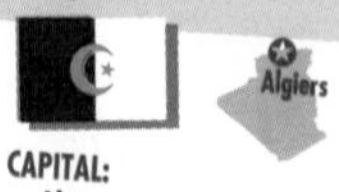
Algiers
LANDMASS: 919,590 sq. mi./ 2,381,738 sq. km
CAPITAL: Algiers
MAJOR LANGUAGE(S): Arabic, Berber
POPULATION: 27,900,000
MONEY: Dinar
MAJOR EXPORT: Petroleum
MAJOR IMPORT: Machinery

Bahrain 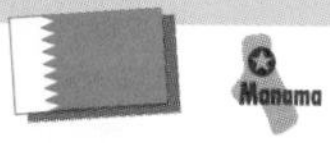
Manama
LANDMASS: 260 sq. mi./ 673 sq. km
CAPITAL: Manama
MAJOR LANGUAGE(S): Arabic, Farsi, Urdu
POPULATION: 600,000
MONEY: Dinar
MAJOR EXPORT: Petroleum Products
MAJOR IMPORT: Petroleum Products

Egypt
Cairo
LANDMASS: 384,340 sq. mi./ 995,441 sq. km
CAPITAL: Cairo
MAJOR LANGUAGE(S): Arabic, English
POPULATION: 58,900,000
MONEY: Pound
MAJOR EXPORT: Petroleum
MAJOR IMPORT: Machinery

Iran 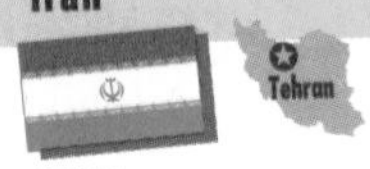
Tehran
LANDMASS: 631,660 sq. mi./ 1,635,999 sq. km
CAPITAL: Tehran
MAJOR LANGUAGE(S): Farsi, Turkish, Kurdish, Arabic
POPULATION: 61,200,000
MONEY: Rial
MAJOR EXPORT: Petroleum
MAJOR IMPORT: Motor Vehicles

Iraq 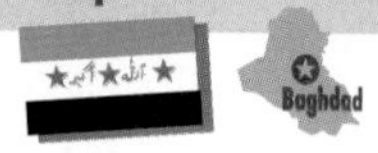
Baghdad
LANDMASS: 168,870 sq. mi./ 437,373 sq. km
CAPITAL: Baghdad
MAJOR LANGUAGE(S): Arabic, Kurdish
POPULATION: 19,900,000
MONEY: Dinar
MAJOR EXPORT: Fuels
MAJOR IMPORT: Machinery

Israel 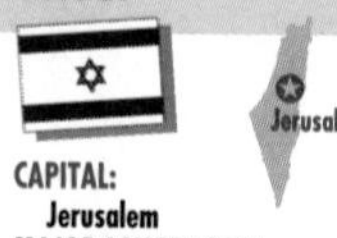
Jerusalem
LANDMASS: 7,850 sq. mi./ 20,332 sq. km
CAPITAL: Jerusalem
MAJOR LANGUAGE(S): Hebrew, Arabic
POPULATION: 5,400,000
MONEY: Shekel
MAJOR EXPORT: Machinery, Polished Diamonds
MAJOR IMPORT: Rough Diamonds

Jordan 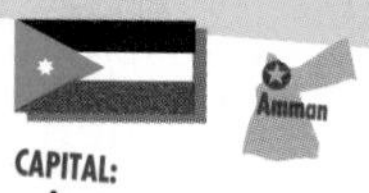
Amman
LANDMASS: 34,340 sq. mi./ 88,941 sq. km
CAPITAL: Amman
MAJOR LANGUAGE(S): Arabic
POPULATION: 4,200,000
MONEY: Dinar
MAJOR EXPORT: Phosphate Fertilizers
MAJOR IMPORT: Food and Live Animals

Kuwait 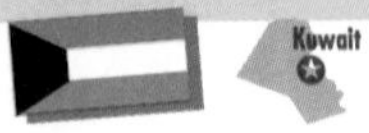
Kuwait
LANDMASS: 6,880 sq. mi./ 17,819 sq. km
CAPITAL: Kuwait
MAJOR LANGUAGE(S): Arabic
POPULATION: 1,300,000
MONEY: Dinar
MAJOR EXPORT: Petroleum
MAJOR IMPORT: Machinery

Lebanon 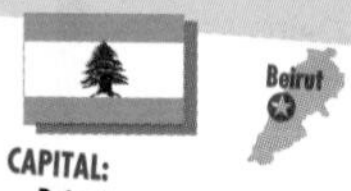
Beirut
LANDMASS: 3,950 sq. mi./ 10,231 sq. km
CAPITAL: Beirut
MAJOR LANGUAGE(S): Arabic, French
POPULATION: 3,600,000
MONEY: Pound
MAJOR EXPORT: Jewelry
MAJOR IMPORT: Machinery

Countries not drawn to scale.

Fun Facts

• **Syria** The ancient city of Palmyra in central Syria was built around a desert oasis. It served as a rest stop for camel caravans on the route between Mediterranean lands and Asia.

• **Saudi Arabia** Makkah and Madinah are Islam's most sacred cities. Non-Muslims are not permitted to enter Makkah. While non-Muslims may enter Madinah, they are not allowed to visit any of the city's holy places.

Libya

Tripoli

CAPITAL: Tripoli
MAJOR LANGUAGE(S): Arabic
POPULATION: 5,100,000
LANDMASS: 679,360 sq. mi./ 1,759,542 sq. km
MONEY: Dinar
MAJOR EXPORT: Petroleum
MAJOR IMPORT: Food

Morocco*

Rabat

CAPITAL: Rabat
MAJOR LANGUAGE(S): Arabic, Berber
POPULATION: 28,600,000
LANDMASS: 172,320 sq. mi./ 446,309 sq. km
MONEY: Dirham
MAJOR EXPORT: Food
MAJOR IMPORT: Crude Oil

Oman

Muscat

CAPITAL: Muscat
MAJOR LANGUAGE(S): Arabic
POPULATION: 1,900,000
LANDMASS: 82,030 sq. mi./ 212,458 sq. km
MONEY: Rial
MAJOR EXPORT: Petroleum
MAJOR IMPORT: Machinery

Qatar

Doha

CAPITAL: Doha
MAJOR LANGUAGE(S): Arabic, English
POPULATION: 500,000
LANDMASS: 4,250 sq. mi./ 11,008 sq. km
MONEY: Riyal
MAJOR EXPORT: Petroleum
MAJOR IMPORT: Machinery

Saudi Arabia

Riyadh

CAPITAL: Riyadh
MAJOR LANGUAGE(S): Arabic
POPULATION: 18,000,000
LANDMASS: 830,000 sq. mi./ 2,149,700 sq. km
MONEY: Riyal
MAJOR EXPORT: Petroleum
MAJOR IMPORT: Machinery

Syria

Damascus

CAPITAL: Damascus
MAJOR LANGUAGE(S): Arabic, Kurdish, Armenian, Turkish
POPULATION: 14,000,000
LANDMASS: 71,070 sq. mi./ 184,071 sq. km
MONEY: Pound
MAJOR EXPORT: Petroleum
MAJOR IMPORT: Food Products

Tunisia

Tunis

CAPITAL: Tunis
MAJOR LANGUAGE(S): Arabic, French
POPULATION: 8,700,000
LANDMASS: 59,980 sq. mi./ 155,348 sq. km
MONEY: Dinar
MAJOR EXPORT: Clothing
MAJOR IMPORT: Textiles

Turkey

Ankara

CAPITAL: Ankara
MAJOR LANGUAGE(S): Turkish, Kurdish, Arabic, Greek
POPULATION: 61,800,000
LANDMASS: 297,150 sq. mi./ 769,619 sq. km
MONEY: Lira
MAJOR EXPORT: Textiles
MAJOR IMPORT: Machinery

United Arab Emirates

Abu Dhabi

CAPITAL: Abu Dhabi
MAJOR LANGUAGE(S): Arabic
POPULATION: 1,700,000
LANDMASS: 32,280 sq. mi./ 83,605 sq. km
MONEY: Dirham
MAJOR EXPORT: Petroleum
MAJOR IMPORT: Machinery

Yemen

San'a

CAPITAL: San'a
MAJOR LANGUAGE(S): Arabic
POPULATION: 12,900,000
LANDMASS: 203,850 sq. mi./ 527,972 sq. km
MONEY: Rial (also the Dinar)
MAJOR EXPORT: Coffee
MAJOR IMPORT: Food and Live Animals

*** Morocco claims the Western Sahara area but other countries do not accept this claim.**

417

Cultural Kaleidoscope

Southwest Asia and North Africa Islamic prayer rugs feature a pointed or arch-shaped pattern. During prayer, Muslims kneel on their rug with the arch pointing toward Makkah.

Curriculum Connection

Mathematics By about 3000 B.C., ancient Egyptian mathematicians had developed a decimal system, or method of counting in groups of 10. They also had formulas for finding area and volume of geometric figures. The Egyptians used these formulas when they built the pyramids.

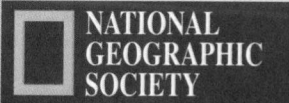

National Geographic Society

These materials are available from Glencoe.

Software

ZipZapMap! World

Teacher Notes

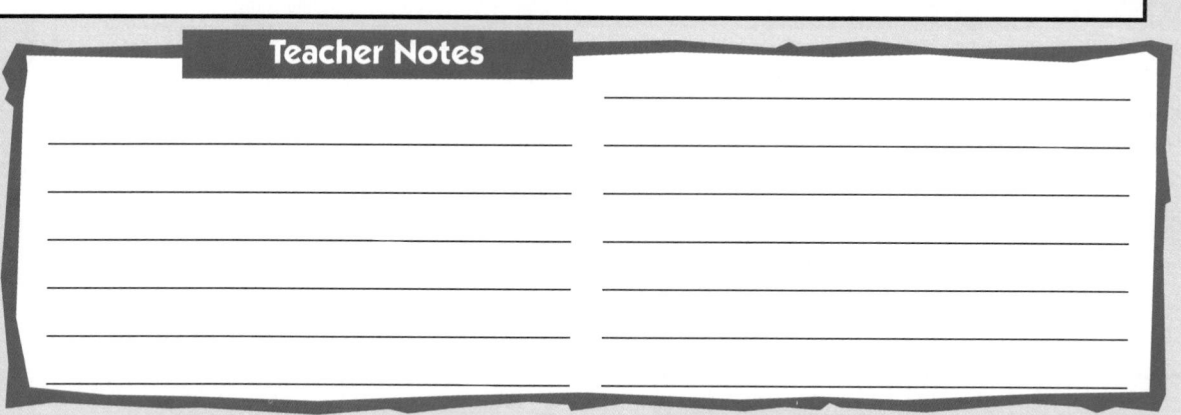

Chapter 16 Planning Guide

Southwest Asia

PERFORMANCE ASSESSMENT ACTIVITY

CREATE A TRADING COMPANY Have students make a matrix of the countries of Southwest Asia, including the resources and products that are present in each. Upon completion, each student should take the role of the owner of a trading company that would like to trade with the nations of the region. Students should decide on a company name, slogan, and products that they would sell or trade with each country. They should also predict three problems their companies might have as they do business in the region, and recommend solutions. Students must make a presentation of their choice of business, explaining their company plans to the various "governments" of each nation as played by groups of students in the class.

POSSIBLE RUBRIC FEATURES: Problem-solving skills, accuracy of content information, analysis and inference skills, oral communication skills, ability to recognize multiple points of view, and impact of presentation on audience

MENTAL MAPPING ACTIVITY

Before teaching Chapter 16, have students point out Southwest Asia on a globe. Have them determine which parts of this region are at the same latitudes as parts of the United States.

TEACHER'S CORNER

NATIONAL GEOGRAPHIC SOCIETY

INDEX TO NATIONAL GEOGRAPHIC MAGAZINE

The following articles may be used for research relating to this chapter:

- "Gaza," by Alexandra Avakian, September 1996.
- "Syria Behind the Mask," by Peter Theroux, July 1996.
- "The Three Faces of Jerusalem," by Alan Mairson, April 1996.
- "Israel's Galilee," by Don Belt, June 1995.
- "New Views of the Holy Land," by Richard Cleave and Technion-Israel Institute of Technology, June 1995.

- "Oman," by Peter Ross Range, May 1995.
- "Water—The Middle East's Critical Resource," by Priit J. Vesilind, May 1993.
- "Struggle of the Kurds," by Christopher Hitchens, August 1992.
- "Who Are the Palestinians?" by Tad Szulc.
- "Retracing the First Crusade," by Tim Severin, September 1989.

NATIONAL GEOGRAPHIC SOCIETY PRODUCTS AVAILABLE FROM GLENCOE

To order the following products for use with this chapter, contact your local Glencoe sales representative or call Glencoe at 1-800-334-7344:

- *STV: World Geography* (Videodisc)
- *Picture Atlas of the World* (CD-ROM)
- *Physical Geography of the World* (Transparencies)
- *ZipZapMap! World* (Software)

- *GeoBee* (Software)
- *Images of the World* (Posters)
- *Eye on the Environment* (Posters)

ADDITIONAL NATIONAL GEOGRAPHIC SOCIETY PRODUCTS

To order the following products for use with this chapter, call National Geographic Society at 1-800-368-2728:

- *Jerusalem: Within These Walls* (Video)
- *Nations of the World Series:* Israel (Video)

- *Physical Geography of the Continents Series:* "Asia." (Video)
- *Portraits of the Continents Series:* "Part III: Europe; Asia; Africa." (Filmstrip)

TEACHER-TO-TEACHER
Curriculum Connection: Earth Science

—from Linda Prieskorn
Clague Middle School, Ann Arbor, MI

WATER, WATER, EVERYWHERE *The purpose of this activity is to have students understand the importance of water in Southwest Asia by drawing up guidelines for the use of water in the area.*

Suggest that students review the **Eye on the Environment** feature on pages 464–465 before they begin the activity. Then have students use atlases as references to draw a map of Southwest Asia. Direct students to mark and label all the rivers in the region. Each time a river crosses an international boundary, have students use a different color to mark the river. Ask students to note how many rivers cross international boundaries and how many countries each of these rivers flows through. Call on volunteers to suggest ways one country might control the flow of a river into other countries.

Point out that when Turkey filled the reservoir behind the Atatürk Dam in 1990, it severely restricted the flow of the Euphrates River into Syria and Iraq. Instruct students to work in small groups to draw up a list of rules on the control of rivers that flow through a number of countries. Have groups write their lists and post them on the bulletin board. As a follow-up, have students discuss the following: How would you feel about these rules if you lived upstream of a dam? How would you feel if you lived downstream of a dam? Why?

BIBLIOGRAPHY

Readings for the Student

Pearson, Robert P., and Leon F. Clark, eds. *Through Middle Eastern Eyes.* New York: Center for International Training and Education, 1993.

Pimlott, John. *Middle East: A Background to the Conflicts.* New York: Gloucester Press, 1991.

Readings for the Teacher

The Cambridge Encyclopedia of the Middle East and North Africa. Boston: Cambridge University Press, 1988.

The Middle East Today: An Atlas of Reproducible Pages. Wellesley, MA: World Eagle, 1993.

Multimedia

Families of the World: Israel. National Geographic Society, 1986. Videocassette/film, 14 minutes.

Middle East: The Desert of God. Coronet/MTI. Videocassette/film, 15 minutes.

KEY TO ABILITY LEVELS

Teaching strategies have been coded for varying learning styles and abilities.

L1 **BASIC** activities for all students

L2 **AVERAGE** activities for average to above-average students

L3 **CHALLENGING** activities for above-average students

LEP **LIMITED ENGLISH PROFICIENCY** activities

MEETING NATIONAL STANDARDS

Geography for Life

All of the 18 standards are demonstrated in Unit 6. The following ones are highlighted in this chapter:

• Standards 1, 2, 3, 4, 6, 9, 10, 12, 13, 14, 15, 16, 18

Use these *Geography: The World and Its People* resources to teach, reinforce, and extend chapter content.

CHAPTER 16 RESOURCES

*Vocabulary Activity 16

Cooperative Learning Activity 16

Workbook Activity 16

Geography Skills Activity 16

GeoLab Activity 16

Chapter Map Activity 16

Performance Assessment Activity 16

*Reteaching Activity 16

Enrichment Activity 16

World Literature Reading Activity 16

Chapter 16 Test, Form A and Form B

Unit Overlay Transparencies 6-0, 6-1, 6-2, 6-3, 6-4, 6-5, 6-6

Political Map Transparency 6; World Cultures Transparency 9

Geoquiz Transparencies 16-1, 16-2

*Chapter 16 Digest Audiocassette, Activity, Test

Vocabulary PuzzleMaker Software

Student Self-Test: A Software Review

Testmaker Software

MindJogger Videoquiz

If time does not permit teaching the entire chapter, summarize using the Chapter 16 Highlights on page 439, and the Chapter 16 English (or Spanish) Audiocassettes. Review students' knowledge using the Glencoe MindJogger Videoquiz. *Also available in Spanish

Use these *Geography: The World and Its People* resources to teach and reinforce section content.

SECTION 1 RESOURCES

- Reproducible Lesson Plan 16-1
- Section Focus Transparency 16-1
- Section Quiz 16-1 (also available in Spanish)

SECTION 2 RESOURCES

- Reproducible Lesson Plan 16-2
- Section Focus Transparency 16-2
- Section Quiz 16-2 (also available in Spanish)

SECTION 3 RESOURCES

- Reproducible Lesson Plan 16-3
- Section Focus Transparency 16-3
- Section Quiz 16-3 (also available in Spanish)

SECTION 4 RESOURCES

- Reproducible Lesson Plan 16-4
- Section Focus Transparency 16-4
- Section Quiz 16-4 (also available in Spanish)

SECTION 5 RESOURCES

- Reproducible Lesson Plan 16-5
- Section Focus Transparency 16-5
- Section Quiz 16-5 (also available in Spanish)

ADDITIONAL RESOURCES FROM GLENCOE

Reproducible Masters

- Glencoe Social Studies Outline Map Book, pages 31, 32
- Foods Around the World, Unit 6
- Facts On File: GEOGRAPHY ON FILE, North Africa and Southwest Asia

World Music: Cultural Traditions
Lesson 5

Transparencies

- National Geographic Society PicturePack Transparencies, Unit 6

Posters

- National Geographic Society: Images of the World, Unit 6

Countries of the World Flashcards
Unit 6

World Crafts Activity Cards Unit 6

World Games Activity Cards
Unit 6

Videodiscs

- Reuters Issues in Geography
- ABCNews InterActive™: In the Holy Land
- National Geographic Society: STV: Water

CD-ROM

- National Geographic Society: Picture Atlas of the World

Performance Assessment

Refer to the Planning Guide on page 418A for a Performance Assessment Activity for this chapter. See the *Performance Assessment Strategies and Activities* booklet for suggestions.

Chapter Objectives

1. **Explain** how Turkey's location affected its development and show how Turkey blends ancient and modern ways.

2. **Describe** how Israel has developed its resources and note how the past affects Israel today.

3. **Compare** the geography of Syria, Lebanon, and Jordan.

4. **Evaluate** the impact of Islam and oil on the lives of the people of the Arabian Peninsula.

5. **Summarize** the major geographical features of Iraq, Iran, and Afghanistan.

 VIDEODISC

Use the Chapter 16 MindJogger Videoquiz to preview chapter content.

MindJogger Videoquiz

Chapter 16
Disc 2 Side B

 Also available in VHS.

418

Chapter 16 Southwest Asia

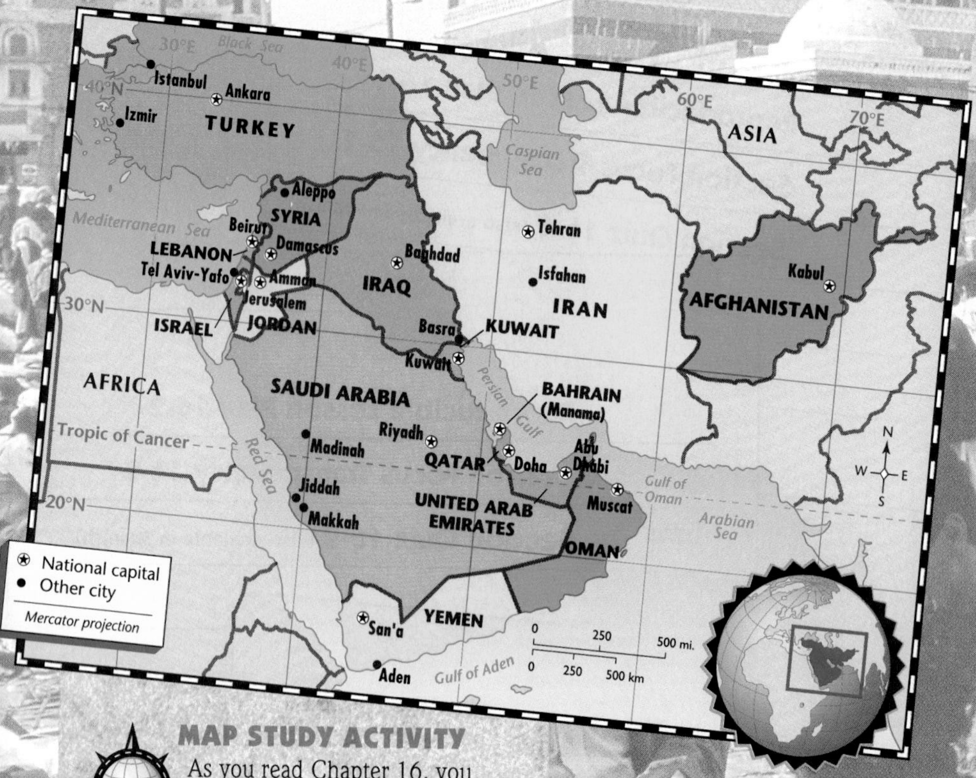

MAP STUDY ACTIVITY

As you read Chapter 16, you will learn about Southwest Asia. The countries in this region lie in the southwest corner of Asia.

1. **How many countries make up Southwest Asia?**

2. **What are the two largest countries?**

3. **Which country connects Europe to Asia?**

418

MAP STUDY ACTIVITY

Answers
1. 15
2. Iran, Saudi Arabia
3. Turkey

Map Skills Practice
Reading a Map What Southwest Asian countries border the Mediterranean Sea? *(Israel, Lebanon, Syria, Turkey)*

SECTION 1 ·
Turkey

PREVIEW

Words to Know
- mosque
- migrate

Places to Locate
- Turkish Straits
- Anatolian Plateau
- Pontic Mountains
- Taurus Mountains
- Istanbul
- Ankara

Read to Learn . . .
1. how Turkey's location has affected its history and development.
2. how Turkey blends its ancient heritage with modern ways.

Turkey, one of the largest countries in Southwest Asia.

Flocks of grazing sheep and goats are common sights in Turkey and other Southwest Asian countries. This shepherd watches a flock of sheep in

A little larger than the state of Texas, Turkey has a unique location—it lies on two continents: Europe and Asia. Because of its location, Turkey has links to both Asian and European cultures.

Location
The Land

The European and Asian parts of Turkey are separated by three important waterways—the Bosporus (BAHS•puh•ruhs), the Sea of Marmara (MAHR•muh•ruh), and the Dardanelles (DAHR•duhn•EHLZ). Together, these waterways are known as the Turkish Straits. The Turkish Straits join the Black Sea and the Mediterranean Sea. Find these waterways on the map on page A20.

The Land and Climate Highlands and plateaus cover much of Turkey. The heart of Turkey is the Anatolian (A•nuh•TOH•lee•uhn) Plateau. Mountain ranges surround this broad sweep of dry highlands. The Pontic Mountains border the plateau on the north. The Taurus Mountains tower over it on the south.

Grassy plains cover the northern part of Turkey along the Black Sea. On the western coast, you find broad, fertile river valleys running inland from the Aegean Sea. In the south, coastal plains extend along the Mediterranean Sea.

CHAPTER 16

 419

Classroom Resources for Section 1

 REPRODUCIBLE MASTERS
Reproducible Lesson Plan 16-1
Guided Reading Activity 16-1
Geography Skills Activity 16
Cooperative Learning
  Activity 16
Section Quiz 16-1

 TRANSPARENCIES
Section Focus
  Transparency 16-1
Geoquiz Transparencies 16-1,
  16-2

MULTIMEDIA
- Vocabulary PuzzleMaker Software
- Testmaker

Meeting National Standards
Geography for Life The following standards are highlighted in this section:
- Standards 1, 4, 6, 9, 10, 12

FOCUS

Section Objectives
1. **Explain** how Turkey's location has affected its history and development.
2. **Discuss** how Turkey blends its ancient heritage with modern ways.

Bellringer Motivational Activity

Prior to taking roll at the beginning of the class period, project Section Focus Transparency 16-1 and have students answer the activity questions. Discuss students' responses. **L1**

Vocabulary Pre-check

Display pictures of mosques. Point out the *minarets,* or towers, from which criers call Muslims to prayer. Explain that a mosque is usually the most important building in a Muslim city. **L1 LEP**

Use the Vocabulary PuzzleMaker Software to create a crossword puzzle. **L1**

TEACH

Guided Practice
Diagram Help students create a web diagram illustrating the characteristics of Turkey's population. In the
(continued)

419

center circle of the web, write "62 million people." Radiating from the center, draw spokes of the web with circles representing characteristics such as languages, religions, occupations, population centers. Have students fill in circles with text information. **L2**

MAP STUDY

Answer
Anatolian Plateau

Map Skills Practice
Reading a Map What two bodies of water are connected by the Gulf of Aden? *(Red Sea, Arabian Sea)*

Independent Practice

📁 Guided Reading Activity 16-1 **L1**

ASSESS

Check for Understanding

Assign Section 1 Review as homework or an in-class activity.

Meeting Lesson Objectives

Each objective below is tested by the questions that follow it in parentheses.

1. **Explain** how Turkey's location has affected its history and development. (2, 3, 5)
2. **Discuss** how Turkey blends its ancient heritage with modern ways. (1, 4)

420

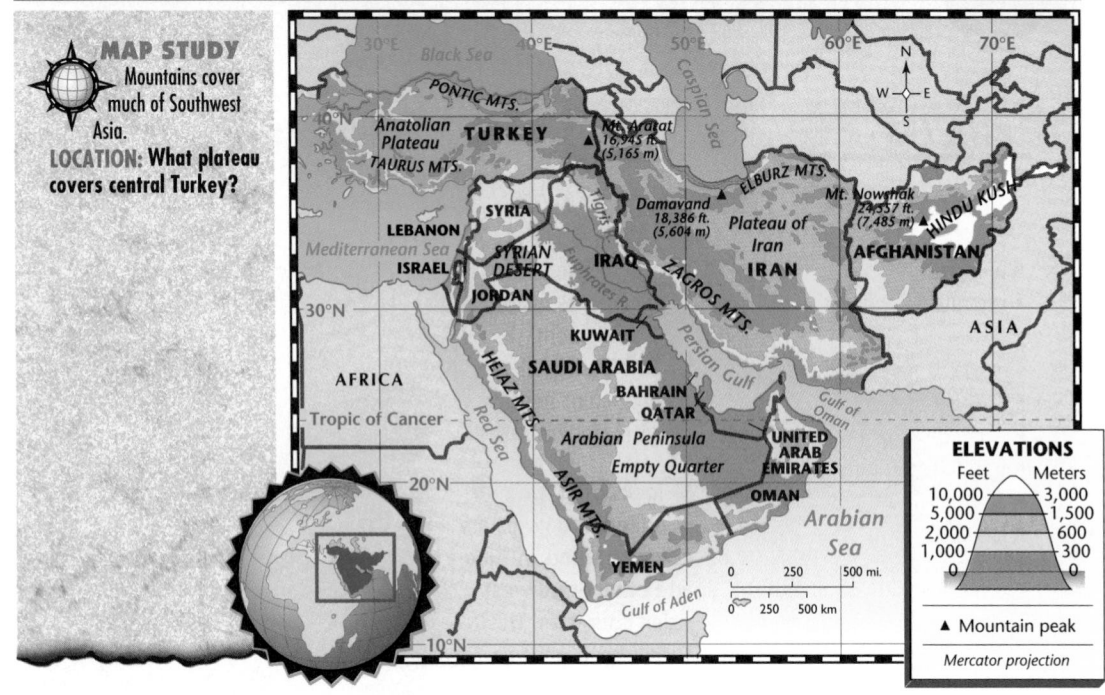

SOUTHWEST ASIA: Physical

MAP STUDY Mountains cover much of Southwest Asia.
LOCATION: What plateau covers central Turkey?

ELEVATIONS
Feet	Meters
10,000	3,000
5,000	1,500
2,000	600
1,000	300
0	0

▲ Mountain peak

Mercator projection

The map on page 424 shows you that Turkey's climate varies throughout the country. If you lived on the Anatolian Plateau, you would experience the hot, dry summers and cold, snowy winters of the steppe climate. The coastal areas enjoy a Mediterranean climate—hot, dry summers and mild, rainy winters.

Human/Environment Interaction
The Economy

Agriculture plays an important role in Turkey's economy. Wheat and livestock are the country's leading farm products. The best farmlands are found in the coastal areas. In recent years, Turkish industry has grown tremendously, especially in the making of textiles. The map on page 429 shows that Turkey is rich in mineral deposits, including iron ore, coal, and copper.

Region
The People

Most of Turkey's 62 million people live in the northern part of the Anatolian Plateau, on coastal plains, or in valleys. Turkish is the major language, but Arabic, Greek, and Kurdish are also spoken. About 98 percent of the Turkish people are Muslims. They follow Islam, a religion that teaches belief in one God. Islam is the major faith of most of Southwest Asia. Turkey also has small groups of Christians and Jews.

420

UNIT 6

Cooperative Learning Activity

Organize students into small groups and assign one of the following physical features of Turkey to each group: mountains, plateaus, plains, rivers, straits. Have each group locate its assigned features on a map of Turkey and explain how the feature influences economic activity, transportation, and trade. **L2**

City and Country About 60 percent of Turkey's people live in cities or towns. Ahmed ul-Hamid, a 14-year-old Turkish student, lives among the 7.6 million people of Istanbul (IHS•tuhn•BOOL), Turkey's largest city. Istanbul is the only city in the world located on two continents. Ahmed has the opportunity to visit Istanbul's beautiful palaces, museums, and **mosques**, or Muslim places of worship. Only once has Ahmed visited Turkey's capital and second-largest city, Ankara (ANG•kuh•ruh). It has about 2.8 million people and is located on the Anatolian Plateau.

Roughly 40 percent of Turkey's people live on farms or in small villages. Since the mid-1900s, hundreds of thousands of people have left their villages to seek work in the cities. The cities, however, can't supply enough jobs for everyone. This results in many searching for work in other parts of Southwest Asia and in western Europe and Australia.

Influences of the Past Most of Turkey's people are descendants of an Asian people called Turks who **migrated**, or moved, to the Anatolian Plateau during the A.D. 900s. One group of Turks—the Ottomans—established a Muslim empire centered in Turkey.

During World War I, the Ottomans were defeated. Kemal Atatürk, a military hero, became the first president of the new country—the modern republic of Turkey. Atatürk introduced many political and social changes to the country.

Turkey soon began to consider itself European as well as Asian. Modern Turkish people, however, remain proud of their Muslim faith. In 1996 the Muslim party became the major political group in Turkey's government.

Istanbul, Turkey

Domed roofs and graceful towers, called minarets, rise above the many mosques of Istanbul.
PLACE: What is the major religion of Turkey?

SECTION 1 REVIEW

REVIEWING TERMS AND FACTS

1. Define the following: mosque, migrate.
2. LOCATION What makes Turkey's location unique?
3. PLACE What areas of Turkey are best for farming?
4. MOVEMENT Where do Turkish workers move to find work?

 MAP STUDY ACTIVITIES

 5. Look at the physical map on page 420. What feature dominates central Turkey?
6. Turn to the political map on page 418. About how far is Ankara from Istanbul?

More About the Illustration

Answer to Caption
Islam

Place Istanbul's architecture reflects the influence of many cultures, including Roman, Greek, and Persian.

Evaluate

Section 1 Quiz **L1**

Use the Testmaker to create a customized quiz for Section 1. **L1**

Reteach

Have students write one question for each subheading in this section. Then have them work in pairs, using their questions to review the material. **L1**

Enrich

Have students investigate the history of Istanbul. Suggest that they present their findings in an illustrated report. **L3**

CLOSE

Have students write a paragraph explaining how Turkey's location has influenced its development. **L2**

Answers to Section 1 Review

1. All vocabulary words are defined in the Glossary.
2. It lies on two continents—Europe and Asia.
3. Anatolian Plateau, coastal plains, valleys
4. Turkish cities, other parts of Southwest Asia, western Europe, Australia
5. Anatolian Plateau
6. about 250 miles (about 400 km)

BUILDING GEOGRAPHY SKILLS

BUILDING GEOGRAPHY SKILLS

TEACH

Have students consider the following question: "Why can't all countries set their clocks to the same time?" After students offer reasons, provide the following explanation: "The earth rotates 360° in 24 hours; therefore, when it is day on one side, it is night on the other. To clarify time relationships among places, the earth has been divided into 24 international time zones."

Have students read the skill on page 422. Then ask: "If you travel west, will it become earlier or later?" *(earlier)* "What happens if you cross the International Date Line traveling east?" *(a day is lost)* **L1**

Additional Skills Practice

1. You leave Chicago at 8 P.M. on an eight-hour flight to Paris, France. What time will you arrive in Paris? *(10 A.M. the following morning)*

2. If you leave Chicago on Monday and travel westward, on what day would you arrive in Sydney, Australia? *(Tuesday)*

📁 Geography Skills Activity 16

📁 Building Skills in Geography, Unit 1, Lesson 10

Reading a Time Zone Map

Earth's surface is divided into 24 time zones. The 0° line of longitude—the Prime Meridian—is the starting point for figuring out time around the world.

To read a time zone map, follow these steps:

- Locate a place where you know what time it is and select another place where you wish to know the time.

- Notice the time zones you cross between these places.

- If the second place lies east of the first, add an hour for each time zone. If it lies west, subtract an hour for each zone. If you must cross the International Date Line—the 180° meridian—add or subtract one day.

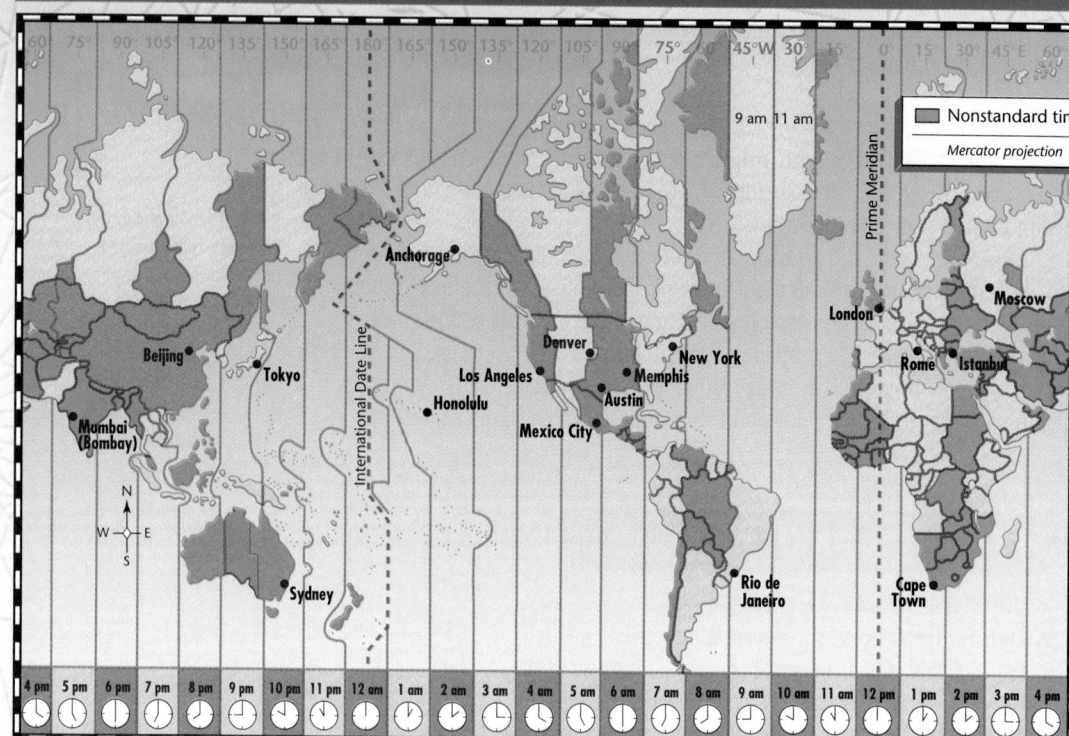

World Time Zones

Geography Skills Practice

1. If it is 7 A.M. in New York City, New York, what time is it in Istanbul, Turkey?

2. If it is 6 A.M. on Wednesday in Mumbai, India, what day and time is it in Rio de Janeiro, Brazil?

422

Answers to Geography Skills Practice

1. 2 P.M.
2. 10 P.M., Tuesday

LESSON PLAN
Chapter 16, Section 2

PREVIEW

Words to Know
- potash
- phosphate
- monotheism
- Holocaust

Places to Locate
- Mediterranean Sea
- Sea of Galilee
- Dead Sea
- Negev
- Tel Aviv

Read to Learn . . .
1. how Israelis have developed their resources.
2. what groups of people live in Israel.
3. how the past affects Israel today.

srael is the only Jewish nation in the world. Established as a nation in 1948, Israel celebrates its Independence Day just as Americans do—with parades and flag-waving.

srael lies at the eastern end of the Mediterranean Sea. It is 256 miles (412 km) long from north to south and only 68 miles (109 km) wide from east to west. An hour's drive takes you from the eastern boundary of the country almost to the Mediterranean Sea.

Place
The Land

Israel's landscape includes plains, deserts, and highlands. In Israel's far north lie the mountains of Galilee. East of these mountains is a plateau called the Golan Heights. The Judean (ju•DEE•uhn) Hills lie south.

The Dead Sea lies between Israel and Jordan. At 1,312 feet (400 m) *below* sea level, the rim of the Dead Sea is the lowest place on the earth! It is also the saltiest body of water in the world—about 9 times as salty as the ocean. A desert landform, the Negev (NEH•gehv) in southern Israel, covers almost one-half of the country. It is a triangular area of dry hills, valleys, and plains. Not all of southern Israel is desert, however. A narrow fertile plain—only 20 miles

---

Meeting National Standards
Geography for Life The following standards are highlighted in this section:
- Standards 1, 4, 6, 9, 10, 12, 13, 14, 18

FOCUS

Section Objectives
1. **Explain** how Israelis have developed their resources.
2. **Identify** the groups of people that live in Israel.
3. **Discuss** how the past affects Israel today.

Bellringer Motivational Activity
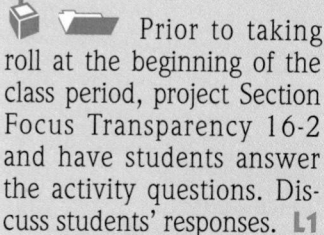
Prior to taking roll at the beginning of the class period, project Section Focus Transparency 16-2 and have students answer the activity questions. Discuss students' responses. **L1**

Vocabulary Pre-check
Have students study the **Words to Know** list and think of ways to remember correct spellings. This may include dividing words into syllables, or finding smaller words within each word. **L1** **LEP**

MULTICULTURAL PERSPECTIVE
Cultural Heritage
Jews come from two distinct cultural backgrounds: the *Sephardim,* whose culture originated and developed in the Mediterranean region, and the *Ashkenazim,* whose culture originated and developed in Northern Europe.

Classroom Resources for Section 2

 REPRODUCIBLE MASTERS
Reproducible Lesson Plan 16-2
Guided Reading Activity 16-2
Section Quiz 16-2

 TRANSPARENCIES
Section Focus
  Transparency 16-2
Political Map Transparency 6

MULTIMEDIA
 ABCNews InterActive™:
  In the Holy Land
 Testmaker

423

MAP STUDY

Answer
desert

Map Skills Practice
Reading a Map Which countries of Southwest Asia are on or below the Tropic of Cancer? *(Saudi Arabia, United Arab Emirates, Oman, Yemen)*

TEACH

Guided Practice

Compare Write these headings on the chalkboard: *Modern Israel* and *Ancient Palestine*. Have students examine the text and photographs in this section to find items that might be classified under each heading. Call on volunteers to enter their items under the appropriate heading on the chalkboard. Discuss ways that modern Israel reflects its ancient heritage. **L2**

Independent Practice

📁 Guided Reading Activity 16-2 **L1**

ABCNEWS
INTERACTIVE™

These materials are available from Glencoe.

 VIDEODISC

In the Holy Land

Three Religions
Side 1 Chapter 9

424

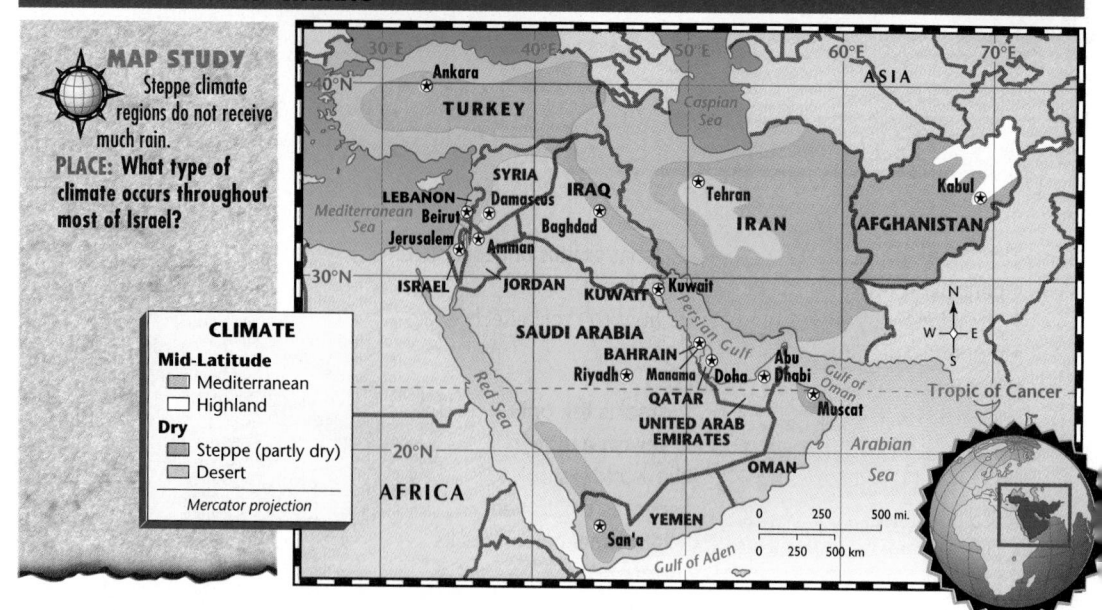

SOUTHWEST ASIA: Climate

MAP STUDY
Steppe climate regions do not receive much rain.
PLACE: What type of climate occurs throughout most of Israel?

CLIMATE
Mid-Latitude
☐ Mediterranean
☐ Highland
Dry
☐ Steppe (partly dry)
☐ Desert
Mercator projection

(33 km) wide at its widest point—lies along Israel's Mediterranean coast. To the east, the Jordan River cuts through the floor of a long, narrow valley. The longest of Israel's few rivers, the Jordan flows south into the Dead Sea.

The Climate As you can see on the map above, Israel's climate varies from north to south. Northern Israel has a Mediterranean climate with hot, dry summers and mild winters. About 40 inches (100 cm) of rain fall in the north each year. Southern Israel has a desert climate, with summer temperatures soaring to 120°F (49°C) or higher. Less than 1 inch of rain (2.5 cm) falls in the south.

Place
The Economy

Israel has limited farmland and natural resources. Israelis, however, make good use of their few resources and have built a strong economy. A developing system of transportation and communication supports this economy.

Mining and Manufacturing Israel is the most industrialized country in Southwest Asia. Its factories produce food products, clothing, chemicals, building materials, electronic appliances, and machinery. Diamond cutting is also a major Israeli industry. Tel Aviv is Israel's largest manufacturing center.

Mining is important to Israel's economy. The Dead Sea area is rich in deposits of **potash**, a type of mineral salt. Potash and other minerals support a growing chemical industry in Israel. The Negev area is also a source of copper and **phosphate**, a mineral used in making fertilizer.

(424)

Cooperative Learning Activity

Organize students into six groups and assign one of the following topics to each group: physical geography, climate, mineral resources, agricultural products, population density, historical events. Provide each group with an outline map of Israel. Instruct groups to create a map that illustrates their assigned topic. Remind groups that maps should have a title and a key. Call on group representatives to present their completed maps. Then have students discuss similarities and differences among the maps. **L2**

Agriculture The Israelis have drained swamps, built irrigation systems, and used fertilizers to improve their soil. Israel's major farming area, which produces citrus fruits such as oranges, grapefruits, and lemons, is the coastal plain. Citrus fruits are Israel's chief farm export. Farming also occurs in parts of the Negev but only with the help of irrigation.

Place

The People

A single Israeli law—The Law of Return—increased the country's population more than any other factor. Passed in 1950, the law states that settlement in Israel is open to all Jews from anywhere in the world. Today Israel has about 5.4 million people. About 80 percent of Israel's population are Jews, but Israel's people also include Muslims and Christian Arabs. Most of Israel's Arabs are part of a group called the Palestinians. Both Israeli Jews and Arabs trace their roots to groups that lived in the area centuries ago.

Because of their country's industrial growth, about 90 percent of Israelis live in urban areas. The largest cities are Tel Aviv and the neighboring cities of Jerusalem, Yafo (yah•FOH), and Haifa (HY•fuh).

Jerusalem is a holy city for Jews, Christians, and Muslims, whose religions began in Southwest Asia centuries ago. All three groups practice **monotheism**, or a belief in one God.

A Special Birthday

If you see a look of pride on Shamir Mazar's face, it comes from his many days of preparation for his bar mitzvah. A bar mitzvah is a special ceremony for Jewish boys when they turn 13. A similar ceremony called a bat mitzvah is held for girls at the same age. At this time young people are accepted as members of the Jewish community. Dressed in a traditional prayer shawl and cap, Shamir carried the scrolls of the Torah, the Jewish holy book. As part of the ceremony, he read a passage from the scrolls.

Jerusalem, Israel

The Muslim Dome of the Rock and the Jewish Western Wall are holy sites in Jerusalem's old city.
PLACE: Jerusalem is considered a holy city for followers of what three religions?

425

LESSON PLAN
Chapter 16, Section 2

ASSESS

Check for Understanding

Assign Section 2 Review as homework or an in-class activity.

Meeting Lesson Objectives

Each objective below is tested by the questions that follow it in parentheses.

1. **Explain** how Israelis have developed their resources. (1, 2, 3)
2. **Identify** the groups of people that live in Israel. (1)
3. **Discuss** how the past affects Israel today. (4)

More About the Illustration

Answer to Caption
Christianity, Islam, Judaism

Place The Western Wall or "Wailing Wall" is the only remaining wall of a Jewish temple built in 516 B.C. A large rock inside the Golden Dome is said to be the site from which Muhammad rose into heaven.

Meeting Special Needs Activity

Attention Deficit Disorder Students who have difficulty keeping their focus on a topic may benefit from the following strategy. Ask a question related to the first subhead. Set a time limit in which students should find the answer in the text. Continue asking one question for each subhead and setting a specific limit for response time. **L1**

More About the Illustration

Answer to Caption
They fought a series of wars in which Israel gained Arab land.

 Region Because their ancestors settled in Palestine 4,000 years ago, Jewish people have always considered the area their homeland.

Evaluate

Section 2 Quiz L1

Use the Testmaker to create a customized quiz for Section 2. L1

Reteach

Have students write five cause-and-effect statements on section content. Have volunteers read either the "cause" or the "effect" half of their statements. As each cause or effect is read, call on students to complete the statement. L1

Enrich

Have students research and report on the Holocaust. Then lead a discussion on how this tragedy led to the creation of Israel. L3

CLOSE

Ask students to imagine that they are visiting Israel. Have them write a postcard to a friend describing the major features of Israel's geography. L1

Meeting of Leaders

Benjamin Netanyahu (right) became Israel's leader in 1996. He firmly defended Israeli interests but was willing to meet with Palestinian leader Yasir Arafat (left).
REGION: What happened to Israel and the Arab countries between 1948 and 1974?

Early History Many Israeli Jews today come from such places as Europe, North Africa, and the Americas. They trace their origins, however, to the Hebrews who set up a kingdom with its capital at Jerusalem in about 1000 B.C. By 922 B.C. the Hebrew kingdom had split into two states—Israel and Judah. The people of Judah were known as Jews. Israel and Judah were eventually destroyed by invaders.

Over time, the area that is now Israel was ruled by a series of peoples—Greeks, Romans, Byzantines, Arabs, and Ottoman Turks. Under the Romans, Israel and Judah became the country of Palestine. In the late 1800s, the people in Palestine included Arabs and a small community of Jews. At that time many Jews from Europe moved to Palestine. These Jewish settlers, known as Zionists, wanted to set up a homeland for Jews.

Modern Israel After World War I, Palestine came under British control. During World War II, millions of Jews in Europe were killed by German Nazis. This mass slaughter known as the **Holocaust** brought worldwide attention to the Zionist cause. In 1947 the United Nations voted to divide Palestine into a Jewish and an Arab state. In 1948 the Jews of Palestine proclaimed an independent republic called the State of Israel. The Arabs of Palestine, however, rejected this action.

A series of Arab-Israeli wars in which Israel gained Arab land were fought between 1948 and 1974. A 1978 treaty between Israel and Egypt was a first step toward peace. Agreements made in 1993 and 1994 between Israel and Palestinian Arab leaders and between Israel and Jordan also gave hope that peace may come to the region.

SECTION 2 REVIEW

REVIEWING TERMS AND FACTS
1. **Define the following:** potash, phosphate, monotheism, Holocaust.
2. **LOCATION** Where is most of Israel's industry located?
3. **PLACE** What is Israel's major crop?
4. **MOVEMENT** Why do people around the world visit Jerusalem?

 MAP STUDY ACTIVITIES
5. Turn to the map on page 424. What body of water borders Israel on the west?
6. Look at the same map on page 424. What climates are found in Israel?

Answers to Section 2 Review
1. All vocabulary words are defined in the Glossary.
2. around Tel Aviv
3. citrus fruit
4. It is considered a holy place by Christians, Jews, and Muslims.
5. Mediterranean Sea
6. Mediterranean and desert climates

MAKING CONNECTIONS

IT'S TIME FOR DINNER

| MATH | SCIENCE | HISTORY | LITERATURE | TECHNOLOGY |

Your friend Bashir from Southwest Asia has invited you to dinner. Some of the foods you're served look familiar. The table holds dates and other fruit, a bowl of rice, and the pita bread your own mother buys to make sandwiches. Whether you are eating Israeli matzo bread or Lebanese pita bread, you are enjoying the foods of Southwest Asia.

ANCIENT ORIGINS The foods of this region date to ancient times. Olives—a key ingredient in the olive oil used in the region's foods—have always grown well in a Mediterranean climate. About 5,000 years ago the Phoenicians taught the Greeks to burn olive oil in their clay lamps. The Greeks quickly learned to cook with the oil. Date palms were also cultivated in the region as early as 400 B.C. Wheat was grown in the Fertile Crescent by the earliest known civilizations. Oranges and spices came to the region from India in the 1200s.

SHARED REGIONAL FOODS The foods common to all of Southwest Asia include dishes made from wheat, fruits and vegetables, and flavorful spices. Dishes prepared

A family meal in Jordan

with eggplant, chickpeas and other legumes, celery, tomatoes, onions, chicken, and lamb appear in most countries. A generous dash of saffron, cinnamon, cardamom, cumin, or cilantro adds flavor. Fruits such as dates, raisins, lemons, figs, pomegranates, and apricots are also popular, especially as desserts.

Lack of refrigeration in some places led people to develop a taste for yogurt—fermented camel's milk—rather than fresh milk products. In your friend's home you might be offered fruit juices or tea to end your meal.

Making the Connection

1. What are some of the oldest foods grown in the Southwest Asian region?
2. Of the foods described here, which have you tasted? How did you like them?

MAKING CONNECTIONS

TEACH

Take a hand vote to discover how many students have eaten the foods mentioned in this feature. Write the results on the chalkboard. Then encourage students to discuss how foods from distant regions, such as Southwest Asia, find their way into the American diet.

More About . . .

Yogurt Many Southwest Asians believe that yogurt has extraordinary qualities. They say that yogurt lengthens life; helps the complexion; cures such ailments as sunburn, ulcers, and malaria; and encourages spiritual well-being.

Foods Around the World, Unit 6

Answers to Making the Connection

1. olives, dates, wheat, oranges, spices
2. Answers will vary. Encourage students to offer reasons why they liked or disliked certain foods.

SECTION 3
Syria, Lebanon, and Jordan

Meeting National Standards

Geography for Life The following standards are highlighted in this section:
• Standards 1, 4, 6, 9, 10, 12, 13, 15

FOCUS

Section Objectives

1. **Identify** the landforms that make up Syria.
2. **Explain** why Lebanon was torn by a fierce civil war.
3. **Discuss** how Jordan is governed.

Bellringer Motivational Activity

Prior to taking roll at the beginning of the class period, project Section Focus Transparency 16-3 and have students answer the activity questions. Discuss students' responses. **L1**

Vocabulary Pre-check

Have students find the section's **Words to Know** in the text. Have them read the sentences in which the words are used. Then have students substitute another word with a similar meaning for each of the **Words to Know**. **L1 LEP**

TEACH

Guided Practice

Chart Create a chart on the chalkboard with *Syria, Lebanon,* and *Jordan* as column headings. Down the left side of the chart, list the
(continued)

PREVIEW

Words to Know
• Bedouin
• civil war
• constitutional monarchy

Places to Locate
• Syria
• Lebanon
• Syrian Desert
• Damascus
• Beirut
• Jordan
• Jordan River
• Amman

Read to Learn . . .
1. what landforms make up Syria.
2. why Lebanon was torn by a fierce civil war.
3. how Jordan is governed.

From the sight of modern buildings in busy Amman, Jordan, you might not guess that the city is more than 3,500 years old. Today Amman is the capital and major economic center of Jordan, a small country that lies east of Israel in Southwest Asia.

Syria, Lebanon, and Jordan share a stretch of fertile land in the middle of Southwest Asia's mountains and deserts. This area is called the Fertile Crescent, because it is shaped like a half-moon. The Fertile Crescent runs from the Mediterranean coast in the west to the Persian Gulf in the east.

Place

Syria

Syria lies at the eastern end of the Mediterranean Sea and has been a center of commerce and trade for centuries. Throughout its early history, Syria was a part of many empires, but in 1946 it became an independent republic.

The Land and Economy Syria's land includes fertile coastal plains and valleys along the Mediterranean Sea. Inland mountains running north and south keep moist sea winds from reaching the eastern part of Syria. The map on page 420 shows that the vast Syrian Desert covers this region.

Agriculture is Syria's major economic activity. Syrian farmers raise mostly cotton and wheat in the rich soil of mountain valleys and coastal plains. In

428

Classroom Resources for Section 3

REPRODUCIBLE MASTERS
Reproducible Lesson Plan 16-3
Guided Reading Activity 16-3
Section Quiz 16-3

TRANSPARENCIES
Section Focus
   Transparency 16-3
Unit Overlay Transparencies
   6-0, 6-1, 6-2, 6-3, 6-4, 6-5, 6-6

MULTIMEDIA
Testmaker

ZipZapMap! World

SOUTHWEST ASIA: Land Use and Resources

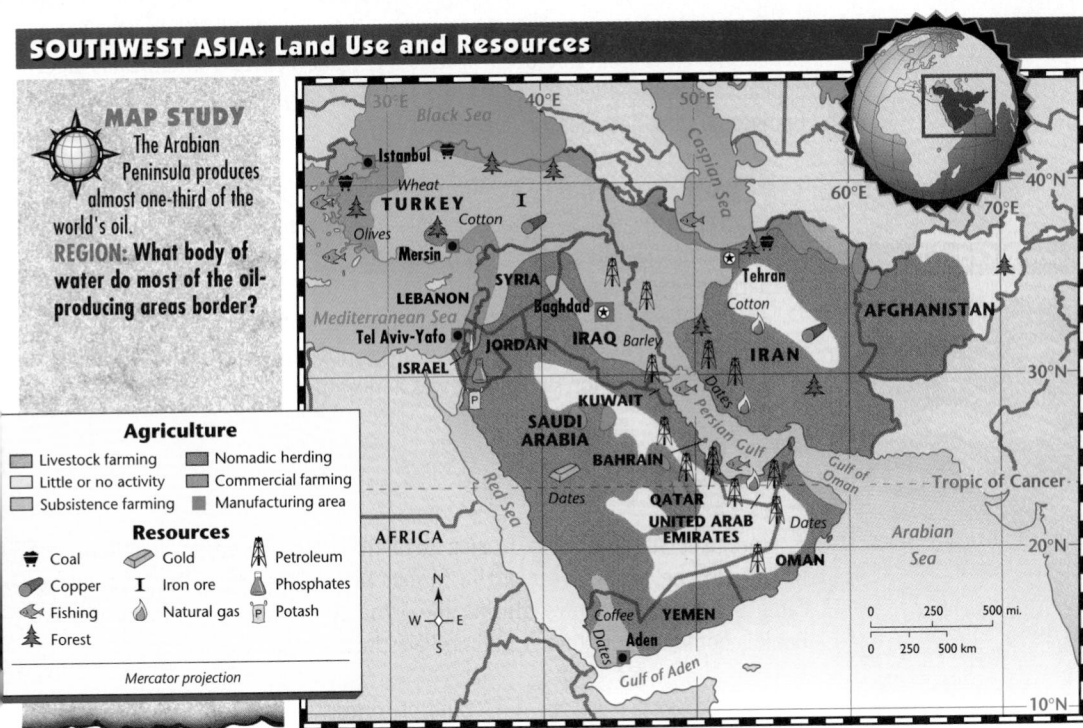

MAP STUDY
The Arabian Peninsula produces almost one-third of the world's oil.
REGION: What body of water do most of the oil-producing areas border?

Agriculture
- Livestock farming
- Little or no activity
- Subsistence farming
- Nomadic herding
- Commercial farming
- Manufacturing area

Resources
- Coal
- Copper
- Fishing
- Forest
- Gold
- Iron ore
- Natural gas
- Petroleum
- Phosphates
- Potash

Mercator projection

headings *Landforms, Rivers, Crops, Industries, Cities,* and *Religions.* Have students copy and complete the chart. **L1** **LEP**

MAP STUDY

Answer
Persian Gulf

Map Skills Practice
Reading a Map What is Afghanistan's main agricultural activity? *(nomadic herding)*

Independent Practice

Guided Reading Activity 16-3 **L1**

ASSESS

Check for Understanding
Assign Section 3 Review as homework or an in-class activity.

---

many dry areas, farmers rely on irrigation to grow crops. Syria's primary industry is textile making. Syrian cloth and fabrics have been highly valued since ancient times. The country also produces cement, chemicals, and glass.

The People About half of Syria's 14 million people live in rural areas, and a few are **Bedouins**, or people who move through the deserts. The word *bedouin* is Arabic for "desert dweller." The other half of the population live in cities. The city of Damascus, Syria's capital, is home to more than 1.5 million people. It was founded more than 5,000 years ago as a trading center.

Place

Lebanon

Lebanon, one of the smallest countries of Southwest Asia, lies west of Syria on the eastern coast of the Mediterranean Sea. Its land area is slightly smaller than the land area of the state of Connecticut.

The Land and Economy Lebanon consists of plains and mountains. Sandy beaches and a narrow plain run along the Mediterranean coast. Rugged mountains rise east of the plain and along Lebanon's eastern border. A fertile valley lies between the two mountain ranges. People living on the coast enjoy a mild, humid climate. Inland areas receive less rainfall and have lower winter and higher summer temperatures.

CHAPTER 16 429

Meeting Lesson Objectives

Each objective below is tested by the questions that follow it in parentheses.

1. **Identify** the landforms that make up Syria. (5)
2. **Explain** why Lebanon was torn by a fierce civil war. (3)
3. **Discuss** how Jordan is governed. (4)

Cooperative Learning Activity

Organize students into three groups and assign each group one of the countries discussed in Section 3. Instruct the groups to create a study guide for their assigned country. Make sure each group member makes a copy of the guide. Then have the class separate into groups of three, consisting of one member from each of the original groups. Each group member is responsible for teaching the other two about the country he or she originally studied. **L2**

More About the Illustration

Answer to Caption
the slopes of hills or mountains

Human/ Environment Interaction Cedar trees were once plentiful in Lebanon. As they were cut down, however, no new trees were planted to replace them. Today, only a single grove of the ancient trees—located in the northern mountains—remains.

Evaluate

 Section 3 Quiz **L1**

Use the Testmaker to create a customized quiz for Section 3. **L1**

Reteach

Ask students to write a one-paragraph geographical description for each of the countries discussed in this section. **L1**

Enrich

Have students investigate the major features of Damascus, Beirut, or Amman. Suggest that they present their findings in the form of a travelogue. **L2**

CLOSE

Ask students to note similarities and differences among Syria, Lebanon, and Jordan. **L2**

430

Service industries such as banking are Lebanon's major source of income. Lebanon's chief manufactured products include cement, chemicals, electric appliances, and textiles. If you were a Lebanese farmer, you might grow apples, oranges, grapes, potatoes, or olives.

The People Almost 60 percent of Lebanon's 3.6 million people are Muslims, while most of the rest are Christians. About 86 percent of Lebanon's people live in urban areas. Beirut (bay•ROOT), located on the coast, is the nation's capital and largest city.

From 1975 to 1991, groups of Muslims and Christians in Lebanon fought a **civil war,** or a conflict among different groups within a country. Many lives were lost and the nation's economy was almost destroyed.

Place
Jordan

Jordan is slightly larger in size than the state of Kentucky. Before 1967 Jordan also held land west of the Jordan River. This area—the West Bank— came under Israeli control during the 1967 Arab-Israeli war. Today the West Bank is home to more than 1 million Palestinian Arabs.

The Land and Economy A land of contrasts, Jordan reaches from the fertile Jordan River valley in the west to dry, rugged country in the east. The northern and western parts of Jordan have mild, rainy winters and dry, hot summers. Southern and eastern parts of the country have a hot, dry climate year-round.

Most workers in Jordan are employed in agriculture and manufacturing. The leading manufactured goods are cement, chemicals, and petroleum products. Jordan's farmers in the Jordan River valley grow citrus fruits, tomatoes, and melons.

The People Most of Jordan's 4.2 million people are Arab Muslims and Christians. They include about 1 million Palestinian Arabs who fled Israel or the West Bank during the Arab-Israeli wars. Amman, with 1.2 million people, is the leading city and capital. Jordan is a **constitutional monarchy,** a form of government in which the monarch shares power with elected officials.

Ancient Cedars

The stately cedar tree is Lebanon's national symbol and appears on its flag, coins, and stamps.
LOCATION: How would you describe the landscape where these trees grow?

SECTION 3 REVIEW

REVIEWING TERMS AND FACTS
1. **Define the following:** Bedouin, civil war, constitutional monarchy.
2. **PLACE** What city in Syria is considered one of the oldest cities in the world?
3. **PLACE** How has civil war affected Lebanon?

 MAP STUDY ACTIVITIES
4. Look at the map on page 429. What is the primary economic activity of Syria?
5. Look at the same map on page 429. What countries border Lebanon?

Answers to Section 3 Review
1. All vocabulary words are defined in the Glossary.
2. Damascus
3. The civil war has caused the loss of many lives and damaged Lebanon's economy.
4. agriculture
5. Syria, Israel

The Arabian Peninsula

PREVIEW

Words to Know
- oasis
- *hajj*

Places to Locate
- Saudi Arabia
- Kuwait
- Bahrain
- United Arab Emirates
- Qatar
- Yemen
- Oman
- Makkah
- Riyadh

Read to Learn . . .
1. why Saudi Arabia is important to the world's Muslims.
2. how oil affects the lives of the people of the Arabian Peninsula.

The hot desert sun rises and sets over the Arabian Peninsula. Under these windswept dunes, oil constantly flows through the country's network of pipelines.

The Arabian Peninsula is like a giant platform that tilts toward the east and the north. Its highest elevations are in the south. The mostly desert land in the north slopes toward flat, sandy beaches along the Persian Gulf. The country of Saudi Arabia takes up about 80 percent of the Peninsula.

Place
Saudi Arabia

Saudi Arabia, the largest country in Southwest Asia, is about the size of the eastern half of the United States. Saudi Arabia holds a major share of the world's oil. The graph on page 432 compares the amount of oil that Southwest Asia pumps to that of other oil-producing nations.

The Land Vast deserts cover Saudi Arabia. The largest desert is the Empty Quarter in the southern part of the country. The Empty Quarter has mountains of sand that reach heights of more than 1,000 feet (305 m). In the west, highlands stretch along the Red Sea. Valleys among these highlands provide fertile farmland. Eastern Saudi Arabia is a coastal plain along the Persian Gulf.

CHAPTER 16

431

Meeting National Standards

Geography for Life The following standards are highlighted in this section:
- Standards 1, 4, 6, 9, 10, 12, 16

FOCUS

Section Objectives
1. **Explain** why Saudi Arabia is important to the world's Muslims.
2. **Discuss** how oil affects the lives of the people of the Arabian Peninsula.

Bellringer Motivational Activity

Prior to taking roll at the beginning of the class period, project Section Focus Transparency 16-4 and have students answer the activity questions. Discuss students' responses. L1

Vocabulary Pre-check
Have students discover the importance of the *hajj* to the Islamic faith. L1 **LEP**

TEACH

Guided Practice
Location Encourage students to make up a riddle on the location of each of the countries on the Arabian Peninsula. Have volunteers read their riddles and call on the class to identify the countries. L2

Classroom Resources for Section 4

 **REPRODUCIBLE MASTERS**
Reproducible Lesson Plan 16-4
Guided Reading Activity 16-4
Section Quiz 16-4

 TRANSPARENCIES
Section Focus
Transparency 16-4

MULTIMEDIA
 Testmaker

 ZipZapMap! World

Independent Practice

📁 Guided Reading Activity 16-4

More About the Illustration

Answer to Caption
oil exports

Place In Saudi Arabia, women must wear a veil in public. In addition, they are not allowed to drive cars and they must have the permission of a male relative to travel.

ASSESS

Check for Understanding

Assign Section 4 Review as homework or an in-class activity.

Meeting Lesson Objectives

Each objective below is tested by the questions that follow it in parentheses.

1. **Explain** why Saudi Arabia is important to the world's Muslims. (1)
2. **Discuss** how oil affects the lives of the people of the Arabian Peninsula. (4)

GRAPHIC STUDY

Answer
33.1 percent

Graphic Skills Practice
Reading a Graph What percentage of the world's oil reserves exist in North America? *(3.1 percent)*

World Oil Producers

GRAPHIC STUDY

Southwest Asia has more oil than all other regions of the world combined.
REGION: What is the total percentage of oil found in all other regions of the world?

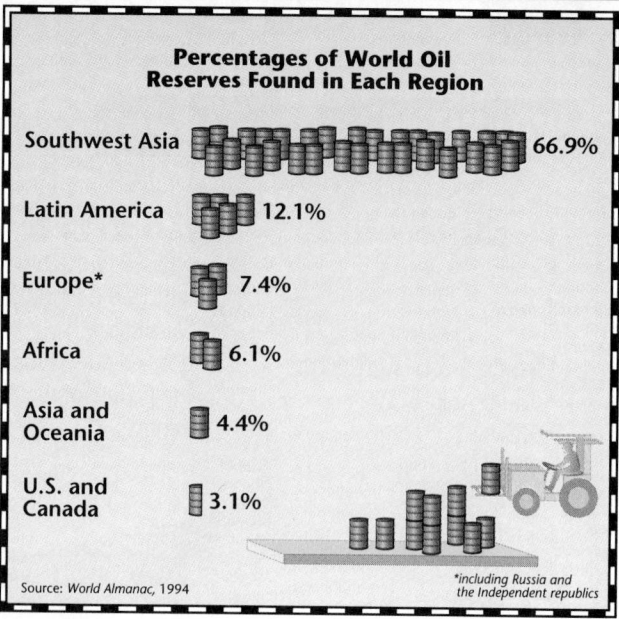

Percentages of World Oil Reserves Found in Each Region

Region	Percentage
Southwest Asia	66.9%
Latin America	12.1%
Europe*	7.4%
Africa	6.1%
Asia and Oceania	4.4%
U.S. and Canada	3.1%

Source: *World Almanac*, 1994

*including Russia and the Independent republics

The Economy The economy of Saudi Arabia is based on oil. Money from oil exports has built schools, hospitals, roads, and airports. Agriculture is less important than oil because of the limited amount of water and farmland. To meet the need for freshwater, the country's engineers have built industrial plants to remove salt from the seawater of the Persian Gulf and the Red Sea.

The People The people of Saudi Arabia once were divided into numerous family groups. In 1932 a monarchy led by the Saud family unified the country. The Saud family still rules Saudi Arabia today.

Most of the 18 million Saudi Arabians live in towns and villages. These are found along the oil-rich Persian Gulf coast or around **oases**—lush, green places in dry areas that have enough water for crops to grow. Some of the oases have grown into large cities such as Riyadh (ree•YAHD).

The Islamic religion affects almost all parts of life in Saudi Arabia—from government to the everyday lives of the people. The holy city of Makkah (MAH•kah) was very important in the life of Muhammad (moh•HA•muhd), the man who brought Islam to the Arabs in the A.D. 600s. Today hundreds of thousands of Muslims from around the world visit Makkah each year. Each Muslim tries to make a *hajj,* or religious journey to Makkah, at least once.

Saudi Women at Work

A female doctor in Saudi Arabia examines her patient.
PLACE: Income from what resource has helped build Saudi Arabia's hospitals?

(432)

UNIT 6

Cooperative Learning Activity

Organize students into groups of five. Inform groups that their task is to create an illustrated wall chart showing the major features of the geography of the Arabian Peninsula. Ensure that each group member is assigned an activity— research, art and design, writing captions, constructing the chart, and so on. Have groups display and discuss their completed wall charts.
L2

Place
Other Arabian Peninsula Nations

Saudi Arabia shares the Arabian Peninsula with six other Arab Muslim countries. They are Kuwait (ku•WAYT), Bahrain (bah•RAYN), Qatar (KAH•tuhr), the United Arab Emirates, Oman (oh•MAHN), and Yemen (YEH•muhn).

Persian Gulf States Kuwait, Bahrain, Qatar, and the United Arab Emirates are located along the Persian Gulf. Beneath their flat deserts and offshore areas lie vast deposits of oil. The Persian Gulf states have used oil profits to build prosperous economies. Their people—for the most part—enjoy a high standard of living with excellent health care, education, and housing. Many workers from other countries have settled in the Persian Gulf countries to work in the modern cities and oil fields.

Oman and Yemen At the southern end of the Arabian Peninsula are the countries of Oman and Yemen. Oman is largely desert, but its bare land yields oil—the basis of the country's economy. Most of the people of Oman live in rural villages. The oil industry, however, has drawn many rural dwellers as well as foreigners to Oman's growing urban areas.

The country of Yemen is made up of a narrow coastal plain and inland mountains. It has almost no natural mineral resources, and most of the people are farmers or herders of sheep and cattle. They live in the high fertile interior where the capital, San'a (sa•NAH), is located. Farther south lies Aden (AH•duhn), a major port for ships traveling between the Arabian and Red seas.

The Kaaba, the most sacred Islamic shrine, is in the courtyard of Makkah's Grand Mosque.
PLACE: Who brought the Islamic religion to the Arabs?

SECTION 4 REVIEW

REVIEWING TERMS AND FACTS
1. **Define the following:** oasis, *hajj.*
2. **PLACE** What country takes up most of the Arabian Peninsula?
3. **LOCATION** Where is the Empty Quarter located?
4. **MOVEMENT** Why is the port of Aden important to Yemen?

GRAPHIC STUDY ACTIVITIES
5. Turn to the graph on page 432. What region comes closest to matching Southwest Asia's oil production?
6. Look at the graph on page 432. What two regions together produce about as much oil as Europe does?

Answers to Section 4 Review
1. All vocabulary words are defined in the Glossary.
2. Saudi Arabia
3. in the southern part of the Arabian Peninsula
4. Aden is a stopping point for shipping moving between the Arabian and Red seas.
5. Latin America
6. Asia and Oceania, North America

LESSON PLAN
Chapter 16, Section 4

More About the Illustration
Answer to Caption
Muhammad

 **Place** When Muslims around the world kneel in prayer, they face in the direction of the Kaaba in Makkah.

Evaluate
Section 4 Quiz **L1**

Use the Testmaker to create a customized quiz for Section 4. **L1**

Reteach
Display an outline map of the Arabian Peninsula. Point to the locations of countries or major physical features. As you point to each location, call on a volunteer to name it. **L1**

Enrich
Have students read stories from the *Arabian Nights* to share with the class. Encourage students to discuss the ways these stories reflect the physical and human characteristics of the Arabian Peninsula. **L2**

CLOSE
Have students discuss what might happen to the economies of the countries of the Arabian Peninsula when the region's oil reserves are exhausted. **L2**

SECTION 5 · Iraq, Iran, and Afghanistan

Meeting National Standards

Geography for Life The following standards are highlighted in this section:
• Standards 1, 2, 4, 6, 9, 10, 12, 13, 16, 18

FOCUS

Section Objectives

1. **Explain** why the Tigris and Euphrates rivers are important to the people of Iraq.
2. **Describe** how a religious government came to power in Iran.

Bellringer Motivational Activity

Prior to taking roll at the beginning of the class period, project Section Focus Transparency 16-5 and have students answer the activity questions. Discuss students' responses. **L1**

Vocabulary Pre-check

List the **Words to Know** on the chalkboard. Have students find the definitions of these terms in a dictionary. Then direct students to use each term correctly in a sentence. **L1 LEP**

PREVIEW

Words to Know
• alluvial plain
• shah
• ethnic group

Places to Locate
• Iraq
• Baghdad
• Iran
• Caspian Sea
• Tehran
• Afghanistan
• Hindu Kush Mountains
• Kabul

Read to Learn . . .
1. why the Tigris and Euphrates rivers are important to the people of Iraq.
2. how a religious government came to power in Iran.

P oems, stories, and songs describe the ancient beauty of Baghdad—the heart of Iraq. Today Baghdad boasts modern buildings as well as old neighborhoods and busy marketplaces.

B aghdad, with a population of about 5.5 million, is one of the largest cities in Southwest Asia. It is the capital of Iraq, a country in the northern part of Southwest Asia. The other countries in this region are Iran and Afghanistan.

Location

Iraq

Iraq lies at the head of the Persian Gulf and north of the Arabian Peninsula. Some of the world's oldest civilizations developed there between the Tigris and Euphrates rivers in an area called Mesopotamia. As a nation, however, modern Iraq has been in existence only since the early 1900s. Today it is one of the important oil-producing countries of Southwest Asia.

The Land The Tigris and Euphrates rivers are the major geographic features of Iraq. They flow through Iraq's northern highlands and central plain before joining to enter the Persian Gulf. West of the Tigris-Euphrates area, the landscape is mostly desert. What types of climates can you find in Iraq on the map on page 424?

434

UNIT 6

Classroom Resources for Section 5

REPRODUCIBLE MASTERS

Reproducible Lesson Plan 16-5
Guided Reading Activity 16-5
Critical Thinking Skills
   Activity 16-5
Reteaching Activity 16
Enrichment Activity 16
Section Quiz 16-5

 TRANSPARENCIES

Section Focus
   Transparency 16-5
Geoquiz Transparencies 16-1,
16-2

 MULTIMEDIA

Reuters Issues in
   Geography
Testmaker

MAP STUDY
Iraq's land includes large desert areas that are mostly unpopulated.
HUMAN/ENVIRONMENT INTERACTION: Which Southwest Asian country has the largest uninhabited areas?

CITIES
■ City with more than 5,000,000 people
● City with 1,000,000 to 5,000,000 people
○ City with 500,000 to 1,000,000 people

Persons per	
sq. mi.	sq. km
Uninhabited	Uninhabited
2-25	1-10
2-60	1-25
60-125	25-50
125-250	50-100
Over 250	Over 100

Mercator projection

MAP STUDY

Answer
Saudi Arabia

Map Skills Practice
Reading a Map What are the two largest cities in Southwest Asia?
(Tehran, Istanbul)

The Economy Iraq's economy is based on oil and agriculture as you can see on the map on page 429. Oil is the country's major export. Manufacturing also has developed in recent years, with Iraqi factories making food products, cloth, soap, and leather goods.

Most farming in Iraq takes place near the Tigris and Euphrates rivers. Find this area on the map on page 420. The plain between the rivers is called an **alluvial plain** because it is built up by rich fertile soil left by river floods. Iraq's farmers grow crops such as barley, dates, grapes, melons, citrus fruits, and wheat.

The People The land lined with date palms along the Tigris and Euphrates rivers is home to most of Iraq's 20 million people. About 70 percent live in urban areas such as Baghdad. Muslim Arabs make up the largest group in Iraq's population. A Muslim people, Kurds, are the second-largest group. The Kurds have their own culture and live in the mountains of northeast Iraq. Many Kurds in Iraq and in neighboring Turkey want to form their own independent country.

Past and Present Throughout the centuries, many empires have risen and fallen on the Tigris-Euphrates plain. From the A.D. 700s to 1200s, Baghdad was the center of a large Muslim empire that made many advances in the arts and sciences. Modern Iraq, with its wealth in oil, seeks to become a leading power in Southwest Asia. Over the past few years, Iraq has fought wars with some of its neighbors to gain more territory.

In 1990 Iraqi armies invaded Kuwait, Iraq's southern neighbor. In early 1991 a United Nations force led by the United States launched air and missile attacks on Iraq and its forces in Kuwait. The war against Iraq became known as

CHAPTER 16

435

TEACH

Guided Practice
Human/Environment Interaction Challenge students to suggest examples of the impact of climate and landscape on economic activities, housing, clothing, and recreation in Iraq, Iran, and Afghanistan. Note responses on the chalkboard. Ask students to use the information on the chalkboard to write a paragraph explaining the link between physical geography and ways of life. **L2**

CURRICULUM CONNECTION

History The body of water created by the joining of the Tigris and Euphrates is called the Shatt-al-Arab. It marks the southernmost section of the border between Iraq and Iran, and the two countries have long struggled over control of the waterway. In 1980, the two countries went to war over the issue. After eight years of bitter conflict, however, the issue remained unresolved.

Cooperative Learning Activity

Organize students into three groups and assign one of the countries discussed in this section to each group. Each student should write five facts about the country assigned to his or her group. Then group members should compare their lists and create a "Fact File" of the most important items. Provide time for the groups to share their Fact Files with the rest of the class. **L2**

Independent Practice

 Guided Reading Activity 16-5 **L1**

 Workbook Activity 16 **L1**

GLENCOE
TECHNOLOGY

 **VIDEODISC**

Reuters Issues in Geography

Chapter 5
North Africa and Southwest Asia: The Influence of Oil and Water
Disc 1 Side A

?WHAT IN THE WORLD

Muslim Art

The Quran, the holy book of Islam, forbids drawing humans and animals. Muslim artists in Southwest Asia focus instead on painting beautiful patterns and designs, which are found in buildings, jewelry, and metal objects. Artists using calligraphy, or the art of fine writing, decorate the words of the Quran and other important writings.

the Persian Gulf War. Iraqi forces were forced to withdraw from Kuwait, but not before they set afire hundreds of Kuwait's oil wells. Oil spilling from the wells created huge, explosive lakes of petroleum and fires and smoke that threatened much of the Persian Gulf's plant and animal life.

Since the conflict, Iraq's people have faced many hardships. Iraq's government continues to threaten to invade its neighbors. In response many foreign nations, including the United States, refuse to trade with Iraq. This has severely damaged Iraq's economy, which depends on selling oil to other parts of the world.

Location
Iran

Iran lies east of Iraq and northeast of the Arabian Peninsula. Zapha, a 14-year-old schoolgirl in modern Iran, knows she lives in one of the world's oldest countries. In school she learned that her country even had a different name at one time—Persia. What has not changed is the land of Iran itself.

The Land Iran is about the size of the state of Alaska. A central, dry plateau stretches across most of Iran. Two vast mountain ranges—the Elburz and the Zagros—surround most of this plateau. Coastal plains lie outside the mountains along the shore of the Caspian Sea in the north and along the Persian Gulf coast in the south.

Iran has a variety of climates. Mountainous areas experience severely cold winters and mild summers. Most of the central plateau has a dry climate, with cold winter temperatures and mild to hot summers.

A Science Lesson in an Iranian School

Iranian students and their teacher work in an all-male science class.
PLACE: How is this science classroom similar to science classrooms in the United States?

(436)

Meeting Special Needs Activity

Visual Learners Some students may rely on photographs and other graphics to aid understanding. Ask these students to state what they learn from each visual in this section. Then help them correlate their statements to information in the text. Guide visual learners to understand how pictures and graphics reinforce and/or extend the text. **L1**

The Economy Iran is one of the world's leading oil-producing countries. A major part of its income results from the sale of oil to foreign countries. In recent years, however, the falling price of oil on world markets has brought less money into Iran. Other Iranian industries produce a variety of goods, including leather goods, machine tools, and tobacco products.

As in much of Southwest Asia, water is scarce in Iran. Less than 12 percent of Iran's land can be used for farming. In dry areas, you see Iranian farmers bringing water to their fields from wells or through underground tunnels. Iran's chief agricultural products are wheat, barley, corn, cotton, dates, nuts, tea, and tobacco.

Tehran, Iran

Iran's capital and largest city, Tehran, lies at the base of the majestic Elburz Mountains.
MOVEMENT: Why are many Iranians moving from rural areas to the cities?

The People Today about 61 million people live in Iran, most of whom speak Persian, or Farsi, a language that is different from Arabic. About 98 percent of Iranians are Muslims who belong to a group within Islam known as Shiites (SHEE•YTES). Most of the world's Muslims belong to another group known as Sunnis (SU•NEES).

About 57 percent of Iran's people live in urban areas. In recent years towns and cities have grown tremendously because many people from rural areas have moved to urban areas in search of jobs. Find Tehran, Iran's capital and largest city, on the map on page 435. It has a population of about 10 million.

Iran was part of the powerful Persian Empire about 3,500 years ago. The descendants of ancient Persians make up the majority of Iran's population today. For hundreds of years Iran was ruled by kings known as **shahs.** Mohammed Reza Pahlavi (PA•luh•vee) became shah during World War II. In the 1960s and 1970s, he used money from oil sales to promote industry and to build Iran's military forces. The Shah, however, was accused of ruling harshly, and many Iranians turned against him. In 1979 a revolution forced the Shah out of Iran. In place of the Shah's government, the Iranians set up a government run by religious leaders. Iran's Muslim government has placed strict laws on the country, and many non-Muslim customs are now forbidden.

437

Critical Thinking Activity

Drawing Conclusions Have students list all the ways that religion influences life in Iraq, Iran, and Afghanistan. They may use information gathered from the text, the illustrations, or reference books. Record students' responses on the chalkboard. Then ask students to draw conclusions about the importance of religion in this region. **L2**

LESSON PLAN
Chapter 16, Section 5

More About the Illustration
Answer to Caption
to find jobs

GEOGRAPHIC THEMES

Place Tehran's modern section has many new office and apartment buildings. Its old section has mosques and bazaars that were built hundreds of years ago.

Cultural Kaleidoscope

Iran Many men in Iran keep physically fit by practicing traditional weight-lifting exercises and gymnastics. They train at athletic clubs called *zurkhanehs,* or "houses of strength."

ASSESS

Check for Understanding
Assign Section 5 Review as homework or an in-class activity.

Meeting Lesson Objectives
Each objective below is tested by the questions that follow it in parentheses.

1. **Explain** why the Tigris and Euphrates rivers are important to the people of Iraq. (1)
2. **Describe** how a religious government came to power in Iran. (3)

Section 5 Quiz **L1**

 Use the Testmaker to create a customized quiz for Section 5. **L1**

Chapter 16 Test Form A and/or Form B **L1**

Reteach

Reteaching Activity 16 **L1**

Spanish Reteaching Activity 16 **L1**

Enrich

Enrichment Activity 16 **L1**

CLOSE

Mental Mapping Activity

Provide each student with a large sheet of white paper and map pencils. Inform students that this will not be a graded activity. Have students draw, freehand from memory, a map of Southwest Asia that includes the following labeled countries and bodies of water: Turkey, Israel, Syria, Lebanon, Jordan, Saudi Arabia, Kuwait, Bahrain, United Arab Emirates, Qatar, Yemen, Oman, Iraq, Iran, Afghanistan, Black Sea, Caspian Sea, Persian Gulf, Red Sea, Tigris River, and Euphrates River.

Place

Afghanistan

Afghanistan is the easternmost country in Southwest Asia. It is a land of rugged landscapes and has no seacoast. Because of these geographic features, Afghanistan has had difficulty developing its resources.

The Land The Hindu Kush range covers most of Afghanistan. It has peaks as high as 25,000 feet (7,620 m). The Khyber (KY•buhr) Pass, which cuts through the Hindu Kush, for centuries has been used as a major trade route linking Southwest Asia with other parts of Asia. Rolling grasslands and low deserts make up the nonmountainous areas of Afghanistan. Highland and desert climates prevail in most of the country.

The Economy Afghanistan's major economic activity is farming. You see such crops as wheat, barley, corn, cotton, nuts, fruits, and vegetables growing there. Afghanistan is rich in minerals, but it has little industry. A few mills produce textiles, and skilled craftspeople produce beautiful rugs and jewelry in their homes.

The People The 18 million people of Afghanistan are divided into about 20 different **ethnic groups,** with each group having its own ancestry, language, and culture. The two largest ethnic groups are the Pushtuns (PUHSH•TOONZ) and the Tajiks (tah•JIHKS). Afghanistan's people practice Islam and live in the fertile valleys of the Hindu Kush range. The capital city, Kabul, is located in one of these valleys.

Afghanistan has had a long and troubled history. From ancient times to the present, a series of foreign powers have won or tried to win control of the country. The country also has been torn apart by continuing civil wars among its many ethnic groups.

SECTION 5 REVIEW

REVIEWING TERMS AND FACTS
1. **Define the following:** alluvial plain, shah, ethnic group.
2. **HUMAN/ENVIRONMENT INTERACTION** How did the Persian Gulf War affect the environment of the Persian Gulf area?
3. **PLACE** What type of government rules Iran today?
4. **PLACE** What is the major landform of Afghanistan?

 MAP STUDY ACTIVITIES
5. Using the map on page 435, name the large body of water that lies southeast of Iraq and south of Iran.
6. Look at the map on page 435. Around what body of water do most Iranians live?

UNIT 6

Answers to Section 5 Review

1. All vocabulary words are defined in the Glossary.
2. Iraqi soldiers set afire many oil wells, causing destruction and oil spillage that threatened plant and animal life in the Persian Gulf area.
3. a theocracy of Muslim religious leaders
4. mountains, most notably the Hindu Kush range
5. Persian Gulf
6. Caspian Sea

Chapter 16 Highlights

Important Things to Know About Southwest Asia

SECTION 1 TURKEY

- Turkey is located in both Europe and Asia.
- Highlands and plateaus make up most of Turkey's land area.
- Most of Turkey's people live in cities and towns.

SECTION 2 ISRAEL

- Israel is the only Jewish nation in the world.
- Despite its small size, Israel has a well-developed economy.
- Israel and its Arab neighbors are moving toward peace after a long period of tension and warfare.

SECTION 3 SYRIA, LEBANON, AND JORDAN

- Syria has a coastal plain, mountains, and deserts.
- Lebanon is one of the smallest countries of Southwest Asia.
- Jordan's farmers grow crops in the rich Jordan River valley.

SECTION 4 THE ARABIAN PENINSULA

- Oil-rich Saudi Arabia is the largest country of Southwest Asia.
- Most of Saudi Arabia's people live in coastal areas.
- Persian Gulf countries have strong economies based on oil.

SECTION 5 IRAQ, IRAN, AND AFGHANISTAN

- Iraq has recently fought wars to increase its power in Southwest Asia.
- Iran is governed by Muslim religious leaders.
- Afghanistan's land area is largely mountainous.

A market in San'a, Yemen ▶

Using the Chapter 16 Highlights

- Use the Chapter 16 Highlights to preview, review, condense, or reteach the chapter.

Preview/Review

 Vocabulary Puzzle-Maker Software reinforces the terms used in Chapter 16.

Condense

Have students read the Chapter 16 Highlights. Spanish Chapter Highlights also are available.

🎧 📁 Chapter 16 Digest Audiocassettes and Activity

📁 Chapter 16 Digest Audiocassette Test

Spanish Chapter Digest Audiocassettes, Activities, and Tests also are available.

📁 Guided Reading Activities

Reteach

📁 📁 Reteaching Activity 16. Spanish Reteaching activities also are available.

MAP STUDY

Place Make copies of page 32 of the Outline Map Resource Book and distribute them to the class. Have students label on the map the countries of Southwest Asia. Then have them mark the region's major centers of population. Call on volunteers to suggest a title for the map.

Extra Credit Project

Time Line Have students work in groups to create a time line illustrating major events in the history of one of the countries in this chapter. Provide an opportunity for groups to compare their time lines. **L2**

Chapter 16 Review and Activities

ANSWERS

Reviewing Key Terms

1. H
2. G
3. C
4. F
5. D
6. I
7. E
8. A
9. B

Mental Mapping Activity

This exercise helps
students to visualize the
countries they have been
studying and understand the
relationships among various
points located in these countries. All attempts at freehand mapping should be
accepted.

REVIEWING KEY TERMS

Match the numbered
terms in Column A
with their definitions
in Column B.

A
1. potash
2. Bedouin
3. *hajj*
4. shah
5. ethnic group
6. monotheism
7. oasis
8. alluvial plain
9. phosphate

B
A. area of fertile soil left by river floods
B. mineral used in making fertilizer
C. Muslim pilgrimage to Makkah
D. group having a common language and culture
E. area in dry places that has enough water to grow crops
F. Iranian king
G. people who move from place to place herding animals
H. type of mineral salt
I. belief in one God

Mental Mapping Activity

On a separate sheet of
paper, draw a freehand
map of the Arabian Pen-
insula. Label the following
on your map:
• Persian Gulf
• Kuwait
• Saudi Arabia
• Yemen
• Red Sea

REVIEWING THE MAIN IDEAS

Section 1
1. **LOCATION** In what part of Turkey is the Anatolian Plateau?
2. **PLACE** What is a major industrial product of Turkey?

Section 2
3. **LOCATION** What river flows into the Dead Sea?
4. **HUMAN/ENVIRONMENT INTERACTION** How have Israelis made their land suitable for farming?

Section 3
5. **PLACE** What kind of climate does Lebanon have?
6. **LOCATION** Where is Jordan's major farming region?

Section 4
7. **PLACE** What is the major industry of Saudi Arabia?
8. **LOCATION** What two nations are located at the southern end of the Arabian Peninsula?

Section 5
9. **PLACE** What major rivers flow through Iraq?
10. **LOCATION** What Southwest Asian country is located east of Iran?

Reviewing the Main Ideas

Section 1
1. the central part of Asian Turkey
2. textiles

Section 2
3. Jordan River
4. They have used water to irrigate the land.

Section 3
5. Mediterranean climate
6. Jordan River Valley

Section 4
7. oil production
8. Oman, Yemen

CRITICAL THINKING ACTIVITIES

1. **Making Inferences** From what you know about a desert climate, what could you expect to be true about farming in the Negev region of Israel?
2. **Determining Cause and Effect** How has the discovery of oil changed the lives of people living on the Arabian Peninsula?

GeoJournal Writing Activity

In this chapter on Southwest Asia, you were introduced to several important cities. Go back through the chapter to see how these cities are similar and how they are different. Write a short story or letter describing a trip to two of these cities.

COOPERATIVE LEARNING ACTIVITY

Working in a small group, select a country of Southwest Asia and research how the people have successfully used their environment. Each person in your group can read about a different part of daily life and then share findings with the group. Prepare charts, storyboards, and posters that will illustrate your presentation to the rest of the class.

PLACE LOCATION ACTIVITY: SOUTHWEST ASIA

Match the letters on the map with the places in Southwest Asia. Write your answers on a separate sheet of paper.

1. Zagros Mountains
2. Arabian Sea
3. Istanbul
4. Persian Gulf
5. Afghanistan
6. Tel Aviv
7. Euphrates River
8. Makkah
9. Tehran
10. Saudi Arabia

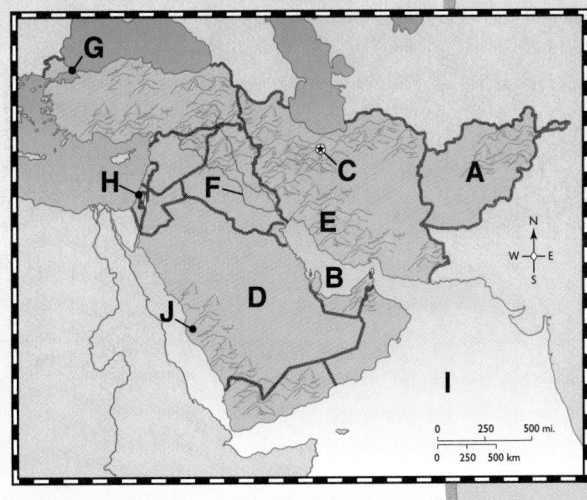

Cooperative Learning Activity

Remind students that humans relate to the environment in two ways. They adjust their activities to suit the environment, or they adjust the environment to suit their activities. Encourage students to make their presentations as colorful and dramatic as possible. Display the visuals in the classroom and invite other classes to share them.

GeoJournal Writing Activity

Encourage students to share their short stories or letters with the rest of the class.

Place Location Activity

1. E
2. I
3. G
4. B
5. A
6. H
7. F
8. J
9. C
10. D

Chapter Bonus Test Question

This question may be used for extra credit on the chapter test.

Some people have referred to Southwest Asia as a "cultural crossroads." What about the region's location supports this view? *(Southwest Asia is located at the juncture of Asia and Europe and of Asia and Africa.)*

Section 5

9. Tigris and Euphrates rivers
10. Afghanistan

Critical Thinking Activities

1. The land is dry and requires human intervention, such as irrigation, for crops to grow.
2. It has brought prosperity and high standards of living.

GeoLab
ACTIVITY

FOCUS

Remind students that as the earth revolves, the sun appears to move from east to west across the sky. As the sun "moves," the shadows it produces change. Ask students what kind of shadow would be cast by a tall building in the early hours of daylight. *(long)* What kind of shadow would the building cast at noon? *(short or no shadow)* What would happen to the shadow as the afternoon progresses? *(it would lengthen)* Conclude by mentioning that sundials employ changing shadows to measure time.

TEACH

Point out that sundials, regardless of their design, have two basic parts: the **gnomon,** the part that casts the shadow, and the **plane,** the face of the dial. Have students identify the gnomon *(pencil)* and the plane *(paper plate)* of the sundial they are asked to construct in the GeoLab Activity.

Then inform students that sundials were used in Southwest Asia—in Babylonia—at least 4,000 years ago. Call on volunteers to suggest how the climate of the region might have encouraged the development of the sundial. *(The sun shines much of the time throughout most of the region.)* **L1**

From the classroom of Eleanor Bloom, Bloom-Carroll Schools, Lithopolis, Ohio

SUNDIALS: Shadow Time

Background

The Chinese used water clocks to measure time more than 3,000 years ago. Egyptians also made water clocks, in which water spilled from one vessel to another. Early civilizations in Southwest Asia and North Africa measured time, too, but instead of water they used the sun as their timekeeper. Sundials measure the shadows an object casts as the sun moves from east to west during the day. Find out more about how these early civilizations measured time by making your own sundial.

Modern Sundial

In the 1300s Henry de Vick invented a clock that contained many of the parts that modern clocks have. Atomic clocks are the world's most accurate timepieces today, gaining or losing only a few seconds every 100,000 years!

A white or light-colored paper plate makes the best base for your sundial.

Materials

- 1 empty thread spool
- 1 ruler
- 1 paper plate
- 1 marker
- glue
- 1 new, sharpened pencil

Curriculum Connection Activity

Art Point out that, like many other practical objects, sundials have been decorated and transformed into works of art. Today, in fact, sundials are popular as ornaments. Display pictures of decorated sundials for the class. Encourage students to use these pictures as models for decorating their sundials. Some students may wish to construct sundials out of sturdier materials, such as wood or metal. Display the finished sundials in the classroom and invite other classes to view them. **L1**

What To Do

A. Glue the spool to the center of the paper plate.

B. Stand the pencil, with the point up, in the top hole of the spool.

C. Draw an arrow on the paper plate.

D. Place your sundial in a sunny window where it gets sunlight all day long. Make sure the arrow is pointing south.

E. Every hour, trace a line along the shadow's line. Use pictograms or Roman numerals to mark the hours.

Lab Activity Report

1. Are the shadow lines you made an equal distance apart?

2. As you made your shadow lines, did the same thing happen every hour?

3. What are the advantages and disadvantages of using a sundial to measure time?

4. Drawing Conclusions Does the space between the shadow lines have anything to do with telling time? Explain.

Go A Step Further

Activity

Invent a new way to tell time. Think of using everyday items and materials that you usually do not associate with telling time. Write a news report explaining your new invention.

ASSESS

Have students answer the **Lab Activity Report** questions on page 443.

CLOSE

Encourage students to complete the **Go A Step Further** writing activity. Suggest that they illustrate their news reports with blueprints or working models of their inventions.

The world's largest sundial is located in Jaipur, India. Built in 1724, this sundial has a gnomon more than 100 feet (30.5 m) high and the plane covers an area of nearly 1 acre (0.4 hectares).

Answers to Lab Activity Report

1. No; The shadow lines are closer together when the sun is high in the sky. The lines are further apart when the sun is lower in the morning and evening.

2. No; The sun cast shorter or no shadows when it was directly overhead. The sun cast longer shadows in the morning and evening.

3. Students should point out that sundials have to be placed in a location where they get sunshine all day long and at night you would not be able to tell the time using the sundial. It is pretty accurate, however, if used when full sunlight is available.

4. Answers will vary, but students should realize that the space between lines represents one hour.

Cultural
HERITAGE:
SOUTHWEST ASIA AND NORTH AFRIC

FOCUS

Before students study this feature, ask them what they know about the architecture, arts, crafts, music, and so on of Southwest Asia and North Africa. List responses on the chalkboard. Have students review and adjust the list as they work through the feature.

TEACH

Review with students the various entries in the feature. Call on volunteers to describe in their own words what is shown in each illustration. Then ask students what, based on the feature entries, appears to be the major characteristic of culture in the region. *(Most students will note that many of the entries are related to religion.)* **L1**

CURRICULUM CONNECTION

Art Marc Chagall created these stained-glass windows for the Haddassah-Hebrew University Hospital synagogue in Jerusalem. Completed in 1961, the windows contain no human forms. (Jewish law forbids the display of human images in houses of worship.) Chagall told the story of the original 12 tribes of Israel with various symbols and images of animals.

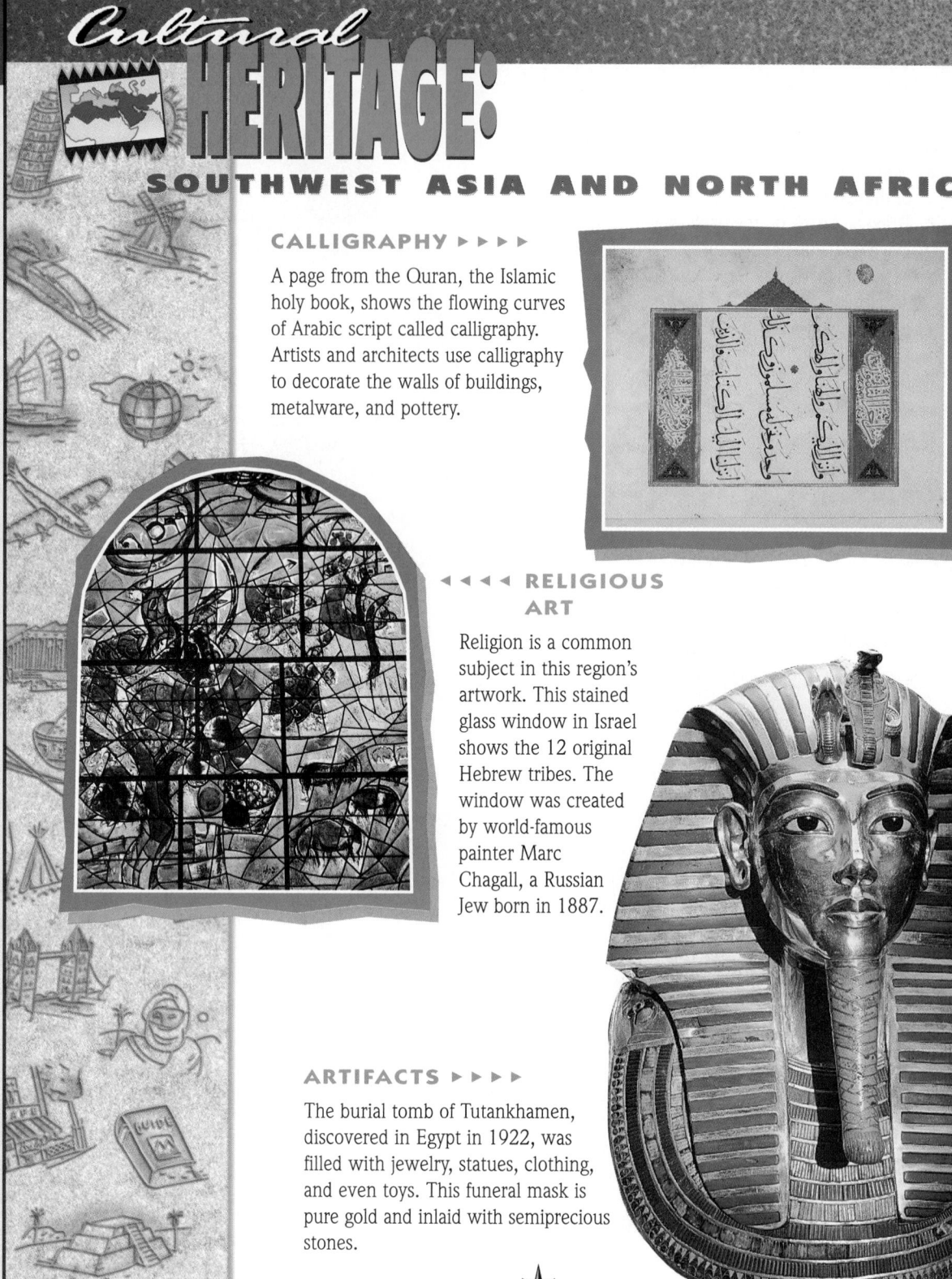

CALLIGRAPHY ▶ ▶ ▶ ▶

A page from the Quran, the Islamic holy book, shows the flowing curves of Arabic script called calligraphy. Artists and architects use calligraphy to decorate the walls of buildings, metalware, and pottery.

◀ ◀ ◀ ◀ RELIGIOUS ART

Religion is a common subject in this region's artwork. This stained glass window in Israel shows the 12 original Hebrew tribes. The window was created by world-famous painter Marc Chagall, a Russian Jew born in 1887.

ARTIFACTS ▶ ▶ ▶ ▶

The burial tomb of Tutankhamen, discovered in Egypt in 1922, was filled with jewelry, statues, clothing, and even toys. This funeral mask is pure gold and inlaid with semiprecious stones.

444

More About . . .

Architecture Although mosques dominate the architecture of most Islamic cities, other buildings, even common dwellings, are also of great architectural interest. In most Southwest Asian and North African cities, the typical home is the courtyard house. These houses are built closely together, and families live in very near proximity to their neighbors. Certain architectural techniques, however, promote a considerable amount of privacy. The doors of courtyard houses, for example, usually are not positioned directly across from one another. Windows are small and positioned above eye level to discourage people from looking in. And the entryways to the courtyards are angled to prevent people from the street seeing inside.

◄ ◄ ◄ ◄ **ARCHITECTURE**

A blue-domed mosque in Isfahan, Iran, shows typical calligraphy and geometric patterns. Most mosques have a *mihrab*—an arch or other structure that marks the wall closest to the holy city of Makkah.

Geography and the Humanities

📁 World Literature Reading 6

🌎 World Music: Cultural Traditions, Lesson 5

🎲 World Cultures Transparencies 9, 10

📕 Focus on World Art Print 13

MUSIC ▲ ▲ ▲ ▲

An Egyptian musician draws a bow across this instrument called a *kamanja*. The music of North Africa and Southwest Asia is most often played on stringed instruments, flutes, drums, and tambourines.

ASSESS

Have students answer the **Appreciating Culture** questions on page 445.

CLOSE

Encourage students to discuss the impact of religion on the cultural life of Southwest Asia and North Africa.

JEWELRY ▲ ▲ ▲ ▲

This elaborate head covering is called a *gargush*. It is made by knitting tiny silver wires together with multicolored beads. Some Jewish women of Yemen still wear this traditional headdress.

APPRECIATING CULTURE

1. What does the mask of Tutankhamen tell us about the king and the time he lived?

2. In what ways is calligraphy used as art in our country?

Answers to Appreciating Culture

1. Answers will vary, but may include: the king's appearance, the kinds of clothes he wore, the materials available at the time for artists to work with, the skill level of Egyptian artisans.

2. Answers will vary, but may include: in advertising, on greeting cards, on clothing, or any other acceptable uses.

PERFORMANCE ASSESSMENT ACTIVITY

PLAN A MUSEUM Place students in groups of three to five. Have them take the roles of museum curators who have received a grant of money from the governments of North Africa to fund a wing of the museum dedicated to learning about their area of the world. Have students develop a map or floorplan poster showing their design for the wing—including symbols and a key indicating the items they would like to represent each country in the region. Each student in the group should take responsibility for one or two countries in the region. These plans should be presented to another group of students who will take the role of the museum's board of directors. For individual accountability, each student should write a reflection telling the decision-making process they went through in order to make their contribution to the product.

POSSIBLE RUBRIC FEATURES:
Map skills, accuracy of content information, decision-making process skills, clarity of written reflection

MENTAL MAPPING ACTIVITY

Before teaching Chapter 17, point out the countries of North Africa on a political map. Have students create acrostics to help remember the locations and/or sizes of the countries relative to one another. For example: <u>E</u>at <u>L</u>ess <u>T</u>han <u>A</u> <u>M</u>ouse provides a means of remembering the countries east to west (Egypt, Libya, Tunisia, Algeria, Morocco).

TEACHER'S CORNER

NATIONAL GEOGRAPHIC SOCIETY

INDEX TO NATIONAL GEOGRAPHIC MAGAZINE

The following articles may be used for research relating to this chapter:

- "Morocco," by Erla Zwingle, October 1996.
- "Africa's Dinosaur Castaways," by Paul C. Sereno, June 1996.
- "Egypt's Old Kingdom," by David Roberts, January 1995.
- "Water—The Middle East's Critical Resource," by Priit J. Vesilind, May 1993.

- "Cairo—Clamorous Heart of Egypt," by Peter Theroux, April 1993.
- "Oasis of Art in the Sahara," by Henri Lhote, August 1987.
- "Morocco's Ancient City of Fez," by Harvey Arden, March 1986.
- "Journey Up the Nile," by Robert Caputo, May 1985.
- "Tunisia: Sea, Sand, Success," by Mike Edwards, February 1980.

NATIONAL GEOGRAPHIC SOCIETY PRODUCTS AVAILABLE FROM GLENCOE

To order the following products for use with this chapter, contact your local Glencoe sales representative or call Glencoe at 1-800-334-7344:

- *STV: World Geography* (Videodisc)
- *Picture Atlas of the World* (CD-ROM)
- *Physical Geography of the World* (Transparencies)
- *ZipZapMap! World* (Software)

- *GeoBee* (Software)
- *Images of the World* (Posters)
- *Eye on the Environment* (Posters)

ADDITIONAL NATIONAL GEOGRAPHIC SOCIETY PRODUCTS

To order the following products for use with this chapter, call National Geographic Society at 1-800-368-2728:

- *Egypt: Quest for Eternity*
- *Nations of the World Series:* Egypt (Video)

- *Physical Geography of the Continents Series:* "Africa." (Video)
- *Portraits of the Continents Series:* "Part III: Europe; Asia; Africa." (Filmstrip)

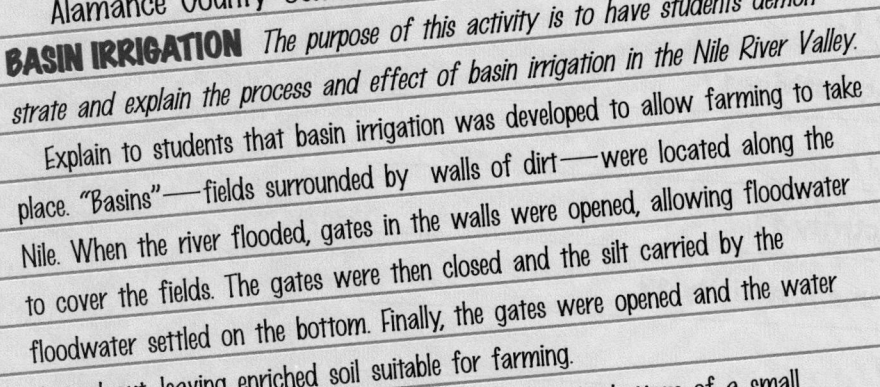

TEACHER-TO-TEACHER
Curriculum Connection: Earth Science

—from Ted Henson,
Alamance County Schools, Graham, NC

BASIN IRRIGATION *The purpose of this activity is to have students demonstrate and explain the process and effect of basin irrigation in the Nile River Valley.*

Explain to students that basin irrigation was developed to allow farming to take place. "Basins"—fields surrounded by walls of dirt—were located along the Nile. When the river flooded, gates in the walls were opened, allowing floodwater to cover the fields. The gates were then closed and the silt carried by the floodwater settled on the bottom. Finally, the gates were opened and the water drained out, leaving enriched soil suitable for farming.

To demonstrate this process, have students cover the bottom of a small aquarium with sand. Then direct them to carefully pour muddy water into the aquarium until it is about half-full. Have students observe the aquarium over a few days. The water will begin to clear as the mud particles settle to the bottom. Finally, have students use a plastic hose to siphon off the water, warning them not to swallow any.

Ask students to write a paragraph telling what they learned about basin irrigation. Have them conclude the paragraph by explaining how basin irrigation affected the lives of people living along the Nile.

BIBLIOGRAPHY

Readings for the Student
Hart, George. *Ancient Egypt: Exploring the Past.* Orlando: Harcourt, Brace and Company, 1989.

Muhammad and the Spread of Islam. London: HarperCollins, 1988.

Readings for the Teacher
Lamb, David. *The Arabs: Journeys Beyond the Mirage.* New York: Random House, 1988.

The Middle East. Washington, D.C.: Congressional Quarterly, 1991.

Multimedia
Nations of the World: Egypt. National Geographic Society, 1987. Videocassette/film, 25 minutes.

The Giant Nile. Blue Bird Films, 1990. Three videocassettes, approximately 60 minutes each.

KEY TO ABILITY LEVELS
Teaching strategies have been coded for varying learning styles and abilities.

L1 **BASIC** activities for all students

L2 **AVERAGE** activities for average to above-average students

L3 **CHALLENGING** activities for above-average students

LEP **LIMITED ENGLISH PROFICIENCY** activities

MEETING NATIONAL STANDARDS

Geography for Life

All of the 18 standards are demonstrated in Unit 6. The following ones are highlighted in this chapter:
- Standards 1, 2, 3, 4, 6, 7, 9, 10, 12, 14, 15, 17

Use these *Geography: The World and Its People* resources to teach, reinforce, and extend chapter content.

CHAPTER 17 RESOURCES

*Vocabulary Activity 17

Cooperative Learning Activity 17

Workbook Activity 17

Geography Skills Activity 17

Performance Assessment Activity 17

GeoLab Activity 17

Chapter Map Activity 17

*Reteaching Activity 17

Chapter 17 Test Form A and Form B

*Chapter 17 Digest Audiocassette, Activity, Test

Unit Overlay Transparencies 6-0, 6-1, 6-2, 6-3, 6-4, 6-5, 6-6

Geography Handbook Transparency 22

Political Map Transparency 6; World Cultures Transparency 10

Geoquiz Transparencies 17-1, 17-2

Vocabulary PuzzleMaker Software

Student Self-Test: A Software Review

Testmaker Software

MindJogger Videoquiz

Focus on World Art Print 7

If time does not permit teaching the entire chapter, summarize using the Chapter 17 Highlights on page 614, and the Chapter 17 English (or Spanish) Audiocassettes. Review students' knowledge using the Glencoe MindJogger Videoquiz. * Also available in Spanish

Use these *Geography: The World and Its People* resources to teach and reinforce section content.

SECTION 1 RESOURCES

Reproducible Lesson Plan 17-1

Section Focus Transparency 17-1

Guided Reading Activity 17-1

Section Quiz 17-1 (also available in Spanish)

SECTION 2 RESOURCES

Reproducible Lesson Plan 17-2

Section Focus Transparency 17-2

Guided Reading Activity 17-2

Section Quiz 17-2 (also available in Spanish)

ADDITIONAL RESOURCES FROM GLENCOE

 Reproducible Masters

- Foods Around the World, Unit 6
- Facts On File: GEOGRAPHY ON FILE, North Africa and Southeast Asia

Workbooks

- Building Skills in Geography, Unit 2, Lesson 9

World Music: Cultural Traditions
Lesson 5

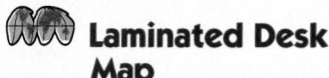 Transparencies

- National Geographic Society PicturePack, Unit 6

Laminated Desk Map

Posters

- National Geographic Society: Eye on the Environment, Unit 6

 World Games Activity Cards Unit 6

Videodiscs

- Geography and the Environment: The Infinite Voyage
- Reuters Issues in Geography
- ABCNews InterActive™: In the Holy Land
- National Geographic Society: STV: World Geography, Vol. 2
- National Geographic Society: GTV: Planetary Manager

 CD-ROM

- National Geographic Society: Picture Atlas of the World

Performance Assessment

Refer to the Planning Guide on page 446A for a Performance Assessment Activity for this chapter. See the *Performance Assessment Strategies and Activities* booklet for suggestions.

Chapter Objectives

1. **Explain** the importance of the Nile River to Egypt's people and analyze the environmental impact of the Aswan High Dam.
2. **Examine** ways of life in North Africa.

GLENCOE
TECHNOLOGY

 VIDEODISC

Use the Chapter 17 MindJogger Videoquiz to preview chapter content.

MindJogger Videoquiz

Chapter 17
Disc 3 Side A

 Also available in VHS.

NATIONAL GEOGRAPHIC SOCIETY

These materials are available from Glencoe.

 POSTERS

Images of the World

Unit 6: Southwest Asia and North Africa

446

Chapter 17 North Africa

MAP STUDY ACTIVITY

As you read Chapter 17, you will learn about Egypt and other countries in North Africa.
1. **What countries are considered part of North Africa?**
2. **What are the two largest countries?**

446

MAP STUDY ACTIVITY

Answers
1. Morocco, Algeria, Libya, Tunisia, Egypt, Western Sahara
2. Algeria and Libya

Map Skills Practice
Reading a Map What is the capital of Morocco? *(Rabat)*

SECTION 1 ·············· Egypt

PREVIEW

Words to Know
- silt
- delta
- hydroelectric power
- consumer goods
- *fellahin*
- bazaar
- hieroglyph

- Arabian Desert
- Cairo
- Alexandria

Places to Locate
- Nile River
- Sinai Peninsula
- Suez Canal
- Libyan Desert

Read to Learn . . .
1. why the Nile River is important to Egypt's people.
2. how the Aswan High Dam has affected Egypt's environment.

You live along the lush, green banks of the longest river in the world. Where are you living? You live along the Nile River in the North African country of Egypt.

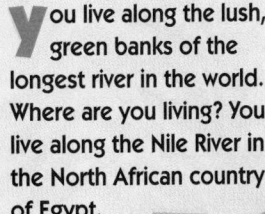

For centuries the people of Egypt have called their country the "gift of the Nile." The river gives them water, fertile soil, and transportation. Egypt receives little rain, and almost all of it is desert. Deserts also cover most of the lands of Egypt's neighbors in North Africa—Libya, Tunisia, Algeria, and Morocco. Egypt has the most people of any country in the region. Look at the graph on page 451 to see how Egypt's population compares with the population of the United States.

Place
The Land

Egypt has three major land areas. They are the Nile River valley, the Sinai (SY•NY) Peninsula, and desert areas. Most of Egypt has a desert climate, with hot summers and mild winters.

Nile River Valley The lifeline of Egypt is the Nile River, which supplies 85 percent of the country's water. The Nile River begins in east Africa. It flows north through the countries of Sudan and Egypt to the Mediterranean Sea. Look

Meeting National Standards
Geography for Life The following standards are highlighted in this section:
- Standards 1, 4, 6, 7, 9, 10, 12, 14, 15, 17

FOCUS

Section Objectives
1. **Determine** why the Nile River is important to Egypt's people.
2. **Analyze** the impact of the Aswan High Dam on Egypt's environment.

Bellringer Motivational Activity
Prior to taking roll at the beginning of the class period, project Section Focus Transparency 17-1 and have students answer the activity questions. Discuss students' responses. **L1**

Vocabulary Pre-check
Tell students that the word *delta* derives from the Greek alphabet letter *delta,* which is shaped like a triangle. Have students locate the word in the text. Then ask them to draw a diagram of a river delta. **L1 LEP**

Use the Vocabulary PuzzleMaker Software to create a crossword puzzle. **L1**

447

TEACH

Guided Practice

Describe Ask students to suggest words that might be used in a description of a desert. List responses on the chalkboard. Encourage students to use the words listed on the chalkboard to write a description of Egypt's desert lands. Call on volunteers to read their descriptions to the class. **L1** **LEP**

▼ Cooperative Learning Activity 17 **L1**

Compare On the chalkboard draw a two-column chart titled "Life in Egypt." Head the two columns *Urban* and *Rural.* Call on students to suggest entries for the chart. Encourage them to consider topics such as housing, ways of earning a living, and transportation. Make entries on the chart as they are suggested. Then have volunteers use the information in the chart to role-play urban or rural Egyptians. Other members of the class might ask the role-players questions about life in Egypt. **L2**

NORTH AFRICA: Physical

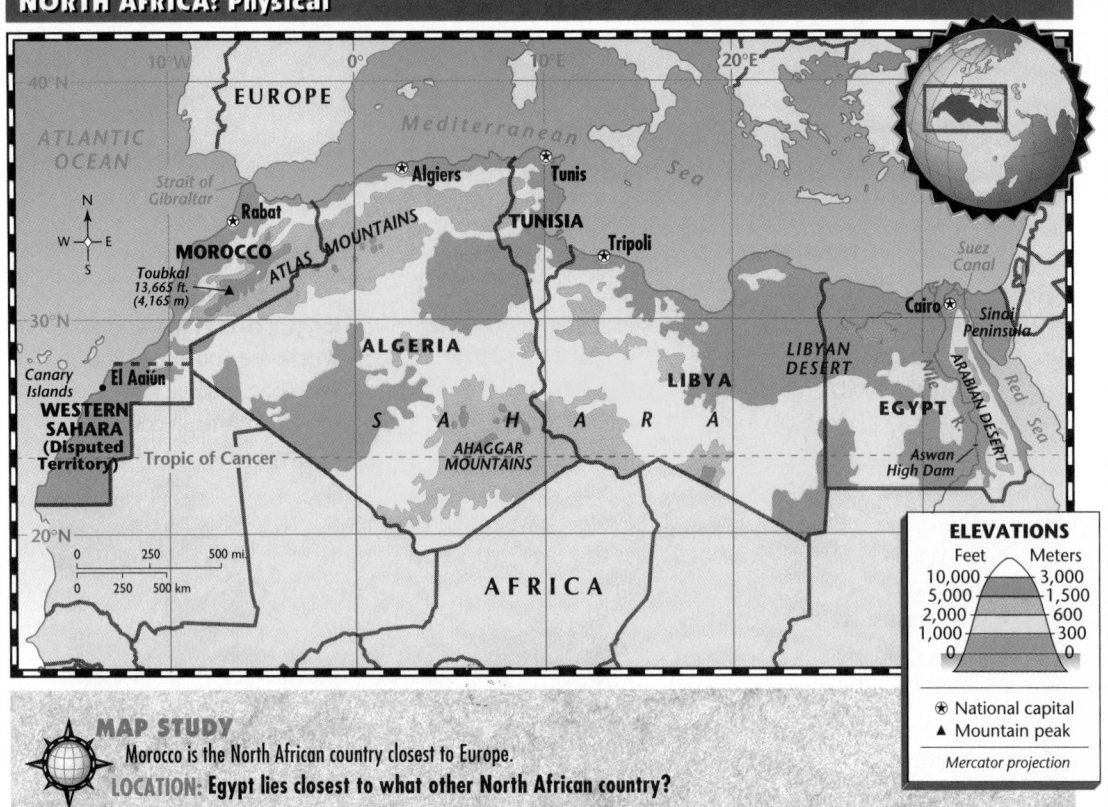

MAP STUDY
Morocco is the North African country closest to Europe.
LOCATION: Egypt lies closest to what other North African country?

at the chart on page 453 to see how the Nile compares in length with other major world rivers.

Imagine you are looking down on Egypt from an airplane flying from south to north. Below you the Nile River and its banks appear as a narrow, green ribbon cutting through the vast desert.

Flood Control The Nile River valley has rich soils formed by **silt,** or particles of earth deposited by the river. In ancient times, floodwaters left heavy deposits of silt every summer. Today dams control the river's flow, and flooding no longer occurs regularly.

As you continue your flight, you can see the Nile River spread out into a broad wedge of farmland about 150 miles (240 km) from the Mediterranean Sea. You are viewing a **delta,** or triangle-shaped area of land at a river's mouth.

Sinai Peninsula The Sinai Peninsula lies southeast of the Nile Delta. This section of Egypt is not in Africa but is actually part of Southwest Asia. If you look at the map above, you can see why this area is a major crossroads between these two continents.

Cooperative Learning Activity

Organize students into three groups and provide each group with an outline map of Egypt. Have one group create a population density map, another group create a map showing major agricultural products, and the third group create a map showing major industrial centers and products. Display the completed maps. Then ask students to discuss how each of these special-purpose maps illustrates the importance of the Nile River to Egypt. **L2**

A human-made waterway called the Suez Canal separates the Sinai Peninsula from the rest of Egypt. Egyptians and Europeans built this important canal in the mid-1860s to allow ships to pass between the Red and the Mediterranean seas.

Deserts From the airplane you can see that most of Egypt is desert. West of the Nile River lies the Libyan (LIH•bee•uhn) Desert. It covers two-thirds of Egypt's land area. Oases are found throughout the Libyan Desert. East of the Nile River and west of the Red Sea is the Arabian Desert.

The Libyan and Arabian deserts are parts of the Sahara—one of the largest deserts in the world. *Sahara* is from the Arabic word meaning "desert." The Sahara—about the size of the United States—runs from Egypt westward through North Africa to the Atlantic Ocean.

Human/Environment Interaction
The Economy

Egypt has a developing economy that has grown considerably in recent years. Since the 1950s government and business leaders have started many new industries. Agriculture, however, remains Egypt's main economic activity.

Agriculture Only about 4 percent of Egypt's land is used for farming. The map on page 450 shows that Egypt's best farmland lies in the fertile Nile River valley. Most Egyptian farmers tend small plots and grow cotton, dates, vegetables, sugarcane, and wheat. They use irrigation, modern tools, and farming methods developed by their ancestors. Cotton is Egypt's leading agricultural export.

Aswan High Dam Until the 1960s Egyptian farmers could plant only once a year—after the Nile River finished flooding in the summer. In 1968 engineers completed the construction of the Aswan High Dam. It holds back the Nile's floodwater. Now Egyptian farmers plant crops two or three times a year.

The Aswan High Dam brings problems as well as blessings. Why? It now stops much of the rich silt flow that fertilized Egypt's fields each year. Egypt's farmers must rely on expensive artificial fertilizers. The dam also restricts the flow of water into the river valley and the Mediterranean Sea. Salty seawater now moves farther inland into the delta.

Industry The Aswan High Dam provides **hydroelectric power,** or electric energy from moving water, to Egypt's growing industries. The largest industrial centers are the capital city of Cairo and the seaport of Alexandria. Egyptian factories there produce mainly food products, textiles, and **consumer goods,** or household goods, clothing, and shoes that people buy.

Harvesting Cotton in Egypt

An Egyptian woman gathers cotton, Egypt's most valuable crop.
LOCATION: Where do you think most of Egypt's cotton fields are located?

LESSON PLAN
Chapter 17, Section 1

Independent Practice

Guided Reading Activity 17-1 **L1**

More About the Illustration
Answer to Caption
in the Nile River valley

GEOGRAPHIC THEMES

Movement Egypt produces high-quality cotton with long fibers that are very strong. During the American Civil War, Egypt's cotton was in special demand. The ports of cotton-growing states in the South were blockaded and very little cotton could be exported from the United States.

NATIONAL GEOGRAPHIC SOCIETY

These materials are available from Glencoe.

 VIDEODISC

STV: World Geography Vol. 2: Africa and Europe

Egypt
Side 1, Frames 10481-10676

Meeting Special Needs Activity

Language Delayed Students with decoding problems may skip over proper nouns. Have these students point out the place names in this section. Then help them use the respellings, which appear in parentheses after the words, as a guide to pronunciation. **L1**

MULTICULTURAL PERSPECTIVE

Culturally Speaking

Egyptians often use the phrase *Ma'alesh,* meaning "Don't worry," or "Never mind." This term reflects their relaxed attitude and lack of concern with unimportant matters.

About one-fifth of Earth's surface is desert. Most deserts lie between 15° and 35° north latitude or between 15° and 35° south latitude. Only about 10 to 20 percent of most deserts are covered by sand.

Answer
phosphates and petroleum

Map Skills Practice
Reading a Map In which area of Egypt does the most farming take place? *(around the Nile River and its delta)*

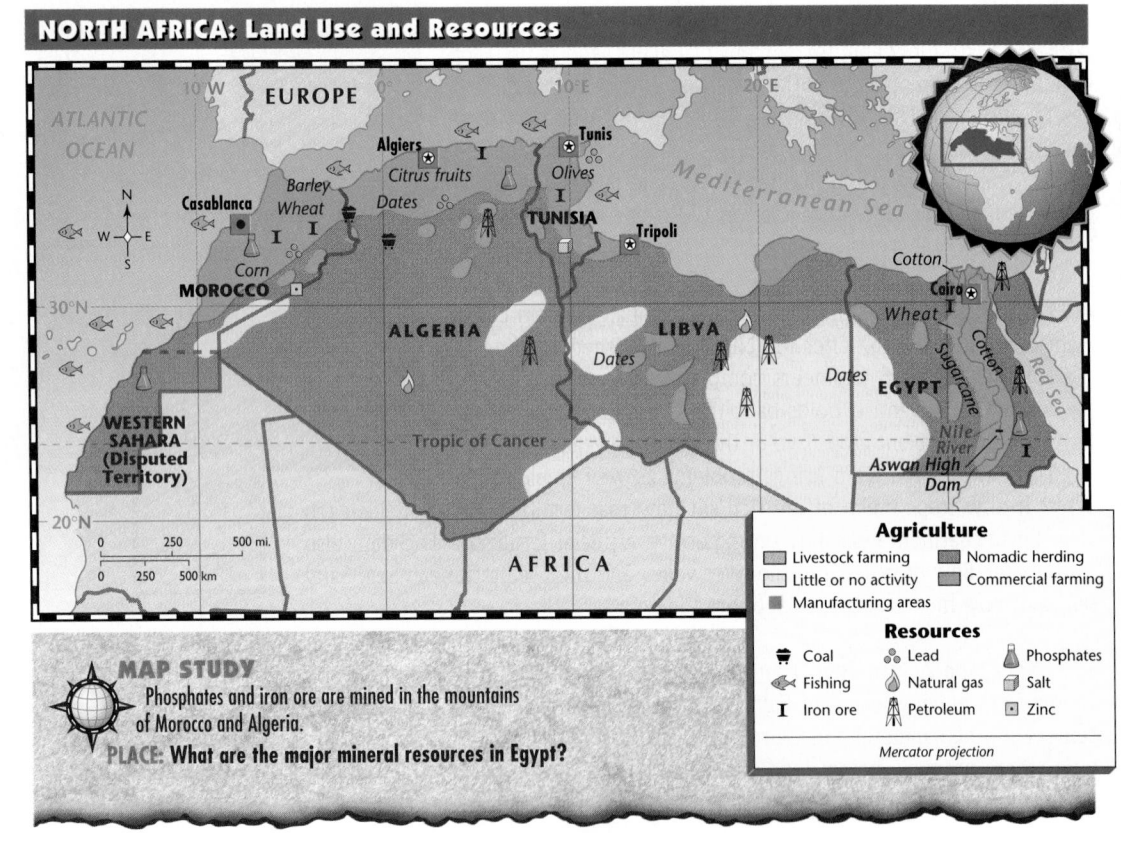

Egyptian Music

Can you imagine keeping an orchestra in your home? In ancient Egypt, some wealthy households had their own household orchestras. Ancient Egyptians also put musical instruments in tombs. They believed that this would guarantee them entertainment in their lives after death.

Oil ranks as the country's most important natural resource and a major export. The map below shows that many of Egypt's oil wells are in and around the Red Sea. Another growing industry is tourism.

Place
The People

Most Egyptians live within 20 miles (33 km) of the Nile River. This means that 99 percent of Egypt's population live on only 3.5 percent of the land. Look at the population density map on page 457. You can see that the Nile River valley is one of the most densely populated areas in the Southwest Asia/North Africa region. It is also one of the most densely populated areas in the world.

Rural Life About 55 percent of Egypt's people live in rural areas. Most rural Egyptians are farmers called *fellahin* (FEHL•uh•HEEN). They live in villages and work on small plots of land rented from landowners. Many *fellahin* raise only

NORTH AFRICA: Land Use and Resources

Agriculture
- Livestock farming
- Little or no activity
- Manufacturing areas
- Nomadic herding
- Commercial farming

Resources
- Coal
- Fishing
- Iron ore
- Lead
- Natural gas
- Petroleum
- Phosphates
- Salt
- Zinc

Mercator projection

MAP STUDY
Phosphates and iron ore are mined in the mountains of Morocco and Algeria.
PLACE: What are the major mineral resources in Egypt?

Critical Thinking Activity

Drawing Conclusions Have students make a chart comparing the positive and negative effects of the Aswan High Dam on life in Egypt. Then ask them to draw conclusions on whether the dam has done more harm or more good. Call on volunteers to defend their conclusions. **L2**

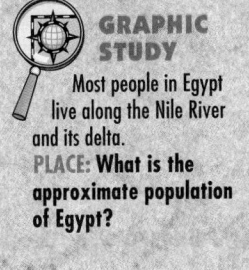

GRAPHIC STUDY

Most people in Egypt live along the Nile River and its delta.
PLACE: What is the approximate population of Egypt?

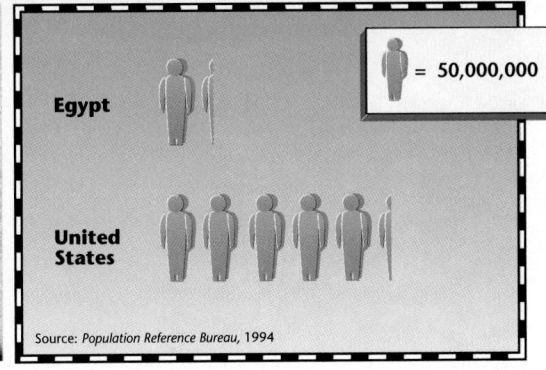

= 50,000,000

Egypt

United States

Source: *Population Reference Bureau, 1994*

enough crops to feed their families. Any food left over is sold in towns at a **bazaar,** or marketplace.

City Life Egypt's large cities offer a different kind of life. Skilled and unskilled workers make up the largest groups of Egyptians in cities. Cairo is a huge and rapidly growing city, with more than 11 million people. It is the largest city in Egypt as well as in Africa. Cairo's mosques, schools, and universities make it a leading center of the Muslim world.

Why is Cairo's population increasing? First, Egypt is a country with a high birthrate. Second, many *fellahin* from rural areas have moved into Cairo to find work. As Cairo's population grows, the city cannot provide enough living space for all of its people. If you were a student in one of Cairo's schools, you might attend crowded classes on a split shift.

Influences of the Past What do you think of when you hear the word "Egypt"? Pyramids? Pharaohs and kings? Egypt has a long and fascinating history. One of the ancient world's most advanced civilizations developed along the Nile River. Ancient Egyptians used **hieroglyphs,** or picture symbols, for writing. One of the first civilizations to make paper from the reedlike papyrus plant, the Egyptians also created a calendar to keep track of the growing season. They became skilled in building such lasting monuments as the Great Sphinx and the Pyramids.

In later centuries, Egypt was ruled by outsiders—Romans, Persians, Arabs, Turks, and the British. Under Arab rule, Islam—its followers are called Muslims—and the Arabic language came to Egypt. Both Islam and Arabic have been an important part of Egyptian life ever since. Today almost 90 percent of Egyptians are Muslims and about 10 percent are Christians.

Government Egypt became a republic in 1953. Gamal Abdel Nasser was president of Egypt from 1954 to 1970. Under Nasser, Egypt became the most powerful country in the Arab world. Nasser, supported by other Arab countries,

ASSESS

Check for Understanding

Assign Section 1 Review as homework or an in-class activity.

Meeting Lesson Objectives

Each objective below is tested by the questions that follow it in parentheses.

1. **Determine** why the Nile River is important to Egypt's people. (1, 2, 4)
2. **Analyze** the impact of the Aswan High Dam on Egypt's environment. (1, 3, 5)

Evaluate

 Section 1 Quiz **L1**

Use the Testmaker to create a customized quiz for Section 1. **L1**

More About . . .

Aswan High Dam The Aswan High Dam took 10 years to build at a cost of some $1 billion. The dam is more than 2 miles (3.2 km) long and rises some 364 feet (111 meters) above the bed of the Nile River. Irrigation provided by the dam brought about 900,000 acres (360,000 hectares) into use as arable land. Much of this area formerly was desert. In addition, the hydroelectricity generated at the Aswan High Dam provides for about 25 percent of Egypt's power needs.

More About the Illustration

Answer to Caption

because many *fellahin* have come from the country to Cairo in search of work

Place The lifestyle of the middle- and upper-class residents of Cairo is very different from that of its poorer residents. Many residents who cannot afford to rent apartments or own homes have moved into some of Cairo's ancient cemeteries. They use the large tombs as homes or build small houses in these areas known as "Cities of the Dead."

Reteach

Have students use the section's headings and subheadings to create a section outline. **L1**

Enrich

Have students research and report on the history of the Suez Canal. They might include a time line of important events in their reports. **L3**

CLOSE

Encourage students to discuss the following statement: Since the construction of the Aswan High Dam, Egypt is no longer the "gift of the Nile."

Cairo, Egypt

Wide avenues wind through modern Cairo.
PLACE: Why is Cairo so densely populated?

opposed Israel. He also angered many Europeans by taking control of the British-run Suez Canal, which was vital to European trade.

To regain control of the canal, Britain, France, and Israel invaded Egypt in 1956. The United Nations persuaded the Europeans and the Israelis to withdraw their troops. Tensions between the countries remained, however. In the 1970s Egypt and Israel decided to settle their differences. In recent years, Egypt has also strengthened its ties with the United States.

SECTION 1 REVIEW

REVIEWING TERMS AND FACTS

1. **Define the following:** silt, delta, hydroelectric power, consumer goods, *fellahin,* bazaar, hieroglyph.
2. **PLACE** Why is Egypt called "the gift of the Nile"?
3. **HUMAN/ENVIRONMENT INTERACTION** How has the Aswan High Dam affected the lives of Egyptians?
4. **LOCATION** Where is Cairo located?

MAP STUDY ACTIVITIES

5. Turn to the physical map of North Africa on page 448. Where is the Aswan High Dam located?
6. Look at the political map of North Africa on page 446. (a) What direction is Alexandria from Cairo? (b) About how many miles (km) separate the two cities?

452

UNIT 6

Answers to Section 1 Review

1. All vocabulary words are defined in the Glossary.
2. The Nile River provides the water and fertile soils on which Egyptians depend for survival in a desert environment.
3. It stores water for irrigation, allowing farmers to plant more than one crop a year. However, it has blocked silt that once fertilized Egypt's field, so farmers have to buy expensive fertilizers.
4. on the low-lying flood plain of the Nile River south of the Nile Delta
5. in southern Egypt
6. **(a)** northwest; **(b)** about 100 miles (160 km)

BUILDING GEOGRAPHY SKILLS

Using a Table or Chart

Charts and tables can make it easier for you to understand how facts and figures relate to each other. In a chart or table, similar kinds of information are arranged down columns and across rows. A label at the beginning of each column or row explains the information that you will find there. Charts and tables make it simple to compare groups of facts.

To use a table or chart, follow these steps:

• Read the chart title and all labels on the columns and rows.

• Look for similarities and differences among the facts presented in the table.

• Draw conclusions about the subject of the table.

Longest Rivers of the World

RIVER	CONTINENT	LENGTH miles (km)
Nile	Africa	4,160 (6,693)
Amazon	South America	4,000 (6,436)
Chang Jiang	Asia	3,964 (6,378)
Huang He	Asia	3,395 (5,463)
Ob-Irtysh	Asia	3,362 (5,409)
Amur	Asia	2,744 (4,415)
Lena	Asia	2,734 (4,399)
Zaire	Africa	2,718 (4,373)
Mackenzie	North America	2,635 (4,240)
Mekong	Asia	2,600 (4,180)
Niger	Africa	2,590 (4,167)
Yenisey	Asia	2,543 (4,091)
Paraná	South America	2,485 (3,998)
Mississippi	North America	2,340 (3,765)

Source: 1994 *World Almanac*

Geography Skills Practice

1. What kind of information appears in each column?

2. What rivers listed in the chart are in Africa? How long is each of these rivers?

3. What conclusion can you draw about the locations of the world's longest rivers?

CHAPTER 17

453

Other North African Countries

PREVIEW

FOCUS

Section Objectives

1. **Identify** the natural resources found in North Africa.
2. **Explain** why most people in North Africa live in coastal areas.
3. **Describe** how urban and rural life differs in North Africa.

Bellringer Motivational Activity

 Prior to taking roll at the beginning of the class period, project Section Focus Transparency 17-2 and have students answer the activity questions. Discuss students' responses. **L1**

Vocabulary Pre-check

Have students create "Pictionary" clues for the **Words to Know** in this section. Then have them work in pairs to identify the terms from each other's clues. **L1**
LEP

Words to Know
• dictatorship
• erg
• *casbah*

Places to Locate
• Sahara
• Libya
• Tunisia
• Atlas Mountains
• Maghreb
• Algeria
• Morocco
• Casablanca

Read to Learn . . .

1. what natural resources are found in North Africa.
2. why most people in North Africa live in coastal areas.
3. how urban and rural life differ in North Africa.

The busy sounds of this modern oil-refining plant echo across the Sahara in the country of Libya. You probably think the Sahara is nothing but miles and miles of shifting sand. The Sahara covers major oil reserves.

Like Egypt, the other countries of North Africa—Libya, Tunisia, Algeria, and Morocco—have established economies based on oil and other natural resources found in the Sahara. Look at the map on page 448. You see that the vast stretches of the Sahara lie to the south of North Africa, and the deep blue waters of the Mediterranean Sea lie to the north.

Place

Libya

Libya is a huge country about one-fifth the size of the United States. Until the mid-1900s Libya was a poor country with few natural resources. The discovery of oil in 1959 brought Libya great wealth.

The Land More than 90 percent desert, Libya is one of the world's driest countries. Its need for water increases as its population grows. In 1996 the Libyan government opened a long irrigation pipeline to bring water to coastal areas from aquifers in the south.

Classroom Resources for Section 2

 **REPRODUCIBLE MASTERS**

Reproducible Lesson Plan 17-2
Guided Reading Activity 17-2
Reteaching Activity 17
Enrichment Activity 17
Section Quiz 17-2

 TRANSPARENCIES

Section Focus Transparency 17-2
Unit Overlay Transparencies 6-0, 6-1, 6-2, 6-3, 6-4, 6-5, 6-6

MULTIMEDIA

 National Geographic Society: Picture Atlas of the World

 Reuters Issues in Geography

 Testmaker

The Economy Although most of Libya's land is too dry for farming, Libya is the richest country in North Africa. It earns an enormous income from its oil exports. A Libyan oil worker earns more in one month than could be earned in one year of farming. The government uses profits from oil to improve farming and to build new schools, houses, and hospitals.

The People The 5 million people of Libya are of mixed Arab and Berber ancestry. The Berbers were the first people known to live in North Africa. During the A.D. 600s, the Arabs brought Islam and the Arabic language to North Africa, including Libya.

If you lived in Libya today, you would probably live in an urban area along the Mediterranean coast, as about 70 percent of Libyans do. Most live in two modern cities—Tripoli, which is the capital, and Benghazi.

For many centuries Libya was a part of either Muslim or European empires. It finally became an independent country in 1951 under a monarchy. In 1969 a military officer named Muammar al-Qaddhafi (kuh•DAH•fee) came to power. He set up a **dictatorship,** or a government under the control of one all-powerful leader.

NORTH AFRICA: Climate

Map labels: EUROPE, ASIA, Mediterranean Sea, Algiers, Tunis, TUNISIA, Rabat, MOROCCO, Tripoli, WESTERN SAHARA (Disputed Territory), El Aaiún, ALGERIA, LIBYA, Cairo, EGYPT, Red Sea, Tropic of Cancer, AFRICA

CLIMATE
Dry
☐ Desert
☐ Steppe (partly dry)
Mid-Latitude
☐ Mediterranean

Mercator projection

40°N, 30°N, 20°N, 10°W, 0°, 10°E, 20°E
0 250 500 mi.
0 250 500 km

MAP STUDY
Most of North Africa has hot, dry desert climates.
PLACE: What entire country in eastern North Africa has a desert climate?

Cooperative Learning Activity

Organize students into four groups to study the Sahara. Assign one of the following topics to each group: landscapes, plants, animals, people. Have groups research their topic and share their findings. Then have each group contribute to a desert diorama illustrating the environment and ways of life in the Sahara. **L2**

TEACH

Guided Practice
Identifying Similarities
Challenge students to identify characteristics of the countries of North Africa. Note their responses on the chalkboard. Then have students determine which countries share common characteristics. (For example, four of the countries border the Mediterranean Sea.) Finally, have students create diagrams or other graphics to illustrate their findings on shared characteristics. **L2**

CURRICULUM CONNECTION

Earth Science More than 10,000 years ago, the Sahara had a much wetter climate. What today is a barren desert was then a region of grasslands and forests that supported elephants, giraffes, and many other animals.

Answer
Egypt
Map Skills Practice
Reading a Map What is the climate in Libya's capital city? *(steppe)*

Independent Practice

 Guided Reading Activity 17-2 **L1**

More About the Illustration

Answer to Caption

Morocco, Algeria, and Tunisia

Place Dense forests cover some of the northern slopes of the Atlas Mountains, but the southern slopes are bare with only scattered scrub vegetation. On the higher plateaus are fields of *esparto,* a type of grass that is used to make paper and rope.

GLENCOE
TECHNOLOGY

 **VIDEODISC**

Reuters Issues in Geography

**Chapter 5
North Africa and Southwest Asia:
The Influence of Oil and Water**
Disc 1 Side A

Shepherding in the Atlas Mountains

Shepherds tend their flocks on the rocky slopes of North Africa's Atlas Mountains.
LOCATION: Through what three countries do the Atlas Mountains extend?

Region
Tunisia

Tunisia and two other western North African countries—Algeria and Morocco—form a region known as the Maghreb (MAH•gruhb). *Maghreb* means "the West" in Arabic. These three countries form the westernmost region of the Arab-Islamic world.

The Land About the size of the state of Missouri, Tunisia is North Africa's smallest country. Like other countries of North Africa, Tunisia includes large areas of the Sahara.

Two branches of the rugged Atlas mountain range sweep into the northern part of Tunisia. The Atlas Mountains stretch from western Morocco to northeastern Tunisia.

The Economy Tunisia depends on mining and agriculture for its income. It exports phosphates and oil. If you were a Tunisian farmer, you would probably grow wheat, barley, olives, and grapes—the country's major farm products. Most of Tunisia's farms are located along the fertile east coast. As the map on page 455 shows, this coastal area has a mild Mediterranean climate that favors the growing of crops.

The People Almost all of Tunisia's 9 million people are of Arab and Berber ancestry and speak Arabic. Look at the population density map on page 457.

Meeting Special Needs Activity

Inefficient Readers Some students may have trouble reading the different forms of a word that are used to denote a country and the people who live there. Have these students make a list of the five countries in North Africa. Next to each country's name, have students write the name of its people. (For example, Libya—Libyans.) Point out that the word that refers to people can also be used as an adjective. (For example, Libyan Desert.) **L1**

Notice that most Tunisians live in the northern and eastern areas of the country. About 60 percent of Tunisia's people live in urban areas. Tunis, with a population of about 1 million, is the capital and largest city.

In ancient times the city of Carthage arose in the area that is now Tunisia. Carthage fought unsuccessfully with Rome for control of the Mediterranean world. In modern times France ruled Tunisia as a colony. Although Tunisia became an independent republic in 1956, French influences can still be seen.

Place
Algeria

Crossing Algeria is no quick trip. Algeria is the largest country in the region of Southwest Asia and North Africa. It is more than three times bigger than the state of Texas.

The Land The landscape of Algeria varies from plains to mountains to deserts. Along the Mediterranean coast is a narrow strip of land known as the Tell. The word *tell* is an Arabic term meaning "hill." The hills and plains of the

Muslims gather at a mosque for daily prayers.
REGION: What is the major religion of Tunisia?

More About the Illustration
Answer to Caption
Islam

Place According to Islamic teachings, Muslims must pray facing Makkah five times a day. A *muezzin*, or crier, calls Muslims to prayer. Before each prayer session, Muslims wash their face, hands, and feet.

Cultural Kaleidoscope

Tunisia When Tunisians enter a store or office, they greet the owner and workers. To neglect greeting someone upon meeting is considered very rude.

NORTH AFRICA: Population Density

CITIES
- City with more than 5,000,000 people
- City with 1,000,000 to 5,000,000 people
- City with 500,000 to 1,000,000 people

Persons per	
sq. mi.	sq. km
Uninhabited	Uninhabited
Under 2	Under 1
2-60	1-25
60-125	25-50
125-250	50-100
Over 250	Over 100

Mercator projection

EUROPE — Tunis — Oran — Algiers — Rabat — Casablanca — MOROCCO — TUNISIA — Tripoli — Mediterranean Sea — Alexandria — Giza — Cairo — WESTERN SAHARA (Disputed Territory) — Tropic of Cancer — ALGERIA — LIBYA — EGYPT — Red Sea — AFRICA

0 — 250 — 500 mi.
0 — 250 — 500 km

MAP STUDY
Most people in North Africa live near the Mediterranean coast and in river valley areas where water is available.
HUMAN/ENVIRONMENT INTERACTION: What North African country has the highest population density ?

MAP STUDY

Answer
Egypt

Map Skills Practice
Reading a Map What city in North Africa has a population of more than 5 million people? *(Cairo, Egypt)*

Critical Thinking Activity

Analyzing Information Discuss the growth of tourism in North Africa. Have students find information in the text about the countries that have become popular tourist destinations. Then ask students to analyze possible reasons why tourism is a growing industry in some North African countries but not in others. Call on volunteers to share their ideas with the rest of the class. **L2**

Cultural Kaleidoscope

Algeria Algerians are complimented if a guest leaves some food on the plate at the end of a meal. This indicates that the host has provided more than enough to eat.

ASSESS

Check for Understanding

Assign Section 2 Review as homework or an in-class activity.

Meeting Lesson Objectives

Each objective below is tested by the questions that follow it in parentheses.

1. **Identify** the natural resources found in North Africa. (4, 6)
2. **Explain** why most people in North Africa live in coastal areas. (2)
3. **Describe** how urban and rural ways of life differ in North Africa. (2)

Teen Scene

It's Festival Time!

Amineh and Jamila are wearing their nicest *caftans* and head scarves. Tonight, the Berber folklore festival will begin. Dancers, singers, and acrobats from all over Morocco come to entertain the large crowds. The 15-day folklore festival takes place every year in May. Amineh and Jamila like to share in all the activities, but their favorite part is meeting new people from other Berber villages.

Tell contain Algeria's best farmland. A Mediterranean climate with hot, dry summers and mild, rainy winters helps crops grow.

South of the Tell are the Atlas Mountains. In Algeria the Atlas Mountains form two major chains with an average height of about 7,000 feet (2,134 m). Another mountain range known as the Ahaggar (uh•HAH•guhr) lies in southern Algeria. Between the Atlas and Ahaggar mountains are parts of the Sahara known as **ergs,** huge areas of shifting sand dunes. You find a hot, dry climate in most parts of Algeria south of the Tell.

The Economy Agriculture, manufacturing, and mining are the basis of Algeria's developing economy. Algeria has large deposits of natural gas and petroleum. These two natural resources make up 30 percent of the country's income.

The People About 28 million people live in Algeria. Like Libya and Tunisia, Algeria has a population of mixed Arab and Berber ancestry. Most Algerians practice Islam and speak Arabic. People in the countryside raise livestock or farm small pieces of land. Algiers, the capital and largest city, is the home of about 1.5 million people.

What is it like to be a teenager living in Algiers? Fatimah Malek is an Algerian teen who enjoys visiting the old areas of Algiers called **casbahs.** In these sections, shops, mosques, bazaars, and homes line narrow streets. Fatimah lives in a new section of Algiers that has broad streets and modern office and apartment buildings.

In Algerian history class Fatimah has learned about her country's independent spirit. Algeria came under French rule in the 1800s. The Algerians never accepted French control of their country, however. In 1962 Algeria became an independent republic.

Location
Morocco

The country of Morocco lies in the northwestern corner of Africa. Its long seacoast is bordered by the Mediterranean Sea on the north and the Atlantic Ocean on the west. The northern tip of Morocco meets the Strait of Gibraltar.

The Land Morocco's landforms are similar to those of neighboring Algeria. A fertile coastal plain borders the Mediterranean Sea and the Atlantic Ocean. This

More About . . .

French Influence in North Africa France has had a lasting influence on three countries of North Africa. French is the principal foreign language spoken in Algeria. In Tunisia and Morocco, French is widely used, especially in government, business, and the media. In Morocco, for example, daily newspapers are published in both Arabic and French. Trade ties between France and these North African countries remain strong. And the vast majority of Algerians, Moroccans, and Tunisians who live abroad make their home in France.

area has a Mediterranean climate of hot, dry summers and mild, rainy winters. Farther inland rise the Atlas Mountains. You can see from the map on page 448 that the Sahara lies east and south of the Atlas Mountains.

The Economy The major economic activities of Morocco are agriculture, mining, fishing, and tourism. Morocco's farms produce citrus fruits, potatoes, olives, sugar beets, and wheat. The country's mines yield many resources, such as phosphates, iron ore, lead, and natural gas. Morocco leads the world in the export of phosphate rock and is a leading producer of phosphates. Tourism ranks as one of the country's major service industries.

The People Morocco has a population of about 29 million—more people than any other North African country except Egypt. Most Moroccans are Muslims of mixed Arab and Berber ancestry.

About half of Morocco's population live in rural areas. Some live in villages and farm the land; others are herders in desert areas. The rest of the population live in cities. Rabat, with a population of about 500,000, is Morocco's capital.

The country's largest city and port is Casablanca (KA•suh• BLANG•kuh), home to more than 3 million people. Both Rabat and Casablanca are located on the Atlantic coast. Moroccan cities face the challenge of rapidly growing populations.

History From the A.D. 1000s to the early 1900s, Morocco was a Muslim kingdom that ruled much of northwestern Africa. France and Spain, however, gained control of Morocco's affairs at the beginning of this century. In 1956 Morocco became an independent kingdom once again. At this time, the Moroccan government laid claim to Western Sahara, a large stretch of African territory south of Morocco. Many countries, however, do not recognize this claim.

SECTION 2 REVIEW

REVIEWING TERMS AND FACTS
1. **Define the following:** dictatorship, erg, *casbah*.
2. **PLACE** Where do most of Libya's people live?
3. **LOCATION** What body of water lies north and east of Tunisia?
4. **MOVEMENT** What is Morocco's major export?

MAP STUDY ACTIVITIES
5. Look at the population map of North Africa on page 457. What is the population density around Algiers, Algeria?
6. Using the climate map on page 455, name the North African countries that have both steppe and Mediterranean climates.

Evaluate

 Section 2 Quiz **L1**

 Use the Testmaker to create a customized quiz for Section 2. **L1**

Chapter 17 Test Form A and/or Form B **L1**

Reteach

Reteaching Activity 17 **L1**

Enrich

Enrichment Activity 17 **L3**

CLOSE

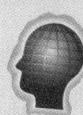 **Mental Mapping Activity**

Provide each student with a large sheet of white paper and map pencils. Inform students that this will not be a graded activity. Have students draw, freehand from memory, a map of North Africa that includes the following labeled countries and bodies of water: Egypt, Libya, Tunisia, Algeria, Morocco, Nile River, Mediterranean Sea.

459

Answers to Section 2 Review
1. All vocabulary words are defined in the Glossary.
2. along the Mediterranean coast
3. Mediterranean Sea
4. phosphate rock
5. 125–250 persons per square mile (50–100 people per sq. km)
6. Libya, Tunisia, Algeria, and Morocco

MAKING CONNECTIONS

TEACH

Ask students to share what they know about the mathematical branches of algebra and trigonometry.

Then inform students that in this feature they will learn that both subjects were developed by Arab scholars. **L1**

More About . . .

Muslim Math and Art

Islamic law forbade the artistic rendering of humans or animals. Muslims made use of mathematics to fulfill their desire for artistic expression. Muslim artists covered the surfaces of buildings and objects with elaborate decorations composed of geometric patterns. They used only a compass and a straight edge to make their designs.

Art historians believe that the Arabs refined and used the compass as no preceding culture had done, for the circle was the foundation for all Islamic designs.

Muslim Contributions to Math

| MATH | SCIENCE | HISTORY | LITERATURE | TECHNOLOGY |

By A.D. 750 Arabs united by the religion of Islam had created a huge empire. Their empire's territory stretched across Southwest Asia and North Africa. The Arabs came into contact with Persian, Indian, Christian, and Jewish scholars living within their empire.

From the A.D. 700s to the 1200s, the city of Baghdad was the Muslim empire's center of learning. The Arabs not only learned Persian astronomy, Indian mathematics, and Greek science, but also added new knowledge to these subjects. Indeed, Muslim mathematicians gave the Western world two kinds of mathematics: algebra and trigonometry.

ALGEBRA About A.D. 825 al-Khwarizmi (ahl•KWAHR•eez•mee), a teacher of mathematics in Baghdad, wrote a book titled *Kitab al-jabr wa al-muqabalah*. His work was a presentation of what was known about algebra at the time. The Arab mathematician improved and expanded algebra, the mathematics that uses symbols to solve complex mathematical problems. Al-Khwarizmi—and readers of his book—invented so many things that scholars consider algebra an Arab creation. Even the name *algebra* comes from part of the title of al-Khwarizmi's book—*al-jabr wa*.

TRIGONOMETRY During the A.D. 900s Muslim mathematicians developed trigonometry, a type of mathematics that deals with the measurements of triangles. They studied and described the spheres in which the sun, moon, and planets were thought to move. Their studies led them to invent the basic ratios, or comparisons, that allow engineers and scientists to measure triangles and distances.

Astronomers and Scientists in Istanbul

Making the Connection

1. From whom did the Arabs learn basic science and mathematics?

2. What contributions did Muslim mathematicians make to Western culture?

460

Answers to Making the Connection

1. Persian, Indian, Christian, and Jewish scholars

2. They developed two new kinds of mathematics: algebra and trigonometry.

Important Things To Know About North Africa

SECTION 1 EGYPT

- The Libyan Desert covers two-thirds of Egypt's land area.

- The Aswan High Dam holds back the Nile's floodwater. The dam provides hydroelectric power to Egypt's growing industries.

- Most of Egypt's people live near the Nile River and depend on it for water and transportation.

- About 55 percent of Egypt's people live in rural areas and farm small plots of land.

- Cairo, Egypt's capital, is the largest city in Africa and a leading center of the Muslim world.

- One of the ancient world's most advanced civilizations developed in the Nile River valley.

- Today Egypt is a republic. Almost 90 percent of Egyptians are Muslims and about 10 percent are Christians.

SECTION 2 OTHER NORTH AFRICAN COUNTRIES

- North Africa also includes the countries of Libya, Tunisia, Algeria, and Morocco. The people of all four countries are of mixed Arab and Berber ancestry.

- The landscape of North Africa is mostly desert or mountains.

- The region's people live on fertile coastal plains or near oases.

- Oil, natural gas, and phosphate rock are among the most important natural resources of North Africa.

- All countries of the region border on the Mediterranean Sea or Atlantic Ocean. This affects climate.

A *casbah* near Duarzazate, Morocco ▶

Extra Credit Project

Documentary Have students research how the scarcity of water affects life in Egypt and North Africa. Suggest that they present their findings in the form of a mini-documentary, consisting of a bulletin-board display of pictures, graphs, charts, diagrams, and articles. L2

461

Chapter **Review and Activities**
17

ANSWERS

Reviewing Key Terms

1. E
2. C
3. H
4. A
5. I
6. G
7. B
8. D
9. F

 Mental Mapping Activity

This exercise helps students to visualize the countries they have been studying and understand the relationships among various points located in these countries. All attempts at freehand mapping should be accepted.

REVIEWING KEY TERMS

Match the numbered terms in Column A with their definitions in Column B.

A
1. delta
2. *fellahin*
3. hydroelectric power
4. hieroglyphs
5. bazaar
6. consumer goods
7. dictatorship
8. erg
9. *casbah*

B
A. picture symbols
B. government ruled by one all-powerful leader
C. Egyptian farmers
D. sand dune
E. broad swampy triangle where a river runs into an ocean or sea
F. old section of a North African city
G. household products, clothes, and shoes that people buy
H. energy produced from moving water
I. marketplace

Mental Mapping Activity

On a separate sheet of paper, draw a freehand map of North Africa. Label the following on your map:

• Egypt
• Nile River
• Cairo
• Libya
• Sahara
• Atlas Mountains

REVIEWING THE MAIN IDEAS

Section 1
1. **LOCATION** Where is Egypt located?
2. **HUMAN/ENVIRONMENT INTERACTION** Why did the Egyptians build the Aswan High Dam?
3. **PLACE** In what direction does the Nile River flow?

Section 2
4. **LOCATION** What countries border Libya?
5. **PLACE** What landforms are found in the Sahara?
6. **REGION** What region includes Tunisia, Morocco, and Algeria?
7. **HUMAN/ENVIRONMENT INTERACTION** How does Morocco's location help it have a prosperous fishing industry?

UNIT 6

Reviewing the Main Ideas

Section 1

1. northeastern Africa
2. to provide water for multiple harvests and to generate hydroelectric power
3. south to north

Section 2

4. Tunisia and Algeria
5. huge areas of shifting sand dunes known as ergs
6. Maghreb
7. It has greater access to the sea, with coastlines on both the Mediterranean Sea and the Atlantic Ocean.

CRITICAL THINKING ACTIVITIES

1. **Determining Cause and Effect** Why do you think an early Egyptian civilization developed in a river valley?
2. **Analyzing Information** How have the people of North Africa adapted to their desert environment?

GeoJournal Writing Activity

Imagine that you are taking a boat trip along the Nile River. Write a daily diary for seven days, describing what you see and telling about your experiences.

COOPERATIVE LEARNING ACTIVITY

Work in small groups to discover what life forms can exist in the deserts of North Africa. Different groups can research the plants and animals that exist there. Collect photographs, posters, or real objects to represent what you discover. Put all information together to make a visual presentation of "Life in the Desert."

PLACE LOCATION ACTIVITY: NORTH AFRICA

Match the letters on the map with the places and physical features of North Africa. Write your answers on a separate sheet of paper.

1. Tripoli
2. Sinai Peninsula
3. Tunis
4. Atlas Mountains
5. Libya
6. Strait of Gibraltar
7. Mediterranean Sea
8. Morocco
9. Cairo

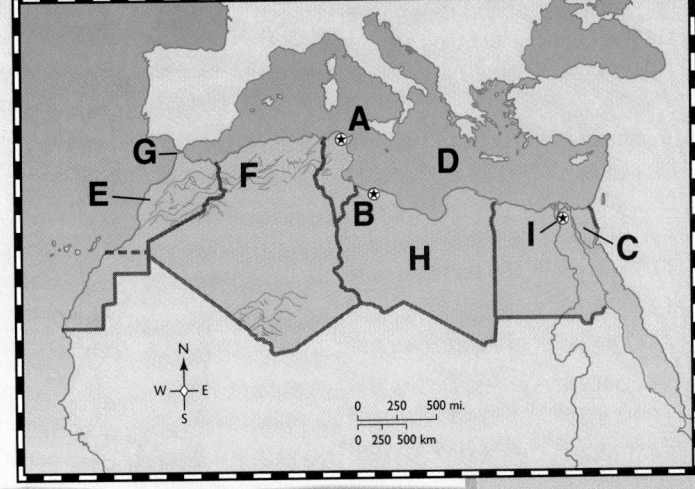

Cooperative Learning Activity

Suggest that students write captions for the various visual materials they collect. These captions should focus on particular characteristics that enable the life forms to survive in the harsh desert climate. Display the visuals in the classroom and invite other classes to share them.

 **GeoJournal Writing Activity**

Encourage students to share their diary entries with the rest of the class.

Place Location Activity

1. B
2. C
3. A
4. F
5. H
6. G
7. D
8. E
9. I

Chapter Bonus Test Question

This question may be used for extra credit on the chapter test.

You find yourself in a North African country. By asking questions, you discover the following:

1. This country has a population of 5 million people.
2. About 70 percent of the population lives along the Mediterranean coast.
3. This country is the richest in North Africa.
4. More than 90 percent desert, this country is one of the world's driest countries.

Where are you? (*Libya*)

Critical Thinking Activities

1. A river valley provided water and fertile soil for growing crops; the river also aided transportation and communication.
2. The people have used natural resources found in the desert, such as oil, to build their economies. They have used various kinds of technology to obtain scarce water and control the water supply.

463

FOCUS

Write the following heading on the chalkboard: "Water, water everywhere." Ask students to identify various sources of water. Guide them by offering such examples as oceans, lakes, rivers, glaciers, and so on. List their responses under the heading. Then point out that water covers some 70 percent of Earth's surface.

TEACH

Ask students to speculate on how Earth could have a shortage of water when it occurs in such abundance. Explain that less than 1 percent of the world's water is fresh and readily accessible. Then point out that this water is distributed very unevenly across the earth and remind students that Southwest Asia and North Africa is one of the world's driest regions. **L1**

NATIONAL GEOGRAPHIC SOCIETY

EYE ON THE ENVIRONMENT

Water: A Precious Resource

PROBLEM

Most of the usable water in **Southwest Asia and North Africa** comes from aquifers and from three river basins: the Jordan, the Tigris-Euphrates, and the Nile. Despite these large river systems, water is scarce. Each of the rivers and its tributaries flows through several countries—and each country's needs are growing. Drought, industrialization, irrigation needs, and mushrooming populations—all strain the limited water supply.

SOLUTIONS

- To resolve water-rights issues, Israel is negotiating with Jordan and the Palestinians over rights to the Jordan River and West Bank aquifers.

- To promote conservation, Tunisia is introducing higher prices for water used for irrigation.

- A recycling project in Tel Aviv, Israel, generates enough recycled wastewater to cultivate 20,000 acres (8,100 hectares) of farmland.

- Advances in technology have lowered the cost of converting seawater to freshwater.

- Cooperative measures, such as Turkey's plans for a Peace Pipeline, may carry water from water-rich countries to water-poor countries.

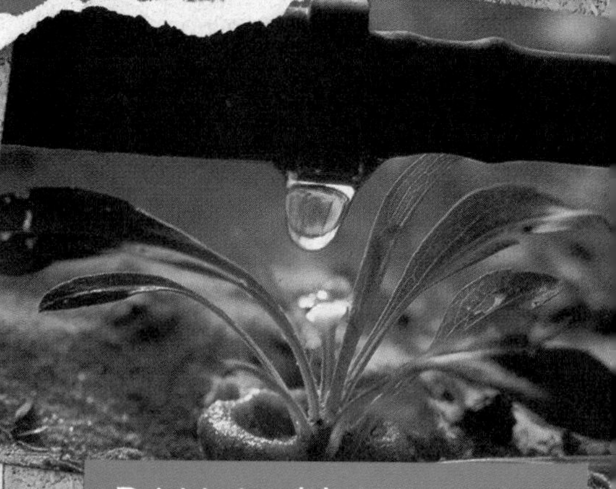

Drip irrigation, which minimizes water loss, delivers a precise drink of water to a seedling in the desert. Drip irrigation may use 25 percent less water than sprinkler systems.

WATER FACT BANK

- By the year 2015, most of the nations of North Africa will not have enough water to meet basic human needs.

- Crops growing in the desert require huge amounts of water. It takes 3 cubic meters of water (780 gallons) to produce about 2 pounds (1 kilogram) of wheat.

- Turkey's Atatürk Dam, on the Euphrates River, holds back water that once reached other nations downstream. Soon, Syria's share of the Euphrates will shrink by 40 percent, and Iraq's share by 80 percent.

- Saudi Arabia leads the world in producing freshwater from salt water. Its 22 desalination plants produce 30 percent of all the desalinated water in the world.

464

UNIT 6

More About the Issues

Human/Environment Interaction
The earliest recorded use of irrigation occurred in Mesopotamia in Southwest Asia and North Africa. Today, irrigation remains a major use of freshwater in the region. Problems related to traditional methods of irrigation continue to plague the region's farmers. Because the region's climate is so hot and dry, surface irrigation systems lose a high percentage of water to evaporation. High rates of evaporation lead to the accumulation of certain minerals in the soil. This can make the land unsuitable for farming. To address such problems, many governments in the region have encouraged the use of new irrigation technology.

> Source of life in a dry land, the Jordan River winds through green orchards and gray fields of cotton on a kibbutz, an Israeli collective farm.

TEEN TRIBUTE

Hamidreza Modaberi, age 12, lives in the desert country of Iran. In a letter to an environmental publication, she wrote: "I dreamed the whole world had changed. All these dry lands had been turned into a beautiful nature full of trees, full of rivers ... war and bloodshed had ended.... I wish the world would be like my dream."

TAKE A CLOSER LOOK

1 What sources provide Southwest Asia and North Africa with water?

2 What human activities have caused water shortage problems in this region?

good planets are hard to find

WHAT CAN YOU DO?

🍃 Put "Water Watch" posters around your school giving tips about conserving water.

🍃 Improve your own water-saving habits and avoid using products that are polluters.

🍃 Write a letter to the editor of your local newspaper to protest industrial pollution in your community.

🍃 Educate yourself – find out where your water comes from and what is being done to protect that source.

🍃 Fix that dripping faucet!

465

All's well—as long as the well water flows. This Tunisian farmer bought a pump with a government loan to lift well water to his thirsty fields. Tunisian farms have been irrigated since Roman times.

Global Issues

Interdependence Establishing a fair system of access to freshwater is a worldwide problem. Even within the United States, some western states are engaged in legal battles over water rights.

GLENCOE
TECHNOLOGY

🔘 **VIDEODISC**

Reuters Issues in Geography

Chapter 5
North Africa and Southwest Asia:
The Influence of Oil and Water
Disc 1 Side A

ASSESS

Have students answer the **Take A Closer Look** questions on page 465.

CLOSE

Encourage students to gather information on the water supply in their community. Suggest that they look for answers to the following questions: What is the source of the community's drinking water? How is this water purified and transported to homes in the community? What problems confront the community's water supply? Consider inviting a representative of the local water utility to give a brief presentation and answer students' questions.

* For an additional regional case study, use the following:

◣ Environmental Case Study 6 **L1**

OUT OF TIME?

If time does not permit teaching each chapter, you may use the Chapter Highlights, Audiocassettes, and their corresponding activities and tests.

GLENCOE
TECHNOLOGY

 **VIDEODISC**

Reuters Issues in Geography

Chapter 6
Sub-Saharan Africa: Preserving Its Legacy
Disc 1 Side A

interNET
CONNECTION

For information on the countries of Africa, check the following address:

World Wide Web
http://www.gorp.com/gorp/location/africa/africa.htm

Learn about Africa's geography, culture, politics, and history. Contact the following address:

World Wide Web
http://www.africaonline.com/

Unit 7 — Africa South of the Sahara

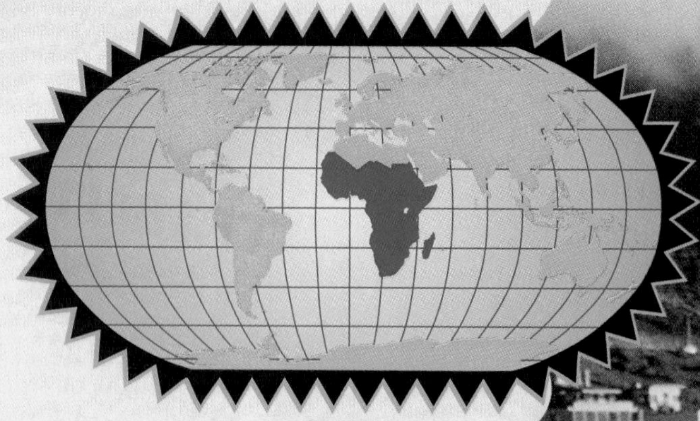

What Makes Africa South of the Sahara a Region?

Most people share...

- a location almost entirely in the tropics.
- the world's fastest-growing and youngest population.
- challenges to the environment, especially natural resources and wildlife.
- a struggle to improve the quality of life.

To find out more about Africa south of the Sahara, see the Unit Atlas on pages 468–483.

Cape of Good Hope, Republic of South Africa

466

About the Illustration

The Cape of Good Hope is a peninsula located at the southern tip of Africa. The first European to sight the cape was Portuguese explorer Bartholomeu Dias in 1488. Vasco da Gama, another Portuguese explorer, in 1497 became the first European to sail round the cape. He continued sailing eastward in search of a trade route to India. Cape Town, at the head of the peninsula, was the first white settlement in the area. The Dutch East India Company founded the colony in 1652.

 **Movement** Why would Europeans be interested in finding a trade route to India by sailing around Africa? *(They did not have control of the overland routes.)*

NATIONAL GEOGRAPHIC SOCIETY

These materials are available from Glencoe.

 CD-ROM

Picture Atlas of the World

 SOFTWARE

ZipZapMap! World

 TRANSPARENCIES

PicturePack Transparencies, Unit 7

GeoJournal Activity

Have students collect newspaper and/or news magazine articles about Africa south of the Sahara. Instruct them to include the source and date for each article they bring to class. Use the articles to create a reference corner in the classroom that students may consult to research the "treasures" of Africa south of the Sahara. **L2**

- This journal activity provides the basis for the "GeoJournal Writing Activity" exercise in the Chapter Review.
- The GeoJournal may be used as an integral part of Performance Assessment.

GeoJournal Activity

As you read this unit, think about why many geographers call Africa an unopened treasure chest. In your journal, choose the three most important treasures you think Africa has to share with the rest of the world.

467

Location Activity

Display Political Map Transparency 7 and ask students the following questions: What body of water borders Africa on the west? *(Atlantic Ocean)* On the east? *(Indian Ocean)* What major river flows through this region to the Mediterranean Sea? *(Nile River)* Which countries are islands? *(Madagascar, Comoros, Mauritius, Seychelles, São Tomé and Príncipe, Cape Verde Islands)* **L1**

IMAGES OF THE WORLD

UNIT 7 ATLAS

NATIONAL GEOGRAPHIC SOCIETY

These materials are available from Glencoe.

 CD-ROM

- **Picture Atlas of the World**
- **PictureShow: Geology**

 VIDEODISC

STV: World Geography Vol. 2: Africa and Europe

Africa
Side 1, Frames 00001-49739

 POSTERS

Images of the World

Unit 7: Africa South of the Sahara

 TRANSPARENCIES

PicturePack Transparencies, Unit 7

NATIONAL GEOGRAPHIC SOCIETY

IMAGES of the WORLD

1. African elephant, Kenya
2. Woman in Tombouctou, Mali
3. Victoria Falls, Zimbabwe
4. Shoppers at a market, Madagascar
5. Mosque at Omdurman, Sudan
6. Storefronts, Kenya
7. Oil company supervisor, off Gabon

All photos viewed against a village in the Tibesti Mountains, Chad.

468

Fun Facts

- **Chad** The Tibesti Mountains cover an area of more than 50,000 square miles (129,500 sq. km) and reach heights of over 11,000 feet (3,353 m). One of the world's most rugged and inaccessible places, the Tibesti Mountains have long served as a refuge for desert bandits.

- **Zambia and Zimbabwe** Victoria Falls, on the border between Zambia and Zimbabwe, was named for Queen Victoria of Britain by the explorer and missionary, David Livingstone. The traditional name used by the African people of the area is far more descriptive. They called the falls *Mosi oa Tunya,* or "the smoke that thunders."

**Geography
and the Humanities**

World Literature
Reading 7

World Music:
Cultural Traditions,
Lesson 6

World Cultures
Transparencies 11, 12

Focus on World Art
Print 12

MULTICULTURAL PERSPECTIVE

Cultural Diversity

The Masai of Kenya are tall, slender nomads. They move from place to place in search of grazing land for their animals. Since the Masai depend on animals for their livelihood, wealth is judged by the number of animals a person owns.

469

Teacher Notes

Using the Unit Atlas

These features and activities may be used as an introduction to the unit or as teaching tools throughout the course of the unit.

 Unit Atlas Activity 7
L1

 Unit Overlay Transparencies 7-0, 7-1, 7-2, 7-3, 7-4, 7-5, 7-6

Home Involvement Activity 7 L1

Environmental Case Study 7 L1

Countries of the World Flashcards, Unit 7

World Games Activity Cards, Unit 7

More About the Illustration

Answer to Caption
more than 2,000 feet (610 m) above sea level

Place Burkina Faso's economy is heavily dependent on the raising of cattle, goats, and sheep. In fact, about 50 percent of the country's export income comes from the sale of livestock.

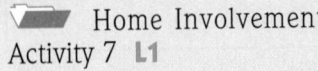

Regional Focus

Africa south of the Sahara covers the central and southern parts of the continent. With the Equator crossing its middle, Africa south of the Sahara is the only region in the world that lies almost entirely in the tropics.

Region

The Land

The region of Africa south of the Sahara is about 9.5 million square miles (24.6 sq. km) in size. Its center is mostly made up of large, rolling plateaus. Narrow coastal plains run along the Atlantic and Indian oceans.

Plateaus The region's plateaus rise from west to east like a stairway. Because of these plateaus, Africa has a higher elevation than any other continent in the world. The map on page 474 shows you that the average elevation in Africa is more than 2,000 feet (610 m) above sea level. Separating the plateaus are steep cliffs. Rivers often spill over the cliffs in thundering waterfalls. The cliffs and waterfalls make land and river travel difficult in many parts of Africa.

Mountains and Valleys Even with its high elevation, Africa south of the Sahara has few tall mountains. Wind and water wore away many of them over millions of years. At 19,340 feet (5,895 m) high, Kilimanjaro is the tallest peak in Africa.

Crossing the plateaus is the Great Rift Valley. This long, narrow break in the earth's surface runs more than 3,000 miles (4,800 km) through eastern Africa. Movements of the earth's crust formed the Great Rift Valley millions of years ago. A chain of deep lakes lies in the Great Rift Valley near the Equator.

Rivers Four great rivers slice through Africa: the Nile, the Zaire, the Niger, and the Zambezi. The largest river system in Africa south of the Sahara is the Zaire. It twists and turns for almost 2,800 miles (4,505 km) through the region. A number of small rivers flow into the Zaire and the other three large rivers. Africa's major river systems are important means of transportation that link inland areas with oceans.

A Herder in Burkina Faso

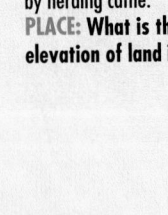

Most people in the country of Burkina Faso make their living by herding cattle.
PLACE: What is the average elevation of land in Africa?

 470

Fun Facts

- **Lesotho** This country has no forests and is subject to severe soil erosion. The government sponsors projects to help the environment. Tree Planting Day is an official holiday celebrated on March 21.

- **Rwanda** Rwanda's oral literary tradition consists of myths, fables, folktales, poetry, and proverbs. Since relatively few Rwandans can read and write, stories are passed from one generation to another by storytellers.

Region
Climate and Vegetation

Africa south of the Sahara has many types of climate and vegetation. Differences in elevation create much of this variety. If you travel in Africa toward the Equator from either the north or the south, you will find four main climate regions. They are desert, steppe, tropical savanna, and tropical rain forest. Only a few areas—the highlands in eastern Africa and the southern tip of Africa—have moderate climates.

Deserts More of Africa is covered by deserts than any other continent. This occurs because many high plateaus block rain clouds from reaching inland areas. In the north you find the Sahara; in the south, the Namib and Kalahari deserts. Africa's desert areas often have daytime temperatures that soar to 120°F (50°C). At night temperatures drop to about 50°F (10°C) or lower. Scattered cactuses and thorny shrubs are the only plants found in the deserts.

Steppe Around Africa's deserts stretch steppe areas, or partly dry grasslands. Directly south of the Sahara is a steppe area known as the Sahel. Because of drought and human activity, the Sahel is gradually becoming a desert. This has brought famine and hardship to people living in the area.

Tropical Savanna Much of Africa south of the Sahara is made up of savannas—tropical grasslands with scattered trees. The savannas have wet and dry seasons. A wide variety of animal life lives in the savannas. African governments have set aside many savanna areas for the protection of endangered animals such as elephants and lions.

Tropical Rain Forest Steamy rain forests cover land near the Equator. This climate area is hot, and rain falls daily. Africa's rain forests have many hardwood trees that provide timber. Because so many trees have been cut down, many rain forest areas are disappearing.

Human/Environment Interaction
The Economy

The economy of Africa south of the Sahara depends mainly on farming. Another important economic activity is the mining and export of mineral resources.

Volcanoes National Park, Rwanda

Volcanic peaks tower over the vast plateaus of northwestern Rwanda.
HUMAN/ENVIRONMENT INTERACTION: What do Africa's rain forests provide?

More About the Illustration

Answer to Caption
timber

 **Location** These volcanic peaks are located in Volcanoes National Park in Rwanda's Virunga Mountains. Before Rwanda's bloody civil war in 1993, this park and others attracted tourists from all over the world.

Cultural Kaleidoscope

Benin Animism is a traditional religion practiced in Benin and elsewhere in West Africa. The religion is based on a belief that all things in nature have spirits.

Fun Facts

• **Kenya** Nairobi, the capital of Kenya, was originally the site of a water hole. It was called *Enkare Nairobi*, which means "cold water."

• **Namibia** The Namib Desert has some of the largest sand dunes in the world. In the southern part of the desert, dunes reach heights of 660 feet (200 m) and may spread up to a mile (1.6 km) wide.

471

FactsOnFile

Use the reproducible masters from **GEOGRAPHY ON FILE,** *Sub-Saharan Africa,* and **AFRICAN HISTORY ON FILE** to enrich unit content.

More About the Illustration

Answer to Caption
more than 542 million

Location Johannesburg is a modern, thriving city that has profited from its location near vast goldfields. It is located in northeast South Africa in the Witwatersrand, the most productive goldfield in the world. Some tunnels from gold mines in the Witwatersrand extend under the city.

Johannesburg, South Africa

Johannesburg is South Africa's largest urban area and most important industrial and commercial center.
PLACE: About how many people live in Africa south of the Sahara?

Agriculture Some areas of Africa have fertile soil and plenty of rain. Many other areas do not have these advantages. Africans have used their environment to meet their needs. Farming and herding are the leading economic activities. Some Africans grow only enough food to feed their families. Others work on large plantations that raise crops for export. Africa leads the world in the production of cacao, cassava, cloves, and sweet potatoes. It is also a major producer of coffee, bananas, cotton, peanuts, rubber, tea, and sugar.

Minerals and Manufacturing Africans also work in mining. In central and southern Africa, miners work some of the largest deposits of minerals in the world. These resources include gold, diamonds, and copper. Oil is found in western and central Africa.

In spite of many resources, manufacturing plays only a small role in Africa's economy. In the past, European countries ruled most of the continent. Today many countries in Africa south of the Sahara are working to build industries and improve transportation. Few paved roads exist outside of urban areas. Railroads and airplanes are best for traveling long distances in the region.

Region
The People

South of the Sahara, Africa is divided into about 50 countries. Each of these countries has many ethnic groups, languages, religions, and customs.

Population More than 542 million people live in this region, which is growing rapidly in population. Because so much of Africa is desert or partly dry grasslands, most people live along the coasts. Some African governments have encouraged their people to move from the crowded coasts to inland areas. Three countries—Côte d'Ivoire, Nigeria, and Tanzania—have moved or will move their capitals from coastal cities to newly built inland cities.

About 73 percent of Africans south of the Sahara live in rural villages. Economic hardships and the desire for a better life, however, are drawing many people to the cities. African cities are among the world's fastest-growing urban areas. The three largest cities are Lagos in Nigeria, Kinshasa in Zaire, and Abidjan in Côte d'Ivoire.

Culture People in this region are devoted to family life. Households often are made up of extended families—grandparents and other relatives as well as parents and children. Africans also have close ties to their ethnic groups. In many countries, loyalty to an ethnic group is more important than loyalty to the national government.

472

UNIT 7

Fun Facts

- **Côte d'Ivoire** French sailors came to this region in the late 1400s in search of ivory. They are responsible for the name, which means "Ivory Coast."

- **South Africa** South Africa produces about one-third of the world's gold.

- **Namibia** About 100,000 people in Namibia and other parts of southwest Africa speak Khoisan languages. These languages are not related to any others spoken in Africa. Many words in Khoisan are expressed with unusual "click" sounds.

History Early Africa saw the rise and fall of powerful empires, kingdoms, and city-states. Many of these became important centers of trade, learning, and culture. From the 1500s to the 1800s, Europeans explored the continent, enslaved many Africans, and sent them to other parts of the world. By 1914 Europeans had divided nearly all of Africa among themselves. They set up borders without regard to the different peoples living in the area.

Praying in Asmara, Eritrea

Worshipers in Eritrea gather to observe the Easter holiday.
PLACE: What percentage of Africans live in rural villages?

In many parts of Africa, however, Africans resisted European rule. In the 1950s and 1960s, most of the European colonies finally became independent African countries. Descendants of European settlers, however, continued to rule the country of South Africa. They withheld many rights from black Africans and other non-Europeans. In the early 1990s, South Africa finally became a democracy based on equal rights for all citizens.

The Arts Africans have developed many art forms. African artists are known for their decorative cloth weavings, wooden sculptures, metal jewelry, and pottery. Through the centuries African storytellers have passed down myths, legends, history, and poems from early ancestors. African authors today blend these traditions with modern writing styles. African music has influenced popular music throughout the world.

Technology Activity

Using a CD-ROM
Use the National Geographic CD-ROM "Picture Atlas of the World" to research an African nation south of the Sahara. Assume that you are an exchange student from Africa who has been living with an American family and going to an American school for three months. Use the Collector's button to create your own file containing notes that you write to explain to your American family ways in which your home country is like the United States.

REVIEW AND ACTIVITIES

1. **REGION** What four major rivers cross Africa south of the Sahara?
2. **HUMAN/ENVIRONMENT INTERACTION** Why is the Sahel becoming a desert?
3. **HUMAN/ENVIRONMENT INTERACTION** What are the region's two major economic activities?
4. **MOVEMENT** Why are Africans moving to cities from the countryside?
5. **PLACE** How has South Africa changed since the early 1990s?

473

Answers to Review and Activities

1. Nile, Niger, Zaire, Zambezi
2. because of drought and human activity
3. farming and exporting mineral resources
4. They face economic hardships in the country and are attracted by the better opportunities available in the cities.
5. It has become a democracy with full rights for all citizens. In their first free elections, South Africans elected a black South African president.

LESSON PLAN
Unit 7 Atlas

Applying Geographic Themes

Region About 90 percent of Africa lies within the tropics. It has the largest tropical region of any continent. How many African countries lie on the Equator? *(six— Gabon, Congo, Zaire, Uganda, Kenya, Somalia)*

NATIONAL GEOGRAPHIC SOCIETY

These materials are available from Glencoe.

 CD-ROM

• **Picture Atlas of the World**
• **PictureShow: Geology**

 SOFTWARE

ZipZapMap! World

MAP STUDY

Answers
1. Mali, Niger, Chad
2. east central Africa

Map Skills Practice
Reading a Map Which three countries border on Lake Victoria? *(Uganda, Kenya, Tanzania)* Into what ocean does the Zambezi River flow? *(Indian Ocean)*

UNIT 7 ATLAS
Physical Geography

AFRICA SOUTH OF THE SAHARA: Physical

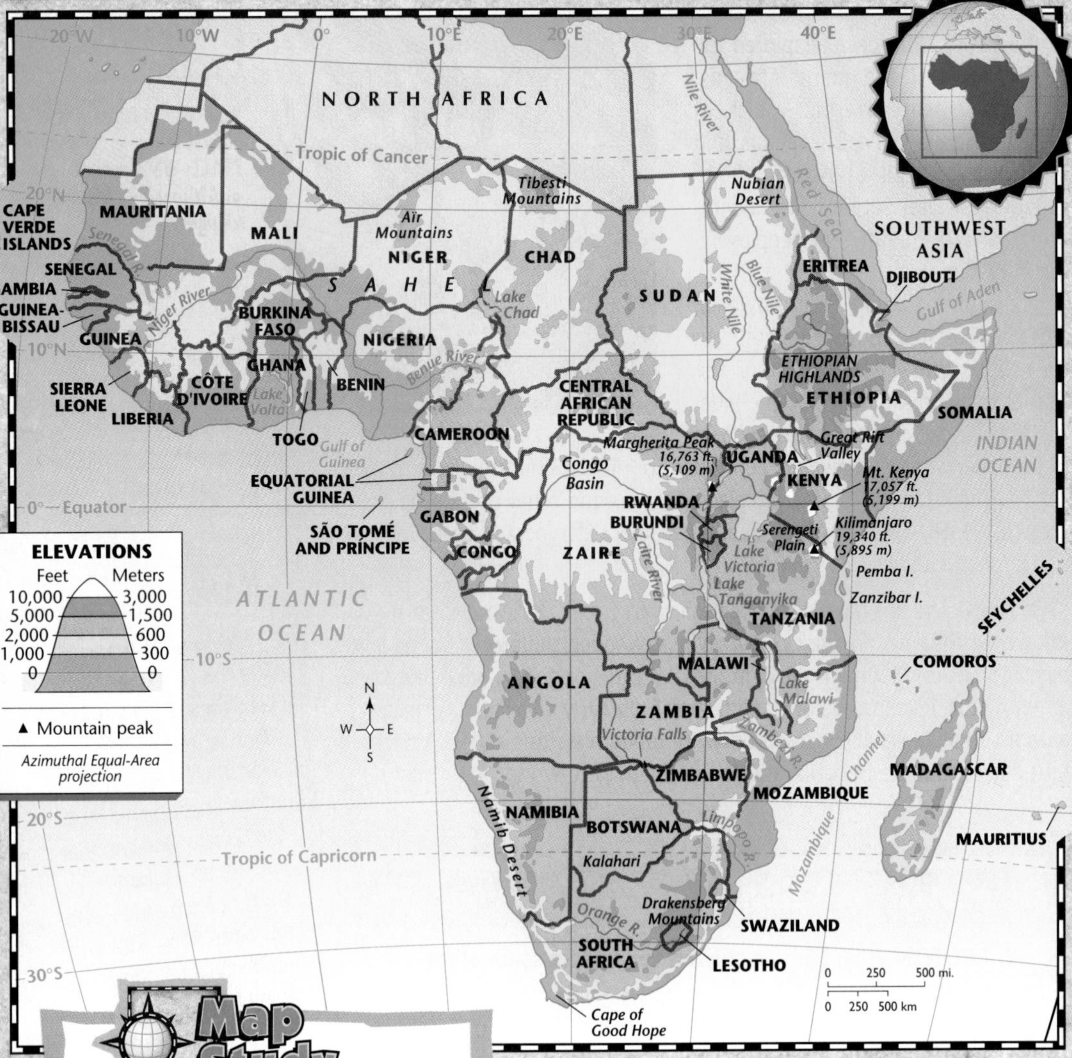

(Map of Africa showing physical geography, countries, elevations, and features including NORTH AFRICA, MAURITANIA, MALI, NIGER, CHAD, SUDAN, ERITREA, DJIBOUTI, ETHIOPIA, SOMALIA, CAPE VERDE ISLANDS, SENEGAL, GAMBIA, GUINEA-BISSAU, GUINEA, SIERRA LEONE, LIBERIA, CÔTE D'IVOIRE, BURKINA FASO, GHANA, TOGO, BENIN, NIGERIA, CAMEROON, CENTRAL AFRICAN REPUBLIC, EQUATORIAL GUINEA, SÃO TOMÉ AND PRÍNCIPE, GABON, CONGO, ZAIRE, UGANDA, RWANDA, BURUNDI, KENYA, TANZANIA, MALAWI, ANGOLA, ZAMBIA, ZIMBABWE, MOZAMBIQUE, MADAGASCAR, NAMIBIA, BOTSWANA, SWAZILAND, LESOTHO, SOUTH AFRICA, SEYCHELLES, COMOROS, MAURITIUS)

ELEVATIONS

Feet	Meters
10,000	3,000
5,000	1,500
2,000	600
1,000	300

▲ Mountain peak

Azimuthal Equal-Area projection

Margherita Peak 16,763 ft. (5,109 m)
Mt. Kenya 17,057 ft. (5,199 m)
Kilimanjaro 19,340 ft. (5,895 m)

0 250 500 mi.
0 250 500 km

Map Study

1. **PLACE** Name three countries crossed by the Sahel.
2. **REGION** In what part of Africa are the three highest mountain peaks located?

Atlas Activity

Place Have students work individually or in pairs to describe the physical geography of Africa south of the Sahara. Have them choose a region shown on the map on page 474. Then ask them to write at least five adjectives that describe the region's physical features. Call on students to identify and describe their selected regions. **L1**

ELEVATION PROFILE: Africa

18,000 ft.	
	Mt. Kenya
14,000 ft.	4,267 m
	Great Rift Valley
10,000 ft.	3,048 m
	Ruwenzori Mountains
Atlantic Ocean	Lake Victoria Indian Ocean
5,000 ft.	1,524 m
	Congo Basin
2,000 ft.	610 m
1,000 ft.	305 m
Sea level	Sea level

⟵ West to East at 0° latitude ⟶

Source: *Goode's World Atlas*, 19th edition

GeoFacts

Highest point: Kilimanjaro (Tanzania) 19,340 ft. (5,895 m) high

Lowest point: Lake Assal (Djibouti) 512 ft. (156 m) below sea level

Longest river: Nile River 4,160 mi. (6,693 km) long

Largest lake: Lake Victoria (Uganda/Tanzania) 26,828 sq. mi. (69,485 sq. km)

Largest desert: Sahara 3,500,000 sq. mi. (9,065,000 sq. km)

AFRICA SOUTH OF THE SAHARA AND THE UNITED STATES: Land Comparison

Graphic Study

1. PLACE What is the elevation of Lake Victoria?

2. REGION About how many times larger than the continental United States is Africa south of the Sahara?

(475)

GLENCOE
TECHNOLOGY

 **VIDEODISC**

The Infinite Voyage: The Keepers of Eden

Chapter 8
Preserves of Endangered Species: San Diego and Kenya
Disc 1 Side A

FactsOnFile

Use the reproducible masters from **GEOGRAPHY ON FILE**, *Sub-Saharan Africa*, and **AFRICAN HISTORY ON FILE** to enrich unit content.

GRAPHIC STUDY

Answers

1. about 3,000 feet (914 m)
2. about three times larger

Graphic Skills Practice
Reading a Diagram
How do elevations in the east of the area shown on the profile compare to those in the west? *(elevations in the east are much higher)*

Teacher Notes

475

LESSON PLAN
Unit 7 Atlas

UNIT 7 ATLAS
Cultural Geography

Applying Geographic Themes

Place Most countries of Africa south of the Sahara have a coastline. Which country has coastline on two oceans? *(South Africa)* Which countries are landlocked? *(Mali, Burkina Faso, Niger, Chad, Ethiopia, Central African Republic, Uganda, Rwanda, Burundi, Malawi, Zambia, Zimbabwe, Botswana, Swaziland, and Lesotho)*

NATIONAL GEOGRAPHIC SOCIETY

These materials are available from Glencoe.

VIDEODISC

STV: World Geography
Vol. 2: Africa and Europe

|||||||||||||||||

Africa
Side 1, Frames 00001-49739

SOFTWARE

ZipZapMap! World

AFRICA SOUTH OF THE SAHARA: Political

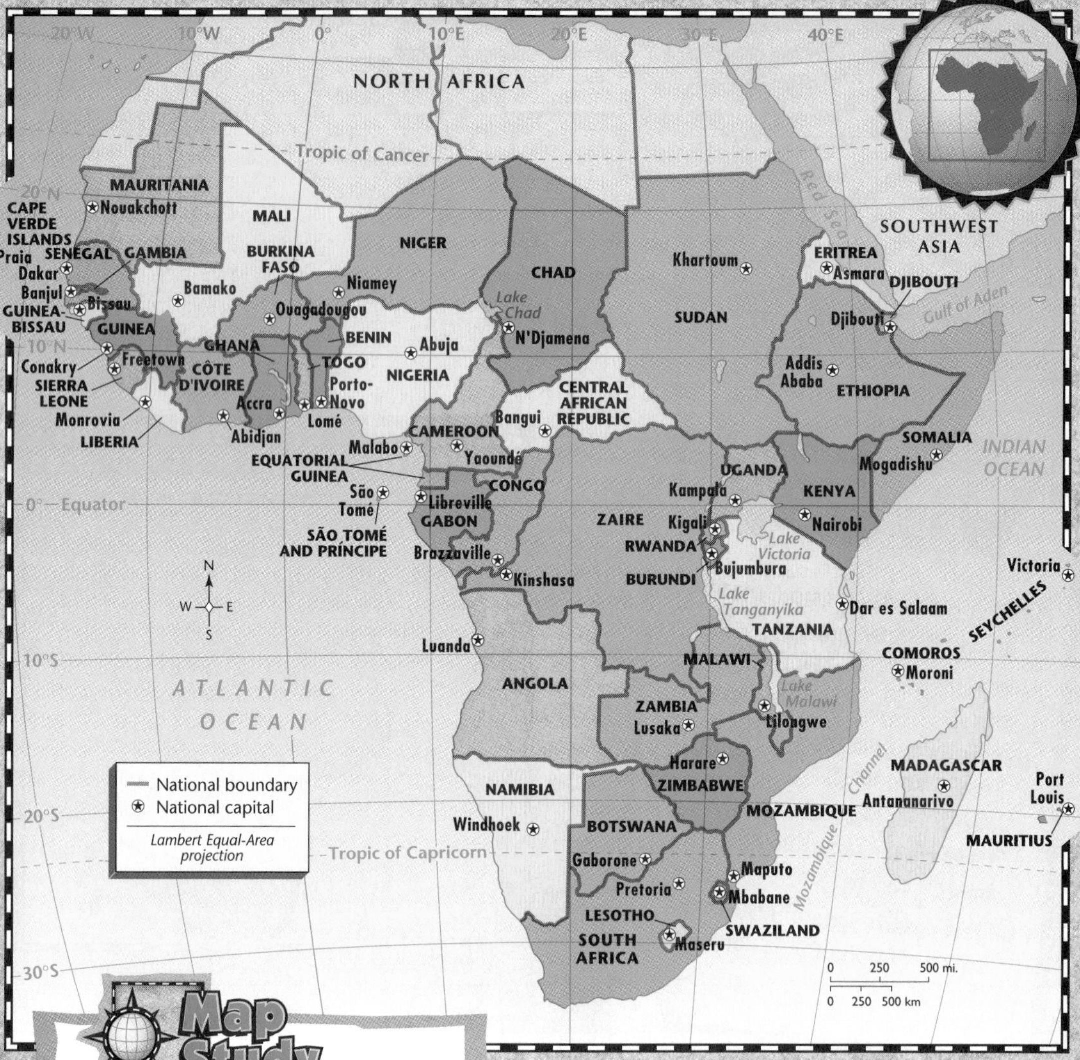

[Map of Africa South of the Sahara showing countries, capitals, and geographic features including:]

NORTH AFRICA
Tropic of Cancer
MAURITANIA — ⊛Nouakchott
CAPE VERDE ISLANDS — ⊛Praia
MALI
NIGER
CHAD
Khartoum⊛
SOUTHWEST ASIA
ERITREA ⊛Asmara
DJIBOUTI
SUDAN
⊛Djibouti
Gulf of Aden
Red Sea
SENEGAL ⊛GAMBIA
BURKINA FASO
Bamako⊛
Niamey⊛
Ouagadougou⊛
N'Djamena⊛
Addis Ababa⊛ ETHIOPIA
Dakar⊛
Banjul⊛
GUINEA-BISSAU ⊛Bissau
GUINEA
GHANA
BENIN
Abuja⊛
CENTRAL AFRICAN REPUBLIC
SOMALIA
Mogadishu⊛
INDIAN OCEAN
Conakry⊛ ⊛Freetown
CÔTE D'IVOIRE
TOGO
NIGERIA
Bangui⊛
SIERRA LEONE
Accra⊛
Porto-Novo⊛
Lomé⊛
Monrovia⊛
Abidjan⊛
LIBERIA
CAMEROON
Yaoundé⊛
UGANDA
Kampala⊛
KENYA
⊛Nairobi
EQUATORIAL GUINEA
Malabo⊛
CONGO
ZAIRE
Kigali⊛
RWANDA
Victoria⊛
São Tomé⊛
Libreville⊛
GABON
Lake Victoria
BURUNDI
Bujumbura⊛
SEYCHELLES
SÃO TOMÉ AND PRÍNCIPE
Brazzaville⊛
⊛Kinshasa
Lake Tanganyika
⊛Dar es Salaam
TANZANIA
Equator
Luanda⊛
COMOROS
⊛Moroni
ATLANTIC OCEAN
ANGOLA
MALAWI
ZAMBIA
Lusaka⊛
Lilongwe⊛
Lake Malawi
MADAGASCAR
⊛Antananarivo
Port Louis⊛
NAMIBIA
Harare⊛
ZIMBABWE
MOZAMBIQUE
MAURITIUS
Windhoek⊛
BOTSWANA
Tropic of Capricorn
Gaborone⊛
Pretoria⊛
Maputo⊛
Mbabane⊛
LESOTHO
SWAZILAND
SOUTH AFRICA
⊛Maseru
Mozambique Channel

— National boundary
⊛ National capital
Lambert Equal-Area projection

0   250   500 mi.
0   250   500 km

Map Study

1. **LOCATION** What country is totally surrounded by South Africa?
2. **PLACE** What is Africa's largest island country?

MAP STUDY

Answers
1. Lesotho
2. Madagascar

Map Skills Practice
Reading a Map Which countries lie on the Tropic of Capricorn? *(Namibia, Botswana, South Africa, Mozambique, Madagascar)* What is the capital of Somalia? *(Mogadishu)*

Atlas Activity

Location Help students develop a mental map of Africa south of the Sahara by playing a location game. Assign one or more countries to each student. Instruct students to write a riddle, consisting of 3-5 clues, about the location of their country or countries. *(Example: This country is east of Gabon. It borders the Central African* Republic. It is west of Zaire. It has a coastline on the Atlantic Ocean. Answer—Congo)* Have each student read clues, one at a time, until a classmate can identify the country. You might use the riddles for team competition. Award points based on the number of clues a team uses to correctly identify a country. **L1**

AFRICA SOUTH OF THE SAHARA:
Selected Rural and Urban Populations

	Rural	Urban
WEST AFRICA		
Niger	85%	15%
Cape Verde	56%	44%
CENTRAL AFRICA		
Angola	72%	28%
Central African Republic	53%	47%
EAST AFRICA		
Rwanda	95%	5%
Djibouti	23%	77%
SOUTHERN AFRICA		
Lesotho	81%	19%
South Africa	43%	57%

Source: *Population Reference Bureau, 1994*

Geography and the Humanities

World Literature Reading 7

World Music: Cultural Traditions, Lesson 6

World Cultures Transparencies 11, 12

Focus on World Art Print 12

GeoFacts

Biggest country (land area): Sudan 967,500 sq. mi. (2,505,813 sq. km)

Smallest country (land area): São Tomé and Príncipe 370 sq. mi. (958 sq. km)

Largest city (population): Lagos (1995) 9,799,000; (2000 projected) 12,528,000

Highest population density: Mauritius 1,549 people per sq. mi. (598 people per sq. km)

Lowest population density: Namibia 5 people per sq. mi. (2 people per sq. km)

COMPARING POPULATION:
Africa South of the Sahara and the United States

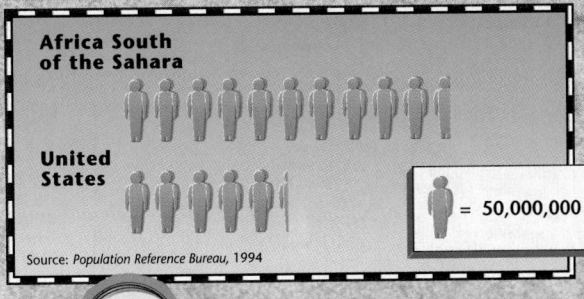

Africa South of the Sahara

United States

= 50,000,000

Source: *Population Reference Bureau, 1994*

Graphic Study

1. **PLACE** About how many people live in Africa south of the Sahara?
2. **REGION** What countries in Africa south of the Sahara have a greater urban population than a rural population?

GRAPHIC STUDY

Answers
1. about 542 million people
2. Djibouti, South Africa

Graphic Skills Practice
Reading a Graph How many more people live in Africa South of the Sahara than in the United States? *(about 280 million)*

477

Teacher Notes

UNIT 7 ATLAS

Countries at a Glance

Angola 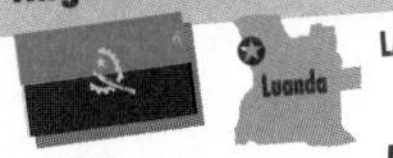
CAPITAL:
Luanda
MAJOR LANGUAGE(S):
Portuguese, Bantu languages
POPULATION:
11,200,000
LANDMASS:
481,350 sq. mi./
1,246,697 sq. km
MONEY:
New Kwanza
MAJOR EXPORT:
Mineral Fuels
MAJOR IMPORT:
Transport Equipment

Benin 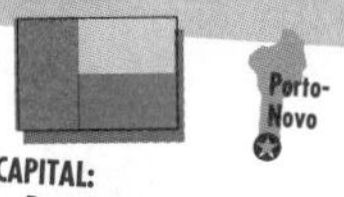
CAPITAL:
Porto-Novo
MAJOR LANGUAGE(S):
French, Fon, Yoruba, Somba
POPULATION:
5,300,000
LANDMASS:
42,710 sq. mi./
110,619 sq. km
MONEY:
CFA Franc
MAJOR EXPORT:
Cotton
MAJOR IMPORT:
Yarn and Fabric

Botswana 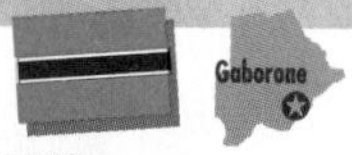
CAPITAL:
Gaborone
MAJOR LANGUAGE(S):
English, Setswana
POPULATION:
1,400,000
LANDMASS:
218,810 sq. mi./
566,718 sq. km
MONEY:
Pula
MAJOR EXPORT:
Diamonds
MAJOR IMPORT:
Transport Equipment

Burkina Faso 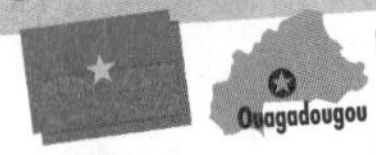
CAPITAL:
Ouagadougou
MAJOR LANGUAGE(S):
French, Sudanic native languages
POPULATION:
10,100,000
LANDMASS:
105,710 sq. mi./
273,789 sq. km
MONEY:
CFA Franc*
MAJOR EXPORT:
Cotton
MAJOR IMPORT:
Machinery

Burundi 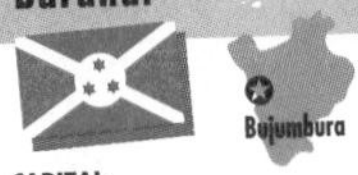
CAPITAL:
Bujumbura
MAJOR LANGUAGE(S):
Rundi, French
POPULATION:
6,000,000
LANDMASS:
9,900 sq. mi./
25,641 sq. km
MONEY:
Burundi Franc
MAJOR EXPORT:
Coffee
MAJOR IMPORT:
Machinery

Cameroon 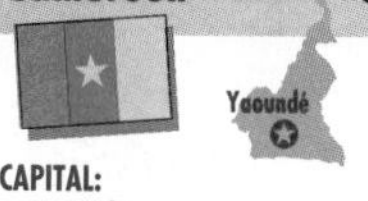
CAPITAL:
Yaoundé
MAJOR LANGUAGE(S):
English, French
POPULATION:
13,100,000
LANDMASS:
179,690 sq. mi./
465,397 sq. km
MONEY:
CFA Franc*
MAJOR EXPORT:
Petroleum
MAJOR IMPORT:
Machinery

Cape Verde Islands 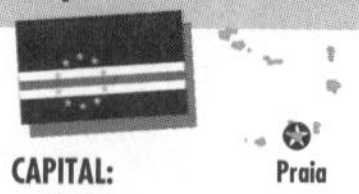
CAPITAL:
Praia
MAJOR LANGUAGE(S):
Portuguese
POPULATION:
400,000
LANDMASS:
1,560 sq. mi./
4,040 sq. km
MONEY:
Escudo
MAJOR EXPORT:
Bananas
MAJOR IMPORT:
Food

Central African Republic 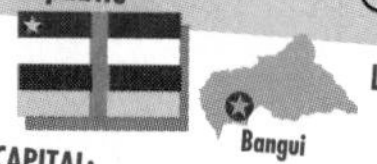
CAPITAL:
Bangui
MAJOR LANGUAGE(S):
French, local dialects
POPULATION:
3,100,000
LANDMASS:
240,530 sq. mi./
622,973 sq. km
MONEY:
CFA Franc*
MAJOR EXPORT:
Diamonds
MAJOR IMPORT:
Food Products

Chad 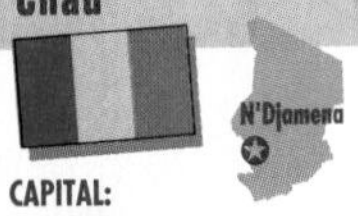
CAPITAL:
N'Djamena
MAJOR LANGUAGE(S):
French, Arabic
POPULATION:
6,500,000
LANDMASS:
486,180 sq. mi./
1,259,206 sq. km
MONEY:
CFA Franc*
MAJOR EXPORT:
Cotton
MAJOR IMPORT:
Petroleum Products

Comoros 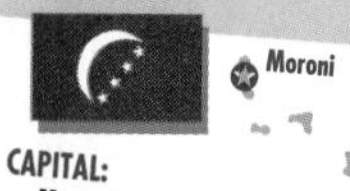
CAPITAL:
Moroni
MAJOR LANGUAGE(S):
Arabic, French
POPULATION:
500,000
LANDMASS:
860 sq. mi./
2,227 sq. km
MONEY:
CFA Franc*
MAJOR EXPORT:
Vanilla
MAJOR IMPORT:
Rice

Countries not drawn to scale.

UNIT 7

Facts

• **Ethiopia** According to tradition, Ethiopians are descendants of the Biblical rulers Solomon and the Queen of Sheba.

• **Cameroon** Telephones are rare and mail service is unreliable in Cameroon. People communicate by *radio trattoir*, or "pavement radio." This is a system of passing news by verbal relay.

Congo

CAPITAL:
Brazzaville
MAJOR LANGUAGE(S):
French, Kongo, Teke
POPULATION:
2,400,000
LANDMASS:
131,850 sq. mi./
341,492 sq. km
MONEY:
CFA Franc*
MAJOR EXPORT:
Petroleum
MAJOR IMPORT:
Machinery

Côte d'Ivoire

ADMINISTRATIVE CAPITAL:
Abidjan
MAJOR LANGUAGE(S):
French, Akan, Kru, Voltaic
POPULATION:
13,900,000
LANDMASS:
122,780 sq. mi./
318,000 sq. km
MONEY:
CFA Franc*
MAJOR EXPORT:
Food Products
MAJOR IMPORT:
Petroleum

Djibouti

CAPITAL:
Djibouti
MAJOR LANGUAGE(S):
French, Arabic
POPULATION:
600,000
LANDMASS:
8,950 sq. mi./
23,181 sq. km
MONEY:
Franc
MAJOR EXPORT:
Live Animals
MAJOR IMPORT:
Food and Beverages

Equatorial Guinea

CAPITAL:
Malabo
MAJOR LANGUAGE(S):
Spanish, Fang, Bubi
POPULATION:
400,000
LANDMASS:
10,830 sq. mi./
28,050 sq. km
MONEY:
CFA Franc*
MAJOR EXPORT:
Food and Live Animals
MAJOR IMPORT:
Food and Beverages

Eritrea

CAPITAL:
Asmara
MAJOR LANGUAGE(S):
Native Languages
POPULATION:
3,500,000
LANDMASS:
48,260 sq. mi./
124,993 sq. km
MONEY:
Ethiopian Birr
MAJOR EXPORT:
Not Available
MAJOR IMPORT:
Not Available

Ethiopia

CAPITAL:
Addis Ababa
MAJOR LANGUAGE(S):
Amharic, Tigre, Oromo
POPULATION:
55,200,000
LANDMASS:
376,830 sq. mi./
975,990 sq. km
MONEY:
Birr
MAJOR EXPORT:
Coffee
MAJOR IMPORT:
Machinery

Gabon

CAPITAL:
Libreville
MAJOR LANGUAGE(S):
French, Bantu dialects
POPULATION:
1,100,000
LANDMASS:
99,490 sq. mi./
257,679 sq. km
MONEY:
CFA Franc*
MAJOR EXPORT:
Petroleum
MAJOR IMPORT:
Machinery

Gambia

CAPITAL:
Banjul
MAJOR LANGUAGE(S):
English, Mandinka, Wolof
POPULATION:
1,100,000
LANDMASS:
3,860 sq. mi./
9,997 sq. km
MONEY:
Dalasi
MAJOR EXPORT:
Peanuts
MAJOR IMPORT:
Food

Ghana

CAPITAL:
Accra
MAJOR LANGUAGE(S):
English, Akan, Mossi, Ewe
POPULATION:
16,900,000
LANDMASS:
88,810 sq. mi./
230,018 sq. km
MONEY:
Cedi
MAJOR EXPORT:
Cocoa
MAJOR IMPORT:
Machinery

Guinea

CAPITAL:
Conakry
MAJOR LANGUAGE(S):
French, Peul, Mande
POPULATION:
6,400,000
LANDMASS:
94,930 sq. mi./
245,869 sq. km
MONEY:
Franc
MAJOR EXPORT:
Bauxite
MAJOR IMPORT:
Petroleum

* **Communauté Financière Africaine
(African Financial Community)**

479

Cultural Kaleidoscope

Cape Verde The *Funáná* is lively dance music with a strong beat. It was forbidden in Cape Verde before independence. The music survived, however, and it is now enjoyed throughout the country.

Teacher Notes

LESSON PLAN
Unit 7 Atlas

Geography and the Humanities

📖 World Literature Reading 7

🌎 World Music: Cultural Traditions, Lesson 6

📦 World Cultures Transparencies 11, 12

📕 Focus on World Art Print 12

Cultural Kaleidoscope

Ghana In Ghana, it is impolite and defiant for a child to look an adult in the eye.

Global 🍲 Gourmet

Ethiopia Many Ethiopians use *injera*—a pancake-shaped bread—to scoop up *wat*—a spicy stew.

UNIT 7 ATLAS
Countries at a Glance

Guinea-Bissau
CAPITAL: Bissau
MAJOR LANGUAGE(S): Portuguese, Crioulo, native languages
POPULATION: 1,100,000
LANDMASS: 10,860 sq. mi./ 28,127 sq. km
MONEY: Peso
MAJOR EXPORT: Cashews
MAJOR IMPORT: Transport Equipment

Malawi
CAPITAL: Lilongwe
MAJOR LANGUAGE(S): English, Chewa, Lomwe, Yao
POPULATION: 9,500,000
LANDMASS: 36,320 sq. mi./ 94,069 sq. km
MONEY: Kwacha
MAJOR EXPORT: Tobacco
MAJOR IMPORT: Machinery

Kenya
CAPITAL: Nairobi
MAJOR LANGUAGE(S): Swahili, English, Kikuyu, Luhya
POPULATION: 27,000,000
LANDMASS: 219,960 sq. mi./ 569,696 sq. km
MONEY: Shilling
MAJOR EXPORT: Tea
MAJOR IMPORT: Machinery

Mali
CAPITAL: Bamako
MAJOR LANGUAGE(S): French, Bambara, Senufo
POPULATION: 9,100,000
LANDMASS: 471,120 sq. mi./ 1,220,201 sq. km
MONEY: CFA Franc*
MAJOR EXPORT: Cotton Products
MAJOR IMPORT: Machinery

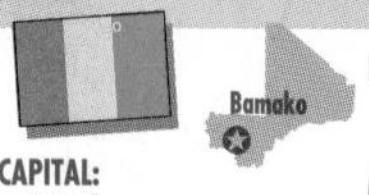

Lesotho
CAPITAL: Maseru
MAJOR LANGUAGE(S): English, Sotho
POPULATION: 1,900,000
LANDMASS: 11,720 sq. mi./ 30,355 sq. km
MONEY: Loti
MAJOR EXPORT: Machinery
MAJOR IMPORT: Clothing

Mauritania
CAPITAL: Nouakchott
MAJOR LANGUAGE(S): Arabic, French, Hassanya Arabic
POPULATION: 2,300,000
LANDMASS: 395,840 sq. mi./ 1,025,226 sq. km
MONEY: Ouguiya
MAJOR EXPORT: Fish
MAJOR IMPORT: Machinery

Liberia
CAPITAL: Monrovia
MAJOR LANGUAGE(S): English, native dialects
POPULATION: 2,900,000
LANDMASS: 37,190 sq. mi./ 96,322 sq. km
MONEY: Dollar
MAJOR EXPORT: Iron Ore
MAJOR IMPORT: Machinery

Mauritius
CAPITAL: Port Louis
MAJOR LANGUAGE(S): English, French Creole, Bhojpuri
POPULATION: 1,100,000
LANDMASS: 710 sq. mi./ 1,839 sq. km
MONEY: Rupee
MAJOR EXPORT: Clothing and Textiles
MAJOR IMPORT: Machinery

Madagascar
CAPITAL: Antananarivo
MAJOR LANGUAGE(S): Malagasy, French
POPULATION: 13,700,000
LANDMASS: 224,530 sq. mi./ 581,533 sq. km
MONEY: Franc
MAJOR EXPORT: Coffee
MAJOR IMPORT: Machinery

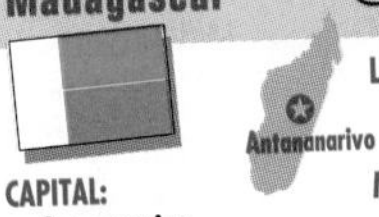

Mozambique
CAPITAL: Maputo
MAJOR LANGUAGE(S): Portuguese, Makua, Malawl
POPULATION: 15,800,000
LANDMASS: 302,740 sq. mi./ 784,097 sq. km
MONEY: Metical
MAJOR EXPORT: Shrimp
MAJOR IMPORT: Food

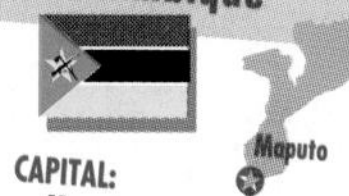

Countries not drawn to scale.

480

Fun Facts

- **Niger** Students in Niger get their teacher's attention by snapping their fingers rather than raising their hands.

- **Guinea** At family celebrations in Guinea, *griots*, or traditional singers, are hired to sing about individual guests. Each song includes the person's name and something about his or her appearance or character.

- **Madagascar** During the 1600s and 1700s Madagascar was used as a base by pirates. Among them was the notorious Scottish pirate known as Captain Kidd.

Namibia

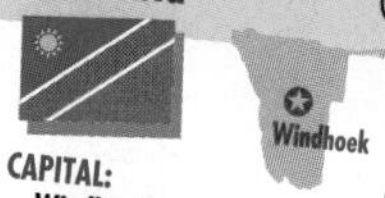

CAPITAL:
Windhoek
MAJOR LANGUAGE(S):
Afrikaans, English
POPULATION:
1,600,000

LANDMASS:
317,870 sq. mi./
823,283 sq. km
MONEY:
South African Rand
MAJOR EXPORT:
Minerals
MAJOR IMPORT:
Petroleum Products

Niger

CAPITAL:
Niamey
MAJOR LANGUAGE(S):
French, Hausa, Fulani
POPULATION:
8,800,000

LANDMASS:
489,070 sq. mi./
1,266,691 sq. km
MONEY:
CFA Franc*
MAJOR EXPORT:
Uranium
MAJOR IMPORT:
Machinery

Nigeria

CAPITAL:
Abuja
MAJOR LANGUAGE(S):
English, Hausa,
Yoruba, Ibo
POPULATION:
98,100,000

LANDMASS:
351,650 sq. mi./
910,774 sq. km
MONEY:
Naira
MAJOR EXPORT:
Petroleum
MAJOR IMPORT:
Machinery

Rwanda

CAPITAL:
Kigali
MAJOR LANGUAGE(S):
French, Rwanda
POPULATION:
7,700,000

LANDMASS:
9,530 sq. mi./
24,683 sq. km
MONEY:
Franc
MAJOR EXPORT:
Coffee and Tea
MAJOR IMPORT:
Machinery

São Tomé and Príncipe

CAPITAL:
São Tomé
MAJOR LANGUAGE(S):
Portuguese
POPULATION:
100,000

LANDMASS:
370 sq. mi./
958 sq. km
MONEY:
Dobra
MAJOR EXPORT:
Cocoa
MAJOR IMPORT:
Food

Senegal

CAPITAL:
Dakar
MAJOR LANGUAGE(S):
French, Wolof, Serer, Peul
POPULATION:
8,200,000

LANDMASS:
74,340 sq. mi./
192,541 sq. km
MONEY:
CFA Franc*
MAJOR EXPORT:
Peanut Oil
MAJOR IMPORT:
Machinery

Seychelles

CAPITAL:
Victoria
MAJOR LANGUAGE(S):
Portuguese
POPULATION:
400,000

LANDMASS:
1,560 sq. mi./
4,040 sq. km
MONEY:
Escudo
MAJOR EXPORT:
Bananas
MAJOR IMPORT:
Food

Sierra Leone

CAPITAL:
Freetown
MAJOR LANGUAGE(S):
English, native languages
POPULATION:
4,600,000

LANDMASS:
27,650 sq. mi./
71,614 sq. km
MONEY:
Leone
MAJOR EXPORT:
Rutile
MAJOR IMPORT:
Food and Live Animals

Somalia

CAPITAL:
Mogadishu
MAJOR LANGUAGE(S):
Somali, Arabic
POPULATION:
9,800,000

LANDMASS:
242,220 sq. mi./
627,350 sq. km
MONEY:
Shilling
MAJOR EXPORT:
Live Animals
MAJOR IMPORT:
Petroleum

South Africa

ADMINISTRATIVE CAPITAL:
Pretoria
MAJOR LANGUAGE(S):
Afrikaans, English, Nguni
POPULATION:
41,200,000

LANDMASS:
471,440 sq. mi./
1,221,030 sq. km
MONEY:
Rand
MAJOR EXPORT:
Gold
MAJOR IMPORT:
Machinery

* **Communauté Financière Africaine
(African Financial Community)**

481

MULTICULTURAL PERSPECTIVE

Cultural Heritage

The flag of Mauritania reflects the country's position as a link between North Africa and West Africa. The color green and the star and crescent honor Mauritania's connection to Islamic North Africa. The color yellow signifies its ties to the African nations south of the Sahara.

Nigeria In general, Nigerians do not eat much meat. Their diet consists mainly of yams, corn, beans, and rice. Nigerians usually cook their food in palm or peanut oil and often spice it with hot red pepper.

Teacher Notes

LESSON PLAN
Unit 7 Atlas

UNIT 7 ATLAS

Countries at a Glance

Global Gourmet

Sudan Most Sudanese eat little meat. Their main dish is *ful*—beans cooked in oil. The national drink is *karkadai*, which is made from hibiscus plants.

MULTICULTURAL PERSPECTIVE

Culturally Speaking

In the 1970s, Zaire's President Mobutu Sese Seko tried to reduce European influence in his country with an "Africanization" campaign. He ordered the people to adopt names of African origin, and he banned the wearing of European-style clothes. The ban on European clothing styles was not lifted until 1990.

Cultural Kaleidoscope

Mali Malians never use the left hand to offer or accept food or money. When shaking hands, a Malian shows special respect by touching his or her own right elbow with the fingers of the left hand.

Sudan

Khartoum

CAPITAL:
Khartoum
MAJOR LANGUAGE(S):
Arabic, Dinka, Nubian, Nuer
POPULATION:
28,200,000

LANDMASS:
967,500 sq. mi./
2,505,813 sq. km
MONEY:
Dinar
MAJOR EXPORT:
Cotton
MAJOR IMPORT:
Machinery

Swaziland

Mbabane

CAPITAL:
Mbabane
MAJOR LANGUAGE(S):
Swazi, English
POPULATION:
800,000

LANDMASS:
6,640 sq. mi./
17,198 sq. km
MONEY:
Lilangeni
MAJOR EXPORT:
Sugar
MAJOR IMPORT:
Machinery

Tanzania

Dar es Salaam

CAPITAL:
Dar es Salaam (seat of govt.)
Dodoma (capital designate)
MAJOR LANGUAGE(S):
Swahili, English
POPULATION:
29,800,000

LANDMASS:
342,100 sq. mi./
886,039 sq. km
MONEY:
Shilling
MAJOR EXPORT:
Coffee
MAJOR IMPORT:
Machinery

Togo

Lomé

CAPITAL:
Lomé
MAJOR LANGUAGE(S):
French, Gur & Kwa languages
POPULATION:
4,300,000

LANDMASS:
21,000 sq. mi./
54,390 sq. km
MONEY:
CFA Franc*
MAJOR EXPORT:
Calcium Phosphates
MAJOR IMPORT:
Machinery

Uganda

Kampala

CAPITAL:
Kampala
MAJOR LANGUAGE(S):
English, Luganda, Swahili
POPULATION:
19,800,000

LANDMASS:
77,050 sq. mi./
199,560 sq. km
MONEY:
Shilling
MAJOR EXPORT:
Coffee
MAJOR IMPORT:
Sugar

Zaire

Kinshasa

CAPITAL:
Kinshasa
MAJOR LANGUAGE(S):
French, Kongo, Luba, Mongo
POPULATION:
42,500,000

LANDMASS:
875,520 sq. mi./
2,267,597 sq. km
MONEY:
Zaire
MAJOR EXPORT:
Copper
MAJOR IMPORT:
Machinery

Zambia

Lusaka

CAPITAL:
Lusaka
MAJOR LANGUAGE(S):
English, Bantu dialects
POPULATION:
9,100,000

LANDMASS:
287,020 sq. mi./
743,382 sq. km
MONEY:
Kwacha
MAJOR EXPORT:
Copper
MAJOR IMPORT:
Machinery

Zimbabwe

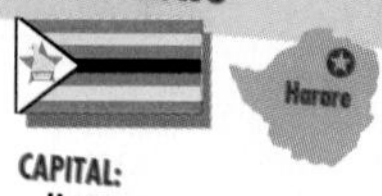

Harare

CAPITAL:
Harare
MAJOR LANGUAGE(S):
English, Shona, Sinde bele
POPULATION:
11,200,000

LANDMASS:
149,290 sq. mi./
386,661 sq. km
MONEY:
Dollar
MAJOR EXPORT:
Tobacco
MAJOR IMPORT:
Machinery

Countries not drawn to scale.

* Communauté Financière Africaine (African Financial Community)

Fun Facts

- **Namibia** The Welwitschia is a plant that grows in the Namib Desert. Its trunk spreads to a width of more than 5 feet (1.5 m). Two giant green leaves grow out from the trunk and split into ribbonlike shreds. The plants may live as long as 2,000 years!

- **Rwanda** Basket weaving is an important craft in Rwanda. Weavers use fibers from banana plants to make waterproof baskets. A family's social standing may be determined by the number and quality of baskets it owns.

The Sahel

A dust storm kicks up a huge wall of sand on the Sahel in northern Burkina Faso.

Bringing in the Nets in Accra, Ghana

A fisher pulls in his fishing nets from a harbor in Accra, Ghana's capital.

Drilling for Gold in South Africa

South African gold miners work with large drills and dynamite to extract gold ore.

483

More About the Illustration

Human/ Environment Interaction In addition to natural soil erosion caused by wind and rain, desertification caused by overgrazing and poor farming techniques is creating vast stretches of unusable land in the Sahel.

More About the Illustration

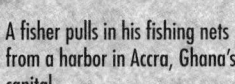
Human/ Environment Interaction Ghana's coastal waters in the Gulf of Guinea supply about 400,000 tons of fish each year. Even with this supply, the country must still import much of its food, including fish and meat.

More About the Illustration

Human/ Environment Interaction South Africa produces about one-third of all the gold in the world. Gold and other valuable minerals, such as diamonds, have contributed greatly to the country's wealth and economic development.

Teacher Notes

West Africa

PERFORMANCE ASSESSMENT ACTIVITY

WRITING A FABLE Point out that the telling of fables—tales that teach a moral or lesson—plays an important part in traditional West African culture. Encourage students to research West African fables. *The Intimate Folklore of Africa* by Alta Jablow and *The Black Cloth: A Collection of African Folktales* by Bernard Biblin Dadié are possible sources. Have students work in small groups to brainstorm possible "moral dilemmas" for fables set in West Africa. Ask groups also to list possible characters—animal or human—and locations for the fables. Then have students work individually to write one fable using ideas developed in the brainstorming session. Suggest that they illustrate their fables with sketches. Fables should include at least five references to the geography of the area. Collect all the fables and bind them together in the form of a book. Display the book in the classroom and invite other classes to view it.

POSSIBLE RUBRIC FEATURES:
Creative thinking, composition skills, analysis skills, content information, collaborative skills, application skills

MENTAL MAPPING ACTIVITY

Before teaching Chapter 18, point out the countries of West Africa on a globe. Have students note the location of the countries relative to one another, relative to the coast of Africa, and relative to the Equator.

TEACHER'S CORNER

NATIONAL GEOGRAPHIC SOCIETY

INDEX TO NATIONAL GEOGRAPHIC MAGAZINE

The following articles may be used for research relating to this chapter:

- "African Gold," by Carol Beckwith and Angela Fisher, October 1996.
- "Africa's Dinosaur Castaways," by Paul C. Sereno, June 1996.
- "The African Roots of Voodoo," by Carol Beckwith and Angela Fisher, August 1995.
- "Fantasy Coffins of Ghana," by Carol Beckwith and Angela Fisher, September 1994.
- "The Cruelest Commerce: African Slave Trade," by Colin Palmer, September 1992.

- "Below the Cliff of Tombs: Mali's Dogon," by David Roberts, October 1990.
- "Africa's Sahel: The Stricken Land," by William S. Ellis, August 1987.
- "Tsetse-Fly of the Deadly Sleep," by Georg Gerster, December 1986.
- "Finding West Africa's Oldest City," by Susan and Roderick McIntosh, September 1982.
- "The Desert: An Age-old Challenge Grows," November 1979.

NATIONAL GEOGRAPHIC SOCIETY PRODUCTS AVAILABLE FROM GLENCOE

To order the following products for use with this chapter, contact your local Glencoe sales representative or call Glencoe at 1-800-334-7344:

- *STV: World Geography* (Videodisc)
- *Picture Atlas of the World* (CD-ROM)
- *Physical Geography of the World* (Transparencies)
- *ZipZapMap! World* (Software)

- *GeoBee* (Software)
- *Images of the World* (Posters)
- *Eye on the Environment* (Posters)

ADDITIONAL NATIONAL GEOGRAPHIC SOCIETY PRODUCTS

To order the following products for use with this chapter, call National Geographic Society at 1-800-368-2728:

- *Lions of the African Night* (Video)
- *Physical Geography of the Continents Series:* "Africa." (Video)

- *Portraits of the Continents Series:* "Part III: Europe; Asia; Africa." (Filmstrip)

TEACHER-TO-TEACHER
Curriculum Connection: Social Studies

—from Debora F. Bittner
John Marshall School, San Antonio, TX

EXPORT CONNECTIONS FOR WEST AFRICA *The purpose of this activity is to help students see spatial connections among countries by constructing an export map for West Africa.*

Draw a large outline map of Africa on a sheet of butcher paper or poster board. Draw the boundaries of the West African countries and add country labels. Display the map prominently in the classroom. Then, using sources such as *Rand McNally World Facts & Maps: Concise International Review*, create a list of West African exports. Organize students into six groups. Assign equal numbers of West African exports to three groups. Ask these groups to research which West African countries export their assigned products. Have groups select symbols for their products and enter these symbols in the appropriate countries on the map. Assign equal numbers of West African countries to the other three groups. Direct these groups to research the leading trade partners of each of their assigned countries and to decide upon a way to enter this information on the map. Remind all groups that symbols will need to be explained in a map key.

As a follow-up activity, have students note the product exported by most West African countries and West Africa's three leading trade partners.

MEETING NATIONAL STANDARDS

Geography for Life

All of the 18 standards are demonstrated in Unit 7. The following ones are highlighted in this chapter:

• Standards 3, 5, 8, 9, 10, 13, 17

KEY TO ABILITY LEVELS

Teaching strategies have been coded for varying learning styles and abilities.

L1 **BASIC** activities for all students

L2 **AVERAGE** activities for average to above-average students

L3 **CHALLENGING** activities for above-average students

LEP **LIMITED ENGLISH PROFICIENCY** activities

BIBLIOGRAPHY

Readings for the Student

Koslow, Philip. *Centuries of Greatness, 750–1900: The West African Kingdoms.* New York: Chelsea House, 1995.

Murray, Jocelyn. *Africa: Cultural Atlas for Young People.* New York: Facts on File, 1990.

Ramsay, F. Jeffress. *Africa: Global Studies,* 6th ed. Guilford, Conn.: Dushkin, 1995.

Readings for the Teacher

Africa Today: A Reproducible Atlas. Wellesley, Mass.: World Eagle, 1994.

Grove, A.T. *The Changing Geography of Africa.* New York: Oxford University Press, 1990.

Jackson, John G. *Introduction to African Civilizations.* New York: Citadel, 1990.

Map Skills and Outlines: Africa. St. Louis: Milliken, 1991. Transparencies and reproducible worksheets.

Multimedia

Africa: Continent of Contrasts. Huntsville, Tex.: Educational Video Network, 1994. Videocassette, 34 minutes.

Africa: Western Region. Chicago: Encyclopedia Britannica, 1993. Videocassette, approximately 20 minutes.

Global Studies With Country Databases: Africa. Toronto, Canada: WorldView Software, 1994. Apple 5.25" disk or IBM 3.5" disk.

Nigeria. VIDEA, 1991. Videocassette, 28 minutes.

Chapter 18

Resource Organizer
West Africa

CHAPTER 18 RESOURCES

*Vocabulary Activity 18

Cooperative Learning Activity 18

Chapter Map Activity 18

Geography Skills Activity 18

GeoLab Activity 18

Critical Thinking Skills Activity 18

Performance Assessment Activity 18

*Reteaching Activity 18

Enrichment Activity 18

Environmental Case Study 7

Chapter 18 Test, Form A and Form B

*Chapter 18 Digest Audiocassette, Activity, Test

Unit Overlay Transparencies 7-0, 7-1, 7-2, 7-3, 7-4, 7-5, 7-6

Political Map Transparency 7

Geoquiz Transparencies 18-1, 18-2

Vocabulary PuzzleMaker Software

Student Self-Test: A Software Review

Testmaker Software

MindJogger Videoquiz

Use these Geography: The World and Its People resources to teach, reinforce, and extend chapter content.

If time does not permit teaching the entire chapter, summarize using the Chapter 18 Highlights on page 499, and the Chapter 18 English (or Spanish) Audiocassettes. Review students' knowledge using the Glencoe MindJogger Videoquiz. *Also available in Spanish

*Use these **Geography: The World and Its People** resources to teach and reinforce section content.*

SECTION 1 RESOURCES

- Reproducible Lesson Plan 18-1
- Section Focus Transparency 18-1
- Guided Reading Activity 18-1
- Section Quiz 18-1 (also available in Spanish)

SECTION 2 RESOURCES

- Reproducible Lesson Plan 18-2
- Section Focus Transparency 18-2
- Guided Reading Activity 18-2
- Section Quiz 18-2 (also available in Spanish)

SECTION 3 RESOURCES

- Reproducible Lesson Plan 18-3
- Section Focus Transparency 18-3
- Guided Reading Activity 18-3
- Section Quiz 18-3 (also available in Spanish)

ADDITIONAL RESOURCES FROM GLENCOE

Reproducible Masters

- Glencoe Social Studies Outline Map Book, pages 36, 37
- Facts On File: GEOGRAPHY ON FILE, Sub-Saharan Africa; AFRICA HISTORY ON FILE
- Foods Around the World, Unit 7

World Music: Cultural Traditions

Lesson 6

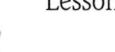

Transparencies

- National Geographic Society PicturePack Transparencies, Unit 7

Laminated Desk Map

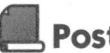

Posters

- National Geographic Society: Images of the World, Unit 7

World Games Activity Cards

Unit 7

Videodiscs

- Reuters Issues in Geography
- National Geographic Society: STV: World Geography, Vol. 2

CD-ROM

- National Geographic Society: Picture Atlas of the World
- National Geographic Society PictureShow: Earth's Endangered Environments

Software

- National Geographic Society: ZipZapMap! World

Performance Assessment

Refer to the Planning Guide on page 484A for a Performance Assessment Activity for this chapter. See the *Performance Assessment Strategies and Activities* booklet for suggestions.

Chapter Objectives

1. **Identify** the physical features of the countries of West Africa.
2. **Relate** West Africa's natural resources to its economic development.
3. **Describe** the impact of historical influences on West Africa's cultures.

GLENCOE
TECHNOLOGY

 **VIDEODISC**

Use the Chapter 18 MindJogger Videoquiz to preview chapter content.

MindJogger Videoquiz

Chapter 18
Disc 3 Side A

 Also available in VHS.

Chapter 18 West Africa

ATLANTIC OCEAN

Tropic of Cancer

NORTH AFRICA

30°N

CAPE VERDE ISLANDS
• Nouakchott
MAURITANIA
MALI
20°N

• Praia   Dakar
SENEGAL   • Bamako
NIGER
CHAD

Banjul
GAMBIA   • Bissau
BURKINA FASO
• Niamey
Lake Chad

GUINEA-BISSAU   GUINEA
• Ouagadougou
BENIN
• Abuja
• N'Djamena

Conakry   • Freetown
CÔTE D'IVOIRE   GHANA   TOGO
NIGERIA
10°N

SIERRA LEONE
• Lomé

Monrovia   • Accra   Porto-Novo
LIBERIA   Abidjan
Gulf of Guinea

Equator

CENTRAL AFRICA
0°

- - - Disputed boundary
— National boundary
⊛ National capital
• Other city

Lambert Equal-Area projection

N
W — E
S

0   250   500 mi.
0   250   500 km

20°W   10°W   0°   10°E   20°E

 MAP STUDY ACTIVITY

In this chapter you will read about Nigeria and the other countries of West Africa.
1. **How many countries make up West Africa?**
2. **What West African country is made up of islands?**
3. **What body of water do the coastal countries border?**

MAP STUDY ACTIVITY

Answers
1. 17
2. Cape Verde Islands
3. Atlantic Ocean or Gulf of Guinea

Map Skills Practice
Reading a Map What is the capital of Nigeria? *(Abuja)* Which countries border Ghana? *(Côte d'Ivoire, Burkina Faso, Togo)*

Nigeria

PREVIEW

Words to Know
- mangrove
- savanna
- harmattan
- cacao
- ethnic group
- compound

Places to Locate
- Nigeria
- Gulf of Guinea
- Niger River
- Lagos
- Abuja

Read to Learn . . .
1. what landforms are found in Nigeria.
2. what mineral supports Nigeria's economy.
3. why Nigeria has many ethnic groups.

Alaska! The Niger River is a major "highway" for those who live along it.

The mighty Niger River flows about 2,600 miles (4,180 km) through the western part of Africa. It is the third-longest river in Africa and drains an area larger than the state of

The West African country of Nigeria takes its name from the Niger River, which flows through western and central Nigeria. One of the largest nations in Africa, Nigeria covers 351,650 square miles (910,774 sq. km). It is more than twice the size of the state of California.

Place
The Land

Imagine taking a trip through Nigeria from south to north. Nigeria has a long coastline on the Gulf of Guinea. The map on page 484 shows you that this gulf is part of the Atlantic Ocean. Along Nigeria's coast, the land is laced with many rivers and creeks and covered with mangrove swamps. A **mangrove** is a tropical tree with roots extending above and beneath the water.

As you travel north, the land becomes a vast tropical rain forest. Small villages and farms appear in only a few clearings. The forests gradually thin into woodlands and savannas in central Nigeria. **Savannas** are tropical grasslands with scattered trees. Highlands and plateaus also blanket this area. Farther north, a stretch of partly dry grasslands lies on the edge of the vast Sahara.

CHAPTER 18

485

Section Objectives
1. **Describe** the landforms that are found in Nigeria.
2. **Identify** the mineral that supports Nigeria's economy.
3. **Explain** why Nigeria has many ethnic groups.

Bellringer Motivational Activity

Prior to taking roll at the beginning of the class period, project Section Focus Transparency 18-1 and have students answer the activity questions. Discuss students' responses. L1

Vocabulary Pre-check

Have students find the definition of *cacao*. Then ask them what similar word they might use for hot chocolate. *(cocoa)* Point out that a spelling mistake, probably made by English exporters, resulted in the different words. L1 LEP

Use the Vocabulary PuzzleMaker Software to create a crossword puzzle. L1

Classroom Resources for Section 1

 **REPRODUCIBLE MASTERS**

Reproducible Lesson Plan 18-1
Guided Reading Activity 18-1
Vocabulary Activity 18
Geography Skills Activity 18
Critical Thinking Skills
  Activity 18
Section Quiz 18-1

TRANSPARENCIES

Section Focus
  Transparency 18-1
Political Map Transparency 7

MULTIMEDIA

 Vocabulary PuzzleMaker
Software

 Testmaker

MAP STUDY

Answer
Gulf of Guinea

Map Skills Practice

Reading a Map Where are the highest elevations found in Nigeria? *(north central Nigeria)*

TEACH

Guided Practice

Flow Chart Have students discuss how history affected ways of life in Nigeria. Suggest that they begin with the arrival of Europeans in the late 1400s and end with British rule in the 1900s. Ask students to pay close attention to how European influences affected ethnic groups, religions, and languages in Nigeria. Then have students summarize the discussion in the form of a flow chart. **L1**

Independent Practice

Guided Reading Activity 18-1 **L1**

MULTICULTURAL PERSPECTIVE

Culturally Speaking
More than 250 languages are spoken in Nigeria. Although English is the official language, fewer than half of all Nigerians speak it fluently.

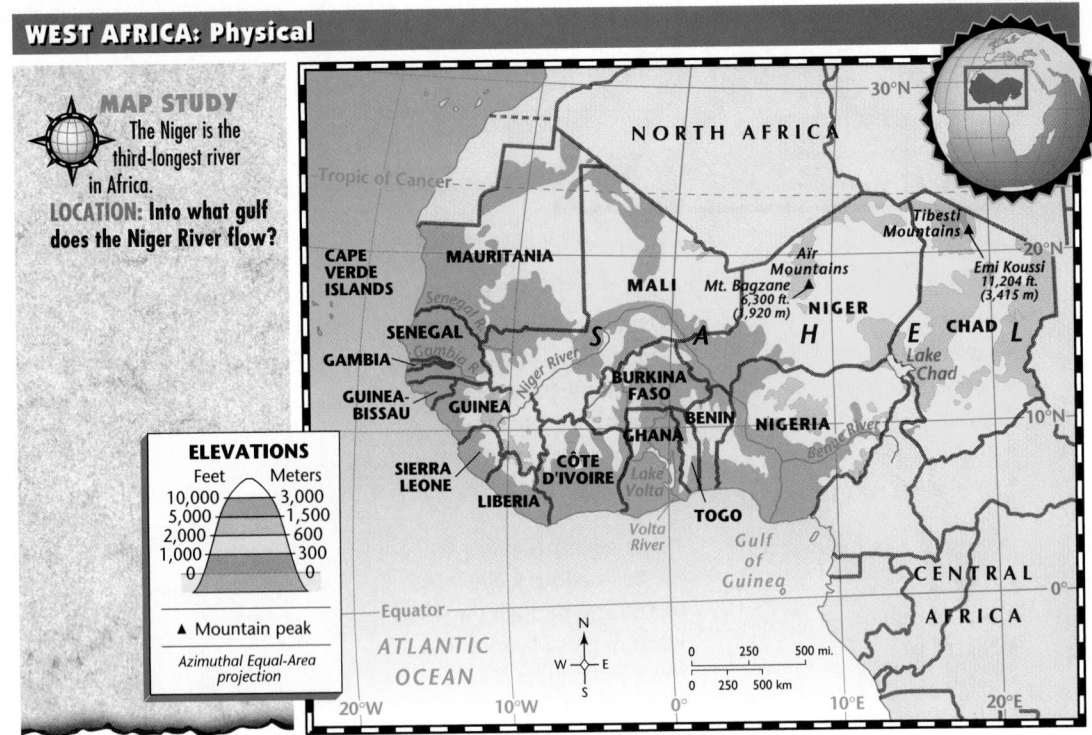

WEST AFRICA: Physical

MAP STUDY
The Niger is the third-longest river in Africa.
LOCATION: Into what gulf does the Niger River flow?

ELEVATIONS

Feet	Meters
10,000	3,000
5,000	1,500
2,000	600
1,000	300

▲ Mountain peak

Azimuthal Equal-Area projection

The Climate The map on page 491 shows you that Nigeria has several climate regions. Most of the country, however, is tropical with high average temperatures. Tropical areas in the south receive plenty of rainfall year-round. Steppe regions to the north have a distinct dry season. In the winter months, a dry, dusty wind called the **harmattan** blows south from the Sahara.

Human/Environment Interaction
The Economy

Nigeria has a developing economy based on mining and farming. It is one of the world's major oil-producing countries and one of the few industrialized African countries. In recent years oil has become the leading source of income for the government. The Nigerian government has used money from oil exports to build factories that turn out textiles, food products, and chemicals. Coal and tin deposits also help boost the country's industrial growth.

Boom and Bust In the 1980s Nigeria's oil-rich economy faced hard times. World oil prices fell, and Nigeria's income declined. Nigerians also realized that, in focusing on industry, they had neglected agriculture. Many farmers had left their farms in search of jobs in the cities. Food production fell so much that Nigeria had to import food. Today Nigeria is using income from oil to improve farming and to raise more food products.

Cooperative Learning Activity

Organize students into five groups and assign one of the following topics to each group: physical features, climate, agriculture, minerals, population distribution. Provide an outline map of Nigeria for each group. Instruct group members to create a map illustrating their assigned topic. Then have groups compare their maps and formulate generalizations about relationships between Nigeria's physical geography and cultural geography. **L1**

Agriculture If you like chocolate, you may have Nigeria to thank. Nigeria's major crops include peanuts, cotton, rubber, palm products, and **cacao.** Beans from the cacao, or cocoa, pod are used to make chocolate. Despite their country's mineral riches, most Nigerians earn their livings as farmers. More than one-half of Nigeria's land is good for farming. Some Nigerian farmers grow subsistence crops, or crops only for their own use. Others produce cash crops, or crops to sell for export.

Place
The People

Like many African countries, Nigeria has a high birthrate and a rapidly growing population. About 98 million people live in Nigeria—more people than in any other country in Africa.

Ethnic Groups Nigeria's people belong to more than 250 ethnic groups. An **ethnic group** is a group of people that has a common language, culture, and history. The four largest ethnic groups are the Hausa (HOW•suh), Fulani (FOO•LAH•nee), Yoruba (YAWR•uh•buh), and Ibo (EE•boh).

Nigerians speak many different African languages. They use English, however, in business and government affairs. About half of Nigeria's people are Muslims and another 40 percent are Christians. The remaining 10 percent practice traditional African religions.

Lagos, Nigeria

Lagos is Nigeria's chief port and commercial center.
PLACE: What language do most Nigerians use when conducting business?

LESSON PLAN
Chapter 18, Section 1

More About the Illustration
Answer to Caption
English

GRAPHIC THEMES

Location The city of Lagos spreads across four islands in the Gulf of Guinea and spills over onto mainland Africa. Lagos is Nigeria's major industrial center and was, until 1991, the country's capital.

ASSESS

Check for Understanding
Assign Section 1 Review as homework or an in-class activity.

Meeting Lesson Objectives
Each objective below is tested by the questions that follow it in parentheses.
1. **Describe** the landforms that are found in Nigeria. (2, 6)
2. **Identify** the mineral that supports Nigeria's economy. (3)
3. **Explain** why Nigeria has many ethnic groups. (1)

Leading Cocoa-Producing Countries

GRAPHIC STUDY
Half of the world's top cocoa-producing countries are in Africa south of the Sahara.
PLACE: What African country leads the world in cocoa production?

Cocoa Beans Produced in One Year

Thousands of metric tons: 0, 100, 200, 300, 400, 500, 600, 700, 800
Thousands of short tons: 0, 110, 220, 331, 441, 551, 661, 771, 882

Côte d'Ivoire, Brazil, Ghana, Malaysia, Indonesia, Nigeria

COCOA XXX = 200,000 metric tons

Source: FAO Production Yearbook, 1992

GRAPHIC STUDY
Answer
Côte d'Ivoire

Graphic Skills Practice
Reading a Graph How much tonnage of cocoa beans is produced by Nigeria each year? *(about 145,000 short tons, or about 132,000 metric tons)*

Meeting Special Needs Activity

Reading Disability Students with reading problems sometimes have difficulty locating information necessary to answer questions. Point out that some answers are "right there," or stated in the text. Other questions require the reader to "think and search," or put together information from different parts of the text. Help students differentiate these two question/answer relationships (QAR) in the subheading "The Land."

Ask, "Where is Nigeria's coastline?" The answer—on the Gulf of Guinea—is "right there." Then ask, "What kinds of landforms would you see on a trip from southern to northern Nigeria?" The answer—mangrove swamps, rain forests, savannas, highlands, plateaus, dry grasslands—requires students to "think and search" for the information in the next two paragraphs. **L1**

CURRICULUM CONNECTION

History In 1967, the Ibo seceded from Nigeria, setting up their own country, Biafra. A bitter civil war followed, which claimed the lives of a million people. After more than two years of fighting, the Biafrans surrendered and agreed to rejoin Nigeria.

Evaluate

Section 1 Quiz **L1**

Use the Testmaker to create a customized quiz for Section 1. **L1**

Reteach

Have students work individually or in small groups to create a country profile of Nigeria. They should summarize facts under the following headings: *Landscapes, Climates, Resources and Industry, Agriculture,* and *People.* **L1**

Enrich

Have students create an exhibit of Nigerian arts. They might photocopy pictures of traditional art objects, or create masks or cloth in Nigerian styles and colors. **L3**

CLOSE

Have students write at least three questions they would like to ask Elizabeth Tofa about her way of life. Then have them write how their lives differ from that of a teenager in rural Nigeria.

488

Influences of the Past The history of Nigeria reaches farther back than that of almost any other in Africa. About 500 B.C., the Nok people hammered iron into tools and made baked clay sculptures. During the following centuries, powerful city-states and kingdoms arose throughout the region and became centers of trade and the arts. Territories in the north came under the influence of Islamic cultures from North Africa. Peoples in the south developed cultures based on traditional African religions.

Europeans looking for gold and slave labor arrived in Africa in the 1400s. They set up trading forts in Nigeria as early as 1498. By the early 1900s, the British controlled the area that is now Nigeria. The many ethnic groups there resisted British rule. On October 1, 1960, the colony of Nigeria finally became an independent nation. The ethnic differences that intensified during the time of European rule later erupted in civil war. Today Nigeria is under harsh military rule. In response, many foreign countries refuse to trade with Nigeria.

Ways of Life About 84 percent of Nigeria's population live in rural villages. Fourteen-year-old Elizabeth Tofa lives in a small village in western Nigeria. Like most rural Nigerians, Elizabeth and her family live in a **compound,** or a group of houses surrounded by walls. Every four days there is a market in her village. Run by women, markets are important in Nigeria and the rest of Africa. They not only provide a variety of goods but also a chance for friends to meet. Elizabeth, her mother, and her sisters sell meat, palm oil, cloth, yams, and nuts.

Nigeria has several large cities bustling with activity. Lagos, located on the coast, is the largest city and commercial center. Other cities—such as Ibadan (ih•BAH•duhn) and Abuja (ah•BOO•jah), the national capital—lie inland.

Nigerians take pride in their art. They make elaborate wooden masks, metal sculptures, and colorful cloth. Playing drums, horns, and other instruments, musicians create rhythms with two or more patterns going at once. In the past, Nigerians passed on stories, proverbs, and riddles by word of mouth from one generation to the next. During the mid-1900s, however, many Nigerian authors began to publish novels, stories, and poetry. In 1986 Nigerian writer Wole Soyinka (WAH•lay shaw•YIHNG•kah) became the first African to win the Nobel Prize for literature.

SECTION 1 REVIEW

REVIEWING TERMS AND FACTS

1. **Define the following:** mangrove, savanna, harmattan, cacao, ethnic group, compound.
2. **PLACE** How does Nigeria's landscape change from south to north?
3. **HUMAN/ENVIRONMENT INTERACTION** What is Nigeria's most important mineral resource?
4. **PLACE** Why are village markets important in Nigeria?

MAP STUDY ACTIVITIES

5. Turn to the political map on page 484. What is Nigeria's capital?
6. Look at the physical map on page 486. Most of northern Nigeria lies at what elevation?

488

UNIT 7

Answers to Section 1 Review

1. All vocabulary words are defined in the Glossary.
2. Coastal areas are lowlands with mangrove swamps, rivers, and creeks; thick rain forests cover central Nigeria. Farther north, the land is covered with savanna. The very northern area is partly dry grassland on the edge of the Sahara.
3. oil
4. Markets provide a variety of goods and enable people to meet.
5. Abuja
6. 1,000 to 2,000 feet (300 to 600 m)

BUILDING GEOGRAPHY SKILLS

Interpreting a Diagram

Have you ever tried to put something together without directions or even a picture to follow? A diagram can be very helpful in cases like this. A *diagram* is a drawing that shows how something works or how its parts fit together.

In most diagrams, labels point out parts or steps in a process. Labels such as words, letters, colors, or numbers identify different parts. Arrows sometimes show the order in which steps follow one another or the direction of an object's movement.

To interpret a diagram, apply the following steps:

- Read the title or caption to find out what the diagram shows.
- Read all labels to determine their meanings.
- Look for arrows that show movement or the order of the steps to be completed.

Sahara Landforms

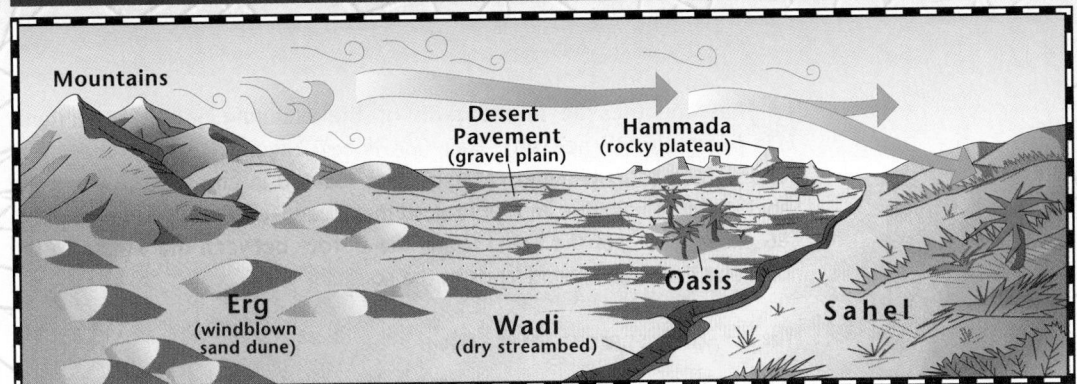

Mountains · Desert Pavement (gravel plain) · Hammada (rocky plateau) · Erg (windblown sand dune) · Wadi (dry streambed) · Oasis · Sahel

Geography Skills Practice

1. What is the title of this diagram?

2. What labels are used?

3. What do the arrows in the diagram show?

4. According to the diagram, how is the Sahel landscape different from the desert landscape?

CHAPTER 18

 489

Answers to Geography Skills Practice

1. Sahara Landforms
2. mountains, erg, desert pavement, hammada, oasis, wadi, Sahel
3. the movement of deserts into grassland areas
4. Sahel landscape includes some vegetation. Desert landscape, except for oases, lacks vegetation.

TEACH

Have students imagine that they have bought a new desk. The desk arrives from the store packaged in a box labeled "Assembly Required." Ask them how they will figure out how to assemble the desk. Point out that such packages usually include instructions for assembly, and that these instructions often take the form of diagrams. Then explain that a diagram is a drawing that shows the steps in a process, or the way parts fit together.

Next, show students examples of diagrams from cookbooks, household repair manuals, sewing patterns, and so on. Once students have familiarized themselves with the many different kinds of diagrams, have them read the skill and complete the practice questions. **L1**

Additional Skills Practice

1. When reading a diagram, how do you identify the parts of the object? *(by viewing the labels)*
2. How can you determine the sequence of steps shown in a diagram? *(numbers or letters in sequential order, arrows showing direction)*

▶ Geography Skills Activity 18

SECTION 2
The Sahel Countries

Meeting National Standards

Geography for Life The following standards are highlighted in this section:
• Standards 4, 7, 14

FOCUS

Section Objectives

1. **Show** where the Sahel is located.
2. **Explain** why Sahel grasslands are turning into desert areas.
3. **Describe** how people in the Sahel countries live.

Bellringer Motivational Activity

Prior to taking roll at the beginning of the class period, project Section Focus Transparency 18-2 and have students answer the activity questions. Discuss students' responses. **L1**

Vocabulary Pre-check

Have students define the two vocabulary terms. Then ask them to formulate a sentence that shows a cause/effect relationship between the terms. **L1 LEP**

TEACH

Guided Practice

Display Political Map Transparency 7. Have students locate the Sahel countries relative to the Sahara.

(continued)

PREVIEW

Words to Know
• drought
• desertification

Places to Locate
• the Sahel
• Niger River
• Mauritania
• Mali
• Niger
• Burkina Faso
• Chad

Read to Learn . . .
1. where the Sahel is located.
2. why Sahel grasslands are turning into desert areas.
3. how people in the Sahel countries live.

Slowly but surely, the desert is creeping into grassy inland areas of West Africa north of Nigeria. Over the past 100 years, a stretch of the Sahara more than 93 miles (150 km) wide has swallowed parts of countries in West Africa. This growing desert is like an invading army slowly taking over the countries.

What countries lie in the path of the creeping Sahara? Mauritania (MAWR•uh•TAY•nee•uh), Mali (MAH•lee), Niger (NY•juhr), Chad, and Burkina Faso (bur•KEE•nuh FAH•soh) are located in an area known as the Sahel. *Sahel* comes from an Arabic word that means "border." The map on page 486 shows you that the Sahel forms the border between the Sahara to the north and the fertile, humid lands to the south.

Human/Environment Interaction
The Growing Desert

The Sahel receives little rainfall, so only short grasses and small trees are able to grow here. As you might guess, this type of vegetation is good for grazing animals. Most people in the Sahel are livestock herders. Their flocks, unfortunately, have overgrazed the land, stripping areas so bare that plant life does not grow back. Without plants, the bare soil is blown away by winds.

To make matters worse, the Sahel has entered a period of **drought**. A drought is an extreme shortage of water. Droughts are common in the Sahel,

490

UNIT 7

Classroom Resources for Section 2

REPRODUCIBLE MASTERS
Reproducible Lesson Plan 18-2
Guided Reading Activity 18-2
Chapter Map Activity 18
Cooperative Learning Activity 18
GeoLab Activity 18
Section Quiz 18-2

TRANSPARENCIES
Section Focus Transparency 18-2
Unit Overlay Transparencies 7-0, 7-1, 7-2

MULTIMEDIA
 Testmaker

 National Geographic Society: ZipZapMap! World

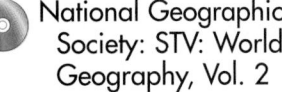 National Geographic Society: STV: World Geography, Vol. 2

MAP STUDY

West Africa's climates are affected by the Sahara, the Atlantic Ocean, and closeness to the Equator.

PLACE: What is the major climate type of the countries in the northern part of West Africa?

CLIMATE

Tropical
- Tropical rain forest
- Tropical savanna

Dry
- Steppe (partly dry)
- Desert

Azimuthal Equal-Area projection

Then ask them to predict possible problems these countries face because of their location. Record predictions and have students check their accuracy after completing study of Section 2. **L1**

MAP STUDY

Answer
desert

Map Skills Practice
Reading a Map Which countries of the Sahel have tropical climate zones? *(Mali, Burkina Faso)*

Independent Practice

Guided Reading Activity 18-2 **L1**

where dry and wet periods regularly follow each other. The latest period of drought began in the 1980s. Since then rivers have dried up, crops have failed, and millions of cattle and other animals have died. Thousands of people have died of hunger. Millions more have fled to fertile southern areas in search of food.

Over the years, both overgrazing and drought have ruined once-productive areas of the Sahel. Many grassland areas have become desert—a change called **desertification.** Scientists believe that the increasing human use of land in the Sahel will only increase the spread of desert. You can read more about desertification south of the Sahara on page 568.

Place
People of the Sahel

The countries of the Sahel are large in size but have small populations. Look at the population density map on page 496. It shows you that most people live in the southern areas of these countries. Can you guess why? Rivers flow here, and the land can be farmed or grazed. Even these areas do not have enough water and fertile land to support large numbers of people, however.

WHAT IN THE WORLD?

The Wet Sahara?
The Sahara was not always a desert. Many caves in Mali have paintings that date back 10,000 years. The paintings show hippos, rhinos, giraffes, crocodiles, and even fish. These drawings suggest that at least the southern part of the Sahara had a wet climate in the distant past.

ASSESS

Check for Understanding
Assign Section 2 Review as homework or an in-class activity.

Meeting Lesson Objectives
Each objective below is tested by the questions that follow it in parentheses.

1. **Show** where the Sahel is located. (2)
2. **Explain** why Sahel grasslands are turning into desert areas. (1, 3)
3. **Describe** how people in the Sahel countries live. (4)

CHAPTER 18 491

Cooperative Learning Activity

Organize students into five groups and assign one of the Sahel countries to each group. Have group members research and prepare a mini-lesson about their assigned country. Topics the lesson should cover include landscapes, re-sources, economic activities, and cultures. Then form new groups consisting of one member from each original group. Have each member in the new groups teach the other group members about the country he or she researched. **L1**

More About the Illustration

Answer to Caption
the mosque

Place Djenné was founded in the thirteenth century and later became a center of Muslim learning and trade. The Great Mosque, completed in 1907, is the latest of many mosques that have occupied this site since the city's founding.

Evaluate

📁 Section 2 Quiz **L1**

🔘 Use the Testmaker to create a customized quiz for Section 2. **L1**

Reteach

Review the causes and effects of desertification in the Sahel. Have students create pictures or graphic organizers to illustrate these causes and effects. **L1**

Enrich

Have students imagine they are consultants to the Sahel countries. Ask them to draw up a plan that offers solutions to the problem of desertification. **L3**

CLOSE

Encourage students to compose newspaper headlines that summarize the greatest challenges facing the Sahel countries today.

Djenné, Mali

Villagers set up shop at a street market outside the Great Mosque in Djenné.
MOVEMENT: How is the Arab influence in Mali shown in this photo?

Influences of the Past You may be surprised to learn that the Sahel was an early trading and cultural area. At a time when the climate was favorable, three great African empires—Ghana (GAH•nuh), Mali, and Songhai (SONG•hy)—arose in the Sahel. From the A.D. 300s to the 1500s, these empires controlled the trade in gold, salt, and other resources between West Africa and the Arab lands of the Mediterranean. The city of Tombouctou (tohn•book•too) on the Niger River was a leading center of learning, drawing students from Europe, Asia, and Africa.

During the 1800s, the countries of the Sahel came under French rule. They finally won their independence in 1960.

A Rural Population Most people in the Sahel live in small towns and villages. They raise only enough food for their own use. Porridges made with sorghum, millet, and other grains are popular dishes.

For years, nomadic groups such as the Tuareg (TWAH•REHG) and Fulani crossed northern desert areas with herds of camels, cattle, goats, and sheep. The recent droughts, however, have forced many of them to give up their traditional way of life and move to the towns and villages.

The people of the Sahel practice a mix of African, Arab, and European traditions. Most follow the religion of Islam. They speak Arabic as well as a variety of African languages. French is also spoken in such cities as Nouakchott (nu•AHK•SHAHT) in Mauritania, Bamako (BAH•muh•KOH) in Mali, and Niamey (nee• AH•MAY) in Niger.

SECTION 2 REVIEW

REVIEWING TERMS AND FACTS
1. **Define the following:** drought, desertification.
2. **LOCATION** Where is the Sahel located?
3. **HUMAN/ENVIRONMENT INTERACTION** How have humans affected the spread of desert in the Sahel?
4. **PLACE** How do most people make their livings in the Sahel countries?

 MAP STUDY ACTIVITIES
5. Look at the climate map on page 491. What climates do you find in the Sahel countries?

Answers to Section 2 Review
1. All vocabulary words are defined in the Glossary.
2. The Sahel is located between the Sahara in the north and a lush, fertile area in the south.
3. Humans have overgrazed and overfarmed the land.
4. They herd livestock or farm the land.
5. desert and steppe climates; southern areas of Mali and Burkina Faso have tropical savanna zones

MAKING CONNECTIONS

THE INTERPRETERS · BY WOLE SOYINKA

| MATH | SCIENCE | HISTORY | LITERATURE | TECHNOLOGY |

Wole Soyinka of Nigeria is perhaps Africa's best-known playwright and writer. In this excerpt from *The Interpreters,* friends talk as they journey in their canoe. They discuss how the landscape makes them feel.

Two paddles clove [moved through] the still water of the creek, and the canoe trailed behind it a silent groove, between gnarled tears of mangrove; it was dead air, and they came to a spot where an old rusted cannon showed above the water. . . . The paddlers slowed down and held the boat against the cannon. Egbo put his hand in the water and dropped his eyes down the brackish stillness, down the dark depths to its bed of mud.

. . . The canoe began to move off.

. . . From the receding cannon a quizzical crab emerged, seemed to stretch its claws in the sun and slipped over the edge, making a soft hole in the water. The mangrove arches spread seemingly endless and Kola broke the silence saying, 'Mangrove depresses me.'

'Me too,' said Egbo. . . . I remember when I was in Oshogbo I loved [the] grove and would lie there for hours listening at the edge of the water. It has a quality of this part of the creeks, peaceful and comforting. . . .' He trailed his hand in the water as he went, pulling up lettuce and plaiting [braiding] the long whitened root-strands.

Mangrove Trees

'That was only a phase of course, but I truly yearned for the dark. I loved life to be still, mysterious. I took my books down there to read, during the holidays. But later, I began to go further, down towards the old suspension bridge where the water ran freely, over rocks and white sand. And there was sunshine. . . . It was so different from the grove where depth swamped me. . . .

From Wole Soyinka's *The Interpreters.* Copyright © 1965 by Wole Soyinka, New York, N.Y.: Africana Publishing Corporation.

Making the Connection

1. What type of physical environment surrounds Kola and Egbo?
2. How do they feel when they are in this area?

493

Answers to Making the Connection

1. a swampy area of mangrove trees
2. The area depresses them.

TEACH

Display photographs or slides of mangrove swamps. Alternatively, show the videocassette *World Climate and Landscape Regions: Rainy Tropics* (Educational Video Network), concentrating on the section that deals with swamp vegetation. Call on volunteers to describe the landscape shown in the visuals. Note their responses on the chalkboard under the heading *Description.* Call on other volunteers to suggest the mood or feelings this type of landscape generates. Note these responses on the chalkboard under the heading *Feelings.* Then have students read the feature to see how many words they suggested appear in the excerpt. **L1**

More About . . .

Literature While Wole Soyinka has written novels, poetry, and criticism, he is best known as a playwright. In his plays, he combines western forms of drama with the words, music, and dance of traditional African festivals. Soyinka is an outspoken champion of human rights, and he often incorporates political themes in his work. This sometimes has created problems for Soyinka. His support of Biafran secession in the late 1960s, for example, earned him a two-year jail term. And Soyinka's constant criticism of Nigeria's military government led to his exile in 1994.

SECTION 3
Coastal Countries

FOCUS

Section Objectives

1. **Identify** the climates found in coastal West Africa.
2. **Explain** how people earn their livings in this region.
3. **Describe** how coastal West African countries won their independence.

Bellringer
Motivational Activity

Prior to taking roll at the beginning of the class period, project Section Focus Transparency 18-3 and have students answer the activity questions. Discuss students' responses. **L1**

Vocabulary Pre-check

Have students define the term *cassava*. Point out that cassava is grown in the southern United States. It is used to make tapioca, which is an ingredient in many puddings. **L1 LEP**

PREVIEW

Words to Know
• cassava
• bauxite

Places to Locate
• Senegal
• Dakar
• Cape Verde Islands
• Liberia
• Côte d'Ivoire
• Abidjan
• Ghana

Read to Learn . . .
1. what climates are found in coastal West Africa.
2. how people earn their livings in this region.
3. how coastal West African countries won their independence.

Do you like peanuts? The next time you bite into a peanut butter sandwich, think about West Africa. Workers in the West African country of Gambia dry peanuts to send to other countries. Gambia began exporting peanuts in the 1800s. Today it depends on this one crop for about 95 percent of its income.

Besides Nigeria and the Sahel countries, West Africa includes 11 coastal countries. One—Cape Verde—is a group of 15 islands in the Atlantic Ocean. The other countries spread along West Africa's Atlantic bend, north of the Equator, from Senegal in the west along the coast to Benin. Find these countries on the map on page 484.

Human/Environment Interaction
The Land

Coastal lowlands sweep along the Atlantic shore of West Africa's coastal countries. Sandy beaches, rain forests, and thick mangrove swamps cover most of this area. Inland from the coast lie highland areas with grasses and trees. Several major rivers—the Senegal, Gambia, Volta, and Niger—flow from these highlands to the coast. You won't see many ships on these rivers. Why? Rapids and shallows prevent ships from traveling far into inland areas.

The Climate Because they border the ocean, the coastal countries of West Africa receive plenty of rainfall. Warm currents in the Gulf of Guinea create a

494

UNIT 7

moist, tropical climate in the coastal lowlands year-round. Some coastal areas receive more than 170 inches (432 cm) of rain a year! Farther inland, rainfall is not as plentiful. These highland areas have a tropical savanna climate with dry and wet seasons.

The Economy Because of the wet climate, most people in coastal West Africa are farmers. On small plots of land, farmers grow yams, corn, rice, cassava, and other foods for their families. **Cassava** is a plant root that looks like a potato. It is used to make flour for bread.

West Africa also has large plantations that grow coffee, rubber, cacao, palm oil, and kola nuts for export. Can you guess what the kola nut is used for? This sweet-tasting nut is used in making soft drinks. Until industries become more developed in the region, these cash crops will continue to be the major source of income.

Region
The People

West Africa's coastal countries have a long and rich history. In early times, powerful kingdoms such as the Ashanti and Benin ruled the region. These kingdoms were centers of trade, learning, and the arts.

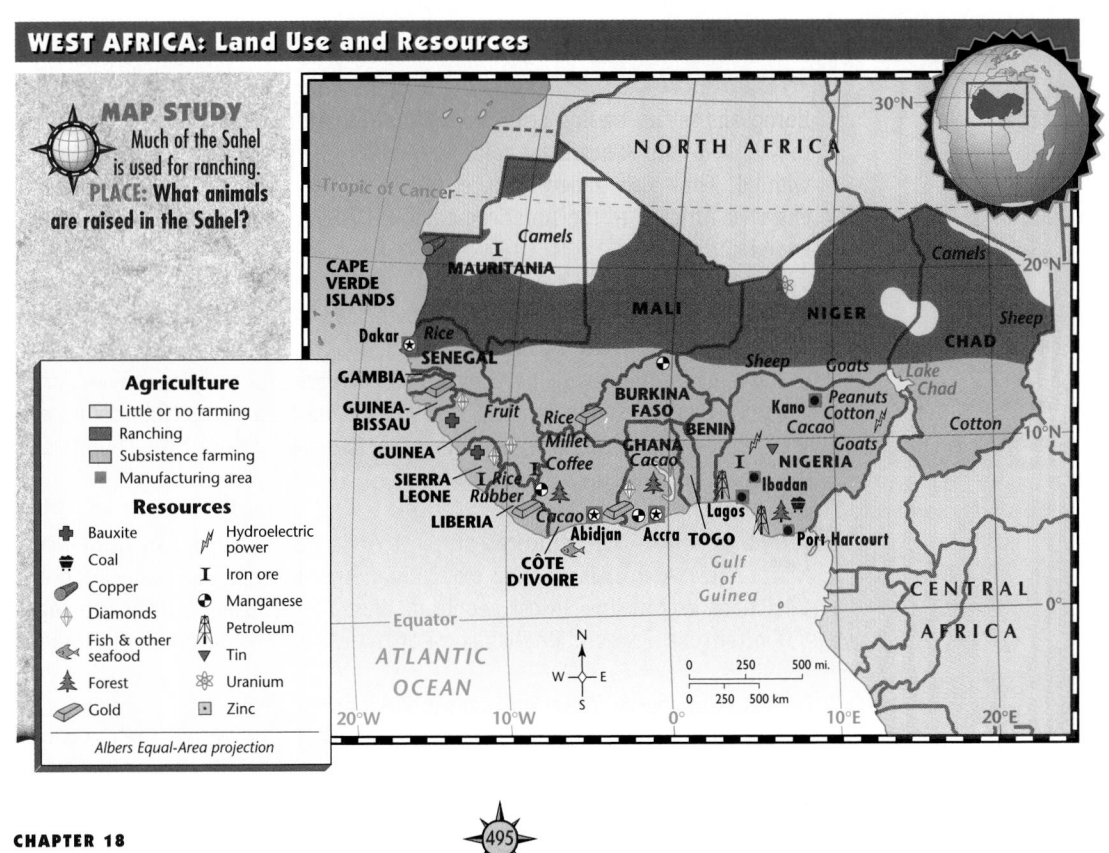

WEST AFRICA: Land Use and Resources

MAP STUDY
Much of the Sahel is used for ranching.
PLACE: What animals are raised in the Sahel?

Agriculture
- Little or no farming
- Ranching
- Subsistence farming
- Manufacturing area

Resources
- Bauxite
- Coal
- Copper
- Diamonds
- Fish & other seafood
- Forest
- Gold
- Hydroelectric power
- Iron ore
- Manganese
- Petroleum
- Tin
- Uranium
- Zinc

Albers Equal-Area projection

Cooperative Learning Activity

Organize students into five groups and assign each group two of the mainland countries of coastal West Africa. Have groups prepare illustrated profiles of their assigned countries. Suggest that groups include information such as physical features, resources, economic activities, population statistics, and cultural features in their profiles. Allow time for groups to share and discuss their profiles. **L1**

TEACH

Guided Practice
Making Comparisons
Have students compare urban and rural ways of life in the coastal countries of West Africa. Then ask students to discuss why rural people are more likely to preserve traditions than are urban people. *(Urban people may have more contact with other cultures; their jobs, homes, and ways of life tend to weaken their ties to tradition.)* **L1**

MULTICULTURAL PERSPECTIVE
Cultural Diffusion
French missionaries brought Christianity to Côte d'Ivoire in the 1600s. Today, Christianity remains one of the country's major religions. The church of Our Lady of Peace in Yamoussoukro is the largest Christian place of worship in Africa.

MAP STUDY

Answer
sheep, goats, camels
Map Skills Practice
Reading a Map What crops are grown in Côte d'Ivoire? *(cacao, coffee, millet)*

MAP STUDY

Answer
Mauritania, Mali, Niger, Chad

Map Skills Practice
Reading a Map What four West African cities have populations of more than one million?
(Abidjan, Dakar, Ibadan, Lagos)

Independent Practice

Guided Reading Activity 18-3 **L1**

Cultural Kaleidoscope

Ghana On special occasions, Ghanaians wear clothing made from kente cloth. The cotton fabric is handwoven in narrow strips then sewn together. Each pattern tells a story or represents an event. The weaving of kente cloth dates back some 250 years.

NATIONAL GEOGRAPHIC SOCIETY

These materials are available from Glencoe.

VIDEODISC

STV: World Geography Vol. 2: Africa and Europe

West Africa
Side 1, Frames 15035-18799

WEST AFRICA: Population Density

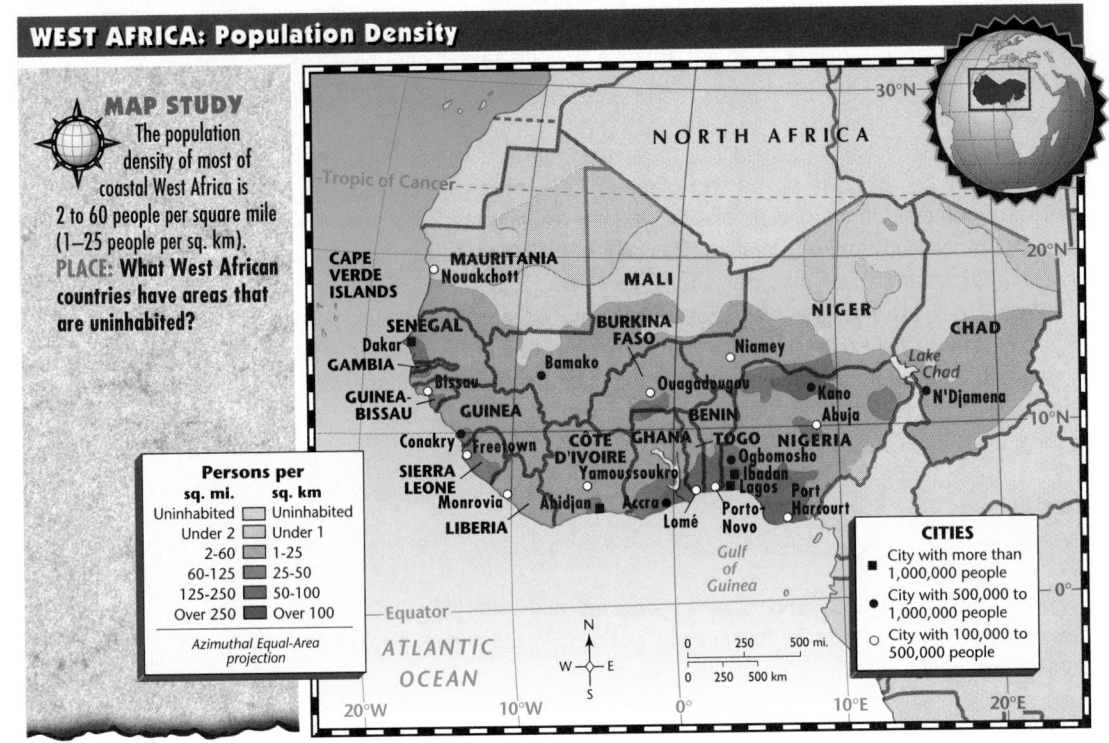

MAP STUDY
The population density of most of coastal West Africa is 2 to 60 people per square mile (1–25 people per sq. km).
PLACE: What West African countries have areas that are uninhabited?

Persons per

sq. mi.	sq. km
Uninhabited	Uninhabited
Under 2	Under 1
2-60	1-25
60-125	25-50
125-250	50-100
Over 250	Over 100

Azimuthal Equal-Area projection

CITIES
- ■ City with more than 1,000,000 people
- ● City with 500,000 to 1,000,000 people
- ○ City with 100,000 to 500,000 people

The Arrival of Europeans From the late 1400s to the early 1800s, Europeans set up trading posts along the West African coast. From these posts they traded with Africans for gold, ivory, and other goods that people in Europe wanted. They also traded enslaved Africans. The traders shipped millions of enslaved Africans to the American colonies. This trade in human beings finally ended in the 1800s.

Toward Freedom The British, French, and Portuguese eventually set up colonies in coastal West Africa. They brought Christianity and European culture to the region. After several decades of European rule, the peoples of West Africa began to demand independence. In 1957 Ghana became the first country in West Africa to become independent. By the late 1970s, all of the countries in the region had won their independence.

Ways of Life The cultures of West Africa show European influences. Many West Africans practice traditional African religions, but some are Christians or Muslims. Most people in West Africa speak African languages but often use French, English, or Portuguese in business or government activities.

Cities in coastal West Africa are modern and growing. If you visited a city like Ghana's capital—Accra (uh•KRAH)—or Côte d'Ivoire's capital—Abidjan (A•bih•JAHN)—you would find downtown areas with modern government and

496

Meeting Special Needs Activity

Language Delayed Help students discover different ways that cause/effect relationships are expressed in the text. For example, ask a student to read the last two sentences under "The Land" on page 494. Point out the cause/effect relationship: cause—rapids and shallows; effect—ships cannot sail to inland areas. Then have students find an example of a cause/effect relationship under the subhead "The Climate." *(Cause—the coastal countries of West Africa border the ocean; effect—these countries receive plenty of rainfall)* **L1**

office buildings. You would also see people dressed in business clothes or in traditional African clothing and wide streets with busy traffic. In rural areas, people live in small villages. Many village homes have mud walls and straw or metal roofs. People here follow many of their ancestors' traditions.

Region
Country Profiles

The 11 countries of coastal West Africa have many things in common. Each country, however, has certain features that make it stand out from the others.

Senegal, Gambia, and Guinea These countries all depend on farming as their major source of income. The map on page 495, however, shows that Guinea is rich in **bauxite**—a mineral from which we make aluminum—and other mineral resources.

Senegal has more city dwellers than either Gambia or Guinea. About 40 percent of Senegal's people live in urban areas. Dakar (DA•KAHR), Senegal's capital and largest city, also is a major port on the Atlantic Ocean.

Guinea-Bissau and Cape Verde Guinea-Bissau lies between Senegal and Guinea. The islands of Cape Verde are located about 375 miles (600 km) offshore in the Atlantic Ocean. Guinea-Bissau and Cape Verde were Portuguese colonies until they won independence in 1975. Portuguese influences are widespread in the languages and cultures of both countries. Most of the people here make their livings by farming or fishing.

Liberia and Sierra Leone At the "bend" of Africa's west coast are Liberia and Sierra Leone. Liberia is the only West African country that was never a colony. African Americans freed from slavery founded Liberia in 1822. Monrovia, Liberia's capital and largest city, was named for James Monroe—American president when Liberia was founded. Since 1989, fighting among Liberian political groups has cost many lives and left thousands homeless.

Like Liberia, Sierra Leone was established as a home for people freed from enslavement. The British ruled Sierra Leone from 1787 until 1961, when Sierra Leone became independent. Most of the country's land is used for agriculture, but diamonds and other minerals also provide wealth.

Côte d'Ivoire and Ghana Côte d'Ivoire was a French colony until winning its independence in 1960. Many ethnic groups live in Côte d'Ivoire's rural

Village Life

Iyabo Ayinde lives in a small village outside Abidjan. The people in her village are related, and each family has its own group of dwellings. Her family's homes have cone-shaped thatched roofs. Villagers share a small plot of land where they raise corn and yams. Today after school, Iyabo will meet her cousins in the central square of her village to plan a trip to Abidjan. Iyabo's cousins want to go to the art museum. Iyabo has a better idea. "Let's go to the ice skating rink!" she says.

ASSESS

Check for Understanding

Assign Section 3 Review as homework or an in-class activity.

Meeting Lesson Objectives

Each objective below is tested by the questions that follow it in parentheses.

1. **Identify** the climates found in coastal West Africa. (5)
2. **Explain** how people earn their livings in this region. (3)
3. **Describe** how coastal West African countries won their independence. (4)

Critical Thinking Activity

Drawing Conclusions Ask students to note how European contact influenced West African ways of life. Then have them categorize these effects as *positive* or *negative*. Provide time for students to discuss and justify their categorizations. L2

 Section 3 Quiz **L1**

 Use the Testmaker to create a customized quiz for Section 3. **L1**

 Chapter 18 Test Form A and/or Form B **L1**

Reteach

 Reteaching Activity 18 **L1**

Enrich

 Enrichment Activity 18 **L1**

CLOSE

 **Mental Mapping Activity**

Provide each student with a large sheet of white paper and map pencils. Inform students that this will not be a graded activity. Have students draw, freehand from memory, a map of West Africa that includes the following labeled countries: Nigeria, Mauritania, Mali, Niger, Chad, Burkina Faso, Senegal, Gambia, Guinea, Guinea-Bissau, Cape Verde, Liberia, Sierra Leone, Ghana, Côte d'Ivoire, Togo, and Benin.

More About the Illustration

Answer to Caption
France

 **Place** The population of Abidjan has grown dramatically in recent years. In 1960, Abidjan's population numbered less than 200,000 people. Today, however, almost 2 million people live in the city.

498

Abidjan, Côte d'Ivoire

The modern port of Abidjan handles many of the imports and exports of West Africa. **PLACE: What European country controlled Côte d'Ivoire until 1960?**

villages. They cultivate coffee, cacao, and forest products for export. In recent years some farmers have moved from their villages to find factory jobs in Abidjan and other cities.

Ghana—like Côte d'Ivoire—produces cacao and forest products for export. Ghana's people belong to about 100 different ethnic groups. The Ashanti and the Fante are the largest groups. About 34 percent of Ghanaians live in cities. Accra, located on the coast, is the capital and largest city.

Togo and Benin The countries of Togo and Benin lie between Ghana and Nigeria. The map on page 484 shows you that both countries are long and narrow. Togo and Benin—once French colonies—became independent in 1960. Many ethnic groups whose dress, language, and other ways of life differ a great deal make up the cultures of Togo and Benin.

SECTION 3 REVIEW

REVIEWING TERMS AND FACTS

1. Define the following: cassava, bauxite.
2. LOCATION Where is Cape Verde located?
3. HUMAN/ENVIRONMENT INTERACTION What food crops are grown in coastal West Africa?
4. REGION How did the arrival of Europeans influence life in coastal West Africa?

 MAP STUDY ACTIVITIES

5. Study the population density map on page 496. Where do most people in coastal West Africa live?

UNIT 7

Answers to Section 3 Review

1. All vocabulary words are defined in the Glossary.
2. Cape Verde is located in the Atlantic Ocean about 375 miles (600 km) from West Africa.
3. yams, corn, rice, cassava
4. The Europeans set up trading posts in West Africa. They traded for gold, ivory, and enslaved people. Millions of Africans were transported to the Western Hemisphere in the slave trade. Eventually, the Europeans set up colonies, bringing Christianity and Western culture to the region.

5. in the coastal lowlands

Chapter 18 Highlights

Important Things to Know About West Africa

SECTION 1 NIGERIA

- Nigeria's major landforms are coastal lowlands, savanna highlands, and partly dry grasslands.
- Most Nigerians farm the land, but oil is Nigeria's major export.
- Many different ethnic groups live in Nigeria.
- Nigeria is the most highly populated African country.

SECTION 2 THE SAHEL COUNTRIES

- The Sahel countries are Mauritania, Mali, Niger, Chad, and Burkina Faso.
- The Sahel lies between the Sahara in the north and fertile, humid lands in the south.
- Most people in the Sahel countries are herders or farmers.
- Desertification has destroyed much of the Sahel's most productive land.

SECTION 3 COASTAL COUNTRIES

- The 11 countries that make up coastal West Africa are Senegal, Gambia, Guinea, Guinea-Bissau, Cape Verde, Liberia, Sierra Leone, Ghana, Côte d'Ivoire, Togo, and Benin.
- Because of the wet climate, farming is the major economic activity in coastal West Africa.
- Most people in the coastal countries of West Africa live in the coastal lowlands.
- People freed from slavery helped establish the countries of Liberia and Sierra Leone.

Rain forest in Liberia ▶

Using the Chapter 18 Highlights

- Use the Chapter 18 Highlights to preview, review, condense, or reteach the chapter.

Preview/Review

Vocabulary Puzzle-Maker Software reinforces the terms used in Chapter 18.

Condense

Have students read the Chapter 18 Highlights. Spanish Chapter Highlights also are available.

Chapter 18 Digest Audiocassettes and Activity

Chapter 18 Digest Audiocassette Test

Spanish Chapter Digest Audiocassettes, Activities, and Tests also are available.

Guided Reading Activities

Reteach

Reteaching Activity 18. Spanish Reteaching activities also are available.

MAP STUDY

Location Distribute copies of page 37 of the Outline Map Resource Book. Have students locate and label the countries of West Africa and their capitals. Suggest that students use the **Unit Atlas** to find this information.

Extra Credit Project

Musical Heritage Have students research and report on the music of Africa south of the Sahara. Have them find out how the strong rhythms from this music influenced the following styles of music in the United States: Spirituals, jazz, rock 'n' roll. Students might illustrate their reports by playing recordings of different styles of American music that express African rhythms. **L3**

499

Chapter 18 Review and Activities

REVIEWING KEY TERMS

Match the numbered terms in Column A with their definitions in Column B.

A
1. cassava
2. savanna
3. drought
4. mangrove
5. ethnic group
6. cacao
7. desertification
8. harmattan
9. bauxite

B
A. change of grasslands to desert
B. dusty wind blowing south from the Sahara
C. plant root used in making flour
D. people with the same language, culture, and history
E. tropical grassland with scattered trees
F. extreme shortage of water
G. mineral from which aluminum is made
H. beans used in making chocolate
I. tropical tree with roots above and below water

ANSWERS

Reviewing Key Terms

1. C
2. E
3. F
4. I
5. D
6. H
7. A
8. B
9. G

Mental Mapping Activity

On a separate sheet of paper, draw a freehand map of West Africa. Label the following items on your map:
- Gulf of Guinea
- Atlantic Ocean
- Lagos
- Senegal River
- Gambia
- Sahara

Mental Mapping Activity

This exercise helps students to visualize the countries and geographic features they have been studying and understand the relationships among them. All attempts at freehand mapping should be accepted.

REVIEWING THE MAIN IDEAS

Section 1
1. PLACE What river is important to Nigeria?
2. HUMAN/ENVIRONMENT INTERACTION What challenges has Nigeria faced because of its dependence on oil?
3. MOVEMENT Why are Nigeria's cities growing?

Section 2
4. HUMAN/ENVIRONMENT INTERACTION Why is farming difficult in the Sahel?
5. PLACE Why was the city of Tombouctou important?
6. REGION What is the major religion in the Sahel countries?
7. HUMAN/ENVIRONMENT INTERACTION What are the major food products of the Sahel?

Section 3
8. REGION Why does the coastal strip in coastal West Africa receive plenty of rainfall?
9. HUMAN/ENVIRONMENT INTERACTION What cash crops are raised in coastal West Africa?
10. PLACE What West African country was the first to win independence?

Reviewing the Main Ideas
Section 1
1. Niger River
2. Falling world oil prices have resulted in decreased income for Nigeria.
3. Many Nigerians have moved from rural areas to the cities in the hope of finding jobs.

Section 2
4. because the area receives little rainfall
5. It was a leading center of learning, drawing students from elsewhere in Africa, Asia, and Europe.
6. Islam
7. grains, such as sorghum and millet

CRITICAL THINKING ACTIVITIES

1. **Drawing Conclusions** Why do you think people in many areas of the Sahel have followed a nomadic way of life?
2. **Determining Cause and Effect** How has location on the coast affected the history and culture of coastal West Africa?

GeoJournal Writing Activity

Imagine that you are on a journey down the Niger River. Write several entries in a diary describing how the landscape changes as you travel down the river from inland areas to the coast.

COOPERATIVE LEARNING ACTIVITY

Work in a group of three to learn more about the arts of West Africa. As a group, choose one country in West Africa. Each member in the group will then research that country's sculpture, music, or dance. After your research is complete, share your information with your group. Brainstorm ways to present your country's art to the class as a whole.

PLACE LOCATION ACTIVITY: WEST AFRICA

Match the letters on the map with the places and physical features of West Africa. Write your answers on a separate sheet of paper.

1. Côte d'Ivoire
2. Monrovia
3. Tibesti Mountains
4. Senegal
5. Mali
6. Gulf of Guinea
7. Lake Chad
8. Niger River
9. Cape Verde Islands
10. Lagos

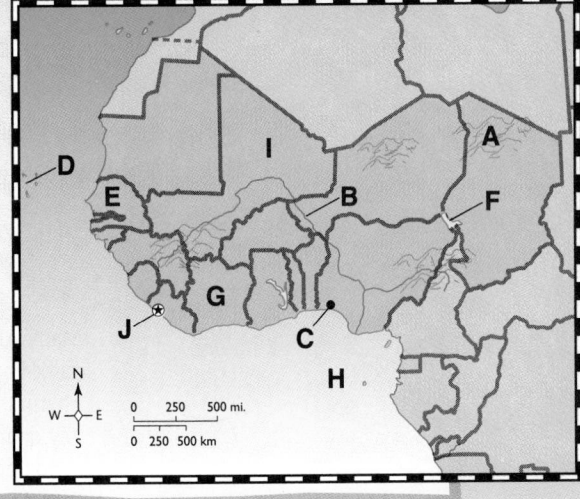

Cooperative Learning Activity

Suggest that groups make multimedia presentations, mixing visuals, the spoken word, and musical recordings.

GeoJournal Writing Activity

Suggest that students read the excerpt from Wole Soyinka's *The Interpreters* on page 493 before completing this activity. Encourage students to share their journal entries with the rest of the class.

Place Location Activity

1. G
2. J
3. A
4. E
5. I
6. H
7. F
8. B
9. D
10. C

Chapter Bonus Test Question

This question may be used for extra credit on the chapter test.

The Organization of Petroleum Exporting Countries (OPEC) announces changes in oil prices. Which West African nation would be directly affected by this announcement? *(Nigeria)*

Section 3
8. Warm currents in the Gulf of Guinea create a moist, tropical climate in the coastal lowlands.
9. coffee, rubber, cacao, palm oil, kola nuts
10. Ghana

Critical Thinking Activities

1. Because the Sahel receives a limited amount of rainfall, the people have to move their herds to where there is adequate grassland.

2. The coastal lowlands had fertile soils and adequate rainfall. This made it suitable for agriculture, and many people settled there. When the Europeans came, they made the coastal lowlands their base for exploration and control of the region. Today, the coastal cities are centers of trade, government, and culture.

Cultural
HERITAGE:
AFRICA SOUTH OF THE SAHARA

FOCUS

Ask students how national boundaries are usually determined. *(Possible responses: long-term occupation of an area by cultural groups, wars, treaties)* Then display a political map of Africa. Tell students that most of the national boundaries in Africa were determined by a group of leaders at a meeting in Berlin in the mid-1880s.

TEACH

Continue this discussion by telling students that the Berlin meeting included representatives from Belgium, France, Germany, Portugal, Spain, and Great Britain. Competing land claims threatened to explode into conflict among these European colonial powers. Therefore, they held the Berlin meeting to divide Africa among themselves. No Africans participated in the conference. The new country borders were drawn without respect to ethnic boundaries. As a result, many ethnic groups were split between countries, while other groups were forced to join the same country.

Guide students in a discussion on how this boundary-making process disrupted hundreds of cultures and led to many of the challenges that face Africa today, such as ethnic conflict and unstable national governments. **L2**

◄ ◄ ◄ ◄ MASKS

Many of Africa's ethnic groups can be identified by the types of masks they create. Masks are usually worn by African dancers in social and religious ceremonies.

SCULPTURE ▶ ▶ ▶ ▶

Europeans often paid African artists to create intricate carvings for their personal collections. Skilled ivory carvers created this salt container from a single elephant tusk in the 1600s.

◄ ◄ ◄ ◄ DANCE

Many African ethnic groups use dance to mark important events in their lives. The rhythms in African music are reproduced by dancers' bodies. Their heads may be moving to one rhythm, their shoulders to another, and their arms and feet to yet another.

502

More About . . .

Art Masks are common among the peoples of western and central Africa, where mask-making is considered a fine art. Masks are, however, created for specific purposes and have to be made according to strict traditional requirements. Masks play important roles in various religious and social celebrations. Most masks represent spirits that are part of religious beliefs, while others represent ancestors. Point out to students that Africa is home to hundreds of distinct cultures, each with its own heritage. Masks are unique to individual cultures and, thus, have a wide variety of meanings.

◄ ◄ ◄ ◄ ARCHITECTURE

Much of the architecture in Africa reflects the influence of the Arab world. This modern mosque with its towering minarets is located in Nigeria.

MUSIC ► ► ► ►

African music can have as many as 5 to 12 different rhythms. The complicated rhythms are often produced on drums, horns, and trumpets made from natural materials. Jazz and other music popular in the United States have been greatly influenced by African music.

JEWELRY ▲ ▲ ▲ ▲

The gold trade in the Ashanti empire in the 1600s brought great riches to the Ashanti kings. Many kings even had their own goldsmiths. This breast ornament was made by twisting delicate gold wire into lacelike patterns.

CHAPTER 18

503

APPRECIATING CULTURE

1. How would you describe this African mask to someone who had not seen it?

2. What are some characteristics of African music?

3. Why do you think gold was a symbol of power and status in the Ashanti empire?

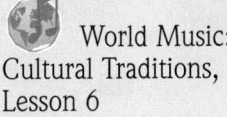

Geography and the Humanities

📁 World Literature Reading 7

🌎 World Music: Cultural Traditions, Lesson 6

📦 World Cultures Transparencies 11, 12

📕 Focus on World Art Print 12

ASSESS

Have students answer the **Appreciating Culture** questions on page 503.

CLOSE

Have students make a list of African countries and the ethnic groups that live within each. Encourage students to use this information to make statements about the difference between cultural and political boundaries.

Answers to Appreciating Culture

1. Answers will vary.

2. It is very rhythmic; can have multiple rhythms; often incorporates drums, horns, and trumpets made of natural materials.

3. Answers will vary. Most students will note that gold brought the Ashanti great wealth, so the possession of gold would give the owner power and status.

POSSIBLE RUBRIC FEATURES: Content information, analysis skills, concept attainment, appropriate advertisement design

PERFORMANCE ASSESSMENT ACTIVITY

"HELP WANTED" ADS Have students assume the role of copy editors for a classified section of an international newsletter. Ask them to develop "Help Wanted" ads for the types of jobs available in Central Africa. Encourage students to look at the classified section of a local newspaper or large urban newspaper in order to determine the size, type of information included, and style of text used in such advertisements. Suggest that students develop 10 ads, pointing out that all ads should include geographical information about Central Africa. Students might combine their ads to form one large classified section for a bulletin-board display.

MENTAL MAPPING ACTIVITY

Before teaching Chapter 19, point out the Central African countries on a globe. Have students note the relative sizes of the countries. Also have them determine which countries are landlocked and which have coastlines.

TEACHER'S CORNER

NATIONAL GEOGRAPHIC SOCIETY

INDEX TO NATIONAL GEOGRAPHIC MAGAZINE

The following articles may be used for research relating to this chapter:

- "Ndoki—Last Place on Earth," by Douglas Chadwick, July 1995.
- "Bonobos, Chimpanzees With a Difference," by Eugene Linden, March 1992.
- "Lifeline for a Nation—Zaire River," by Robert Caputo, November 1991.
- "The Efe: Archers of the African Rain Forest," by Robert C. Bailey, November 1989.
- "Silent Death from Cameroon's Killer Lake," by Curt Stager, September 1987.

NATIONAL GEOGRAPHIC SOCIETY PRODUCTS AVAILABLE FROM GLENCOE

To order the following products for use with this chapter, contact your local Glencoe sales representative or call Glencoe at 1-800-334-7344:

- *STV: World Geography* (Videodisc)
- *Picture Atlas of the World* (CD-ROM)
- *Physical Geography of the World* (Transparencies)
- *ZipZapMap! World* (Software)
- *GeoBee* (Software)
- *Images of the World* (Posters)
- *Eye on the Environment* (Posters)

ADDITIONAL NATIONAL GEOGRAPHIC SOCIETY PRODUCTS

To order the following products for use with this chapter, call National Geographic Society at 1-800-368-2728:

- *African Wildlife* (Video)
- *Lions of the African Night* (Video)
- *Physical Geography of the Continents Series:* "Africa." (Video)
- *Portraits of the Continents Series:* "Part III: Europe; Asia; Africa." (Filmstrip)

TEACHER-TO-TEACHER
Curriculum Connection: Social Studies, Art

—from Susie Gerschutz
Norton Middle School, Columbus, OH

FLAG FORMATIONS The purpose of this activity is to help students understand how history, politics, and beliefs play a role in the design of a nation's flag. Organize students into seven groups and assign each group one of the countries of Central Africa. Direct groups to undertake research on the flag of their assigned country. Then provide groups with a sheet of butcher paper. Have each group make a color sketch of their assigned country's flag in the top half of their sheet of paper. In the bottom half of the paper, have each group list the meanings of the colors and symbols incorporated in the flag. Display the flag information sheets around the classroom and encourage students to discuss the following questions: What are the most frequently used colors in the flags? What are the most frequently used symbols? What themes, if any, are there in the meanings of the flags?

As a follow-up activity, some students might like to investigate the flags of other countries in Africa south of the Sahara and create similar flag information sheets. Other students might like to design a flag for the school, explaining the meaning of the colors and the symbols used.

MEETING NATIONAL STANDARDS

Geography for Life

All of the 18 standards are demonstrated in Unit 7. The following ones are highlighted in this chapter:
• Standards 4, 6, 9, 15, 16, 18

KEY TO ABILITY LEVELS

Teaching strategies have been coded for varying learning styles and abilities.

L1   **BASIC** activities for all students

L2   **AVERAGE** activities for average to above-average students

L3   **CHALLENGING** activities for above-average students

LEP   **LIMITED ENGLISH PROFICIENCY** activities

BIBLIOGRAPHY

Readings for the Student

Murray, Jocelyn. *Africa: Cultural Atlas for Young People.* New York: Facts On File, 1990.

Ramsay, F. Jeffress. *Africa: Global Studies,* 6th ed. Guilford, Conn.: Dushkin, 1995.

Readings for the Teacher

Africa Today: A Reproducible Atlas. Wellesley, Mass.: World Eagle, 1994.

Grove, A.T. *The Changing Geography of Africa.* New York: Oxford University Press, 1990.

Jackson, John G. *Introduction to African Civilizations.* New York: Citadel, 1990.

Map Skills and Outlines: Africa. St. Louis: Milliken, 1991. Transparencies and reproducible worksheets.

Multimedia

Africa: Continent of Contrasts. Huntsville, Tex.: Educational Video Network, 1994. Videocassette, 34 minutes.

Africa: Central and Eastern Regions. Chicago: Encyclopedia Britannica, 1993. Videocassette, approximately 20 minutes.

Global Studies With Country Databases: Africa. Toronto, Canada: WorldView Software, 1994. Apple 5.25" disk or IBM 3.5" disk.

Chapter 19 Resource Organizer
Central Africa

Use these *Geography: The World and Its People* resources to teach, reinforce, and extend chapter content.

CHAPTER 19 RESOURCES

- *Vocabulary Activity 19
- Cooperative Learning Activity 19
- Workbook Activity 19
- Geography Skills Activity 19
- GeoLab Activity 19
- Performance Assessment Activity 19
- *Reteaching Activity 19
- Enrichment Activity 19
- Chapter 19 Test, Form A and Form B
- *Chapter 19 Digest Audiocassette, Activity, Test
- Unit Overlay Transparencies 7-0, 7-1, 7-2, 7-3, 7-4, 7-5, 7-6
- Political Map Transparency 7; World Cultures Transparency 12
- Geoquiz Transparencies 19-1, 19-2
- Vocabulary PuzzleMaker Software
- Student Self-Test: A Software Review
- Testmaker Software
- MindJogger Videoquiz
- Focus on World Art Print 12

If time does not permit teaching the entire chapter, summarize using the Chapter 19 Highlights on page 517, and the Chapter 19 English (or Spanish) Audiocassettes. Review students' knowledge using the Glencoe MindJogger Videoquiz. * *Also available in Spanish*

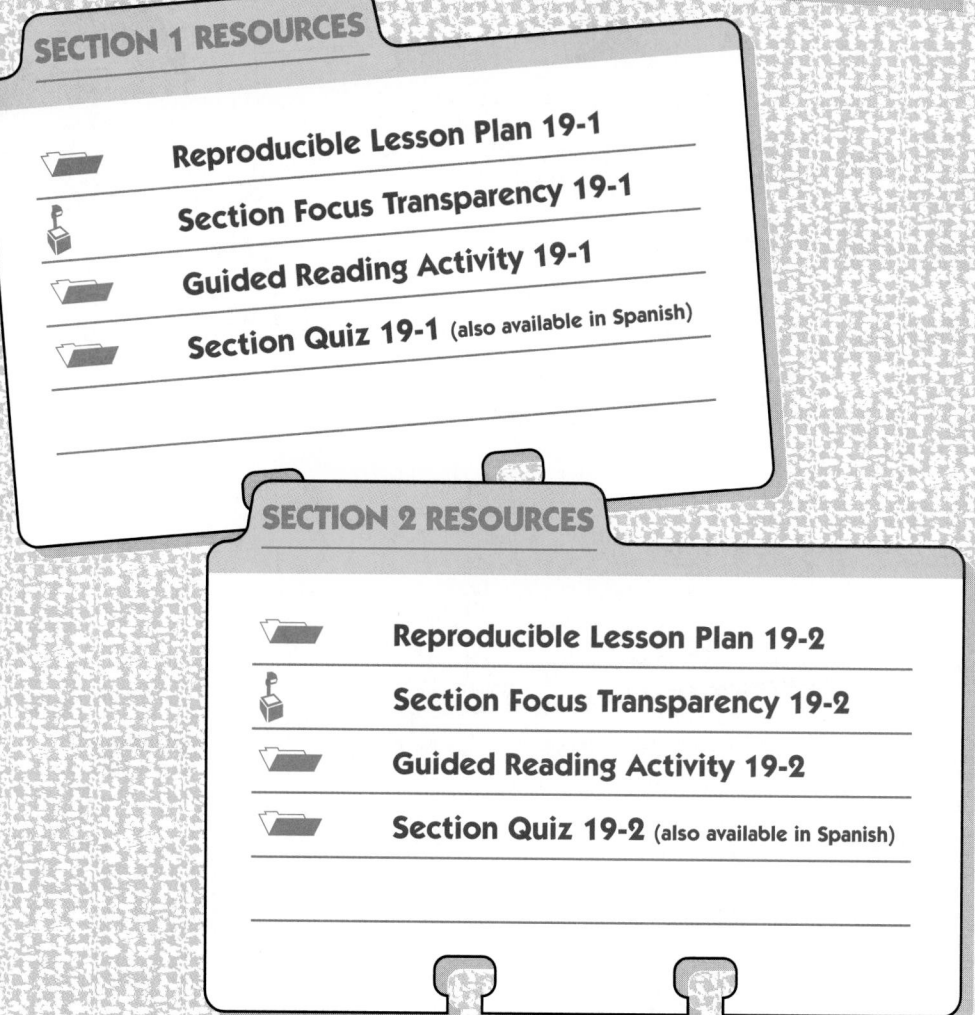

Use these *Geography: The World and Its People* resources to teach and reinforce section content.

SECTION 1 RESOURCES

Reproducible Lesson Plan 19-1

Section Focus Transparency 19-1

Guided Reading Activity 19-1

Section Quiz 19-1 (also available in Spanish)

SECTION 2 RESOURCES

Reproducible Lesson Plan 19-2

Section Focus Transparency 19-2

Guided Reading Activity 19-2

Section Quiz 19-2 (also available in Spanish)

ADDITIONAL RESOURCES FROM GLENCOE

 **Reproducible Masters**

- Glencoe Social Studies Outline Map Book, pages 36, 37

- Facts On File: GEOGRAPHY ON FILE, Sub-Saharan Africa; AFRICAN HISTORY ON FILE

- Foods Around the World, Unit 7

Workbook

- Building Skills in Geography, Unit 2, Chapter 5

 World Music: Cultural Traditions Lesson 6

Transparencies

- National Geographic Society PicturePack Transparencies, Unit 7

 Videodiscs

- Geography and the Environment: The Infinite Voyage

- Reuters Issues in Geography

- National Geographic Society: STV: World Geography, Vol. 2

 CD-ROM

- National Geographic Society: Picture Atlas of the World

- National Geographic Society PictureShow: Earth's Endangered Environments

 **Software**

- National Geographic Society: ZipZapMap! World

Performance Assessment

Refer to the Planning Guide on page 504A for a Performance Assessment Activity for this chapter. See the *Performance Assessment Strategies and Activities* booklet for suggestions.

Chapter Objectives

1. **Examine** the physical and cultural geography of Zaire.

2. **Compare** the geographic features of the other Central African countries.

GLENCOE
TECHNOLOGY

 **VIDEODISC**

Use the Chapter 19 MindJogger Videoquiz to preview chapter content.

MindJogger Videoquiz

Chapter 19
Disc 3 Side A

Also available in VHS.

504

Chapter 19 Central Africa

Map labels:
WEST AFRICA
10°N
Garoua
CAMEROON
CENTRAL AFRICAN REPUBLIC
Malabo
Douala
Bangui
EQUATORIAL GUINEA
Yaoundé
SÃO TOMÉ AND PRÍNCIPE
São Tomé
Libreville
GABON
CONGO
Kisangani
Brazzaville
ZAIRE
Equator
Pointe-Noire
Kinshasa
Cabinda (ANGOLA)
Kikwit
ATLANTIC OCEAN
10°S
Kolwezi
SOUTHERN AFRICA
10°E   20°E   30°E
0°

N W E S

0   150   300 mi.
0   150   300 km

— National boundary
⊛ National capital
● Other city

Azimuthal Equal-Area projection

MAP STUDY ACTIVITY

As you read Chapter 19, you will learn about Zaire and the other countries in Central Africa.

1. **How many countries are in Central Africa?**
2. **What is the largest country in Central Africa?**
3. **What is the capital of Zaire?**

504

 MAP STUDY ACTIVITY

Answers
1. seven
2. Zaire
3. Kinshasa

Map Skills Practice
Reading a Map What island country is part of Central Africa? *(São Tomé and Príncipe)* Which country outside the Central African region governs Cabinda? *(Angola)*

SECTION 1 · Zaire

PREVIEW

Words to Know
- canopy
- hydroelectricity

Places to Locate
- Zaire River
- Kinshasa
- Lake Tanganyika

Read to Learn . . .
1. why the Zaire River is a highway for Zaire's people.
2. about Zaire's potential wealth.
3. how large movements of people have shaped Zaire's history.

This lush countryside lies in Zaire, one of the countries in the region of Central Africa. Water is abundant in the rainy and green stretches of this region.

Central Africa's many rivers are a source of life for the people of the region. Africa's second-longest river—the Zaire—flows through the country of the same name. Zaire is located on the Equator in the very heart of Africa. A large country, Zaire is bordered by nine other African countries.

Place
The Land

A burning torch on a field of green marks the flag waving above the third-largest country in Africa—Zaire. Zaire covers an area of 875,520 square miles (2,267,597 sq. km)—about one-fourth the size of the United States. Only Sudan and Algeria are larger. Zaire encompasses many landscapes, including rain forests, savannas, and highlands. Its only coastline is a 25-mile (41-km) strip on the Atlantic Ocean.

Rain Forests Vast tropical rain forests cover about one-third of Zaire. It is one of the largest rain forest areas in the world. The treetops of the rain forests form a **canopy,** or umbrella-like forest covering. This canopy is so thick that sunlight seldom reaches parts of the forest floor, which is almost clear of plants.

CHAPTER 19 505

Classroom Resources for Section 1

 REPRODUCIBLE MASTERS

Reproducible Lesson Plan 19-1
Vocabulary Activity 19
Cooperative Learning
 Activity 19
Geography Skills Activity 19
Building Skills in Geography,
 Unit 2, Lesson 5
Section Quiz 19-1

TRANSPARENCIES

Section Focus
 Transparency 19-1
World Cultures Transparency 12

MULTIMEDIA

 Vocabulary PuzzleMaker
 Software

 Testmaker

National Geographic
 Society: STV: World
 Geography, Vol. 2

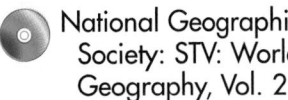 Geography and
 the Environment:
 The Infinite Voyage

FOCUS

Section Objectives
1. **Discover** why the Zaire River is a highway for Zaire's people.
2. **Discuss** Zaire's potential wealth.
3. **Explain** how large movements of people have shaped Zaire's history.

Bellringer Motivational Activity

Prior to taking roll at the beginning of the class period, project Section Focus Transparency 19-1 and have students answer the activity questions. Discuss students' responses. **L1**

Vocabulary Pre-check

Read several different meanings of the word *canopy* from a dictionary. Then have students find the term in the text and discuss why rain forest treetops are called a canopy. *(They form a covering for the forest.)* **L1** **LEP**

Use the Vocabulary PuzzleMaker Software to create a crossword puzzle. **L1**

MAP STUDY

Answer
north, then west,
then south

Map Skills Practice
Reading a Map What
lake forms part of Zaire's
southeastern border?
(Lake Tanganyika)

TEACH

Guided Practice

Display Political Map
Transparency 7. Have stu-
dents analyze the size and
location of Zaire relative to
other countries in the region. Ask them to speculate
on how Zaire's large size,
central location, and short
coastline might have af-
fected its history and economic development. Ask
students to record responses
in their notebooks. Have
them check the accuracy of
these responses as they
study Section 1. **L1**

CURRICULUM CONNECTION

Earth Science The Zaire
River is Africa's second-
longest river. Draining almost all of west Central
Africa, the Zaire carries a
huge volume of water.
Only the Amazon River
in South America carries
more.

CENTRAL AFRICA: Physical

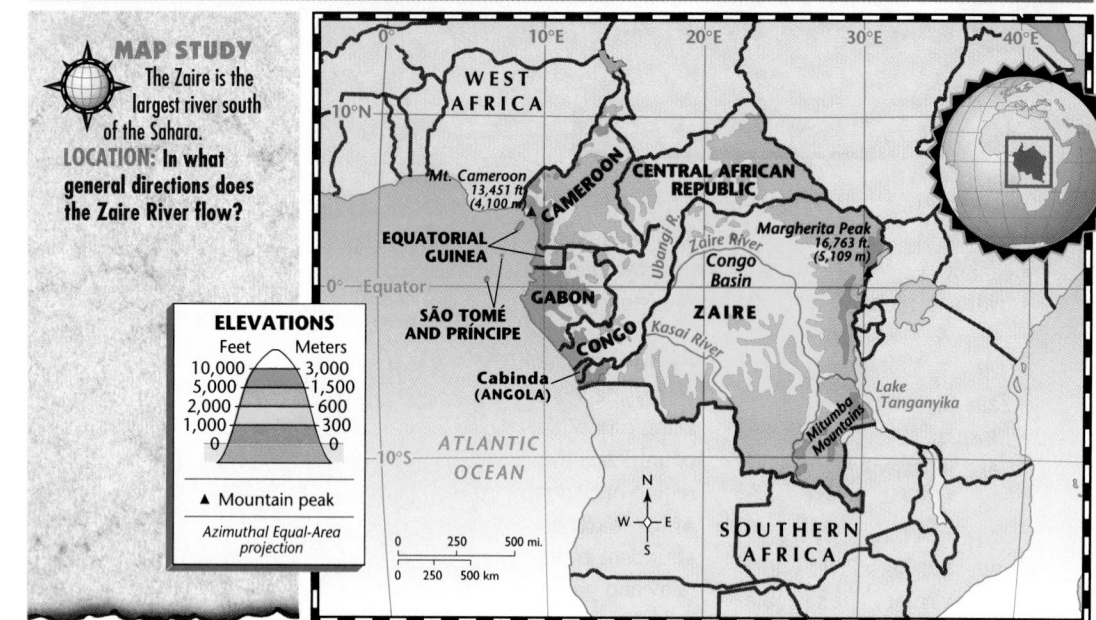

MAP STUDY
The Zaire is the largest river south of the Sahara.
LOCATION: In what general directions does the Zaire River flow?

ELEVATIONS

Feet	Meters
10,000	3,000
5,000	1,500
2,000	600
1,000	300
0	0

▲ Mountain peak

Azimuthal Equal-Area projection

Many monkeys, snakes, birds, and other small animals spend their entire lives in the canopy.

More than 750 different kinds of trees grow in Zaire's rain forests. Mahog-any, ebony, and other woods prized by furniture makers are harvested in abun-dance. Like South America's Amazon rain forest, Central Africa's rain forests are being destroyed at a fast rate.

Highlands and Savannas Spectacular mountains rise along Zaire's east-ern border. Savannas cross the northern and southern parts of the country. These vast open grasslands are home to some of Zaire's most valuable natural resources—its wildlife. Antelopes, leopards, lions, rhinoceroses, zebras, and giraffes roam freely on gov-ernment preserves. An animal similar to a giraffe—the okapi—is found only in Zaire and is the country's national symbol.

WHAT IN THE WORLD?

That's a Strong River!
The Zaire River doesn't stop when it meets the Atlantic Ocean. The river's current carries water about 100 miles (161 km) into the ocean. The force of the current has gouged a 4,000-foot (1,200-m) canyon into the ocean floor!

Rivers and Lakes The mighty Zaire River—above 2,800 miles (4,505 km) long—courses its way through the rain forests of Central Africa to the Atlantic Ocean. Many smaller rivers branch out from the Zaire River.

One of the world's longest rivers, the Zaire River is the country's highway for trade and travel. Dugout canoes, steamers carrying passengers to various port cities, and cargo ships travel the river to the Atlantic Ocean.

506

Cooperative Learning Activity

Organize students into five groups and assign one of the following topics to each group: phys-ical features, climate, resources, agriculture, or cities. Have group members study the material about Zaire that pertains to their assigned topic. Then form new groups of five students, consisting of one member of each of the origi-nal groups. Have members of the new groups combine their knowledge to write a country profile about Zaire. **L1**

In Zaire's eastern mountainous region, you will find several beautiful blue lakes—Lake Albert, Lake Edward, Lake Kivu, and Lake Tanganyika (TAN•guh•NYEE•kuh). Lake Tanganyika is the world's longest freshwater lake and the second-deepest, after Lake Baikal in Russia.

The Climate Because of its location on the Equator, Zaire generally has a tropical climate. Rain forest areas are hot and humid all year. Average daytime temperatures here reach about 90°F (32°C). Heavy rainstorms bring 80 inches (203 cm) or more of rain each year. Savanna and highland areas are cooler and drier.

Human/Environment Interaction

The Economy

Zaire has a developing economy and the potential to become a wealthy country. The map on page 512 shows you Zaire's mineral resources. Mining is Zaire's main economic activity. Zaire is a world leader in copper production. Industrial diamonds—used in precision tools—rank second in importance. Zaire mines more industrial diamonds than any other country in Africa.

Energy Sources Zaire has oil and natural gas deposits, but rivers provide the main source of energy. Experts believe that Zaire's rivers have the ability to produce about 13 percent of the world's **hydroelectricity,** electricity created by

Cruising Along the Zaire River

A ferry shuttles passengers and vehicles across the Zaire River.
HUMAN/ENVIRONMENT INTERACTION: What is hydroelectricity?

Leading Diamond-Producing Countries

GRAPHIC STUDY

Three of the world's top diamond-producing countries are in Africa south of the Sahara.
PLACE: What two countries in Africa produce the most diamonds?

Diamonds Produced in One Year

Thousands of carats

Kilograms

Australia   Russia   Botswana   Zaire   South Africa

Country

◆ = 5,000 carats
(2,204 pounds)

Source: *Minerals Yearbook*, 1992

Meeting Special Needs Activity

Spatial Disability Have students with spatial orientation problems practice location skills using a map of Central Africa. Have them identify countries that border Zaire. Then have them state the location of each country relative to Zaire. Finally, have them make up questions about Central African countries based on location. Provide time for students to use their questions to quiz one another. **L1**

More About the Illustration

Answer to Caption
electricity generated by moving water

GEOGRAPHIC THEMES

Movement The Zaire River and its tributaries are navigable for more than 7,000 miles (11,263 km) within the borders of Zaire. When ships come to rapids or waterfalls, cargo is moved from the ships to railroad cars. It is then transported further up river, where it can be loaded back onto ships.

Independent Practice

▱ Guided Reading Activity 19-1 **L1**

MULTICULTURAL PERSPECTIVE

Culturally Speaking

Time does not have the same importance in Zaire as it does in Western cultures. The Lingala language has only one word for both "yesterday" and "tomorrow." The meaning must be determined from context.

GRAPHIC STUDY

Answer
Botswana, Zaire

Graphic Skills Practice
Reading a Graph About how many kilograms of diamonds does Zaire produce in one year? *(about 3,000 kg)*

NATIONAL GEOGRAPHIC SOCIETY

These materials are available from Glencoe.

 VIDEODISC

STV: World Geography Vol. 2: Africa and Europe

Zaire River
Side 1, Frames 23880-24199

ASSESS

Check for Understanding

Assign Section 1 Review as homework or an in-class activity.

Meeting Lesson Objectives

Each objective below is tested by the questions that follow it in parentheses.

1. **Discover** why the Zaire River is a highway for Zaire's people. (2)
2. **Discuss** Zaire's potential wealth. (3)
3. **Explain** how large movements of people have shaped Zaire's history. (4)

 MAP STUDY

Answer
Zaire

Map Skills Practice
Reading a Map What are the two main types of climate found in Central Africa? *(tropical rain forest, tropical savanna)*

508

---

moving water. The Zaire River carries more water than any other river in the world except the Amazon. About 10 million gallons (38 million l) of water rush down the Zaire River every second! No wonder the name *Zaire* comes from a word meaning "big river."

Farming Despite its rich mineral and energy resources, most people in Zaire make a living by subsistence farming. Farmers grow crops such as corn, rice, and cassava for their families. Coffee, cotton, and palm oil are grown to sell. Along with farming, the people of Zaire also hunt wild animals in the forests and fish in the rivers, sometimes with traps woven from vines.

Movement
The People

Zaire has a population of almost 43 million. Like other African countries, Zaire is made up of different ethnic groups with their own separate languages. At times, tensions between ethnic groups have led to fighting. In recent years, Zaire's government has helped reduce ethnic divisions and has given the people a greater sense of unity.

Influences of the Past Groups of people began moving into Zaire from other parts of Africa about 2,000 years ago. Before the A.D.1400s, several powerful kingdoms developed in the savanna area south of the rain forests. The largest kingdom was the Kongo, which ruled much of western Zaire.

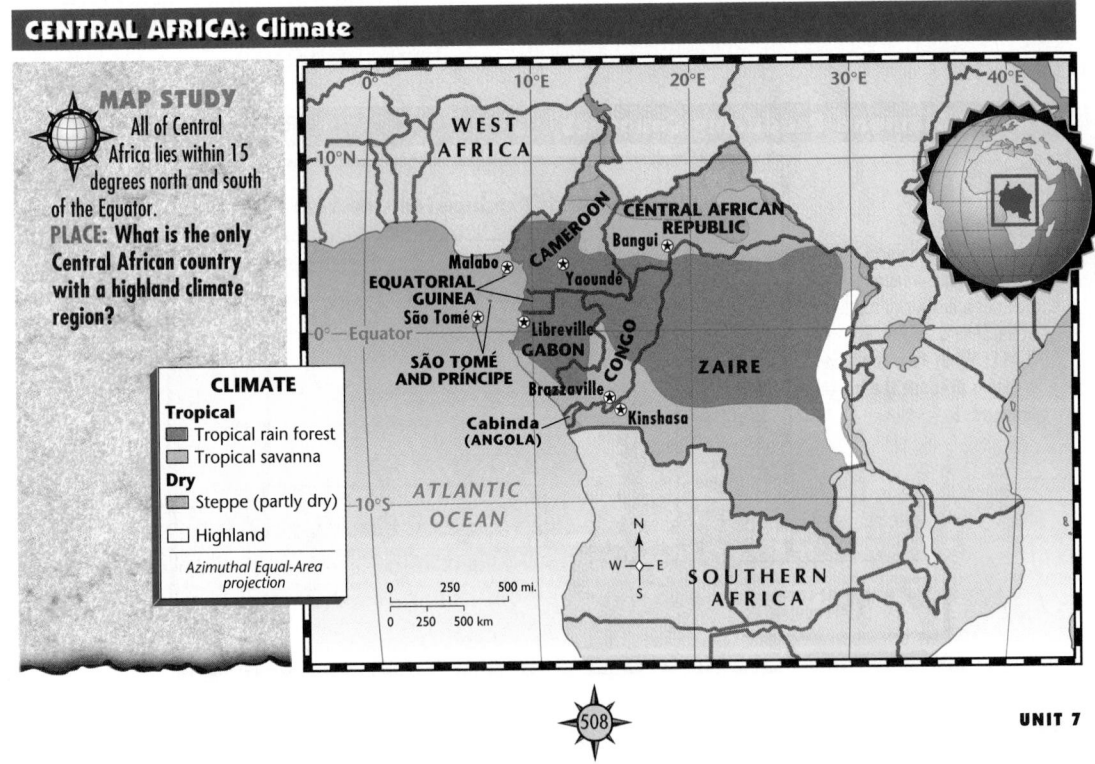

CENTRAL AFRICA: Climate

MAP STUDY
All of Central Africa lies within 15 degrees north and south of the Equator.
PLACE: What is the only Central African country with a highland climate region?

CLIMATE
Tropical
- Tropical rain forest
- Tropical savanna
Dry
- Steppe (partly dry)
- Highland

Azimuthal Equal-Area projection

0   250   500 mi.
0 250 500 km

508

Critical Thinking Activity

Drawing Conclusions Have students list obstacles to Zaire's economic development. Next, have students identify and list Zaire's major resources. Then encourage students to use this information to draw conclusions about the most serious economic challenges that Zaire faces and about the resources that might help Zaire overcome these challenges. **L2**

In the late 1400s, Portuguese and other European traders arrived in Central Africa. During the next 300 years, they enslaved thousands of Central Africans, shipping them to the Americas.

In 1878 Belgium's King Leopold II took over Zaire. He considered the entire region his own plantation and treated the people harshly. When these outrages were publicized in 1908, the Belgian government took control of the colony and named it the Belgian Congo. In 1960 the Belgian Congo became an independent nation known as Congo. Nine years later, the country changed its name to Zaire.

Ways of Life Zaire's culture is largely African with European influences. More than 75 percent of Zaire's people are Christians. Most of these are Roman Catholics. Other people in the country practice traditional African religions.

About 60 percent of Zaire's people live in rural villages. Since the 1960s, however, many Zairians have migrated to cities in search of jobs. This movement of people has turned Kinshasa, the capital, into a city of about 3.9 million people. Kinshasa is a major port on the Zaire River.

Fifteen-year-old Kilundu Basese lives in Kinshasa. During his free time, Kilundu strolls along Kinshasa's streets lined with trees, elegant shops, and outdoor cafés. Many areas of the city are crowded with new residents looking for work. In recent years Zaire's government has tried to improve housing, education, and health care for poorer people.

Market Day

Masika Mbwana lives in a village on the banks of the Zaire River. This morning, with other girls from her village, she will walk to the market in a nearby town. Masika's family grows its own food. Any extra food is gathered to sell at the market. Today Masika has a basket of cassava roots to sell. The meal made from ground cassava is mixed with corn or rice and is the main food of most people in Zaire.

SECTION 1 REVIEW

REVIEWING TERMS AND FACTS

1. Define the following: canopy, hydroelectricity.

2. PLACE Why is the Zaire River important to Zaire's people?

3. HUMAN/ENVIRONMENT INTERACTION Zaire leads the world in the production of what mineral resource?

4. MOVEMENT How was early Zaire settled?

MAP STUDY ACTIVITIES

5. Turn to the physical map on page 506. What large body of water forms part of Zaire's southeast boundary?

6. Look at the climate map on page 508. What climates are found in Zaire?

Evaluate

Section 1 Quiz **L1**

Use the Testmaker to create a customized quiz for Section 1. **L1**

Reteach

Make photocopies of the section and cut the paragraphs apart, omitting the subheadings. Give small groups of students a set of section pieces. Have groups put the paragraphs in order and write headings for the subsections. **L1**

Enrich

Have students research the animals of Zaire. Ask them to find pictures or make drawings of the animals. They might share their work with the class in the form of an oral report or a captioned pictorial display. **L3**

CLOSE

Have students take an imaginary trip along the Zaire River and describe the landscapes they would see. They might write journal entries describing their journey.

Answers to Section 1 Review

1. All vocabulary words are defined in the Glossary.

2. It is the country's major highway for trade and travel.

3. copper

4. About 2,000 years ago, groups of people from other parts of Africa began to move into what today is Zaire.

5. Lake Tanganyika

6. tropical rain forest, tropical savanna, highland, and several small areas of steppe

BUILDING GEOGRAPHY SKILLS

TEACH

Count the numbers of students in the class who have birthdays in the following periods: January through March, April through June, July through September, and October through December. List results on the chalkboard. Then ask students to present the information on the chalkboard in the form of a bar graph. If students need assistance, suggest that they read the introduction to this skill feature.

When students have completed their graphs, ask them what presenting the information in this fashion allows them to do. *(to visually compare the numbers of students born at various times during the year)* Then direct students to carefully read the skill and complete the practice questions. **L1**

Additional Skills Practice

1. Which countries have a literacy rate above 50 percent? *(Cameroon, Congo, Gabon, São Tomé and Príncipe, Zaire)*
2. Which countries have a literacy rate of 50 percent or lower? *(Angola, Central African Republic, Chad, Equatorial Guinea)*

▱ Geography Skills Activity 19

▱ Building Skills in Geography, Unit 2, Lesson 5

Reading a Bar Graph

You want to compare your running time in the 100-yard dash to that of the other sprinters on your team. Putting your running times on a bar graph would allow you to visually compare all the runners' times at once.

A *bar graph* presents numerical information in a visual way. Bars of various lengths stand for different quantities. Bars may be drawn vertically—up and down—or horizontally—left to right. Labels along the axes, or the left side and bottom of the graph, explain what the bars represent. Some bar graphs show changes over time. Others compare quantities during the same time period, but in different locations. To read a bar graph, apply these steps:

• Read the title to find out what the graph is about.
• Study the information on both axes to figure out what the bars represent.
• Compare the lengths of the bars to draw conclusions about the graph's topic.

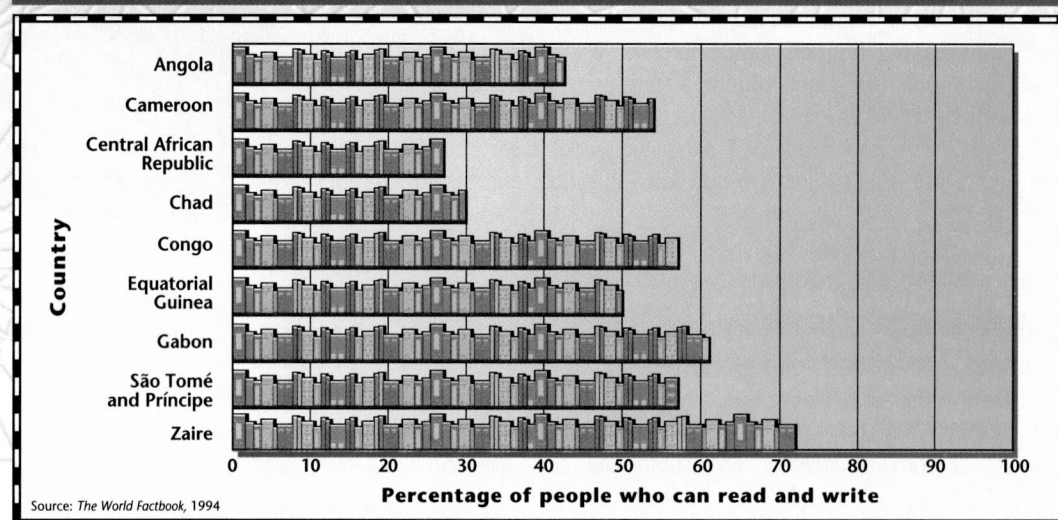

Literacy Rate in Selected African Countries

Source: *The World Factbook*, 1994

Geography Skills Practice

1. Which countries are compared on this graph?
2. What quantities appear on the horizontal axis? The vertical axis?
3. Which country has the highest literacy rate? Which has the lowest?

Answers to Geography Skills Practice

1. Angola, Cameroon, Central African Republic, Chad, Congo, Equatorial Guinea, Gabon, São Tomé and Príncipe, Zaire
2. percentage of people who can read and write; countries
3. Zaire; Central African Republic

Other Countries of Central Africa

PREVIEW

Words to Know
- basin
- tsetse fly

Places to Locate
- Central African Republic
- Cameroon
- Gabon
- Congo
- Equatorial Guinea
- São Tomé and Príncipe

Read to Learn . . .
1. where the Central African Republic and Cameroon are located.
2. what resources are found in Gabon and Congo.
3. how people in Equatorial Guinea and São Tomé and Príncipe make their livings.

The sound of buzz saws rings through Gabon's rain forests. These rain forests hold vast resources, including the ebony trees shown here.

Ebony and mahogany hardwoods provide Central Africa with one of its most valuable sources of income.

In addition to Zaire, Central Africa includes Gabon (ga•BOHN), the Central African Republic, Cameroon, Congo, Equatorial Guinea, and São Tomé (SOWN tuh•MAY) and Príncipe (PRIHN•suh•puh). All of these countries are only beginning to develop their natural resources.

Location
The Central African Republic and Cameroon

The Central African Republic and Cameroon became independent from France in 1960. Both countries are working to develop their economies. Cameroon, with a coastline on the Gulf of Guinea, enjoys greater natural advantages than the landlocked Central African Republic.

The Land The Central African Republic lies deep in the middle of Africa, just north of the Equator. Most of the country lies on a vast plateau bordered on both sides by basins. A **basin,** you will remember, is a broad flat valley. Savannas cover most of the plateau, but tropical rain forests are found in the

 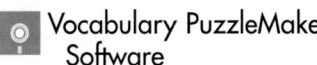
511

MAP STUDY

Answer
petroleum, manganese, uranium

Map Skills Practice
Reading a Map What city in Cameroon is a major manufacturing area? *(Douala)*

TEACH

Guided Practice

Predicting Direct students' attention to maps in Chapter 19 as well as those in the Unit 7 Atlas. Discuss what kinds of information they can learn about Central African countries. *(location, physical features, resources, climate)* Ask students to use this information from maps to predict which Central African countries have the greatest economic potential. Call on several volunteers to defend their answers. **L2**

MULTICULTURAL PERSPECTIVE

Culturally Speaking

The name *Cameroon* comes from *camaroes,* the Portuguese word for "shrimp." Portuguese explorers found small crayfish that looked like shrimp along the coast of present-day Cameroon.

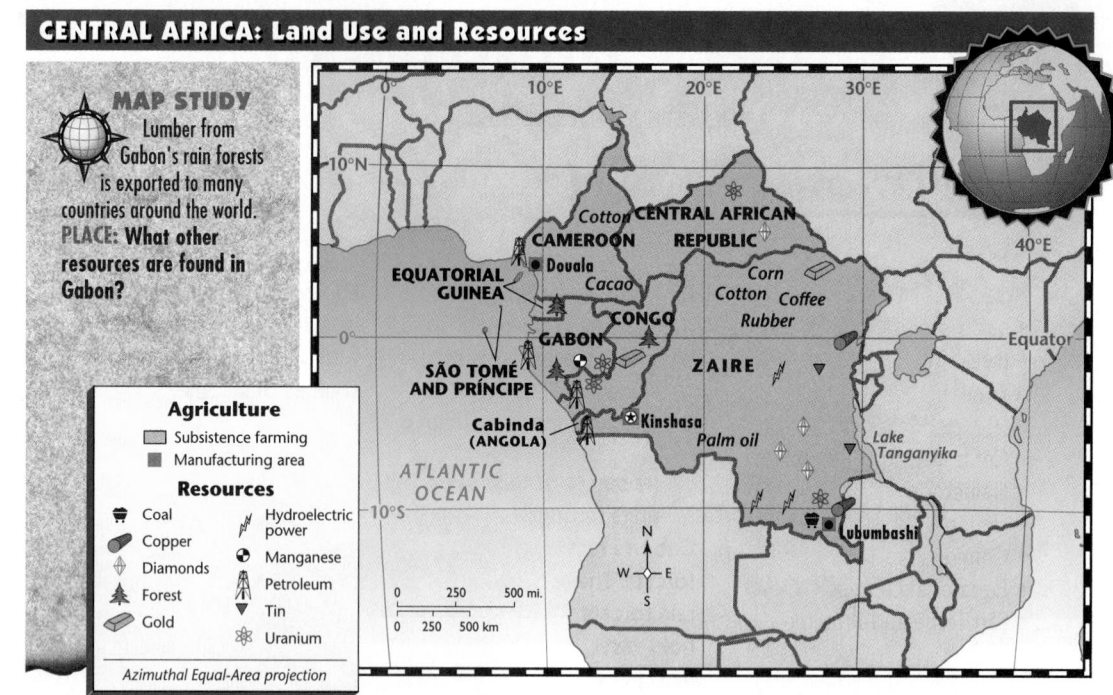

CENTRAL AFRICA: Land Use and Resources

MAP STUDY
Lumber from Gabon's rain forests is exported to many countries around the world.
PLACE: What other resources are found in Gabon?

Agriculture
☐ Subsistence farming
■ Manufacturing area

Resources
- Coal
- Copper
- Diamonds
- Forest
- Gold
- Hydroelectric power
- Manganese
- Petroleum
- Tin
- Uranium

Azimuthal Equal-Area projection

southwestern part of the country. The map on page 508 shows you that the Central African Republic has a tropical savanna climate and a steppe climate.

Cameroon lies on the west coast of Central Africa. In the south hot, humid lowlands stretch along the Gulf of Guinea. The central part of Cameroon is a forested plateau with cooler temperatures and less rainfall. To the north lies a partly dry grasslands area. Mountains and hills run along Cameroon's western border.

The Economy Most people in the Central African Republic and Cameroon depend on farming for a living. In the Central African Republic, most farmers grow only enough food to feed their families. A few large plantations raise coffee, cotton, and rubber for export. Farmers in Cameroon raise cassava, corn, millet, and yams. The chief cash crops are bananas, cacao, coffee, cotton, and peanuts.

The Central African Republic and Cameroon are only beginning to industrialize. Cameroon, however, has had greater success in this effort. It has coastal ports and many natural resources such as petroleum, bauxite, and forest products. With no seaports and limited resources, the Central African Republic can claim only diamond mining as an important industry.

Some people in these two countries herd livestock. They raise their animals in regions that are safe from tsetse flies. The bite of the **tsetse fly** causes a deadly disease called sleeping sickness in cattle. Tsetse flies can also spread this disease to humans.

512

Cooperative Learning Activity

Organize students into three groups and assign each group one of the following subdivisions of this section: the Central African Republic and Cameroon, Gabon and Congo, or the Island Countries. Have group members prepare summaries, maps, charts, and graphs pertaining to their assigned countries. Then ask each group to teach the rest of the class about its assigned region. **L2**

The People Most of the people in the Central African Republic and Cameroon live in rural areas. In recent years, however, many people from the countryside have moved to cities such as Yaoundé (yown•DAY) in Cameroon and Bangui (bahn•gee) in the Central African Republic in search of work.

Many languages are spoken in these cities. The people of the Central African Republic and Cameroon belong to many ethnic groups. In both countries, most people follow traditional African religions. Smaller numbers are Christians or Muslims.

Place
Gabon and Congo

Gabon and Congo—once French colonies—became independent countries in 1960. Both have many natural resources and excellent waterways. The map on page 508 shows you that Gabon and Congo lie on the Equator along the west coast of Central Africa. Because of this location, both countries have hot, humid climates.

The Land Although they share the same type of climate, Gabon and Congo have different landforms. In Gabon, palm-lined beaches and swamps run along the Gulf of Guinea. As you move inland, you see that the land rises to become rolling hills and low mountains. Thick rain forests crossed by rivers cover most of Gabon.

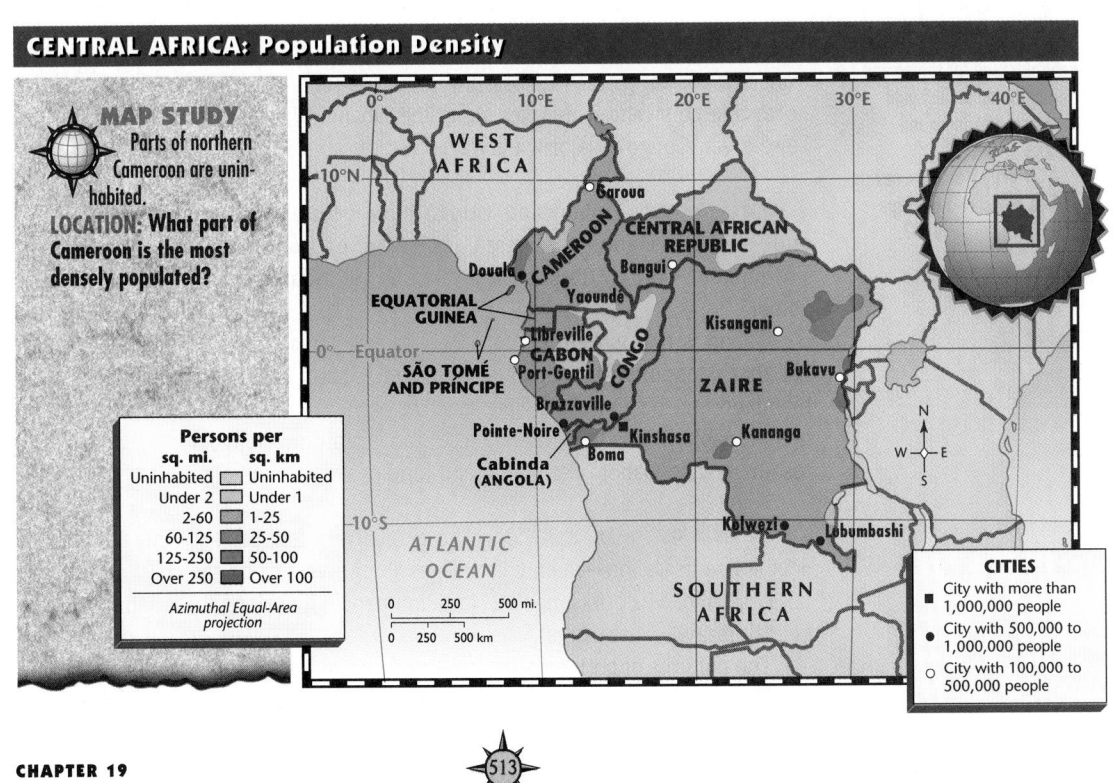

CENTRAL AFRICA: Population Density

MAP STUDY

Parts of northern Cameroon are uninhabited.

LOCATION: What part of Cameroon is the most densely populated?

Persons per

sq. mi.	sq. km
Uninhabited	Uninhabited
Under 2	Under 1
2-60	1-25
60-125	25-50
125-250	50-100
Over 250	Over 100

Azimuthal Equal-Area projection

CITIES

■ City with more than 1,000,000 people
● City with 500,000 to 1,000,000 people
○ City with 100,000 to 500,000 people

Meeting Special Needs Activity

Reading Disability Review the question-answer relationships (QAR) described as "right there" and "think and search." (see p. 487) Then add another QAR called "on your own," in which answers are not stated in text. These questions require students to do research. Refer students to the last paragraph under "The Economy" on page 512. Ask, "What are the symptoms of sleeping sickness?" Point out that the answer is not in the text. Have volunteers use references to find the answer. **L1**

LESSON PLAN
Chapter 19, Section 2

Independent Practice

 Guided Reading Activity 19-2 **L1**

NATIONAL GEOGRAPHIC SOCIETY

These materials are available from Glencoe.

 VIDEODISC

STV: World Geography Vol. 2: Africa and Europe

Central Africa
Side 1, Frames 26591-30935

Strange but TRUE!

In August 1986, a cloud of carbon dioxide gas rose from the waters of Lake Nios in western Cameroon. As the gas drifted across the land, it suffocated more than 1,700 people. No one knows what caused the release of gas from the lake. Some scientists believe it was the result of volcanic activity on the lake floor.

MAP STUDY

Answer
northwestern coast

Map Skills Practice
Reading a Map What is the only city in Central Africa with a population of more than one million people? *(Kinshasa)*

More About the Illustration

Answer to Caption
because it consists of volcanic ash

Movement During the 1500s, São Tomé and Príncipe became a center of the slave trade. Enslaved people were sent from mainland Africa to the islands. There, they were loaded on ships and transported to the Americas.

ASSESS

Check for Understanding

Assign Section 2 Review as homework or an in-class activity.

Meeting Lesson Objectives

Each objective below is tested by the questions that follow it in parentheses.

1. **Note** where the Central African Republic and Cameroon are located. (2)
2. **Identify** what resources are found in Gabon and Congo. (3, 5)
3. **Describe** how people in Equatorial Guinea and São Tomé and Príncipe make their livings. (4)

Boats are used for fishing and transportation in the island country of São Tomé and Príncipe.
PLACE: Why does the nation of São Tomé and Príncipe have such fertile soil?

In Congo, a low treeless plain stretches along the Atlantic coast. Just inland from this coastal plain, you will find low mountain ranges and plateaus. To the north a large swampy area lies along the Ubangi River. Farther south flows the Zaire River, Congo's major waterway.

The Economy Gabon and Congo have economies based on agriculture and mining. Gabon's farmers cultivate cacao and coffee for export. Throughout Congo subsistence farmers grow food crops such as cassava and yams.

Gabon is rich in natural resources. Its rain forests provide high-quality lumber that is exported throughout the world. Gabon also has oil reserves on the mainland and off the coast. The map on page 512 shows you that Gabon also has valuable deposits of manganese and uranium. Congo has fewer natural resources than Gabon. Petroleum and lumber are its main exports.

The People Gabon and Congo have relatively few people. Only 1.1 million people live in Gabon—mainly in villages along rivers or on the coast at the capital, Libreville. Most of Congo's 2.4 million people live along the Atlantic coast or near Brazzaville, the capital.

The cultures of both countries are made up of many ethnic groups. In Gabon the Fang group controls the national government. Two groups—the Kongo and

514

UNIT 7

Critical Thinking Activity

Analyzing Information Point out that the economies of most Central African countries are based on agriculture. Have students analyze reasons why farming is a primary activity in countries with developing economies. *(Developing countries have few resources or they have not developed their resources.)* Then have students speculate on the changes that must take place to encourage economic development in Central African countries. *(Answers might include: more education, job training, financial assistance from developed countries.)* **L2**

the Bateke (bah•TEH•keh)—make up most of Congo's population. Most Kongo farm for a living. The Bateke are primarily hunters and fishers.

Place
The Island Countries

The map on page 504 shows you that Equatorial Guinea and São Tomé and Príncipe are island nations. Equatorial Guinea includes mainland territory on the west coast of Africa, plus five offshore islands. São Tomé and Príncipe lies in the Gulf of Guinea, about 180 miles (290 km) west of the Central African coast. This country consists of two main islands and several tiny islands.

Equatorial Guinea Equatorial Guinea was a Spanish colony until 1968. Today the country is home to about 400,000 people from different ethnic groups. Most of Equatorial Guinea's people live on the mainland. They live in rural areas where they grow bananas and coffee. They also harvest timber from the country's thick rain forests. Equatorial Guinea's largest island, Bioko (bee•OH•koh), lies in the Gulf of Guinea. Farmers grow cacao in Bioko's rich volcanic soil. Malabo, the nation's capital and largest city, is on Bioko.

São Tomé and Príncipe At one time a Portuguese colony, the island country of São Tomé and Príncipe became independent in 1975. The Portuguese first settled the islands in 1470. At that time, the islands had no humans living on them. Today São Tomé and Príncipe has about 100,000 people. Most are of mixed African and Portuguese ancestry. Nearly all of the population live on São Tomé, the largest island.

São Tomé and Príncipe lies on a chain of inactive volcanoes, which no longer erupt. Over the years, volcanic ash formed deep layers of fertile soil. Today plantation workers make use of this rich soil to grow bananas, cacao, coconuts, and coffee for export.

SECTION 2 REVIEW

REVIEWING TERMS AND FACTS
1. Define the following: basin, tsetse fly.
2. LOCATION What Central African country is landlocked?
3. HUMAN/ENVIRONMENT INTERACTION What is Gabon's most important natural resource?
4. HUMAN/ENVIRONMENT INTERACTION How do most people in Equatorial Guinea and São Tomé and Príncipe make their livings?

MAP STUDY ACTIVITIES

5. Turn to the land use map on page 512. What natural resources are found in Congo?
6. Look at the population density map on page 513. What is the population density of most of Central Africa?

Answers to Section 2 Review

1. All vocabulary words are defined in the Glossary.
2. Central African Republic
3. lumber
4. by farming
5. petroleum, uranium, gold, forests
6. 2–60 persons per sq. mile (1–25 persons per sq. km)

Evaluate

Section 2 Quiz **L1**

 Use the Testmaker to create a customized quiz for Section 2. **L1**

Chapter 19 Test Form A and/or Form B **L1**

Reteach

Reteaching Activity 19 **L1**

Spanish Reteaching Activity 19 **L1**

Enrich

Enrichment Activity 19 **L3**

CLOSE

Mental Mapping Activity

Provide each student with a large sheet of white paper and map pencils. Inform students that this will not be a graded activity. Have students draw, freehand from memory, a map of Central Africa that includes the following labeled countries: Zaire, Central African Republic, Cameroon, Gabon, Congo, Equatorial Guinea, and São Tomé and Príncipe.

MAKING CONNECTIONS

TEACH

Ask students to define the term *migration.* Note their responses on the chalkboard. Then ask students to suggest reasons why people might migrate. List these responses on the chalkboard. As students read the feature, have them compare the reasons for the Bantu migrations with their suggestions. L1

More About . . .

Cultural Diffusion Indonesian sailors venturing eastward in search of trade introduced the Asian banana to coastal settlements in East Africa. Of much better quality than its African relative, the Asian banana quickly spread deep into Central Africa. This further enhanced the food supply and greatly added to population pressures among the Bantu.

THE BANTU MIGRATIONS

| MATH | SCIENCE | HISTORY | LITERATURE | TECHNOLOGY |

Around 500 B.C., a people known as the Bantu lived in West Africa. Bantu, meaning "the people," is the name of a major African language group. The Bantu lived simply—farming, hunting, and fishing—probably in the Benue Valley of Nigeria. Within 500 years the Bantu had learned to mine and to work metals.

POPULATION GROWTH The development of metalworking allowed Bantu farmers to make iron tips for their tools and Bantu hunters to fashion weapons. These advancements made farming easier and hunting more successful. As food supplies increased, more people could be supported. The population in Bantu settlements thrived and grew.

By about 2,000 years ago, the Bantu population had increased so much that most settlements could no longer feed all their members. As a result, small groups split off to search for new land.

CENTURIES OF MOVEMENT Anthropologists have come to believe that Bantu-speaking groups gradually spread throughout central, eastern, and southern Africa. As they pushed into new areas, the Bantu shared their farming and metalworking knowledge—

as well as their language—with other African groups.

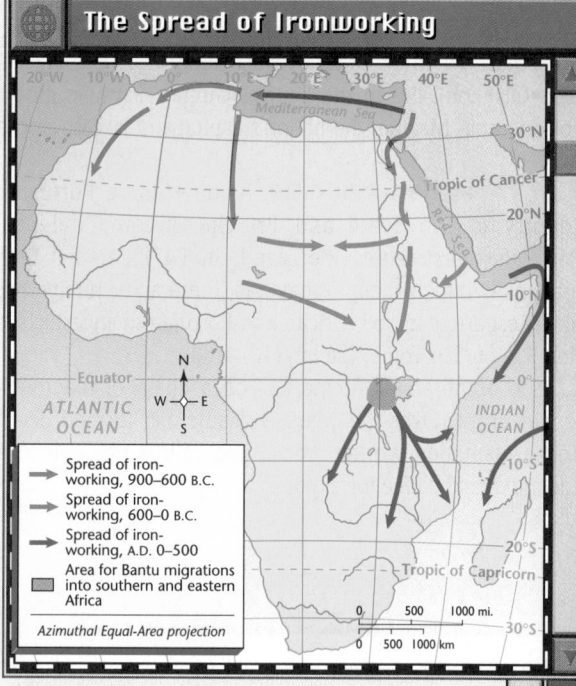

The Spread of Ironworking

→ Spread of ironworking, 900–600 B.C.
→ Spread of ironworking, 600–0 B.C.
→ Spread of ironworking, A.D. 0–500
▨ Area for Bantu migrations into southern and eastern Africa

Azimuthal Equal-Area projection

0   500   1000 mi.
0   500 1000 km

By A.D. 1500, nearly 15 centuries of migration had produced hundreds of ethnic groups, each culturally different from the others. Today between 60 and 80 million African people speak Bantu languages.

Making the Connection

1. Why did Bantu-speaking settlements increase in population?
2. To where did Bantu groups migrate?

—516—

Answers to Making the Connection

1. The use of iron allowed the Bantu to make better tools and weapons. This, in turn, allowed them to increase their food supply.
2. to the central, eastern, and southern areas of Africa

Chapter 19 Highlights

Important Things to Know About Central Africa

SECTION 1 ZAIRE

- Zaire's land area is about one-fourth the size of the United States.
- The Zaire River is Zaire's major "highway."
- Zaire has one of the largest tropical rain forest areas in the world.
- Vast savannas are home to Zaire's most valuable natural resources—antelopes, leopards, lions, zebras, giraffes, and other wildlife.
- Because of its location on the Equator, Zaire generally has a tropical climate.
- Many important mineral resources, including copper and industrial diamonds, are found in Zaire.
- Zaire's 43 million people belong to many different ethnic groups.

SECTION 2 OTHER COUNTRIES OF CENTRAL AFRICA

- The Central African Republic's landlocked location makes economic development difficult.
- Cameroon's coastal location has boosted its economic growth.
- Gabon and Congo export lumber and petroleum.
- Most of the people in Equatorial Guinea live on the mainland where they farm and harvest wood from the rain forests.
- São Tomé and Príncipe is an island nation with rich volcanic soil good for farming.

A market in Zaire ▶

Using the Chapter 19 Highlights

- Use the Chapter 19 Highlights to preview, review, condense, or reteach the chapter.

Preview/Review

 Vocabulary Puzzle-Maker Software reinforces the terms used in Chapter 19.

Condense

Have students read the Chapter 19 Highlights. Spanish Chapter Highlights also are available.

Chapter 19 Digest Audiocassettes and Activity

Chapter 19 Digest Audiocassette Test

Spanish Chapter Digest Audiocassettes, Activities, and Tests also are available.

Guided Reading Activities

Reteach

Reteaching Activity 19. Spanish Reteaching activities also are available.

MAP STUDY

Human/Environment Interaction

Copy and distribute page 39 of the Outline Map Resource Book. Have students label the countries of Central Africa. Then have them locate and mark with symbols and/or colors the main resources and forms of economic activity in these countries.

Extra Credit Project

Create an Encyclopedia Have students work in small groups to create illustrated encyclopedias for Central Africa. Offer the following as a sample entry: **T** is for *Tanganyika* . . . a lake in southeastern Central Africa on the border between Zaire and Tanzania. Lake Tanganyika is the world's longest freshwater lake and the second-deepest, after Lake Baikal in Russia.

Point out that this entry was taken directly from the chapter. Students will have to consult almanacs and encyclopedias for items beginning with certain letters. Display finished encyclopedias for other geography classes to view and use. **L3**

517

Chapter 19 Review and Activities

ANSWERS

Reviewing Key Terms

1. C
2. B
3. D
4. A

Mental Mapping Activity

This exercise helps students to visualize the countries and geographic features they have been studying and understand the relationships among them. All attempts at freehand mapping should be accepted.

REVIEWING KEY TERMS

Match the numbered terms in Column A with their definitions in Column B.

A
1. hydroelectricity
2. tsetse fly
3. basin
4. canopy

B
A. forest covering
B. insect that spreads tropical diseases to humans and animals
C. power developed from the force of moving water
D. broad, flat valley

Mental Mapping Activity

On a separate piece of paper, draw a freehand map of Central Africa. Label the following items on your map:
- Zaire
- Zaire River
- Gulf of Guinea
- Gabon
- Central African Republic
- Cameroon
- Congo

REVIEWING THE MAIN IDEAS

Section 1

1. PLACE What landscapes are found in Zaire?
2. HUMAN/ENVIRONMENT INTERACTION What challenges to the environment does Zaire face?
3. PLACE How do Zaire's many ethnic groups affect its national unity?
4. REGION Why are the savannas of Zaire important?

Section 2

5. LOCATION Describe how the locations of the Central African Republic and Cameroon have influenced their economic development.
6. PLACE What mineral resources are found in Gabon?
7. MOVEMENT What city in Congo has become home to many of the country's people?
8. PLACE What Central African nation is made up entirely of islands?

Reviewing the Main Ideas
Section 1

1. rain forests, savannas, and highlands
2. Its rain forests are quickly being destroyed.
3. Sometimes, ethnic differences have led to conflict. Recently, the government of Zaire has taken steps to reduce ethnic tensions and build a greater sense of national unity.
4. They provide grass for Zaire's most valuable resource—wildlife.

Section 2

5. The Central African Republic is landlocked and has few resources, so its development has been relatively slow. Cameroon has a coastline, good seaports, and many resources. Its development has been more rapid.
6. forests, petroleum, manganese, uranium
7. Brazzaville
8. São Tomé and Príncipe

CRITICAL THINKING ACTIVITIES

1. **Determining Cause and Effect** How do you think the presence of thick rain forests has affected the development of Central Africa?

2. **Drawing Conclusions** As the nations of Central Africa make use of their resources, what challenges do they face in developing their economies?

GeoJournal Writing Activity

Imagine that you work in a large corporation. You have been sent by your company to open a branch office in Kinshasa, Zaire. Write a letter to your employer back in the United States describing how life in Zaire is different from and similar to life in the United States.

COOPERATIVE LEARNING ACTIVITY

Work in groups of three to learn about environmental challenges facing one of the Central African countries. Then create a documentary showing your findings. Each group should choose one country and do research for a documentary. Your findings should be presented in storyboard form—showing still pictures and dialogue that go together to form the documentary. Research as a group, then assign the following individual roles: (a) author, (b) editor, and (c) critic.

PLACE LOCATION ACTIVITY: CENTRAL AFRICA

Match the letters on the map with the places and physical features of Central Africa. Write your answers on a separate sheet of paper.

1. Congo
2. Lake Tanganyika
3. Libreville
4. Cameroon
5. São Tomé and Príncipe
6. Kinshasa
7. Zaire River

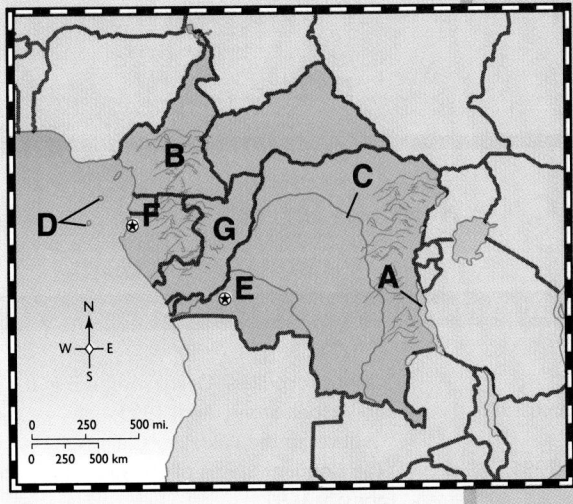

Cooperative Learning Activity

As groups present their storyboards, have the rest of the class consider the other areas of the world that face similar environmental challenges.

GeoJournal Writing Activity

Encourage students to share their letters with the rest of the class.

Place Location Activity

1. G
2. A
3. F
4. B
5. D
6. E
7. C

Chapter Bonus Test Question

This question may be used for extra credit on the chapter test.

If you traveled on the Zaire River from its source to the Atlantic Ocean, how many times would you cross the Equator? *(twice)*

Critical Thinking Activities

1. Answers will vary. Some students will suggest that the bountiful resources of the rain forests helped development by providing Central African countries with important export industries, such as the lumber industry. Others might add that the humid tropical climate encouraged the cultivation of such crops as cacao. The rain forests' very bounty may have encouraged dependence on one-crop economies in the region. Further, the rain forests have hindered communications and travel in the region.

2. Answers will vary, but might include: depletion of the rain forests and other non-renewable resources, development of new trade ties with foreign countries, improving education and training of population.

PERFORMANCE ASSESSMENT ACTIVITY

AWARDS FOR EAST AFRICA Organize students into small groups to act as awards committees. Groups should design awards to give to different countries, cities, or physical features in East Africa. These awards might include "Best Source of a Major River" for Victoria Falls in Uganda, or "The Country Most Likely to Alter Archaeologists' Opinions" for Tanzania. Students should create at least five awards, each of a different type (certificate, trophy, and so on) with appropriate illustrations. A brief explanation should accompany each award, showing in more detail the importance of the country, city, or physical feature and why the award was given.

Students may want to extend the task by holding an awards show, where they choose a winner for each of the different categories.

POSSIBLE RUBRIC FEATURES: Analysis and evaluation skills, design skills, creative thinking, content information, decision-making skills

MENTAL MAPPING ACTIVITY

Before teaching Chapter 20, point out the countries of East Africa on a globe. Have students note the location of these countries relative to the countries they studied in Chapters 18 and 19. Also point out the bodies of water that border East Africa—the Red Sea, the Gulf of Aden, and the Indian Ocean.

TEACHER'S CORNER

NATIONAL GEOGRAPHIC SOCIETY

INDEX TO NATIONAL GEOGRAPHIC MAGAZINE

The following articles may be used for research relating to this chapter:

- "Eritrea Wins the Peace," by Charles E. Cobb, Jr., June 1996.
- "Face-to-Face With Lucy's Family," by Donald C. Johanson, March 1996.
- "Jane Goodall," by Peter Miller, December 1995.
- "The Mountain Gorillas of Africa," by Paul F. Salopek, October 1995.
- "The Dawn of Humans," by Meave Leakey, September 1995.
- "Tragedy Stalks the Horn of Africa," by Robert Caputo, August 1993.

- "Giant Crocodiles—Deadly Ambush in the Serengeti," by Mark Deeble and Victoria Stone, April 1993.
- "Captives in the Wild," by Craig Packer, April 1992.
- "The Eloquent Surma of Ethiopia," by Carol Beckwith and Angela Fisher, February 1991.
- "Africa's Great Rift," by Curt Stager, May 1990.

NATIONAL GEOGRAPHIC SOCIETY PRODUCTS AVAILABLE FROM GLENCOE

To order the following products for use with this chapter, contact your local Glencoe sales representative or call Glencoe at 1-800-334-7344:

- *STV: World Geography* (Videodisc)
- *Picture Atlas of the World* (CD-ROM)
- *Physical Geography of the World* (Transparencies)
- *ZipZapMap! World* (Software)

- *GeoBee* (Software)
- *Images of the World* (Posters)
- *Eye on the Environment* (Posters)

ADDITIONAL NATIONAL GEOGRAPHIC SOCIETY PRODUCTS

To order the following products for use with this chapter, call National Geographic Society at 1-800-368-2728:

- *Africa's Animal Oasis* (Video)
- *Lions of the African Night* (Video)
- *Among the Wild Chimpanzees* (Video)
- *Jane Goodall: My Life with the Chimpanzees* (Video)
- *Search for the Great Apes* (Video)

- *Serengeti Diary* (Video)
- *Eternal Enemies: Lions and Hyenas* (Video)
- *Zebra: Patterns in the Grass* (Video)
- *Physical Geography of the Continents Series: "Africa."* (Video)

TEACHER-TO-TEACHER
Curriculum Connection: Language Arts

—from Susie Gerschutz
Norton Middle School, Columbus, OH

CODES OF COMMUNICATION *The purpose of this activity is to help students appreciate the challenges to communication and national unity when different groups within a country speak different languages.*

Inform students that while Swahili and English are the official languages of Kenya, most of the country's more than 40 different ethnic groups speak their own regional languages. Then organize students into several small groups. Encourage groups to develop their own alphabets. Some might use an alphabet where the letters are switched—*B* becomes *A, C* becomes *B,* and so on. Others might use numerals for some or all of the letters. Have each group use its new alphabet to write on a sheet of notepaper three statements about the geography of Kenya. Have groups exchange papers and attempt to translate the statements they have received. If this proves difficult, have groups exchange copies of the new alphabets. Call on groups to read out their translations.

Have students discuss the following: What difficulties did you face in sharing information about Kenya? What special challenges does a multi-language nation such as Kenya face? What would you do to meet such challenges? Why might some Kenyans be reluctant to speak English? What new problems might be created by forcing all Kenyans to speak Swahili?

MEETING NATIONAL STANDARDS

Geography for Life

All of the 18 standards are demonstrated in Unit 7. The following ones are highlighted in this chapter:

• Standards 4, 5, 7, 8, 17, 18

KEY TO ABILITY LEVELS

Teaching strategies have been coded for varying learning styles and abilities.

L1 **BASIC** activities for all students

L2 **AVERAGE** activities for average to above-average students

L3 **CHALLENGING** activities for above-average students

LEP **LIMITED ENGLISH PROFICIENCY** activities

BIBLIOGRAPHY

Readings for the Student

Bechky, Allen. *Adventuring in East Africa.* San Francisco: Sierra Club, 1990.

Heritage Library of African Peoples: East Africa. New York: Rosen, 1994.

Jefferson, Margo, and Elliott P. Skinner. *Roots of Time: A Portrait of African Life and Culture.* Trenton, N.J.: African World Press, 1990.

Ricciuti Edward. *Somalia: A Crisis of Famine and War.* Brookfield, Conn.: Millbrook Press, 1993.

Readings for the Teacher

Africa Today: A Reproducible Atlas. Wellesley, Mass.: World Eagle, 1994.

Clark, Leon E. *Through African Eyes, Volume 1: The Past—The Road to Independence,* revised ed. New York: Center for International Training and Education, 1991.

Creative Activities for Teaching About Africa. Stockton, Calif.: Stevens & Shea, 1990.

Map Skills and Outlines: Africa. St. Louis: Milliken, 1991. Transparencies and reproducible worksheets.

Multimedia

Africa: Central and Eastern Regions. Chicago: Encyclopedia Britannica, 1993. Videocassette, approximately 20 minutes.

Global Studies With Country Databases: Africa. Toronto, Canada: WorldView Software, 1994. Apple 5.25" disk or IBM 3.5" disk.

Reflections of Elephants. Washington, D.C.: National Geographic Society, 1994. Videocassette, 59 minutes.

Use these *Geography: The World and Its People* resources to teach, reinforce, and extend chapter content.

CHAPTER 20 RESOURCES

- *Vocabulary Activity 20

- Cooperative Learning Activity 20

- Workbook Activity 20

- Geography Skills Activity 20

- Chapter Map Activity 20

- GeoLab Activity 20

- Critical Thinking Skills Activity 20

- Performance Assessment Activity 20

- *Reteaching Activity 20

- Enrichment Activity 20

- Chapter 20 Test, Form A and Form B

- *Chapter 20 Digest Audiocassette, Activity, Test

- Unit Overlay Transparencies 7-0, 7-1, 7-2, 7-3, 7-4, 7-5, 7-6

- Political Map Transparency 7; World Cultures Transparency 11

- Geoquiz Transparencies 20-1, 20-2

- Vocabulary PuzzleMaker Software

- Student Self-Test: A Software Review

- Testmaker Software

- MindJogger Videoquiz

If time does not permit teaching the entire chapter, summarize using the Chapter 20 Highlights on page 539, and the Chapter 20 English (or Spanish) Audiocassettes. Review students' knowledge using the Glencoe MindJogger Videoquiz. *Also available in Spanish*

Use these *Geography: The World and Its People* resources to teach and reinforce section content.

SECTION 1 RESOURCES

Reproducible Lesson Plan 20-1

Section Focus Transparency 20-1

Guided Reading Activity 20-1

Section Quiz 20-1 (also available in Spanish)

SECTION 2 RESOURCES

Reproducible Lesson Plan 20-2

Section Focus Transparency 20-2

Guided Reading Activity 20-2

Section Quiz 20-2 (also available in Spanish)

SECTION 3 RESOURCES

Reproducible Lesson Plan 20-3

Section Focus Transparency 20-3

Guided Reading Activity 20-3

Section Quiz 20-3 (also available in Spanish)

SECTION 4 RESOURCES

Reproducible Lesson Plan 20-4

Section Focus Transparency 20-4

Guided Reading Activity 20-4

Section Quiz 20-4 (also available in Spanish)

ADDITIONAL RESOURCES FROM GLENCOE

Reproducible Masters

- Glencoe Social Studies Outline Map Book, pages 36, 37

- Facts On File, GEOGRAPHY ON FILE, Sub-Saharan Africa; AFRICAN HISTORY ON FILE

- Foods Around the World, Unit 7

Workbook

- Building Skills in Geography, Unit 2, Chapter 6

World Music: Cultural Traditions
Lesson 6

Transparencies

- National Geographic Society PicturePack Transparencies, Unit 7

World Games Activity Cards
Unit 7

Countries of the World Flashcards
Unit 7

Videodiscs

- Reuters Issues in Geography

- National Geographic Society: STV: World Geography, Vol. 2

CD-ROM

- National Geographic Society: Picture Atlas of the World

Software

- National Geographic Society: ZipZapMap! World

Performance Assessment

Refer to the Planning Guide on page 520A for a Performance Assessment Activity for this chapter. See the *Performance Assessment Strategies and Activities* booklet for suggestions.

Chapter Objectives

1. **Discuss** the physical and cultural geography of Kenya.
2. **Review** the cultural geography of Tanzania.
3. **Describe** climate, economic activities, and ethnic tensions in inland East Africa.
4. **Outline** the geography of the Horn of Africa.

GLENCOE TECHNOLOGY

VIDEODISC

Use the Chapter 20 MindJogger Videoquiz to preview chapter content.

MindJogger Videoquiz

Chapter 20
Disc 3 Side A

 Also available in VHS.

Chapter 20 East Africa

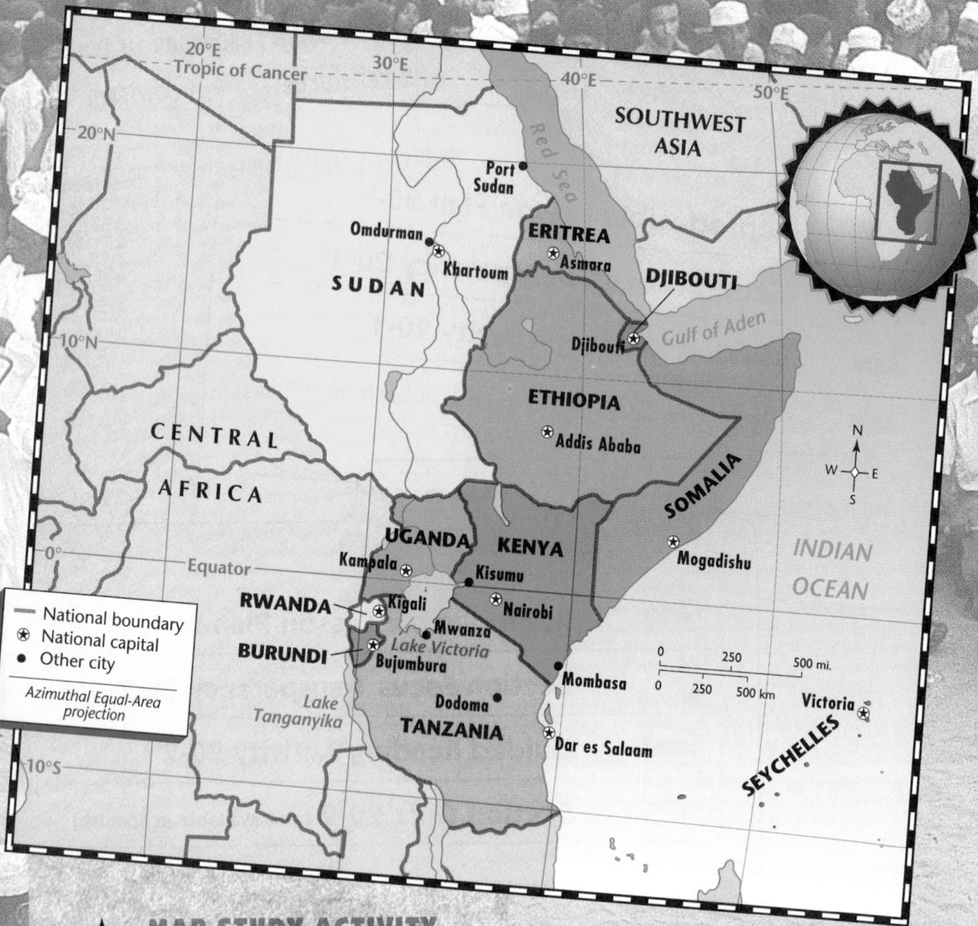

National boundary
⚿ National capital
• Other city

Azimuthal Equal-Area projection

MAP STUDY ACTIVITY

As you read Chapter 20, you will learn about Kenya and the other countries in East Africa.

1. What East African countries border the Red Sea?
2. What is the largest East African country?
3. What country east of the African mainland is made up of islands?

520

MAP STUDY ACTIVITY

Answers
1. Sudan, Eritrea
2. Sudan
3. Seychelles

Map Skills Practice
Reading a Map Which countries' capitals are located on the shores of the Indian Ocean? (*Somalia, Tanzania*) What is the capital of Kenya? (*Nairobi*)

LESSON PLAN
Chapter 20, Section 1

PREVIEW

Words to Know
- coral
- reef
- fault
- escarpment
- poacher

Places to Locate
- Kenya
- Great Rift Valley
- Indian Ocean
- Nairobi
- Mombasa

Read to Learn . . .
1. what landforms are found in Kenya.
2. why most Kenyans live in highland areas.
3. what languages are spoken in Kenya.

The Great Rift Valley cuts a deep gash through East Africa. In places, the valley's sides are about 1 mile (1.6 km) high, and the valley floor is more than 50 miles (80 km) wide. The Great Rift Valley was formed when two of the plates that make up the earth's crust moved apart millions of years ago.

Meeting National Standards

Geography for Life The following standards are highlighted in this section:
- Standards 8, 17, 18

FOCUS

Section Objectives

1. **Describe** the landforms that are found in Kenya.
2. **Explain** why most Kenyans live in highland areas.
3. **Identify** the languages that are spoken in Kenya.

Bellringer Motivational Activity

 Prior to taking roll at the beginning of the class period, project Section Focus Transparency 20-1 and have students answer the activity questions. Discuss students' responses. L1

Vocabulary Pre-check

Ask students to define the terms listed in **Words to Know.** Then have them find the terms in the text to determine how many of their definitions are accurate. L1 LEP

 Use the Vocabulary PuzzleMaker Software to create a crossword puzzle. L1

The Great Rift Valley is only one of East Africa's many geographic features. The region of East Africa includes 11 countries. Find these countries on the map on page 520. East Africa begins in the north with the country of Sudan. The region then runs along the Red Sea and the Indian Ocean to Tanzania (TAN•zuh•NEE•uh) in the south. About 1,000 miles (1,600 km) east of the African mainland lies the island country of Seychelles (say•SHEHLZ). On the mainland, Kenya lies close to the center of East Africa.

Region
The Land

Kenya's land area of 219,960 square miles (569,696 sq. km) makes it slightly smaller than the state of Texas. Like Texas's landforms, Kenya's landforms include coastal areas, plains, and highlands.

The Coast The blue Indian Ocean borders Kenya on the east. Kenya's long coastline has stretches of white beaches lined with palm trees. Not far offshore is a coral reef. **Coral** is a hard, rocklike material made of the skeletons of small

CHAPTER 20

521

Classroom Resources for Section 1

📁 REPRODUCIBLE MASTERS

Reproducible Lesson Plan 20-1
Guided Reading Activity 20-1
Workbook Activity 20
Vocabulary Activity 20
Geography Skills Activity 20
Building Skills in Geography,
  Unit 2, Lesson 6
Section Quiz 20-1

📽 TRANSPARENCIES

Section Focus
  Transparency 20-1
Political Map Transparency 7
World Cultures Transparency 11

MULTIMEDIA

💿 Vocabulary PuzzleMaker
  Software

💿 Testmaker

🔵 National Geographic
  Society: STV: World
  Geography, Vol. 2

MAP STUDY

Answer
Ethiopia, Kenya, Tanzania

Map Skills Practice
Reading a Map In which East African country is the Nubian Desert located? *(Sudan)*

TEACH

Guided Practice

Graphic Organizer Have students identify the major economic activities in Kenya. *(tourism; agriculture—subsistence farming and cash crops; industries)* Then have them create a graphic organizer to show the various sub-divisions of each economic activity. **L1**

NATIONAL GEOGRAPHIC SOCIETY

These materials are available from Glencoe.

 **VIDEODISC**

STV: World Geography
Vol. 2: Africa and Europe

East Africa
Side 1, Frames 18834-26590

EAST AFRICA: Physical

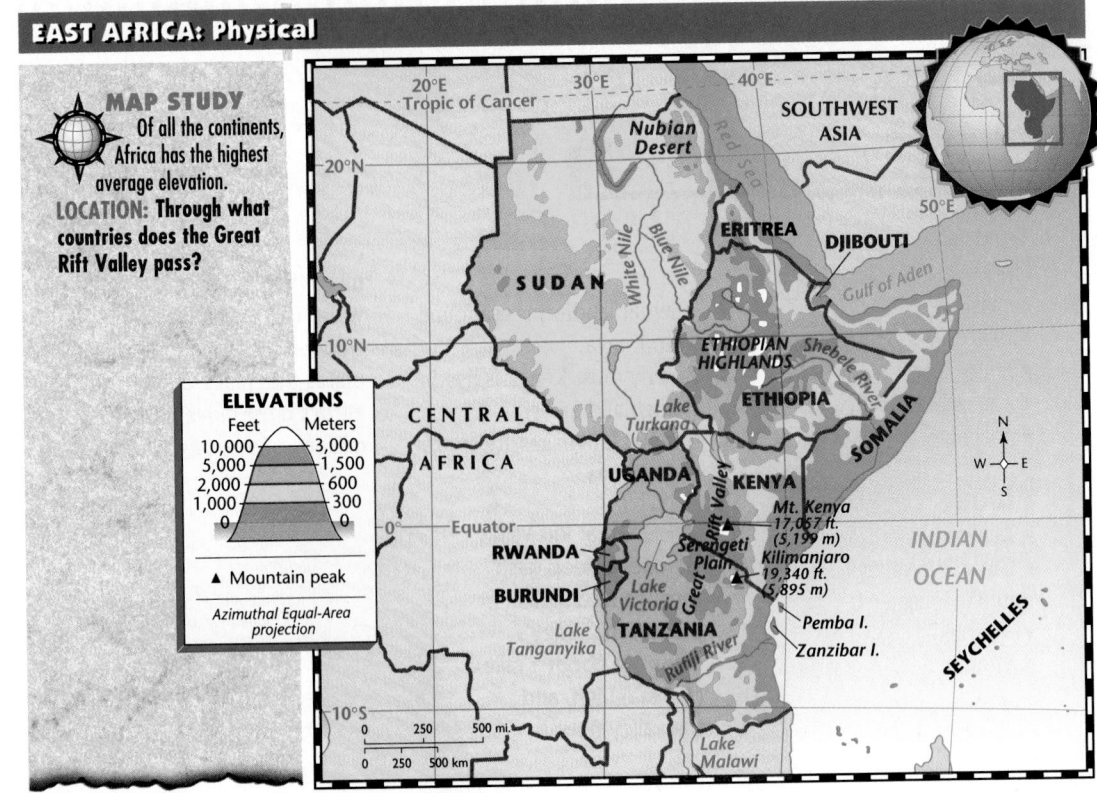

MAP STUDY Of all the continents, Africa has the highest average elevation.
LOCATION: Through what countries does the Great Rift Valley pass?

ELEVATIONS

Feet	Meters
10,000	3,000
5,000	1,500
2,000	600
1,000	300
0	0

▲ Mountain peak

Azimuthal Equal-Area projection

---

sea animals. A **reef** is a narrow ridge of coral, rock, or sand at or near the water's surface. Beaches near Kenya's coral reef area draw tourists from around the world.

Plains If you traveled about 20 miles (32 km) inland, you would find a vast plains area covering about three-fourths of Kenya. Bushes, shrubs, and low thorn trees thrive on the plains. Except for groups of nomads who herd their cattle through the region, few people make their homes here. Many animals, however, roam these vast plains. These include antelopes, water buffaloes, elephants, giraffes, lions, zebras, and other animals.

Highlands Southwestern Kenya is a highlands area made up of mountains, valleys, and plateaus. Forests and grasslands sweep through the highlands. This region, which boasts Kenya's most fertile soil, is home to about 75 percent of Kenya's people.

Look at the map above. You see that the Great Rift Valley **fault**, or crack in the earth, divides Kenya's highlands into eastern and western areas. Mountain ranges and **escarpments**, or steep cliffs, tower over both sides of the valley. Mount Kenya—Kenya's highest peak—rises 17,057 feet (5,199 m) in the eastern highlands. Many lakes are found in the western highlands.

522

Cooperative Learning Activity

Organize students into three groups and assign one of the following regions of Kenya to each group: coast, plains, or highlands. Remind students that tourism is important to Kenya's economy. Have groups design a travel brochure to attract tourists to their assigned regions. Call on group representatives to present and discuss the travel brochures. **L2**

The Climate The map on page 528 shows you that the Equator passes through the middle of Kenya. Almost all of Kenya has a savanna or steppe climate with hot or warm temperatures. The coast is hot and humid year-round. This area gets enough rain to support a few small rain forests. The plains region is the driest part of Kenya and has a desert climate.

In the highlands, the climate is mild with plenty of rainfall. Temperatures here average about 67°F (19°C), and rainfall may reach about 50 inches (130 cm) each year. The mild climate and fertile soil make the highlands Kenya's most important farming area.

Human/Environment Interaction
The Economy

Kenya has a developing economy based on free enterprise. Its chief economic activity is farming. About 50 percent of Kenya's farm products are subsistence food crops, the most important of which is corn. Kenyans grind corn into a porridge and mix it with vegetables to make a stew. Other subsistence food crops include bananas, beans, and cassava.

Cash Crops The other half of Kenya's farm products are cash crops cultivated for export. The country's leading cash crops—coffee and tea—are Kenya's main source of income. You can find both of these products growing on the slopes of the highlands.

Industries Although Kenya has no major mineral deposits to develop, the government has encouraged the growth of manufacturing in recent years. Kenya's chief factory-made products are cement, chemicals, light machinery, and household appliances.

Tourism adds to Kenya's treasury. Thousands of tourists visit Kenya each year to take trips called safaris. On safaris, visitors tour through national parks to see the country's spectacular wildlife in natural surroundings. The government has set up these parks to protect endangered animals from **poachers,** or people who hunt and kill animals illegally.

Place
The People

Harambee, which means "pulling together," is an important word in Kenya. Kenyans come from many different ethnic groups and speak many languages. Since their independence, Kenyans have tried to strengthen their country by working together. *Harambee,* however, is being tested as ethnic disputes become more frequent.

Tsavo National Park, Kenya

Elephants and other animals roam the vast plains in Kenya's parks and game reserves.
REGION: What is the climate of Kenya's plains region?

Independent Practice

Guided Reading Activity 20-1 **L1**

More About the Illustration

Answer to Caption
desert

Human/ Environment Interaction National parks and game reserves are Kenya's biggest tourist attraction. Tourism employs some 40,000 Kenyans and brings some $200 million into the economy each year.

MULTICULTURAL PERSPECTIVE

Cultural Heritage
The family has the highest value in Kenyan society. Individuals are expected to sacrifice personal interests for the good of the family. And a person who fails to keep close ties with his or her extended family is considered rebellious.

Cultural Kaleidoscope

Kenya Soccer is Kenya's most popular team sport. Kenya is best known, however, for its many Olympic track champions. The first, and perhaps greatest, of these was Kipchoge Keino, who won gold medals at the 1968 and 1972 Olympic Games.

Meeting Special Needs Activity

Language Delayed Students with language problems tend to skip over place names and other words that they find difficult to pronounce. Help them practice using the phonetic respellings that appear in parentheses to pronounce unfamiliar words in Section 1. **L1**

523

ASSESS

Check for Understanding

Assign Section 1 Review as homework or an in-class activity.

Meeting Lesson Objectives

Each objective below is tested by the questions that follow it in parentheses.

1. **Describe** the landforms that are found in Kenya. (1, 5)
2. **Explain** why most Kenyans live in highland areas. (2)
3. **Identify** the languages that are spoken in Kenya. (4)

Evaluate

 Section 1 Quiz **L1**

Use the Testmaker to create a customized quiz for Section 1. **L1**

 **MAP STUDY**

Answer
Sukama, Chaga
Map Skills Practice
Reading a Map Which is the major ethnic group in Kenya? *(Kikuyu)*

Impact of the Past Scientists have found some of the earliest-known remains of humans in the Great Rift Valley. About 3,000 years ago, people from other parts of Africa began to move into Kenya. These people farmed and herded animals in the highlands. They became the ancestors of today's Kenyans.

In the A.D. 700s, Arabs from Southwest Asia set up trading settlements along Kenya's coast. Some of these Arabs married Africans. From these marriages came a new people, culture, and language known as Swahili.

In the late 1800s, Kenya came under British control. Many British people moved to Kenya's highlands because of the mild climate and fertile soil. They took land away from Africans and established their own farms on that land. By the 1940s Kenya's African population had organized to reclaim their land. They opposed the British until Kenya finally won its independence in 1963.

After independence, Kenya became the political and economic leader of East Africa. Since the late 1970s, Kenya has struggled to overcome ethnic conflicts and to build a strong and lasting democracy.

Kenyans Today Kenya's population of about 27 million people is growing rapidly. Kenya has one of the world's fastest population growth rates—almost 3.5 percent a year. If the population keeps increasing at this rate, it will double in about 21 years! One of the biggest challenges Kenya faces is providing enough food and jobs for its people.

Major African Ethnic Groups

MAP STUDY There are more than 100 different ethnic groups in Tanzania.
MOVEMENT: What are two major ethnic groups in Tanzania?

Critical Thinking Activity

Predicting Consequences Point out that Kenya has one of the world's highest population growth rates. Ask students to predict the consequences of a rapidly growing population. Students should consider the impact of population growth on food supplies and employment. Remind students that current farming practices in Kenya mostly produce subsistence crops. **L2**

About 40 different ethnic groups are found in Kenya. The map on page 524 shows several of the many ethnic groups of Africa. The largest of Kenya's ethnic groups is the Kikuyu (kee•KOO• yoo), who live mostly in the highlands. What other major groups are found in East Africa?

Although Kenya's many ethnic groups speak many different African languages, Swahili—a blend of African languages and Arabic—is the most widely spoken. About 65 percent of Kenya's population are Christians, while another 25 percent of the people practice traditional African religions.

Most Kenyans—about 75 percent—live in rural villages. Only 25 percent of the population live in cities such as Nairobi (ny•ROH•bee)—Kenya's modern capital—and the Indian Ocean port of Mombasa (mahm•BAH•suh). Find these cities on the map on page 520.

In recent years, many Kenyans have moved from the countryside to the cities. Lucy Ombasa, a Nairobi teenager, and her family recently moved to the city from their village in western Kenya. Their village specialized in stone carvings of animals, fish, and birds. Lucy's parents and sister now sell carvings to tourist shops in the Kenyan capital.

Nairobi, Kenya

Kenya's capital and largest city has many parks and modern office buildings.
MOVEMENT: Why do you think many rural Kenyans have moved to cities?

SECTION 1 REVIEW

REVIEWING TERMS AND FACTS

1. Define the following: coral, reef, fault, escarpment, poacher.
2. LOCATION Where is Kenya's most important farming area?
3. HUMAN/ENVIRONMENT INTERACTION What draws tourists to Kenya?
4. PLACE What is the most widely spoken language in Kenya?

MAP STUDY ACTIVITIES
5. Turn to the physical map on page 522. What large lake lies entirely within Kenya?
6. Look at the political map on page 520. About how far is Kenya's capital from the Ugandan border?

Answers to Section 1 Review

1. All vocabulary words are defined in the Glossary.
2. the highlands
3. the chance to see African wildlife in its natural surroundings
4. Swahili
5. Lake Turkana
6. about 180 miles (290 km)

More About the Illustration
Answer to Caption
to find jobs

Place Nairobi is a modern commercial center, yet it includes a national park within its borders where lions, zebras, and other wild animals roam freely.

Reteach
Have students make up one question for each subheading in Section 1. Then organize students into teams and have teams use their questions to quiz one another. **L1**

Enrich
Encourage students to research and report on the Masai of Kenya. Suggest that they make the following question the focus of their reports: How have these people maintained their traditional ways of life in a changing world? **L3**

CLOSE

Have students work in groups to create posters illustrating the concept of *harambee.* Display and discuss students' work.

TEACH

Present the following scenario to students: You want to see if you are increasing the distance you cover in the 30-minute run you take every other day. You measure your distances for the next 10 running sessions. Your results are 2.0 miles, 2.0 miles, 2.25 miles, 2.5 miles, 2.25 miles, 2.5 miles, 3.0 miles, 2.5 miles, 2.5 miles, and 3.0 miles.

Direct students to chart the information on a line graph. Remind them to plot time intervals on the horizontal axis and the distance run on the vertical axis. When students have completed the line graph, ask them what trend, if any, the graph shows. *(Overall, there was a steady increase in distance covered.)* Have students read the skill and complete the practice questions. **L1**

Additional Skills Practice

1. In what years did factory-related exports exceed farming-related exports? *(1979–1981)*

2. If the next entry on the graph for factory-related exports was just below 40 percent, where do you think the entry for farming-related exports would be? *(just above 60 percent)*

📁 Geography Skills Activity 20

📁 Building Skills in Geography, Unit 2, Lesson 6

526

BUILDING GEOGRAPHY SKILLS

Reading a Line Graph

A **line graph** is a good way to show how things change. On a line graph, time intervals appear on the bottom of the graph—the horizontal axis. The information being compared usually appears on the left side of the graph—the vertical axis. Dots or other symbols mark specific quantities. These dots are connected with a line to show relationships and trends. To read a line graph, apply these steps:

- Read the title to find out the subject of the graph.
- Familiarize yourself with the information on the horizontal axis.
- Familiarize yourself with the information on the vertical axis.
- Examine where the dots are placed on the graph.
- Determine what the lines or curves mean.

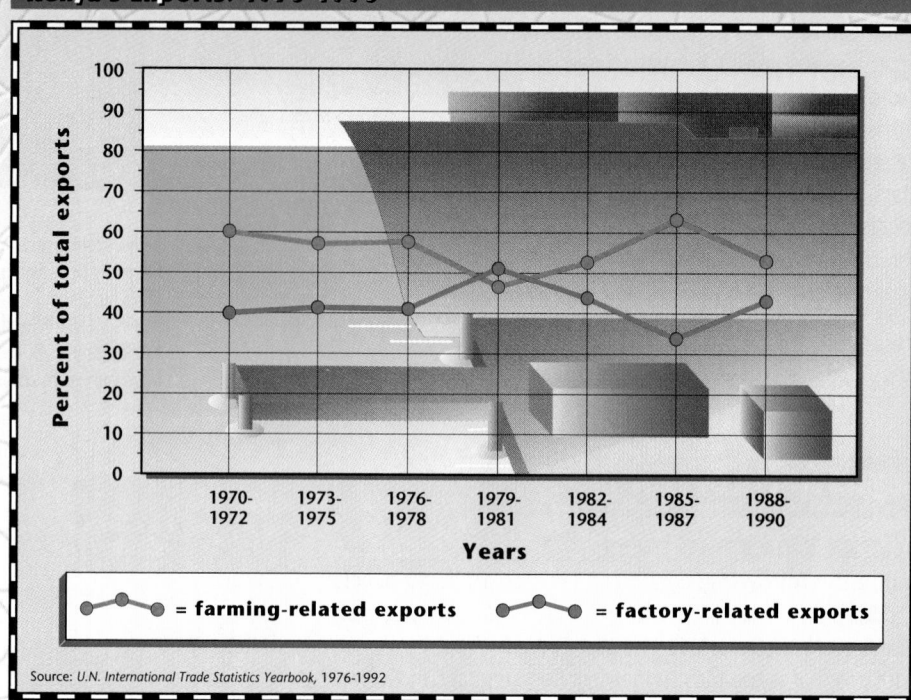

Kenya's Exports: 1970-1990

Source: U.N. International Trade Statistics Yearbook, 1976-1992

Geography Skills Practice

1. What is measured on this graph? Over what time period?

2. In what years did Kenya have the fewest farming-related exports?

3. In what years did Kenya have the most factory-related exports?

Answers to Geography Skills Practice

1. Kenya's exports from 1970 to 1990
2. 1979–1981
3. 1979–1981

PREVIEW

Words to Know
- sisal
- cloves

Places to Locate
- Tanzania
- Lake Victoria
- Lake Tanganyika
- Dar es Salaam
- Zanzibar
- Dodoma

Read to Learn . . .
1. how many ethnic groups make up Tanzania's population.
2. how most people in Tanzania earn a living.

The gleaming, snow-capped peak of Kilimanjaro towers 19,340 feet (5,895 m) over grassy plains in northern Tanzania. For centuries, Africans have gazed at Africa's tallest mountain with awe and wonder. An old African story says that God built Kilimanjaro as a throne from which to view the world.

Kilimanjaro is probably the best-known sight in Tanzania. One of the largest countries in East Africa, Tanzania has a land area of 342,100 square miles (886,039 sq. km). Look at the map on page 520. You see that Tanzania lies south of Kenya and below the Equator.

Place
The Land

Most of Tanzania lies on the mainland of East Africa. The country also includes several small coral islands just off the coast, in the Indian Ocean. The largest of these islands is called Zanzibar.

Coast, Plateaus, and Lakes Like the coast of Kenya, Tanzania's coastline boasts white beaches and palm trees. As you travel inland, the country's elevation rises gradually from humid lowlands to partially dry plateaus. Huge grasslands with patches of trees and shrubs cover the plateaus. To the north, near the Kenyan border, lies a mountainous area that includes Kilimanjaro.

Much of western Tanzania is part of the Great Rift Valley. A number of lakes lie in this area, including Lake Victoria and Lake Tanganyika. Lake Victoria is the

CHAPTER 20

527

LESSON PLAN
Chapter 20, Section 2

Meeting National Standards
Geography for Life The following standards are highlighted in this section:
- Standards 4, 10, 17

FOCUS

Section Objectives
1. **Show** how many ethnic groups make up Tanzania's population.
2. **Explain** how most Tanzanians earn a living.

Bellringer Motivational Activity

 Prior to taking roll at the beginning of the class period, project Section Focus Transparency 20-2 and have students answer the activity questions. Discuss students' responses. **L1**

Vocabulary Pre-check

Provide a few whole cloves for students to examine. Have them describe the fragrance. Then point out that the word *cloves* comes from a French word meaning "nail." **L1 LEP**

NATIONAL GEOGRAPHIC SOCIETY

These materials are available from Glencoe.

 VIDEODISC

STV: World Geography Vol.2: Africa and Europe

Lake Tanganyika
Side 1, Frames 24756-25114

Classroom Resources for Section 2

 REPRODUCIBLE MASTERS

Reproducible Lesson Plan 20-2
Guided Reading Activity 20-2
Chapter Map Activity 20
Vocabulary Activity 20
GeoLab Activity 20
Section Quiz 20-2

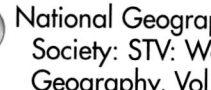 TRANSPARENCIES

Section Focus
  Transparency 20-2
Political Map Transparency 7
Unit Overlay Transparencies
  7-0, 7-4, 7-5

MULTIMEDIA

Testmaker

National Geographic Society: STV: World Geography, Vol. 2

TEACH

Guided Practice

 Display Unit Overlay Transparencies 7-0, 7-4, and 7-5. Call on volunteers to identify and locate Tanzania's major physical features. Ask students to speculate on Tanzania's economic activities based on its physical features. Have students check the accuracy of their speculations as they read the section. **L1**

Independent Practice

Guided Reading Activity 20-2 **L1**

MAP STUDY

Answer

mostly tropical savanna with small areas of highland and steppe climates

Map Skills Practice

Reading a Map What climate type is found in central Kenya? *(desert)*

ASSESS

Check for Understanding

Assign Section 2 Review as homework or an in-class activity.

largest lake in Africa. Lake Tanganyika's floor is the deepest point on the African continent. Unusual fish found nowhere else in the world live in Lake Tanganyika's deep, dark waters.

Wildlife Like Kenya, Tanzania has many kinds of animals. The Tanzanian government has set aside thousands of square miles to protect its wildlife. Serengeti National Park covers about 5,600 square miles (14,500 sq. km). It is home to many lions and huge herds of antelopes and zebras. During the dry season, thousands of animals roam the plains in search of water.

Place

The Economy and People

Tanzania has a developing economy based on agriculture. Service industries are growing, but manufacturing plants are still small. Tanzania is rich in mineral resources such as gold and diamonds. These riches, however, have not yet been developed.

Most Tanzanians raise livestock or farm small plots of land. Farmers grow only enough bananas, cassava, corn, millet, and rice to feed their families. Large, government-run farms grow many of the crops that Tanzania exports. These cash crops include coffee, cotton, tea, and tobacco.

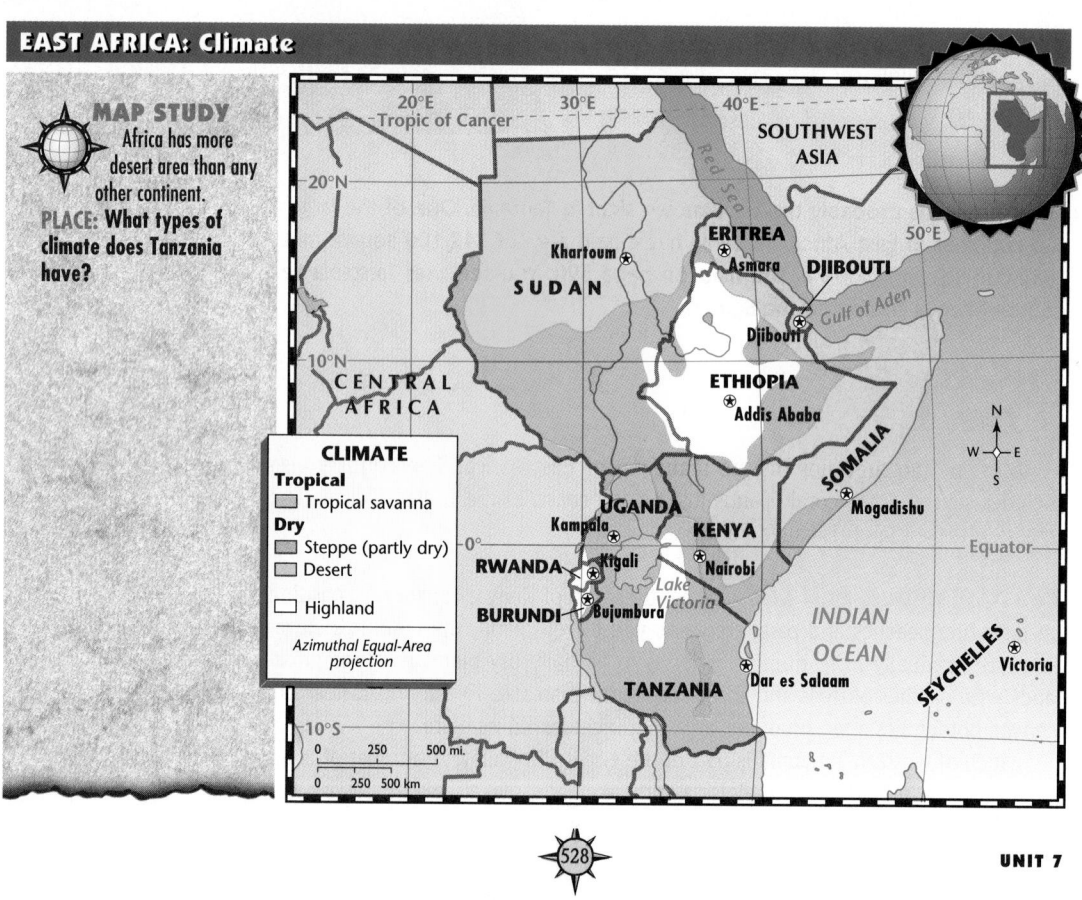

EAST AFRICA: Climate

MAP STUDY Africa has more desert area than any other continent. **PLACE: What types of climate does Tanzania have?**

CLIMATE

Tropical
- Tropical savanna

Dry
- Steppe (partly dry)
- Desert

- Highland

Azimuthal Equal-Area projection

528

Cooperative Learning Activity

Organize students into groups to create commemorative stamps for Tanzania's "world records." Subjects include the following: World's Largest Producer of Sisal, World's Leading Producer of Cloves, Largest African Lake (Victoria), Deepest Point on the African Continent (Lake Tanganyika), Highest Point on the African Continent (Kilimanjaro). Call on groups to present and discuss their stamps. **L2**

Tanzania is the world's largest producer of **sisal,** a plant fiber used in making rope and twine. Farmers on the island of Zanzibar raise the world's largest supply of **cloves,** a spice made from the buds of clove trees.

National Unity About 30 million people live in Tanzania. Most of the population belong to one of about 120 African ethnic groups. A small number of the country's people are Asians or Europeans. Although each ethnic group has its own language, many people speak Swahili, the national language. The population is about evenly divided among Christian, Muslim, and traditional African religions. Because there are so many ethnic groups and religions in Tanzania, no single group controls the country. This ethnic and religious balance has helped Tanzania achieve national unity.

The Past and Present Scientists have found the remains of some of the earliest human settlements in Tanzania. By about A.D. 500, Bantu-speaking peoples had settled in this area. Nearly 600 years later, Arabs from Southwest Asia set up major trading centers on Zanzibar and other islands. The region that is now Tanzania came under European control beginning in the early 1500s. After periods of Portuguese and German rule, the area came under British control following World War I. In the early 1960s, the Tanzanians finally won their independence as the United Republic of Tanzania.

Today about 80 percent of Tanzania's people live in rural villages. Dar es Salaam, on the Indian Ocean, is Tanzania's capital, largest city, and major port. The central area of Tanzania has few people. In an effort to encourage people to move there, the Tanzanian government is planning to build a new national capital—Dodoma—in the area.

Teen Scene

Olympic Hopeful

Sadiki Abasi hopes to be in the Olympics someday. Soccer is his favorite hobby, but his real talent is running. Sadiki is competing with long-distance runners from all over Tanzania for a spot on the national team. He thinks he's got a chance. "I started running with my uncle when I was seven," Sadiki says. "I try to run at least 100 miles a week to stay competitive." Sadiki's uncle, a former national champion, is Sadiki's coach.

SECTION 2 REVIEW

REVIEWING TERMS AND FACTS

1. **Define the following:** sisal, cloves.
2. **LOCATION** In what area of Tanzania is Kilimanjaro located?
3. **PLACE** What agricultural products does Tanzania export?
4. **MOVEMENT** Why will Tanzania move its capital from Dar es Salaam to Dodoma?

 MAP STUDY ACTIVITIES

5. Turn to the climate map on page 528. What type of climate is found in southern Tanzania?

LESSON PLAN
Chapter 20, Section 2

Meeting Lesson Objectives

Each objective below is tested by the questions that follow it in parentheses.

1. **Show** how many ethnic groups make up Tanzania's population. (4)
2. **Explain** how most Tanzanians earn a living. (3)

Evaluate

Section 1 Quiz **L1**

Use the Testmaker to create a customized quiz for Section 2. **L1**

Reteach

Have students write newspaper headlines that summarize the main ideas of this section. **L1**

Enrich

Have students prepare oral reports about the Serengeti National Park or the Selous Game Reserve in Tanzania. **L3**

CLOSE

Ask students to explain why Tanzania would be of interest to the following scientists: geologist, archaeologist, anthropologist, zoologist.

MAKING CONNECTIONS

Ask students to suggest why some people refer to Africa as the "birthplace of humankind." *(Some of the oldest human remains have been found there.)* Then point out that the work of the Leakey family established that humankind had its origins in Africa. **L1**

More About . . .

Archaeology By the mid-1960s, Louis and Mary Leakey felt they had uncovered—literally—enough fossil evidence to prove that humans had existed at least 2 million years ago, and possibly as far back as 14 million years. Later evidence discovered by Louis—who died in 1972—and his son Richard also supports the theory that at least three kinds of early humans and near-humans existed side by side. The Leakeys theorized that the near-human strains became extinct between 2 and 4 million years ago, leaving only the true humans to evolve into *Homo sapiens.*

A REMARKABLE FIND

MATH | **SCIENCE** | **HISTORY** | **LITERATURE** | **TECHNOLOGY**

On July 17, 1959, archaeologist Mary Leakey discovered fossils of teeth buried in the ground. An expert on prehistoric bones, she had never seen teeth quite like these before. Hopeful and excited, Mary Leakey ran to tell her husband, Louis, also an archaeologist.

EARLY DIGS Louis and Mary Leakey worked at the Olduvai (OHL•duh•vy) Gorge in northern Tanzania. The Gorge—a 25-mile-long (41km) canyon—had proven to be a rich treasure of human and animal fossils. The Leakeys had been digging at various places along the Gorge since the early 1930s. They had already unearthed thousands of remains.

In 1948 Mary Leakey had found a fossilized skull and other bones. These remains were between 25 and 40 million years old. Putting the bones together, the Leakeys produced a creature that had human-like jaws and walked upright on hind legs. It was the first significant evidence that the human race may have started in Africa rather than in Asia, as most archaeologists believed. As the Leakeys ran back to excavate the teeth on that July day in 1959, they thought the teeth might provide further evidence of humankind's beginnings in Africa.

The Leakeys at Olduvai Gorge

TWICE AS OLD With dental picks and brushes, the Leakeys slowly uncovered the teeth, which turned out to be part of an almost-complete human skull. When dated, the skull proved to be about 1.75 million years old. It was twice as old as scientists had judged the human race to be.

The skull and other Leakey finds forced archaeologists to change their search for human origins from Asia to Africa. In addition, scientists were forced to hypothesize that the development of human beings took much longer than previously thought.

Making the Connection

1. Where is Olduvai Gorge?
2. How did the Leakeys' discoveries change scientists' thinking about human origins?

530

Answers to Making the Connection

1. northern Tanzania
2. The Leakeys' work persuaded scientists that humankind had its origins in Africa rather than Asia and that the development of human beings took much longer than previously thought.

SECTION 3 · Inland East Africa

PREVIEW

Words to Know
- autonomy
- watershed
- civil war
- refugee

Places to Locate
- Uganda
- Rwanda
- Burundi
- Lake Victoria
- Nile River
- Kampala

Read to Learn . . .
1. why inland East Africa has a mild climate.
2. what farm products are grown in inland East Africa.
3. how conflict between ethnic groups has divided the people of inland East Africa.

The sparkling waters of Lake Victoria

cover 26,828 square miles (69,485 sq. km). About half of this magnificent lake lies inside the landlocked East African country of Uganda. Large lakes are also found in two other inland countries of East Africa—Rwanda and Burundi.

Great lakes, snow-covered mountains, deep tropical valleys, and thundering rivers are all found in Uganda, Rwanda, and Burundi. Each of these three countries of inland East Africa is also home to many different peoples.

Place

Uganda

Uganda lies in the highlands region of East Africa. Uganda's 77,050 square miles (199,560 sq. km) make it slightly smaller than Oregon. A plateau covers most of northern Uganda. In the center of the country is a large area of marshes and lakes. Here, waters flowing out of Lake Victoria gather to form a source of the Nile River. In southern Uganda, thick forests spread across the landscape. To the east and west, mountains form Uganda's borders with Kenya and Zaire.

The Equator crosses southern Uganda. The country's generally high elevation, however, keeps temperatures mild. In most areas, at least 40 inches (100 cm) of rain falls each year. Only a few northern areas receive less than 20 inches (51 cm) of rain.

CHAPTER 20

531

Meeting National Standards
Geography for Life The following standards are highlighted in this section:
- Standards 4, 13, 15

FOCUS

Section Objectives
1. **Explain** why inland East Africa has a mild climate.
2. **List** the farm products grown in inland East Africa.
3. **Examine** how conflict between ethnic groups has divided the people of inland East Africa.

Bellringer Motivational Activity

 Prior to taking roll at the beginning of the class period, project Section Focus Transparency 20-3 and have students answer the activity questions. Discuss students' responses. **L1**

Vocabulary Pre-check

Write the definitions of the vocabulary terms on the chalkboard. Challenge students to match each term with its definition. Then have them check their answers by finding the defined terms in the text. **L1 LEP**

Classroom Resources for Section 3

⬗ REPRODUCIBLE MASTERS
Reproducible Lesson Plan 20-3
Guided Reading Activity 20-3
Cooperative Learning
  Activity 20
Critical Thinking Skills
  Activity 20
Section Quiz 20-3

⬗ TRANSPARENCIES
Section Focus
  Transparency 20-3
Political Map Transparency 7
Unit Overlay Transparencies
  7-0, 7-4, 7-5, 7-6

MULTIMEDIA
 Testmaker

National Geographic
  Society: ZipZapMap!
  World

Answer
subsistence farming

Map Skills Practice
Reading a Map Name two crops grown in Uganda. *(tea, coffee)*

TEACH

Guided Practice

Display Unit Overlay Transparencies 7-0, 7-4, 7-5, and 7-6. Have students locate Uganda, Rwanda, and Burundi. Ask them to hypothesize about why these countries are grouped as a region. *(Possible answers: location, similar features, all landlocked, similar economic activities, cultural ties.)* L1

Independent Practice

Guided Reading Activity 20-3 L1

CURRICULUM CONNECTION

Life Science Coffee is a major export of inland East Africa. On land up to about 4,500 feet (1,370 m) above sea level, farmers raise *robusta*—used to make instant coffee. In altitudes between 4,500 and 6,000 feet (1,370 and 1,800 m), farmers raise *arabica*—coffee used for regular brewing.

EAST AFRICA: Land Use and Resources

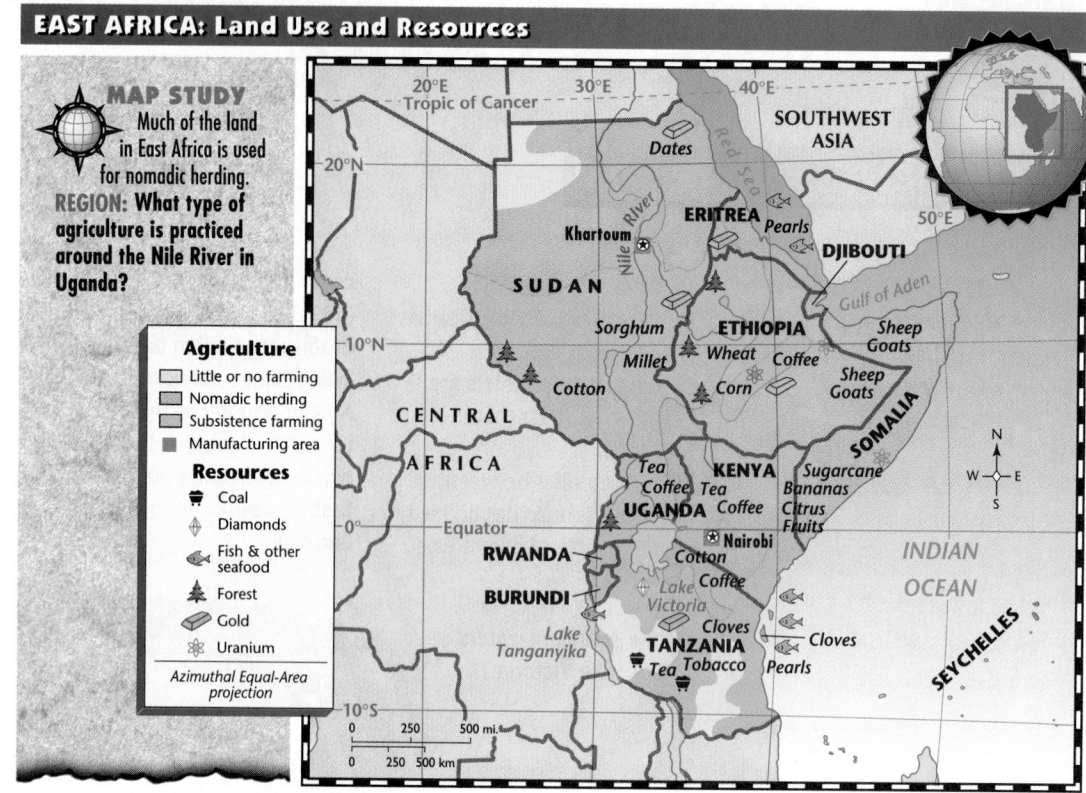

MAP STUDY
Much of the land in East Africa is used for nomadic herding.
REGION: What type of agriculture is practiced around the Nile River in Uganda?

Agriculture
☐ Little or no farming
▨ Nomadic herding
▨ Subsistence farming
▪ Manufacturing area

Resources
⛏ Coal
◈ Diamonds
🐟 Fish & other seafood
🌲 Forest
▬ Gold
❄ Uranium

Azimuthal Equal-Area projection

The Economy Agriculture is the most important economic activity in Uganda. Fertile soil and a mild climate help crops to grow well. The most productive areas lie along the western and northern sides of Lake Victoria and in the western highlands.

Uganda's farmers grow bananas, beans, cassava, corn, and sweet potatoes to feed their families. They sell any extra food at local markets. Coffee is Uganda's leading export crop. Cotton, sugarcane, and tea are valuable exports, too. These cash crops support Uganda's economy.

The People Uganda has about 20 million people. You will find almost two-thirds of Uganda's people living in the fertile south. There you will find Kampala, the capital, on the shores of Lake Victoria.

The people of Uganda belong to more than 20 different ethnic groups. The Ganda are the largest and most prosperous group. Until 1967 the Ganda enjoyed local **autonomy,** or self-government. Today they are part of a united Uganda.

For most of this century, the British ruled Uganda. After Uganda won its independence in 1962, fighting broke out between ethnic groups in the north and the south. Ethnic conflict, plus a period of cruel dictatorship, hurt Uganda throughout much of the 1980s. Ugandans are still rebuilding their country after this time of suffering and warfare.

Cooperative Learning Activity

Organize students into three groups and assign one of the following countries to each group: Uganda, Rwanda, or Burundi. Have group members prepare a number of maps for their assigned country. Maps should indicate physical features, climate, major cities, population density, and land use and resources. Have groups display their maps and discuss the geographical similarities and differences among the countries of inland East Africa. L1

Place
Rwanda and Burundi

Rwanda and Burundi are located deep in inland East Africa. Although these two countries lie just south of the Equator, high altitudes give them a mild climate. They also lie on the ridge that separates the Nile and Zaire **watersheds.** A watershed is an area drained by a river. To the west of the ridge, waters eventually run into the Zaire River watershed. To the east, they run into the Nile watershed.

The Economy Most people in Rwanda and Burundi are farmers. They grow tea, coffee, beans, bananas, cotton, and grains on gently rolling plateaus. In both countries, some people raise livestock. Along lakes, such as Lake Tanganyika and Lake Kivu, fishing is an important economic activity. Some tin mining takes place in mountainous areas, but neither country has valuable minerals.

Because Burundi and Rwanda are landlocked, they have trouble getting their exports to foreign buyers. There are few paved roads and no railroads. Most goods must be transported by road to Lake Tanganyika, where boats take them to Tanzania and Zaire. Another route is by dirt road to Tanzania and then by rail to Dar es Salaam.

Ethnic Conflict Burundi and Rwanda are two of the smallest and most crowded nations in Africa. Rwanda, for example, has an average of 805 persons per square mile (2,085 persons per sq. km). This density is about 13 times as

WHAT IN THE WORLD?

Glaciers in Africa?

"Impossible!" thought British geographers in the mid-1800s. They were proved wrong in 1889, when British explorer Henry Morton Stanley saw glaciers in the Ruwenzori (ROO•uhn•ZOHR•ee) Mountains. The snowcapped Ruwenzori range lies between Uganda and Zaire. Even harder to believe is the fact that the Nile—the river that flows through steamy rain forests and dry deserts—begins as a trickle in these frozen cliffs.

Volcanoes National Park, Rwanda

A small village sits in the shadow of one of several inactive volcanoes in Rwanda.
MOVEMENT: How does the lack of a coastline affect the economy of Rwanda?

ASSESS

Check for Understanding

Assign Section 3 Review as homework or an in-class activity.

Meeting Lesson Objectives

Each objective blow is tested by the questions that follow it in parentheses.

1. **Explain** why inland East Africa has a mild climate. (2)
2. **List** the farm products grown in inland East Africa. (3, 5)
3. **Examine** how conflict between ethnic groups has divided the people of inland East Africa. (4)

More About the Illustration

Answer to Caption
it is difficult to get goods to foreign buyers

Human/ Environment Interaction Rwanda reserves a greater percentage of its land for national parks than any other country in Africa. Volcanoes National Park was the first wildlife park on the continent. It is the home of the mountain gorillas, made famous by the movie *Gorillas in the Mist.*

CHAPTER 20 533

Meeting Special Needs Activity

Visual Learners Help students who learn best visually relate the text to a map of inland East Africa. Display a physical map of East Africa. Have students read parts of the text that describe locations of features. For example, "In southern Uganda, thick forests spread across the landscape. To the east and west, mountains form Uganda's borders with Kenya and Zaire." Have students locate these areas on the map. Continue in this manner with features of Rwanda and Burundi. **L1**

More About the Illustration

Answer to Caption

farming

 Human/ Environment Interaction Tea packers harvest an average of 40 pounds (18 kg) of leaves a day. When processed, this will make about 10 pounds (4.5kg) of tea.

Evaluate

Section 3 Quiz **L1**

Use the Testmaker to create a customized quiz for Section 3. **L1**

Reteach

Have students create a chart to compare Uganda, Rwanda, and Burundi. Bases of comparison might include physical features, economic activities, and ethnic groups. **L1**

Enrich

Have students find out about the research conducted by Dian Fossey among the mountain gorillas of Rwanda. **L3**

CLOSE

Have students bring to class newspaper and/or magazine articles dealing with refugees in inland East Africa. Discuss the causes and effects of migrations of people in this region.

534

A Field of Tea in Ijenda, Burundi

Tea pickers gather leaves in the highland plantations of Burundi.
PLACE: What is the occupation of most people in Burundi?

great as Africa's population density as a whole! Less than 10 percent of these people live in towns. Most live in villages surrounded by their fields.

The majority of the population in Rwanda and Burundi belong to two ethnic groups—the Hutu and the Tutsi. The Hutu are the largest group in both countries. The Tutsi, however, have controlled the countries' governments and economies. Since Rwanda's and Burundi's independence from Belgium in 1962, the Hutu have tried to gain part of this control.

In 1994 **civil war,** or fighting within a country, broke out in Rwanda between the Hutu and the Tutsi. Hundreds of thousands of people were killed. More than 2 million people, fearful of losing their lives, fled from their homes. These refugees settled in huge camps on the border of Zaire and neighboring countries. A **refugee** is a person who must flee his or her home and seek safety elsewhere. In 1995 conflict between the Hutu and the Tutsi spread to Burundi.

SECTION 3 REVIEW

REVIEWING TERMS AND FACTS

1. Define the following: autonomy, watershed, civil war, refugee.
2. PLACE How does elevation affect climate in inland East Africa?
3. HUMAN/ENVIRONMENT INTERACTION What is Uganda's most important economic activity?
4. PLACE What two major ethnic groups live in Rwanda and Burundi?

 MAP STUDY ACTIVITIES
5. Turn to the land use map of East Africa on page 532. What type of farming takes place in Rwanda and Burundi?

Answers to Section 3 Review

1. All vocabulary words are defined in the Glossary.
2. The area's elevation moderates the climate.
3. farming
4. Hutu and Tutsi
5. subsistence farming

The Horn of Africa

LESSON PLAN
Chapter 20, Section 4

PREVIEW

Words to Know
- drought
- clan

Places to Locate
- Sudan
- Ethiopia
- Eritrea
- Somalia
- Djibouti
- Gulf of Aden
- Indian Ocean

Read to Learn . . .
1. what ethnic groups live in Sudan.
2. why Ethiopia has been independent for thousands of years.
3. how civil war has affected countries in the Horn of Africa.

of Khartoum (kahr•TOOM), the capital of Sudan. Yet water flows nearby. Two branches of the Nile River meet at Khartoum. As in Egypt, the Nile River is the giver of life to people in Sudan and other countries of northeastern Africa.

Fierce desert winds often blow through the streets

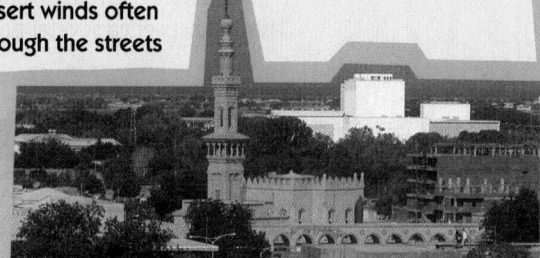

A large area of northeastern Africa is called the Horn of Africa. The map on page 520 shows you how this region got its name. The land is shaped like a horn as it juts out into the Indian Ocean. The countries that lie in the Horn of Africa region are Sudan, Ethiopia, Eritrea (EHR•uh•TREE•uh), Somalia, and Djibouti (juh•BOO•tee).

Place
Sudan

Sudan is the largest country in Africa. Its 967,500 square miles (2,505,813 sq. km) cover an area about one-third the size of the contiguous United States.

Northern Sudan is mostly a desert of bare rocks or sand dunes. Grassy plains cross the central part of Sudan. Here, two great branches of the Nile River—the Blue Nile and the White Nile—meet. This area is the most fertile part of the country. To the south, Sudan has humid tropical rain forests and swamps.

Sudan is one of the world's leading producers of cotton. Most of Sudan's 28 million people live along the Nile River or one of its branches. They use water from the Nile to irrigate huge fields of cotton. Some work on large farms owned by the government.

CHAPTER 20

535

Meeting National Standards
Geography for Life The following standards are highlighted in this section:
- Standards 10, 13, 17

FOCUS

Section Objectives
1. **Identify** the ethnic groups that live in Sudan.
2. **Discuss** why Ethiopia has been independent for thousands of years.
3. **Describe** how civil war has affected countries in the Horn of Africa.

Bellringer Motivational Activity
  Prior to taking roll at the beginning of the class period, project Section Focus Transparency 20-4 and have students answer the activity questions. Discuss students' responses. **L1**

Vocabulary Pre-check
Pronounce the word *drought.* Help new speakers of English by pointing out how the same combination of letters has different sounds in the following words: *drought/thought; through/though/trough.*
L1 LEP

Classroom Resources for Section 4

 REPRODUCIBLE MASTERS
Reproducible Lesson Plan 20-4
Guided Reading Activity 20-4
Performance Assessment
  Activity 20
Reteaching Activity 20
Enrichment Activity 20
Section Quiz 20-4

 TRANSPARENCIES
Section Focus
  Transparency 20-4
Political Map Transparency 7
Geoquiz Transparencies 20-1,
  20-2

MULTIMEDIA
 Testmaker

National Geographic
  Society: Picture Atlas
  of the World

MULTICULTURAL PERSPECTIVE

Culturally Speaking

The name of the country now called Sudan has varied throughout history. Ancient Egyptians called it Kush. The Greeks considered it part of Abyssinia, a name that reflected its mixed ethnic heritage. To the Romans, it was Nubia. The Arabs called it *bilad as sudan,* or "land of the blacks."

TEACH

Guided Practice

Display Political Map Transparency 7. Have students compare the five countries of the Horn of Africa. Have them identify the following: Which country is the largest? *(Sudan)* the smallest? *(Djibouti)* Which has the longest coastline? *(Somalia)* Which border the Red Sea? *(Sudan, Eritrea)* **L1**

Independent Practice

Guided Reading Activity 20-4 **L1**

MAP STUDY

Answer

along the Nile River from the northern border to just south of Khartoum

Map Skills Practice

Reading a Map What is the largest city in the Horn of Africa? *(Addis Ababa)*

The People If you visited Sudan, you would find that the people differ greatly from north to south. Most people who live in the northern two-thirds of the country are Muslim Arabs. People who live in the south come from several different African ethnic groups. They are Christians or practice traditional African religions.

After its independence in 1956, Sudan was torn by civil war between the north and the south. The fighting disrupted food production and caused widespread hunger. In the early 1990s, a **drought**—an extended dry period—occurred in Sudan and caused even more suffering.

Place

Ethiopia

To the east of Sudan lies Ethiopia. If you look at the map on page 522, you will see that Ethiopia is a rugged, mountainous country with a high plateau. The Great Rift Valley cuts through this plateau, forming deep river gorges and sparkling waterfalls. Mild temperatures and fertile soil make areas of the plateau excellent for farming. Ethiopia's farmers, however, cannot always depend on regular rainfall. Drought often occurs, destroying crops and bringing famine. This has happened several times since the 1970s.

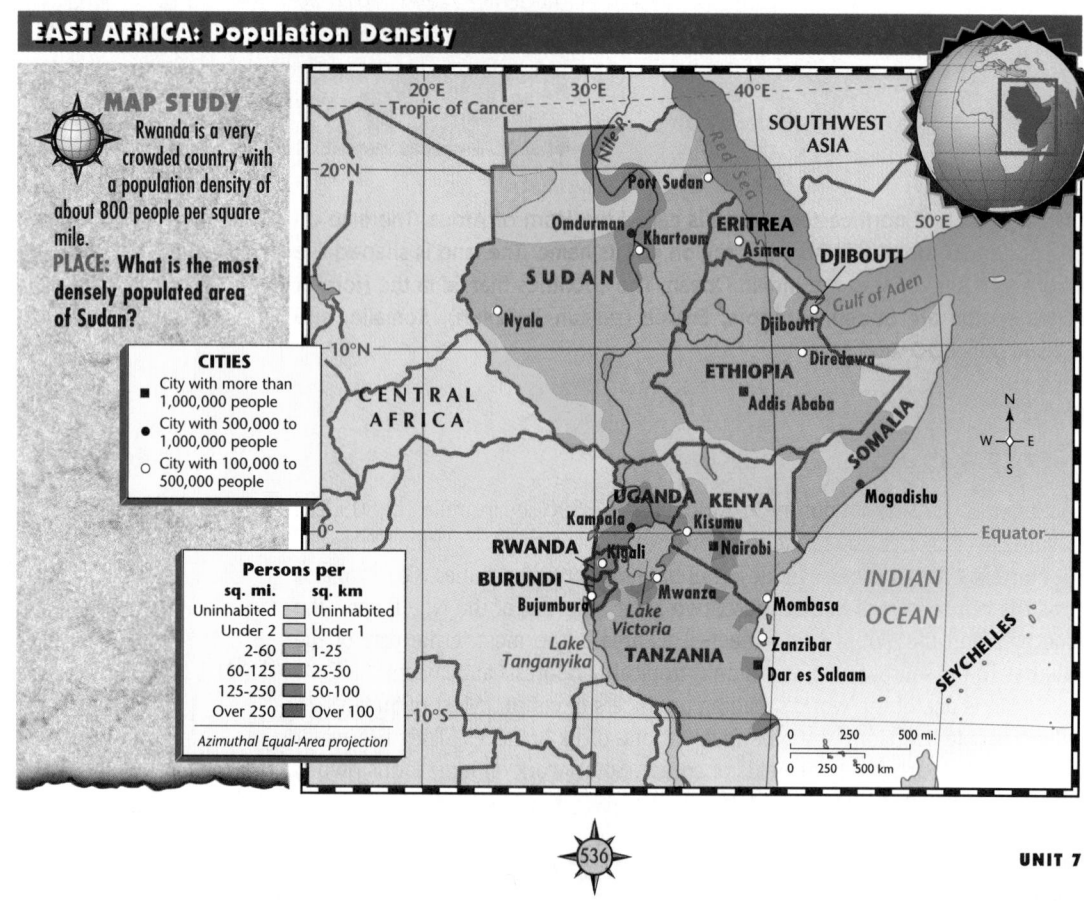

EAST AFRICA: Population Density

MAP STUDY Rwanda is a very crowded country with a population density of about 800 people per square mile.

PLACE: What is the most densely populated area of Sudan?

CITIES
- City with more than 1,000,000 people
- City with 500,000 to 1,000,000 people
- City with 100,000 to 500,000 people

Persons per sq. mi.	sq. km
Uninhabited	Uninhabited
Under 2	Under 1
2-60	1-25
60-125	25-50
125-250	50-100
Over 250	Over 100

Azimuthal Equal-Area projection

UNIT 7

Cooperative Learning Activity

Organize students into five groups and assign each group one of the following countries: Sudan, Ethiopia, Eritrea, Somalia, or Djibouti. Have groups study the people of their assigned countries. Their investigations should include the following characteristics: ethnic groups, population density, standard of living, and challenges they face. Have groups discuss and compare the results of their investigations. **L1**

Most Ethiopians live on the high plateau and grow wheat, corn, and other grains on subsistence farms. Some farmers grow coffee to sell. Others also raise livestock.

The People Ethiopia—one of the world's oldest countries—sent government representatives to meet Egyptian pharaohs centuries ago. Ethiopia's mountains kept it isolated and independent for thousands of years. During most of its history, emperors and empresses ruled. In 1974, military leaders overthrew the last emperor. Today Ethiopia is struggling to build a democracy.

About 55 million people live in Ethiopia. Only 15 percent of them live in urban areas. Addis Ababa (A•duhs A•buh•buh) is Ethiopia's capital and largest city. About 70 different languages are spoken in Ethiopia. Amharic, similar to Hebrew and Arabic, is the official language. Most Ethiopians are either Christians or Muslims. A smaller number practice traditional African religions.

Place

Eritrea

Ethiopia is one of the oldest countries in Africa, and Eritrea is the newest! In 1993, after 30 years of war, Eritrea won its independence from Ethiopia.

Look at the map on page 522. It shows you that Eritrea lies on the Red Sea. Along this Red Sea coast runs a wide plain—one of the hottest and driest areas in Africa. As you travel inland, you will discover that mountains cover the rest of Eritrea.

Most of Eritrea's 3.5 million people farm the land, although farming is uncertain in this dry land. Eritrea's long war with Ethiopia has also made farming difficult. Soldiers cut down the forests, which led to soil erosion. Now that peace has returned, Eritreans are rebuilding their country.

The walls and ceiling of this church tell the story of Ethiopian Christianity *(left)*. Many residents of Ethiopia's capital live and work in modern buildings *(right)*.
PLACE: What are the two main religions of Ethiopia?

More About the Illustration

Answer to Caption
Islam, Christianity

GEOGRAPHIC THEMES **Movement** Some 40 percent of Ethiopians belong to the Ethiopian Orthodox Christian Church. This church is different from the Roman Catholic and Eastern Orthodox churches in that it has adapted many Jewish customs. Worshippers observe the Sabbath on Saturday, cantors chant the liturgy, and there is an emphasis on unclean and clean foods.

ASSESS

Check for Understanding
Assign Section 4 Review as homework or an in-class activity.

Meeting Lesson Objectives
Each objective below is tested by the questions that follow it in parentheses.

1. **Identify** the ethnic groups that live in Sudan. (2)
2. **Discuss** why Ethiopia has been independent for thousands of years. (3)
3. **Describe** how civil war has affected countries in the Horn of Africa. (3)

Meeting Special Needs Activity

Memory Disability Use the following exercise to help students who have difficulty recalling facts: Photocopy paragraphs of Section 4 in which each country in the Horn of Africa is introduced. Black out the names of the countries. Have students read the paragraphs and identify the countries from context. L1

Evaluate

 Section 4 Quiz **L1**

Use the Testmaker to create a customized quiz for Section 4. **L1**

Chapter 20 Test Form A and/or Form B **L1**

Reteach

Reteaching Activity 20 **L1**

Enrich

Enrichment Activity 20 **L3**

More About the Illustration

Answer to Caption
Ethiopia

Human/ Environment Interaction About 80 percent of Eritreans are farmers or herders. Most struggle to make a living. Years of civil war, drought, and the movement of refugees have damaged the best farm and pasture lands.

Mental Mapping Activity

Provide each student with a large sheet of white paper and map pencils. Inform students that this will not be a graded activity. Have students draw, freehand from memory, a map of East Africa that includes the following labeled countries: Kenya, Tanzania, Uganda, Rwanda, Burundi, Sudan, Ethiopia, Eritrea, Somalia, and Djibouti.

538

The Central Mosque in Asmara, Eritrea

Asmara is one of the few large cities in the tiny country of Eritrea.
PLACE: What country was Eritrea a part of until 1993?

Place
Somalia

Like Eritrea, Somalia is a hot, dry country. Grassy plains cover most of Somalia. About half of the country's 10 million people are nomads. Southern Somalia, however, has rivers that provide water for irrigation. Farmers there grow sugarcane and citrus fruits.

The People Nearly all Somalis are Muslims and speak either Somali or Arabic. Yet they are deeply divided. They belong to **clans,** groups of people related to one another. In the late 1980s, clan disputes led to civil war. When a drought struck a few years later, hundreds of thousands of Somalis starved. Other countries sent food, but the fierce fighting kept much of the food from reaching the starving people.

In 1992 the United States and other countries sent troops to protect food supplies. The troops withdrew in 1995 after the worst of the famine ended and roads were reopened.

Place
Djibouti

Djibouti is a tiny land wedged between Eritrea and Somalia. Djibouti's 600,000 people are mostly Muslims who, in the past, lived a nomadic life. Now many of them have settled permanently in the capital city, also called Djibouti.

Almost all of the country's income derives from shipping. Farming is difficult on its dry lands. The city of Djibouti on the Gulf of Aden is an important port. A railroad line connecting it to Ethiopia makes Djibouti an important outlet for Ethiopia's products.

SECTION 4 REVIEW

REVIEWING TERMS AND FACTS

1. **Define the following:** drought, clan.
2. **PLACE** Where do most of Sudan's people live?
3. **PLACE** What is the newest country in Africa?
4. **HUMAN/ENVIRONMENT INTERACTION** What challenges do farmers face in the Horn of Africa?

 MAP STUDY ACTIVITIES

5. According to the population density map on page 536, what country in the Horn of Africa is the most densely populated?

 (538)

Answers to Section 4 Review

1. All vocabulary words are defined in the Glossary.
2. along the Nile River and its branches
3. Eritrea
4. The climate is hot and dry, there are frequent droughts, and in some places soil erosion has occurred.
5. Burundi

Chapter 20 Highlights

Important Things to Know About East Africa

SECTION 1 KENYA

- Kenya has coastal lowlands as well as inland plains and highlands.
- The Great Rift Valley is a huge gash in the earth that runs through most of East Africa.
- Two important economic activities in Kenya are farming and tourism.
- Coffee and tea exports are Kenya's main source of income.
- Kenya's population of about 27 million is growing rapidly.

SECTION 2 TANZANIA

- Tanzania is made up of a mainland area and a group of offshore islands.
- Kilimanjaro, Africa's tallest mountain, lies in the northern area of Tanzania.
- Serengeti National Park in Tanzania is home to many species, including lions, antelopes, and elephants.
- Tanzania's population consists of about 120 ethnic groups.

SECTION 3 INLAND EAST AFRICA

- Uganda has a fertile plateau that is good for farming.
- Uganda, Rwanda, and Burundi lie close to the Equator, but their generally high elevation keeps their temperatures mild.
- The two main ethnic groups in Rwanda and Burundi are the Hutu and the Tutsi.

SECTION 4 THE HORN OF AFRICA

- Most of Sudan's people live along the Nile River and farm for a living.
- Ethiopia has been an independent country for thousands of years.
- Eritrea, Somalia, and Djibouti are hot, dry countries where farming is difficult.

Goma cane dancers in Kenya ▶

Extra Credit Project

Documentary Have students work in small groups to research one of the challenges facing East Africa today—drought and famine in the Horn of Africa or the civil war in Rwanda, for example. Have groups present their findings in documentary-style—a script accompanied by appropriate images. **L3**

CHAPTER HIGHLIGHTS

Using the Chapter 20 Highlights

- Use the Chapter 20 Highlights to preview, review, condense, or reteach the chapter.

Preview/Review

Vocabulary Puzzle-Maker Software reinforces the terms used in Chapter 20.

Condense

Have students read the Chapter 20 Highlights. Spanish Chapter Highlights also are available.

Chapter 20 Digest Audiocassettes and Activity

Chapter 20 Digest Audiocassette Test

Spanish Chapter Digest Audiocassettes, Activities, and Tests also are available.

Guided Reading Activities

Reteach

Reteaching Activity 20. Spanish Reteaching activities also are available.

MAP STUDY

Location Copy and distribute page 37 of the Outline Map Resource Book. Have students locate and label the countries of East Africa. Then have them locate and mark all the places listed in this chapter's **Section Previews.**

Chapter 20 Review and Activities

ANSWERS

Reviewing Key Terms

1. J
2. A
3. H
4. B
5. D
6. F
7. I
8. G
9. E
10. C

Mental Mapping Activity

This exercise helps students to visualize the countries and geographic features they have been studying and understand the relationships among them. All attempts at freehand mapping should be accepted.

REVIEWING KEY TERMS

Match the numbered terms in Column A with their definitions in Column B.

A
1. clan
2. poacher
3. autonomy
4. watershed
5. refugee
6. fault
7. civil war
8. cloves
9. reef
10. sisal

B
A. person who kills animals illegally
B. ridge dividing the flow of waters
C. plant fiber used to make rope
D. person who flees home for safety elsewhere
E. sea ridge rising above or below the water's surface
F. crack in the earth
G. spice made from dried flower buds
H. self-government
I. conflict within a country
J. large group of related people

Mental Mapping Activity

On a separate sheet of paper, draw a freehand map of East Africa. Label the following items on your map:

- Kenya
- Lake Tanganyika
- Rwanda
- Nairobi
- Eritrea
- Indian Ocean
- Somalia

REVIEWING THE MAIN IDEAS

Section 1
1. PLACE What is the Great Rift Valley?
2. MOVEMENT What group of people set up trading settlements along East Africa's coast?
3. HUMAN/ENVIRONMENT INTERACTION What major challenge does Kenya face today?

Section 2
4. PLACE What is the largest lake in Africa?
5. PLACE Where are cloves grown in Tanzania?

Section 3
6. LOCATION In what part of Uganda do most Ugandans live?
7. MOVEMENT Why did more than 2 million refugees flee Rwanda during the early 1990s?

Section 4
8. PLACE How do cultures differ in northern Sudan and southern Sudan?
9. MOVEMENT Why were troops sent to Somalia in the early 1990s?
10. LOCATION How does Djibouti's location affect its economy?

540

Reviewing the Main Ideas

Section 1

1. a deep gash in the earth's surface caused by plate movements millions of years ago—it runs through central East Africa and into Asia
2. Arabs from Southwest Asia
3. a rapidly growing population that needs jobs and food

Section 2

4. Lake Victoria
5. on the island of Zanzibar

Section 3

6. the southern part of the country
7. because of a civil war between the Hutu and the Tutsi

CRITICAL THINKING ACTIVITIES

1. **Identifying Central Issues** How do ethnic differences create problems for the countries of the East African region?
2. **Predicting Consequences** What effect do you think moving from the countryside to cities has on the way people live in East Africa?

GeoJournal Writing Activity

Imagine you have been sent to Somalia, Sudan, or Rwanda as part of a relief agency such as the American Red Cross, CARE, or Doctors Without Borders. Write a short article for your home-town newspaper describing your experience.

COOPERATIVE LEARNING ACTIVITY

Work in a group of four to learn about an ethnic group in East Africa. Choose one ethnic group, then have each student research one of the following topics: (a) family life; (b) education; (c) way of earning a living; or (d) the arts. After your research is complete, share your information with your group. Brainstorm ways to present your ethnic group to the entire class.

PLACE LOCATION ACTIVITY: EAST AFRICA

Match the letters on the map with the places and physical features of East Africa. Write your answers on a separate sheet of paper.

1. Zanzibar
2. Tanzania
3. Lake Victoria
4. Nile River
5. Mogadishu
6. Red Sea
7. Khartoum
8. Uganda
9. Ethiopia

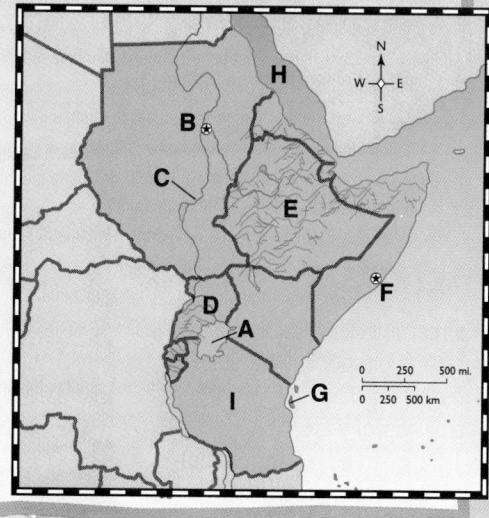

need for increased social services. Further, this movement often leads to the destruction of traditional ways of life.

Cooperative Learning Activity

Encourage students to make their presentations as colorful and dramatic as possible. Display any visuals or written work in the classroom and invite other classes to share them.

 GeoJournal Writing Activity

Interested students might like to interview a representative from a relief agency before undertaking this activity.

Place Location Activity

1. G
2. I
3. A
4. C
5. F
6. H
7. B
8. D
9. E

Chapter Bonus Test Question

This question may be used for extra credit on the chapter test.

Why is the Great Rift Valley an area of many volcanoes, lakes, and highlands? *(It was created by the movement of plates that make up the earth's crust.)*

Section 4

8. People in the north are Muslim Arabs; people in the south belong to several different African ethnic groups and practice either Christianity or traditional African religions.
9. to protect food supplies going to starving and sick Somalis
10. Djibouti's location on the Gulf of Aden makes the country an important shipping center.

Critical Thinking Activities

1. Answers will vary. Most students will note that ethnic differences often have led to open conflict and civil war. Further, governments in East Africa have to spend a great deal of time, money, and energy encouraging a sense of national identity among the many different ethnic groups.
2. Answers will vary. Most students will note that it leads to overcrowding in cities and the

GeoLab
ACTIVITY

FOCUS

Ask students how each of the following are shown on a map, and sketch their responses on the board: a river *(a line, usually blue in color)*; a lake *(an outline, area within colored blue)*; a road *(a line)*; forests *(a symbol or color, usually green).* Then ask students how a mountain or slope is depicted. Tell them that mountains and other changes in elevation are often shown by *contour lines.*

TEACH

Make sure students understand the two basic facts about contour lines: 1) that every point on a given contour line is at the same elevation above sea level; and 2) that there is a constant difference in elevation, called the *contour interval,* between contour lines. Then give students practice in reading contour lines by playing a contour matching game. Draw profiles of several hills of different shapes on the board. Have students draw rough contour maps to depict each hill. Next, draw contour maps on the board and have students create the profiles. Then assign partners and have students take turns matching various contours and hills. **L2**

From the classroom of
Carole Mayrose
Clay City, Indiana

LANDFORMS: Contour Mappi

Background

Africa south of the Sahara rises from west to east in a series of plateaus. How can you show plateaus or a rise in elevation on a map? In this activity you will create a contour map by drawing lines that follow the shape of a landform at different elevations.

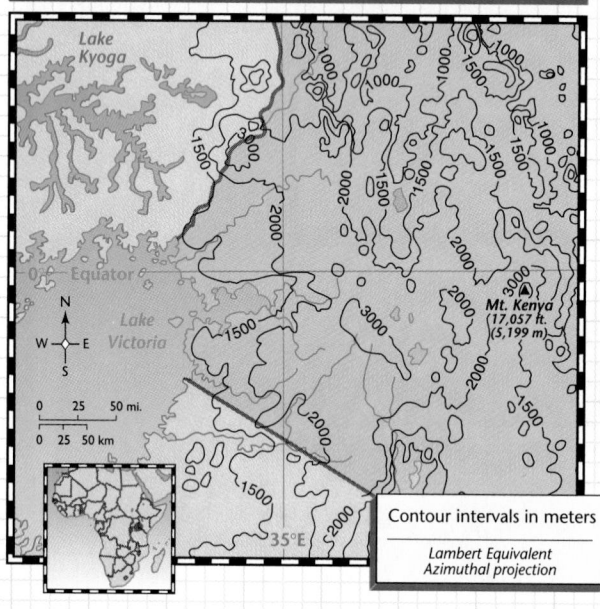

WEST KENYA: Contour Map

Contour intervals in meters

Lambert Equivalent
Azimuthal projection

Most of the African continent is a plateau, with the exception of the Great Rift Valley, which slices through eastern Africa. This valley has sides up to 1 mile (1.6 km) high and a floor up to 50 miles (80 km) wide. The Great Rift Valley is like a huge scar on the earth—and can be seen from space.

Materials

- a metric ruler
- 1 clear plastic box and lid
- modeling clay
- a beaker of water
- 1 transparency (clear plastic)
- a transparency marker
- tape

Curriculum Connection Activity

Government The United States federal government makes more maps than any other organization in the world. About 40 federal agencies, from the Agricultural Stabilization and Conservation Service to the National Aeronautic and Space Administration, have produced hundreds of thousands of different maps and distributed hundreds of millions of copies. The United States Geological Survey (USGS) makes topographic maps, which use contour lines and symbols. Have students contact the USGS to obtain a catalog of topographic maps for your area. Then help them order small-scale maps of your community. **L1**

What To Do

A. Using the ruler and the transparency marker, measure and mark 1-cm lines up the side of the box. (The bottom of the box will be zero elevation, or sea level.)

B. Tape the transparency to the top of the box lid.

C. With the modeling clay, mold a landform with mountains, plateaus, and valleys.

D. Place the model in the box.

E. Using the beaker, pour water into the box to a height of 1 cm. Place the lid on the box.

F. Looking down through the transparency on top of the box, use the marker to trace the top of the water line as it surrounds the landform.

G. Using the scale 1 cm=5 feet, mark the elevation on the line.

H. Repeat steps D through F, adding water to the next 1-cm level and tracing until you have mapped the landform by means of contour lines.

I. Transfer the tracing of the contours of the landform onto paper.

Lab Activity Report

1. What is the difference in elevation between each contour line on your map?

2. What is the average elevation of your landform?

3. Making Comparisons How would the number of contour lines on an area of steep mountains compare with the number on an area of flat plains?

Go A Step Further

Activity
Describe the kind of landform you think occurs where the contour lines are close together. What kind of landform occurs where the lines are far apart?

543

GeoLab
ACTIVITY

ASSESS

Have students answer the **Lab Activity Report** questions on page 543.

CLOSE

Encourage students to complete the **Go A Step Further** writing activity. Use maps of your area obtained from the USGS for a field trip to give students first-hand knowledge of how real physical features appear on contour maps.

Go A Step Further
Answer
mountain; flat plains area

CURRICULUM CONNECTION

Sports The sport of orienteering involves the use of compasses and topographic maps. Competitors use these tools to race across unfamiliar territory, locating landmarks. The successful competitor reads contour lines to determine the most efficient routes.

South Africa and Its Neighbors

PERFORMANCE ASSESSMENT ACTIVITY

WORLD'S FAIR EXHIBIT Have students work in groups to plan a booth for southern Africa at the next World's Fair. Groups first should conduct preliminary research to discover what items countries exhibit at the World's Fair. They should then apply this information to southern Africa. Have groups draw up designs for their booth and lists of items they want to exhibit. Each item should be accompanied by a brief explanation of its connection to southern Africa. Designs and listed items should illustrate the five themes of geography. Groups might present their work in the form of illustrated brochures, or they might like to construct small booths with photographs, sketches, or actual examples of their listed items. Invite members of other classes, school officials, and parents to view the brochures or booths.

POSSIBLE RUBRIC FEATURES: Accuracy of content information, concept attainment for the five themes, collaborative skills, analysis and evaluation skills, decision-making skills, organization skills

MENTAL MAPPING ACTIVITY

Before teaching Chapter 21, point out the countries of southern Africa on a globe. Have students find other countries at the same latitudes. *(parts of Australia and South America)*

TEACHER'S CORNER

NATIONAL GEOGRAPHIC SOCIETY

INDEX TO NATIONAL GEOGRAPHIC MAGAZINE

The following articles may be used for research relating to this chapter:

- "A Place for Parks," by Douglas H. Chadwick, July 1996.
- "Lions of Darkness," by Dereck Joubert, August 1994.
- "The Twilight of Apartheid," by Charles E. Cobb, Jr., February 1993.
- "A Gathering of Waters and Wildlife," Photo essay by Frans Lanting, December 1990.
- "Okavango Delta: Old Africa's Last Refuge," by Douglas B. Lee, December 1990.
- "Botswana, the Adopted Land," by Arthur Zich, December 1990.
- "The Plant Hunters: A Portrait of the Missouri Botanical Garden (including Expeditions to Madagascar)," by Boyd Gibbons, August 1990.
- "Emeralds," by Fred Ward, July 1990.
- "Tsetse-Fly of the Deadly Sleep," by Georg Gerster, December 1986.
- "After Rhodesia, a Nation Named Zimbabwe," by Charles E. Cobb, Jr., November 1981.

NATIONAL GEOGRAPHIC SOCIETY PRODUCTS AVAILABLE FROM GLENCOE

To order the following products for use with this chapter, contact your local Glencoe sales representative or call Glencoe at 1-800-334-7344:

- *STV: World Geography* (Videodisc)
- *Picture Atlas of the World* (CD-ROM)
- *Physical Geography of the World* (Transparencies)
- *ZipZapMap! World* (Software)
- *GeoBee* (Software)
- *Images of the World* (Posters)
- *Eye on the Environment* (Posters)

ADDITIONAL NATIONAL GEOGRAPHIC SOCIETY PRODUCTS

To order the following products for use with this chapter, call National Geographic Society at 1-800-368-2728:

- *Bushmen of the Kalahari* (Video)
- *Physical Geography of the Continents Series: "Africa."* (Video)
- *The Rhino War* (Video)
- *Creatures of the Namib Desert* (Video)
- *Africa: Wilds of Madagascar* (Video)
- *Lions of the African Night* (Video)
- *African Wildlife* (Video)

TEACHER-TO-TEACHER
Curriculum Connection: Earth Science

—from Eleanor Bloom
Bloom-Carroll Schools, Lithopolis, OH

PIN A PRODUCT ON THE MAP *The purpose of this activity is to help students develop an awareness of the economic activities of South Africa.*

List the following on the chalkboard, under the title *South Africa's Major Products:* **Agriculture**—corn, wheat, potatoes, citrus fruits, sugarcane, tobacco, dairy products, beef cattle, wool; **Manufacturing**—iron and steel, other metals, automobiles, machinery, chemicals, processed foods, beverages, textiles and clothing; **Mining**—gold, diamonds, platinum, copper, uranium, iron ore, coal, limestone.

Assign each student one or more of the listed products, making certain all products are assigned. Draw a large outline map of South Africa on a sheet of butcher paper and attach it to the bulletin board. Have students draw symbols for their products on separate sheets of paper. Below the symbol, have students write three pieces of information about the product. Direct students to attach their sheets to the bulletin board around the map. Then have them run lead lines from the sheets to the locations on the map where products are farmed, manufactured, or mined.

In a follow-up discussion, have students identify South Africa's major agricultural, manufacturing, and mining areas.

MEETING NATIONAL STANDARDS

Geography for Life

All of the 18 standards are demonstrated in Unit 7. The following ones are highlighted in this chapter:

• Standards 4, 6, 11, 13, 16

KEY TO ABILITY LEVELS

Teaching strategies have been coded for varying learning styles and abilities.

L1   **BASIC** activities for all students

L2   **AVERAGE** activities for average to above-average students

L3   **CHALLENGING** activities for above-average students

LEP   **LIMITED ENGLISH PROFICIENCY** activities

BIBLIOGRAPHY

Readings for the Student

Meisel, Jacqueline Drobis. *South Africa at the Crossroads.* Brookfield, Conn.: Millbrook Press, 1994.

Murray, Jocelyn. *Africa: Cultural Atlas for Young People.* New York: Facts On File, 1990.

"South Africa." *Faces: The Magazine About People,* January 1991.

Readings for the Teacher

Africa Today: A Reproducible Atlas. Wellesley, Mass.: World Eagle, 1994.

Creative Activities for Teaching About Africa. Stockton, Calif.: Stevens & Shea, 1990.

Map Skills and Outlines: Africa. St. Louis: Milliken, 1991. Transparencies and reproducible worksheets.

"Teaching About the New South Africa." *Social Education,* Vol. 59 No. 2, February 1995.

Multimedia

Africa Recovery. New York: United Nations, 1990. Videocassette, 15 minutes.

Africa: Southern Region. Chicago: Encyclopedia Britannica, 1993. Videocassette, approximately 20 minutes.

Global Studies With Country Databases: Africa. Toronto, Canada: WorldView Software, 1994. Apple 5.25" disk or IBM 3.5" disk.

They Come in Peace: A New Democratic South Africa. Niles Ill.: United Learning, 1994. Videocassette, 27 minutes.

Use these *Geography: The World and Its People* resources to teach, reinforce, and extend chapter content.

CHAPTER 21 RESOURCES

- *Vocabulary Activity 21
- Cooperative Learning Activity 21
- Workbook Activity 21
- Chapter Map Activity 21
- Geography Skills Activity 21
- GeoLab Activity 21
- Critical Thinking Skills Activity 21
- Performance Assessment Activity 21
- *Reteaching Activity 21
- Enrichment Activity 21
- Environmental Case Study 7
- Chapter 21 Test, Form A and Form B
- *Chapter 21 Digest Audiocassette, Activity, Test
- Unit Overlay Transparencies 7-0, 7-1, 7-2, 7-3, 7-4, 7-5, 7-6
- Political Map Transparency 7
- Geoquiz Transparencies 21-1, 21-2
- Vocabulary PuzzleMaker Software
- Student Self-Test: A Software Review
- Testmaker Software
- MindJogger Videoquiz

If time does not permit teaching the entire chapter, summarize using the Chapter 21 Highlights on page 565, and the Chapter 21 English (or Spanish) Audiocassettes. Review students' knowledge using the Glencoe MindJogger Videoquiz. *Also available in Spanish

Use these *Geography: The World and Its People* resources to teach and reinforce **section** content.

SECTION 1 RESOURCES

Reproducible Lesson Plan 21-1

Section Focus Transparency 21-1

Guided Reading Activity 21-1

Section Quiz 21-1 (also available in Spanish)

SECTION 2 RESOURCES

Reproducible Lesson Plan 21-2

Section Focus Transparency 21-2

Guided Reading Activity 21-2

Section Quiz 21-2 (also available in Spanish)

SECTION 3 RESOURCES

Reproducible Lesson Plan 21-3

Section Focus Transparency 21-3

Guided Reading Activity 21-3

Section Quiz 21-3 (also available in Spanish)

SECTION 4 RESOURCES

Reproducible Lesson Plan 21-4

Section Focus Transparency 21-4

Guided Reading Activity 21-4

Section Quiz 21-4 (also available in Spanish)

ADDITIONAL RESOURCES FROM GLENCOE

Reproducible Masters

- Glencoe Social Studies Outline Map Book, pages 36, 37

- Facts On File, GEOGRAPHY ON FILE, Sub-Saharan Africa; AFRICAN HISTORY ON FILE

- Foods Around the World, Unit 7

World Music: Cultural Traditions
Lesson 6

Transparencies

- National Geographic Society PicturePack Transparencies, Unit 7

Posters

- National Geographic Society: Eye on the Environment, Unit 7

Videodiscs

- Reuters Issues in Geography

- National Geographic Society: STV: World Geography, Vol. 2

- National Geographic Society: GTV: Planetary Manager

CD-ROM

- National Geographic Society: Picture Atlas of the World

Software

- National Geographic Society: ZipZapMap! World

Reference Books

- Picture Atlas of Our World

Performance Assessment

Refer to the Planning Guide on page 544A for a Performance Assessment Activity for this chapter. See the *Performance Assessment Strategies and Activities* booklet for suggestions.

Chapter Objectives

1. **Review** the geography of South Africa.
2. **Outline** the geographic features of the Atlantic countries of southern Africa.
3. **Compare** and contrast the countries of Zambia, Zimbabwe, Malawi, and Botswana.
4. **Discuss** the geography of the Indian Ocean countries of southern Africa.

GLENCOE TECHNOLOGY

VIDEODISC

Use the Chapter 21 MindJogger Videoquiz to preview chapter content.

MindJogger Videoquiz

Chapter 21
Disc 3 Side B

Also available in VHS.

Chapter 21 South Africa and Its Neighbors

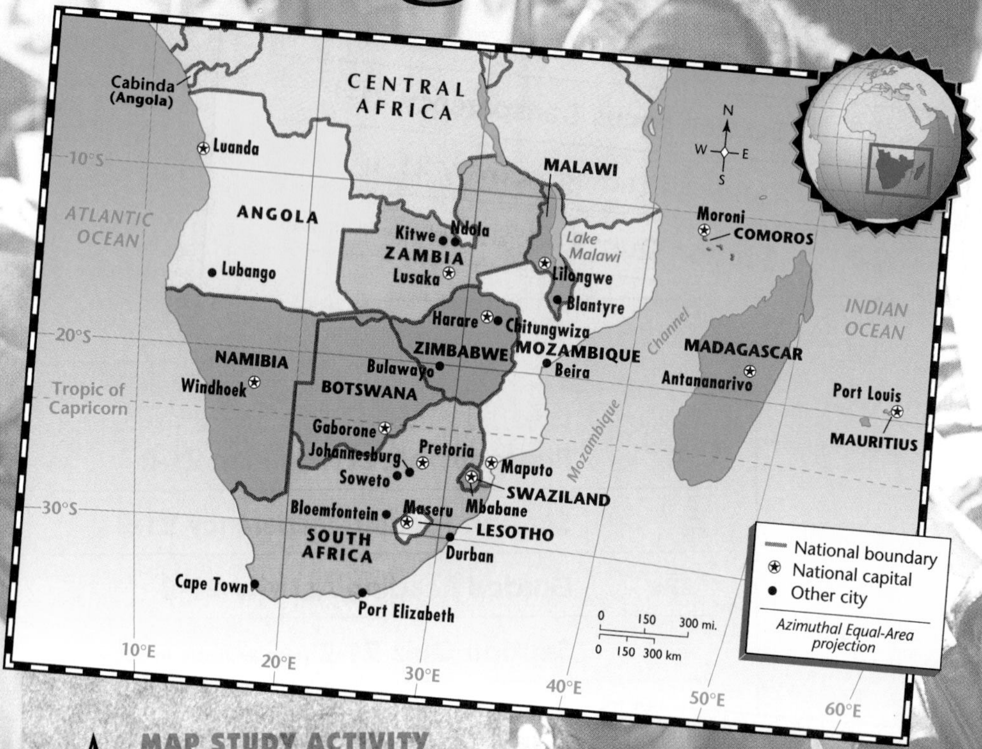

MAP STUDY ACTIVITY

As you read Chapter 21, you will learn about South Africa and its neighbors.

1. **What three island countries lie off Africa's southeastern coast?**
2. **What country forms the southern tip of the African continent?**
3. **Through what countries does the Tropic of Capricorn pass?**

544

MAP STUDY ACTIVITY

Answers

1. Comoros, Madagascar, Mauritius
2. South Africa
3. Namibia, Botswana, South Africa, Mozambique, Madagascar

Map Skills Practice

Reading a Map What is the capital of Zimbabwe? *(Harare)* At what latitude is the city of Durban located? *(30°S)*

Republic of South Africa

LESSON PLAN
Chapter 21, Section 1

PREVIEW

Words to Know
- enclave
- high veld
- escarpment
- apartheid
- township

Places to Locate
- Republic of South Africa
- Cape Town
- Drakensberg Mountains
- Johannesburg
- Durban
- Pretoria

Read to Learn . . .
1. what landscapes are found in South Africa.
2. what mineral resources South Africa has.
3. what changes have occurred recently in South Africa's government.

This huge natural wonder is called Table Mountain. It looms over the city of Cape Town—a major port in the Republic of South Africa.

The Republic of South Africa spreads across the southern end of Africa. It is a land of breathtaking scenery and great mineral wealth. It is also a land of great change. In recent years, South Africa's people have gone through many changes in their lives and their government.

Region
The Land

South Africa sprawls across 471,440 square miles (1,221,030 sq. km)—an area nearly three times the size of California. South Africa has many different landscapes: winding coastlines, tall mountains, deep valleys, and a high plateau. The map on page 544 shows you that South Africa's large land area also swallows up two small independent African nations—Lesotho (luh•SOH•toh) and Swaziland. These are **enclaves,** or small countries surrounded or nearly surrounded by a larger country.

Coasts The west coast of South Africa borders the Atlantic Ocean, and the south and east coasts border the Indian Ocean. Northwest along the Atlantic

---

Meeting National Standards

Geography for Life The following standards are highlighted in this section:
- Standards 10, 13, 17

FOCUS

Section Objectives
1. **Describe** the landscapes that are found in South Africa.
2. **Identify** the mineral resources that South Africa possesses.
3. **Discuss** the changes that occurred recently in South Africa's government.

Bellringer Motivational Activity

Prior to taking roll at the beginning of the class period, project Section Focus Transparency 21-1 and have students answer the activity questions. Discuss students' responses. **L1**

Vocabulary Pre-check

Tell students that *veld* is a Dutch word for "open grasslands." South Africa's high veld is one of the world's major prairies. Others include the American Midwest, Canada's Prairie Provinces, and Argentina's pampas. **L1 LEP**

Use the Vocabulary PuzzleMaker Software to create a crossword puzzle. **L1**

Classroom Resources for Section 1

REPRODUCIBLE MASTERS

Reproducible Lesson Plan 21-1
Guided Reading Activity 21-1
Vocabulary Activity 21
Geography Skills Activity 21
Section Quiz 21-1

**TRANSPARENCIES**

Section Focus
    Transparency 21-1
Political Map Transparency 7

MULTIMEDIA

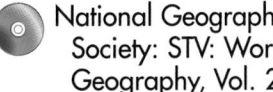Vocabulary PuzzleMaker Software

Testmaker

National Geographic Society: STV: World Geography, Vol. 2

545

TEACH

Guided Practice

Drawing Conclusions
Draw a four-column chart on the chalkboard, using the following as column headings: *The Land, The Economy, Influences of the Past, The People.* Call on volunteers to come to the chalkboard and enter in the appropriate columns information on the Republic of South Africa. Then ask students to use information in the chart to write a paragraph explaining why South Africa has the most developed economy in Africa. **L2**

CURRICULUM CONNECTION

Earth Science Because of its elevation, Lesotho does not have a tropical climate like that of the surrounding region. Also because of its elevation, Lesotho is free of many diseases common in other parts of Africa.

Answer
Drakensberg Mountains

Map Skills Practice
Reading a Map What is the name of the dry, flat land in southern South Africa? *(Great Karroo)*

---

coast stretches the vast Namib Desert. Farther south lies the Cape of Good Hope, the southernmost point of Africa. A group of long, narrow mountain ranges runs through this southwestern area. Between the ranges lies a dry, flat land called the Great Karroo. Farther east, along the Indian Ocean, are high cliffs, sandy beaches, rolling hills, and a lowland plain.

A Plateau and Mountains A large plateau spreads through the center of South Africa. It ranges from about 2,000 to 8,000 feet (610 to 2,440 m) above sea level. Part of the South African plateau is made up of flat, grass-covered plains called the **high veld.** Isolated rocky hills rise as high as 100 feet (30 m) above the surrounding land.

A group of mountains and cliffs circle the plateau and divide it from the coastal areas. They are known as the Great Escarpment. An **escarpment,** you will recall, is a steep cliff or slope that divides high and low ground. The Great Escarpment rises to its highest elevations in the eastern Drakensberg Mountains. Many peaks in this range are more than 10,000 feet (3,000 m) high.

The Climate South Africa lies south of the Equator. Its seasons are opposite those in the Northern Hemisphere. When it is winter north of the Equator, it is summer south of the Equator. In winter the South African plateau is cool and sunny—with some rainfall—during the day. Temperatures sometimes drop below freezing at night. Summers are mild because of the high elevation.

Cape Town—a major port city—has a Mediterranean climate of cool, rainy winters and hot, often dry summers. Along the eastern coast warm winds from the Indian Ocean bring a humid subtropical climate.

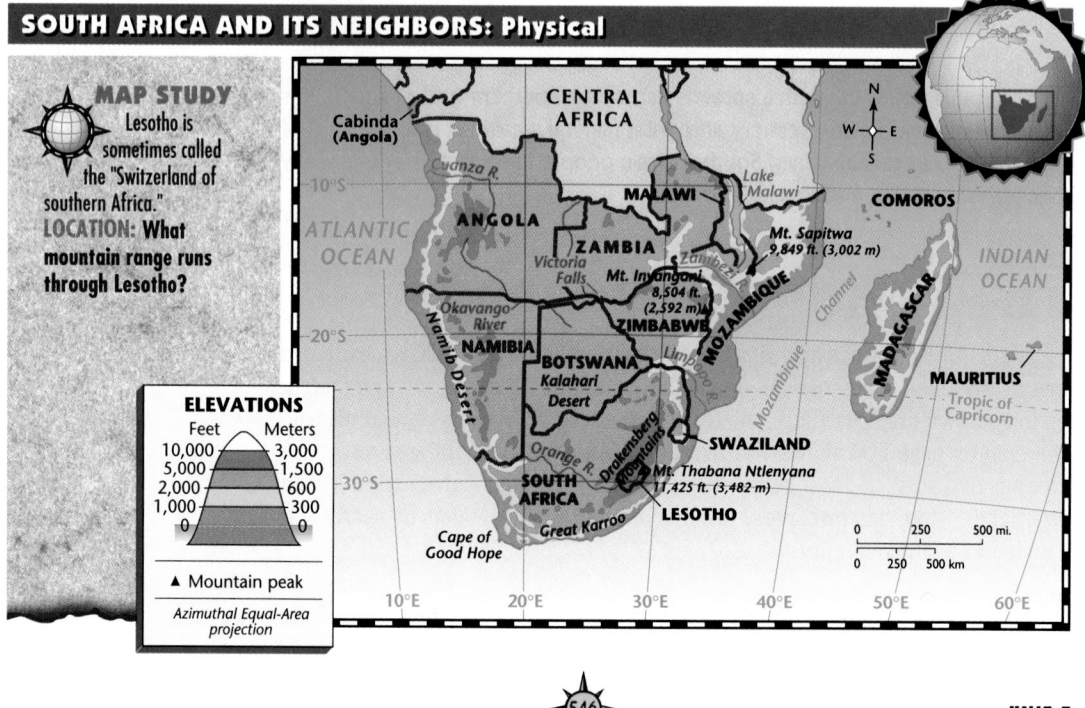

SOUTH AFRICA AND ITS NEIGHBORS: Physical

MAP STUDY
Lesotho is sometimes called the "Switzerland of southern Africa."
LOCATION: What mountain range runs through Lesotho?

ELEVATIONS
Feet	Meters
10,000	3,000
5,000	1,500
2,000	600
1,000	300
0	0

▲ Mountain peak

Azimuthal Equal-Area projection

Cooperative Learning Activity

Organize students into small groups to research the history of South Africa. Assign a time period or a topic to each group. Groups should research how their assigned time periods or topics affected life in South Africa. Then groups should determine a way to share their information with the class. Possible methods of presentation include illustrated time lines, skits, oral reports, and captioned drawings. **L2**

Human/Environment Interaction
The Economy

South Africa has the most developed economy in Africa. Its workers mine one-half of Africa's minerals and produce two-fifths of Africa's manufactured goods. South Africa is also the continent's largest exporter of farm products.

Not all South African workers take part in this prosperous economy. Workers in modern industries and businesses live in urban areas. Yet many people in rural South Africa are poor and depend on subsistence farming.

Land of Gold and Diamonds In terms of mineral wealth, South Africa is one of the richest countries in the world. South Africa produces about one-third of all the gold mined in the world each year. The Witwatersrand (WIHT•WAW•tuhrz•RAND)—an area around the city of Johannesburg—holds the world's largest and richest goldfield. South Africa also has the world's largest deposits of diamonds.

Manufacturing and Farming South Africa's industrial workers produce many manufactured goods. Many of the products South Africans use are made in their own factories. South Africa also exports metal products, chemicals, clothing, and processed foods.

Most of South Africa is either too dry or too hilly to farm. On the land that is available, South Africa's farmers grow enough food for the country's needs and for export. Among the crops they cultivate are corn, fruits, potatoes, and wheat. Herding sheep and livestock is a major economic activity on the plateau.

Many people raise livestock in South Africa's high veld *(left)* and in the small enclave of Swaziland *(right)*.
PLACE: What is an enclave?

WHAT IN THE WORLD

The Biggest Diamond
The largest diamond ever found was dug from Premier Mine near the city of Pretoria in South Africa. It was about the size of a person's fist and weighed 3,106 carats—about 1.4 pounds (.64 kg). When it was cut, it made 9 large jewels and more than 100 smaller ones. The largest jewel became known as the Star of Africa.

Meeting Special Needs Activity

Learning Disabled Use a globe and a light source to demonstrate why seasons in the Southern Hemisphere are the opposite of those in the Northern Hemisphere. Tell students that the light represents the sun. Point out that when the North Pole is tilted toward the sun, the Northern Hemisphere has summer. When the North Pole is tilted away from the sun, the Southern Hemisphere has summer. Have students move the globe around the light to demonstrate the positions of the earth relative to the sun that produce autumn and spring in each hemisphere. **L1**

LESSON PLAN
Chapter 21, Section 1

More About the Illustration
Answer to Caption
small country surrounded or nearly surrounded by a larger country

GEOGRAPHIC THEMES

Place Cattle are a symbol of wealth and status in rural Swaziland. When a Swazi man gets married, for example, his family gives cattle to the bride's family. The transaction makes the marriage official.

Independent Practice

Guided Reading Activity 21-1 **L1**

Strange but TRUE!

South Africa's diamond mines were discovered in 1867. A farmer's child found a "pretty pebble" near the banks of the Orange River. The pebble turned out to be a diamond worth $2,500!

Cultural Kaleidoscope

Lesotho As many as 250,000 men at a time may be gone from home to work in South Africa's mines or factories. As a result, women in Lesotho do most of the farming and make many day-to-day family decisions.

547

MULTICULTURAL PERSPECTIVE

Culturally Speaking

Under the apartheid system, the South African government classified people into four groups—white, black, colored (mixed), and Asian.

NATIONAL GEOGRAPHIC SOCIETY

These materials are available from Glencoe.

VIDEODISC

STV: World Geography Vol. 2: Africa and Europe

Southern Africa: The South
Side 1, Frames 42895-48595

MAP STUDY

Answer
1975

Map Skills Practice
Reading a Map Which southern African nation gained its independence most recently? *(Namibia)*

Place
Influences of the Past

About 42 million people live in South Africa. African ethnic groups make up 74 percent of South Africa's population. Africans of European origin represent 14 percent, and Asians 3 percent. People of mixed European, African, and Asian backgrounds make up the remaining 9 percent.

Africans Most Africans in South Africa trace their ancestry to Bantu-speaking peoples that settled throughout Africa between A.D. 100 and 1000. The largest ethnic groups in South Africa today are the Sotho, Zulu, and Xhosa (KOH•suh).

Europeans In the 1600s the Dutch became the first Europeans to settle in South Africa. Over time German and French settlers joined them. Together these European groups came to speak their own language—Afrikaans (A•frih•KAHNS). The Afrikaners pushed Africans off the best land and set up farms.

The British settled in South Africa in the early 1800s. By the end of the century, diamonds and gold drew many British people to South Africa. War broke out between the British and the Afrikaners in 1899. After three years of fighting, the British won the conflict.

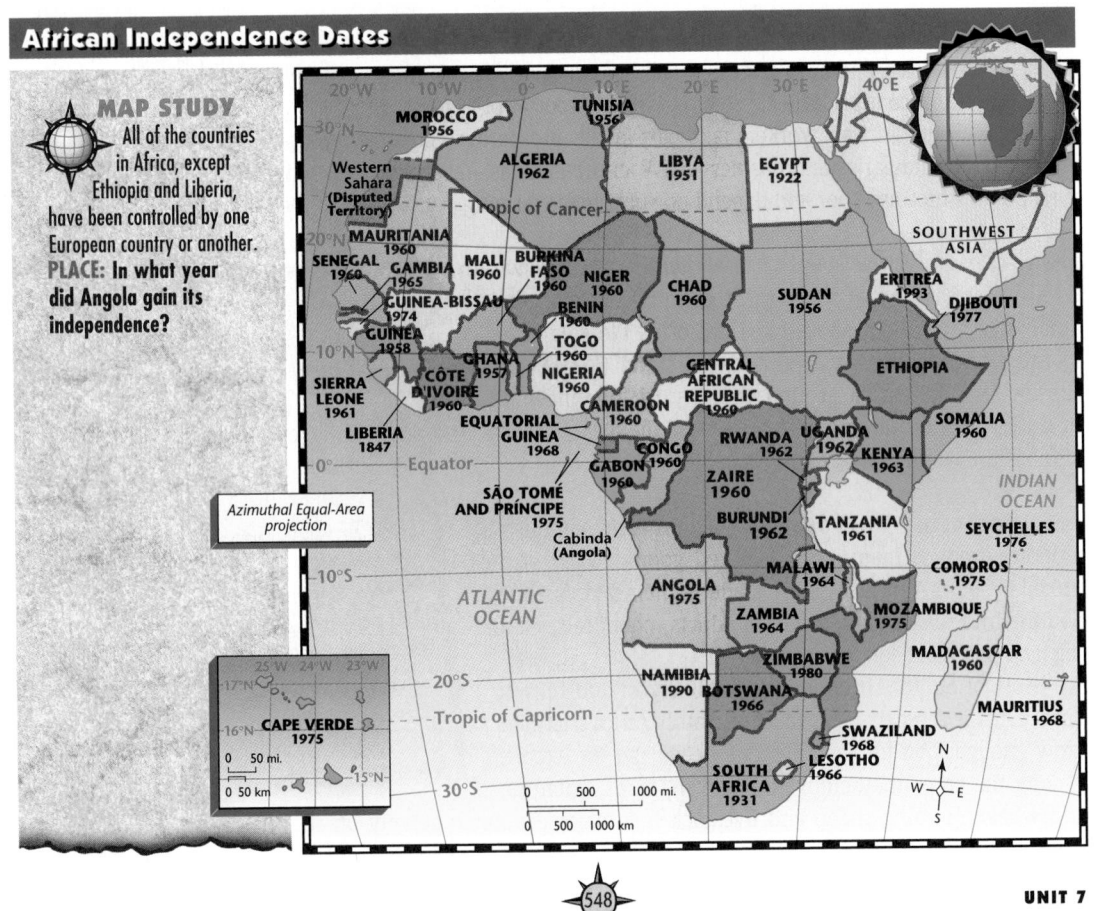

African Independence Dates

MAP STUDY All of the countries in Africa, except Ethiopia and Liberia, have been controlled by one European country or another. **PLACE: In what year did Angola gain its independence?**

Azimuthal Equal-Area projection

Critical Thinking Activity

Drawing Conclusions Have students consider whether being an enclave, such as Swaziland and Lesotho, has more advantages or disadvantages. *(Possible advantages: has its own government, people can work in South Africa, may have protection from outside influences; Possible disadvantages: lacks access to other countries except through South Africa)* Have students with differing opinions present their views. Then have the class determine which arguments have the most merit. **L2**

A New Nation In 1910 Afrikaner and British territories united into one country—the Union of South Africa—within the British Empire. Black South Africans founded the African National Congress in 1912 in hopes of gaining power. In 1931 the United Kingdom gave South Africa full independence.

In 1948 the Afrikaners set up **apartheid**—"apartness"—or practices that separated South Africans of different ethnic groups. For example, laws forced black South Africans to live in separate areas and attend different schools from European South Africans. People of non-European background could not vote. They had virtually no political rights enjoyed by those of European ancestry.

For more than 40 years, people inside and outside South Africa protested against the practice of apartheid. In 1991, the South African government finally agreed to end apartheid. In April 1994, South Africa had its first-ever election in which people of all ethnic groups could vote. South Africans elected their first black African president, Nelson Mandela.

In 1994, for the first time, South Africa's elections were open to all races *(left)*. Nelson Mandela became the country's first black president *(right)*.
PLACE: What group of people were favored under apartheid?

More About the Illustration

Answer to Caption
white South Africans

Place Shortly after receiving his law degree in 1942, Nelson Mandela helped establish the first black law firm in South Africa. At about the same time, he joined the African National Congress (ANC), an organization dedicated to winning freedom for black South Africans. As a result of his work for the ANC, he was imprisoned for life in 1962. After his release in 1990, Mandela played a major role in leading South Africa toward democracy.

COMPARING POPULATION:
South Africa and the United States

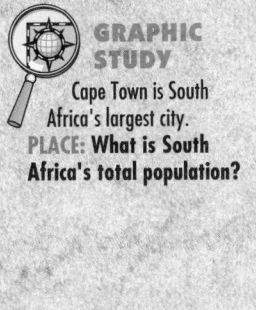

GRAPHIC STUDY
Cape Town is South Africa's largest city.
PLACE: What is South Africa's total population?

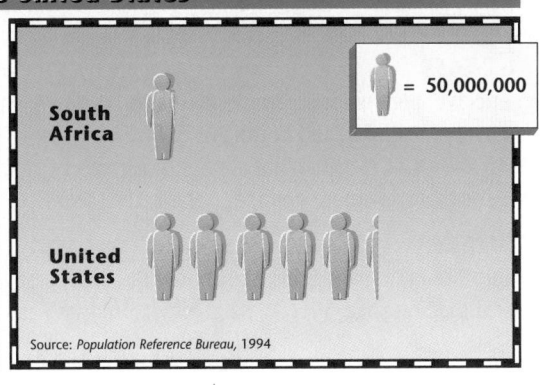

☻ = 50,000,000

South Africa

United States

Source: *Population Reference Bureau, 1994*

ASSESS

Check for Understanding

Assign Section 1 Review as homework or an in-class activity.

GRAPHIC STUDY

Answer
42 million

Graphic Skills Practice
Reading a Graph About how many more people live in the U.S. than in South Africa? *(about 218 million)*

Global Issues Activity

Economics Have students investigate ways that other countries brought pressure on South Africa to abolish apartheid. They should research the widespread economic boycott of South Africa during the 1980s. They also should investigate the role of the United Nations in opposing South Africa's apartheid policies. **L2**

549

Meeting Lesson Objectives

Each objectives below is tested by the questions that follow it in parentheses.

1. **Describe** the landscapes that are found in South Africa. (2)
2. **Identify** the mineral resources that South Africa possesses. (3)
3. **Discuss** the changes that occurred recently in South Africa's government. (4)

Evaluate

 Section 1 Quiz **L1**

Use the Testmaker to create a customized quiz for Section 1. **L1**

Reteach

Have students work in pairs or small groups to study and become "experts" on an assigned part of Section 1. Then have each "expert" pair or group answer questions from the rest of the class. **L1**

Enrich

Ask students to find out how gold and/or diamonds are mined. Have students share their findings with the class. **L3**

CLOSE

Have students explore the effects of apartheid by answering the following questions: How did government policies regulate education, employment, transportation, and neighborhoods? How did apartheid violate human rights?

550

Teen Scene

Creating History

Mosi Faraji feels lucky to be living in South Africa right now. Someday he will tell his children about what it was like to live under apartheid. Now he is allowed to attend schools that were once all-white. The history books in his classes finally include the story of his nation's black people. Many restaurants and theaters that served only whites are now open to everyone. Even sports teams can now be mixed.

Place
The People

About 57 percent of South Africans live in urban areas. The map on page 562 shows you that South Africa has five large cities with more than 500,000 people. These cities are Johannesburg, Cape Town, Durban, Soweto and Pretoria. Cape Town and Durban are port cities. Johannesburg is an inland industrial and commercial center. North of Johannesburg is Pretoria, which is one of the national capitals along with Cape Town and Bloemfontein (BLOOM•fuhn•TAYN).

One of the challenges facing South Africa is to develop a better standard of living for all of its people. Most European South Africans live in modern homes or apartments, own cars, and enjoy foods similar to those common in the United States and Europe. Most African, Asian, and mixed-group South Africans, however, have a much harder life. Many live in crowded **townships,** or neighborhoods outside cities. Others live in rural areas where the land is difficult to farm.

The arts in South Africa reflect the country's long struggle for justice and equality. In recent years, African musicians have combined traditional African dance rhythms and modern rock music. This style of music has become popular in many countries, especially the United States. You may be familiar with South African groups such as Mahlathini and the Mahotella Queens and Ladysmith Black Mambazo.

SECTION 1 REVIEW

REVIEWING TERMS AND FACTS

1. **Define the following:** enclave, high veld, escarpment, apartheid, township.
2. **PLACE** What landform covers most of inland South Africa?
3. **HUMAN/ENVIRONMENT INTERACTION** What are South Africa's two main mineral resources?
4. **PLACE** How did South Africa's government change in the early 1990s?

 MAP STUDY ACTIVITIES

5. Look at the political map on page 544. About how far is Cape Town from Pretoria?
6. Turn to the physical map on page 546. At what elevation does Lesotho lie?

Answers to Section 1 Review

1. All vocabulary words are defined in the Glossary.
2. high plateau
3. gold, diamonds
4. a white-controlled government based on apartheid gave way to a democratic government based on equality for all South Africans
5. about 700 miles (1,127 km)
6. between 5,000 and 10,000 feet (1,500 to 3,000 m)

BUILDING GEOGRAPHY SKILLS

Reading a Time Line

A time line is a good way to show the order of events. Dates at the beginning and the end of a time line mark a *time span,* or the amount of time the line covers. Events listed on the time line all took place within its time span.

Most time lines are divided into sections representing equal time intervals. For example, a time line showing 1,000 years might be divided into ten 100-year sections. Each event appears on the time line beside the date when the event took place. In reading a time line, apply these steps:

• Find the dates on the opposite ends of the time line to know the total time span. Also note the intervals.

• Study the order of events.

• Analyze relationships among events or look for trends.

Key Events in Modern South Africa

1910: Union of South Africa formed

1931: South Africa becomes independent

1961: South Africa leaves Commonwealth of Nations

1990-1991: Apartheid laws ended

1900 1910 1920 1930 1940 1950 1960 1970 1980 1990 2000

1912: African National Congress founded

1948: Apartheid laws begin

1984: Africans protest new constitution that excludes them

1994: Nelson Mandela elected president

Geography Skills Practice

1. What time span does this time line represent?

2. How many years does each section represent?

3. Was the African National Congress formed before or after South Africa became independent?

4. About how long were the apartheid laws in force?

TEACH

On the chalkboard, draw a time line spanning 1,000 years divided into 100-year intervals. Have students identify the time span the time line covers. Point out the intervals and have students note the number of years intervals represent. Then add 5 to 10 events to the time line at their appropriate points.

Have students study the time line. Direct them to create a similar time line covering their own lives. Suggest that they list five or six important events. Then have students read the skill and complete the practice questions. **L1**

Additional Skills Practice

1. How many years after South African independence did the apartheid laws begin? *(17 years)*

2. Which came first, the formation of the Union of South Africa or South African independence? *(formation of the Union of South Africa)*

📁 Geography Skills Activity 21

Answers to Geography Skills Practice

1. 100 years

2. 10 years

3. the ANC was formed before South Africa became independent

4. 43 years

FOCUS

Section Objectives

1. **Identify** the landforms that Angola and Namibia share.
2. **Explain** why Portuguese is the official language of Angola.
3. **Discuss** why Namibia has difficulty feeding its people.

Bellringer Motivational Activity

 Prior to taking roll at the beginning of the class period, project Section Focus Transparency 21-2 and have students answer the activity questions. Discuss students' responses. **L1**

Vocabulary Pre-check

Have students find the meaning of *exclave* in the text. Help them locate Cabinda, Swaziland and Lesotho on the map on page 544. Have students use these examples to write a comparative definition of the terms *exclave* and *enclave*. **L1** **LEP**

TEACH

Guided Practice

Chart Help students make a chart comparing Namibia and Angola. Bases of comparison include size, landscapes, climates, economic

(continued)

552

SECTION 2
Atlantic Countries

Words to Know
• exclave

Places to Locate
• Namibia
• Angola
• Windhoek
• Luanda
• Namib Desert
• Kalahari Desert

Read to Learn . . .
1. what landforms Angola and Namibia share.
2. why Portuguese is the official language of Angola.
3. why Namibia has difficulty feeding its people.

Namib Desert stretches north to south about 1,200 miles (1,931 km) through southwestern Africa.

You would not want to get a flat tire on the vast Namib Desert! This empty stretch of sand and rock is one of the loneliest places on the earth. The

The Namib Desert extends most of the length of the country of Namibia. In addition to desert, Namibia and its northern neighbor—Angola—have long coastlines on the Atlantic Ocean. For this reason, they are known as southern Africa's Atlantic countries.

Place
Angola

Angola is a huge, nearly square mass of land. It covers 481,350 square miles (1,246,697 sq. km)—an area larger than Texas and California put together. Angola also includes a tiny exclave called Cabinda. An **exclave** is a tiny area of a country that is separated from the main part. The map on page 544 shows you that Cabinda is separated from Angola.

The Land Most of Angola is part of the same huge inland plateau that sweeps through the Republic of South Africa. Many rivers cross this area in Angola. Some flow into the Zaire River in the north; others flow into the Atlantic Ocean or the Indian Ocean.

552

UNIT 7

Classroom Resources for Section 2

 REPRODUCIBLE MASTERS
Reproducible Lesson Plan 21-2
Guided Reading Activity 21-2
Cooperative Learning Activity 21
Chapter Map Activity 21
Section Quiz 21-2

 **TRANSPARENCIES**
Section Focus Transparency 21-2
Political Map Transparency 7

MULTIMEDIA
◎ Testmaker

◎ National Geographic Society: ZipZapMap! World

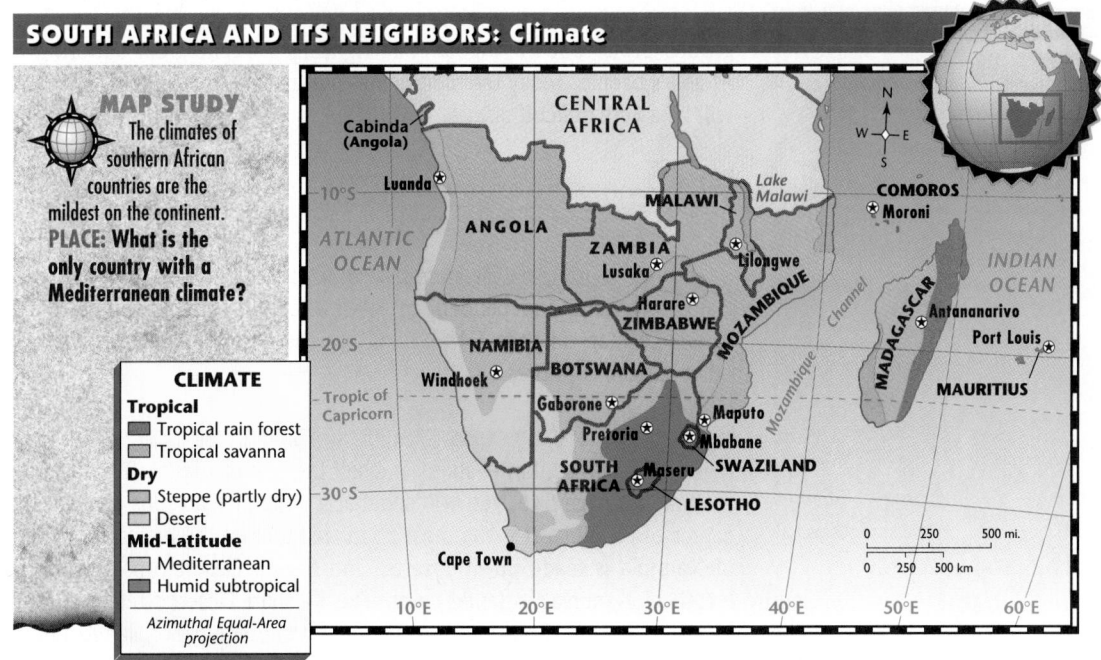

MAP STUDY
The climates of southern African countries are the mildest on the continent.
PLACE: What is the only country with a Mediterranean climate?

CLIMATE
Tropical
☐ Tropical rain forest
☐ Tropical savanna
Dry
☐ Steppe (partly dry)
☐ Desert
Mid-Latitude
☐ Mediterranean
☐ Humid subtropical

Azimuthal Equal-Area projection

LESSON PLAN
Chapter 21, Section 2

activities, and historical influences. You might draw the outline of the chart on the chalkboard and have students copy and complete it in their notebooks. **L1**

MAP STUDY

Answer
South Africa
Map Skills Practice
Reading a Map What is the major climate type found in northern Namibia? *(steppe)*

Hilly grasslands cover northern Angola. The southern part of the country, however, is a rocky desert. A low strip of land winds along the 900-mile (1,448-km) Atlantic coastline. This lowland area has little natural vegetation, except for a few rain forests in the north.

If you traveled across Angola, you would experience three kinds of climates. The map above shows you that coastal areas have a steppe or desert climate. The inland plateau has a tropical savanna climate of wet and dry seasons. This area receives enough rainfall for farming.

The Economy Angola's major economic activity is agriculture. Nearly 75 percent of Angola's 11.2 million people live in the countryside. Subsistence farmers there grow corn, cotton, and sugarcane. Coffee is the leading export crop. Some Angolans also herd cattle, sheep, and goats.

Oil and mining, however, provide most of Angola's income. Look at the map on page 558. It shows you that Angola is rich in diamonds, iron ore, manganese, and petroleum. Oil is the country's leading export. Most of the oil deposits are located off the shore of Cabinda.

Influences of the Past Most of Angola's people are Africans who belong to several ethnic groups. They trace their ancestry to the ancient Bantu-speaking peoples. In the 1400s the Kongo kingdom ruled a large part of northern Angola.

From the 1500s until its independence in 1975, Angola was a colony of Portugal. Portuguese influence still remains strong in Angola. Although African languages are spoken, Portuguese is the official language. Almost 70 percent of Angolans practice the Roman Catholic faith brought to Angola by Portuguese missionaries.

CHAPTER 21

 553

Independent Practice

📁 Guided Reading Activity 21-2 **L1**

ASSESS

Check for Understanding

Assign Section 2 Review as homework or an in-class activity.

Meeting Lesson Objectives

Each objective below is tested by the questions that follow it in parentheses.

1. **Identify** the landforms that Angola and Namibia share. (4)
2. **Explain** why Portuguese is the official language of Angola. (3)
3. **Discuss** reasons why Namibia has difficulty feeding its people. (5)

Cooperative Learning Activity

Have students work in small groups to create dioramas or other displays illustrating the Namib Desert environment. Each group might prepare its own display. Or you might assign the following topics to groups: landscapes, animals, or plants. Then have groups combine their work to create a single classroom display. **L1**

Evaluate

⬦ Section 2 Review **L1**

⊙ Use the Testmaker to create a customized quiz for Section 2. **L1**

Reteach

Have students each make five fact cards on the Atlantic countries of southern Africa. Shuffle the cards. Have a student draw a card and identify the country it refers to. Continue until all the cards have been drawn. **L1**

Enrich

Ask students to investigate the events leading to independence for Angola and Namibia. Then have them present their findings in a newspaper article. **L3**

CLOSE

Have students compare the Atlantic countries with another area in Africa that they have studied. Ask them to identify factors that account for similarities and differences among regions.

Luanda, Angola

Angola's capital and major port is located on the country's long coastline.
LOCATION: What is the name of Angola's exclave?

After independence, civil war among different political and ethnic groups broke out in Angola. The fighting brought great suffering to Angola's people. Today the fighting has diminished, but rebuilding Angola will be a very difficult task.

Place
Namibia

South of Angola lies Namibia, one of Africa's newest countries. Namibia became independent in 1990 after 75 years of South African rule. The map on page 548 shows you the dates of independence for other African countries.

A Plateau and Deserts Namibia has a land area of 317,870 square miles (823,283 sq. km)—about one-half the size of Alaska. If you look at the map on page 546, you will see that a large plateau runs through the center of Namibia. It is the most populated area of the country. The rest of Namibia is made up of deserts. The Namib Desert runs almost the entire length of Namibia's Atlantic coast. The Kalahari Desert stretches across the southeastern part of the country. As you might guess, most of Namibia has a hot, dry climate.

The Economy Namibia's economy depends on the export of minerals. The country has rich deposits of diamonds, manganese, copper, silver, and zinc. Many Namibians work in the country's mines.

Because of the desert environment, Namibia has difficulty feeding its people. Farmers in the plateau region receive barely enough rainfall to grow corn. Most Namibians are herders who raise cattle, goats, and sheep.

Land of Few People Only 1.6 million people live in Namibia—one of the most sparsely populated countries in Africa. Most Namibians belong to African ethnic groups. A small number are of European ancestry. English and Afrikaans are the official languages, but most Namibians speak African languages.

SECTION 2 REVIEW

REVIEWING TERMS AND FACTS
1. **Define the following:** exclave.
2. **HUMAN/ENVIRONMENT INTERACTION** What is Angola's most important mineral export?
3. **PLACE** What European country once ruled Angola?
4. **LOCATION** Where do most Namibians live?

MAP STUDY ACTIVITIES
5. Turn to the climate map on page 553. What climate regions are found in Angola and Namibia?

Answers to Section 2 Review
1. All vocabulary words are defined in the Glossary.
2. oil
3. Portugal
4. on the central plateau
5. desert, steppe, and tropical savanna in Angola; desert and steppe in Namibia

MAKING CONNECTIONS

GREAT ZIMBABWE

| MATH | SCIENCE | HISTORY | LITERATURE | TECHNOLOGY |

Long before Europeans arrived on Africa's shores, inland African kingdoms were wealthy from the gold trade. Between A.D. 1000 and 1500, a Bantu-speaking people—the Shona—built nearly 300 stone-walled fortresses throughout southern Africa. The largest fortress was called Great Zimbabwe (zihm•BAH•bwee)—meaning "house of stone." It was the religious and trading center of the Shona kingdom.

A GIGANTIC CAPITAL Great Zimbabwe included fortress walls, temples, market-places, and homes spread out on a fertile, gold-rich plateau south of the Zambezi River. The city was the capital of the Shona kingdom and displayed the wealth and power of its kings. Its shape was oval, and its buildings were made of stone. This made the city stand out from the traditional circular mud-and-thatch villages.

The most impressive part of the city was an area called the Great Enclosure. The Great Enclosure's outer wall was 16.5 feet (5 m) thick and 32 feet (9.8 m) high. It was built of 900,000 granite stones fitted together without mortar. Enclosed within this wall, a maze of interior walls and hidden passages pro-

tected the circular house of the king, the Great Temple, and other religious buildings.

Remains of Zimbabwe Fortress

INTERNATIONAL TRADE During the A.D. 1400s, Shona kings controlled the trade routes between inland goldfields and seaports on the Indian Ocean coast. Civil wars and attacks from European invaders brought an end to the empire. Archaeologists in the late 1800s concluded that Great Zimbabwe, however, enjoyed a profitable trade with overseas lands. In Great Zimbabwe's ruins, they uncovered articles from India, China, and Asia.

Making the Connection

1. Why did Great Zimbabwe become so powerful?
2. With whom did Great Zimbabwe trade?

555

Answers to Making the Connection

1. because the Shona controlled the trade routes between the inland goldfields and seaports on the Indian Ocean coast
2. India, China, Asia

Meeting National Standards
Geography for Life The following standards are highlighted in this section:
• Standards 8, 11, 17

FOCUS

Section Objectives

1. **Identify** Zambia's most important natural resource.
2. **Illustrate** how Zimbabwe's recent history is similar to South Africa's.
3. **Explain** why most of Botswana's people live in the eastern part of the country.

Bellringer Motivational Activity

Prior to taking roll at the beginning of the class period, project Section Focus Transparency 21-3 and have students answer the activity questions. Discuss students' responses. **L1**

Vocabulary Pre-check

Review the meanings of terms such as *farm belt* and *corn belt*. Then ask students what they think *copper belt* means. Have them check their answer by finding the word in the text. **L1 LEP**

SECTION 3
Inland Southern Africa

PREVIEW

Words to Know
• copper belt
• sorghum

Places to Locate
• Zambia
• Malawi
• Zimbabwe
• Botswana
• Zambezi River
• Harare

Read to Learn . . .
1. what Zambia's most important natural resource is.
2. how Zimbabwe's recent history is similar to South Africa's.
3. why most of Botswana's people live in the eastern part of the country.

So much mist and noise rise from this waterfall that people in the area call it the "smoke that thunders." You are looking at the waters of the Zambezi River as they plunge 350 feet (107 m) to the bottom of a deep gorge. They form Victoria Falls, one of Africa's great natural wonders.

The Zambezi (zam•BEE•zee) River is one of southern Africa's longest rivers. For part of its journey, the Zambezi forms the border between two inland countries in southern Africa—Zambia and Zimbabwe. Two other countries—Malawi (muh•LAH•wee) and Botswana (baht•SWAH•nuh)—are also located in this region.

Place
Zambia

Zambia is a landlocked country of 287,020 square miles (743,382 sq. km). Zambia—slightly larger than the state of Texas—lies near the Equator. Zambia's high elevation gives most of the country a mild climate.

The map on page 546 shows you that a high plateau covers much of Zambia. Several rivers, including the Zambezi, cross the country. Kariba Dam—one of Africa's largest hydroelectric projects—spans the Zambezi River.

The Economy Zambia is one of the world's largest producers of copper. This resource provides more than 80 percent of Zambia's income. You will find a

---

Classroom Resources for Section 3

 REPRODUCIBLE MASTERS
Reproducible Lesson Plan 21-3
Guided Reading Activity 21-3
GeoLab Activity 21
Critical Thinking Skills Activity 21
Section Quiz 21-3

 TRANSPARENCIES
Section Focus Transparency 21-3
Unit Overlay Transparencies 7-0, 7-4, 7-5

MULTIMEDIA
 Testmaker

National Geographic Society: ZipZapMap! World

 National Geographic Society: GTV: Planetary Manager

copper belt, a large area of copper mines, in northern Zambia near the border with Zaire.

Because copper is so valuable, Zambians have paid less attention to developing agriculture. As a result, Zambia must import much of its food. Railroads link landlocked Zambia to ports in neighboring countries, such as Mozambique, Angola, and Tanzania.

The People Zambia's 9.1 million people belong to more than 70 different ethnic groups. They speak 8 major African languages. Once a British colony, Zambia became independent in 1964. Because of the British influence, many Zambians today also speak English.

About half of Zambia's population live in urban areas. Lusaka is the capital and largest city. Many urban dwellers work in mining and service industries. The rest of Zambia's people live in rural villages. They grow corn and other subsistence food crops for their families.

Place
Malawi

East of Zambia lies Malawi, a small, narrow country of only 36,320 square miles (94,069 sq. km). In some places, Malawi is no more than 50 miles (81 km) wide! Green plains and savanna grasslands cover western Malawi. Part of the Great Rift Valley runs through the country from north to south. In the middle of the valley lies beautiful Lake Malawi. High plateaus and tall mountains rise on both sides of the lake. Malawi lies in the tropics, but its plateaus and mountains give it a mild climate.

The Economy Because of its mountains, only one-third of the land in Malawi is suitable for farming. Agriculture, however, is the country's major economic activity. Some of Malawi's farmers work on large commercial farms that grow tea for export. Others cultivate sugarcane, corn, cotton, peanuts, and **sorghum** for local use on small family plots. Sorghum is a tall grass with seeds like corn. People in Malawi also fish and herd livestock.

The People Look at the map on page 562. It shows you that Malawi is one of the most densely populated countries in Africa. Malawi has about 260 people per square mile (100 people per sq. km). Most of its 9.5 million people belong to many different African ethnic groups and live in small villages.

All in a Day's Work

Fish from Lake Malawi are an important part of the diet of local residents.
REGION: Why does the tropical country of Malawi have a mild climate?

Cooperative Learning Activity

Organize students into four groups and assign one of the following countries to each group: Zambia, Malawi, Zimbabwe, or Botswana. Have groups determine what they consider to be the 10 most significant points of information about their assigned countries.

Groups may use other references in addition to this textbook as sources. Call on each group to present its 10 information points. Have individual group members explain reasons for selecting various points of information. **L2**

TEACH

Guided Practice

Project Unit Overlay Transparencies 7-0, 7-4, 7-5, and have students locate southern Africa's inland countries and study their physical features. Then have students study the **Countries at a Glance** listings on pages 478 to 481 to find population figures for each of the four inland countries. Call on a volunteer to list the countries and their populations on the chalkboard. Then ask students to synthesize the information and determine which countries are most densely populated and where most people live. Students can check the accuracy of their answers as they study Section 3. **L2**

More About the Illustration

Answer to Caption
because of its high elevation

Location
Lake Malawi stretches about 400 miles (644 km) along Malawi's western border. The lake occupies about 20 percent of the country's total area.

Global Gourmet

Zambia *Nsima* is the national food of Zambia. It is made from cornmeal, prepared as a dough or thick porridge. *Nsima* may be eaten with fish, meat, and/or vegetables.

Independent Practice

 Guided Reading Activity 21-3 L1

NATIONAL GEOGRAPHIC SOCIETY

These materials are available from Glencoe.

💿 **VIDEODISC**

GTV: Planetary Manager

Kyle National Park, Zimbabwe
Any side, Frame 47506

MAP STUDY

Answer
Zambia, Zimbabwe

Map Skills Practice
Reading a Map What crop is grown in southern Zimbabwe? *(coffee)*

Place
Zimbabwe

South of Zambia lies the country of Zimbabwe. With 149,290 square miles (386,661 sq. km), Zimbabwe is a little larger than the state of Montana. Like other countries in southern Africa, most of Zimbabwe occupies a high plateau. North and south of the plateau are lowlands. The Zambezi River crosses the northern lowlands. The Limpopo River winds through the southern lowlands.

Most of Zimbabwe has a tropical savanna climate of wet and dry seasons. High elevations, however, keep temperatures cool and pleasant in many parts of the country.

The Economy Gold, asbestos, and copper are three of the most important minerals mined in Zimbabwe. Although mining provides most of Zimbabwe's income, a majority of its people are farmers. About half of Zimbabwe's land is fertile enough for farming. Farmers grow corn and herd cattle and livestock. Some people also work on large commercial farms that grow coffee and tobacco for export.

The People Zimbabwe has about 11.2 million people. Most of the people belong to two African ethnic groups—the Shona and the Ndebele (ehn•duh•BEH•leh). The country takes its name from the famous trading center—Great Zimbabwe—built by the Shona in the A.D. 1400s.

In the 1890s the British began to rule the area that is now Zimbabwe. As in South Africa, Europeans ran the government and owned all the best farmland.

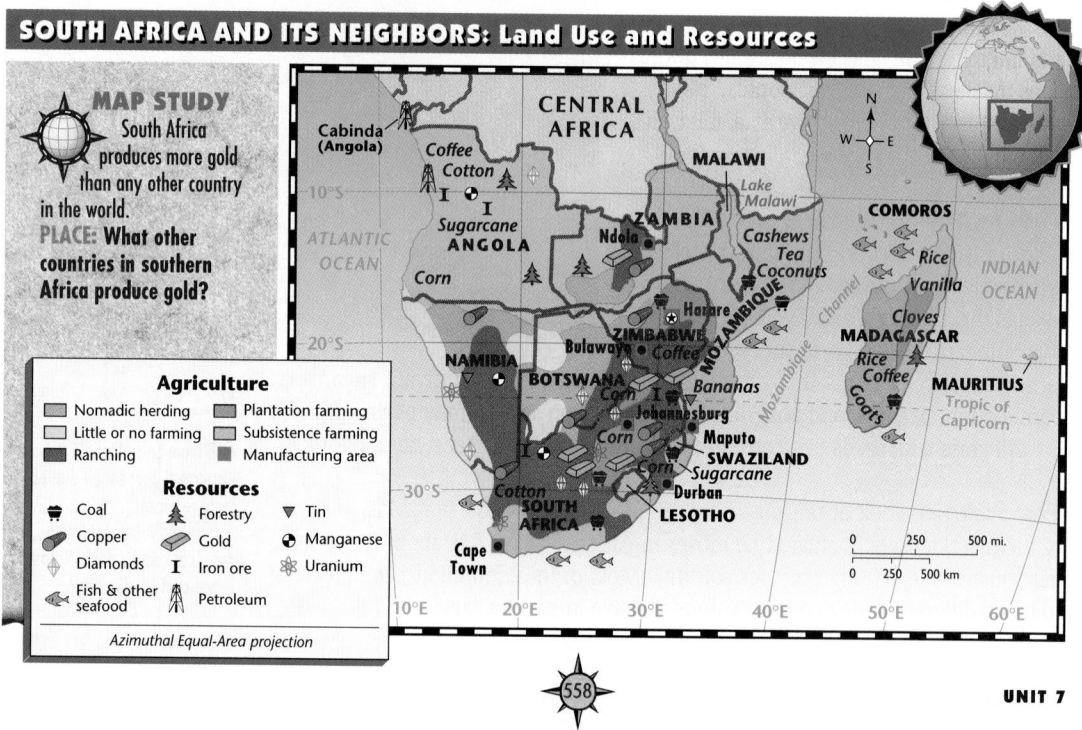

SOUTH AFRICA AND ITS NEIGHBORS: Land Use and Resources

MAP STUDY
South Africa produces more gold than any other country in the world.
PLACE: What other countries in southern Africa produce gold?

Azimuthal Equal-Area projection

558

UNIT 7

Meeting Special Needs Activity

Memory Delayed Have students who have difficulty recalling facts create a "memory game." Provide index cards and markers. Instruct each student to make at least one pair of cards for a country of inland southern Africa. On one card, students should print the name of the country. On the other card, students should write a fact about the country, such as its capital city or major product. Students should also draw a small picture or symbol on both cards to facilitate the game. (Note that one side of each card remains blank.) Shuffle all cards and place them face down on a table. Students play "memory" by turning over two cards. If the cards match, the player may keep them. If they do not match, they are returned to their place on the table. The object of the game is to find matching cards by remembering where they are. L1

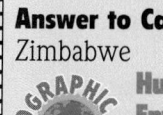

The Kariba Dam harnesses the power of the mighty Zambezi River.
LOCATION: The Zambezi forms the border between Zambia and what other country?

More About the Illustration

Answer to Caption
Zimbabwe

Human/ Environment Interaction The Kariba Dam, on Zambia's southern border, stands some 420 feet (128 m) high. Lake Kariba, the body of water created by the dam, measures more than 2,000 square miles (5,180 sq. km) in area. Hydroelectric power plants at the dam provide much of the electricity for Zambia's copper mines.

They refused to share power with the Africans. The Africans, in turn, resisted European rule. Peace talks finally led to a settlement. In 1980 free elections brought an independent African government to power.

Today about 75 percent of the population live in rural villages. A growing number, however, are moving to cities to find factory jobs. The largest city is Harare, the national capital. One of Harare's residents is fifteen-year-old Moses Katiyo. He and his family live in a crowded township in the suburbs. When school is not in session, Moses and his older brother take the train to Bulawayo (BOO•luh•WAY•oh), a city in the south. From there they often visit the Khami Ruins. These are the remains of stone buildings built hundreds of years ago by Moses' Bantu-speaking ancestors.

Place
Botswana

Botswana is the most isolated country in inland southern Africa. Desert covers much of its land area of 218,810 square miles (566,718 sq. km). If you look at the map on page 546, you will see that the vast Kalahari Desert spreads over southwestern Botswana. The Kalahari Desert is hot and dry, but parts of it have some vegetation. In eastern Botswana the land rises to a savanna region of grasses, bushes, and trees. The Okavango River flows through northwestern Botswana. There it forms one of the largest swamp areas in the world.

CHAPTER 21

559

ASSESS

Check for Understanding

Assign Section 3 Review as homework or an in-class activity.

Meeting Lesson Objectives

Each objective below is tested by the questions that follow it in parentheses.

1. **Identify** Zambia's most important resource. (2)
3. **Explain** why most of Botswana's people live in the eastern part of the country. (4)

Critical Thinking Activity

Making Comparisons/Drawing Conclusions
Have students compare the economies of the four inland countries of southern Africa. Then ask them to conclude from this information which country has the greatest potential for economic development. Call on students to support their choices with information from the text. **L2**

More About the Illustration

Answer to Caption
mining

Place Zimbabwe was formerly the British colony of Southern Rhodesia. Its neighbor, Zambia, was known as Northern Rhodesia before independence. Both these colonies were named for their founder, Cecil Rhodes.

Evaluate

Section 3 Quiz **L1**

Use the Testmaker to create a customized quiz for Section 3. **L1**

Reteach

Have students work in pairs to review the material in this section. Partners should take turns asking and answering questions based on the text. **L1**

Enrich

Challenge students to discover the origin of the names of various physical and cultural features in inland southern Africa. Suggest that they find the origins of at least 10 names. **L3**

CLOSE

Have students imagine the four countries of inland southern Africa decided to form an alliance. Encourage students to design a symbol or logo—based on the area's geography—for the new alliance.

560

Celebrating National Day in Harare, Zimbabwe

On National Day in Zimbabwe, a flame is lit to mark the country's independence from the United Kingdom.
PLACE: What industry generates the most income for Zimbabwe?

The Economy Like its neighbors, Botswana is rich in mineral resources. Among its treasures are diamonds, copper, and uranium. Mining, however, provides jobs for only a small number of Botswana's people. Most raise livestock or farm the land, but growing crops is difficult in Botswana. During the 1980s a severe drought brought hardships to many of Botswana's farmers. To earn a living, many young people work in the Republic of South Africa for several months a year.

The People The map on page 562 shows you that Botswana has few people for its large size. Most of the 1.4 million people live in eastern Botswana, where the land is most fertile. About 75 percent of the people live in rural areas, but thousands of them move to the cities each year. Gaborone is the capital and largest city.

SECTION 3 REVIEW

REVIEWING TERMS AND FACTS
1. **Define the following:** copper belt, sorghum.
2. **HUMAN/ENVIRONMENT INTERACTION** What is Zambia's most abundant natural resource?
3. **PLACE** What lake lies in Malawi's Great Rift Valley?
4. **REGION** Why do many young people from Botswana go to the Republic of South Africa?

MAP STUDY ACTIVITIES
5. Look at the land use map on page 558. What are the major resources of Zambia?

560

Answers to Section 3 Review
1. All vocabulary words are defined in the Glossary.
2. copper
3. Lake Malawi
4. to find jobs, because work is scarce in Botswana
5. forests, copper, gold

Indian Ocean Countries

PREVIEW

Words to Know
• slash-and-burn farming

Places to Locate
• Mozambique
• Madagascar
• Comoros
• Mauritius
• Indian Ocean

Read to Learn . . .
1. why Mozambique's ports are so valuable.

2. why Madagascar has unusual plants and animals.
3. how the islands of Comoros and Mauritius were formed.

Modern buildings rise along the tree-lined streets of Maputo, capital of the country of Mozambique (MOH•zuhm•BEEK). Maputo is a port city near the Indian Ocean. It provides a link to the outside world for the landlocked countries of southern Africa.

Meeting National Standards

Geography for Life The following standards are highlighted in this section:
• Standards 3, 8, 12

FOCUS

Section Objectives
1. **Explain** why Mozambique's ports are so valuable.
2. **Determine** why Madagascar has unusual plants and animals.
3. **Describe** how the islands of Comoros and Mauritius were formed.

Bellringer Motivational Activity

 Prior to taking roll at the beginning of the class period, project Section Focus Transparency 21-4 and have students answer the activity questions. Discuss students' responses. **L1**

Vocabulary Pre-check

Have students look up the meaning of *slash-and-burn farming.* Then ask them why this practice is sometimes called "shifting cultivation." *(Farmers plant in different areas each season.)* Point out that the ashes from burned vegetation helps fertilize the soil so that it can be planted again. **L1 LEP**

Mozambique is a large country in southern Africa that borders the Indian Ocean. Three island countries—Madagascar (MA•duh•GAS•kuhr), Comoros (KAH•muh•ROHS), and Mauritius (maw•RIH•shuhs)—also form part of southern Africa's Indian Ocean region.

Place
Mozambique

The map on page 544 shows you that Y-shaped Mozambique stretches along the Indian Ocean. Its landscapes include sandy lowlands, high plateaus, and tall mountains. Sand dunes, swamps, and fine natural harbors line Mozambique's Indian Ocean coast. In the center of the country lies a flat plain covered with grasses and tropical forests. Beyond the plain, the land slowly rises to form high plateaus and mountains along Mozambique's border with Malawi.

A number of rivers cross the northern and central parts of Mozambique. The most important river is the Zambezi. The Cabora Bassa Dam on the Zambezi provides electric power for much of the country. The map on page 553 shows you that most of Mozambique has a tropical savanna climate with wet and dry seasons. Rainfall is heaviest in the north. Much of the southwest is dry.

CHAPTER 21  561

Classroom Resources for Section 4

 REPRODUCIBLE MASTERS

Reproducible Lesson Plan 21-4
Guided Reading Activity 21-4
Performance Assessment Activity 21
Reteaching Activity 21
Enrichment Activity 21
Section Quiz 21-4

TRANSPARENCIES

Section Focus Transparency 21-4
Geoquiz Transparencies 21-1, 21-2

MULTIMEDIA
Testmaker

National Geographic Society: ZipZapMap! World

National Geographic Society: World Geography, Vol. 2

MAP STUDY

Answer

125–250 persons per sq. mile (50–100 persons per sq. km)

Map Skills Practice

Reading a Map Where in Madagascar is population density the heaviest? *(in the central highlands around Antananarivo)*

TEACH

Guided Practice

Display Political Map Transparency 7. Have students locate Mozambique, Madagascar, Comoros, and Mauritius. State the following facts from the text: "Mozambique's major source of income comes from its seaports." "Madagascar has many plants and animals that are not found elsewhere on the earth." "Most people on Comoros speak Arabic and Swahili and practice Islam." "Providing food, jobs, and housing for more people will be a major challenge for Mauritius in the years ahead." Have students use information on the map to hypothesize about reasons why the statements above are true. **L2**

Independent Practice

Guided Reading Activity 21-4 **L1**

SOUTH AFRICA AND ITS NEIGHBORS: Population Density

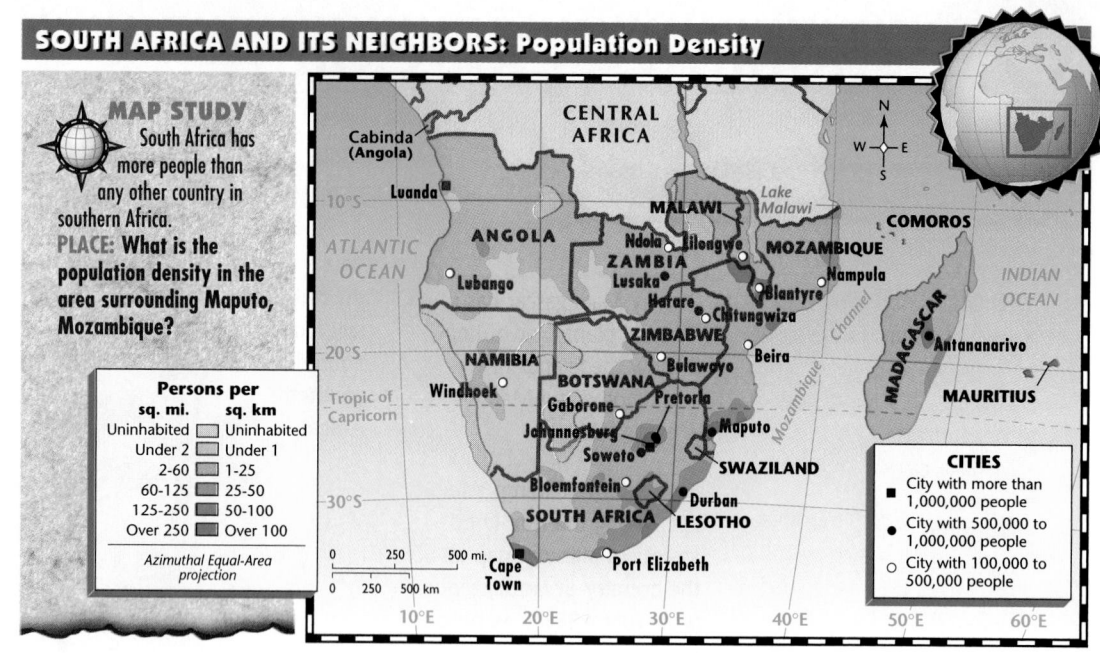

MAP STUDY South Africa has more people than any other country in southern Africa.
PLACE: What is the population density in the area surrounding Maputo, Mozambique?

Persons per	
sq. mi.	**sq. km**
Uninhabited	Uninhabited
Under 2	Under 1
2-60	1-25
60-125	25-50
125-250	50-100
Over 250	Over 100

Azimuthal Equal-Area projection

CITIES
- City with more than 1,000,000 people
- City with 500,000 to 1,000,000 people
- City with 100,000 to 500,000 people

The Economy Most people in Mozambique are farmers. Some of them practice **slash-and-burn farming.** This means they cut and burn forest trees to clear areas for planting. Mozambique's major crops are cashews, coconuts, tea, sugarcane, and bananas. Fishing provides an income for people living along the Indian Ocean.

Mozambique's major source of income, however, comes from its seaports. South Africa, Zimbabwe, Swaziland, and Malawi pay Mozambique to use docks at Maputo and other ports.

During the 1980s and early 1990s, a fierce civil war slowed industrial growth in Mozambique. Although the fighting has ended, the country's economy still cannot provide enough jobs. In order to earn their livings, many people from Mozambique work in the Republic of South Africa.

The People About 16 million people live in Mozambique. Nearly all of them belong to African ethnic groups. Most live in the southern part of the country around Maputo, the capital. Portugal governed Mozambique from the early 1500s until 1975, when Mozambique became independent. Portuguese is still the official language. Most of the people, however, speak African languages.

Place

Madagascar

Madagascar is an island nation in the Indian Ocean. Look at the map on page 546. It shows you that Madagascar lies about 250 miles (400 km) southeast of the African mainland. Cool highland areas cross the middle of

562

Cooperative Learning Activity

Organize students into four groups and assign one of the following countries to each group: Mozambique, Madagascar, Comoros, or Mauritius. Instruct students to choose or create one or more images to represent their as- signed country. For example, a lemur might represent Madagascar. Call on groups to display their images and explain the significance of each one. **L1**

Madagascar. Most coastal areas have warm, humid plains and fertile river valleys. Partly dry grasslands cover southern Madagascar.

In the past, Madagascar's island location kept it isolated from other parts of the world. Today Madagascar has many plants and animals that are not found elsewhere on the earth. One of these unusual animals is the lemur, which looks like a monkey. Lemurs have large eyes that appear to glow at night as they leap from tree to tree in the forests.

The Economy Agriculture is the chief economic activity in Madagascar. Farmers grow food crops—mainly rice—and herd cattle. Coffee is the leading export. If you like vanilla ice cream, thank Madagascar! It produces most of the world's vanilla beans.

The People About 14 million people live in Madagascar. Most trace their ancestry to Southeast Asian and African groups that settled the island centuries ago. Malagasy, the major language, is similar to languages spoken in Southeast Asia. French is also spoken in Madagascar's cities. Madagascar was a colony of France from the 1890s until it gained its independence in 1960.

Place
Small Island Countries

Far from Africa in the Indian Ocean are two other island republics—Comoros and Mauritius. The people of both countries have many different backgrounds.

Comoros Comoros is a group of four mountainous islands that lie between mainland Africa and Madagascar. Volcanoes formed all four islands thousands of years ago. Thick tropical forests cover Comoros today. At one time a French territory, Comoros became independent in 1975.

Most of Comoros's 500,000 people are farmers. They grow rice, cassava, corn, vanilla, sugarcane, and coffee. The people of Comoros are a mixture of Arabs, Africans, and people from Madagascar. They speak Arabic and Swahili, and most practice the religion of Islam.

Mauritius The island republic of Mauritius lies about 500 miles (800 km) east of Madagascar in the Indian Ocean. Mauritius includes one major island and a number of smaller islands, which are territories. Mauritius gained its freedom from the United Kingdom in 1968.

Madagascar's Wildlife

The ring-tailed lemur is one of the many unusual species of animals that live in Madagascar.
LOCATION: Why does Madagascar have so many unusual plants and animals?

LESSON PLAN
Chapter 21, Section 4

More About the Illustration
Answer to Caption
Because of Madagascar's isolation, many species that have developed there are found nowhere else.

Place The Mauritian kestrel, the echo parakeet, and the pink pigeon—three of the world's rarest birds—live on Mauritius.

ASSESS

Check for Understanding
Assign Section 4 Review as homework or an in-class activity.

Meeting Lesson Objectives
Each objective below is tested by the questions that follow it in parentheses.
1. **Explain** why Mozambique's ports are so valuable. (2)
2. **Determine** why Madagascar has unusual plants and animals. (3)

Evaluate
🗀 Section 4 Quiz **L1**

◉ Use the Testmaker to create a customized quiz for Section 4. **L1**

Meeting Special Needs Activity

Learning Disabled Some students may have difficulty relating facts to countries. Suggest that they make a fact sheet for each country in this section. As they read, they should write down important facts about landscapes, economies, and populations. When they have finished, they should underline the most important points. Then have each student work with a partner to summarize key facts about each country. **L1**

CLOSE

Mental Mapping Activity

Provide each student with a large sheet of white paper and map pencils. Inform students that this will not be a graded activity. Have students draw, freehand from memory, a map of southern Africa that includes the following labeled countries: Republic of South Africa, Lesotho, Swaziland, Namibia, Angola, Zambia, Malawi, Zimbabwe, Botswana, Mozambique, Madagascar, Comoros, and Mauritius.

Antananarivo, Madagascar

Horse-drawn wagons share the streets with cars in Madagascar's capital.
PLACE: What flavoring is Madagascar known for?

Like Comoros, the islands of Mauritius were formed by volcanoes. Bare, black peaks rise sharply above green fields, and palm-dotted white beaches line the coasts. The climate is subtropical. Because Mauritius has few natural resources, its economy depends on agriculture. Sugar and sugar products are the island country's major exports.

Mauritius has about 1.2 million people who come from many different backgrounds. About 70 percent are descendants of settlers from India. The rest are of African, European, or Chinese ancestry. The population of Mauritius is increasing rapidly. Providing food, jobs, and housing for more people will be a major challenge for Mauritius in the years ahead.

SECTION 4 REVIEW

REVIEWING TERMS AND FACTS

1. Define the following: slash-and-burn farming.
2. HUMAN/ENVIRONMENT INTERACTION What is Mozambique's major source of income?
3. LOCATION Where is Madagascar located?
4. MOVEMENT What major groups settled in Madagascar?

 MAP STUDY ACTIVITIES
5. According to the population density map on page 562, where are Mozambique's most densely populated areas?

Answers to Section 4 Review

1. All vocabulary words are defined in the Glossary.
2. payments from other countries for the use of its seaports
3. about 250 miles (400 km) southeast of the African mainland
4. Southeast Asians and Africans
5. on the coast, especially in the south

Chapter 21 Highlights

Important Things to Know About South Africa and Its Neighbors

SECTION 1 REPUBLIC OF SOUTH AFRICA

- Most of inland South Africa is a high plateau.
- Because of its mineral resources, South Africa has the most developed economy in Africa.
- In recent years, South Africa has given equality to citizens of all ethnic groups and is struggling to become a democracy.

SECTION 2 ATLANTIC COUNTRIES

- Angola has large deposits of oil, diamonds, and iron ore.
- Two major deserts cross Namibia. This harsh landscape limits agriculture in Nambia.

SECTION 3 INLAND SOUTHERN AFRICA

- Zambia is one of the world's largest producers of copper.
- Tiny, mountainous Malawi has a large lake running almost the entire length of its eastern border.
- Zimbabwe has many mineral resources and good farmland.
- Desert covers a large part of Botswana.

SECTION 4 INDIAN OCEAN COUNTRIES

- Mozambique's ports on the Indian Ocean are used by landlocked countries in southern Africa.
- Madagascar's people trace their ancestry to African and Southeast Asian settlers.

Celebrating South Africa National Independence Day ▶

CHAPTER 21

Extra Credit Project

Campaign Speech Have students imagine that they are running for political office in one of the countries studied in this chapter. Have them identify the country's problems and pre- pare a campaign speech outlining their pro- posed solutions. Provide time for volunteers to deliver their speeches to the class. L2

CHAPTER HIGHLIGHTS

Using the Chapter 21 Highlights

- Use the Chapter 21 High- lights to preview, review, condense, or reteach the chapter.

Preview/Review

Vocabulary Puzzle- Maker Software reinforces the terms used in Chapter 21.

Condense

Have students read the Chapter 21 Highlights. Spanish Chapter Highlights also are available.

Chapter 21 Di- gest Audiocassettes and Ac- tivity

Chapter 21 Digest Audiocassette Test

Spanish Chapter Digest Au- diocassettes, Activities, and Tests also are available.

Guided Reading Ac- tivities

Reteach

Reteaching Activity 21. Spanish Reteach- ing activities also are avail- able.

MAP STUDY

Place Copy and distribute page 37 of the Outline Map Resource Book. Have students mark on the map the countries of southern Africa. Then have them mark the region's major centers of population. Call on volunteers to sug- gest a title for the map.

Chapter 21 Review and Activities

REVIEWING KEY TERMS

Match the numbered terms in Column A with their definitions in Column B.

A
1. exclave
2. copper belt
3. high veld
4. sorghum
5. enclave
6. township
7. escarpment
8. apartheid

B
A. cliff that divides high and low ground
B. flat, grass-covered plains in South Africa
C. official system of racial separation in South Africa
D. crowded African neighborhood outside a city
E. large area of copper mines
F. part of a country separated from the main part
G. tall grass with seeds like corn
H. small country surrounded by another country

ANSWERS

Reviewing Key Terms

1. F
2. E
3. B
4. G
5. H
6. D
7. A
8. C

Mental Mapping Activity

This exercise helps students to visualize the countries and geographic features they have been studying and understand the relationships among them. All attempts at freehand mapping should be accepted.

Mental Mapping Activity

On a separate piece of paper, draw a freehand map of South Africa and its neighbors. Label the following items on your map:
- South Africa
- Indian Ocean
- Angola
- Zambia
- Zimbabwe
- Johannesburg

REVIEWING THE MAIN IDEAS

Section 1
1. PLACE What oceans border South Africa?
2. MOVEMENT How did the coming of Europeans affect life for Africans in South Africa?

Section 2
3. LOCATION Describe Namibia's coast.
4. HUMAN/ENVIRONMENT INTERACTION What is Angola's main agricultural export?

Section 3
5. HUMAN/ENVIRONMENT INTERACTION What is the chief food crop of Zambia?
6. PLACE How does the population density of Malawi compare with populations in other African countries?
7. MOVEMENT What two African ethnic groups are found in Zimbabwe?

Section 4
8. LOCATION Why has industrial development been slow in Mozambique?
9. REGION How were Comoros and Mauritius formed?

Reviewing the Main Ideas

Section 1

1. Atlantic and Pacific oceans
2. The Europeans pushed Africans off the best land. Later they passed apartheid laws that separated racial and ethnic groups and barred nonwhites from participating in government.

Section 2

3. it consists of desert
4. coffee

Section 3

5. corn
6. It is one of the most densely populated countries in Africa.
7. Shona, Ndebele

CRITICAL THINKING ACTIVITIES

1. Formulating Questions Write three questions you would ask a South African about his or her way of life and recent changes there.

2. Identifying Central Issues What are some of the challenges faced by countries in southern Africa as they try to develop their economies?

GeoJournal Writing Activity

Choose a country in southern Africa and imagine that you are a travel writer visiting the place. Write a short article about the country. You may want to describe its major natural attractions or its cities, or give interesting information about its people.

COOPERATIVE LEARNING ACTIVITY

Work in small groups to learn more about democracy in southern Africa today. Each group will choose one country to research. Groups may want to divide the research among their members according to the following topics: (a) form of government before independence; (b) present form of government; or (c) current political challenges faced by the government. As a group, create a skit, play, or news broadcast to share your information with the class.

PLACE LOCATION ACTIVITY: SOUTH AFRICA AND ITS NEIGHBORS

Match the letters on the map with the places of South Africa and its neighbors. Write your answers on a separate sheet of paper.

1. Mauritius
2. Lake Malawi
3. Drakensberg Mountains
4. Botswana
5. Cape Town
6. Luanda
7. Kalahari Desert
8. Zambezi River
9. Madagascar
10. Mozambique

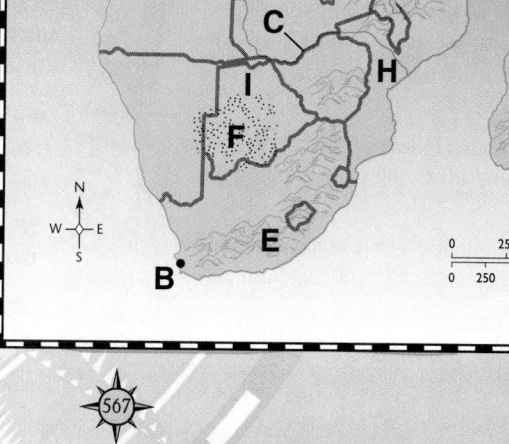

Cooperative Learning Activity

Have groups point out the location of their selected country on a map of Africa before they make their presentations.

GeoJournal Writing Activity

Encourage students to illustrate their articles with maps, photographs, or sketches.

Place Location Activity

1. J
2. A
3. E
4. I
5. B
6. G
7. F
8. C
9. D
10. H

Chapter Bonus Test Question

This question may be used for extra credit on the chapter test.

More people live in the eastern section of southern Africa than in the western section. What accounts for this population distribution? *(The western section has dry climates that do not support agriculture.)*

Section 4

8. Civil war has prevented economic growth there.

9. They were formed by volcanic activity.

Critical Thinking Activities

1. Questions will vary. Possible questions include: How has your life changed since the end of apartheid? What actions taken by the government do you approve of? What actions do you think the government should take? What are your hopes for the future?

2. Answers will vary, but might include that much of the land is dry and there are frequent droughts; civil wars have disrupted farming; farmers lack modern farming equipment and techniques.

These materials are available from Glencoe.

 POSTERS

Eye on the Environment, Unit 7

FOCUS

Write the words *static* and *dynamic* on the board and guide students in developing a definition for each *(static: unchanging; dynamic: constantly changing)*. Ask students which term applies to the earth, and have them offer examples of how the earth is a dynamic system. Record their examples on the board. Add *desertification* to the list. Make sure students understand it as the process by which deserts are made, especially as a result of human activities.

TEACH

Emphasize to students that deserts occur, expand, and retract naturally, but that they have expanded dramatically recently as a direct result of human activities. Then organize the class into two work groups. Have one group create a graphic organizer about desertification. The organizer should show the causes, consequences, and possible solutions for the problem. Have the group create a world map showing where desertification is occurring. Combine the two visuals in a bulletin-board display. **L1**

NATIONAL GEOGRAPHIC SOCIETY

EYE ON THE ENVIRONMENT

Africa South of the Sahara

DESERTIFICATION

Camels drink at a well in Niger. Ironically, use of wells and water holes can contribute to desertification as animals trample and kill plants when they drink.

PROBLEM

Desertification, or the "making of deserts," is occurring in parts of Africa south of the Sahara. Poor conservation methods lead to soil erosion and dried-up wells and water holes. Trees and shrubs are cut down and used as fuel. Vegetation dies out as herds of goats and cattle overgraze fragile lands. Dry winds and fierce sandstorms blow away the naked topsoil and cover once-productive fields with sheets of sand. Marching dunes bury fields, roads, and entire villages. And so the desert advances.

SOLUTIONS

● Farmers in some areas confine grazing animals to fenced areas and bring food to the animals.

● To help save rapidly dwindling forests, Mali, Niger, and Burkina Faso have started national campaigns to place more efficient clay or metal cookstoves in every household.

● In Niger, more than 435 miles of windbreaks have been planted to slow advancing sand and protect fragile seedlings.

● In Ethiopia, communities are planting hundreds of trees, and in Kenya, people are terracing farmland to control erosion.

DESERTIFICATION FACT BANK

🖋 Thirty-four percent of Africa's land is at risk from desertification.

🖋 The Sahel, a fragile belt of grasses and forests that stretches 3,000 miles across Africa, has lost 30 percent of its trees in the last 20 years.

🖋 Once 40 percent of Ethiopia was forested; in 1990 less than 4 percent of the forests remained.

🖋 In Mali, the Sahara has spread more than 400 miles in 20 years.

🖋 During the last 20 years, many villages in the Sahel have lost as much as one-half of their farmland to desertification.

UNIT 7

More About the Issues

GEOGRAPHIC THEMES

Place The popular image of a desert is of a vast and lifeless expanse of sand dunes. Actually, sand covers only 10 to 20 percent of most deserts. Most of the Sahara, for example, is made up of gravel-covered plains and rocky plateaus. Another mistaken idea is that the desert is a lifeless place. Deserts are home to a wide variety of plants that have adapted to the dry conditions through extensive root systems, small leaves, and other ways. Animals, too, from insects to birds to some large mammals, live in desert regions. The deserts of the world are diverse.

Dwellings nestle in a sandy area of trees and shrubs in Kenya. Drought and overuse could easily turn a semiarid region into desert.

TEEN TRIBUTE

Eleven-year-old George Otieno lives on a farm in Kenya. George's family could not grow enough food to feed themselves because the soil on their farm was so poor. Then at school George learned that planting trees can help reverse desertification. George planted several hundred seedlings he received from CARE, an environmental relief group. The trees restored nutrients to the soil—and improved the family's harvest and their standard of living.

environmental activities

TAKE A CLOSER LOOK

1 What is desertification?

2 What are the main causes of desertification in Africa south of the Sahara?

BE NICE TO TREES

WHAT CAN YOU DO?

• Adopt a piece of land in your own neighborhood. Clean it up, plant flowers, and put up a No Littering sign.

• Conserve water—cut your shower water use by half by a using a low-flow shower head.

• Volunteer at your local zoo. Help with programs to save endangered species.

• Support agencies that help African countries to reforest empty land.

Men on camels herd cattle in Chad. As livestock strip away vegetation, soil blows away, and bare patches become desert.

569

ASSESS

Have students answer the **Take A Closer Look** questions on page 569.

CLOSE

Discuss with students the **What Can You Do?** question. You might want to have students explain in a paragraph why people who live far away from desert frontiers should be concerned about desertification.

*For an additional case study, use the following:

▼ Environmental Case Study 7 **L1**

Cultural Kaleidoscope

North Africa The name *Sahara* is an Arabic word that literally means "desert."

Answers to Take A Closer Look

1. the "making of deserts"
2. Cutting down trees and shrubs and overgrazing the land leaves the topsoil naked. Dry winds and fierce sandstorms blow away the topsoil and cover the once-productive land with sheets of sand.

Global Issues

Resources About 15 percent of the earth's land is desert, and about another 15 percent is semidesert. Expansion of these already large areas is a worldwide problem. Other dramatic examples of desertification include the Gobi and the Thar Desert in Asia.

0:00 OUT OF TIME?

If time does not permit teaching each chapter, you may use the Chapter Highlights, Audiocassettes, and their corresponding activities and tests.

GLENCOE
TECHNOLOGY

 VIDEODISC

Reuters Issues in Geography

**Chapter 7
South Asia: Preserving the Himalayas**
Disc 1 Side A

Southeast Asia: Patterns of Migration
Disc 1 Side A

inteNET
CONNECTION

Explore a variety of topics about China, Taiwan, Korea, Japan, Singapore, Malaysia, Thailand, and the Philippines on the Asian Net.

World Wide Web
http://www.asiannet.com/

The latest news about India is available at:

World Wide Web
http://www.indiaxs.com/
HOMEPG1.HTM

What's going on in Pakistan? Contact:

http://www.scsu-cs.ctstateu.edu/~memon/pak.html

Unit 8 Asia

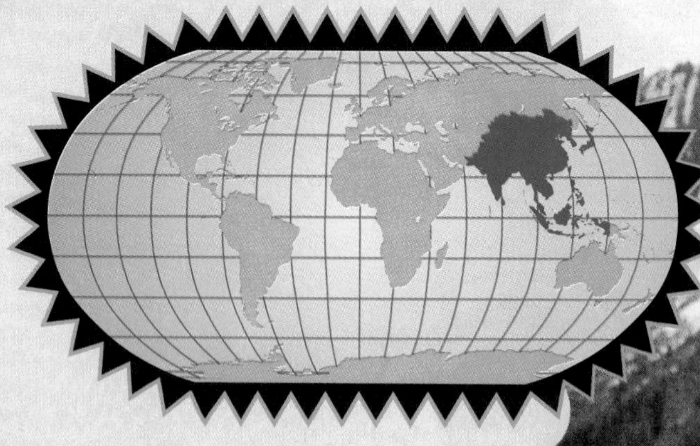

What Makes Asia a Region?

The people share . . .

• high population densities.

• deadly natural hazards—typhoons, floods, volcanic activity.

• expanding economies.

• rich natural resources.

• ancient religions and cultures.

To find out more about Asia, see the Unit Atlas on pages 572–585.

Himalayas, India

570

About the Illustration

The Himalayas, the highest mountains in the world, actually are made up of three ranges that run very close together. The highest peak in the Himalayas—and in the world—is Mount Everest, which rises to 29,028 feet (8,848 m). K2, in the Karakoram Range, is the world's second-highest peak. These two ranges, along with the Hindu Kush, separate the Indian subcontinent from the rest of Asia.

 GEOGRAPHIC THEMES

Human/Environment Interaction
Climbing Everest is the ultimate challenge for mountaineers. Only two expeditions are allowed on the mountain at the same time, however, and each expedition must take a different route to the top. Why do you think these restrictions have been placed on climbers? *(to limit the environmental damage to this area)*

NATIONAL GEOGRAPHIC SOCIETY

These materials are available from Glencoe.

 CD-ROM

Picture Atlas of the World

 SOFTWARE

ZipZapMap! World

 TRANSPARENCIES

PicturePack Transparencies, Unit 8

GeoJournal Activity

Display news articles about Asia. Each day that the class works on Unit 8, have a student select an article. Ask the students to review each article for information on the listed categories. **L1**

• This journal activity provides the basis for the "GeoJournal Writing Activity" exercise in the Chapter Review.

• The GeoJournal may be used as an integral part of Performance Assessment.

GeoJournal Activity

As you read about Asia, follow news reports about the region either on television or in newspapers and magazines. Organize your information in your journal under the following categories: Environment, Culture and Daily Life, and Economy.

571

Location Activity

Display Political Map Transparency 8 and ask students the following questions: What continents lie closest to Asia? *(Europe and Africa)* What countries border India? *(Pakistan, China, Nepal, and Bangladesh)* Which of the countries that lie totally in Asia covers the most area? *(China)* What oceans border Asia to the south and east? *(Indian and Pacific oceans)*

NATIONAL GEOGRAPHIC SOCIETY

These materials are available from Glencoe.

CD-ROM

Picture Atlas of the World

POSTERS

Images of the World
Unit 8: Asia

TRANSPARENCIES

PicturePack Transparencies, Unit 8

Strange but TRUE!

Asia covers more square miles than the combined area of Europe, Australia, and North America.

NATIONAL GEOGRAPHIC SOCIETY

IMAGES *of the* WORLD

1. Horse breeder, Mongolia
2. Traffic in the heart of the city, Hong Kong
3. Harvesting coffee, Indonesia
4. Mosque in a rainstorm, Indonesia
5. Girl holding the Quran, Indonesia
6. Father and son readying their boat, Indonesia
7. Pilgrims bathing in the Ganges River, India

All photos viewed against snow-covered peaks in the Karakoram Range, Pakistan.

572

Fun Facts

• **The Koreas** The Korean Peninsula is about the same size as Florida.

• **Bangladesh** The average age of the Bangladeshi population is 16.

• **Indonesia** This island country is made up of the summits of a mountain chain that is mostly underwater.

Geography and the Humanities

📁 World Literature Reading 8

🌐 World Music: Cultural Traditions, Lessons 7, 8, 9

📠 World Cultures Transparencies 13, 14, 15, 16, 17, 18

📖 Focus on World Art Print 10

Cultural Kaleidoscope

India In the ancient Indian language of Sanskrit, *Himalayas* means "Abode of Snow."

Global Gourmet

China A mandarin was a public official in Imperial China. Over time, the term came to be used to describe the food eaten by "mandarins," as foreigners called the Chinese upper classes.

Teacher Notes

LESSON PLAN
Unit 8 Atlas

Using the Unit Atlas

These features and activities may be used as an introduction to the unit or as teaching tools throughout the course of the unit.

📁 Unit Atlas Activity 8
L1

📦 📁 Unit Overlay Transparencies 8-0, 8-1, 8-2, 8-3, 8-4, 8-5, 8-6 **L1**

📁 Home Involvement Activity 8 **L1**

📁 Environmental Case Study 8 **L1**

🗂 Countries of the World Flashcards, Unit 8

✂ World Crafts Activity Cards, Unit 8

More About the Illustration

Answer to Caption
Taklimakan

Region The Gobi is home to numerous animal herders who traverse the desert in search of grazing land for their sheep, goats, and cattle. Raising animals and the processing of animal products is the main source of income for the people of the Gobi.

Regional Focus

Asia is both a continent and a region. The *continent* of Asia covers the eastern part of the large landmass called Eurasia. The *region* of Asia is a smaller part of the Asian continent. As a region, Asia reaches from the rugged mountains of Pakistan in the west to the volcanic islands of the Philippines in the east. From north to south, the region of Asia stretches from the cool highlands of northeastern China to the tropical islands of Indonesia.

Region
The Land

The Gobi

The Gobi stretches across southern Mongolia and northern China. It is colder and farther north than any other desert in the world.
PLACE: What other desert lies in central Asia?

The region of Asia covers about 7.8 million square miles (20.2 million sq. km). Look at the physical map on page 578. Asia's winding coastlines touch the Indian and Pacific oceans as well as many seas. Within Asia's vast land area are mountains, deserts, plains, and great rivers.

Mountains Rugged mountain ranges sweep through central Asia. The best known are the Himalayas. Their highest peak—Mount Everest—is the world's tallest mountain. Mountains also cross northeastern China, the Korean Peninsula, and Southeast Asia. Off Asia's coasts you will find mountainous groups of islands, such as Japan, the Philippines, and Indonesia. They lie in the Ring of Fire, a Pacific area known for its earthquakes and active volcanoes.

Deserts and Plains Large deserts—the Taklimakan (tah•kluh•muh•KAHN) and the Gobi—stretch across inland parts of central Asia. Fertile plains lie in

574

UNIT 8

Facts

- **Asia** Three out of every five people in the world live on the continent of Asia.

- **China/Mongolia** *Gobi* is Mongolian for "place without water."

northern India, eastern China, and mainland Southeast Asia. Many of Asia's rivers flow through these plains to the ocean or sea. The most important rivers are the Indus in Pakistan; the Ganges and Brahmaputra in India; the Mekong in Southeast Asia; and the Chang Jiang and Huang He in China.

Region
Climate and Vegetation

Because of its vast size, Asia has diverse climates. They include cold highlands and hot deserts in north and central Asia, mild climates in the east, and tropical climates in the south.

Monsoons Winds called monsoons affect climate in much of Asia. In winter, monsoons from the north bring cool, dry weather. In summer, the winds reverse direction and blow from the seas that lie south of Asia. The summer monsoons bear hot, humid weather. Wet monsoons often bring heavy rains and floods.

Vegetation Dry areas in central Asia have little plant life, except for grasses. These grasses provide food for livestock. You will find many trees and plants in eastern Asia, which gets plenty of rain. The warm, wet climate of Southeast Asia favors tropical vegetation. Many Southeast Asian exports are hardwoods from rain forests.

Place
The Economy

Asia is made up of 23 countries. Economies differ from country to country, although agriculture is the major economic activity. Important crops include rice, wheat, cotton, rubber, and tea. Because of Asia's mountainous areas and large populations, farmers try to produce as much food as they can on small areas of land, including terraced mountain slopes.

Mining and Industry Asia is rich in mineral resources. Coal, iron ore, manganese, and mica come from India and China. Southeast Asia is a major supplier of tin, oil, and bauxite. Most of Asia's manufacturing takes place in Japan, South Korea, Taiwan, China, and India. Factories in these countries produce cars, electronic equipment, ships, and textiles. In other parts of Asia, industry is less developed. Governments in these areas, however, are trying to create new industries and improve old ones.

Transportation Industrial growth in Asia has led to better means of transportation. Roads and railroads link major urban areas. In Asian cities you can see modern buses, trains, and cars. In rural areas people often travel by foot or use carts pulled by animals. People living along rivers rely on small boats for traveling and transporting goods.

575

Mohenjo-Daro, Pakistan

Ruins of one of the world's first civilizations were discovered in the Indus River Valley in Pakistan.
PLACE: What forms of transportation might have been used by these early people?

 NATIONAL GEOGRAPHIC SOCIETY

These materials are available from Glencoe.

 CD-ROM

Picture Atlas of the World

 SOFTWARE

ZipZapMap! World

POSTERS

Eye on the Environment, Unit 8

More About the Illustration
Answer to Caption
river travel

 **Human/ Environment Interaction** The Indus Valley civilization flourished between 4,500 and 3,500 years ago. Mohenjo-Daro was one of the great cities of this civilization. The city was well-planned. Most buildings, for example, were constructed on mud-brick platforms to protect them from flooding when the Indus overflowed its banks.

 **Fun Facts**

• **Indonesia** About 1 in every 15 cups of coffee consumed around the world and about 1 in every 20 cups of tea come from crops grown on plantations in Indonesia.

• **China/Nepal** Mount Everest is named for Sir George Everest, a British colonial official who surveyed the Himalayas in the 1860s. The Tibetans call Everest *Chomolungma*, which means "Goddess Mother of the World."

ASIA

UNIT 8 ATLAS

Movement
The People

About 3.1 billion people live in the region of Asia. China, India, Indonesia, Bangladesh, and Japan are among the world's most heavily populated countries.

Population People are not evenly distributed throughout Asia. Few Asians live in the mountains or desert areas. Most of the region's people make their homes in river or mountain valleys or near seacoasts. There they fish, farm, or work in factories. Among the most crowded areas of Asia are Bangladesh, eastern China, northern India, southern Japan, and the island of Java in Indonesia.

Only about 32 percent of the region's people live in urban areas. Asia's cities, however, are rapidly growing. Many are overcrowded, and some show great contrasts of wealth and poverty. Among the most important cities are Mumbai (Bombay) and Calcutta in India; Beijing and Shanghai in China; Tokyo and Osaka in Japan; Seoul in South Korea; and Singapore, Bangkok, and Manila.

Ethnic Groups and Religions Japan and the Koreas each have one main ethnic group. China also has one, but many small non-Chinese groups live on the edges of the country. India and Southeast Asia, however, have many ethnic groups with different languages and cultures.

Anuradhapura, Sri Lanka

The graceful domes of Buddhist temples are a common sight in the island country of Sri Lanka. **PLACE: What religion is practiced by more Asians than any other religion?**

Hinduism is the primary religion of the region. Hindus generally live in India, where the religion began thousands of years ago. The next largest religion—Islam—is Asia's most widespread. Islam began in Southwest Asia and later spread to India, western China, and Southeast Asia. Buddhism, which started in India, is now a major religion of Southeast Asia. It also has followers in China, Korea, and Japan. Smaller numbers of Asians practice Christianity or local religions.

History Some of the world's oldest civilizations arose in Asia. Until the A.D. 1500s, Asia was more advanced than Europe in its cultural and technological development. Early Asians founded cities, set up powerful governments, carved out trade routes, and farmed the land. They developed writing systems and created great works of literature. Among other inventions, Asians developed paper, the magnetic compass, and the first printed books.

576

UNIT 8

Facts

• **Pakistan** K2 was so named because it was the second peak in the Karakoram Range measured in a survey conducted in the 1850s.

• **Southeast Asia** China, Taiwan, Vietnam, Malaysia, and the Philippines have all laid claim to the Spratly Islands in the South China Sea. These countries are not interested in what is on this group of some 400 desolate, uninhabited coral islands, but what is underneath—vast deposits of oil and natural gas.

Europeans arrived in Asia around 1500. By the 1800s many parts of the region had fallen under European control. The Europeans brought Western ideas about industry and science to Asia. In the late 1800s, Japan led Asia in accepting modern ways and soon became the world's leading Asian power. World War II led to Japan's defeat, but it also ended Europe's hold on Asia. Nearly all of the Asian lands ruled by foreigners became independent in the mid-1900s.

From the late 1940s to the 1970s, Asia was caught in the worldwide struggle between communist and noncommunist countries. Today the countries of Asia have many different forms of government. Communist governments rule in China, Vietnam, and North Korea. Military leaders govern Myanmar and Indonesia. Kings reign in Nepal, Bhutan, and Thailand. Among Asia's democratic countries are India, Japan, and the Philippines.

The Arts Religion has strongly influenced the traditional arts of Asia. Buddhists created gold-roofed temples, cave paintings, and sculptures. Muslims built domed mosques, palaces, and fortresses. Hindus erected stone temples decorated with figures of gods and goddesses. Chinese, Korean, and Japanese artists are famous for their painted porcelain and scroll paintings. Indians and Southeast Asians have created unique forms of dance and music.

The Beijing Opera in China

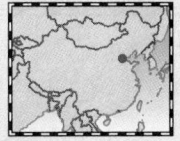
Operas attract huge audiences in China.
PLACE: What are Chinese, Korean, and Japanese artists famous for?

Technology Activity

Using a Spreadsheet
Create a spreadsheet that shows the (A) land area, (B) population, and (C) population density of each of these 10 Asian cities: Mumbai (Bombay), Calcutta, Shanghai, Beijing, Tokyo, Osaka, Seoul, Bangkok, Manila, and Singapore. Enter this formula to calculate population density for each city: Divide (B) by (A) to get (C). Use your program's chart capability to display your spreadsheet data. Share with the class.

REVIEW AND ACTIVITIES

1. **REGION** What is the Ring of Fire?
2. **REGION** How do monsoons affect Asia's climate?
3. **HUMAN/ENVIRONMENT INTERACTION** Name five Asian countries where manufacturing takes place.
4. **LOCATION** Why are some areas of Asia sparsely populated and other areas crowded?
5. **PLACE** What faith is Asia's most widespread religion?
6. **REGION** Name three Asian countries that have communist governments.

577

FactsOnFile

Use the reproducible masters from **MAPS ON FILE I,** *Asia,* to enrich unit content.

More About the Illustration

Answer to Caption painted porcelain and scroll paintings

Place The Beijing opera is one of the most popular forms of dramatic entertainment in China. Operatic performers combine acting, singing, and dancing in plays that focus on Chinese history and folklore.

Answers to Review and Activities

1. It is a Pacific area known for its earthquakes and active volcanoes.
2. Monsoons from the north bring dry, cool weather in the winter; in summer, the monsoons blow across the seas to the south, bringing hot, humid weather.
3. Japan, South Korea, Taiwan, China, and India
4. Many areas of Asia have mountains or deserts; most Asians live in mountain or river valleys and along the coasts.
5. Islam
6. China, North Korea, Vietnam

Physical Geography

Applying Geographic Themes

Location The Indian Ocean covers slightly less area than the Atlantic and holds 20 percent of the world's water. What continent lies to the north of the Indian Ocean? *(Asia)* What mainland country lies between the Arabian Sea and the Bay of Bengal? *(India)*

These materials are available from Glencoe.

 CD-ROM

Picture Atlas of the World

 SOFTWARE

ZipZapMap! World

MAP STUDY

Answers
1. Gobi
2. Himalayas

Map Skills Practice

Reading a Map What island country lies about 750 miles *(1,207 km)* east of Vietnam? *(Philippines)* What is the shortest distance between mainland China and the island of Taiwan? *(about 100 miles, or 161 km)*

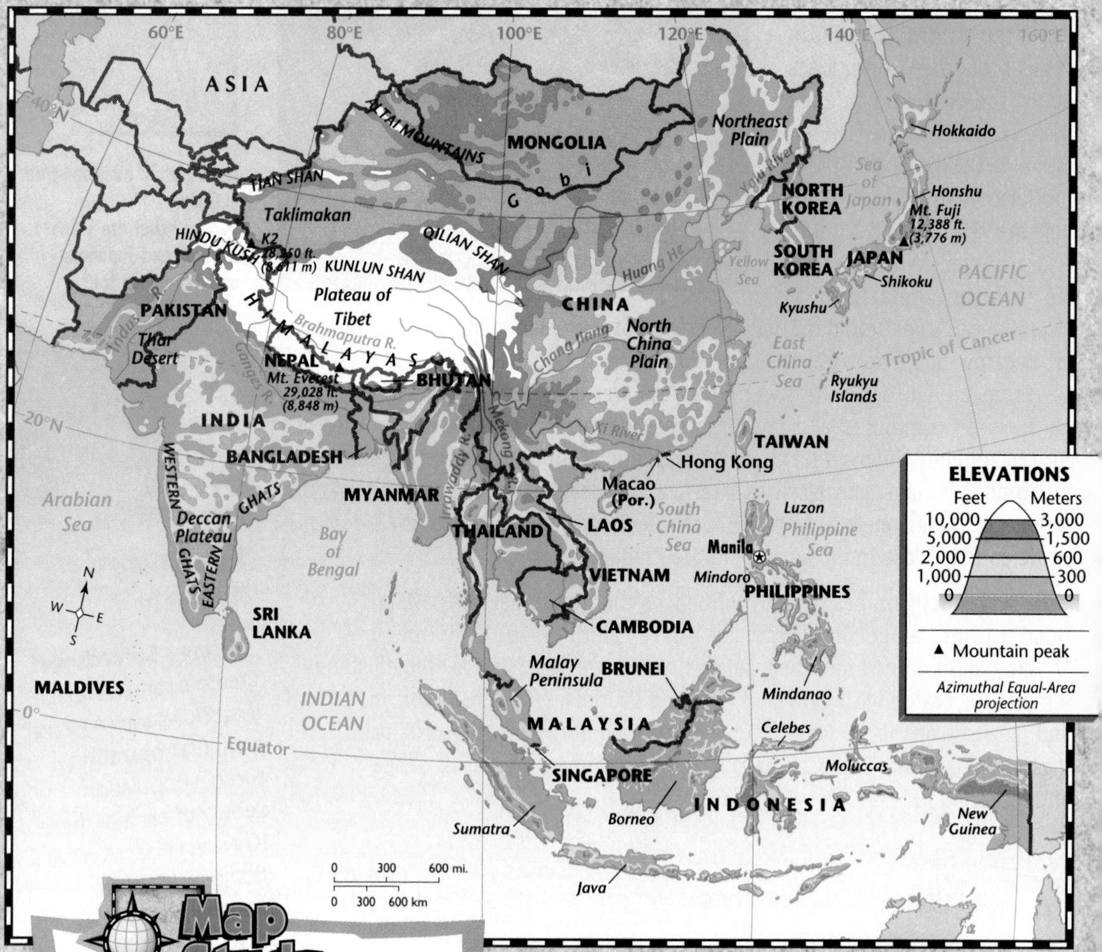

Map Study

1. **LOCATION** What desert extends across northern China and southern Mongolia?
2. **PLACE** What mountain range forms a natural border between India and China?

Atlas Activity

Movement Organize students into pairs and have each pair map a route between two places, such as central China and southeastern Pakistan. Ask the pairs to describe their route to the class, using directions and referring to physical features. Also have them point out deserts, mountains, and other barriers along their route, and hypothesize about how travelers can best overcome or circumvent these barriers. **L1**

ELEVATION PROFILE: South Asia

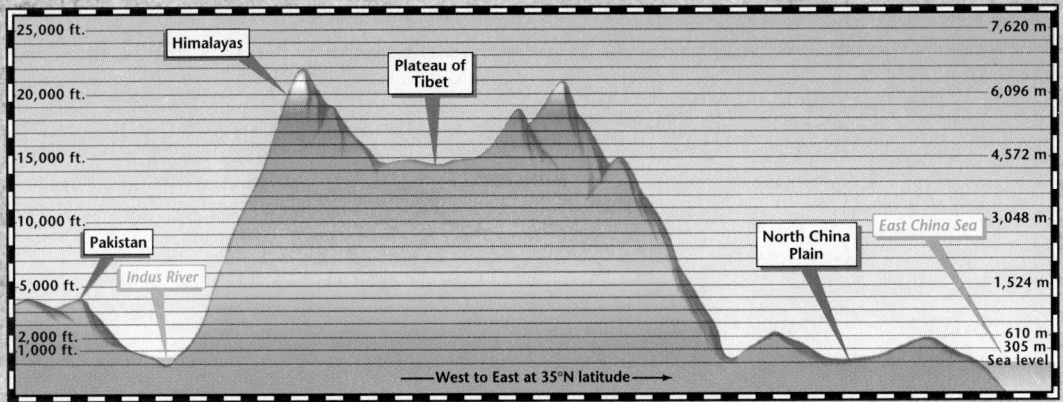

25,000 ft.	7,620 m
20,000 ft.	6,096 m
15,000 ft.	4,572 m
10,000 ft.	3,048 m
5,000 ft.	1,524 m
2,000 ft.	610 m
1,000 ft.	305 m
	Sea level

Himalayas
Plateau of Tibet
Pakistan
Indus River
North China Plain
East China Sea

← West to East at 35°N latitude →

Source: *Goode's World Atlas,* 19th edition

GeoFacts

Highest point: Mount Everest (Nepal-Tibet) 29,028 ft. (8,848 m) high

Longest river: Chang Jiang (China) 3,964 mi. (6,378 km) long

Largest desert: Gobi (Mongolia and China) 500,000 sq. mi. (1,300,000 sq. km)

Highest waterfall: Jog Falls (Gersoppa Falls) 830 ft. (253 m) high

ASIA AND THE UNITED STATES: Land Comparison

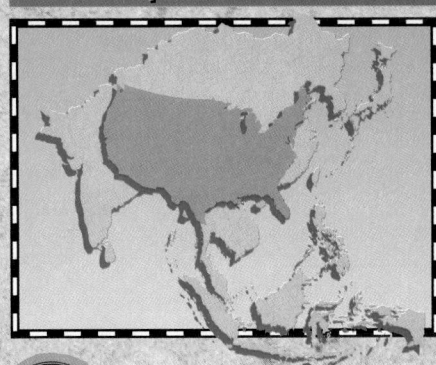

Graphic Study

1. **PLACE** What is the average elevation of the Himalayas?
2. **PLACE** What is the difference in elevation between the Plateau of Tibet and the North China Plain?

Teacher Notes

_____   _____

_____   _____

_____   _____

_____   _____

_____   _____

GLENCOE TECHNOLOGY

 VIDEODISC

The Infinite Voyage: To the Edge of the Earth

**Chapter 2
Exploring Tibet: The Gateway of Exchange**
Disc 1 Side B

The Infinite Voyage: Living With Disaster

**Chapter 10
Adapting to the Nature of Volcanoes: Sakurajima**
Disc 3 Side A

FactsOnFile

Use the reproducible masters from **MAPS ON FILE I,** *Asia,* to enrich unit content.

GRAPHIC STUDY

Answers
1. more than 20,000 feet (6,096 m)
2. about 14,000 feet (4,267 m)

Graphic Skills Practice
Reading a Graph If you traveled eastward from the Himalayas, would you be moving toward the Indus River or the Plateau of Tibet? *(Plateau of Tibet)*

LESSON PLAN
Unit 8 Atlas

UNIT 8 ATLAS

Cultural Geography

Applying Geographic Themes

Place Because Japan is a small island country, it has limited natural resources and farmland. How do these facts help explain why Tokyo is Japan's most important and populous city? *(Tokyo is on the coast and, therefore, can serve as a trading port and a fishing center.)* What other island countries in Asia have their capitals near the coast? *(Taiwan, Brunei, Indonesia, Sri Lanka, the Maldives, and the Philippines)*

NATIONAL GEOGRAPHIC SOCIETY

These materials are available from Glencoe.

VIDEODISC

STV: World Geography Vol. 1: Asia and Australia

Asia
Side 1, Frames 00002-47133

SOFTWARE

ZipZapMap! World

ASIA (map title)

Map Study

1. **LOCATION** What country lies at the southern tip of mainland Malaysia?
2. **PLACE** What country borders northwestern India?

580

UNIT 8

MAP STUDY

Answers
1. Singapore
2. Pakistan

Map Skills Practice
Reading a Map What is the capital of South Korea? *(Seoul)* What is the capital of Vietnam? *(Hanoi)*

Atlas Activity

Region Assign each student a city on the map. Have the students use political maps of North and South America to find a city in the Western Hemisphere along the same line of latitude as their assigned city in Asia. Then tell the students to use physical and climate maps in the text to compare the landform and climate regions of the two cities. Instruct them to present their comparisons in the form of a chart. **L1**

COMPARING POPULATION: Asia and the United States

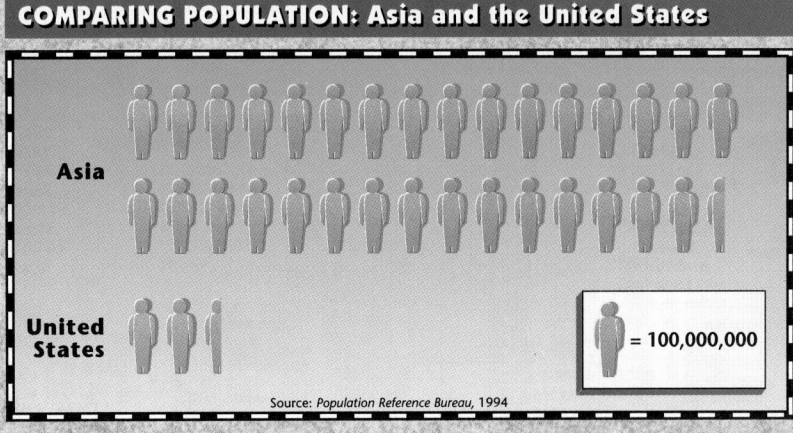

Asia

United States

 = 100,000,000

Source: *Population Reference Bureau, 1994*

GeoFacts

Biggest country (land area): China 3,600,930 sq. mi. (9,326,409 sq. km)

Smallest country (land area): Maldives 120 sq. mi. (311 sq. km)

Largest urban area (population): Tokyo (1995) 28,447,000; (2000 projected) 29,971,000

Highest population density: Singapore 12,083 people per sq. mi. (4,662 people per sq. km)

Lowest population density: Mongolia 4 people per sq. mi. (2 people per sq. km)

Asian Population

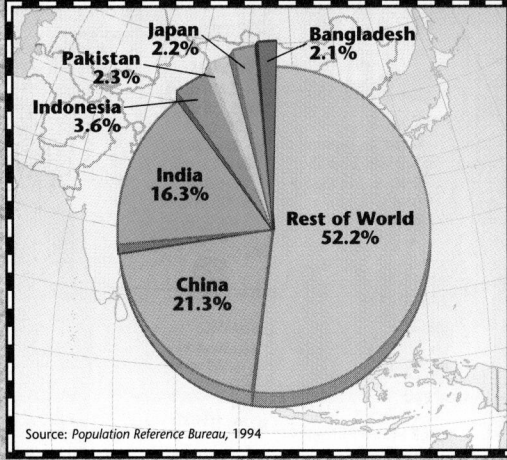

Japan 2.2%
Pakistan 2.3%
Bangladesh 2.1%
Indonesia 3.6%
India 16.3%
Rest of World 52.2%
China 21.3%

Source: *Population Reference Bureau, 1994*

Graphic Study

1. **PLACE** About how many times more people live in Asia than live in the United States?
2. **REGION** What two Asian countries have the largest populations?

Geography and the Humanities

World Literature Reading 8

World Music: Cultural Traditions, Lessons 7, 8, 9

World Cultures Transparencies 13, 14, 15, 16, 17, 18

Focus on World Art Print 10

Strange but TRUE!

By the year 2000, the city of Calcutta, India, will have a greater population than the nation of Canada.

GRAPHIC STUDY

Answers
1. about 11 times more people live in Asia
2. China and India

Graphic Skills Practice

Reading a Graph What is the population of Asia? *(about 3 billion people)* What percentage of the world's people live in the six Asian countries shown in the circle graph? *(47.8 percent)*

Teacher Notes

LESSON PLAN
Unit 8 Atlas

UNIT 8 ATLAS

Countries at a Glance

Bangladesh
CAPITAL: Dhaka
MAJOR LANGUAGE(S): Bengali, Chakma, Magh
POPULATION: 116,600,000
LANDMASS: 50,260 sq. mi./ 130,173 sq. km
MONEY: Taka
MAJOR EXPORT: Garments
MAJOR IMPORT: Textile Yarn

Bhutan
CAPITAL: Thimphu
MAJOR LANGUAGE(S): Dzongkha, Nepali and Tibetan dialects
POPULATION: 800,000
LANDMASS: 18,150 sq. mi./ 47,009 sq. km
MONEY: Ngultrum
MAJOR EXPORT: Electricity
MAJOR IMPORT: Petroleum

Brunei
CAPITAL: Bandar Seri Begawan
MAJOR LANGUAGE(S): Malay, English, Chinese
POPULATION: 300,000
LANDMASS: 2,030 sq. mi./ 5,258 sq. km
MONEY: Brunei Dollar
MAJOR EXPORT: Petroleum
MAJOR IMPORT: Machinery

Cambodia
CAPITAL: Phnom Penh
MAJOR LANGUAGE(S): Khmer, French
POPULATION: 10,300,000
LANDMASS: 68,150 sq. mi./ 176,509 sq. km
MONEY: Riel
MAJOR EXPORT: Rubber
MAJOR IMPORT: Machinery

China
CAPITAL: Beijing
MAJOR LANGUAGE(S): Mandarin, Yue, Wu Hakka, Xiang
POPULATION: 1,192,000,000
LANDMASS: 3,600,930 sq. mi./ 9,326,409 sq. km
MONEY: Yuan
MAJOR EXPORT: Textile Products
MAJOR IMPORT: Machinery

India
CAPITAL: New Delhi
MAJOR LANGUAGE(S): Hindi, English
POPULATION: 919,900,000
LANDMASS: 1,147,950 sq. mi./ 2,973,191 sq. km
MONEY: Rupee
MAJOR EXPORT: Diamonds
MAJOR IMPORT: Fuels

Indonesia
CAPITAL: Jakarta
MAJOR LANGUAGE(S): Bahasa Indonesia (Malay), Javanese
POPULATION: 199,700,000
LANDMASS: 705,190 sq. mi./ 1,826,442 sq. km
MONEY: Rupiah
MAJOR EXPORT: Petroleum
MAJOR IMPORT: Machinery

Japan
CAPITAL: Tokyo
MAJOR LANGUAGE(S): Japanese
POPULATION: 125,000,000
LANDMASS: 145,370 sq. mi./ 376,508 sq. km
MONEY: Yen
MAJOR EXPORT: Motor Vehicles
MAJOR IMPORT: Fuels

Laos
CAPITAL: Vientiane
MAJOR LANGUAGE(S): Lao
POPULATION: 4,700,000
LANDMASS: 89,110 sq. mi./ 230,795 sq. km
MONEY: New Kip
MAJOR EXPORT: Wood
MAJOR IMPORT: Food Products

Malaysia
CAPITAL: Kuala Lumpur
MAJOR LANGUAGE(S): Malay, English, Chinese, Indian languages
POPULATION: 19,500,000
LANDMASS: 126,850 sq. mi./ 328,542 sq. km
MONEY: Ringgit
MAJOR EXPORT: Machinery
MAJOR IMPORT: Machinery

Countries not drawn to scale.

582

UNIT 8

Fun Facts

- **Bhutan** This Himalayan country is called "Land of the Thunder Dragon" because of its violent storms.

- **Brunei** With a net worth of $31 billion, the Sultan of Brunei is the wealthiest person in the world.

- **Maldives** The highest natural elevation on these islands is about 15 feet (5 m) above sea level.

- **Mongolia** Livestock outnumbers people 12 to 1 in Mongolia, and the horse is the principal means of transportation.

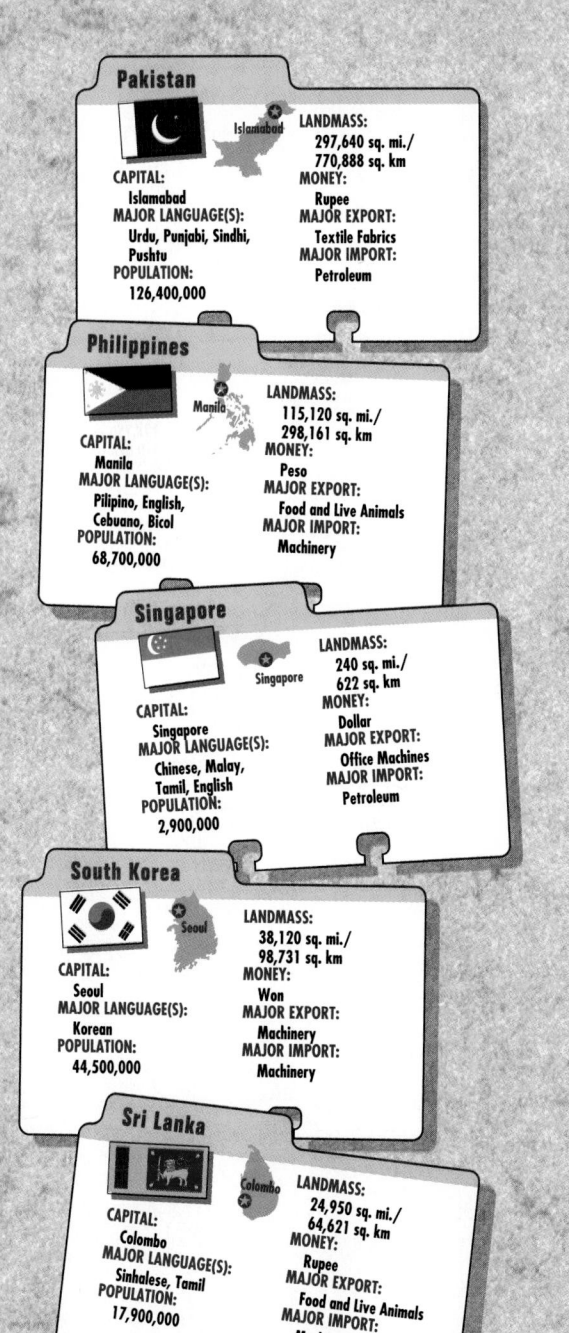

Maldives

CAPITAL:
Male
MAJOR LANGUAGE(S):
Divehi (Sinhalese dialect)
POPULATION:
252,000

LANDMASS:
120 sq. mi./
311 sq. km
MONEY:
Rufiyaa
MAJOR EXPORT:
Clothing
MAJOR IMPORT:
Food and Beverages

Mongolia

CAPITAL:
Ulan Bator
MAJOR LANGUAGE(S):
Khalkha, Mongolian
POPULATION:
2,400,000

LANDMASS:
604,830 sq. mi./
1,566,510 sq. km
MONEY:
Tugrik
MAJOR EXPORT:
Minerals and Metals
MAJOR IMPORT:
Machinery

Myanmar

CAPITAL:
Yangon
MAJOR LANGUAGE(S):
Burmese, Karen, Shan
POPULATION:
45,400,000

LANDMASS:
253,880 sq. mi./
657,549 sq. km
MONEY:
Kyat
MAJOR EXPORT:
Agricultural Products
MAJOR IMPORT:
Machinery

Nepal

CAPITAL:
Kathmandu
MAJOR LANGUAGE(S):
Nepali
POPULATION:
22,100,000

LANDMASS:
52,820 sq. mi./
136,804 sq. km
MONEY:
Rupee
MAJOR EXPORT:
Food and Live Animals
MAJOR IMPORT:
Machinery

North Korea

CAPITAL:
Pyongyang
MAJOR LANGUAGE(S):
Korean
POPULATION:
23,100,000

LANDMASS:
46,490 sq. mi./
120,409 sq. km
MONEY:
Won
MAJOR EXPORT:
Minerals
MAJOR IMPORT:
Petroleum

Pakistan

CAPITAL:
Islamabad
MAJOR LANGUAGE(S):
Urdu, Punjabi, Sindhi,
Pushtu
POPULATION:
126,400,000

LANDMASS:
297,640 sq. mi./
770,888 sq. km
MONEY:
Rupee
MAJOR EXPORT:
Textile Fabrics
MAJOR IMPORT:
Petroleum

Philippines

CAPITAL:
Manila
MAJOR LANGUAGE(S):
Pilipino, English,
Cebuano, Bicol
POPULATION:
68,700,000

LANDMASS:
115,120 sq. mi./
298,161 sq. km
MONEY:
Peso
MAJOR EXPORT:
Food and Live Animals
MAJOR IMPORT:
Machinery

Singapore

CAPITAL:
Singapore
MAJOR LANGUAGE(S):
Chinese, Malay,
Tamil, English
POPULATION:
2,900,000

LANDMASS:
240 sq. mi./
622 sq. km
MONEY:
Dollar
MAJOR EXPORT:
Office Machines
MAJOR IMPORT:
Petroleum

South Korea

CAPITAL:
Seoul
MAJOR LANGUAGE(S):
Korean
POPULATION:
44,500,000

LANDMASS:
38,120 sq. mi./
98,731 sq. km
MONEY:
Won
MAJOR EXPORT:
Machinery
MAJOR IMPORT:
Machinery

Sri Lanka

CAPITAL:
Colombo
MAJOR LANGUAGE(S):
Sinhalese, Tamil
POPULATION:
17,900,000

LANDMASS:
24,950 sq. mi./
64,621 sq. km
MONEY:
Rupee
MAJOR EXPORT:
Food and Live Animals
MAJOR IMPORT:
Machinery

583

NATIONAL GEOGRAPHIC SOCIETY

These materials are available from Glencoe.

 VIDEODISC

STV: World Geography Vol. 1: Asia and Australia

Interior and North Asia
Side 1, Frames 13631-17712

South and Southeast Asia
Side 1, Frames 27589-42384

Cultural Kaleidoscope

Bangladesh
Bangladeshis prize literature so highly that touching a book with one's shoe is considered an insult.

Teacher Notes

LESSON PLAN
Unit 8 Atlas

Countries at a Glance

More About the Illustration

Place This hillside monastery, called *Tashichodzong*, was built in 1641 and houses Bhutan's government offices. It is also the center of Bhutan's dominant Buddhist sect. Bhutan has a hereditary monarchy, but the king shares some of his power with a council of ministers and an assembly. Bhutan relies on India for its defense, and India also handles Bhutan's foreign affairs.

Cultural Kaleidoscope

Malaysia Malays remove their shoes and sunglasses before entering a mosque or house.

More About the Illustration

Region The Guilin Hills are limestone peaks in southeastern China near the city of Guilin. Some of the hills stand alone and reach as high as 600 feet. The Chinese compare their shape to dragons' teeth.

584

Taiwan

Taipei

LANDMASS:
13,900 sq. mi./
36,001 sq. km
MONEY:
New Taiwan Dollar
MAJOR EXPORT:
Machinery
MAJOR IMPORT:
Machinery

CAPITAL:
Taipei
MAJOR LANGUAGE(S):
Mandarin Chinese,
Taiwanese, Hakka dialects
POPULATION:
21,100,000

Thailand

Bangkok

LANDMASS:
197,250 sq. mi./
510,878 sq. km
MONEY:
Baht
MAJOR EXPORT:
Machinery
MAJOR IMPORT:
Machinery

CAPITAL:
Bangkok
MAJOR LANGUAGE(S):
Thai, Chinese, Malay,
regional dialects
POPULATION:
59,400,000

Vietnam

Hanoi

LANDMASS:
125,670 sq. mi./
325,485 sq. km
MONEY:
Dong
MAJOR EXPORT:
Fuels
MAJOR IMPORT:
Machinery

CAPITAL:
Hanoi
MAJOR LANGUAGE(S):
Vietnamese, Chinese
POPULATION:
73,100,000

Thimphu, Bhutan

Bhutan's king rules from this fortified hillside monastery in Bhutan's capital.

Guangxi Province, China

These unusual cone-shaped hills in southern China were formed by erosion and weathering.

Countries not drawn to scale.

584

Fun Facts

- **Philippines** The Philippines is a major movie producer; only three other countries in the world make more films.

- **South Korea** The martial art known as tae kwon do originated in Korea.

- **Thailand** Thais always honor their king and his image, including his picture on a stamp. A Thai wets the stamp with a damp sponge rather than show disrespect by licking it.

Angkor Wat, Cambodia

This elaborate stone temple was built almost 900 years ago to honor a Hindu king.

More About the Illustration

GEOGRAPHIC THEMES

Human/ Environment Interaction Angkor Wat was built in the 1100s when the kings of Cambodia were worshipped as gods. The walls are nearly half a mile in length. The building was designed to symbolize the Hindu view of the universe. The large tower in the center represents the home of the Hindu gods. The surrounding walls were considered to be the mountains that enclosed the world. The large moat represented the ocean beyond the mountains.

World's Tallest Mountains

GRAPHIC STUDY

Kilimanjaro is the highest mountain peak in Africa.
PLACE: What is the highest mountain peak in the world?

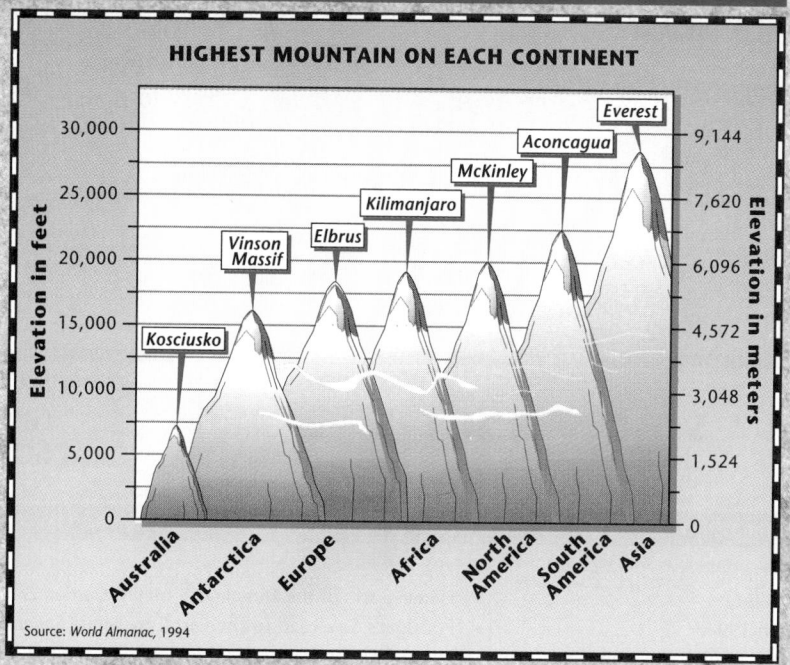

HIGHEST MOUNTAIN ON EACH CONTINENT

Elevation in feet: 30,000 / 25,000 / 20,000 / 15,000 / 10,000 / 5,000 / 0

Elevation in meters: 9,144 / 7,620 / 6,096 / 4,572 / 3,048 / 1,524 / 0

Kosciusko — Australia
Vinson Massif — Antarctica
Elbrus — Europe
Kilimanjaro — Africa
McKinley — North America
Aconcagua — South America
Everest — Asia

Source: World Almanac, 1994

Global Gourmet

Korea Cooks in North and South Korea try to serve meals that include all the following traditional colors: red, green, yellow, white, and black.

GRAPHIC STUDY

Answer
Mount Everest

Graphic Skills Practice
Reading a Chart What is the difference in elevation between Mount Everest and the highest peak in North America? *(about 9,000 feet, or 2,740 m)*

585

Teacher Notes

PERFORMANCE ASSESSMENT ACTIVITY

A GARAGE SALE Have students examine ads in the classified sections of the newspaper to determine the types of things sold at garage sales. Then ask students to draw up a list of items that might be found at a garage sale held by a family who had spent many years living in the Indian subcontinent. The list should include clothing, furniture, religious art or icons, souvenirs, magazines, and so on. Point out that the items on the list also should represent all of the five themes of geography. Encourage students to share and compare their lists.

POSSIBLE RUBRIC FEATURES: Accuracy of content information, concept attainment for the five themes, prioritization skills, application skills, ability to recognize relationships

MENTAL MAPPING ACTIVITY

Before teaching Chapter 22, point out the countries of South Asia on a globe to students. Have students identify the hemispheres in which South Asia lies and where the countries are located relative to the United States.

TEACHER'S CORNER

NATIONAL GEOGRAPHIC SOCIETY

INDEX TO NATIONAL GEOGRAPHIC MAGAZINE

The following articles may be used for research relating to this chapter:

- "Storming the Tower," by Todd Skinner, April 1996.
- "High Road to Hunza," by John McCarry, March 1994.
- "Himalayan Caravans," by Eric Valli and Diane Summers, December 1993.
- "Wandering With India's Rabari," by Robyn Davidson, September 1993.
- "Bangladesh: When the Water Comes," by Charles E. Cobb, Jr., June 1993.

- "Gatekeepers of the Himalaya," by Jim Carrier, December 1992.
- "Portugal's Sea Road to the East," by Merle Severy, November 1992.
- "India's Wildlife Dilemma," by Geoffrey C. Ward, May 1992.
- "Bhutan, Kingdom in the Clouds," by Bruce W. Bunting, May 1991.
- "Elephants—Out of Time, Out of Space," by Douglas Chadwick, May 1991.

NATIONAL GEOGRAPHIC SOCIETY PRODUCTS AVAILABLE FROM GLENCOE

To order the following products for use with this chapter, contact your local Glencoe sales representative or call Glencoe at 1-800-334-7344:

- *STV: World Geography* (Videodisc)
- *Picture Atlas of the World* (CD-ROM)
- *Ancient Civilizations: China and India* (CD-ROM)
- *Physical Geography of the World* (Transparencies)

- *ZipZapMap! World* (Software)
- *GeoBee* (Software)
- *Images of the World* (Posters)
- *Eye on the Environment* (Posters)

ADDITIONAL NATIONAL GEOGRAPHIC SOCIETY PRODUCTS

To order the following products for use with this chapter, call National Geographic Society at 1-800-368-2728:

- *Return to Everest* (Video)
- *Himalayan River Run* (Video)
- *The Great Indian Railway* (Video)

- *Man-Eaters of India* (Video)
- *Physical Geography of the Continents Series: "Asia."* (Video)

TEACHER-TO-TEACHER
Curriculum Connection: Global Studies, Home Economics

—from Jack Bransford,
Boston Middle School, LaPorte, IN

INDIAN FOOD DAY The purpose of this activity is to have students learn about some aspects of Indian culture by having them plan and prepare an Indian meal.

Organize students into four groups and inform them that their task is to prepare an Indian meal. Assign one of the following dishes to each group: an appetizer, a main course, a side dish, and a dessert. Direct groups to use library resources to find a recipe for their assigned dish. Review the recipes to make sure they are appropriate in terms of cost, availability of ingredients, and preparation and cooking time. When all groups have an approved recipe, set a date for "Indian Food Day." Also, direct groups to provide items such as paper plates, plastic cutlery, napkins, and so on.

On Indian Food Day, have groups use facilities in the Home Economics classroom to prepare their dishes. Make sure there is enough of each dish for every student to have a small portion. As students are sampling the food, call on a representative of each group to describe the ingredients and explain the preparation techniques of his or her dish. In the next class period, lead students in a discussion of Indian foods and cooking methods.

MEETING NATIONAL STANDARDS

Geography for Life

All of the 18 standards are demonstrated in Unit 8. The following ones are highlighted in this chapter:
• Standards 1, 2, 3, 4, 6, 7, 9, 10, 12, 15, 16

KEY TO ABILITY LEVELS

Teaching strategies have been coded for varying learning styles and abilities.

L1 **BASIC** activities for all students

L2 **AVERAGE** activities for average to above-average students

L3 **CHALLENGING** activities for above-average students

LEP **LIMITED ENGLISH PROFICIENCY** activities

BIBLIOGRAPHY

Readings for the Student
Das, Prodeepta. *Inside India.* New York: Franklin Watts, 1990.

Ganeri, Anita. *The Indian Subcontinent.* New York: Franklin Watts, 1994.

Kalman, Bobbie. *India: The Lands, Peoples, and Cultures* Series. New York: Crabtree, 1990. Three 32-page paperbacks.

Norton, James K. *India and South Asia: Global Studies, 2nd ed.* Guilford, Conn.: Dushkin, 1995.

Readings for the Teacher
Robinson, Francis, ed. *The Cambridge Encyclopedia of India, Pakistan, Bangladesh, Sri Lanka, Nepal, Bhutan, and the Maldives.* New York: Cambridge University Press, 1989.

Creative Activities for Teaching About India. Stockton, Calif.: Stevens & Shea, 1988. Reproducible activity sheets.

Johnson, Donald J., and Jean E. Johnson, eds. *Through Indian Eyes: The Living Tradition,* revised ed. New York: Center for International Training and Education, 1992.

Multimedia
Assignment: India. Boston: Christian Science Monitor Video, 1993. Videocassette, 60 minutes.

Exploring the Himalayas, Nepal, and Kashmir. Chicago: Questar, 1990. Videocassette, 60 minutes.

Use these *Geography: The World and Its People* resources to teach, reinforce, and extend chapter content.

CHAPTER 22 RESOURCES

- *Vocabulary Activity 22
- Cooperative Learning Activity 22
- Workbook Activity 22
- Chapter Map Activity 22
- Geography Skills Activity 22
- GeoLab Activity 22
- Critical Thinking Skills Activity 22
- Performance Assessment Activity 22
- *Reteaching Activity 22
- Enrichment Activity 22
- Chapter 22 Test, Form A and Form B
- *Chapter 22 Digest Audiocassette, Activity, Test
- Unit Overlay Transparencies 8-0, 8-1, 8-2, 8-3, 8-4, 8-5, 8-6
- Political Map Transparency 8; World Cultures Transparencies 13, 14
- Geoquiz Transparencies 22-1, 22-2
- Vocabulary PuzzleMaker Software
- Student Self-Test: A Software Review
- Testmaker Software
- MindJogger Videoquiz

If time does not permit teaching the entire chapter, summarize using the Chapter 22 Highlights on page 605, and the Chapter 22 English (or Spanish) Audiocassettes. Review students' knowledge using the Glencoe MindJogger Videoquiz. *Also available in Spanish*

Use these *Geography: The World and Its People* resources to teach and reinforce section content.

SECTION 1 RESOURCES

Reproducible Lesson Plan 22-1

Section Focus Transparency 22-1

Guided Reading Activity 22-1

Section Quiz 22-1 (also available in Spanish)

SECTION 2 RESOURCES

Reproducible Lesson Plan 22-2

Section Focus Transparency 22-2

Guided Reading Activity 22-2

Section Quiz 22-2 (also available in Spanish)

SECTION 3 RESOURCES

Reproducible Lesson Plan 22-3

Section Focus Transparency 22-3

Guided Reading Activity 22-3

Section Quiz 22-3 (also available in Spanish)

SECTION 4 RESOURCES

Reproducible Lesson Plan 22-4

Section Focus Transparency 22-4

Guided Reading Activity 22-4

Section Quiz 22-4 (also available in Spanish)

ADDITIONAL RESOURCES FROM GLENCOE

 Reproducible Masters

- Facts On File, MAPS ON FILE I, Asia

- Foods Around the World, Unit 8

 **Workbook**

- Building Skills in Geography, Unit 2, Lesson 8

World Music: Cultural Traditions Lesson 7

 **Transparencies**

- National Geographic Society PicturePack Transparencies, Unit 8

Posters

- National Geographic Society: Images of the World, Unit 8

World Games Activity Cards Unit 8

 Videodiscs

- Geography and the Environment: The Infinite Voyage

- Reuters Issues in Geography

- National Geographic Society: STV: World Geography, Vol. 1

 CD-ROM

- National Geographic Society: Picture Atlas of the World

 **Software**

- National Geographic Society: ZipZapMap! World

586D

Performance Assessment

Refer to the Planning Guide on page 586A for a Performance Assessment Activity for this chapter. See the *Performance Assessment Strategies and Activities* booklet for suggestions.

Chapter Objectives

1. **Describe** the land, climate, and religions of India.
2. **Compare** the economies and cultures of Pakistan and Bangladesh.
3. **Explain** how the landscapes and religions of Nepal and Bhutan affect their people.
4. **Specify** the locations, landforms, and products of Sri Lanka and the Maldives.

GLENCOE
TECHNOLOGY

VIDEODISC

Use the Chapter 22 MindJogger Videoquiz to preview chapter content.

MindJogger Videoquiz

Chapter 22
Disc 3 Side B

 Also available in VHS.

586

Chapter 22 South Asia

MAP STUDY ACTIVITY

As you read Chapter 22, you will learn about India and the other countries in South Asia.

1. What is the largest country in South Asia?
2. What two island countries lie off the coast of southern India?
3. What is the capital of India?

586

MAP STUDY ACTIVITY

Answers
1. India
2. Sri Lanka, the Maldives
3. New Delhi

Map Skills Practice
Reading a Map What South Asian countries lie completely to the north of the Tropic of Cancer? *(Pakistan, Nepal, Bhutan)* Which South Asian capital city lies farthest south? *(Male, the Maldives)*

SECTION 1 India

PREVIEW

Words to Know
- subcontinent
- monsoon
- jute
- cottage industry

Places to Locate
- India
- Himalayas
- Ganges River
- Deccan Plateau
- Calcutta
- Mumbai (Bombay)
- New Delhi

Read to Learn . . .
1. what a subcontinent is.
2. how seasonal winds affect India's climate.
3. what religions India's people follow.

Hindus—the largest religious group in India—consider the Ganges a holy river. About 1 million Hindus from all areas of India pray and bathe in this river each year.

This is the Ganges River in the South Asian country of India.

India and several other countries—Pakistan, Bangladesh (BAHN•gluh•DEHSH), Nepal, Bhutan, Sri Lanka, and the Maldives—make up the area known as South Asia. South Asia is often called a **subcontinent,** or a landmass that is like a continent, only smaller. The map on page 588 shows you that towering mountains in the north separate South Asia from the rest of the continent of Asia. Bordering South Asia are three large bodies of water: the Indian Ocean, the Arabian Sea, and the Bay of Bengal.

Location
The Land

India takes up about 75 percent of the land area of South Asia. India covers 1,147,950 square miles (2,973,191 sq. km)—about one-third the area of the United States. The map on page 588 shows you that mountains run along three sides and through the middle of the country. Plateaus and desert plains run through the rest of the country.

Mountains Two huge walls of mountains—the Karakoram (KAR•uh•KOHR•uhm) Range and the Himalayas (HIH•muh•LAY•uhs)—form India's

CHAPTER 22

587

FOCUS

Section Objectives
1. **Define** the term subcontinent.
2. **Explain** how seasonal winds affect India's climate.
3. **Name** religions that India's people follow.

Bellringer
Motivational Activity
 Prior to taking roll at the beginning of the class period, project Section Focus Transparency 22-1 and have students answer the activity questions. Discuss students' responses. L1

Vocabulary Pre-check
Direct students to locate the **Words to Know** terms in the text. Then have students use the terms in sentences to demonstrate that they understand the meanings. L1 LEP

Use the Vocabulary PuzzleMaker Software to create a crossword puzzle. L1

Meeting National Standards
Geography for Life The following standards are highlighted in this section:
- Standards 1, 2, 7, 9, 10, 16

Classroom Resources for Section 1

 REPRODUCIBLE MASTERS
Reproducible Lesson Plan 22-1
Guided Reading Activity 22-1
Geography Skills Activity 22
Building Skills in Geography, Unit 2, Lesson 8
Workbook Activity 22
Section Quiz 22-1

TRANSPARENCIES
Section Focus Transparency 22-1
World Cultures Transparencies 13, 14
Unit Overlay Transparencies 8-0, 8-4, 8-5

MULTIMEDIA
 Vocabulary PuzzleMaker Software

Testmaker

National Geographic Society: STV: World Geography, Vol. 1

TEACH

Guided Practice

Display Unit Overlay Transparencies 8-0, 8-4, and 8-5. Have students show with a marker how the Himalayas block cold winds and the direction the monsoon winds blow during the summer and winter. Then have students write a brief paragraph explaining the impact of the Himalayas on climate in the Indian subcontinent. **L1**

MAP STUDY

Answer
Brahmaputra, Ganges, Narmada, Indus, Mahanadi, Godavari, Krishna

Map Skills Practice
Reading a Map What mountain ranges can be found in the southernmost part of India?
(Western Ghats, Eastern Ghats)

NATIONAL GEOGRAPHIC SOCIETY

These materials are available from Glencoe.

VIDEODISC

STV: World Geography Vol. 1: Asia and Australia

Karakoram Range, Kashmir
Any side, Frame 50252

northern border. The Himalayas—made up of several ranges—stretch more than 1,500 miles (2,414 km) across northern South Asia. They are the tallest mountains in the world. Their snowcapped peaks average more than 5 miles (8 km) in height!

In the center of India lies another mountain range—the Vindhya (VIHN•dyuh) Mountains. These low mountains divide India in half. They have helped create two very different cultures in northern India and southern India.

Two chains of hills and mountains called the Eastern Ghats (GAHTS) and Western Ghats edge the southern coasts of India. These mountain chains lie just inland from the Bay of Bengal and the Arabian Sea.

Plains and Plateaus Sweeping through northern India between the Himalaya and Vindhya ranges is the Ganges Plain. It boasts some of the most fertile soil in the country. Look at the map below. You see that the Ganges River begins in the Himalayas and flows through the Ganges Plain to the Bay of Bengal. At the western edge of the Ganges Plain lies the Thar (TAHR) Desert. It is about 500 miles (804 km) long and 275 miles (442 km) wide.

The Deccan Plateau is located south of the Vindhya Mountains. This triangle-shaped landmass makes up the southern two-thirds of India. Forests, fertile farmland, and rich deposits of minerals make the Deccan Plateau a valuable region.

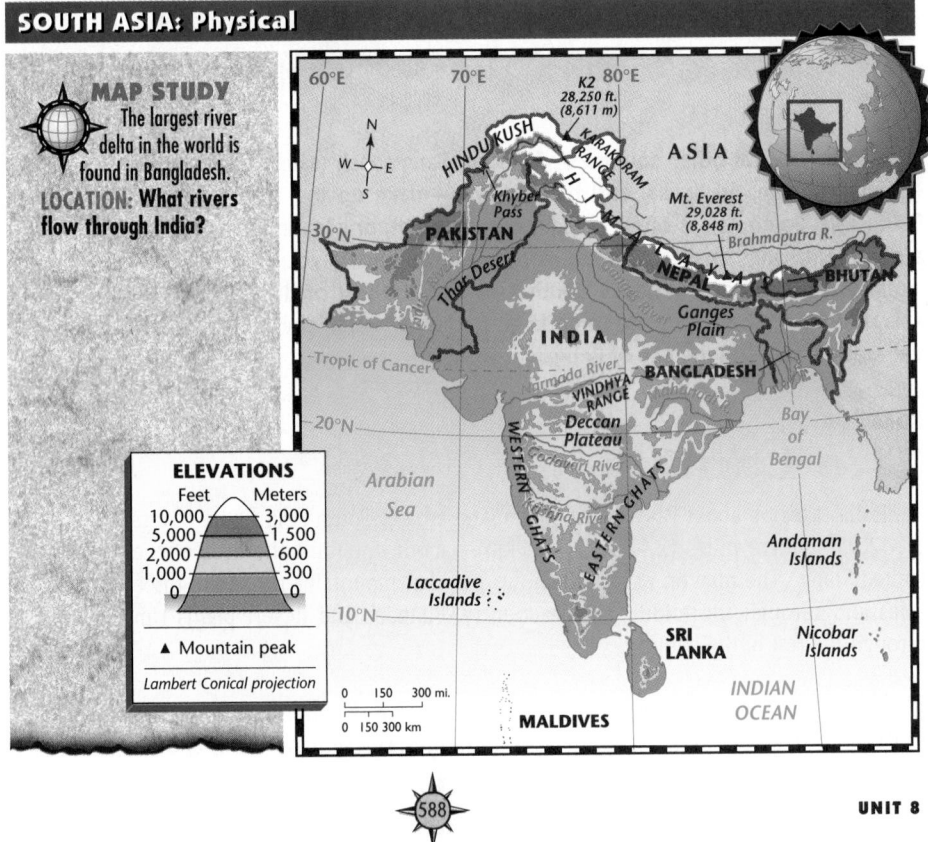

SOUTH ASIA: Physical

MAP STUDY
The largest river delta in the world is found in Bangladesh.
LOCATION: What rivers flow through India?

ELEVATIONS
Feet	Meters
10,000	3,000
5,000	1,500
2,000	600
1,000	300
0	0

▲ Mountain peak

Lambert Conical projection

Cooperative Learning Activity

Organize the students into groups and assign each group one of the following Hindu festivals: Diwali, Durga Puja or Dassehra, Holi, Jagannath, Makara-Sankranti, Pongal, Ramlila, and Vaisakhi. Ask groups to find out what their festival commemorates and how Hindus celebrate it. Refer them to encyclopedias and other sources, such as *The Aquarian Dictionary of Festivals* by J. C. Cooper. Have groups share their findings with the class in a skit or an oral report. **L1**

Independent Practice

 Guided Reading Activity 22-1

Climate Most places in India are warm or hot most of the year. The people of India can thank the Himalayas for their warm climate. These mountains block cold northern air from entering India. The map on page 595 shows you the climate regions of India.

The **monsoons,** or seasonal winds, are another important influence on India's climate. Most of India has three seasons—cool, hot, and rainy. During the cool season—November through February—and the hot season—March through April—monsoon winds from the north bring dry air. During the rainy season—May through October—monsoon winds reverse direction and bring moist air from the Indian Ocean.

Human/Environment Interaction

The Economy

Agriculture and industry are equally important to India's economy. The government of India has set up plans to increase the production of farm and industrial goods. The goal is to develop India's resources and to improve the standard of living for India's people.

Agriculture Most of India's best farmland lies in the Ganges Plain and the Deccan Plateau. In both places you will see many farmers work small plots of land. India is the world's second-largest producer of rice. Other important crops are tea, sugarcane, wheat, barley, cotton, and jute. **Jute** is a plant fiber used in making rope, twine, and burlap bags.

Industry and Mining Huge factories in India turn out cotton textiles and produce iron and steel. Oil and sugar refineries loom over the industrial landscape

The Thar Desert is one of the most sparsely populated areas of India *(left)*. In contrast, Calcutta is one of the most crowded cities in the world *(right)*.
LOCATION: In what part of India is Calcutta located?

More About the Illustration

Answer to Caption
northeastern

GEOGRAPHIC THEMES

Movement Overcrowding and poverty are serious problems in Calcutta. Many Calcuttans live in slum areas called *bustees.* Most homes in the bustees have neither electricity nor running water. Other Calcuttans are even worse off, making their homes on the city's streets. Poor sanitation and malnutrition make the city's population very susceptible to disease.

Strange but TRUE!

The world's wettest place is Chjerrapunji, India, which receives, on average, 457 inches (1,161 cm) of rain a year.

Meeting Special Needs Activity

Visual Learning Disability Visual processing problems often impair a student's ability to recall words by sight, and this affects reading and comprehension. Confusion between similar words is common because of weaknesses in visual attention to detail or visual memory problems. In private sessions, have students read aloud the paragraphs under "Agriculture" and "Industry." Note if they read "must" for "most" and "were" for "where." Encourage students to reread and to self-correct their mistakes. L1

Answer

about 900 million people

Graphic Skills Practice

Reading a Graph About how many more people live in India than in the United States? *(about 650 million)*

MULTICULTURAL PERSPECTIVE

Culturally Speaking

Many Hindu women have a red dot on their foreheads. This mark, which is called a *bindi,* signifies that the woman's husband is alive.

India Wealthy Indians honor special guests by decorating rice dishes with thin sheets of silver or gold leaf.

COMPARING POPULATION: India and the United States

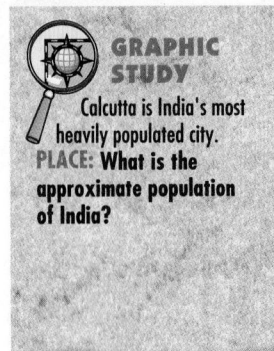

GRAPHIC STUDY

Calcutta is India's most heavily populated city. **PLACE: What is the approximate population of India?**

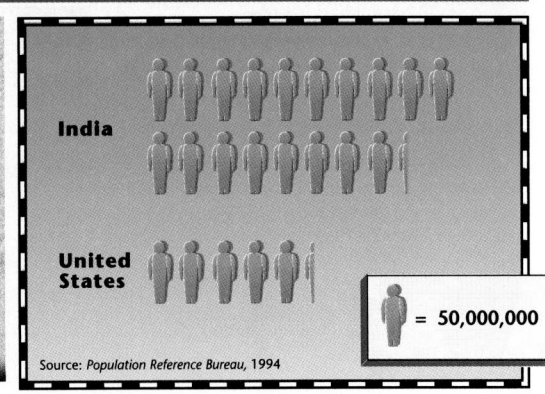

India

United States

= 50,000,000

Source: *Population Reference Bureau,* 1994

of many cities. Other factories produce locomotives, cars, cement, and chemical products. India's mica mines produce much of the world's supply of this mineral.

Cottage industries are also the source of many Indian products. A **cottage industry** is based in a rural village where family members use their own equipment. Goods produced in cottage industries include cotton cloth, silk cloth, rugs, leather products, and metalware.

Human/Environment Interaction
The People

India has more than 919 million people. It is the world's second-largest country in population. Only China is larger. The graph above shows you how India's population compares with that of the United States. The people of India speak 14 major languages and more than 1,000 other languages and dialects. Hindi is India's official language, but English is commonly spoken in government and business.

Influences of the Past About 4,000 years ago, the first Indian civilization built well-planned cities in present-day Pakistan. About 1500 B.C. warriors known as Aryans (AR•ee•uhnz) entered the subcontinent from central Asia. They set up kingdoms in northern India, forcing many of the earlier Indian peoples to move southward. The Aryans brought the religious teachings of Hinduism and the traditional system of social groups to India's culture.

Many other groups came later to the area that is now India. Beginning in the A.D. 700s, Muslims from Southwest Asia brought Islam to India. In the 1500s they founded the Mogul (MOH•guhl) Empire, which lasted more than 200 years.

The British ruled large areas of India from the 1700s to the mid-1900s. They built roads, railroads, and seaports. They also made large profits from the plantations, mines, and factories they set up in India. An Indian leader named Mohandas Gandhi (moh•HAHN•das GAHN•dee) led a movement that brought

590

Critical Thinking Activity

Making Generalizations Display a transparency comparing India and the Indian state of Kerala:

Per Capita Domestic Product in $US (1991) India—225; Kerala—200
Infant Mortality per 1,000 Live Births (1992) India—83; Kerala—17
Women's Life Expectancy in Years (1990) India—59; Kerala—74

Percentage of Girl Drop-outs, Grades 1–5 (1988) India—50; Kerala—0
Literacy Rate Among Women (1991) India—34; Kerala—87

Ask students to use this information to write a generalization about life in India as compared to life in the state of Kerala. **L2**

India independence from the United Kingdom in 1947. Since independence India has been a democracy.

Religion About 80 percent of India's people are Hindus. Hindus believe that all living things have souls that belong to one eternal spirit. After the body dies, the soul is reborn and returns to the earth. This process is repeated many times until the soul reaches perfection in a higher state of existence.

Islam also claims many followers in India. India's 100 million Muslims form one of the largest Muslim populations in the world. Other religions in India are Buddhism, Sikhism (SEE•KIH•zuhm), Jainism (JY•NIH•zuhm), and Christianity.

The Arts Religion has influenced the arts of India. Ancient Hindu builders constructed temples with tall, elaborately carved towers. Hindu writers left stories, poems, and legends. Hindu artists developed dances and music that are still performed today.

Muslims also have added to India's artistic heritage. One of the finest Muslim buildings in India is the Taj Mahal. A ruler in the mid-1600s built the Taj Mahal as a monument to a beloved wife. By the 1800s European influences affected the arts of India. Today Indian arts reflect a blend of both East and West, old and new.

Where's Mumbai?

In 1996 Bombay became known as Mumbai to English speakers. The word "Mumbai" comes from the name of a Hindu goddess.

Movie-Mania in New Delhi, India

A large billboard advertises a popular movie—one of hundreds produced every year in India.
PLACE: What other forms of entertainment are popular in India?

More About the Illustration

Answer to Caption sports, such as soccer and rugby, and religious festivals

GEOGRAPHIC THEMES

Place Movies are a big business in India. Mumbai, the center of the movie industry, is sometimes called the Hollywood of India. The typical Indian movie focuses on fantasy and escape, and that seems to be what the audience wants. An average of 5 billion tickets are sold to Indian movie-goers each year.

ASSESS

Check for Understanding

Assign Section 1 Review as homework or an in-class activity.

Meeting Lesson Objectives

Each objective below is tested by the questions that follow it in parentheses.

1. **Define** the term subcontinent. (1)
2. **Explain** how seasonal winds affect India's climate. (1, 2)

More About . . .

Sikhism Sikhism, which combines elements of the Hindu and Muslim faiths, was founded by Guru Nanak, who lived between 1469 and 1539. Sikh festivals, or *melas,* correspond with Hindu festivals. Sikhs, however, differ from Hindus in that they reject the caste system, the priesthood, pilgrimages, begging, and bathing in sacred streams. Sikhs accept the equality of men and women and a belief in one God. Their Holy Book is the *Granth Sahib* and their holiest place is the Golden Temple at Amritsar in the state of Punjab.

Section 1 Quiz **L1**

Use the Testmaker to create a customized quiz for Section 1. **L1**

Reteach

On the chalkboard, write the subheadings from the section and the numbers one through five under each subheading. Have students list a detail related to the subheading next to each number as they reread the section. **L1**

Enrich

Ask interested students to research the life of Mohandas Gandhi and to watch the movie *Gandhi*. Then have the students evaluate the factual accuracy of the movie. **L3**

CLOSE

Have students write to Rajinder Swarup comparing life in the United States to life in India.

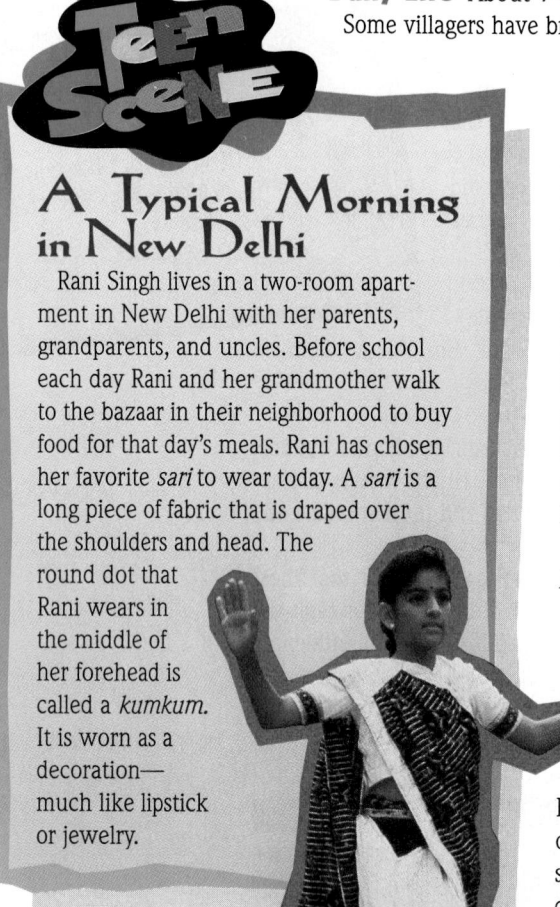

Teen Scene

A Typical Morning in New Delhi

Rani Singh lives in a two-room apartment in New Delhi with her parents, grandparents, and uncles. Before school each day Rani and her grandmother walk to the bazaar in their neighborhood to buy food for that day's meals. Rani has chosen her favorite *sari* to wear today. A *sari* is a long piece of fabric that is draped over the shoulders and head. The round dot that Rani wears in the middle of her forehead is called a *kumkum*. It is worn as a decoration— much like lipstick or jewelry.

Daily Life About 74 percent of India's people live in rural villages. Some villagers have brick homes. Others live in mud-and-straw shelters. Both men and women work in nearby fields. The Indian government tries to provide electricity, drinking water, better schools, and paved roads for many villages. Life is still difficult, however, for many Indian villagers. In recent years large numbers of people from the countryside have moved to urban areas.

India's cities today are very crowded. Delhi, Calcutta, and Mumbai are among the cities in India that have more than 5 million people each. Modern high-rise apartment and office buildings rise above the skyline. Next to these modern buildings stand slum areas of poverty. Bicycles, carts, animals, and people fill the narrow streets lined by small family-owned shops.

Fifteen-year-old Rajinder Swarup lives in Mumbai. Like many Indians, he wears a mix of traditional and modern clothing. He favors light, loose clothes because of India's hot climate. Like you, Rajinder loves going to the movies. Filled with action, adventure, and romance, movies are the most popular form of entertainment in India today. India makes more movies than any other country in the world. Rajinder also enjoys sports such as rugby and soccer. His favorite holiday is Diwali—the festival of lights. It is a Hindu festival marking the coming of winter and the victory of good over evil.

SECTION 1 REVIEW

REVIEWING TERMS AND FACTS

1. **Define the following:** subcontinent, monsoon, jute, cottage industry.
2. **LOCATION** What two mountain ranges lie along India's northern border?
3. **PLACE** What kinds of landforms are found in India?
4. **HUMAN/ENVIRONMENT INTERACTION** What is India's leading food crop?

 MAP STUDY ACTIVITIES

5. Look at the physical map on page 588. What three major bodies of water border India?
6. Turn to the political map on page 586. What is the capital of India?

Answers to Section 1 Review

1. All vocabulary words are defined in the Glossary.
2. Karakoram Range and Himalayas
3. mountains, plateaus, plains, deserts
4. rice
5. Arabian Sea, Bay of Bengal, Indian Ocean
6. New Delhi

BUILDING GEOGRAPHY SKILLS

Reading a Circle Graph

Have you ever watched someone dish out pieces of pie? When the pie is cut evenly, everybody gets the same size slice. If one slice is cut a little larger, however, someone else gets a smaller piece.

A circle graph is like a sliced pie. Often it is even called a pie chart. In a circle graph, the complete circle represents a whole group—or 100 percent. The circle is divided into "slices," or wedge-shaped sections representing parts of the whole.

Suppose, for example, that 25 percent of your friends watch five hours of television per day. On a circle graph representing hours of TV watching, 25 percent of the circle would represent friends who watch five hours. To read a circle graph, follow these steps:

- Read the title to find out what the subject is.
- Study the labels or key to determine what the parts or "slices" represent.
- Compare the parts to draw conclusions about the subject.

Religions of South Asia

RELIGION	NUMBER OF FOLLOWERS
Hinduism	755,240,000
Islam	324,657,000
Christianity	21,210,000
Buddhism	20,310,000
Sikhism	17,000,000
Other	10,290,000

Sikhism: 1%
Buddhism 2%
Christianity 2%
Other 1%
Islam 28%
Hinduism 66%

Source: *Encyclopedia Britannica Book of the Year*, 1994

Geography Skills Practice

1. Which religion in Asia has the most followers?

2. What percentage of South Asians are Muslim?

3. What is the combined percentage of Buddhist and Christian followers?

CHAPTER 22

593

TEACH

Conduct a class survey by counting numbers of students who arrive at school 1) on foot, 2) by bicycle, 3) on a school bus, 4) on a public bus, or 5) by car. (Have students choose the method they use most often.) Note the results on the chalkboard and call on volunteers to convert these numbers into percentages. Next, draw a circle on the chalkboard and explain that it represents the whole class, or 100 percent. Divide the circle into sections to represent the percentages of the subgroups. Have students note how the circle graph illustrates the relationship of the parts to the whole. Then have students read the skill and complete the practice questions. **L1**

Additional Skills Practice

1. What is the combined percentage of those who follow Hinduism and Islam? *(94 percent)*

2. How many more South Asians follow Christianity than follow Sikhism? *(4,210,000)*

📁 Geography Skills Activity 22

📁 Building Skills in Geography, Unit 2, Lesson 8

Answers to Geography Skills Practice

1. Hinduism
2. 28 percent
3. 4 percent

FOCUS

Section Objectives

1. **Locate** Pakistan and Bangladesh.
2. **Contrast** how people in Pakistan and Bangladesh earn their livings.
3. **Interpret** how Islam influences the cultures of Pakistan and Bangladesh.

Bellringer Motivational Activity

Prior to taking roll at the beginning of the class period, project Section Focus Transparency 22-2 and have students answer the activity questions. Discuss students' responses. **L1**

Vocabulary Pre-check

Write each vocabulary term on the chalkboard and give its definition on a slip of paper to a student. Ask the student to pantomime the definition for the class. Allow students a specified time to guess each word. **L1**
LEP

SECTION 2 Muslim South Asia

PREVIEW

Words to Know
• tributary
• delta
• cyclone
• teak

Places to Locate
• Pakistan
• Bangladesh
• Hindu Kush
• Brahmaputra River
• Indus River
• Karachi
• Islamabad
• Dhaka

Read to Learn . . .
1. where Pakistan and Bangladesh are located.
2. how people in Pakistan and Bangladesh earn their livings.
3. how Islam influences the cultures of Pakistan and Bangladesh.

This beautiful domed mosque soars above the city sky-

line of Lahore, Pakistan. About 1,000 years old, Lahore boasts many buildings that reflect the religion of Islam.

Two countries in South Asia—Pakistan and Bangladesh—are largely Muslim. Although they share a religion, the peoples of these two countries have very different cultures. They also are separated from each other by more than 1,000 miles (1,610 km).

Location
Pakistan

Pakistan lies northwest of India. Its 297,640 square miles (770,888 sq. km) make it about the size of Texas. Pakistan is an independent country today because of religion. Its Muslim population did not want to be part of largely Hindu India. When British rule of India ended in 1947, the western and eastern parts of India became Pakistan. In 1971 the eastern section became Bangladesh.

Land and Climate Snow-topped mountains, high plateaus, fertile plains, and sandy deserts make up Pakistan. In the north you will find the Hindu Kush. Mountain passes cut through these rugged peaks. Centuries ago, the Khyber Pass was an important passage for people entering into South Asia from the north.

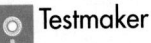

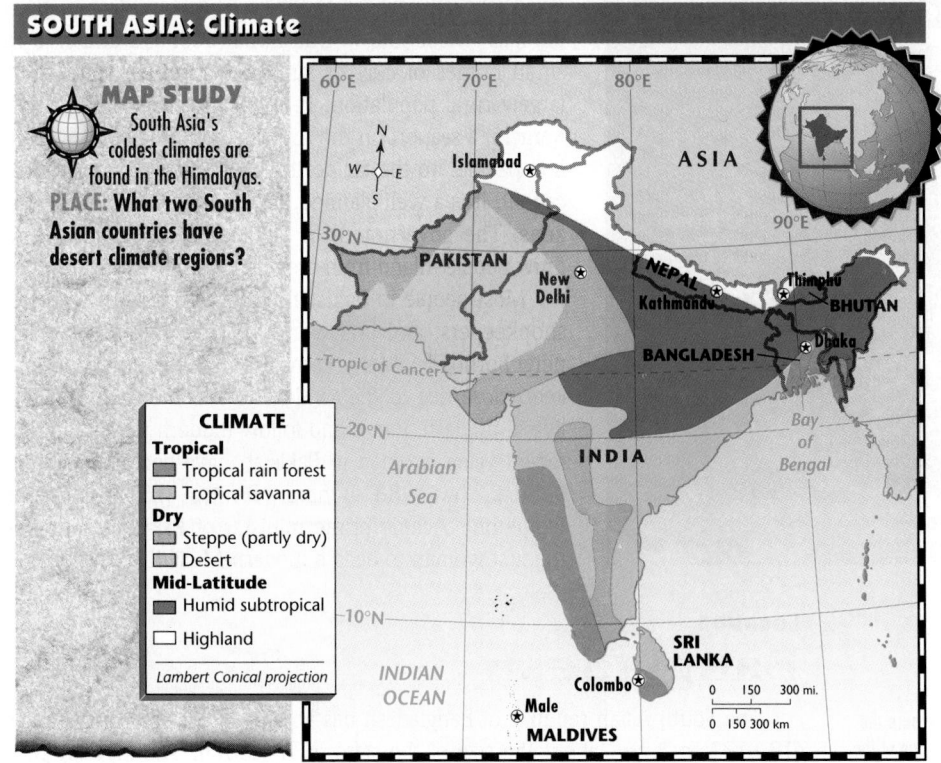

MAP STUDY
South Asia's coldest climates are found in the Himalayas.
PLACE: What two South Asian countries have desert climate regions?

CLIMATE
Tropical
Tropical rain forest
Tropical savanna
Dry
Steppe (partly dry)
Desert
Mid-Latitude
Humid subtropical
Highland

Lambert Conical projection

MAP STUDY

Answer
Pakistan and India
Map Skills Practice
Reading a Map What are the two major climate zones found in Bangladesh? *(tropical rain forest, humid subtropical)*

TEACH

Guided Practice

Making Comparisons
Display a transparency with two blank circle graphs. Use markers to show the percentage of Pakistanis who follow various religions on one graph and the percentage of Bangladeshis who follow Islam. Title the graphs. Have students use circle graphs to make comparisons of Pakistan and Bangladesh. One comparison might be urban and rural populations. L1

Independent Practice

Guided Reading Activity 22-2 L1

Cultural Kaleidoscope

Pakistan More than 40 percent of Pakistan's population is younger than age 15.

Plains cover most of eastern Pakistan. They are rich in fertile soil deposited by rivers. The major river system running through Pakistan's plains is the Indus River and its tributaries. A **tributary** is a small river that flows into a larger river or body of water. West of the Indus River valley, the land rises to form a plateau. A large part of this plateau is dry and rocky with little vegetation. Another vast barren area—the Thar Desert—sweeps east of the Indus River valley.

Pakistan has desert and steppe climates, with hot summers and cool winters. Rainfall in most regions is less than 10 inches (25 cm) a year. As in India, mountains in northern Pakistan block cold air from central Asia.

The Economy Many Pakistanis earn their livings by farming. Farmers use irrigation to grow wheat, cotton, and corn. They also raise cattle, goats, and sheep. Pakistan's factories manufacture cotton textiles for export. Other manufactured goods include cement, chemicals, fertilizer, and steel. Many craftworkers participate in cottage industries making metalware, pottery, and carpets.

The People Although 97 percent of Pakistanis are Muslim, the country's population includes many different ethnic and language groups. Among the major languages of Pakistan are Punjabi and Urdu. English is widely spoken in government and business.

CHAPTER 22

595

Cooperative Learning Activity

On separate slips of paper, copy the following incomplete statements: 1) If mountains did not border Pakistan to the north . . . 2) If British rule in India never ended . . . 3) If the Indus River did not run through Pakistan . . . 4) If the Brahmaputra and Ganges rivers did flow through Bangladesh . . . 5) If farmers in Bangladesh had modern tools . . . 6) If most Bangladeshis were Hindus instead of Muslims . . .

Organize the students into six groups and give each group a slip of paper. Allow a specified time for the groups to discuss the consequences of the condition on their assigned slip. Tell them to base their discussion on facts from the section. Then direct the groups to complete their statement and to share it with the rest of the class. L1

Place Chapati, a flatbread made from wheat flour, is eaten at almost every meal in Pakistan. It is baked in a *tandoor,* or clay oven. It is served as a side dish with rice and meat or vegetables prepared in a spicy sauce.

ASSESS

Check for Understanding

Assign Section 2 Review as homework or an in-class activity.

Meeting Lesson Objectives

Each objective below is tested by the questions that follow it in parentheses.

1. **Locate** Pakistan and Bangladesh. (2, 3, 5)
2. **Contrast** how people in Pakistan and Bangladesh earn their livings. (3)
3. **Interpret** how Islam influences the cultures of Pakistan and Bangladesh. (4)

Cultural Kaleidoscope

Bangladesh When someone does a favor for a Bangladeshi, the custom is to return the favor instead of saying thanks.

Working Together in Gulmit, Pakistan

Pakistani women prepare the evening meal while one of the men spins wool.
PLACE: What is the major religion of Pakistan?

About 72 percent of Pakistan's people live in rural villages. Most follow traditional customs and live in small homes of clay or sun-dried mud. In spite of its largely rural population, Pakistan has many large cities. Karachi, a seaport on the Arabian Sea, has about 5.3 million people. To the far north lies Islamabad, the national capital. It is a well-planned, modern city of 200,000 citizens. The government of Pakistan built Islamabad to draw people inland from crowded coastal areas.

Most people in Pakistan's cities are factory workers, shopkeepers, and craftworkers. They live in crowded neighborhoods. Many come from the countryside and keep close ties to their villages. Wealthier city dwellers live in modern homes and follow modern ways of life. In recent years, women in Pakistan have become active in politics. In 1988 Benazir Bhutto (BEHN•uh•zihr BOO•toh) became prime minister of Pakistan. She was the first woman to head a modern Muslim country.

Location
Bangladesh

The South Asian republic of Bangladesh has an area of 50,260 square miles (130,173 sq. km)—about the size of the state of Wisconsin. The map on page 586 shows that Bangladesh is nearly surrounded by India. Although Bangladesh is an Islamic nation, it shares many cultural features with eastern India.

From 1947 to 1971, Bangladesh was part of Pakistan, but separated from the rest of Pakistan by India. In 1971 Bangladesh gained its independence from Pakistan after a nine-month civil war.

The Land Flat, low plains cover most of Bangladesh. Two major rivers—the Brahmaputra (BRAH•muh•POO•truh) and the Ganges—flow through these plains. Find these rivers on the map on page 588. The people of Bangladesh depend on the rivers for transportation and for farming. The rivers often overflow their banks, leaving fertile soil. These deposits have formed a **delta,** or land formed by mud and sand at the mouth of a river. The largest river delta in the world has formed in Bangladesh where the Brahmaputra and Ganges rivers flow into the Indian Ocean.

Bangladesh has tropical and humid subtropical climates. As in India, the monsoons affect Bangladesh. During the summer monsoons, **cyclones** may cause flooding. A cyclone is an intense storm with high winds and heavy rains. Cyclones kill many people and animals and ruin crops.

The Economy Bangladesh's economy depends on farming. Farmers there raise rice, sugarcane, and jute in the wet, fertile soil. Bangladesh produces the world's largest supply of jute. Its thick forests provide **teak,** a wood used for

596

Almost 80 percent of the land in Bangladesh is part of a large delta.

HUMAN/ENVIRONMENT INTERACTION: What is the advantage and one disadvantage of living in a delta area?

shipbuilding and fine furniture. Although it has natural riches, Bangladesh is a struggling country. Its farmers have few modern tools and use outdated farming methods. The country often suffers from disastrous floods and food shortages.

The People Bangladesh—with about 117 million people—is one of the world's most densely populated countries. It has about 2,320 people per square mile (6,000 people per sq. km). About 86 percent of Bangladesh's people live in rural areas. Their homes are made of bamboo with thatched roofs. In recent years, many people have moved to crowded neighborhoods in urban areas to find work in factories. Their most common choice is Dhaka (DA•kuh), Bangladesh's capital and major port.

Most of Bangladesh's people speak the Bengali language, which is also spoken in neighboring India. About 87 percent of the people in Bangladesh are Muslim. Muslim influences are strong in the country's art, literature, and music.

SECTION 2 REVIEW

REVIEWING TERMS AND FACTS
1. **Define the following:** tributary, delta, cyclone, teak.
2. **LOCATION** Describe the landforms of Pakistan.
3. **PLACE** How have rivers affected life in Bangladesh?
4. **REGION** What is the major religion of Pakistan and Bangladesh?

MAP STUDY ACTIVITIES
5. Turn to the climate map on page 595. What are the three major climate regions of Pakistan?

Answers to Section 2 Review
1. All vocabulary words are defined in the Glossary.
2. Pakistan has deserts, mountains, plains, and plateaus.
3. The people of Bangladesh depend on rivers for transportation and farming.
4. Islam
5. steppe, desert, highland

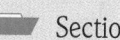
597

MAKING CONNECTIONS

TEACH

Ask students to recall other mathematicians they have studied during this course. If necessary, refer them back to the **Making Connections** feature on page 460. Point out that the Arabs used and popularized many of the mathematical ideas developed by Indian scholars. L1

More About . . .

Mathematics The Hindu scholar Aryabhata, who lived in the A.D. 400s, made many contributions to the study of astronomy and mathematics. He suggested that the earth was round, rotated on an axis, and revolved around the sun. He also calculated that *pi* had a value of 3.1416. *Pi* is the number that represents the ratio of a circle's circumference to its diameter.

MAKING IT COUNT

| MATH | SCIENCE | HISTORY | LITERATURE | TECHNOLOGY |

Have you ever stopped to think where the number system you use came from? Centuries ago, Hindu scholars in India made great discoveries in mathematics.

NUMBER SYSTEM As early as 300 B.C., Hindu scholars came up with a number system based on 10. This number system used just 9 symbols, or numerals: 1, 2, 3, 4, 5, 6, 7, 8, 9. The strength of their number system lay in the idea of *place value*. The same symbol could represent different amounts, depending on its place in a number. For example, the symbol "3" represents 3 ones in the number 13, and 30 ones in the number 31. The Hindu scholars discovered that an infinitely large number could be represented using only 9 symbols.

ZERO About A.D. 600, Hindu thinkers invented a tenth symbol, *sunya,* which means "empty." This symbol, which became known as zero, was used to fill an empty place in a number. Until the invention of *sunya,* Hindus wrote a word to show the place value of each numeral above 99. For example, the number we would write as 306 was written "3 *sata* 6."

Zero ended the need for written place names, so "3 *sata* 6" became "306."

From Zero to Microchips

SPREAD OF NUMBERS For several centuries, only well-educated Hindus used the number system. Later, Arabs from Southwest Asia began using the Hindu system and found it much easier than their own. They adopted the system and eventually brought it to Europe. There it became known as the Arabic-Hindu number system. It replaced the long, clumsy system of numerals that had been used by the Romans.

Making the Connection

1. Hindu scholars developed 9 symbols and what special new symbol as part of their number system?
2. What people later used the Hindu number system?

Answers to Making the Connection

1. zero
2. Arabs and Europeans

The Himalayan Countries

PREVIEW

Words to Know
• dzong

Places to Locate
• Nepal
• Bhutan
• Mount Everest
• Kathmandu
• Thimphu

Read to Learn . . .
1. why Nepal and Bhutan were historically isolated.
2. how the people of Nepal and Bhutan earn their livings.
3. what influence religion has on the Himalayan countries.

I n the Himalayas, north and east of India, you will find two small kingdoms—Nepal and Bhutan. The people of these

Himalayan countries visit street markets like this one every day to shop for food and visit with their neighbors.

Nepal and Bhutan both lie hidden among the towering peaks of the Himalayas. For centuries their mountain location kept them isolated. Today, airplanes and roads open up Nepal and Bhutan to the rest of the world.

Place
Nepal

A well-known myth of the Himalayas tells about a gigantic, apelike beast that roams the mountainous wilderness of Nepal. This "abominable snowman" seems almost too large for the country of Nepal, a small kingdom of only 52,820 square miles (136,804 sq. km)—an area about the size of North Carolina.

The Land The map on page 588 shows you that the Himalayas cover about 80 percent of Nepal's land area. The Himalayas are actually three mountain chains running side by side. Steep river valleys cut through the ice and snow of these mountains. You will find 8 of the 10 highest mountains in the world in the Himalayas of Nepal. Mount Everest is the highest peak at 29,028 feet (8,848 m).

Hills and valleys lie south of the Himalayas. Thick forests of trees and bamboo grasses grow in this region. A flat fertile river plain runs along Nepal's

CHAPTER 22

599

Classroom Resources for Section 3

 REPRODUCIBLE MASTERS
Reproducible Lesson Plan 22-3
Guided Reading Activity 22-3
Cooperative Learning Activity 22
Vocabulary Activity 22
Chapter Map Activity 22
Section Quiz 22-3

 TRANSPARENCIES
Section Focus Transparency 22-3
Geoquiz Transparencies 22-1, 22-2

MULTIMEDIA
Testmaker

National Geographic Society: Picture Atlas of the World

Meeting National Standards

Geography for Life The following standards are highlighted in this section:
• Standards 3, 4, 6, 16

FOCUS

Section Objectives
1. **Understand** why Nepal and Bhutan were historically isolated.
2. **State** how the people of Nepal and Bhutan earn their livings.
3. **Establish** what influence religion has on the Himalayan countries.

Bellringer Motivational Activity

Prior to taking roll at the beginning of the class period, project Section Focus Transparency 22-3 and have students answer the activity questions. Discuss students' responses. **L1**

Vocabulary Pre-check
Have students compare dzongs in Bhutan to religious education centers in their community. **L1** **LEP**

TEACH

Guided Practice
Applying Information Explain that the temperature drops 4°F for every 1,000 feet of elevation. Then ask students to calculate the temperature at the top of Mount Everest if the temperature at sea level is 80°F (26.7°C). *(−36°F, or −38°C)* **L1**

ASSESS

Check for Understanding

Assign Section 3 Review as homework or an in-class activity.

Meeting Lesson Objectives

Each objective below is tested by the questions that follow it in parentheses.

1. **Understand** why Nepal and Bhutan were historically isolated. (2)

2. **State** how the people of Nepal and Bhutan earn their livings. (4)

3. **Establish** what influence religion has on the Himalayan countries. (1)

MAP STUDY

Answer
hydroelectric power

Map Skills Practice
Reading a Map What kinds of agriculture are practiced in Nepal?
(nomadic herding, subsistence farming)

---

southern border with India. The plain includes farmland, rain forests, and swamps. Animals such as tigers and elephants roam these forests.

Differences in elevation affect Nepal's climate. Mountainous areas have long, harsh winters and short, cool summers. Valley areas have a cool climate with heavy summer rains. The plains in the south have a humid subtropical climate.

The Economy Nepal's economy depends almost entirely on farming. The farmers of Nepal grow enough wheat, barley, rice, and potatoes to feed themselves and their families. At markets farmers trade surplus crops for other items.

With few railroads or roads, Nepal carries on limited trade with the outside world. Herbs, jute, rice, spices, and wheat are exported to India. In return Nepal imports gasoline, machinery, and textiles. The country's rugged mountains, which attract thousands of climbers and hikers each year, have helped build a growing tourist industry.

The People Nepal has a population of 22.1 million. Most of its people are related to peoples in northern India and in Tibet to the north. One group in Nepal—the Sherpa—are known for their skills as mountain guides. About 92 percent of Nepal's people live in rural villages. A growing number of Nepal's people live in Kathmandu, the capital and largest city. Most Nepalese follow a form of Hinduism mixed with Buddhist practices.

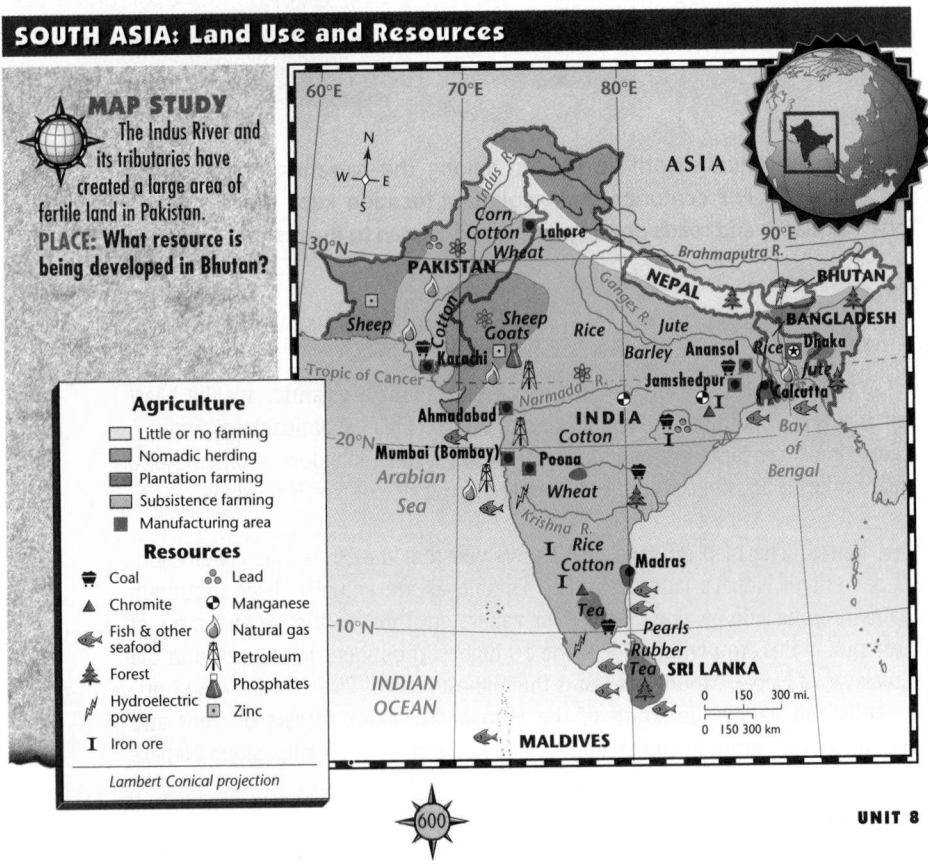

SOUTH ASIA: Land Use and Resources

MAP STUDY
The Indus River and its tributaries have created a large area of fertile land in Pakistan.
PLACE: What resource is being developed in Bhutan?

Agriculture
- ☐ Little or no farming
- ▨ Nomadic herding
- ■ Plantation farming
- ☐ Subsistence farming
- ■ Manufacturing area

Resources
- 🜨 Coal
- ▲ Chromite
- 🐟 Fish & other seafood
- 🌲 Forest
- ⚡ Hydroelectric power
- I Iron ore
- ⚫ Lead
- ● Manganese
- Natural gas
- Petroleum
- Phosphates
- Zinc

Lambert Conical projection

Cooperative Learning Activity

Have two pairs of students write a script for guides on tour buses—the first bus will cross Nepal, while the second will cross Bhutan. Suggest that writers describe their assigned country's landforms, climate, plants, animals, and people. Encourage the writers to find facts in the text and other sources to give their descriptions "local color." Also remind them to add fun facts and humor. Then line up chairs as if they were seats on a bus and direct the students in the class to take their seats. Ask the first pair of guides to stand at the head of the "bus" and to conduct the tour of Nepal. Then have the second pair lead the class on the tour of Bhutan. Encourage the "tourists" to ask questions. **L1**

Place
Bhutan

East of Nepal lies the even smaller kingdom of Bhutan. The map on page 586 shows you that a small part of India separates Bhutan from Nepal. Bhutan has a land area of about 18,150 square miles (47,009 sq. km)—about the size of New Hampshire and Vermont put together.

The Land and Economy As in Nepal, the Himalayas are the major landform of Bhutan. They rise more than 24,000 feet (7,320 m) in many places. Mountainous areas of the country have snow and ice all year. In the foothills of the Himalayas the climate is mild. Thick forests cover much of this area. To the south along Bhutan's border with India lies an area of subtropical plains and river valleys.

Most of Bhutan's people are subsistence farmers. They live in the fertile mountain valleys and grow barley, rice, and wheat. In the severe climate of the mountain areas people herd cattle and yaks, or longhaired oxen. With India's help, Bhutan is developing its economy. It has set up commercial farms and built hydroelectric plants.

The People Bhutan has about 800,000 people. Most of them speak the Dzongkha dialect. Many live in rural villages that dot valleys and plains in the southern part of the country. Thimphu, the national capital, is in this area.

Bhutan once was called the "hidden holy land." Bhutan's people are not as isolated as they used to be. They are deeply loyal to the Buddhist religion, which teaches that people can find peace from life's troubles by living simply, doing good deeds, and praying. An Indian holy man named Siddhartha Gautama first preached Buddhism in India in the 500s B.C. Buddhism later spread eastward and northward to other parts of Asia. In Bhutan, Buddhist centers of prayer and study called **dzongs** have shaped the country's art and culture.

The Sherpas of Nepal

The Sherpas of northern Nepal serve as guides in the Himalayas.
LOCATION: What is the highest peak in the Himalayas?

SECTION 3 REVIEW

REVIEWING TERMS AND FACTS
1. Define the following: dzong.
2. PLACE What climate areas are found in Nepal?
3. LOCATION Where do most people in Nepal and Bhutan live?

MAP STUDY ACTIVITIES
4. Turn to the land use map on page 600. What natural resource is found in eastern Nepal?

Answers to Section 3 Review
1. All vocabulary words are defined in the Glossary.
2. highland, humid subtropical
3. in the southern parts of the countries
4. forest

LESSON PLAN
Chapter 22, Section 3

More About the Illustration

Answer to Caption
Mount Everest

Human/ Environment Interaction The Sherpas are farmers by tradition. Since the 1950s, however, more and more Sherpas have made a living by serving as guides and porters for the many climbing expeditions to the Himalayas.

Evaluate

Section 3 Quiz **L1**

Use the Testmaker to create a customized quiz for Section 3. **L1**

Reteach

Have students make up questions based on the section subheadings. Encourage students to quiz each other with their questions. **L1**

Enrich

Ask students to research the major expeditions to Mount Everest and to arrange their findings chronologically on an illustrated time line. **L3**

CLOSE

Write HIMALAYAN COUNTRIES vertically on the chalkboard. Have students create an acrostic by writing a different geography-related term from the section for each letter.

SECTION 4 Island Countries

FOCUS

Section Objectives

1. **Find** where Sri Lanka and the Maldives are located.
2. **Identify** the major products grown in Sri Lanka.
3. **Describe** what formed the islands of the Maldives.

Bellringer Motivational Activity

Prior to taking roll at the beginning of the class period, project Section Focus Transparency 22-4 and have students answer the activity questions. Discuss students' responses. **L1**

Vocabulary Pre-check

Pose this riddle: Is a lagoon inside an atoll or is an atoll inside a lagoon? Have students read the section to find the answer. **L1** **LEP**

TEACH

Guided Practice

Classifying Information On the chalkboard, draw a chart with the following column headings: Natural Resources, Agricultural Products, and Manufactured Goods. Have students read the text concerning Sri Lanka's economy on page 603. Call on volunteers to enter Sri Lanka's products in the appropriate column. **L1**

PREVIEW

Words to Know
• atoll
• lagoon

Places to Locate
• Sri Lanka
• Maldives
• Colombo

Read to Learn . . .
1. where Sri Lanka and the Maldives are located.
2. what major products are grown in Sri Lanka.
3. what formed the islands of the Maldives.

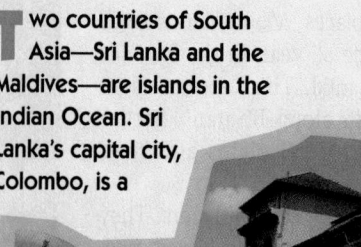

Two countries of South Asia—Sri Lanka and the Maldives—are islands in the Indian Ocean. Sri Lanka's capital city, Colombo, is a major seaport. Most of its people are Buddhists.

Sri Lanka and the Maldives lie south of India in the Indian Ocean. The sea affects the ways in which the peoples of both island countries live and earn their livings.

Place

Sri Lanka

Pear-shaped Sri Lanka lies about 20 miles (32 km) off the southeastern coast of India. Covering 24,950 square miles (64,621 sq. km), Sri Lanka is about the size of West Virginia. Sri Lanka, located on an important ocean route between Africa and Asia, became a natural stopping place for traders. Beginning in the 1500s, Sri Lanka—then known as Ceylon—came under the control of European countries. The British ruled the island from 1802 to 1948, when it became independent once again. In 1972 Ceylon took the name Sri Lanka, an ancient term meaning "brilliant land."

Sri Lanka is a land of white beaches and thick forests. Much of the country along the coast is rolling lowlands. Highlands cover the center. Rivers flow down the low mountains, providing irrigation for crops.

602

Classroom Resources for Section 4

REPRODUCIBLE MASTERS
Reproducible Lesson Plan 22-4
Guided Reading Activity 22-4
Performance Assessment
  Activity 22
Reteaching Activity 22
Enrichment Activity 22
Section Quiz 22-4

TRANSPARENCIES
Section Focus
  Transparency 22-4
Political Map Transparency 8

MULTIMEDIA
Testmaker

National Geographic
  Society: ZipZapMap!
  World

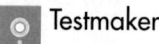

The Economy Farming is Sri Lanka's major economic activity. Many farmers grow rice and other food crops in lowland areas. Water buffaloes help to plow their fields. In higher elevations, tea, rubber, and coconuts grow on large plantations. Sri Lanka is one of the world's leading producers of tea and rubber.

Sri Lanka, rich in natural resources, is famous for its sapphires, rubies, and other gemstones. Sri Lanka's forests contain many valuable woods such as ebony and satinwood, and a variety of birds and animals. To protect wildlife, the government of Sri Lanka has set aside land for national parks.

Sri Lanka is developing its industrial economy. Factories in Sri Lanka's cities produce textiles, fertilizers, cement, leather products, and wood products for export. Colombo, the capital city, is a bustling port on the country's west coast.

The People About 18 million people live in Sri Lanka. They belong to two major ethnic groups, the Sinhalese (SIHN•guh•LEEZ) and the Tamils (TA•muhlz). Forming about 70 percent of the population, the Sinhalese live in the southern and western parts of the island. They speak Sinhalese and are Buddhists. The Tamils make up about 20 percent of the population. They live in the north and east, speak Tamil, and are Hindus. Some Tamils have fought to set up a separate country in northern Sri Lanka.

Whatever their ethnic background, people in Sri Lanka value their history and the arts—still visible in the remains of ancient cities. These remains hold

Independent Practice

📁 Guided Reading Activity 22-4 **L1**

ASSESS

Check for Understanding

Assign Section 4 Review as homework or an in-class activity.

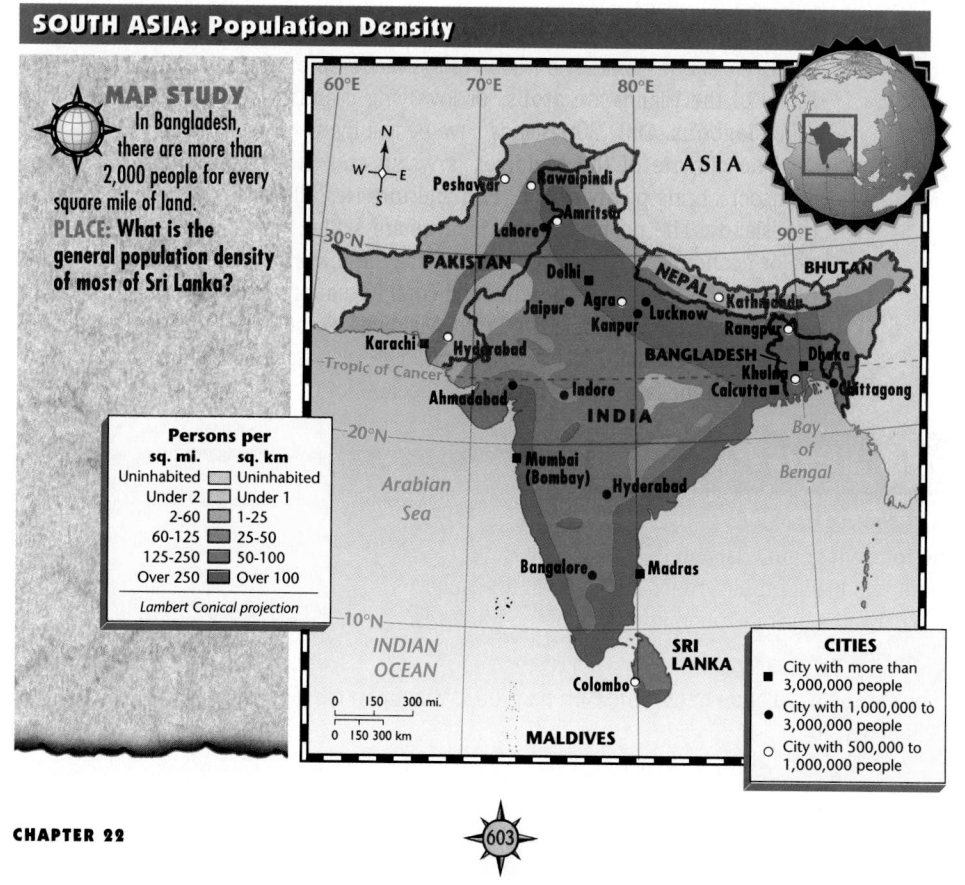

SOUTH ASIA: Population Density

MAP STUDY
In Bangladesh, there are more than 2,000 people for every square mile of land.
PLACE: What is the general population density of most of Sri Lanka?

Persons per

sq. mi.	sq. km
Uninhabited	Uninhabited
Under 2	Under 1
2-60	1-25
60-125	25-50
125-250	50-100
Over 250	Over 100

Lambert Conical projection

0   150   300 mi.
0 150 300 km

CITIES
■ City with more than 3,000,000 people
● City with 1,000,000 to 3,000,000 people
○ City with 500,000 to 1,000,000 people

603

Meeting Lesson Objectives

Each objective below is tested by the questions that follow it in parentheses.

1. **Find** where Sri Lanka and the Maldives are located. (4)
3. **Describe** what formed the islands of the Maldives. (1)

MAP STUDY

Answer
125–250 persons per square mile, or 50–100 persons per sq. km
Map Skills Practice
Reading a Map Which part of Sri Lanka is most densely populated, the north or the south? *(south)*

Cooperative Learning Activity

Organize students into a number of groups and inform them that their task is to create an illustrated pamphlet titled *The Story of Tea in Sri Lanka.* Point out that the purpose of the pamphlet is to teach students in the school's lower grades about this important Sri Lankan product.

Suggest that group members share various tasks—research, writing, artwork, design, and so on—among themselves. Display finished pamphlets in the library or the classroom resource center and encourage lower grade classes to study them. **L2**

Evaluate

 Section 4 Quiz **L1**

Use the Testmaker to create a customized quiz for Section 4. **L1**

Chapter 22 Test Form A and/or Form B **L1**

Reteach

 Reteaching Activity 22 **L1**

Enrich

 Enrichment Activity 22 **L3**

CLOSE

 **Mental Mapping Activity**

Provide each student with a large sheet of white paper and map pencils. Inform students that this will not be a graded activity. Have them draw, freehand from memory, an outline map of South Asia, Sri Lanka, and the Maldives that includes the following labeled cities and bodies of water: Colombo, Male, Indian Ocean, Bay of Bengal, and Arabian Sea.

Island Countries

Workers harvest tea leaves on large plantations in Sri Lanka's highlands *(left)*. Fish are the most important food and export in the Maldives *(right)*. **LOCATION: Which of these two island countries is closer to India?**

many old palaces, temples, and statues of Buddha. Sri Lankan craftspeople also take pride in their carved wooden masks, brass work, and handmade cloth.

Region

Maldives

About 370 miles (595 km) south of India lie the Maldives, the smallest country of South Asia. The Maldives are made up of about 1,200 coral islands. Many of the islands are **atolls,** or low-lying islands that circle pools of water called **lagoons.** Only 200 islands have people living on them.

The climate of the Maldives is warm and humid throughout the year. Monsoons bring plenty of rain. Fish and tortoises fill the tropical waters around the islands. The people of the Maldives are skilled sailors, and fishing is their major economic activity. Fish and rice are the major foods. In recent years, the islands' palm-lined sandy beaches and coral formations have attracted tourists.

About 250,000 people live in the Maldives. Some 30,000 make their homes in Male, the capital. Most are Muslims. The islands, which came under British rule during the late 1800s, became independent in 1965.

SECTION 4 REVIEW

REVIEWING TERMS AND FACTS
1. **Define the following:** atoll, lagoon.
2. **PLACE** What two major ethnic groups are found in Sri Lanka?
3. **REGION** What is the major economic activity of the Maldives?

 MAP STUDY ACTIVITIES
4. Look at the population density map on page 603. What part of Sri Lanka has the highest population density?

 604

UNIT 8

Answers to Section 4 Review
1. All vocabulary words are defined in the Glossary.
2. Sinhalese and Tamil
3. fishing
4. the southern coast

Important Things to Know About South Asia

SECTION 1 INDIA

- India is the largest country of South Asia in size and population.
- The Himalayas and the monsoons, or seasonal winds, affect India's climate.
- India's economy is based on farming and industry.
- Although the peoples of India belong to many religious and language groups, most are Hindus.

SECTION 2 MUSLIM SOUTH ASIA

- Islam has shaped the cultures of Pakistan and Bangladesh.
- Pakistan is made up of mountains, deserts, and fertile river valleys.
- The Ganges and Brahmaputra rivers form a fertile delta area in Bangladesh.
- Monsoons bring plenty of rain to Bangladesh's low, fertile plains.

SECTION 3 THE HIMALAYAN COUNTRIES

- The Himalayas are the major landform of Nepal and Bhutan.
- Most people in Nepal and Bhutan are farmers or herders. They live in the southern parts of their countries.
- The Hindu and Buddhist religions have shaped the cultures of Nepal and Bhutan.

SECTION 4 ISLAND COUNTRIES

- Sri Lanka and the Maldives are island republics in the Indian Ocean.
- Sri Lanka is one of the world's major tea-producing countries.
- The Maldives are made up of about 1,200 coral islands. Only 200 are inhabited.

Festival market in Rajasthan, India ▶

CHAPTER HIGHLIGHTS

Using the Chapter 22 Highlights

- Use the Chapter 22 Highlights to preview, review, condense, or reteach the chapter.

Preview/Review

Vocabulary Puzzle-Maker Software reinforces the terms used in Chapter 22.

Condense

Have students read the Chapter 22 Highlights. Spanish Chapter Highlights also are available.

Chapter 22 Digest Audiocassettes and Activity

Chapter 22 Digest Audiocassette Test

Spanish Chapter Digest Audiocassettes, Activities, and Tests also are available.

Guided Reading Activities

Reteach

Reteaching Activity 22. Spanish Reteaching activities also are available.

MAP STUDY

Place Copy and distribute page 34 from the Outline Map Resource Book. Have students label India, Pakistan, Bangladesh, Nepal, Bhutan, Sri Lanka, and the Maldives. Then have them draw symbols to show the countries' natural resources. Remind them to add a key that explains the symbols.

Extra Credit Project

Model Encourage students to research and recreate in miniature a tropical hardwood forest of Bangladesh, Bhutan, or Sri Lanka. Encourage students to reproduce the forest's flora and fauna in clay, paper, and cardboard, and to make the model as detailed as possible. **L3**

Chapter 22 Review and Activities

ANSWERS

Reviewing Key Terms

1. G
2. C
3. I
4. B
5. A
6. H
7. J
8. E
9. F
10. D

 Mental Mapping Activity

This exercise helps students to visualize the countries and geographic features they have been studying and understand the relationships among them. All attempts at freehand mapping should be accepted.

REVIEWING KEY TERMS

Match the numbered terms in Column A with their definitions in Column B.

A
1. cottage industry
2. tributary
3. jute
4. atoll
5. subcontinent
6. monsoon
7. lagoon
8. delta
9. teak
10. cyclone

B
A. landmass like a continent, only smaller
B. low-lying island circling a pool of water
C. small river that flows into a larger river
D. heavy rainstorm with high winds
E. land formed by mud and sand deposited at the mouth of a river
F. wood used for shipbuilding and fine furniture
G. where family members use their own equipment to create a product
H. seasonal wind
I. plant fiber used in making burlap bags
J. pool of water circled by an island

Mental Mapping Activity

On a separate piece of paper, draw a freehand map of South Asia. Label the following items on your map:

- Calcutta
- Arabian Sea
- Hindu Kush
- Bay of Bengal
- Bhutan
- Deccan Plateau
- Indian Ocean

REVIEWING THE MAIN IDEAS

Section 1
1. LOCATION Where is the Deccan Plateau located?
2. REGION What major river flows through northern India?
3. PLACE How would you describe life in a typical Indian city?

Section 2
4. PLACE What large city in Pakistan lies on the coast of the Arabian Sea?
5. HUMAN/ENVIRONMENT INTERACTION What has helped increase food production in Pakistan?
6. MOVEMENT What is the major means of transportation in Bangladesh?

Section 3
7. PLACE What climates are found in Nepal?
8. PLACE Why was Bhutan once called the "hidden holy land"?

Section 4
9. HUMAN/ENVIRONMENT INTERACTION What agricultural products are grown on plantations in Sri Lanka?
10. PLACE What is the major religion of the Maldives?

Reviewing the Main Ideas

Section 1

1. It is found in south central India between the Vindhya Range and the Ghats.
2. Ganges River
3. Indian cities are very crowded; they are a mix of rich and poor, of modern and old.

Section 2

4. Karachi
5. use of irrigation
6. rivers

Section 3

7. highland, moderate, and tropical climates
8. It is loyal to Buddhism and was, for many years, isolated from the rest of the world by mountains.

CRITICAL THINKING ACTIVITIES

1. Drawing Conclusions Why do you think the description "brilliant island" applies to Sri Lanka?

2. Cause and Effect How have the environments of Nepal and Bhutan affected the way the countries' people live?

GeoJournal Writing Activity

Imagine what it would be like to visit the Himalayas. Write about your experiences to a friend. Describe the types of transportation that you might use and the scenes that you might see on your journey.

COOPERATIVE LEARNING ACTIVITY

Work in groups of three to learn more about the major religions of South Asia: Hinduism, Islam, and Buddhism. Choose one of the religions and research how it has affected one of the following areas: (a) the arts; (b) diet; and (c) family life. After your research is complete, share your information with your group. As a group, prepare a chart and a poster that present the group's findings.

PLACE LOCATION ACTIVITY: SOUTH ASIA

Match the letters on the map with the places and physical features of South Asia. Write your answers on a separate sheet of paper.

1. Karachi
2. Himalayas
3. New Delhi
4. Ganges River
5. Mumbai (Bombay)
6. Bangladesh
7. Thar Desert
8. Sri Lanka
9. Indus River
10. Western Ghats

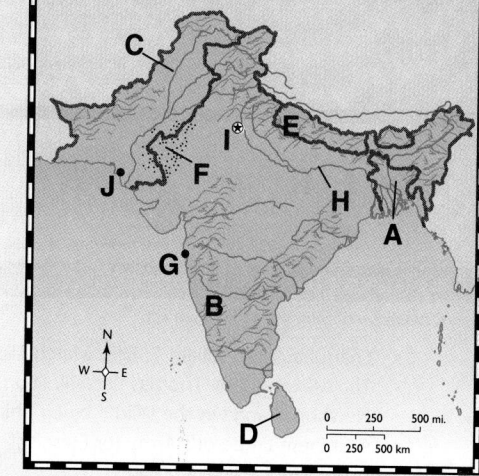

Cooperative Learning Activity

Tell students to copy art influenced by the major religions and to display the copies on their posters. Also suggest that they find recipes of dishes eaten for religious reasons and distribute copies of the recipes to the rest of the class.

 **GeoJournal Writing Activity**

Encourage students to compare their journeys.

Place Location Activity

1. J
2. E
3. I
4. H
5. G
6. A
7. F
8. D
9. C
10. B

 Chapter Bonus Test Question

This question may be used for extra credit on the chapter test.

What accounts for India having a dry season and a wet season? *(The monsoon winds, which blow from the land in the dry season and from the sea in the wet season.)*

Section 4

9. tea, rubber, coconuts
10. Islam

Critical Thinking Activities

1. Answers will vary, but might mention the scenic beauty and mild climate of the island, as well as its rich cultural heritage.

2. The Himalayas have, until recently, so isolated Nepal and Bhutan that these countries had little contact with the outside world. Farming land is limited because of the high mountains. Transportation is difficult, but groups such as the Sherpa have developed mountain skills and can carry heavy loads at high altitudes.

PERFORMANCE ASSESSMENT ACTIVITY

AMERICAN BUSINESS IN CHINA Point out that a number of American businesses have opened franchises or divisions on the Chinese mainland. Have students assume the roles of experts on American business opportunities in China. Direct students to work in small groups to research several large U.S. corporations. Then have groups select three corporations to recommend for expansion to China. Choices should be based on the economic needs of China. Encourage groups to develop brief presentations to be made either to the corporations or to Chinese government officials. Presentations should identify best locations on the Chinese mainland for the new venture and should be accompanied by appropriate illustrations. Call on groups to make their presentations to the class.

POSSIBLE RUBRIC FEATURES:
Accuracy of content information, ability to recognize cause/effect relationships, decision-making skills, research skills, visual quality of illustrative materials, collaborative skills

MENTAL MAPPING ACTIVITY

Before teaching Chapter 23, point out China on a globe to students. Have them identify the hemispheres in which China is located and where it is located relative to the United States.

TEACHER'S CORNER

NATIONAL GEOGRAPHIC SOCIETY

INDEX TO NATIONAL GEOGRAPHIC MAGAZINE

The following articles may be used for research relating to this chapter:

- "China's Terra-cotta Warriors," by O. Louis Mazzatenta, October 1996.
- "Dinosaurs of the Gobi," by Donovan Webster, July 1996.
- "Pilgrimage to Buddhist Caves," by Reza, April 1996.
- "Xinjiang," by Thomas B. Allen, March 1996.
- "The Silk Road's Lost World," by Thomas B. Allen, March 1996.
- "New Hope for China's Pandas," by Pan Wenshi, February 1995.
- "Shanghai," by William S. Ellis, March 1994.
- "The Mekong," by Thomas O'Neill, February 1993.
- "Newborn Panda in the Wild," by Lü Zhi, February 1993.
- "A Chinese Emperor's Army for Eternity," by O. Louis Mazzatenta, August 1992.

NATIONAL GEOGRAPHIC SOCIETY PRODUCTS AVAILABLE FROM GLENCOE

To order the following products for use with this chapter, contact your local Glencoe sales representative or call Glencoe at 1-800-334-7344:

- *STV: World Geography* (Videodisc)
- *Picture Atlas of the World* (CD-ROM)
- *Ancient Civilizations: China and India* (CD-ROM)
- *Physical Geography of the World* (Transparencies)
- *ZipZapMap! World* (Software)
- *GeoBee* (Software)
- *Images of the World* (Posters)
- *Eye on the Environment* (Posters)

ADDITIONAL NATIONAL GEOGRAPHIC SOCIETY PRODUCTS

To order the following products for use with this chapter, call National Geographic Society at 1-800-368-2728:

- *China: Sichuan Province* (Video)
- *China: Beyond the Clouds* (Video)
- *Hong Kong: A Family Portrait* (Video)
- *Save the Panda* (Video)
- *Secrets of the Wild Panda* (Video)
- *Physical Geography of the Continents Series:* "Asia." (Video)
- *The People's Republic of China,* "Introducing China Today," "Living and Working in China Today," "Chinese Arts and Culture." (Filmstrip)

TEACHER-TO-TEACHER
Curriculum Connection: Earth Science

—from Ted Henson
Alamance County Schools, Graham, NC

TERRACE FIELDS *The purpose of this activity is to help students understand terrace farming in southern China by having them construct a model of hillside terrace fields.*

Inform students that the farmers of southern China increase the amount of land available for cultivation by building terraces on the mountain slopes.

Demonstrate the construction of hillside terraces in the following fashion: create mountains by crumpling up sheets of newspaper and taping them to a piece of board to form peaks and slopes. Use clay to construct terraces on several slopes. Using strips of paper and wallpaper paste, cover the entire model with papier maché. After the model has dried, paint the various areas appropriate colors, making sure to paint the tops of the terrace soil brown. Place "plants" on the terraces using toothpicks and crepe paper. If time allows, organize students into several groups and have groups build their own models of hillside terraces.

As a follow-up activity, have students discuss the following questions: How does the process of terracing increase the amount of land available for cultivation? Would the mountain slopes be suitable for cultivation without terracing? Why or why not? How has terrace farming helped to improve the lives of the people of southern China?

MEETING NATIONAL STANDARDS

Geography for Life

All of the 18 standards are demonstrated in Unit 8. The following ones are highlighted in this chapter:

- Standards 1, 2, 3, 4, 5, 6, 9, 10, 11, 12, 13, 14, 15, 16, 17, 18

KEY TO ABILITY LEVELS

Teaching strategies have been coded for varying learning styles and abilities.

L1 **BASIC** activities for all students

L2 **AVERAGE** activities for average to above-average students

L3 **CHALLENGING** activities for above-average students

LEP **LIMITED ENGLISH PROFICIENCY** activities

BIBLIOGRAPHY

Readings for the Student

Carter, Alden R. *China Past—China Future.* New York: Franklin Watts, 1994.

Kalman, Bobbie. *China: The Lands, Peoples, and Cultures Series.* New York: Crabtree, 1989. Three 32-page paperbacks.

Keeler, Stephen. *Passport to China.* New York: Franklin Watts, 1994.

Ogden, Suzanne. *China: Global Studies,* 6th ed. Guilford, Conn.: Dushkin, 1995.

Readings for the Teacher

China: A Teaching Workbook, revised ed. New York: East Asian Curriculum Project at Columbia University, 1991.

Creative Activities for Teaching About China. Stockton, Calif.: Stevens & Shea, 1988. Reproducible activity sheets.

Seybolt, Peter J., ed. *Through Chinese Eyes: The Living Tradition,* revised ed., New York: Center for International Training and Education, 1988.

Multimedia

Assignment: China. Boston: Christian Science Monitor Video, 1993. Videocassette, 60 minutes.

China: Sichuan Province. Washington, D.C.: National Geographic Society, 1992. Videocassette, 25 minutes.

Discovering China and Tibet, revised ed., New York: Video Visits, 1990. Videocassette, 52 minutes.

Use these *Geography: The World and Its People* resources to teach, reinforce, and extend chapter content.

CHAPTER 23 RESOURCES

📁 *Vocabulary Activity 23

📁 Cooperative Learning Activity 23

📁 Workbook Activity 23

📁 Chapter Map Activity 23

📁 Geography Skills Activity 23

📁 GeoLab Activity 23

📁 Critical Thinking Skills Activity 23

📁 Performance Assessment Activity 23

📁 *Reteaching Activity 23

📁 Enrichment Activity 23

📁 Chapter 23 Test, Form A and Form B

🎧 📁 *Chapter 23 Digest Audiocassette, Activity, Test

🖋 Unit Overlay Transparencies 8-0, 8-1, 8-2, 8-3, 8-4, 8-5, 8-6

🖋 Political Map Transparency 8; World Cultures Transparency 15

🖋 Geoquiz Transparencies 23-1, 23-2

💿 Vocabulary PuzzleMaker Software

💿 Student Self-Test: A Software Review

💿 Testmaker Software

💿 MindJogger Videoquiz

If time does not permit teaching the entire chapter, summarize using the Chapter 23 Highlights on page 627, and the Chapter 23 English (or Spanish) Audiocassettes. Review students' knowledge using the Glencoe MindJogger Videoquiz. *Also available in Spanish

Use these *Geography: The World and Its People* resources to teach and reinforce *section* content.

SECTION 1 RESOURCES

Reproducible Lesson Plan 23-1

Section Focus Transparency 23-1

Guided Reading Activity 23-1

Section Quiz 23-1 (also available in Spanish)

SECTION 2 RESOURCES

Reproducible Lesson Plan 23-2

Section Focus Transparency 23-2

Guided Reading Activity 23-2

Section Quiz 23-2 (also available in Spanish)

SECTION 3 RESOURCES

Reproducible Lesson Plan 23-3

Section Focus Transparency 23-3

Guided Reading Activity 23-3

Section Quiz 23-3 (also available in Spanish)

SECTION 4 RESOURCES

Reproducible Lesson Plan 23-4

Section Focus Transparency 23-4

Guided Reading Activity 23-4

Section Quiz 23-4 (also available in Spanish)

ADDITIONAL RESOURCES FROM GLENCOE

Reproducible Masters

- Glencoe Social Studies Outline Map Book, page 34
- Foods Around the World, Unit 9

Workbook

- Building Skills in Geography, Unit 2, Lesson 6

World Music: Cultural Traditions
Lesson 8

Transparencies

- National Geographic Society PicturePack Transparencies, Unit 8

Posters

- National Geographic Society: Images of the World, Unit 8

World Games Activity Cards
Unit 8

Videodiscs

- Geography and the Environment: The Infinite Voyage
- Reuters Issues in Geography
- National Geographic Society: STV: World Geography, Vol. 1

CD-ROM

- National Geographic Society: Picture Atlas of the World

Software

- National Geographic Society: ZipZapMap! World

Performance Assessment

Refer to the Planning Guide on page 608A for a Performance Assessment Activity for this chapter. See the *Performance Assessment Strategies and Activities* booklet for suggestions.

Chapter Objectives

1. **Identify** the landforms, populous areas, and climates of China.
2. **Describe** the past and present Chinese economies.
3. **Evaluate** the influences on Chinese culture and ways of life.
4. **Outline** important details about the economies of Mongolia, Taiwan, and Hong Kong.

GLENCOE
TECHNOLOGY

VIDEODISC

Use the Chapter 23 MindJogger Videoquiz to preview chapter content.

MindJogger Videoquiz

Chapter 23
Disc 3 Side B

  Also available in VHS.

608

Chapter 23 China

MAP STUDY ACTIVITY

As you read Chapter 23, you will learn about China and its neighbors.

1. **What area is a European colony?**
2. **What is the capital of China?**
3. **What is the capital of Taiwan?**

National boundary
⊛ National capital
● Other city
Conic projection

608

 MAP STUDY ACTIVITY

Answers
1. Macao
2. Beijing
3. Taipei

Map Skills Practice
Reading a Map What is the capital of Mongolia? *(Ulan Bator)* What country borders China and Mongolia on the north? *(Russia)*

SECTION 1

The Land

PREVIEW

Words to Know
- loess
- dike
- typhoon

Places to Locate
- China
- Plateau of Tibet
- Kunlun Shan
- Taklimakan
- Gobi
- Huang He
- Chang Jiang
- Xi

Read to Learn . . .
1. why the Plateau of Tibet is called the "Roof of the World."
2. where most of China's people live.
3. what climates are found in China.

The Great Wall of China is one of the wonders of the world. The longest structure ever built, the wall stretches about 4,000 miles (6,437 km). The Chinese built the wall centuries ago to keep out invaders.

The People's Republic of China lies in the eastern part of Asia. It is a vast country that covers more than 3,600,930 square miles (9,326,409 sq. km). China is the world's third-largest country in area, after Russia and Canada.

Region
Landscapes

If you travel by train across China, you will be amazed by the many different landscapes and climates. Look at the map on page 610. You see that China has tall mountains, vast deserts, fertile plains, and mighty rivers.

Mountains Rugged mountains cover about one-third of China. They lie in the western part of the country. The Himalayan range sweeps across the south on the border between China and Nepal. The center of China's mountainous area holds the largest plateau in the world—the Plateau of Tibet. It is called the "Roof of the World" because its average height is about 13,000 feet (4,000 m) above sea level. Scattered shrubs and grasses cover the countryside. The eastern edge of the plateau is home to pandas, golden monkeys, and other rare animals.

CHAPTER 23

609

Meeting National Standards
Geography for Life The following standards are highlighted in this section:
- Standards 1, 2, 4, 15

FOCUS

Section Objectives
1. **Explain** why the Plateau of Tibet is called the "Roof of the World."
2. **Describe** where most of China's people live.
3. **Name** climates found in China.

Bellringer Motivational Activity

 Prior to taking roll at the beginning of the class period, project Section Focus Transparency 23-1 and have students answer the activity questions. Discuss students' responses. **L1**

Vocabulary Pre-check
Pronounce *loess* (LEHS) for the students and have them find its definition in the text before reading the rest of the section. **L1 LEP**

Use the Vocabulary PuzzleMaker Software to create a crossword puzzle. **L1**

Classroom Resources for Section 1

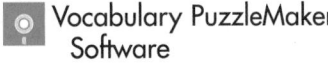

REPRODUCIBLE MASTERS
Reproducible Lesson Plan 23-1
Guided Reading Activity 23-1
Geography Skills Activity 23
Building Skills in Geography, Unit 2, Lesson 6
Section Quiz 23-1

TRANSPARENCIES
Section Focus Transparency 23-1
Political Map Transparency 8

MULTIMEDIA
Vocabulary PuzzleMaker Software
Testmaker

TEACH

Guided Practice

Understanding Information On the chalkboard, draw an outline map of China. As students read through the section, call on volunteers to come to the chalkboard and locate and label China's major landforms and bodies of water on the map. **L1**

Independent Practice

 Guided Reading Activity 23-1 **L1**

MAP STUDY

Answer
Huang He

Map Skills Practice
Reading a Map What desert spreads across parts of China and Mongolia? *(Gobi)*

Deserts In the north of China, you notice that the mountain ranges circl desert basins. One of these deserts is the Taklimakan. It is an isolated regio with some of the world's highest temperatures. Sandstorms there may last fo days and create huge, shifting dunes.

Farther east lies another desert, called the Gobi. The Gobi is made up c rock and gravel instead of sand. Temperatures there can rise to 110°F (43°C during summer days. Because the Gobi is far north and at a high altitude, i temperatures during winter nights can drop as low as –30°F (–34°C)!

Plains and Highlands Plains cover the eastern part of China. Look at th map below. You see that these plains run along the coasts of the South Chin and East China seas. This area has fertile land and is home to almost 90 percer of China's people.

The Northeast Plain lies in the center of Manchuria, a region in northea China. To the east of the plain lies a heavily forested, hilly area near China's bo der with Korea. To the south is the wide, flat North China Plain. Other plair areas extend farther south near the coast.

Inland in the southeast, the land changes from plains to green highland: This region is one of the most scenic areas in China. Many tourists come to se its many scattered limestone hills that rise 100 to 600 feet (30 to 182 m).

Rivers Three of China's major rivers—the Huang He (HWAHNG HUH), th Chang Jiang (CHAHNG jee•AHNG), and the Xi (SHEE)—flow through th

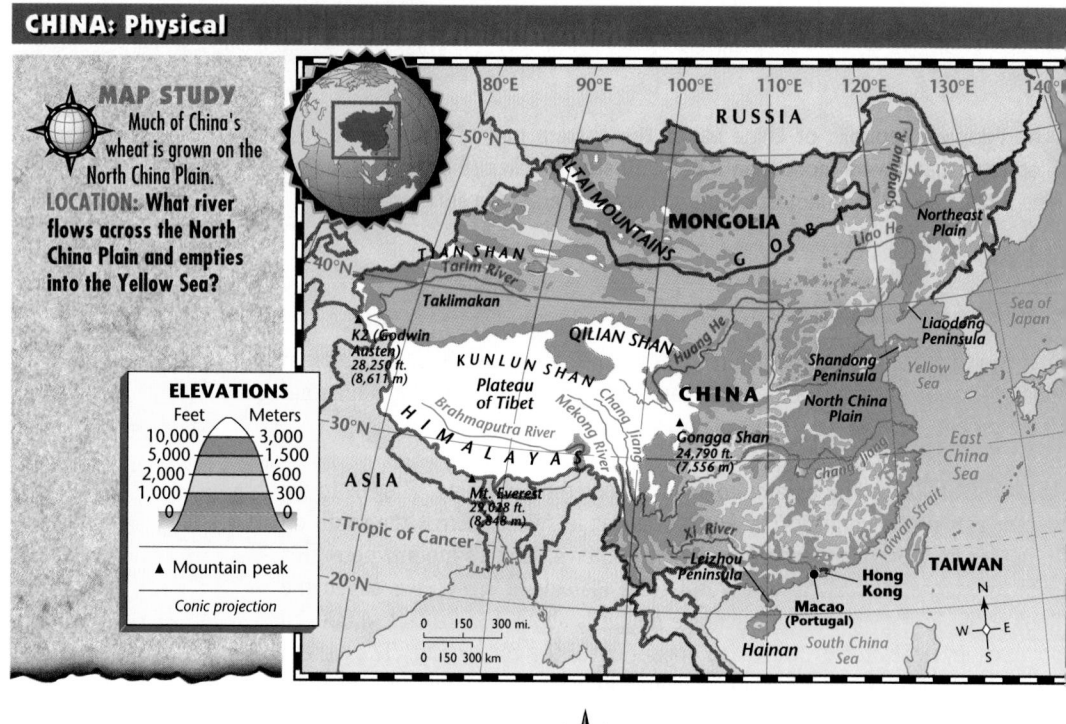

CHINA: Physical

MAP STUDY
Much of China's wheat is grown on the North China Plain.
LOCATION: What river flows across the North China Plain and empties into the Yellow Sea?

ELEVATIONS
Feet	Meters
10,000	3,000
5,000	1,500
2,000	600
1,000	300
0	0

▲ Mountain peak

Conic projection

610

UNIT

Cooperative Learning Activity

Organize students into several groups. Have groups construct wall-size maps of China showing the seven climate regions of China. Then direct groups to use encyclopedias and atlases, as well as the text, to decide what kinds of clothing would be most appropriate for each of the seven climate regions. Have group members draw summer and winter wardrobes for the various climate regions, using clothing shown in pictures in research materials as models. Direct groups to display clothing pictures around their wall maps, running lead lines from the pictures to the appropriate climate regions on the maps. **L1**

CHINA: Climate

LESSON PLAN
Chapter 23, Section 1

MAP STUDY
In southeastern China and northern Taiwan the summers are warm or hot.
PLACE: What type of climate does southeastern China have?

CLIMATE

High Latitude
- ◼ Subarctic
- ☐ Highland

Dry
- ◼ Steppe (partly dry)
- ◻ Desert

Mid-Latitude
- ◼ Humid continental
- ◼ Humid subtropical

Tropical
- ◼ Tropical rain forest

Conic projection

0 150 300 mi.
0 150 300 km

MAP STUDY

Answer
humid subtropical

Map Skills Practice
Reading a Map If you traveled north along 105°N longitude through Mongolia, what climate regions would you pass through? *(desert, steppe, subarctic)*

ASSESS

Check for Understanding
Assign Section 1 Review as homework or an in-class activity.

Meeting Lesson Objectives
Each objective below is tested by the questions that follow it in parentheses.

1. **Explain** why the Plateau of Tibet is called the "Roof of the World." (2)
3. **Name** climates found in China. (4, 6)

---

plains and southern highlands. They serve as important transportation routes and as a source of soil. How? For centuries, floodwaters have deposited rich soil to form flat river basins.

The Huang He basin is an important farming area. It is thickly covered with **loess,** a fertile, yellow-gray soil deposited by wind and water. The name *Huang He* means "yellow river," referring to the large amounts of loess carried by the river.

The rivers of China have brought disasters as well as blessings. The Chinese, for example, call the Huang He "China's sorrow." In the past, its flooding cost hundreds of thousands of lives and caused much damage. To control floods, the Chinese have built dams and **dikes,** or high banks of soil, along the sides of this river. You can read more about the Huang He and ways of preventing soil erosion in the GeoLab on page 630.

Region
Climate

Like the United States, China has many different climate regions. The map above shows you the seven climate regions of China. Location, elevation, and wind currents affect the type of climate found in any particular area of the country. Southeastern China has a humid subtropical climate with hot, humid summers. The northeast has a humid continental climate with cold winters. In desert areas to the northwest, summers are hot and dry and winters are cold

Evaluate

 Section 1 Quiz L1

 Use the Testmaker to create a customized quiz for Section 1. L1

Meeting Special Needs Activity

Language Deficiency Students with language problems often have difficulty seeing differences in similar words. The objectives for this section require students to explain, describe, and name. Point out that explaining means giving reasons or interpreting; describing usually requires listing characteristics; and naming usually involves labeling people, places, or things. Provide students opportunities for explaining, describing, and naming in the context of normal lesson discussion. L1

Human/ Environment Interaction Even though only 13 percent of its land is suitable for farming, China has been able to produce enough food to meet its people's needs. China grows more rice, cotton, pears, and potatoes than any other country in the world.

Reteach

Distribute copies of a skeleton outline with the section's main headings as main topics. Tell students to complete the outline with subheadings and details from the section. **L1**

Enrich

Ask interested students to research rare animals found in China. Have them present their findings in a classroom display. **L3**

CLOSE

Have students imagine they are astronauts training a powerful telescope on China. Encourage them to describe the physical features they see.

Rice fields are a common sight in warm, humid southeastern China *(left)*. Beijing has a humid continental climate with cold, snowy winters *(right)*.
LOCATION: What are hurricanes in the Pacific Ocean called?

and windy. In the southwestern part of China, the high Plateau of Tibet has bitterly cold winters and cool summers.

Monsoons greatly affect China's climates. In winter cold, dry air flows from central Asia across China. In summer the monsoons blow in from the sea, bringing warm, moist air. The summer monsoons often bring typhoons to coastal areas in the south. **Typhoons**—called hurricanes in the Atlantic Ocean—are tropical storms with strong winds and heavy rains.

SECTION 1 REVIEW

REVIEWING TERMS AND FACTS

1. **Define the following:** loess, dike, typhoon.

2. **LOCATION** Where is the Plateau of Tibet located?

3. **PLACE** How would you describe the landscape of southeastern China?

4. **REGION** How do monsoons affect China's climates?

MAP STUDY ACTIVITIES

5. Look at the physical map on page 610. What three major rivers flow through China?

6. Study the climate map on page 611. What are the three major climate regions of western China?

Answers to Section 1 Review

1. All vocabulary words are defined in the Glossary.
2. in southwestern China
3. The landscape changes from flat plains to green highlands. It is one of the most scenic areas of China.
4. Winter monsoons bring cold, dry air from central Asia, while summer monsoons bring warm, moist air from the sea.
5. Huang He, Chang Jiang, Xi River
6. highland, desert, steppe

BUILDING GEOGRAPHY SKILLS

Comparing Two or More Graphs

Comparing two or more graphs often gives you a better understanding of a topic. When comparing graphs, first see how they are related. Are they on the same topic? Are they the same kind of graph? Are the same units of measurement used? Then look for similarities and differences in the information on the graphs. To compare two or more graphs, apply these steps:

• Read the title and labels on each graph to see how they are related.

• Look for similarities and differences in the information given on each graph.

• Draw conclusions based on the comparison.

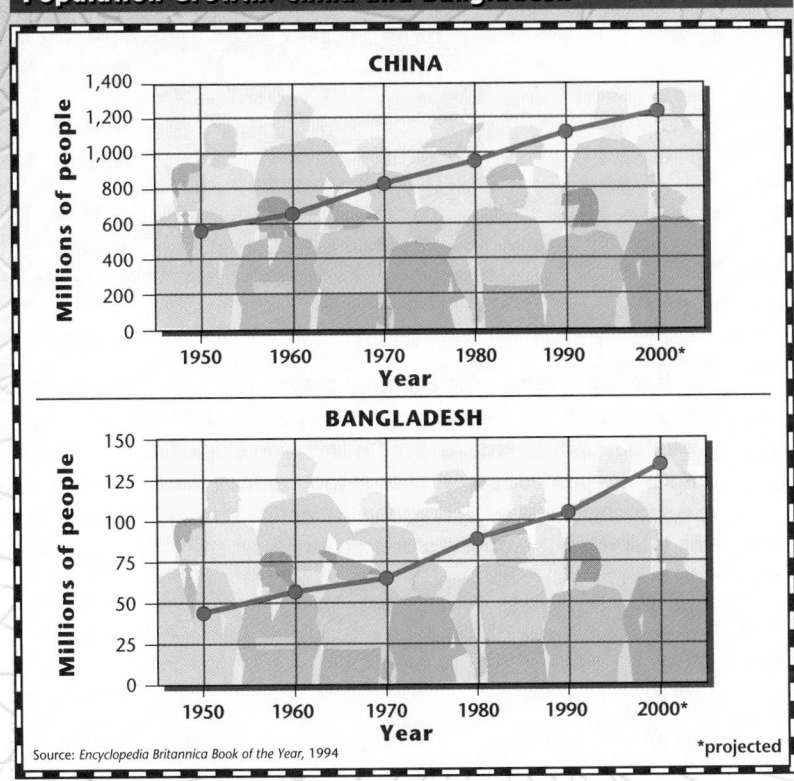

Population Growth: China and Bangladesh

CHINA

BANGLADESH

Source: *Encyclopedia Britannica Book of the Year, 1994*

*projected

Geography Skills Practice

1. What is the subject of the graphs?
2. What was Bangladesh's population in 1960?
3. By how much did China's population increase from 1980 to 1990?
4. What conclusions can you draw from the information on these graphs?

CHAPTER 23

TEACH

Divide the class into two groups. Give the first group the following information: Student A does the following homework hours over a five-week period—Week 1, 5 hours; Week 2, 10 hours; Week 3, 12 hours; Week 4, 10 hours; Week 5, 15 hours. Give the second group the following information: Student B does the following homework hours over a five-week period—Week 1, 8 hours; Week 2, 8 hours; Week 3, 14 hours; Week 4, 8 hours; Week 5, 12 hours. Have students work individually to plot the data on a line graph. Then direct students to pair off, one student from each group per pair, and compare their graphs.
L2

Additional Skills Practice

1. How are the two graphs in the skill feature similar? *(Both are line graphs showing population growth over the same period.)*

2. How are the two graphs different? *(different countries, China and Bangladesh)*

Geography Skills Activity 23

Building Skills in Geography, Unit 2, Lesson 6

Answers to Geography Skills Practice

1. population growth in China and Bangladesh
2. about 58 million people
3. nearly 200 million
4. Both countries show a marked increase in population, but Bangladesh's population increased at a greater rate than China's.

FOCUS

Section Objectives

1. **Explain** how most Chinese earn their livings.
2. **Identify** the three economic regions of China.
3. **Summarize** how China's economy has changed in recent years.

Bellringer Motivational Activity

Prior to taking roll at the beginning of the class period, project Section Focus Transparency 23-2 and have students answer the activity questions. Discuss students' responses. **L1**

Vocabulary Pre-check

Pass around an incandescent light bulb and draw students' attention to the tungsten filament inside. Then point out that tungsten is also used in spark plugs, cutting tools, and x-ray tubes. **L1** **LEP**

SECTION 2
The Economy

PREVIEW

Words to Know
• invest
• consumer goods
• tungsten
• terraced field

Places to Locate
• Shanghai
• Beijing
• Wuhan
• Guangzhou

Read to Learn . . .
1. how most Chinese earn their livings.
2. what the three economic regions of China are.
3. how China's economy has changed in recent years.

Shanghai is China's largest city and its major port. The name *Shanghai* means "on the water." Shanghai is a leader in trade and business. Because Shanghai is a port city, it has many contacts with the rest of the world.

China has a developing economy based on farming and industry. About 60 percent of the Chinese people farm the land. Because there are so many mountains and deserts, farmland is limited in China. Chinese farmers, however, are able to produce almost enough food to feed the entire country. China also has growing industries. It is among the world's ten leading countries in terms of the total value of products produced each year.

Region
A New Economy

Since 1949 China has been a communist country. Under communism, the government—not individuals or businesses—decides what crops are grown, what products are made, and what prices are charged for both. In recent years China's government has made many changes in the economy. Why? It hopes to make China a modern, industrialized nation. It also wants to increase trade with the United States and other Western countries. Without completely giving up communism, the government has allowed some features of free enterprise to develop in China.

614

UNIT 8

Classroom Resources for Section 2

 REPRODUCIBLE MASTERS

Reproducible Lesson Plan 23-2
Guided Reading Activity 23-2
Workbook Activity 23
Chapter Map Activity 23
Section Quiz 23-2

 **TRANSPARENCIES**

Section Focus Transparency 23-2
Political Map Transparency 8

MULTIMEDIA

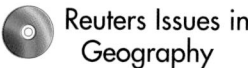 Testmaker

Reuters Issues in Geography

Making a Living Since the 1980s, China has increased its ties with foreign countries. Eager to learn about new business methods, it has asked other countries to **invest,** or put money, in Chinese businesses. Many businesses in China are now jointly owned by Chinese and foreign businesspeople.

Instead of telling people what jobs they should have, the Chinese government now wants individuals to choose where to work and even to start their own businesses. Shop owners, dentists, and barbers are among the people competing for customers in China's major cities.

Changes have also come to China's countryside. Farmers once spent most of their time working on government farms. Now they grow extra crops on private plots of land. Farmers can keep the money they make from selling these crops.

Because of all these changes, farming, manufacturing, and trade are booming in China. Some Chinese now enjoy a good standard of living. They can afford **consumer goods,** or products such as television sets, cars, and motorcycles. Not everyone, however, finds it easy to adjust to the new economy. Many Chinese are finding that the prices of food and goods are rising faster than their incomes. Some Chinese are getting very rich, while others remain poor.

China has not developed modern ways of dealing with industrial wastes. Many factories dump poisonous chemicals into China's rivers. Factories powered by coal are another cause of pollution. When the coal is burned, it sends out a thick, black smoke. Breathing this smoke is a health hazard. There are few laws to control pollution. As China continues to industrialize, it will need to deal with these environmental problems.

CHINA: Land Use and Resources

MAP STUDY
Because of its location, Hong Kong is a center of world trade and manufacturing.
PLACE: What are three major manufacturing areas along China's east coast?

Resources
- ✚ Bauxite
- 🗦 Coal
- Copper
- 🐟 Fish & other seafood
- 🌲 Forest
- ⚡ Hydroelectric power
- I Iron ore
- ⚙ Lead
- ◐ Manganese
- ⚒ Petroleum
- ▽ Tin
- ▨ Tungsten
- ▣ Zinc

Conic projection

Agriculture
- ☐ Little or no farming
- ☐ Nomadic herding
- ☐ Commercial farming
- ☐ Subsistence farming
- ■ Manufacturing area

Cooperative Learning Activity

Organize students into several groups and provide each group with a copy of the following chart:

PRODUCTS OF CHINA
	North	South	West
Mining			
Farming			
Manufacturing			
Other			

Have group members work together to complete the chart. Then ask groups to imagine they are Chinese officials who want to encourage foreign investment in China. Have groups select one economic region and, using information in the chart, create a four-page illustrated pamphlet encouraging foreign businesses to invest in the region. Call on groups to present and discuss their pamphlets. **L2**

TEACH

Guided Practice

Analyzing Information
Direct students to write the headings *Early Communist China* and *Communist China Today* on separate pages in their notebooks. As students read the text under "A New Economy" on pages 614–615, have them list descriptions of the past or present Chinese economy under the appropriate heading in their notebooks. **L1**

Independent Practice

📁 Guided Reading Activity 23-2 **L1**

Deforestation has contributed to soil erosion and other environmental problems in China. To help prevent further damage, the Chinese are planting a shelterbelt of trees called the Great Green Wall of China.

MAP STUDY

Answer
Shanghai, Hangzhou, Guangzhou

Map Skills Practice
Reading a Map In what part of China is rice grown? *(southeastern China)*

Economic Regions

The physical geography of China influences its economy. Many geographers divide China into three economic regions: the north, the south, and the west.

The North The north region includes the plains and highlands of northeastern China. It has a variety of economic activities, including manufacturing, mining, farming, and commercial fishing. Because of the partly dry, often cold climate, farmers in the region grow hearty grains such as wheat, cotton, and soybeans. In isolated grassland areas, nomads herd livestock.

The map on page 615 shows you that the north is rich in natural resources such as coal, petroleum, iron ore, and tungsten. **Tungsten** is a metal used in electrical equipment. China is a world leader in the mining of coal and iron ore. A number of large urban areas in the north manufacture many goods. Workers in these industrial centers produce textiles, chemicals, electronic equipment, farm machinery, airplanes, and other metal goods. Beijing (BAY•JIHNG), China's capital and major industrial city, is located in the north.

Fourteen-year-old Lai Sau Chun has lived in Beijing all of her life. She explains that, like many Chinese cities, Beijing is a blend of old and new. Many factories have been built near the city, but Sau Chun prefers the old Beijing. She takes visitors to the Imperial Palace in the heart of the city. Chinese rulers built it in the early 1400s. The palace was called "the forbidden city" because only the ruler's family and government officials were allowed to enter.

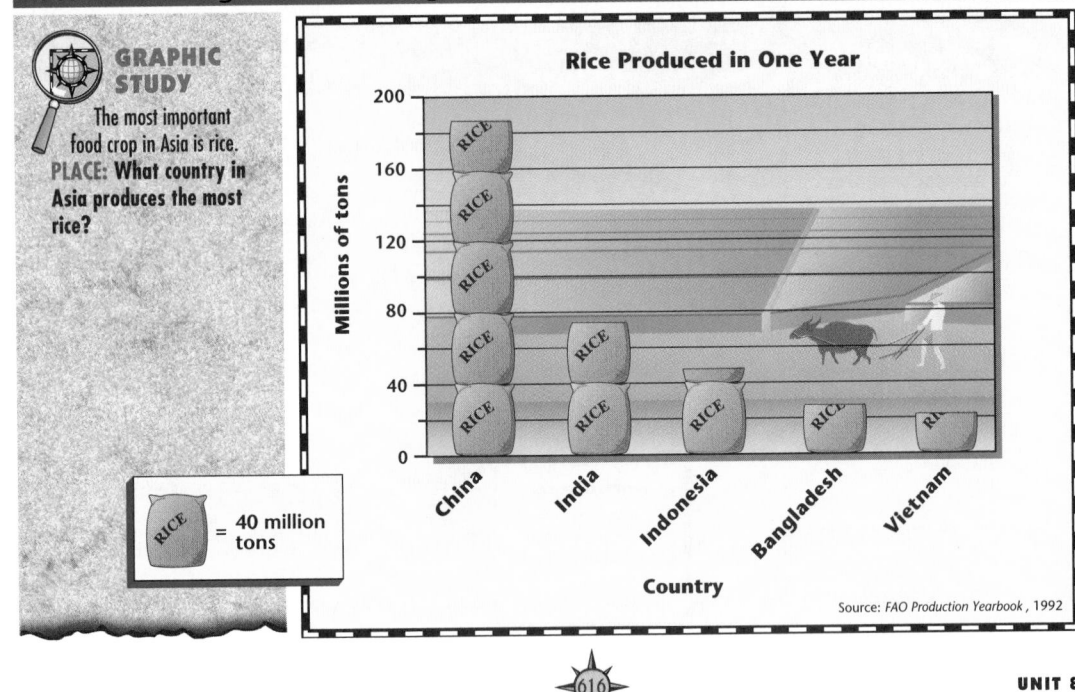

World's Leading Rice-Producing Countries

GRAPHIC STUDY
The most important food crop in Asia is rice.
PLACE: What country in Asia produces the most rice?

Rice Produced in One Year

Millions of tons

RICE = 40 million tons

Country: China, India, Indonesia, Bangladesh, Vietnam

Source: *FAO Production Yearbook*, 1992

Meeting Special Needs Activity

Language Delayed To help students with language concept delays, discuss the meaning of "developing economy" and "industrialized nation" in class. Read aloud the second-to-last paragraph on page 615 and ask why China is experiencing a gap between rich and poor. Then read aloud the last paragraph on page 615 and have students define health hazard and explain how it is related to pollution in China. **L1**

The South The south region includes the southern half of eastern China. It has fertile soil, a humid climate, and a long growing season. If you travel through hilly areas, you will see farmers growing crops on terraced fields. A **terraced field** has strips of land cut out of a hillside like stair steps. Rice is the south's major crop. Farmers also grow tea, jute, silk, fruits, and vegetables.

In addition to fertile soil, the south is rich in bauxite, iron ore, tin, and other minerals. It also has many urban manufacturing areas. Some cities, such as Shanghai, are located on or near the coast. Others, such as Wuhan (WOO•HAHN) or Guangzhou (GWAHNG•JOH), are on rivers. Workers in these industrial cities make ships, machinery, textiles, and electrical equipment.

The West The west region takes up the part of China that is covered by mountains, deserts, and grasslands. The dry and cold Plateau of Tibet is not very good for farming. It provides only limited grazing for hardy animals such as yaks. In low-lying fertile areas, farmers are able to grow cotton, wheat, and other food crops. The west is also rich in petroleum, coal, and iron ore.

A worker puts the finishing touches on a vase in a Beijing factory *(left)*. A sheepherder tends his flock in northwestern China *(right)*.
PLACE: What natural resources are found in northern China?

SECTION 2 REVIEW

REVIEWING TERMS AND FACTS
1. **Define the following:** invest, consumer goods, tungsten, terraced field.
2. **PLACE** How do most Chinese earn their livings?
3. **HUMAN/ENVIRONMENT INTERACTION** How has China's economy been changing?
4. **REGION** What region is the major rice-producing area of China?

MAP STUDY ACTIVITIES
5. Look at the map on page 615. What minerals are found in northeastern China?
6. Where do you think oranges and other tropical crops grow in China?

Evaluate

Section 2 Quiz **L1**

Use the Testmaker to create a customized quiz for Section 2. **L1**

Reteach

Assign a paragraph from the section to each student. Direct students to summarize their assigned paragraphs in a sentence. Copy the sentences, omitting key words, and distribute copies to the class. Have the students fill in the missing words. **L1**

Enrich

Have students research and report on recent economic developments in China. **L3**

CLOSE

Encourage students to write Lai Sau Chun asking her questions about life in Beijing.

Answers to Section 2 Review
1. All vocabulary words are defined in the Glossary.
2. by farming the land
3. Since the 1980s, the Chinese government has increased trade with foreign countries and encouraged foreign investment in the economy. In addition, the government has encouraged people to start their own businesses and allowed farmers to sell surplus crops for a profit.
4. the south
5. coal, iron ore, petroleum, copper, lead, zinc, tungsten, manganese
6. the south

TEACH

Ask students to scan the chapter for photographs of landscapes that Wang Wei might have described in his poetry. Encourage students to write a poem, in the style of Wang Wei, on one of the landscapes they select. **L2**

More About . . .

Wang Wei Scholars consider Wang Wei, who lived from A.D. 699–759, one of China's greatest poets. He wrote many four-line poems that celebrated scenes of nature. His poems also emphasized the qualities of quiet and contemplation, showing the influence of Buddhism. Wang Wei was a painter as well as a poet, and he is often credited with founding the Southern school of Chinese landscape painting.

CHINESE NATURE POETRY

| MATH | SCIENCE | HISTORY | LITERATURE | TECHNOLOGY |

Chinese poets aimed to create a certain mood and atmosphere in their works. Many of their poems describe brief moments of deep feelings, such as those expressed in these poems by Wang Wei.

NORTH HILL

North Hill stands out above the lake
Against thick evergreens gleams startlingly
    a vermilion [red-orange] gate.
Below, South River zig-zags toward the horizon,
Glistening, here and there, beyond the treetops of the
    blue forest.

SAILING ON THE RIVER TO CH'ING-HO

Sailing on the great river
The gathered waters reach to the very rim of heaven.
Suddenly the sky-high waves part,
Disclosing country of a thousand and cities of ten
    thousand roofs.
And sailing further, still other villages and
    market-towns appear
Mid mulberry trees and growing flax.
Looking back towards my native village . . .
The vast waters have joined the clouds.

Chinese Painting

From *Poems by Wang Wei*, translated by Chang Yin-nan and Lewis C. Walmsley. Copyright © 1958 by Charles E. Tuttle Co., Inc., Tokyo, Japan. Reprinted by permission.

Making the Connection

1. What geographic features are mentioned in these Chinese nature poems?
2. How does the poet describe the environment?

UNIT 8

Answers to Making the Connection

1. hills, rivers, forests
2. in terms of beauty, color, and tranquility

SECTION 3
The People

Meeting National Standards
Geography for Life The following standards are highlighted in this section:
• Standards 6, 9, 10, 12, 13

PREVIEW

Words to Know
• dynasty
• calligraphy
• pagoda

Places to Locate
• Taiwan
• Beijing

Read to Learn . . .
1. what groups influenced the culture of China.
2. how city life differs from country life in China.
3. what arts and recreation the Chinese enjoy.

Along city streets in China, you will see many young people dressed as you are—in jeans. Western-style clothing is very popular. People are prepared to pay high prices for this clothing.

Crowded city streets are common in China. The country has more people than any other place in the world. About 1.2 billion people—one-fourth of the world's population—make China their home. The graph on page 620 shows you how China's population compares with that of the United States.

About 94 percent of China's people are Han Chinese. They have a unique culture, although they speak different dialects of the same Chinese language. The remaining 6 percent belong to 55 other ethnic groups. Most of these groups, such as the Tibetans, live in the western part of the country. For years the Tibetans have struggled to keep their culture and win their independence from China.

Place
Influences of the Past

Chinese civilization is more than 4,000 years old. For centuries **dynasties,** or ruling families, governed China. Each dynasty was made up of a line of rulers from the same family. Under the dynasties, China set up a strong government and enlarged its borders.

CHAPTER 23

619

FOCUS

Section Objectives
1. **Recall** what groups influenced the culture of China.
2. **Cite** how city life differs from country life in China.
3. **Enumerate** the arts and recreation the Chinese enjoy.

Bellringer Motivational Activity

 Prior to taking roll at the beginning of the class period, project Section Focus Transparency 23-3 and have students answer the activity questions. Discuss students' responses. **L1**

Vocabulary Pre-check
Have students read the definitions of the terms in the text. Then challenge students to use all three terms correctly in one sentence. **L1 LEP**

Classroom Resources for Section 3

REPRODUCIBLE MASTERS
Reproducible Lesson Plan 23-3
Guided Reading Activity 23-3
Vocabulary Activity 23
GeoLab Activity 23
Section Quiz 23-3

TRANSPARENCIES
Section Focus
  Transparency 23-3
World Cultures Transparency 15

MULTIMEDIA
 Testmaker

National Geographic
  Society: Picture Atlas of
  the World

GRAPHIC STUDY

Answer
about 1.2 billion people

Graphic Skills Practice
Reading a Graph What is the difference in population between China and the United States? *(about 930 million people)*

TEACH

Guided Practice

Interpreting Information
On a transparency, draw three circle graphs. Shade 25 percent of the first, 94 percent of the second, and 75 percent of the third. Have students create titles for the graphs based on information in the section. *(Graph 1: China's Share of the World's Population; Graph 2: Percentage of China's People Who Are Han Chinese; Graph 3: Percentage of Chinese Who Live in Rural Areas)* **L1**

More About the Illustration

Answer to Caption
Most will suggest that the government saw the demonstrators as a threat to its authority and wanted to make clear that such protests would not be tolerated.

Place Estimates suggest that the death toll in the Tiananmen Square protest was more than 1,000. At least 10,000 people were later arrested for taking part in the demonstrations. Many were summarily executed.

620

COMPARING POPULATION: China and the United States

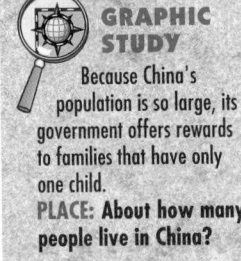

GRAPHIC STUDY

Because China's population is so large, its government offers rewards to families that have only one child.
PLACE: About how many people live in China?

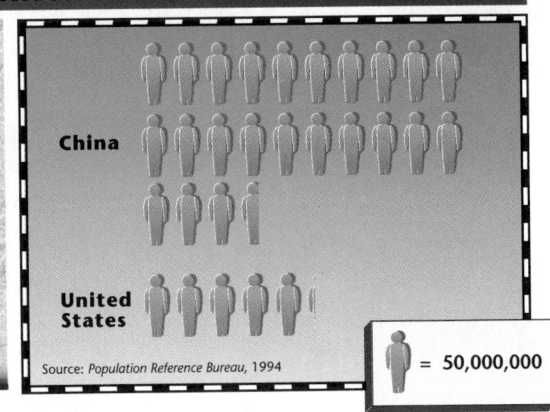

China

United States

Source: *Population Reference Bureau,* 1994

= 50,000,000

Ideas and Inventions The early Chinese developed a rich civilization. They believed that learning was a key to good behavior. A philosopher named Konfuzi, or Confucius, taught that people should be polite, honest, brave, and wise. Children were to obey their parents, and every person was to respect the elderly and obey the country's rulers. For more than 2,000 years, the Chinese followed the teachings of Confucius. They blended these teachings with those of Buddhism, which reached China from India about A.D. 100.

Did you know that the early Chinese were using paper and ink before people in other parts of the world? Early Chinese developed a number of outstanding inventions. Besides paper and ink, these inventions include the clock, the compass, the first printed book, and fireworks.

Modern China The early Chinese tried to avoid contact with other parts of the world. In the 1700s and 1800s, however, European countries forced the Chinese to trade with them and tried to make the Chinese accept Western ways. This angered the Chinese.

In 1912 a national uprising overthrew the last dynasty, and China became a republic. Soon civil war divided the country. A military leader named Chiang Kai-shek (jee•AHNG KY•SHEHK) headed the government. Chiang's forces fought against the Chinese Communists, led by Mao Zedong (MOW ZUH•DUNG). In 1949 the Communists won the civil war, driving Chiang's government to the offshore island of Taiwan.

Mainland China became the communist People's Republic of China. The Communists wanted to make China a modern industrial country. But the Communists refused to allow much freedom to the Chinese people. In 1989 thousands of Chinese students staged a protest in Beijing's Tiananmen Square. They demanded free elections and the right to speak and act freely. The Chinese government answered their protests with gunfire, killing many and putting others in prison. Human rights issues continue to be a problem in China's relations with the United States.

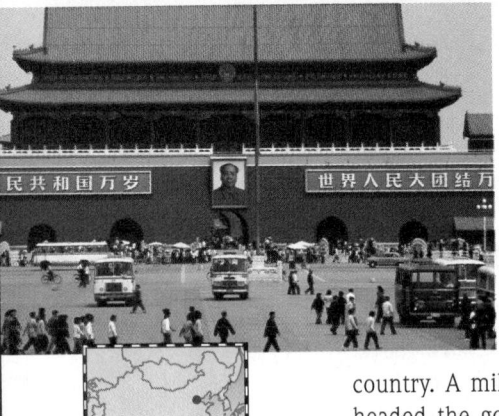

Tiananmen Square, Beijing

民共和国万岁   世界人民大团结万

In 1989 hundreds of students were killed here when they protested against the Chinese government.
PLACE: Why do you think the Chinese government would kill protesting students?

620

Cooperative Learning Activity

On the chalkboard, copy the following sayings of Confucius: 1) Have no friends not equal to yourself. 2) When you have faults, do not fear to abandon them. 3) Learning without thought is labor lost; thought without learning is perilous. 4) The cautious seldom err.

Organize the students into four groups and assign each group a saying. Have groups discuss their sayings and, after a specified time, reach a consensus about what the sayings mean. Ask groups to present their interpretations of the sayings in poster form. Call on groups to display and discuss their posters. **L1**

Rural Life About 75 percent of China's people live in rural areas. The map below shows you that most people are crowded into the fertile river valleys of eastern China. Families in the countryside work very hard in their fields. They often use simple hand tools because farming equipment is too expensive.

Village life, though, has improved in recent years. Most rural families live in three- or four-room houses. They have enough food and some modern appliances. Many villages have community centers where people can watch movies, read books, and play sports such as table tennis and basketball.

City Life More than 230 million Chinese—about 20 percent of the population—live in cities. China's cities are growing rapidly as people from the countryside move to cities hoping to find work. Living conditions are crowded, but most homes and apartments have heat, electricity, and running water. Many people now earn enough money to buy extra clothes and television sets. They also have more leisure time to attend concerts or Chinese operas, walk in parks, visit zoos, or talk with friends.

Region
The Arts

China is famous for its traditional arts. Chinese craftworkers make bronze bowls, jade jewelry, glazed pottery, and fine porcelain. The Chinese are also known for their painting, sculpture, and architecture.

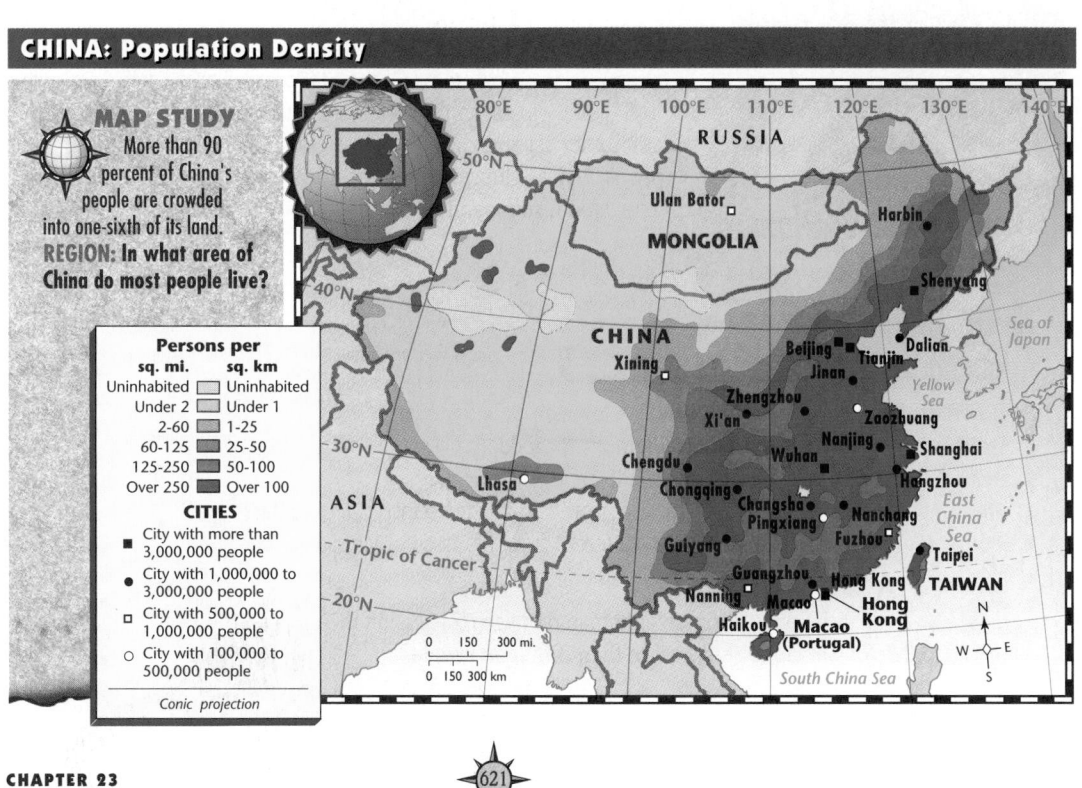

CHINA: Population Density

MAP STUDY More than 90 percent of China's people are crowded into one-sixth of its land.
REGION: In what area of China do most people live?

Persons per

sq. mi.	sq. km
Uninhabited	Uninhabited
Under 2	Under 1
2–60	1–25
60–125	25–50
125–250	50–100
Over 250	Over 100

CITIES
■ City with more than 3,000,000 people
● City with 1,000,000 to 3,000,000 people
□ City with 500,000 to 1,000,000 people
○ City with 100,000 to 500,000 people

Conic projection

Independent Practice

📁 Guided Reading Activity 23-3 **L1**

ASSESS

Check for Understanding

Assign Section 3 Review as homework or an in-class activity.

Meeting Lesson Objectives

Each objective below is tested by the questions that follow it in parentheses.

1. **Recall** what groups influenced the culture of China. (2)
2. **Cite** how city life differs from country life in China. (3, 4)
3. **Enumerate** the arts and recreation the Chinese enjoy. (4)

MAP STUDY

Answer
eastern and southeastern China

Map Skills Practice
Reading a Map What geographic feature would you find in the uninhabited areas shown on the map? *(desert)*

Meeting Special Needs Activity

Language Delayed Give students who have difficulty asking questions of their teachers an opportunity to practice formulating questions. Organize the class into small groups to develop questions. Have students focus on the text under the heading "Influences of the Past." Reward the groups that formulate the most varied questions. **L1**

621

Art The Chinese love of nature has influenced painting and poetry in China. Chinese artists paint on long panels of paper or silk. Artwork often shows scenes of mountains, rivers, and forests. Artists attempt to portray the harmony between nature and the human spirit.

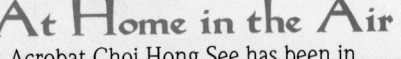

At Home in the Air

Acrobat Choi Hong See has been in training since she was three years old. Her family performs in Shanghai's Acrobatics Theater. She can walk on high wires, do triple somersaults, and balance on a human pyramid. Acrobatics have been popular in China for more than 2,000 years. Many acrobatic groups are supported by the government and are often invited to perform in Europe or the United States.

Many Chinese paintings include a verse or poem written in **calligraphy,** the art of beautiful writing. Chinese writing is different from the print you are reading right now. It uses characters that represent words or ideas instead of sounds. There are more than 50,000 Chinese characters, but the average person recognizes only about 8,000.

Chinese pottery is famous for its painted decoration. The Chinese developed the world's first porcelain centuries ago. Porcelain is made from coal dust and a fine, white clay. Porcelain vases from early Chinese history are considered priceless today.

Architecture Most buildings in Chinese cities are modern in style, but a few traditional buildings stand. Some have large tiled roofs with edges that curve gracefully upward. Others are Buddhist temples with many-storied towers called **pagodas.** Large statues of Buddha are found in these buildings.

Foods Cooking differs greatly from region to region in China. In coastal areas fish, crab, and shrimp dishes are common, while central China is famous for its spicy dishes. Most Chinese, however, eat very simply. A typical Chinese meal includes vegetables with bits of meat or seafood, soup, and rice or noodles.

SECTION 3 REVIEW

REVIEWING TERMS AND FACTS
1. **Define the following:** dynasty, calligraphy, pagoda.
2. **PLACE** How was early China governed?
3. **MOVEMENT** Why have many Chinese moved from the countryside to cities?
4. **REGION** How is life in urban China different from life in rural China?

 MAP STUDY ACTIVITIES
5. Look at the population density map on page 621. Name the cities in the People's Republic of China that have more than 3 million people.

 622

Answers to Section 3 Review
1. All vocabulary words are defined in the Glossary.
2. Early China was governed by dynasties.
3. Standards of living are better in the cities.
4. Most rural families live in three- or four-room houses that have few amenities. They work very hard on farms, often using simple hand tools. Many families make just enough to get by. City dwellers live in crowded conditions, but most apartments have modern amenities. Many city dwellers earn enough to buy luxury items.
5. Shanghai, Beijing, Shenyang, Tianjin, Wuhan

SECTION 4 · China's Neighbors

PREVIEW

Words to Know
- empire
- yurt
- high-technology industry

Places to Locate
- Mongolia
- Taiwan
- Hong Kong
- Macao

Read to Learn . . .
1. how people live and earn their livings in Mongolia.

2. what products are made in Taiwan.
3. why Hong Kong is one of the world's busiest ports.

North of China lies Mongolia. If you flew to Mongolia, you would arrive in Ulan Bator (OO•LAHN BAH•TAWR), the capital and largest city. Ulan Bator began as a Buddhist community in the early 1600s. Today it is a modern cultural and industrial center.

Mongolia is one of several territories that border or lie close to the People's Republic of China. The others are the island of Taiwan and the mainland areas of Hong Kong and Macao (muh•KOW). Throughout their history, the peoples of these territories have had close ties to China. Mongolia once had a communist government similar to China's. Many of Taiwan's people are Chinese who left the mainland and set up a noncommunist Chinese government on Taiwan. Macao, once part of China, is a European colony that will soon return to Chinese rule.

Place
Mongolia

Mongolia lies in east central Asia between China and Russia. The map on page 608 shows you that Mongolia is a landlocked country. Its land area of 604,830 square miles (1,566,510 sq. km) is more than twice the size of Texas.

Centuries ago, Mongolia was the home of an Asian people called the Mongols. During the 1200s powerful Mongol armies carved out the largest land empire in history. An **empire** is a collection of territories under one ruler. The

---

Meeting National Standards

Geography for Life The following standards are highlighted in this section:
- Standards 3, 5, 10, 11, 12, 13, 17, 18

FOCUS

Section Objectives
1. **Examine** how people live and earn their livings in Mongolia.
2. **Classify** products made in Taiwan.
3. **Point** out why Hong Kong is one of the world's busiest ports.

Bellringer Motivational Activity

 Prior to taking roll at the beginning of the class period, project Section Focus Transparency 23-4 and have students answer the activity questions. Discuss students' responses. **L1**

Vocabulary Pre-check

Display pictures of a yurt and a computer and have students choose the vocabulary term that they think identifies each picture. Point out that the pictures illustrate the contrasts found in China's neighbors. **L1** **LEP**

Classroom Resources for Section 4

 REPRODUCIBLE MASTERS

Reproducible Lesson Plan 23-4
Guided Reading Activity 23-4
Reteaching Activity 23
Enrichment Activity 23
Performance Assessment Activity 23
Section Quiz 23-4

TRANSPARENCIES

Section Focus Transparency 23-4
Geoquiz Transparencies 23-1, 23-2

MULTIMEDIA

 Testmaker

National Geographic Society: ZipZapMap! World

Strange but TRUE!

Less than 1 percent of the land in Mongolia is suitable for farming.

TEACH

Guided Practice
Organizing Information
Ask students to arrange the territories discussed in the section according to: 1) area from largest to smallest; *(Mongolia, Taiwan, Hong Kong, Macao)* 2) population from greatest to smallest; *(Taiwan, Hong Kong, Mongolia, Macao)* and 3) population density from most dense to least dense. *(Macao, Hong Kong, Taiwan, Mongolia)* **L1**

Independent Practice

Guided Reading Activity 22-4 **L1**

Electronics Factory

Workers assemble electronic circuit boards in a Taiwan factory.
MOVEMENT: Why do the Republic of China and mainland China have different governments?

Mongol Empire reached from China to eastern Europe. Today Mongolia is smaller in size, but its people still honor their Mongol past.

The Land Rugged mountains and high plateaus cover most of Mongolia. In the southeast, river basins dot the mountainous landscape. The bleak desert landscape of the Gobi spreads through central and southeastern Mongolia. Most of the country receives only small amounts of rain, and temperatures are usually very cold or very hot.

The Economy Mongolia's economy relies on the raising of livestock. Many Mongolians herd sheep, goats, cattle, or camels on grasslands that cover large areas of the country. For centuries the people of Mongolia have been famous for their skills in raising and riding horses. In limited areas, farmers grow wheat, barley, and oats.

The People Nearly all of Mongolia's 2.4 million people are Mongols who speak the Mongol language. About 57 percent of them live in urban areas such as Ulan Bator. Mongolians in the countryside live on farms. A few still follow the nomadic way of their ancestors. These herder-nomads live in **yurts,** or large circle-shaped tents made of animal skins.

Mongolians traditionally followed a form of Buddhism. When Mongolia was under a rigid communist government from 1924 to 1990, however, the Communists discouraged religious practice. Now that Mongolia is independent, people are free to observe their Buddhist beliefs again.

Place
Taiwan

About 100 miles (161 km) off the southeast coast of China lies the island country of Taiwan. In 1949 the Chinese Communists conquered mainland China. The defeated Chinese government of Chiang Kai-shek moved to Taiwan and established the Republic of China. Today Taiwan is still governed separately from the communist-ruled mainland.

The Land Thickly forested mountains run through the center of Taiwan. A flat plain along the western coast is home to most of the island's people. Like southeastern China, Taiwan has a humid subtropical climate of mild winters and hot, rainy summers.

The Economy Taiwan has one of the world's most prosperous economies. Taiwan's wealth rests largely on high-technology industries, manufacturing, and trade with foreign countries. **High-technology industries** produce computers and other kinds of electronic equipment. Taiwanese workers produce many different products, including computers, clothing, furniture, ships, radios, and televisions. You have probably seen goods from Taiwan sold in stores in your community.

Taiwan's mountains limit available farmland. Farmers often terrace hills to provide more land for growing rice. Other crops grown in Taiwan include bananas, citrus fruits, corn, peanuts, sugar, tea, and vegetables.

The People About 75 percent of Taiwan's 21.1 million people live in urban areas. Taipei is the capital and largest city. Skyscrapers and busy streets are major features of Taipei, which has about 7 million people.

Taiwan is a modern country, but many of its people still follow Chinese traditions. About 95 percent of the people are Chinese and speak Chinese dialects. Buddhism is the leading religion of Taiwan.

WHAT IN THE WORLD?

Growing Land?
You've heard of growing crops, but have you heard of growing land? Hong Kong has many people, and there is no place to put them all. Therefore, workers are blasting the mountains on Hong Kong's Kowloon Peninsula. They dump the rock and dirt from the blastings onto the shore of the South China Sea—growing new land.

Place
Hong Kong and Macao

Big changes are in store for Hong Kong and Macao, which lie on the southeastern coast of China. Both are made up of hilly peninsulas and groups of islands. The United Kingdom ruled Hong Kong until 1997, when it returned to Chinese rule. Although Portugal governs Macao, China and Portugal have agreed that Macao will revert to China in 1999.

Busy Port Hong Kong has a land area of about 410 square miles (1,061 sq. km). Natural harbors make Hong Kong one of the world's busiest ports and a major center of trade and business. It is also one of the world's most crowded places. About 6 million people live in Hong Kong. This means that the population density is about 14,500 people per square mile (5,600 per sq. km).

Nearly all the people of Hong Kong are Chinese. They live mostly in urban areas in crowded, high-rise apartments. Others make their homes on boats in harbors. Many of Hong Kong's people work in manufacturing, banking, government, foreign trade, and tourism.

Hong Kong Souvenirs

A gift shop in Hong Kong sells souvenirs commemorating the territory's return to China.
PLACE: What country ruled Hong Kong before 1997?

ASSESS

Check for Understanding
Assign Section 4 Review as homework or an in-class activity.

Meeting Lesson Objectives
Each objective below is tested by the questions that follow it in parentheses.
1. **Examine** how people live and earn their livings in Mongolia. (3)
3. **Point** out why Hong Kong is one of the world's busiest ports. (6)

Evaluate
Section 4 Quiz **L1**

Use the Testmaker to create a customized quiz for Section 4. **L1**

Chapter 23 Test Form A and/or Form B **L1**

More About the Illustration
Answer to Caption
the United Kingdom

Location
Hong Kong is made up of the city of Kowloon on the mainland and more than 200 islands. Hong Kong is known for its fine harbors.

Meeting Special Needs Activity

Reading Disability The presentation of background knowledge, or "schema," greatly aids reading comprehension. A schema is developed from the reader's current knowledge, then extended with new information from the text and illustrations. Have students with reading comprehension problems read only the introductory paragraphs on page 623. Compare this information with another region students have studied with a similar past and present, such as South Asia. After this orientation, have students read the rest of section 4. **L1**

More About the Illustration

Answer to Caption
Answers will vary.

Place At Hindu weddings, the bride dresses in either a pink or red embroidered sari. The delicate designs on the bride's forehead serve as an announcement that she is now a married woman. The groom wears a traditional turban, often decorated with jewelry or feathers.

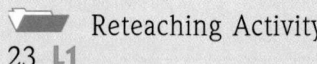

Reteach

 Reteaching Activity 23 **L1**

Enrich

Enrichment Activity 23 **L3**

CLOSE

Mental Mapping Activity

Provide each student with a large sheet of white paper and map pencils. Inform students that this will not be a graded activity. Have students draw, freehand from memory, a map of China, Mongolia, Hong Kong, and Macao. Tell them to label China and its neighbors and to shade and label the Gobi. Also have them locate and label Beijing, Ulan Bator, and Taipei.

Hindu Wedding in Hong Kong

At a Hindu wedding, the bride and groom dress in traditional clothing and sit in front of a sacred fire.
PLACE: How is this wedding different from or similar to the weddings you have attended?

A Bit of Portugal About 40 miles (64 km) west of Hong Kong lies the territory of Macao. The map on page 608 shows you that Macao covers about 6 square miles (16 sq. km). It has about 302,000 people, most of whom are Chinese. Most of the territory's food is imported from China. Macao's economy is based on industry, trade, and tourism.

Macao has an interesting mix of Chinese and Portuguese cultures. In the older sections of Macao, you will find Roman Catholic churches and old European-looking houses next to Chinese temples and Chinese-owned shops. Across town high-rise apartments and modern office buildings crowd the skyline. There is much new construction here in anticipation of the political change in 1999.

SECTION 4 REVIEW

REVIEWING TERMS AND FACTS
1. **Define the following:** empire, yurt, high-technology industry.
2. **PLACE** What desert area crosses Mongolia?
3. **HUMAN/ENVIRONMENT INTERACTION** What is the major economic activity of Mongolia?
4. **PLACE** Where do most of Taiwan's people live?

 MAP STUDY ACTIVITIES
   5. Look at the climate map on page 611. What is the major climate region of Mongolia?
   6. Turn to the physical map on page 610. What body of water touches Taiwan, Hong Kong, and Macao?

626

UNIT 8

Answers to Section 4 Review

1. All vocabulary words are defined in the Glossary.
2. Gobi
3. raising livestock
4. in urban areas
5. steppe
6. South China Sea

Chapter 23 Highlights

Important Things to Know About China

SECTION 1 THE LAND

- China has more people than any other country in the world. It is the world's third-largest country in area, after Russia and Canada.
- Three important rivers—the Huang He, Chang Jiang, and Xi—flow through fertile river valleys in eastern China.
- The western part of China is largely mountains and deserts.
- China's climate is affected by monsoons.

SECTION 2 THE ECONOMY

- Modern China's communist leaders have moved China's economy toward free enterprise.
- Most of China's people are farmers. Rice is the major food crop in the south.
- China's largest urban manufacturing areas lie near rivers or on the coast.

SECTION 3 THE PEOPLE

- China has one of the world's oldest civilizations.
- The early Chinese passed on many inventions to the rest of the world, including paper, ink, porcelain, and the first printed book.

SECTION 4 CHINA'S NEIGHBORS

- Mongolia has a landscape of rugged mountains, plateaus, and desert.
- Taiwan has a prosperous economy based on high-technology industries, manufacturing, and trade.
- Hong Kong, once under British rule, became part of China in 1997. Macao is a European colony but will return to Chinese rule in 1999.

Tiananmen Square, Beijing ▶

Using the Chapter 23 Highlights

- Use the Chapter 23 Highlights to preview, review, condense, or reteach the chapter.

Preview/Review

Vocabulary Puzzle-Maker Software reinforces the terms used in Chapter 23.

Condense

Have students read the Chapter 23 Highlights. Spanish Chapter Highlights also are available.

Chapter 23 Digest Audiocassettes and Activity

Chapter 23 Digest Audiocassette Test

Spanish Chapter Digest Audiocassettes, Activities, and Tests also are available.

Guided Reading Activities

Reteach

Reteaching Activity 23. Spanish Reteaching activities also are available.

MAP STUDY

Location Copy and distribute page 34 from the Outline Map Resource Book. Have students label China, Mongolia, Taiwan, Hong Kong, and Macao.

Extra Credit Project

Trivia Game Have each student make up questions about the geography of East Asia based on facts from the chapter. Ask students to write each question on one side of an index card and the answer on the other side. Store the cards in a box. Use the cards to quiz the students on the chapter. Inform students that they will earn a point for each correct answer and that the student with the most points becomes the trivia champion. **L1**

627

Chapter 23 Review and Activities

ANSWERS

Reviewing Key Terms

1. F
2. G
3. H
4. A
5. D
6. I
7. C
8. E
9. J
10. B

Mental Mapping Activity

This exercise helps students to visualize the countries and geographic features they have been studying and understand the relationships among them. All attempts at freehand mapping should be accepted.

REVIEWING KEY TERMS

Match the numbered terms in Column A with their definitions in Column B.

A
1. loess
2. typhoon
3. dike
4. terraced field
5. invest
6. tungsten
7. yurt
8. high-technology industry
9. consumer goods
10. calligraphy

B
A. level strip of land cut out of a hillside
B. the art of beautiful writing
C. tent made of animal skins
D. to put money in
E. production of computers and electronic equipment
F. yellow-gray soil deposited by the wind
G. tropical storm with strong winds and heavy rains
H. bank of earth built to prevent floods
I. metal used in electrical equipment
J. products such as television sets, cars, and motorcycles

Mental Mapping Activity

On a separate sheet of paper, draw a freehand map of China and its neighbors. Label the following items on your map:
- Beijing
- Chang Jiang
- Gobi
- Taiwan
- Ulan Bator

REVIEWING THE MAIN IDEAS

Section 1
1. **REGION** How does China compare in size to other countries?
2. **PLACE** Why was the Huang He called "China's sorrow"?
3. **PLACE** What climate is found in northeastern China?

Section 2
4. **MOVEMENT** Why has China made changes in its economy?
5. **HUMAN/ENVIRONMENT INTERACTION** How has the growth of industry affected China's economy?
6. **PLACE** What economic activities take place in western China?

Section 3
7. **PLACE** What kind of government has China had since 1949?
8. **REGION** What percent of China's people live in rural areas?
9. **PLACE** On what ideas did the ancient Chinese base their way of living?

Section 4
10. **PLACE** What religion is practiced in Mongolia?
11. **PLACE** Why does Taiwan have limited farmland?
12. **PLACE** What will happen to Macao in 1999?

Reviewing the Main Ideas

Section 1

1. It is the third-largest in size, after Russia and Canada.
2. The Huang He often flooded, causing death and widespread destruction.
3. humid continental

Section 2

4. to become a modern, industrialized country

5. It has raised the standard of living of some Chinese, but now there is a growing gap between the rich and the poor. Also, it has put extra burdens on the economy by increasing pollution.
6. herding, farming, mining

Section 3

7. a communist government
8. about 75 percent
9. the teachings of Confucius mixed with Buddhism

CRITICAL THINKING ACTIVITIES

1. **Determining Cause and Effect** How have recent changes in China's economy affected the Chinese people?
2. **Drawing Conclusions** How have landscape and climate affected where the Chinese people live?

GeoJournal Writing Activity

Imagine that you are a reporter traveling on one of the major rivers of China. Write a news story about the river and the people who live along it.

COOPERATIVE LEARNING ACTIVITY

Work in a group of three to learn about and create the flags of China, Mongolia, and Taiwan. Each group will choose the flag of one country, and each group member will select one of the following to research: (a) flag symbols and their meanings; (b) flag colors and their meanings; or (c) the historical events related to the creation of the flags. After your research is complete, prepare a group report describing the symbols, colors, and history of the flag. Create a sample flag to present to the rest of the class.

PLACE LOCATION ACTIVITY: CHINA

Match the letters on the map with the places and physical features of China and its neighbors. Write your answers on a separate sheet of paper.

1. Huang He
2. Mongolia
3. Taklimakan
4. Beijing
5. Hong Kong
6. Taipei
7. Shanghai
8. South China Sea
9. Tian Shan
10. Himalayas

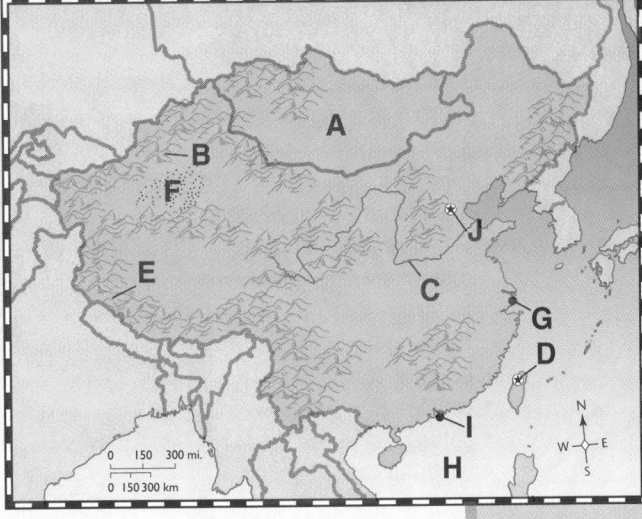

Cooperative Learning Activity

Suggest that students reproduce the flags on large sheets of paper and use dowel rods as flag poles. Display the flags near the reports in the classroom.

GeoJournal Writing Activity

Encourage students to act like on-the-scene reporters and read their news stories to the class.

Place Location Activity

1. C
2. A
3. F
4. J
5. I
6. D
7. G
8. H
9. B
10. E

Chapter Bonus Test Question

This question may be used for extra credit on the chapter test.

The Huang He is known as "China's sorrow" because of the damage it does when it floods. But how has this flooding helped farming in China? *(When the river floods, it deposits fertile soil on the land.)*

Section 4

10. Buddhism
11. because the central area of the country consists of thickly forested mountains
12. It will switch to mainland Chinese rule.

Critical Thinking Activities

1. Some Chinese now enjoy a higher standard of living and can afford to buy consumer goods. For some, however, prices are rising faster than wages. These people are falling behind in terms of buying power, and there is a growing gap between the rich and the poor.
2. Most Chinese settled in the eastern part of the country, where the land is fertile, the climate is relatively mild, and there is good access to rivers and the sea. In contrast, western China has few people because of its isolation, cold climate, and mountain and desert landscapes.

GeoLab
ACTIVITY

FOCUS

Ask students how rain can be both the best friend and the worst enemy of a farmer. Guide them to the understanding that rainfall, although necessary for growing crops, can also damage the land. In fact, the erosion from heavy rainfall is such a serious problem that it threatens the food supply in parts of Asia.

TEACH

Write *topsoil* on the board. With student input, create a concept map that answers *What? Where? When?* and *Why?* questions. The map might include the following answers: *What?* Topsoil is the topmost, fertile layer of earth. *Where?* Topsoil is located at the surface of the earth, usually just several inches deep. Because it is thin and fine, it is easily washed away. *When?* Topsoil is created over decades and centuries, as other substances break down. *Why?* Topsoil is important because it contains substances and bacteria essential for crop growth. **L2**

From the classroom of Don Mendenhall, Coleman Junior High, Van Buren, Arkansas

EROSION: Saving the Soil

Background

Because of the enormous amount of rainfall in many Asian countries, growing crops has become very difficult. Too much rain erodes topsoil and results in the destruction of crops. In this activity, you will try various ways to solve the problem of erosion.

Terraced Fields

Believe it or NOT!

Monsoon rains drench Asia every year, bringing much-needed water. But they also cause serious flooding. In 1887 the Huang He overflowed its banks because of heavy rains and the spring thaw. About 1 million people died as a result—the greatest flood disaster in history.

Materials

- 5 aluminum 8- or 9-inch pans
- leaves or grass clippings
- 500-ml beaker
- water
- ruler
- newspaper
- pebbles
- potting soil
- watering can
- dishpan

Curriculum Connection Activity

Mathematics Explain to students that quantitative data is often presented in a visual form to make it more easy to analyze. Encourage students to translate the results of their experiments—runoff water and soil measurements—into an appropriate chart or graph. If possible, have students use computers to generate their graphic materials. **L3**

What To Do

A. Fill each pan almost to the rim with soil. Pat the soil until the surface is firm and flat. Soak the soil by pouring 100 ml of water into each pan.

B. In one pan, cover the surface with a layer of grass clippings or leaves.

C. In the second pan, lay newspaper strips across the surface.

D. In the third pan, build a terrace by forming two small walls of pebbles.

E. In the fourth pan, use your finger to make curved grooves across the surface. This represents contour plowing.

F. Set the fifth pan aside.

G. Now follow this procedure with all five pans:

H. Measure 200 ml of water into the watering can.

I. Hold the pan so that one side dips slightly into the dishpan.

J. Slowly pour the water onto the soil at the top of the pan. Wait until the excess water runs across the soil and into the dishpan. **Note:** Be sure to hold the terracing and contour plowing pans so the water runs across the pebble walls and the grooves.

K. After repeating step J for each pan, pour the water and soil that ends up in the dishpan into the beaker. Measure the amount and record the results. When the soil settles to the bottom, measure and record its height in the beaker.

Lab Activity Report

1. Which pan had the least erosion?

2. Which pan had the most erosion?

3. **Drawing Conclusions** What did you learn about controlling erosion?

Go A Step Further

Activity

Experiment with different types of soils to see if soil composition—as well as soil covering—affects erosion. Mix potting soil with sand, clay, or flour. Which mixture best prevents erosion?

ASSESS

Have students answer the Lab Activity Report questions on page 631.

CLOSE

Encourage students to complete the **Go A Step Further** writing activity. Consider inviting a farmer or other qualified individual to explain how soil erosion is controlled in your region.

Go A Step Further
Answer

Results of experiments will vary.

Strange but TRUE!

Topsoil that took thousands of years to develop can be washed away in just one growing season of a few months.

Answers to Lab Activity Report

1. Answers will vary, but contour plowing should show some of the best results.

2. Results will vary, but all "treated" pans should show less erosion than the fifth pan.

3. Students should observe that some methods are more effective than others.

PERFORMANCE ASSESSMENT ACTIVITY

CREATING A RÉSUMÉ Show students the format for a professional résumé. Then have them develop a résumé for Japan in application for the position of a leading world economic power. Suggest that they categorize Japan's qualities according to the five themes of geography. After students have completed their résumés, have them compose a cover letter of introduction briefly expressing reasons why Japan is qualified for the job opening.

POSSIBLE RUBRIC FEATURES: Accuracy of content information, concept attainment, organization skills, application and analysis skills, composition skills

MENTAL MAPPING ACTIVITY
Before teaching Chapter 24, point out Japan and the Korean Peninsula on a globe to students. Have students identify the hemispheres in which Japan and the Korean Peninsula are located and where the countries are located relative to the United States.

TEACHER'S CORNER

NATIONAL GEOGRAPHIC SOCIETY

INDEX TO NATIONAL GEOGRAPHIC MAGAZINE

The following articles may be used for research relating to this chapter:

- "Kuril Islands," by Charles E. Cobb, Jr., October 1996.
- "The Great Tokyo Fish Market," by T.R. Reid, November 1995.
- "Geisha," by Jodi Cobb, October 1995.
- "Hiroshima," by Jodi Cobb, August 1995.
- "Kobe Wakes to a Nightmare," by T.R. Reid, July 1995.

- "Inner Japan," by Patrick Smith, September 1994.
- "Kyushu," by Tracy Dahlby, January 1994.
- "Japan's Sun Rises Over the Pacific," by Arthur Zich, November 1991.
- "The South Koreans," by Boyd Gibbons, August 1988.
- "Kyongju, Where Korea Began," by Cathy Newman, August 1988.

NATIONAL GEOGRAPHIC SOCIETY PRODUCTS AVAILABLE FROM GLENCOE

To order the following products for use with this chapter, contact your local Glencoe sales representative or call Glencoe at 1-800-334-7344:

- *STV: World Geography* (Videodisc)
- *Picture Atlas of the World* (CD-ROM)
- *Physical Geography of the World* (Transparencies)
- *ZipZapMap! World* (Software)

- *GeoBee* (Software)
- *Images of the World* (Posters)
- *Eye on the Environment* (Posters)

ADDITIONAL NATIONAL GEOGRAPHIC SOCIETY PRODUCTS

To order the following products for use with this chapter, call National Geographic Society at 1-800-368-2728:

- *Democratic Government Series:* "Japan." (Video)
- *Nations of the World Series:* "Japan." (Video)

- *Living Treasures of Japan* (Video)
- *Physical Geography of the Continents Series:* "Asia." (Video)

TEACHER-TO-TEACHER

Curriculum Connection: History, Language Arts

—from Doris Knoch
South Adams School, Berne, IN

DRAWING "PARALLELS" The purpose of this activity is to help students gain an understanding of the division of North and South Korea by developing a scenario for a division elsewhere on the 38th parallel.

Have students use library resources to research and develop a time line of the major events that led to the division of Korea along the 38th parallel. Have students add to their time lines a brief paragraph discussing the results of that division.

On a globe, trace the 38th parallel with your finger, moving eastward from Korea to the United States. Invite students to turn to the atlas at the front of this textbook and "take a walk" along the 38th parallel across the United States. Have students imagine a "parallel scenario"—a division of the United States along the 38th parallel. Encourage them to write a paragraph explaining what geographic features, cultural differences, or historical developments or events might have led to such a division.

Call on volunteers to display their time lines and read their "parallel scenarios." Then lead the class in a discussion of why such a division occurred in Korea but would have been improbable in the United States.

MEETING NATIONAL STANDARDS

Geography for Life

All of the 18 standards are demonstrated in Unit 8. The following ones are highlighted in this chapter:

- Standards 1, 2, 7, 9, 11, 12, 13, 14, 15, 16, 17, 18

KEY TO ABILITY LEVELS

Teaching strategies have been coded for varying learning styles and abilities.

L1 **BASIC** activities for all students

L2 **AVERAGE** activities for average to above-average students

L3 **CHALLENGING** activities for above-average students

LEP **LIMITED ENGLISH PROFICIENCY** activities

BIBLIOGRAPHY

Readings for the Student

Bunce, Vincent. *Japan: People and Places.* New York: Franklin Watts, 1994.

Kalman, Bobbie. *Japan: The Lands, Peoples, and Cultures Series.* New York: Crabtree, 1989. Three 32-page paperbacks.

Tames, Richard. *Passport to Japan.* New York: Franklin Watts, 1994.

Warshaw, Steven. *Japan Emerges,* revised ed. Emeryville: Diablo, 1993.

Readings for the Teacher

Creative Ideas for Teaching About Japan. Stockton, Calif.: Stevens & Shea, 1988. Reproducible activity sheets.

Minear, Richard H. *Through Japanese Eyes,* 3rd revised ed. New York: Center for International Training and Education, 1994.

Modern Japan: An Idea Book for K-12 Teachers. Bloomington, Ind.: Social Studies Development Center, 1992.

Multimedia

Discover Korea: Geography and Industry. New York: Asia Society, 1988. Videocassette, 25 minutes.

Exotic Japan, revised ed. Mountain View, Calif.: Pixonix, 1994. CD-ROM for Macintosh.

Japan. Washington, D.C.: National Geographic Society, 1990. Videocassette, 25 minutes.

Chapter 24

Resource Organizer

Japan and the Koreas

Use these *Geography: The World and Its People* resources to teach, reinforce, and extend chapter content.

CHAPTER 24 RESOURCES

*Vocabulary Activity 24

Cooperative Learning Activity 24

Workbook Activity 24

Chapter Map Activity 24

Geography Skills Activity 24

GeoLab Activity 24

Critical Thinking Skills Activity 24

Performance Assessment Activity 24

*Reteaching Activity 24

Enrichment Activity 24

Chapter 24 Test, Form A and Form B

*Chapter 24 Digest Audiocassette, Activity, Test

Unit Overlay Transparencies 8-0, 8-1, 8-2, 8-3, 8-4, 8-5, 8-6

Political Map Transparency 8; World Cultures Transparency 16

Geoquiz Transparencies 24-1, 24-2

Vocabulary PuzzleMaker Software

Student Self-Test: A Software Review

Testmaker Software

MindJogger Videoquiz

Focus on World Art Print 10

If time does not permit teaching the entire chapter, summarize using the Chapter 24 Highlights on page 649, and the Chapter 24 English (or Spanish) Audiocassettes. Review student's knowledge using the Glencoe MindJogger Videoquiz. * *Also available in Spanish*

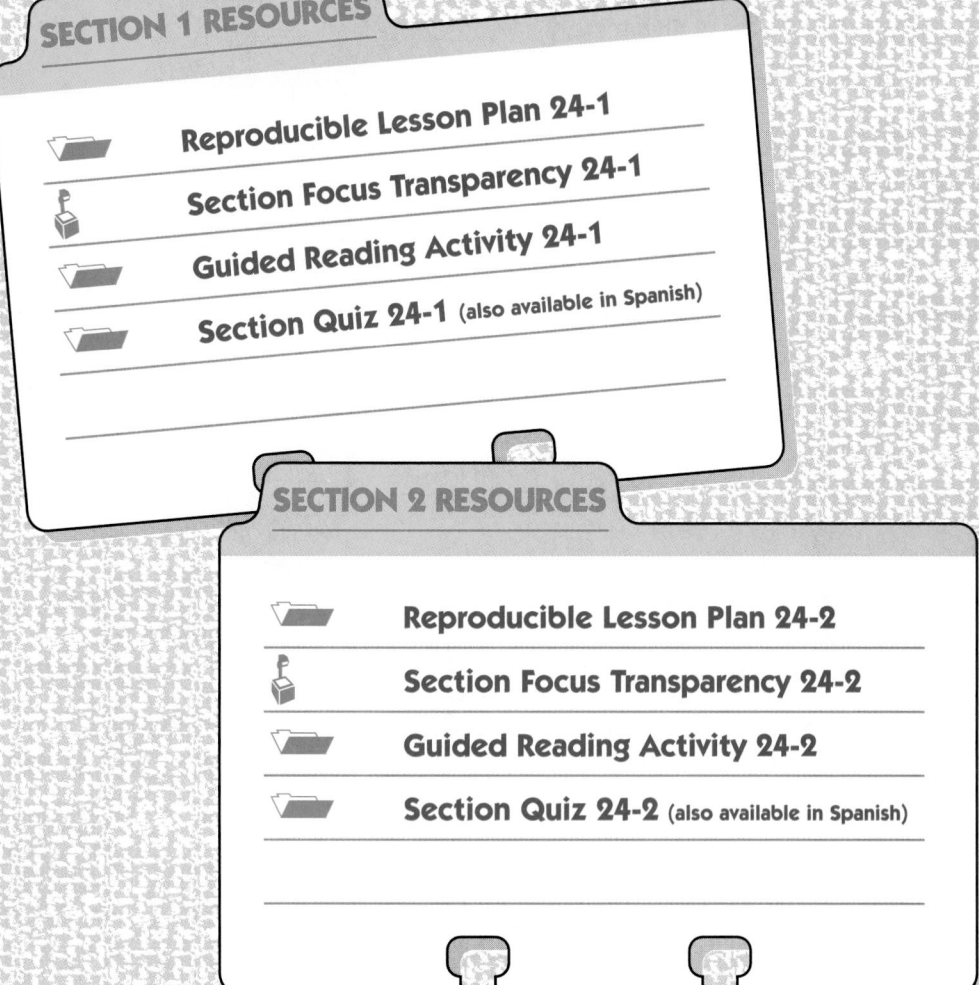

Use these *Geography: The World and Its People* resources to teach and reinforce section content.

SECTION 1 RESOURCES

Reproducible Lesson Plan 24-1

Section Focus Transparency 24-1

Guided Reading Activity 24-1

Section Quiz 24-1 (also available in Spanish)

SECTION 2 RESOURCES

Reproducible Lesson Plan 24-2

Section Focus Transparency 24-2

Guided Reading Activity 24-2

Section Quiz 24-2 (also available in Spanish)

ADDITIONAL RESOURCES FROM GLENCOE

 **Reproducible Masters**

- Glencoe Social Studies Outline Map Book, pages 34, 35
- Foods Around the World, Unit 9
- Facts On File: GEOGRAPHY ON FILE, East Asia

 World Music: Cultural Traditions Lesson 8

 **Transparencies**

- National Geographic Society PicturePack Transparencies, Unit 8

 **Posters**

- National Geographic Society: Images of the World, Unit 8

 **Countries of the World Flashcards** Unit 8

 **World Crafts Activity Cards** Unit 8

 **Videodiscs**

- Geography and the Environment: The Infinite Voyage
- National Geographic Society: STV: World Geography, Vol. 1

CD-ROM

- National Geographic Society: Picture Atlas of the World

Software

- National Geographic Society: ZipZapMap! World

Performance Assessment

Refer to the Planning Guide on page 632A for a Performance Assessment Activity for this chapter. See the *Performance Assessment Strategies and Activities* booklet for suggestions.

Chapter Objectives

1. **Describe** the location, economy, and culture of Japan.
2. **Explain** why and how life in North Korea differs from life in South Korea.

GLENCOE TECHNOLOGY

 **VIDEODISC**

Use the Chapter 24 MindJogger Videoquiz to preview chapter content.

MindJogger Videoquiz

Chapter 24
Disc 3 Side B

 Also available in VHS.

Chapter 24 Japan and the Koreas

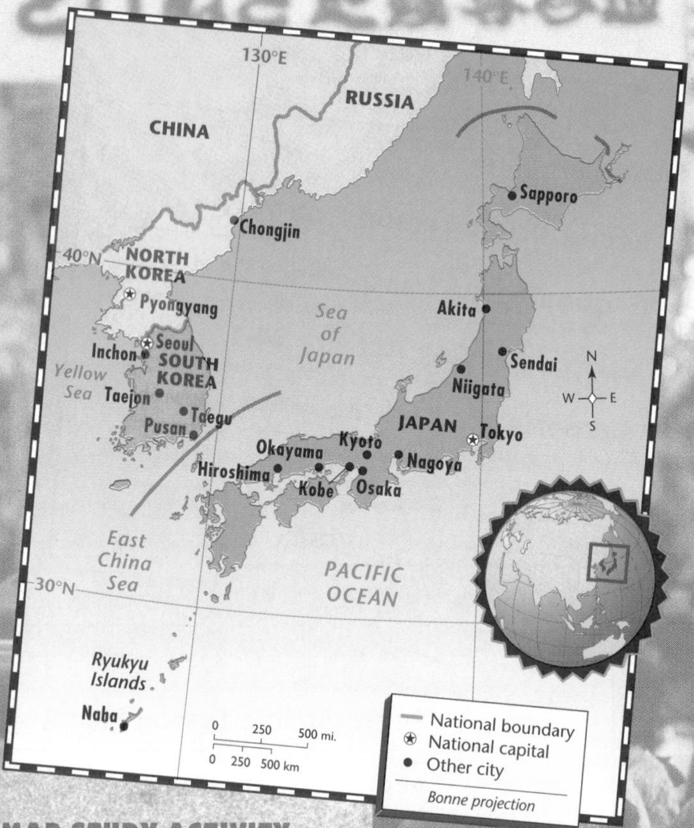

National boundary
National capital
Other city

Bonne projection

MAP STUDY ACTIVITY

As you read Chapter 24, you will learn about Japan and the Koreas.

1. **What is the capital of North Korea?**
2. **What sea lies between Japan and North Korea?**
3. **What is Japan's capital?**

MAP STUDY ACTIVITY

Answers
1. Pyongyang
2. Sea of Japan
3. Tokyo

Map Skills Practice
Reading a Map Which capital is closer to the border between the Koreas, North Korea's or South Korea's? *(South Korea's)* Is Sapporo located north or south of Tokyo? *(north)*

PREVIEW

Words to Know
- archipelago
- tsunami
- intensive cultivation
- clan
- samurai
- shogun
- megalopolis

Places to Locate
- Japan
- Tokyo
- Honshu
- Hokkaido
- Kyushu
- Shikoku

Read to Learn . . .
1. where Japan is located.
2. why Japan has a strong economy.
3. what religions influenced the culture of Japan.

On a clear day in Tokyo, Japan, you can see

beautiful Mount Fuji. Mount Fuji, Japan's highest peak, is an inactive volcano. During the summer thousands of Japanese make their way to the top of the mountain.

Meeting National Standards

Geography for Life The following standards are highlighted in this section:
- Standards 1, 2, 7, 11, 12, 14, 15, 17

FOCUS

Section Objectives
1. **Describe** where Japan is located.
2. **Investigate** why Japan has a strong economy.
3. **Identify** what religions influenced the culture of Japan.

Bellringer Motivational Activity

Prior to taking roll at the beginning of the class period, project Section Focus Transparency 24-1 and have students answer the activity questions. Discuss students' responses. **L1**

Vocabulary Pre-check

Display a transparency showing a map of the world and point out examples of archipelagoes, such as Japan, the Philippines, and Hawaii. Ask students to formulate a definition of *archipelago* based on these examples. **L1 LEP**

Use the Vocabulary PuzzleMaker Software to create a crossword puzzle. **L1**

Love of nature is a major part of the Japanese way of life. The sun, the mountains, and the sea have always been important to the people of Japan. The map on page 632 shows you that Japan is an **archipelago,** or a chain of islands. The Japanese islands form a curve off the coast of eastern Asia between the Sea of Japan and the Pacific Ocean. Four main islands make up Japan. They are Hokkaido (hah•KY•doh), Honshu, Shikoku (shih•KOH•koo), and Kyushu (kee•OO•shoo). Thousands of smaller islands are also part of Japan.

Location
The Land

Japan has a land area of 145,370 square miles (376,508 sq. km)—slightly smaller than California. The Japanese islands are the peaks of a great mountain range rising 20,000 to 30,000 feet (6,000 to 9,000 m) from the floor of the Pacific Ocean. They lie in the Ring of Fire—an area surrounding the Pacific Ocean where the earth's crust often shifts. As a result, earthquakes and volcanic eruptions are common in Japan and other parts of the Ring of Fire.

Classroom Resources for Section 1

REPRODUCIBLE MASTERS

Reproducible Lesson Plan 24-1
Guided Reading Activity 24-1
Vocabulary Activity 24
Workbook Activity 24
GeoLab Activity 24
Chapter Map Activity 24
Section Quiz 24-1

TRANSPARENCIES

Section Focus
    Transparency 24-1
Political Map Transparency 8
Unit Overlay Transparencies
    8-0, 8-4, 8-5

MULTIMEDIA

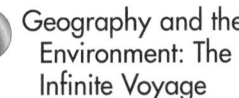 Vocabulary PuzzleMaker Software

Testmaker

Geography and the Environment: The Infinite Voyage

MAP STUDY

Answer
Hokkaido, Honshu, Shikoku, Kyushu

Map Skills Practice
Reading a Map What body of water separates Shikoku and Kyushu from Korea? *(Korea Strait)*

TEACH

Guided Practice

Understanding Information Distribute copies of the following:

My country consists of thousands of islands that are really the peaks of a massive underwater mountain range. It is located in the Ring of Fire, and my country is regularly jolted by earthquakes and volcanic eruptions. The ocean plays an important part in everybody's life, for no location in the country is further than 70 miles (113 km) from the sea.

Ask students to identify the country described in the paragraph. *(Japan)* Have them continue the description by adding information on Japan's physical features, climate, economy, history, people, and culture. **L1**

GLENCOE
TECHNOLOGY

The Infinite Voyage:
Living with Disaster

Chapter 5
Earthquakes: A Turbulence Beneath the Earth
Disc 3 Side A

634

JAPAN AND THE KOREAS: Physical

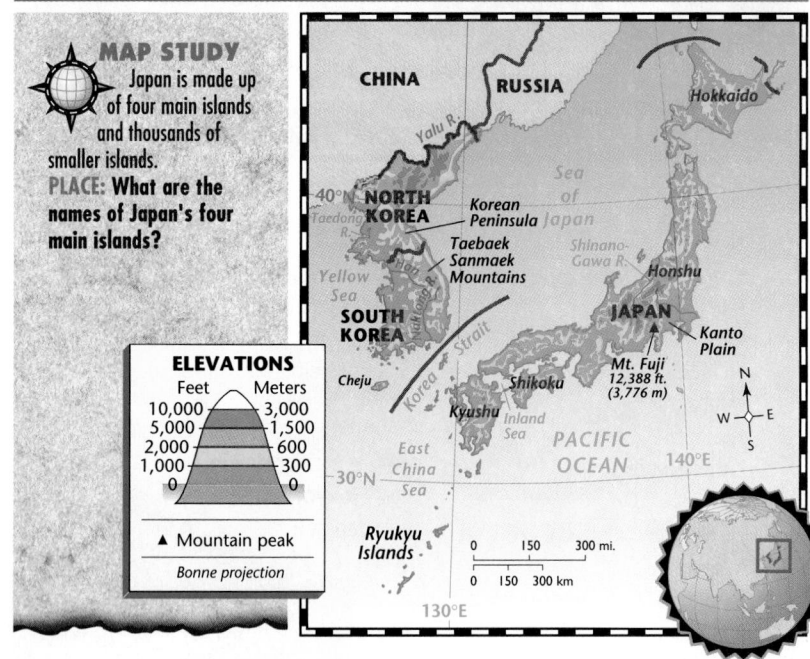

MAP STUDY Japan is made up of four main islands and thousands of smaller islands.
PLACE: What are the names of Japan's four main islands?

ELEVATIONS

Feet	Meters
10,000	3,000
5,000	1,500
2,000	600
1,000	300
0	0

▲ Mountain peak

Bonne projection

Earthquakes Japan experiences about 5,000 earthquakes a year. Most are minor and cause little damage. But every few years, severe earthquakes shake the land. One of the worst earthquakes took place in 1995, near the port city of Kobe (KOH•bee). Thousands of people died, and damage was estimated at $50 billion to $70 billion. People in Japan and other areas of the Ring of Fire also have to face **tsunamis.** These huge sea waves caused by undersea earthquakes are very destructive along Japan's Pacific coast.

Mountains The physical map above shows you that rugged mountains and steep hills cover most of Japan. Rocky gorges and narrow valleys cut through the mountains. Thick forests grow on the lower slopes. Swift, short rivers flow through the mountains in thundering waterfalls before running to the sea.

Sea and Plains No part of Japan is more than 70 miles (113 km) from the sea. Coastlines are rough and jagged. In bay areas along the coasts, you will find many fine harbors and ports. One of Japan's most important seacoast areas lies along a body of water called the Inland Sea. This sea winds its way among the islands of Honshu, Shikoku, and Kyushu.

Narrowly squeezed between the seacoast and the mountains are the plains areas of Japan. The largest is the Kanto Plain, on the east coast of Honshu where Tokyo and Yokohama are located. You will find most of Japan's cities, farms, and industries in the coastal plains.

Cooperative Learning Activity

Organize students into several groups. Direct groups to create posters, collages, or annotated and illustrated wall maps titled *Life in the Ring of Fire.* Visual materials should illustrate the incidence of earthquakes and volcanic activity in Japan. Suggest that groups include information such as the most destructive earthquakes, volcanic eruptions, and tsunamis; first-person accounts of such disasters; and the steps the Japanese have taken and are taking to reduce the impact of these events. Have groups display their visual materials around the room. **L1**

The Climate Ocean and wind currents affect Japan's climate. Look at the map below. You see that climate differs in northern and southern Japan. Hokkaido and northern Honshu have a humid continental climate. Cold ocean and wind currents from the Arctic area and Siberia result in cold winters here. Southern Honshu and the other two major islands have a humid subtropical climate. This climate is caused by a warm current flowing north from the Pacific Ocean.

Human/Environment Interaction
The Economy

Although Japan's land area is relatively small, the country is an economic giant. Japan is one of the world's major industrial powers. It makes use of every resource—people, technology, land, and sea. The Japanese people have one of the highest standards of living in the world.

Farming Farmland in Japan is very limited. Most of Japan's farms are small—about three acres (one ha). Japanese farmers, however, raise about 60 percent of their country's food on privately owned plots of land. How do they do this? Japanese farmers use fertilizers and modern machinery to produce high crop yields. They also practice **intensive cultivation** in which farmers grow crops on

Inland Sea, Japan

The Inland Sea is an important waterway linking the Pacific Ocean and the Sea of Japan.
PLACE: What is a tsunami?

JAPAN AND THE KOREAS: Climate

MAP STUDY
In northern Japan, the winters are cold and snowy.
REGION: What type of climate is found in northern Japan and all of North Korea?

CHINA
RUSSIA
Sapporo
Chongjin
NORTH KOREA
Sea of Japan
40°N
Pyongyang
Yellow Sea
Seoul
SOUTH KOREA
JAPAN
Tokyo
Kobe
Osaka
East China Sea
PACIFIC OCEAN
140°E
30°N
Ryukyu Islands

CLIMATE
Mid-Latitude
Humid continental
Humid subtropical
Bonne projection

0   150   300 mi.
0   150   300 km

130°E

CHAPTER 24   635

Independent Practice

Guided Reading Activity 24-1 **L1**

More About the Illustration

Answer to Caption
a huge sea wave produced by underwater earthquakes

GEOGRAPHIC THEMES

Place Hundreds of small islands dot Japan's Inland Sea. The mountains on these islands act as a wind break, creating very calm waters for such a large body of water.

MAP STUDY

Answer
humid continental

Map Skills Practice
Reading a Map Where in Japan would you find a humid subtropical climate? *(southern Japan)*

Meeting Special Needs Activity

Learning Disability Students who have learning disabilities may benefit from having a specific purpose for listening. After students have read the subsection headed "The Land," lead a discussion about the material. Tell students to listen for words or phrases from the text. Tell them to write each word or phrase that also appeared in the text. Use the following terms: Ring of Fire, earthquakes, volcanic eruptions, tsunamis, Kanto Plain, humid continental climate, and humid subtropical climate. Give extra credit to any student who lists all the terms. **L1**

CURRICULUM CONNECTION

Earth Science Geologists believe that the basin beneath the Inland Sea was formed by faulting, or the cracking of the earth's crust. This faulting split apart a single piece of land to form three of the four main islands of Japan.

GRAPHIC STUDY

Answer

by about 9.9 million cars

Graphic Skills Practice
Reading a Graph How many cars were produced in Japan in 1990? *(about 9.9 million)*

Japan The Japanese harvest six kinds of edible seaweed. One kind, *nori,* is rich in vitamins A, D, and B12, while another kind, *kombu,* is rich in iodine and vitamin C. Seaweed plays many roles in the Japanese diet, such as a garnish, a wrapper for fish, and an ingredient in soup stock and snack foods.

every available piece of land. If you visit Japan, you will see farmers growing rice on terraces cut into the hillsides. You will also see rows of crops growing between buildings and highways. Using these methods, Japanese farmers are able to plant and harvest three crops a year in the warmer regions of the country. They grow rice, wheat, fruit, vegetables, and tea.

The Japanese also rely on the sea for food. People who fish in Japan catch salmon, tuna, and snapper. Huge catches make Japan one of the world's leading fishing countries.

Cars Made in Japan

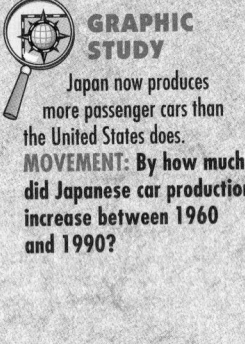

GRAPHIC STUDY

Japan now produces more passenger cars than the United States does.
MOVEMENT: By how much did Japanese car production increase between 1960 and 1990?

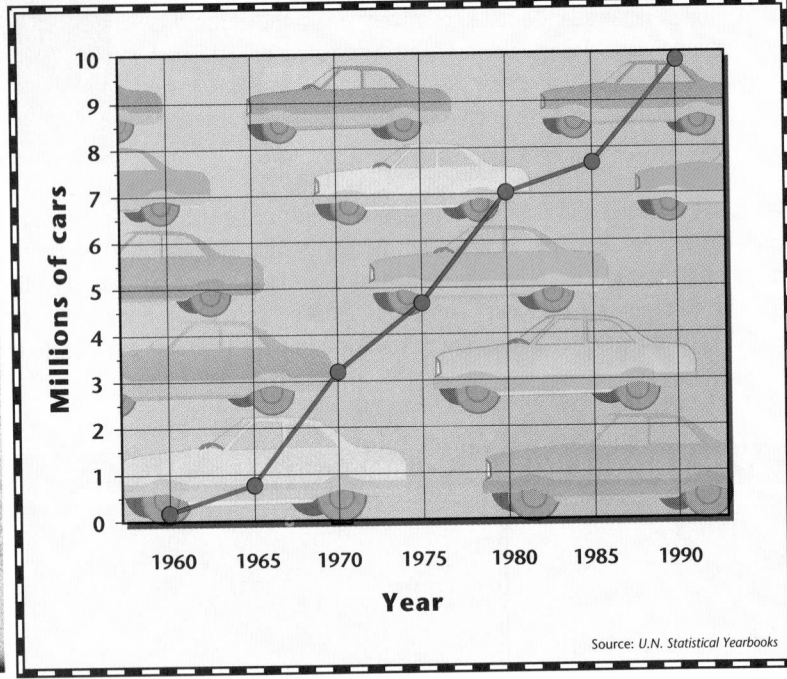

Source: *U.N. Statistical Yearbooks*

Manufacturing Japan is known around the world for the variety and quality of its manufactured goods. It is a leader in manufacturing because of its use of high technology, including robots. Japan produces and exports steel, cars, ships, computers, cameras, televisions, and textiles. The graph above shows you how much Japan has increased its production of cars since 1960. You probably own a piece of electronic equipment—TV, camera, or VCR—that was made in Japan or has Japanese parts.

In addition to the latest technology, Japan's economy has prospered because of its highly skilled workers. The Japanese people value education, hard work, and cooperation. Because of Japan's industrial success, many countries look to Japan as a model for improving their own economies.

Critical Thinking Activity

Analyzing Information On the chalkboard, copy the following Japanese proverb: *Failure is the source of success.* Then ask students to analyze the information under "Environment," "Early Japan," and "Modern Japan" for examples that confirm the proverb. For each example, direct students to find passages that describe a failure and the success that came from that failure. Encourage students to share and compare their examples and passages. **L3**

Trade Japan has limited mineral resources and must import oil, coal, iron ore, and other materials for manufacturing. In return, Japan exports manufactured goods to other countries. The chart on page 648 compares Japan's trade with that of other Asian countries in the Pacific Rim. The term "Pacific Rim" refers to the rapidly developing Asian countries that border the Pacific Ocean.

Environment Japan's industrial growth has created many environmental challenges. By the 1970s, waste products from factories had polluted the air and water. The Japanese government urged industries to prevent pollution. Today Japan's pollution-control laws are among the strictest in the world. Japan also leads the world in the development of advanced public transportation systems. You can read about Japan's famous bullet trains on page 641.

Movement
Influences of the Past

The heritage of the Japanese people is long and rich. The Japanese trace their ancestry to various **clans,** or groups of related families, that lived and ruled in Japan as early as the late A.D. 400s.

Early Japan The Japanese developed close ties with the Chinese on the Asian mainland. Ruled by emperors, Japan modeled its society on the Chinese way of life. The Japanese also borrowed the Chinese system of writing and accepted the Buddhist religion brought by Chinese missionaries.

Beginning in the 790s, the Japanese emperors' power declined. From the late 1100s to the 1860s, Japan was ruled by the **samurai**—the warrior class—and their leaders—the **shoguns.** Like China, Japan worked to keep out foreign countries that wanted to trade with them. In 1853 the United States government sent a fleet headed by Commodore Matthew Perry to Japan to demand trading privileges. On threat of war, the Japanese opened their country to the rest of the world.

Modern Japan In the late 1800s, Japan began to change into a modern country. Japanese leaders used European and American ideas to improve education and to set up industries. By the early 1900s, Japan had become the world's leading Asian power.

In the 1930s Japan needed more resources for its growing population. It began taking over land in China. In 1941 the Japanese attacked the American naval base at Pearl Harbor in Hawaii and began the Pacific conflict in World War II. The fighting between Japan and the United States and other countries ended in 1945 with Japan's defeat. After the war, many of Japan's cities lay in ruins.

Exercising at the Office

Japanese office workers in Kyoto stand for their morning exercises.
MOVEMENT: How can Japan produce so many goods when it has so few natural resources?

Cultural Kaleidoscope

Japan Japan produces more garbage than any other country in the world except the United States.

More About the Illustration

Answer to Caption
it imports large quantities of mineral resources

Place A popular Japanese proverb states that "the nail that sticks out gets hammered down." Indeed, individualism is frowned upon in Japan. From an early age, Japanese children are taught that cooperation and loyalty to the group are the most important qualities. The ideas of cooperation, loyalty, and belonging are encouraged at work. Workers wear company pins, sing company songs, and attend company social functions. Some businesses have employees wear similar clothing or uniforms to ensure that no one is treated differently.

More About . . .

Shinto All Japanese belong to the State Shinto—Japan's official religion, or "way of the gods." Shinto has no founder or sacred book. Its followers worship numerous gods, emperors, heroes, and ancestors. State Shinto demands loyalty to the emperor, who supposedly descends from the Sun Goddess. Another division of Shinto, Shrine Shinto, centers its rites around state-supported shrines. At these shrines, priests pray for peace, good harvests, and prosperity for all. Festivals held at these shrines incorporate many Buddhist practices.

Global Gourmet

Japan A dish of raw fish is called *sashimi*. A dish of cooked or raw fish served with rice and vinegar is called *sushi*. If the cook adds a vegetable to *sushi* or substitutes vegetables for the fish, the dish is called *norimaki*.

Cultural Kaleidoscope

Japan On March 3, city girls dress in their best outfits to celebrate the Dolls Festival. They place their dolls on shelves and offer them peach blossoms and *saké*. The festival is derived from a rural ceremony in which paper figures are set afloat to take evil and disease away.

GRAPHIC STUDY

Answer
about 125 million people

Graphic Skills Practice
Reading a Graph About how many times larger is the population of the United States than that of Japan? *(about two and one-half times larger)*

Paper Art

People in Japan make an art—called origami—out of folding paper. From paper they create decorative objects such as animals, fish, or flowers. There are about 100 traditional origami patterns. Each year new patterns are developed.

The Japanese economy was near collapse. With help from the United States, Japan became a democracy and rebuilt its economy.

Movement

The People

Although only the size of California, Japan has about one-half the population of the entire United States. Look at the graph below to compare Japan's population of 125 million people with that of the United States. About 77 percent of Japan's people are crowded into urban areas on the coastal plains. More than one-third of these city dwellers live in four large cities located on the plains of Honshu. These four cities are Tokyo, Yokohama, Osaka, and Nagoya. They form a **megalopolis,** a huge supercity made up of several large cities and the smaller cities near them.

City Life Japanese cities are very modern. They are crowded with tall office buildings, busy streets, and freeways. Homes are small and close to one another. In the suburbs, you will find small apartment buildings and wooden houses with tiny gardens.

In Japan's cities, the past remains very much a part of urban life. Along the narrow side streets of most cities, small shops sell traditional items. Parks, gardens, and religious shrines are found close to modern neighborhoods. In Tokyo and other cities, it is common to see a person dressed in a traditional garment known as a kimono walking with another person wearing a T-shirt and jeans!

Country Life Only 23 percent of Japan's people live in rural areas. These people generally earn less than city dwellers, but their standard of living has improved in recent years. Most have a small house, a TV, and modern kitchen

COMPARING POPULATION: Japan and the United States

GRAPHIC STUDY
Almost 30 million people live in Tokyo, the capital of Japan.
PLACE: What is the approximate population of Japan?

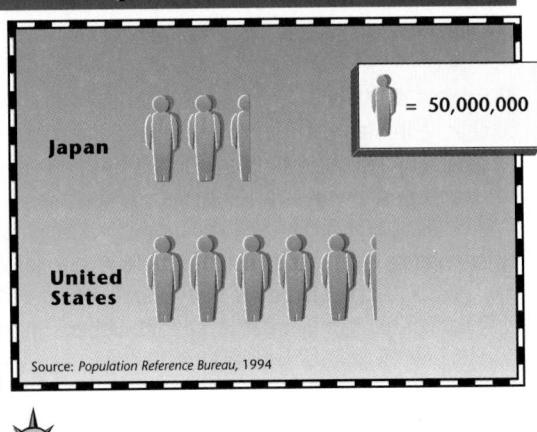

= 50,000,000

Japan

United States

Source: *Population Reference Bureau, 1994*

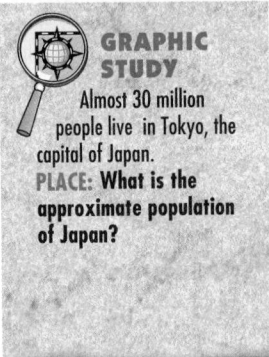

Global Issues Activity

To Protect or Not To Protect? Present the following conservation issue to the class: "Japanese along the coasts traditionally have hunted dolphins and porpoises and sold the meat locally. When the International Whaling Commission (IWC) declared a moratorium on commercial whaling, the killing of dolphins and porpoises in Japan's waters increased. Almost 50,000 were reported killed in Japanese waters during 1988. Also in 1988, a large publicity campaign encouraged the Japanese to eat more dolphin meat. Many conservationists believe that the Japanese whaling industry is exploiting dolphin and porpoise populations for profit just as they over-exploited the whale population." Ask students to play the role of mediators and negotiate an agreement between conservationists and the Japanese whaling industry. **L3**

appliances. In some rural areas, a small farmer might have three jobs. For example, a person might farm one day, fish or cut timber another day, and work occasionally in a factory.

Family Life In traditional Japan the family was the center of one's life. Each family member had to obey certain rules. Grandparents, parents, and children all lived in one house. In recent years much has changed. Family ties remain strong, but each family member is allowed more freedom. Today many Japanese also live in small family groups that consist only of parents and children.

Fifteen-year-old Michiko Wakita lives with her parents and older brother Naohiro in Tokyo. Michiko, like most Japanese teenagers, studies hard six days a week in school. After graduation from high school, she plans to study marine biology at a local university.

Despite Japan's strong emphasis on education, life is not all work for Michiko and other Japanese young people. She enjoys rock music, modern fashions, television, and movies. Her brother Naohiro likes motorcycles and comic books. When Michiko isn't watching game shows on television, Naohiro watches baseball and sumo, a Japanese form of wrestling. Judo is also popular. It is a form of martial arts that developed from ancient Asian fighting skills. Today martial arts are practiced for self-defense and for exercise. They are also very popular in the United States.

Place
Culture

Over the centuries the Japanese have developed many customs and art forms to express their delight in nature. They have passed on some of these traditions to the rest of the world. At the same time, the Japanese have accepted cultural practices from foreign countries. Many of Japan's people, for example, enjoy Western movies, hamburgers, and rock music.

Religion Japan has two major religions—Shinto and Buddhism. Shinto began in Japan centuries ago. It teaches a respect for nature, a love of simple things, and a concern for cleanliness and good manners. Buddhism came to Japan from China. It teaches respect for nature and the need for inner peace. Many Japanese practice both religions.

CHAPTER 24 639

Making the Grade

It's time for exams in Japan, and all eyes are turned toward middle school students. To get into high school, students must take very difficult entrance exams. Kin Toshiro has been studying every day and weekend for months. Exams are the main topic of conversation in his neighborhood and city. Even television shows give advice on how and what to study. After the exam, Kin and his mother will go to his school with hundreds of others to wait for the grades to be posted on the bulletin board. If Kin fails, he will have to go to a "cram school" and take the test again next year.

Critical Thinking Activity

Expressing Problems Clearly Ask students to list cultural practices that the Japanese have accepted from foreign countries. *(Examples might include: Western fast food, rock music, and television programs and movies.)* Direct students to reread the sentences about Japanese religious beliefs on page 639. Then have students express why combining certain aspects of Western culture with Japanese religion may create cultural problems in Japan. *(Answers will vary. Some students might suggest that violence depicted in movies and television programs contradicts the Japanese desire for inner peace.)* **L3**

MULTICULTURAL PERSPECTIVE
Cultural Diffusion

Tea came from China to Japan around A.D. 800. It quickly became a beverage of the rich, although the poor occasionally used it as a medicine. In the 1400s, however, the practice of drinking tea became very popular throughout Japan. This was due, in part, to the Japanese court's adoption of the Buddhist tea ceremony.

ASSESS

Check for Understanding

Assign Section 1 Review as homework or an in-class activity.

Meeting Lesson Objectives

Each objective below is tested by the questions that follow it in parentheses.

1. **Describe** where Japan is located. (5)
2. **Investigate** why Japan has a strong economy. (3)

Cultural Kaleidoscope

Japan The average life span is longer in Japan than anywhere else in the world. A Japanese man can expect to live 75.8 years and a Japanese woman 81.8 years.

Evaluate

 Section 1 Quiz **L1**

Use the Testmaker to create a customized quiz for Section 1. **L1**

Reteach

Have students work in pairs to quiz each other on the content of Section 1. **L1**

Enrich

Suggest that students find and follow directions for building a traditional Japanese kite shaped like a box, animal, or dragon. **L3**

CLOSE

Have students sketch super heroes, such as sumo wrestlers or judo masters, for a comic book that Naohiro Wakita might enjoy.

Japanese Recreation

The people of Japan love baseball *(left)*. A colorful float used for puppet shows is part of a festival parade in Japan *(right)*. **PLACE: What are some other popular pastimes in Japan?**

The Arts You may sense a peaceful feeling of nature in Japanese painting, music, dance, and literature. Many paintings show scenes of the countryside and include verses in calligraphy. The Japanese enjoy traditional and modern forms of drama. Since the 1600s Japanese theatergoers have attended the historical plays of the Kabuki theater. Kabuki plays are set on colorful stages with actors wearing brilliantly colored costumes.

Japanese literature includes poetry and novels. Many scholars believe that the world's first novel came from Japan. This novel is called *The Tale of Genji* and was written by a noblewoman of the emperor's court about A.D. 1000. Haiku (HY•koo) is a well-known type of Japanese poetry that is written according to a very specific formula. A haiku has only 3 lines and 17 syllables.

SECTION 1 REVIEW

REVIEWING TERMS AND FACTS

1. **Define the following:** archipelago, tsunami, intensive cultivation, clan, samurai, shogun, megalopolis.
2. **PLACE** What is the major landform of Japan?
3. **HUMAN/ENVIRONMENT INTERACTION** What are the main economic activities that take place in Japan?
4. **MOVEMENT** Where do most Japanese live?

 MAP STUDY ACTIVITIES
5. Look at the physical map on page 634. About how many miles is Japan from north to south?
6. Study the climate map on page 635. What are Japan's two climate regions?

Answers to Section 1 Review

1. All vocabulary words are defined in the Glossary.
2. mountains
3. manufacturing, farming, fishing, trade
4. in the coastal plains
5. about 2,400 miles (3,800 km)
6. humid continental, humid subtropical

MAKING CONNECTIONS

JAPANESE BULLET TRAINS

| MATH | SCIENCE | HISTORY | LITERATURE | **TECHNOLOGY** |

What travels almost as fast as a speeding bullet? Japan's famous bullet trains first whizzed across the Japanese countryside in 1964. The trains completely changed ground travel in Japan. They also became the model for public transport vehicles in other densely populated countries.

FROM TOKYO TO OSAKA The first blue-and-ivory bullet-nosed train raced at 125 miles per hour (242 km per hour) between Tokyo and Osaka. These two cities lie 320 miles (515 km) apart. The new train gained instant popularity. The number of people riding the train daily grew by 300 percent during its first five years.

THE BULLETS INCREASE The bullet train became the first Japanese train to run at a profit. It was so successful that other trains were built in different parts of Japan. About 260 bullet trains run daily now, covering a distance equal to circling the earth three times. Lines stretch from Tokyo to cities north and south. Eventually a bullet train will run under the sea through a tunnel to Hokkaido, the northernmost Japanese island.

Speeding Along

MEETING CHALLENGES The bullet trains carry 400,000 passengers every day—135 million each year—at speeds up to 170 miles per hour (274 km per hour). They have a 99 percent on-time record, and have never had a fatality.

The trains help the Japanese successfully overcome several challenges. They streamline transportation in tiny, crowded Japan. Most Japanese now prefer the comfort and speed of the trains to cars. As a result, the popular trains help to keep Japan's overseas oil purchases under control.

Making the Connection

1. How fast and how far do bullet trains travel daily?
2. What problems in Japan do the bullet trains help solve?

641

The Two Koreas

FOCUS

Section Objectives

1. **Recognize** where the Korean Peninsula is located.
2. **State** why the Koreas are divided.
3. **Comprehend** how life in South Korea differs from life in North Korea.

Bellringer Motivational Activity

Prior to taking roll at the beginning of the class period, project Section Focus Transparency 24-2 and have students answer the activity questions. Discuss students' responses. **L1**

Vocabulary Pre-check

Have students find the meaning of *anthracite* in the text. Then explain that although coal is a low-cost fuel, burning it contributes to acid rain and the greenhouse effect. Add that hard coal pollutes less than soft coal. **L1** **LEP**

PREVIEW

Words to Know
• anthracite

Places to Locate
• North Korea
• South Korea
• Seoul
• Pyongyang

Read to Learn . . .
1. where the Korean Peninsula is located.
2. why the Koreas are divided.
3. how life in South Korea differs from life in North Korea.

Forests—often shrouded in mist—cover much of Korea. You feel a sense of peace and calm as you look at the mountainous Korean landscape. Because of its geography, Korea was once known as "land of the morning calm."

In modern times Korea has been anything but calm. Today it is divided into two nations—communist North Korea and noncommunist South Korea. The two governments of Korea are bitter enemies, although efforts are being made to bring them together.

Region

A Divided Country

Although Korea is now divided, it has a long history as a united country. The land of Korea lies on the Korean Peninsula. This peninsula juts out from northern China, between the Sea of Japan and the Yellow Sea.

The Korean Past The Koreans trace their ancestry to people who settled the Korean Peninsula thousands of years ago. From A.D. 668 to 935, a Korean kingdom called Silla united much of the peninsula for the first time under one government. Under Silla rulers, Korea enjoyed a period of cultural and

642

UNIT 8

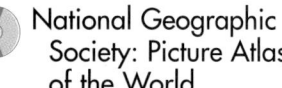

scientific advances. For example, the Koreans built one of the world's earliest astronomical observatories in the A.D. 600s.

For centuries the Korean Peninsula was a bridge between Japan and the mainland of Asia. Trade and ideas went back and forth. China and Japan, however, wanted to control Korea. China ruled part of Korea from the 100s B.C. until the early A.D. 300s. Japan later governed Korea from 1910 until the end of World War II in 1945.

Division of Korea After Japan's defeat in World War II, troops from the communist Soviet Union took over the northern half of Korea. American troops occupied the southern half. Korea eventually divided along the 38th parallel. A noncommunist government ruled in what became South Korea, while a communist government governed what became North Korea.

The Korean War In 1950 North Korean armies attacked South Korea. Their plan was to bring all of Korea under communist rule—a plan that led to the Korean War. Noncommunist United Nations countries, led by the United States, rushed to aid South Korea. China's communist leaders eventually sent troops to help North Korea. The fighting finally ended in 1953—without a peace treaty or a victory for either side. By the 1960s, two separate countries with their own ways of life had developed in the Korean Peninsula. Today the United States continues to encourage peaceful relations between the two Koreas.

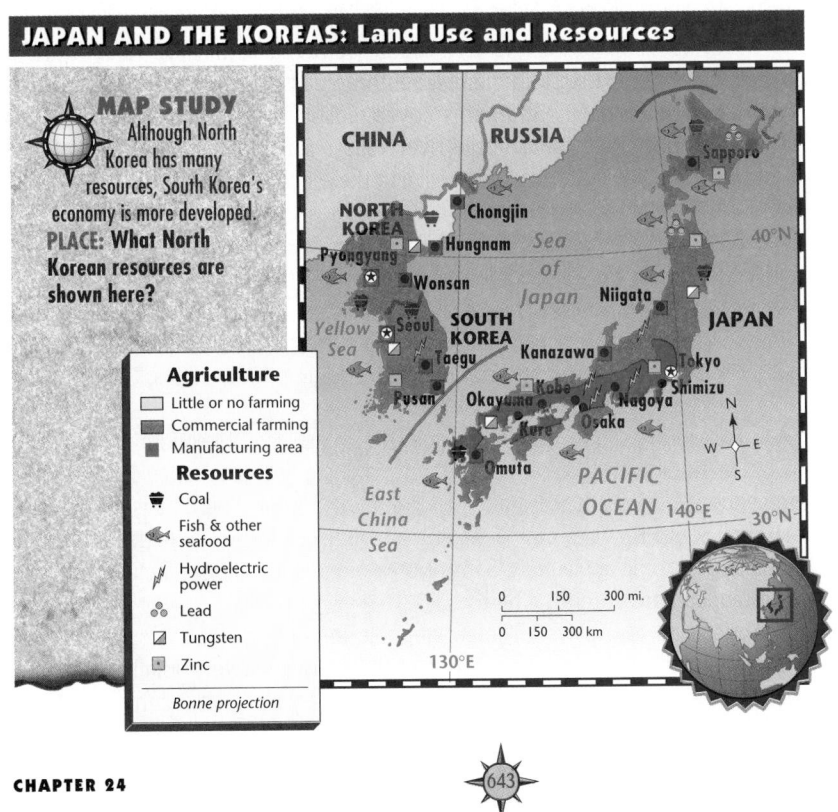

JAPAN AND THE KOREAS: Land Use and Resources

MAP STUDY
Although North Korea has many resources, South Korea's economy is more developed.
PLACE: What North Korean resources are shown here?

Agriculture
- ☐ Little or no farming
- Commercial farming
- ■ Manufacturing area

Resources
- Coal
- Fish & other seafood
- Hydroelectric power
- Lead
- Tungsten
- Zinc

Bonne projection

TEACH

Guided Practice

Time Line Draw a line across the chalkboard. At the left end of the line, write *100 B.C.* Label the right end of the line *A.D. 2000.* As students read the section, have them note the dates and events mentioned. When all students have completed the reading, call on volunteers to come to the chalkboard and add dates and events to the time line. **L1**

Cultural Kaleidoscope

Korea In A.D. 935, a kingdom called *Koryo* took control of much of the Korean Peninsula. The modern name of *Korea* is derived from the name of that ancient kingdom.

MAP STUDY

Answers
coal, tungsten, zinc, fish and other seafood

Map Skills Practice
Reading a Map In which part of Korea is hydroelectric power important? *(South Korea)*

Cooperative Learning Activity

Ask students to write questions and answers based on information from the section. Collect the questions and answers and use them to play Korean Squares. Seat nine students, or "squares," at the front of the room. Give each "square" a cardboard with a large *X* on one side and a large *O* on the other. Organize the rest of the class into Team X and Team O. Ask a question and have a player in Team X choose a "square" to answer it. Tell the player to agree or disagree with the square's answer. If the player chooses correctly, have the square hold up the card with the *X* facing outward. If the player chooses incorrectly, give the correct answer and then ask a question of a player in Team O and give them a chance to get an *O.* Continue until three *X*'s or *O*'s are lined up horizontally, vertically, or diagonally in the squares' hands. **L1**

Independent Practice

 Guided Reading Activity 24-2 **L1**

More About the Illustration

Answer to Caption

ships, cars, textiles, computers, electronic appliances

 **Place** Seoul is a very modern city located in the northwestern part of the country near the border with North Korea. About 1 person in 30 in the city owns a car. Those who do not own cars can use the efficient railroad system or buses. In contrast, people who live in rural areas rely on bicycles for transportation.

NATIONAL GEOGRAPHIC SOCIETY

These materials are available from Glencoe.

 CD-ROM

Picture Atlas of the World

 TRANSPARENCIES

PicturePack Transparencies, Unit 8

Seoul, South Korea

South Korea's modern capital city is home to almost one-fourth of its people.
PLACE: What are some of South Korea's most important exports?

Place
South Korea

South Korea lies at the southern end of the Korean Peninsula. About 38,120 square miles (98,731 sq. km) in area, South Korea is slightly larger than Indiana. The Taebaek Sanmaek Mountains cover most of central and eastern South Korea. Plains with hills and fertile river valleys spread along the southern and western coasts. Most South Koreans live in these coastal areas.

The Climate Fierce winds howl through the streets as rain pelts the windows and bends trees. To the people of South Korea, this is as common as North American spring rains. South Korea has a climate affected by monsoon winds. During the summer, a monsoon from the south brings hot, humid weather. In winter, a monsoon blows in from the north, bringing cold weather.

The Economy South Korea has one of Asia's strongest economies. In recent years manufacturing, high-technology, and service industries have grown tremendously. Many of these privately run industries make products for overseas markets. South Korea is now a leading exporter of ships, cars, textiles, computers, and electronic appliances. South Korea's mineral resources include iron ore, tungsten, and **anthracite,** a hard coal.

Agriculture also plays an important role in South Korea's economy. South Korean farmers own their land. Most of their farms are very small—about 2.7 acres (1.1 ha) in size. The major crops are rice, apples, barley, Chinese cabbage,

 644

UNIT 8

---

onions, and potatoes. Many farmers add to their income by fishing. The map on page 643 shows you that South Korea, like Japan, is a major fishing country.

The People The people of the two Koreas belong to the same Korean ethnic group. South Korea has a population of nearly 45 million people. About 74 percent of South Koreans live in cities and towns in the coastal plains. Since the mid-1950s, South Korea's capital of Seoul has grown into a busy, modern city of more than 15 million people.

Most people in South Korea's cities live in tall apartment buildings and modern homes. Many own cars, but they also rely on buses, subways, and trains to get from one place to another. In rural areas many people live in small, one-story houses made of bricks or concrete blocks. Some rural dwellers still follow traditional ways. A large number of South Koreans have also emigrated to the United States since the end of the Korean War.

Religion and the Arts South Koreans enjoy freedom of religion. Although most practice Buddhism or the teachings of Confucius, some are Christians. The traditional arts of Korea were influenced by Chinese religion and culture. The Koreans, however, have developed their own culture. The most widely practiced art forms are music, poetry, pottery, sculpture, and painting.

If you visit South Korea, you can see examples of traditional Korean art and architecture. In Seoul stand ancient palaces that were modeled after the Imperial Palace in Beijing, China. Ancient Buddhist temples dot the hills and valleys of the Korean countryside. Within these temples are beautifully sculpted figures of Buddha in stone, iron, and gold. One of the great achievements of early Koreans was well-crafted pottery. Even today, Korean potters still make bowls and dishes that are admired throughout the world.

Place
North Korea

Communist North Korea lies at the northern end of the Korean Peninsula. Slightly larger than South Korea, North Korea covers 46,490 square miles (120,409 sq. km). Rugged mountains run through the center of North Korea. A group of narrow valleys separate these ranges. Plains and lowlands run along the western and eastern coasts.

Towering Buddha

Statues of Buddha are common sights at temples in South Korea.
MOVEMENT: What two neighboring countries have had a major influence on the history and culture of Korea?

LESSON PLAN
Chapter 24, Section 2

More About the Illustration

Answer to Caption
China and Japan

Movement
About 20 percent of South Koreans practice Buddhism. Many Koreans, even though they may be Buddhist or Christian, follow at least some of the teachings of Confucius. One Confucian practice most Koreans adhere to is the honoring of ancestors.

Cultural Kaleidoscope

Korea Although North and South Koreans have fought no major battles since the 1953 truce, minor incidents have occurred. Tunnels have been discovered running under the border defenses. They probably were used by raiding parties to gain access and escape unseen.

Critical Thinking Activity

Identifying Central Issues Tell students that many people in North and South Korea would like to reunite their countries. Add that in 1990 travel between the two countries was allowed for the first time since 1953, and that North and South Koreans already co-sponsor sporting and cultural events. Appoint students to research the political differences between the countries and to present the central issues that still divide them in reports to the class. Suggest that students document their reports with information from encyclopedias, current affairs magazines, and newspapers. **L3**

ASSESS

Check for Understanding

Assign Section 2 Review as homework or an in-class activity.

Meeting Lesson Objectives

Each objective below is tested by the questions that follow it in parentheses.

1. **Recognize** where the Korean Peninsula is located. (2)
2. **State** why the Koreas are divided. (3)
3. **Comprehend** how life in South Korea differs from life in North Korea. (4, 5, 6)

MAP STUDY

Answers
the western areas

Map Skills Practice
Reading a Map Based on information on the map, which of the Koreas is more urbanized? Why? *(South Korea, because it has more large cities.)*

Evaluate

 Section 2 Quiz **L1**

Use the Testmaker to create a customized quiz for Section 2. **L1**

The Climate Monsoons also affect climate in North Korea. Most of North Korea has hot summers and cold, snowy winters. Because the central mountains block the full effects of the winter monsoon, the east coast generally has warmer winters than the rest of the country.

The Economy Like South Korea, North Korea has an industrial economy. With more minerals than the south, the northern part of the peninsula was the major industrial area for many years. The map on page 643 shows you that North Korea has minerals such as anthracite, tungsten, and zinc. Since the 1950s, however, South Korea has surpassed North Korea in industrial growth because of its dynamic, free enterprise economy.

In North Korea the communist government owns and runs factories, businesses, and farms. Under the government's direction, North Korean industrial workers produce chemicals, iron and steel, machinery, and textiles. Most farmers work on government-run farms. They grow rice, potatoes, corn, and vegetables and raise livestock.

The People North Korea has about 23.1 million people. About 60 percent of the population live in urban areas along the coasts and river valleys. Pyongyang is North Korea's capital and largest city. Largely rebuilt since the Korean War, Pyongyang has many modern buildings and monuments to communist leaders.

Under communist rule North Koreans have had to put aside much of their traditional culture in favor of communist ways. For example, the government

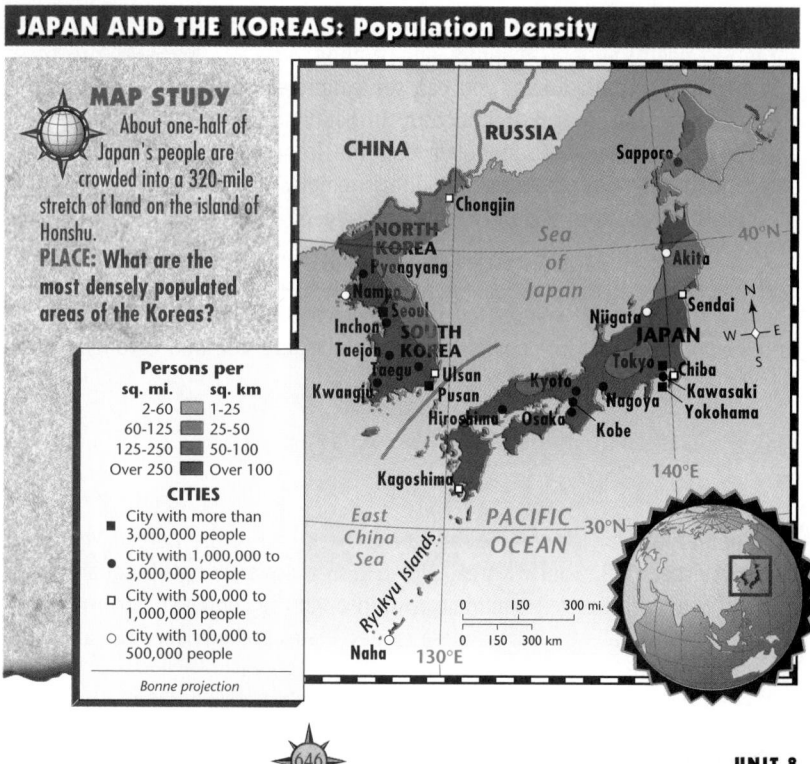

JAPAN AND THE KOREAS: Population Density

MAP STUDY
About one-half of Japan's people are crowded into a 320-mile stretch of land on the island of Honshu.
PLACE: What are the most densely populated areas of the Koreas?

Persons per	
sq. mi.	sq. km
2-60	1-25
60-125	25-50
125-250	50-100
Over 250	Over 100

CITIES
■ City with more than 3,000,000 people
● City with 1,000,000 to 3,000,000 people
□ City with 500,000 to 1,000,000 people
○ City with 100,000 to 500,000 people

Bonne projection

646

More About . . .

Korean Food Korean cooks balance their meals by following the rule of Five Flavors—salt, sweet, sour, bitter, and hot. Soy sauce and bean paste supply the salty flavor; honey, sugar, and sweet potatoes add sweetness; vinegar provides the sourness; ginger contributes bitterness; and mustard and chili peppers create heat. Garlic, green onions, sesame seeds, cinnamon, and eggs are some of the other ingredients found in many Korean dishes. An everyday dinner consists of three to seven of these dishes. For example, a typical menu might include chicken soup, beef and vegetable rolls, whole steamed fish in vegetable sauce, *kimchi* (spicy pickled cabbage), and rice pudding with fruit, nuts, and honey.

Pyongyang, North Korea

A modern hotel is surrounded by office buildings and apartments in North Korea's capital. **PLACE: Before Korea was divided, why was most industry located in North Korea?**

discourages the practice of religion, although many people still hold on to their beliefs. The government also places the needs of the communist system over the needs of individuals or families. During the 1990s North Korea's communist leadership became unstable. The death of leader Kim Il Sung caused a power struggle.

North Koreans still enjoy traditional Korean foods. Rice is the basic food item. One of the most popular side dishes in all of Korea is *kimchi,* a highly spiced blend of Chinese cabbage, onions, and other vegetables.

SECTION 2 REVIEW

REVIEWING TERMS AND FACTS
1. **Define the following:** anthracite.
2. **LOCATION** What body of water borders Korea on the east?
3. **MOVEMENT** How did Korea's location on a peninsula affect its history?
4. **HUMAN/ENVIRONMENT INTERACTION** What farm products are grown in South Korea?

MAP STUDY ACTIVITIES
5. Look at the land use map on page 643. What are three of South Korea's large manufacturing areas?
6. Turn to the population density map on page 646. What are the two most densely populated South Korean cities shown?

Answers to Section 2 Review
1. All vocabulary words are defined in the Glossary.
2. Sea of Japan
3. Its location served as a bridge between Japan and mainland Asia. Trade and ideas went back and forth through Korea. China and Japan, however, wanted to control Korea.

Each country ruled all or part of the peninsula at certain times.
4. rice, apples, barley, Chinese cabbage, onions, and potatoes
5. Pusan, Taegu, Seoul
6. Pusan, Seoul

LESSON PLAN
Chapter 24, Section 2

Chapter 24 Test Form A and/or Form B **L1**

Reteach

Reteaching Activity 24 **L1**

Enrich

Enrichment Activity 24 **L3**

More About the Illustration
Answer to Caption
most mineral resources were located in the north

Place North Korea's Communist government controls the social and cultural lives of the people. It decides what entertainment is acceptable and prohibits any social or cultural activity that conflicts with government policy.

CLOSE

Mental Mapping Activity

Provide each student with a large sheet of white paper and map pencils. Inform students that this will not be a graded activity. Have students draw, freehand from memory, a map of the Korean Peninsula that includes the following labeled countries, cities, and bodies of water: North Korea, South Korea, Seoul, Pyongyang, Sea of Japan, and Yellow Sea.

647

BUILDING GEOGRAPHY SKILLS

BUILDING GEOGRAPHY SKILLS

TEACH

Direct students to carefully study the map and the graph in the skill feature. Then guide them through the four steps for using a map and graph. As you read each step, call on a volunteer to demonstrate it. For example, for the first step, the volunteer would read out the title of the map/graph. When you have worked through all the steps, have students read the skill again and complete the practice questions. **L1**

Additional Skills Practice

1. How do maps and graphs show geographic information differently? *(Maps show locations; graphs show numeric information.)*
2. Why is this area called the Pacific Rim? *(All countries are located on the Pacific Ocean.)*

Geography Skills Activity 24

Using a Map and a Graph

A map and a graph both show you a picture. They can show geographic information but in different ways. Maps show locations, while graphs display information—usually data in numbers—about those locations. Comparing maps and graphs can help you understand more about a location. Comparing also helps show relationships between locations. In using a map and a graph, apply these steps:

• Read the map/graph title to see how the two are related.

• On the map, locate places included in the graph.

• Use information on the graph to draw conclusions about the places on the map.

• Use the map to help explain the information on the graph.

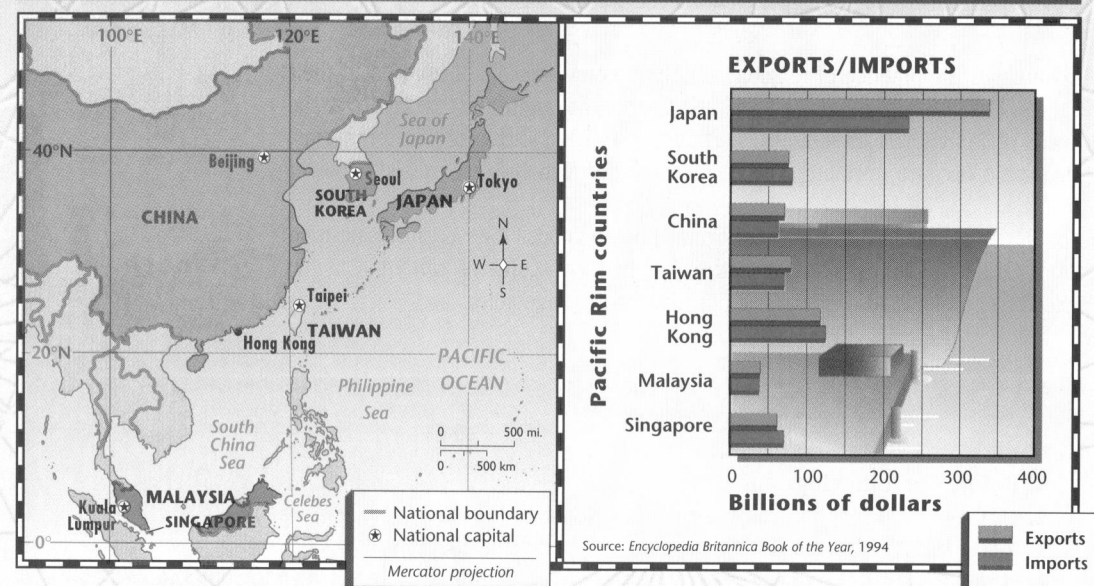

ASIA'S PACIFIC RIM: Exports/Imports

Source: *Encyclopedia Britannica Book of the Year, 1994*

Geography Skills Practice

1. How are the map and the graph related?
2. Which country shown on the map and listed on the graph has the largest land area?
3. Which country has the most exports and imports?
4. How does the physical geography of this region explain the trading activity shown on the graph?

Answers to Geography Skills Practice

1. The graph includes export/import data for the seven Pacific Rim countries shown in the map.
2. China
3. Japan
4. All the countries are located on the Pacific Ocean, an outlet for trade. Also, except for China, their relatively small land areas have increased their dependence on trade.

Chapter 24 Highlights

Important Things to Know About Japan and the Koreas

SECTION 1 JAPAN

- Japan is made up of four main islands and thousands of smaller islands.
- Japan lies on the Pacific Ring of Fire and experiences many earthquakes.
- Japan's strong economy is based on manufacturing and trade.
- Japan—a densely crowded country—holds a population almost half the size of that of the United States.
- The Japanese follow modern ways of life but keep many of their traditions.
- The religions of Shinto and Buddhism have influenced the Japanese arts.

SECTION 2 THE TWO KOREAS

- The Korean Peninsula is divided between communist North Korea and noncommunist South Korea.
- After World War II, communist troops from the Soviet Union took over the northern half of Korea. American troops occupied the southern half. Korea eventually divided along the 38th parallel.
- Both Koreas have inland mountains and coastal plains.
- Monsoon winds affect the climate of the Koreas, bringing hot weather in the summer and cold weather in winter.
- South Korea has a booming free enterprise economy that exports many industrial goods to other countries.
- A communist government controls North Korea's economy.
- South Korea has about 45 million people; North Korea, about 23 million.
- Most South Koreans practice Buddhism or the teachings of Confucius. Under communism, North Koreans are discouraged from practicing religion.

Tokyo, Japan ▶

Extra Credit Project

Oral Interviews Ask interested students to interview someone who served in South Korea during the Korean War or after. Suggest that students ask about the interviewees' impressions of the land and the people. Encourage students to tape their interviews and to play the tapes for the class. **L1**

Using the Chapter 24 Highlights

- Use the Chapter 24 Highlights to preview, review, condense, or reteach the chapter.

Preview/Review

 Vocabulary Puzzle-Maker Software reinforces the terms used in Chapter 24.

Condense

Have students read the Chapter 24 Highlights. Spanish Chapter Highlights also are available.

Chapter 24 Digest Audiocassettes and Activity

Chapter 24 Digest Audiocassette Test

Spanish Chapter Digest Audiocassettes, Activities, and Tests also are available.

Guided Reading Activities

Reteach

Reteaching Activity 24. Spanish Reteaching activities also are available.

MAP STUDY

Region Copy and distribute page 35 from the Outline Map Resource Book. Have students label North Korea, South Korea, and the countries in East Asia that influenced their history—China and Japan. Have students include in the labels the dates that these countries ruled or influenced the Korean Peninsula.

Chapter 24 Review and Activities

REVIEWING KEY TERMS

Match the numbered terms in Column A with their definitions in Column B.

A
1. megalopolis
2. anthracite
3. shogun
4. samurai
5. intensive cultivation
6. archipelago
7. tsunami
8. clan

B
A. an urban area made up of several large cities
B. the warrior class of early Japan
C. huge sea wave created by an undersea earthquake
D. farming nearly every area of land
E. a group of islands
F. hard coal
G. Japanese military leader
H. group of related families

ANSWERS

Reviewing Key Terms

1. A
2. F
3. G
4. B
5. D
6. E
7. C
8. H

Mental Mapping Activity

This exercise helps students to visualize the countries and geographic features they have been studying and understand the relationships among them. All attempts at freehand mapping should be accepted.

Mental Mapping Activity

On a separate piece of paper, draw a freehand map of Japan and the two Koreas. Label the following items on your map:

- Honshu
- North Korea
- Pacific Ocean
- Kyushu
- Kobe
- Tokyo

REVIEWING THE MAIN IDEAS

Section 1

1. **REGION** What is the Ring of Fire?
2. **PLACE** Why does Hokkaido have cold winters?
3. **HUMAN/ENVIRONMENT INTERACTION** Why are Japanese farmers able to produce high crop yields?
4. **HUMAN/ENVIRONMENT INTERACTION** How has Japan tried to improve its environment?
5. **PLACE** What do the religions of Shinto and Buddhism teach?

Section 2

6. **REGION** Why was Korea divided after World War II?
7. **REGION** How do monsoons affect climate in the two Koreas?
8. **PLACE** What is South Korea's capital and largest city?
9. **LOCATION** Where is most mineral wealth on the Korean Peninsula found?
10. **PLACE** How many North Koreans live in urban areas?

650

UNIT 8

Reviewing the Main Ideas
Section 1

1. an area of volcanic and earthquake activity around the Pacific Ocean
2. The cold Arctic current brings cold temperatures.
3. They use fertilizers and modern farm equipment and practice intensive agriculture.
4. The government has passed strict pollution-control laws.

5. Shinto teaches a respect for nature, a love of simple things, and a concern for cleanliness and good manners. Buddhism teaches respect for nature and the need for inner peace.

Section 2

6. After the war, Soviet troops occupied the north, while American troops took control in the south. The Soviets set up a communist government in the north, while a non-communist government ruled in the south.

CRITICAL THINKING ACTIVITIES

1. **Drawing Conclusions** Why do other countries want to model their economies after Japan's economy?
2. **Making Comparisons** How do ways of life differ in North Korea and South Korea?

GeoJournal Writing Activity

A haiku is a three-line poem, usually about nature and human emotions. The traditional haiku requires 17 syllables—5 in the first line, 7 in the second line, and 5 in the third line. In your journal, write your own haiku in which you poetically sketch a scene from nature, such as a snowfall or a sunrise.

COOPERATIVE LEARNING ACTIVITY

The United States is an important trading partner of Japan and South Korea. Work in groups of three to find out what products from Japan and South Korea the members of your group use. Look at home and school for products made in these two countries. Make a list of the products and where they were made. As a group, prepare a list of your findings and share the list with the class.

PLACE LOCATION ACTIVITY: JAPAN AND THE TWO KOREAS

Match the letters on the map with the places and physical features of Japan and the two Koreas. Write your answers on a separate sheet of paper.

1. Seoul
2. Honshu
3. Taebaek Sanmaek Mountains
4. Hokkaido
5. East China Sea
6. Inland Sea
7. Sapporo
8. Tokyo
9. Pyongyang
10. Sea of Japan

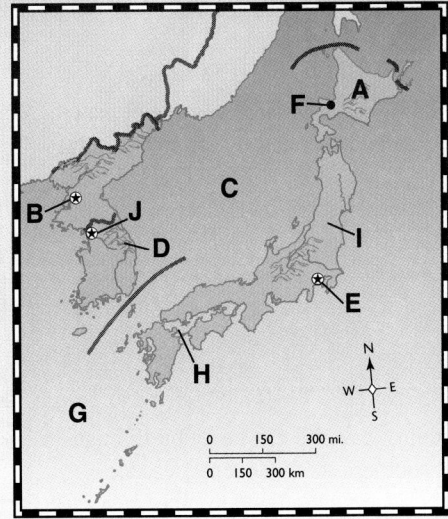

Encourage students to bring in examples of Japanese and South Korean products, such as radios and other small electrical appliances. Display the sample products when groups present their lists.

GeoJournal Writing Activity

Students might like to collect their poems and bind them in a book. Display the haiku collection for other classes to read.

Place Location Activity

1. J
2. I
3. D
4. A
5. G
6. H
7. F
8. E
9. B
10. C

Chapter Bonus Test Question

This question may be used for extra credit on the chapter test.

How do ocean currents affect Japan's climate? *(A cold current flows south from the Arctic, cooling northern Japan; a warm current flows north from the Pacific Ocean, warming southern Japan.)*

7. Summer—monsoons blowing from the south bring hot, humid weather. Winter—monsoons blow from the north, bringing cold weather.
8. Seoul
9. in North Korea
10. about 60 percent

Critical Thinking Activities

1. Answers should mention Japanese business efficiency; a skilled, hardworking labor force, a strong sense of employee loyalty, and the use of modern technology.
2. North Korea and South Korea share a common geography, history, language, and cultural background. North Korea has a communist government that runs the economy and has attempted to replace traditional Korean culture with communist ways. South Korea has a non-communist government and a free enterprise economy.

651

Cultural HERITAGE: ASIA

MUSIC ▶ ▶ ▶ ▶

The music of India is played mostly on guitarlike stringed instruments and drums. Ancient melodies, called *ragas,* sound very different from Western music because they use a different scale.

ARTIFACTS ▲ ▲ ▲ ▲

This tiny stone carving was one of many discovered at Mohenjo-Daro in the Indus River valley. At least 4,000 years old, it was probably used by traders as a stamp or seal to mark their merchandise.

ARCHITECTURE ▶ ▶ ▶ ▶

Buddhist pagodas can be found throughout southern Asia. Most pagodas are eight-sided with an odd number of stories. Each story has an umbrella-shaped roof made of tiles.

Cultural HERITAGE

FOCUS

Have a student volunteer remind the class of the definition of *culture.* Then ask students whether there is such a thing as one "Asian culture." Point out that the continent of Asia is home to many distinct ways of life.

TEACH

Display Political Map Transparency 8. Tell students that they can think about Asian culture on three levels. On the highest level, they can think of all Asia as a cultural region. As you mention this point, draw a circle on the transparency that encompasses all of Asia. On the middle level, they can think of two cultural subregions within Asia: the Chinese subregion (China, Japan, and Vietnam) and the Indian subregion (South Asia). Draw circles around the areas mentioned. Finally, on the smallest level, they can think of specific cultural groups within the subregions. Call on volunteers to name several examples. *(Japanese culture, Indian culture, and so on)* L1

More About . . .

Architecture One of the most recognized buildings in the world, the Taj Mahal, is also among the most beautiful. Shah Jahan, the builder, spared no expense in creating this "dream in marble" as a tribute to his wife, Mumtaz Mahal. He commissioned engineers and architects from all over the Muslim world to draw up designs for the tomb. The raw materials came from far and wide, too. Teams of elephants dragged the white marble some 100 miles (161 km) across country. The precious metals and stones used to decorate the structure came from as far afield as Russia, China, Persia, and Afghanistan. Many of the craftworkers who constructed the tomb came from equally distant locations.

◄ ◄ ◄ THEATER

Actors with wigs, painted faces, and elaborate costumes perform in Japan's popular Kabuki theaters. Kabuki performers act out historic events or real-life experiences in a lively, dramatic way. All Kabuki actors are men who begin their training at a very young age.

ISLAMIC ARCHITECTURE ► ► ► ►

The magnificent Taj Mahal was built by an Indian ruler in the 1600s as a tomb and monument to his beloved wife. The white marble dome and minarets reflect the Islamic influence in India at the time. It took 20,000 laborers almost 20 years to complete this elaborate structure.

◄ ◄ ◄ DECORATIVE ART

These colorful, hand-painted umbrellas are an art form in Thailand. The umbrellas are made from paper and bamboo and decorated with colorful flowers, dragons, and birds.

APPRECIATING CULTURE

1. Compare and contrast the architecture of the Taj Mahal and the Buddhist pagoda.
2. The animal seal from Mohenjo-Daro was probably used as a stamp or trademark to indicate ownership of goods. Design your own personal trademark.

653

Answers to Appreciating Culture

1. Answers will vary, but most students will contrast the angular lines of the pagoda with the smooth, curving lines of the Taj Mahal.
2. Designs will vary. Encourage students to include their trademarks on papers they submit for grading.

PERFORMANCE ASSESSMENT ACTIVITY

SCOUTING LOCATIONS FOR MOVIES Ask students to brainstorm a list of television series and action-adventure movies such as those featuring Indiana Jones, James Bond, or Rambo. Encourage students to assume the role of an assistant director for a similar type of movie. Tell them they have been set the task of selecting a Southeast Asia location for the movie. Students should decide in which country they would locate the movie and give examples of how the geographic features of that country could be woven into the adventure plot. These examples may take the form of snippets of script or "scouting" notes on the location. Students should add a summary paragraph in which they explain the process they followed in selecting their locations and writing their examples.

POSSIBLE RUBRIC FEATURES: Accuracy of content information, ability to recognize relationships, composition skills, decision-making skills

MENTAL MAPPING ACTIVITY

Before teaching Chapter 25, point out Myanmar, Thailand, Laos, Cambodia, Vietnam, Indonesia, Malaysia, Singapore, Brunei, and the Philippines on a globe to students. Have them identify the hemispheres in which the Southeast Asian nations are located and where the countries are located relative to the United States.

TEACHER'S CORNER

NATIONAL GEOGRAPHIC SOCIETY

INDEX TO NATIONAL GEOGRAPHIC MAGAZINE

The following articles may be used for research relating to this chapter:

- "Irian Jaya," by Thomas O'Neill, February 1996.
- "Irian Jaya's People of the Trees," by George Steinmetz, February 1996.
- "Thailand," by Noel Grove, February 1996.
- "Burma, the Richest of Poor Countries," by Joel L. Swerdlow, July 1995.
- "The New Saigon," by Tracy Dahlby, April 1995.

- "The Tale of the *San Diego*," by Franck Goddio, July 1994.
- "The Mekong," by Thomas O'Neill, February 1993.
- "Volcanoes: Crucibles of Creation," by Noel Grove, December 1992.
- "Vietnam: The Hard Road to Peace," by Peter T. White, November 1989.
- "Indonesia: Two Worlds, Time Apart," by Arthur Zich, January 1989.

NATIONAL GEOGRAPHIC SOCIETY PRODUCTS AVAILABLE FROM GLENCOE

To order the following products for use with this chapter, contact your local Glencoe sales representative or call Glencoe at 1-800-334-7344:

- *STV: World Geography* (Videodisc)
- *Picture Atlas of the World* (CD-ROM)
- *Physical Geography of the World* (Transparencies)
- *ZipZapMap! World* (Software)

- *GeoBee* (Software)
- *Images of the World* (Posters)
- *Eye on the Environment* (Posters)

ADDITIONAL NATIONAL GEOGRAPHIC SOCIETY PRODUCTS

To order the following products for use with this chapter, call National Geographic Society at 1-800-368-2728:

- *Physical Geography of the Continents Series:* "Asia." (Video)
- *Changing Faces of Communism:* "Vietnam." (Video)

- *Bali: Masterpiece of the Gods* (Video)
- *Born of Fire* (Video)

TEACHER-TO-TEACHER
Curriculum Connection: Art, Language Arts

—from Doris Knoch
South Adams School, Berne, IN

SIGHTSEEING IN SOUTHEAST ASIA *The purpose of this activity is to help students gain a better understanding of Southeast Asia by creating travel brochures for the region.*

Have students work in small groups to create travel brochures for Southeast Asia. Groups must select three countries to visit. For each country they should note major cities and other areas of interest and add any "fun facts" they find in the course of their research.

Point out that brochures should include a master map of Southeast Asia with countries and capital cities labeled. The map should also show the itinerary through their three selected countries. Distances between stops and local climatic conditions might also be included in annotations on the map.

Each of the three group members should create an insert page for one of their three countries. The insert should contain information on topics such as religious and ethnic groups, customs and traditions, food, clothing, government, sports, and entertainment.

Display finished brochures and provide students time to review them. Then have students discuss which country in Southeast Asia they would most like to visit and why.

BIBLIOGRAPHY

Readings for the Student

Detzer, David. *An Asian Tragedy: America and Vietnam.* Brookfield, Conn.: Millbrook Press, 1992.

Parks, Carl. *Southeast Asia Handbook.* Chicago: Moon Publishing, 1990.

Warshaw, Steven. *Southeast Asia Emerges,* revised ed. Emeryville: Diablo, 1990.

Readings for the Teacher

Asia Today: An Atlas of Reproducible Pages, revised ed. Wellesley, Mass.: World Eagle, 1991.

Southeast Asia. Stockton, Calif.: Stevens & Shea, 1994. Reproducible activity sheets.

Multimedia

Assignment: Southeast Asia. Boston: Christian Monitor Video, 1993. Videocassette, 60 minutes.

Vietnam. Washington, D.C.: National Geographic Society, 1990. Videocassette, 25 minutes.

MEETING NATIONAL STANDARDS

Geography for Life

All of the 18 standards are demonstrated in Unit 8. The following ones are highlighted in this chapter:

• Standards 1, 2, 3, 4, 6, 9, 10, 12, 13, 15, 16

KEY TO ABILITY LEVELS

Teaching strategies have been coded for varying learning styles and abilities.

L1 **BASIC** activities for all students

L2 **AVERAGE** activities for average to above-average students

L3 **CHALLENGING** activities for above-average students

LEP **LIMITED ENGLISH PROFICIENCY** activities

Chapter 25 Resource Organizer

Southeast Asia

Use these *Geography: The World and Its People* resources to teach, reinforce, and extend chapter content.

CHAPTER 25 RESOURCES

- *Vocabulary Activity 25
- Cooperative Learning Activity 25
- Workbook Activity 25
- Chapter Map Activity 25
- Geography Skills Activity 25
- GeoLab Activity 25
- Performance Assessment Activity 25
- *Reteaching Activity 25
- Enrichment Activity 25
- Environmental Case Study 8
- Chapter 25 Test, Form A and Form B
- *Chapter 25 Digest Audiocassette, Activity, Test
- Unit Overlay Transparencies 8-0, 8-1, 8-2, 8-3, 8-4, 8-5, 8-6
- Political Map Transparency 8; World Cultures Transparencies 17, 18
- Geoquiz Transparencies 25-1, 25-2
- Vocabulary PuzzleMaker Software
- Student Self-Test: A Software Review
- Testmaker Software
- MindJogger Videoquiz

*If time does not permit teaching the entire chapter, summarize using the Chapter 25 Highlights on page 669, and the Chapter 25 English (or Spanish) Audiocassettes. Review student's knowledge using the Glencoe MindJogger Videoquiz. * Also available in Spanish*

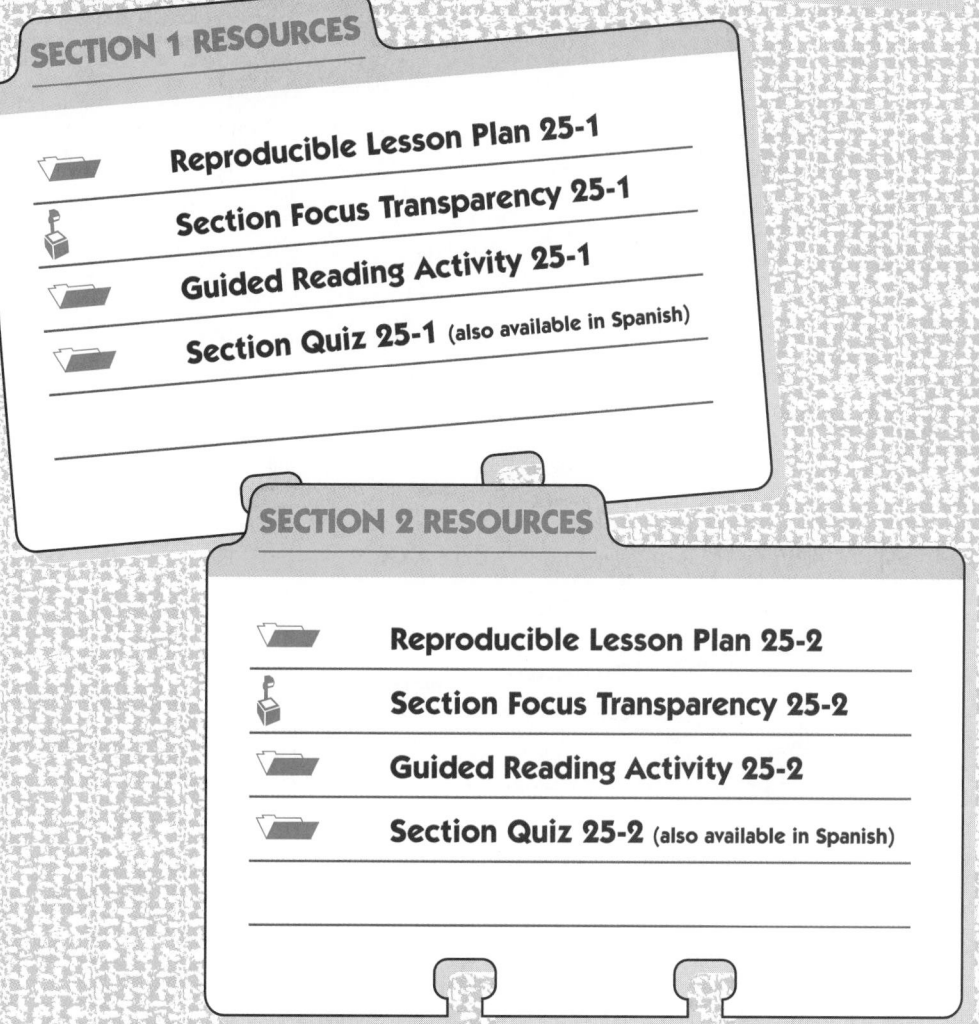

Use these *Geography: The World and Its People* resources to teach and reinforce section content.

SECTION 1 RESOURCES

Reproducible Lesson Plan 25-1

Section Focus Transparency 25-1

Guided Reading Activity 25-1

Section Quiz 25-1 (also available in Spanish)

SECTION 2 RESOURCES

Reproducible Lesson Plan 25-2

Section Focus Transparency 25-2

Guided Reading Activity 25-2

Section Quiz 25-2 (also available in Spanish)

ADDITIONAL RESOURCES FROM GLENCOE

 Reproducible Masters

- Foods Around the World, Unit 10

- Facts On File: GEOGRAPHY ON FILE, Southeast Asia

Workbook

- Building Skills in Geography, Unit 2, Chapter 3

 **World Music: Cultural Traditions**
Lesson 9

 Transparencies

- National Geographic Society PicturePack Transparencies, Unit 8

 **Posters**

- National Geographic Society: Eye on the Environment, Unit 8

 Countries of the World Flashcards
Unit 8

Videodiscs

- Reuters Issues in Geography

- National Geographic Society: STV: World Geography, Vol. 1

- National Geographic Society: GTV: Planetary Manager

CD-ROM

- National Geographic Society: Picture Atlas of the World

 **Software**

- National Geographic Society: ZipZapMap! World

Performance Assessment

Refer to the Planning Guide on page 654A for a Performance Assessment Activity for this chapter. See the *Performance Assessment Strategies and Activities* booklet for suggestions.

Chapter Objectives

1. **Describe** the location, economies, and resources of island countries in Southeast Asia.

2. **Examine** how physical and human influences affect life in island countries of Southeast Asia.

GLENCOE
TECHNOLOGY

 **VIDEODISC**

Use the Chapter 25 MindJogger Videoquiz to preview chapter content.

MindJogger Videoquiz

Chapter 25
Disc 4 Side A

 Also available in VHS.

Chapter 25 Southeast Asia

SOUTHEAST ASIA: Political

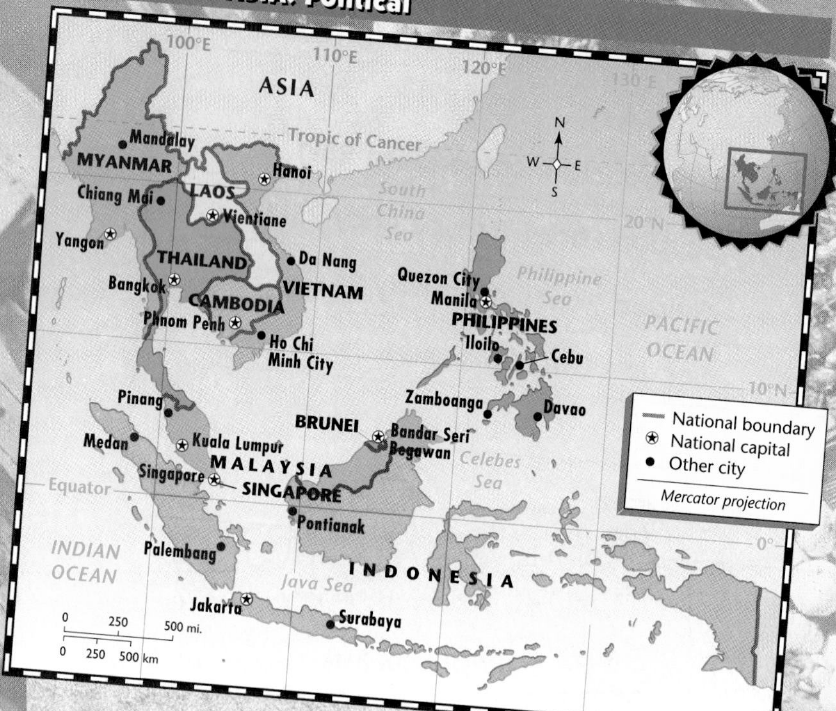

Map showing Southeast Asian countries: Myanmar, Laos, Thailand, Cambodia, Vietnam, Philippines, Brunei, Malaysia, Singapore, Indonesia, with cities including Mandalay, Hanoi, Chiang Mai, Vientiane, Yangon, Da Nang, Bangkok, Phnom Penh, Ho Chi Minh City, Quezon City, Manila, Iloilo, Cebu, Zamboanga, Davao, Pinang, Kuala Lumpur, Bandar Seri Begawan, Medan, Singapore, Pontianak, Palembang, Jakarta, Surabaya

Legend:
— National boundary
⊛ National capital
● Other city
Mercator projection

 MAP STUDY ACTIVITY

As you read Chapter 25, you will learn about Southeast Asia.

1. What countries in Southeast Asia border China?
2. What Southeast Asian countries are made up of islands?
3. What three countries share one island?

654

MAP STUDY ACTIVITY

Answers
1. Myanmar, Laos, Vietnam
2. Philippines, Malaysia, Indonesia
3. Brunei, Malaysia, Indonesia

Map Skills Practice
Reading a Map What is the capital of the Philippines? *(Manila)* What ocean washes the western edge of Southeast Asia? *(Indian Ocean)*

Mainland Southeast Asia

PREVIEW

Words to Know
- alluvial plain
- deforestation
- delta

Places to Locate
- Myanmar
- Thailand
- Laos
- Cambodia
- Vietnam
- Bangkok
- Mekong River

Read to Learn . . .
1. what countries make up mainland Southeast Asia.
2. what minerals are found in mainland Southeast Asia.
3. why the economies of Laos, Cambodia, and Vietnam are not fully developed.

Imagine visiting a village deep in the forests of

Myanmar, a country in Southeast Asia. You would see that many village homes are built on poles. Homes are built this way to protect the occupants from floods and wild animals.

South of China and east of India lies a region known as Southeast Asia. The peninsulas and thousands of islands that make up this region cover an area of about 1,750,000 square miles (4,536,000 sq. km). Including its seas, Southeast Asia spreads over an area as big as the contiguous United States. On the peninsulas of mainland Southeast Asia, you will find the countries of Myanmar, Thailand, Laos, Cambodia, Vietnam, and part of Malaysia.

Location
Myanmar

Southeast Asia's northernmost country is Myanmar. Once called Burma, Myanmar for many years was part of British India. It became an independent republic in 1948. Since independence, military leaders have turned Myanmar into a socialist country. The government runs the economy and forbids any criticism of its policies. It also limits Myanmar's contacts with the outside world.

The Land and Economy Myanmar's 253,880 square miles (657,549 sq. km) make it about the size of Texas. Rugged mountains run through the eastern

FOCUS

Section Objectives
1. **Identify** countries that make up mainland Southeast Asia.
2. **Itemize** the minerals found in mainland Southeast Asia.
3. **Cite** why the economies of Laos, Cambodia, and Vietnam are not fully developed.

Bellringer Motivational Activity

Prior to taking roll at the beginning of the class period, project Section Focus Transparency 25-1 and have students answer the activity questions. Discuss students' responses. **L1**

Vocabulary Pre-check

Tell students that *alluvial* derives from a Latin word meaning "washed up on." Have them read the definition for *alluvial plain* in the section to find out what is being washed up onto the plain. *(soil deposits)* **L1**
LEP

Use the Vocabulary PuzzleMaker Software to create a crossword puzzle. **L1**

Classroom Resources for Section 1

 **REPRODUCIBLE MASTERS**
Reproducible Lesson Plan 25-1
Guided Reading Activity 25-1
Geography Skills Activity 25
Building Skills in Geography, Unit 2, Lesson 3
Workbook Activity 25
Section Quiz 25-1

 TRANSPARENCIES
Section Focus Transparency 25-1
Political Map Transparency 8
World Cultures Transparency 17

MULTIMEDIA
 Vocabulary PuzzleMaker Software
Testmaker

 National Geographic Society: STV: World Geography, Vol. 1

MAP STUDY

Answer
South China Sea

Map Skills Practice
Reading a Map Which river flows through Laos, Cambodia, and Vietnam? *(Mekong River)*

TEACH

Guided Practice

Applying Information Review with students the discussion of climate regions on pages 45–48 of this text, paying special attention to the impact of elevation and latitude on climate. Call on volunteers to come to the chalkboard and write brief descriptions of each climate region. Before the students read the section, direct them to the physical map on page 656. Tell them to note the latitudes and landforms of Myanmar, Thailand, Laos, Cambodia, and Vietnam, and to predict which climate types might be found in each country. Have them read the section and study the climate map on page 657 to verify their predictions. **L1**

Independent Practice

Guided Reading Activity 25-1 **L1**

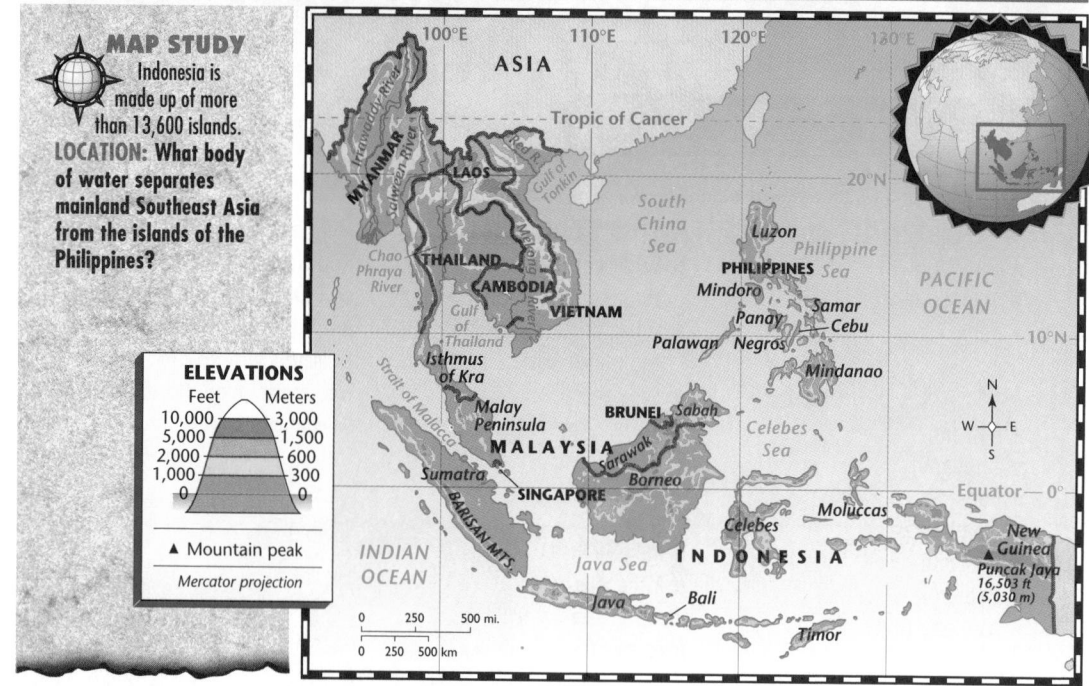

SOUTHEAST ASIA: Physical

MAP STUDY
Indonesia is made up of more than 13,600 islands.
LOCATION: What body of water separates mainland Southeast Asia from the islands of the Philippines?

ELEVATIONS

Feet	Meters
10,000	3,000
5,000	1,500
2,000	600
1,000	300
0	0

▲ Mountain peak

Mercator projection

and western parts of the country. Between the mountain ranges flow two major rivers—the Irrawaddy (IHR•uh•WAH•dee) and the Salween. The land along these rivers contains Myanmar's most fertile soil.

Southern Myanmar has tropical climates. Monsoon winds from the Indian Ocean bring rain during Myanmar's summers. Dry monsoons from the north occur during winter. As is typical in high elevations, temperatures are cooler in Myanmar's northern mountains.

Myanmar's economy is based on farming. Rice is the major food crop. Some farmers work their fields with tractors, but most rely on plows drawn by water buffalo. Forests cover large areas of Myanmar and provide about 80 percent of the world's teakwood. Mines yield minerals such as tin, copper, and tungsten. Gemstones—rubies, sapphires, and jade—also come from Myanmar.

The People About 75 percent of Myanmar's nearly 46 million people live in small villages. They build their homes on poles above the ground to protect themselves from floods and wild animals. The most densely populated area of the country is the fertile Irrawaddy River valley.

Many rural people in Myanmar follow traditional ways. In cities you will notice a blend of old and new. The capital and largest city, Yangon, is famous for both its modern university and its gold-covered Buddhist temples. Buddhism is the major religion of Myanmar.

Cooperative Learning Activity

Organize students into five committees and assign each committee one of the countries discussed in this section. Tell the committees that they represent a group of American investors and that their task is to: 1) decide on a profitable investment in their assigned country, and 2) list the advantages and disadvantages of investing in that country. Suggest that groups review the text on the countries' economies to find ideas for investments, and review the countries' land, climate, people, and governments to find advantages and disadvantages. Encourage each committee to discuss the facts from the section, to appoint a secretary to record the discussion, and to put together a report for investors based on the secretary's notes. Have committees share their reports; then ask the class to select the wisest investment. **L1**

Place
Thailand

Southeast of Myanmar lies Thailand. The map on page 654 shows you that Thailand looks like a flower on a stem. The "flower" is the northern part of Thailand, located on the mainland. The "stem" is a very narrow strip of Thailand that runs southward to the Malay Peninsula.

Once called Siam, Thailand means "land of the free." It is the only Southeast Asian country that has never been a European colony. The people of Thailand have skillfully blended modern ways with traditional practices and a loyalty to Thailand's royal family.

The Land and Economy Mountains, plateaus, and river valleys spread through northern Thailand. In the central part of the country, the Chao Phraya (chow•PRY•uh) River has formed a large alluvial plain. An **alluvial plain** is a land area built up by a river's soil deposits.

Because of its fertile soil, central Thailand has most of the country's people, farms, and industries. Farther south the thin strip of Thailand along the Malay Peninsula boasts thick forests, rubber trees, and mineral wealth.

The map below shows you that Thailand has tropical and subtropical climates. Monsoon winds bring dry and wet seasons. Rainfall is heaviest in the south and southeast. Thailand is an agricultural country. Thai farmers grow rice

LESSON PLAN
Chapter 25, Section 1

CURRICULUM CONNECTION

History During World War II, the Japanese forced some 330,000 prisoners of war to build a 250-mile (402-km) railroad between Burma and Thailand. Starvation, disease, and the sheer brutality of their guards claimed the lives of thousands of these prisoners. The story of the Burma Railway was the subject of the Academy-Award-winning film, *The Bridge on the River Kwai* (1957).

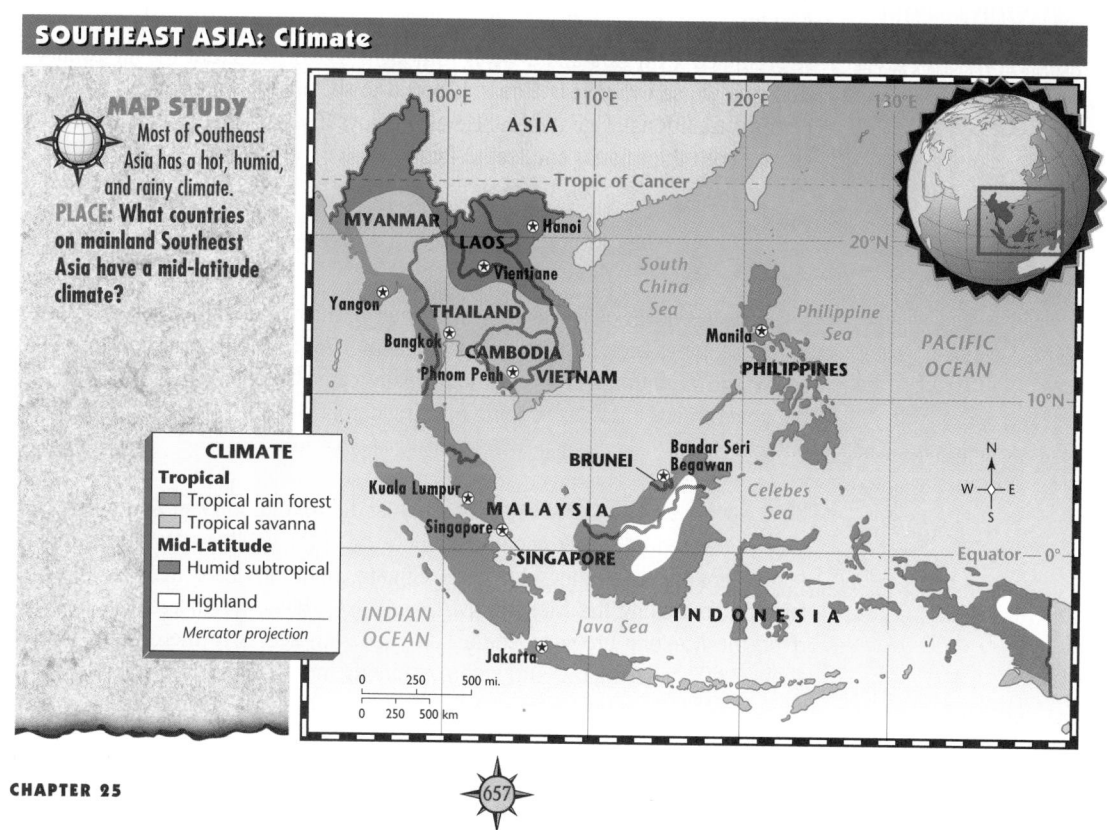

SOUTHEAST ASIA: Climate

MAP STUDY
Most of Southeast Asia has a hot, humid, and rainy climate.
PLACE: What countries on mainland Southeast Asia have a mid-latitude climate?

CLIMATE
Tropical
- Tropical rain forest
- Tropical savanna

Mid-Latitude
- Humid subtropical
- Highland

Mercator projection

0   250   500 mi.
0   250 500 km

MAP STUDY

Answer
Myanmar, Thailand, Laos, Vietnam

Map Skills Practice
Reading a Map In one word, describe the climate found in Cambodia. *(tropical)*

NATIONAL GEOGRAPHIC SOCIETY

These materials are available from Glencoe.

 VIDEODISC

STV: World Geography Vol. 1: Asia and Australia

South and Southeast Asia
Side 1, Frames 27589-42384

Meeting Special Needs Activity

Reading Disability Many reading comprehension questions require students to restate what they have read. Model a restating strategy for students. Have them read the first sentence of the second paragraph under the subheading "The People" on page 658. Demonstrate restating: "Thailand's population is mostly rural." Have students practice this skill by restating other sentences from the section. **L1**

More About the Illustration

Answer to Caption
about 20 percent

GEOGRAPHIC THEMES

Human/ Environment Interaction Within and around Bangkok are hundreds of miles of canals called *klongs*. These canals connect with the Chao Phraya River, which runs through the city. The *klongs* always teem with local boat traffic. Many boats are water busses. Some serve as small stores, selling food and other goods. Still others are used to deliver mail.

Cultural Kaleidoscope

Cambodia A popular attraction in Angkor is Angkor Wat, an 800-year-old temple. Huge sandstone carvings cover the inner wall of the first level of this 3-tier building. The carvings depict scenes from wars and sacred stories.

Bangkok, Thailand

Traffic on the crowded streets of Thailand's capital has created an air pollution problem. **HUMAN/ENVIRONMENT INTERACTION: What percentage of Thailand's people live in cities?**

as their major crop. Plantations in the south provide rubber for export. Teak and other wood products come from Thailand's forests. In recent years developers have cut down so many forests that many Thai people fear the complete loss of one of their country's most valuable resources. The government now tries to limit **deforestation,** or the widespread cutting of trees.

Thailand is rich in mineral resources. It is one of the world's leading exporters of tin and gemstones. Most of Thailand's manufacturing is located around Bangkok, the capital and largest city. Thai workers there produce textiles, cement, and paper products.

The People Most of Thailand's 59.4 million people belong to the Thai ethnic group and follow the Buddhist religion. Hundreds of Buddhist temples called *wats* dot cities and the countryside. Bangkok alone has more than 300 Buddhist temples! Every morning Buddhist monks, or holy men, carrying small bowls and wearing yellow-orange robes walk among the people to receive food offerings. Buddhism inspires many Thai festivals celebrated with parades and colorful flowers.

About 80 percent of Thailand's people live in rural villages. Bangkok, however, is growing rapidly as more and more people move into the city from rural areas. Today Bangkok has about 6 million people. Thirteen-year-old Chuan Soonsiri notices the contrasts of life in Bangkok. He points out beautiful Buddhist temples and royal palaces next to modern skyscrapers and crowded streets filled with cars and people. Although he worries about the air pollution that gets worse every year, Chuan is proud to see dancers in elaborate costumes perform traditional Thai dances. He also enjoys Thai boxing, in which the boxers use both their hands and feet to fight.

Place
Laos

Laos lies north of Thailand in the center of mainland Southeast Asia. Once a French colony, Laos became independent in 1953. A civil war soon tore Laos apart, and a communist government finally came to power in 1975. Recently the government has relaxed some of its rigid policies and has opened Laos to the outside world.

The Land and Economy The only landlocked country in Southeast Asia, Laos is covered by rugged mountains. Because of the cooling effect of the mountains, Laos has a humid subtropical climate. Southern Laos includes a fertile farming area along the Mekong (MAY•KAWNG) River. Southeast Asia's longest river, the Mekong provides landlocked Laos with access to the ocean. It flows about 2,600 miles (4,180 km) before emptying into the South China Sea.

UNIT 8

Critical Thinking Activity

Synthesizing Information Recent changes in Vietnam have created social problems. The use of child labor, child abuse, juvenile delinquency, and the number of homeless children have all increased. The government is ill-equipped to fix these problems, for the following reasons:

- outdated laws concerning children
- no child welfare organizations
- few and poorly trained social workers

- police and prison officials who lack the skills to deal with children

After participating in the 1990 World Summit for Children, Vietnamese leaders drew up a national plan of action for children. Have students suggest measures that they think should be part of that plan, such as the government should pass a law that bans the hiring of children under the age of 15. **L2**

Laos's communist government runs the country's economy, which is mostly agricultural. Farmers grow fruits, vegetables, and rice along the fertile banks of the Mekong River. Industry is still largely undeveloped because of isolation and years of civil war.

The People About 80 percent of Laos's 4.7 million people live in villages and towns along the Mekong River. You find Vientiane (vyehn•TYAHN), the largest city and capital, near the river. The communist government discourages religion, but most of Laos's people remain Buddhists.

Place
Cambodia

South of Thailand and Laos is Cambodia. About 1,000 years ago, Cambodia was the center of a large empire that ruled much of Southeast Asia. Today you can still see the remains of Angkor, the ancient empire's capital. In modern times Cambodia was under French rule, finally becoming independent in 1953.

Cambodians have faced almost constant warfare since the 1960s. A communist government took control of the country in the mid-1970s, and the people suffered great hardships. More than 1 million Cambodians died. Many fled overseas to the United States and other countries. In 1993 a democratic government finally came to power in Cambodia.

The Land and Economy Low mountains, lakes, and forested plains stretch across Cambodia. The Mekong River and a number of smaller rivers crisscross the country, providing water, fertile soil, and waterways for transportation. The map on page 657 shows you that Cambodia has tropical climates.

For many years Cambodia was a rich farming country that exported rice and rubber. By the 1980s its economy was in ruins because of war and harsh communist rule. Today the new democratic government is trying to rebuild the economy. Rice and soybeans are the major crops.

The People Most of Cambodia's 10 million people belong to the Khmer (kuh•MEHR) ethnic group. They live in rural villages and farm the land. Only about 10 percent live in cities such as the capital, Phnom Penh (NAHM PEHN). Buddhism is the major religion of Cambodia.

Teen Scene

Happy New Year!

In Thailand the New Year is celebrated in April, one of the hottest months of the year. Vina Charas and her sister Uma look forward to the celebration. This year they will be in their village parade. After the parade, the fun begins. On New Year's Day, people drench anyone and everyone with buckets of water. The tradition started many years ago by people hoping for rain for the crops. Young people splash anyone who comes near. "No one seems to mind because it's so hot!" Vina says.

LESSON PLAN
Chapter 25, Section 1

ASSESS

Check for Understanding
Assign Section 1 Review as homework or an in-class activity.

Meeting Lesson Objectives
Each objective below is tested by the questions that follow it in parentheses.

1. **Identify** countries that make up mainland Southeast Asia. (2, 5)

MULTICULTURAL PERSPECTIVE

Cultural Heritage
Hanoi was founded in A.D. 1010 and is Southeast Asia's oldest capital. Between 1887 and 1954, Hanoi served as the capital of French Indochina and as the capital of North Vietnam while the country was divided.

Evaluate

Section 1 Quiz **L1**

Use the Testmaker to create a customized quiz for Section 1. **L1**

More About . . .

Dress in Southeast Asia The orange robes of Buddhist monks may set them apart from the many Southeast Asians in Western-style dress, but monks are not the only ones who wear traditional clothing. Laotian women often wear handwoven, multicolor calf-length sarongs with their Western-style blouses. Thai women also wear sarongs, and often wear long or short wrap skirts around the house. Thais of Chinese descent prefer jackets over loose, calf-length pants. Cambodian men and women wear sarongs that hang to the ankles. A Cambodian woman may also don a small, colorful hat or a *krama*—a large scarf that can serve as a hat, blanket, or baby sling. For special occasions in Vietnam, women wear the *ao dai*—a dress with a long front and back panels over satin pants.

Place Ho Chi Minh City was formerly called Saigon and served as the capital of South Vietnam when the country was divided. In 1976, when the country was united under communist rule, the city was renamed for Ho Chi Minh, the revolutionary leader and first president of North Vietnam.

Reteach

Have students summarize the section by writing a brief paragraph about each of the countries of mainland Southeast Asia. **L1**

Enrich

Have students research and write reports on the Vietnam War or the civil war in Cambodia. Call on volunteers to present and discuss their reports. **L3**

CLOSE

Ask students to list places of interest, sports, and other forms of entertainment in their community that they think Chuan Soonsiri might like to experience.

Ho Chi Minh City, Vietnam

Rush-hour traffic in Vietnam's largest city means curb-to-curb mopeds.
PLACE: What is the capital of Vietnam?

Place
Vietnam

Vietnam lies along the east coast of mainland Southeast Asia. A few decades ago, Americans talked about Vietnam every day. United States forces were fighting a war there. Today Vietnam is a communist country, but its leaders are cautiously opening Vietnam to Western ideas and practices.

The Land and Economy Vietnam has a long coastline bordering the South China Sea and the Gulf of Thailand. Rugged mountains and high plateaus cover most of Vietnam. The rest of the country is made up of coastal lowlands and river deltas. A **delta** is an area of land formed by soil deposits at the mouth of a river. Two important river deltas lie at opposite ends of Vietnam. One is the Red River delta in the north; the other is the Mekong River delta in the south. Vietnam has tropical and subtropical climates. Monsoons bring wet and dry seasons.

Vietnam's communist government runs the country's economy. Coal, zinc, and other mineral resources are plentiful in north and central Vietnam. Most of Vietnam's industries are located there. Years of warfare have kept Vietnam's industries from fully developing. The country is rich in fertile soil, and agriculture is the major economic activity.

The People About 85 percent of Vietnam's 73 million people belong to the Vietnamese ethnic group. The rest are Chinese, Cambodians, and other Asian ethnic groups. Most Vietnamese are crowded along the coasts and river deltas. The largest cities are Ho Chi Minh (HOH chee MIHN) City in the Mekong delta, and Hanoi, the capital, in the north. The most widespread religion is Buddhism.

SECTION 1 REVIEW

REVIEWING TERMS AND FACTS
1. **Define the following:** alluvial plain, deforestation, delta.
2. **LOCATION** Where is Myanmar located?
3. **REGION** How do monsoons affect climate in mainland Southeast Asia?
4. **PLACE** What major food crop is grown in mainland Southeast Asia?

MAP STUDY ACTIVITIES

5. Look at the physical map on page 656. What river flows through Cambodia?
6. Study the map on page 657. What are the major climates of mainland Southeast Asia?

Answers to Section 1 Review
1. All vocabulary words are defined in the Glossary.
2. Myanmar is the northernmost country of Southeast Asia, located just to the east of India and Bangladesh.
3. The monsoons bring wet and dry seasons.
4. rice
5. Mekong River
6. humid subtropical, tropical savanna, tropical rain forest

BUILDING GEOGRAPHY SKILLS

Reading a Contour Map

Hills and valleys have highs and lows. How do you show this on a flat map? One way is with a contour map. Contour maps use lines, called *isolines,* to outline the shape—or contour—of the landscape. Each isoline connects all points that are at the same elevation. If you walked along one isoline, you would always be at the same height above sea level.

Where isolines are far apart, the land rises gradually. Where they are close together, the land rises steeply. To read a contour map, follow these steps:

- Name the area shown on the map.
- Read the numbers on the isolines to find out how much the elevation increases with each line.
- Identify areas of steep and level terrain.

BORNEO: Contour Map

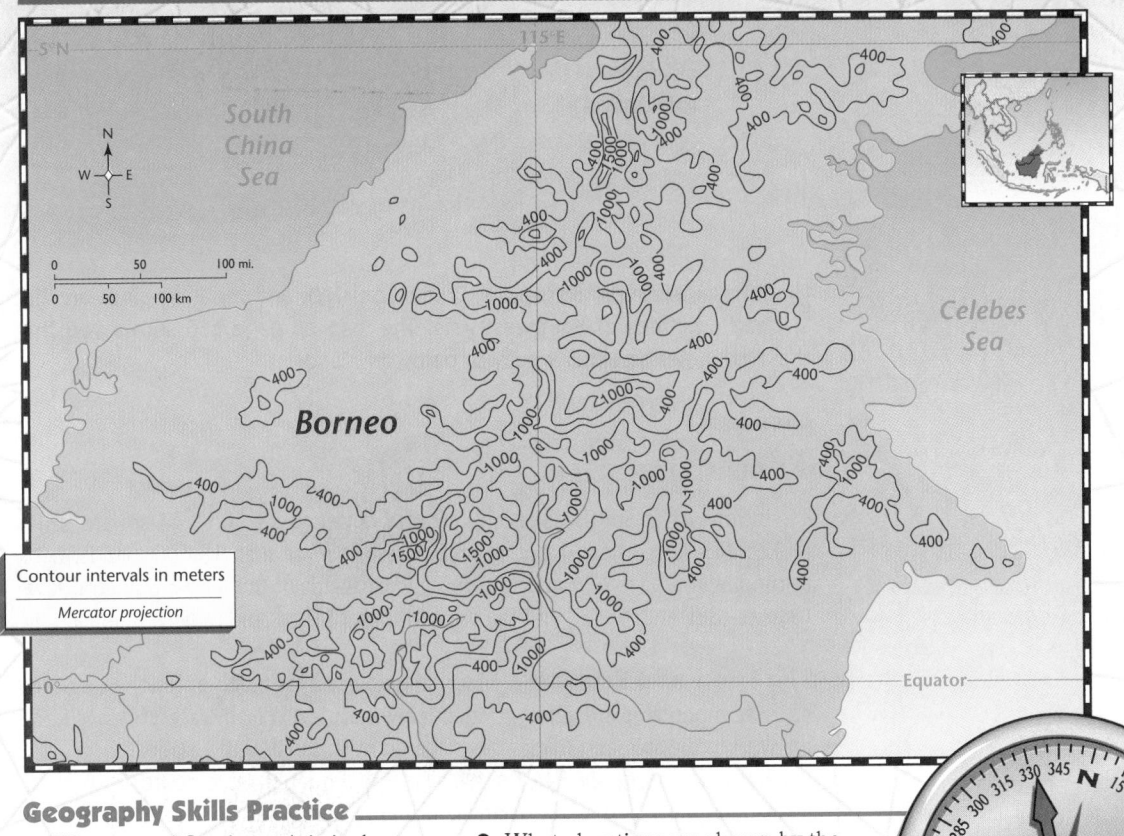

Contour intervals in meters

Mercator projection

Geography Skills Practice

1. What area of Southeast Asia is shown on the map?

2. What elevations are shown by the isolines?

CHAPTER 25

TEACH

On the chalkboard, draw a simple trail map with two routes leading from a town to a lake. One route should be short and direct, the other should be long and circuitous. Ask students which route they would take if they were hiking to the lake. Most will select the shorter route. Then draw contour lines that show that the shorter route goes over a mountain and the longer route is on flat land. Ask students again which route they would chose. Then direct students to read the skill and complete the practice questions. **L1**

Additional Skills Practice

1. What are isolines? *(lines that connect all the points at the same elevation)*
2. What kind of slope would you expect to see if isolines are far apart? *(a gradual slope)*

Geography Skills Activity 25

Building Skills in Geography, Unit 2, Lesson 3

Answers to Geography Skills Practice

1. Borneo
2. 400 meters, 1,000 meters, 1,500 meters

Island Southeast Asia

Meeting National Standards

Geography for Life The following standards are highlighted in this section:
• Standards 2, 4, 6, 9, 10, 13, 15

FOCUS

Section Objectives

1. **Distinguish** among groups of people that live in island Southeast Asia.
2. **Analyze** why people in Malaysia, Singapore, and Brunei have high standards of living.

Bellringer Motivational Activity

Prior to taking roll at the beginning of the class period, project Section Focus Transparency 25-2 and have students answer the activity questions. Discuss students' responses. **L1**

Vocabulary Pre-check

Write *cassava* and *abaca* on the chalkboard, and explain that the terms are names of plants. Then have students speculate how the plants are used. Note their speculations under the terms. As students read the section, have them check which, if any, of their speculations are correct. **L1** **LEP**

PREVIEW

Words to Know
• cassava
• free port
• abaca

Places to Locate
• Indonesia
• Malaysia
• Singapore
• Brunei
• Philippines
• Jakarta
• Manila
• Java

Read to Learn . . .
1. what groups of people live in island Southeast Asia.
2. why people in Malaysia, Singapore, and Brunei have high standards of living.

This volcano is part of a chain on the island of Java, one of thousands of islands that make up the country of Indonesia. The islands of Southeast Asia lie along the Pacific Ocean's Ring of Fire. People living here constantly face the threat of volcanic activity and earthquakes.

Indonesia, Malaysia, Singapore, Brunei (bru•NY), and the Philippines are the island countries of Southeast Asia. The map on page 656 shows you that these countries lie either wholly or partly on islands.

Location

Indonesia

Indonesia is Southeast Asia's largest country. Its 705,190 square miles (1,826,443 sq. km) consist of an archipelago of more than 13,600 islands. Look at the map on page 656 and find the major islands of Indonesia: Sumatra, Java, Celebes (SEH•luh•BEEZ), and the western part of the island of New Guinea.

The Land and Economy Mountains tower over much of Indonesia. Many of these mountains are active or inactive volcanoes. Volcanic ash provides farmers with fertile soil, and thick forests spread over much of the islands.

Because Indonesia lies on or near the Equator, its climate is tropical. Monsoons blow across the islands, bringing a wet season and a dry season. Indonesia's major economic activities are farming, mining, and manufacturing. Indonesian farmers grow rice, coffee, spices, and **cassava,** a plant root used in

662

UNIT 8

Classroom Resources for Section 2

 REPRODUCIBLE MASTERS

Reproducible Lesson Plan 25-2
Guided Reading Activity 25-2
Chapter Map Activity 25
Cooperative Learning Activity 25
Reteaching Activity 25
Enrichment Activity 25
Section Quiz 25-2

TRANSPARENCIES

Section Focus Transparency 25-2
World Cultures Transparency 18
Geoquiz Transparencies 25-1, 25-2

MULTIMEDIA

Testmaker

National Geographic Society: ZipZapMap! World

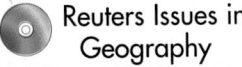 Reuters Issues in Geography

making flour. Forests supply teak and other hardwoods. Indonesia also mines tin, petroleum, nickel, and other minerals. The graph below shows that Indonesia is one of the world's leading producers of tin.

Low labor costs attract manufacturing and processing industries to Indonesia from all over the world. You find most of Indonesia's industries on the island of Java. Factories there churn out steel products, chemicals, textiles, and cement.

The People Indonesia has about 200 million people—the largest population of any country in Southeast Asia. Its rapidly growing population has made it the country with the fourth-highest population density in the world. About 60 percent of Indonesia's people live on the island of Java. There you will find Jakarta (juh•KAHR•tuh), Indonesia's capital and largest city. Jakarta has modern buildings and streets crowded with cars and bicycles.

Most Indonesians belong to the Malay ethnic group. They are divided into about 250 to 300 smaller groups with their own languages and dialects. The official language, Bahasa Indonesia, is taught in schools. Some urban Indonesians also understand the Dutch language. Why? The Dutch ruled Indonesia from the 1600s to the mid-1900s. In 1945 Indonesia declared itself an independent republic.

Islam—brought to the islands some 500 years ago by Arab traders—is the major religion of the country. Some Indonesians are Hindus or Christians, or they practice traditional religions. In past centuries, cultural influences from China and India affected Indonesian ways of life. Today European and American influences are also widespread.

WHAT IN THE WORLD?

Waxing Cloth?
Indonesia and Malaysia are known for a striking art form called *batik*—a Javanese word meaning "drop." To create a *batik* pattern, wax is first dripped onto fabric. When the fabric is dyed, the dye reaches only the fabric not covered with wax. After the dye has dried, the fabric is boiled to remove the wax. A beautiful pattern is revealed.

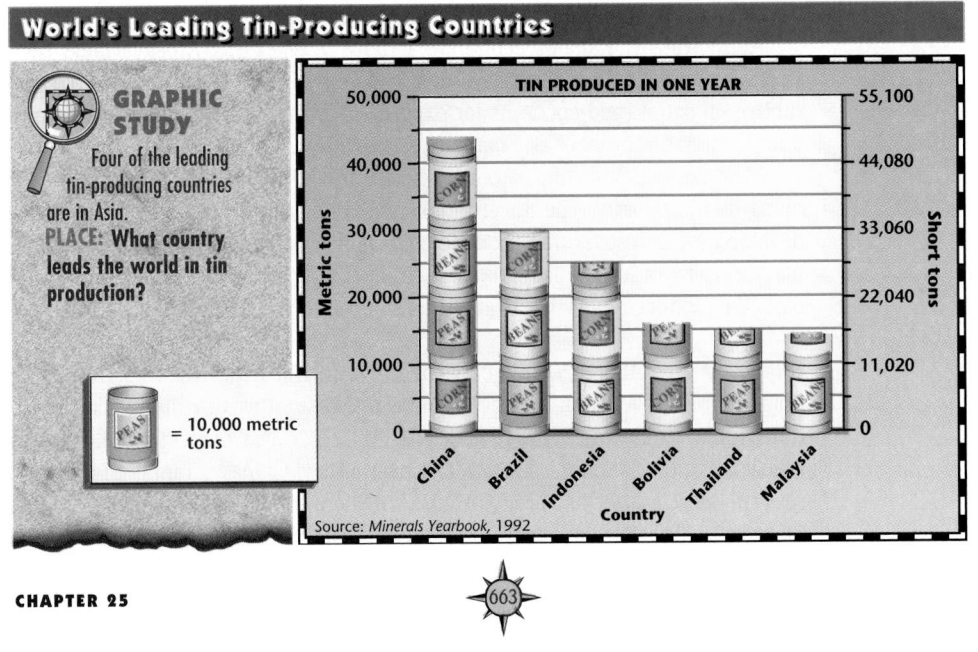

World's Leading Tin-Producing Countries

GRAPHIC STUDY
Four of the leading tin-producing countries are in Asia.
PLACE: What country leads the world in tin production?

= 10,000 metric tons

TIN PRODUCED IN ONE YEAR

Metric tons: 50,000 / 40,000 / 30,000 / 20,000 / 10,000 / 0
Short tons: 55,100 / 44,080 / 33,060 / 22,040 / 11,020 / 0

Country: China, Brazil, Indonesia, Bolivia, Thailand, Malaysia

Source: *Minerals Yearbook*, 1992

LESSON PLAN
Chapter 25, Section 2

TEACH

Guided Practice
Making Generalizations
Divide the chalkboard into five columns and label the columns *Indonesia, Malaysia, Singapore, Brunei,* and *The Philippines.* As students read the information on Indonesia, have volunteers list each country's major geographical features—mountains and volcanoes; fertile soil; tropical climate; farming, mining, and agricultural; rapidly growing population; mostly Malay; Islam; and so on—in the first column. Have them do the same for the other four countries discussed in the section. Then ask the students to make generalizations about the island countries of Southeast Asia based on the information in the chart. **L1**

Cultural Kaleidoscope

Indonesia More Muslims live in Indonesia than in any other country.

GRAPHIC STUDY
Answer
China
Graphic Skills Practice
Reading a Graph
Which is the leading Southeast Asian tin producer? *(Indonesia)*

Cooperative Learning Activity

Organize students into five groups and assign each group one of the Southeast Asian island countries. Have members of each group study the way of life in their assigned country. Then instruct groups to prepare a presentation about their country to share with the class. Presentations may consist of skits, panel discussions, or graphics. **L2**

Independent Practice

 Guided Reading Activity 25-2 **L1**

Brunei has an average humidity of 82 percent all year long.

Global Gourmet

Malaysia The durian is the best of all fruits, as far as Malaysians are concerned. It has a green, spiny rind and beautifully flavored, soft pulp. However, it gives off a rather strong and—some say—disagreeable odor.

Bali and Malaysia

Young boys join in a dance at a Bali religious festival *(left)*. A Malaysian rubber worker taps a rubber tree *(right)*.
PLACE: What is the most important economic activity in Malaysia?

Place
Malaysia

North of Indonesia is Malaysia, one of Southeast Asia's most prosperous countries. Malaysia has two separate areas. One is the mainland part, called Malaya, on the Malay Peninsula. The other is made up of two territories— Sarawak (suh•RAH•wahk) and Sabah (SAH•buh)—on the northern edge of the island of Borneo.

The Land and Economy Mountains cover the center of Malaya. Low-lying plains are found east and west of these mountains. In Sarawak and Sabah, coastal swamps, rain forests, and rugged mountains make up most of the landscape. The map on page 657 shows you that both parts of Malaysia have a tropical rain forest climate. They have high temperatures and plenty of rain.

Agriculture is a major economic activity in Malaysia. Most farmers own small plots of land and raise rice, fruits, and vegetables. Plantation owners grow rubber, oil palms, and coconuts for export. Malaysia leads the world in the production of natural rubber and palm oil.

Malaysia is rich in tin, bauxite, petroleum, and iron ore. It is rapidly developing its oil industry. The government uses money from tin and oil exports to develop new industries that produce consumer goods. Malaysians perform manufacturing tasks for industries across the globe. For example, many computer parts are produced in Malaysia and then shipped to the United States.

The People About 20 million people live in Malaysia. Most belong to the Malay ethnic group, but a large Chinese population lives in the cities. People from India, Pakistan, Bangladesh, and Sri Lanka also make their homes in Malaysia. In the marketplaces you can hear Malay, Chinese, Tamil, and English spoken. Most Malaysians are Muslims, but there are large numbers of Hindus, Buddhists, Christians, and followers of traditional Chinese religions.

Meeting Special Needs Activity

Poor Learners Most students are familiar with the study system "SQ3R" (Survey, Question, Read, Recite, Review), but those with reading and learning problems often have difficulty *surveying* information. One task involved in surveying is predicting what kind of information will be given in a piece of text. Guide students to recognize that the first paragraph under each main head makes general statements and the rest of the subsection gives detailed information. **L1**

The people of Malaysia are about evenly divided between those living in cities and those living in rural areas. In rural villages many people live in thatched-roof homes built on posts a few feet off the ground. Most people in the cities live in high-rise apartments. Kuala Lumpur (KWAH•luh LUM•PUR), located on the Malay Peninsula, is the capital and largest city.

Movement
Singapore

Singapore is made up of 58 islands off the southern shore of the Malay Peninsula. With only 240 square miles (622 sq. km), Singapore is the smallest country in Southeast Asia. The city of Singapore takes up much of the largest island. In the early 1800s, Singapore was covered with rain forests and swamps. Today it is one of the world's leading commercial and trading centers.

The Land and Economy The central part of Singapore is hilly; coastal areas are flat. Urban centers with modern buildings, streets, and highways have replaced much of the rain forest. Like Malaysia, Singapore has a hot, humid climate.

Singapore's economy relies on trade and manufacturing. The city of Singapore has one of the world's busiest harbors. It is a **free port,** a place where goods can be loaded, stored, and shipped again without payment of import taxes. Singapore's workers add to the country's wealth by making industrial goods for export.

MAP STUDY

Answer
Kuala Lumpur

Map Skills Practice
Reading a Map In which part of Indonesia are tea and coffee grown? *(western Indonesia)*

MULTICULTURAL PERSPECTIVE

Culturally Speaking
The name *Singapore* is derived from the Sanskrit *Singa Pur,* which means "city of the lion." This name was probably given to the area by Sumatrans who settled there in the 1200s.

GLENCOE
TECHNOLOGY

Reuters Issues in Geography

Chapter 9
Southeast Asia: Patterns of Migration
Disc 1 Side A

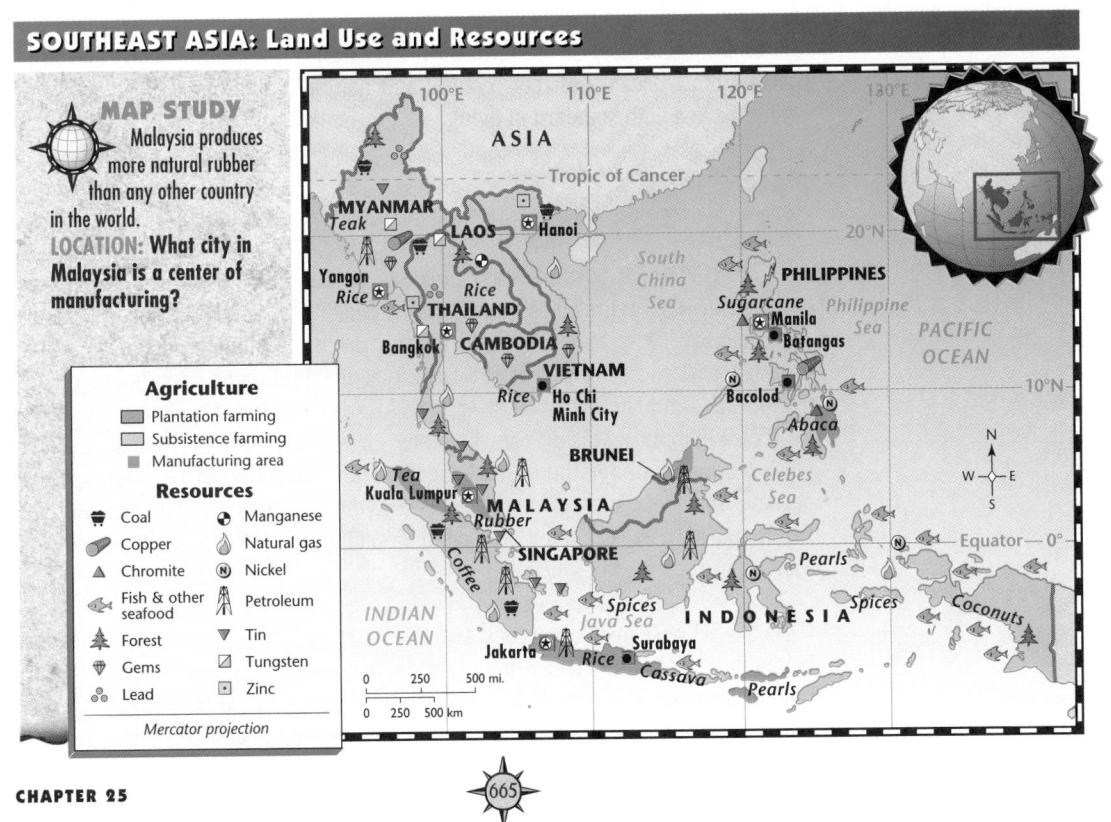

SOUTHEAST ASIA: Land Use and Resources

MAP STUDY
Malaysia produces more natural rubber than any other country in the world.
LOCATION: What city in Malaysia is a center of manufacturing?

Agriculture
- Plantation farming
- Subsistence farming
- Manufacturing area

Resources
- Coal
- Copper
- Chromite
- Fish & other seafood
- Forest
- Gems
- Lead
- Manganese
- Natural gas
- Nickel
- Petroleum
- Tin
- Tungsten
- Zinc

0 250 500 mi.
0 250 500 km
Mercator projection

CHAPTER 25

665

Identifying Central Issues Point out that many people in the West have criticized some Southeast Asian countries for their record on human rights. These critics have cited numerous incidents of imprisonment without trial, unduly harsh punishments for minor crimes, and violations of freedom of speech to back up their charges. Some Southeast Asian leaders have responded by saying that human rights are part of the "Western value system" and not part of Asian thinking. Ask students to discuss the following: Is freedom of speech a universal human right, or is it a "Western value" that other cultures can accept or reject? **L2**

More About the Illustration

Answer to Caption
Goods are not subject to import taxes when they are unloaded and stored for reshipment.

Place Social behavior is strictly regulated in Singapore. People are fined for smoking or spitting in public, for hogging the road, and even for not flushing public toilets. In 1991, chewing gum in public was outlawed because of the messy litter that discarded gum creates. Convicted gum-chewers face up to a year in prison or a heavy fine.

ASSESS

Check for Understanding

Assign Section 2 Review as homework or an in-class activity.

Meeting Lesson Objectives

Each objective below is tested by the questions that follow it in parentheses.

1. **Distinguish** among groups of people that live in island Southeast Asia. (2, 3, 4)
2. **Analyze** why people in Malaysia, Singapore, and Brunei have high standards of living. (3)

The Skyline of Singapore, Singapore

Singapore, the largest port in Southeast Asia, has been described as one of the cleanest and safest cities in the world.
MOVEMENT: Why is Singapore called a *free port?*

The People Most of Singapore's nearly 3 million people are Chinese, but Malays, Indians, and Europeans also live there. Singapore's people enjoy a high standard of living. More than 90 percent of them live in the capital city of Singapore.

Place
Brunei

On the northern coast of Borneo lies another small nation—Brunei. Brunei's 2,030 square miles (5,258 sq. km) make it about the size of Delaware. The map on page 654 shows you that Brunei borders the South China Sea. Malaysia divides Brunei into two parts. The western part has low plains and coastal swamps. The eastern part has forested hills and mountains. All of Brunei experiences a hot and humid tropical climate.

The Economy and People Oil and natural gas are Brunei's main exports and sources of income. With money from these resources, the government of Brunei is building new industries. Farming and fishing are other important economic activities. Brunei's farmers grow rice, pepper, coconuts, and fruit.

Most of Brunei's 300,000 people are either Malays or Chinese. Malay, Chinese, and English are the most commonly spoken languages. Islam is the leading religion. About 58 percent of the people of Brunei live in cities. The capital and largest city is Bandar Seri Begawan (BUHN•duhr sehr•EE buh•GAH•wuhn). Because of their rich oil and natural gas resources, the people of Brunei have a high standard of living.

Place
The Philippines

East of Vietnam in the South China Sea lies the Philippines. The Philippines is an archipelago with more than 7,000 islands. Only about 900 of the islands are inhabited. Only 11 islands hold 95 percent of the Filipino people. The map on page 656 shows you that the largest of the Philippine islands are Luzon (loo•ZAHN) and Mindanao (MIHN•duh•NAH•oh).

The Philippines was a Spanish colony for more than 300 years. In the late 1800s, it came under the rule of the United States. In 1946 the Philippines became an independent democracy. Twenty-five years later, Filipino President Ferdinand Marcos set up a dictatorship. A national uprising in 1986 overthrew Marcos and restored democracy.

The Land Volcanic mountains and forests spread over the landscape of the Philippines. Because of steep mountain slopes, only one-third of the land is suitable for farming. The climate of the Philippines is tropical, with wet and dry seasons. Typhoons often cause flooding and much loss of life and property.

666

UNIT 8

More About . . .

Southeast Asia's Rubber Industry Rubber trees are not indigenous to Southeast Asia. They were introduced to the region during the 1870s by British government officials. At the time, the only source for natural rubber was the South American rain forests. The supply was limited and quite expensive, however. Growing rubber trees on organized plantations, British officials thought, would cut costs and greatly increase

supply. In 1876, Henry Wickham, working for the British, collected some 70,000 rubber-tree seeds from Brazil. He transported them to England, where they were planted in hothouses. Later, rubber-tree seedlings were sent to Malaysia and other British colonies in Asia. By the 1890s several British plantations were producing natural rubber. Today, Southeast Asia leads the world in natural rubber production.

SOUTHEAST ASIA: Population Density

MAP STUDY
Indonesia ranks fourth in world population after China, India, and the United States.
PLACE: What Indonesian island has the highest population density?

Persons per

sq. mi.	sq. km
Under 2	Under 1
2-60	1-25
60-125	25-50
125-250	50-100
Over 250	Over 100

Mercator projection

ASIA

Tropic of Cancer

Mandalay
MYANMAR Hanoi
Sittwe **LAOS** Haiphong
Chiang Mai Vientiane
Yangon Da Nang
THAILAND South China Sea
Bangkok
CAMBODIA
Phnom Penh **VIETNAM**
Ho Chi Minh City
Quezon City *Philippine Sea*
Manila **PHILIPPINES**
Pinang Cebu
Davao
BRUNEI *PACIFIC OCEAN*
Medan *Celebes Sea*
Kuala Lumpur
MALAYSIA
Singapore **SINGAPORE**
Pontianak
Palembang
Jakarta *Java Sea* **I N D O N E S I A** Jayapura
Bandung Semarang Ujung Pandang
Surabaya

INDIAN OCEAN

Equator — 0°

0   250   500 mi.
0   250   500 km

CITIES

- City with more than 1,000,000 people
- City with 500,000 to 1,000,000 people
- City with 100,000 to 500,000 people

The Economy and People Agriculture, mining, and industry are the major economic activities in the Philippines. Filipino farmers harvest rice, sugarcane, pineapples, and **abaca,** a plant fiber used to make rope. Filipino workers produce textiles, chemicals, electronic products, and other manufactured goods.

About 69 million people live in the Philippines. Most Filipinos live on the island of Luzon. Manila, the capital and largest city, is located on Luzon and is one of the great commercial centers of Asia.

The Filipinos are Malay in ethnic background. Their culture, however, blends Malay, Spanish, and American influences. About 90 percent of Filipinos practice the Roman Catholic faith, brought to the islands by Spanish missionaries. American influences show up in the products sold along busy city streets.

SECTION 2 REVIEW

REVIEWING TERMS AND FACTS

1. **Define the following:** cassava, free port, abaca.
2. **PLACE** What are the four major islands of Indonesia?
3. **REGION** What are Singapore's two important economic activities?

MAP STUDY ACTIVITIES

4. Look at the population density map above. What two cities in the Philippines have more than 1 million people?

Answers to Section 2 Review

1. All vocabulary words are defined in the Glossary.
2. Sumatra, Celebes, New Guinea, Java
3. trade, manufacturing
4. Quezon City, Manila

MAP STUDY
Answer
Java
Map Skills Practice
Reading a Map Where on the Malay Peninsula is population density the heaviest? *(southwest coast)*

Evaluate

Section 2 Quiz L1

Use the Testmaker to create a customized quiz for Section 2. L1

Chapter 25 Test Form A and/or Form B

Reteach

Reteaching Activity 25 L1

Enrich

Enrichment Activity 25 L3

CLOSE

Mental Mapping Activity

Provide each student with a large sheet of white paper and map pencils. Inform students that this will not be a graded activity. Have students draw, freehand from memory, a map of Southeast Asia that includes the following labeled countries and bodies of water: Indonesia, Malaysia, Singapore, Brunei, the Philippines, South China Sea, Indian Ocean, and Pacific Ocean.

667

TEACH

Ask students to name the gemstone they think the most beautiful and precious. Most students will name diamonds. Point out, however, that rubies and sapphires as beautiful as and, in some cases, more precious than diamonds can be found in several Southeast Asian countries. **L1**

More About . . .

Rubies, Sapphires, and Jade

- In the Middle Ages, rubies were a symbol of power and romance.
- Ancient Persian legends told of a giant sapphire on which the world stands. The sky is blue, the legends contend, because it reflects the color of the sapphire.
- Spanish conquistadors believed the green stones they found in the Americas could cure kidney diseases, so they called them *kidney stones,* or *piedras de hijadas.* The word jade is derived from *hijadas.* In Europe the name *jade* was applied to the material of the same color and hardness that was imported from China.

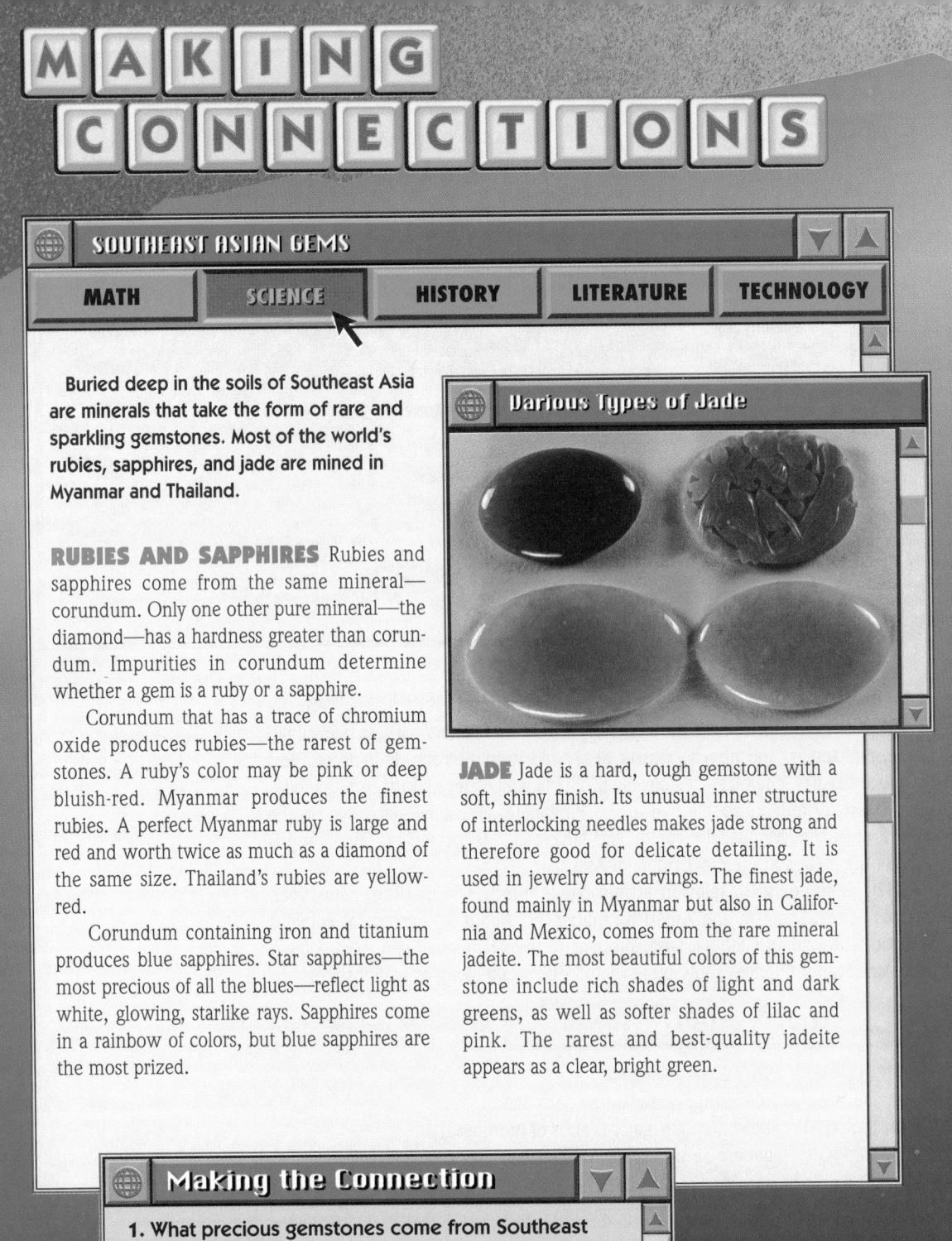

MAKING CONNECTIONS

SOUTHEAST ASIAN GEMS

| MATH | SCIENCE | HISTORY | LITERATURE | TECHNOLOGY |

Buried deep in the soils of Southeast Asia are minerals that take the form of rare and sparkling gemstones. Most of the world's rubies, sapphires, and jade are mined in Myanmar and Thailand.

RUBIES AND SAPPHIRES Rubies and sapphires come from the same mineral—corundum. Only one other pure mineral—the diamond—has a hardness greater than corundum. Impurities in corundum determine whether a gem is a ruby or a sapphire.

Corundum that has a trace of chromium oxide produces rubies—the rarest of gemstones. A ruby's color may be pink or deep bluish-red. Myanmar produces the finest rubies. A perfect Myanmar ruby is large and red and worth twice as much as a diamond of the same size. Thailand's rubies are yellow-red.

Corundum containing iron and titanium produces blue sapphires. Star sapphires—the most precious of all the blues—reflect light as white, glowing, starlike rays. Sapphires come in a rainbow of colors, but blue sapphires are the most prized.

Various Types of Jade

JADE Jade is a hard, tough gemstone with a soft, shiny finish. Its unusual inner structure of interlocking needles makes jade strong and therefore good for delicate detailing. It is used in jewelry and carvings. The finest jade, found mainly in Myanmar but also in California and Mexico, comes from the rare mineral jadeite. The most beautiful colors of this gemstone include rich shades of light and dark greens, as well as softer shades of lilac and pink. The rarest and best-quality jadeite appears as a clear, bright green.

Making the Connection

1. What precious gemstones come from Southeast Asia?
2. Which Southeast Asian countries produce the gems?

668

UNIT 8

Answers to Making the Connection

1. rubies, sapphires, jade
2. Myanmar, Thailand

Chapter 25 Highlights

Important Things to Know About Southeast Asia

SECTION 1 MAINLAND SOUTHEAST ASIA

- Mainland Southeast Asia includes the countries of Myanmar, Thailand, Laos, Cambodia, and Vietnam.
- Plentiful water supplies and fertile soil enable farmers to grow rice as a major food crop.
- Most of this area of Southeast Asia has tropical and subtropical climates.
- Most mainland Southeast Asians live in rural villages, but cities are growing in population.
- Most people in mainland Southeast Asia practice Buddhism.

SECTION 2 ISLAND SOUTHEAST ASIA

- The islands of Southeast Asia lie along the Pacific Ocean's Ring of Fire. Their people face the threat of volcanic activity and earthquakes.
- Indonesia is the largest country in Southeast Asia in area and population.
- The city of Singapore is one of the world's busiest trading and commercial centers.
- The people of Malaysia come from many different ethnic backgrounds.
- The culture of the Philippines shows Malay, Spanish, and American influences.

Floating market in Bangkok, Thailand ▶

Extra Credit Project

Music Have students research the history of an instrument or type of music played in Southeast Asia. Suggest that students use public library resources to find a tape of the music or instrument. Have students report orally about the music and use the tapes to illustrate points in their reports. **L2**

Using the Chapter 25 Highlights

- Use the Chapter 25 Highlights to preview, review, condense, or reteach the chapter.

Preview/Review

Vocabulary Puzzle-Maker Software reinforces the terms used in Chapter 25.

Condense

Have students read the Chapter 25 Highlights. Spanish Chapter Highlights also are available.

Chapter 25 Digest Audiocassettes and Activity

Chapter 25 Digest Audiocassette Test

Spanish Chapter Digest Audiocassettes, Activities, and Tests also are available.

Guided Reading Activities

Reteach

Reteaching Activity 25. Spanish Reteaching activities also are available.

MAP STUDY

Place Copy and distribute page 34 from the Outline Map Resource Book. Have students locate and label the countries and major urban centers of Southeast Asia, including Bangkok, Hanoi, Singapore, Jakarta, and Manila.

Chapter 25 Review and Activities

Chapter
25
Review and Activities

GLENCOE TECHNOLOGY

 **VIDEODISC**

Use the Chapter 25 MindJogger Videoquiz to review students' knowledge before administering the Chapter 25 Test.

MindJogger Videoquiz

Chapter 25
Disc 4 Side A

Also available in VHS.

ANSWERS

Reviewing Key Terms

1. F
2. E
3. B
4. D
5. A
6. C

Mental Mapping Activity

This exercise helps students to visualize the countries and geographic features they have been studying and understand the relationships among them. All attempts at freehand mapping should be accepted.

REVIEWING KEY TERMS

Match the numbered terms in Column A with their definitions in Column B.

A
1. cassava
2. deforestation
3. abaca
4. delta
5. alluvial plain
6. free port

B
A. land area formed by a river's soil deposits
B. plant fiber used in making rope
C. place where goods are unloaded, stored, and reshipped without import taxes
D. land formed by soil deposits at a river's mouth
E. widespread cutting of trees
F. plant root used in making flour

Mental Mapping Activity

On a separate piece of paper, draw a freehand map of Southeast Asia. Label the following items on your map:
- Java
- Kuala Lumpur
- South China Sea
- Thailand
- Bangkok
- Vietnam

REVIEWING THE MAIN IDEAS

Section 1
1. PLACE What is the capital of Myanmar?
2. PLACE Which Southeast Asian country was never a colony?
3. HUMAN/ENVIRONMENT INTERACTION What two major agricultural products are grown in Thailand?
4. HUMAN/ENVIRONMENT INTERACTION Where do most people live in Myanmar and Thailand?
5. MOVEMENT Why did many people flee Cambodia during the 1970s?

Section 2
6. PLACE What is the major religion of Indonesia and Malaysia?
7. HUMAN/ENVIRONMENT INTERACTION Why is only one-third of the Philippines suitable for farming?
8. MOVEMENT Why is Singapore important to the economies of Southeast Asia?
9. MOVEMENT Why do the people of Brunei have a high standard of living?
10. REGION How are Indonesia and the Philippines alike geographically?

670

UNIT 8

Reviewing the Main Ideas

Section 1
1. Yangon
2. Thailand
3. rice, rubber
4. in river valleys and coastal areas
5. to escape harsh rule by the Communist government

Section 2
6. Islam
7. because the landscape is mountainous
8. because it is a major trade and manufacturing center
9. because of the country's profitable oil industry
10. both are archipelagos consisting of many islands

CRITICAL THINKING ACTIVITIES

1. **Drawing Conclusions** Why is the Mekong River important to the people of Southeast Asia?
2. **Predicting Outcomes** Experts believe that Brunei has enough oil reserves to last until 2018. What might happen to the country's economy and standard of living at that time?

GeoJournal Writing Activity

Imagine you are a news reporter on assignment in Southeast Asia to report on an active volcano. Write about the volcanic activity as if you were an eyewitness.

COOPERATIVE LEARNING ACTIVITY

Work in a group of three to learn about one of the countries of Southeast Asia. Each group should select one country. Then each member should choose one of the following topics to research: (a) the arts; (b) holidays and festivals; or (c) forms of government. After your research is complete, share your information with your group. As a group, prepare a written report with illustrations or photos that illustrate the group's findings.

PLACE LOCATION ACTIVITY: SOUTHEAST ASIA

Match the letters on the map with the places and physical features of Southeast Asia. Write your answers on a separate sheet of paper.

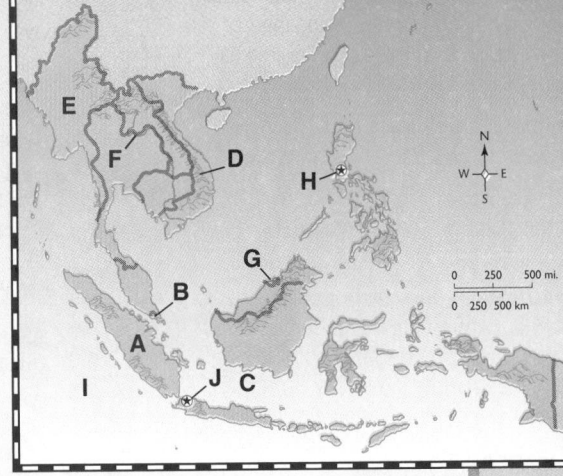

1. Myanmar
2. Mekong River
3. Singapore
4. Jakarta
5. Sumatra
6. Java Sea
7. Vietnam
8. Brunei
9. Manila
10. Indian Ocean

Cooperative Learning Activity

Ask groups to use food from their country as well as drawings and photos to illustrate their findings. Suggest that they search their grocery stores for Asian foods. Have students sample the food.

 GeoJournal Writing Activity

Have students simulate and tape the sounds of a volcano erupting. Then play the tape as background sound while volunteers share their journal entries with the class.

Place Location Activity

1. E
2. F
3. B
4. J
5. A
6. C
7. D
8. G
9. H
10. I

Chapter Bonus Test Question

This question may be used for extra credit on the chapter test.

Why are there so many volcanoes on the islands of Southeast Asia? *(because the area is part of the Ring of Fire)*

Critical Thinking Activities

1. Answers will vary, but should note that the river is an important source of water and is a major means of transportation. Fertile land along the river banks supports farming.

2. Answers will vary, but might point out that Brunei will not have the income to develop new economic endeavors, and the people will have to accept a lower standard of living. On the other hand, industries currently being set up with oil money might enable Brunei to continue its prosperity.

FOCUS

Make sure students understand what the term *species* means, and ask them to guess how many plant and animal species there are. Record their responses on the board. Then tell them that scientists have identified about 1.5 million species so far, and that they estimate that there are as many as 10 to 50 million species worldwide.

TEACH

Ask students why species extinction is a cause for concern when there are so many millions of species. Help them understand the consequences of extinction by creating a consequence wheel on the board. Write "_____ becomes extinct" and circle it. Choose different Asian species to complete the statement. For each, have students suggest consequences of the extinction and record them in smaller circles surrounding the main circle. Draw an arrow from the central circle to each consequence. Have students identify consequences such as disruption of food chains and ecosystems and the loss of natural beauty. **L2**

672

**NATIONAL
GEOGRAPHIC
SOCIETY**

EYE ON THE ENVIRONMENT

Asia

HABITAT LOSS

PROBLEM

Over the last 100 years, Asia has suffered greater destruction of habitats than any other continent. Asian countries have drained wetlands, felled forests, and plowed under grasslands to feed growing populations and boost their economies. As the demand for resources increases, species disappear at an alarming rate. Scientists estimate that every few hours one plant or animal species in Asia becomes extinct.

SOLUTIONS

● India has created more than 370 national parks and preserves since 1960 to protect endangered wildlife such as the tiger. Since 1970 the population of Indian tigers in the wild has doubled.

● Countries in Asia have passed tough laws protecting wildlife. In China, poachers are executed.

● China has established 12 reserves for its giant pandas over the past few decades and plans to create 14 additional protected areas.

● In Indonesia, villagers now learn how to use the forests without destroying them. Some locals are hired as park rangers and tourist guides; others make money raising and selling butterflies or growing forest crops of nuts and palm leaves.

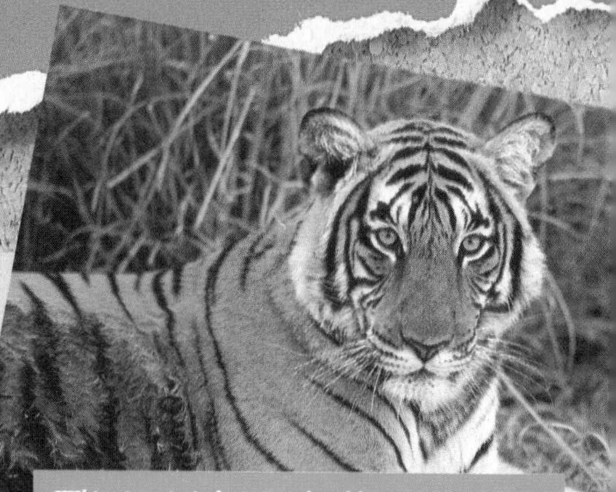

672

This tiger in India's Ranthambhore National Park is part of the success story of Project Tiger. Since the effort to protect tigers was launched in 1973, the number of tigers living in the park's forest has grown from around 14 to more than 40.

HABITAT LOSS FACT BANK

🌿 In 1900, 40,000 tigers lived in the wild in India alone; today about 6,000 survive, mostly in Asia.

🌿 In 1900, 40 percent of India was forested; today, only 15 percent is forested—and the number of endangered species has increased tenfold.

🌿 Indonesia has 210 endangered plant and animal species—more than any other nation in the world.

🌿 Javan and Sumatran rhinos are among the world's most endangered species. Fewer than 800 animals survive.

🌿 An estimated 1,200 giant pandas survive in bamboo forests in China. Poachers kill 40 or more pandas a year.

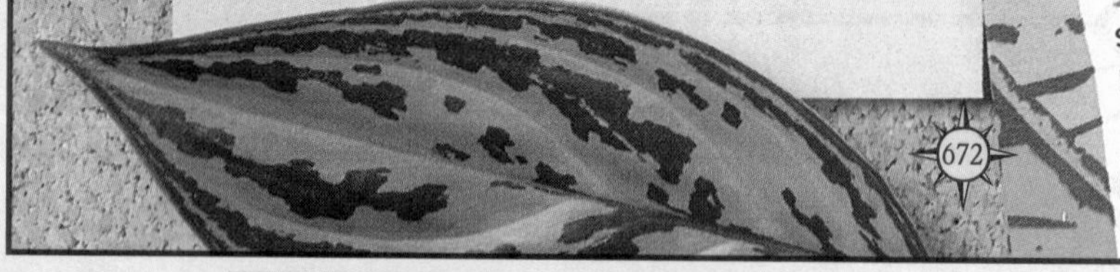

UNIT 8

More About the Issues

Human/Environment Interaction The rapid growth of the population of one species—the human—is perhaps the root cause of the loss of habitat for other species. An expanding human population requires more space, consumes more resources, entails more economic activities, and generates more waste. In 1900, about 1.65 billion people inhabited the planet. Today, the population is about 5.6 billion. Estimates suggest that the world population will exceed 6 billion by the year 2000. The human population of Asia, the continent with the greatest habitat loss, is currently about 3.1 billion. It is forecast to increase by nearly 50 percent in the next two decades to more than 4.5 billion, putting even greater strains on already pressured habitats.

O il palms advance on the wilderness. Hundreds of thousands of acres of Malaysia's rain forests have been lost to commercial projects such as this one on a Malaysian plantation.

NATIONAL GEOGRAPHIC SOCIETY

These materials are available from Glencoe.

GTV: Planetary Manager

Malaysian lacewing butterfly
Any side, Frame 47682

TEEN TRIBUTE

Maria Martinez, a student at Boston English High School in Boston, Massachusetts, wanted to do something to save endangered Asian wildlife. So she got a job. Maria works two to three hours a day at the World Society for the Protection of Animals. She sends out petitions that call for the end of inhumane treatment of endangered Asiatic black bears in China. Maria also answers letters from other young people who want to help.

environmental activities

T hese men in Foshan, China, carry bamboo poles for sale in local markets. Pandas and villagers often compete for bamboo.

TAKE A CLOSER LOOK

1 What has happened to Asia's habitats in the last 100 years?

2 What growing demand has caused the most destruction of Asia's plants and animals?

Save the Ancient Forests

WHAT CAN YOU DO?

🍃 Talk to your parents. Ask them to vote to keep our national parks and nature preserves.

🍃 Save a tree—carry your own shopping bag. The average American uses up seven trees a year.

🍃 Stuff envelopes or help a group raise money to protect an endangered animal.

🍃 Make a bird feeder. Spread a pinecone with peanut butter and hang it from a tree.

ASSESS

Have students answer the **Take A Closer Look** questions on page 673.

CLOSE

Discuss with students the **What Can You Do?** question. Encourage students to identify local habitats and determine what species live in them. Have them share what they learn with the class.

* For an additional case study, use the following:

📁 Environmental Case Study 8 **L1**

Answers to Take A Closer Look

1. They have suffered greater destruction than habitats on any other continent.
2. demand for resources

Global Issues

Wildlife One Asian species threatened by habitat loss is the Komodo dragon of Indonesia. It is the largest lizard in the world. Some specimens are more than 10 feet long and weigh 300 pounds—and they do resemble the mythical dragon.

0:00 OUT OF TIME?

If time does not permit teaching each chapter, you may use the Chapter Highlights, Audiocassettes, and their corresponding activities and tests.

GLENCOE
TECHNOLOGY

 VIDEODISC

Reuters Issues in Geography

**Chapter 10
Antarctica: The Ozone Hole**
Disc 1 Side 1

interNET
CONNECTION

Locate teacher/student materials and other resources on Antarctica at the following site:

World Wide Web
http://quest.arc.nasa.gov/antarctica/index.html

Find leads to information on Australia, New Zealand, and other Pacific countries at:

World Wide Web
http://sunsite.anu.edu.au/spin/wwwvl-pacific/index.html

Unit 9 Australia, Oceania, and Antarctica

What Makes Australia, Oceania, and Antarctica a Region?

Most people share . . .

- a location mostly south of the Equator.
- a location near or in the Pacific Ocean.
- animals and vegetation found nowhere else in the world.
- specialized economies.

To find out more about Australia, Oceania, and Antarctica, see the Unit Atlas on pages 676–687.

Milford Sound, New Zealand

-674-

About the Illustration

Milford Sound, on New Zealand's South Island, is one of the many fjords that open into the Tasman Sea. Some of New Zealand's most beautiful scenery is found in this area. The fjords are surrounded by the lush, green slopes of the country's Southern Alps.

 **Location** What other area of the world is known for its fjords? *(Scandinavia)*

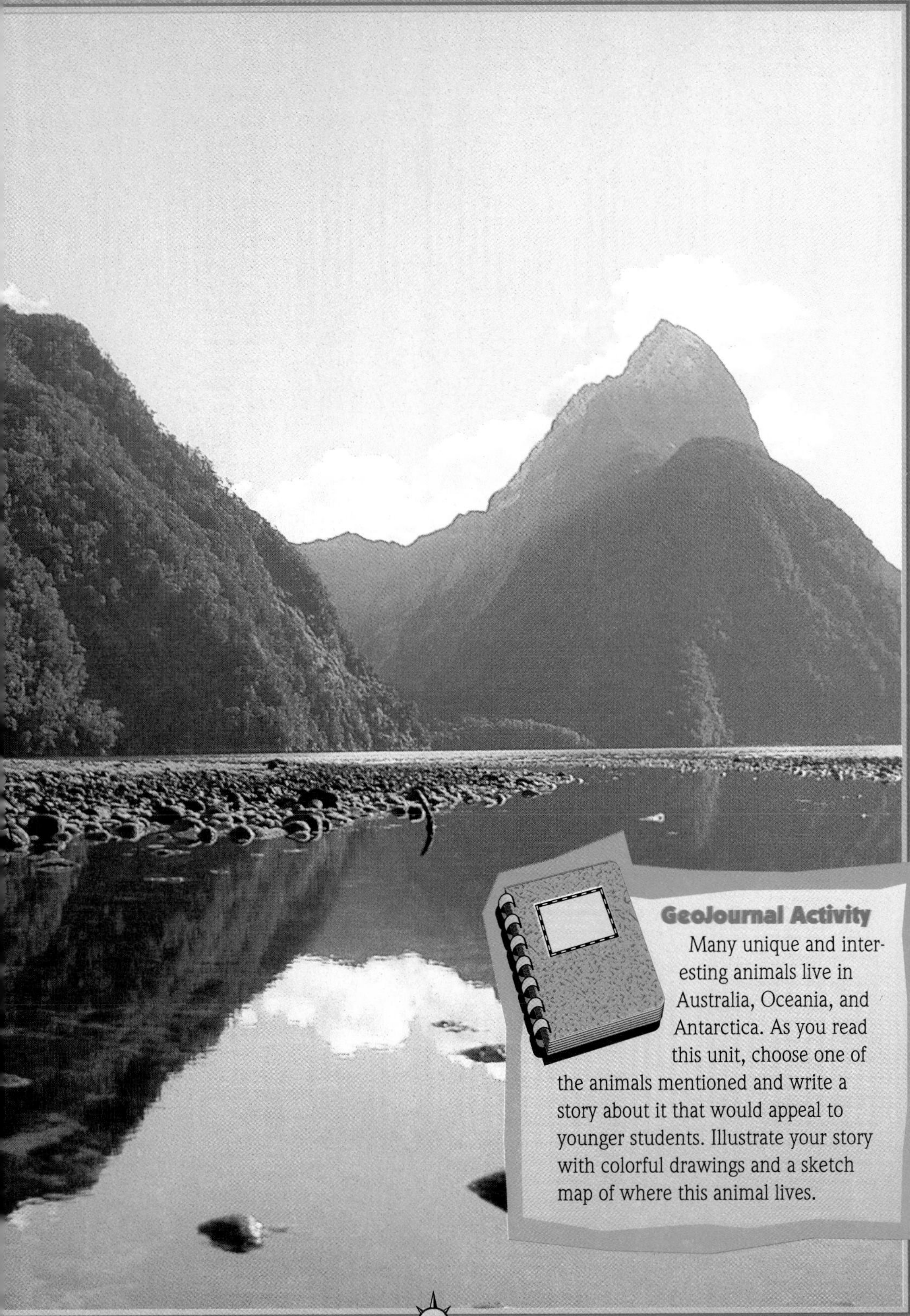

GeoJournal Activity

Encourage students to search children's books about nature to find information about the habitats of their selected animals. Remind them to depict their animals' habitats in the drawings for their stories. L1

- This journal activity provides the basis for the "GeoJournal Writing Activity" exercise in the Chapter Review.
- The GeoJournal may be used as an integral part of Performance Assessment.

GeoJournal Activity

Many unique and interesting animals live in Australia, Oceania, and Antarctica. As you read this unit, choose one of the animals mentioned and write a story about it that would appeal to younger students. Illustrate your story with colorful drawings and a sketch map of where this animal lives.

675

Location Activity

Display Political Map Transparency 9 and ask students the following questions: In what hemisphere do Australia, Oceania, and Antarctica lie? *(Southern Hemisphere)* Australia has a warmer climate than Antarctica. Why? *(because Australia lies closer to the Equator)* What major island country lies southeast of Australia? *(New Zealand)* What land features do Australia and New Zealand share? *(coastal lowlands and interior mountains)* Where does most of Oceania lie in relation to Australia? *(to the north and east)* L1

IMAGES OF THE WORLD

UNIT 9 ATLAS

NATIONAL GEOGRAPHIC SOCIETY

These materials are available from Glencoe.

 CD-ROM

Picture Atlas of the World

 POSTERS

Images of the World

Unit 9: Australia, Oceania, and Antarctica

TRANSPARENCIES

PicturePack Transparencies, Unit 9

Some sheep stations in Australia are larger than the state of Rhode Island.

NATIONAL GEOGRAPHIC SOCIETY

IMAGES of the WORLD

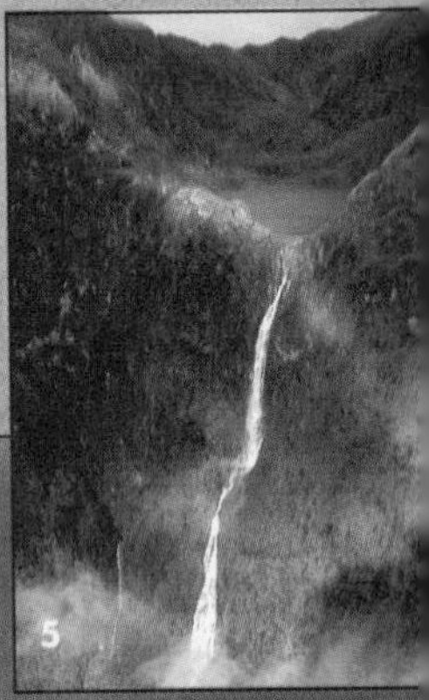

Facts

• **Tonga** Tongans have always considered obesity as a sign of good health, especially among their leaders. The *Guinness Book of World Records* listed their 462-pound (210-kg) King Tapou as the world's heaviest monarch. Recently, however, the government has started a national weight-awareness campaign, and King Tapou has shed 182 pounds (83 kg).

Papua New Guinea Traditional musicians play the *kundu*—an hourglass-shaped drum covered with lizard skin.

1. Kangaroo and joey, Australia
2. Three cowboys, Australia
3. Chinstrap penguins, Antarctica
4. Stone faces, Easter Island
5. Sutherland Falls, New Zealand
6. Boab tree, Western Australia
7. Traditional fishing, Vanuatu

All photos viewed against the soaring roofline of Sydney's Opera House, Australia.

Global  Gourmet

Fiji Islanders prepare many foods in *lolo,* or coconut milk.

Geography and the Humanities

World Literature Reading 9

World Music: Cultural Traditions, Lesson 10

World Cultures Transparencies 19, 20

Teacher Notes

UNIT 9 ATLAS

Regional Focus

Using the Unit Atlas

These features and activities may be used as an introduction to the unit or as teaching tools throughout the course of the unit.

📁 Unit Atlas Activity 9
L1

🔲 📁 Unit Overlay Transparencies 9-0, 9-1, 9-2, 9-3, 9-4, 9-5, 9-6

📁 Home Involvement Activity 9 **L1**

📁 Environmental Case Study 9 **L1**

🗂 Countries of the World Flashcards, Unit 9

The region of Australia, Oceania, and Antarctica lies in the Pacific Ocean southeast of Asia. It includes two continents—Australia and Antarctica. Australia is the only place on the earth that is both a continent *and* a country. Antarctica is the world's coldest and iciest continent. This region also takes in Oceania—an area of thousands of islands in the Pacific Ocean.

Region
The Land

The region of Australia, Oceania, and Antarctica lies mostly south of the Equator. It spreads out across millions of square miles of the Pacific Ocean. Long distances and rugged landscapes have kept many parts of the region isolated from each other and the rest of the world.

Kangaroo Crossing?

NEXT 8 km

In some parts of Australia, kangaroo populations are so large that signs are posted to warn motorists.
HUMAN/ENVIRONMENT INTERACTION: In what ways do you think humans could be a threat to the kangaroo?

Australia A vast spread of dry, flat land makes up most of Australia. Through the central part of the country, however, runs a thick ribbon of pastureland. Underground pools of water allow farming and ranching to take place in this area. Hills and mountains stretch along Australia's east coast. A thin coastal strip faces the Pacific Ocean.

Oceania Oceania is made up of different kinds of Pacific islands. Some of the islands are mountainous. They are known for volcanic and earthquake activity. Others are massive formations of rock that rise from the ocean floor. Still other islands are low-lying, ring-shaped coral islands.

Antarctica Antarctica spreads over the southern end of the globe. It lies under an enormous sheet of ice. Mountains run through Antarctica and divide it into two parts. To the east is a high, flat plateau. A landmass largely below sea level lies to the west.

More About the Illustration

Answer to Caption
automobile accidents, hunting them, destroying their habitats through development or pollution

GEOGRAPHIC THEMES

Place
Australian wild dogs, called *dingoes*, and humans are the only predators of kangaroos. Australia has strict laws to protect kangaroos from hunters. In some parts of the country, hunting is allowed only when kangaroo populations become too large.

Fun Facts

• **Polynesia** Rapanui—or Easter Island—is the most remote of the Polynesian islands. This tiny island of 63 square miles (163 sq. km) lies some 1,200 miles (1,931 km) from its nearest Polynesian neighbor. It is famous for a number of stone statues of heads. Some of these imposing sculptures are as tall as 40 feet (12 m) and weigh as much as 50 tons (45 t).

Place
Climate and Vegetation

Location and wind and ocean currents affect the climates of Australia, Oceania, and Antarctica. These account for the extreme temperatures in the region. Because of centuries of isolation, Australia, Oceania, and Antarctica have animals and vegetation found nowhere else in the world.

Australia Australia's thin eastern coastal strip receives rain from the Pacific Ocean. It enjoys mild climates year-round. Mountains in the area, however, block the Pacific moisture from reaching inland areas. For this reason, most of central and western Australia has dry or partly dry climates.

Oceania Oceania has mostly tropical climates. Warm days follow each other in an almost unbroken chain. Most of the region has wet and dry seasons. During the wet season, days are constantly rainy and humid. In the dry season, rainfall is low, and a brilliant blue sky and ocean blend into one endless horizon. Tropical plants, such as coconut palms, cover many of the islands.

Antarctica Antarctica is one of the coldest places on the earth. It is also one of the highest and driest of the continents. As air rises across Antarctica, moisture is lost. This dryness, in turn, makes the air colder. During the long winter, temperatures may fall as low as −100°F (−73°C).

Along the Antarctic Peninsula in the north, temperatures are warmer. There, you will be surprised by the vegetation that breaks through the endless white of the continent's interior. Mosses grow on the rocks, and algae tints the ice itself with red, green, or yellow.

Human/Environment Interaction
The Economy

Agriculture is the major economic activity of Oceania. Because of great distances and high development costs, manufacturing is limited to Australia and the island country of New Zealand. Mining is also important to Australia and

Emperor Penguins

Penguins thrive in the Southern Hemisphere where icy Antarctic currents flow.
REGION: What do you think makes up the main portion of a penguin's diet?

NATIONAL GEOGRAPHIC SOCIETY

 CD-ROM
Picture Atlas of the World

 SOFTWARE
ZipZapMap! World

POSTERS
Eye on the Environment, Unit 9

More About the Illustration

Answer to Caption
fish

Place Penguins are found in the colder ocean waters of the Southern Hemisphere. They spend most of their time in the water hunting for fish and other food. On land, they gather in large colonies called *rookeries* to raise their young. Some rookeries have a million or more birds.

The Nullarbor Plain, a vast, dry, treeless plateau, extends some 400 miles (644 km) along the central southern coast of Australia. Its name is Latin for "no tree."

Fun Facts

- **Tahiti** The famous mutiny on the *Bounty* took place during a voyage from Tahiti to the West Indies. The *Bounty* was carrying breadfruit plants, which the British government hoped would thrive in the islands and provide a cheap source of food for enslaved people.

- **Australia** Australia has a large variety of pouched mammals, or *marsupials,* such as the kangaroo and the koala. They are different from other mammals because their young are born before they are fully developed. The young mature in the mother's pouch, getting their food from nipples within the pouch. The only marsupial in the United States is the opossum.

FactsOnFile

Use the reproducible masters from **MAPS ON FILE I**, *Australia and Oceania,* to enrich unit content.

More About the Illustration

Answer to Caption
large, deep-water harbor, shelter from the open sea

Location
Suva is Fiji's main seaport and business center. People and goods are constantly moving in and out of this busy port. Suva is also the political center of Fiji. UN agencies and the embassies of the United States, France, and China are all located there.

Cultural Kaleidoscope

Australia More Greek-speaking people live in the city of Melbourne than in any other city outside of Greece.

Antarctica. In recent years, modern means of transportation have increased trade between Australia, Oceania, and Antarctica and the rest of the world.

Australia Only a small area of Australia is good for farming. Australian farmers use the limited land and water resources well. Australia has many cattle and sheep ranches. It is a world leader in the export of beef, lamb, and wool. Australians also mine uranium, bauxite, iron ore, copper, nickel, and gold. Factories produce cars, machinery, food products, and textiles.

Oceania In much of Oceania, poor soil limits the growing of crops for export. Most people raise only enough food to feed themselves. Some larger islands, however, have rich volcanic soil and abundant rain. Farmers there grow cash crops—tropical fruits, sugar, coffee, and coconut products. New Zealand, located between Oceania and Australia, is both an agricultural and manufacturing country. It exports mainly meat, wool, and dairy products. Fishing is an important industry throughout the South Pacific area.

Docking at Suva, Fiji

Fiji's capital and largest city is a busy seaport.
PLACE: What physical features do you think are necessary for a city to be a seaport?

Antarctica Because of its harsh climate, Antarctica produces no farm crops. Its icy coastal waters, however, are a rich source of seafood. The continent is believed to be rich in minerals. To preserve Antarctica as the world's last wilderness, many nations have agreed—for now—not to mine Antarctica's mineral wealth.

Movement
People

Many different groups of people have settled Australia, Oceania, and Antarctica. Today the region is a blend of European, traditional Pacific, and Asian cultures.

Population In spite of its vast size, the region has only about 28 million people. Nearly 18 million are in Australia, and most of them live in cities along the coasts. These cities include Sydney and Brisbane on the eastern coast, Melbourne and Adelaide on the southern coast, and Perth on the western coast. Another 3.5 million people live in New Zealand, which also has large urban populations. These urban areas include Auckland, Wellington, and Christchurch.

The rest of the South Pacific islands have smaller urban populations. Many Pacific islanders, especially young people, leave their villages and head to the

---

Fun Facts

• **New Zealand** The first European settlers in New Zealand were sealers and whalers. They set up stations on both islands to take advantage of the abundance of seals and whales in the waters of the South Pacific.

• **Australia** The *woomera* was one of many useful tools employed by the Aborigines. This wooden rod with a hooked end was used to launch spears a great distance. The Australian government named its rocket and missile testing range *Woomera* after the Aborigine spear thrower.

Six Months of Sunshine

Can you sleep while the sun is shining? If you lived in Antarctica you would have to. Daylight there lasts for six months at a time—not the 12 hours most parts of the world experience. How can this be? The earth's axis is tilted as it orbits around the sun. The South Pole faces the sun for six months, causing half a year of constant daylight. During the next six months the South Pole faces away from the sun, giving Antarctica six months of darkness.

region's major cities to find work. Although Antarctica has no permanent population, multinational groups of scientists live there for brief periods to carry out research.

History The first humans to settle the South Pacific region probably came from Southeast Asia and then spread eastward. In Australia, the original settlers were the Aborigines; in New Zealand, they were the Maori. They lived by fishing, farming, and hunting. In the late 1700s and early 1800s, the British began to settle and rule Australia and New Zealand. European countries and Japan struggled throughout the late 1800s and early 1900s for control of Oceania.

In the early 1900s, Australia and New Zealand became independent countries. Other South Pacific areas won their independence in the years following World War II. Today most countries in the region have democratic forms of government.

The Arts Traditional Pacific cultures created many art forms still practiced today. For example, Aborigines paint and Maori carve beautiful wooden masks and figures. European settlers brought traditions from their homelands. In recent years traditional Pacific and European ways have blended to form a new culture unique to the region.

REVIEW AND ACTIVITIES

1. **LOCATION** Where are many of Australia's ranches located?
2. **REGION** What manufacturing countries are found in the region?
3. **PLACE** Why do most people in Australia live in coastal cities?
4. **REGION** What is an important industry throughout the South Pacific region?
5. **PLACE** What is unique about Antarctica's population?

Technology Activity

Using the Internet Choose one geographic area—Australia, Antarctica, or Oceania. Log on to the Internet and use a search engine such as InfoSeek to find one of these sites: Glacier, Australian Environment, or Oceania—The New Country. Check out one of the sites to find three facts you did not know about the geographic area you chose. Share these with a classmate, explaining why you think they are interesting.

681

Geography and the Humanities

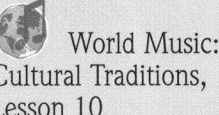   World Literature Reading 9

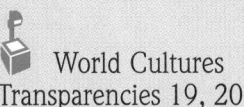   World Music: Cultural Traditions, Lesson 10

   World Cultures Transparencies 19, 20

NATIONAL GEOGRAPHIC SOCIETY

These materials are available from Glencoe.

 **VIDEODISC**

STV: World Geography Vol. 1: Asia and Australia

Australia
Side 2, Frames 00002-44834

STV: World Geography Vol. 3: South America and Antarctica

Antarctica
Side 2, Frames 00001-44754

Answers to Review and Activities

1. They are located inland in a pastureland area in central Australia.
2. Australia and New Zealand
3. Inland Australia is very dry and hot; coastal areas have a mild climate.
4. fishing
5. It is not permanent; multinational groups of scientists live there for brief periods to carry out research.

LESSON PLAN
Unit 9 Atlas

UNIT 9 ATLAS
Physical Geography

 **Applying Geographic Themes**

Movement Ancestors of the Aborigines arrived in Australia during the Ice Age before rising seas cut off the land passage between Australia and Asia. The Aborigines remained isolated from the rest of the world until the arrival of the Dutch in the 1600s. What means of transportation did the Dutch use to reach Australia? *(ship)*

These materials are available from Glencoe.

 SOFTWARE

ZipZapMap! World

 POSTERS

Eye on the Environment, Unit 9

 **MAP STUDY**

Answers
1. Great Barrier Reef
2. about 98 percent

Map Skills Practice
Reading a Map In what country are the Southern Alps located? *(New Zealand)* Use the map scale to determine the distance between mainland Australia and New Zealand. *(about 1,200 miles; 1,930 km)*

AUSTRALIA, OCEANIA, AND ANTARCTICA: Physical

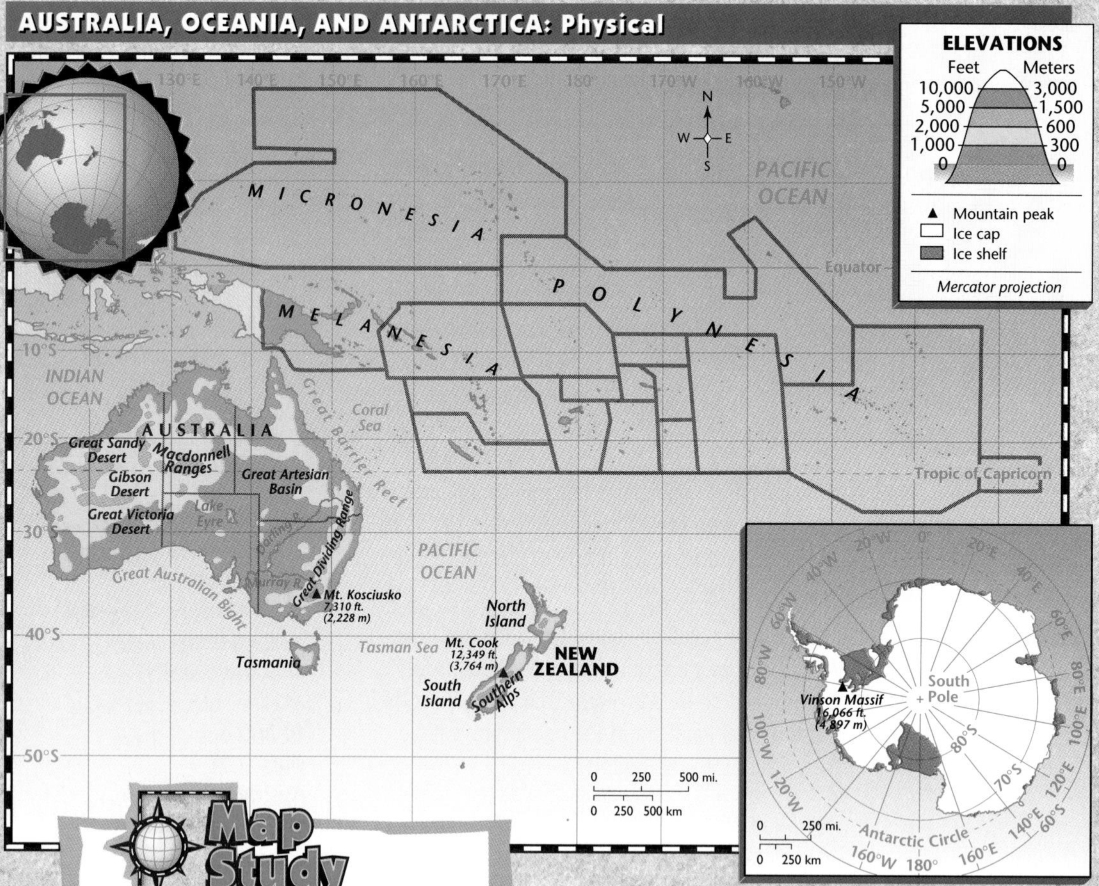

ELEVATIONS

Feet	Meters
10,000	3,000
5,000	1,500
2,000	600
1,000	300
0	0

▲ Mountain peak
☐ Ice cap
▨ Ice shelf

Mercator projection

PACIFIC OCEAN

MICRONESIA

MELANESIA

POLYNESIA

Equator

INDIAN OCEAN

10°S

Coral Sea

Tropic of Capricorn

AUSTRALIA

Great Sandy Desert
Macdonnell Ranges
Gibson Desert
Great Artesian Basin
Great Victoria Desert
Lake Eyre
Great Dividing Range
Great Barrier Reef

20°S

30°S

▲ Mt. Kosciusko 7,310 ft. (2,228 m)

Great Australian Bight
Murray R.

PACIFIC OCEAN

North Island

40°S

Tasmania
Tasman Sea
Mt. Cook 12,349 ft. (3,764 m)
NEW ZEALAND
South Island
Southern Alps

50°S

| 0 | 250 | 500 mi. |
| 0 | 250 | 500 km |

Vinson Massif 16,066 ft. (4,897 m)
South Pole
Antarctic Circle

| 0 | 250 mi. |
| 0 | 250 km |

 **Map Study**

1. **LOCATION** What reef lies off the northeastern coast of Australia?
2. **PLACE** About how much of Antarctica is covered by ice cap?

UNIT 9

Atlas Activity

Location On the chalkboard, write the names of islands in Oceania, and assign each island to a different student. Have each student determine the absolute location of his or her island by finding its latitude and longitude. Then ask each student to express the island's relative location by figuring out its distance and direction from three or four other places. Direct students to list the absolute and relative locations on the chalkboard below the name of their islands. **L1**

ELEVATION PROFILE: Australia

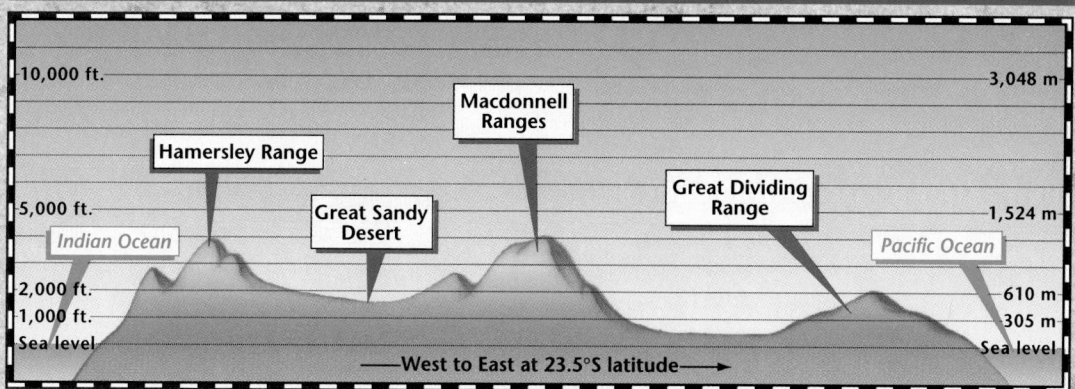

| 10,000 ft. | | | | 3,048 m |

Macdonnell Ranges

Hamersley Range

Great Sandy Desert

Great Dividing Range

| 5,000 ft. | | 1,524 m |

Indian Ocean

Pacific Ocean

2,000 ft.		610 m
1,000 ft.		305 m
Sea level		Sea level

←—— West to East at 23.5°S latitude ——→

Source: *Goode's World Atlas*, 19th edition

GeoFacts

Highest point: Vinson Massif (Antarctica) 16,066 ft. (4,897 m) high

Lowest point: Lake Eyre (Australia) 52 ft. (16 m) below sea level

Longest river: Murray-Darling (Australia) 2,310 mi. (3,717 km) long

Largest lake: Lake Eyre (Australia) 3,600 sq. mi. (9,324 sq. km)

Largest waterfall: Sutherland Falls (New Zealand) 1,904 ft. (580 m)

Largest desert: Gibson Desert (Australia) 120,000 sq. mi. (310,800 sq. km)

AUSTRALIA/OCEANIA/ANTARCTICA AND THE UNITED STATES: Land Comparison

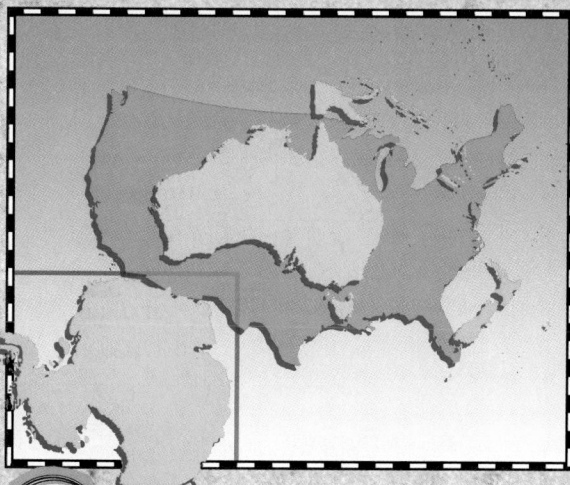

Graphic Study

1. **PLACE** What is the highest point in the Australia/Oceania/Antarctica region?
2. **PLACE** How would you describe the size of the United States compared with the combined sizes of Australia, Oceania, and Antarctica?

GLENCOE TECHNOLOGY

  VIDEODISC

The Infinite Voyage: Secrets of a Frozen World

Chapter 6 The Antarctica Peninsula: Pack Ice and Life Cycles Disc 2 Side B

Reuters Issues in Geography

Chapter 10 Antarctica: The Ozone Hole Disc 1 Side A

FactsOnFile

Use the reproducible masters from **MAPS ON FILE I,** *Australia and Oceania,* to enrich unit content.

GRAPHIC STUDY

Answers
1. Vinson Massif
2. about one-third smaller

Graphic Skills Practice
Reading a Graph What is the elevation of Australia's Great Sandy Desert? *(about 1,750 feet; 533 m)* Is the elevation profile north or south of the Equator? *(south)*

Teacher Notes

_____
_____
_____
_____
_____
_____
_____

Applying Geographic Themes

Place Some 80 percent of all Australians live in a crescent-shaped stretch of coast about 1,000 miles (1,609 km) long. Compare the map on this page to the maps on pages 682 and 688. Based on these maps, where do you think this crescent lies? *(in the southeast corner of Australia)*

NATIONAL GEOGRAPHIC SOCIETY

These materials are available from Glencoe.

VIDEODISC

STV: World Geography

Vol. 1: Asia and Australia

Australia
Side 2, Frames 00002-44834

STV: World Geography

Vol. 3: South America and Antarctica

Antarctica
Side 2, Frames 00001-44754

MAP STUDY

Answers
1. Canberra
2. Tarawa

Map Skills Practice
Reading a Map What is the capital of Papua New Guinea? *(Port Moresby)*

AUSTRALIA, OCEANIA, AND ANTARCTICA

UNIT 9 ATLAS

Cultural Geography

AUSTRALIA AND OCEANIA: Political

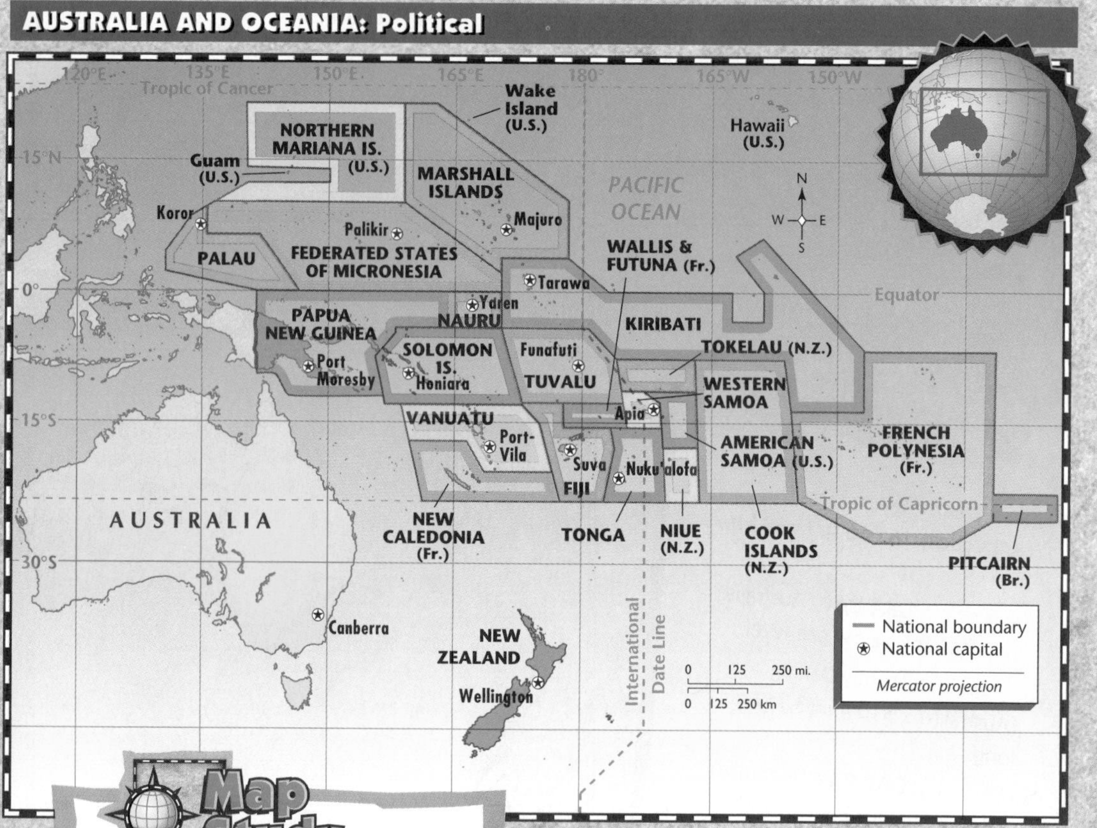

Tropic of Cancer — 120°E — 135°E — 150°E — 165°E — 180 — 165°W — 150°W

Wake Island (U.S.)

Hawaii (U.S.)

NORTHERN MARIANA IS. (U.S.)

Guam (U.S.)

MARSHALL ISLANDS

PACIFIC OCEAN

Koror ☆ ★ Majuro

Palikir ☆

PALAU

FEDERATED STATES OF MICRONESIA

★ Tarawa

WALLIS & FUTUNA (Fr.)

Equator

☆ Yaren

NAURU

KIRIBATI

PAPUA NEW GUINEA

Port ☆ Moresby

SOLOMON IS.

☆ Honiara

Funafuti ☆

TUVALU

TOKELAU (N.Z.)

WESTERN SAMOA

VANUATU

Port- ☆ Vila

Apia ★

AMERICAN SAMOA (U.S.)

FRENCH POLYNESIA (Fr.)

Suva ★ ☆ Nuku'alofa

AUSTRALIA

FIJI

NEW CALEDONIA (Fr.)

TONGA

NIUE (N.Z.)

COOK ISLANDS (N.Z.)

Tropic of Capricorn

PITCAIRN (Br.)

☆ Canberra

NEW ZEALAND

International Date Line

☆ Wellington

— National boundary
★ National capital

0 — 125 — 250 mi.
0 — 125 — 250 km

Mercator projection

Map Study

1. **PLACE** What city is Australia's national capital?
2. **LOCATION** Which Pacific island capital lies closest to both the Equator and the International Date Line?

Atlas Activity

Place Explain that Western Australia covers one-third of the country and has the largest area but the lowest population density of the six states. Have students compare the maps on pages 682 and 684. Ask them why they think Western Australia is sparsely populated. *(Most students will suggest that the deserts of Western Australia discourage settlement.)* Then, based on their comparison of the maps, have students list the factors that they think encourage or discourage settlement in Australia's other states. *(encourage settlement: level land for farming; presence of lakes, rivers, and harbors; proximity to big cities; discourage settlement: mountainous terrain; lack of water sources; isolation from population centers.)* **L1**

COMPARING POPULATION:
Australia/Oceania and the United States

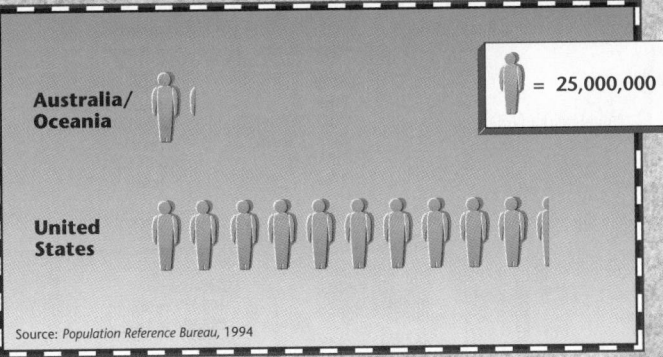

= 25,000,000

Australia/Oceania

United States

Source: *Population Reference Bureau, 1994*

Geography and the Humanities

World Literature Reading 9

World Music: Cultural Traditions, Lesson 10

World Cultures Transparencies 19, 20

GeoFacts

Biggest country (land area): Australia 2,941,290 sq. mi. (7,617,941 sq. km)

Smallest country (land area): Nauru 8 sq. mi. (21 sq. km)

Largest city (population): Sydney (1995) 3,619,000; (2000 projected) 3,708,000

Highest population density: Marshall Islands 1,429 people per sq. mi. (552 people per sq. km)

Lowest population density: Australia 6 people per sq. mi. (2 people per sq. km)

AUSTRALIA: Population Growth

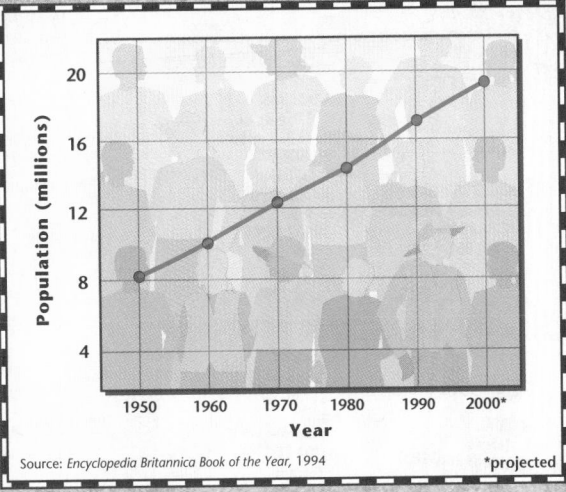

Source: *Encyclopedia Britannica Book of the Year, 1994*

*projected

Graphic Study

1. **REGION** What is the combined population of Australia and Oceania?
2. **PLACE** How much did Australia's population grow between 1950 and 1990?

GRAPHIC STUDY

Answers
1. about 28 million people
2. about 9 million people

Graphic Skills Practice
Reading a Graph How does the population of Australia and Oceania compare to that of the United States? *(United States has about 230 million more people.)* What is Australia's population expected to be in the year 2000? *(about 19.5 million people)*

685

Teacher Notes

UNIT 9 ATLAS

Countries at a Glance

Cultural Kaleidoscope

Australia People in Australia consider themselves equals and are wary of people who put on "airs and graces." A taxi driver, for example, might take offense if a person traveling alone chooses to ride in the back seat of the cab.

Global Gourmet

Kiribati Mealtimes on the islands of Kiribati occur whenever fresh fish is available, although some islanders believe fish for breakfast makes people lazy.

Geography and the Humanities

World Literature Reading 9

World Music: Cultural Traditions, Lesson 10

World Cultures Transparencies 19, 20

Australia
CAPITAL: Canberra
MAJOR LANGUAGE(S): English, Aboriginal languages
POPULATION: 17,800,000
LANDMASS: 2,941,290 sq. mi./ 7,617,941 sq. km
MONEY: Australian Dollar
MAJOR EXPORT: Crude Oil
MAJOR IMPORT: Machinery

FEDERATED STATES OF Micronesia
CAPITAL: Palikir
MAJOR LANGUAGE(S): English
POPULATION: 100,000
LANDMASS: 270 sq. mi./ 699 sq. km
MONEY: U.S. Dollar
MAJOR EXPORT: Copra
MAJOR IMPORT: Food and Beverages

Fiji
CAPITAL: Suva
MAJOR LANGUAGE(S): English, Fijian, Hindi
POPULATION: 800,000
LANDMASS: 7,050 sq. mi./ 18,260 sq. km
MONEY: Fiji Dollar
MAJOR EXPORT: Sugar
MAJOR IMPORT: Machinery

Kiribati
CAPITAL: Tarawa
MAJOR LANGUAGE(S): English, Gilbertese
POPULATION: 76,900
LANDMASS: 313 sq. mi./ 811 sq. km
MONEY: Australian Dollar
MAJOR EXPORT: Copra
MAJOR IMPORT: Machinery

Marshall Islands
CAPITAL: Majuro
MAJOR LANGUAGE(S): English, Marshallese, Japanese
POPULATION: 100,000
LANDMASS: 70 sq. mi./ 181 sq. km
MONEY: U.S. Dollar
MAJOR EXPORT: Crude Coconut Oil
MAJOR IMPORT: Food and Live Animals

Nauru
CAPITAL: Yaren
MAJOR LANGUAGE(S): Nauruan, English
POPULATION: 9,460
LANDMASS: 8 sq. mi./ 21 sq. km
MONEY: Australian Dollar
MAJOR EXPORT: Phosphates
MAJOR IMPORT: Food

New Zealand
CAPITAL: Wellington
MAJOR LANGUAGE(S): English, Maori
POPULATION: 3,500,000
LANDMASS: 103,470 sq. mi./ 267,987 sq. km
MONEY: New Zealand Dollar
MAJOR EXPORT: Food and Live Animals
MAJOR IMPORT: Machinery

Palau
CAPITAL: Koror
MAJOR LANGUAGE(S): English, Palauan
POPULATION: 15,122
LANDMASS: 177 sq. mi./ 458 sq. km
MONEY: U.S. Dollar
MAJOR EXPORT: Tuna
MAJOR IMPORT: Food

Papua New Guinea
CAPITAL: Port Moresby
MAJOR LANGUAGE(S): English, Melanesian, Papuan languages
POPULATION: 4,000,000
LANDMASS: 174,850 sq. mi./ 452,862 sq. km
MONEY: Kina
MAJOR EXPORT: Gold
MAJOR IMPORT: Machinery

Solomon Islands
CAPITAL: Honiara
MAJOR LANGUAGE(S): English, Papuan, Melanesian, Polynesian languages
POPULATION: 400,000
LANDMASS: 10,810 sq. mi./ 27,998 sq. km
MONEY: Solomon Islands Dollar
MAJOR EXPORT: Fish Products
MAJOR IMPORT: Machinery

Countries not drawn to scale.

UNIT 9

Fun Facts

• **Antarctica** The largest land animal on the continent is an insect less than one-tenth of an inch (2.5 mm) in length.

• **Fiji** In January, village women play a game called "kick the orange" to welcome the New Year.

Tonga

CAPITAL:
Nuku'alofa
MAJOR LANGUAGE(S):
Tongan, English
POPULATION:
99,100

LANDMASS:
301 sq. mi./
780 sq. km
MONEY:
Pa'anga
MAJOR EXPORT:
Squash
MAJOR IMPORT:
Food and Live Animals

Nuku'alofa

Tuvalu

CAPITAL:
Funafuti
MAJOR LANGUAGE(S):
Tuvaluan, English
POPULATION:
9,500

LANDMASS:
9 sq. mi./
23 sq. km
MONEY:
Australian Dollar
MAJOR EXPORT:
Copra
MAJOR IMPORT:
Food

Funafuti

Vanuatu

CAPITAL:
Port-Vila
MAJOR LANGUAGE(S):
Bislama, French, English
POPULATION:
200,000

LANDMASS:
4,710 sq. mi./
12,199 sq. km
MONEY:
Vatu
MAJOR EXPORT:
Copra
MAJOR IMPORT:
Machinery, Food

Port-Vila

Western Samoa

CAPITAL:
Apia
MAJOR LANGUAGE(S):
Samoan, English
POPULATION:
200,000

LANDMASS:
1,090 sq. mi./
2,823 sq. km
MONEY:
Tala
MAJOR EXPORT:
Coconut Oil
MAJOR IMPORT:
Food

Apia

Delivering Supplies to Antarctica

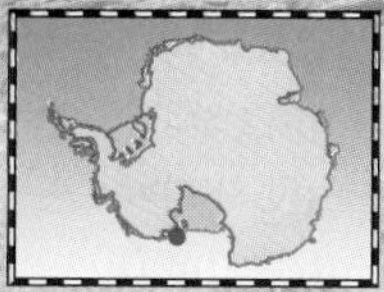

A Coast Guard icebreaker cuts through the ice to deliver passengers and supplies to a science station in Antarctica.

Overlooking Sydney, Australia

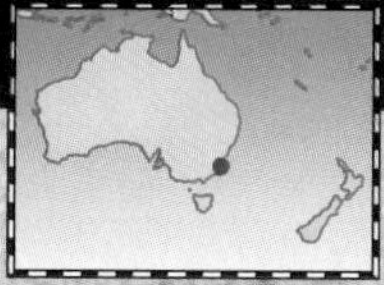

A bridge over Sydney Harbor carries traffic from the main business district to the suburbs.

Teacher Notes

NATIONAL GEOGRAPHIC SOCIETY

These materials are available from Glencoe.

 **SOFTWARE**

ZipZapMap! World

 Strange but TRUE!

Averaging 8,000 feet (2,438 m) in elevation, Antarctica is higher than any other continent.

Cultural Kaleidoscope

Marshall Islands The Marshall Islanders' greeting *Iokwe* (YAH-quay) is similar to the Hawaiians' *Aloha*. It can mean "hello," "good-bye," "love," or "like."

CURRICULUM CONNECTION

Earth Science Some coral reefs were formed almost 600 million years ago and make up the earth's oldest ecosystems—areas where organisms interact with air, nutrients, and water.

687

PERFORMANCE ASSESSMENT ACTIVITY

CREATING A SCRAPBOOK Have students imagine that they have spent a summer in Australia and New Zealand as part of a geography course they are taking for credit. During their visit, they learned about the landforms, climate, cities, people, resources, and tourist attractions of the two countries. Have students create a scrapbook of their time in the area, including such simulated items as pictures, ticket stubs, newspaper headlines, matchbook covers, postcards, menu items, and so on. Entries in the scrapbook should show the students' knowledge of the most important cultural and physical features of the area and include captions that relate the items in some manner to the five geographic themes.

Students may share their scrapbooks by forming groups and passing their books within their sharing group. Then have students write a paragraph on how they decided which items to include in their scrapbooks. Suggest that they conclude their paragraphs with a discussion on the items they saw in other scrapbooks that they would now include in their own product.

POSSIBLE RUBRIC FEATURES:

Planning skills, decision-making skills, accuracy of content information, evaluation and prioritization skills, creative thinking, visual quality of product

MENTAL MAPPING ACTIVITY

Before teaching Chapter 26, point out Australia and New Zealand on a globe to students. Have them identify the hemisphere in which Australia and New Zealand are located and where the countries are located relative to the United States.

TEACHER'S CORNER

NATIONAL GEOGRAPHIC SOCIETY

INDEX TO NATIONAL GEOGRAPHIC MAGAZINE

The following articles may be used for research relating to this chapter:

- "Cape York Peninsula," by Cathy Newman, June 1996.
- "Saltwater Crocodiles," by David Doubilet, June 1996.
- "Koalas—Out on a Limb," by Oliver Payne, April 1995.
- "Australian Wildflowers," by Cary Wolinsky, January 1995.
- "The Simpson Outback," by Jane Vessels, April 1992.
- "Australia's Magnificent Pearls," by David Doubilet, December 1991.
- "Wetas—New Zealand's Insect Giants," by Mark W. Moffett, November 1991.
- "Journey into Dreamtime," by Harvey Arden, January 1991.
- "The Sea Beyond the Outback," by Rodney Fox, January 1991.
- "New Zealand's Magic Waters," by David Doubilet, October 1989.

NATIONAL GEOGRAPHIC SOCIETY PRODUCTS AVAILABLE FROM GLENCOE

To order the following products for use with this chapter, contact your local Glencoe sales representative or call Glencoe at 1-800-334-7344:

- *STV: World Geography* (Videodisc)
- *Picture Atlas of the World* (CD-ROM)
- *Physical Geography of the World* (Transparencies)
- *ZipZapMap! World* (Software)
- *GeoBee* (Software)
- *Images of the World* (Posters)
- *Eye on the Environment* (Posters)

ADDITIONAL NATIONAL GEOGRAPHIC SOCIETY PRODUCTS

To order the following products for use with this chapter, call National Geographic Society at 1-800-368-2728:

- *Physical Geography of the Continents Series:* "Australia." (Video)
- *Nations of the World Series:* "Australia." (Video)
- *Australia's Aborigines* (Video)
- *Valley of the Kangaroos* (Video)

TEACHER-TO-TEACHER
Curriculum Connection: Math

—from Sherry Henderson,
Northbrook School, Houston, TX

MEASURING AUSTRALIA *The purpose of this activity is to help students develop an understanding of region size by making a rough computation of the measurements of Australia and comparing these figures to those of the United States.*

Ask students: Which landmass is bigger, the continental United States or Australia? Most students probably will select the United States. Mention that most people have a fixed perception of Australia as being much smaller than the United States. Then ask students to plan two trips across Australia. The first should run roughly east to west—Brisbane to Perth, for example. The second should run roughly north to south—from Cape York Peninsula to Melbourne is one possibility. Have students use a map with a scale so that they may compute the distance, in both miles and kilometers, that they will travel on their two trips. Ask students to plan and measure similar east-west and north-south trips across the United States. Then have them compare the measurements for the two countries.

Encourage students to reconsider their perceptions of Australia's size by asking the following questions: Which country is larger? By how much?

MEETING NATIONAL STANDARDS

Geography for Life

All of the 18 standards are demonstrated in Unit 9. The following ones are highlighted in this chapter:

- Standards 1, 2, 3, 5, 8, 10, 11, 12, 13, 14, 15, 16, 17, 18

KEY TO ABILITY LEVELS

Teaching strategies have been coded for varying learning styles and abilities.

L1 **BASIC** activities for all students

L2 **AVERAGE** activities for average to above-average students

L3 **CHALLENGING** activities for above-average students

LEP **LIMITED ENGLISH PROFICIENCY** activities

BIBLIOGRAPHY

Readings for the Student

Gouck, Maura M. *The Great Barrier Reef.* Plymouth, Minn.: Child's World, 1993.

New Zealand in Pictures. Minneapolis: Lerner Publications, 1990.

Newman, Graeme, and Tamsin Newman. *Hippocrene Companion Guide to Australia.* New York: Hippocrene Books, 1992.

Readings for the Teacher

Dennis, Anthony. *Ticket to Ride: A Rail Journey Around Australia.* New York: Prentice-Hall, 1990.

Keneally, Thomas, Patsy Adam-Smith, and Robyn Davidson. *Australia: Beyond the Dreamtime.* New York: Facts On File, 1987.

Theroux, Paul. *The Happy Isles of Oceania: Paddling the Pacific.* New York: G.P. Putnam's Sons, 1992.

Multimedia

The Australian Way of Life. Chatsworth, Calif.: AIMS Media, 1989. Videocassette, 21 minutes.

Australia's Aborigines. Washington, D.C.: National Geographic Society, 1989. Videocassette, 60 minutes.

The Pacific World. New York: Educational Design, nd. Videocassette, 45 minutes.

Touring New Zealand. Chicago: Questar, 1989. Videocassette, 60 minutes.

CHAPTER 26 RESOURCES

CHAPTER 26 RESOURCES

📁 *Vocabulary Activity 26

📁 Cooperative Learning Activity 26

📁 Workbook Activity 26

📁 Chapter Map Activity 26

📁 Geography Skills Activity 26

📁 GeoLab Activity 26

📁 Critical Thinking Skills Activity 26

📁 Performance Assessment Activity 26

📁 *Reteaching Activity 26

📁 Enrichment Activity 26

📁 Chapter 26 Test, Form A and Form B

🎧 📁 *Chapter 26 Digest Audiocassette, Activity, Test

🕹 Unit Overlay Transparencies 9-0, 9-1, 9-2, 9-3, 9-4, 9-5, 9-6

🕹 Political Map Transparency 9; World Cultures Transparencies 19, 20

🕹 Geoquiz Transparencies 26-1, 26-2

💿 Vocabulary PuzzleMaker Software

💿 Student Self-Test: A Software Review

💿 Testmaker Software

💿 MindJogger Videoquiz

> Use these *Geography: The World and Its People* resources to teach, reinforce, and extend chapter content.

*If time does not permit teaching the entire chapter, summarize using the Chapter 26 Highlights on page 703, and the Chapter 26 English (or Spanish) Audiocassettes. Review student's knowledge using the Glencoe MindJogger Videoquiz. * Also available in Spanish*

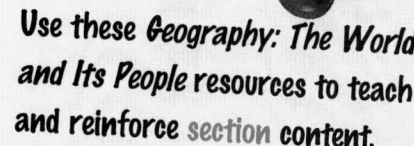

Use these Geography: The World and Its People resources to teach and reinforce section content.

SECTION 1 RESOURCES

Reproducible Lesson Plan 26-1

Section Focus Transparency 26-1

Guided Reading Activity 26-1

Section Quiz 26-1 (also available in Spanish)

SECTION 2 RESOURCES

Reproducible Lesson Plan 26-2

Section Focus Transparency 26-2

Guided Reading Activity 26-2

Section Quiz 26-2 (also available in Spanish)

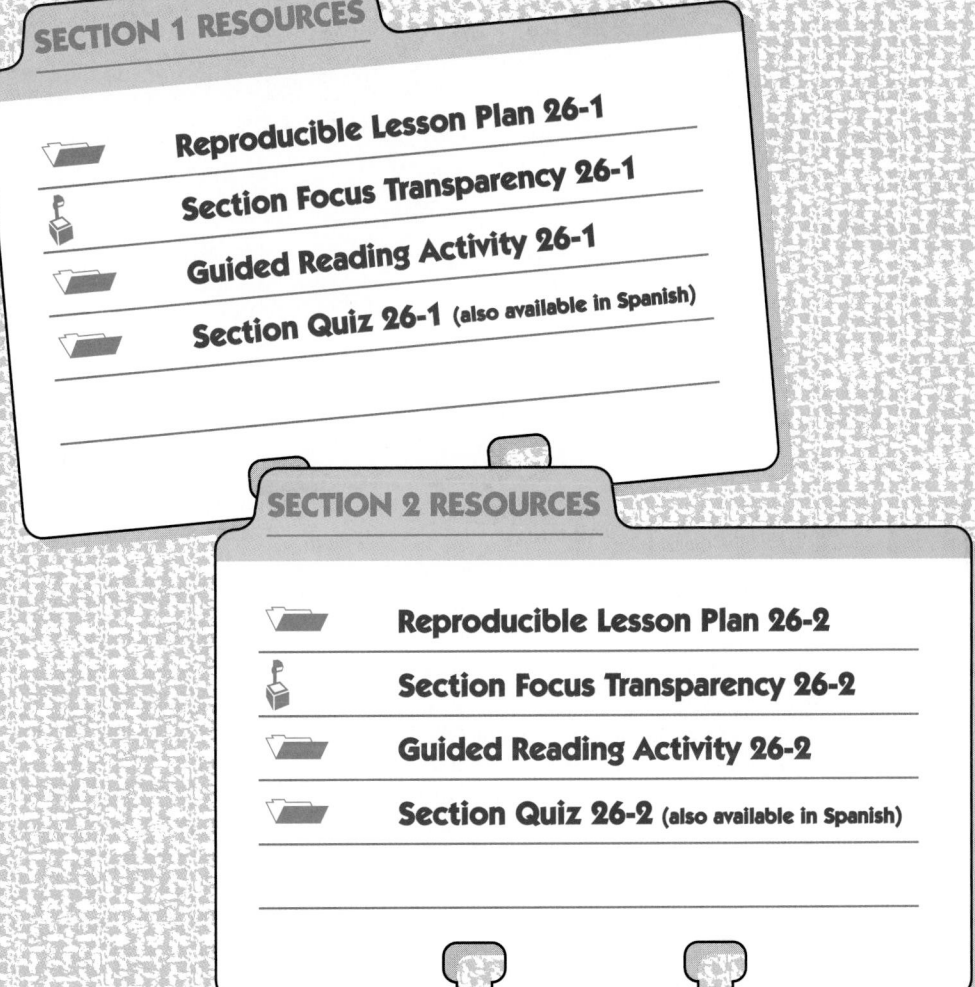

ADDITIONAL RESOURCES FROM GLENCOE

Reproducible Masters

- Glencoe Social Studies Outline Map Book, page 38
- Facts On File: MAPS ON FILE I, Australia and Oceania
- Foods Around the World, Unit 11

World Music: Cultural Traditions
Lesson 10

Transparencies

- National Geographic Society PicturePack Transparencies, Unit 9

Posters

- National Geographic Society: Images of the World, Unit 9

Countries of the World Flashcards
Unit 9

World Games Activity Cards
Unit 9

Videodiscs

- Geography and the Environment: The Infinite Voyage
- National Geographic Society: STV: World Geography, Vol. 1

CD-ROM

- National Geographic Society: Picture Atlas of the World

Software

- National Geographic Society: ZipZapMap! World

Performance Assessment

Refer to the Planning Guide on page 688A for a Performance Assessment Activity for this chapter. See the *Performance Assessment Strategies and Activities* booklet for suggestions.

Chapter Objectives

1. **Examine** the geography and economy of Australia.

2. **Investigate** how the geography of New Zealand affects its people and its relations with other countries.

GLENCOE
TECHNOLOGY

 VIDEODISC

Use the Chapter 26 MindJogger Videoquiz to preview chapter content.

MindJogger Videoquiz

Chapter 26
Disc 4 Side A

Also available in VHS.

Chapter 26 Australia and New Zealand

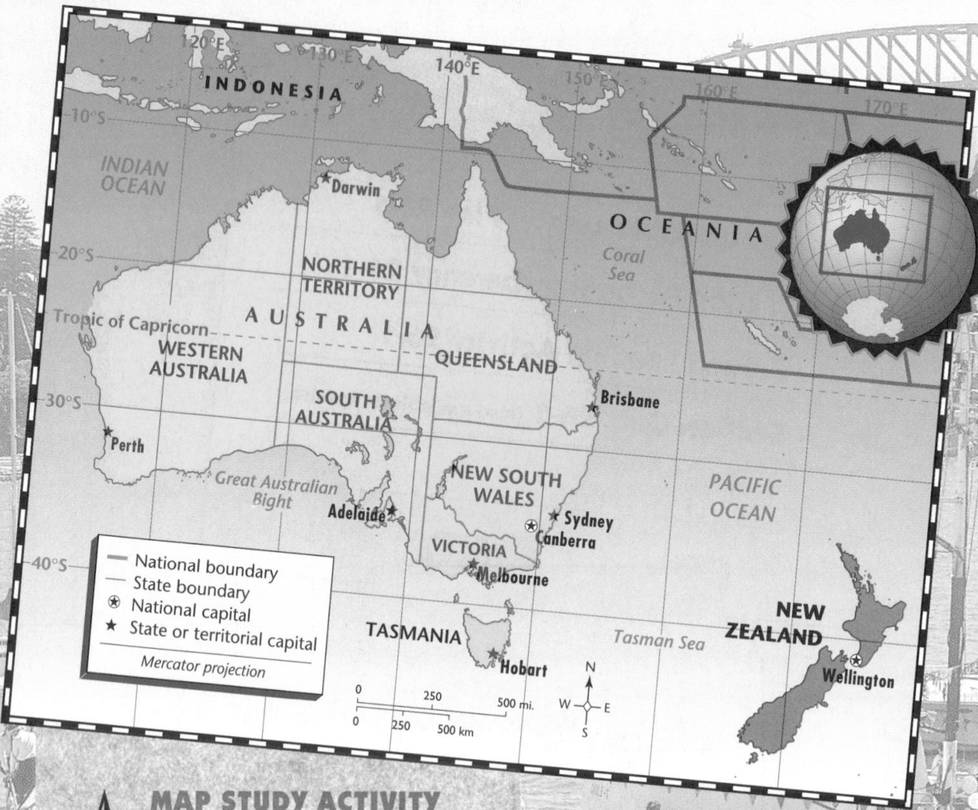

MAP STUDY ACTIVITY

In this chapter you will read about Australia and New Zealand, which are completely surrounded by water.

1. **What is the national capital of Australia?**
2. **What sea separates Australia and New Zealand?**

 MAP STUDY ACTIVITY

Answers
1. Canberra
2. Tasman Sea

Map Skills Practice
Reading a Map What large island lies south of Australia? *(Tasmania)*

Meeting National Standards

Geography for Life The following standards are highlighted in this section:
• Standards 1, 2, 5, 10, 11, 12, 14, 16, 17, 18

PREVIEW

Words to Know
• coral
• outback
• station
• marsupial
• bush

Places to Locate
• Australia
• Great Barrier Reef
• Great Dividing Range
• Western Plateau
• Canberra
• Sydney
• Melbourne

Read to Learn . . .
1. why most Australians live in coastal areas.
2. why Australia has a strong economy.
3. what groups have influenced Australia's culture.

"I started the morning routine—boil the tea, pack the gear, saddle the camels—and head south once more. . . . At Areyonga I filled my drinking-water bag with rainwater and set off for Ayers Rock. . . ." Explorer Robyn Davidson describes her recent trek across the desert of Australia. Ayers Rock—a huge red island of stone rising 1,000 feet (305 m)—is a landmark to all who travel through central Australia.

Australia lies far south on the globe between the Indian Ocean and the Pacific Ocean. People often call Australia the Land Down Under because it is located entirely within the Southern Hemisphere.

Location
The Land

Surrounded by water, Australia is a country that is a continent all by itself! Why isn't it called an island? It is too large. Instead geographers describe it as a continent—the smallest one in the world. The world's largest island, Greenland, has only about one-third of Australia's land area. Australia is the world's sixth-largest country—about the size of the continental United States.

Not only is Australia the smallest continent—it is also the oldest. Australia does not have a wide variety of physical features. It is made up of mostly deserts and dry grasslands, with a few low mountains and coastal plains. Winds and time have worn the land into the flattest and lowest continent in the world.

CHAPTER 26

689

FOCUS

Section Objectives
1. **Analyze** why most Australians live in coastal areas.
2. **Discuss** why Australia has a strong economy.
3. **List** the groups that have influenced Australia's culture.

Bellringer Motivational Activity

 Prior to taking roll at the beginning of the class period, project Section Focus Transparency 26-1 and have students answer the activity questions. Discuss students' responses. **L1**

Vocabulary Pre-check

Ask students to illustrate familiar meanings of *station* by using the word in several sentences. Then have them scan the section to find another meaning of *station* used in Australia. **L1 LEP**

Use the Vocabulary PuzzleMaker Software to create a crossword puzzle. **L1**

Classroom Resources for Section 1

REPRODUCIBLE MASTERS
Reproducible Lesson Plan 26-1
Guided Reading Activity 26-1
Workbook Activity 26
Vocabulary Activity 26
Chapter Map Activity 26
Geography Skills Activity 26
Section Quiz 26-1

TRANSPARENCIES
Section Focus Transparency 26-1
Political Map Transparency 9
World Cultures Transparencies 19, 20

MULTIMEDIA
Vocabulary PuzzleMaker Software
Testmaker

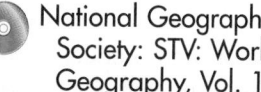 National Geographic Society: STV: World Geography, Vol. 1

MAP STUDY

Answer

less than 1,000 feet (300 m)

Map Skills Practice

Reading a Map What mountain range separates Australia's southeast coast from the rest of the country? (*Great Dividing Range*)

TEACH

Guided Practice

Making Connections On the chalkboard, copy the headings *CROPS, MANUFACTURED GOODS, MINERALS,* and *OTHER PRODUCTS.* Have students list items from the section under the correct headings. Then have them hypothesize about how farmers, manufacturers, miners, and ranchers in Australia depend on one another. Tell them to cite passages from the section to support their hypotheses. **L1**

NATIONAL GEOGRAPHIC SOCIETY

These materials are available from Glencoe.

 VIDEODISC

STV: World Geography

Vol. 1: Asia and Australia

Australian Outback
Side 2, Frames 05001-15553

690

AUSTRALIA AND NEW ZEALAND: Physical

MAP STUDY
Australia's landforms offer great contrasts.
PLACE: At what elevation is most of the Great Artesian Basin?

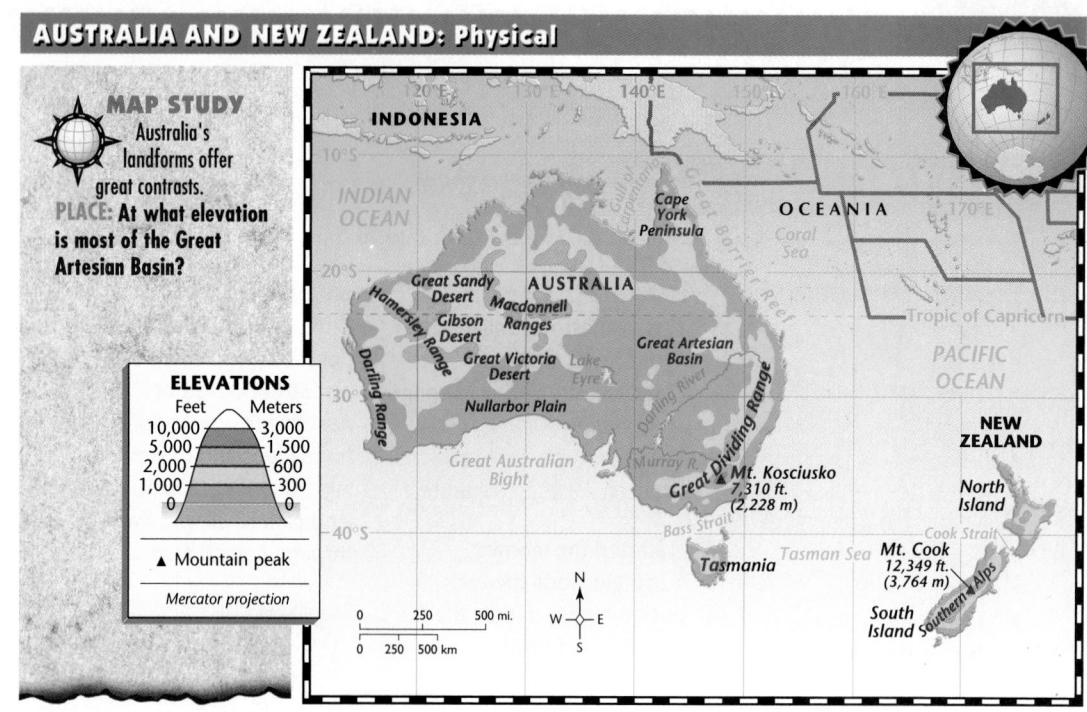

ELEVATIONS

Feet	Meters
10,000	3,000
5,000	1,500
2,000	600
1,000	300
0	0

▲ Mountain peak

Mercator projection

Off Australia's northeastern coast lies one of the world's great natural wonders—the Great Barrier Reef. It is a 1,250-mile (2,011 km) chain of colorful **coral** formations and islands rising out of blue-green Pacific waters. Coral is a hard, rocklike material made of the skeletons of small sea animals. To Australia's southeast lies the small island of Tasmania.

The Great Dividing Range The Great Dividing Range stretches along Australia's Pacific coast. It divides the flow of Australia's rivers—eastward toward the Pacific Ocean or westward to inland areas. Two of Australia's major rivers—the Murray and the Darling—begin in the Great Dividing Range and flow westward into central Australia.

In Australia's southeastern corner, you find a narrow coastal plain. This fertile flatland boasts Australia's best farmland. You can see on the map on page 700 that it also includes some of Australia's most densely populated areas. The southern end of the Great Dividing Range hugs this narrow plain. In this area rise the peaks of the Australian Alps, Australia's highest mountains. At 7,310 feet (2,228 m), Mount Kosciusko (KAH•zee•UHS•koh) is Australia's highest peak.

The Central Lowlands As you move inland, the Great Dividing Range levels off into pastureland known as the Central Lowlands. This and other inland parts of Australia are called the **outback**. Partly dry grasslands cover much of the Central Lowlands. Here in the outback, mining towns and huge **stations**—Australian sheep and livestock ranches—are scattered over a vast area.

UNIT 9

Cooperative Learning Activity

Organize students into a number of small groups. Give each group the following list of minerals found in Australia: bauxite, lead, coal, gold, iron ore, uranium, silver, zinc, copper, tin, tungsten, diamonds, natural gas, and petroleum. Have groups research to find two products that use each mineral. Have group members draw a picture of each product on index cards, labeling each drawing with the name of the mineral and the product. Then ask groups to make one card with a picture of a platypus. Inform the groups that they can use the cards to play a game by following the rules of "Old Maid"—players divide the cards, pair up pictures of products made from the same mineral, and draw cards from the player on their left. The player left with the unmatched platypus card loses. Allow time in class for groups of students to play "Platypus." **L1**

A resource prized above all others, water is scarce or lacking across 75 percent of the country. An unusual feature of the Central Lowlands is the Great Artesian Basin, where water is found deep underground. The basin is like a huge pail that catches water from eastern mountain rivers. Ranchers drill wells to make the underground water available for irrigating grazing land.

The Western Plateau Picture a carpet of sand twice as large as Alaska, Texas, California, and New Mexico combined. That's about the size of Australia's Western Plateau. Travel across the plateau, which covers about 75 percent of Australia, is mostly by airplane.

The map on page 690 shows you the huge deserts that spread across much of the Western Plateau. This part of the outback, however, is not a wasteland—it is rich in mineral wealth. To the south, a narrow coastal plain is home to almost all of the Western Plateau's farms and people.

The Climate Look at Australia's climate regions on the map on page 692. You can see that most of the country is dry with desert or partly desert grassland. But there are several other climate regions. Northern Australia—the part of the country closest to the Equator—has a tropical climate with rainy and dry seasons and year-round hot or warm temperatures. Coastal areas in the south and southeast have warm summers and mild winters. Only the Australian Alps and the southern island of Tasmania have winter temperatures that fall below freezing for more than a day.

Animals and Plants Unless you visit a zoo, you probably don't see kangaroos or koalas near your home. Kangaroos and koalas are native only to Australia. These animals and about 120 similar kinds are called **marsupials**, or mammals that carry their young in a pouch. Why do they live only in Australia?

For millions of years, Australia was isolated from other continents. This separation led to a very special wildlife population. Dingoes, wombats, kookaburras, emus, bandicoots, and dugongs are just some of the animals found in Australia. One particularly unusual Australian animal—the platypus—lays eggs but has a furry body, a large bill, and webbed feet. Rare plants such as the eucalyptus are also native to Australia. The long, leathery leaves of the eucalyptus are the only food a koala will eat.

A Busy Station in the Outback

Sprawling cattle and sheep ranches, called stations, dot the landscape of Australia's outback.
LOCATION: Where in Australia do you find the outback?

CHAPTER 26

Independent Practice

📁 Guided Reading Activity 26-1 **L1**

📁 Workbook Activity 26 **L1**

More About the Illustration

Answer to Caption
the interior

Movement Some stations are more than 100 miles (161 km) from the nearest town. Because of this, some wealthy ranchers own airplanes. Others make the long trip to town only when it is absolutely necessary to get supplies.

Although eucalyptus trees are native only to Australia, many other countries use them for reforestation programs.

Meeting Special Needs Activity

Language Delayed Students with difficulty categorizing information will find it helpful to use a framework for comparing climate zones. Develop categories for these comparisons such as temperature, precipitation, and vegetation. Direct students in searching the section for information that might be included in these categories. Point out that they will need to look in other sources to find some information needed to complete this task. **L1**

MAP STUDY

Answer

southeastern corner and Tasmania

Map Skills Practice

Reading a Map What type of climate is found on Australia's northern coast? *(tropical savanna)*

Australia's sunny climate has its drawbacks. Sunburn and sunstroke are major health hazards, and Australians have one of the highest incidences of skin cancer in the world.

NATIONAL GEOGRAPHIC SOCIETY

These materials are available from Glencoe.

 VIDEODISC

STV: World Geography

Vol. 1: Asia and Australia

Gems from Coober Pedy; Australia

Any side, Frame 50292

AUSTRALIA AND NEW ZEALAND: Climate

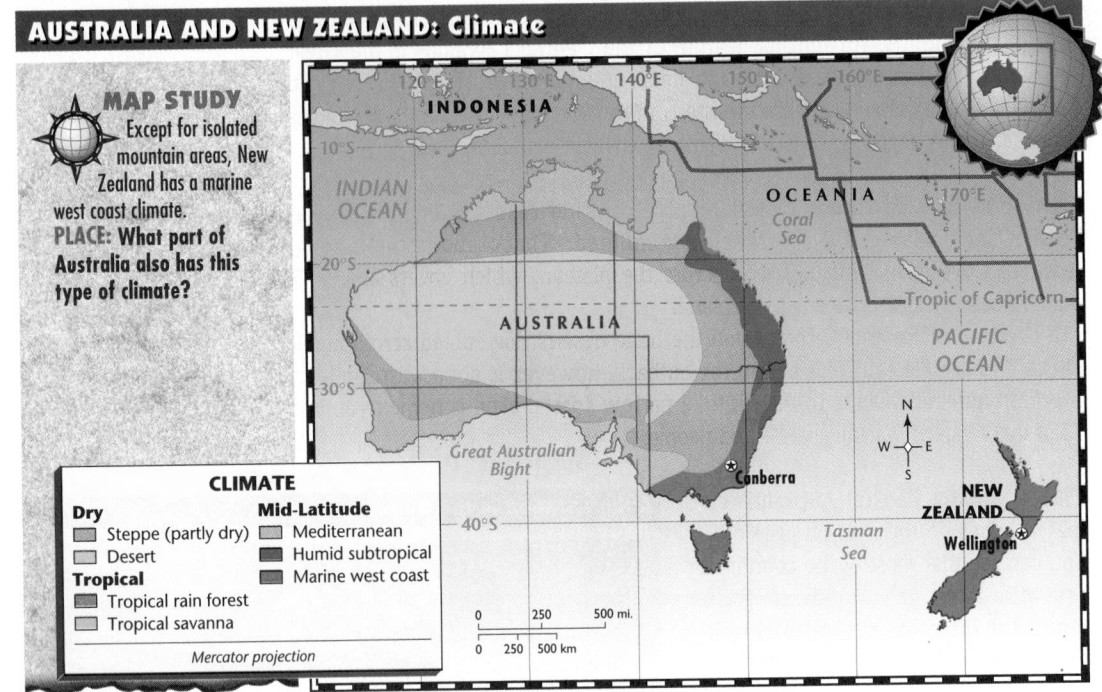

MAP STUDY Except for isolated mountain areas, New Zealand has a marine west coast climate. **PLACE: What part of Australia also has this type of climate?**

CLIMATE

Dry
- ☐ Steppe (partly dry)
- ☐ Desert

Tropical
- ☐ Tropical rain forest
- ☐ Tropical savanna

Mid-Latitude
- ☐ Mediterranean
- ☐ Humid subtropical
- ☐ Marine west coast

Mercator projection

0 250 500 mi.
0 250 500 km

Movement

The Economy

A treasure chest overflowing with mineral resources—that's Australia. The continent's strong economy relies on exporting these resources. In the past the United Kingdom ranked as Australia's most important trading partner. Today Australia trades more with Japan and the United States.

Many of the products that are manufactured in the country meet local needs. Australian factories produce cars, trains, computers, and consumer goods. Tourism is also an important industry. Many tourists flock to Australia each year, and the country's service industries thrive on the tourist trade.

Agriculture Australia's dry climate limits farming. Only about 10 percent of Australia's land is good for growing crops. Farmers rely on machines and irrigation to improve crop yields. Wheat is the main crop. Other grains, sugarcane, fruits, and vegetables are also grown. Most of Australia's land is set aside for grazing. Ranchers raise cattle for beef and sheep for meat and wool. The graph on page 693 shows that Australia is the world's leading wool-producing country.

Mining When gold was discovered in Australia in the mid-1800s, thousands flocked to the outback. But gold is only one of the country's many mineral resources. Australia is the world's leading producer of bauxite and lead. The map on page 698 shows you the location of Australia's coal, iron ore, uranium, silver, opal, and zinc mines.

 692

UNIT 9

Critical Thinking Activity

Making Generalizations Read aloud the following passage, concerning the Aborigines, from the writings of Captain James Cook:

They are far happier than we Europeans, being wholly unacquainted not only with the superfluous but the necessary conveniences so much sought after in Europe; . . . The earth and sea of their own accord furnishes them with all things necessary for life.

Then ask students to make generalizations about European and Aboriginal societies of the 1700s based on the passage. *(Possible answers are that Europeans were preoccupied with material possessions and that the Aboriginal lifestyle was very simple.)* **L2**

Place
The People

For its huge area, Australia has very few people—only 18 million. Most Australians live in scattered urban areas along the coasts. Australia has long needed more skilled workers to develop its open land and boost its economy. To meet this goal, the Australian government has encouraged people from other countries to settle in Australia.

Past and Present A small part of the continent's current population were the first Australians—the Aborigines (A•buh•RIHJ•neez). The Aborigines came to Australia from Asia about 30,000 to 40,000 years ago. They moved throughout the continent hunting animals and gathering plants. The Aborigines developed a unique culture. Ancient cave paintings show the work of their artists.

The voyages of Captain James Cook provided the British with a claim to Australia in the late 1700s. At first the United Kingdom used Australia as a colony for prisoners. Then other British emigrants settled in Australia. These settlers took land from the Aborigines and forced them far into the outback. Many Aborigines died of European diseases.

In 1901 the British colonies became states and united to form the independent Commonwealth of Australia. Australia is divided into seven states as shown on the map on page 688. Australians created a British-style parliamentary

WHAT IN THE WORLD?

On Target

Hunters around the world could take a lesson from Australia's Aborigines. They invented a hunting tool called the boomerang. Shaped like a bent wing, the wooden boomerang—thrown by a hunter—sails out toward its target. If the boomerang misses, it returns to land near the person who threw it!

Leading Wool-Producing Countries

GRAPHIC STUDY

Some Australian sheep ranches are so large they have their own post offices. **HUMAN/ENVIRONMENT INTERACTION: About how many more pounds of wool would Argentina have to produce to equal what Australia produces?**

🐑 = 300 million pounds

Millions of pounds

| | 0 | 300 | 600 | 900 | 1,200 | 1,500 | 1,800 | 2,100 |

Australia
Russia*
New Zealand
China
Argentina

Millions of kilograms

| 0 | 136 | 273 | 409 | 545 | 681 | 817 | 953 |

*1988 figures before breakup of Soviet Union   Source: *World Book*, 1994

More About . . .

Strine The Australian form of English not only includes British terms but also many terms peculiar to Australia. *Strine,* for example, is what Australians call their form of English. The word *Strine* is a reference to the Australian practice of running syllables—and words—together when they speak. *Strine,* then, is how many people say "Australian." *Apples* is Strine for "under control," *barney* or *blue* means "an argument," *cackleberry* means "egg," *chook* means "chicken," a *flyer* is a "swiftly-moving kangaroo," and *Noah's Ark* means "a shark." Good guides to Strine include Danielle Martin's *Australians Say G'Day* and K. J. Condren's *The Ridgy Didge Guidebook to Australia.*

LESSON PLAN
Chapter 26, Section 1

MULTICULTURAL PERSPECTIVE

Cultural Heritage

Today, only about 206,000 Aborigines remain in Australia, accounting for around 1 percent of the population. Some live in settlements and follow a more traditional lifestyle than those who have moved to the cities.

ASSESS

Check for Understanding

Assign Section 1 Review as homework or an in-class activity.

Meeting Lesson Objectives

Each objective below is tested by the questions that follow it in parentheses.

1. **Analyze** why most Australians live in coastal areas. (2, 5, 6)
2. **Discuss** why Australia has a strong economy. (3)
3. **List** the groups that have influenced Australia's culture. (4)

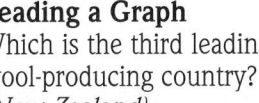

GRAPHIC STUDY

Answer
1,700 million pounds (771 million kg)

Graphic Skills Practice
Reading a Graph
Which is the third leading wool-producing country? *(New Zealand)*

Evaluate

Section Quiz 26-1
L1

Use the Testmaker to create a customized quiz for Section 1. L1

Reteach

Outline the section, scramble the parts of the outline, and distribute copies of the scrambled outline. Have students reread the section to help them rearrange the outline in the correct order. L1

Enrich

Have students prepare an illustrated report on the arrival of the first European settlers—convicts—in Australia in 1789. L3

CLOSE

Tell students that many young Australians who live in remote areas do not attend schools. Rather, they participate in "schools of the air," talking with their teachers by two-way radios. Encourage students to imagine they "attend" a school of the air. Have them write a conversation they might have with a geography teacher over the two-way radio.

Teen Scene

Winter in June

It's June in Australia—time to get out the skis! Because the seasons in Australia are opposite of those in the Northern Hemisphere, Paul Noone and his friends plan their winter activities from June to September. Paul and his family go to Snowy Mountains to ski. This area lies just south of Canberra in one of Australia's many national parks. From the chairlift, Paul can see Australia's highest peak, Mount Kosciusko.

democracy and preserved much of their British heritage. Australia has developed its own national character, however. The "new" Australia comes from the diversity of its people. Many present-day Australians have Italian, Greek, Slavic, Aboriginal, or Chinese backgrounds.

City Life About 85 percent of Australia's population live in cities. Sydney and Melbourne—on the southeastern coast—are the largest cities. With populations of more than 3 million, both are lively cultural and commercial centers. Australia's next largest cities—Adelaide, Brisbane, and Perth—also are located on the coast. Canberra, the national capital, lies a relatively short distance inland from Sydney and Melbourne. With about 280,000 people, Canberra was developed through a government plan designed to draw people away from the coasts and into the outback.

Rural Life Only about 15 percent of Australians live in rural areas known as the **bush**. Many rural people live on stations that raise sheep or cattle. Thirteen-year-old Alice Adams lives on a sheep station in the Central Lowlands. She and her family live in a large, comfortable farmhouse. Working on the station is a family effort. Alice and other rural Australians sometimes find life in the bush very lonely. The Adamses' station covers more than 300 square miles (700 sq. km). The nearest settlement is about 50 miles (81 km) away. A trip to the market or shopping center is not quick. Alice and her family travel two hours on unpaved roads to reach a distant rural town.

SECTION 1 REVIEW

REVIEWING TERMS AND FACTS

1. **Define the following:** coral, outback, station, marsupial, bush.
2. **PLACE** What is the landscape of the central three-fourths of the country like?
3. **PLACE** What important agricultural products does Australia produce?
4. **MOVEMENT** Who were the first people to settle Australia?

MAP STUDY ACTIVITIES

5. Compare the maps on pages 688 and 690. What large Australian city lies closest to the Great Barrier Reef?
6. Turn to the climate map on page 692. What area of Australia enjoys a Mediterranean climate?

694

Answers to Section 1 Review

1. All vocabulary words are defined in the Glossary.
2. It is an empty stretch of land that is made up mostly of deserts.
3. wool, wheat, sugarcane, beef
4. Aborigines
5. Brisbane
6. southwest corner, south central coast

BUILDING GEOGRAPHY SKILLS

Analyzing a LANDSAT Image

Look carefully at the photograph below. What does it show? A LANDSAT satellite carrying cameras took this picture of southern Florida from space. The cameras record millions of energy waves invisible to the human eye. Computers then change this information into pictures of the earth's surface.

In LANDSAT images, the colors red or pink mean healthy vegetation. Deep blue shows areas of clear water, while light blue shows unclear water. These images help to pinpoint polluted areas. To interpret a LANDSAT image, apply these steps:

- Name the area shown in the image.
- Point out the natural and human-made features.
- Look for areas of healthy vegetation and clean water, and areas showing signs of pollution.

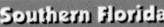
Southern Florida

Geography Skills Practice

1. What geographic area is shown in this image?

2. Where do you see the healthiest vegetation?

3. Where in the picture do you see water that might be polluted or muddy?

TEACH

Point out that different shades of color on a LAND-SAT image indicate different things. Purple, a mixture of red and blue, shows the location of forest areas. Some shades of red indicate such cultural features as roads and cities. Other shades show where certain kinds of minerals are located. Ask students for what purposes they think LANDSAT images might be used. *(to locate minerals, to note how such resources as forests are used, to forecast crop harvests, to note the growth of cities, and so on)* Then direct students to read the skill and complete the practice questions.

Additional Skills Practice

1. LANDSAT images are sometimes called "false-color images." Why do you think this is so? *(Because they do not show the natural colors of objects.)*

2. How would a murky, polluted river appear on a LANDSAT image? *(It would be shown as a light blue color.)*

📁 Geography Skills Activity 26

Answers to Geography Skills Practice

1. southern Florida

2. in the southeastern and northernmost areas of the image

3. in the coastal areas

SECTION 2
New Zealand

FOCUS

Section Objectives

1. **Recognize** the two large islands that make up most of New Zealand.
2. **Enumerate** the products that New Zealand sends to other countries.
3. **Discuss** how New Zealanders spend their leisure time.

Bellringer Motivational Activity

 Prior to taking roll at the beginning of the class period, project Section Focus Transparency 26-2 and have students answer the activity questions. Discuss students' responses. **L1**

Vocabulary Pre-check

Explain that *geo-* comes from a Greek word for "earth" and that *-thermal* comes from a Greek word for "heat." Ask students to speculate on the meaning of *geothermal power*. Have students check their speculations against the definition in the section. **L1 LEP**

PREVIEW

Words to Know
• *manuka*
• fjord
• geothermal power

Places to Locate
• New Zealand
• North Island
• South Island
• Wellington
• Auckland
• Southern Alps
• Canterbury Plains

Read to Learn . . .
1. what two large islands make up most of New Zealand.
2. what products New Zealand exports to other countries.
3. how New Zealanders spend their leisure time.

What country has more than 15 times as many sheep as people? New Zealand is the answer. Rolling green hills provide excellent grazing land for some 55 million sheep. New Zealand is one of the world's leading producers of lamb and wool.

New Zealand is an island country in the southwest Pacific Ocean. It lies about 1,200 miles (1,930 km) southeast of its nearest neighbor, Australia. For centuries, distance and isolation separated New Zealand from other parts of the world. Today it is a modern Pacific nation with one of the highest standards of living in the world.

Human/Environment Interaction
The Land

If you were shown two photographs—one of a desert and one of a lush green landscape—you could probably guess which one was Australia and which New Zealand. Although New Zealand lies relatively close to Australia, its landscape is very different from that of its larger neighbor. With 103,470 square miles (267,987 sq. km), New Zealand is about the size of the state of Colorado. The map on page 690 shows you that New Zealand is a long, narrow country. Anywhere you stand in New Zealand, you are not more than 80 miles (129 km) from the sea. New Zealand includes two main islands—North Island and South Island—as well as many smaller islands.

 696

UNIT 9

North Island A large volcanic plateau forms the center of North Island. Three active volcanoes and the perfectly shaped cone of inactive Mount Egmont rise above the island's fertile plains. Small shrubs called *manuka* grow well in the plateau's volcanic soil.

Farther north on the plateau you see geysers spewing hot water as high as 100 feet (30 m) in the air. Nearby flow icy cold streams. Engineers use natural steam from the geysers and other volcanic activity to produce electricity for New Zealand's industries, farms, and urban areas.

Fertile lowlands, forested hills, and sandy beaches surround the North Island's volcanic plateau. On the plateau's slopes, sheep and cattle graze. Fruits and vegetables are grown on the coastal lowlands. Seaside resorts offer tourists deep-sea fishing for marlin, shark, and tuna.

South Island A massive mountain chain called the Southern Alps runs along South Island's west coast. Snowcapped Mount Cook, the highest peak in New Zealand, soars 12,349 feet (3,764 m) in the Southern Alps.

This mountainous region reveals some of New Zealand's most beautiful scenery. Glaciers lie on mountain slopes high above thick, green forests and sparkling lakes. Along the southwestern coast, long narrow fingers of the sea called **fjords** cut into the land. New Zealand's fjords are similar to those in Norway.

From the Southern Alps to the east coast stretch the Canterbury Plains. They form New Zealand's largest area of flat or nearly flat land. Farmers grow grains and ranchers raise sheep here. The map on page 698 shows you the kinds of products New Zealand produces.

Animals and Plants As in Australia, animals and plants in New Zealand are found nowhere else. New Zealanders take pride in this fact. They chose the unusual kiwi (KEE•wee)—a long-beaked bird that cannot fly—as their national symbol. You may have eaten a fruit called a kiwi. This fruit—really a Chinese gooseberry—originated in New Zealand and was named for the kiwi bird.

Giant Australian kauri trees once covered all of North Island. European settlers cut down much of the kauri forest to build homes and ships. Today the kauri trees are protected. One of them is more than 2,000 years old!

The Climate New Zealand has a mild, wet climate. Compare the map on page 692 to the map on page 94. You see that New Zealand lies in a marine west coast region similar to the Pacific Northwest coast of the United States.

Thermal Springs in Rotorua, New Zealand

"Big Splash" is one of the many spectacular geysers created by volcanic activity in northern New Zealand.
HUMAN/ENVIRONMENT INTERACTION: How are New Zealand's engineers using geysers as a source of energy?

More About the Illustration

Answer to Caption
They are harnessing the steam to produce electricity.

Place The city of Rotorua is in the center of New Zealand's thermal region. It is sometimes called the "Sulfur City" because the smell that spews from underground is so strong. Boiling mud pools, hot springs, and geysers dot the city, and it is not unusual to see steam rising from sewer grates and storm drains.

TEACH

Guided Practice

Compare Write the following on the chalkboard:

	North	South
Features		
Crops		
Other products		

Have students make a comparison chart of North Island and South Island by filling in the information. *(North: Features—manuka, plateau, volcanoes, and geysers; Crops—fruits and vegetables; Other products—sheep, cattle, marlin, shark, tuna. South: Features—Southern Alps, glaciers, forests, lakes, and fjords; Crops—grains; Other products—sheep)* **L1**

Cooperative Learning Activity

Organize students into several small groups. Have groups research the wool-making process and prepare a report illustrating the different steps—shearing, sorting, cleaning, carding, roving, and spinning. Suggest that they include illustrations of the differences between woolen and worsted fabrics and between fabrics made from sheep wool and goat, camel, llama, alpaca, and vicuña wool. Invite groups to present their reports. Some might like to further illustrate their reports by displaying articles of clothing made from wool. **L2**

In the Canterbury Plains area of South Island, sheep outnumber the human population by a ratio of 20 to 1.

MULTICULTURAL PERSPECTIVE
Culturally Speaking

Most New Zealanders refer to themselves as *Kiwis,* after the symbol of their country. Maoris call white New Zealanders *Pakeha*—a Maori word meaning "fair skinned."

Answer

bauxite, coal, petroleum, gold, silver, iron ore

Map Skills Practice

Reading a Map Where in New Zealand is hydro-electric power generated? *(South Island)*

Ocean breezes from the west warm the land in winter and cool it in summer. As in the Pacific Northwest, rain falls throughout the year in New Zealand.

Because New Zealand lies south of the Equator, its seasons are the opposite of those in the Northern Hemisphere. July is New Zealand's coolest month, while January and February are the warmest months. Temperatures get cooler as you move from the north to the south.

Movement
The Economy

About 55 percent of New Zealand's land—its greatest resource—is cropland or pastureland. With its mild climate and fertile soil, New Zealand supports a thriving farm economy and exports many farm products. In recent years, manufacturing and tourism also have become important sources of income for New Zealand.

Farming A return to the use of natural fibers in clothing has boosted New Zealand's economy. A large percentage of the world's wool comes from New Zealand ranches. Its farmers and ranchers also produce dairy products and meat. The major agricultural exports of New Zealand are wool, lamb, butter, and cheese. Wheat, barley, oats, apples, and kiwifruit are also grown there.

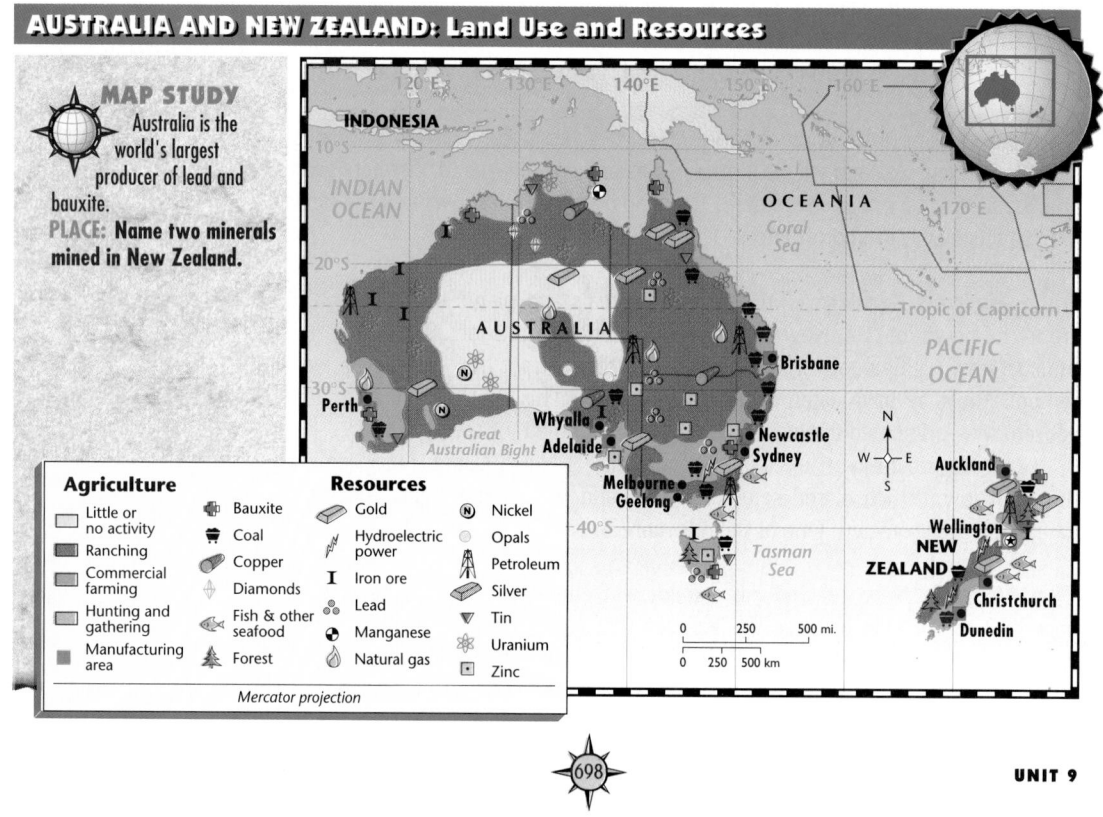

AUSTRALIA AND NEW ZEALAND: Land Use and Resources

MAP STUDY Australia is the world's largest producer of lead and bauxite.
PLACE: Name two minerals mined in New Zealand.

Agriculture	Resources	
Little or no activity	🔷 Bauxite	🟨 Gold	Ⓝ Nickel
Ranching	⬛ Coal	⚡ Hydroelectric power	○ Opals
Commercial farming	Copper	I Iron ore	🛢 Petroleum
Hunting and gathering	Diamonds	Lead	🟪 Silver
Manufacturing area	🐟 Fish & other seafood	Manganese	▽ Tin
	🌲 Forest	💧 Natural gas	☢ Uranium
			⊡ Zinc

Mercator projection

Meeting Special Needs Activity

Reading Disability Practice a focused reading exercise with students who experience reading comprehension problems. Write the following questions on the chalkboard: How did the Maoris travel from the Pacific islands to New Zealand? Why did the British decide to make New Zealand part of their empire? What "firsts" occurred in New Zealand? Have students consider these questions as they read the subsection titled "The People." **L1**

Mining and Manufacturing New Zealand has only a few minerals but is rich in **geothermal power**, or electric power produced from natural steam. The major source of power, however, is hydroelectricity.

In recent years, New Zealand has developed its industrial economy. Its factories now produce cars, furniture and other wood products, clothing, and electronic equipment. Most manufacturing, however, involves making food products such as butter, milk, and cheese.

A Maori wood-carver practices his skill *(left)*. The British sport of lawn bowling is popular with New Zealanders of European descent *(right)*.
PLACE: What percentage of New Zealand's population is Maori?

Place
The People

Picture this scene: The year is about A.D. 1300. Small canoes paddled by a Polynesian people called the Maori reach the New Zealand shore. The lush, green landscape is a welcome sight. A quick hunt produces a large mao bird and a lizard for dinner. The Maori decide to stay.

The arrival of the first people in New Zealand could have happened just this way. Today about 9 percent of the population are Maori (MOWR•ee). Most of the rest of New Zealand's 3.5 million people trace their ancestry to British settlers who came to the islands in the 1800s. Immigrants still come to New Zealand from the United Kingdom and other countries.

Influences of the Past The Maori are believed to have arrived in New Zealand about 700 years ago from Pacific islands far to the northeast. They developed skills in farming, weaving, fishing, and bird hunting and became well-known for their artistic wood carvings.

Geologically, North Island and South Island could be considered tiny continents because they lie atop separate tectonic plates.

Global Gourmet

New Zealand New Zealanders eat vegemite, made from yeast extract. They spread it on bread as Americans spread peanut butter.

CHAPTER 26 (699)

Critical Thinking Activity

Drawing Conclusions Inform students that New Zealand does not allow nuclear-armed or even nuclear-powered vessels to use its port facilities. Then ask them to draw a conclusion, based on this information and information in the section, about the probable amount of energy generated by nuclear power plants in New Zealand. Have them explain their answer. *(None, because New Zealand is strongly antinuclear and has other sources of energy, such as geothermal power and hydroelectricity.)* **L1**

ASSESS

Check for Understanding

Assign Section 2 Review as homework or an in-class activity.

Meeting Lesson Objectives

Each objective below is tested by the questions that follow it in parentheses.

1. **Recognize** the two large islands that make up most of New Zealand. (2, 6)
2. **Enumerate** the products that New Zealand sends to other countries. (5)
3. **Discuss** how New Zealanders spend their leisure time. (4)

Evaluate

Section 2 Quiz **L1**

Use the Testmaker to create a customized quiz for Section 2. **L1**

Chapter 26 Test Form A and/or Form B **L1**

MAP STUDY

Answer
2–60 persons per sq. mile (1–25 persons per sq. km)

Map Skills Practice
Reading a Map What area of New Zealand has the lowest population density? *(southwest coast of South Island)*

A few European explorers first came to New Zealand in the mid-1600s. Almost 200 years passed before European settlers—most of them British—arrived in large numbers. Then the British government decided to add New Zealand to its empire. In 1840 British officials signed a treaty with Maori leaders. The Maori agreed to accept British rule in return for the right to keep their lands. British settlers, however, eventually moved into Maori territory. Although the Maori fought to keep their lands, the British eventually defeated them.

New Zealand is a country of "firsts." In 1893 it became the first country in the world to give women the right to vote. New Zealand was also among the first countries to provide government help to the elderly, the sick, and those without jobs.

In 1907 New Zealand became independent of the United Kingdom. A parliamentary democracy, the New Zealand government works to improve housing, job opportunities, and education for all of its citizens, including Maori. Many Maori also want the government to return lands taken from them under British rule. In 1994 the New Zealand government paid one Maori group millions of dollars for land taken by the British.

Way of Life About 85 percent of New Zealanders live in urban areas. You can see where these cities are located on the map on page 698. The largest cities are Auckland (AW•kluhnd), a major port, and Wellington, the national capital. Both lie on North Island where three-fourths of New Zealanders live. Two smaller cities—Christchurch and Dunedin—are manufacturing centers on South Island.

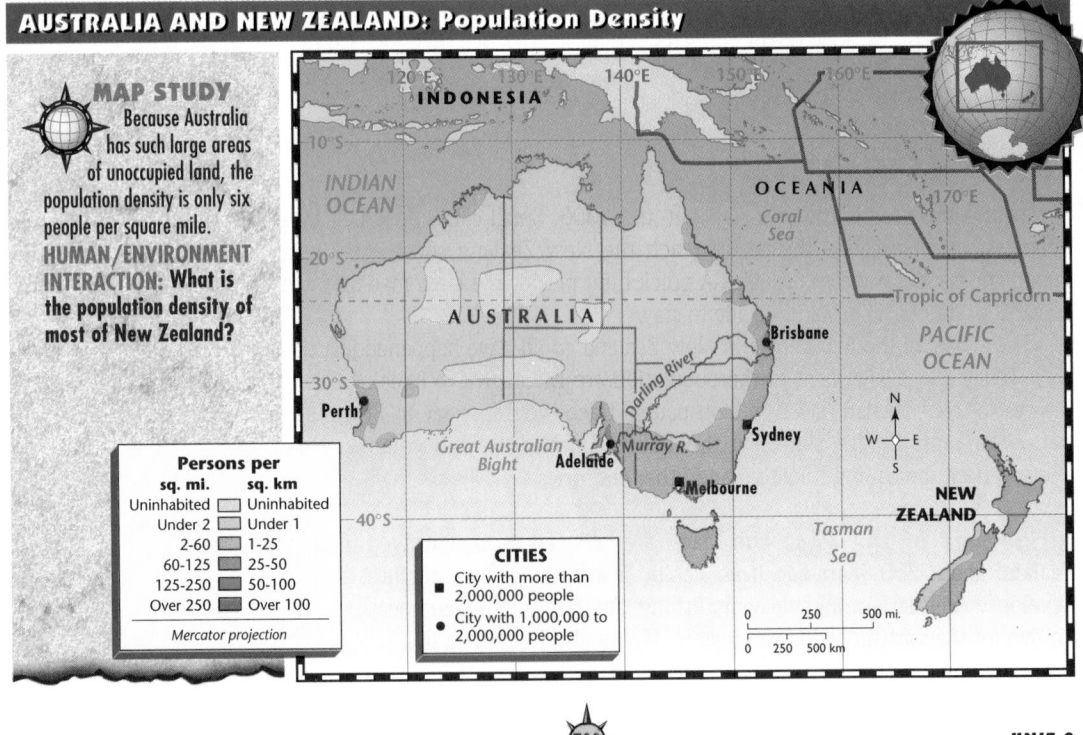

AUSTRALIA AND NEW ZEALAND: Population Density

MAP STUDY
Because Australia has such large areas of unoccupied land, the population density is only six people per square mile. **HUMAN/ENVIRONMENT INTERACTION:** What is the population density of most of New Zealand?

Persons per

sq. mi.	sq. km
Uninhabited	Uninhabited
Under 2	Under 1
2–60	1–25
60–125	25–50
125–250	50–100
Over 250	Over 100

Mercator projection

CITIES
■ City with more than 2,000,000 people
● City with 1,000,000 to 2,000,000 people

0   250   500 mi.
0 250 500 km

More About . . .

The Treaty of Waitangi February 6, or Waitangi Day, is a national holiday that commemorates the treaty in which the Maoris agreed to accept British rule and the British recognized the Maoris as the landowners of New Zealand. Hone Heke Pokai was the first Maori chief to sign the treaty and also one of the first to declare war on the British in 1844 when he felt they had violated their agreement. The British defeated Hone Heke Pokai and his followers in 1846, forcing the Maoris to honor the treaty. Interestingly, the Maoris have since used the treaty to reclaim land that was misappropriated by white settlers.

New Zealanders love outdoor activities. In New Zealand's mild climate, they enjoy camping, hiking, hunting, boating, and mountain climbing in any season. The people of New Zealand also enjoy sports such as cricket, a British game played with a bat and ball, and rugby, a British form of football.

Many New Zealanders still observe British customs. Yet immigrants from Asia, the Pacific, and various European nations have brought cultural diversity to New Zealand. Today, the growing Maori population also receive more recognition than they did in the past.

New Zealand is rapidly becoming a modern Pacific nation that blends European and Maori ways. This mix is evident in the life of 15-year-old Ruth Erutoe, a Maori teenager who lives in a suburb of Auckland. Ruth and her family listen to Italian opera and admire Kiri Te Kanawa (KEE•ree teh KAN• ah•wah), the world-famous opera star from New Zealand. The Erutoes also remain close to their Maori heritage. For weddings and other special occasions, they go to a *marae*, or a Maori meetinghouse. Some meetinghouses are simple buildings. Others are decorated with elaborate wood carvings. Turn to page 708 to see an example of a Maori wood carving.

Bustling Auckland, New Zealand

Auckland is New Zealand's largest city, chief seaport, and industrial center.
PLACE: What is the capital of New Zealand?

Turn to page 708 to see an example of a Maori wood carving.

SECTION 2 REVIEW

REVIEWING TERMS AND FACTS

1. **Define the following:** *manuka*, fjord, geothermal power.
2. **REGION** What physical feature lies in the center of North Island?
3. **MOVEMENT** From where did the Maori come?
4. **LOCATION** Where is the city of Wellington in relation to Auckland?

MAP STUDY ACTIVITIES

5. Look at the land use map on page 698. What mineral resources does New Zealand have?
6. Turn to the population density map on page 700. Where do the fewest people of New Zealand live?

Answers to Section 2 Review

1. All vocabulary words are defined in the Glossary.
2. a large volcanic plateau
3. from Pacific islands far to the northeast of New Zealand
4. Wellington is located at the southern end of North Island, some 250 miles (402 km) south of Auckland.
5. bauxite, coal, petroleum, gold, silver, iron ore
6. southwest coast of South Island

More About the Illustration

Answer to Caption
Wellington

GEOGRAPHIC THEMES

Place When New Zealand was established as a separate British colony in 1841, Russell, a small town in northern North Island, was chosen as its capital. The capital was moved to Auckland in the mid-1840s, and to Wellington in 1865.

Reteach

 Reteaching Activity 26 **L1**

Enrich

Enrichment Activity 26 **L1**

CLOSE

Mental Mapping Activity

Provide each student with a large sheet of white paper and map pencils. Inform students that they will not be graded on this activity. Have students draw, freehand from memory, a map of New Zealand that includes North Island, South Island, Mount Egmont, the Southern Alps, Auckland, and Wellington.

MAKING CONNECTIONS

TEACH

Ask students to identify the major characteristics of the kangaroo and the koala. Note their suggestions on the chalkboard. Circle the word *pouch*, or write it on the chalkboard if students do not identify this feature, and point out that because both animals have pouches they are classified as *marsupials*. **L1**

More About . . .

Life Science Getting a young kangaroo in the mother's pouch is an acrobatic feat. The mother stands still, while the young kangaroo tumbles inside the pouch head first. Once inside, it does a forward somersault, ending curled in a ball, head—and sometimes its hind legs and tail tip—poking out of the pouch.

Cultural Kaleidoscope

Australia Australians call a male kangaroo a *boomer,* a female kangaroo a *blue flier,* and any young kangaroo a *joey.*

AUSTRALIA'S AMAZING CREATURES

| MATH | SCIENCE | HISTORY | LITERATURE | TECHNOLOGY |

Australia is home to more than 120 kinds of marsupials—mammals whose young grow and mature inside a pouch on the mother's belly. Wallabies, wombats, and other marsupials are unfamiliar to people outside Australia. Kangaroos and koala bears, however, are known throughout the world.

KANGAROOS No matter what their size or where they live, all kangaroos have one thing in common: big hind feet. Either red or gray, kangaroos bound along at about 20 miles (32 km) per hour. Some may race up to 40 miles (65 km) per hour. In a single jump a kangaroo can hop 10 feet (3 m) high and cover a distance of 45 feet (14 m)!

Kangaroos also have big ears. Sensitive instruments, a kangaroo's ears are never still. One ear points forward, the other backward. Female kangaroos have front pouches that open at the top to hold their joeys, or infants. These big pockets provide nourishment and protection for several months.

KOALAS Despite their name, koalas are not bears. Like their kangaroo cousins, female koalas protect and nourish their young in pouches on their bellies. A female koala's pouch, however, opens at the bottom. Strong muscles keep the pouch shut and the young koalas safe inside.

Koala Mom and Joey

Koalas make their homes in the eucalyptus trees of southeastern Australia. Quiet, calm, and sleepy, the cuddly-looking animals spend at least 20 hours a day dozing high in a favorite tree. Koalas are fussy eaters. They eat nothing but eucalyptus leaves—about 2.5 pounds (1.2 kg) a day. The leaves also provide the animals with all the moisture they need.

Making the Connection

1. About how many kinds of marsupials live in Australia?
2. How are kangaroos and koalas alike? How are they different?

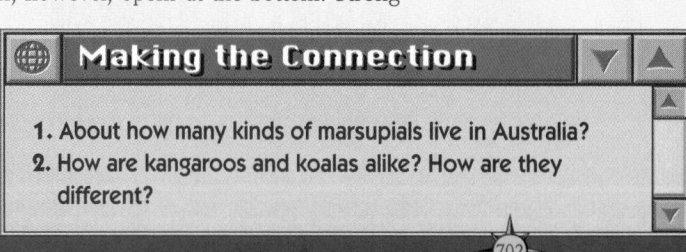

702

Answers to Making Connections

1. about 120
2. Kangaroos and koalas both carry their young in pouches; however, the kangaroo's opens at the top while the koala's opens at the bottom. The kangaroo travels great distances on the ground by jumping. The koala spends most of its time in eucalyptus trees, dozing and eating.

Important Things to Know About Australia and New Zealand

SECTION 1 AUSTRALIA

- Australia is both a country and the world's smallest continent.
- Deserts and grasslands cover most of the country west of the Great Dividing Range.
- Australia's economy relies mostly on farming and mining.
- Australia is the world's leading wool-producing country. It also leads in the production of bauxite and lead.
- The first Australians were the Aborigines. Others are of European and Asian descent.
- Most Australians live in cities. Sydney, Melbourne, Adelaide, Brisbane, and Perth are the largest.

SECTION 2 NEW ZEALAND

- New Zealand includes two major islands—North Island and South Island—and many smaller islands.
- New Zealand has a mild, wet climate.
- New Zealand's agricultural exports—especially wool—are the mainstays of its economy.
- New Zealand gets most of its electricity from geothermal energy.
- The Maori, a Polynesian ethnic group from the South Pacific, were the first people to settle New Zealand.
- New Zealand's culture at one time was mainly influenced by British traditions. Now it is a mix of many cultures, including those of the Maori and other immigrants.

Sydney, Australia ▶

CHAPTER 26   703

Extra Credit Project

Skit Ask students to undertake research and then write scripts that dramatize important events in the history of Australia or New Zealand, such as the arrival of the First Fleet or the signing of the Treaty of Waitangi. Encourage students to work in small groups to make dramatic presentations of their scripts.   L1

Chapter 26 Review and Activities

REVIEWING KEY TERMS

Match the numbered terms in Column A with their definitions in Column B.

A

1. *manuka*
2. bush
3. geothermal power
4. outback
5. marsupial
6. fjord
7. coral
8. station

B

A. inlet of the sea that cuts into the land
B. hard material made of the skeletons of small sea animals
C. electricity produced from natural steam
D. Australian sheep or cattle ranch
E. vast inland areas of Australia
F. small shrubs that grow on New Zealand's volcanic soil
G. Australian word for rural areas
H. mammal that carries its young in a pouch

ANSWERS

Reviewing Key Terms

1. F
2. G
3. C
4. E
5. H
6. A
7. B
8. D

 Mental Mapping Activity

This exercise helps students to visualize the countries they have been studying and understand the relationships among various points located in these countries. All attempts at freehand mapping should be accepted.

Mental Mapping Activity

On a separate piece of paper, draw a freehand map of Australia and New Zealand. Label the following items on your map:
- Sydney
- Melbourne
- Auckland
- North Island
- South Island
- Wellington

REVIEWING THE MAIN IDEAS

Section 1

1. **LOCATION** Why is Australia called the Land Down Under?
2. **LOCATION** Where is Australia's tropical climate region located?
3. **MOVEMENT** What two countries are Australia's major trading partners?
4. **PLACE** Why was the city of Canberra built?

Section 2

5. **LOCATION** Where is New Zealand located in relation to Australia?
6. **PLACE** What large mountain range runs down the western coast of South Island?
7. **MOVEMENT** What are New Zealand's major agricultural exports?
8. **HUMAN/ENVIRONMENT INTERACTION** What leisure activities do New Zealanders enjoy that are made possible by the country's climate?

Reviewing the Main Ideas

Section 1

1. because it is located entirely in the Southern Hemisphere
2. in the northern parts of the country
3. Japan, United States
4. It was built as a planned city to draw settlers inland from the heavily populated coastal areas.

Section 2

5. New Zealand is located 1,200 miles (1,930 km) southeast of Australia.
6. Southern Alps
7. wool, lamb, butter, cheese
8. camping, hiking, hunting, boating, and mountain climbing

CRITICAL THINKING ACTIVITIES

1. **Making Comparisons** How does the climate of Australia compare with that of New Zealand?
2. **Making Inferences** Why do you think most Australians live in coastal cities?

GeoJournal Writing Activity

Imagine that you are visiting Australia. Using the map on page 688, prepare a list of places to see—one place for each state of the country. Write an itinerary describing the places you want to visit, what you hope to see there, and how many miles (km) you will travel from place to place.

COOPERATIVE LEARNING ACTIVITY

Work in a group of three to learn more about the people and cultures of Australia and New Zealand. Each group will choose a country. Then each member will select *one* of the following topics to research: (a) leisure activities; (b) arts and crafts; or (c) music and dance. After your research is complete, share your information with the rest of your group. As a group, prepare a written report, a poster, or a travel brochure that presents your group's findings.

PLACE LOCATION ACTIVITY: AUSTRALIA AND NEW ZEALAND

Match the letters on the map with the places and physical features of Australia and New Zealand. Write your answers on a separate sheet of paper.

1. Brisbane
2. Great Barrier Reef
3. Wellington
4. Darling River
5. Canberra
6. Great Dividing Range
7. Auckland
8. Southern Alps
9. Perth
10. Tasmania

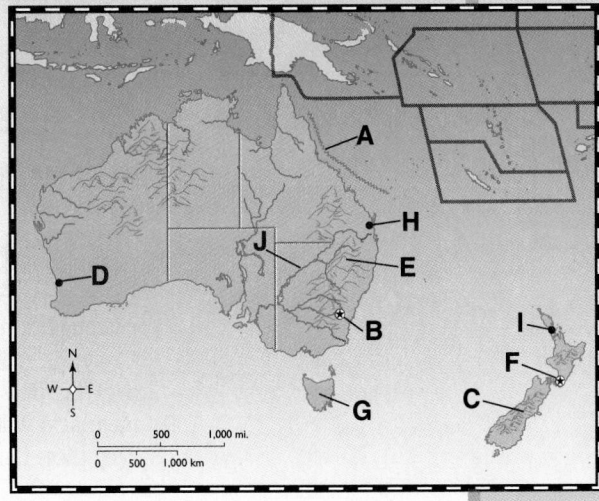

Encourage students who choose arts and crafts to look for pictures of Aboriginal art, such as the 20,000-year-old paintings found at Australia's Kakadu National Park, and Maori art, such as masks and wood carvings. Suggest that all students include a variety of illustrations in their reports, posters, and brochures.

GeoJournal Writing Activity
Encourage students to compare their itineraries with those of other class members.

Place Location Activity
1. H
2. A
3. F
4. J
5. B
6. E
7. I
8. C
9. D
10. G

Chapter Bonus Test Question

This question may be used for extra credit on the chapter test.

Prevailing westerly winds bring heavier rain totals to the west coast of New Zealand's South Island than to its east coast. Why? *(The east coast lies in the rain shadow of the Southern Alps.)*

Critical Thinking Activities

1. Australia generally has a dry climate, with tropical, humid subtropical, Mediterranean, and marine west coast climates in coastal areas; whereas all of New Zealand has a marine west coast climate.

2. The coastal areas have moderate climates and fertile soils. The interior of the country is dry and much of the area has desert climates.

705

GeoLab
ACTIVITY

FOCUS

Call on students to indicate on a globe the locations of the world's glaciers. Then point out that more than 6 million square miles (15.5 million sq. km) of Earth's land surface is permanently glaciated.

TEACH

Before assigning the experiment, organize students into three groups. Direct one group to create a large wall map of the world showing the locations of glaciers. Have the second group create a large diagram showing a cross section of a glacier. Mention that the diagram should include the following labels: *head wall, bergschrund, firn field,* and *snout.* Direct the third group to locate pictures of glaciated landscapes. Have group members write captions identifying the features shown in the pictures. Display student work around the classroom. L1

The world's largest glacier is Lambert Glacier in Antarctica. It is about 40 miles (64 km) wide and 250 miles (402 km) long.

From the classroom of Rebecca Revis, Sylvan Hills Junior High School, Sherwood, Arkansas

GLACIERS: Earth's Scouring Pad

Background

Antarctica, the world's coldest and iciest region, has mountains, valleys, and lowlands underneath its ice cap. Like the Arctic Circle, the Antarctic Circle also has many glaciers. Glaciers and rivers erode the land and change it a great deal. In this activity, you'll observe how glaciers and rivers change the earth's surface.

Glacier-created valley, New Zealand

Believe it or NOT!

Glaciers cover about one-tenth of the earth's land and hold about 85 percent of all freshwater. The power of glaciers is enormous. Consider that the Great Lakes were formed by a glacier scouring out a river valley.

Materials

- water
- sand, gravel, and clay
- stream table with sand
- lamp with reflector
- ruler

(NOTE: You will be using electrical equipment near water in this GeoLab. Please keep these items away from one another.)

Curriculum Connection Activity

Earth Science Many different factors combine to determine how a glacier flows. One of the basic elements of glacial movement, however, is *basal slip.* This involves the glacier sliding over melted water underneath it. The water melts because the glacier's great weight exerts enormous pressure at its base. Have students demonstrate this in a simple experiment. Provide them with three small trays or tin plates and two ice cubes. Direct them to place the ice cubes on separate trays or plates. Then have them take the third tray or plate and place it on top of one of the ice cubes, applying increasing pressure to the top tray or plate. Ask students to compare the rate of melting of the two ice cubes. *(weighted cube melts quicker)* L1

What To Do

A. Set up the stream table and lamp as shown.

B. Make an ice block by mixing water with sand, gravel, and clay in a container that measures about 2 inches (5 cm) by 8 inches (20 cm) by 1 inch (2 cm). Then freeze this container.

C. Make a V-shaped river channel. Measure and record its width and depth. Draw a sketch that includes these measurements.

D. Place the ice block, to act as a moving glacier, at the upper end of the stream table.

E. Gently push the glacier along the river channel until it is under the light, halfway between the top and bottom of the stream table.

F. Turn on the light and allow the ice to melt. Observe and record what happens.

G. Measure and record the width and depth of the glacial channel. Draw a sketch of the channel and include these measurements in your journal.

GeoLab
ACTIVITY

NATIONAL GEOGRAPHIC SOCIETY

These materials are available from Glencoe.

 VIDEODISC

STV: World Geography Vol. 3: South America and Antarctica

Ice-layered glacier moves toward the ocean
Any side, Frame 50364

Lab Activity Report

1. How can you figure out the direction from which a glacier traveled?

2. How can you tell how far down the valley the glacier traveled?

3. How do glaciers affect the surface over which they move?

4. **Drawing Conclusions** How can you identify land that was once covered by a glacier?

Go A Step Further!

Activity
Before your "glacier" melts completely, take one small part of it and rub it over a piece of wood. What happens in the wood's surface? How do glaciers make similar patterns in rocks?

ASSESS

Have students answer the **Lab Activity Report** questions on page 707.

CLOSE

Encourage students to complete the **Go A Step Further** writing activity. When students have completed the activity, tell them that geographers often use working models to simulate geographic events.

Go A Step Further
Answer
Wood becomes scratched; rocks have grooves or scratches in their surfaces.

Answers to Lab Activity Report

1. The area over which the glacier has moved will probably be smoother and have steeper sides than the section that has not eroded.

2. The channel will be U-shaped to the point where the glacier stopped, and V-shaped past that point. Small deposits will form at the end of the glacier.

3. Glaciers erode the surface like a bulldozer, leaving U-shaped valleys with steep sides and flat bottoms.

4. Glaciers leave behind characteristic deposits and patterns of erosion.

FOCUS

Draw students' attention to the photographs of Sydney Opera House and the thatched buildings. Have them note similarities and differences between the two structures. Record student observations on the chalkboard in the form of a Venn diagram. Emphasize the commonality of the two structures, despite their vast differences. Point out that such commonality often underlies apparent differences in cultures.

TEACH

Inform students that New Guinea is among the most culturally diverse regions in the world. For instance, more than 700 different languages are spoken on the island. Ask students how differences in language reflect differences in culture. *(A common language is a basic feature of a culture.)* **L1**

Geography and the Humanities

📁 World Literature Reading 9

🌐 World Music: Cultural Traditions, Lesson 10

🕹 World Cultures Transparencies 19, 20

Cultural **HERITAGE:**
AUSTRALIA AND OCEANIA

ARCHITECTURE ▶ ▶ ▶ ▶

The opera house in Sydney's harbor is a landmark recognized throughout the world. The overlapping white shells of its roofline were designed to imitate the shapes of boat sails in the harbor.

◀ ◀ ◀ ◀ WOOD CARVING

The Maori of New Zealand are expert wood-carvers. Many of their intricately textured carvings portray ancestors. The carvings decorate homes and community meetinghouses.

PAINTING ▶ ▶ ▶ ▶

Nature plays an important role in the lives of the Australian Aborigines, and it is reflected in their art. This painting, done on bark, shows many animal, plants, and landscape features.

708

UNIT 9

More About . . .

History Boomerangs were developed by many peoples in prehistoric times, but this tool is most often connected with the Aborigines of Australia. Indeed, *boomerang* is an Aborigine word. Original boomerangs were not designed to return to the thrower. Since they were weapons used for hunting and, later, warfare, they were designed to hit a target. A skilled thrower with a high-quality boomerang could hit a moving target at 100 yards (91 m). Aboriginal boomerangs used for warfare were quite impressive. Highly decorated and measuring some 3 feet (0.9 m) in length, they flew at incredible speeds.

◄ ◄ ◄ ◄ **MASKS**

This mask, with its large eyes and bright colors, represents the face of an owl. The people of Papua, New Guinea, make animal masks to use in traditional dances.

THATCHED BUILDINGS ▶ ▶ ▶ ▶

New Guinea builders use the trunks of palm trees to build the frames for their beautiful spirit houses. The walls and roof are woven from palm leaves. Only men are allowed in these buildings, which hold each village's religious relics.

BOOMERANGS ▲ ▲ ▲ ▲

Australian boomerangs come in many shapes and sizes. The Aborigines use boomerangs mostly for hunting. Some are carved or painted for use in religious ceremonies. The small boomerangs are toys for children.

CHAPTER 26

709

APPRECIATING CULTURE

1. What natural materials do the Aborigines, Maori, and people of New Guinea need to create their artwork and buildings?

2. How would you describe the Sydney Opera House to someone who has not seen it?

Cultural Kaleidoscope

Oceania Many cultures in Oceania practice the art of body tattooing. (The word *tattoo* comes from a Tahitian term.) Tattoos often are very elaborate, involving intricate spiral patterns. Traditionally, the people of Oceania have considered a beautiful tattoo a sign of good breeding.

ASSESS

Have students answer the **Appreciating Culture** questions on page 709.

CLOSE

Have students imagine that they are young members of one of Oceania's many cultures. Have them write a letter to a friend in the United States comparing their culture with American culture.

Answers to Appreciating Culture

1. Answers will vary. Most students will identify such materials as different kinds of wood; tree bark; palm leaves; and leaves, berries, or roots ground into powder to make paint.

2. Answers will vary. Most descriptions will make the comparison between the roof sections and ships' sails.

Oceania and Antarctica

PERFORMANCE ASSESSMENT ACTIVITY

A PUPPET SHOW Organize students into groups of four. Have groups assume the roles of local public librarians who put on a summer program for vacationing elementary school students. Have groups develop and present a puppet show for this program. The show should inform the audience about the geographic importance of Oceania and Antarctica. Character puppets should take the form of people or animals native to the area. Set a time limit for the show, so that group members will have to prioritize the information to be presented. Call on groups to present their puppet shows to the rest of the class, ensuring that all group members take part in the presentation.

POSSIBLE RUBRIC FEATURES: Oral presentation skills, planning skills, collaborative skills, accuracy of content information, decision-making skills

MENTAL MAPPING ACTIVITY

Before teaching Chapter 27, point out Oceania and Antarctica on a globe. Have students identify the hemisphere in which Oceania and Antarctica are located and where they are located relative to the United States.

TEACHER'S CORNER

NATIONAL GEOGRAPHIC SOCIETY

INDEX TO NATIONAL GEOGRAPHIC MAGAZINE

The following articles may be used for research relating to this chapter:

- "Exploring Antarctic Ice," by Jane Ellen Stevens, May 1996.
- "Emperor Penguins," by Glenn Oeland, March 1996.
- "The Two Worlds of Fiji," by Roger Vaughan, October 1995.
- "Gray Reef Sharks," by Bill Curtsinger, January 1995.
- "Return to Hunstein Forest," by Edie Bakker, February 1994.
- "Reclaiming a Lost Antarctic Base," by Michael Parfit, March 1993.

- "Six Across Antarctica: Into the Teeth of the Ice," by Will Steger, November 1990.
- "Ghosts of War in the South Pacific," by Peter Benchley, April 1988.
- "In the Far Pacific: At the Birth of Nations," by Carolyn Bennett Patterson, October 1986.
- "The Two Samoas, Still Coming of Age," by Robert Booth, October 1985.

NATIONAL GEOGRAPHIC SOCIETY PRODUCTS AVAILABLE FROM GLENCOE

To order the following products for use with this chapter, contact your local Glencoe sales representative or call Glencoe at 1-800-334-7344:

- *STV: World Geography* (Videodisc)
- *Picture Atlas of the World* (CD-ROM)
- *Physical Geography of the World* (Transparencies)
- *ZipZapMap! World* (Software)

- *GeoBee* (Software)
- *Images of the World* (Posters)
- *Eye on the Environment* (Posters)

ADDITIONAL NATIONAL GEOGRAPHIC SOCIETY PRODUCTS

To order the following products for use with this chapter, call National Geographic Society at 1-800-368-2728:

- *Lost Fleet of Guadalcanal* (Video)

- *Antarctic Wildlife Adventure* (Video)

TEACHER-TO-TEACHER
Curriculum Connection: Language Arts, Earth Science

—from Patricia K. Bosh
Mifflin International Middle School, Columbus, OH

THE OCEAN FLOOR *The purpose of this activity is to have students understand and visualize ocean landforms by creating a landforms dictionary, an ocean-floor diagram, and an ocean-floor map.*

Oceania is a huge region. As its name suggests, much of the area Oceania covers is ocean—the Pacific Ocean. What lay at the depths of this vast body of water remained a mystery until sonic depth recorders and satellite imaging outlined its topography. Ask students to create an ocean topography dictionary, using such terms as *abyssal plain, basin, continental shelf, continental slope, fracture zone, guyot, ridge, seamount, trench, and trough.* Suggest that students accompany each entry with pronunciation instructions and a definition. When dictionaries have been completed, organize students into four or six groups. Direct half the groups to use their dictionaries and other resources to construct a three-dimensional diagram of the ocean floor. Have the rest of the groups use their dictionaries and other resources to create a relief map of the Pacific Ocean. Call on groups to display and discuss their diagrams and maps. Then have students compare ocean landforms to landforms found on the earth's surface.

MEETING NATIONAL STANDARDS

Geography for Life

All of the 18 standards are demonstrated in Unit 9. The following ones are highlighted in this chapter:

- Standards 1, 2, 3, 5, 7, 8, 10, 11, 14, 15, 16

KEY TO ABILITY LEVELS

Teaching strategies have been coded for varying learning styles and abilities.

L1 **BASIC** activities for all students

L2 **AVERAGE** activities for average to above-average students

L3 **CHALLENGING** activities for above-average students

LEP **LIMITED ENGLISH PROFICIENCY** activities

BIBLIOGRAPHY

Readings for the Student

Hermes, Jules. *The Children of Micronesia.* Minneapolis: Carolrhoda Books, 1994.

Lyle, Garry. *Pacific Islands.* New York: Chelsea House, 1988.

Pringle, Laurence. *Antarctica.* New York: Simon & Schuster, 1992.

Readings for the Teacher

Laws, Richard. *Antarctica: The Last Frontier.* New York: State Mutual Books, 1990.

Stewart, John. *Antarctica: An Encyclopedia.* Jefferson, Mo.: McFarland & Co., 1990.

Theroux, Paul. *The Happy Isles of Oceania: Paddling the Pacific.* New York: G.P. Putnam's Sons, 1992.

Multimedia

Antarctica. Washington, D.C.: National Geographic Society, 1991. Videocassette, 25 minutes.

Glaciers: Ice on the Move. Washington, D.C.: National Geographic Society, 1994. Videocassette, 25 minutes.

The Pacific World. New York: Educational Design, nd. Videocassette, 45 minutes.

Use these *Geography: The World and Its People* resources to teach, reinforce, and extend chapter content.

CHAPTER 27 RESOURCES

- *Vocabulary Activity 27
- Cooperative Learning Activity 27
- Workbook Activity 27
- Geography Skills Activity 27
- GeoLab Activity 27
- Chapter Map Activity 27
- Critical Thinking Skills Activity 27
- Performance Assessment Activity 27
- *Reteaching Activity 27
- Enrichment Activity 27
- Environmental Case Study 9
- Chapter 27 Test Form A and Form B
- *Chapter 27 Digest Audiocassette, Activity, Test
- Unit Overlay Transparencies 9-0, 9-1, 9-2, 9-3, 9-4, 9-5, 9-6
- Political Map Transparency 9
- Geoquiz Transparencies 27-1, 27-2
- Vocabulary PuzzleMaker Software
- Student Self-Test: A Software Review
- Testmaker Software
- MindJogger Videoquiz

If time does not permit teaching the entire chapter, summarize using the Chapter 27 Highlights on page 723, and the Chapter 27 English (or Spanish) Audiocassettes. Review student's knowledge using the Glencoe MindJogger Videoquiz. * *Also available in Spanish*

Use these *Geography: The World and Its People* resources to teach and reinforce section content.

SECTION 1 RESOURCES

Reproducible Lesson Plan 27-1

Section Focus Transparency 27-1

Guided Reading Activity 27-1

Section Quiz 27-1 (also available in Spanish)

SECTION 2 RESOURCES

Reproducible Lesson Plan 27-2

Section Focus Transparency 27-2

Guided Reading Activity 27-2

Section Quiz 27-2 (also available in Spanish)

710D

Performance Assessment

Refer to the Planning Guide on page 710A for a Performance Assessment Activity for this chapter. See the *Performance Assessment Strategies and Activities* booklet for suggestions.

Chapter Objectives

1. **Identify** places, people, and occupations in Oceania.
2. **Describe** the location, climate, and scientific importance of Antarctica.

GLENCOE
TECHNOLOGY

VIDEODISC

Use the Chapter 27 MindJogger Videoquiz to preview chapter content.

MindJogger Videoquiz

Chapter 27
Disc 4 Side A

 Also available in VHS.

Chapter 27 Oceania and Antarctica

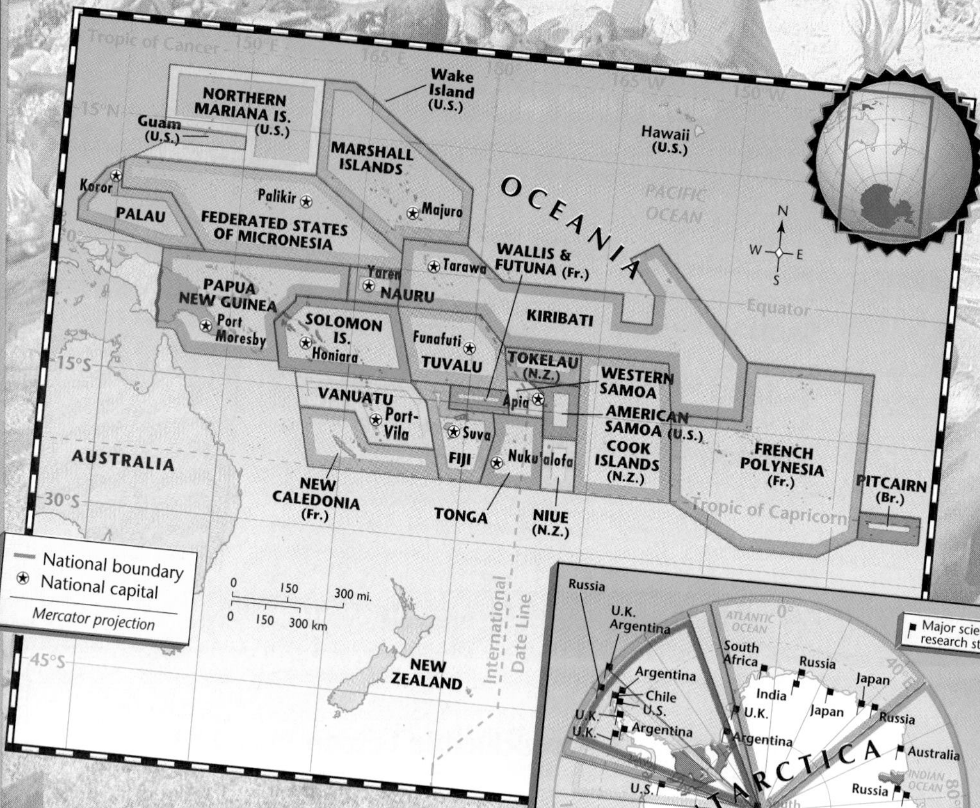

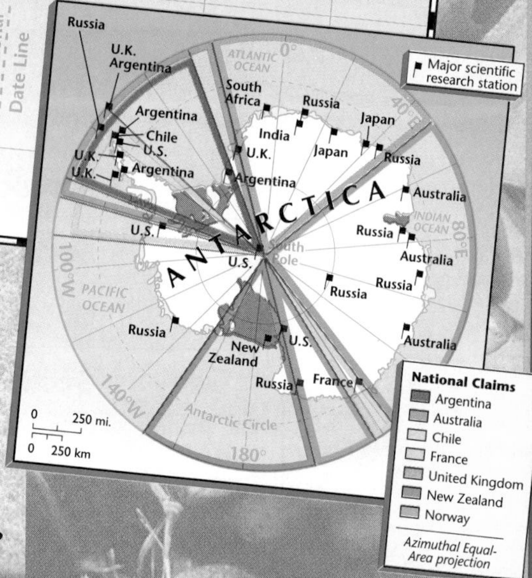

MAP STUDY ACTIVITY

You are ending your geographic "tour" of the world by learning about Oceania and Antarctica.

1. What is the capital of Papua New Guinea?
2. How many research stations does the United States have in Antarctica?

710

MAP STUDY ACTIVITY

Answers
1. Port Moresby
2. four

Map Skills Practice
Reading a Map What three island areas in Oceania are territories of France? *(New Caledonia, Wallis and Futuna, French Polynesia)*

SECTION 1

Oceania

Meeting National Standards

Geography for Life The following standards are highlighted in this section:
• Standards 1, 2, 5, 7, 8, 10, 11, 14, 15

PREVIEW

Words to Know
• continental island
• copra
• high island
• low island
• atoll
• phosphate
• trust territory

Places to Locate
• Melanesia
• Micronesia
• Polynesia

Read to Learn . . .
1. what geographic areas make up Oceania.
2. how land and climate affect the way people in Oceania earn their livings.
3. what groups have settled in Oceania.

What do you picture when you hear the word "paradise"? You might see one of the more than 30,000 tropical islands that dot the blue waters of the Pacific Ocean. Some of the islands, like New Guinea shown here, are mountainous.

The Pacific island region is called Oceania. Great distances separate this region from other parts of the world. The islands of Oceania—about 25,000 of them—are spread out across 70 million square miles (181.3 million sq. km). Because of Oceania's vast distances, geographers group the islands into three main regions: Melanesia, Micronesia, and Polynesia.

Location
Melanesia

North and east of Australia lie the islands of Melanesia (MEH•luh•NEE•zhuh). Because of their dense vegetation, Melanesia's islands are often called the "black islands." Melanesia's land and ocean area combined almost match the land area of the continental United States. The largest country in size and population is Papua New Guinea (PA•pyuh•wuh noo GIH•nee). It lies on the eastern half of the island of New Guinea.

Southeast of Papua New Guinea are three other independent island countries: the Solomon Islands, Fiji, and Vanuatu (VAN•WAH•TOO). Near these countries is New Caledonia, a group of islands ruled by France.

CHAPTER 27

711

FOCUS

Section Objectives
1. **Identify** the geographic areas that make up Oceania.
2. **Summarize** how land and climate affect the way people in Oceania earn their livings.
3. **Recall** what groups have settled in Oceania.

Bellringer Motivational Activity

 Prior to taking roll at the beginning of the class period, project Section Focus Transparency 27-1 and have students answer the activity questions. Discuss students' responses. **L1**

Vocabulary Pre-check
Turn the meanings of the vocabulary terms in the section into questions. For example: What is dry coconut meat called? *(copra)* Write the questions on the chalkboard and have students match each question with a term. Tell them to read the section to find out if their matches are correct. **L1**
LEP

Use the Vocabulary PuzzleMaker Software to create a crossword puzzle. **L1**

Classroom Resources for Section 1

 REPRODUCIBLE MASTERS
Reproducible Lesson Plan 27-1
Guided Reading Activity 27-1
Vocabulary Activity 27
Cooperative Learning
  Activity 27
Workbook Activity 27
Geography Skills Activity 27
Section Quiz 27-1

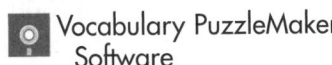 **TRANSPARENCIES**
Section Focus
  Transparency 27-1
Political Map Transparency 9
Unit Overlay Transparency 9-0

MULTIMEDIA
 Vocabulary PuzzleMaker
  Software
Testmaker

 National Geographic
  Society: STV: Biodiversity

711

TEACH

Guided Practice

Display Political Map Transparency 9 and write the headings *Melanesia, Micronesia,* and *Polynesia* on the chalkboard. As students read about each island group in the section, point to it on the transparency. Call on students to list the major geographic features of each group under the appropriate heading on the chalkboard. **L1**

Independent Practice

Guided Reading Activity 27-1 **L1**

Cultural Kaleidoscope

Papua New Guinea One of the most important crops grown in the highlands of Papua New Guinea is the sweet potato. This plant, native to the Western Hemisphere, came to the island by way of Europe and Asia.

MAP STUDY

Answer
Micronesia

Map Skills Practice
Reading a Map Which island group lies closest to Australia? *(Melanesia)*

The Climate and Economy Melanesia consists mostly of continental islands. A **continental island** is formed by chunks of land split off from a larger continent or through the erosion of a link of land that once connected the island to the mainland. Rugged mountains and thick rain forests cover the continental islands of Melanesia. Strips of fertile plains hug island coastlines.

Thermometers record little change in temperatures in Melanesia year-round. The map on page 720 shows you that most of Melanesia has a tropical climate. The temperature along island coasts seldom falls below 70°F (21°C) or rises above 80°F (27°C).

Most Melanesians are subsistence farmers, growing only enough to feed their families. Papua New Guinea and Fiji, however, cultivate agricultural products for export. Farmers in Papua New Guinea grow cacao, tea, and coffee. Those in Fiji raise sugarcane. A major money-making product of Melanesia—and the rest of Oceania—is **copra,** or dried coconut meat. Countries around the world use coconut oil from copra to make margarine, soap, and other products.

Some Melanesian islands, such as Papua New Guinea and New Caledonia, hold rich mineral resources. Rugged mountains and vast shipping distances, however, make it costly to develop mining and other industries.

The Past and Present Melanesians share many different cultures and languages. More than 700 languages are spoken in Papua New Guinea alone! To ease communication, English is the language of business and government throughout most of Melanesia. From the 1800s to the mid-1900s, Western countries ruled most of Melanesia. Many island countries today are a blend of Western and Melanesian cultures.

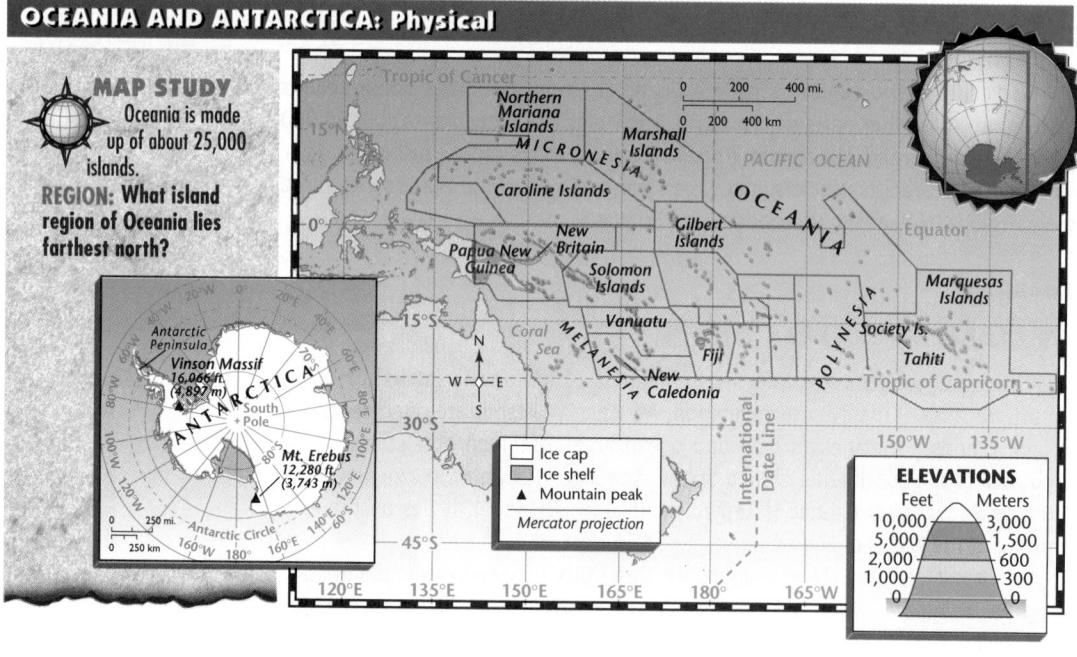

OCEANIA AND ANTARCTICA: Physical

MAP STUDY Oceania is made up of about 25,000 islands.
REGION: What island region of Oceania lies farthest north?

ELEVATIONS
Feet	Meters
10,000	3,000
5,000	1,500
2,000	600
1,000	300
0	0

Cooperative Learning Activity

Tell the class that storytelling dances are a tradition on the islands of Kiribati. The dancers wear knee-length skirts of coconut fronds or pandanus leaves. Males tie belts of braided hair around their waist and crisscross their chest with bright bands. Women wear belts of seashells and garlands of flowers in their hair.

Organize the class into groups and have each group choose a well-known myth or fairy tale. Tell them to choreograph a dance that conveys their chosen story. Suggest that one member in each group act as a narrator to interpret the moves in the dance. Have groups make costumes with yarn, paints, paper bags, and construction paper. Allow time for each group to perform its dance for the rest of the class. **L1**

More Melanesians live in small villages like this one in the Solomon Islands *(left)* than in modern cities such as Fiji's capital of Suva *(right)*.
PLACE: How are the lifestyles of rural Melanesians different from those who live in the city?

Many Melanesians still live in small villages in houses made of grass and other natural materials. They keep strong ties to their local group and hold on to traditional customs. Only a small number of Melanesians live in cities. Port Moresby in Papua New Guinea and Suva in Fiji are among the largest urban areas. Many people in these cities live in modern houses and have jobs in business and government.

Region
Micronesia

North of Melanesia lies another island region—Micronesia. Look at the map on page 712 to see the contrasts between the two regions. The map shows you that the islands of Micronesia are scattered across a vast ocean area. The independent countries of Micronesia are the Federated States of Micronesia, the Marshall Islands, Palau (puh•LOW), Nauru (nah•OO•roo), and Kiribati (KIHR•uh•BAS). The Northern Mariana Islands and the island of Guam are territories linked to the United States.

High and Low Islands Rich soil, ribbonlike waterfalls, and thick green vegetation blanket Micronesia's **high islands**. Volcanic activity formed most of these rugged islands many centuries ago. The rest of Micronesia includes **low islands** usually made of coral, and have little vegetation. Skeletons of millions of tiny sea animals formed the low islands over thousands of years. Most of the low islands are **atolls**, or ring-shaped coral reefs that surround small lagoons. You often will find atolls built up on the rims of underwater volcanoes that no longer erupt. The chart on page 714 shows you how a typical atoll is formed.

The Climate and Economy Micronesia has a tropical climate. Daily temperatures average about 80°F (27°C) year-round, and rain is plentiful across the region. Tropical breezes flowing over these islands are not always gentle,

CHAPTER 27 713

Meeting Special Needs Activity

Pronouncing Difficult Words Some students may have difficulty pronouncing the place names in this section, such as Papua New Guinea, Melanesia, Micronesia, and Polynesia. You may wish to sound out the correct pronun- ciations of these places. You also may direct students to the Gazetteer in the back of their textbooks for pronunciations of other places mentioned in the text. **L1**

Phosphate sales have been so lucrative for the government of Nauru that it does not tax its people to raise revenues.

Marshall Islands Popular foods in this part of Micronesia include breadfruit baked on coals, raw fish, and fried banana pancakes.

Answer
because the landform on which the atoll develops is close to the surface
Graphic Skills Practice
Reading a Diagram
What eventually happens to a volcanic island surrounded by an atoll? *(it subsides into the water)*

however. From July to October, fierce storms called typhoons sometimes strike the islands, causing loss of life and much destruction.

Subsistence farming is the major economic activity on Micronesia's high islands. Farmers there grow food crops in the fertile soil and raise cattle, sheep, and goats. On some high islands, copra is a major export. People on the low coral islands rely on the sea for food.

Most Micronesian island countries depend on aid from the United States, Australia, and Europe. With this aid they are building roads, ports, and airfields so that industry can develop in the future. The only exception is the tiny island republic of Nauru. For years the 9,500 people of Nauru have prospered from the mining of **phosphate,** a mineral used in making fertilizers. With Nauru's phosphate deposits running out, however, the people are looking for other sources of income.

The Past and Present Southeast Asians first settled Micronesia about 4,000 years ago. European explorers, traders, and missionaries settled on the Micronesian islands during the 1700s and early 1800s. By the early 1900s, many European countries, the United States, and Japan owned colonies in this region.

During World War II, the United States and Japan fought a number of bloody battles on Micronesian islands. Why? Look at the map on page A30 to see the islands' location in relation to Japan and the United States. Following World War II, most of Micronesia was turned over to the United States as **trust territories**. These territories were under temporary United States control. Since the 1970s, many of the Micronesian islands have become independent.

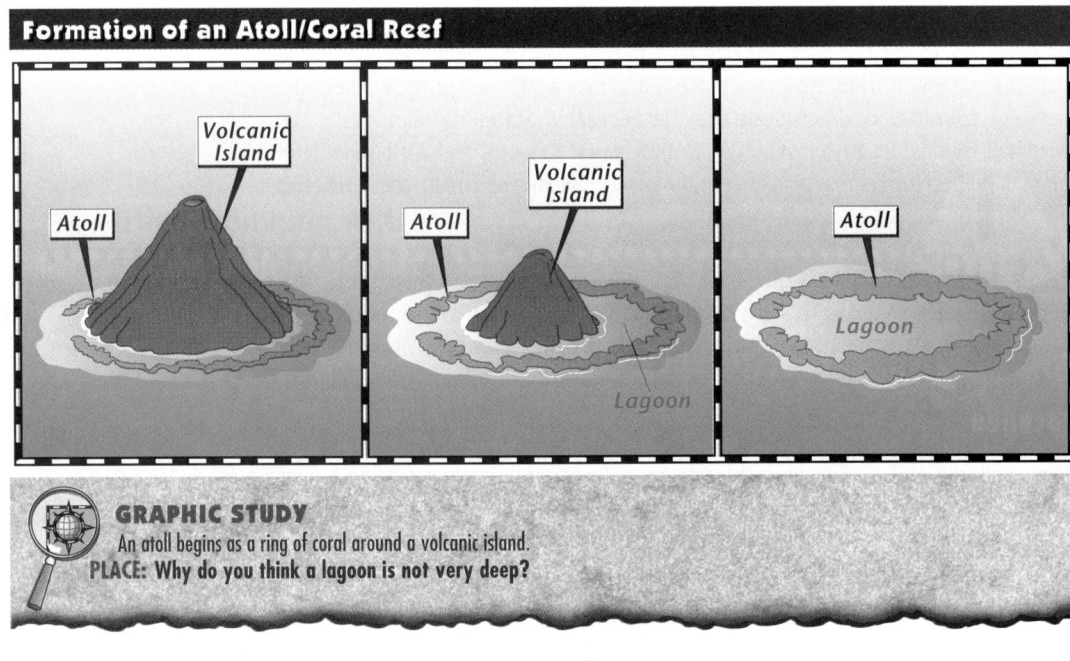

Formation of an Atoll/Coral Reef

GRAPHIC STUDY
An atoll begins as a ring of coral around a volcanic island.
PLACE: Why do you think a lagoon is not very deep?

Critical Thinking Activity

Determining Cause and Effect Remind the students that Melanesians have many different cultures and languages, and that linguists have identified hundreds of languages on Papua New Guinea alone. Point out that isolation of groups of people often results in this kind of diversity. Read aloud the part of the section that explains how the ocean isolates Melanesia from the rest of the world. Suggest that the ocean also isolates the groups on different islands from one another and helps to create the diversity of cultures and languages among the islands. Have students reread the descriptions of Papua New Guinea. Then ask them what might isolate groups from one another on Papua New Guinea and help to create a diversity of languages there. *(mountains)* **L1**

Today Micronesia's 400,000 people speak 11 languages and several dialects. Most of them live in villages headed by local chiefs. In recent years many young people have left the villages to find jobs in towns.

Place
Polynesia

East of Melanesia and Micronesia lie the tropical islands often pictured on travel brochures—the islands that make up Polynesia. The map on page 712 shows you that Polynesia is the largest of the three Pacific island regions. Three independent countries—Western Samoa, Tonga, and Tuvalu—are found in Polynesia. Other island groups are under French rule and are known as French Polynesia. Tahiti, Polynesia's largest island, is part of this French territory.

The Climate and Economy Most Polynesian islands are high volcanic islands, some with tall, rugged mountains. Thick rain forests cover mountain valleys and coastal plains. Other Polynesian islands are of the low coral type. With little soil, the only vegetation is scattered coconut palms. Because most of Polynesia lies in the tropics, the climate is hot and humid.

Most people in Polynesia raise crops or fish for their food. Some farmers raise coconuts and tropical fruits for export. Tourism is the fastest-growing industry of Polynesia. Each year thousands of visitors come by air or sea to its emerald-green mountains and white, palm-lined beaches. New roads, hotels, shops, and restaurants serve these tourists.

The Past and Present Settlers came later to Polynesia than to the other two Pacific regions. Can you imagine why? Look at the map on page 710. The first Polynesians were probably Melanesians or Micronesians who sailed to the islands in canoes.

During the late 1800s, the stronger European powers divided up Polynesia among themselves. Military bases were built on these well-located islands in the 1900s. They were perfect refueling stops for airplanes crossing the Pacific. Beginning in the 1960s, some Polynesian territories chose independence while others decided to remain under Western rule.

Today about 580,000 people live in Polynesia. Modern influences have affected the islands, yet some Polynesians live traditionally in villages. Women in Samoa and Tonga still make tapa cloth. They first strip the inner bark from paper

Island Life

Kaulana Pau lives on a tiny atoll surrounded by ocean. The atoll is in the Caroline Islands in Micronesia. Kaulana has visited some of the cities on the larger islands and plans to go the University of Guam someday. But when asked if she would like to move to a city, she politely answers, "No, thank you." The ocean is a part of Kaulana's life. "It gives us food, it's our highway to other islands, and it's how my family earns a living," she says. Kaulana's family takes the fish they catch to a processing plant on one of the larger islands. There it is either dried or canned for export.

Strange but TRUE!

French Polynesia stretches over an area as great as Europe, but its total land area is little larger than Rhode Island.

ASSESS

Check for Understanding

Assign Section 1 Review as homework or an in-class activity.

Meeting Lesson Objectives

Each objective below is tested by the questions that follow it in parentheses.

1. **Identify** the geographic areas that make up Oceania. (1, 2, 3, 6)
2. **Summarize** how land and climate affect the way people in Oceania earn their livings. (4)
3. **Recall** what groups have settled in Oceania. (4)

Evaluate

 Section Quiz 27-1
L1

Use the Testmaker to create a customized quiz for Section 1. L1

More About . . .

Coconut On most Pacific islands, coconut is a staple. The I-Kiribati people, for example, eat coconut with fish, grate it into tea, and use its milk to sweeten breadfruit soup. For many islanders, coconut sap provides a rich source of vitamin C. Twice a day, young boys on Kiribati cut *toddy*, or gather coconut sap. The *toddy* is boiled over a low heat to make a thick, sweet molasses used in place of sugar to sweeten drinks. Sometimes the *toddy* is made into hard candy.

Answer
under 2 persons per sq. mile (under 1 person per sq. km)

Map Skills Practice
Reading a Map What is the only city in Oceania with a population greater than 100,000 people? *(Port Moresby)*

Reteach

Assign each student a paragraph from the section to sum up in one sentence. Arrange their sentences in a summary of the section. Distribute copies of the summary and encourage students to retain them for review purposes. **L1**

Enrich

Have students research and report on the famous mutiny on the *Bounty.* **L3**

CLOSE

Have students imagine that they are reporters for a television show that discusses lifestyles around the world. Tell them to draft questions for an interview with Henri Moeino about life in Tahiti.

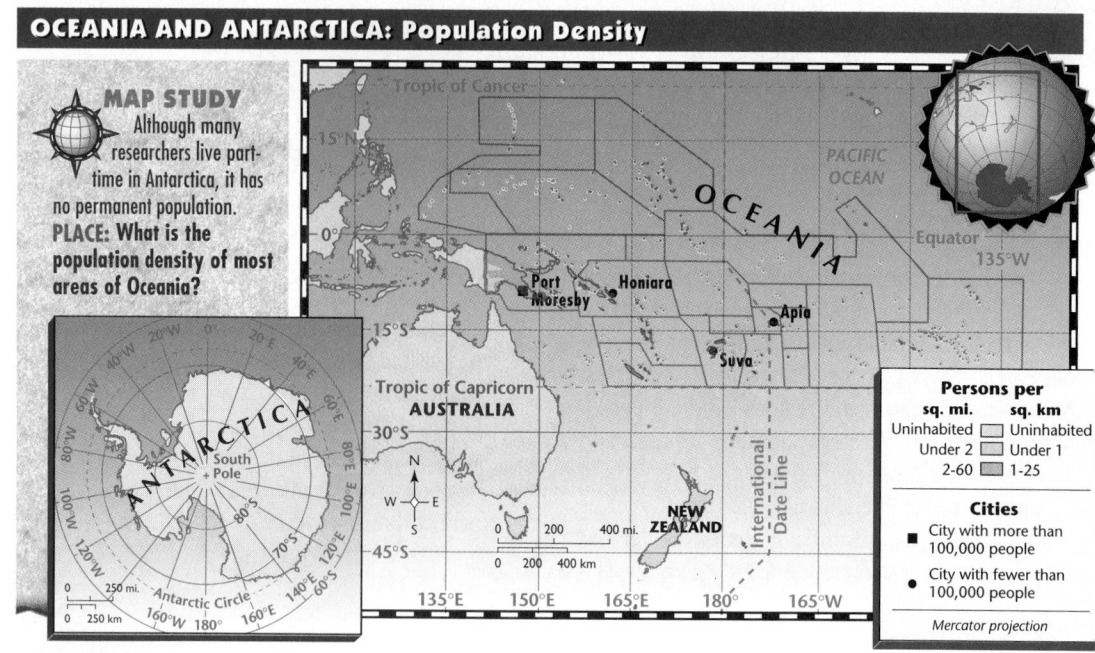

OCEANIA AND ANTARCTICA: Population Density

MAP STUDY
Although many researchers live part-time in Antarctica, it has no permanent population.
PLACE: What is the population density of most areas of Oceania?

Persons per

sq. mi.		sq. km
Uninhabited		Uninhabited
Under 2		Under 1
2-60		1-25

Cities
■ City with more than 100,000 people
• City with fewer than 100,000 people

Mercator projection

mulberry trees. Then they soak the bark and beat it with wooden clubs to form the cloth which is used to make clothing.

An increasing number of Polynesians live in cities or towns. Thirteen-year-old Henri Moeino lives in Papeete (PAH•pee•AY•tee), the largest city and port on the island of Tahiti. Henri points out that Papeete and its surrounding area are home to more than 70,000 people. He and other Tahitians belong to an interesting mix of peoples. Henri has Polynesian and French ancestry. Like most urban Polynesians, Henri follows both European and American ways.

SECTION 1 REVIEW

REVIEWING TERMS AND FACTS

1. Define: continental island, copra, high island, low island, atoll, phosphate, trust territory.

2. REGION What are the three island regions of Oceania?

3. PLACE How do the high and low islands in Micronesia differ in appearance?

4. HUMAN/ENVIRONMENT INTERACTION How do most people in Oceania make a living?

MAP STUDY ACTIVITIES

5. Turn to the political map on page 710. What is the capital of Fiji?

6. Look at the physical map on page 712. What is the largest island in Oceania?

UNIT 9

Answers to Section 1 Review

1. All vocabulary words are defined in the Glossary.
2. Melanesia, Micronesia, Polynesia
3. High islands have rugged mountains with thick rain forests, and fertile plains run along their coasts. Low islands barely rise above the water and have little vegetation.
4. Most people are subsistence farmers.
5. Suva
6. Papua New Guinea

BUILDING GEOGRAPHY SKILLS

Analyzing a Photograph

You've probably heard the saying, "A picture is worth a thousand words." You can learn much about the geography and culture of a place from a photograph. The key is to examine clues in the photo. Because photographs show only a small "slice" of reality, however, they can be misleading. For example, a picture of the wealthiest part of a city does not accurately show the living standards of all of the city's people. To accurately interpret a photograph, follow these steps:

- Read the caption.
- Name the landforms, climate, vegetation, cultural features (clothing and housing styles, transportation, technology), and human activities in the photograph.
- Think about what may be excluded from this view.

Drying Copra on the Solomon Islands

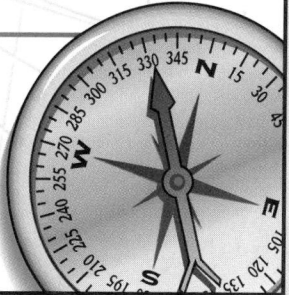

A woman collects dried coconuts, called copra, which will be pressed to produce coconut oil. About 30 coconuts are needed to make 1 gallon of oil. The oil is then used to make soaps, margarine, detergents, cosmetics, and many other products.

Geography Skills Practice

1. What is the main topic of this photograph?

2. What natural features are included in the photograph?

3. What information about culture is shown?

4. What information do you think is left out of the photograph?

CHAPTER 27

717

Answers to Geography Skills Practice

1. a woman collecting copra

2. tropical vegetation and copra

3. clothing, how people make their living

4. whether she is working alone or with other people, how big the work space is, and so on

Meeting National Standards

Geography for Life The following standards are highlighted in this section:
• Standards 3, 7, 13, 16

FOCUS

Section Objectives

1. **Point out** where Antarctica is located.
2. **Understand** why Antarctica is called a polar desert.
3. **Relate** how Antarctica is like a scientific laboratory.

Bellringer Motivational Activity

Prior to taking roll at the beginning of the class period, project Section Focus Transparency 27-2 and have students answer the activity questions. Discuss students' responses. **L1**

Vocabulary Pre-check

Have students scan the section to find the terms listed in **Words to Know.** Have them use context clues to define the terms. **L1** **LEP**

NATIONAL GEOGRAPHIC SOCIETY

These materials are available from Glencoe.

VIDEODISC

STV: World Geography Vol. 3: South America and Antarctica

Antarctic ice cap
Any side, Frame 50472

PREVIEW

Words to Know
• rookery
• ice shelf
• crevasse
• krill
• ozone

Places to Locate
• Antarctica
• Transantarctic Mountains
• South Pole

Read to Learn . . .
1. where Antarctica is located.
2. why Antarctica is called a polar desert.
3. how Antarctica is like a scientific laboratory.

Who doesn't love penguins? Their shiny "tuxedos" and waddling walk have fascinated people for years. Scientists study the habits of penguins, the largest group of marine animals in

Antarctica. For example, studying where they build their nests—or **rookeries**—gives scientists more clues to understanding the vast continent of Antarctica.

Antarctica—surrounded by icy ocean water—sits on the southern end of the earth. It covers about 5,400,000 square miles (14,000,000 sq. km). It is larger in size than either Europe or Australia. Mysterious Antarctica is the least explored of all the continents.

Location

Landforms

Picture Antarctica—a rich, green land covered by forests and lush plants. Does this description match your mental picture of the continent? Fossils discovered there tell scientists that millions of years ago Antarctica's landscape was inhabited by dinosaurs and small mammals.

Today a huge ice cap buries nearly 98 percent of Antarctica's land area. In many places, the ice cap is 1.5 to 2 miles (2.4 to 3.2 km) thick—about the height of 10 tall skyscrapers. This sea of ice holds about 70 percent of the world's freshwater. Where the ice cap spreads beyond the land and over the ocean, it forms an **ice shelf.**

Classroom Resources for Section 2

 REPRODUCIBLE MASTERS

Reproducible Lesson Plan 27-2
Guided Reading Activity 27-2
GeoLab Activity 27
Chapter Map Activity 27
Workbook Activity 27
Reteaching Activity 27
Enrichment Activity 27
Section Quiz 27-2

 **TRANSPARENCIES**

Section Focus
  Transparency 27-2
Political Map Transparency 9
Geoquiz Transparencies 27-1,
  27-2

MULTIMEDIA

 Testmaker

National Geographic
  Society: STV: World
  Geography, Vol. 3

Icy Coasts The ice cap is heavy. In some areas, it forms **crevasses**, or cracks, more than 100 feet (30 m) deep. As glaciers meet the sea, large chunks of ice break off to form icebergs. Antarctica borders the Atlantic, Pacific, and Indian oceans. As these oceans reach Antarctica, they become cooler and less salty, which allows them to freeze. Some of the fish in these waters have a special antifreeze—a protein in their blood that keeps it from freezing.

Landforms Beneath the packed layers of the ice cap lie highlands, valleys, and lowlands—the same features you would find on other continents. A long mountain range called the Transantarctic Mountains crosses Antarctica and splits it in two. The highest peak on the continent—Vinson Massif—rises here to 16,066 feet (4,897 m). The Transantarctic Mountains sweep out into the Antarctic Peninsula, which reaches within 600 miles (965 km) of South America's Cape Horn.

A high, flat plateau covers the area east of the Transantarctic Mountains. The earth's southernmost point, the South Pole, lies on the plateau at the center of Antarctica. West of the mountains is a group of low islands buried under layers and layers of ice. On Ross Island rises the peak of Mount Erebus. At 12,220 feet (3,743 m), it is Antarctica's most active volcano.

Place
Climate

Now that you have a mental picture of Antarctica's ice cap, think about this: Antarctica receives so little precipitation that it is the world's largest, coldest desert! Inland Antarctica receives no rain and hardly any new snow each year.

A scientist lowers into a crevasse near his research station. **HUMAN/ENVIRONMENT INTERACTION: What may happen to Antarctica's environment if too many people work there?**

More About the Illustration

Answer to Caption
trash and pollution problems; damage to habitats of plants and animals; minerals may be depleted

Human/ Environment Interaction The United States' McMurdo Station is the largest research facility in Antarctica. It has its own airstrip and helicopter pad and, during the summer months, a staff of about 1,000 scientists.

OCEANIA AND ANTARCTICA: Land Use and Resources

MAP STUDY
Many countries have research stations in Antarctica, but no single nation controls it.
PLACE: What resource has been discovered near the coasts of Antarctica?

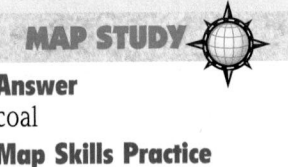

ANTARCTICA
South Pole

Tropic of Cancer
15°N
0°
PACIFIC OCEAN
OCEANIA
Equator
135°W
15°S
AUSTRALIA
30°S
Tropic of Capricorn
International Date Line
NEW ZEALAND
45°S
135°E 150°E 165°E 180° 165°W

0   200   400 mi.
0   200   400 km

Agriculture
Hunting and gathering
Commercial farming
Little or no activity

Resources
Coal
Gold
Nickel
Manganese
Phosphates

Mercator projection

CHAPTER 27

719

TEACH

Guided Practice

Making Predictions Before students read the section, have them predict the kinds of animals, plants, and resources that might be found in Antarctica. Write their predictions on the chalkboard. As students work through the section, have them check the correct predictions on the chalkboard and cross out the incorrect ones. **L1**

Independent Practice

Guided Reading Activity 27-2 **L1**

MAP STUDY

Answer
coal

Map Skills Practice
Reading a Map Where in Antarctica is most coal found? *(in the east)*

Cooperative Learning Activity

Have students hypothesize about whether bacteria can grow in Antarctica. Write hypotheses on the chalkboard. Organize students into groups and direct groups to perform the following experiment: 1) Pour a cup of milk in each of two pint-size jars and close each jar. 2) Place one jar in a refrigerator; the other in the classroom. 3) Check the milk in each jar daily. After seven days, have groups display their jars. Warm milk should be lumpy and sour, cold milk should still look and smell fresh. Mention that bacteria sours milk. Ask students to draw a conclusion about cold temperatures and bacteria growth based on the experiment. *(Cold temperatures slow down bacteria growth.)* Although milk eventually will sour even in a refrigerator, temperatures are so much lower in Antarctica that food keeps there indefinitely. **L1**

MAP STUDY

Answer
ice cap

Map Skills Practice

Reading a Map What does Antarctica's climate have in common with desert climates?
(It is very dry.)

ASSESS

Check for Understanding

Assign Section 2 Review as homework or an in-class activity.

Meeting Lesson Objectives

Each objective below is tested by the questions that follow it in parentheses.

1. **Point out** where Antarctica is located. (2)
2. **Understand** why Antarctica is called a polar desert. (3, 6)
3. **Relate** how Antarctica is like a scientific laboratory. (4)

Antarctica's ice cap is so large that if it were to melt, the water released would cause sea level to rise some 200 feet (60 m) worldwide. This would leave cities such as Boston, Miami, New Orleans, and Houston under the sea.

720

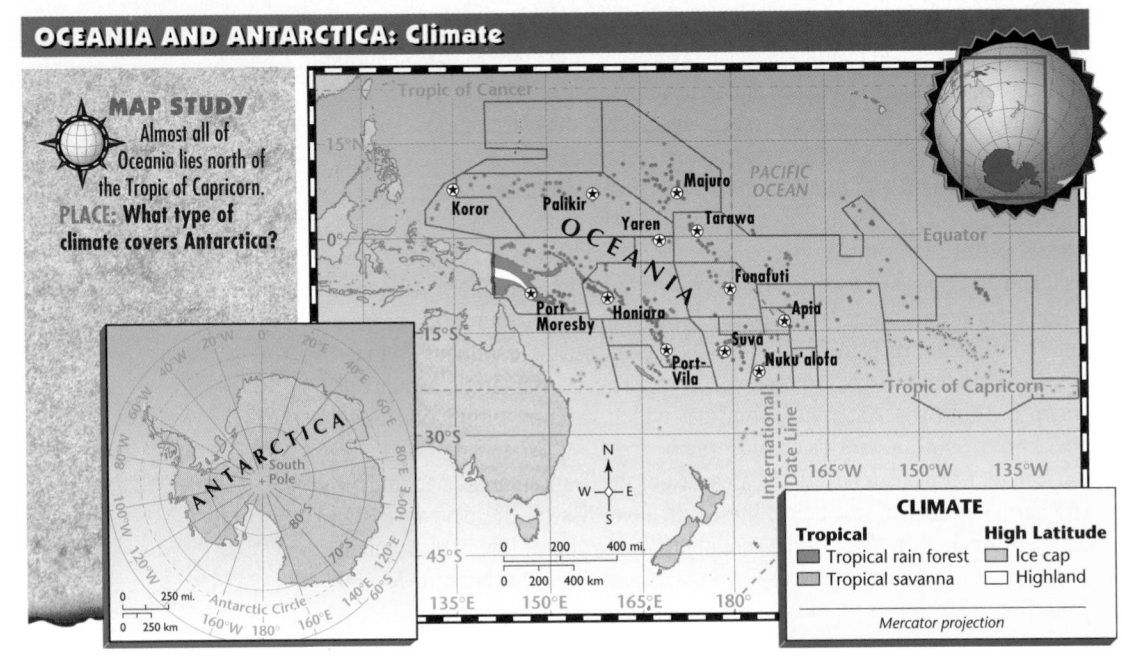

OCEANIA AND ANTARCTICA: Climate

MAP STUDY Almost all of Oceania lies north of the Tropic of Capricorn.
PLACE: What type of climate covers Antarctica?

CLIMATE

Tropical
- Tropical rain forest
- Tropical savanna

High Latitude
- Ice cap
- Highland

Mercator projection

The map above shows you that Antarctica has a polar ice cap climate. Imagine summer in a place where temperatures may fall as low as −30°F (−34.5°C) and climb only to 32°F (0°C)! Antarctic summers last from December through February. Winter temperatures along the coasts fall to −40°F (−40°C), and in inland areas to a low of −100°F (−73°C).

Region

Resources

Scientists believe that the Antarctic ice hides a treasure chest of minerals. The mining of resources on this continent is a source of international disagreement, however.

Mineral Resources Antarctica has major deposits of coal, and lesser amounts of copper, gold, iron ore, manganese, and zinc hidden underneath its ice cap. Petroleum may lie offshore. None of Antarctica's mineral resources have been developed. Many people believe that mining these resources would harm Antarctica's fragile environment.

Plants and Animals What is the largest permanent land animal in Antarctica? Most people would say the penguin. They would be incorrect, however. Penguins are marine animals, not land animals. Today the life-form considered to be the largest permanent land animal is *Beligica antarctica,* a wingless insect about one-tenth of an inch long.

Meeting Special Needs Activity

Language Delayed Help students with language concept problems to categorize information. Direct their attention to the following heading: "Explorers and Scientists." After they read the material under the heading, have students think of an alternate title, such as "The World Shares Antarctica." Direct them to do the same exercise with other headings in the section. **L1**

Only a few small plants and insects can survive in Antarctica's cold, dry, inland areas. Most plants on the continent are algae and mosses that grow along the coast on rocky surfaces. The waters and coasts around Antarctica, however, are home to penguins, fish, whales, and many kinds of flying birds. A small shrimplike animal called **krill** is a key source of food for other animals in Antarctica.

Human/Environment Interaction
Exploration and Research

Because of its harsh climate, Antarctica is the only continent that has no permanent human settlement. Its largely unspoiled environment has made it a favorable place for scientific research.

Operation Highjump

In 1946 the United States carried out the largest exploration of Antarctica up to that time. Captain Richard E. Byrd led 4,700 people in Operation Highjump. With 13 ships and 23 airplanes, Byrd and his group found more than 25 new islands and photographed 1,400 miles (2,300 km) of uncharted coastline.

Explorers and Scientists Europeans first sighted Antarctica during the early 1800s. In 1911 explorers competed with one another to reach the South Pole. You can read about this exciting race on page 722. The discovery of the South Pole opened the rest of Antarctica for exploration.

By the 1960s scientists from 12 countries had set up research centers on Antarctica. The map on page 710 shows you the major research centers. To preserve Antarctica as a peaceful scientific research lab, 12 countries signed the Antarctic Treaty in 1959. In the early 1990s, a number of countries agreed to keep Antarctica free of mining for 50 years.

Much research in Antarctica deals with **ozone,** a form of oxygen in the atmosphere. You learned in Chapter 3 that ozone protects all living things on the earth from certain harmful rays of the sun. In the 1980s scientists discovered a weakening or "hole" in the ozone layer above Antarctica. Studying this layer will help scientists learn more about possible changes in the world's climates.

SECTION 2 REVIEW

REVIEWING TERMS AND FACTS
1. **Define the following:** rookery, ice shelf, crevasse, ozone.
2. **LOCATION** What oceans surround Antarctica?
3. **PLACE** What kind of climate does Antarctica have?
4. **HUMAN/ENVIRONMENT INTERACTION** Why was the 1959 Antarctic Treaty important?

MAP STUDY ACTIVITIES
5. Look at the map on page 719. What major mineral resource lies in Antarctica?
6. Turn to the climate map on page 720. How do the climates of Oceania and Antarctica differ?

LESSON PLAN
Chapter 27, Section 2

Evaluate
 Section 2 Quiz **L1**

Use the Testmaker to create a customized quiz for Section 2. **L1**

Chapter 27 Test Form A and/or Form B **L1**

Reteach
Reteaching Activity 27 **L1**

Enrich
 Enrichment Activity 27 **L3**

CLOSE

 Mental Mapping Activity

Have students draw, freehand from memory, a view of Antarctica from the South Pole. Have them include and label the South Pole, the Antarctic Circle, and the 60°S and 80°S lines of latitude.

TEACH

Have students scan Section 2 of Chapter 27 and pages 674 to 687 for photographs and descriptions of Antarctica. Ask students to speculate on why Amundsen, Scott, and others were so enthusiastic to explore such a place. Then have them discuss under what circumstances they would be willing to travel to Antarctica. **L1**

More About . . .

Polar Exploration Roald Amundsen did not originally intend to explore Antarctica. His plan was to pinpoint the geographical North Pole. However, Robert Peary and Matthew Henson of the United States accomplished that feat in 1909, before Amundsen had set off for the Arctic. Amundsen quickly changed his plan, setting his sights on the South Pole. When Amundsen arrived in Antarctica, some of Scott's team visited him. While the British and the Norwegians were wary of one another, the meeting was friendly.

Better organized and equipped, and far more aggressive in approach, Amundsen's team got off to a quick start and reached their goal first. Plagued by poor planning, Scott's expedition ran into problems early on and never recovered. Scott and his men died just 11 miles (18 km) from one of their supply bases.

MAKING CONNECTIONS

THE RACE TO THE SOUTH POLE

| MATH | SCIENCE | HISTORY | LITERATURE | TECHNOLOGY |

The big news in the fall of 1911 was the race between Norway's Roald Amundsen and Britain's Robert Scott. The two seasoned explorers took on the challenge of finding the exact location of the South Pole.

READY, SET . . . Arriving at Ross Island, Scott and his crew set up headquarters at Cape Evans. Their equipment included 21 ponies and 3 tractor-like, motorized sleds.

Amundsen and his team pitched camp at the extreme western edge of the Ross Ice Shelf. To make the journey easier, Amundsen brought trained sled dogs and special skis.

. . . GO!! Amundsen's crew and dogsleds pulled out on October 19. Their expedition cut a new route over the Great Ice Barrier and through Antarctica's great central plateau. During the 1,400-mile (2,253-km) trip, the men left supplies in seven well-marked places. They reached the South Pole on December 14. There they put up a small tent on the Pole's exact location and left a letter inside for Scott. The team arrived back at their original campsite on January 25, 1912, in good health and high spirits.

On November 1, Scott and his 15-member team headed for the central plateau on a well-known route across Beardmore Glacier. The trip was a disaster. Their motorized sleds bogged down in soft snow, and supplies ran low. Eleven men and all the ponies died. Only Scott and 4 others made it to the South Pole on January 17, 1912. On the return trip to Cape Evans, however, the men were unable to find their poorly marked supply sites in the vast windy Antarctic wastes. All died from hunger and cold.

Scott's Expedition

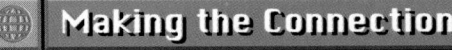

Making the Connection

1. What transportation equipment did each team bring?
2. How did Antarctica's geography and climate defeat Scott?

722

Answers to Making the Connection

1. Scott brought 21 ponies and 3 motorized sleds; Amundsen brought trained sled dogs and specialized skis.
2. The motorized sleds bogged down in the soft snow; the weather was too cold for the ponies; Scott and his men did not mark their supply deposits well, and could not find them on the trip back because the icy wastes had no distinguishing features.

Important Things to Know About Oceania and Antarctica

SECTION 1 OCEANIA

- Oceania is made up of three regions—Melanesia, Micronesia, and Polynesia.
- Oceania has three kinds of islands: continental islands separated from continents, high islands formed by volcanoes, and low islands made from coral reefs.
- Most islands of Polynesia were trust territories at one time. Today many are independent.
- The cultures of these island groups are a mixture of local traditions, the cultures of European settlers, and modern influences.
- Farming, fishing, and tourism are the major economic activities of Oceania.
- The largest island region in Oceania by size is Polynesia.

SECTION 2 ANTARCTICA

- A huge ice cap covers most of Antarctica.
- The earth's southernmost point, the South Pole, lies near the center of Antarctica.
- Potential uses of Antarctica's resources include mining of minerals and transforming ice into freshwater that could be used by the rest of the world.
- Animals such as penguins, fish, whales, birds, and krill make their homes in the icy waters and on the coasts of Antarctica.
- Antarctica is the only continent with no permanent human settlement.
- Antarctica is a major center of international scientific research.

Port Ucla, Vanuatu ▶

Extra Credit Project

Catalog Suggest that interested students research clothing worn in Oceania and compile pictures and descriptions of pieces of the clothing in a catalog. Furnish actual clothing catalogs as models. Display the students' completed catalogs in the classroom. **L1**

Using the Chapter 27 Highlights

- Use the Chapter 27 Highlights to preview, review, condense, or reteach the chapter.

Preview/Review

Vocabulary Puzzle-Maker Software reinforces the terms used in Chapter 27.

Condense

Have students read the Chapter 27 Highlights. Spanish Chapter Highlights also are available.

Chapter 27 Digest Audiocassettes and Activity

Chapter 27 Digest Audiocassette Test

Spanish Chapter Digest Audiocassettes, Activities, and Tests also are available.

Guided Reading Activities

Reteach

Reteaching Activity 27. Spanish Reteaching activities also are available.

MAP STUDY

Location Copy and distribute outline maps of Oceania. Have students use broken lines to divide the Pacific into Melanesia, Micronesia, and Polynesia. Tell students to label the three island groups and then locate and label the following: Tropic of Capricorn, Papua New Guinea, the Solomon Islands, Fiji, and Tahiti.

Chapter 27 Review and Activities

REVIEWING KEY TERMS

Match the numbered terms in Column A with their definitions in Column B.

A

1. krill
2. atoll
3. crevasse
4. high island
5. copra
6. low island
7. continental island
8. phosphate

B

A. large island thrust up from the ocean
B. large break in the Antarctic ice
C. mineral used to make fertilizer
D. shrimplike animal found in Antarctic waters
E. island formed from coral reefs
F. coral reef surrounding a bay
G. island made from volcanoes
H. dried coconut meat

ANSWERS

Reviewing Key Terms

1. D
2. F
3. B
4. G
5. H
6. E
7. A
8. C

Mental Mapping Activity

On a separate sheet of paper, draw a freehand map of Antarctica. Label the following:
- Atlantic Ocean
- Pacific Ocean
- Antarctic Peninsula
- South Pole
- Antarctic Circle

Mental Mapping Activity

This exercise helps students to visualize the countries and physical features they have been studying and understand the relationships among them. All attempts at freehand mapping should be accepted.

REVIEWING THE MAIN IDEAS

Section 1

1. **LOCATION** Where is Oceania located?
2. **REGION** What is the major climate found in Melanesia, Micronesia, and Polynesia?
3. **PLACE** What mineral is mined in the island republic of Nauru?
4. **MOVEMENT** Why do people in Polynesia leave their islands?

Section 2

5. **LOCATION** What lies under the ice cap of Antarctica?
6. **PLACE** What is summer like in Antarctica?
7. **PLACE** What kinds of plants grow in Antarctica?
8. **REGION** Why is Antarctica considered a good place for scientific research?

Reviewing the Main Ideas

Section 1

1. in the Pacific Ocean northeast of Australia
2. tropical
3. phosphate
4. to find jobs elsewhere

Section 2

5. highlands, lowlands, and valleys—the same landforms that can be found on other continents
6. Summer, which lasts from December to February, is very cold. Temperatures rarely rise above 32°F (0°C) and may fall as low as −30°F (−34.5°C).

CRITICAL THINKING ACTIVITIES

1. **Drawing Conclusions** Why is tourism a growing industry in Oceania?
2. **Predicting Consequences** What might have happened to Antarctica if the Antarctic treaties had not been signed?

GeoJournal Writing Activity

Imagine that you are a scientist working in Antarctica. Write a letter home describing your work and the work of other scientists at your research station.

COOPERATIVE LEARNING ACTIVITY

Work in a group of three to learn more about one of the island countries of Oceania. Each group will choose one country from Melanesia, Micronesia, or Polynesia. Then each group member will select one topic to research for a group essay entitled: "One Day of My Life in _____." The essay should describe what a person's daily activities in that location might be like. Put your group's information together and write your essay. Add illustrations if possible. Share your essays with the rest of the class.

PLACE LOCATION ACTIVITY: OCEANIA AND ANTARCTICA

Match the letters on the map with the places and features of Oceania and Antarctica. Write your answers on a separate sheet of paper.

1. Port Moresby
2. Solomon Islands
3. French Polynesia
4. South Pole
5. Fiji
6. Federated States of Micronesia
7. Antarctic Peninsula
8. Pacific Ocean

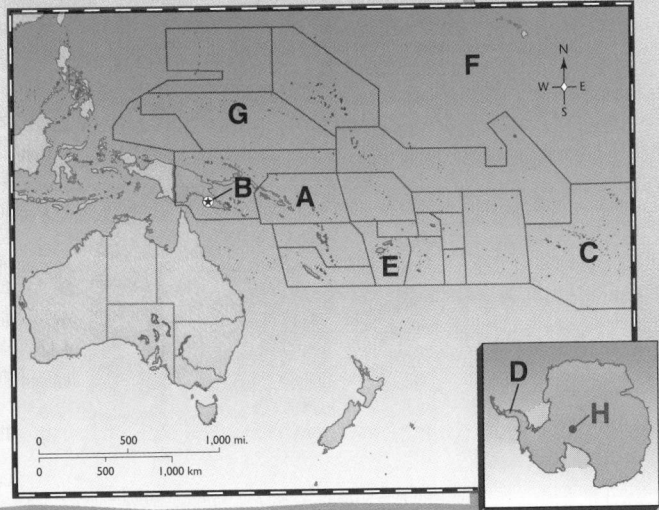

Cooperative Learning Activity

Display the essays under three headings—*Melanesia, Micronesia,* and *Polynesia.* Have artistic students create a tropical backdrop for the display, or find photographs of tropical landscapes to intersperse among the reports.

GeoJournal Writing Activity

Encourage students to exchange and answer one another's letters.

Place Location Activity

1. B
2. A
3. C
4. H
5. E
6. G
7. D
8. F

Chapter Bonus Test Question

This question may be used for extra credit on the chapter test.

You join the crew of a ship sailing from Tasmania to Tahiti. In which direction will you travel? *(northeast)*

7. small plants, such as algae and mosses
8. Its largely unspoiled environment has made Antarctica a favorable place for scientific research.

Critical Thinking Activities

1. Answers will vary, but should mention the islands' natural beauty, tropical climate, and friendly people.
2. Answers will vary. Most will focus on the overdevelopment of the continent, with the side effects of depletion of natural resources and pollution.

NATIONAL
GEOGRAPHIC
SOCIETY

These materials are available from Glencoe.

📕 **POSTERS**

**Eye on the Environment,
Unit 9**

FOCUS

Ask students to identify large structures built by humans—the Great Wall of China, the pyramids in Egypt, and skyscrapers in the United States, for example. List their responses on the chalkboard. Then ask students to identify the single largest structure ever built. Inform them that this structure is the Great Barrier Reef, which is made of countless billions of coral. Compare the size of the reef to the largest of the structures listed on the chalkboard, emphasizing the reef as an example of the awesome power that exists in the natural world.

TEACH

Display Unit Overlay Transparency 9–0 to help students locate the Great Barrier Reef. Next, write the following on the chalkboard: Australia, Continental Shelf, Lagoon, Great Barrier Reef, Coral Sea. Then ask students to construct a cross section of the Great Barrier Reef, using the items in the list as labels. Call on volunteers to display and discuss their cross sections. **L1**

NATIONAL GEOGRAPHIC SOCIETY
EYE ON THE ENVIRONMENT

Great Barrier Reef
TROUBLE DOWN UNDER

PROBLEM

Australia's Great Barrier Reef, the world's largest coral reef, is threatened by human activity. Pollution and commercial fishing endanger the reef. The possibility of a major oil spill—hundreds of oil tankers pass near the reef each year—looms as a constant worry. But tourism poses the most serious threat to the reef's health. And every year more than *2 million* people visit the reef!

SOLUTIONS

● To protect the reef and its treasures, the government of Australia works with environmental organizations around the world.

● New laws restrict where and how people may fish, dive, sail, snorkel, or build near the reef.

● Oil tankers, other large ships, and ships carrying hazardous cargo must carry a specially trained pilot to help navigate the reef.

● Specific areas of the reef are set aside for certain tourist activities. Other areas are off-limits except to scientists.

A leopard moray eel pokes its head out of coral. The Great Barrier Reef supports fish, whales, turtles, and countless other sea creatures.

CORAL REEF FACT BANK

🐟 Coral reefs, formed by the accumulated skeletons of coral, are "fish factories" crucial to the world's food chain.

🐟 The Great Barrier Reef covers about 135,000 square miles (350,000 sq. km). It is 1,250 miles (2,011 km) long and contains 2,600 separate reefs and about 300 islands.

🐟 It is home to more than 1,500 species of fish, 400 types of coral, 4,000 species of mollusks, and 22 types of whales.

🐟 The Great Barrier Reef—the largest structure built by living creatures anywhere in the world—is clearly visible from space.

726

UNIT 9

More About the Issues

Human/Environment Interaction The Great Barrier Reef has been under construction for perhaps as much as 30 million years. In contrast, the serious damage from human activity has taken just a few decades. The Australian government has taken a number of steps to arrest this dam-

age. In 1975, for example, the government declared the reef a national park. The park authority's plan of setting aside specific areas for specific uses has become the model for other countries who are attempting to preserve their coral reefs.

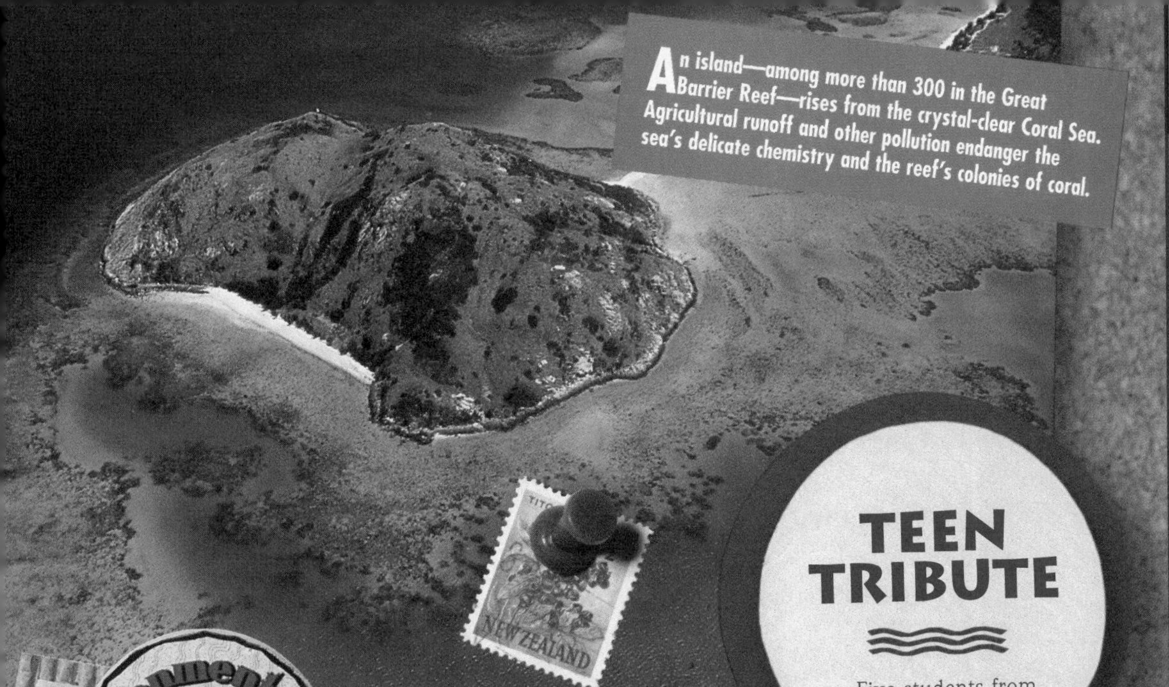

An island—among more than 300 in the Great Barrier Reef—rises from the crystal-clear Coral Sea. Agricultural runoff and other pollution endanger the sea's delicate chemistry and the reef's colonies of coral.

These materials are available from Glencoe.

⊙ **VIDEODISC**

STV: World Geography Vol. 1: Asia and Australia

Coral; Great Barrier Reef, Australia
Any side, Frame 50432

TEEN TRIBUTE

Five students from Pioneer State High School, in Mackay, Australia, set up a self-guided snorkeling tour of the Great Barrier Reef that keeps tourists off the coral. The students studied an area and then laid out an underwater trail. By carefully choosing the path and placing information markers on the trail, the students help protect the reef—and educate sightseers.

WHAT CAN YOU DO?

🌿 Write to oil companies, encouraging them to be watchful when shipping oil.

🌿 Leave nature the way you find it whenever hiking or camping in state and national parks.

🌿 Find out about a natural area in your community or state that people are trying to protect. Join in!

🌿 Support the International Green Cross—an environmental protection group organized at the recent Earth Summit.

TAKE A CLOSER LOOK

1 Where is the Great Barrier Reef located?

2 What human activities have threatened the reef?

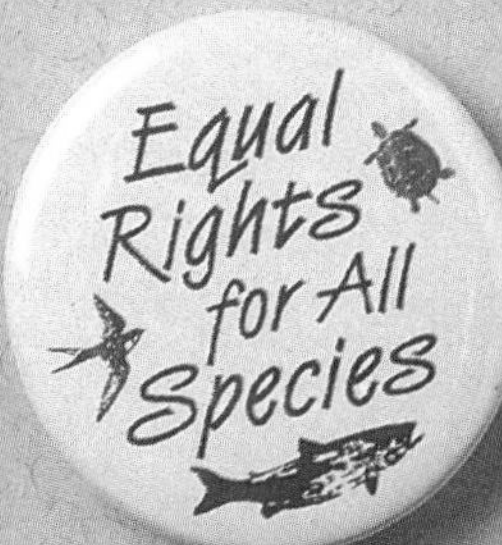

Equal Rights for All Species

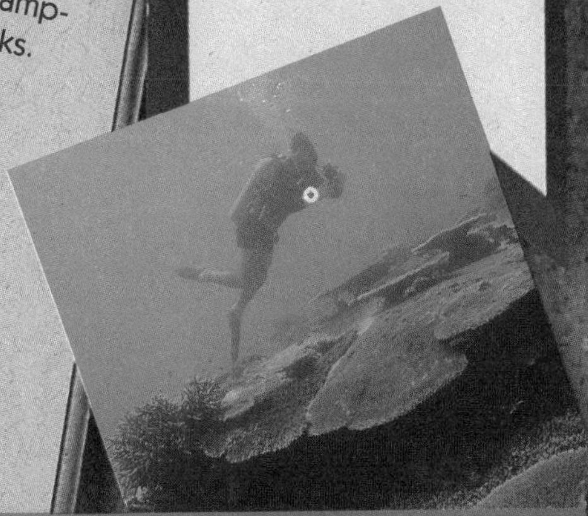

727

A diver explores the reef, eyeing table coral and other wonders in this enchanted undersea garden that stretches for more than a thousand miles.

ASSESS

Have students answer the **Take A Closer Look** questions on page 727.

CLOSE

Discuss with students the **What Can You Do?** question. Then have students debate whether tourism should be banned in unique and fragile environments such as the Great Barrier Reef.

*For an additional regional case study, use the following:

📁 Environmental Case Study 9

Answers to Take A Closer Look
1. in the Coral Sea off the northeast coast of Australia
2. tourism, commercial fishing, pollution

Global Issues

Tourism Worldwide, there are more than 60,000 miles (96,558 km) of coral reefs lining the coastlines of some 109 nations. Many reefs are now threatened with destruction brought about by human invasion.

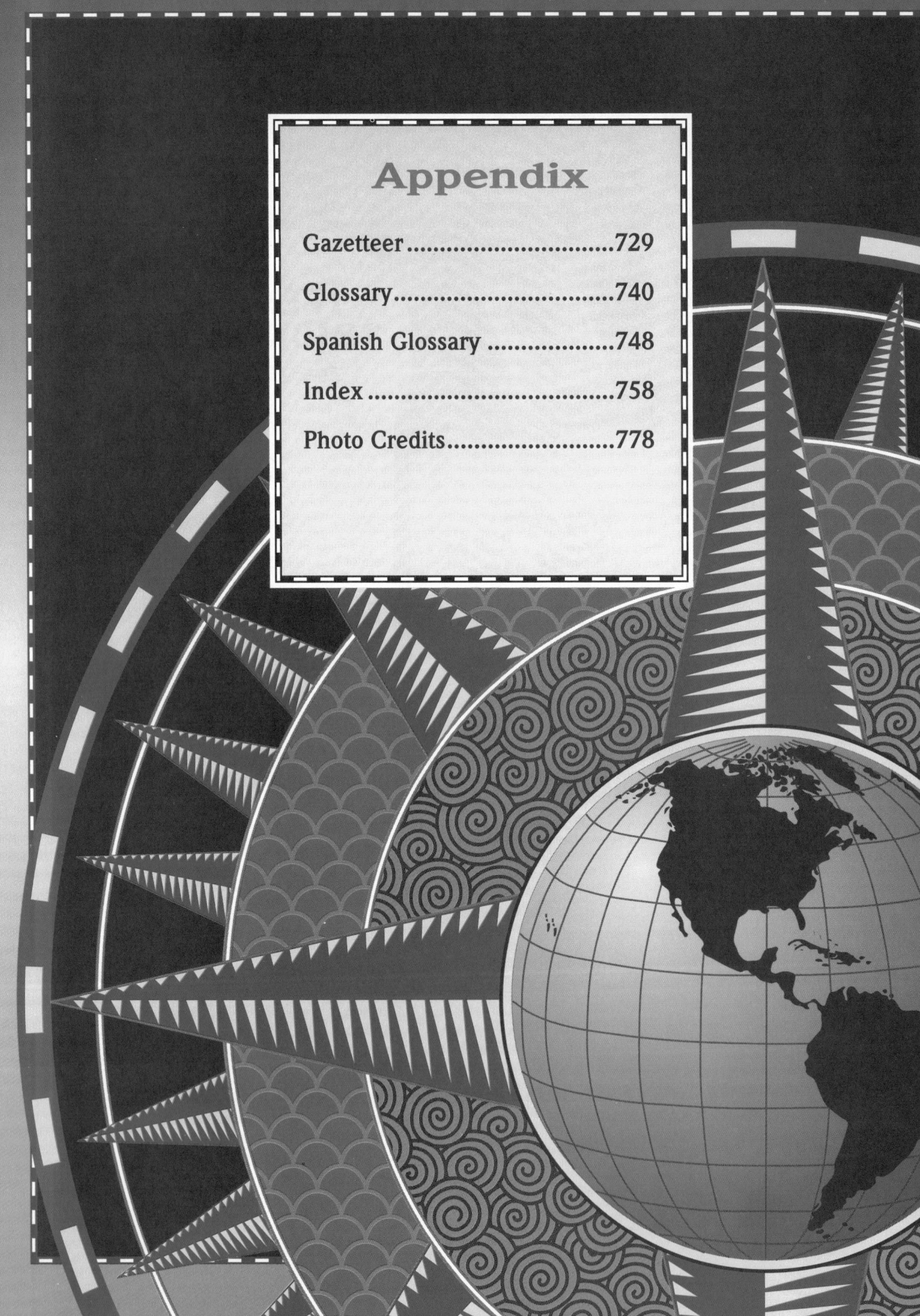

Appendix

Gazetteer

A Gazetteer (GAZ•uh•TIR) is a geographic index or dictionary. It shows latitude and longitude for cities and certain other places. Latitude and longitude are shown in this way: 48°N 2°E, or 48 degrees north latitude and two degrees east longitude. This Gazetteer lists most of the world's largest independent countries, their capitals, and several important geographic features. The page numbers tell where each entry can be found on a map in this book. As an aid to pronunciation, most entries are spelled phonetically.

A

Abidjan [A•bih•JAHN] Capital and port of Côte d'Ivoire, Africa. 5°N 4°W (p. 476)

Abu Dhabi [AH•boo DAH•bee] Capital of the United Arab Emirates, on the Persian Gulf. 24°N 54°E (p. 414)

Abuja [ah•BOO•jah] Capital of Nigeria. 8°N 9°E (p. 476)

Accra [uh•KRAH] Port city and capital of Ghana. 6°N 0° longitude (p. 476)

Addis Ababa [A•duhs A•buh•buh] Capital of Ethiopia. 9°N 39°E (p. 476)

Adriatic [AY•dree•A•tihk] **Sea** Arm of the Mediterranean Sea between the Balkan Peninsula and Italy. 44°N 14°E (p. 246)

Afghanistan [af•GA•nuh•STAN] Country in central Asia, west of Pakistan. 33°N 63°E (p. 414)

Albania [al•BAY•nee•uh] Country on the east coast of the Adriatic Sea, south of Serbia and Montenegro. 42°N 20°E (p. 248)

Algeria [al•JIHR•ee•uh] Country in North Africa. 29°N 1°E (p. 414)

Algiers [al•JIHRZ] Capital of Algeria. 37°N 3°E (p. 414)

Almaty [al•MAH•tee] Capital of Kazakstan. 43°N 77°E (p. 354)

Alps [ALPS] Mountain system extending east to west through central Europe. 46°N 9°E (p. 246)

Amazon [A•muh•ZAHN] **River** Largest river in the world by volume and second-largest in length. 2°S 53°W (p. 144)

Amman [ah•MAHN] Capital of Jordan. 32°N 36°E (p. 414)

Amsterdam [AMP•stuhr•DAM] Capital of the Netherlands. 52°N 5°E (p. 248)

Andes [AN•DEEZ] Mountain system extending north to south along the western side of South America. 13°S 75°W (p. 144)

Andorra [an•DAWR•uh] Small country in southern Europe, between France and Spain. 43°N 2°E (p. 248)

Angola [ang•GOH•luh] Country in Africa, south of Zaire. 14°S 16°E (p. 476)

Ankara [ANG•kuh•ruh] Capital of Turkey. 40°N 33°E (p. 414)

Antananarivo [AN•tuh•NA•nuh•REE•VOH] Capital of Madagascar. 19°S 48°E (p. 476)

Arabian [uh•RAY•bee•uhn] **Peninsula** Large peninsula of southwest Asia, extending north to south into the Arabian Sea. 28°N 40°E (p. 412)

Argentina [AHR•juhn•TEE•nuh] Country in the southern part of South America, east of Chile on the Atlantic Ocean. 36°S 67°W (p. 146)

Armenia [ahr•MEE•nee•uh] Southeastern European country between the Black Sea and the Caspian Sea. 40°N 45°E (p. 354)

Ashkhabad [ASH•kuh•BAD] Capital of Turkmenistan. 38°N 58°E (p. 354)

Asmara [az•MAHR•uh] Capital of Eritrea. 16°N 39°E (p. 476)

Asunción [uh•SOONT•see•OHN] Capital of Paraguay. 25°S 58°W (p. 146)

Athens [A•thuhnz] Capital and largest city in Greece. 38°N 24°E (p. 248)

Australia [aw•STRAYL•yuh] Country and continent southeast of Asia. 23°S 135°E (p. 682)

Austria [AWS•tree•uh] Country in central Europe, east of Switzerland and south of Germany and the Czech Republic. 47°N 12°E (p. 248)

Azerbaijan [A•zuhr•BY•JAHN] European-Asian country on the Caspian Sea. 40°N 47°E (p. 354)

B

Baghdad [BAG•DAD] Capital of Iraq. 33°N 44°E (p. 414)

Bahama [buh•HAH•muh] **Islands** Country made up of many islands, between Cuba and the United States. 23°N 74°W (p. 145)

Baku [bah•KOO] Port city and capital of Azerbaijan. 40°N 50°E (p. 354)

Balkan [BAWL•kuhn] **Peninsula** Peninsula in southeastern Europe, mostly covered by the country of Greece. 42°N 20°E (p. 246)

Baltic [BAWL•tihk] **Sea** Sea in northern Europe that is an arm of the Atlantic Ocean connected to the North Sea. 55°N 17°E (p. 246)

Bamako [BAH•muh•KOH] Capital of Mali. 13°N 8°W (p. 476)

Bangkok [BANG•KAHK] Capital of Thailand. 14°N 100°E (p. 580)

Bangladesh [BAHNG•gluh•DEHSH] Country in South Asia, bordered by India and Myanmar. 24°N 90°E (p. 580)

Bangui [BAHNG•GEE] Capital of the Central African Republic. 4°N 19°E (p. 476)

Banjul [BAHN•JOOL] Port city and capital of Gambia. 13°N 17°W (p. 476)

Barbados [bahr•BAY•DOHS] Island country between the Atlantic Ocean and the Caribbean Sea. 14°N 59°W (p. 146)

Beijing [BAY•JIHNG] Cultural and industrial center and capital of China. 40°N 116°E (p. 580)

Beirut [bay•ROOT] Capital of Lebanon. 34°N 36°E (p. 414)

Belarus [BEE•luh•ROOS] Eastern European country west of Russia and east of Poland. 54°N 28°E (p. 354)

Belgium [BEHL•juhm] Country in northwestern Europe, south of the Netherlands. 51°N 3°E (p. 248)

Belgrade [BEHL•GRAYD] Capital of Yugoslavia (Serbia). 45°N 21°E (p. 248)

Belize [buh•LEEZ] Country in Central America, east of Guatemala. 18°N 89°W (p. 146)

Belmopan [BEHL•moh•PAN] Capital of Belize. 17°N 89°W (p. 146)

Benin [buh•NIHN] Country in western Africa. 8°N 2°E (p. 476)

Berlin [BUHR•LIHN] Capital of Germany. 53°N 13°E (p. 248)

Bern [BUHRN] Capital of Switzerland. 47°N 7°E (p. 248)

Bhutan [boo•TAHN] Country in the eastern Himalayas, northeast of India. 27°N 91°E (p. 580)

Bishkek [bihsh•KEHK] Largest city and capital of Kyrgyzstan. 43°N 75°E (p. 354)

Bissau [bih•SOW] Capital of Guinea-Bissau. 12°N 16°W (p. 476)

Black Sea Large sea between Europe and Asia. 43°N 32°E (p. 352)

Bogotá [BOH•guh•TAH] Capital of Colombia. 5°N 74°W (p. 146)

Bolivia [buh•LIH•vee•uh] Country in the central part of South America, north of Argentina. 17°S 64°W (p. 146)

Bosnia-Herzegovina [BAHZ•nee•uh HEHRT•suh•goh•VEE•nuh] Southeastern European country between Yugoslavia and Croatia. 44°N 18°E (p. 248)

Botswana [baht•SWAH•nuh] Country in Africa, north of the Republic of South Africa. 22°S 23°E (p. 476)

Brasília [bruh•ZIHL•yuh] Capital of Brazil. 16°S 48°W (p. 146)

Bratislava [BRA•tuh•SLAH•vuh] Capital and largest city of Slovakia. 48°N 17°E (p. 248)

Brazil [bruh•ZIHL] Largest country in South America. 9°S 53°W (p. 146)

Brazzaville [BRA•zuh•VIHL] Port city and capital of the Congo Republic in west central Africa. 4°S 15°E (p. 476)

Brunei [bru•NY) Country in Southwest Asia, on northern coast of the island of Borneo. 5°N 114°E (p. 578)

Brussels [BRUH•suhlz] Capital of Belgium. 51°N 4°E (p. 248)

Bucharest [BOO•kuh•REHST] Capital of Romania. 44°N 26°E (p. 248)

Budapest [BOO•duh•PEHST] Capital of Hungary. 48°N 19°E (p. 248)

Buenos Aires [BWAY•nuhs AR•eez] Capital of Argentina, on the Río de la Plata. 34°S 58°W (p. 146)

Bujumbura [BOO•juhm•BUR•uh] Capital of Burundi in eastern Africa near Lake Tanganyika. 3°S 29°E (p. 476)

Bulgaria [BUHL•GAR•ee•uh] Country in southeastern Europe, south of Romania. 42°N 24°E (p. 248)

Burkina Faso [bur•KEE•nuh FAH•soh] Country in western Africa, south of Mali. 12°N 3°E (p. 476)

Burundi [bu•ROON•dee] Country in central Africa at the northern end of Lake Tanganyika. 3°S 30°E (p. 476)

730

C

Cairo [KY•ROH] Capital of Egypt, on the Nile River. 31°N 32°E (p. 414)

Cambodia [kam•BOH•dee•uh] Country in Southeast Asia, south of Thailand. 12°N 104°E (p. 580)

Cameroon [KA•muh•ROON] Country in west Africa, on the northeast shore of the Gulf of Guinea. 6°N 11°E (p. 476)

Canada [KA•nuh•duh] Northernmost country in North America. 50°N 100°W (p. 84)

Canberra [KAN•buh•ruh] Capital of Australia. 35°S 149°E (p. 684)

Cape Town Port city and legislative capital of the Republic of South Africa. 34°S 18°E (p. 544)

Caracas [kuh•RAH•kuhs] Capital of Venezuela. 11°N 67°W (p. 146)

Caribbean [KAR•uh•BEE•uhn] **Sea** Part of the Atlantic Ocean, bordered by the West Indies, South America, and Central America. 15°N 76°W (p. 144)

Caspian [KAS•pee•uhn] **Sea** Salt lake between Europe and Asia that is the world's largest inland body of water. 40°N 52°E (p. 352)

Caucasus [KAW•kuh•suhs] **Mountains** Mountain range in the southern part of the former Soviet Union. 43°N 42°E (p. 352)

Central African Republic Country in central Africa, south of Chad. 8°N 21°E (p. 476)

Chad [CHAD] Country in central Africa, west of Sudan. 18°N 19°E (p. 476)

Chang [CHANG] **River** Principal river of China that rises in Tibet and flows into the East China Sea near Shanghai. 31°N 117°E (p. 578)

Chile [CHIH•lee] South American country, west of Argentina. 35°S 72°W (p. 146)

China [CHY•nuh] Country in eastern and central Asia, known officially as the People's Republic of China. 37°N 93°E (p. 580)

Chisinau [KEE•shih•NOW] Largest city and capital of Moldova. 47°N 29°E (p. 354)

Colombia [kuh•LUHM•bee•uh] Country in South America, west of Venezuela. 4°N 73°W (p. 146)

Colombo [kuh•LUHM•BOH] Capital of Sri Lanka. 7°N 80°E (p. 580)

Comoros [KAH•muhr•ROHS] **Islands** Small island country in Indian Ocean between the island of Madagascar and the southeast African mainland. 13°S 43°E (p. 476)

Conakry [KAH•nuh•kree] Capital of Guinea. 10°N 14°W (p. 476)

Congo [KAHNG•GOH] **Republic** Country in equatorial Africa, south of the Central African Republic. 3°S 14°E (p. 476)

Copenhagen [KOH•puhn•HAY•guhn] Seaport, largest city, and capital of Denmark. 56°N 12°E (p. 248)

Costa Rica [KAHS•tuh REE•kuh] Central American country, south of Nicaragua. 11°N 85°W (p. 146)

Côte d'Ivoire [KOHT dee•VWAHR] West African country, south of Mali. 8°N 7°W (p. 476)

Croatia [kroh•AY•shuh] Southeastern European country on the Adriatic Sea. 46°N 16°E (p. 248)

Cuba [KYOO•buh] Island in the West Indies. 22°N 79°W (p. 146)

Cyprus [SY•pruhs] Island country in the eastern Mediterranean Sea, south of Turkey. 35°N 31°E (p. 248)

Czech [CHEHK] **Republic** Central European country south of Germany and Poland. 50°N 15°E (p. 248)

D

Dakar [DA•KAHR] Port city and capital of Senegal. 15°N 17°W (p. 476)

Damascus [duh•MAS•kuhs] Capital of Syria. 34°N 36°E (p. 414)

Dar es Salaam [DAHR•ehs suh•LAHM] Capital of Tanzania. 7°S 39°E (p. 520)

Denmark [DEHN•MAHRK] Country in northwestern Europe, between the Baltic and North Seas. 56°N 9°E (p. 248)

Dhaka [DA•kuh] Capital of Bangladesh. 24°N 90°E (p. 580)

Djibouti [juh•BOO•tee] Country in east Africa, on the Gulf of Aden. 12°N 43°E (p. 476)

Dodoma [DOH•doh•mah] Future capital of Tanzania. 7°S 36°E (p. 476)

Doha [DOH•HAH] Capital of Qatar. 25°N 51°E (p. 414)

Dominican [duh•MIH•nih•kuhn] **Republic** Country in the West Indies on the eastern part of Hispaniola Island. 19°N 71°W (p. 174)

GAZETTEER

Dublin [DUH•bluhn] Port city and capital of Ireland. 53°N 6°W (p. 248)

Dushanbe [doo•SHAM•buh] Largest city and capital of Tajikistan north of Pakistan. 39°N 69°E (p. 354)

E

Ecuador [EH•kwuh•DAWR] Country in South America, south of Colombia. 0° latitude 79°W (p. 146)

Egypt [EE•juhpt] Country in northern Africa on the Mediterranean Sea. 27°N 27°E (p. 414)

El Salvador [el SAL•vuh•DAWR] Country in Central America, southwest of Honduras. 14°N 89°W (p. 146)

Equatorial Guinea [EE•kwuh•TOHR•ee•uhl GIH•nee] Country in western Africa, south of Cameroon. 2°N 8°E (p. 476)

Eritrea [EHR•uh•TREE•uh] Country in eastern Africa on the Red Sea, north of Ethiopia. 17°N 39°E (p. 476)

Estonia [eh•STOH•nee•uh] Northern European country on the Baltic Sea, north of Latvia. 59°N 25°E (p. 248)

Ethiopia [EE•thee•OH•pee•uh] Country in eastern Africa, north of Somalia and Kenya. 8°N 38°E (p. 476)

Euphrates [yu•FRAY•TEEZ] **River** River in southwestern Asia that flows through Syria and Iraq and joins the Tigris River. 36°N 40°E (p. 412)

F

Fiji [FEE•JEE] Country comprised of an island group in the southwest Pacific Ocean. 19°S 175°E (p. 684)

Finland [FIHN•luhnd] Country in northern Europe, east of Sweden. 63°N 26°E (p. 248)

France [FRANTS] Country in western Europe, south of the English Channel. 47°N 1°E (p. 248)

Freetown [FREE•TOWN] Port city and capital of Sierra Leone, in western Africa. 9°N 13°W (p. 476)

French Guiana [gee•AH•nah] French-owned territory in northern South America. 5°N 53°W (p. 146)

G

Gabon [ga•BOHN] Country in western Africa on the Atlantic Ocean, west of Congo. 0° latitude 12°E (p. 476)

Gaborone [GAH•buh•ROH•NAY] Capital of Botswana, in southern Africa. 24°S 26°E (p. 476)

Gambia [GAM•bee•uh] Country in western Africa, along the Gambia River. 13°N 16°W (p. 476)

Georgetown [JAWRJ•TOWN] Capital of Guyana. 8°N 58°W (p. 146)

Georgia [JAWR•juh] Asian-European country bordering the Black Sea, south of Russia. 42°N 43°E (p. 354)

Germany [JUHR•muh•nee] Country in north central Europe, officially called the Federal Republic of Germany. 52°N 10°E (p. 248)

Ghana [GAH•nuh] Country in western Africa on the Gulf of Guinea, east of Côte d'Ivoire. 8°N 2°W (p. 476)

Great Plains The continental slope extending through the United States and Canada. 45°N 104°W (p. 82)

Greece [GREES] Country in southern Europe, mostly on the Balkan Peninsula. 39°N 22°E (p. 246)

Greenland [GREEN•luhnd] Island in northwestern Atlantic Ocean and the largest island in the world. 74°N 40°W (p. 84)

Guatemala [GWAH•tuh•MAH•luh] Country in Central America, south of Mexico. 16°N 92°W (p. 146)

Guatemala City Capital of Guatemala and the largest city in Central America. 15°N 91°W (p. 146)

Guinea [GIH•nee] West African country on the Atlantic coast north of Sierra Leone. 11°N 12°W (p. 476)

Guinea-Bissau [GIH•nee bih•SOW] West African country on the Atlantic coast. 12°N 20°W (p. 476)

Gulf of Mexico Gulf on the southeast coast of North America. 25°N 94°W (p. 82)

Guyana [gy•A•nuh] South American country on the Atlantic coast, between Venezuela and Suriname. 8°N 59°W (p. 146)

GEOGRAPHY: The World and Its People

H

Haiti [HAY•tee] Country on Hispaniola Island in the West Indies. 19°N 72°W (p. 174)

Hanoi [ha•NOY] Capital of Vietnam. 21°N 106°E (p. 580)

Harare [huh•RAH•RAY] Capital of Zimbabwe. 18°S 31°E (p. 476)

Havana [huh•VA•nuh] Seaport, capital city of Cuba, and largest city of the West Indies. 23°N 82°W (p. 146)

Helsinki [HEHL•SIHNG•kee] Capital of Finland. 60°N 24°E (p. 248)

Himalayas [HI•muh•LAY•uhs] Mountain range in South Asia, bordering the Indian subcontinent on the north. 30°N 85°E (p. 578)

Honduras [hahn•DUR•uhs] Central American country, on the Caribbean Sea. 15°N 88°W (p. 146)

Hong Kong [HAHNG KAHNG] Administrative district and port in southern China. 22°N 115°E (p. 580)

Huang He [HWAHNG•HE] River in north central and eastern China, also known as the Yellow River. 35°N 114°E (p. 578)

Hungary [HUHNG•guh•ree] Central European country, south of Slovakia. 47°N 18°E (p. 248)

I

Iberian [y•BIHR•ee•uhn] **Peninsula** Peninsula in southwest Europe, occupied by Spain and Portugal. 41°N 1°W (p. 246)

Iceland [YS•luhnd] Island country between the North Atlantic and the Arctic oceans. 65°N 20°W (p. 248)

India [IHN•dee•uh] South Asian country, south of China and Nepal. 23°N 78°E (p. 580)

Indonesia [IHN•duh•NEE•zhuh] Group of islands that forms the Southeast Asian country known as the Republic of Indonesia. 5°S 119°E (p. 580)

Indus [IHN•duhs] **River** River in Asia that rises in Tibet and flows through Pakistan to the Arabian Sea. 27°N 68°E (p. 578)

Iran [ih•RAHN] Southwest Asian country that was formerly named Persia, east of Iraq. 31°N 54°E (p. 414)

Iraq [ih•RAHK] Southwest Asian country, south of Turkey. 32°N 43°E (p. 414)

Ireland [YR•luhnd] Island west of England, occupied by the Republic of Ireland and by Northern Ireland. 54°N 8°W (p. 248)

Islamabad [ihs•LAH•muh•BAHD] Capital of Pakistan. 34°N 73°E (p. 580)

Israel [IHZ•ree•uhl] Country in southwest Asia, south of Lebanon. 33°N 34°E (p. 414)

Italy [IH•tuhl•ee] Southern European country, south of Switzerland and east of France. 44°N 11°E (p. 248)

J

Jakarta [juh•KAHR•tuh] Capital of Indonesia. 6°S 107°E (p. 580)

Jamaica [juh•MAY•kuh] Island country in the West Indies. 18°N 78°W (p. 146)

Japan [juh•PAN] Country in east Asia, consisting of the four large islands of Hokkaido, Honshu, Shikoku, and Kyushu, plus thousands of small islands. 37°N 134°E (p. 580)

Jerusalem [juh•ROO•suh•luhm] Capital of Israel and a holy city for Christians, Jews, and Muslims. 32°N 35°E (p. 414)

Jordan [JAWR•duhn] Country in southwest Asia, south of Syria. 30°N 38°E (p. 414)

K

Kabul [KAH•buhl] Capital of Afghanistan. 35°N 69°E (p. 414)

Kampala [kahm•PAH•luh] Capital of Uganda. 0° latitude 32°E (p. 476)

Kathmandu [KAT•MAN•DOO] Capital of Nepal. 28°N 85°E (p. 580)

Kazakstan [KA•ZAK•STAN] Large Asian country south of Russia, bordering the Caspian Sea. 48°N 59°E (p. 354)

Kenya [KEH•nyuh] Country in eastern Africa, south of Ethiopia. 1°N 37°E (p. 476)

Khartoum [kahr•TOOM] Capital of Sudan. 16°N 33°E (p. 476)

Kiev [KEE•ehf] Capital of Ukraine. 50°N 31°E (p. 354)

GAZETTEER

Kigali [kih•GAH•lee] Capital of Rwanda, in central Africa. 2°S 30°E (p. 476)

Kingston [KIHNG•stuhn] Capital of Jamaica. 18°N 77°W (p. 146)

Kinshasa [kihn•SHAH•suh] Capital of Zaire. 4°S 15°E (p. 476)

Kuala Lumpur [KWAH•luh LUM•PUR] Capital of Malaysia. 3°N 102°E (p. 580)

Kuwait [ku•WAYT] Country between Saudi Arabia and Iraq, on the Persian Gulf. 29°N 48°E (p. 414)

Kyrgyzstan [KIHR•gih•STAN] Small central Asian country on China's western border. 41°N 75°E (p. 354)

L

Lagos [LAY•GAHS] Port city of Nigeria. 6°N 3°E (p. 495)

Laos [LOWS] Southeast Asian country, south of China and west of Vietnam. 20°N 102°E (p. 580)

La Paz [luh PAHZ] The administrative capital of Bolivia, and the highest capital in the world. 17°S 68°W (p. 146)

Latvia [LAT•vee•uh] Northeastern European country on the Baltic Sea, west of Russia. 57°N 25°E (p. 248)

Lebanon [LEH•buh•nuhn] Country on the Mediterranean Sea, south of Syria. 34°N 34°E (p. 414)

Lesotho [luh•SOH•TOH] Country in south Africa, within the borders of the Republic of South Africa. 30°S 28°E (p. 476)

Liberia [ly•BIHR•ee•uh] West African country, south of Guinea. 7°N 10°W (p. 476)

Libreville [LEE•bruh•VIHL] Port city and capital of Gabon. 1°N 9°E (p. 476)

Libya [LIH•bee•uh] North African country on the Mediterranean Sea, west of Egypt. 28°N 15°E (p. 414)

Liechtenstein [LIHK•tuhn•SHTYN] Small country in central Europe, between Switzerland and Austria. 47°N 10°E (p. 248)

Lilongwe [lih•LAWNG•way] Capital of Malawi, in southeastern Africa. 14°S 34°E (p. 476)

Lima [LEE•muh] Capital of Peru. 12°S 77°W (p. 146)

Lisbon [LIHZ•buhn] Port city and capital of Portugal. 39°N 9°W (p. 248)

Lithuania [LIH•thuh•WAY•nee•uh] Northeastern European country on the Baltic Sea, west of Belarus. 56°N 24°E (p. 248)

Ljubljana [lee•OO•blee•AH•nuh] Largest city and capital of Slovenia. 46°N 14°E (p. 248)

Lomé [loh•MAY] Port city and capital of Togo in Africa. 6°N 1°E (p. 476)

London [LUHN•duhn] Capital of the United Kingdom, on the Thames River. 52°N 0° longitude (p. 248)

Luanda [lu•AN•duh] Port city and capital of Angola. 9°S 13°E (p. 476)

Lusaka [loo•SAH•kuh] Capital of Zambia. 15°S 28°E (p. 476)

Luxembourg [LUHK•suhm•BUHRG] Small European country between France, Belgium, and Germany. 50°N 7°E (p. 248)

M

Macao [muh•KOW] Portuguese-ruled territory in southern China. 22°N 113°E (p. 580)

Macedonia [MA•suh•DOH•nee•uh] Southeastern European country north of Greece. 42°N 22°E (p. 248). Macedonia also refers to a geographic region covering northern Greece, the country Macedonia, and part of Bulgaria.

Madagascar [MA•duh•GAS•kuhr] Island in the Indian Ocean off the southeast coast of Africa. 18°S 43°E (p. 476)

Madrid [muh•DRIHD] Capital of Spain. 40°N 4°W (p. 248)

Malabo [mah•LAH•BOH] Capital of Equatorial Guinea. 4°N 9°E (p. 476)

Malawi [muh•LAH•wee] Southeastern African country, south of Tanzania and east of Zambia. 11°S 34°E (p. 476)

Malaysia [muh•LAY•zhuh] Federation of states in Southeast Asia on the Malay Peninsula and the island of Borneo. 4°N 101°E (p. 578)

Maldive [MAWL•DEEV] **Islands** Island country in the Indian Ocean near South Asia. 5°N 42°E (p. 580)

Mali [MAH•lee] Country in western Africa, east of Mauritania and south of Algeria. 16°N 0° longitude (p. 476)

Managua [muh•NAH•gwuh] Capital of Nicaragua. 12°N 86°W (p. 146)

Manila [muh•NIH•luh] Port city and capital of the Republic of the Philippines. 15°N 121°E (p. 580)

Maseru [MA•suh•ROO] Capital of Lesotho, in southern Africa. 29°S 27°E (p. 476)

Mauritania [MAWR•uh•TAY•nee•uh] Western African country, north of Senegal. 20°N 14°W (p. 476)

Mauritius [maw•RIH•shuhs] Small island country in the Indian Ocean east of Madagascar. 21°S 58°E (p. 476)

Mbabane [EHM•bah•BAH•nay] Capital of Swaziland, in southeastern Africa. 26°S 31°E (p. 476)

Mediterranean [MEH•duh•tuh•RAY•nee•uhn] **Sea** Large inland sea surrounded by Europe, Asia, and Africa. 36°N 13°E (p. 246)

Mekong [MAY•KAWNG] **River** River in Southeast Asia that rises in Tibet and empties into the South China Sea. 18°N 104°E (p. 578)

Mexico [MEHK•sih•KOH] Country in North America, south of the United States. 24°N 104°W (p. 146)

Mexico City Capital and most populous city of Mexico. 19°N 99°W (p. 146)

Minsk [MIHNTSK] Capital of Belarus. 54°N 28°E (p. 354)

Mississippi [MIH•suh•SIH•pee] **River** Large river system in the central United States that flows southward into the Gulf of Mexico. 32°N 92°W (p. 82)

Mogadishu [MAH•guh•DIH•SHOO] Major seaport and capital of Somalia, in eastern Africa. 2°N 45°E (p. 476)

Moldova [mahl•DOH•vuh] Small European country between Ukraine and Romania. 48°N 28°E (p. 354)

Monaco [MAH•nuh•KOH] Small country in southern Europe, on the French Mediterranean coast. 44°N 8°E (p. 248)

Mongolia [mahn•GOH•lee•uh] Country in Asia between Russia and China 46°N 100°E (p. 580)

Monrovia [MUHN•ROH•vee•uh] Major seaport and capital of Liberia, in western Africa. 6°N 11°W (p. 476)

Montevideo [MAHN•tuh•vuh•DAY•OH] Seaport and capital of Uruguay in South America. 35°S 56°W (p. 146)

Morocco (muh•RAH•KOH) Country in northwestern Africa on the Mediterranean Sea and the Atlantic Ocean. 32°N 7°W (p. 414)

Moscow (MAHS•KOH) Capital and largest city of Russia. 56°N 38°E (p. 354)

Mount Everest (EHV•ruhst) Highest mountain in the world, in the Himalayas between Nepal and Tibet. 28°N 87°E (p. 578)

Mozambique (MOH•zuhm•BEEK) Country in southeastern Africa, south of Tanzania. 20°S 34°E (p. 476)

Muscat (MUHS•KAT) Seaport and capital of Oman. 23°N 59°E (p. 414)

Myanmar (MYAHN•MAHR) Country in Southeast Asia, south of China and India, formerly called Burma. 21°N 95°E (p. 580)

N

Nairobi [ny•ROH•bee] Capital of Kenya. 1°S 37°E (p. 476)

Namibia [nuh•MIH•bee•uh] Country in southwestern Africa, south of Angola on the Atlantic Ocean. 20°S 16°E (p. 476)

Nassau [NA•SAW] Capital of the Bahamas. 25°N 77°W (p. 146)

N'Djamena [EN•juh•MAY•nuh] Capital of Chad. 12°N 15°E (p. 476)

Nepal [nuh•PAWL] Mountain country between India and China. 29°N 83°E (p. 580)

Netherlands [NEH•thuhr•lundz] Western European country on the North Sea, north of Belgium. 53°N 4°E (p. 248)

New Delhi [NOO DEH•lee] Capital of India. 29°N 77°E (p. 580)

New Zealand [NOO ZEE•luhnd] Major island country in the South Pacific, southeast of Australia. 42°S 175°E (p. 684)

Niamey [nee•AH•may] Capital and commercial center of Niger, in western Africa. 14°N 2°E (p. 476)

Nicaragua [NI•kuh•RAH•gwuh] Country in Central America, south of Honduras. 13°N 86°W (p. 146)

Nicosia [NIH•kuh•SEE•uh] Capital of Cyprus. 35°N 33°E (p. 248)

Niger [NY•juhr] Landlocked country in western Africa, north of Nigeria. 18°N 9°E (p. 476)

Nigeria [ny•JIHR•ee•uh] Country in western Africa, south of Niger. 9°N 7°E (p. 476)

Nile [NYL] **River** Longest river in the world, flowing north and east through eastern Africa. 19°N 33°E (p. 412)

North Korea [kuh•REE•uh] Asian country in the northernmost part of the Korean Peninsula. 40°N 127°E (p. 580)

Norway [NAWR•WAY] Country on the Scandinavian Peninsula. 64°N 11°E (p. 248)

Nouakchott [noo•AHK•SHAHT] Capital of Mauritania. 18°N 16°W (p. 476)

O

Oman [oh•MAHN] Country on the Arabian Sea and the Gulf of Oman, east of Saudi Arabia. 20°N 58°E (p. 414)

Oslo [AHZ•loh] Capital and largest city of Norway. 60°N 11°E (p. 248)

Ottawa [AH•tuh•wuh] Capital of Canada. 45°N 76°W (p. 84)

Ouagadougou [WAH•guh•DOO•goo] Capital of Burkina Faso, in western Africa. 12°N 2°W (p. 476)

P

Pakistan [PA•kih•STAN] South Asian country on the Arabian Sea, northwest of India. 28°N 68°E (p. 580)

Palau [pah•LOW] Island country in the Pacific Ocean. 7°N 135°E (p. 684)

Panama [PA•nuh•MAH] Country in Central America, on the Isthmus of Panama. 9°N 81°W (p. 146)

Panamá Capital city of Panama. 9°N 79°W (p. 146)

Papua New Guinea [PA•pyuh•wuh NOO GIH•nee] Independent island country in the Pacific Ocean north of Australia. 7°S 142°E (p. 684)

Paraguay [PAR•uh•GWY] Country in South America, north of Argentina. 24°S 57°W (p. 146)

Paramaribo [PAIR•uh•MAIR•uh•BOH] Port city and capital of Suriname. 6°N 55°W (p. 146)

Paris [PAIR•uhs] River port and capital of France. 49°N 2°E (p. 248)

Persian [PUHR•zhuhn] **Gulf** Arm of the Arabian Sea between Iran and Saudi Arabia. 28°N 51°E (p. 412)

Peru [puh•ROO] Country in South America, south of Ecuador and Colombia. 10°S 75°W (p. 146)

Philippines [FIH•luh•PEENZ] Country in the Pacific Ocean, southeast of Asia. 14°N 125°E (p. 580)

Phnom Penh [NAWM•PEN] Capital of Cambodia. 12°N 106°E (p. 580)

Poland [POH•luhnd] Country on the Baltic Sea in eastern Europe. 52°N 18°E (p. 248)

Port-au-Prince [POHRT•oh•PRINTS] Port city and capital of Haiti. 19°N 72°W (p. 146)

Port Moresby [MOHRZ•bee] Port city and capital of Papua New Guinea. 10°S 147°E (p. 684)

Port of Spain [SPAYN] Capital of Trinidad and Tobago in the West Indies. 11°N 62°W (p. 146)

Porto-Novo [POHR•tuh•NOH•voh] Port city and capital of Benin, in western Africa. 7°N 3°E (p. 476)

Portugal [POHR•chih•guhl] Country on the Iberian Peninsula, south and west of Spain. 39°N 8°W (p. 246)

Prague [PRAHG] Capital of the Czech Republic. 51°N 15°E (p. 248)

Pretoria [prih•TOHR•ee•uh] Capital of South Africa. 26°S 28°E (p. 476)

Puerto Rico [PWEHR•tuh•REE•koh] Island in the Caribbean Sea; U.S. Commonwealth. 19°N 67°W (p. 146)

Pyongyang [pee•AWNG•YAHNG] Capital of North Korea. 39°N 126°E (p. 580)

Q

Qatar [KAH•tuhr] Country on the southwestern shore of the Persian Gulf. 25°N 53°E (p. 414)

Quito [KEE•toh] Capital of Ecuador. 0° latitude 79°W (p. 146)

R

Rabat [ruh•BAHT] Capital of Morocco. 34°N 7°W (p. 414)

Reykjavík [RAY•kyuh•VIK] Capital of Iceland. 64°N 22°W (p. 248)

Rhine [RYN[**River** River in western Europe that flows into the North Sea. 51°N 7°E (p. 246)

Riga [REE•guh] Capital and largest city of Latvia. 57°N 24°E (p. 248)

Rio Grande [REE•OH•GRAND] River that forms the boundary between the United States and Mexico. 30°N 103°W (p. 82)

Riyadh [ree•YAHD] Capital of Saudi Arabia. 25°N 47°E (p. 414)

Rocky Mountains Mountain system in western North America. 50°N 114°W (p. 82)

Romania [roo•MAY•nee•uh] Country in eastern Europe, east of Hungary. 46°N 23°E (p. 248)

Rome [ROHM] Capital of Italy. 42°N 13°E (p. 248)

Russia [RUH•shuh] Largest country in the world, covering parts of Europe and Asia. 60°N 90°E (p. 354)

Rwanda [roo•AHN•duh] Country in Africa, south of Uganda. 2°S 30°E (p. 476)

S

Sahara [suh•HAIR•uh] Desert region in northern Africa that is the largest desert in the world. 24°N 2°W (p. 412)

Saint Lawrence [LAWR•uhns] **River** River that flows from Lake Ontario to the Atlantic Ocean and forms part of the boundary between the United States and Canada. 48°N 70°W (p. 82)

San'a [sa•NAH] Capital of Yemen. 15°N 44°E (p. 414)

San José [SAN•uh•ZAY] Capital of Costa Rica. 10°N 84°W (p. 146)

San Marino [SAN•mah•REE•noh] Small European country, located in the Italian Peninsula. 44°N 13°E (p. 248)

San Salvador [san•SAL•vuh•DAWR] Capital and industrial center of El Salvador. 14°N 89°W (p. 146)

Santiago [SAN•tee•AH•goh] Capital and major industrial center of Chile. 33°S 71°W (p. 146)

Santo Domingo [SAN•tuh•duh•MIN•goh] Capital of the Dominican Republic. 19°N 70°W (p. 146)

São Tomé and Príncipe [SOWN•tuh•MAY PRIN•suh•puh] Small island country in Gulf of Guinea off the coast of Central Africa. 1°N 7°E (p. 474)

Sarajevo [SAR•uh•YAY•voh] Capital of Bosnia-Herzegovina. 43°N 18°E (p. 248)

Saudi Arabia [SOW•dee uh•RAY•bee•uh] Country on the Arabian Peninsula. 23°N 46°E (p. 412)

Senegal [SEH•nih•GAWL] Country on the coast of western Africa, on the Atlantic Ocean. 15°N 14°W (p. 476)

Seoul [SOHL] Capital of South Korea. 38°N 127°E (p. 580)

Serbia [SUHR•bee•uh] European country south of Hungary. 44°N 21°E (p. 248)

Seychelles [say•SHELZ] Small island country in the Indian Ocean near East Africa. 6°S 56°E (p. 476)

Sierra Leone [see•EHR•uh•lee•OHN] Country in western Africa, south of Guinea. 8°N 12°W (p. 476)

Singapore [SIHNG•uh•POHR] Multi-island country in Southeast Asia near tip of Malay Peninsula. 2°N 104°E (p. 578)

Skopje [SKAW•PYAY] Capital of the country of Macedonia, in southeastern Europe. 42°N 21°E (p. 248)

Slovakia [sloh•VAH•kee•uh] Central European country south of Poland. 49°N 19°E (p. 248)

Slovenia [sloh•VEE•nee•uh] Small central European country on the Adriatic Sea, south of Austria. 46°N 15°E (p. 248)

Sofia [SOH•fee•uh] Capital of Bulgaria. 43°N 23°E (p. 248)

Solomon [SAW•lah•mahn] **Islands** Island country in the Pacific Ocean, northeast of Australia. 7°S 160°E (p. 684)

Somalia [soh•MAH•lee•uh] Country in Africa, on the Gulf of Aden and the Indian Ocean. 3°N 45°E (p. 476)

South Africa [A•frih•kuh] Country at the southern tip of Africa. 28°S 25°E (p. 476)

South Korea [kuh•REE•uh] Country in Asia on the Korean Peninsula between the Yellow Sea and the Sea of Japan. 36°N 128°E (p. 580)

Spain [SPAYN] Country on the Iberian Peninsula. 40°N 4°W (p. 246)

Sri Lanka [sree•LAHNG•kuh] Country in the Indian Ocean south of India, formerly called Ceylon. 9°N 83°E (p. 580)

Stockholm [STAHK•HOHLM] Capital of Sweden. 59°N 18°E (p. 248)

Sucre [SOO•kray] Constitutional capital of Bolivia. 19°S 65°W (p. 146)

Sudan [soo•DAN] Northeast African country on the Red Sea. 14°N 28°E (p. 476)

Suriname [SUHR•uh•NAH•muh] South American country on the Atlantic Ocean between Guyana and French Guiana. 4°N 56°W (p. 146)

Suva [SOO•vuh] Port city and capital of Fiji. 18°S 177°E (p. 684)

Swaziland [SWAH•zee•LAND] South African country west of Mozambique, almost entirely within the Republic of South Africa. 27°S 32°E (p. 476)

Sweden [SWEE•duhn] Northern European country on the eastern side of the Scandinavian Peninsula. 60°N 14°E (p. 246)

Switzerland [SWIT•suhr•luhnd] European country in the Alps, south of Germany. 47°N 8°E (p. 246)

Syria [SIHR•ee•uh] Country in Asia on the east side of the Mediterranean Sea. 35°N 37°E (p. 414)

T

Taipei [TY•PAY] Capital of Taiwan. 25°N 122°E (p. 580)

Taiwan [TY•WAHN] Island country off the southeast coast of China, and the seat of the Chinese Nationalist government. 24°N 122°E (p. 580)

Tajikistan [tah•JIH•kih•STAN] Central Asian country east of Turkmenistan. 39°N 70°E (p. 354)

Tallinn [TA•luhn] Largest city and capital of Estonia. 59°N 25°E (p. 248)

Tanzania [TAN•zuh•NEE•uh] East African country on the coast of the Indian Ocean, south of Uganda and Kenya. 7°S 34°E (p. 476)

Tashkent [tash•KENT] Capital of Uzbekistan and a major industrial center. 41°N 69°E (p. 354)

Tbilisi [tuh•BEE•luh•see] Capital of the Republic of Georgia. 42°N 45°E (p. 354)

Tegucigalpa [tuh•GOO•suh•GAL•puh] Capital of Honduras. 14°N 87°W (p. 146)

Tehran [TAY•RAN] Capital of Iran. 36°N 52°E (p. 414)

Thailand [TY•LAND] Southeast Asian country south of Myanmar. 17°N 101°E (p. 580)

Thimphu [THIHM•boo] Capital of Bhutan. 28°N 90°E (p. 580)

Tigris [TY•gruhs] **River** River in southeast Turkey and Iraq that merges with the Euphrates River. 35°N 44°E (p. 412)

Tiranë [tih•RAH•nuh] Capital of Albania. 42°N 20°E (p. 248)

Togo [TOH•goh] West African country between Benin and Ghana, on the Gulf of Guinea. 8°N 1°E (p. 474)

Tokyo [TOH•kee•OH] Capital of Japan. 36°N 140°E (p. 580)

Trinidad and Tobago [TRIH•nih•DAD tuh•BAY•goh] Island country between the Atlantic Ocean and the Caribbean Sea, near Venezuela. 11°N 61°W (p. 146)

Tripoli [TRIH•puh•lee] Capital of Libya. 33°N 13°E (p. 414)

Tunis [TOO•nuhs] Port city and capital of Tunisia. 37°N 10°E (p. 414)

Tunisia [too•NEE•zhuh] North African country on the Mediterranean Sea between Libya and Algeria. 35°N 10°E (p. 414)

Turkey [TUHR•kee] Country in southeastern Europe and western Asia. 39°N 32°E (p. 414)

Turkmenistan [TUHRK•MEH•nuh•STAN] Central Asian country on the Caspian Sea. 41°N 56°E (p. 354)

U

Uganda [oo•GAN•duh] East African country south of Sudan. 2°N 32°E (p. 476)

Ukraine [yoo•KRAYN] Large eastern European country west of Russia, on the Black Sea. 49°N 30°E (p. 354)

Ulan Bator [OO•LAHN•BAH•TAWR] Capital of Mongolia. 48°N 107°E (p. 580)

United Arab Emirates [ih•MIR•uhts] Country made up of seven states on the eastern side of the Arabian Peninsula. 24°N 54°E (p. 414)

United Kingdom Country in western Europe made up of England, Scotland, Wales, and Northern Ireland. 57°N 2°W (p. 248)

United States of America Country in North America made up of 50 states, mostly between Canada and Mexico. 38°N 110°W (p. 84)

Uruguay [UR•uh•GWY] South American country, south of Brazil on the Atlantic Ocean. 33°S 56°W (p. 146)

Uzbekistan [OOZ•BEH•kih•STAN] Central Asian country south of Kazakstan, on the Caspian Sea. 42°N 60°E (p. 354)

V

Vanuatu [VAN•WAH•TOO] Country made up of islands in the Pacific Ocean, east of Australia. 17°S 170°W (p. 684)

Vatican [VA•tih•kuhn] **City** Headquarters of the Roman Catholic Church, located in the city of Rome in Italy. 42°N 13°E (p. 253)

Venezuela [VEH•nuh•ZWAY•luh] South American country on the Caribbean Sea, between Colombia and Guyana. 8°N 65°W (p. 146)

Vienna [vee•EH•nuh] Capital of Austria. 48°N 16°E (p. 248)

Vientiane [VYEHN•TYAHN] Capital of Laos. 18°N 103°E (p. 580)

Vietnam [vee•EHT•NAHM] Southeast Asian country, east of Laos and Cambodia. 18°N 107°E (p. 580)

Virgin Islands Island territory of the United States, east of Puerto Rico in the Caribbean Sea. 18°N 65°W (p. 146)

W

Warsaw [WAWR•SAW] Capital of Poland. 52°N 21°E (p. 248)

Washington, D.C. Capital of the United States, in the District of Columbia. 39°N 77°W (p. 84)

Wellington [WEH•lihng•tuhn] Capital of New Zealand. 41°S 175°E (p. 684)

West Indies [IHN•deez] Islands in the Caribbean Sea, between North America and South America. 19°N 79°W (p. 144)

Windhoek [VIHNT•HOOK] Capital of Namibia, in southwestern Africa. 22°S 17°E (p. 476)

Y

Yamoussoukro [YAH•muh•SOO•kroh] Second capital of Côte d'Ivoire, in western Africa. 7°N 6°W (p. 496)

Yangon [YAHN•GOHN] Capital of Myanmar. 17°N 96°E (p. 580)

Yaoundé [yown•DAY] Capital of Cameroon, in western Africa. 4°N 12°E (p. 476)

Yemen [YEH•muhn] Country on the Arabian Peninsula, south of Saudi Arabia on the Gulf of Aden and the Red Sea. 15°N 46°E (p. 414)

Yerevan [YEHR•uh•VAHN] Largest city and capital of Armenia. 40°N 44°E (p. 354)

Z

Zagreb [ZAH•GREHB] Largest city and capital of Croatia. 46°N 16°E (p. 248)

Zaire [zah•IHR] African country on the Equator, north of Zambia and Angola. 1°S 22°E (p. 476)

Zambia [ZAM•bee•uh] Country in south central Africa, south of Zaire and Tanzania. 14°S 24°E (p. 476)

Zimbabwe [zim•BAH•bwee] Country in south central Africa. 18°S 30°E (p. 476)

Glossary

A

abaca [a•buh•KAH] plant fiber used to make rope (p. 671)

absolute location the exact position of a place on the earth's surface (pp. 5, 21)

acid rain precipitation in which water carries large amounts of chemicals, especially sulfuric acid (pp. 68, 100, 283)

adobe [uh•DOH•bee] sun-dried clay brick (p. 166)

alluvial [uh•LOO•vee•uhl] **plain** plain built up from soil deposited by a river (pp. 435, 657)

altiplano [AL•tih•PLAH•NOH] high plateau region of the Andes (p. 222)

altitude [AL•tuh•TOOD] height above sea level (pp. 156, 202)

anthracite [AN•thruh•SYT] hard coal (p. 644)

apartheid [uh•PAHR•TAYT] South African legal restrictions and practices separating racial and ethnic groups, word means "apartness" (p. 549)

aquifer [AH•kwuh•fuhr] underground rock layer that stores large amounts of water (p. 39)

archipelago [AHR•kuh•PEH•luh•GOH] large group or chain of islands (pp. 183, 633)

atmosphere thick cushion of gases surrounding the earth, made up mainly of the gases nitrogen and oxygen (p. 32)

atoll [A•TAWL] ring-shaped coral reef or island surrounding a small bay (pp. 604, 713)

autobahn [AW•toh•BAHN] a superhighway in Germany (p. 284)

autonomy [aw•TAH•nuh•mee] self-government (p. 532)

axis (plural, **axes**) the horizontal (bottom) or vertical (side) line of measurement on a graph (p. 13)

B

bar graph graph in which vertical or horizontal bars represent quantities (p. 13)

basin broad flat lowland area surrounded by higher land (pp. 155, 195, 511)

bauxite [BAWK•SYT] mineral ore from which aluminum is taken (pp. 178, 497)

bazaar a marketplace (p. 451)

Bedouin [BEH•duh•wuhn] member of the nomadic desert peoples of North Africa and Southwest Asia (p. 429)

bilingual having or speaking two languages (p. 129)

birthrate number of children born each year for every 1,000 people (p. 62)

bog low-lying marshy land (pp. 261, 323)

buffer state small country located between larger, often hostile, states (p. 208)

bush Australian term for remote rural areas (p. 694)

C

cacao [kuh•KOW] tropical tree whose seeds are used to make cocoa and chocolate (p. 487)

calligraphy [kuh•LIH•gruh•fee] the art of beautiful handwriting (p. 622)

canopy topmost layer of a rain forest, which shades the forest floor (p. 505)

campesino [KAM•puh•SEE•noh] farmer in Latin America (p. 220)

cardinal directions the basic directions on the earth: north, south, east, west (p. 7)

casbah [KAZ•BAH] old area of cities with narrow streets and small shops in North Africa (p. 458)

cash crop crop grown to be sold, often for export (p. 220)

cassava [kuh•SAH•vuh] plant whose roots are ground into flour and eaten; in tapioca pudding, for example (pp. 210, 495, 663)

caudillo [kaw•DEE•yoh] in Latin American history, a military dictator (p. 203)

chart graphic way of presenting information clearly (p. 17)

chicle juice of the sapodilla tree, used in making chewing gum (p. 177)

circle graph round or pie-shaped graph showing how a whole is divided (p. 14)

city-state in history, an independent city and the lands around it (p. 306)

civil war conflict involving different groups within one country (pp. 430, 534)

civilization highly developed culture, usually with organized religion and laws (p. 56)

clan group of people or families related to one another (pp. 538, 637)

climate usual pattern of weather events in an area over a long period of time (pp. 15, 41)

climate region a broad area of earth with the same climate (p. 12)

climograph combination bar and line graph giving information about temperature and precipitation (p. 15)

cloves spice made from flower buds of clove trees (p. 529)

cold war era from late 1940s to early 1990s in which the United States and Soviet Union competed for world influence (p. 371)

colony overseas settlement made by a parent country (pp. 103, 187)

command economy economic system in which the government owns land, resources, and means of production and makes all economic decisions (p. 364)

commonwealth term for a republic or state governed by the people; the official title of Puerto Rico (p. 189)

communism authoritarian political system in which the central government controls the economy and society (pp. 187, 282, 319, 364)

compass rose device drawn on maps to show the directions (p. 7)

compound in rural Africa, a group of houses surrounded by a wall (p. 488)

condensation process in which water vapor changes into a liquid form (p. 38)

constitutional monarchy form of government headed by a monarch but with most power given to an elected legislature (pp. 259, 430)

consumer goods products for personal use, such as clothing and household goods (pp. 336, 366, 449, 615)

contiguous referring to areas that touch or share a boundary (p. 89)

continent one of the major land areas of the earth (p. 30)

continental island land originally part of a continent, separated by rising water or by geological activity (p. 712)

contour line on a contour map, line connecting all points at the same elevation (p. 9)

cooperative in Cuba, a farm owned and operated by the government (p. 188)

copper belt in Zambia, region of rich copper mines (p. 557)

copra dried inner meat of coconuts (p. 712)

coral rocklike material formed of the skeletons of small sea animals (pp. 521, 690)

coral reef low-lying ocean ridge made of coral (p. 93)

cordillera [KAW•duhl•YEHR•uh] group of mountain chains that run side by side (pp. 118, 217)

core central layer of the earth, probably composed of hot iron and nickel, solid in the inner core and molten in the outer (p. 29)

cottage industry home- or village-based industry in which family members supply their own equipment to make such goods as cloth and metalware (p. 590)

crevasse [krih•VAS] deep crack in an ice cap or glacier (p. 718)

crust outer layer of the earth (p. 29)

cultural diffusion process by which knowledge and skills spread from one area to another (p. 56)

culture way of life of a group of people who share similar beliefs and customs, including language and religion (p. 55)

currents moving streams of water, warm or cold, in the oceans; also streams of air (p. 42)

cyclone intense storm system with heavy rain and high winds blowing in a circular pattern (p. 596)

czar [ZAHR] title of the former emperors of Russia (p. 370)

D

death rate number of deaths each year for every 1,000 people (p. 62)

deforestation loss of forests due to widespread cutting of trees (p. 658)

delta triangle-shaped area at a river's mouth, formed of mud and sand deposited by water (pp. 448, 596, 660)

demographer scientist who studies population (p. 62)

desertification process in which grasslands become drier and desert areas expand (p. 491)

developed country country that is industrialized rather than agricultural (p. 62)

developing country country in the process of becoming industrialized (p. 62)

diagram drawing that shows steps in a process or parts of an object (p. 16)

dialect local form of a language (pp. 188, 284, 300)

dictatorship government under the control of a single all-powerful leader (p. 455)

dike high bank of soil or concrete built to hold back the water of a river or ocean (p. 611)

drought [DROWT] extended period of extreme dryness and water shortages (pp. 490, 537)

dry farming method of wheat farming in dry areas that conserves moisture by deep plowing (p. 98)

dynasty a family of rulers (p. 619)

dzong Buddhist center for prayer and study (p. 601)

E

earthquake violent jolting or shaking of the earth caused by movement of rocks along a fault (p. 30)

elevation height above sea level (pp. 9, 48, 310)

elevation profile cutaway diagram showing changes in elevation of land (p. 9)

emigrate to move from one's native country to another country (pp. 60, 312)

empire group of lands or states under the rule of one nation or ruler (pp. 223, 623)

enclave small nation or distinct region located inside a larger country (p. 545)

environment natural surroundings (p. 23)

equinox day in March and September when the sun's rays are directly overhead at the Equator, making day and night of equal length (p. 27)

erg desert region of shifting sand dunes (p. 458)

erosion wearing away of the earth's surface, mainly by water, wind, or ice (p. 31)

escarpment steep cliff separating two fairly flat land surfaces, one higher than the other (pp. 196, 522, 546)

estancia [ehs•TAHN•syah] a large ranch in Argentina (p. 230)

ethnic group people who share a common cultural background, including ancestry and language (pp. 104, 371, 438, 487)

evaporation process in which water from oceans, lakes, and streams is turned into a gas or vapor by the sun's heat (p. 38)

exclave small part of a nation separated from the main part of the country (p. 552)

F

famine lack of food, affecting a large number of people (p. 63)

farm belt region of the midwestern United States with flat land and fertile soil (p. 98)

fault crack in the rocks of the earth's crust along which movement occurs (pp. 30, 391, 522)

favela [fuh•VEH•luh] in Brazil, term used for urban slum areas (p. 199)

fellahin [fehl•uh•HEEN] Egyptian farmers and farm workers (p. 450)

fjord [FYOHRD] long, narrow, steep-sided inlet from the sea (pp. 266, 697)

flow chart type of diagram that shows direction or movement (p. 16)

food processing industry in which foods are prepared and packaged for sale (p. 393)

fossil fuels group of nonrenewable mineral resources—coal, oil, natural gas—formed in the earth's crust from plant and animal remains (pp. 66, 124)

free port area where goods can be loaded, stored, and reshipped without payment of import-export taxes (p. 666)

free enterprise basis of a market economy in which people start and run businesses to make a profit with little government intervention (p. 58)

free enterprise system economic system based on free enterprise (see above) (pp. 96, 365)

G

galaxy huge system in the universe including millions of stars (p. 25)

gaucho in southern South America, a cowhand on horseback (pp. 208, 230)

geography the study of the earth and its land, water, and plant and animal life (p. 21)

geothermal power electric power produced by natural underground sources of steam (p. 699)

geyser hot spring that spouts steam and hot water through a crack in the earth (p. 269)

glasnost [GLAZ•nohst] Russian term for "openness," used for President Gorbachev's policy in late 1980s (p. 371)

great circle route ship or airplane route following a great circle, the shortest distance between two points on the earth (p. 6)

grid system network of imaginary lines on the earth's surface, formed by the crisscrossing patterns of the lines of latitude and longitude (pp. 5, 22)

groundwater water stored in rock layers beneath the earth's surface (p. 39)

H

hajj religious journey made by Muslims to Makkah (p. 432)

harmattan [HAHR•muh•TAN] dry wind that blows southward from the Sahara in winter (p. 486)

heavy industry production of industrial goods such as machinery and military equipment (p. 365)

hemisphere one-half of the globe; the Equator divides the earth into Northern and Southern hemispheres; the Prime Meridian divides it into Eastern and Western hemispheres (pp. 4, 22)

hieroglyphs ancient form of Egyptian writing using pictographs (p. 451)

high island in Micronesia, a mountainous volcanic island (p. 713)

high-technology industry factory that produces computers or other electronic equipment (p. 624)

high veld flat, grassy plateau region of southern Africa (p. 546)

Holocaust [HOH•luh•KAWST] term for the mass imprisonment and slaughter of European Jews and others by German Nazis during World War II (pp. 284, 426)

humid continental climate climate in continental interiors, with cold winters and short, hot summers (p. 47)

humid subtropical climate warm, mild, rainy mid-latitude climate (p. 47)

hurricane fierce tropical storm system with high-speed winds, formed over warm oceans (pp. 43, 177)

hydroelectricity/hydroelectric power electric power generated by falling water (pp. 68, 202, 449, 507)

I

ice shelf part of an ice cap that extends from the land over the ocean (p. 718)

immigrant person who moves from one place to make a permanent home in another (p. 102)

industrialized describing a country in which industry has replaced farming as the main economic activity (p. 161)

intensive cultivation farming method using all available land, producing several crops a year (p. 635)

interdependent referring to countries or people that rely on one another (p. 100)

intermediate direction any direction between the cardinal directions, such as southeast or northwest (p. 7)

invest to put money into a company in return for a share of its profits (pp. 329, 615)

isthmus [IHS•muhs] narrow piece of land connecting two larger pieces of land (pp. 32, 181)

J-K

jute [JOOT] plant fiber used in making rope and burlap bags (p. 589)

key on a map, an explanation of the symbols used (p. 7)

krill small shrimplike shellfish, source of food for many sea animals (p. 721)

L

ladino in Guatemala, a person who speaks Spanish and follows a Spanish-American, not Native American, lifestyle (p. 180)

lagoon pool of water surrounded by reefs or sandbars (p. 604)

land bridge narrow strip of land that joins two larger landmasses (p. 153)

landlocked describing a country that has no land on a sea or ocean (p. 207)

language family group of languages that comes from a common ancestor (p. 57)

latitude location north or south of the Equator, measured by imaginary lines (parallels) numbered in degrees north or south (pp. 5, 22, 155)

GLOSSARY

leap year year with 366 days to account for the extra one-fourth day in Earth's revolution around the sun (p. 26)

light industry production of consumer goods, such as food products and household goods (pp. 188, 365)

line graph graph in which one or more lines connect dots representing changing quantities (p. 14)

literacy rate percentage of adults in a society who can read and write (pp. 58, 181)

llanos [LAH•nohs] large, grassy plains regions of Latin America (pp. 202, 218)

loch [LAHK] long, narrow bay with mountains on either side; also, a lake (p. 256)

loess [LEHS] fine-grained fertile soil deposited by the wind (pp. 283, 611)

longitude location east or west of the Prime Meridian, measured by imaginary lines (meridians) numbered in degrees east or west (pp. 5, 22)

low island in Micronesia, a low-lying island formed of coral (p. 713)

M

magma [MAG•muh] melted rock within the earth's mantle (p. 30)

mainland the main landmass of a country, as contrasted with nearby islands (p. 309)

mangrove tropical tree that grows in swampy land, it has roots both above and below the water (p. 485)

mantle middle layer of the earth, composed of thick hot rock (p. 29)

manuka shrub typical of volcanic soil in New Zealand (p. 697)

maquiladoras [mah•KEE•luh•DOH•rahz] factories that assemble parts shipped in from other countries, producing automobiles, stereos, etc. (p. 161)

marine west coast climate coastal climate with mild winters and cool summers (p. 47)

marsupial type of mammal that carries its young in a pouch as the infants mature (p. 691)

Mediterranean climate mild mid-latitude coastal climate with rainy winters and hot, dry summers (p. 47)

megalopolis [MEH•guh•LAH•puh•luhs] area in which neighboring urban areas blend into one "super city" (pp. 91, 638)

mestizo [meh•STEE•zoh] person of mixed Native American and European ancestry (p. 165)

migrate to move to another place (p. 421)

mobile moving from place to place (p. 105)

monotheism belief in one God (p. 425)

monsoon seasonal winds that bring rain to parts of Asia (pp. 43, 589)

moor in the United Kingdom, a treeless, windy highland area (p. 256)

mosque [MAHSK] place of worship for followers of Islam (pp. 338, 421)

movement one of the geographic themes, describing how people from different places interact (p. 23)

multicultural referring to something that includes many different cultures (p. 104)

multinational firm a company that has offices in several countries and does business in several countries (p. 289)

N

national park public lands set aside for recreation and wilderness protection (p. 106)

natural resource anything from the natural environment that people use to meet their needs (p. 65)

nature preserve land that is set aside by a government to protect plants and wildlife (pp. 332, 387)

navigable [NA•vih•guh•buhl] describing a body of water wide and deep enough for ships to pass (pp. 222, 278)

neutrality policy of refusing to take sides in international disputes and wars (p. 290)

newsprint type of paper used for printing newspapers (p. 124)

nomads people who move from place to place, often with herds of sheep or other animals (p. 396)

nonrenewable resources metals and other minerals that cannot be replaced once they are used up (p. 66)

nutrients minerals in soil that supply growing plants with food (p. 177)

O

oasis (plural, **oases**) place with water and green vegetation surrounded by desert (pp. 398, 432)

oil shale layered rock that contains oil (p. 320)

orbit elliptical path that a planet follows in revolving around the sun (p. 25)

outback remote inland regions of Australia (p. 690)

ozone form of oxygen (O_3) in the atmosphere (pp. 72, 721)

P

pagoda many-storied tower built as a temple or shrine (p. 622)

parliamentary democracy form of government in which voters elect representatives to a law-making body (parliament), which then chooses a leader, the prime minister (pp. 128, 259)

peat partly decayed plant matter, found in bogs, used for fuel and fertilizer (p. 261)

peninsula piece of land surrounded by water on three sides (pp. 32, 153)

permafrost permanently frozen lower layers of soil in Arctic regions (pp. 47, 362)

pesticide chemicals used to kill insects and other pests (p. 67)

phosphate mineral salt containing phosphorus, used in fertilizers (pp. 424, 714)

pictograph graph in which small symbols represent quantities (p. 15)

place one of the geographic themes, describing the typical characteristics that distinguish one place from another (p. 22)

plantation large farm on which a single cash crop is raised (pp. 161, 177)

plate tectonics theory in geology that the earth's crust is made up of huge, moving plates of rock (p. 30)

plateau flat landform whose surface is raised above the surrounding land, with a steep cliff on one side (pp. 32, 297)

poacher person who hunts and kills animals illegally (p. 523)

polder [POHL•duhr] in the Netherlands, an area of land reclaimed from the sea (p. 289)

pollution putting impure or poisonous substances into land, water, or air (p. 67)

pope title of the head of the Roman Catholic Church (p. 326)

population density average number of people living in a square mile or square kilometer (pp. 12, 61)

potash [PAHT•ASH] mineral salt, often used in fertilizers (p. 424)

prairie rolling, inland grassland area with fertile soil (p. 118)

precipitation water that falls to the earth as rain, snow, or sleet (p. 38)

Prime Meridian Line of 0° longitude from which east and west locations are measured, runs through Greenwich, United Kingdom (p. 22)

prime minister government leader chosen in a parliamentary democracy by members of parliament (p. 128)

projection in mapmaking, a way of drawing the round Earth on a flat surface (p. 2)

Q

quinoa [KEEN•WAH] cereal grain grown in Bolivia (p. 226)

R

rain forest dense forest in tropical regions with heavy, year-round rainfall (p. 45)

rain shadow area on the inland side of mountain ranges, where little rain falls (p. 43)

reef narrow ridge of coral, rock, or sand near the surface of water (p. 522)

refugee person who has had to flee to another country for safety from disaster or danger (pp. 60, 534)

region one of the geographic themes, defining parts of the earth that share common characteristics (p. 23)

relative location the position of a place on the earth's surface in relation to another place (p. 22)

relief differences in height in a landscape; how flat or rugged the surface is (p. 9)

renewable resource resource that is replaced naturally or can be grown quickly (p. 66)

GLOSSARY

republic government without a monarch, in which people elect important officials (pp. 103, 280)

revolution for a planet, one complete trip around the sun (p. 26); in politics, a sudden radical change in government (p. 103)

rookery nesting place for large numbers of birds or marine animals (p. 718)

rural relating to the countryside, not the city (p. 91)

S

samurai class of warrior-nobles in feudal Japan (p. 637)

savanna tropical grassland with scattered trees, usually with wet and dry seasons (pp. 46, 485)

scale the relationship between distance on a map and actual distance on the earth (p. 8)

scale bar on a map, a divided line showing the map scale, usually in feet, miles, or kilometers (p. 8)

selva thick tropical forests of the Amazon basin in Brazil (p. 195)

serfs farm laborers bound to the land they worked (p. 370)

service industry business that provides services to people rather than producing goods (pp. 67, 97, 159, 310, 333)

shah title held by kings of Iran, formerly Persia (p. 437)

shogun military ruler or dictator of feudal Japan (p. 637)

silt particles of soil carried and deposited by running water (p. 448)

sirocco [shuh•RAH•koh] hot, dry wind that blows across southern Europe from North Africa (p. 304)

sisal [SY•suhl] plant fiber used to make rope (pp. 197, 529)

slash-and-burn farming farming techniques in which areas of forest are cleared for farmland by burning (p. 562)

smog fog mixed with smoke and chemicals (p. 162)

socialism economic system in which government sets economic goals and may own some businesses (p. 58)

sodium nitrate mineral used in making fertilizer and explosives (p. 227)

solar energy power produced by the heat of the sun (p. 68)

solar system group of planets and other bodies that revolve around the sun (p. 25)

solstice day in June and December when the sun is directly overhead at the Tropic of Cancer (23 1/2° N) or Tropic of Capricorn (23 1/2° S), marking the beginning of summer or winter (p. 26)

sorghum tall grass used as grain and to make syrup (p. 557)

standard of living measure of quality of life based on income and material possessions (p. 58)

station Australian term for a sheep or cattle ranch (p. 690)

steppe dry treeless grasslands, often found at the edges of deserts (pp. 48, 384)

strait narrow body of water lying between two pieces of land (pp. 32, 419)

subcontinent large landmass that is part of another continent but distinct from it, such as South Asia (p. 587)

subsistence farm/farming farm that produces only enough to support a family's needs (pp. 67, 161)

suburbs smaller communities surrounding a central city (p. 311)

T

table graphic way of organizing and presenting statistics or facts (p. 17)

taiga [TY•guh] huge, subarctic evergreen forests in northern Europe, Asia, and North America (p. 361)

tannin substance from tree bark used in turning hides into leather (p. 229)

teak tropical wood used in furniture and ship-building (p. 597)

terraced field hillside field cut in steplike strips to hold water (p. 617)

textile fabrics and clothing (p. 299)

timberline elevation above which trees cannot grow (p. 48)

township in South Africa, term for some settlements outside cities (p. 550)

tributary small stream or river that flows into a larger river (p. 595)

tropics region of the world located between the Tropics of Cancer and Capricorn (about 30° North and 30° South), with a generally hot climate (p. 41)

trust territory region placed under control of a another country by international agreement (p. 714)

tsetse fly insect whose bite causes sleeping sickness (p. 512)

tsunami [tsoo•NAH•mee] huge sea wave caused by an earthquake on the ocean floor (pp. 30, 634)

tundra broad, dry, treeless plain in the high latitudes (pp. 47, 361)

tungsten hard, grayish metallic element used in electrical equipment (p. 616)

typhoon a hurricane, or tropical storm system, that forms in the Pacific Ocean (pp. 43, 612)

U-V

urban related to a city or densely populated area (p. 91)

urbanization tendency of a country's people to move from rural areas to cities (p. 62)

vegetation plant life in a region (p. 37)

W-X-Y-Z

water cycle process by which the earth's water moves from the oceans to the air to the land and back to the oceans (p. 37)

water vapor water in the form of a gas (p. 37)

watershed high ridge from which rivers of an area flow in different directions (pp. 290, 533)

weather changes in temperature and precipitation over a short period of time (p. 41)

weathering process by which surface rocks are broken down into smaller pieces by water, chemicals, or frost (p. 31)

welfare state country in which government money is used to provide needy people with health care, unemployment benefits, and so on (pp. 208, 268)

yurt large, round tents of animal skin, used by nomads in Mongolia (p. 624)

GLOSSARY

Spanish Glossary

A

abaca [a•buh•KAH]/**abacá** fibra de una planta la cual es usada para hacer soga (p. 671)

absolute location/localización absoluta la posición exacta de un lugar en la superficie de la Tierra (pp. 5, 21)

acid rain/lluvia ácida precipitación la cual posee grandes cantidades de químicos, especialmente ácido sulfúrico (pp. 68, 100, 283)

adobe [uh•DOH•bee]/**adobe** ladrillo secado al Sol (p. 166)

alluvial [uh•LOO•vee•uhl] **plain/plano aluvial** plano que fue creado por la tierra acumulada de un río (pp. 435, 657)

altiplano [AL•tih•PLAH•NOH]/**altiplano** región altiplana de los Andes (p. 222)

altitude [AL•tuh•TOOD]/**altitud** altura sobre el nivel del mar (pp. 156, 202)

anthracite [AN•thruh•SYT]/**antracita** carbón duro (p. 644)

apartheid [uh•PAHR•TAYT]/**apartheid** restricciones y prácticas legales sudafricanas que segregan a los grupos raciales y étnicos, la palabra significa "aparte" (p. 549)

aquifer [AH•kwuh•fuhr]/**acuífer** capa de tierra subterránea que contiene grandes cantidades de agua (p. 39)

archipelago [AHR•kuh•PEH•luh•GOH]/**archipiélago** grupo grande o cadenas de islas (pp. 183, 633)

atmosphere/atmósfera cojín espeso de gases que rodean a la Tierra, formados principalmente de los gases nitrógeno y oxígeno (p. 32)

atoll [A•TAWL]/**atolón** arrecife o isla coralina de forma anular que rodea a una laguna pequeña (pp. 604, 713)

autobahn [AW•toh•BAHN]/*autobahn* una autopista en Alemania (p. 284)

autonomy [aw•TAH•nuh•mee]/**autonomía** auto gobierno (p. 532)

axis (plural, **axes**)/**axis** la línea o eje horizontal o vertical de medida usado en una gráfica (p. 13)

B

bar graph/gráfica de franjas gráfica en la cual las franjas verticales u horizontales representan cantidades (p. 13)

basin/cuenca extensa área de tierra baja plana, la cual está rodeada por tierra más alta (pp. 155, 195, 511)

bauxite [BAWK•SYT]/**bauxita** mineral metálico del cual se saca el aluminio (pp. 178, 497)

bazaar/bazar un mercado público (p. 451)

Bedouin [BEH•duh•wuhn]/**beduino** miembro de los nómadas del desierto del África del norte y del sudoeste de Asia (p. 429)

bilingual/bilingüe persona que habla dos idiomas (p. 129)

birthrate/índice de natalidad el número de niños que nace cada año por cada 1,000 personas (p. 62)

bog/pantano área de tierra baja y pantanosa (pp. 261, 323)

buffer state/estado intermedio pequeño país que sirve de valla entre dos naciones rivales (p. 208)

bush/área remota término australiano que designa a las áreas rurales remotas (p. 694)

C

cacao [kuh•KOW]/**cacao** árbol tropical cuyas semillas se emplean para hacer cacao y chocolate (p. 487)

calligraphy [kuh•LIH•gruh•fee]/**caligrafía** el arte de escribir con letra bonita (p. 622)

canopy/bóveda capa superior de un bosque tropical, la cual resguarda el suelo del bosque (p. 505)

campesino [KAM•puh•SEE•noh]/**campesino** granjero de Latinoamérica (p. 220)

cardinal directions/puntos cardinales las direcciones básicas: el norte, sur, este y oeste (p. 7)

casbah [KAZ•BAH]/**casbah** la parte vieja de las ciudades en África del norte donde hay calles estrechas y mercados pequeños (p. 458)

cash crop/cosecha al contado cosecha que se cultiva para la venta, a menudo con fines de exportación (p. 220)

cassava [kuh•SAH•vuh]/**yuca** planta cuyas raíces se transforman en harina y se comen, tal como en el pudín de tapioca, por ejemplo (pp. 210, 495, 663)

caudillo [kaw•DEE•yoh]/**caudillo** un dictador militar en la historia de Latinoamérica (p. 203)

chart/diagrama forma gráfica de representar datos claramente (p. 17)

chicle/chicle jugo del árbol de zapotillo, el cual se usa en la producción de la goma de mascar (p. 177)

circle graph/gráfica circular gráfica redonda, la cual muestra cómo se divide un conjunto (p. 14)

city-state/ciudad estado en la historia, consiste de una ciudad independiente y las tierras a su alrededor (p. 306)

civil war/guerra civil conflicto que incluye a diferentes grupos de ciudadanos en un mismo país (pp. 430, 534)

civilization/civilización cultura altamente desarrollada, usualmente posee la religión organizada y leyes (p. 56)

clan/clan grupo de personas o familias relacionadas unas a las otras (pp. 538, 637)

cloves/clavos de especia especia que se produce de los capullos de las flores del clavero (p. 529)

climate/clima los patronos normales del clima de un área a través de un largo período de tiempo (pp. 15, 41)

climate region/región climática un área extensa del mundo con el mismo clima (p. 12)

climograph/climograma una combinación de gráfica lineal y de franjas, la cual provee información sobre la temperatura y precipitación (p. 15)

cold war/la guerra fría época desde los finales del 1940 al comienzo de los 1990 en la cual los EE. UU. y la Unión Soviética compitieron por la influencia mundial (p. 371)

colony/colonia colonización extranjera hecha por un país matriarcal (pp. 103, 187)

command economy/economía autoritaria sistema económico en el cual el gobierno toma posesión de las propiedades, los recursos y medios de producción, y hace todas las decisiones económicas (p. 364)

commonwealth/estado libre asociado término usado para designar a una república o un estado gobernado por la gente; el título oficial de Puerto Rico (p. 189)

communism/comunismo sistema político autoritario en el cual el gobierno central controla a la economía y a la sociedad (pp. 187, 282, 319, 364)

compass rose/rosa de los vientos emblema que aparece dibujado en los mapas para señalar direcciones (p. 7)

compound/campamento grupo de viviendas rodeadas por una muralla en el África rural (p. 488)

condensation/condensación paso del vapor de agua al estado líquido (p. 38)

constitutional monarchy/monarquía constitucional forma de gobierno encabezado por un monarca pero donde el mayor control es otorgado a la legislatura electa (pp. 259, 430)

consumer goods/bienes del consumidor productos para el consumo personal, tales como la ropa y los productos caseros (pp. 336, 366, 449, 615)

contiguous/contiguo se refiere a las áreas adyacentes o que comparten una frontera (p. 89)

continent/continente una de las áreas de población más grandes de la Tierra (p. 30)

continental island/isla continental tierra que fue originalmente parte de un continente, separada por agua creciente o por alguna actividad geológica (p. 712)

contour line/curva de nivel en un mapa topográfico, la línea que une a todos los puntos en la misma elevación (p. 9)

cooperative/cooperativa en Cuba, es una finca controlada y operada por el gobierno (p. 188)

copper belt/región cobreña en Zambia, región de valiosas minas de cobre (p. 557)

copra/copra médula del coco seco (p. 712)

coral/coral material rocoso formado de los esqueletos de animales marinos pequeños (pp. 521, 690)

coral reef/arrecife de coral arrecife que yace en un nivel bajo del océano y está hecho de coral (p. 93)

cordillera [KAW•duhl•YEHR•uh]**/cordillera** grupo de montañas enlazadas entre si (pp. 118, 217)

core/núcleo capa central de la Tierra, probablemente compuesta de hierro caliente y níquel, es sólida en su núcleo interno y fundida en su exterior (p. 29)

cottage industry/industria autosuficiente industria casera o iniciada en un pueblo donde los miembros de la familia suplen su propio equipo de trabajo para producir bienes tal como las telas y efectos de metal (p. 590)

SPANISH GLOSSARY

crevasse [krih•VAS]/**grieta** rajadura profunda en una capa de hielo o glaciar (p. 718)

crust/corteza la capa externa de la Tierra (p. 29)

cultural diffusion/difusión cultural el proceso por el cual los conocimientos y las destrezas se diseminan de un área a la otra (p. 56)

culture/cultura forma de vida de un grupo de gente que comparten costumbres y creencias similares, tanto como un mismo idioma y religión (p. 55)

currents/corrientes corrientes de agua, caliente o fría, en los océanos (p. 42)

cyclone/ciclón tormenta intensa que produce fuertes lluvias y vientos veloces que giran en forma circular (p. 596)

czar [ZAHR]/**zar** título de los emperadores antiguos de Rusia (p. 370)

D

death rate/índice de mortalidad número de muertes que ocurren cada año por cada 1,000 personas (p. 62)

deforestation/deforestación la pérdida de los bosques debido a la extensa práctica de cortar los árboles (p. 658)

delta/delta área triangular en la boca de un río, la cual se forma debido al lodo y la arena depositadas por el agua (pp. 448, 596, 660)

demographer/demógrafo científico que estudia la población (p. 62)

desertification/desiertificación proceso por el cual los prados se secan y las áreas desiertas aumentan (p. 491)

developed country/país desarrollado país que es industrial en vez de ser agricultural (p. 62)

developing country/país en vías de desarrollo país en proceso de industrialización (p. 62)

diagram/diagrama dibujo que muestra los pasos de un proceso o las partes de un objeto (p. 16)

dialect/dialecto variante regional de un idioma (pp. 188, 284, 300)

dictatorship/dictadura un gobierno que está bajo el control de un sólo líder que posee todo el poder (p. 455)

dike/dique banco alto hecho de tierra o concreto, el cual es construido para contener el agua de un río u océano (p. 611)

drought [DROWT]/**sequía** un largo período de extrema sequía y escasez de agua (pp. 490, 537)

dry farming/cultivo seco método de cultivar el trigo en áreas secas, el cual conserva humedad por medio del arado profundo (p. 98)

dynasty/dinastía una familia de gobernantes (p. 619)

dzong/dzong centro budista para el rezo y el estudio (p. 601)

E

earthquake/terremoto estremecimiento violento o temblor de la tierra causado por el movimiento de rocas en una falla (p. 30)

elevation/elevación la altura sobre el nivel del mar (pp. 9, 48, 310)

elevation profile/perfil de elevación diagrama transversal que muestra los cambios en la elevación de la Tierra (p. 9)

emigrate/emigrar trasladarse de su país nativo a otro país (pp. 60, 312)

empire/imperio grupo de países o estados bajo el mando de una nación o un gobernante (pp. 223, 623)

enclave/enclave pequeña nación o región específica establecida dentro de un país más grande (p. 545)

environment/medio ambiente el ambiente natural (p. 23)

equinox/equinoccio día en los meses de marzo y septiembre en que los rayos del Sol caen directamente sobre el ecuador, causando que el día y la noche tengan la misma duración (p. 27)

erg/ergio región desértica con dunas de arena movedizas (p. 458)

erosion/erosión desgaste de la superficie de la Tierra, principalmente por el agua, el viento o el hielo (p. 31)

escarpment/escarpadura acantilado alto que separa a dos superficies de terreno plano, una más alta que la otra (pp. 196, 522, 546)

estancia [ehs•TAHN•syah]/**estancia** rancho grande en Argentina (p. 230)

ethnic group/grupo étnico personas que comparten una historia cultural común, tanto como la misma raza e idioma (pp. 104, 371, 438, 487)

evaporation/evaporación el proceso por el cual

el agua de los océanos, lagos y arroyos se transforma en un gas o vapor por el calor del Sol (p. 38)

exclave la parte pequeña de una nación que está separada del área principal del país (p. 552)

F

famine/hambre escasez de alimentos que afecta a un gran número de personas (p. 63)

farm belt/zona de cultivo la región del centro occidental de los EE. UU. donde el terreno es plano y la tierra es fértil (p. 98)

fault/falla fractura de las rocas en la corteza de la Tierra, a lo largo de la cual ocurren desplazamientos (pp. 30, 391, 522)

favela [fuh•VEH•luh]*/favela* en el Brasil, término que se refiere a los barrios urbanos pobres (p. 199)

fellahin [fehl•uh•HEEN]*/fellahin* granjeros egipcios y trabajadores de la granja (p. 450)

fjord (FYOHRD)**/fiordo** entradas que provienen del mar, las cuales son largas, estrechas y de lados profundos (pp. 266, 697)

flow chart/diagrama de progreso tipo de diagrama que muestra dirección o movimiento (p. 16)

food processing/proceso de comidas la industria en la que las comidas son preparadas y empaquetadas para la venta (p. 393)

fossil fuels/combustibles fósiles grupo de recursos minerales no renovables — el carbón, el petróleo, el gas natural — formados en la corteza de la Tierra de los restos de plantas y animales (pp. 66, 124)

free port/puerto libre el área donde los artículos pueden ser cargados, almacenados y reembarcados sin pagar impuestos de importación-exportación (p. 666)

free enterprise/empresa libre la base de una economía de mercados en la cual las personas comienzan y manejan sus propios negocios para lograr una ganancia con poca intervención del gobierno (p. 58)

free enterprise system/sistema libre de empresa sistema económico basado en la libre empresa (ver definición anterior) (pp. 96, 365)

G

galaxy/galaxia sistema enorme en el universo, el cual incluye a millones de estrellas (p. 25)

*gaucho/***gaucho** en la parte sur de América Latina, es un ganadero a caballo (pp. 208, 230)

geography/geografía el estudio de la Tierra y sus regiones, su agua y su vida vegetal y animal (p. 21)

geothermal power/energía geotérmica energía eléctrica producida por fuentes de vapor naturales subterráneas (p. 699)

geyser/géiser manantial caliente que arroja vapor y agua caliente a través de una rajadura en la Tierra (p. 269)

glasnost [GLAZ•nohst]*/glasnost* término ruso que significa "política abierta"; lema usado por Gorbachev en su política a finales de los 1980 (p. 371)

great circle route/ruta del gran círculo ruta tomada por un barco o avión en la cual éste hace un círculo grande, la distancia más corta entre dos puntos de la Tierra (p. 6)

grid system/sistema de cuadrícula red de líneas imaginarias sobre la superficie de la Tierra, formadas cuando las líneas de latitud y longitud se cruzan, creando patrones (pp. 5, 22)

groundwater/agua subterránea el agua que está almacenada en las capas de roca bajo la superficie de la Tierra (p. 39)

H

hajj/hajj viaje religioso que los musulmanes hacen al Makkah (p. 432)

harmattan [HAHR•muh•TAN]**/harmatán** viento seco que sopla hacia el sur desde el Sahara en el invierno (p. 486)

heavy industry/industria pesada producción de bienes industriales tales como las maquinarias y equipo militar (p. 365)

hemisphere/hemisferio una mitad del globo terráqueo; el ecuador divide a la Tierra en los hemisferios norte y sur; el primer meridiano lo divide en los hemisferios oriental y occidental (pp. 4, 22)

SPANISH GLOSSARY

hieroglyphs/jeroglífico estilo antiguo de escritura egipcia, el cual usa pictografías (p. 451)

high island/isla montañosa es una isla volcánica montañosa en Micronesia (p. 713)

high-technology industry/industria de tecnología avanzada fábrica que produce computadoras u otros equipos electrónicos (p. 624)

high veld/estepa región altiplana y herbosa de Sudáfrica (p. 546)

Holocaust [HOH•luh•KAWST]**/Holocausto** término usado para describir el encarcelamiento y matanza de judíos europeos por los nazi alemanes durante la Segunda Guerra Mundial (pp. 284, 426)

humid continental climate/clima húmedo continental el clima en el interior del continente, con inviernos fríos y veranos cortos y calientes (p. 47)

humid subtropical climate/clima húmedo subtropical el clima cálido, templado y lluvioso de la altitud media (p. 47)

hurricane/huracán tormenta tropical devastadora, con fuertes vientos, la cual se forma en las aguas cálidas del océano (pp. 43, 177)

hydroelectric power (hydroelectricity)/ energía hidroeléctrica (hidroelectricidad) energía eléctrica generada por el agua que cae (pp. 68, 202, 449, 507)

I

ice shelf/capa de hielo parte de una capa espesa de hielo que se extiende desde la tierra hasta el océano (p. 718)

immigrant/inmigrante persona que se muda de un lugar y forma un hogar permanente en otro lugar (p. 102)

industrialized/industrializado término que describe a un país en el que la industria ha reemplazado al cultivo como actividad económica principal (p. 161)

intensive cultivation/cultivo intenso método de cultivo que utiliza toda la tierra disponible para producir varias cosechas al año (p. 635)

interdependent/interdependiente se refiere a los países o a personas que dependen unos de otros (p. 100)

intermediate direction/dirección intermedia cualquier dirección que está entre los puntos cardinales, tal como el sudeste o el noreste (p. 7)

invest/invertir poner dinero en una compañía a cambio de recibir un porcentaje de sus ganancias (pp. 329, 615)

isthmus [IHS•muhs]**/istmo** lengua de tierra que une a dos partes más grandes (pp. 32, 181)

J-K

jute [JOOT]**/yute** fibra de una planta, se usa para hacer cordeles y telas de saco (p. 589)

key/clave la explicación de los símbolos usados en un mapa (p. 7)

krill/crustáceo marisco pequeño parecido al camarón, el cual es fuente de alimento para muchos animales marinos (p. 721)

L

*ladino/***ladino** en Guatemala, se refiere a una persona que habla español y que sigue el estilo de vida español-americano, no el nativo americano (p. 180)

lagoon/laguna cuerpo de agua rodeado de arrecifes o bancos de arena (p. 604)

land bridge/puente terrestre pedazo estrecho de tierra que une a dos áreas mayores (p. 153)

landlocked/rodeado de tierra término que describe a un país cuya tierra no tiene salida al mar o al océano (p. 207)

language family/familia de idiomas grupo de idiomas que proviene de la misma lengua predecesora (p. 57)

latitude/latitud distancia al norte o sur del ecuador, la cual se mide con líneas imaginarias (paralelos) contadas en grados al norte o sur (pp. 5, 22, 155)

leap year/año bisiesto año que tiene 366 días, establecido para compensar por la fracción adicional de un cuarto de día que se produce cuando la Tierra gira alrededor del Sol (p. 26)

light industry/industria liviana producción de bienes al consumidor, tales como productos comestibles y del hogar (pp. 188, 365)

752

line graph/gráfica lineal gráfica en la cual una o más líneas conectan los puntos que representan cantidades variantes (p. 14)

literacy rate/índice de alfabetización porcentaje de adultos en una sociedad que pueden leer y escribir (pp. 58, 181)

llanos [LAH•nohs]**/llanos** las amplias regiones llanas y verdosas de Latinoamérica (pp. 202, 218)

loch [LAHK]**/ensenada** bahía larga y estrecha con montañas a cada lado; también, un lago (p. 256)

loess [LEHS]**/loes** tierra fértil de grano fino, depositada por el viento (pp. 283, 611)

longitude/longitud distancia al este u oeste del primer meridiano, la cual se mide con líneas imaginarias (meridianas) contadas en grados al este u oeste (pp. 5, 22)

low island/isla baja es una isla en Micronesia situada a nivel bajo y formada de coral (p. 713)

M

magma [MAG•muh]**/magma** roca derretida situada dentro de la capa intermedia de la Tierra (p. 30)

mainland/territorio continental el área principal y de más volumen de tierra que tiene un país, en contraste con las islas adyacentes (p. 309)

mangrove/mangle árbol tropical que crece en tierra pantanosa, tiene raíces tanto sobre como bajo la superficie del agua (p. 485)

mantle/capa capa intermedia de la Tierra, compuesta de roca espesa y caliente (p. 29)

manuka/manuka arbusto típico de la tierra volcánica de Nueva Zelandia (p. 697)

maquiladoras [mah•KEE•luh•DOH•rahz]**/ maquiladoras** factorías que ensamblan las partes recibidas de otros países para producir automóviles, sistemas de sonido estereofónicos, etc. (p. 161)

marine west coast climate/clima marino de la costa occidental clima costeño, el cual tiene inviernos templados y veranos frescos (p. 47)

marsupial/marsupial tipo de mamífero que lleva a sus crías en una bolsa abdominal donde terminan su desarrollo (p. 691)

Mediterranean climate/clima mediterráneo clima templado costeño a mitad de latitud, el cual tiene inviernos lluviosos y veranos calientes y secos (p. 47)

megalopolis/la megalópolis región en la que las áreas urbanas adyacentes están integradas, formando así una "gran ciudad" (pp. 91, 638)

mestizo [meh•STEE•zoh] persona nacida de padres procedentes de las razas americana nativa y europea (p. 165)

migrate/migrar mudarse a otro lugar (p. 421)

mobile/móvil mudarse de lugar a lugar (p. 105)

monotheism/monoteísmo la creencia en un sólo Dios (p. 425)

monsoon/monzón vientos estacionales que traen la lluvia a ciertas partes de Asia (pp. 43, 589)

moor/páramo un terreno alto y sin árboles pero ventoso en el Reino Unido (p. 256)

mosque [MAHSK]**/mezquita** lugar de devoción de los que siguen la religión del Islam (pp. 338, 421)

movement/movimiento uno de los temas geográficos que describen cómo las personas de diferentes lugares se comunican (p. 23)

multicultural/multicultural se refiere a un asunto que incluye a muchas culturas diferentes (p. 104)

multinational firm/empresa multinacional una compañía que tiene oficinas y comercia en varios países (p. 289)

N

national park/parque nacional terrenos públicos que se reservan para el recreo de los visitantes al área y para la protección de su estado virgen (p. 106)

natural resource/recurso natural cualquier cosa que proviene del medio ambiente natural y que las personas usan para satisfacer sus necesidades (p. 65)

nature preserve/santuario terreno reservado por el gobierno para proteger a las plantas y a la fauna silvestre (pp. 332, 387)

navigable/navegable término que describe a un cuerpo de agua lo suficientemente ancho y

SPANISH GLOSSARY

profundo para permitir el paso a los barcos
(pp. 222, 278)

neutrality/neutralidad   la política que rechaza la preferencia de una u otra posición en lo relativo a desacuerdos y guerras internacionales (p. 290)

newsprint/papel de periódico   tipo de papel usado para imprimir los periódicos (p. 124)

nomads/nómadas   personas que se mudan de un lugar a otro, a menudo se llevan consigo a sus rebaños de ovejas u otros animales (p. 396)

nonrenewable resources/recursos no renovables   metales y otros minerales que no pueden ser reemplazados después de ser utilizados (p. 66)

nutrients/nutrientes   los minerales de la tierra que suplen alimento a las plantas en crecimiento (p. 177)

O

oasis (plural, **oases**)**/oasis**   lugar donde el agua y la vegetación verdosa está rodeada por un desierto (pp. 398, 432)

oil shale/esquisto   roca arcillosa en capas, la cual contiene aceite (p. 320)

orbit/órbita   trayectoria elíptica que sigue un planeta cuando da vueltas alrededor del Sol (p. 25)

outback/llanura desértica   regiones distantes del interior de Australia (p. 690)

ozone/ozono   forma de oxígeno (O_3) en la atmósfera (pp. 72, 721)

P

pagoda/pagoda   una torre de muchos pisos, la cual fue construida para servir de templo o lugar de adoración (p. 622)

parliamentary democracy/democracia parlamentaria   estilo de gobierno en el cual los votantes eligen a representantes para formar parte de un cuerpo creador de leyes (parlamento), el cual después escoge a un líder, el primer ministro (pp. 128, 259)

peat/turba   materia de las plantas en un estado parcialmente descompuesto, se encuentra en pantanos y es usada como combustible y fertilizante (p. 261)

peninsula/península   pedazo de tierra rodeado por agua en tres de sus lados (pp. 32, 153)

permafrost/permagel   capas inferiores de la tierra en las regiones árticas, las cuales están permanentemente congeladas (pp. 47, 362)

pesticide/insecticida   químicos usados para matar a los insectos y a otros animales dañinos (p. 67)

phosphate/fosfato   sal mineral que contiene fósforo, usado en fertilizantes (pp. 424, 714)

pictograph/pictografía   gráfica en la que los símbolos pequeños representan cantidades (p. 15)

place/lugar   uno de los temas geográficos, el cual describe las características típicas que distinguen a un lugar de otro (p. 22)

plantation/hacienda   granja de gran tamaño donde se cultiva sólo una cosecha al contado (pp. 161, 177)

plate tectonics/tectónicas de lámina   teoría en el campo de geología la cual dice que la corteza de la Tierra está compuesta de enormes láminas movedizas de roca (p. 30)

plateau/meseta   forma terrestre plana cuya superficie está por encima de la tierra que le rodea, con una colina muy alta en uno de sus lados (pp. 32, 297)

poacher/cazador furtivo   persona que caza y mata a los animales ilegalmente (p. 523)

polder [POHL•duhr]**/pólder**   área de tierra en los Países Bajos, la cual fue recuperada del mar (p. 289)

pollution/contaminación   el arrojar substancias impuras o venenosas en la tierra, agua o aire (p. 67)

pope/papa   título del cabecilla de la Iglesia Católica Romana (p. 326)

population density/densidad de la población   número promedio de personas que viven en una milla cuadrada o kilómetro cuadrado (pp. 12, 61)

potash [PAHT•ASH]**/potasio**   sal mineral; usada a menudo en fertilizantes (p. 424)

prairie/pradera   terreno ondulado de pastoreo, se encuentra en áreas interiores con tierra fértil (p. 118)

precipitation/precipitación   agua que cae sobre la Tierra en forma de lluvia, nieve o aguanieve (p. 38)

Prime Meridian/primer meridiano la línea de longitud de 0° desde la cual se miden las distancias al este y al oeste, atraviesa a Greenwich, Reino Unido (p. 22)

prime minister/primer ministro líder gubernamental el cual es seleccionado en una democracia parlamentaria por los miembros del parlamento (p. 128)

projection/proyección en la cartografía, es una manera de dibujar la redondez de la Tierra en una superficie plana (p. 2)

Q

quinoa [KEEN•WAH]/**quinoa** grano de cereal que se crece en Bolivia (p. 226)

R

rain forest/bosque tropical bosque espeso en las regiones tropicales, donde llueve mucho todo el año (p. 45)

rain shadow/sombra de lluvia región en la parte interior de las cordilleras de montañas, donde cae muy poca lluvia (p. 43)

reef/arrecife hilera estrecha de coral, roca o arena cerca de la superficie del agua (p. 522)

refugee/refugiado persona que ha tenido que escapar a otro país para protegerse de un desastre o de peligro (pp. 60, 534)

region/región uno de los temas geográficos, el cual define a las partes de la Tierra que comparten características comunes (p. 23)

relative location/ubicación relativa la ubicación de un lugar en la superficie de la Tierra con relación a otro lugar (p. 22)

relief/relieve diferencias en la altura de un terreno; cuán plano o desigual es su superficie (p. 9)

renewable resource/recurso renovable recurso que puede ser reemplazado naturalmente o que se puede desarrollar con rapidez (p. 66)

republic/república un gobierno sin monarca, en el cual las personas eligen a los oficiales importantes (pp. 103, 280)

revolution/revolución al referirse a un planeta, es un recorrido completo alrededor del Sol (p. 26); en la política, es un cambio radical y repentino de gobierno (p. 103)

rookery/criadero nido para un gran número de pájaros o animales marinos (p. 718)

rural/rural relativo al campo, no a la ciudad (p. 91)

S

samurai/samurai clase de guerreros aristocráticos en el Japón feudal (p. 637)

savanna/sabana terreno de pastoreo tropical con árboles dispersos, generalmente tiene épocas lluviosas y secas (pp. 46, 485)

scale/escala la relación entre la representación de una distancia en un mapa y la distancia verdadera en la Tierra (p. 8)

scale bar/barra de escala la línea dividida en un mapa, la cual muestra la escala del mapa, generalmente en pies, millas o kilómetros (p. 8)

selva/**selva** bosque tropical espeso en la cuenca del Amazonas de Brasil (p. 195)

serfs/siervos labradores vinculados a la tierra en la que han trabajado (p. 370)

service industry/industria de servicio comercio que provee un servicio a las personas en vez de producir bienes para el consumidor (pp. 67, 97, 159, 310, 333)

shah/sha título de los reyes de Irán, país anteriormente llamado Persia (p. 437)

shogun/shogún mandatario militar o dictador del Japón feudal (p. 637)

silt/sedimento partículas de tierra que son arrastradas y depositadas por el agua (p. 448)

sirocco [shuh•RAH•koh]/**siroco** viento caliente y seco que sopla a través de Europa desde el África del Norte (p. 304)

sisal [SY•suhl]/**sisal** fibra de una planta, la cual se usa para hacer cordeles (pp. 197, 529)

slash-and-burn farming/agricultura de tala y quemado técnicas agrícolas en las cuales ciertas áreas forestales son despejadas para el cultivo por medio de incendios (p. 562)

smog/smog humo mezclado con niebla y productos químicos (p. 162)

socialism/socialismo sistema económico en el que el gobierno establece las metas económicas y puede poseer algunos comercios (p. 58)

SPANISH GLOSSARY

sodium nitrate/nitrato de sodio mineral que se usa al preparar los abonos y en explosivos (p. 227)

solar energy/energía solar la energía producida por el calor del Sol (p. 68)

solar system/sistema solar grupo de planetas y otros cuerpos celestes que dan vuelta alrededor del Sol (p. 25)

solstice/solsticio día en junio y diciembre cuando el Sol está directamente sobre el Trópico de Cáncer (23 1/2° N) o el Trópico de Capricornio ((23 1/2° S), lo que marca el comienzo del verano o el invierno (p. 26)

sorghum/melaza hierba alta que se usa como un grano y también para hacer almíbar (p. 557)

standard of living/nivel de vida término que mide la calidad de vida de una persona con base en su ingreso y posesiones materiales (p. 58)

station/estación término australiano que se refiere a un rancho de ovejas o ganado (p. 690)

steppe/estepa tierras secas y sin árboles utilizadas para el pastoreo, a menudo se encuentran a los bordes de los desiertos (pp. 48, 384)

strait/desfiladero estrecho cuerpo de agua que yace entre dos pedazos de tierra (pp. 32, 419)

subcontinent/subcontinente gran área de tierra que forma parte de otro continente, pero es distinto a éste, tal como el Asia del Sur (p. 587)

subsistence farm (farming)/granja de subsistencia/cultivo granja que sólo produce lo suficiente para satisfacer las necesidades de una familia (pp. 67, 161)

suburbs/suburbios comunidades más pequeñas que rodean a una ciudad central (p. 311)

T

table/tabla forma gráfica de organizar y representar datos o estadísticas (p. 17)

taiga [TY•guh]**/taiga** enormes selvas subárticas de árboles siempre verdes en el norte de Europa, en Asia y América del Norte (p. 361)

tannin/tanino substancia que proviene de la corteza de los árboles y que se usa para convertir las pieles de animales en cuero (p. 229)

teak/teca madera tropical que se usa en la construcción de muebles y barcos (p. 597)

terraced field/campo abancalado campo donde la ladera ha sido cortada en franjas parecidas a escaleras para contener el agua (p. 617)

textile/textil telas y ropa (p. 299)

timberline/altura límite límite de elevación sobre el cual los árboles no pueden crecer (p. 48)

township/municipio término usado para describir a varios pueblos a las afueras de las ciudades en el África del Sur (p. 550)

tributary/tributario arroyo pequeño o río que desemboca en un río más grande (p. 595)

tropics/tropical región del mundo ubicada entre los Trópicos de Cáncer y Capricornio (alrededor de 30° al norte y 30° al sur), generalmente es de clima caliente (p. 41)

trust territory/territorio bajo administración fiduciaria región que está bajo el control de otro país debido a un pacto internacional (p. 714)

tsetse fly/mosca tse-tsé insecto cuya picada causa la enfermedad del sueño (p. 512)

tsunami [tsoo•NAH•mee]**/tsunami** ola gigantesca de mar causada por un terremoto en el suelo del océano (pp. 30, 634)

tundra una llanura extensa, seca y sin árboles en las latitudes altas (pp. 47, 361)

tungsten/tungsteno elemento metálico duro y de color gris, usado en los equipos eléctricos (p. 616)

typhoon/tifón un huracán o tormenta tropical, el cual se forma en el Océano Pacífico (pp. 43, 612)

U-V

urban/urbano relacionado a una ciudad o área densamente poblada (p. 91)

urbanization/urbanización la tendencia de la gente de un país a mudarse de áreas rurales a las ciudades (p. 62)

vegetation/vegetación la vida vegetal de una región (p. 37)

W-X-Y-Z

water cycle/ciclo de agua proceso por el cual el agua de la Tierra se mueve de los océanos al aire, a la tierra y vuelve otra vez a los océanos (p. 37)

water vapor/vapor de agua agua en forma de gas (p. 37)

watershed/cuenca arrecife alto del cual los ríos de un área fluyen en direcciones diferentes (pp. 290, 533)

weather/clima cambios de temperatura y de precipitación que ocurren en un período corto de tiempo (p. 41)

weathering/acción corrosiva proceso donde las rocas en la superficie se desmoronan en trozos más pequeños debido a la acción corrosiva del agua, los elementos químicos o la congelación (p. 31)

welfare state/estado benefactor país donde se usa el dinero del gobierno para proveer cuidado de la salud, beneficios de desempleo y otros servicios a las personas necesitadas (pp. 208, 268)

yurt/yurta tiendas de campaña grandes y redondas, las cuales están hechas de piel de animales y son usadas por los nómadas en Mongolia (p. 624)

SPANISH GLOSSARY

Index

people, 576–77; population, *g581,*
61, 576; population by country,
g581; population distribution, 576;
religion, 576; resources, 575; trans-
portation, 575; vegetation, 575.
See also China. *See also* South
Asia. *See also* Southeast Asia. *See
also* Southwest Asia
Asir Mountains, 408
Asmara, Eritrea, *p473, m476, p538*
Assal, Lake, 475
assembly plants. *See maquiladoras*
astronomers; Muslim, *ptg460*
astronomy, 327
Asunción, Paraguay, 210
Aswan High Dam, 449
Atatürk, Kemal, 421
Atatürk Dam, 464
Athena, goddess, 308
Athens, Greece, 244, *m248,* 308;
arts, 245; early government, 312;
location, 309
Atlanta, Georgia, 98
Atlantic Coastal Plain, 90–91
Atlantic Ocean, 39, 90, 243, 498
Atlantis, 312
Atlas Mountains, *m412, p456,* 408;
Algeria, 458; Morocco, 459;
Tunisia, 456
atmosphere, 32
atolls, *p30, d714,* 604, 713
Attica, Plain of, 309
Auckland, New Zealand, 680, 700,
p701
Australia, *m684,* 678–81, 688–94,
714; agriculture, *m698,* 692; architecture, 708; area, 685; cities, 694;
climate, *m692,* 679, 691; culture,
708–09; diamond mining, *c507;*
economy, 679–80, 692; elevations,
p683; facts, 686; history, 681,
693–94; land, 678, 689–91; landforms, *m682, m690;* location, 689;
minerals, 692; people, 680–81,
693–94; population, *g685,* 680,
693; population density, *m700,*
685; population growth, *g685;*
resources, *m698;* rural life, 694;
seasons, 694; states, *m687;* vegetation, 691; water, 691; wildlife, 691,
702
Australian Alps, 690, 691
Austria, *m276,* 290, 291–92; agriculture, *m288;* climate, *m284;* facts,
250; landforms, *m278;* population
density, *m292;* resources, *m288*
Austrian Empire, 291
authoritarian governments, 56
autobahns, 284
automobile industry; in the U.S.
Midwest, 99; Mercedes-Benz, *p58;*
in Japan, *g636*
autonomy; defined, 532
axes; on graphs, 13, 510
axis; of the earth, 26
Ayers Rock, *p689*
Azerbaijan, *m354, m382,* 348, 391,
394; agriculture, *m393;* climate,
m388; facts, 356; immigration,
392; landforms, *m384;* membership
in Commonwealth of Independent

States, 393; population, *c355;* pop-
ulation density, *m396;* resources,
m393
Azeri people, 394
Azov Sea, 390
Aztec, 142, 154, 165
Azteca Stadium, Mexico City, 168

B

Bad Lands, *p101,* 101
Baghdad, Iraq, *m414, p434,* 411,
434, 435
Bahamas, *m145;* facts, 148; location,
183
Bahasa Indonesian language, 663
Bahrain, *m418,* 433; area, 415; facts,
416; population density, 415
Baikal, Lake, *m360,* 348, 353, 362,
507
Baja California, *m158,* 153, 160
Baku, Azerbaijan, *m354,* 394
Balaton, Lake, 329
Balearic Islands, *m298*
Balkan Countries, 335–38; coun-
tries of, 335
Balkan Mountains, 337
Balkan Peninsula, *m246, p335,*
308, 338
ballet, *p374,* 373
Ballet Folklórico of Mexico, 215,
p215
Baltic republics, 319–21
Baltic Sea, *m246,* 267, 319, 359
Bamako, Mali, *m476,* 492
bamboo, *p673*
bananas, 142, 219
Bandar Seri Begawan, Brunei, 666
bandura, p380, 385
Banff National Park, Canada, 118
Bangkok, Thailand, *m580,* 576, 658,
p658
Bangladesh, *m586, p597,* 576, 587,
596–97; agriculture, *m600;* climate,
m595; landforms, *m588;* population
density, *m603;* population growth,
g613; resources, *m600*
banking industry, centers of, 291
Bantu migrations, *m516,* 516
Bantu-speaking people; in Angola,
553; early trade, 516, 555; in
South Africa, 548; in Tanzania, 529
Barbados, *m146;* facts, 148; population density, 147
Barcelona, Spain, 301
bar mitzvah, 425, *p425*
baroque architecture, *p215*
Barranquilla, Colombia, 220
baseball, *p106;* Japan, *p640;* Mexico,
168
basins, 155, 195, 511. *See also* valleys
Basque people, 300
Bastille Day, 279, *p279*
Bateke people, 515
batik, 663
bat mitzvah, 425
bauxite, 178, 497, 692

Bavarian Alps, *p283,* 285. *See also*
Alps
bays, *d31. See also* lochs
bazaars, 451. *See also* markets
Beardmore Glacier, 722
Bedouins, 429
Beijing, China, *m580,* 576, 616, 645
Beijing Opera, *p577*
Beirut, Lebanon, *m414,* 430
Belarus, *m354, m382,* 348, 387–88;
agriculture, *m393,* 350; climate,
m388; facts, 356; landforms,
m384; pollution, 402; population,
c355; population density, *m396;*
resources, *m393*
Belgian Congo, 509
Belgium, *m276,* 287–88; agriculture,
m288; climate, *m284;* colonies,
509, 534; facts, 250; landforms,
m278; population density, *m292;*
resources, *m288*
beliefs. *See* culture
Beligica antarctica, 720
Belize, *m174,* 175; agriculture,
m178; climate, *m184;* facts, 148;
landforms, *m176;* population, 180;
population density, *m186;*
resources, *m178*
Belovezha Forest, 387
Benelux countries, 287–89; coun-
tries of, 287
Bengal, Bay of, 587
Bengali language, 597
Benghazi, Libya, 455
Benin, *m484,* 494, 498; agriculture,
m495; climate, *m491;* facts, 478;
landforms, *m486;* population den-
sity, *m496;* resources, *m495*
Benin kingdom, 495
Benin people, 495
Benue Valley, 516
Berber people, Algeria, 458; Libya,
455; Morocco, 459; Tunisia, 456
Bergen, Norway, *p267*
Bering Sea, 361
Berlin, Germany, *m248,* 244, 285
Berlin, Massachusetts, *m40*
Berlin Wall, *p282, p285,* 285
Bern, Switzerland, *m248,* 291
Bhutan, *m586, 577,* 587, 599, 601;
agriculture, *m600;* climate, *m595;*
landforms, *m588;* population den-
sity, *m603;* resources, *m600*
Bhutto, Benazir, 596
bicycling, 280
bilingualism, 129
Bioko, Equatorial Guinea, 515
birthrate, 62-63
Bishkek, Kyrgyzstan, *m354,* 397
Bismarck, Otto von, 284
bison, *p77-78;* European, 387
Black Forest, 283
Black Rock Desert, 93
Black Sea, *m352, p347,* 242, 349;
and Bulgaria, 337; and the Danube
River, 328; and the Dnieper River,
384; and Georgia, 392; and Russia,
359, 390; and Southwest Asia, 408
Blanc, Mont, 247, 278
Bloemfontein, South Africa, 550
Blue Nile River, 535

C

GEOGRAPHY: The World and Its People

D

G

L

M

INDEX

N

Q

INDEX

U

W

X

y

z

Photo Credits

(r) Matt Meadows; **317,** Matt Meadows; **318,** ©B. Bisson/Sygma; **319,** TRIP/W. Jacobs; **321,** TRIP/T. Noorits; **323,** ©Superstock, Inc.; **325, 326,** James P. Blair/National Geographic Society; **327,** The Granger Collection; **328,** ©Superstock, Inc; **329,** TRIP/M. Barlow; **331,** ©Superstock, Inc.; **332,** James L. Stanfield/National Geographic Society; **333,** ©Peter Turnley/Black Star; **335,** TRIP/I. Wellbelove; **338,** James L. Stanfield/National Geographic Society; **339,** ©B. Bisson/Sygma; **342,** James P. Blair/National Geographic Society; **342-343,** (background) Aaron Haupt; **343,** (t) James L. Stanfield/National Geographic Society, (b) Albert Moldvay/National Geographic Society; **344-345,** George F. Mobley/National Geographic Society; **346,** Dean Conger/National Geographic Society; **346-347,** (c) Bruce Dale/National Geographic Society, (background) Bruce Dale/National Geographic Society; **347,** Steve Raymer/National Geographic Society; **348, 349,** Novosti/Sovfoto/Eastfoto; **350,** (t) Bruce Dale/National Geographic Society, (b) Dean Conger/National Geographic Society; **357,** (t) ©Shinichi Kanno/FPG International, (b) Steve Raymer/National Geographic Society; **358,** Dean Conger/National Geographic Society; **359,** Sovfoto/Eastfoto; **362,** Steve Raymer/National Geographic Society; **364,** ©Bill Swersey/Liaison International; **365,** ©Superstock, Inc.; **369,** ©G. Pinkhassov/Magnum Photos, Inc.; **370,** ©De Keerle/Sygma; **373,** (l) Steve Raymer/National Geographic Society, (r) ©Dave Bartruff/Artistry International; **374,** Steve Raymer/National Geographic Society; **375,** Dean Conger/National Geographic Society; **378,** (l) Wolfgang Kaehler, (r) Matt Meadows; **379,** Matt Meadows; **380,** (t) ©Superstock, Inc., (c) ©TRIP/S. Pozharskij, (b) Tate Gallery, London/Art Resource; **381,** (t) Dean Conger/National Geographic Society, (bl) ©Forbes Magazine Collection/Superstock, Inc., (b) George F. Mobley/National Geographic Society; **382,** Steve Raymer/National Geographic Society; **383,** Sovfoto/Eastfoto; **387,** ©Jeremy Hartley/Panos Pictures; **389,** ©TRIP/V. Kolpakov; **390,** G. Romantsova from Sovfoto/Eastfoto; **391, 392,** George F. Mobley/National Geographic Society; **394,** ©TRIP/V. Slapinia; **395,** Buddy Mays/Travel Stock; **397,** Sovfoto/Eastfoto; **398,** Tass/Sovfoto/Eastfoto; **399,** Steve Raymer/National Geographic Society; **402,** George F. Mobley/National Geographic Society; **402-403,** (background) Aaron Haupt; **403,** (t) William H. Bond/National Geographic Society, (br) Steve Raymer/National Geographic Society; **404-405,** Anne Krumbhaar/National Geographic Society; **406,** (l) James L. Stanfield/National Geographic Society, (b) Steve Raymer/National Geographic Society; **406-407,** (t,background) David Alan Harvey/National Geographic Society, (c) James L. Stanfield/National Geographic Society; **407,** (t,br) Jodi Cobb/National Geographic Society, (bl) Winfield Parks/National Geographic Society; **408,** ©Nik Wheeler; **409,** Robert Harding Picture Library; **410,** (l) James L. Stanfield/National Geographic Society, (r) Robert Harding Picture Library; **411,** ©Superstock, Inc.; **418, 419,** James L. Stanfield/National Geographic Society; **421,** file photo; **423,** ©Richard T. Nowitz; **425,** (t) Aaron Haupt, (b) Jodi Cobb/National Geographic Society; **426,** Sygma; **427,** ©Katrina Thomas/Photo Researchers, Inc.; **428, 430,** ©Superstock, Inc.; **431,** TRIP Photographic; **432,** ©J. Langevin/Sygma; **433,** ©Nabeel Turner/Tony Stone Images; **434,** ©Christina Dameyer/Photo 20-20; **436,** James P. Blair/National Geographic Society; **437,** ©Bernard Sioberstein/FPG International; **439,** James L. Stanfield/National Geographic Society; **442,** (l) ©Stephen Marks/The Image Bank, (r) Matt Meadows; **443,** Matt Meadows; **444,** (t) Victoria & Albert Museum/Art Resource, (c) D.W. Funt/Art Resource, (b) Erich Lessing/Archaeological Museum, Cairo, Egypt/Art Resource; **445,** (t) ©Superstock, Inc., (c) ©Josef Beck/FPG International, (b) ©Richard T. Nowitz; **446,** ©Murray & Associates/Tony Stone Images; **447,** Leo de Wys, Inc./©Fridmar Damm; **449,** ©F. Lazi/FPG International; **452,** Robert Harding Picture Library; **454,** ©Superstock, Inc.; **456,** Thomas J. Abercrombie/National Geographic Society; **457, 458,** ©Dave Bartruff/Artistry International; **460,** The Granger Collection; **461,** ©Murray & Associates/Tony Stone Images; **464,** James L. Stanfield/National Geographic Society; **464-465,** (background) Aaron Haupt; **465,** (t) James L. Stanfield/National Geographic Society, (b) David Alan Harvey/National Geographic Society;

466-467, Walter Meayers Edwards/National Geographic Society; **468,** (l) George F. Mobley/National Geographic Society, (br) Steve Raymer/National Geographic Society; **468-469,** (c) James L. Stanfield/National Geographic Society, (background) Thomas J. Abercrombie/National Geographic Society; **469,** (tl) James L. Stanfield/National Geographic Society, (tr) Steve Raymer/National Geographic Society, (cr,br) George F. Mobley/National Geographic Society; **470,** Jeremy Hartly/Panos Pictures; **471,** ©Gerry Ellis Nature Photography; **472,** N. Durrell McKenna/Panos Pictures; **473,** Neil Cooper/Panos Pictures; **483,** (tl) Jeremy Hartley/Panos Pictures, (tr) Betty Press/Panos Pictures, (bl) Jason Lauré; **484,** Panos Pictures; **485,** Bruce Paton/Panos Pictures; **487,** Betty Press/Panos Pictures; **490,** Jeremy Hartley/Panos Pictures; **492,** Marcus Rose/Panos Pictures; **493,** Sara Leigh/Panos Pictures; **494,** **497,** Dave G. Houser; **498,** Ron Giling/Panos Pictures; **499,** Panos Pictures; **502,** (t,c) Boltin Picture Library, (b) ©Travelpix/FPG International; **503,** (t) ©C.M. Hardt/Liaison International, (bl) Boltin Picture Library, (br) ©Superstock, Inc.; **504,** Nick Robinson/Panos Pictures; **505,** Robert Harding Picture Library; **507,** Jason Lauré; **509,** Dave G. Houser; **511,** ©Marco Corsetti/FPG International; **514,** Martin Adler/Panos Pictures; **517,** Nick Robinson/Panos Pictures; **520,** ©Gerald Cubitt/FPG International; **521,** Emory Kristof/National Geographic Society; **523,** Joseph J. Scherschel/National Geographic Society; **525,** Trygue Bolstad/Panos Pictures; **527,** Bruce Dale/National Geographic Society; **529,** Jason Lauré; **530,** Robert F. Sisson/National Geographic Society; **531,** Sarah Leen/National Geographic Society; **533,** Emory Kristof/National Geographic Society; **534,** Marc Schlossman/Panos Pictures; **535,** TRIP/Helen Rogers; **537,** (l) James P. Blair/National Geographic Society, (r) Liba Taylor/Panos Pictures; **538,** Betty Press/Panos Pictures; **539,** ©Gerald Cubitt/FPG International; **542, 543,** Matt Meadows; **544,** Johann Van Tonder/Images of Africa Photobank; **545,** James P. Blair/National Geographic Society; **547,** (l) Volkmar Wentzel/National Geographic Society, (r) Neil Cooper/Panos Pictures; **549,** (l) Liaison International, (r) Jason Lauré; **550,** David Atchinon-Jones/Robert Harding Picture Library; **552,** Rob Cousins/Panos Pictures; **554,** Bruce Paton/Panos Pictures; **555,** Walter Meayers Edwards/National Geographic Society; **556,** James L. Stanfield/National Geographic Society; **557,** Trygve Bolstad/Panos Pictures; **559,** Gary John Norman/Panos Pictures; **560,** James L. Stanfield/National Geographic Society; **561,** Mark Schlossman/Panos Pictures; **563,** Alan Carey; **564,** Steve Raymer/National Geographic Society; **565,** Johann Van Tonder/Images of Africa Photobank; **568,** James L. Stanfield/National Geographic Society; **568-569,** (background) Aaron Haupt; **569,** (t) Emory Kristof/National Geographic Society, (b) Gordon W. Gahan/National Geographic Society; **570-571,** Thomas J. Abercrombie/National Geographic Society; **572,** (l) Dean Conger/National Geographic Society, (br) George F. Mobley/National Geographic Society; **572-573,** (c) Jodi Cobb/National Geographic Society, (background) George F. Mobley/National Geographic Society; **573,** (tl) Sam Abell/National Geographic Society, (tr,cl) James L. Stanfield/National Geographic Society, (br) David Alan Harvey/National Geographic Society; **574,** Dean Conger/National Geographic Society; **575,** James L. Stanfield/National Geographic Society; **576,** Dallas & John Heaton/Westlight; **577,** Dean Conger/National Geographic Society; **584,** (t) ©Superstock, Inc., (b) James P. Blair/National Geographic Society; **585,** ©Josef Beck/FPG International; **586,** ©N. Shah/Superstock, Inc.; **587,** George F. Mobley/National Geographic Society; **589,** (l) James L. Stanfield/National Geographic Society, (r) George F. Mobley/National Geographic Society; **591,** Steve Raymer/National Geographic Society; **594,** James L. Stanfield/National Geographic Society; **596,** George F. Mobley/National Geographic Society; **597,** Dick Durrance II/National Geographic Society; **598,** Steve Raymer/National Geographic Society; **599,** ©Naomi Duguid/Asia Access; **601,** Paul von Stroheim/Westlight; **602,** ©Travelpix/FPG International, Inc.; **604,** (l) ©Superstock, Inc., (r) James L. Stanfield/National Geographic Society; **605,** ©N. Shah/Superstock, Inc.; **608, 609,** ©Superstock, Inc.; **612,** (l) Bruce Dale/National